ALGAE *and* ELEMENT CYCLING *in* WETLANDS

Jan Vymazal

Duke University
School of the Environment
Duke Wetland Center
Durham, North Carolina

and

Ecology and Use of Wetlands
Prague, Czech Republic

LEWIS PUBLISHERS
Boca Raton   Ann Arbor   London   Tokyo

Library of Congress Cataloging-in-Publication Data

Vymazal, Jan.
  Algae and element cycling in wetlands / by Jan Vymazal.
    p.   cm.
  Includes bibliographical references (p.   ) and index.
  ISBN 0-87371-899-2
    1. Algae—Ecophysiology. 2. Algae. 3. Wetland plants. 4. Wetlands. 5. Biogeochemical cycles. I. Title.
QK565.V86   1994
589.3—dc20   94-37620
                                                              CIP

No claim to original U.S. Government works
International Standard Book Number 0-87371-899-2
Library of Congress Card Number 94-37620
Printed in the United States of America 1 2 3 4 5 6 7 8 9 0
Printed on acid-free paper

TO MY WIFE EVA AND SON ONDŘEJ

Jan Vymazal lives in Prague, Czech Republic. Dr. Vymazal received his MSc and PhD degrees from the Prague Institute of Chemical Technology in 1980 and 1985, respectively. His thesis dealt with eutrophication and algal growth potential tests and his dissertation concerned the use of periphyton for nutrient and heavy metals removal from waters. Since 1985 he has worked in the Water Research Institute in Prague. His research interest focused on the use of algae and constructed wetlands for wastewater treatment. During the years 1991–1993 Dr. Vymazal worked as visiting professor at Duke University Wetland Center in North Carolina where he studied algal and macrophyte communities in the Florida Everglades. Since 1993 he has been working as a private research scientist and consultant in Prague. He cooperates with Botanical Institute of Academy of Sciences of the Czech Republic on two long-term projects dealing with wetland ecology and the use of constructed wetlands for wastewater treatment.

Dr. Vymazal holds memberships in the International Association of Water Quality, the Society of Wetland Scientists, the Phycological Society of America, the International Wildlife Federation and the Czech Algological, Limnological and Microbiological Societies. He has published more than 120 journal articles and participated in more than 50 scientific reports.

Algae and wetlands—what do they have in common? Both of them are ubiquitous, present in all climatic regions in the world, both provide extremely important functions in the environment. Unfortunately most people underestimate their value and are often more aware of their negative aspects, which occur, but definitely do not top the beneficial aspects. In addition, both algae and wetlands are difficult to define and classify. Algae are present in all wetlands and sometimes they can play an important role in wetland structure and function, but so far they have received little attention from the wetland scientists, especially in element cycling evaluation.

The primary aim of this book is to bring general information about algae to wetland scientists and I hope that readers will treat this book accordingly. On the other hand, the book may bring useful information about both wetlands and algae also for phycologists and many other scientists. Wetland and especially phycological (algological) literature is extremely vast and it is not possible to cover all aspects of algal physiology, biochemistry, biology, cytology, ultrastructure, morphology, genetics or ecology in detail. I am aware of the fact that a generalization of large subjects, as algae and wetlands, can lead to oversimplification and therefore I included as many references as possible to which the reader is referred to find further and detailed information. This choice, however, was not easy as thousands of papers on both subjects are published every year and I know that many readers will miss papers they do regard to be important, but simply said, the choice had to be made.

Although the book is mostly aimed at algae, in Chapter 3 basic information about wetlands is also included. The information concerns the definition and classification of wetlands and three major wetland features: vegetation, soils and hydrology. In Chapters 4–6 further information about element cycling and presence in wetlands is also included. For elements whose cycling in wetlands has been studied to a larger extent, special chapters are included. For elements which have not been studied in detail in wetlands, information on the particular element in wetland plants is included in the chapter dealing with the element occurrence in the environment. In the Appendix, macrophyte standing crop and productivity and element concentrations, uptake rates, and standing stocks in macrophytes are summarized.

In the text, elements are divided into three major groups: "macronutrients," "micronutrients," and "other elements" according to algal requirements. These, however, are not identical with higher plant requirements and, therefore, e.g. calcium is included in the group of micronutrients. In addition, the position of several elements in algal nutrition is still not fully understood (e.g. nickel, selenium) and, therefore, the presented division is not definite but only reflects the author's approach.

In chapters dealing with elements, the text is divided according to the following pattern: global cycle of the element (for C, N, P and S), the element in the environment (including aquatic chemistry), the element in wetlands (for macronutrients and several micronutrients), the element in algal nutrition and element toxicity and accumulation in algae (for heavy metals). The aquatic chemistry in water is included as the major environment for algae.

There are many people I would like to thank. In particular, I acknowledge those who either read parts of the book or discussed special topics of the book (in alphabetical order): Christopher B. Craft, Frantisek Hindák, Štepán Husák, Jiří Komárek, Jaroslava Komárková, Lubomír Kováčik, Jan Květ, Robert B. Qualls, Eliška Rejmánková, Curtis J. Richardson, Alena Sládečková, Vladimír Sládeček, Jan R. Stevenson. I am grateful to the many publishers and authors who gave permission for the use of their illustrations and tables. The courtesy of V. Sládeček and F. Hindák for using their plates is especially appreciated. The Duke University "Biology-Forestry Library" staff—David Talbert, Tracy Delius and Margaret Wilbur—were very helpful in finding many older references. I also thank C.B. Craft, E. Rejmánková and C.J. Richardson for the opportunity to use their personal libraries. I am grateful to Donald A. Hammer who introduced me to Lewis Publishers and to Lewis Publishers' editorial staff, especially Jon R. Lewis and Sharon Ray, for their tremendous support during the work. I am also grateful to Jana and Robert Preuhsler for allowing the use of their computer for typing the manuscript before I got my own one. Finally, I want to thank my wife for her continuous encourage-ment, help with figures and tiresome reference list organization, and especially for her willingness to let me move thousands of papers and hundreds of books into our tiny flat for nearly a year.

Most of the book was written in Prague, Czech Republic; smaller parts were completed in Durham, North Carolina and Loxahatchee, Florida.

Jan Vymazal

List of symbols and abbreviations used in the text

ADH = alcohol dehydrogenase
ADP = adenosine 5′-diphosphate
AFDM = ash-free dry mass (= determined after ashing in muffle furnace for 2 hours at 475–515°C; older term, still frequently used = AFDW, ash-free dry weight)
AH = allophanate hydrolase
AMP = adenosine 5′-monophosphate
cAMP = cyclic adenosine 3′,5′-monophosphate
APS = adenosine-5′-phosphosulfate
ATP = adenosine 5′-triphosphate
CER = chloroplast endoplasmic reticulum
CF = concentration factor
CMP = cytidine-(3′ or 5′)-monophosphate
CoA = coenzyme A
ø = diameter
DHAP = dihydroxyacetone phosphate
DM = dry mass (= oven-dry biomass, determined mostly at 65–105°C ; older term was dry weight
Note: in the text, biomass is expressed as DM or AFDM only, wet mass IS NOT used
DMSe = dimethyl sulfide
DNA = desoxyribonucleic acid
EC-50$_{96}$ = the metal concentration which reduces population growth by 50% after 96 hours of exposure
EDTA = ethylenediaminetetraacetic acid
Eh = redox potential
EMP = Embden-Meyerhoff-Parnas (pathway)
ET = evapotranspiration
F_d = ferredoxin
FAC = facultative plants
FACU = facultative upland plants
FACV = facultative wetland plants
FAD = flavin adenine dinucleotide
FD = few data
FMN = flavin mononucleotide
FBP = fructose-1,6-biphosphate
GAP = glyceraldehyde-3-phosphate
GDH = glutamic dehydrogenase (EC 1.4.1.4)
GOGAT = glutamate synthase (=glutamine-oxoglutarate aminotransferase, EC 1.4.7.1)
GPP = gross primary productivity
GS = glutamine synthetase (EC 6.3.1.2)
HMP = hexose monophosphate pathway
I = irradiance (PFD)
I_c = photosynthetic light compensation point
I_d = PFD at depth d
I_o = incident radiation (light intensity, PFD)
I_{opt}, I_k = optimum saturating light intensity or PFD (at which $\mu = \mu_{max}$)
K_s, K_L, K_i = half saturation constants for nutrient uptake or growth, irradiance and inhibitor, respectively
K_m = concentration supporting half-maximum velocity of the reaction
$K_{0.5}$ (CO$_2$) = CO$_2$ concentration at which rate of photosynthesis is one half of the maximum
NAD = nicotinamide adenine dinucleotide (= DPN, diphosphopyridine nucleotide)
NADP = nicotinamide adenine dinucleotide phosphate (= TPN, triphosphopyridine nucleotide)
NADPH$_2$ = reduced nicotinamide adenine dinucleotide phosphate (=TPNH$_2$, reduced triphosphopyridine nucleotide)
NPP = net primary productivity
OBL = obligate wetland plants
P_i = inorganic orthophosphate
P_{max} = maximum photosynthetic rate
PAPS = adenosine-3′-phosphoadenosine-5′-phosphosulfate
PAR = photosynthetically active radiation

PCOC = photosynthetic carbon oxidation cycle
PEP = phosphoenolpyruvate
PEPC = phosphoenolpyruvate carboxylase (EC 4.1.1.31)
PEPCK = phosphoenolpyruvate carboxykinase (EC 4.1.1.49)
PFD = photon flux density
PGA = 3-phosphoglyceric acid
PP = particulate phosphorus
PPP = pentose phosphate pathway
PS = photosystem
PSP = paralytic shellfish poisoning
Q = cell nutrient quota; the concentration of nutrient per cell or per unit biomass
Q_o = subsistence quota; lowest Q at which the alga can grow
Q_{max} = amount of nutrient in the cell supporting μ_{max}
Q_p = phosphorus cellular quota
Q_{10} = ratio of rate at T °C to that at T - 10 °C
RNA = ribonucleic acid
RP = reactive phosphorus
RPI = relative preference index
RPPC = reductive pentose phosphate cycle (= Calvin cycle = photosynthetic carbon reduction cycle)
RS-P = reductant-soluble phosphorus
RuBP = ribulose 1,5-biphosphate
RuBisCO = ribulose 1,5-biphosphate carboxylase/oxygenase (EC 4.1.1.39) (= RuBPCO, in older literature RuDPC or RuBPC)
S = concentration of the nutrient
SRP = soluble reactive phosphorus
t = time
T = temperature
TCA = tricarboxylic acid
TP = total phosphorus
TPP = total particulate phosphorus
TRP = total reactive phosphorus
TSP = total soluble phosphorus
UPL = obligate upland plants
U, μ = specific growth rate
U_{max}, μ_{max} = maximum specific growth rate
μ'_{max} = apparent maximum specific growth rate at infinite Q
urea CA = urea carboxylase
V = uptake rate
V_{max} = maximum uptake rate
W_o = dry mass initially present
W_k = dry mass remaining at the end of the period of measurement

CHAPTER 1

Introduction

1.1 ALGAE

The assemblage of primitive plants which are collectively referred to as algae includes a tremendously diverse array of organisms. Algae may range in size from single cells as small as one micrometer to large seaweeds that may grow to over 50 meters. Many of the unicellular forms are motile, and may integrate with the Protozoa.[4643] It is difficult to give an exact number of species as these data differ considerably in the literature. Chapman and Chapman[765] reported 1,800 genera and 29,000 species, Alexopoulos and Bold[55] reported 1,200 genera and 21,000 species.

Algae are ubiquitous, they occur in every kind of water habitat (freshwater, brackish and marine). However, they can be also found in almost every habitable environment on earth—in soils, permanent ice, snow fields, hot springs, and hot and cold deserts.

From their long fossil record it is evident that the prokaryotic algae were the first photosynthetic cellular plants; it is now generally agreed that algae are the group from which all subsequent cryptogamic groups of plants, and ultimately the flowering plants (or Spermatophyta), arose.[4643]

Algae are simply constructed; even the most complex multicellular forms show a low level of differentiation compared with other groups of plants. Yet the relative simplicity of algae is misleading, because even the smallest may exhibit, at the cellular level, a high degree of complexity. Biochemically and physiologically algae are similar in many respects to other plants. They possess the same basic biochemical pathways; all possess chlorophyll-a and have carbohydrate and protein end products comparable to those of higher plants.[4643]

Algae are the major primary producers of organic compounds, and they play a central role as the base of the food chain in aquatic systems. Besides forming the basic food source for these food chains, they also form the oxygen necessary for the metabolism of the consumer organisms.[2745,4643]

Man's uses of algae, particularly marine algae, are far more diverse and economically important than generally realized.[10] They are used as a human food, in agriculture (fertilizers, manure, fodder, aquaculture), medicine, textile, paper and paint industries, chemical extracts from larger marine algae (e.g. alginic acid, carrageenan or agar) are used in the manufacture of food or in cosmetic industry, and diatomaceous earth (deposits of diatom frustules) is widely used as filtration and polishing materials. Algae are also important surface binding agents which reduce erosion and can be used for wastewater treatment.

Algae have played only a minor role as disease agents, but they are significant agents of a variety of toxic (e.g. cyanophyte blooms or dinoflagellate red tides) and nuisance problems (e.g. eutrophication, discoloration of water, clogging filters in drinking water treatment in water-works).

The first written references to algae are found in the ancient Chinese classic Sze Sen (Bible of Poems).[3855] Phycology or algology is the study of algae. The word phycology is derived from the Greek word "phykos," which means "seaweed." The scientific name Algae (singular alga) was used by G. Plinius Secundus (23 B.C.) for herbaceous marine algae.[5091]

The first new age reference to algae can be found in a book, *Methodi herbariae libri tres,* by Adam Zálužanský ze Zálužan (often quoted as Von Zalusian) published in Prague in 1592.[5540] Other early phycological studies include those by Ray,[4014] Dillenius,[1083] Bauhin,[246] Linnaeus,[2871–2] and De Reaumur.[1059]

The advantage of improved microscopy resulted in revolutionary progress in the systematics of cryptograms during the first three quarters of the 19th century when many of the presently known algal genera had been described. Important studies from this era include e.g. those by Stackhouse,[4669] Vaucher,[5142] Lamouroux,[2691–3] Lyngbye,[2988] Greville,[1817] Harvey,[1953–4] and Kützing.[2648–9]

Once the majority of the algae had been described the development of modern systematics became possible. The fast accumulating information about morphology, reproduction and life histories of algae led to the improvement in the classification of algae and simultaneously opened the way for recognition of new algal species (e.g. References 452, 465, 534, 1064–5, 1246, 1511, 1514, 1731, 2256, 2521, 2670, 3549, 3665, 4296, 4353, 4524, 4848, 5272, 5429–31, 5438).

The beginning of the "modern" phase of phycology in the early 1950s is characterized by widespread use of electron microscopy, both scanning and transmission. These modern techniques have greatly improved the knowledge of the structure and functioning of algal cells and their diverse organelles, including an understanding of their micromorphology, cytology, physiology, molecular organization, and genetics.[4643]

The phycological literature is enormous, including journals devoted exclusively to the study of algae (*Phycologia*, the *Journal of Phycology*, the *British Phycological Journal*—now the *European Journal of Phycology, Phykos*, the *Japanese Journal of Phycology*, the *Bulletin de la Société Phycologique de France*, the *Journal of Applied Phycology, Algological Studies* [formerly *Arch. Hydrobiol. Suppl.*]) or to specific groups of algae (e.g. *Bacillaria*, International Journal for Diatom Research, devoted to diatoms) as well as proceedings of regular symposia on various algal groups (e.g. *International Seaweed Symposia, Symposia of International Association for Cyanophyte Research*), monographs or other books. However, it is beyond the scope of this introduction to make a list of basic phycological literature but some publications, because of their quality and importance should be mentioned here. First of all, there are two comprehensive publications on the biochemistry and physiology of algae.[2820,4770] Other important books include those on:

- introduction to algae and general phycology (including determination keys): References 55, 424, 434, 485–6, 797, 898, 1005–7, 1284, 1460, 1462–3, 2098, 2299, 2300, 2490–1, 2745, 2977, 3310, 3879–81, 4190–1, 4215, 4592, 4594, 4643, 4971, 5014, 5091, 5112, 5535;
- biology of algae: References 753, 973, 2885, 4209, 4886;
- ecology: References 769, 2099, 2886, 4210;
- morphology: Reference 425;
- genetics: Reference 2824;
- biochemistry: Reference[4171]
- cultivation and phycological methods: References 421, 601, 636, 1561, 2021, 2818, 2887, 4593, 4695, 4717, 5150.

Useful information on algal ecology and nutrient cycling in the water environment can be found in excellent limnological books by Hutchinson,[2257–8,2260] Wetzel,[5349] or Golterman.[1709] Important papers and books devoted to specific classes of algae are mentioned in Chapter 2.

Algae are found in all wetlands. In some of them, such as the Florida Everglades, they can play a crucial role in wetland biogeochemical cycles. Despite this fact, algae have received little attention in wetland research so far. Nixon and Lee[3447] pointed out that since higher plants form the most conspicuous vegetation component of many wetlands, they also tend to receive the most attention in studies on mass balance, nutrient budget, and inflow-outflow relations. The role of other vegetation, including algae, is often ignored even though the productivity of algae can easily exceed the productivity of higher plants. In some sloughs of the Florida Everglades, the periphyton growing on *Eleocharis* or *Nymphaea* stems or peduncles produces in summer and fall up to six times more biomass than higher plants themselves.[5210]

Wetland literature, with few exceptions,[1205,2009,3834,4393] has not described many detailed studies on algae so far. In fact, many papers on algae in wetlands have appeared in ecological, botanical, limnological and also phycological (e.g. References 871, 976–7, 1879, 4156, 4254, 4831–2, 4834–6, 5445) literature. Most of the studies on algae in wetlands deal with benthic salt marsh communities.

The number of phycological studies in wetlands depends on what we consider to be a wetland (in Chapter 3, the difficulties with wetland definition and classification will be discussed). No matter which definition we use, one example clearly shows how algae are treated by wetland scientists. In the journal *Wetlands,* which is published by the Society of Wetland Scientists and devoted exclusively to wetlands, only two papers[3348,3953] from the total number of 130 contributions, have dealt with algae during the years of 1988 to 1993.

1.2 WETLANDS

Wetlands are ubiquitous and found on every continent except Antarctica and in every climate from the tropics to the tundra. An estimated 6% of the land surface of the world is wetland.[3032,3246] Wetlands are among the most important ecosystems on the Earth. In the great scheme of things, it was the swampy environment of the Carboniferous Period that produced and preserved many of the fossil fuels on which we now depend. On a much shorter time scale, wetlands are valuable as sources, sinks, and transformers of a multitude of chemical, biological, and genetic material.[3246]

Compared with other major natural forms of landscape, wetlands are young and dynamic. Many are physically unstable, changing in a season or even in a single storm. They change as vegetation changes, sediments are laid down or land sinks. The process which formed some wetlands stopped a long time ago, so they may be regarded as fossil landscapes. In other cases, a single, recent event initiated them.[3032]

Wetlands are enigmas as ecosystems, being neither wholly terrestrial nor aquatic, nor easily classified as ecotonal intergrades between the two. Like intertidal zones, prairie-forest borders, and other edges of small or large scale between distinct natural communities, wetlands are intermediate between two extremes of conditions, wet and dry. They possess not only intermediary characteristics—physical, chemical and biological—but in addition, qualities that are uniquely their own and not part of adjacent land or water ecosystems. Neither terrestrial nor aquatic ecologists can claim wetlands with complete comfort, and perhaps because wetlands do not fit conventional classification schemes they have until recently been largely unclaimed, and unstudied.[3697]

For millennia man has settled in or near floodplains or shallow lakes, which provided water and food from the aquatic environment and agriculture. Most primary civilizations emerged on floodplains.[2897] Many examples of paleolithic and neolithic settlements from China, India, and near Eastern and Mediterranean countries occur in such areas.[1620,4870]

Wetlands provide people—directly and indirectly—with an enormous range of goods and services (Figure 1-1): water supply and control (recharge of groundwater aquifers, drinking water, irrigation, flood control, water quality and wastewater treatment), mining (peat, sand, gravel), use of plants (staple food plants, grazing land, timber, paper production, roofing, agriculture, horticulture, fertilizers, fodder), wildlife (e.g. breeding grounds for waterfowl, preservation of flora and fauna), fish and invertebrates (shrimps, crabs, oysters, clams, mussels), integrated systems and aquaculture (e.g. fish cultivation combined with rice production), erosion control, gene pools and diversity, energy (hydro-electric, solar energy, heat pumps, gas, solid and liquid fuel), education and training, recreation and reclamation.[2897,3032,3246] The use of reclaimed natural wetlands for agriculture (including paddies), fish ponds, power, industry, and recreation may turn out often in the long term to be unprofitable.[2897]

Wetlands mean different things to different people. To some they are stretches of waterlogged wastelands that harbor pests and diseases, and should therefore be drained or filled in order to make better use of the land. To others, wetlands are beautiful serene landscapes, usually associated with open waters, and serve as important habitats for waterfowl, fish and other wildlife.[1750]

There has been an explosive growth of knowledge about, and a radical change of attitude toward wetlands during the last few decades. Before the early 1960s wetlands were largely neglected and unappreciated, and were probably the most poorly understood of landscapes and ecosystems, being neither sound land nor good water.[5440] Wetlands are the only ecosystem type that have their own international convention—often known as Ramsar Convention[2294] from its place of adoption in Iran in 1971. Under this convention signatories agree to include and to promote the sound utilization of wetlands.[3032,3034]

Wetland literature published in the last fifteen years is voluminous and it is beyond the scope of this book to refer to most of it but some books and conference proceedings, in my opinion, should be mentioned here as primary sources of information about wetlands:

- general features: References 697, 1278, 1351, 1713, 1755–6, 2754, 3027, 3032, 3246, 3698, 4418, 5325, 5441;
- ecology and management: References 1205, 1735, 1750–1, 2177, 3710, 3699, 4099, 5379–80, 5564;
- creation and restoration of wetlands: References 1889, 2462, 2646;
- specific wetland types: References 885, 959, 1015, 1298, 1535, 1989, 2011, 2683, 2903, 3834, 3898, 4102, 4393, 5119;
- the use of wetlands for wastewater treatment: References 883, 1665, 1888, 3322, 4029, 5009, 5340;
- classification, identification and delineation of wetlands: References 893, 1249, 2992.

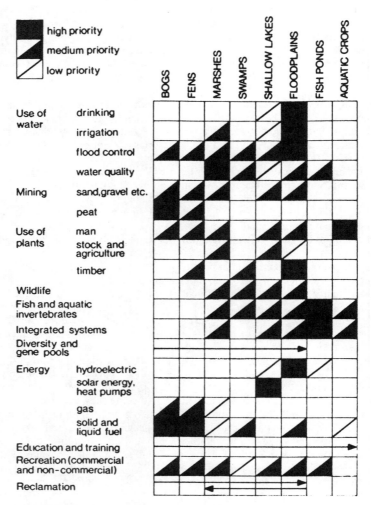

Figure 1-1.   Summary of human uses of wetlands.[2897] (From Löffler, 1990, in *Wetlands and Shallow Continental Water Bodies,* Patten, B.C., Ed., p. 18, reprinted with permission of SPB Academic Publishing bv, The Hague.)

2.1 DEFINITION AND CLASSIFICATION

Algae comprise a very diverse group of organisms which, since the earliest times, defied precise definition.[4643] Perhaps the most striking indication of the diversity of algae is that they are placed in three separate kingdoms (Monera, Protista, Plantae) in a five kingdom concept of classification of organisms.[5395] There is not even an agreement among scientists in this basic classification as e.g. Margulis and Schwartz[3068] placed algae in only two kingdoms (Protoctista and Plantae). Thus it is very difficult to provide an accurate definition of algae.

Bold and Wynne[424] wrote: ". . . The term 'algae' means different things to different people, and even the professional botanist and biologist find algae embarrassingly elusive to define. The reasons for this are that algae share their more obvious characteristics with other plants, while their really unique features are more subtle." Trainor[5014] pointed out that these "simple" plants are not really simple at all.

It is difficult, if not impossible, to decide which classification is "more inclusive," and therefore I include several definitions of algae given by different scientists.

Lee[2745]: "The algae are thallophytes (plants lacking roots, stems and leaves) that have chlorophyll-a as their primary photosynthetic pigment and lack a sterile covering of cell around the reproductive cells. This definition encompasses a number of plant forms that are not necessarily closely related. For example, blue-green algae and prochlorophytes are closer in evolution to the bacteria than to the rest of the algae."

Bold and Wynne[424]: "Algae differ from other chlorophyllous plants in that: (1) In unicellular algae, the organisms themselves may function as gametes* (Figure 2-1a); (2) in some multicellular algae, the gametes may by produced in special unicellular containers or gametangia (Figure 2-1b); or (3) in others, the gametangia are multicellular (Figure 2-1c), every gametangial cell being fertile, that is, producing a gamete. None of these characteristics occurs in liverworts, mosses, and vascular plants. Instead, the multicellular sex organs of many of them are only partially fertile, being covered by sterile cells (Figure 2-1d, e)."

Khan[2491]: "The algae are defined as a group of chlorophyllous plants with uncovered reproductive organs."

It is necessary to mention probably the most comprehensive definition of Fritsch[1511]: "Unless purely artificial limits are drawn, the designation of an alga must involve all holophytic organisms (as well as their numerous colorless derivatives) that fail to reach the level of differentiation of archegoniate plants." The algae thus constitute an heterogenous assemblage of oxygen-producing photosynthetic, non-vascular organisms with unprotected reproductive structures. The definition bridges both pro- and eukaryotic forms.[4643,4497]

The algae are divided into Prokaryota (cells without organized nucleus and lacking membrane-bounded organelles) and Eukaryota (these with organized nucleus).[797,1457] Prokaryota include Cyanophyta and Prochlorophyta (division proposed by Lewin[2826-7]), while Eukaryota include the rest of algal groups. For the comparison of the prokaryotic and eukaryotic cells see e.g. Grant and Long[1880] or Margulis and Schwartz.[3068]

The classification of any group of organisms is subjective, varies with the classifier and reflects his or her emphasis and acceptance or nonacceptance of criteria available at the time of classification. The authors view their own system of classification as tentative and subject to late continuing modification.[424] I agree with this statement but for scientists, who do not follow all algological or phycological literature regularly and algae are not their main concern (e.g. wetland scientists), the result is confusion.

---

*Gamete is a sexually active cell capable of uniting with a compatible cell to form a zygote.[424]

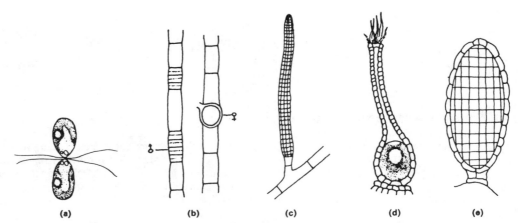

(a)   **(b)**   **(c)**   **(d)**   **(e)**

Figure 2-1.   Characteristics of the sexual reproduction of algae (a)-(c) and nonalgal plants (d), (e). (a) Uniting gametes of a unicellular alga. *Chlamydomonas*. (b) Unicellular gametangia of a filamentous alga, *Oedogonium*. (c) Multicellular gametangium of *Ectocarpus*. Note that every cell is gametogenous. (d) Archegonium and (e) antheridium of a liverwort, a representative nonalgal plant. Sex organs are multicellular and consist of both gametic and sterile (vegetative) cells.[424] (From H.C. Bold and M.J. Wynne, in *Introduction to the Algae*, 2e, © 1985, p. 2, reprinted with permission of Prentice Hall, Englewood Cliffs, New Jersey.)

The early classifications were mostly based on pigmentation and color of the algae. Color as manifestation of the differing pigmentation is still of high importance in classifying the major groups of algae but external shape, presence or absence of flagella, type of stored photosynthetic products, cellular organization, reproduction and life history have become important characteristics for classification. The development of electron microscopy revealed details of flagellar structure, flagella hairs, flagellar root system, and swellings, chloroplast endoplasmic reticulum, thylakoids groupings, phycobilisomes, external scales, pit connections, silification vesicles, theca projectiles, eyespots, and nuclear structure and division. Biochemical analyses clarified the presence and structure of algal pigments, storage products, and cell-wall components.

Until the end of the 18th century the algae were crudely classified, and were grouped with the fungi and lichens. Zálužanský ze Záluzan[5540] included them with fungi, lichens and seaweeds as Musci, under plants "*Ruda et Confusa.*" Bauhin[246] listed under algae *Muscus, Fucus, Conferva* and *Equisetum* (= *Chara*). Ray[4014] describes *Corallinum, Fucus, Muscus* and *Chara*. Swedish botanist Linnaeus[2871-2] describes 108 algal species (mostly marine) which were designated under genera *Fucus, Ulva, Conferva* and *Byssus* (*Chara*). Linnaeus[2872] applied the name Algae for the group of plants and included it with Hepaticeae in his *Species Plantarum*. At the end of the 18th century, algae were mostly named as *Fucus, Corallina, Ulva* and *Conferva*. Jussieu[2392] for the first time classified the algae on the basis of gross morphology in the current sense. Usually *Conferva* included filamentous, *Ulva* membranous and *Fucus* fleshy forms. Stackhouse[4669] broke completely from Linnaeus's criteria of genus and divided *Fucus* into 67 genera. Vaucher[5142] used life history as a basis of taxonomy and recognized three groups of fresh water algae, Conferves, Ulves and Tremelles (*Nostoc* and *Oscillatoria*). Lamouroux[2691-3] divided algae into major taxa (Orders) on the basis of color. C. Agardh[30] created six well defined divisions—Diatomaceae, Nostochineae, Confervoideae, Ulvaceae, Florideae and Fucoideae. Harvey[1953] divided algae into groups based primarily on color—Melanospermae (brown algae), Rhodospermae (red algae), Chlorospermae (green algae), and Diatomaceae (including desmids). This scheme has proved to be the basis of most modern classifications of the algae into divisions.

The second half of the 19th century was an era of upsurge of natural sciences (physics, chemistry, geology), an era when modern philosophical and developmental theories were created. The advancement in these disciplines created new methodological approaches and new evaluation of scientific results.

In 1914, Pascher[3666] proposed a classification which included 7 algal divisions—Chrysophyta, Phaeophyta, Pyrrhophyta, Euglenophyta, Chlorophyta, Charophyta and Rhodophyta. The fast accumulating information about morphology, reproduction and life history of algae gave rise to new classifications.

Classification below the level of division progresses, in descending rank to class (in some divisions to subclass), order, family, genus and species. Similarities between the taxa become greater with progression to the lower levels of classification. The International Code of Botanical Nomenclature laid down the suffixes to be used for the categories of plant classification and the following are those applicable to the algae:[4209]

Taxonomical value	Suffix	Example
division	-phyta	Chlorophyta
class	-phyceae	Chlorophyceae
order	-ales	Chlorococcales
family	-aceae	Scenedesmaceae
genus		*Scenedesmus*
species		*quadricauda*

It is usual, in formal accounts, to indicate the name of the author who describes a genus and species. In case of revision, the name of the first author is listed in parentheses. For *S. quadricauda* it would therefore be *Scenedesmus quadricauda* (Turpin) Brébison.

Bold et al.[426] proposed that "division" should be replaced by "phyllum," but the International Botanical Congress at Sydney (1981) did not approve this proposal. Papenfuss[3639] had pointed out that to use the designation "Chlorophyta," literally "Green algae," for the green algae precluded its use for other members of the plant kingdom with identical pigmentation and storage products. He suggested, therefore, that the names for algal divisions include "phyco," the group being accordingly, Chlorophycophyta, Euglenophycophyta, and so on.[424] Several authors followed this proposal[425,3640,4643] but most authors as well as current phycological journals do not follow these names.

A continuous stream of new information on the fine structure and biochemistry of algae together with the individual approach of authors resulted in a number of different classifications. It is beyond the scope of this book to discuss in detail all classification problems but some of the most important information is listed in Chapter 2.18. I do not want to judge which classification is more correct or more inclusive. I am also not going to create any further classification and add to the confusion which has already existed. Modern classifications vary widely from author to author. Bold and Wynne[424] in their comparative summary found the number of divisions varied from 4 to 15, and the number of classes from 5 to 25. Here are some examples of algal divisions and classes of algae proposed by different authors:

Tilden[4971]
Cyanophyceae
Chlorophyceae
Chrysophyceae
Phaeophyceae
Rhodophyceae

Papenfuss[3640]
Schizophycophyta
  Schizophyceae
Chlorophycophyta
  Chlorophyceae
Charophycophyta
  Charophyceae
Chrysophycophyta
  Xanthophyceae
  Chrysophyceae
  Bacillariophyceae
  Cryptophyceae
Pyrrhophycophyta
  Dinophyceae
  Chloromonadophyceae
Phaeophycophyta
  Phaeophyceae
Euglenophycophyta

Euglenophyceae
Rhodophycophyta
  Rhodophyceae

Christensen[797]
Cyanophyta
  Cyanophyceae
Chlorophyta
  Euglenophyceae
  Loxophyceae
  Prasinophyceae
  Chlorophyceae
Chromophyta
  Xanthophyceae
  Chrysophyceae
  Bacillariophyceae
  Cryptophyceae
  Dinophyceae
  Raphidophyceae
  Haptophyceae
  Craspedophyceae
  Phaeophyceae
Rhodophyta
  Rhodophyceae

Chapman and Chapman[765]
Cyanophyta
  Cyanophyceae
Chlorophyta
  Chlorophyceae
  Prasinophyceae
  Charophyceae
Euglenophyta
  Euglenophyceae
Chloromonadophyta
  Chloromonadophyceae
Xanthophyta
  Xanthophyceae
  Eustigmatophyceae
Bacillariophyta
  Bacillariophyceae
Chrysophyta
  Chrysophyceae
  Haptophyceae
Phaeophyta
  Phaeophyceae
Pyrrhophyta
  Desmophyceae
  Dinophyceae
Cryptophyta

Cryptophyceae
Rhodophyta
Rhodophyceae

Round[4209]
Cyanophyta
Cyanophyceae
Chlorophyta
Chlorophyceae
Oedogoniophyceae
Zygnematophyceae
Bryopsidophyceae
Charophyta
Charophyceae
Prasinophyta
Prasinophyceae
Euglenophyta
Euglenophyceae
Xanthophyta
Xanthophyceae
Chrysophyta
Chrysophyceae
Haptophyta
Haptophyceae
Bacillariophyta
Centrobacillariophyceae
Pennatobacillariophyceae
Dinophyta (= Pyrrophyta)
Desmophyceae
Dinophyceae
Eustigmatophyceae
(no division listed)
Phaeophyta
Phaeophyceae
Cryptophyta
Cryptophyceae
Rhodophyta
Rhodophyceae

Hindák et al.[2098]
Cyanophyta
Cyanophyceae
Euglenophyta
Euglenophyceae
Chlorophyta
Chlorophyceae
Conjugatophyceae
Charophyceae
Chromophyta
Chrysophyceae
Bacillariophyceae
Dinophyceae
Phaeophyceae
Cryptophyceae
Raphidophyceae
Xanthophyceae
Rhodophyta
Rhodophyceae

Bold et al.[425]
Cyanochloronta
Myxophyceae
Chlorophycophyta
Chlorophyceae
Euglenophycophyta

Euglenophyceae
Charophyta
Charophyceae
Chrysophycophyta
Xanthophyceae
Chrysophyceae
Bacillariophyceae
Phaeophycophyta
Phaeophyceae
Pyrrhophycophyta
Dinophyceae
Rhodophycophyta
Rhodophyceae

Khan[2491]
Cyanophyta
Cyanophyceae
Chlorophyta
Chlorophyceae
Oedogoniophyceae
Conjugatophyceae
Charophyta
Charophyceae
Euglenophyta
Euglenophyceae
Xanthophyta
Xanthophyceae
Chrysophyta
Chrysophyceae
Haptophyceae
Bacillariophyta
Bacillariophyceae
Dinophyta
Dinophyceae
Desmophyceae
Phaeophyta
Isogeneratophyceae
Heterogeneratophyceae
Cyclosporophyceae
Cryptophyta
Cryptophyceae
Rhodophyta
Rhodophyceae

Bold and Wynne[424]
Cyanophyta
Cyanophyceae
Prochlorophyta
Chlorophyta
Chlorophyceae
Charophyta
Charophyceae
Phaeophyta
Phaeophyceae
Chrysophyta
Chrysophyceae
Prymnesiophyceae
Xanthophyceae
Eustigmatophyceae
Raphidophyceae
Bacillariophyceae
Pyrrhophyta
Ebriophyceae
Ellobiophyceae
Syndiniophyceae

Dinophyceae
Desmophyceae
Rhodophyta
Rhodophyceae
Cryptophyta

South and Whittick[4643]
Cyanophycophyta
Cyanophyceae
Prochlorophycophyta
Prochlorophyceae
Euglenophycophyta
Euglenophyceae
Chlorophycophyta
Chlorophyceae
Charophyceae
Prasinophyceae
Chromophycophyta
Chrysophyceae
Bacillariophyceae
Dinophyceae
Phaeophyceae
Cryptophyceae
Raphidophyceae
Xanthophyceae
Eustigmatophyceae
Prymnesiophyceae
(= Haptophyceae)
Rhodophycophyta
Rhodophyceae
(Glaucophycophyta)

Lee[2745]
Cyanophyta
Cyanophyceae
Prochlorophyta
Glaucophyta
Rhodophyta
Rhodophyceae
Chlorophyta
Micromonadophyceae
Charophyceae
Ulvophyceae
Chlorophyceae
Euglenophyta
Euglenophyceae
Dinophyta
Dinophyceae
Cryptophyta
Cryptophyceae
Chrysophyta
Chrysophyceae
Synurophyceae
Prymnesiophyta
Prymnesiophyceae
Bacillariophyta
Bacillariophyceae
Xanthophyta
Xanthophyceae
Eustigmatophyta
Eustigmatophyceae
Raphidophyta
Raphidophyceae
Phaeophyta

There are numerous other classifications in the literature (see References 973, 1284, 1511, 1514, 2099, 2534, 3310, 4470, 4497, 4868, 5014, 5091).

For a long time, all algae exhibiting movement, e.g. diatoms, desmids as well as flagellates were regarded as animals. Braun[534] established the algal nature of some flagellates and Klebs[2531] was the first to emphasize the plant-like nature of many of these forms.

The classification and systematics of algae are heavily criticized and undergo a lot of changes nowadays. The discussion during the last International Botanical Congress in Japan (1993) revealed that the present system would probably be "turn over" in the near future. The major changes are based on the ultrastructural observations of flagella bases which means that major changes may be expected in classes with flagellate cells. However, the new classification has not been firmly fixed yet, and therefore, in Chapter 2.18 these changes are not included.

2.2 ALGAL MORPHOLOGY

The plant body of an alga is called thallus. Algal thalli range from small solitary cells to large, complex multicellular structures.[4868] The size of the algae varies from unicells less than 1 μm in diameter (e.g. *Chlorella*) to kelps (brown algae) which may attain a length more than 50 m (e.g. *Macrocystis pyrifera*).

The algal forms can be unicellular or multicellular. Multicellular thalli may be categorized into several main types. Again, there is no consistency in these groups. Morris[3310] listed five types: colonial, aggregations, filamentous, siphoneous and parenchymatous. Khan[2491] also listed five types: colonial, filamentous, siphonaceous, foliose and heterotrichous. Round[4209] listed four types: colonial, filamentous, siphonaceous and parenchymatous. Bold and Wynne[424] distinguished colonial, filamentous, membranous or foliose, tubular, blade- and leaf-like types (Figure 2-2). South and Whittick[4643] divided multicellular forms into three main types: colonial, filamentous and pseudoparenchymatous. Algal morphology is discussed in detail by Round[4209] and South and Whittick,[4643] further information has also been presented by many other authors.[424,2098,2491,2745,3324,4868,5091] The summary of levels of organization in the algae is given in Table 2-1.

2.2.1 Unicellular Organization

The unicellular algae (Figure 2-3) may be a) rhizopodial (amoeboid), b) non-motile coccoid (= protococcoid in Round[4209]), and c) motile flagellate.

Rhizopodial form: cells lack rigid cell walls (although they may be enclosed by other structures, such as loricae) and form cytoplasmic projections known as pseudopodia and rhizopodia. Contractile vacuoles are in the protoplast, stigma is usually found in plastids. They are most abundant in the Chrysophyceae and Xanthophyceae, but also occur in the Dinophyceae and several other classes.

Coccoid form: the coccoid habit refers to those unicells where a non-motile state predominates and where motility is either entirely absent or restricted to reproduction stages. Coccoid forms occur in the majority of algal classes, and in some are the predominant, and in others the only type of organization (Eustigmatophyceae).[4643] Coccoid form is mostly one-nucleus. It varies in shape and the cell wall is always present.

Flagellate form: motile vegetative cells, moving by means of flagella are found in all eukaryotic algal classes except the Rhodophyceae, Bacillariophyceae, Eustigmatophyceae, Phaeophyceae and Charophyceae (the last three, however, possess flagellate gametes and/or zoospores). Flagellate unicells are the principal vegetative phase in the Dinophyceae, Raphidophyceae, Cryptophyceae, Euglenophyceae and Prasinophyceae. Within many classes there is considerable uniformity in the features of the flagellate cells, whether vegetative or reproduction.[4643] Motile cells are commonly spherical, elongate, or ovoid, and round in cross-section. All contents are usually polarized, with the forward (anterior) end of the cell commonly the site of flagella insertion and the location of photoreception organelles and contractile vacuoles.[4643] This type is considered primitive among eukaryotic algae and is believed to have given rise to the other types.[4868]

The chloroplasts, when present, occupy the posterior region or lie along the sides, whereas the nucleus is frequently near the middle of the cell. A cell wall may surround the plasma membrane. Some flagellates have a periplast, and amphisema, or a pellicle of strengthening material inside the plasma membrane. Some flagellates settle on a substrate and exhibit creeping movement with the protrusion of pseudopodia or rhizopodia.[2745]

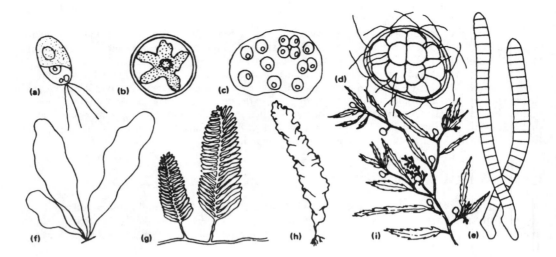

Figure 2-2. Types of algal plant body (diagrammatic). (a) Unicellular, motile. (b) Unicellular, nonmotile. (c) Colonial, noncoenobic. (d) Colonial, coenobic. (e) Filamentous. (f) Membranous or foliar. (g) Tubular, coenocytic. (h) Bladelike, kelp. (i) Leafy axis. (From H.C. Bold and M.J. Wynne, in *Introduction to the Algae,* 2e, © 1985, p. 8, reprinted with permission of Prentice Hall, Englewood Cliffs, New Jersey.)

Photosynthetic flagellates orient or distribute themselves in their environment in response to many different external chemical and physical factors.[1861,2475-6,3482] These responses are important in nature because they help to guide the cell population to its optimum position in the environment.[2476]

2.2.2 Colonial Organization

Colonial types (Figure 2-4) are commonly divided into aggregations (palmelloid, tetrasporal, and dendroid types) and coenobia.

2.2.2.1 Aggregations

This morphological type includes forms in which cells are aggregated into more or less irregular colonial-like masses showing vegetative cell division and not showing the regular differentiated construction of the coenobial forms. Vegetative cell division takes place so that there is an increase in cell number during growth.

Palmelloid form: in flagellated species, a temporary palmelloid stage may occur in which the cells lose their flagella and secrete extensive mucilage. A palmelloid condition is the normal, persistent condition in some species. Cells are surrounded by a sheath of mucilaginous polysaccharides, which may have an indefinite shape or a distinctive form. The cells, separated from each other in the mucilage, retain some of features of flagellated cells, such as basal bodies, eyespots, and contractile vacuoles.[4868] This is the most common type of aggregation.

Dendroid form: the dendroid colony is a variant in which mucilage is produced locally, generally at the base of the cell, resulting in cells with basal mucilage stalks.[2745] These colonies readily revert to the motile conditions.

Tetrasporal form: in most groups of algae, non-motile colonies are found in which the cells are embedded in mucilage; these are known as tetrasporal thalli. In tetrasporal types the motile stages (if present at all) are restricted to the reproductive cells.

Table 2-1. Summary of levels of organization in the algae.[4643] (From South and Whittick, 1987, in *Introduction to Phycology*, p. 93, reprinted with permission of Blackwell Scientific Publications, Oxford and authors.)

	Unicellular			Colonial		Filamentous				Pseudoparenchymatous			
	CD	RH	FL	PA-TE	C	SP	B	H	U	M	P	SC	S
Prokaryote													
Cyanophyceae	X			X		X	X				X		
Prochlorophyceae	X												
Eukaryote													
Rhodophyceae	X						X	X	X	X	X		
Chrysophyceae	X	X	X	X	X		X	X			X		
Prymnesiophyceae			X	X									
Xanthophyceae	X	X	X	X		X	X						X
Eustigmatophyceae	X												
Bacillariophyceae	X			X									
Dinophyceae	X	X	X	X			X						
Phaeophyceae							X	X	X	X	X		
Raphidophyceae			X	X									
Cryptophyceae	X		X	X									
Euglenophyceae			X	X									
Chlorophyceae	X	X	X	X	X	X	X	X	X	X	X	X	X
Charophyceae									X	X		X	
Prasinophyceae	X		X	X									

CD = coccoid, RH = rhizopodial, FL = flagellate, PA-TE = palmelloid or tetrasporal, C = coenobial, SP = simple, B = branched, H = heterotrichous, U = uniaxal, M = multiaxal, P = parenchymatous, SC = siphonocladous, S = siphonous.

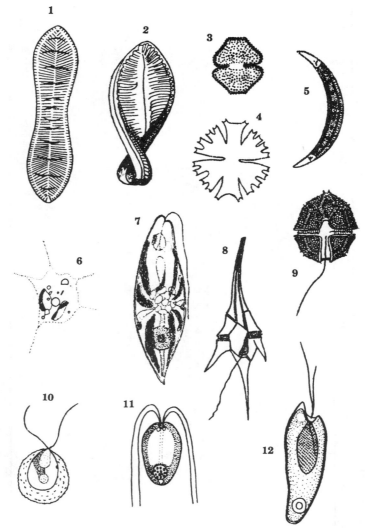

Figure 2-3.   Unicellular organization. <u>Coccoid:</u> 1-*Cymatopleura solea,* 2-*Surirella spiralis,* 3-*Cosmarium turpini,* 4-*Micrasterias crux-melitensis,* 5-*Closterium parvulum;* <u>Rhizopodial:</u> 6-*Rhizochloris stigmatica;* <u>Flagellate:</u> 7-*Euglena viridis,* 8-*Ceratium hirundinella,* 9-*Peridinium biceps* f. *tabulatum,* 10-*Phacotus lenticularis,* 11-*Pyramimonas tetraselmis,* 12-*Cryptomonas* sp. Not to scale.[2098,4559] (1–5 and 7–12 from Sládeček et al., 1973, in *Technical Hydrobiolgy: A Laboratory Manual,* SNTL Praha, reprinted with permission of V. Sládeček; 6 from Hindák et al., 1975, in *Determination Key for Lower Plants,* SPN Bratislava, reprinted with permission of F. Hindák.)

2.2.2.2 Coenobia

Coenobial forms possess no vegetative cell division. In this type of thallus the cells are either embedded in a mucilaginous matrix, or united by a more localized production of mucilage. It is not merely an irregular aggregation of cells but is a well-defined colony with important reproducible features. The coenobium is of constant size and shape for any given species, and the cells show no vegetative divisions.[3310] Thus, the number of cells of a coenobium is determined at its formation and does not increase during growth of the colony.[3310] Colonies range from those with relatively few cells (4–32 in *Gonium*) to those with more (up to 40,000 in *Volvox*).

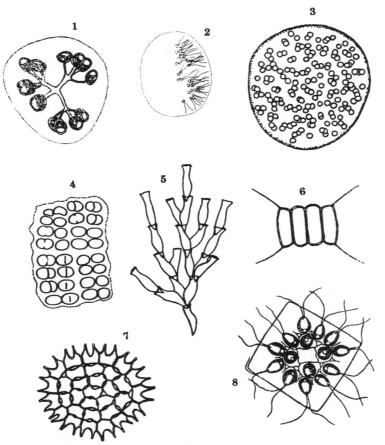

Figure 2-4. Colonial organization. <u>Aggregations</u>: <u>Palmelloid</u>: 1-*Dictyosphaerium pulchellum;* <u>Tetrasporal</u>: 2-*Tetraspora lemmermannii,* 3-*Microcystis grevillei,* 4-*Merismopedia elegans;* <u>Dendroid</u>: 5-*Dinobryon sertularia;* <u>Coenobia</u>: 6-*Scenedesmus quadricauda,* 7-*Pediastrum duplex,* 8-*Gonium pectorale.* Not to scale.[2098,4559] (1, 3–8 from Sládeček et al., 1973, in *Technical Hydrobiology: A Laboratory Manual,* SNTL Praha, reprinted with permission of V. Sládeček; 2 from Hindák et al., 1975, in *Determination Key for Lower Plants,* SPN Bratislava, reprinted with permission of F. Hindák.)

2.2.3 Filamentous Organization

Filamentous forms (Figure 2-5) are characterized by vegetative cell division but unlike the irregular aggregations the cells are arranged in linear rows.[3310] Filaments are either simple (unbranched) or branched.

In filaments, cells are arranged end-to-end with a common cross wall shared by adjacent cells. Cytoplasmic connections may extend through the cross walls.[4868] Filaments can be uniseriate (composed of a single row of cells) or multiseriate (composed of more than one row of cells).[2745]

Simple filaments consist of a single row of cells firmly attached to one another, and with no branching.[4643] It represents the most basic form of multicellular algal thallus.[1511] Simple filaments are found in only a few classes, in the Cyanophyceae the simple filament is the trichome.

There are three principal modifications of *branched filamentous thalli:*[4209,4633]

1. a branched upright filamentous thallus is attached by a simple disc derived from a basal cell,
2. the *heterotrichous* type with a basal attachment system of filaments giving rise to upright branches. Amongst the branched forms, a heterotrichous construction is one which shows a differentiation into a prostrate portion and an erect system,[3310]

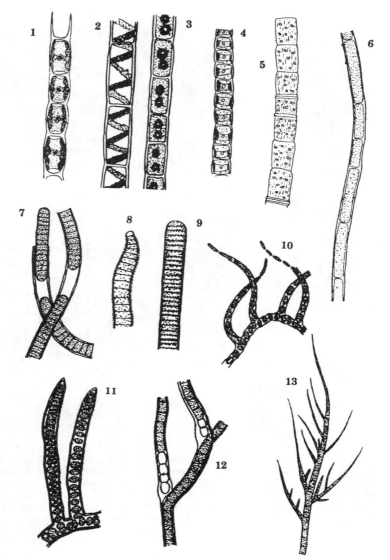

Figure 2-5.   Filamentous organization. <u>Simple</u>: 1-*Microspora amoena*, 2-*Spirogyra* sp., 3-*Zygnema* sp., 4-*Ulothrix zonata*, 5-*Oedogonium* sp., 6-*Rhizoclonium hieroglyphicum*, 7-*Lyngbya martensiana*, 8-*Phormidium favosum*, 9-*Oscillatoria limosa*; <u>Branched</u>: 10-*Mastigocladus laminosus*, 11-*Stigonema ocelatum*, 12-*Tolypothrix tenuis* f. *lanata*; <u>Heterotrichous</u>: 13-*Stigeoclonium flagelliferum*.   Not to scale.[4559,4564] (1–5, 13 from Sládečková et al., 1982; 6–12 from Sládeček et al., 1973, in *Technical Hydrobiology: A Laboratory Manual*, SNTL Praha, reprinted with permission of V. Sládeček.)

3. either single (uniaxal) or numerous (multiaxal) branched filaments become aggregated in a pseudoparenchymatous thallus.

2.2.4 Pseudoparenchymatous Organization

Pseudoparenchymatous thalli (Figure 2-6) develop through the branching of a single (*uniaxal*) or a number (*multiaxal*) of central filaments held together in a common matrix. Alternatively, the division of a filament in two or more planes results in a truly parenchymatous mode of growth.[4643]

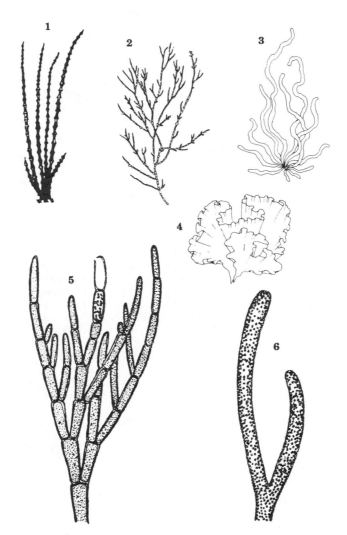

Figure 2-6. Pseudoparenchymatous organization. <u>Uniaxal</u>: 1-*Lemanea fluviatilis;* <u>Multiaxal</u>: 2-*Coral-lina officinalis;* <u>Parenchymatous</u>: 3-*Enteromorpha* sp., 4-*Ulva* sp.; <u>Siphonocladous</u>: 5-*Cladophora glomerata;* <u>Siphonous</u>: 6-*Vaucheria geminata.* Not to scale.[973,2095,4559] (1, 5–6 from Sládeček et al., 1973, in *Technical Hydrobiology: A Laboratory Manual,* SNTL Praha, reprinted with permission of V. Sládeček; 2 from Hillson, *Seaweeds,* University Park: The Pennsylvania State University Press, 1977, p. 139, Copyright 1977 by the Pennsylvania State University, reproduced with permission of the publisher; 3–4 from Darley, 1982, in *Algal Biology: A Physiological Approach,* drawings by Andrew J. Lamp-kin, III., reprinted with permission of Blackwell Scientific Publications and the author.)

Parenchymatous organization occurs when cells of the primary filament divide in all directions; any essentially filamentous structure is thus lost early on.[4643] Growth in the simpler examples is diffuse and may result in leafy thalli of one cell layer or two cell layers. In other simple types a tubular or initially saccate thallus is found. In these leafy or tubular examples there is little cell differentiation, except for the production of rhizoids from the lower cells to form a simple holdfast.[4643]

Growth of filamentous and parenchymatous thalli can be diffuse (all cells capable of division), intercalary (well-defined dividing regions not located terminally), trichothallic (a specialized intercalary meristem at the base of a terminal hair) or apical (one or more well-defined apical cells dividing to give remainder of the thallus).[3310]

Siphonocladous organization[5111] is restricted to members of the Chlorophyceae in which the unbranched or branched filaments are composed of multinucleate (semi-coenocytic) cells.

Siphonous (coenocytic) organization—enlargement and elaboration of the thallus occurs with multiplication of organelles but without septation. This structure is multinucleate and has a large number of chloroplasts. This organization ranges from saccate to uniaxal and multiaxal.[4643] Coenocytic thalli result from repeated nuclear division with relatively few (siphonocladous[5111]) or no (siphonous) cross-walls. Morris[3310] pointed out that it is considered more desirable to refer to such a thallus as acellular and not unicellular.

2.3 BIOLOGICAL FORMS

According to duration and habit of growth, the algae can be classified into the following life forms as suggested by Feldmann.[1328]

I. Annuals

    1. Ephemerophyta—forms found throughout the year, producing several generations in a year from a zygote or spore without a resting period (e.g. *Enteromorpha*).
    2. Echipsiophyta—plants found in obvious state during one part of the year only and existing as microscopic vegetative form during the other period (most Phaeophyceae).
    3. Hypnophyta—algae found in obvious form during one part of the year, existing as cysts or resting stages during the other period (e.g. *Ulothrix, Chara, Porphyra,* desmids).

II. Perennials

    1. Phanerophyta—plants in which complete frond is perennial and erect (*Fucus*)
    2. Chamaephyta—plants in which the entire plant is perennial but grows as a horizontal encrustment (*Lithophyllum*)
    3. Hemiphanerophyta—plants in which only a part of erect frond persists for several years (*Cystoseria*)
    4. Hemicryptophyta—plants in which only the basal portion persists throughout the year when other parts grow as a disc or creeping filaments (*Cladostephus*)

2.4 STRUCTURE OF ALGAL CELLS

There are two basic types of cells in the algae, *prokaryotic* and *eukaryotic*. Prokaryotic cells lack membrane-bounded organelles (plastids, mitochondria, nuclei, Golgi bodies, and flagella) and occur in Cyanophyceae and Prochlorophyceae.[2821,2745] Prokaryotic algae are rather divided into zones, with the nuclear material concentrated centrally (centroplasm) and the photosynthetic lamellae peripherally (the chromoplast).[4643] Apparently all Prokaryota lack steroids,[2785] which are a common feature of nucleate organisms. The remainder of the algae are eukaryotic and have the living material compartmentalized in a number of intracellular membrane-bounded organelles (Figure 2-7). The plasma membrane (plasmalemma) is a living structure responsible for controlling the influx and outflow of substances in the protoplasm. The endoplasmic reticulum comprises a further intracellular membrane system, or cisterna. Other cell inclusions include carbohydrate reserves (often associated with pyrenoids) and a variety of crystalline, lipid or other materials. Locomotion organs, the flagella, are enclosed in the plasma membrane.[2745] For further details on prokaryotic and eukaryotic features see Grant and Long,[1799,1800] Margulis and Schwartz[3068] or Taylor.[4914]

2.4.1 Cell Wall

While many flagellates and algal spores and gametes are described as "naked," the majority of algal cells are covered by one or more bounding layers of relatively inert material (principally carbohydrates), which may or may not be impregnated or layered with inorganic substances such as $CaCO_3$,

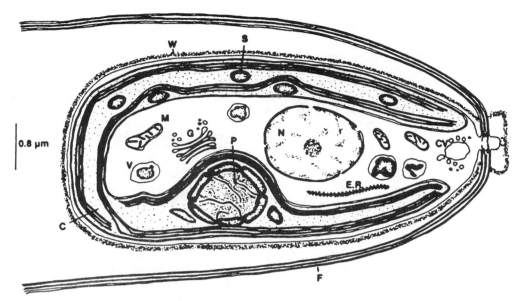

Figure 2-7.   A drawing of a cell of *Chlamydomonas* in the electron microscope showing the organelles present in an eukaryotic algal cell. (C) Chloroplast; (CV) contractile vacuole; (E.R.) endoplasmic reticulum; (F) flagella; (G) Golgi body; (M) mitochondrion; (N) nucleus; (P) pyrenoid; (S) strach; (V) vacuole; (W) wall.[2745] (From Lee, 1989, in *Phycology*, 2e, p. 4, reprinted with permission of Cambridge University Press and the author.)

$MgCO_3$ or silica (SiO_2).[4209,4643]   "Naked" is, in fact, a comparative term used to distinguish the organisms without cellulose or silica walls.[4209] Many of the walls are comparable with those found in higher plants, although in many flagellates, zoospores and gametes the enclosing membrane is merely the outermost layer of the cytoplasm (such as the pellicle, periplast and theca) may be distinctive for a particular group of algae.[4209,4643]

In general, algal cell walls are made up of two components: (1) the inner fibrillar, water-insoluble component, which forms the skeleton of the wall (e.g. cellulose, mannan, xylan), and (2) the amorphous component (e.g. agar, alginic acid, carrageenan, porphyran), soluble in boiling water, which forms a matrix within which fibrillar component is embedded.[2594,2745]

2.4.1.1 Fibrillar components

The most common type of fibrillar component is *cellulose* [β-(1→4)-linked glucan]. Cellulose I is the modification of cellulose which is found in the cell walls of higher plants.[2594] In the algae it was first observed in *Valonia*,[4659] and further observations indicated its presence in other algae. Cellulose II, or regenerated cellulose, is the crystalline modification formed when cellulose is precipitated from a solution; it has been identified in precipitates from neutralized euprammonium extracts of pretreated *Cladophora, Spirogyra, Vaucheria, Mougeotia*, and *Tribonema* filaments.[1492] By boiling in glycerol, cellulose II is converted into cellulose IV, a modification with an X-ray diagram much resembling that of cellulose I, though readily distinguishable therefrom.[2594]

However, a number of other structural compounds can be present, especially in the cell walls of coenocytic green algae, such as mannan or xylans of different polymers.[2745,3007,4172]

Mannan was first obtained from *Porphyra umbilicalis*.[2594] Chemical studies suggested that mannans are essentially linear and contain a predominance of ß-1,4 linked D-mannopyranose residues.[2287,2363,1919]

Some members of the Rhodophyceae and Chlorophyceae contain polymers (*xylans*) based on ß-D-xylose as major polysaccharide constituents. There are of different types and appear to vary in their proportions of 1,3 and 1,4 linkages and in the degree of branching.[3007]

2.4.1.2 Amorphous components

The amorphous mucilaginous components occur in greater amounts in the Phaeophyceae and Rhodophyceae. These compounds include alginic acid, fucoidin, and galactans (agar, carrageenan, porphyran, furcellaran, funoran). The amorphous polysaccharides of the Chlorophyceae are more complex.[2745]

Alginic acid, together with its salts (alginates), is one of the most extensively studied polysaccharide preparations.[3752] Alginates occur in both the intracellular regions and cell walls of the Phaeophyceae[1293] and it is considered that their biological functions are principally of a structural and ion-exchange type. Originally, it was believed that alginic acid was a polymer of ß-D-mannuronic acid only. Subsequently, L-guluronic acid was discovered in variable proportions in hydrolysates of several alginic acid samples.[1355] Evidence was obtained that both units are 1,4 linked[2103,4047] and to some extent, at least, occur in copolymer form, following the isolation of 4-O-ß-D-mannosyl-L-glucose from reduced and hydrolysed alginic acid.[2104]

In addition to alginic acid, water soluble extracts of brown algae usually contain polysaccharides characterized by the presence of sulfate ester and residues of L-fucose (2-deoxyl-L-mannose). Originally the name *fucoidin* (fucoidan) was coined for the fucose containing substances of *Fucus vesiculosus,* but this term tended to be applied to all fucose-rich polysaccharides.[3007] Fucoidin was first isolated by Kylin.[2671] Fucinic acid occurs in the form of its calcium salt, together with that of alginic from which it cannot be completely freed.[2672]

Agar is probably the most familiar of all the red algal polysaccharide preparations.[3007] It is atypical in having a very low sulfate content and for many years it was considered that agar was a mixture of two polysaccharides. Agar was first isolated by Payen[3709] from *Gelidium amansii.* The major component was called agarose, shown to be built up on the basis of a repeating unit of alternating 1,4 linked 3,6-anhydro-α-L-galactopyranose and 1,3 linked ß-D-galactopyranose.[3007] From further results involving fractionation of the agar from *Gelidium amansii* by ion exchange chromatography it has been suggested that these ideas are an oversimplification (for details see Mackie and Preston[3007]).

Carrageenan is the name given to the polysaccharide extracted from closely related members of the family Gigartinaceae, including *Chondrus crispus, Gigartina stellata, G. accicularis, G. pistillata,* and *G. radula.*[3498] Early studies with extractives from *Chondrus* and *G. stellata* indicated that it was a sulfated galactan, yielding 30–40% galactose and 20–30% sulfuric acid on hydrolysis.[3498] Dillon and O'Colla[1088-9] and Percival and Johnston[3751] showed that the galactose residues were 1,3'-linked, each bearing a sulfate group on C-4 and that an unstable hexose was present. Later, Smith and Cook[4583] found that it comprised two components: 40% κ-carrageenan, which can be precipitated from dilute aqueous solutions by KCl, and 60% λ-carrageenan, which remains in solution.

Porphyran is a polysaccharide which appears to be closely related to agarose in its basic structure but it has more variation in monosaccharide composition.[104]

The cell walls of the Cyanophyta and Prochlorophyta are more complicated and are similar to those of bacteria.[2745] Blue-green algae appear to have a typical gram-negative bacterial cell wall, which includes a variety of mucopolysaccharides, mucolipides and amino sugar polymers but few lipopolysaccharides.[1004]

When a wall is present, its chemical constituents vary from one group to another and are sometimes important indications of the taxonomic position of a particular alga.[3310] Further details concerning cell wall composition and function can be found in References 1004, 1118, 2594, 2745, 3007, 3498, 4172, 4209, 4483, 4643, 4959 (see also Chapter 2.14.5).

2.4.2 Plastids

Algal plastids, particularly their color, shape, size, number and distribution within the cells, have been used along with other important characters such as pigments and reserve products in the classification of algae.[378]

The basic type of plastid in the algae is a *chloroplast* (Figure 2-8), a plastid capable of photosynthesis. Chloroplast is the most conspicuous organelle in the algal cell—it occurs in a variety of patterns visible with the light microscope. The form of chloroplast is an important criterion in the class of the green algae. Chromoplast (or chromatophore) is synonymous with chloroplast. Urban and Kalina[5091] define chloroplasts as plastids of algae and higher plants containing chlorophyll-*b* and chromatophores as plastids of algae which do not contain chlorophyll-*b.* Lee[2745] defines chloroplast as a plastid with chlorophyll and chromoplast or chromatophore as a chloroplast with some other color than green.

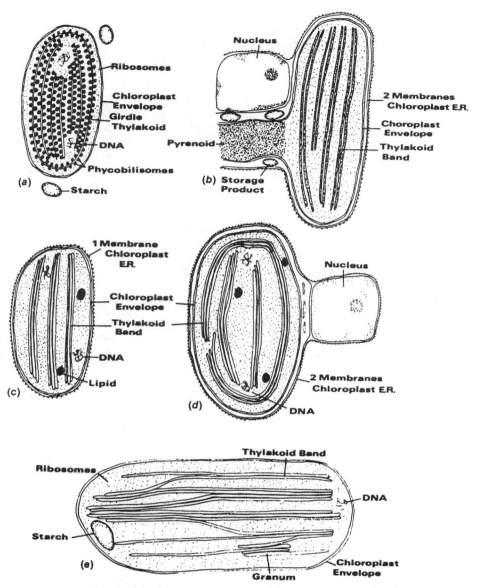

Figure 2-8.   Types of chloroplast structure in eukaryotic algae. (a) One thylakoid per band, no CER (Rhodophyceae). (b) Two thylakoids per band, two membranes of CER (Cryptophyceae). (c) Three thylakoids per band, one membrane of CER (Dinophyceae, Euglenophyceae). (d) Three thylakoids per band, two membranes of CER (Chrysophyceae, Prymnesiophyceae, Bacillariophyceae, Raphidophyceae, Xanthophyceae, Eustigmatophyceae, Phaeophyceae). (e) Two to six thylakoids per band, no CER (Chlorophyceae, Prasinophyceae, Charophyceae).[2745]   (From Lee, 1989, in *Phycology*, 2e, p. 12, reprinted with permission of Cambridge University Press and the author.)

In the Rhodophyta and Chlorophyta, the chloroplasts are bounded by the double membrane of the chloroplast envelope. In the other eukaryotic algae, the chloroplast envelope is surrounded by one or two membranes of *chloroplast endoplasmic reticulum* (CER) which has ribosomes attached to the outer face of the membrane adjacent to the cytoplasm. CER exhibits high enzymatic activity in connection with photosynthesis.[2745] The ground substance of all algal cells is traversed by endoplasmic reticulum. The interconnecting system does not penetrate the chloroplast of pyrenoid.[4209]

The basic structure of the photosynthetic apparatus in a plastid consists of a series of flattened membranous vesicles called *thylakoids* and a surrounding matrix or *stroma*. The thylakoids contain the chlorophylls and are the sites of the photochemical reactions; carbon dioxide fixation occurs in the stroma. In the Cyanophyta and Rhodophyta the thylakoids are usually free from one another (with phycobilisomes containing the phycobiliproteins on the surface of the thylakoids—see Chapter 2.4.3 on Pigments), in other divisions they are grouped in bands.[2745]

The arrangements of the thylakoids in the chloroplasts of various groups of eukaryotic algae (Figure 2-8) can be classified from electron microscopic investigations in five types:[378,2745] 1) no association between individual thylakoids, i.e. thylakoids not banded or stacked, but inter-connexions between neighboring thylakoids may occur, no grana**, CER absent, Rhodophyceae; 2) association of thylakoids into 2-thylakoid bands, no grana, CER present, Cryptophyceae; 3) association of thylakoids into 3-thylakoid bands, no grana, CER present (2 membranes), Xanthophyceae, Chrysophyceae, Bacillariophyceae, Raphidophyceae, Eustigmatophyceae, Prymnesiophyceae, Phaeophyceae; 4) association of thylakoids into 3-thylakoids bands, no grana, CER present (1 membrane), Euglenophyceae, Dinophyceae; and 5) bands of 2-6-many fused thylakoids, grana present, Chlorophyceae, Prasinophyceae, Charophyceae.

A *pyrenoid* is a differentiated region within the chloroplast. It may be embedded within the chloroplast or occur in the form of a projection from the chloroplast.[378] Pyrenoid is composed of polypeptides with enzymatic properties of ribulose-1,5-biphosphate carboxylase (RuBPC) that are capable of fixing carbon dioxide.[4256] Pyrenoids occur in every class of eukaryotic algae and within a class are considered to be a primitive evolutionary characteristic.[2745] They play a part in the synthesis and storage of polysaccharide reserve materials.[1004,2491] In the green algae the pyrenoid is one major site of starch formation, one or more starch grains forming within the chloroplast closely appressed to the surface of the pyrenoid. It has been suggested that the pyrenoid is the region of temporary storage for early products of photosynthesis that, upon overproduction, are converted into starch.[1820] Although in some algae it has been shown that all the starch of the chloroplast arises in anticipation with the pyrenoid, this has not been established for others. Furthermore, pyrenoids are entirely absent from some starch-forming algae. Further details on pyrenoids can be found in Gibbs[1629] or Griffiths.[1820]

Chloroplasts commonly contain small (30–100 nm), spherical lipid droplets between their thylakoids. These lipid droplets serve as pool of lipid reserve for the synthesis and growth of lipoprotein membranes within the chloroplast.[378] Most chloroplasts contain prokaryotic DNA in an area of the chloroplast devoid of 70S ribosomes. The algae can be divided into two general groups according to the distribution of DNA in the plastids:[836] (1) The clumps of DNA (nucleoids) are scattered throughout the plastids; and (2) the DNA occurs in a ring just within the girdle lamella. The Euglenophyceae fit into neither group, showing a variable distribution of chloroplast DNA.[2745]

Many motile algae have groups of tightly packed carotenoid lipid globules that constitute an orange-red *eyespot* or *stigma*. These globules shade the photoreceptor, an area of the plasma membrane or chloroplast envelope containing specialized molecules.[3194] A motile cell may swim toward (positive phototaxis) or away from light (negative phototaxis).[2745]

Bold and Wynne[424] summarized that there are two different suggestions regarding the role of the eyespot in light perception. According to one point of view, the eyespot is the site of light perception, controlling cellular movement as to direction and hence as to phototactic responses. According to the other interpretation, the real site of light perception is thought to be the flagellar swelling, the eyespot functioning as a shading organelle for it (see also Halldal[1881]). Halldal[1881] pointed out that it is certain that the stigma is not the photoreceptor in phototaxis, since both phobo-phototaxis and topo-phototaxis occur in algae lacking this organelle.[1947,2979]

Further information about eyespots can be found in References 308, 1080, 1104, 1107, 1881, 2747–8, 3481. General reviews of algal chloroplasts include those by Bisalputra,[378] Coombs and Greenwood,[875] Dodge,[1107] and Gibbs.[1629–31] Protein and nucleic acid synthesis in chloroplasts have been reviewed by Ellis[1233] and Whatley.[5360]

**Granum (grana) is broadly defined as a stack(s) of thylakoids within a chloroplast, such that the membranes of adjacent thylakoids are fused.[424]

2.4.3 Pigments

Photosynthesis in algae, as in all photosynthetic organisms, depends on the harvesting of light energy and its conversion into chemical energy, specifically into ATP and a reductant, NADPH. The first step in this process is the absorption of light by the photosynthetic pigments.[4643] There are three kinds of photosynthetic pigments in algae: *chlorophylls, carotenoids* and *phycobiliproteins (phycobilins)*.[2745,2820,4770] The algal pigments have been comprehensively reviewed in an excellent book by Rowan.[4219]

2.4.3.1 Chlorophylls

Chlorophylls are the basic pigments involved in light absorption and photochemistry in higher plants, algae, and photosynthetic bacteria.[3173] The photosynthetic algae have chlorophyll in their chloroplasts. Functional chlorophylls consist of four linked pyrrole rings with a Mg atom chelated at the center and a phytol tail. *In vivo* they are conjugated with proteins.[1004,4643,4965] The algae have four types of chlorophyll-*a*, *b*, *c* (c_1 and c_2), and *d* (Figure 2-9).

Chlorophyll-*a* is the primary photosynthetic pigment (the light receptor in photosystem I of the light reaction—see also Chapter 2.12.1) in all photosynthetic algae and ranges approximately from 0.3 to 3.0% of the dry mass.[2745,3944,4225] *In vitro* chlorophyll-*a* is soluble in alcohol, diethyl ether, benzene, and acetone and is insoluble in water.[3173] Whereas chlorophyll-*a* is found in all photosynthetic algae, the other algal chlorophylls have a more limited distribution and function as accessory pigments.[2745,3173]

Chlorophyll-*b* functions photosynthetically as a light-harvesting pigment transferring absorbed light energy to chlorophyll-*a* for primary photochemistry.[3173] Efficiency of energy transfer from chlorophyll-*b* to chlorophyll-*a* in algae is almost 100% both in room and liquid nitrogen temperature, and the efficiency of energy transfer from the various forms of chlorophyll-*a* to the reaction center is also very high.[787–8,1193,5002] Its solubility characteristics are similar to those of chlorophyll-*a*.[4598]

Chlorophyll-*c* probably functions as an accessory pigment to photosystem II and chlorophyll-*c* preparations contain two spectrally distinct components: c_1 and c_2.[2745,3173] The pigment is soluble in either acetone, methanol and ethyl acetate but insoluble in petroleum ether.[4598] Chlorophyll-*d* is a minor component in many Rhodophyceae—its photochemical function is unknown.[2745,3173,3516] It is soluble in ether, acetone, alcohol, and benzene, and very slightly soluble in petroleum ether.[4598]

The ratio of chlorophyll-*a* to chlorophyll-*b* varies from 3:1 to 2:1,[4297] the ratio of chlorophyll-*a* to chlorophyll-*c* ranges from 1.2:1 to 5.5:1.[2323] Mineral nutrition affects a number of growth and metabolic parameters including chlorophyll content of algae.[3173] Predictably, deficiencies of Fe, N, and Mg, essential constituents of haem and chlorophyll, have pronounced effects on chlorophyll synthesis and content.[2515,3529] Within certain limits, the chlorophyll content of numerous algae is inversely proportional to the light intensity during growth.[593,2515,4442]

Detailed information about algal chlorophylls can be found in References 412, 415, 1315, 2324, 3137, 3516, 4219, 4882, 5163.

2.4.3.2 Carotenoids

Carotenoids are yellow, orange or red pigments that usually occur inside the plastid but may be outside in certain cases.[2745] Carotenoids are isoprenoid polyene pigments present in many organisms, but can be synthetized *de novo* only by plants and photosynthetic bacteria.[4643] There are two classes of carotenoids: (1) linear unsaturated hydrocarbons, the *carotens* (Figure 2-10), and (2) their oxygenated derivatives, the *xanthophylls* (Figure 2-11).[2745,3310,4643] The carotens transmit mostly yellow light, xanthophylls transmit more in the orange or red wavelengths.[1004] β-carotene is present in most algae, other carotenes are present in some algae, and since many are unique to particular algal groups, they are important diagnostic features.[3310]

Both chlorophylls and carotenoids are soluble in alcohol, benzene or acetone, but insoluble in water.[2745,3376,4643] For further details on carotenoids the reader is referred to References 1738–40, 3046, 3376, 3955, 4948.

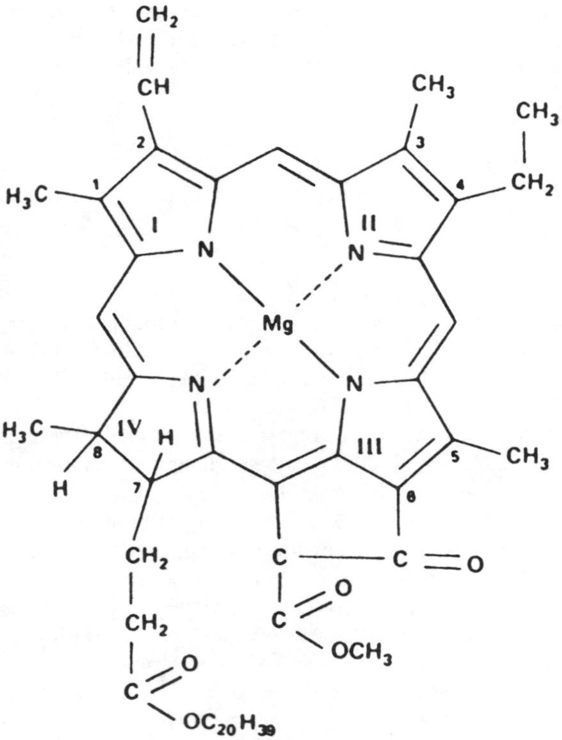

Figure 2-9.   Chemical structures of chlorophylls. Chlorophyll-*a*: as shown; Chlorophyll-*b*: II-3=CHO; Chlorophyll-c_1: IV-7=CH=CHCOOH; double bond at IV-7,8; Chlorophyll-c_2: IV-7=CH=CHCOOH; double bond at IV-7,8; II-4=CH=CH_2; Chlorophyll-*d*: 1-2=CHO.

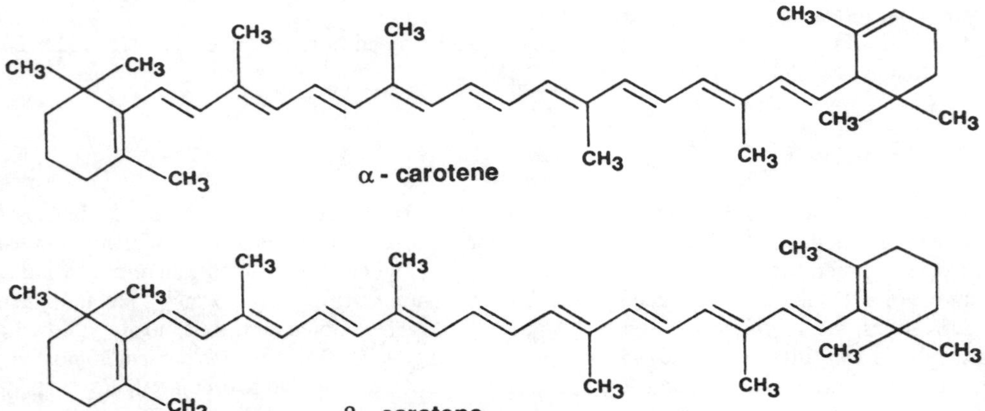

α - **carotene**

β - **carotene**

Figure 2-10.   Chemical structure of major carotens.

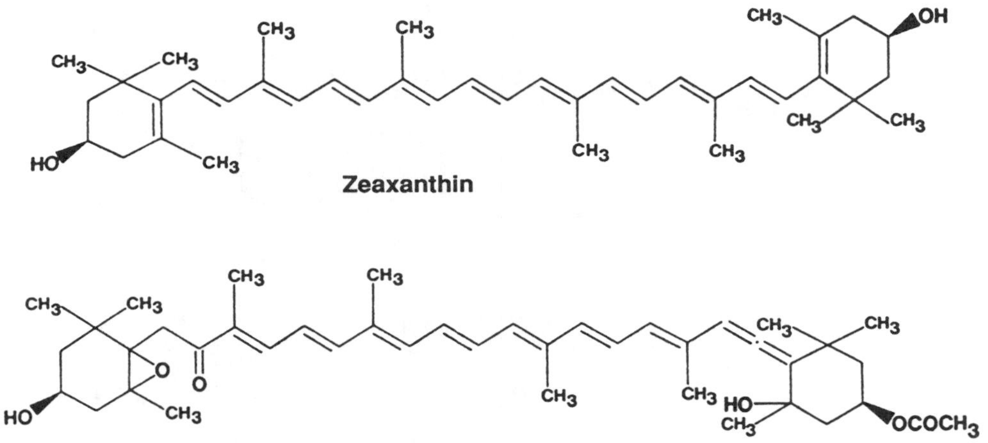

Figure 2-11.   Chemical structure of two xanthophylls.

2.4.3.3 Phycobiliproteins

Phycobiliproteins (phycobilins) are water-soluble red or blue pigments located on (Cyanophyceae, Rhodophyceae) or inside (Cryptophyceae) thylakoids of algal chloroplasts.[1647]

Like the chlorophylls, the phycobilins are linear tetrapyrols, but unlike chlorophylls they lack magnesium and a phytol tail.[1004] They are described as chromoproteins (colored proteins) in which the prosthetic group (nonprotein part of the molecule) or chromophore is a tetrapyrole (bile pigment) known as phycobilin. Because it is difficult to separate the pigment from the apoprotein (protein part of the molecule), the term phycobiliprotein is used (Lee[89]). There are two major types of phycobiliproteins (Figure 2-12): phycocyanin (blue) and phycoerythrin (red); phycocyanin transmits blue light, whereas phycoerythrin transmits red light.[1004,2745]

Phycobiliproteins by themselves do not interact with one another, they must be assembled into proper sequences by means of linker polypeptides. These polypeptides assemble the phycobiliproteins into *phycobilisomes* and attach them to thylakoid membranes.[5314] The linker proteins are basic whereas the phycobiliproteins are acidic; this suggests that electrostatic interactions are important in the assembling of phycobiliproteins.[2745]

The general classification of phycobiliproteins is based on their absorption spectra. Originally, the letters C-, B- and R-designated the source of the pigments, the Cyanophyceae, the Bangiales, and Rhodophyceae other than the Bangiales, respectively. Further investigations showed that this relationship was not entirely valid[3516] and, as stated above, the prefix now refers to pigments with different characteristic absorption spectra.[1740]

The accessory pigments apparently transfer the light energy they absorb to chlorophyll-*a*. Phycobiliproteins constitute an energy-transfer chain through which the incident light energy passes from phycoerythrin or phycoerythrocyanin to chlorophyll-*a* as follows:[1783,1801,2974,3728,4386,4884]

phycoerythrin or phycoerythrocyanin → phycocyanin → allophycocyanin → chlorophyll

In intact cells, the overall efficiency of energy transfer from the phycobilisome to chlorophyll-*a* in the thylakoids exceeds 90%.[3848]

The proportion of one kind of pigment to the other is variable giving a characteristic color to different algal groups. However, the proportion of one type of pigment to the other can vary considerably within changes in the environmental conditions, and it is difficult to justify its use as a taxonomic feature.[3310]

Phycobiliproteins are synthetized in response to low irradiance levels.[56,251,1563,2791,4184] In addition, as a major component of soluble cell protein,[416,831] phycobiliprotein can be an important source of internal nitrogen, acting as a N reserve in blue-green algae[72,5512] and red algae.[375,5151]

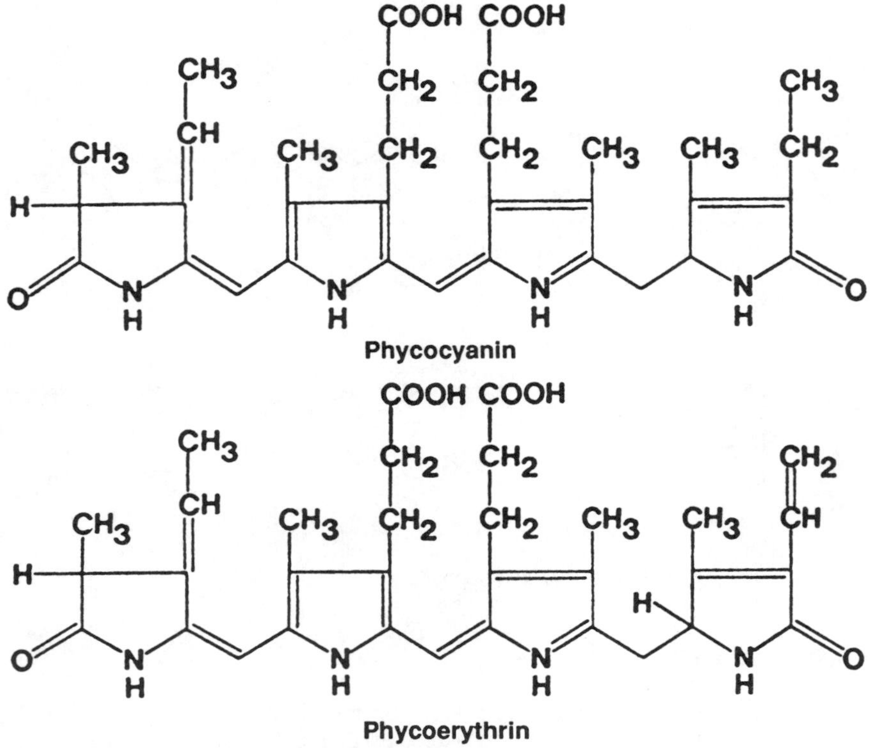

Figure 2-12.   Chemical structure of phycobiliproteins.

Literature on phycobiliprotein composition and phycobilisome structure and function is voluminous (see References 71, 415–16, 606–7, 1527–8, 1560, 1562, 1564, 1648–50, 1740, 3515, 3517–8, 3522–3, 3884, 4884, 5031).

2.4.4 Flagella

Flagella are the locomotory organs which propel the cell through the medium by their beating. The flagella are enclosed in the plasma membrane and have a specific number and orientation of microtubules.[2745] Apart from Cyanophyceae, Prochlorophyceae and Rhodophyceae, flagella are found in all other divisions of algae, and their nature, number and position are important characteristics for the primary classification of the algae.

Among eukaryotic cells, the internal flagellar structure is uniform and consists of two microtubules and 9 outer doublets or microtubules held within the cytoplasmic membrane.[3260] While flagella possess remarkably uniform internal microtubular structure, flagella are highly variable in external morphology, mode of insertion, and root systems.[4643] Algal motile cells can have different arrangements of flagella. If the flagella are of equal length, they are called *isokont flagella*; if they are of unequal length, they are called *anisokont flagella,* and if they form a ring at one end of the cell, they are called *stephanocont flagella. Heterokont* refers to an organism with a hairy and smooth flagellum.[3260]

The flagellar membrane may have no hairs (mastigonemes[***]) on its surface (whiplash or *acronematic flagellum*), or it may have hairs on its surface (tinsel, Flimmergeissel or *pantonematic flagellum*).[2745] Type of flagellum with a single row of mastigonemes or hairs is called *stichonematic flagellum.*[424] Flagellar number, insertion, and organization may be fairly consistent within a division; a summary of common arrangements is given in Figure 2-13.

[***]Mastigoneme is a fine, hairlike appendage of flagella.[424]

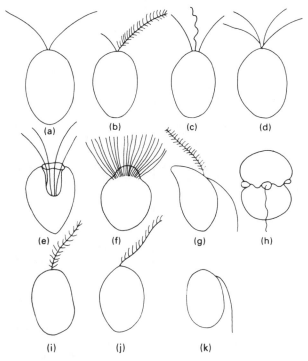

Figure 2-13. Selected flagellar types and arrangements in the algae. (a) Isokont, homodynamic, smooth, apically inserted (e.g. *Chlamydomonas,* Chlorophyceae). (b) Heterokont, unequal, apically inserted (e.g. *Tribonema,* Xanthophyceae). (c) Isokont, homodynamic, apically inserted, with a central haptonema (e.g. *Chrysochromulina,* Prymnesiophyceae). (d) Isokont, homodynamic, apically inserted. (e) Isokont (with scales, not shown), originating from a flagellar pit (e.g. *Pyramimonas,* Prasinophyceae). (f) Multiflagellate, or stephanokont (flagella inserted in a subapical ring) (e.g. *Oedogonium,* Chlorophyceae). (g) Heterokont, heterodynamic, laterally inserted (e.g. *Ectocarpus,* Phaeophyceae). (h) Transverse and trailing, heterokont flagella (hairs not shown), basally inserted (e.g. *Gymnodinium,* Dinophyceae). (i) Single, apical, bilaterally hairy. (j) Single apical, unilaterally hairy. (k) Single, subapically inserted.[4643] (From South and Whittick, 1987, in *Introduction to Phycology,* p. 65, reprinted with permission of Blackwell Scientific Publications, Oxford, and the authors.)

A comprehensive review of algal flagella is given by Moestrup[3259-60] Melkonian,[3188-90] and South and Whittick.[4643] Further information can be found in References 91, 274, 571, 1292, 2190, 2774, 3069, 3192, 3454, 4754.

2.4.5 Golgi Bodies (Dictyosomes)

These are present in all algal cells except the Cyanophyceae and Prochlorophyceae and are fairly easily recognizable in sections under the electron microscope. They may be found in the region of the nucleus (e.g. in *Chlamydomonas*) or associated with the flagellar bases (e.g. in *Chrysochromulina* and *Pedinomonas*), and are composed of stacks of flat vesicles (cisternal elements), the ends of which may be dilated (Round 1973). The Golgi apparatus has a key position in the complex membrane-bound transport system within the cell.[5357] The number of Golgi bodies per cell varies from only one[3048] to ca. 25,000 in the rhizoid apex of *Chara*.[3272]

As well as serving as part of an internal transport system in the cell, Golgi complex is involved in the formation and packaging of substances for extracellular transport (for details see Evans,[1292] for further information see also Chapter 5.1.4). There is also evidence of Golgi involvement in polysaccharide synthesis in larger algae. It seems that materials such as polysaccharides are both synthetized and modified within the Golgi cisternae and vesicles.[1292]

2.4.6 Mitochondria

The mitochondria in algal cells can be one of two types. Mitochondria with flat lamellae cristae associated with those algae that have either phycobiliproteins or chlorophylls-*a* and *b* together.[4756] Mitochondria with tubular cristae occur in groups which never have phycobiliproteins or chlorophylls-*a* and *b* together. All of the photosynthetic groups with tubular cristae have CER.[2745] Mitochondria range in length from less than 1 μm to several μm, are sometimes branched and may be of variable shape. There may be only one mitochondrion per cell, but generally there are more.[1292]

Round[4209] pointed out that mitochondria certainly function as sites of enzyme action in glycolysis, amino acid interconversion and protein synthesis. Mitochondria are functionally connected with respiration.[5091] Cells of the Cyanophyceae do not possess mitochondria.

2.4.7 Nucleus

The nucleus is the principal store of genetic information (the genome) in the cell, and the controlling center for selective expression of the stored information (additional genetic material is contained in the chloroplast and in the mitochondria). An exception occurs in the Cyanophyceae.[1638,2491] There are two basic types of nuclei in the eukaryotic algae and it is possible to divide the algae into two types in regard to nuclear division:[2744] (1) that occurring in the Dinophyceae and Euglenophyceae, and (2) that occurring in the rest of the eukaryotic algae.

The nuclear envelope is frequently in continuity with endoplasmic reticulum, so that there may be free passage from the perinuclear space into the lumen of the endoplasmic reticulum. In some algae, such as members of the Phaeophyceae, the nuclear envelope is characteristically continuous with the envelope of endoplasmic reticulum that surrounds the chloroplast.[479,1291]

Electron micrographs of sections of algae reveal a nuclear membrane consisting of two layers separated by a narrow space and perforated by pores. There is evidence that the nucleus is separated from the cytoplasm, not only by its double membrane, but also by a perinuclear space, which may be bridged at intervals.[1327,4209] The outer membrane mostly bears dense particles, while the inner membrane is smooth.[2491]

Nuclear division (mitotic and meiotic) is almost similar to higher plants.[2491] Mitosis proceeds in many algae as in other plants but some deviations occur; meiosis—few detailed studies have been made, but the data suggest that the process is very similar to that in other organisms.[4209]

2.4.8 Ribosomes

Ribosomes fall into two general classes, based on their sedimentation coefficients in an ultracentrifuge (expressed in Svedberg units or S): (1) the smaller "70S" prokaryotic type of ribosome found in bacteria, Cyanophyceae, Prochlorophyceae, chloroplasts, and mitochondria; and (2) the "80S" eukaryotic type of ribosome found in the cytoplasm of eukaryotic cells outside the chloroplasts and mitochondria.[2745]

2.4.9 Contractile Vacuoles

The ability of algal cells to adjust to changes in the salinity of the medium is an important aspect of the physiology of these cells. In cells with walls, this osmoregulation is accomplished with the aid of turgor pressure, whereas in naked cells it is accomplished by means of contractile vacuoles and/or regulation of the solutes present in the cell.[2745] Contractile vacuoles range in number from one, two, four to many. They lie near the cell surface, often in close proximity to the flagellar bases.[55] A contractile vacuole fills with an aqueous solution (diastole) and then expels the solution outside the cell and contract (systole). The contractile vacuole rhythmically repeats this procedure. If there are two contractile vacuoles, they usually fill and empty alternately. In freshwater algae the contractile vacuole cycle lasts for 4–16 seconds, whereas in marine species the cycle can last for up to 40 seconds.[2745]

2.4.10 Food Storage Products

The initial stages in carbon dioxide fixation are probably the same in all photosynthetic organisms. Thus, the primary products of photosynthesis are the same in all algae. However, the insoluble products which accumulate over a longer period of time are more variable and they afford useful taxonomic criteria. The compounds which are most widespread and most useful in the primary classification of algae are various polysaccharides.[3176] Reviews on storage products include those by Craigie,[906] Frederick,[1477-8] Meeuse,[3176] Percival,[3749] or Percival and McDowell.[3752]

The storage products occurring in algae can be divided broadly as high-molecular weight and low-molecular weight compounds.[2745]

2.4.10.1 High-molecular weight compounds

High-molecular weight compounds are (1) α-1→4 linked glucans (floridean starch, myxophycean starch, "true starch"), (2) β-1→3 linked glucans (laminarin, paramylon, chrysolaminarin), and (3) fructosans and inulin.[2745]

Floridean starch, discovered by Kützing[2648] in 1843, forms the typical photosynthetic product of certain Rhodophyceae. It occurs in the form of layered granules which do not seem to be associated with plastids, and which vary considerably in shape and size, from 0.5 to 25 μm, even within a single alga.[906,3178] Chemical characterization of this starch is given by Percival,[3749] Percival and McDowell,[3752] or Peat and Turey.[3721] Floridean starch is essentially identical with the branched or amylopectin fraction of higher plant starches.[3177-8,3304,4400]

Myxophycean starch. Hough et al.[2212] were the first who succeeded in isolating from an *Oscillatoria* sp. a purified polyglucose with all characteristics of amylopectin.[3776] Within the cell it is not found in the form of clearly discernible granules, but as "undissolved submicroscopic crystals."[2675-6] The morphology of the "granules" varies with the algal source.[747,1634,4151]

True starch is composed of amylose and amylopectin.[3006,3043-4]

Laminarin was first described by Schmiedeberg[4351] in 1885, and has been shown to be ubiquitous in the Phaeophyceae.[3941] It consists of a related group of predominantly β-1→3 linked glucans containing 16 to 31 residues[906] and occurs in two forms. Soluble laminarin dissolves as readily in cold water as in hot water, but the so-called "insoluble" laminarin dissolves in water only when heated.[3176] Laminarin occurs as an oil-like liquid.[2745]

Paramylon, the reserve material typical of the Euglenophyceae, was first described by O.F. Müller.[3338] Paramylon was shown to be a β-1→3 linked glucan by Kreger and Meeuse[2595] in 1952. It occurs as water insoluble, single membrane bound cellular inclusions of variable shapes and dimensions.[906,2745] Kreger and Van der Veer[2596] reported paramylon also in a chrysophyte *Pavlova mesolychnon.*

Chrysolaminarin (leucosin, chrysose) consists of β-1→3 linked D-glucose residues with two 1→6 glycosidic bonds per molecule. Leucosin has more glucose residues per molecule than laminarin.[2745]

Fructosans and *inulin.* Endo[1244] and Kylin[2677] identified free fructose in *Cladophora* (Chlorophyceae) species. The first to notice sphaerocrystals of the inulin type in *Acetabularia* (Chlorophyceae) was Nägeli,[3369] the first to recognize their inulin-like nature was Leitgeb.[2767]

2.4.10.2 Low-molecular weight compounds

Low-molecular weight compounds are (1) sugars (sucrose, trehalose, maltose), (2) glycosides (glycerol glycosides, floridoside, isofloridoside), (3) polyols (mannitol, glycerol, erythriol, ribitol, sorbitol), and (4) cyclitols.[906,2745,2840]

Sucrose mostly occurs in the Chlorophyceae, but it is also present in species from some other algal classes.[238,359,364,906]

Trehalose is the next most frequently identified algal disaccharide (1,1-α-glucosyl-α-glucose). It is found in the Cyanophyceae, Rhodophyceae and Euglenophyceae.[826,906,909]

Maltose occurs in several unrelated species and occasionally may be an important photosynthetic product of some green algae.[2379,3355]

Glycerol glycosides, floridoside (2-O-glycerol-α-D-galactoside[842-4,3938]) and *isofloridoside* (1-O-glycerol-α-D-galactoside[2857]) are widely distributed in the Rhodophyceae, except of Rhodomelaceae where they are largely replaced by Na-monoglycerate and mannitol.[906] Glycosides are also found in the Cryptophyceae and Chrysophyceae.

Mannitol is the most common polyol reported and is conspicuous by its almost universal presence in the Phaeophyceae, and in the Prasinophyceae, where it replaces sucrose as a photosynthetic product.[906] *Free glycerol* occurs fairly widely in algae and is an important photosynthetic product in several zooxanthellae[3357,5020] and in some marine algal species.[907-8] *Sorbitol* is reported from some blue-green algae and sorbitol; *ribitol* and *erythriol* occur in green phycobionts.[906]

Cyclitols (e.g. nyo-Inositol, mytilitol) are known to be present in representatives of five algal classes.[906]

2.5 REPRODUCTION

Although it is convenient to designate reproduction as *asexual* or *sexual,* many life cycles obligately combine both processes. In many unicellular species, cell division, or *binary fission,* is the only mechanism for asexual reproduction. This can be extended to multicellular forms, which may reproduce either by *fragmentation,* or by the producing of specialized multicellular structures termed gemmae (e.g. freshwater red alga *Hildebrandia rivularis*[3444]) or propagules (in brown algae[3163]). These processes are often termed *vegetative reproduction* with *asexual reproduction* being reserved for events which include the formation of specialized unicellular spores.[424]

2.5.1 Vegetative Reproduction

Noncoenobic colonial, filamentous and other types of multicellular algae reproduce by various types of fragmentation (Figure 2-14i), the fragments having the capacity through continuing growth of developing into new individuals. This is sometimes called vegetative reproduction[1511,2491] because the reproductive agent involved is part of the vegetative or somatic plant body of the organism.[424]

Vegetative reproduction through fragmentation is widely reported from all algal divisions in both freshwater and marine habitats and is especially common in filamentous species.[4643] It appears to have an important role in maintaining populations in habitats such as estuaries[3438] or saltmarshes[3466] or at the extremes of geographical ranges.[1096,5397] Bold and Wynne[424] pointed out that it should be noted that fragmentation is not a method of reproduction in coenobic algae. Instead, these undergo *aucolony* formation (Figure 2-14j). An autocolony is a miniature colony produced by a parental colony that it resembles.

In some blue-green algae, filaments break up into motile segments of varying length called *hormogonia* (Figure 2-14i).[3435] *Akinetes* (Figure 2-14k) are vegetative cells that have developed into a sporelike stage with very thick walls and abundant food reserves. Akinetes are resistant to unfavorable environmental conditions.

In some unicellular algae the organism reproduces by cell division. These divisions may be repeated in rapid succession, designated repeated bipartition, to form new individuals like the parent cell. This process is also sometimes called *binary fission* (Figure 2-14a,b,c).[424] In noncoenobic colonial and other multicellulate types of algae, cell division and subsequent enlargement result in growth.

2.5.2 Asexual Reproduction

Asexual reproduction is normally achieved by the formation of spores of various kinds. Spores are cells that germinate without fusing to form new individuals. Asexual reproduction is known from the Chlorophyceae, Prasinophyceae, Rhodophyceae, Xanthophyceae, Eustigmatophyceae, Chrysophyceae, Dinophyceae, Prymnesiophyceae and Phaeophyceae.[4643]

Most groups (except Cyanophyceae and Rhodophyceae) produce *zoospores* (Figure 2-14d,e), which are motile naked unicells. They can be formed in ordinary vegetative cells or in specialized cells called *sporangia.* Zoospores are formed by zoosporogenesis and many of them have eyespots and/or photoreceptors and show a phototactic response.

Nonmotile asexual spores are *aplanospores* (Figure 2-14f) and possess a distinct wall. When they appear identical to the parent cell they are referred to as *autospores* (Figure 2-14g,h) and if they acquire

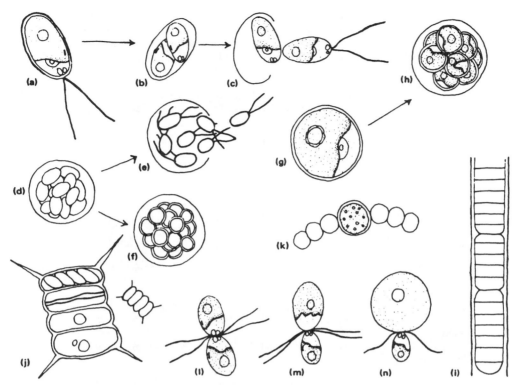

Figure 2-14. Methods of algal reproduction (diagrammatic). (a)-(c) Bipartition or binary fission. (d), (e) Zoospore formation. (f) Aplanospore formation. (g), (h) Autospore formation. (i) Fragmentation or hormogonium formation. (j) Autocolony formation. (k) Akinete formation. (l) Isogamy. (m) Anisogamy. (n) Oogamy.[424] (From H.C. Bold and M.J. Wynne, in *Introduction to the Algae*, 2e, 1985, p. 10, reprinted with permission of Prentice Hall, Englewood Cliffs, New Jersey.)

a thick wall, they are referred to as *hypnospores* (suitable for dormancy). In some unicellular algae the division products (aplanospores and autospores) remain associated for a period within the persistent and enlarging parent cell wall (this phenomena is interpreted as incipient colony formation).[424]

2.5.3 Sexual Reproduction

Sexual reproduction is achieved by fusion of *gametes* to yield a *zygote*. Sexual reproduction generally comes at the end of the growing season, or may be induced by unfavorable or critical changes in the environmental conditions such as nutrition supply, pH of water, light, temperature or oxygen. The great majority of algae produce gametes either from unspecialized vegetative cells or specialized reproductive cells (gametangium: male—*antheridium,* female—*oogonium,* carpogonium in Rhodophyceae).

There are several types of reproduction, depending on the structure and function of the gametes. *Isogamy* (Figure 2-14l) involves the fusion of two morphologically and physiologically similar (identical) gametes. Because the two gametes that fuse to form a zygote look and behave the same way, it is not possible to call them male and female, and therefore the gametes are referred to as plus (+) and minus (-). *Anisogamy* (Figure 2-14m) is fusion between structurally and/or morphologically dissimilar gametes. The larger gamete is the female, and the smaller is the male. *Oogamy* (Figure 2-14n) differs from anisogamy in that large non-motile female gamete (egg or ovum) is not liberated prior to fertilization by a smaller and motile (with the exceptions of Rhodophyta, where spermatia are non-motile) sperm (antherozoid), but is fertilized while still within the oogonium.

The zygote formed by all the above methods of sexual reproduction has an independent existence for a variable length of time. After fertilization some zygotes will accumulate food reserves and yellowish-red oil, and form a thick wall. Then called *zygospores* or *oospores,* they are able to withstand prolonged desiccation, and germinate under favorable conditions. A number of algae produce biflagellate cells which may function either as zoospores or gametes. In a number of other algae in which the gametes differ morphologically from zoospores (e.g. *Ulva, Cladophora*), the gametes may develop without union—that is parthenogenetically—into new individuals.[55]

Sexual reproduction is absent or uncertain in the Prochlorophyceae, Eustigmatophyceae, Raphidophyceae, Cryptophyceae, and Euglenophyceae. It does not occur in the Cyanophyceae, although gene transfer between individuals is well documented.[4643]

2.6 LIFE HISTORIES

Five basic types of life history occur in the algae:[2745]

1. *Haploid*—meiosis is zygotic when it germinates. The zygote is the only diploid part of the life history.
2. *Diploid*—occurs before the formation of gametes. The gametes are the only haploid part of the life cycle and fuse to form the diploid zygote.
3. *Isomorphic* (homologous)—alternation of generations, consisting of the alternation of haploid (gametophytic) plants bearing gametes with structurally identical diploid (sporophytic) plants bearing spores.
4. *Heteromorphic* (anithetic)—alternations of generations, consisting of the alternations of small haploid plants bearing gametes with large diploid plants bearing spores or of large haploid plants alternating with smaller diploid plants.
5. *Triphasic*—life cycle in the red algae, consisting of a haploid gametophyte, a diploid carposporophyte, and a diploid tetrasporophyte.

Detailed studies on life histories of algae have been presented by Lee[2745] and South and Whittick.[4643]

2.7 ECOLOGY

Algae can be broadly classified as: 1) aquatic, 2) aerophytic, 3) soil, 4) thermal, and 5) ice and snow.[1460]

2.7.1 Aquatic Algae

The aquatic environment comprises some 70% of the earth's surface and here the algae are important as primary producers of oxygen and elaborate organic materials, and thus play a critical role in the economy of the seas and freshwaters and the whole environment as well.[4209]

Three general schemes for classification of aquatic microorganisms can be found in the literature.[5322] One scheme, the most recent, is a two component system, as follows:[2260]

- plankton: organisms that swim or float in the water
- benthos: organisms that grow on the bottom of the water

The second and older system makes a distinction within the benthos component:[873,2260,4560]

- plankton: organisms that swim or float in the water
- periphyton: all aquatic organisms that grow on submerged substrates
- benthos: organisms that grow on the bottom of the water

The third scheme, dating around 1918, is also a three-component system:[2258]

- planomenon: plankton, "euplankton" or drifting organisms
- ephaptomenon: organisms attached to a submerged substrate
- rhizomenon: organisms rooted in a solid substrate

2.7.1.1 Phytoplankton

The first use of the term "plankton" is widely attributed to the German biologist Viktor Hensen,[2258] who in the second half of the 19th century, began a series of expeditions to gauge the distribution, abundance and composition of microscopic organisms in the open ocean. According to Hensen's[2037] usage "plankton" included all organic particles "which flow freely and involuntarily in open waters, independent of shores and bottom."

Hensen's "plankton" did not specifically exclude non-living particles, so it is therefore synonymous with "seston" in Kolkwitz's[2557] later terminology, that continues to command common acceptance.[4078] Seston[2557] thus applies to all particulate matter maintained in the pelagic zone: *abioseston*, or *tripton*,[4210] is the non-living fraction (organic + inorganic detritus), and *bioseston* (plankton + neuston + pleuston + nekton) comprises only discrete, living organisms.

Neuston is a term coined by Naumann[3396] for algae living at the interface of water and the atmosphere. In fact, it comprises two groups of organisms, one living on the upper surface of the water film—epineuston, and one attached to the under surface—hyponeuston.[1596,4210] Fott[1460] pointed out that it develops only in perfectly still water, the surface of which is protected from wind. *Pleuston* are big organisms flowing on the water surface and *nekton* are big organisms with movement.[1460]

The phytoplankton may be subdivided. The *euplankton* is the permanent community of the open water, and is by far the most important of the floating communities. Accidental plankters, caught up in water currents or washed into the habitat are termed *pseudoplankton (tychoplankton)*.[4210]

There are two more serious criticisms of Hensen's definition of what we now understand to be "plankton"; but it has taken nearly a century of subsequent investigations to resolve these.[4078]

One is simply that, in general, plankton does not float: there are few planktonic organisms which are consistently buoyant; on the contrary, most are often or always more dense than the water they inhabit.[4078] The specific adaptations of planktonic organisms for pelagic life seem largely directed towards prolonged maintenance in suspension. The second point is that many planktonic organisms are not exclusively confined to the pelagic zone but may spend part, or even most, of their life cycle on the sediments or in other (littoral) habitats. Put another way, many organisms present in open waters are only facultatively planktonic (or meroplanktonic).[4078]

For these reasons, it is perhaps more useful to regard *plankton* as "the community" of plants and animals adapted to suspension in the sea or in fresh waters and which is liable to passive movement by wind and current.[4078] Planktonic organisms are suspended in the water column and lack the means to maintain their position against the current flow, although many of them are capable of limited, local movement with the water mass.[973] The plants and animals are conveniently segregated in the terms phytoplankton and zooplankton, respectively, notwithstanding differences in opinion about where the dividing line is drawn.[4078]

The phytoplankton includes representatives of several groups of algae and bacteria, as well as the infective stages of certain actinomycetes and fungi. Of these, the most conspicuous are undoubtedly the algae, but neither the biomass of bacteria nor the importance of their contribution to the functioning of aquatic ecosystems should be underestimated.[4078]

Phytoplankton encompasses a surprising range of cell size and cell volume from the largest forms visible to the naked eye, e.g. *Volvox* (500–1500 μm) in the freshwater and *Coscinodiscus* species in the ocean to the algae as small as 1 μm in diameter.[4210] Phytoplankton algae are mainly unicellular, though many colonial and filamentous forms occur, especially in fresh waters.[4643] Some examples of phytoplankton algae are given in Figure 2-15.

Phytoplankton occur in virtually all bodies of water. All algal groups except the Rhodophyceae, Charophyceae and Phaeophyceae contribute species to the phytoplankton flora, and in coastal zones the zoospores produced from benthic phaeophytes and chlorophytes may at least temporarily add to the plankton.[4643] The most abundant of net plankton in both fresh and marine waters are diatoms and dinoflagellates; centric diatoms outnumber pennate species in the sea, whereas both are common in lakes. The net phytoplankton in fresh waters will also commonly include colonial green algae, desmids, filamentous and colonial blue-green algae and colonial chrysophytes.[973]

The nanoplankton consists of unicellular blue-green algae, and small, often motile unicells belonging to the Chlorophyceae, Chrysophyceae, Cryptophyceae, and Prymnesiophyceae. Small diatoms (Bacillariophyceae) and dinoflagellates (Dinophyceae) are also included in the nanoplankton.[973]

Since 72% of the earth is covered by ocean, phytoplankton is the most important group of primary producers on earth.[1004] The phytoplankton stands on the base line of food webs in aquatic environments and is in turn dependent on the activities of other microbial organisms, mainly bacteria, which convert

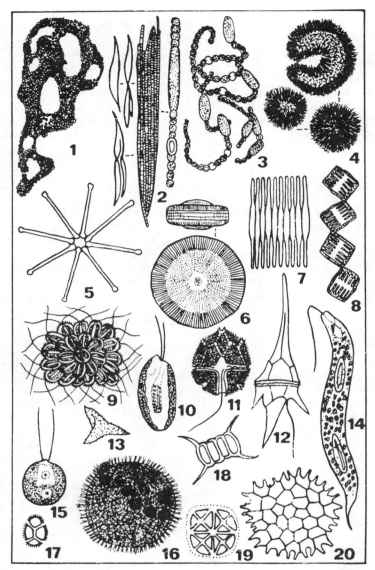

Figure 2-15.   Examples of freshwater phytoplankton. 1-*Microcystis aeruginosa*, 2-*Aphanizonemon flos-aquae*, 3-*Anabaena flos-aquae*, 4-*Gloeotrichia echinulata*, 5-*Asterionella formosa*, 6-*Cyclotella bodanica*, 7-*Fragilaria crotonensis*, 8-*Tabellaria flocculosa*, 9-*Synura peter-senii*, 10-*Cryptomonas ovata*, 11-*Peridinium biceps* f. *tabulatum*, 12-*Ceratium hirundi-nella*, 13-*Goniochloris cochleata*, 14-*Euglena spirogyra*, 15-*Chlamydomonas simplex*, 16-*Volvox aureus*, 17-*Tetrastrum staurogeniaeforme*, 18-*Scenedesmus quadricauda*, 19-*Crucigenia tetrapedia*, 20-*Pediastrum boryanum*. 1–4: Cyanophyceae, 5–8: Bacillario-phyceae, 9: Chrysophyceae (Synurophyceae), 10: Cryptophyceae, 11–12: Dinophyceae, 13: Xanthophyceae, 14: Euglenophyceae, 15–20: Chlorophyceae. Not to scale.[4564] (From Sládečková et al., 1982.)

organic material into the inorganic nutrients required by plants.[437] Phytoplankton is grazed mainly by zooplankton, a group of organisms which is dominated by rotifers and crustaceans, especially cladocer-ans and copepods.[973] (For details see Chapter 2.13.1.3.)

Various authors have introduced prefixes to categorize the phytoplankton according to the individual sizes of the organisms: the terms nannoplankton,[4168] nanoplankton,[5522] ultraplankton,[5343] picoplankton[2745] and μ-algae[1424] have been used to separate the lower size ranges of individuals from the larger ("netplankton") forms. Few workers, however, continue to physically separate the larger algae from quantitative water samples with the result that there is no longer any agreed differentiation between "netplankton" and "nanoplankton." There is some looseness in the utilization of these terms by different authors (Table 2-2).

Reynolds[4078] pointed out that another set of terms relating to size is sometimes encountered. "Microplankton" corresponds roughly with nanoplankton, "mesoplankton" with the netplankton, whilst the meaning of macroplankton (occasionally "megaloplankton" or "megaplankton") embraces those aquatic angiosperms and pteridophytes that float freely on water surfaces (e.g. *Lemna, Wolffia, Hydrocharis, Azolla*). Typical cell concentrations of total phytoplankton range from less than 10 cells mL^{-1} in oligotrophic (nutrient-poor) waters to more than 10,000 cells mL^{-1} during spring blooms and over 50,000 cells mL^{-1} in red tides.[973]

A major environmental factor to which phytoplankton must adapt is the influence of gravity. The adaptations generally tend to increase the buoyancy of the organism. Planktonic species often have a relatively slow rate of sinking or more rarely some buoyancy mechanism. Most cytoplasmic components have densities higher than water (nominally 1.0008 g mL^{-1} for fresh water and 1.027 g mL^{-1} for sea water). The silica walls of diatoms, one of the most successful phytoplankton groups, have a density of 2.6. Only lipids at 0.86 are less dense than water and, while they may contribute to the reduction of cell density, cannot produce neutral buoyancy.[4643] Two factors which increase the buoyancy of algae are the presence of gas vacuoles (e.g. in the Cyanophyceae) and oil globules (e.g. in *Botryococcus,* diatoms and dinoflagellates).[4209] The development of bristles and spines, particularly those that are hollow, also increase the buoyancy.[2491] It is doubtful whether form has much influence on flotation since the range is extremely great; it varies from coccoid forms, through plates (e.g. *Pediastrum*, Figure 2-57, #11), spheres (e.g. *Sphaerocystis*), needles (e.g. *Synedra acus*, Figure 2-49, #24), radiating needles (e.g. *Asterionella*, Figure 2-50, #1 or *Tabellaria*, Figure 2-49, #9, 10) to filaments (e.g. *Melosira*, Figure 2-49, #1–3 *Anabaena*, Figure 2-41, #9, and 12). Long, narrow cells and disc-shaped forms are the two types commonly found in phytoplankton populations.[2491] Motility in the plankton is almost entirely by means of flagella, but it is unlikely that this is an asset, except in very still water.[4209]

Anderson and Sweeney[99] showed that the density of the vacuolar sap of the diatom *Ditylum brightwellii* may be reduced by the selective accumulation of K^+ and Na^+ ions, which replace the heavier divalent ions. The dinoflagellate *Pyrocystis noctiluca* also showed an ability to reduce the concentration of the heavy SO_4^{2-}, Ca^{2+} and Mg^{2+} ions sufficiently to become positively buoyant.[2491]

Literature dealing with phytoplankton is extremely voluminous; for comprehensive reviews see e.g. References 437, 1418, 1709, 1935, 2971, 3312, 3818, 4078, 4210, 4273, 4637, 4804, 5179.

2.7.1.2 Water blooms

The term "water bloom" is widely understood to refer to the accumulation of planktonic blue-green algae at the surface of lakes, reservoirs, or other water bodies. Although the constituent algae are themselves microscopic in size, the intensity of the scums and the rapidity with which they develop are such as to impress the most casual observer: blooms can form and disperse again within a matter of hours.[4080] Water blooms are more frequently observed in freshwater lakes but have also been recorded in the ocean and again blue-green algae (*Trichodesmium*) is usually involved.[1338]

Bloom-formation is generally confined to waters sufficiently rich in dissolved plant nutrients to permit the growth of dense algal populations. In temperate regions, conspicuous blooms develop most frequently during calm weather in summer and autumn. Blooms of blue-green algae occur when populations, which have been growing unnoticed below the surface for several weeks, suddenly float at the surface during periods of calm weather. At lower light intensities turgor pressure declines, gas vesicles (forming continually through self-assembly) increase in number and the alga regains its buoyancy (Figure 2-16). In the tropics they can form at almost any time of the year.[973,4080,4210]

Algal blooms are usually composed of colony-forming, unicellular and filamentous blue-green algae, often belonging to one of the following genera: *Microcystis, Anabaena, Aphanizomenon*, or *Gloeotrichia*[4080] (see also Figures 2-15 and 2-41). Except for the pressure of gas vacuoles, bloom-forming species do not appear to have any special physiological characteristics beyond the general tendency of blue-green algae to favor mildly alkaline, warm and slightly eutrophic waters.[973]

Table 2-2. Phytoplankton - size groups (in μm unless mentioned)

net plankton	macro	micro	nano	ultra	pico	Reference
>50			<50			973
50–70						4209
>64						4643
>150–200			5–60			4210
	>1 mm	60–1,000	5–60	<5		437
	visible	<200	20–200	10–20	0.2–2	2745
	>500	50–500ª	10–50	0.5–10		5349
		60–500ª				2258
		5–600	<50			1460
		>10	5–20	<5		1141
		>20	<20			325
				0.5–3		1942
			<80			3397
			<50			1653
			<30			3308
			<20			1004,5546
			<10			1600,3047

a = net plankton

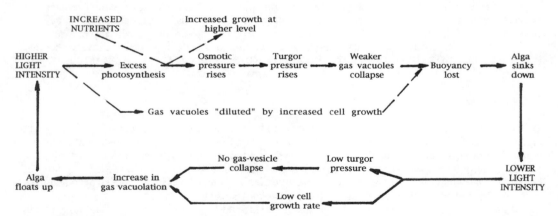

Figure 2-16.   Diagram summarizing the possible sequence of events in buoyancy regulation by plank-tonic blue-green algae (as proposed by Walsby[5247–9] and Dinsdale and Walsby[1091]).[4080] (From Reynolds and Walsby, 1975, in *Biol. Rev.,* Vol. 50, p. 457, reprinted with permission of Cambridge University Press.)

Diatoms, dinoflagellates, green algae and euglenoids are also present in blooms.[424] Excessively polluted or eutrophic habitats are dominated by small green algae and euglenoids which form blooms but not in the sense of the surface phenomena associated with blue-green algae. Periodic imbalances favoring blue-green algae in water bodies have been explained by numerous hypotheses.[2706] Major among these are the role of inorganic carbon availability,[2505,4417] interactions involving cyanophycean inhibition of metabolism and growth in other microbes,[2393-4,24503353,4053,4253] zooplankton predation immunity of blue-green algae,[3849-50] nutrient ratios favoring blue-green algae,[2491,4341,4606] and light availability.[2491] Organic carbon may also make a significant contribution to blue-green algae growth at low light intensities via photoassimilation.[704,1413,2121,2491] Hammer[1891] found that temperature influences the composition and sequence of blue-green algal blooms.

Most planktonic species of blue-green algae are buoyant at some stage of their development whereas planktonic representatives of other major algal groups are typically heavier than water. However, not all blue-green algae form blooms.[4080] Moreover, bloom-forming species may sink in still water under certain conditions.[1557,4523]

Buoyancy is impacted by gas vacuoles in the protoplast of the cell. These structures are found only in prokaryotic organisms (i.e. bacteria and blue-green algae), and with few exceptions, are normally restricted to planktonic species capable of true suspension in water bodies. In planktonic blue-green algae they are more widespread as compared to bacteria and their role in bloom-formation has been well documented.[1395,2966,4073,5250]

In the past, there has been some uncertainty as to whether such concentrations of algae are the result of periods of rapid multiplication of the algae or whether they are produced by a period of more gradual growth followed by an abrupt rise to the surface. In spite of reports of very high rates of growth (e.g. Mackenthum et al.:[3003] 56 mg dry mass per liter in 36 hours), the latter explanation seems more likely.[4073-5,4080]

Reynolds[4075] found concentrations of *Anabaena circinalis* filaments of 16,000 mL^{-1} common in blooms in Shropshire, England. This is equivalent to about 500,000 cells mL^{-1} or dry mass of about 20 mg L^{-1}. Growth of such populations occurs in the water column followed by accumulation at the surface and although this gives the appearance of dominance of blue-green algae, counts often show that other algae are even more abundant in the water column as a whole.[4076-7]

Clasen and Bernhardt[814] found up to 80,000 cells mL^{-1} of *Synura uvella* bloom in the upper shallow parts of the Wahnbach reservoir in Germany in autumn 1978. Utermöhl[5094] calculated a bloom biomass, consisting of filamentous blue-green algae and green alga *Ankistrodesmus falcatus* (Figure 2-58, #15), to be 30 kg ha^{-1} (i.e. 3 g m^{-2}) in the locality near Plön in Germany. Vymazal and Sládečková (unpublished data) found a water bloom from the Fryšták Reservoir in Moravia, dominated by blue-green algae *Aphanizomenon flos-aquae* and *Anabaena flos-aquae* having dry mass as high as 11 g L^{-1}.

Red tides are a coastal marine phenomenon caused not so much by the aggregation due to buoyancy but by rapid cell division stimulated by particular environmental conditions. The algae involved are species of the Dinophyceae usually of the genera *Gonyaulax, Gymnodinium, Prorocentrum* or *Pyrodinium*,[973,4210] which, having red pigments dissolved in their oil globules, impart a reddish tinge to the sea. These algae occur in many oceans but their growth to dangerous proportions tends to occur in tropical and sub-tropical zones; they are dangerous since they produce toxins lethal to fish and man but not to shellfish in which they tend to accumulate.[4210] *Peridinium* blooms have also been observed in lakes.[973]

Among the 5,000 species of extant marine phytoplankton, some 300 species can at times occur in such high numbers that they obviously discolor the surface of the sea (so-called "red tides"), while only 40 or so species have the capacity to produce potent toxins that can find their way through fish and shellfish to humans.[1882]

Hallegraeff[1882] in a very comprehensive review on harmful algal blooms listed different types of harmful algal blooms:

1. species which produce basically harmless water discolorations, however, under exceptional conditions in sheltered bays, blooms can grow so dense that they cause indiscriminate kills of fish and invertebrates due to oxygen depletion (e.g. dinoflagellate *Gymnodinium polygramma* or blue-green alga *Trichodesmium erythraeum*);

2. species which produce potent toxins that can find their way through the food chain to humans, causing a variety of gastrointestinal and neurological illness such as: paralytic shellfish poisoning (e.g. dinoflagellates of genera *Alexandrium* or *Gymnodinium*), diarrhetic shellfish poisoning (e.g. dinoflagellates of genera *Dinophysis* and *Prorocentrum*), amnesic shellfish poisoning (e.g. diatoms of the genus *Nitzschia*), ciguatera fishfood poisoning (various dinoflagellates), neurotoxic shellfish poisoning (e.g. dinoflagellate *Gymnodinium breve*) and cyanophyte toxin poisoning (e.g. blue-green algae *Anabaena flos-aquae* or *Microcystis aeruginosa*);

3. species which are non-toxic to humans, but harmful to fish and invertebrates by damaging or clogging their gills (e.g. some diatoms, dinoflagellates, prymnesiophytes or raphidophytes).

Further details on water blooms can be found e.g. in References 2196, 2535, 2804, 2810, 3618, 5066–7, 5109. For further information about algal toxins see also Chapter 2.15.

2.7.1.3 Benthos

Benthos is composed of attached and bottom-dwelling organisms.[424] Benthic algae grow attached to various substrates and may be classified as:

- epilithic (growing on stones)
- epipelic (attached to mud or sand)
- epiphytic (attached to plants)
- epizoic (attached to animals)

Lee[2745] characterizes benthos as organisms that grow on the bottom of a body of water. He distinguishes three categories:

- lithophytic (attached to the surface of a rock)
- endolithic (alga bores into, or lives inside, a rock—usually a limestone)
- epipelic (living on the surface of mud or sand)

The organisms and communities growing in the benthos have been classified on the growth form of the algae and the following summary gives the main forms:[4210]

- Rhizobenthos: vegetation rooted (by means of rhizoids) in the sediment. Some workers have distinguished associations "rooted" in silt (rhizoopalon) or in sand (rhizopsammon)
- Haptobenthos: organisms adnate to solid surfaces and conveniently divided into
  - *epiphyton* growing on other plants
  - *epilithon* growing on rock surfaces
  - *epipsammon* growing on sand grains
  - *epizoon* growing on animals
- Herpobenthos: community living on or moving through sediments (strictly term refers to organisms creeping in sediments) and it can be subdivided into *epipelon*,[4202] living in the surface of the deposit (mud or sand), *endopelon*, living and moving within muddy sediments, and *endopsammon*, living within sandy sediments.
- Endobenthos: community living on and often boring into solid substrata. The most common example occurs on rock and this is termed *endolithon*.
- Metaphyton: community of algae found loosely associated with the epiphyton but since the species lack attachment organs they can be easily washed out from amongst the epiphyton.[290] The algae are unattached species moved at random by local water movements. It is likely that many species generally considered epiphytic are in reality metaphytic.

The metaphyton is only common in certain types of host and under certain conditions of water chemistry and at a low rate of water flow. It does, however, form a very conspicuous element of the flora in oligotrophic lakes and in bog pools.[4210]

2.7.1.4 Periphyton

The term periphyton is of Russian origin[288–9] and first referred only to organisms growing on objects placed in water by man. Roll[4175] defined periphyton as all sessile organisms on any type of substratum, divisible into two subgroups, epiphyton (= Aufwuchs, sessile organisms on fixed substrata that are not associated with one another) and laison (= Bewuchs, a successively developed stage of periphyton that are thoroughly associated with one another in the phyto-sociological sense). Cooke[873] pointed out that Aufwuchs was first used to describe organisms growing on or attached to a substrate, but not growing into or penetrating the substrate. Later it was used in reference to the development of microcommunities attached to living substrates, and finally, Ruttner[4230] defined Aufwuchs as "all those organisms that are firmly attached to a substratum but do not penetrate into it." Wetzel[5351] pointed out that since the term Aufwuchs is so broad and ambiguous, it is of little meaning and should be abandoned. The term periphyton gradually acquired a broader meaning. Young[5529] defined periphyton as follows: "By periphyton is meant that assemblage of organisms growing upon free surfaces of submerged objects in water, and covering them with a slimy coat. It is that slippery brown or green layer usually found adhering to the surfaces of water plants, wood, stones, or certain other objects immersed in water and may gradually develop from a few tiny gelatinous plants to culminate in a woolly, felted coat that may be slippery, or crusty with contained marl or sand." Now it is generally employed to include all attached microorganisms growing on natural or artificial substrates—periphyton in its broad definition includes all aquatic organisms (microflora) growing on submergent substrates.[873,2260,3500,4560,5178,5349–50]

Sládečková[4560] subdivided periphyton into "true periphyton," immobile organisms attached to the substrate by means of rhizoids, gelatinous stalks, or other holdfast mechanisms, and "pseudoperiphyton," forms that are free-living, mobile, or creeping among or within the true periphyton. Pseudoperiphyton is synonymous with metaphyton.[5349] Wilbert[5425] proposed a small but very useful change: he introduced the Greek term "euperiphyton" instead of "true-periphyton." Together these organisms comprise the periphyton community, and should be treated as such.[5322] In comparison of terms, Hutchinson[2260] indicates that "periphyton" in its broad definition[4560] is synonymous with "haptobenthos." Wetzel and Westlake[5353] suggested that ". . . periphyton includes all of the plant organisms, excluding rooted macrophytes, growing on submerged material in water. Submerged materials include all substrata: sediments, rocks, debris, and living organisms."

Often fragments of this attached microflora can be distributed within the plankton, where it has been defined as "tychoplankton" or "pseudoplankton."[1174,2258,3880,4560]

In the literature, there has been a lot of discussion and controversy about the use of the words "benthos" and "periphyton." All the discussion is based on the difference between two- and three-compartment approaches (see Chapter 2.7.1). In fact, periphyton is usually subdivided into the same categories as benthos (see Chapter 2.7.1.3), i.e. epiphyton, epilithon, epipelon and epizoon and the terms periphyton and benthos have been used interchangeably quite often. Some authors brought even more confusion into the problem by using the term "periphyton" as a noun and "benthic" as an adjective.

Sládečková[4560] in her excellent review on periphyton pointed out that the majority of English hydrobiologists (see References 1134, 2960, 2971, 4202, 4204–5) had preferred the term "benthos" and "benthic organisms" in a wide sense, including "epilithic, epiphytic, epizoic and epipelic communities,"[4202] as an opposite to the term "plankton." But this usage is not in agreement with the original meaning of the word "benthos," which relates directly only to the organisms living on the bottom, i.e., to the epipelic and, sometimes, also to the epilithic community.[4560] Wetzel[5349] pointed out that the term benthos is now nearly uniformly applied to animals associated with substrata.

Round,[4210] on the other hand, criticized the use of the term periphyton: "The term periphyton, phycoperiphyton, and the even more superfluous combination epiphytic periphyton[5353] have crept into the literature as portmanteaux terms to include algae growing on objects placed in the water, and unfortunately have been loosely applied to epiphytic, epipelic and epilithic communities, merely adding confusion, and worse still, imprecision to the subject. I strongly urge that the term periphyton and its variations be dropped or at most be confined to growths on artificial substrata. The term "phycoperiphyton," which has been coined and used for epiphyton e.g. by Foerster and Schlichting[1394] is as superfluous as the term periphyton itself.

In my opinion, the terms "epiphyton" (community growing on the surface of plants), "epipelon" (community growing on sediments and muds), "epilithon" (community growing on the surfaces of rocks and stones), "epizoon" (community growing on surface of animals) and "epipsammon" (community growing on sand particles) are descriptive enough and have the same meaning no matter if they are referred to as periphyton or benthos. In order to avoid controversy these communities could be referred to as "attached" or "sessile" communities. Contrary to Round's opinion,[4210] the terms "periphyton" and "benthos" should be supplemented with an appropriate adjective, e.g. epilithic periphyton or epilithic benthos in order to specify the community type. The community growing on every kind of artificial substrata should be referred to as periphyton only, with an adjective periphytic.

Attached communities are affected by current, which may also influence their response to disturbance. Current influences algal immigration,[2573,4739,4743] emigration,[475,3150] and reproduction via variation in nutrient supply rates,[2203,5166,5390] as well as attachment strength and resistance and community physiognomy.[3150,3769,3823,4060,5012]

Methods for periphyton sampling, measurement, and data interpretation are given by Wetzel and Westlake,[5353] Weitzel,[5322] Wetzel,[5349] and Stevenson and Lowe.[4742] Examples of attached algae are given in Figure 2-17.

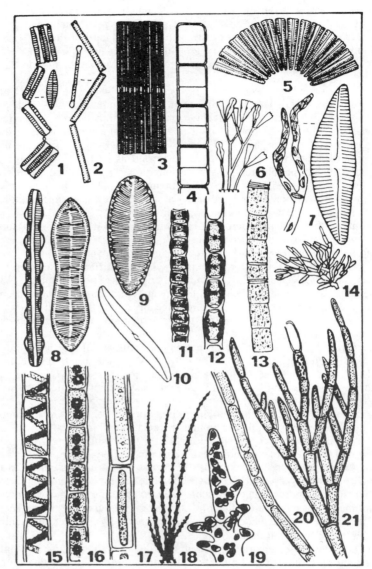

Figure 2-17.   Examples of attached algae (benthos, periphyton). 1-*Diatoma vulgare,* 2-*Diatoma tenue,* 3-*Fragilaria capucina,* 4-*Melosira varians,* 5-*Meridion circulare,* 6-*Gomphonema constrictum,* 7-*Cymbella ventricosa,* 8-*Cymatopleura librilis,* 9-*Surirella ovata,* 10-*Gyrosigma scalproides,* 11-*Ulothrix zonata,* 12-*Microspora amoena,* 13-*Oedogonium* sp., 18-*Lemanea fluviatilis,* 19-*Hydrurus foetidus,* 20-*Rhizoclonium hieroglyphicum,* 21-*Cladophora glomerata.* 1–10: Bacillariophyceae, 11–17 and 20–21: Chlorophyceae, 18: Rhodophyceae, 19: Chrysophyceae.   Not to scale.[4564]   (From Sládečková et al., 1982.)

2.7.1.5 Epipelic algae

The epipelic flora is mainly composed of diatoms, blue-green algae, coccoid Chlorophyceae (Chlorococcales and desmids, especially in fresh waters), and euglenoids. The majority of species are motile (90%), whilst a few are attached by mucilage to sand particles, etc., or lie free on the sediment.[973,4209,4643] Motility is a definite selective advantage in this habitat due to the chance of being buried by the muddy sediments.[973]

Epipelic algae are found on the surface of sediments ranging from fine muds to coarse sands.[4209,4643] There are three distinct components within the algal flora of the sediments:[973,4209] (1) the macroscopic algae rooted in the sediment (e.g. *Chara* and *Nitella*, Figure 2-61); (2) the non-motile often mucilaginous colonial of filamentous masses which rest on the surface of the sediment and are often visible to the naked eye (e.g. *Aphanothece, Spirogyra, Mougeotia,* see Figure 2-58, #11 and 13) will develop near the sediment surface in areas protected from swift current; and (3) the motile microscopic flora forming a film on and in the surface layer of sediment (e.g. innumerable diatoms, desmids, flagellates, filamentous and thalloid Cyanophyceae). This epipelic flora is present on all types of sediment (peat, silt, sand, decaying vegetation, calcium carbonate grains, iron deposits, etc.), extending down to the lower limit of penetration of photosynthetically available light; this is usually to a greater depth to which the macroscopic plants grow. Some of the factors affecting this flora are sediment structure and chemistry, nutrient status of the water, depth and shading, wave action and grazing.[4209]

The association of algae lying loose on the sediments is somewhat arbitrarily distinguished from the actual sediment community merely by its size and lack of motility. In many water bodies, particularly eutrophic waters, it is composed mainly of species of the Cyanophyceae (*Aphanothece stagnina, Nostoc pruniformae*) and some Chlorophyceae, e.g. *Cladophora glomerata* (Figure 2-17, #21) which often occurs here as matted balls of filaments. In oligotrophic waters it is more frequent to find loose mucilaginous masses of other filamentous Chlorophyceae (e.g. *Ulothrix, Microspora, Oedogonium, Spirogyra, Zygnema, Mougeotia,* see Figure 2-17, #11, 12, 13, 15, 16, 17, respectively), and intermingled with these are epiphytes and motile species derived from the sediments. The lack of attachment means that these are often transitory associations and they are frequently very contaminated by causal species.[4209]

In eutrophic water bodies, and especially shallow ponds, the development of epipelic mats of autotrophic algae is limited. They are absent in summer partly because of greater water turbidity due to plankton development, and partly because the bottom is constantly disturbed mainly by the feeding activities of fishes.[3087]

The ecology of epipelic algae is especially difficult to study due to the intimate association of the microalgae with the soft, easily disturbed sediments. Any scraped sample of epipelic algae contains a great deal of sedimentary material which obscures microscopic observations as well as practically any other measurement one would care to make.[973]

Another vexing aspect of epipelic algae is their well-documented, vertical-migration behavior within the upper 0.2–0.4 mm of sediment.[1926] The cells (pennate diatoms, euglenoids and blue-green algae) generally migrate to the sediment surface at dawn and return at sunset, although in freshwater ponds they may retreat from bright, midday light intensities. The diel migratory behavior is thought to regulate by carcadian rhythms of phototaxis and/or geotaxis which may be retained in the laboratory for several days. In intertidal sediments, the migratory response is modified so that the cells appear on the surface during daytime low tides. In most species examined the tidal aspect of the rhythms appears to be maintained by some aspect of the tidal regime (perhaps changes in light intensity, water content of the sediment or pressure) since it disappears rapidly when sediment is brought to the laboratory.[973]

The distribution of epipelic algae appears to be more dependent on the physical and chemical properties of the substrate than is the case for epiphytic flora, although water chemistry and temperature continue to be important variables. Since the depth of the 1% light level may vary from about 0.2 mm in the mud to 3.0 mm in sand, sediment movements which bury the algae can be limiting through their effect on light intensity.[973]

Inorganic nutrients are not considered to be limiting factors for the growth of algae in most epipelic habitats due to the relatively high nutrient concentrations in sediments. It is possible, however, that localized nutrient deficiency could occur during blooms when other parameters (light, temperature) are optimum. Nitrogen limitation of epipelic algae has been demonstrated in salt marshes.[977] It is likely that marsh grasses, which are also N-limited, are competing for available nitrogen.

Literature dealing with epipelic algae: References 23, 976, 3086–7, 3286, 3395, 3558, 3586, 4202, 4210. (For biomass and productivity of epipelic algae see also Tables 2-8 and 2-14, respectively.)

2.7.1.6 Edaphic algae

The edaphic algae[****] (= sediment associated, growing in or on soil, see Chapter 2.7.3) form an important productivity component in some habitats, e.g. temperate salt marshes.[4836] These algae are largely microscopic, extremely diverse in terms of numbers of resident species,[4834] and highly productive in spite of the heavy shading by the vascular plant canopy.[1547,3832,5134,5541] Furthermore, the edaphic algae, unlike the overstory vascular plants, are productive on a year-round basis and represent the major source of newly fixed carbon on the marsh when the latter is dormant.[4836] Ribelin and Collier[4091] reported that 98% of the detrital material exported from a Florida Gulf coast marsh was produced by edaphic algae.

Sullivan[4834] hypothesized that the diversity of salt marsh diatom edaphic communities is directly related to the dynamic nature of the salt marsh habitat. Dominance of the community by a few species is minimal because environmental conditions change rapidly enough to prevent it.

Literature dealing with edaphic communities: References 213, 708–9, 871, 1171, 1232, 1384, 1547, 3832, 4831–4, 5135, 5444–5. (For productivity of edaphic algae see also Table 2-14.)

2.7.1.7 Epiphytic algae

Epiphyton is the assemblage of organisms, including algae, bacteria, fungi and invertebrates which develops on submerged plants.[3325] Some epiphytic organisms are firmly attached to the host plant epidermis while others are only loosely associated,[731,4210,4213] and it is necessary to take care when sampling epiphyton so as to include both components of the community.

Cattaneo and Kalff[731] reported that since tightly attached epiphytes produce a substantial fraction of epiphyte production, and since much of the loose epiphyte biomass is readily lost unless extreme caution is taken during sampling, it is likely that in most of the field studies the role of epiphytes in macrophyte beds has been underestimated.

Epiphytic microflora contribute a major portion of the total production in many shallow aquatic ecosystems.[731,5349] Epiphyte production as a percentage of the total production (macrophytes, epiphytes and phytoplankton) in beds of emergent and submerged plants and mosses varies between 2 and >50%.[62,413,564,2408,2566–7,3793,3984,4627,4798,5355] (See also Tables 2-6, 2-7, 2-8 and 2-14.)

Epiphytic algae generally show little substrate specificity; many epiphytic species are encountered in natural epilithic communities and on artificial substrates. In spite of seeming relative indifference of epiphytic algae to their substrate, the epiphytic habitat has several distinctive attributes. The surface itself has a definite life span. New leaves are colonized as they develop during the growing season resulting in a summer and autumn peak in epiphytic biomass and productivity. The canopy of aquatic macrophytes often creates light-limiting conditions for epiphytic algae.[973] On the other hand, decreases in growth and photosynthetic rates, as well as abundance and occurrence of submersed macrophytes, have been attributed to PAR attenuation by the periphyton complex.[1242,3781,4274,4277]

In their use of nutrients from the sediment (via macrophyte tissue) as well as from the overlying water,[3256] epiphytes can play an important role in nutrient cycling. Much of the physiological research on epiphytic algae has focused on the question of nutrient transfer from rooted, aquatic, vascular plants to their epiphytes. A few studies have demonstrated a transfer of organic carbon, nitrogen and phosphorus from macrophyte to the epiphytic community. Experiments with radio-labelled phosphorus show that this release is small for macrophytes in active growth (3–9%,[730] 15–24%[3256]), though larger proportions (60%) can apparently be obtained by firmly attached epiphytic algae when phosphorus availability in the water phase is extremely low.[3256] The release is probably larger from senescent leaves, but perhaps of little significance because old leaves are subsequently shed.[4281] There is evidence that some rooted aquatic plants act as pumps, transferring phosphorus and other nutrients from the sediment to epiphytes and the water column. The amount of nutrient released, however, is very small.[730]

Interactions between epiphytic algae and their host macrophytes have been subject to controversy.[733,1777,5349] Competing hypotheses differ as to whether (1) the host macrophyte is a neutral sub-

---

[****]The terms edaphic algae or edaphon, in their original meaning, refer to soil algae. The terms themselves do not specify "how wet the soil is." The term "edaphic algae" is now widely used for benthic algal communities in coastal areas where the soil is temporarily (sometimes permanently) flooded. In my opinion, these communities are closer to epipelon than to "classic" edaphon.

strate[729-30] or (2) the host macrophyte influences epiphyton production and community composition by mechanisms independent of morphology.[621,778,3556,4527,5471] Similarities between natural and artificial macrophyte-substrates in community composition, biomass and production of colonizing epiphyton[729,3555] support the former hypothesis. On the other hand, it has been found that epiphyton species composition and abundance were related to the macrophyte-mediated changes in the physico-chemical environment.[3556] The responses of epiphytic and epipelic algae to primary physical, chemical and biotic parameters have been discussed in detail by Wetzel.[5349]

The host algae produce various extracellular products, some of which have been shown to influence both epiphyte attachment and composition.[4480,4527] In *Sargassum*, where the branch tips are free of epiphytes; this is attributed to the production of tannins by the algae.[4480] Algae, especially the filamentous genera, are themselves hosts to secondary epiphytic species.[4210] Chudyba[806-7] found a total of 220 species (176 diatoms) of epiphytic algae on *Cladophora glomerata* in a river. For more information about epiphyton-macrophyte interactions see Chapter 3.3.5.

Experiments showed that periphyton have higher rates of respiration and phosphorus uptake at higher current velocities.[3149,3503,5390-91] Hosseini and Van der Valk[2210] pointed out that metaphyton, though an important component of freshwater marshes, generally have been overlooked by researchers (as are all algal assemblages in wetlands—author's note). It is sometimes difficult to delineate this community from loosely associated epiphyton, but normally they are not sampled when periphyton or phytoplankton are sampled. Nevertheless, metaphyton may attain high densities and have a high annual primary production.[2567,5326] Hosseini and Van der Valk[2210] reported metaphyton production between 19.6 and 150.7 g AFDM m^{-2} year^{-1}.

Further details on epiphytic algae can be found e.g. in References 629, 3280, 3286, 5321, 5349–50.

2.7.1.8 Epilithic algae

The algae of rock surfaces are adapted to swiftly flowing water by being either encrusting or basally attached, but flexible and streaming in the current. The epilithic flora may be contaminated with species filtered out of the water, and where the rate of flow is reduced, silt tends to become trapped amongst the epilithic algae and flora intermediate between the epilithic and epipelic type develops.[4209]

Four groups of algae are found frequently in the epilithic flora: encrusting and filamentous Cyanophyceae, Chlorophyceae and Rhodophyceae and encrusting Bacillariophyceae. It is characteristic of the habitat that relatively pure stands of species develop.[4209]

The epilithic community usually develops in areas where water movements are sufficiently strong to prevent finer sediments from accumulating; streams and rivers and rocky shores are the best studied habitats. The algae found here have various, well-developed attachment mechanisms including: a series of prostrate filaments which support erect, branching filaments in the green alga *Stigeoclonium*; rhizoidal extension of the basal cell in *Cladophora*; a single, relatively undifferentiated, holdfast cell in the unbranched green filamentous *Ulothrix* and *Oedogonium*; and the secretion of a cushion, stalk or tube of gelatinous material by many species of pennate diatoms.[973] PAR attenuation and wave disturbance are primary mechanisms that control the vertical zonation of freshwater epilithic algae.[2117]

The nature of the substrate seems to be less important than other environmental parameters in determining the spatial distribution of algae, although some differences have been noted in the communities developing on different types of rocks.[973]

The type of flora is determined by the chemistry of the water (the chemical nature of the rock surfaces are often correlated with this), the position on the shore and the degree of wave action. Since the substrate is relatively permanent, there is usually a perennial flora but there is little data on its seasonal aspects.[4209]

Further information about epilithic communities can be found e.g. in References 1185, 2539, 3333, 4210, 4434, 5391. (See also Tables 2-6, 2-7, 2-8 and 2-14.)

2.7.1.9 Epizoic algae

Many animals, especially sedentary species with hard shells, are host to numerous common species.[4210] Attached algae, e.g. *Stigeoclonium* grow on the shell of molluscs, whilst on turtles the genera *Basicadia, Dermatophyton* and various *Oscillatoria* species are found. A few algae are reported to grow on fish possibly after injury although the holdfast of the algae is reported to penetrate the bones.[4209]

2.7.2 Aerophytic Algae

Aerophytic (atmospheric) algae live outside the aquatic environment in the air and take up water from the air in the form of rain or high humidity.[1460,4209] Some stands of aerophytic algae are temporarily wet with percolating water and if the source of percolating water is stable and substantial these stands become aquatic. In fact, algae growing on the soil surface are aerophytic, but they are usually classified separately.[1460]

Khan[2491] classifies aerophytic algae as:

- epiphytophytes—algae growing on leaves
- epiphleophytes—algae growing on bark
- epizoophytes—algae growing on animals
- lithophytes—algae growing either on stones or on chalk

The aerial epiphytic habitat is characterized by the absence of nutrients other than those in rain and those obtained by solution of dust or material on or from the host plant tissues. This factor, rather than desiccation, is probably more important in the restriction of the flora to a few species. Desiccation of soils may be equally extreme but here the floras are often rich in species.[4209] Aerophytic algal flora is mostly formed by members of the Cyanophyceae, Bacillariophyceae and Chlorophyceae. Probably the most common aerophytic alga all over the world is *Pleurococcus vulgaris.*[1460]

Rock and stone surfaces which are relatively stable are often coated with algae. On rocks which receive only atmospheric moisture the flora is similar to that on tree bark. However, many rock surfaces receive seepage water from overlying soils and hence a rich nutrient supply is available.[4209] Tree-bark-inhabiting algae have been studied by Cox and Hightower,[899] Graham et al.[1789] and Wylie and Schlichting.[5506]

2.7.3 Soil Algae

The community of all organisms in soil is called edaphon, a floral community is called phytoedaphon.[1460]

Algae are of widespread occurrence in the soils of all continents, from the polar regions to hot deserts.[4689] Soil algae are virtually ubiquitous on or near the soil surface, even in soils which support few higher plants.[973]

The soil flora is selected from a limited range of algae within the Chlorophyceae, Cyanophyceae, Bacillariophyceae and Xanthophyceae, with occasional species from the Euglenophyceae and Rhodophyceae.[973,3202,4210] Most soil algae, however, both in kind and number, are members of the Cyanophyceae and Chlorophyceae.[3202] The soil flora is sufficiently distinct so that it should not merely be regarded as a part of the aquatic flora living in an unfavorable environment.[3202,4210] Metting[3202] listed 38 genera of prokaryotic and 147 genera of eukaryotic algae that include terrestrial species, the majority of which are edaphic.The soil algal flora does not consist of species which are entirely confined to this habitat, many extend their range into other terrestrial habitats, and frequently into small or temporary waters and even into the littoral zones of lakes. However, in these latter habitats they are usually rare in the flora and it is obvious that they attain their greatest growth on the soil surface.[4209] Friedmann et al.[1501] suggested for surface and sub-surface algae the terms epidaphic and endedaphic algae, respectively. The term hypolithic was used for the algae living on the lower surface of stones on soil.

Most algae are probably photoautotrophs. However, a number of soil algae belong to other nutritional categories including photoauxotrophs, colorless chemoheterotrophs or facultative photoheterotrophs.[1157,3202,3657] Feher[1322] identified 685 taxa of soil algae reported from the literature and concluded that it was not possible to determine any geographic distribution, or correlations with soil types. It would appear, however, that alkaline soils support an abundance of cyanophytes, and that these are not found on soils with a pH of less than 5.[565,973,4689] However, green algae appear to be more indifferent to soil pH, and are often found in acidic soils.[565,973,1403,2165] Diatoms are poorly represented in acid soils, though it is probably more precise to say base-deficient rather than acid.[2961] Distribution of soil algae through climatic regions has been reviewed by Starks et al.[4689]

It has been suggested that soil algae may be ecologically significant, especially as pioneer species in the initial stages of succession. They stabilize the soil surface of bare eroded soils against further erosion (especially filamentous blue-green algae which form confluent mucilaginous sheaths) and soil removal by wind, improve infiltration of water into the soil, reduce water loss, improve soil texture by aggregating soil particles, add organic matter to the soil, help solubilize certain minerals and provide a favorable habitat for seed germination.[212,448,973,1382,1802,2816,4210,4447,4461-3] The soil algae appear to prevent loss of soil ammonia and leaching out of nitrates by converting these forms to organic nitrogen.[4447] It is also clear, however, that nitrogen-fixing blue-green algae significantly improve the nitrogen status of soil.[973,1017,1398,1402,2391,3637]

It is generally accepted that the contribution by nitrogen-fixing blue-green algae to the nitrogen pool of natural ecosystems is probably more important than that supported by heterotrophic microorganisms.[4772] The importance of free-living nitrogen-fixing algae was reviewed in depth by several authors.[1417,1434,2390,4446,4768-9,4771-2] The most common factors that affect the distribution of nitrogen-fixing blue-green algae are pH, soil moisture, temperature and light.[4769] Low values of these factors limited nitrogen-fixing in various ecosystems.

Soil algae produce and excrete a variety of substances, including vitamins, organic acids, plypeptides, nucleic acids, polysaccharides, ammonium, amide-N, vitamins and other growth substances and antibiotics.[66,1401,1430,1460,1581,1802,2778,2967,3229,3538,3869,3872,4518]

Even though edaphic algae occupy such important positions in the ecology of terrestrial habitats, they have received relatively little attention from phycologists.[2509] The small amount of information on seasonal patterns in the diversity of soil algae suggests that seasonal succession such as occurs in aquatic systems is not apparent in the soil.[559,2248]

Soil algae grow better in partially dry soil (40–60% of water-holding capacity) than in soil held at field capacity, reminding one of a similar response in some lichens and intertidal seaweeds. Soil algae may remain viable in some type of resting stage for 50–70 years in soils with a water content of 3–10% and up to 10 years in soils with a moisture content of 1% (air dry).[973] It seems likely that algae, found well below the euphotic zone which is the top few millimeters, are washed into the soil during heavy rains and remain viable for many years in this moderated environment, perhaps utilizing the limited supply of usable organic matter in maintenance metabolism.[973] The surface of most soils supports a rich algal flora. Spores and fragments of algae are washed to greater depths and remain viable for long periods, but actively growing algae are probably confined to within a mm or so of the surface.[4210]

Although as many as 10^8 algae per gram of soil have been documented, soils commonly support between 10^3 and 10^4 algae per gram.[3202] Gollerbach[1702] found up to 100,000 algal units (including blue-green algae) in 1 g of the surface soil—it equals about 138 kg of algal biomass per 1 hectare of the soil. Hunt et al.[2249] found up 3.3×10^7 cells of eukaryotic algae per 1 g of soil and up to 8.2×10^5 cells of prokaryotic algae (blue-green algae) per 1 g soil. Soil algae are affected by synthetic pesticides and pollutants. In general, most herbicides, fungicides, and soil fumigates are detrimental to soil algae while most insecticides are not.[3202]

Literature on soil algae: see References 172, 423, 538, 548–553, 563, 672, 674, 707, 1194, 1439, 1441, 1704, 2144, 2249, 2508–9, 2957–8, 2961–3, 2967, 2969, 2997, 3202, 3451, 3469, 3656–7, 3765–6, 4209, 4447, 4689, 4928.

2.7.3.1 Epipsammic algae

A specific community of soil algae lives in sand (psammon). The flora consists in part of coccoid chlorophytes and cyanophytes which occupy depressions in the sediment particles. The most conspicuous component, however, are diatoms, which may be closely adherent to the particles, or attached by short mucilage stalks, in contrast to epipelic flora these are predominantly non-motile genera.[4210] Algae can grow in sand in large quantities resulting in a brown or green color of sand. If organisms in the sand possess their own motion (Dinophyceae, Bacillariophyceae, Euglenophyceae), the sand color can periodically appear and disappear.[2555] According to the amount of water in the sand, Wiszniewski[5465] distinguished "hydropsammon" (community living in wet, periodically flooded sand) and "eupsammon" ("real psammon" forming out of the reach of ground and surface water; the only source of water is rain and aerial humidity).

A host of small diatom species belonging to the genera *Opephora, Fragilaria, Achnanthes,* etc., attach to sand grains and in many lakes, especially in North America, form a very important element of the total benthic flora.[4209]

Literature on epipsammon: see References 2260, 3680, 5349, 5405.

2.7.3.2 Desert soils algae

Cryptogamic crusts, also known as soil algal crusts, microbiotic crusts or rain crusts, cover extensive portions of the arid and semiarid regions of the world.[1382,2344] These crusts consist of water-stable surface soil aggregates held together by algae, fungi, lichens, and mosses.[2344] The filamentous cyanophytes are the most conspicuous and best studied element of these crusts.[2344,4210] Diatoms, coccoid chlorophytes and xanthophytes are also present but have been less studied so far.[2344]

Algae occur in desert soils but may not be obvious macroscopically,[746,1497,1501] although they are important as primary producers.[1500] Friedmann and co-workers[1501] have classified desert soil algae as *endedaphic* (living in soil), *epidaphic* (living on the soil surface), *hypolithic* (on the lower surface of stones on soil) and as rock algae, including *chasmolithic* algae (in rock fissures) and *endolithic* algae (rock penetrating). *Cryptoendolithic* algae passively enter the rock matrix while *euendolithic* algae actively bore into the rock. A review on cryptoendolithic algae has been given by Bell.[297]

The ability to withstand desiccation is quite common amongst the Cyanophyceae so it is not surprising that these are so abundant in such habitats.[4210] When in a dry soil sample, some microalgae can withstand a temperature of 145°C for one hour and up to three months at 40°C.[5013,5015]

2.7.4 Snow and Ice Algae

2.7.4.1 Snow algae

Microscopic plants (algae and fungi) living in snow and ice are called kryoflora; those that do not live in other environments are considered to be kryophilic plants.[1460] Snow algae occur worldwide in areas in which permanent or semipermanent snow banks are found; they are found in high-altitude mountain habitats and polar regions. The most common are species of chlamydomonads (Chlorophyceae), but euglenoids, chrysophytes, dinoflagellates, cryptomonads, cyanophytes, xanthophytes and diatoms have also been reported.[2150] The most extensive investigations on snow algae, aimed mostly at their taxonomy, ecology and geographical distribution, are from Europe,[2554] North America[2150-1] and Japan.[1530]

Snow algae can be found in blooming proportions, coloring the snow green, yellow, orange, pink or red.[2150] The various non-green colors associated with snow chlorophytes are due to astaxanthin (3,3'dihydroxy-4',4'diketo-β-carotene) and other carotenoids which accumulate in vegetative cells and zygotes, probably as a photoprotective mechanism in response to the high light intensities. Nitrogen deficiency and normal maturation process in zygotes may also contribute to carotenoid accumulation.[973]

They appear in late spring or early summer when the air temperature remains above freezing long enough to allow melt-water to form on the surface of the ice crystals. The vegetative cells can survive periodic freezing, but liquid water is required for spore germination and for the development of large populations. Most snow algae belong to one of a few genera of unicellular green algae (*Chlamydomonas, Chlainomonas*).[973] *Chlamydomonas nivalis* is the most frequently recorded alga from permanent snow fields.[973] Most species of snow algae are not known from other habitats, suggesting that they are well adapted to the regions of living in snow.[973]

There is considerable confusion in their taxonomy; many may be stages in the life cycles of other species and many supposed snow chlorococcalean species are known to be zygotic stages of *Chloromonas* spp.[4643] Hoham[2148-9] has suggested that true snow algae have an optimum growth below 10°C. Snow algae may be active even down to 50 cm.[1412,4952]

Further literature on snow algae: see References 358, 1379, 1412, 2147, 2152–4, 2956, 4718, 5320.

2.7.4.2 Marine ice algae

Ice algae occur in the marine pack ice of the polar region, where they grow in the brine channels which develop between the ice crystals and in the loose layer that occurs in the underside of the ice.[2200] Diatoms are the most common species, but dinoflagellates, green flagellates, and cryptophytes are also reported. Most of these algal species are non-planktonic and the assemblage is distinct from the plankton.

Four types of ice microalgal communities have been described based on location within the ice column: (1) surface;[634,3180] (2) interior;[15,2207,2230] (3) bottom congelation;[124,615,2200,3135,3181,3632] and (4) under-water platelet[615,4830] communities. Algae found frozen into pack ice are apparently common in polar regions but may not contribute more than 2% to the total production.[616] The under-ice flora is rather loosely attached and can be easily washed away if the ice is disrupted.[615]

Further information about marine ice algae can be found in References 52, 68, 619, 1572–3, 1769, 1798, 1832–3, 2198–9, 3632–3, 4610.

2.7.5 Algae of Hot Springs

Algae of hot springs are common in active volcanic regions. In addition to elevated temperatures, such waters have 10–15 times higher concentrations of dissolved salts, especially Na^+, K^+, HCO_3^-, SO_4^{2-}, Cl^-, S^{2-}. Others have low pH and may also have elevated concentrations of heavy metals.[4643] Brock[563] suggested that the termophilic species are relict populations from periods early in the earth's history when volcanic conditions were more widespread.

Blue-green algae are abundant in most springs above pH 6 but are completely absent below pH 4 and rarely found below pH 5. Most blue-green algae are also sensitive to soluble sulfide, a frequent component of hot-spring water but there is an evidence that sulfide-tolerant species may actually use S^{2-} as an electron donor for anoxygenic photosynthesis.[725]

Darley[973] reported that the highest temperatures at which blue-green algae are growing is 70–72°C, and Brock[562] gave the upper limit as 73–75°C. Petersen[3763] reported blue-green algae living in hot springs in Kamtschatka up to 77°C. Whitford[5389] reported maximum temperatures for blue-green algae *Synechococcus eximus* and *Oscillatoria filliformis* to be 79°C and 85°C, respectively. The blue-green alga *Synechococcus lividus* has been reported to tolerate a temperature of 74°C.[562,3720] It is blue-green algae alone which survive temperatures above 57°C. The upper temperature limit for eukaryotic algae is 55–57°C for *Cyanidium caldarium*,[1114] some diatoms existing up to 50°C and Chlorophyta only up to 48°C.[4210] Some heterotrophic bacteria are capable of growing in boiling water, the upper limit for fungi is at 60–61°C.[973]

In most cases the algae only occupy the uppermost layer which may vary in color from the typical blue-green to yellow, green or dark brown. The pink, red, orange or pale yellow, multilayered undermat consists of various autotrophic and heterotrophic bacteria.[973] The bacterium *Bacillus coagulans* and the fungus *Dactylaria gallopava* are common components of the mat-like growth associated with *Cyanidium*.[4210]

It appears that heat tolerance in hot-spring algae is a consequence primarily of the singular capacity of their proteins to endure without denaturation abnormally high temperatures, thereby enabling them to flourish in environments which tend to exclude competition from other species.[3075] The indications of an unusual protein structure in heat-resistant algae are in arrangement with similar data from thermophilic bacteria.[3075]

For further details on algae from hot springs see References 247, 562, 566, 723–4, 726, 1309, 1595, 1902, 2304, 2640, 3012, 3174.

2.7.6 Symbiotic Associations

Algae are known to live in intimate associations with a great variety of other organisms including invertebrates, fungi, liverworts and vascular plants. Several diatoms have developed a symbiotic association with a cyanophyte. Since in many cases the partnership is considered to be advantageous to both organisms, mutualism is the preferred term although the more general term symbiosis is often used.[973] The partnership has developed through evolutionary time as two distinct genetic entities, the alga and the "host" have established a harmonious and stable relationship in which the partners are dependent on one another to a greater or lesser extent.[973] Their individual functions have been adjusted to increase the biological fitness of the integrated unit.[973]

2.7.6.1 Endophytic algae

Several algae grow as endophytes or endosymbionts within other plants. Endophyton grows between the cells of other plants or in cavities within plants. Planktonic algae producing mucilage often have other algae growing within the mucilage: a form of endophitism.[1958,4869] The green alga *Chlorochytridium* grows on certain mosses,[4051] blue-green alga *Anabaena azollae* grows within the water fern

Azolla,[2170] and the species of *Anabaena* and/or *Nostoc* live within the hornworth *Anthoceros* (Bryophyta), in the roots of cycads, and in the rhizomes of *Gunnera,* an angiosperm.[424,1172,1821-2,3758,3392,3496,4503,5046]

Molisch[3271] reported that *Nostoc* spp. from liverworts were able to fix atmospheric nitrogen, but the results of his investigation is questionable, since he had no bacteria-free cultures. Nevertheless, some experiments of Winter[5459] indicate that *Nostoc muscorum* from *Gunnera* and *Cycas* possess this property. *Azolla* is able to meet all of its nitrogen requirements through the N_2-fixing activity of the symbiotic *Anabaena*[3759-60] which accounts for approximately 15% of the biomass of the partnership.[973] The alga's role is enhanced by a high heterocyst frequency (20–50%).[973] The N_2-fixing association is, however, unusually resistant to inhibition by NO_3^-. In free-living blue-green algae, nitrogenase activity is lost in a matter of days following the addition of NO_3^- to the medium, while N_2 fixation by the *Anabaena-Azolla* combination is only slightly inhibited after more than a month.[973]

Diatoms of the family Epithemiaceae have developed a symbiotic association with a cyanophyte. These unicellular symbionts are located in the cytoplasm of the diatom cell.[1075] The unidentified bodies were observed in *Rhopalodia* cells more than 100 years ago.[3779] The electron microscopy revealed the cyanophyte nature of these bodies.[1170] In *Rhopalodia* and *Epithemia,* each cell contains from 1 to 10 unicellular nitrogen-fixing endosymbionts.[1170,2532,2926]

2.7.6.2 Lichens

Lichens are the most well-known example of a symbiosis of algae (phycobiont) and fungi (mycobiont). The term phycobiont was coined by Scott.[4377] Lichens grow on soil, rocks and tree trunks and are found in such diverse habitats as deserts, sea-shores, tropical rain forests, mountain peaks and polar regions. Thallus structure varies from crustose forms which are tightly appressed to the substrate, to leafy (foliose) forms and highly branched (fructicose) species. Reindeer moss (a fructicose lichen which grows to a height of 20–30 cm) occurs over vast areas of the arctic tundra and is an important food for reindeer and caribou.[973]

Representatives of 26 genera (for comments see Ahmadjian[42]) of algae enter into lichen associations. The list of genera includes 8 of Cyanophyceae, 17 of Chlorophyceae, and 1 of Xanthophyceae. Many of the algal symbionts are found also in a free-living state. In a majority of associations, perhaps as high as 90%, the phycobiont is one of three genera, namely, *Trebouxia, Trentepohlia,* and *Nostoc.*[42] Although the lichen algae are a heterogenous group, they share at least one common trait, that is, an ability to survive the encroachment of, and subsequent permanent union with, a heterotrophic organisms.[42] The most common phycobionts are members of the Chlorophyceae, especially *Trebouxia,* which has probably the best adaptation of all coccoid green algae to lichen symbiosis.[1460]

The fungal component (mycobiont) constitutes more than 90% of their biomass of most lichens. Nearly 29% of all species of fungi are found as mycobionts in lichens[4584] and most of these belong to the Ascomycetes.

As in algae-invertebrates symbioses, the algal symbiont in lichens releases a large amount of organic carbon (60–80% of the total carbon fixed) which is used by the heterotrophic partner. Blue-green algae phycobionts release glucose, whereas green phycobionts release polyols such as ribitol, erythriol or sorbitol (Figure 2-18).[973] Release of organic carbon by the algae declines dramatically in a matter of hours or days following their isolation from the fungus.[973] The physiology of blue-green phycobionts is considerably modified relative to free-living blue-green algae. In symbionts, nitrogenase is much less sensitive to inhibition by combined nitrogen (ammonia and nitrate) than it is in free-living algae. In symbiosis, the activity of glutamine synthetase in the phycobiont is very low and as a result almost all of the fixed nitrogen is excreted as NH_4^+ and is available for assimilation by the fungus.[973]

Lichens containing heterocystous blue-green algae (usually *Nostoc*) fix N_2 at rates comparable to free-living forms. N_2 fixation is restricted to the heterocyst although it was once thought possible that fungal respiration could lower the O_2 tension in the thallus sufficiently to allow nitrogenase to function in vegetative cells.[973]

For further literature dealing with lichens see e.g. References 41, 43, 272, 1060, 1460, 1496, 1800, 2033–4.

2.7.6.3 Endozoic algae

Some algae live endozoically in various protozoa, coelenterates, mollusc and worms.[49,874,4059] Members of at least 8 algal classes (Cyanophyceae, Prochlorophyceae, Chlorophyceae, Prasinophyceae,

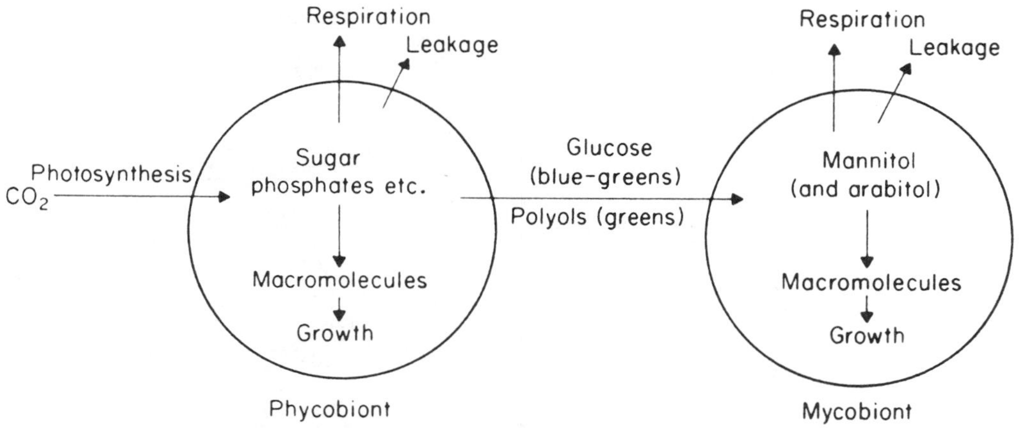

Figure 2-18.   Carbon flow through lichens.[973] (From Darley, 1982, in *Algal Biology: A Physiological Approach,* p. 122, reprinted with permission of Blackwell Scientific Publications, Oxford, and the author.)

Bacillariophyceae, Prymnesiophyceae, Dinophyceae, Rhodophyceae) form symbioses with a wide variety of invertebrates, both freshwater and marine.[187,219,1652,4906] Unicellular algae living within the tissues of invertebrates traditionally have been categorized as either zoochlorellae (green-pigmented symbionts) or zooxanthellae (golden-pigmented symbionts).

Symbiotic endozoic algae are primarily unicellular and live within protozoans or multicellular invertebrates. Algal-invertebrate associations, termed phycozoans by Pardy,[3642] are stable composite organisms having specific characteristics enabling them to exist in niches from which their individual component bionts are excluded.[1652]

Three different algae are involved in the best-studied symbiosis involving invertebrates:[5022]

1. the green prasinophyte flagelatte *Platymonas*, occurs in the marine flatworm, *Convoluta*;
2. the green, non-motile unicell, *Chlorella* (often referred to as zoochlorella) is found in a freshwater coelenterate, *Hydra,* and in the protozoan, *Paramecium*;
3. dinoflagellates, often called zooxanthellae, are common symbionts in the coelenterates, especially tropical sea anemones and corals, and are also found in the giant clam, *Tridacna.*

In the coelenterates, the symbiotic algal cells are localized within the endoderm cells of the host; *Platymonas* occurs in intercellular spaces in its flatworm host. The flatworm *Convoluta roscoffensis* is not holozoic and is entirely dependent for its development on the presence of the alga *Platymonas* within it.[3569,3932] Its symbiont, *Platymonas convolutae,* however, is found free living on the sandy beaches where the worm occurs. Once established in the worm, the alga loses its theca, flagella, and eyespot.[973] The *Platymonas-Convoluta* symbiosis is unusual in that the glutamine and possibly other amino acids are the principal products released by the algae rather than carbohydrates as in other algae-invertebrates associations.[517] These exchanges are illustrated in Figure 2-19.

The occurrence of *Chlorella*-like cells in *Hydra viridis* and *Paramecium bursaria* is well known.[874] *Hydra viridis* is much less dependent on its symbionts than is *Convoluta.* With an ample food supply in the light, green *Hydra* grows as fast as a symbiotic *Hydra*, but green individuals grow faster when food is limited.[973] Literature on endosymbionts of *Hydra* was summarized by Pardy,[3641] Pool[3842] and Muscatine and Neckelmann.[3359] In contrast to *Hydra*, symbiotic *Paramecium bursaria* can grow slowly in the light when depending solely on its symbiont in the absence of bacterial food.[973] Green algae also occur in the freshwater sponges (*Spongilla, Meyenia* and *Tubella*) but little is known of their biology.[4209]

Others, zooxanthellae (members of the Dinophyceae), live in intimate association with corals, where their photosynthetic activity is of primary importance to the reef community.[314]

Another major group of endozoic algae is the marine dinophycean zooxanthellae, the taxonomy and systematics of which are in a state of confusion.[1652] According to older observations from the late 19th

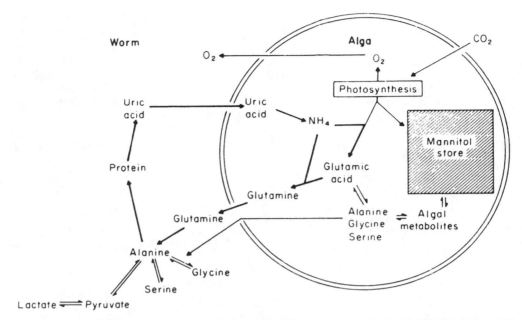

Figure 2-19.   Postulated pathways of nutrient interchanges in the symbiotic association between the green alga *Platymonas convolutae* (now prasinophyte), and the flatworm, *Convoluta roscoffensis*.[517] (From Boyle and Smith, 1975, in *Proc. R. Soc. Lond.,* Vol. B 189, p. 133, reprinted with permission of The Royal Society, London.)

century and early 20th century[1460] zoxanthellae are non-motile stages of cryptomonads, they attain a flagellate-look as soon as they leave the host body. Hovasse and Teissier[2215] proved that zooxanthellae are cell-walled Dinophyceae in a non-motile stage in the host body. Also, McLaughlin and Zahl[3165] and Freudenthal[1490] described zooxanthellae as Dinophyceae. The most common representative of this group is *Symbiodinium microadriaticum*[1490] = *Gymnodinium microadriaticum*[4906] = *Zooxanthella microadriatica*.[2895]

The symbiotic association between *Zooxanthella microadriatica* and various corals and sea anemones is the best studied of the algae-invertebrate symbioses. The relationship is not obligate since both partners can exist independently. As a symbiont the alga is a spherical, encysted cell with a reduced cell wall. A great number of zooxanthellae live in corals forming coral reefs. In 1 mm^3 of a coral reef body there may be up to 30,000 zooxanthellae that form 5% of the coral reef biomass. Their photosynthetic activity is so large that coral reefs are, in fact, autotrophic organisms with a productivity up to 3,500 g C m^{-2} year^{-1}.[1153] Zooxanthellae also significantly influence the precipitation of $CaCO_3$, and thus contribute to coral reef formation.[1758] The symbiotic algae secrete maltose which is utilized by the animal cells.

Further literature on endozoic associations can be found e.g. in References 1482, 1652, 1679, 3165, 3358, 3360, 3569–70, 3642, 4376, 4905–9, 5020–5, 5384.

2.7.7 Parasitic Algae

A few algae lead a parasitic existence. The green alga *Cephaleuros* lives on the leaves, young twigs, and fruits of certain angiosperms. Various colorless or weakly pigmented algae, particularly in the Rhodophyceae, grow on other seaweeds and are assumed to receive organic carbon from their hosts.[2886]

The best known parasitic red alga *Harveyella mirabilis* parasites other red algae.[760,1676,1678,2382,2602] *Harveyella* cells penetrate among the host cells via grazing wounds and cause proliferation of host cells to form pustules, over which the cortex of *Harveyella* forms its reproductive structures. *Harveyella* relies completely on its host for carbon, which it receives as digeneaside and perhaps some amino acids.[2602] Three possible mechanisms have been suggested by Goff[1677] to explain how *Harveyella* induces its host to release organic carbon. Other parasitic algae receive carbon as floridoside[1294] or mannoglycerate;[666] each host-parasite relationship is unique and must be assessed separately.[2886]

The endoparasitic dinoflagellate *Amoebophrya ceratii* occurs in coastal waters of Nova Scotia within cells of two dinoflagellate hosts, a *Scrippsiella* species (probably *S. trochoidea*) and *Dinophysis norvegica*.[1515] Cachon[655] reported that *Amoebophrya ceratii* in an endoparasite within cells of at least 10 different dinoflagellate genera, including *Prorocentrum, Gymnodinium, Gyrodinium, Gonyaulax* and *Ceratium*, in European waters. Taylor[4913] and Nishitani et al.[3442] reported *A. ceratii* along the Pacific Northwest coast within cells of *Gonyaulax catanella, Peridinium trochoideum* (= *Scrippsiella trochoidea*) and *Gymnodinum sanguineum*.

2.8 EVOLUTION AND PHYLOGENY OF ALGAE

The algae are ancient organisms. Seemingly, the most ancient recognizable algae were prokaryotic.[4357] The origins of the prokaryotes lie remarkably close in time to the beginning of the geological record at ca. 3.8 billion years ago.[4643] The first autotrophic organisms, the cyanophytes, made their appearance about halfway through the Archean (3.8 - 2.6 $\times 10^9$ years BP), the earlier part of the Precambrian.[4360] In the latter part of the Precambrian, the Proterozoic (2.6 - 0.58 $\times 10^9$ years BP) they were ubiquitous and representative of most present-day blue-green algae orders have been recorded from the fossil record of this period.[4360] Additional evidence of the early occurrence of blue-green algae is provided by stromatolites, which are rock-like formations of limestone and other materials trapped in layers by colonies of cyanophytes which have been dated as old as 2.8 billion years.[4911] However, some "modern" algae form similar structures, e.g. in Western Australia.[4911]

The development of the nucleated eukaryotic cell type took place probably about 1.4 billion years ago.[4359] Meiosis and eukaryotic sexuality originated at least 800 million years ago,[2141] but may have come into existence well before then, possibly as early as 1.33 billion years ago.[5259] Lee[2745] pointed out that the first eukaryotic algae appeared in a form similar to the Glaucophyta of today, with endosymbiotic blue-green algae instead of chloroplast. The "youngest" classes of algae are probably Euglenophyceae (53.5 million years ago) and Xanthophyceae (25.0 million years ago).[2745]

South and Whittick[6443] summarize that numerous phylogenies involving algae have been proposed.[733,798,884,924,2749,4244,4310,5395] Virtually all utilize biochemical characteristics (e.g. pigmentation) in combination with cytological and morphological features, although some[1108] consider only morphological and ultrastructural characteristics and others[3956] only biochemical features.[4643] Klein and Cronquist[2534] provided an exhaustive comparative account of biochemical, micromorphological, and physiological aspects of algal phylogeny, and constructed a whole array of phyletic schemes based on individual characteristics. Phylogenies may also be constructed taking only a single feature into consideration.[4643]

The general relationship among the classes of algae are shown in Figure 2-20. These are based primarily on similarities in pigments, flagellation, the cell covering, and the principal storage products. All of the classes except for Chrysophyceae appear to represent distinct evolutionary lines, although it should be remembered that that is a largely phenetic***** series.[4643] For further discussion on algal phylogeny see South and Whittick.[4643]

2.9 NUTRIENTS

The criteria of essentiality were set by Arnon and Stout:[158] An element is not considered essential unless (1) a deficiency of it makes it impossible for the plant to complete the vegetative or reproductive stage of its life cycle; (2) such deficiency is specific to the element in question, and can be prevented or corrected only in supplying this element; and (3) the element is directly involved in the nutrition of the plant quite apart from its possible effects in correcting some unfavorable microbiological or chemical condition of the soil or other culture medium.

*****Phenetics is the grouping of organisms into taxa on the basis of estimates of overall similarity, without any initial weighting of characters.[12]

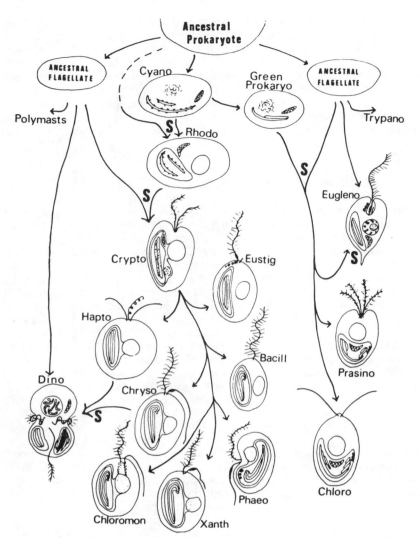

Figure 2-20.   A diagrammatic summary of the phylogenetic relationships among classes of algae. S indicates that a symbiotic event is necessary (Chloromonadophyceae = Raphidophyceae; Charophyceae between Prasinophyceae and Chlorophyceae).[1][110] (From Dodge, 1979, in *Biochemistry and Physiology of the Protozoa,* Hutner, S.A., and Levandarsky, M., Eds., p. 48, reprinted with permission of Academic Press and the author.)

Two methods used to establish the essentiality of a mineral nutrient for growth are:[3427] (1) its omission from the culture solution resulting in a reduction in growth which can be remedied only by returning the nutrient to the plant in a readily available form, and (2) determining a function for the element in the metabolism of the organism. In practice, the first approach is often a prerequisite to the second.

Plant nutrition has been a subject of many reviews.[430,598,1267–8,1290,1646,3199,3963,4749] In the literature, many reviews and publications have also dealt with algal nutrition wholly or at least partly.[490–1,1022,1290,1303–4,1576,1608,1991,2047,2266,2479,2592,3390,3428–9,3529,3806,3923–4,3963,4175,5424]

The elements carbon, hydrogen, oxygen, phosphorus, nitrogen, potassium, calcium, magnesium, sulfur, iron, manganese, copper, zinc, molybdenum, sodium, boron, cobalt, vanadium, silicon, chlorine, and possibly nickel, selenium, bromine, and iodine are required by one or more algal species (Table 2-3). C, H, O, N, P, Mg, Fe, Cu, Mn, Zn, and Mo are considered to be required by all algae and not

Table 2-3. Role of inorganic nutrients in algal metabolism.[1022] (After De Boer, 1981, in *The Biology of Seaweeds,* Lobban, C.S., and Wynne M.J., Eds., p. 361, reprinted with permission of Blackwell Scientific Publications, Oxford)

Element	Probable function	Examples of compounds
Nitrogen	Major metabolic importance as compounds	Amino acids, purines, pyrimidines, porphyrins, amino sugars, amines
Phosphorus	Structural, energy transfer	ATP, GTP, nucleic acids, phospho-phospholipids, coenzymes, coenzyme A, phosphoenolpyruvate
Potassium	Osmotic regulation, pH control, protein conformation and stability	Probably occurs predominantly in ionic form
Calcium	Structural, enzyme activation, ion transport	Calcium alignate, $CaCO_3$
Magnesium	Photosynthetic pigments, enzyme activation, ion transport, ribosome stability	Chlorophyll
Sulfur	Active groups in enzymes and coenzymes, structural	Methionine, cysteine, glutathione, agar, carrageenan, sulfolipids, coenzyme A
Iron	Active groups of porphyrin molecules and enzymes	Ferredoxin, cytochromes, nitrate and nitrite reductases, catalase
Manganese	Electron transport in photosystem II, maintenance of chloroplast membrane structure	Manganin
Copper	Electron transport in photosynthesis, enzymes	Plastocyanin, amine oxidase
Zinc	Enzymes, ribosome structure?	Carbonic andydrase
Molybdenum	Nitrate reduction, ion absorption	Nitrate reductase
Sodium	Enzyme activation, water balance, enzymes	Nitrate reductase
Silicon	Structural	Silica
Chlorine	Photosystem II	Terpenes
Boron	Regulation of C utilization (?), RNA metabolism (?)	Phosphogluconates
Cobalt	Component of vitamin B_{12}	Vitamin B_{12}
Bromine[a]	?	Wide range of halogenated compounds in red algae
Iodine[a]	?	Wide range of halogenated compounds in red algae
Selenium	?	?

[a] Possibly an essential element in some seaweeds.

replaceable even in part by another element.[1022,3530] S, K, and Ca are required by all algae, but can be replaced by other elements. Na, B, Co, V, Si, Cl, Se, and I (and Ni?) are required only by some algae.[3530] Eyster[1304] reported the microinorganic elemental requirements for algae.

Bowen[491] pointed out that it is necessary to keep in mind that only a very small part of the half million species of plants have been investigated with respect to their elemental requirements, and the evidence for essentiality sometimes rests on experiments with a single species.

Essential elements are usually divided into macronutrients and micronutrients, though there is no clear division between these subclasses.[156,490] C, H, O, N, P, S, K, and Mg are generally considered to be macronutrients for algae, while the other essential elements are considered to be micronutrients.[156,1303] Sodium and calcium, however, have been reported as macronutrients for blue-green algae.[156,1303] The macronutrients are required in relatively large quantity and are used generally as building materials. The micronutrients are required in small amounts and are required as co-factors in enzyme systems or are utilized in electron transport systems.[3144,4643] Macronutrients, based on the critical concentration in the culture medium, are required in the range of 10^{-2} M to 10^{-4} M. The quantitative requirements for micronutrients are less than 10^{-5} M.[1303] In addition, there are differences in quantitative requirements within the micronutrient category. The requirement of Fe can be e.g. 1,000 times higher than that of Mo and 100 times higher than that of Co and V. Some authors have used a term "ultramicronutrient" for molybdenum, vanadium and cobalt.[3429]

In Figure 2-21, the idealized diagram showing the yield of an organism as a function of the essential element concentration is given. In fact, every essential element is toxic in higher concentration. For some elements such as N, P or K the optimum range of concentrations is broad, for some elements such as Cu or Zn the range is very narrow.

Eyster et al.[1305] reported that there is a difference in minimal concentrations of elements needed for autotrophic and heterotrophic growth of algae. Elements involved in the photosynthesis (Mg, Fe, Zn, Mn, P) are generally required in higher concentrations for autotrophic growth.

Bowen[490] classified the functions of the essential elements (excluding C, H and O) as:

- electrochemical (Ca, Cl, K, Mg, N, Na, P, S);
- catalytic (Ca, Cl, Co, Cu, Fe, K, Mg, Mn, Mo, Na, P, S, Zn);
- structural (Ba, Ca, F, Fe?, Mg, N, P, S, Si, Sr);
- miscellaneous (B, Br, Cd, Cl, Cu, Fe, I, Se, V, Zn).

Elements with an electrochemical function occur in the cell as free ions at concentrations differing from these of the medium in which the cell lives. Nearly all the essential elements, both major and minor, are believed to have one or more catalytic functions in the cell. Catalysts from living organisms are called enzymes and all known enzymes are proteins, but not all proteins are enzymes.[490]

Bowen[491] classified essential elements functions (other than C, H, O, N, P and S) under the following headings:

1. Inorganic structural materials: Ca, F, Si and rarely Ba, Sr, Mg.
2. Electrochemical and messenger functions of ions: Ca, Cl, K, Mg, Na.
3. Small molecules, including antibiotics (Br, Cl, Cu, F, Fe, I), porphyrins (Co, Cu, Fe, Mg) and miscellaneous (As, B, Ca, Hg(?[******]), Se, Si, V).
4. Large molecules, almost all proteins, some of which are enzymes with catalytic properties (Ca, Cl, Co, Cr?, Cu, Fe, K, Mg, Mn, Mo, Na, Ni, Se, Zn), while others have storage, transport or unknown functions (Ca, Cd, Co, Cu, Fe, Hg(?[******]), I, Mn, Ni, Se, Zn); a few contain more than one metal.
5. Organelles or part organelles, e.g. the nitrogenase enzyme system, mitochondria and chloroplast, in which several metals may function co-operatively (e.g. Fe, Mo, Cu). For detailed discussion about these categories see Bowen.[491]

2.9.1 Pathways of Ion Entry into Cells

The relevance of algal work to the ionic relationships of higher plant cells is sometimes questioned, but the fact remains that work in which it has proved experimentally feasible to study the mechanism of ion transport processes by a series of well-defined specific questions, has largely been confined to algae.[3008] Macrobbie[3008] pointed out that the development of current ideas of ion transport, made possible by modern experimental techniques, has been largely based on algal work. For comprehensive review of ion uptake see Macrobbie[3008] (and references therein). Ions can enter the cell by four pathways:[2886] (1) adsorption, (2) passive transport, (3) facilitated diffusion, and (4) active transport.

2.9.1.1 Adsorption

Ions enter cells by moving across the boundary layer of water surrounding the cell. The route of ion entry into cells is to the cell surface, then passive through the cell wall and plasmalemma into the cytoplasm. Some ions, especially cations, may not reach the plasmalemma because they become

---

[******]author's note

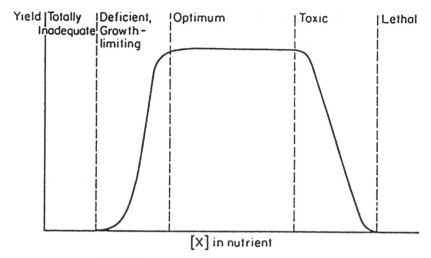

Figure 2-21. Idealized diagram showing the yield of an organism as a function of the concentration of an essential element X in the nutrient supplied.[4601] (From Smith, 1962, reprinted with permission, from the *Annual Reviews of Plant Physiology*, Vol. 13, p. 102, 1962, by Annual Reviews, Inc.)

adsorbed to certain components of the cell wall. Polysaccharides and proteins have sulfate, carboxyl, and phosphate groups from which protons can dissociate, leaving a net negative charge on these compounds in the cell wall. In effect, these macromolecules act as cation exchangers, consequently, large amounts of cation can be adsorbed from the environment.[2886]

2.9.1.2 Passive transport

Algal plasma membranes consist of polar lipid bilayers interspersed with proteins. Nonelectrolytes (uncharged particles) diffuse through membranes with a rate proportional to their solubility in lipid and inversely proportional to their molecular size.[2886] The driving force for passive diffusion of an ion is the difference of electrochemical potential between two points along the path.[1022]

Dissolved gases move more freely than most solutes. As a result, many important gases (e.g. CO_2, NH_3, O_2 and N_2) cross lipid bilayers by dissolving in the lipid portion of the membrane, diffusing to the other lipid-water interface, and dissolving in the aqueous phase on the other side of the membrane. Uncharged molecules such as water and urea are also highly mobile. However, molecules such as NH_3 may be trapped inside cells when they are converted to ions.[2886]

Metabolic energy is not directly required for passive diffusion, however, the maintenance of intact cells and also the maintenance of Donnan potentials, either via macromolecular synthesis or active ion uptake, are dependent on metabolic energy.[1022] The Donnan equilibrium has a small temperature coefficient and is unaffected by oxygen and other gases, but is greatly altered by pH changes which may suppress the ionization of weak acid groups such as carboxyl.[490-1] Donnan equilibria are indeed found for dead tissues, but are wholly inadequate to explain most of the facts about the uptake of ions, such as the selective uptake of K^+ and rejection of Na^+ by most cells.[491] Since passive diffusion occurs without the expenditure of metabolic energy, metabolic inhibitors have no direct effect on diffusive flux. In addition, no carriers or binding sites are involved in diffusion and therefore it is nonsaturable.[2886]

2.9.1.3 Facilitated diffusion

Facilitated diffusion resembles passive diffusion in that transport occurs down an electrochemical gradient,[1022,2886] but frequently the rate of transport by this process is faster. In facilitated diffusion, carriers or enzymes (permease) are thought to bind the ion at the outer membrane surface and to cross the membrane to the inner surface where the bound ion is released.[2886] Although the exact mechanisms of physically binding and moving the solutes through the membrane are still unknown, the carrier concept has found widespread application in the interpretation of experimental observations.[1022] Facilitated diffusion mechanisms have many properties characteristic of enzymatic involvement: (1) they can be saturated, and transport data fit a Michaelis-Menten-like equation; (2) they exhibit stereospecificity with respect to the solute (substrate); and (3) they are susceptible to both competitive and noncompetitive inhibition.[735] However, in contrast to active transport mechanisms, any energy expenditure required for transport must be indirect.[2886]

2.9.1.4 Active transport

The most intriguing transport process of membranes is active transport, by which a material is moved across a membrane against an electrochemical gradient, i.e. from a lower concentration to a higher concentration.[1022] For this reason, active transport has also been termed "uphill" transport.[2886] Active transport is the transfer of ions or molecules across a membrane at rates greater than by the combined rate of free diffusion and facilitated diffusion.[2886] Active transport covers all modes of uptake of ions which require the expenditure of metabolic energy.[491]

Active uptake rates are roughly doubled by raising the temperature from 10°C to 20°C, which would only increase passive diffusion by about 20%.[491] Active uptake is relatively independent of the pH of the nutrient medium over the range 4–9. It does not involve the appearance of an equivalent number of exchanged ions in the medium.[491] Like facilitated diffusion, active transport is selective with respect to various ions and groups of ions. The carriers involved in active transport are thought to be similar if not actually the same carriers involved in facilitated diffusion.[1022] However, active transport systems differ from simple and facilitated diffusion systems in three important ways:[1022] (1) in active transport, energy is expended to facilitate the release of the molecule or ion from the carrier (i.e. uptake is linked with metabolic processes); (2) solutes can be taken up (or eliminated) against electrochemical gradients; and (3) active transport systems are usually directional, and transport a solute either into or out of the cell.

2.9.2 Uptake of Nutrients

Nutrient ions generally enter plant cells by three mechanisms: passive diffusion, facilitated diffusion, and active transport.[2886] Oxygen and carbon dioxide uptake, except at high pH when active transport is necessary for the uptake of bicarbonate at rates sufficient to support net photosynthesis, is by passive diffusion.[3996] Other inorganic nutrients are generally taken up in ionic forms, and this uptake usually involves active transport linked to photosynthesis, which provides the energy necessary for the uptake.[4643] If transport occurs solely by passive diffusion, the transport rate is directly proportional to the external concentration (Figure 2-22a). In contrast, facilitated diffusion and active transport exhibit a saturation of the membrane carriers as the external concentration of the ion increases. The relationship between uptake rate of the ion and its external concentration is generally described by a rectangular hyperbola, similar to the Michaelis-Menten equation for enzyme kinetics.[2886] This approach (Figure 2-22b) was first used by Dugdale[1175] for nutrient uptake kinetics in the growth of phytoplankton and was followed by many others, e.g.:[1047,1257,1264,1896,3351,5005,5552]

$$V = V_{max} \frac{S}{K_s + S} \tag{2.1}$$

where V is the nutrient uptake rate, V_{max} is the maximum uptake rate, S is the concentration of the nutrient, and K_s is the half saturation constant of substrate concentration at which V is $V_{max}/2$. This implies that at low nutrient concentrations the rate of uptake increases rapidly with an increase in external nutrient concentration, but at higher concentrations an increase in concentration adds progressively less to the uptake rate until a high point is reached (V_{max}) at which the uptake is virtually constant and thus independent of the external concentration of the nutrients.[4643]

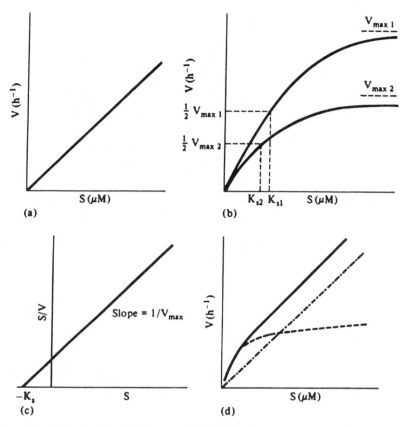

Figure 2-22. Hypothetical plots of nutrient uptake rate (V) and concentration of the limiting nutrient (S), for (a) passive diffusion only, where V is directly proportional to S; (b) facilitated diffusion or active transport in which V_{max} in example 2 is half V_{max} of example 1, resulting in a concomitant decrease in the K_s; (c) linearized plot of the data in (b) to illustrate how the kinetic parameters V_{max} and K_s are determined graphically; (d) passive diffusion plus active transport (—), and active transport (- - -) with the passive diffusion component (- . -) subtracted.[2886] (From Lobban et al., 1985, in *The Physiological Ecology of Seaweeds*, p. 83, reprinted with permission of Cambridge University Press.)

The half-saturation constant reflects enzyme activity at the cell surface. High enzyme activity increases the probability for interaction between enzyme and substrate. Cells with a high enzyme activity will be more efficient in assimilating phosphorus at low concentrations than will cells with low enzyme activity. Efficiency is reflected by the half-saturation constant: a high value of K_s is associated with low enzyme activity and poor uptake efficiency; similarly, low values of K_s indicate high enzyme activity and high uptake efficiency.[181]

Healey[1994] suggested that the slope of the initial part of the hyperbola (the linear portion, V_m/K_s) may be a more useful parameter for comparing the competitive ability of various species for a limiting nutrient. One of the problems in using K_s is that its value is not independent of V_{max} (i.e. when V_{max} decreases the value of K_s will also decrease even though the initial slope of the hyperbola remains the same (Figure 2-22b). The kinetics parameters may be estimated graphically from the rectangular hyperbola, but generally the Michaelis-Menten equation is rearranged to yield a straight line from which K_s and V_{max} can be calculated more accurately by linear regression (Figure 2-22c). The advantages and disadvantages of the three possible linear plots have been examined by Dowd and Riggs;[1135] the S/V vs. S (Woolf plot) or the V/S vs. V plots are generally superior to the 1/S vs. 1/V plot (Lineweaver-Burk equation). In S vs. S/V transformation, K_s is given as the Y-intercept and V_{max} as the slope of the regression equation.[1264]

An enhancement of V_{max} by nutrient deficient phytoplankton is well documented.[785-6,1261,1523,3126,4083] A decline in growth rate and cell quota is usually accompanied by an increase in V_{max}.[1261,4083]

Dugdale[1175] emphasized the ecological advantage given to an alga by the set of physiological adaptations represented by a low half-saturation constant (K) and a high maximum growth rate (μ_{max}). He discussed the question of whether the species with lower μ_{max} values also have lower K that would enable them to compete with species with high specific uptake rate (U_{max}) values, when living in oligotrophic waters. On the other hand, eutrophicated waters should be inhabited mostly by fast-growing species, because, as nutrients are more abundant, ability to take them up at low concentrations is not as important, and natural selection will tend to favor organisms capable of dividing rapidly.[3019] The simplest application of the model, i.e. the comparison of the K and U_{max} values of species competing for the same limiting nutrient, was done soon after Dugdale's paper. Following papers[1257,1259,1264] hypothesized that species involved sequentially in a seasonal succession of dominant species related to declining nutrient concentrations should be ordered by lower and lower K values. Nutrient-based algal competition in theory and experimental studies is discussed in an excellent paper by Maestrini and Bonin.[3019]

The growth rate of an alga will decline if the concentration of the nutrients in the external medium drops below that which will support an uptake rate sufficient to maintain the current growth rate.[4643] The relationship between growth rates and limiting nutrients usually can be described by one or two hyperbolic functions.[973] The Monod equation[3279] which was originally applied to the growth of bacteria in chemostat culture, also describes nutrient limited, steady-state growth of algae:[1158,1835,4974,5127]

$$\mu = \mu_{max} \frac{S}{K_s + S} \qquad (2.2)$$

where μ is the specific growth rate, μ_{max} is the maximum specific growth rate, S is the substrate concentration, and K_s is the half saturation constant for growth.

In other studies it has been observed that growth rate is related to the cellular concentration of the limiting nutrient rather than to its external concentration.[182,367,862,1155-6,1610,1692] This relationship is expressed in the Droop equation:[1155]

$$\mu = \mu'_{max} (1 - Q_o/Q) \qquad (2.3)$$

where Q is the amount of the nutrient in the algal cells, Q_o is the lowest level of Q at which the algae can grow, and μ'_{max} is the growth rate at infinite Q (Figures 2-23 and 2-24). The Droop equation is well established under steady state conditions,[3295,5055] the non-steady state kinetics of nutrient-dependent algal growth and uptake was presented by Grover.[1837] The *critical concentration* is that concentration of a nutrient in the tissue just below the level giving maximal growth[1022] (Figure 2-25). The critical concentrations for a given element vary with algal species involved (see De Boer[1022] for discussion).

Under steady-state conditions both Monod and Droop equations describe nutrient-limited growth equally well because uptake (a function of external nutrient concentration) is in equilibrium with the cell quota of the nutrient.[973] The Droop model offers an advantage that cell quota is more readily measured in the particulate matter following concentration by filtration or centrifugation than the very low external nutrient concentrations.[973,4643] The Droop equation is a theoretical description, in that Q is never infinite and the true maximum growth rate μ_{max} at Q_{max} will be less than the theoretical maximum μ'_{max}; thus both Q_{max} and μ_{max} must be determined experimentally.[4086]

Factors affecting nutrient uptake rates can be divided into three broad categories:[2886] physical, chemical and biological. Physical factors include light (indirect effect through photosynthesis), temperature, water motion and exposure to air.[2886] The most important chemical factors include concentration and ionic form of the element being taken up.[2886] Biological factors that influence uptake rates include the type of tissue, the age of the plant, size of cells, its nutritional history and interplant variability.[1606,1834-5,2886]

Cell size is known to influence many properties of algal cells.[1935,4078] The prevalent viewpoint is that small cells are better competitors for nutrients than larger cells.[1935,3030,4603,5244] This hypothesis is based on observations that small cells have higher nutrient uptake rates per unit biomass,[4602] higher maximal growth rates,[4625-6] and lower susceptibility to transport limitation of nutrient uptake by

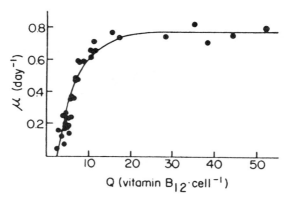

Figure 2-23. Growth rate (μ) as a function of cell quota (Q) of vitamin B₁₂ for *Pavlova (Monochrysis) lutheri.*[1155] (From Droop, 1968, in *J. Mar. Biol. Ass. U.K.,* Vol. 48, p. 699, reprinted with permission of Cambridge University Press.)

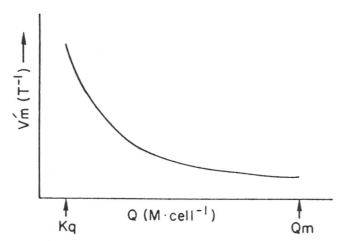

Figure 2-24. Enhanced potential for maximum specific rate of nutrient uptake (V'm) as a function of cell quota (Q) assuming that the maximum rate of uptake per cell (ρ_m) is invariant with Q. K_q and Q_m represent minimum and maximum values of Q for steady-state populations maintained in continuous culture.[3125] (From McCarthy, 1981, in *Physiological Bases of Phytoplankton Ecology,* Platt, T., Ed., p. 220, reprinted with permission of Department of Fisheries and Oceans, Ottawa.)

molecular diffusion[1583] than large cells. Grover[1834] found that for spherical cells, smaller cells are better competitors than large ones. For cells that are very elongated in shape, however, large cells are often better competitors than small ones. The size spectra of benthic and pelagic microalgae have not been compared, but there may be a tendency for benthic microalgae to be larger and heavier.[4281] Affinity for nutrients, however, is also dependent on internal cell concentrations and affinities are often markedly lower for benthic microalgae than phytoplankton because the former are sitting on top of a large nutrient pool and need to invest less in uptake capacity.[4071] Sand-Jensen and Borum[4281] in their excellent review, pointed out that cell requirements of nitrogen and phosphorus relative to carbon are probably not different for the two types of organisms but the photosynthetic capacities and the realized rates of benthic microalgae are usually lower than in phytoplankton communities[23,2406] and accordingly the needs for new supply of nitrogen and phosphorus to growth of benthic organisms are smaller per unit time. Phytoplankton biomass is often strongly nutrient limited in lakes and coastal marine regions[464,5349] and this is rarely the case for microalgae in sediments.[25]

The kinetics of nutrient uptake is broadly discussed by McCarthy,[3125] Morel[3295] and Dugdale et al.[1180]

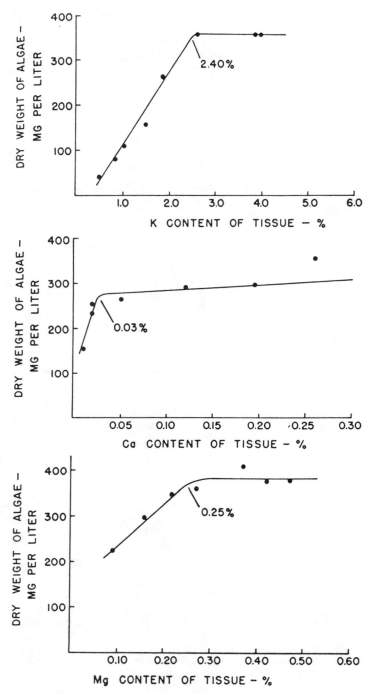

Figure 2-25.   Examples of the curves relating algae yield and element content from which critical concentrations of the essential cations were established.   1. Critical potassium in *Draparnaldia plumosa*.   2. Critical calcium in *Draparnaldia plumosa*.   3. Critical magnesium in *Stigeoclonium tenue*.[1612]   (From Gerloff and Fishbeck, 1969, in *J. Phycol.*, Vol. 5, p. 111, reprinted with permission of the *Journal of Phycology*.)

2.9.2.1 Vitamins and growth regulators

Many algae require vitamins (auxotrophy); most higher plants do not.[3930] The algae stand out among microorganisms for needing only three vitamins.[1152] Vitamin B_{12} and thiamine are required alone or in combination by the majority of the auxotrophic algae, and B_{12} seems required more often than thiamine.[3930,4851] Organisms having a requirement for biotin often require one of these other vitamins.[4851] Biotin so far has been shown to be necessary for a few chrysomonads and dinoflagellates and one euglenoid.[3930] Vitamins are required in very low concentrations (10^{-13} - 10^{-10} mol L^{-1}) for use as cofactors by organisms which otherwise undergo purely autotrophic growth.[4851] Vitamin requirements are scattered among all the major groups of algae. Within a genus there may be species with and without a vitamin requirement.[3924–5] Within a species there may be individual clones with different vitamin requirements.[1843,1914,2805] The review of Provasoli and Carlucci[3930] gives the comprehensive listings of vitamin requirements in algal species studied and the patterns of requirements among algal groups.

Provasoli and Carlucci[3930] pointed out that the need for an exogenously indispensable molecule obviously points to an inability of the organism to synthetize the substance. The physiological impairment is seldom total; often the biosynthesis is only partly blocked at one step on the pathway, as in the familiar phenomenon of leaky mutants studied in microbial genetics. In such cases, the intermediate after the block, if taken up, replaces the vitamin.[3930]

Thiamine was the first vitamin to be identified as a growth factor for algae.[2984–6] Thiamine molecule is composed of a thiazole and a pyrimidine moiety. The thiamine requirement is satisfied by the thiazole moiety alone in some species and by pyrimidine moiety in some species; some species require both moieties, and only *Ochromonas malhamensis*,[2265,2819] and a mutant of *Chlamydomonas reinhardtii*,[1297,2819] require the intact thiamine molecule. Thiamine, as thiamine pyrophosphate, is a cofactor in the decarboxylation of pyruvic and other a-keto acids and in other reactions.[4851]

The isolation of vitamin B_{12} in 1948 was quickly followed by its identification by Hutner and co-workers[2267–8] with hitherto unknown factors needed by the flagellates *Euglena gracilis* and *Ochromonas malhamensis*; and this led to recognition of the prevalence of the requirement for this vitamin among algal flagellates.[1152] Vitamin B_{12} has attracted more interest than thiamine. Vitamin B_{12} (α-5,6-dimethylbenzimidazolyl cyanocobalamin) is one of a family of cobalamines (see also Figure 5-20). The molecule is composed of two major portions, a large corrin nearly planar group, and a nucleotide side chain (α-5,6-dimethylbenzimidazole-D-ribofuranose phosphate).[3930] The minimum cell requirement for vitamin B_{12} by an auxotroph is very small. It is of the order of 2–18 molecules μm^{-3} cell volume computed in batch cultures, almost regardless of the type of microorganism, alga or bacterium.[520,1148,1845] Vitamin B_{12} is unusual as a vitamin in being synthetized primarily by microorganisms. It is important as a vitamin for animals and microorganisms only, and not for higher plants, which lack vitamin B_{12} and contain enzymes using different cofactors for the transformations which utilize this vitamin in those other kingdoms.[2311,4391] Lewin[2819] listed some 65 species requiring an exogenous supply of vitamin B_{12}. For further details see the excellent reviews by Provasoli and Carlucci[3930] and Swift.[4851]

Biotin is required by a much smaller number of algae, all of which require either vitamin B_{12} and/or thiamin.[4209] Biotin is soluble in hot water, sparingly soluble in cold water and is a cofactor in some carboxylations and transcarboxylations. Biotin enzymes are pyruvate carboxylase, catalyzing the conversion of pyruvate to oxaloacetate in mitochondria with ATP, CO_2, manganic ion, magnesium ion, acetyl CoA, and biotin required, and methylcrotonyl CoA carboxylase, catalyzing a carboxylation step in breakdown of leucine to eventually form CoA. Biotin is not a cofactor for enzymes in the major pathways of autotrophic carbon fixation (RuBPC or PEPC).[4851]

The effect of vitamins was first studied quantitatively as all-or-nothing.[4851] However, it became obvious that, at sufficiently low concentration, a graded response in growth rate should be observed, and Ford[1435] found a Michaelis-Menten type of saturation kinetics for *Ochromonas malhamensis*. The half saturation constant for growth is usually very low, as low as 0.14 ng L^{-1} for vitamin B_{12} in *Monochrysis lutheri*.[1154–5] (See also Figure 2-23.) Carlucci,[690] however, reported a thiamine K_s for uptake by *Monochrysis lutheri* as high as 125 ng L^{-1}.

The dissolved vitamins in the sea and in freshwaters have been thought to originate mainly from the activities of bacteria, both in water column and in the sediments.[631–3,3925] However, several algae had been reported to release vitamins in the culture medium.[4,631,1436]

For further details on vitamins and growth regulators see e.g. References 691–2, 1152, 1928, 3301, 3904, 3906, 3924–5, 3927–8, 4951.

2.9.2.2 Effect of nutrient deficiencies

Generally, as expected, a deficiency in any nutrient is reflected in a greatly decreased cellular content of that nutrient. Deficiency of one nutrient is also often accompanied by changes in the cellular content of other elements.[1991]

Perhaps the most general response of algae to nutrient deficiency is a decrease in the content of photosynthetic pigments.[1991] Lowered chlorophyll content has been reported in a variety of algae as a result of deficiencies of nitrogen,[72,4953] phosphorus,[978,1523] sulfur,[978,3866] silicon,[879,5330] magnesium,[978,4585] iron,[978,3559] potassium,[978,3810] or molybdenum.[166] In blue-green algae and red algae, the photosynthetically active biliproteins are also lost in response to various nutrient deficiencies.[72,1229,1982,3559,5106]

A second, rather general response to nutrient deficiency among algae is the accumulation of C-storage compounds.[1991] Usually this is carbohydrate,[72,206,3300,5330] but lipid accumulation has also been reported.[849] Carbohydrate has been found to accumulate early in N deficiency and lipid only in more severe deficient cells.[1,4100,5330] A third general response to nutrient deficiency is a decrease in protein. This has been reported for algae deficient in nitrogen,[5106,5330] phosphorus,[1523,5330] silicon[879,5330] magnesium,[1550] or zinc.[5216]

There are several metabolic responses to nutrient deficiency which, on the basis of present evidence, appear to be rather general among microorganisms.[1991] The most extensively documented of these is the increased ability of the cells to take up the deficient nutrient.[404,1365,4777,4855,5113] This is most easily seen as an increased maximum rate of uptake.[1991] For further details see the excellent review by Healley.[1991]

2.9.2.3 Extracellular products

In general, any environmental condition which inhibits cell multiplication, but still permits photoassimilation to continue results in the release of high proportions of the photoassimilate.[2017]

The production of a great variety of extracellular substances by algae is now well established. It is also clear that such substances often play important roles in algal growth and physiology, as well as in aquatic food chains and ecosystems in general.[2017] Reviews on this subject have been published by Fogg,[1409,1411,1414] and extensive discussion of relevant work in this field may also be found in some other publications.[2,758,1664,5316] The following information is based mostly on an excellent review by Helleburst.[2017]

Carbohydrates—simple and complex polysaccharides are liberated by a large number of taxonomically diverse algae.[66,1844,2017] The amount released may represent a considerable fraction (15–90%) of the photoassimilated carbon of some algae during active growth.[1846,2367,2816,3281,3863] Other algae produce extracellular polysaccharides mainly when they enter stationary growth phase.[1844,3072] Polysaccharide release may become substantial, sometimes showing good correlation with cytological events.[3970,3972] Common constituents of the extracellular polysaccharides and heteropolysaccharides are: glucose, galactose, mannose, rhammose, fucose, arabinose, xylose, and uronic acids.[379,1846,2367,2816,3072,3281] Extracellular carbohydrates are summarized by Helleburst.[2017]

Nitrogenous substances—amino acids and peptides are very common in algal filtrates but in most cases represent only a small fraction of the total extracellular material.[66,1421-2,1664,2013,5292-3] Blue-green algae, on the other hand, liberate a very large portion (14–60%) of their assimilated nitrogenous substances into the medium.[1401,1411,2365,5278] A large portion of the extracellular nitrogenous material released by blue-green algae is in the form of polypeptides, and only small amounts of free amino acids are usually found.[1401,2365,4759,5402,5278]

Organic acids—glycolic acid is the organic acid most commonly liberated by algae.[66,1934,2013,3387,5293-4] The release of glycolate by algae is favored by conditions where CO_2 limits photosynthesis, e.g. high light intensities, low CO_2 concentrations and high pH.[3914,4999,5294,5398] The studies indicate that the amount of glycolate released under favorable conditions varies considerably for different species.[2017] Algae grown anaerobically liberate large quantities of fermentation products including lactate which is most common, acetate, pyruvate and formate.[95,221,2017,2576,4863]

Other compounds released by algae include lipids,[2017,3917,4385,4658] phenolic substances,[861,907,2487,2617,4479] vitamins,[691-2,1409,2817] enzymes,[2,3902-3] organic phosphates,[114,2629,4805] volatile substances,[2290,2403] sex factors,[2139,3668-9] toxins (see Chapter 2.15.1), growth inhibitors and stimulators,[315-6,907,1929-31,3387,4658] and auto-inhibitors and antibiotics (see References 1409 and 2820 for review, see also Chapter 2.15).

2.10 NUTRITIONAL TYPES

Algae can be *autotrophic* (*lithotrophic* or *holophytic*) or *heterotrophic* (*organotrophic*).

If they are autotrophic, they use inorganic compounds as a source of carbon. Autotrophs can be *photoautotrophic* (*photolithotrophic*), using light as a source of energy, or *chemoautotrophic* (*chemolitrotrophic*), oxidizing inorganic compounds for growth.[2745] Some algae, particularly flagellates, are *photoauxotrophic,* requiring a small amount of an organic compound, which does not contribute significantly to the cell carbon. These algae usually require a vitamin, e.g. B_{12}, thiamine or biotin. The term signifies a growth factor requirement.[55,2745,3408] Some algae are *obligately photoautotrophic*—that is, they are unable to utilize organic substances of any kind as a substitute for an inorganic source of any of the required elements.[55]

A number of algae are heterotrophic, either facultatively or obligately so. Several algae in which the photosynthetic mechanism is impaired require an organic carbon source, acetate, for example, for optimal growth. Such organisms are *obligately heterotrophic.* Some of these algae grow anaerobically.[55] A number of photoautotrophic species can live in darkness (some retaining their chlorophyll) as long as they are supplied with organic substances which they can metabolize. Such organisms are *facultatively heterotrophic* (*mixotrophic*).[55,2745] Heterotrophs can be *photoheterotrophs* (*photoorganotrophs*), using light as a source of energy, or *chemoheterotrophs* (*chemoorganotrophs*), oxidizing organic compounds for energy.[2745] Heterotrophic algae may be *phagocytotic* (*holozoic, phagotrophic*), absorbing food particles whole into food vesicles for digestion, or they may be *osmotrophic,* absorbing nutrients in a soluble form through the plasma membrane.[55,2745] If algae live heterotrophically on dead material, they are *saprophytic,* if they live off a live host, they are *parasitic.*[2745]

Probably most algae, including those unable to grow heterotrophically in the dark, are able to incorporate certain organic compounds into cellular material in the light. The terms *photoheterotrophy, photometabolism* and *photoassimilation* have been used for such processes, though the use of the word photoheterotrophy should be restricted to those situations in which the organic substrate serves as the exclusive, or at least the main, source of cell carbon during the growth.[3408]

2.10.1 Photoautotrophy

Photoautotrophs (photolithotrophs) obtain energy by absorption of light, and electrons from the inorganic substrate. Cell carbon is generally obtained by reduction of CO_2; in photosynthetic algae and higher plants, the inorganic reductant is water, which is oxidized to oxygen. Since such plants are generally able to grow in the light in a simple, defined mineral medium with CO_2 as the only source of cellular carbon, they have often been termed photoautotrophs.[3408] Those algae for which this is apparently the only possible nutritional mode are generally called obligate photolithotrophs or obligate photoautotrophs,[967,4152] although further experimentation may, of course, reveal that the photolithotrophy is facultative rather than obligate.[3408] Obligate phototrophy appears to be widespread among diatoms[2794] and dinoflagellates; in most of these cases, other, closely related species are facultative heterotrophs.[967]

2.10.2 Auxotrophy

Auxotrophy, or lack of it, does not relate to sources of energy used by the algae; it appears to be an independent loss of function, sensu Lwoff.[2983] Obligate photoautotrophic species may need vitamins as do some *Synura* spp. and many marine algae. Obligate heterotrophs, such as algae which have lost their photosynthetic apparatus, may not need vitamins, i.e. *Polytoma uvella* and *P. obtusum.*[3930]

Since vitamins are present in all environments, auxotrophy does not strictly correlate with the trophic state of the environment. Thus auxotrophs are found in sewage lagoons and in oligotrophic waters as well.[3930] However, the order of incidence of auxotrophs and autotrophs may be expected to differ either in number of individuals or species. Obviously, in environments perennially rich in vitamins the auxotrophs may thereby have an advantage in avoiding the necessity of synthesizing these vitamins.[3930]

2.10.3 Heterotrophy

Heterotrophy is defined as the utilization of dissolved and particulate organic compounds for growth.[312,1157] Among planktonic algae, the potential for uptake of dissolved compounds is usually limited by low surface/volume ratios of algae compared to planktonic bacteria[2124] and by low concentrations of organic compounds. Algae that grow in mixed assemblages with bacteria, animals and detritus on solid surfaces or in sediments are usually exposed to much higher concentrations of dissolved compounds and heterotrophy may consequently be more important.[3071]

Photosynthetic carbon fixation by algae exposed to reasonable light levels (i.e. above a few percentages of surface irradiance) is usually much more important for growth than uptake of dissolved organic compounds for growth.[312,2124,5167] For algae growing permanently and temporarily at very low light, however, heterotrophy may become important not only for growth but also for the ability to maintain biomass and cellular functions.[3071] Uptake of dissolved organic compounds to cover respiratory maintenance is not heterotrophy according to Droop's definition above but is perhaps more important ecologically than utilization of external compounds for cellular growth.[3071] It is essential for many algae to tolerate low light without losing photosynthetic capacity. In the presence of suitable organic substrates growth and metabolism of photosynthetic algae are often enhanced by light of appropriate intensity.[3802]

Bennett and Hobbie[312] pointed out that algae in nature must always compete with bacteria for dissolved organic compounds. All the evidence indicates that even in sediments and sewage lagoons, the algae can never obtain adequate quantities of substrate to maintain themselves solely by heterotrophy on dissolved organic matter. It may well be, however, that the extra energy these algae obtain in this way is important for survival in extreme environments. The authors also stress the fact that the only algae whose heterotrophy has been well studied have been forms that are relatively easy to grow.

2.10.4 Mixotrophy

Mixotrophy occurs in a few algae which, presumably as a result of impaired capacity to assimilate carbon dioxide in the light, require a supply of organic carbon even for growth in the light.[3408,3903]

True mixotrophy, defined as use of particulate organic matter for cellular growth, has been conclusively demonstrated for only a few species in the genus *Ochromonas*,[1331] and for *Poterioochromonas malhamensis*.[696] Much more widespread is the ability to use a restricted range of organic substances, available at high concentrations, as a dietary supplement or sole C source in the dark.[110] Mixotrophy in algal flagellates has been considered mainly a strategy for gaining C during low light[373,4270] and therefore has been studied in algae that use bacteria as their primary C source. Doddema and Van der Veer[1100] suggested that phagocytosis might permit utilization of particulate organic N and P when inorganic nutrients are in limited supply. Bacteria are therefore a potential source of P when obligate phototrophic algal flagellates are subjected to P limitation, because bacteria are more efficient at sequestering P under these conditions.[533] Thus bacteriovory in obligate phototrophic algal flagellates may be an important strategy for acquiring nutrients during periods of inorganic nutrient limitation.[3488]

2.10.5 Phagotrophy

Phagotrophy is a method of feeding in which particles of food are ingested, digested in food vacuoles, and the digest absorbed through the vacuolar membrane into the cell.[1152] This, of course, is not a nutritional mode characteristic of algae, although it is found in many colorless forms, e.g. *Oxyrrhis marina* as well as in some pigmented ones, e.g. *Ochromonas malhamensis*.[1146,1149,2268,3408,3903-4,4018]

Porter[3852] discussed the role of phagotrophic phytoflagellates in microbial food webs. He found in field studies during a chrysophyte bloom that phytoflagellate grazing exceeded heterotrophic microflagellate grazing and constituted up to 55% of the bacteriovory of all microflagellates, ciliates, rotifers, and crustaceans combined.

2.11 TOXICITY

Wood[5487] classified elements according to their toxicity and availability into three groups: (1) noncritical (e.g. C, N, P, O, Na, Mg, Fe, K, Ca, Si); (2) toxic but very insoluble or rare (e.g. Ti, Ga, Yr, W, Ta, Ru, Re); and (3) very toxic and relatively accessible (mostly heavy metals, e.g. Be, As, Co, Se, Hg, Ni, Cu, Pb, Zn, Ag, Sb, Cd). Some heavy metals in group 3, however, such as copper, zinc, and cobalt, are essential micronutrients and are frequently referred to as trace metals. They may limit algal growth if their concentrations are too low and can be toxic at higher concentrations; frequently the optimum concentration range for growth is narrow (Figure 2-21).

Nieboer and Richardson[3436] developed a scheme for separation of metal and metalloid ions into Class A (oxygen-seeking), Class B (sulfur-seeking) or borderline (intermediate between A and B). Their classification is related to atomic properties such as electronegativity and ionic radius, and solution chemistry of metal ions. Class A ions include the alkali metals and alkaline earths, notably the biologically essential K^+, Na^+, Mg^{2+} and Ca^{2+}. Class B ions, in contrast, include Cu^+, Hg^{2+}, Ag^{2+} and Pb^{2+} which are extremely toxic and for the most part nonessential. Borderline ions include Fe^{2+}, Fe^{3+}, Mn^{2+} and Cu^{2+}, which have biological roles but which may also exhibit toxic effects.

Bowen[490] classified elements according to the concentration needed to produce toxic effects in an otherwise balanced nutrient solution:

1. *very toxic*—toxic effects may be seen at concentrations below 1 mg L^{-1} in the nutrient solution cations (Ag^+, Be^{2+}, Cu^{2+}, Hg^{2+}, Sn^{2+} and possibly also Co^{2+}, Ni^{2+}, Pb^{2+} and the oxygenated anions of Cr^{VI});
2. *moderately toxic*—toxic effects appear at concentrations between 1 and 100 mg L^{-1} in the nutrient solution (simple anions F^- and S^{2-}, the oxygenated anions of As^{III}, As^V, B, Br^V, Cl^V, Mn^{VII}, Mo^{VI}, Sb^{III}, Sb^V, Se^{IV}, Se^{VI}, Te^{IV}, Te^{VI}, V^V and W^{VI}, and the cations Al^{3+}, Ba^{2+}, Bi^{3+}, Cd^{2+}, Cr^{3+}, Fe^{2+}, Mn^{2+}, Tl^+, Zn^{2+} and Zr^{4+});
3. *scarcely toxic*—toxic effects rarely appear except in the absence of a related essential nutrient, or at osmotic pressures exceeding 1 atmosphere (the simple anions Br^-, Cl^-, I^-, the oxygenated anions of Ge, N, P, S^{VI}, Si and Ti, and the cations Ca^{2+}, Cs^+, K^+, Li^+, Mg^{2+}, Rb^+ and Sr^{2+}).

Trace elements are sometimes referred to as heavy metals. The term "heavy metal" has been generally used to describe those metals having atomic numbers greater than iron or having a density greater than 5 g mL^{-1}.[4629] Some authors, however, consider the term "heavy metals" to be unsatisfactory or unacceptable.[3436,4792] In fact, in the literature both terms are frequently used. Heavy metal cycles in water and soil have been reviewed many times; I would like to bring to the reader's attention especially References 78, 1451, 2395, 3287–8, 4261 and 4738.

Heavy metals exert their harmful effects in many ways, although all the major mechanisms of toxicity are consequences of the strong coordinating properties of metal ions.[3497] The effects of heavy metal toxicity in algae may include:

1) an irreversible increase in plasmalemma permeability, leading to the loss of cell solutes, e.g. potassium,[969,1030,1851,2927,3P,3212,3373,3592,3671,4249,4446] or organic osmotica[4045] and changes in cell volume;[796,1269,1456,3948,4947]
2) a reduction in photosynthetic electron transport[1036,3120,4453] and photosynthetic carbon fixation[987] resulting in decreased photosynthesis;[254,399,734,3772,4177,4708–9,4P4957,5467,5497,5563]
3) inhibition of photosynthetic O_2 evolution;[2288,3593,4189]
4) the inhibition of respiratory oxygen consumption;[4153]
5) the disruption of nutrient uptake processes;[1939]
6) enzyme inhibition, due to the displacement of essential metal ions;[490,1036,1358,3962,5103]
7) the inhibition of protein synthesis;[353,2605,3449]
8) formation of stable precipitates or chelates with essential metabolites (e.g. Fe with ATP) and antimetabolic behavior of some anions (e.g. selenate, arsenate);[490,4331,4455,4457]
9) abnormal morphological development;[26,170,2493,3981,4189,4307,4500,4599,4956]
10) the impairment of motility and loss of flagella in certain microalgae;[1035,1336,2045,3377]
11) the degradation of photosynthesis pigments, coupled with reduction in growth[1589,2288,3185,3449,3960] and, in extreme cases, cell mortality;[1336,2045,3614,4947]

In heterocystous, nitrogen-fixing, blue-green algae, toxic levels of heavy metals may reduce nitrogenase activity,[1227,4801] coupled with pronounced heterocyst damage/cell lysis,[3100] and in some cases, increased heterocyst frequency.[4800]

The susceptibility of organisms to a given metal is a function of both the species and clone or population of a given species.[317,1907-8] When a given test organism is considered, differences in the toxicity of a selection of heavy metals can also be observed.[324,4189,4286]

2.11.1 Mechanisms of Heavy Metal Tolerance in Algae

The ability of algae to survive and reproduce in metal-polluted habitats may depend on genetic adaptation over extended time periods by mutation, genetic exchange, selection, etc., or to changes in physiology resulting from metal exposure. "Tolerance" and "resistance" are arbitrary terms and often used interchangeably. However, "resistance" is sometimes used in connection with a direct response to metal exposure, whereas "tolerance" may rely on intrinsic properties of the organism or the physical and chemical nature of the environment.[4046]

Metals generally decrease diversity and productivity. Blue-green algae and diatoms appear generally to be less tolerant than green algae to metals such as copper, zinc, and lead, and metal-contaminated waters often support the filamentous greens *Hormidium, Ulothrix, Mougeotia, Microspora, Zygnema,* and *Stigeoclonium* in flowing waters[1544,4304-5,5403-4] and unicellular chlorococcalean species[4793] in lakes.

Toxicant-resistant populations have been studied both in the laboratory and in the field.[980,1034,2302,2770,3159,4307,4437,4499,4793,5407,5411]

Figure 2-26 summarizes the route of uptake for an algal cell and with this, identifies 7 possible mechanisms or sites where tolerance mechanisms or sites where tolerance mechanisms could operate.[4792] The metal normally has to be in ionic or "free" state for uptake to operate. The ion then has to pass through or become absorbed upon the cell wall. Exclusion could operate at this point if the cell wall or the sheath were able to take up the metal ion and thus prevent its encountering the cell membrane (1). At the membrane, permeability may be controlled, such that the ion does not move into the cell proper, or moves in very slowly (2). Beyond this, mechanisms (3), (4) and (5) refer to the binding of cations to non-sensitive sites inside the cells, and mechanism (6) requires the production of extracellular ligands (normally in response to the metal) which either transport the metal outside or more probably bind external metal and thus prevent further entry. Mechanism (7) is conceived as some type of alternative biochemical pathway or shunt by which the need for the metal-sensitive enzymes is avoided.[4792]

Mechanisms of algal tolerance to heavy metals have been comprehensively summarized by Rai et al.[3962]

2.11.1.1 Extracellular binding and precipitation

The effects of extracellular polypeptides in reducing the toxicity of heavy metals was first shown for the freshwater blue-green alga *Anabaena cylindrica*.[1421] Subsequently, the complexing capacity of extracellular organic ligands, including organic acids and polypeptides, has been demonstrated in other blue-green algae[3159,5114] and in eukaryotic algae.[648,3159] High concentrations of other organic compounds are able to reduce heavy metal toxicity[3962] and may be important in certain natural waters, e.g. near sewage outflows.

Hydroxamic acids may be the only examples of strong metal-complexing agents that are known to be released by algae and to be present in natural waters.[3403-5,3356,4514]

Algae may also produce specific iron-chelating siderophores[3404,3406] which can function as strong copper-complexing agents.[813] In certain cases, metal crystallization may occur on algal surfaces. Thus *Cyanidium caldarium* is able to remove Cu, Ni, Al, and Cr from acidic mine wastes by cell surface precipitation of metal-sulfide microcrystals.[5488] Francke and Hillebrandt[1472] found more than half the copper associated with the cells of filamentous green algae accumulated in the cell wall. Some algae secrete polymers which form large, extracellular aggregates.[968] These are often polysaccharides with anionic properties, and are capable of binding heavy metal cations. Carrageenan in red algae can bind a range of heavy metals, due to its high cation exchange capacity.[5160] The structural components of algal cell walls behave like weakly acidic cation exchangers; in *Scenedesmus obliquus* the cell walls behave like a weakly acidic cation exchanger, though some heavy metals (e.g. Cd) may be accumulated as neutral complexes.[4698] In *Vaucheria* sp. ionic bonding and covalent bonding to cell surface groups (e.g. carboxyl/sulfate and amino/carboxyl groups, respectively) are important extracellular binding processes.[923] The affinity of alginic acid for different metals is as follows:[1973]

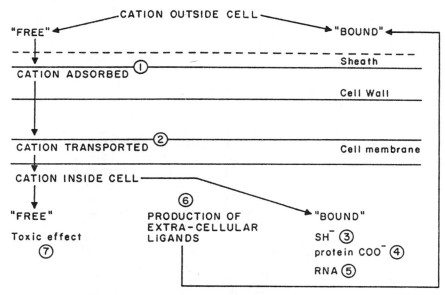

Figure 2-26. Route of uptake for potentially toxic cations. For explanation see text.[4792] (From Stokes, 1983, in *Progress in Phycological Research,* Vol. 2, Round, F.E. and Chapman, D.J., Eds., p. 101, reprinted with permission of Elsevier Science Publishers bv, Amsterdam.)

$$Pb > Cu > Cd > Ba > Sr > Ca > Co > Ni, Zn, Mn > Mg$$

Some microorganisms can synthetize polymers extending from the outer membrane of the cells; these polymers can bind ions from the solution. Protein on the cell surface could be another site for binding.[2495]

The binding capacity for metal ions was shown to be related to algal species composition rather than to total algal biomass or physicochemical parameters.[541] The authors found that most of the binding could be accounted for by certain species of green algae, diatoms, and chrysomonads that usually constituted only a minor fraction of the total algal volume at Heney Lake, Ontario.

2.11.1.2 Impermeability and exclusion

Decreased internal accumulation of a heavy metal has been proposed as a mechanism of tolerance in certain algae.[648,1456] Detoxification of metal ions in the medium or at the cell surface is referred to as an exclusion mechanism since the metal ions do not cross the cell membrane and consequently the uptake of the metal is prevented.[2886] This physical exclusion mechanism may account, in part, for the cotolerance of fouling isolates of *Ectocarpus siliculosus* to elevated levels of Co and Zn.[1875] Some copper-tolerant strains of *Scenedesmus* may also operate an exclusion mechanism, showing a reduced internal accumulation.[3212]

2.11.1.3 Internal detoxification

Previous reports in the literature have demonstrated that heavy metals are often found as deposits in various cellular organelles such as mitochondria,[1782,4500,4820] vacuoles,[2288,4473,4499] chloroplast,[1525,2288,4499] nuclei,[791,2288,3169,3285,4501,4537] polyphosphate bodies,[249,250,2236–8,3951,4472] cell walls,[2288,2338,3948,4470] and even in eyespot.[2288] Evidence of movement from the wall or plasma membrane area to the vacuole or cytoplasm is often reported.[351,589,590,1617,3354,4473] Heavy metals may also exert cytological effects without forming distinct deposits.[679,3948,3951–2,4242,4470,4472,4478,4599] Many of the cytological changes that accompany heavy metal exposure are the result of damage to or alterations in membrane structure, producing swollen organelles or other osmotic disruptions.[4470,4599] Differences in distribution within the cell may be considered to be of great importance, as this allows the algae to reduce the toxic effect.[1272]

Rothstein[4199] has demonstrated that metals interact with sulfhydryl groups in cell membranes to produce -S-metal-S- bridges and Simkiss[4508] has discussed the importance of proteins and lipoproteins in binding with metals as a means of inactivating them within the cell.

It is of interest to speculate on how metals may be sequestered in polyphosphate bodies.[2336] A number of studies have been carried out to determine their composition. In a study using a variety of techniques, Widra[5419] reported that polyphosphate bodies, in bacteria, are composed of P, Mg, lipoprotein and RNA. It can be seen from studies by Doonan et al.,[1120] Adamec and Peverly[17] or Baxter and Jensen[249,250] that polyphosphate bodies contain, or are composed of, three components that may bind metals. Polyphosphate, because of its negative surface charge, could possibly bind positive ions. Lipids may also bind metals; however, they are generally not considered to be a major binding site. Proteins are generally the cellular components that are suggested to sequester heavy metals. Friedberg and Avigad[1494] in a study of isolated bodies from *Micrococcus lysodeikticus* reported that they were composed of 24% protein, 30% lipid and 27% polyphosphate. Small amounts of Na, Mg, Ca, K, Mn and Fe were detected with K, Mg, and Ca present in the largest amounts. Small amounts of carbohydrate and RNA were also detected. Polyphosphates are highly charged anions and thus adhere strongly to cations.[3776] Various metals have been found sequestered in polyphosphate bodies in algae, including Ti in *Anacystis nidulans*,[913] Pb in *Diatoma tenue*,[4472] and *Fragilaria capucina*,[4476] Mg, Ba, Mn, Cd, Co, Cu, Hg, Pb, and Zn in *Plectonema boryanum*,[250,2336-7] and Pb and Zn in *Chlorella saccharophila, Navicula incerta* and *Nitzschia closterium*.[2337] Such sequestering may serve to provide storage sites for essential metals or lessen the activity of toxic metals.[2336-7] Sicko-Goad and Lazinski[4473] pointed out that the movement of species that have sequestered metals in this manner by water mass movement and circulation patterns could lead to the dispersal of metals to areas that are distant from the point source of pollution.

Watanabe et al.[5285] summarize that intracellular concentrations and effects of metals have been shown to depend on the phosphate nutrient status of cells.[243,2650,3275,4472,4474-5,4478] The phosphate sufficient cells are usually less susceptible to metal toxicity than cells which had no phosphate. Twiss and Nalewajko,[5063] based on their results with three strains of the green alga *Scenedesmus acutus,* suggested that polyphosphate plays a passive role in protecting cells from copper. However, with respect to the mechanism of Cu tolerance, polyphosphate appeared to be relatively unimportant because the sensitivity of the Cu-tolerant strains showed less dependence on cellular polyphosphate than did the Cu-sensitive strain. The authors also provided a comprehensive list of literature sources dealing with the association of heavy metals and polyphosphate in algae.

2.11.2 Algal Sensitivity

The relative toxicity of heavy metals to algae has been summarized in several reviews.[3962,4629] The most common overall toxicity sequence is Hg > Cu > Cd > Ag > Pb > Zn,[4046,4629] although the degree of toxicity varies widely with respect to species and environmental conditions (Table 2-4).

It is very difficult to obtain an absolute statement of the toxic effect that a given amount of a given metal will have on a given algal species. Relative values for the toxicities of a variety of metals to a given test organism can be usefully compared, however.[2159]

2.11.3 Accumulation of Heavy Metals

Uptake of heavy metals to produce an internal concentration greater than in the external environment appears widespread in aquatic organisms. Quantitatively, information concerning such accumulation has been obtained both by direct analysis and by studies using radionuclides. The data in the literature are presented in a variety of ways and it is not always easy to compare results.[5408]

Concentration factor (= concentration coefficient, accumulation factor, bioaccumulation factor) defined by Brooks and Rumsby[573] is the ratio of concentration of an element in dry plant biomass and in the water.

Metal accumulation by algae is influenced by a number of abiotic and biotic factors (for details see next chapter) which interact with the organisms and result in a complex of relationships determining metal accumulation giving rise to a number of consequences concerning metabolism, structure, growth and reproduction of the organisms concerned:[2907]

Table 2-4. Orders of metal toxicity on different algal species and populations

Species/population	Order	Reference
Summer phytoplankton Marine	Hg >> Cu > Pb > As^{5+} > Zn = Cd > Ni, Cr, Sb, Se, As^{3+}	2159
	Hg >> Cu = Pb > Cd > Ni = Cr (2 exp.)	2159
Francisco Bay phytoplankton	Hg > Cu >> Cd = Cr, Ni, Pb	2615
Freshwater phytoplankton	Hg > Cr > Cu > Cd > Ni > Zn > Pb	5485
Freshwater phytoplankton	Cd > Zn > Cr	3375
Freshwater phytoplankton	Hg > Cu > Cd > Zn > Pb	1535
Asterionella japonica	Cu > Zn >> Pb > Cd	1359
Chlorella pyrenoidosa 251	Cu > Cr^{6+} > Ni	5483
Chlorella pyrenoidosa	Cu > Pb > Cd > Zn	273
Chlorella saccharophila	Cd > Cu >> Zn >> Pb	3946
Chlorella vulgaris	Cd > Cu > Hg > Zn > Pb	4189
Chlorella sp.	Hg > Ag > Cd > Pb > Cu	1897
Chlorella sp.	Ag > Cd > Cu > Hg > Se > Ni > Co > Ba > Pb	2261
Navicula incerta	Cd > Pb >> Zn > Cu	3949
Nitzschia closterium	Cu > Zn > Cd >> Pb	3950
Nitzschia closterium	Cu > Zn > Co > Mn	4188
Scenedesmus quadricauda	Zn >> Pb > Ni >> Cr (deleterious effect)	3107
	Ni > Cr = Zn >> Pb (lethal effect)	
Scenedesmus quadricauda	Hg > Cu > Ni > Cd > Pb > Zn > Co > Cr	4240
Scenedesmus sp.	Hg > Cd > Pb > Cu > Ni, Zn	545
Scenedesmus sp.	Ag > Cd > Ni, Se > Cu > Ba > Pb	2261
Selenastrum capricornutum	Cu > Ni > Co > Cr > Cd > Zn (-EDTA)	780
	Cd > Ni > Co > Zn > Cr > Cu (+EDTA)	

- abiotic factors: specific traits of metal (affinity to binding site, electronegativity), metal concentration, duration of exposure, concentration of other ions (e.g. Ca, Mg, P, other heavy metals), pH, complexing and chelating agents, redox conditions, temperature, light, turbulence;
- biotic factors: species-specific characteristics (cell wall, mucilage, cellular composition), algal biomass concentration, extracellular products, stage of development, cellular activity.

Consequences:

- growth, reproduction, ultrastructure, metabolism;
- detoxification: cell wall and mucilage, polyphosphate bodies, pinocytosis, metallothioneins;
- specificity: metal specific effect.

Detailed results and other factors can be found in e.g. References 194, 676, 1226, 2495, 2907, 3275-6, 5193, 5195, 5205.

The kinetics of metal uptake by green algae has been studied intensively and found to comprise two stages.[653,984,2495,2770,4369,4538,5026,5200,5205-6] The first stage is very rapid and short-lived, occurring immediately after initial contact with the heavy metal and usually lasting for less than 5 to 10 minutes.[866,1139,1591,1643,2495,2780] This initial phase is thought to be passive (i.e. physical sorption or ion exchange at the cell surface). The subsequent stage is slow and extended, and has been followed for up to 600 hours in some algae.[3169] It may be separated from the fast phase by a lag period[2495,4250] and may be linear[3169] or hyperbolic[866] in nature. The slow stage is possibly active (i.e. related to metabolic activity of the cell).

The accumulation of heavy metals by cell walls of microalgae is often rapid, reversible, and complete within 10 minutes. In many instances, uptake can be described by Freundlich or Langmuir sorption isotherms,[196,1856] confirming a linear equilibrium between the metal concentration in solution and that bound to cell surface/cell walls.[2495] Such uptake is mostly unaffected by metabolic poisons, modest variations in temperature, and/or light/dark treatment.[1537] Passive sorption can be affected by other cations, including H^+, Ca^{2+}, Mg^{2+}, Na^+, and Mn^{2+}, and by other heavy metals.[4250] Other factors supporting the passive adsorption mode of initial uptake, such as uptake of metals in darkness, uptake by killed algae, the possibility to express initial uptake by adsorption isotherms, are discussed e.g. by Vymazal.[5193,5195] In certain cases, the amount of heavy metal passively accumulated by dead cells or by cell wall preparations may be greater than that for living cells,[3040] suggesting that additional binding sites have become available during such experimental treatments. In many marine macroalgae the time period of passive sorption may be significantly extended. These plants often have extremely high concentration ratios for many heavy metals and significant proportions of heavy metal uptake may be attributed to extracellular binding and ion-exchange phenomena.[1974]

The amounts of heavy metals taken by passive mechanisms and bound on the cell surface/cell walls of algae are often quite low, when compared to those accumulated intracellularly by metabolism-dependent processes.[1538] Thus only 20% of the mercury uptake by *Synedra ulna* was due to extraprotoplast sorption.[1524] Similar results have been reported for cadmium in *Eremosphaera viridis*, while *Ankistrodesmus braunii* and *Chlorella vulgaris* showed a greater extraprotoplast component, with over 80% of the accumulated metal being accounted for by cell-wall binding during short-term (5 min.) incubation.[1591] High levels of cell wall-associated Cu have also been reported for *Mougeotia*, *Microspora*, and *Hormidium*, accounting for over 50% of the total cellular copper.[1472]

Although it has been suggested that Cd uptake is via a passive diffusion along a concentration gradient,[3040] it is generally considered, at least partially, an energy-depending process.[4542-3] Active transport has been described for several heavy metals in algae. Campbell and Smith[677] reported active transport of nickel by blue-green alga *Anabaena cylindrica*, active transport of cadmium by another blue-green alga *Anacystis nidulans* is reported by Singh and Yadava.[4520] Hart and Scaife[1943] found that active cadmium uptake in *Chlorella pyrenoidosa* was unaffected by range of divalent cations, but was completely inhibited in darkness, or at 4°C. Hart et al.[1944] found the Cd accumulation by *Chlorella pyrenoidosa* to be light and temperature dependent, indicating an active process. McLean and Williamson[3169] found that cadmium accumulation in the red alga, *Porphyra umbilicalis*, was dependent on illumination and inhibited by the protein synthesis inhibitor, cycloheximide. They concluded that Cd uptake was the result of an ongoing anabolic process and not a consequence of a pH gradient provided by photosynthesis. However, this may be species dependent, as Conway and Williams[866] noted that Cd uptake in the diatom, *Asterionella formosa*, was partially an active process, because of a dependency on illumination, whereas in *Fragilaria crotonensis*, another diatom, it was mainly passive.

For discussion on both the fast and slow phases of heavy metal accumulation by algae see the review by Trevors et al.[5026]

2.11.3.1 Environmental factors affecting toxicity and/or accumulation

A variety of environmental factors are known to modify the toxicity of heavy metals on algae.[676,1226,3962,4792,5193,5195,5205] A very important aspect of the environmental toxicity of heavy metals is specification. The particular state of the metal is dependent on the chemical properties of the metal itself and on the properties of the medium. Numerous studies have demonstrated that biotoxicity and other responses are determined by free ion concentration of the metal.[60,100,103,1245,4325,4840-1] Other factors affecting metal toxicity to algae include organic chelators factors,[60,780,980,1269,1373,1809,2253,2263,2999,3167,3317,3483,4392,4572,4707,4709] pH,[57,676,1643,1854-5,1943,2417,3337,3660,4696,5196,5201,5203,5211,5565] humic substances,[311,1385,3966,5189] particles and complexing agents,[307,2253,4842,5016] presence of other elements and ions such as Ca, Mg, P, C, NO_3^- or other heavy metals,[107,227,525-6,889,941,1535,1643,1908-9,1943-4,2844,3109,3783,3973-4,4301,5407,5411] sulfur compounds and amino acids,[32,440,3484] ionic strength,[1809,3231] cell concentrations,[4707,4711,5196,5530] temperature,[1854,3677,4821] salinity,[350,1450,1491,3037,3342,4821] light intensity,[197,1811,1854,2550,3276] aeration of the medium,[3120] the exchange reactions between suspended sediments and water,[1819,4946] and age of the cells.[660]

Campbell and Stokes,[676] in an excellent review on the relationship between pH and toxicity and availability of metals to aquatic biota divided metals into two groups: those for which a decrease in pH results in a decreased biological response (e.g. Cd, Cu, Zn) and those for which the dominant effect of lower pH is to increase metal availability (e.g. Pb). Rai and co-workers[3962] pointed out that barring a few exceptions[1535,1915-6,4707] where an increase in pH toward the alkaline range has been found to increase the toxicity of copper and zinc to natural populations of algae and to *Chlorella* and *Hormidium rivulare*, all other relevant studies have indicated that heavy metals exert more toxic effect in acidic conditions and that the toxicity of all the metals tested decreases at alkaline pH. This is because of the fact that at acidic pH metals exist in free ionic forms whereas at alkaline pH they tend to precipitate as insoluble carbonates, phosphates, sulfides, oxides or hydroxides.[1450]

Complexing agents have been held responsible for the solubilization of iron and therefore, its greater biological availability.[2809] In contrast, complexing agents are assumed to reduce the biological availability of copper and other heavy metals and minimize their toxic effect.[980,1358,1663,2174,2838,3623,3865]

The synergistic, additive, or antagonistic effects of combinations of metals have been examined. Cu and Zn salts act synergistically and have inhibitory effect on the growth of the diatoms *Amphora coffeaeformis* and *Amphiphora hyalina*.[1484] Thomas[4955] reported Cu and Zn synergic effect on *Thalassiosira aestivalis*. Whitton and Shehata[5411] found that Cd toxicity on *Anacystis nidulans* increased in the presence of Pb and decreased toxicity in the presence of Fe or Zn. Further reports on interactions of metals can be found in References 525, 2261, 3594, 4707 and 5530. The results are not conclusive as the effect of the same metals have been found to act both synergistically and antagonistically depending on species and culture conditions in many experiments (see Rai et al.[3962] for review).

2.12 PHOTOSYNTHESIS AND RESPIRATION

2.12.1 Photosynthesis

Photosynthesis has been reviewed many times (e.g. References 1876–7, 2709, 3199). Photosynthesis can be split into three stages:[4209]

1. Absorption of light energy by the photosynthetic pigments located in the chloroplasts.
2. The transference of this energy, partly to pyrophosphate bond energy of ATP (adenosine triphospahate), in a process of "photosynthetic phosphorylation" and partly to oxidation/reduction energy, in which $NADPH_2$ (reduced nicotinamide adenine dinucleotide phosphate) is formed and oxygen released. Photosynthetic phosphorylation is a term used to describe the conversion of light energy into the energy-rich pyrophosphate bonds of adenosine triphosphate by isolated chloroplast, without the aid of other cellular constituents (Arnon et al.[164] and reference therein).
3. The assimilation of carbon in a series of dark reactions utilizing the reducing power of $NADP_2$ and the phosphorylating action of ATP.

Algae have evolved various pigments for the purpose of light absorption: chlorophylls, carotenoids and phycobiliproteins (see Chapter 2.4.3). These bulk pigments, called such as they are present in large quantities, provide the algae with *antannae* to capture the light energy.[1781] The antenna pigment molecules are arranged in a manner such that the energy of the electron returning to its ground state is not emitted as fluorescence, and ultimately to the photosystems traps. The shorter the wavelength of the light absorbed, the more accessory pigments involved.[4643] There are two photosystems, PS I and PS II, each with their light-absorbing trap, P_{700} and P_{680}, respectively:[1781]

most carotenes → some chl b_{650} → chl a_{670} → chl a_{680} → chl $a_{685-705}$ P 700
most xanthophylls → chl b_{650} → chl a_{670} → chl a_{680} P 680

In 1944, Emerson and co-workers[1241] wrote ". . . We feel, therefore, that there is now sufficient evidence to show that it is quite possible that the function of the light energy in photosynthesis is the formation of 'energy-rich' phosphate."

A scheme for the second stage has been proposed by Arnon[157] (Figure 2-27, with an original caption). Two *cyclic photophosphorylation* pathways were discovered, catalyzed either by vitamin K or by riboflavin phosphate together with nicotinamide adenine dinucleotide phosphate (NADP = TPN, triphosphopyridine nucleotide). The only product of these cyclic pathways is ATP and neither oxygen nor a reductant is formed; light energy is converted by this process into chemical energy in the form of "high energy" pyrophosphate bonds.

The evolution of oxygen is associated with a *non-cyclic photophosphorylation,* in which the hydrogen ion or water is used to form reduced nicotinamide adenine dinucleotide phosphate ($NADPH_2$ = $TPNH_2$, reduced triphosphopyridine nucleotide) and the OH^- moiety yields molecular oxygen and donates electrons via the cytochrome chain to chlorophyll.

The overall equations for these reactions are:[157,164-5]

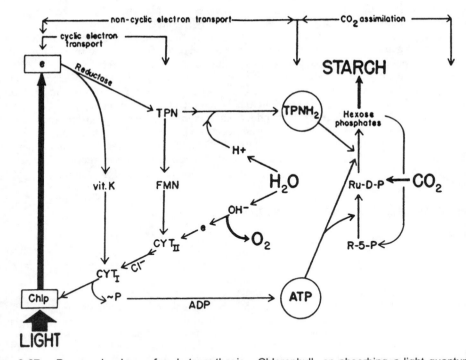

Figure 2-27.   Proposed scheme for photosynthesis. Chlorophyll, on absorbing a light quantum, becomes "excited" and expels electrons which have been raised to a high-energy level. In
both green plants and photosynthetic bacteria the energy thus acquired from the absorbed light is converted into the pyrophosphate bond-energy of adenosine triphosphate
(ATP) during the "downhill" return of electrons to chlorophyll by a cyclic electron transport
system of the vitamin K type. In green plants there is a second cyclic electron transport
system of the riboflavin phosphate (FMN) type for forming ATP and also an "open" non-
cyclic mechanism which transports electrons from chlorophyll to triphosphopyridine
nucleotide (TPN). In the non-cyclic mechanism the electrons accepted by TPN are
replaced by those donated to chlorophyll by OH⁻ ions, with a resulting evolution of oxygen.
Cytochrome (cyt) components are visualized in the electron transport chain of the cyclic
and non-cyclic mechanisms. It is suggested that photosynthetic bacteria use light energy
only for the formation of ATP by the vitamin K type of cyclic photophosphorylation. Green
plants, however, require both the cyclic and non-cyclic electron transport systems for the
formation of assimilatory power, that is, ATP and reduced triphosphopyridine nucleotide
(TPNH₂). These two substances are then used for assimilation of CO_2 to the level of
sugar and starch in a series of dark reactions involving the regeneration of ribose-5-
phosphate (R-5-P) and ribulose diphosphate (Ru-D-P).[157] (From Arnon, 1959, reprinted
with permission from *Nature,* Vol. 184, p. 14. Copyright (1959) Macmillan Magazines
Limited, London.)

$$ADP + P \overset{light}{\rightarrow} ATP \text{ (cyclic photophosphorylation)} \tag{2.4}$$

$$2\ NADP + 2\ H_2O^* + 2\ ADP + 2\ P \overset{light}{\rightarrow} 2\ NADPH_2 + O_2^* + 2\ ATP \tag{2.5}$$
$$\text{(non-cyclic photophosphorylation)}$$

The photosynthetic conversion of light energy into chemical energy is described by the Hill and
Bendall "Z" model (concept Hill and Beardall[2078]) (Figure 2-28). The electron donor to photosystem
II is water:[774,1877,5543]

$$2\ H_2O \rightarrow O_2 + 4\ H^+ + 4\ e^- \tag{2.6}$$

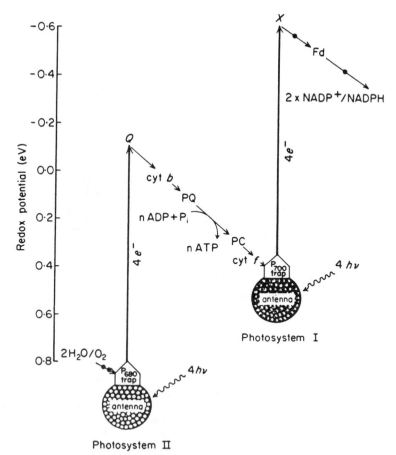

Figure 2-28.   A modified Hill-Bendall or "Z" scheme of photosynthesis. The bold vertical arrows represent the photochemical or "light" reactions: all others, "dark" reactions. Only non-cyclic photophosphorylation is shown here. Q and X are primary electron acceptors for PS II and PS I, respectively. Cyt *b* (cytochrome *b*), cyt *f*, PQ (plastoquinone), PC (plastocyanin) and Fd (ferredoxin) are electron carriers. The large dots represent proposed but unknown electron carriers.[3971] (From Ramus, 1981, in *The Biology of Seaweeds*, Lobban, C.S., and Wynne, M.J., Eds., p. 462, reprinted with permission of Blackwell Scientific Publications, Oxford.)

and the ultimate electron acceptor in photosystem I is oxidized NADP with adenosine triphosphate (ATP) produced in the PS I reactions.[3971] The standard redox potential for the H_2O/O_2 couple is 0.82 eV and the $NADP^+/NADPH$ couple is -0.32 eV, giving a standard potential difference of 1.14 eV per electron, a considerable amount of reducing and oxidizing power.[3971] Since there are two photosystems, in theory the evolution of one mole of O_2 require 8 mole quanta, four for each photoact, but in practice requires more (10±2). Non-cycling electron transport adds 0.5–0.75 molecule of ATP per electron by a phosphorylating couple between PS II and PS I. The actual yield for 10±2 mole quanta captured is 2–3 mole ATP and 2 mole NADPH. The free-energy gradients produced at the reaction centers by charge separation are thus transduced and stored as chemical energy in the form of ATP and NADPH to be used by the cell for CO_2 fixation.[3971]

The assimilation of carbon dioxide in the dark, stage 3, utilizing the energy and reducing power of ATP and $NADPH_2$ formed in stage 2, is represented by the following equation:[157]

$$CO_2 + 2\ NADPH_2 + n\ ATP \xrightarrow{light} (CH_2O) + H_2O + 2\ NADP + n\ ADP + n\ P \qquad (2.7)$$

The sum of equations (2.4), (2.5) and (2.7) is the well known basic equation (2.8) of photosynthesis:

$$CO_2 + 2\ H_2O^* \overset{\text{light}}{\rightarrow} (CH_2O) + O_2^* + H_2O \qquad (2.8)$$

The scheme for the light reactions in photosynthesis together with CO_2 assimilation has been presented by Zehnder[5543] (Figure 2-29).

2.12.1.1 Alternative hydrogen donors

In the above scheme for normal photosynthesis, water acts as the hydrogen donor, but some algae are capable of utilizing elementary hydrogen and others hydrogen sulfide. Gaffron[1539-41] discovered about 50 years ago, that two of the most common green algae, *Scenedesmus* and *Ankistrodesmus* (*Raphidium*), can reduce carbon dioxide with molecular hydrogen in low light intensity, after incubation in hydrogen for a few hours in the dark, and suggested the presence of hydrogenase in these algae.[781] *Scenedesmus* and *Raphidium* incubated in the dark with hydrogen will absorb this gas and when transferred to light will reduce carbon dioxide without release of oxygen; instead an equivalent amount of hydrogen is absorbed. This photosynthetic reaction, discovered and named "photoreduction" by Gaffron,[1540] can be described by the equation:

$$CO_2 + 2\ H_2 \rightarrow (CH_2O) + H_2O \qquad (2.9)$$

This is an adaptive process, not normally found in nature, and is dependent on the presence of a latent hydrogenase enzyme, which is activated under dark anaerobic conditions. Hydrogenase is present in certain autotrophic, heterotrophic, anaerobic and aerobic bacteria and in members of most major groups of algae[2473] (e.g. *Scenedesmus, Ankistrodesmus, Raphidium, Chlamydomonas, Ulva, Ascophyllum, Porphyra, Porphyridium, Synechococcus, Synechocystis*).[2471,2473,4209] The length of time taken to adapt is very variable, short or non-existing in *Chlamydomonas* to several hours for *Ascophyllum*.[1487,4209] The most widely accepted concept[167,4681] is that photoreduction involves photosystem I only, which produces ATP by means of cyclic photophosphorylation, whereas the hydrogen activated by hydrogenase serve for CO_2 reduction via ferredoxin and NADP. For further information see References 1542, 2473, 3944 and 4663.

The utilization of hydrogen sulphide as a hydrogen donor by *Oscillatoria* results in sulfur deposition in place of oxygen evolution according to the equation:[4209]

$$CO_2 + 2\ H_2S \rightarrow (CH_2O) + H_2O + 2\ S \qquad (2.10)$$

2.12.2 Respiration

2.12.2.1 Dark respiration

Whereas photosynthesis consumes CO_2 and produces O_2, respiration uses O_2 and releases CO_2. The respiratory coefficient (R.Q.) is defined as the ratio of CO_2 produced to O_2 consumed.[1625] The major features of dark respiration in algae are essentially identical with those found in other aerobic plants.[4010] The glycolytic or EMP (Embden-Meyerhoff-Parnas) pathway runs parallel with the hexose monophosphate shunt (HMP), while the tricarboxylic acid (TCA) or Krebs cycle is in series with these processes.[3999,4643]

The location of these processes in the cell is essentially the same as in the higher plants. The enzymes of the EMP and HMP pathways lie in the cytosol, but also occur in the chloroplasts, which contain starch.[4010] The TCA cycle enzymes are located in the mitochondrial matrix, while the oxidative phosphorylation processes are associated with the inner mitochondrial membrane. Differences of location are found in the prokaryotic Cyanophyceae, where the EMP, HMP, and TCA enzymes are found in the cytosol, and oxidative phosphorylation is associated with membranes.[4643]

Considerable amounts of organic carbon provided by photosynthesis are reconverted to CO_2 through the EMP pathway of glycolysis and further aerobic degradation.[1625,2601] The EMP pathway functions

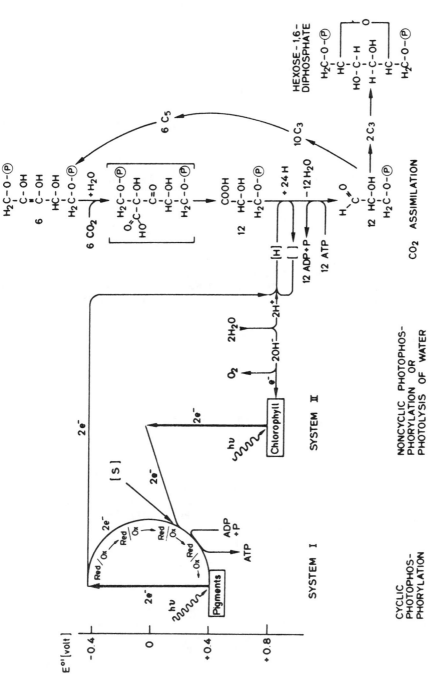

Figure 2-29. Scheme for the light reactions in photosynthesis and the CO_2 assimilation over the "Calvin-Bassham cycle." In green plants and algae the photosystem has two "short-wavelength" light reaction centers. The chlorophyll-*a* of system II absorbs below 680 nm and in systems I at 700 nm. In photosynthetic bacteria the pigments (bacteriochlorophyll-*a*) in systems I are most active at the "long-wavelength" of 890 nm. [S] stands for electron donors other than water. [H] and [ ] mean the electron and proton carrier NADP⁺ in its reduced and oxidized forms, respectively.[5543]   (From Zehnder, 1982, in *The Handbook of Environmental Chemistry*, Vol. 1, Part B, Hutzinger, O., Ed., p. 95, reprinted with permission of Springer Verlag and the author.)

under either anaerobic or aerobic conditions, which affect only the fate of the initially formed pyruvate. Under anaerobic conditions pyruvate is converted to lactate or ethanol, and no further generation of ATP is possible. Under aerobic conditions, however, part of the pyruvate enters the TCA cycle, where it is oxidized to carbon dioxide. Electrons generated during the functioning of the TCA cycle are transferred to oxygen by the respiratory electron-transport chain, thereby producing more ATP. The reactions of the EMP pathway are thus the following:[3408]

Anaerobic or aerobic conditions

$$C_6H_{12}O_6 + 2 \ ADP + 4 \ NAD \rightarrow 2 \ CH_3.CO.CO_2H + 2 \ ATP + 4 \ NADH \qquad (2.11)$$

Anaerobic

$$2 \ CH_3.CO.CO_2H + 4 \ NADH \rightarrow 2 \ CH_3.CH(OH).CO_2H + 4 \ NAD \qquad (2.12)$$

Aerobic

$$2 \ CH_3.CO.CO_2H + 6 \ H_2O + 20 \ NAD \rightarrow 6 \ CO_2 + 20 \ NADH \qquad (2.13)$$

$$24 \ NADH \rightarrow 24 \ NAD + 36 \ ATP \qquad (2.14)$$

The pentose phosphate pathway (PPP), also called the hexose monophosphate pathway (HMP), may, like the EMP pathway, function under either aerobic or anaerobic conditions.[3408] Both share the initial formation of glucose-6-phosphate. In the PPP, however, this is dehydrogenated to 6-phosphogluconolactone and then further oxidatively decarboxylated to carbon dioxide and ribose phosphate. Thereafter the anaerobic and aerobic pathways diverge. Under anaerobic conditions the final products are ethanol and lactate, whereas under aerobic conditions the ribose is eventually oxidized to CO_2. The summary is as follows:[3408]

Anaerobic conditions

$$C_6H_{12}O_6 + ADP \rightarrow C_2H_5OH + CH_3.CH(OH).CO_2H + ATP + CO_2 \qquad (2.15)$$

Aerobic conditions

$$C_6H_{12}O_6 + 6 \ H_2O + ATP \rightarrow 6 \ CO_2 + 24 \ NADH \qquad (2.16)$$

$$NADH \rightarrow 24 \ NAD + 36 \ ATP \qquad (2.17)$$

The relative contribution of the EMP and pentose phosphate pathways has been discussed by Neilson and Lewin.[3408] Determination of the contribution of dark respiration to a photosynthetizing cell is not an easy task. For comparative purposes the respiratory capacity of an algal cell may be expressed as the ratio of the maximum production of carbon dioxide in the dark ($u_{r\,max}$) to the maximum achieved growth rate (u_g). For phototrophically grown cells the ratio may be as low as 0.3–0.5, while for comparable heterotrophically grown cells the ratio is 0.5–1.2.[4643]

2.12.2.2 Photorespiration

Warburg's description[5264] of O_2 inhibition of photosynthesis by *Chlorella* has been referred to as the "Warburg O_2 effect." This phenomenon is photorespiration, in which O_2 inhibition of net photosynthesis is one of the prime effects.[4996] Investigators have used the term photorespiration instead of the Warburg O_2 effect for this overall process occurring in both higher plants and algae.[4996]

Photorespiration is defined as a light dependent O_2 uptake and CO_2 release that occurs in photosynthetic tissues, and it has been extensively investigated with higher plants.[1965,2307,4118,4995,5549–50] In higher plants, photorespiration and the C_2 photosynthetic carbon oxidation cycle (PCOC), or glycolate pathway, are initiated by the oxygenase activity of the bifunctional enzyme RuBP carboxylase/oxygenase (RuBisCO = RuBPCO).[2912,3514] Tolbert[4996] pointed out that this phenomenon in also present and important in algae. Unlike the case with photosynthetic carbon fixation, in many green algae the C_2

PCOC is not to be exactly the same as that in higher plants. The major difference is in the enzyme responsible for oxidation of glycolate to glyoxylate.[4645] In higher plants, this conversion is catalyzed by a flavoprotein enzyme, glycolate oxidase, located in the peroxisome.[4995] In many algae, however this enzyme cannot be found, but a different, nonflavoprotein enzyme, glycolate dehydrogenase, appears to catalyze this oxidation.[4996] Rather than in the peroxisome, glycolate dehydrogenase is apparently located in the mitochondrion.[287,4667]

In C_3 higher plants, photosynthesis, measured as net CO_2 uptake, is inhibited at atmospheric O_2 (21%) relative to low O_2 (1–2%) concentrations. This O_2 inhibition of photosynthesis has two components: direct, competitive inhibition of RuBisCO by the alternate substrate O_2, and indirect inhibition by CO_2 release at the glycine decarboxylase step of the C_2 PCOC.[3514] In general, unicellular green algae exhibit little or no inhibition of photosynthesis by 21% O_2 if they are grown with air concentrations of CO_2.[360,384,3440,4441,4646] On the other hand, when grown at elevated CO_2 concentrations (CO_2-enriched), the same algae exhibit substantial inhibition of photosynthesis by 21% O_2.[1976,3440,4441]

The extent to which photorespiration occurs in algae is a matter of controversy. Photorespiration arises because RuBP carboxylase can also act as an oxygenase, binding O_2 rather than CO_2 to RuBP, and the process includes the metabolic pathway followed by the C_2 portion of RuBP. The carboxylase activity is competitively inhibited by O_2, so that photorespiration is more pronounced at high O_2 or low CO_2 concentrations.[2886] Freshwater algae, with metabolic pathways similar to C_3 plants, show gas exchange characteristics of C_4 plants: little external manifestation of photorespiration. This may be due to a biophysical mechanism for concentrating CO_2 at the chloroplast.[263,3316] A potential indicator of photorespiration is the CO_2 compensation point, an equilibrium maintained between the rate of photosynthetic CO_2 fixation and the rate of CO_2 lost from photorespiration as well as dark respiration. This compensation is nearly zero for C_4 plants, because of an unusually efficient CO_2 fixation, and in the range of 30–70 ppm for the majority of C_3 plants.[2601] Classic C_4 plants show no measurable rates of photorespiration. This is mainly due to the high affinity of PEPC (phosphoenolpyruvate carboxylase) for CO_2 ($K = 7 \times 10^{-6}$ M), which is far greater than that of RuBPC (K about 450×10^{-6} M), resulting in a very low CO_2 compensation point.[2601]

Photorespiration can be explained metabolically by glycolate biosynthesis in the chloroplast followed by its metabolism, which in leaves occurs in peroxisomes and mitochondria.[4996] The total sequence has also been referred to as the glycolate pathway.[4994–5]

Glycolate is either oxidized with molecular O_2 to form H_2O_2 and glyoxylate via glycolate oxidase (as in higher plants) or may be oxidized without O_2 to form glyoxylate via glycolate dehydrogenase (as in the unicellular algae[4996]). Further enzymatic steps lead to the formation of other important intermediates such as glycine and serine.[2601]

These reactions occurring in unicellular green algae are similar to those in higher plants with a few modifications, particularly in the enzymatic oxidation and fate of glycolate. Photorespiration differs from mitochondrial respiration in that it does not occur in the dark, does not conserve energy as ATP, and does not utilize substrates of the tricarboxylic acid cycle.[4996] It is now commonly accepted that glycolate is a product of photorespiration.[2307]

The unicellular green algae can accumulate inorganic carbon through active transport, although it is not clear whether CO_2 or HCO_3^- is transported, nor is it clear across which membrane the transport occurs. The active transport of inorganic carbon results in a substantial increase in the intracellular CO_2 concentration, which inhibits RuBP oxygenase activity, suppressing photorespiration. This inorganic carbon transport is also inducible, apparently being expressed only when inorganic carbon concentrations are low or inorganic carbon is limiting for growth.[4645]

Recently, photorespiration in algae and blue-green algae has been reviewed exhaustively by Splading,[4645] Colman[850] and Beardall.[259]

The summary of algal metabolism has been given by Lobban et al.[2886] (Figure 2-30).

2.12.2.3 Anaerobic respiration (fermentation)

Gibbs[1626] defines the term fermentation as the degradation of a carbohydrate molecule into two more smaller molecules by biological processes not requiring molecular oxygen. In general, the products of fermentation are carbon dioxide, ethanol, lactic acid, and other organic acids. Gibbs[1626] pointed out that it has been common to equate the terms fermentation and glycolysis. The latter process is defined by most reviewers as the degradation of any carbohydrate by fermentative fission into pyruvic acid. The fate of the keto acids is generally not specified.

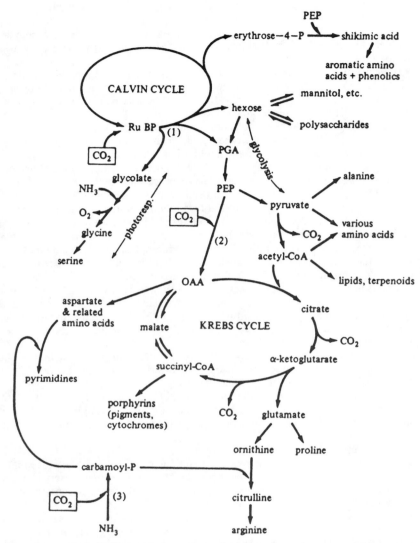

Figure 2-30. Summary of algal metabolism. Enzymes: (1) RuBP carboxylase; (2) PEP carboxylase or carboxykinase; (3) carbamoyl-phosphate-synthetase.[2886] (From Lobban et al., 1985, in *The Physiological Ecology of Seaweeds,* p. 116, reprinted with permission of Cambridge University Press.)

Acid fermentation has been demonstrated in a number of algae.[4209] Glucose may be metabolized under either aerobic and anaerobic conditions. Although fermentation of glucose has been demonstrated in several algae,[1626] there is known only one putative example of heterotrophic growth of an alga, a symbiotic blue-green alga *Nostoc,* under anaerobic conditions.[2122]

2.12.3 Photosynthesis vs. Light Intensity

The response of photosynthetic rate to light intensity is plotted in P versus I (irradiance) curves, shown diagrammatically in Figure 2-31. Photosynthetic rate increases linearly with light intensity until

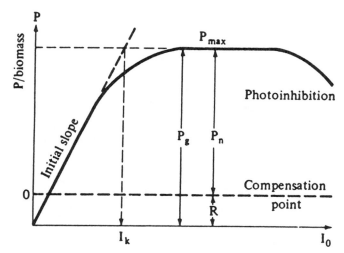

Figure 2-31.   Light saturation curve for photosynthesis. P = photosynthesis, P_{max} = maximum photosynthesis, P_g = gross photosynthesis, P_n = net photosynthesis, R = respiration, I_o = incident photon flux density (PFD),[3971] I_k = saturating PFD. (From Ramus, 1981, in *The Biology of Seaweeds*, Lobban, C.S., and Wynne, M.J., Eds., p. 466, reprinted with permission of Blackwell Scientific Publications, Oxford.)

it approaches a plateau at saturating light intensities.[973] Very high light intensities are often inhibitory.[3364] However, the shape of light-photosynthetic curves and of light-growth curves is markedly affected by temperature,[4410,4571,4633] and by the other factors such as salinity[3130] and nutrient level.[3011]

Impairment of photosynthesis and photoautotrophic growth by supra-optimal intensities of visible light is well known from laboratory[4633] and field studies.[1689,2968,4888] Belay and Fogg[293] pointed out that inhibition of phytoplankton photosynthesis occurs generally near lake and sea surfaces at times when the incident visible radiation exceeds about 0.1 cal cm^{-2} min^{-1} ~ 17 klux or about 385 μE m^{-2} s^{-1} of photosynthetically active light.[1689,2326,2972,3078,4166,4879,4885-6]

The so-called surface inhibition observed in pelagic communities seems to be due to damage by ultraviolet radiation.[1236,1348,4705] The cause of photoinhibition include photooxidation of components of the photosynthetic apparatus: carotenoids may function to mitigate this problem, allowing some seaweeds to photosynthetize in full sunlight.[2886] High light intensities may also inhibit respiration of actively photosynthesizing cells.[576] Such photoinhibition is greatest in blue light, and the size of the effects depends in *Chlorella* on respiratory activity prior to illumination.[4132] Blue light leads to a destruction of cytochromes, especially cytochrome a_3.[1250]

The classic curvilinear plot relating photosynthesis to PFD (photon flux density) (P v. I_o plot) (Figure 2-31) contains considerable information about the photosynthetic apparatus.[3971] At low PFD photosynthetic rate is limited by the light reactions of photosynthesis, i.e. by photochemical reactions involving the number of available quanta and the quantum-absorbing capacity of the cells and is linearly related to incident light.[973,3971] This linear region (the initial slope) is the apparent efficiency of the photosynthetic apparatus. True efficiency, or quantum yield, is necessarily derived from light absorbed data and is directly defined by the slope if this line.[3971] The slope of the initial part of the curve is largely independent of temperature and cell metabolism.[973] At saturating light intensities, the maximum photosynthetic rate (P_{max}), also called photosynthetic capacity, is a function of the dark reactions of photosynthesis and becomes independent of light. Because the reactions are enzymatic in nature, temperature as well as other factors which influence cell metabolism (e.g. nutrient levels) will affect P_{max}. The light intensity (I_k) at which the initial slope and P_{max} intersect is a convenient reference point on the curve and is usually taken as the intensity at which light saturation occurs.[973] At supersaturating PFD ($I_o \gg I_k$) photosynthesis declines, i.e. photoinhibits, the causes for which are likely manifold.[3971] A comprehensive review on the role of light in photosynthesis has been given by Ramus[3971] who also presented model light saturating curves normalizing photosynthesis to biomass (Figure 2-32) and chlorophyll-*a* (Figure 2-33).

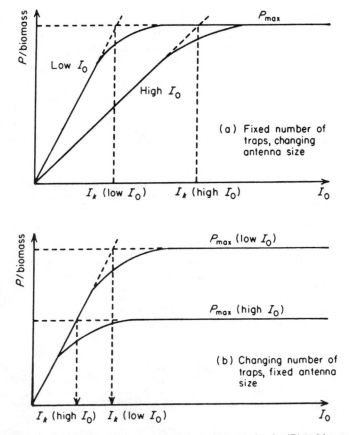

Figure 2-32.   Model light saturation curves normalizing photosynthesis (P) to biomass and comparing two "strategies" (a and b) for the adjustment of the photosynthetic unit to extreme (low and high I_o) light environments. I_o = incident PFD, I_k = saturating PFD, P_{max} = maximum photosynthesis.[3971] (From Ramus, 1981, in *The Biology of Seaweeds,* Lobban, C.S., and Wynne, M.J., Eds., p. 486, reprinted with permission of Blackwell Scientific Publications, Oxford.)

Growth rate increases with increasing irradiance up to an optimum or saturating light intensity (I_{opt}).[821] As irradiance exceeds this optimum, growth rate either plateaus or drops off,[869,1936,3011,3715-6] depending upon the proximity of temperature to T_{opt}. At low temperatures, algal growth is inhibited by high light intensities,[4571,4631,4633] but as temperature increases, higher light intensity is required for optimum growth rate.[4571,4631] Steele[4701] proposed an empirical relation between growth rate and irradiance:

$$\mu = \mu_{max} \cdot 1/1_{opt} \cdot \exp(1 - 1/1_{opt}) \tag{2.18}$$

where μ is observed growth rate at light intensity (PFD) I; μ_{max} is maximum growth rate; and I_{opt} is the light intensity (PFD) at which $\mu = \mu_m$. Assuming that the two parameters μ_m and I_{opt} both vary with temperature, Steele's equation describes the complex response to irradiance and temperature outlined.[821] The Steele's empirical model were modified by Platt and Jassby[3819] and Platt et al.[3820] and has been discussed by Ojala[3522-3] recently.

Brown and Richardson[593] pointed out that results from many algal species studied show that the overall influence of light diminishes as the organism becomes less dependent upon it for production of metabolic substrates. This increase in independence follows the progression: obligate photoautotrophs—facultative photoautotrophs—facultative heterotrophs—obligate heterotrophs.

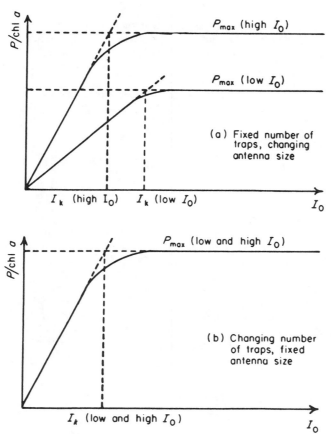

Figure 2-33. Model light saturation curves normalizing photosynthesis (P) to chlorophyll-*a* and comparing two "strategies" (a and b) for the adjustment of the photosynthetic unit to extreme (low and high I_o) light environments. I_o = incident PFD, i_k = saturating PFD, P_{max} = maximum photosynthesis.[3971] (From Ramus, 1981, in *The Biology of Seaweeds,* Lobban, C.S., and Wynne, M.J., Eds., 487, reprinted with permission of Blackwell Scientific Publications, Oxford.)

Respiration rates are generally 5–10% of the light-saturated rate of photosynthesis. At some very low light intensities, the rate of respiration will equal the rate of gross photosynthesis and the cell is unable to grow (net photosynthesis zero). The light intensity at which this *compensation point* occurs is the compensation intensity (I_c) (Figure 2-31).[973] The compensation points for most plants occur around dawn and dusk, but vary with the species.[12] The compensation depth is the point in the water column at which the compensation intensity occurs. The compensation depth also defines the lower limit of the euphotic zone. As a rule of thumb for field work, the compensation depth is set where the light intensity is 1% of the surface intensity; it is located at approximately two to three times the depth of Secchi disc visibility.[973]

Since blue-green wavelengths dominate low-light environments in natural aquatic systems, another selective adaptation would be a shift in a pigment ratio to maximize the absorption of blue-green light, i.e. increased accessory pigment/chlorophyll ratios at depths. Complementary chromatic adaptation is well known in the blue-green algae.[416]

2.13 PRIMARY PRODUCTION

The term *primary production* (or *productivity*) refers to the amount of organic matter present in a unit area per specified time. The *gross primary production* (GPP) is the total amount produced. Some of this is utilized by plants through respiration or is lost through leakage, death, or grazing. The *standing crop* (*biomass*) is the actual biomass present at the time of sampling and may be considered the *net primary production* (NPP). Thus the standing crop may be low and may not reflect the growth of a phytoplankton population because of high grazing pressures.[1004] Kvet[2656] estimated the turnover coefficient (net primary productivity: maximum seasonal biomass) to be as high as 400–650 for phytoplankton.

A first quantitative approach of the relationship between primary production and nutrients can be derived from a mean algal elemental composition:[1720] C = 50% of the dry mass; N = 8–13% of the dry mass; P = 0.5–1.0% of the dry mass; and chlorophyll-*a* = 0.5–1.5% of the dry mass. As a chemical reaction the primary production can be described as follows:[1720]

$$\text{Light} + 5\ CO_2 + 2\ H_2O + NH_3 \rightarrow C_5H_7NO_2 + 5\ O_2 \tag{2.19}$$

$$\text{Light} + 5\ CO_2 + 3\ H_2O + HNO_3 \rightarrow C_5H_7NO_2 + 7\ O_2 \tag{2.20}$$

Adding phosphate to this formulation gives $C_5H_7NO_2P_{1/30}$ as a first approximation of mean algal chemical composition. The formula was thought to give a reasonable mean algal composition by participants in workshops during the International Biological Programme.[1720] It suggests less precision than the "Redfield" ratio, which was derived by Redfield[4040] from data of the English Channel and which suggests very precise ratios indeed: $C_{106}H_{263}O_{110}N_{16}P_1$.

In comparison with macrophytes, considerably less general estimates of algal productivity and biomass have been reported in the literature. The reason is the immense variability in algal morphology: while some planktonic species are only several μm in size, some brown algae can reach more than 60 m in size. Therefore, some estimates are possible only within a certain narrow group of algae. Whittaker and Likens[5396] gave the following ranges for marine environments:

Algal beds and reefs: NPP: 500–4,000 (mean 2,500) g m^{-2} yr^{-1}
Biomass: 40–4,000 (mean 2,000) g m^{-2}
Open oceans: NPP: 2–400 (mean 125) g m^{-2} yr^{-1}
Biomass: 0–5 (mean 3) g m^{-2}

Květ[2656] estimated freshwater phytoplankton biomass in the range from 1 to 2.3 g m^{-2} and productivity from 600 to 2,400 g m^{-2} yr^{-1}.

In comparison to macrophyte biomass and production (see Chapter 3.3.2), productivity and biomass of algae are expressed on much more bases. Standing crop of attached algae may be, similar to macrophytes, expressed in g DM per m^2 (Tables 2-5 and 2-6). However, the standing crop of algae is more often expressed in g AFDM per m^{-2} (Table 2-7), mg chlorophyll-*a* per m^{-2} (Table 2-8), cells number per cm^{-2} (Table 2-9) or as biovolume (in μL) per cm^{-2} (Table 2-10). Standing crop of phytoplankton is mostly expressed as chlorophyll-*a* per m^{-2} (Table 2-11). Productivity of algae is only rarely expressed on dry mass basis (mostly for macroscopic species) (Tables 2-12 and 2-13). The productivity of attached algal communities and seaweeds is more often expressed on carbon basis (Tables 2-14 and 2-15), AFDM basis (Table 2-16), chlorophyll-*a* basis (Table 2-17) or oxygen basis (Table 2-18). The phytoplankton productivity is almost entirely expressed on carbon basis (Table 2-19). The productivity is mostly measured and expressed on a day or an hour basis, the annual production is mostly estimated. In order to compare different units, some authors listed the following relationships:

Hodgson:[2130] 1 mg C = 3.663 mg CO_2
Littler:[2874] 1 mg O_2 = 0.3 mg C
Strickland:[4803] 1 g O_2 = 0.31 g C = 0.78 g glucose = 0.62 g AFDM
For other relationships see Westlake.[5336]

The energy content of algae has been reported as follows:

Larcher:[2709]   planktonic algae: 4,600–4,900 cal/g DM (19.3–20.5 kJ/g)
seaweeds: 4,400–4,500 cal/g DM (18.4–18.9 kJ/g)
Cummins and Wuycheck:[932] green algae: 4,780 cal/g AFDM

Table 2-5. Standing crop of algae (g DM m^{-2})

Species	Standing crop	Locality	Reference
		Freshwater	
Oedogonium sp.	4.6–8.2	Lake Wingra, Wisconsin	126
Oedogonium sp.	10	Lake Wingra, Wisconsin	3141
Cladophora sp.	1.7–18.2	England, fertilizer exps.	962
filamentous algae[a]	33.1–83	Ponds in South Czech Republic	420
Chara sp.	110	Lake Lawrence, Michigan	4096
Chara vulgaris	127	Opatovický pond, Czech Republic	1207
Cladophora glomerata	17.2–144	Colorado River, Arizona	5093
Nitella opaca	170	Lake Thingvallavatn, Iceland	2407
Mougeotia spp.	200	Caplecleugh Low Level, England	3703
Cladophora glomerata	100–214	Lake Erie, Michigan, Ohio	2909
Chara sp.	150–220	Artificial medium	2716
Cladophora sp.	30–272	Lake Krageholmssjön, Sweden	3333
Chara sp.	326–764	nameless pond, David, California	5343
filamentous algae[b]	25–891	Nesyt Pont, Czech Republic	2843
Enteromorpha intestinalis	900	Opatovický pond, Czech Republic	2567
Chara sp.	2,194	Michigan Lake	4975
		Marine	
Hypnea musciformis	32–190	Brazil, coast	4329

[a] Cladophora rivularis, Hydrodictyon reticulatum
[b] Spirogyra sp., Oedogonium sp., Cladophora fracta, Enteromorpha intestinalis

The ash content of the algal biomass varies widely among species. For planktonic species, the ash content has been reported in the range of 1.36 to 46.2% DM, the highest values being recorded in diatoms.[1604,3235,4657,5275] The higher ash values and even wider variations are commonly recorded in attached algal communities as a result of a presence of sediment particles which are "trapped" within the community. In the literature values from 5% DM up to more than 87% DM have been reported with values > 50% DM commonly occurring.[720,1220,2116,3150,4412,5299]

2.13.1 Factors Affecting Primary Productivity

Primary productivity is mainly influenced by nutrients, light, temperature, grazing and successional cycles (and their interactions).[443,473,1004,1092,1258,2089,2689,3002,3209,3489,4725–6,4744,4788]

2.13.1.1 Light

The effect of changes in light intensity on plants have been studied more than those of any other parameter since Oltman's[3548] investigation. Light is considered to be the principal factor limiting algal productivity[2089,3238] and there are three aspects of the role of light: a) diurnal changes, b) light intensity, and c) light quality.

Photosynthesis in the algae occurs mainly in the visible range of the spectrum between 400 nm and 700 nm, and photomorphogenic responses may extend the physiological effect of light into the far red from 700 to 760 nm.[4643] In fact, violet photons (400 nm) are nearly twice as energetic as red photons (700 nm).[3971] The dependence of the quantum yield of algal photosynthesis on wave length of light was presented by Emerson and Lewis in 1943.[1240]

Table 2-6. Standing crop of attached algal communities (g DM m^{-2})

Standing crop	Community type	Locality	Exposure (d)[a]	Reference
0.33–1.24	Periphyton[b]	Sodon Lake, Michigan	10	3421
1.3–6.1	Periphyton	Ditch, The Netherlands		2523
6.8	Epiphyton	Douglas Lake, Michigan		5529
2.5–7.0	Epilithon	Arakawa River, Japan		2546
0.55–9.98	Periphyton	Columbia River, Washington		937
2.5–18	Epiphyton	Lake Globokoye, USSR		175
25	Epilithon	Logan River, Utah		3134
0.3–34.9	Periphyton	Florida Everglades	82	5212
61.4	Periphyton	Sedlice Reservoir, Czech Republic	313	4557
33–86	Epilithon	Two lochs in Scotland		1389
4.7–93.9	Periphyton	Little Miami River, Ohio	13–27	5299
111.5	Epipelon	Silver Spring, Florida		3502
20.7–124	Periphyton	Artificial outdoor stream, Czech Republic	17	5198
1.35–143.6	Periphyton	Ohio River, Ohio	13–27	5299
47.5–148.9	Periphyton	Artificial indoor stream		3151
164	Periphyton	WWTP[c] effluent, Vermont	22	1000
181	Periphyton	Artificial stream		4170
119–182	Epilithon	Lake Balaton, Hungary		1329
187	Periphyton	Artificial indoor stream		3155
188	Epiphyton	Silver Spring, Florida		3502
94–233	Periphyton	Artificial indoor stream	45–66	3149
9–403	Periphyton	Artificial indoor stream		3150
100–500	Epilithon	Lake Superior		1167
593	Periphyton	Artificial indoor stream		3153
58.6–1,180	Epiphyton	Florida Everglades		5210

[a] Where available
[b] Periphyton stands for an attached community on any kind of artificial substratum
[c] Wastewater treatment plant

Phycologists are interested principally in photon flux density (PFD), the number of quanta (= photons in the light range of the electromagnetic spectrum). Such quantum irradiance measurements are given in SI units as micromoles meter^{-2} second^{-1};[2282] the identical, but non-SI, units of microeinsteins meter^{-2} second^{-1} are frequently used.[2282,2976,4643]

Irradiance may also be measured and expressed energetically in units of watts meter^{-2}. For monochromatic light, irradiance and PFD may be readily interconverted as:

$$1\,\mu\text{mol m}^{-2}\text{ s}^{-1} = 1\ \mu\text{E m}^{-2}\text{ s}^{-1} = 6.02 \times 10^{17} \text{ quanta m}^{-2}\text{ s}^{-1} = (119.7/\lambda)\ \text{W m}^{-2} \qquad (2.21)$$

where λ = the wavelength of the light in nm.[2282,2976,4643]

For light of broader spectral range the conversion from irradiance to photosynthetic PFD requires the integration of the PFD obtained over the spectral range of interest. Although it is generally not possible to relate directly illuminances measured in lux to measures of either light energy or photon density, a convenient rule-of-thumb conversion factor emerges from comparisons for daylight and for several artificial light fields:[2976,3142,3293]

Table 2-7. Standing crop of attached algal communities (g AFDM m^{-2})[a]

Standing crop	Community type	Locality	Exposure (d)	Reference
0.095	Periphyton	Artificial outdoor flow-through, Kentucky	28	5044
0.11–0.53	Periphyton	Sodon Lake, Michigan	10	3421
0.2–0.66	Periphyton	Walker Branch stream, Tennessee	60	1237
0.81	Periphyton	Walker Branch stream, Tennessee		1238
0.69–1.79	Periphyton	Fox Creek, California	31	2089
1.1–2.3	Periphyton	Falls Lake, Washington	28	720
2.94	Periphyton	Alkali Lake, Washington	189	720
1.1–3.0	Periphyton	Rangraun Stream, Malaysia	21–40	2106
3.04	Periphyton	White Oak Creek, Tennessee		3412
3.5	Periphyton	Red Cedar River, Michigan	7	1839
0.31–3.51	Periphyton	Otter Creek, Oklahoma	21	4412
1–3.8	Periphyton	Red Cedar River, Michigan	18	2507
1.3–4.0	Periphyton	Gombak River, Malaysia		380
0.24–4.03	Periphyton	Columbia River, Washington	14	937
1.3–4.3	Periphyton	Barnwell Creek, California	31	2089
5.2	Periphyton	Stream, Tennessee	35	2485
5.2	Periphyton	L. Maarsseveen, The Netherlands, summer		3203
2.7–5.7	Periphyton	Stream, Finland	14–42	1235
6.6	Periphyton	Carnation Creek, Canada, BC		4787
7.09	Periphyton	Outdoor troughs		809
8.8	Epilithon	Mountain streams, Oregon		2987
2.7–10.3	Periphyton	Lake Superior		1469
1.3–11.6	Periphyton	Artificial stream, inorg. enrich.	56	1220
11.1	Periphyton	Outdoor troughs		2438
12	Periphyton	Artificial indoor stream	34	2485
12.7 (mean)	Periphyton	Sedlice Reservoir, Czech Republic	313	4557
6.3–13.2	Periphyton	Hart's Run, Kentucky	26	2169
2.7–14	Periphyton	Artificial stream, org. enrich.	56	1220
14.9	Periphyton	Outdoor troughs, enrichment		3074
15	Epilithon	Stream, England		3073
2.1–15.3	Epilithon	Swedish Rivers		3332
2.4–16.7	Epilithon	New Zealand rivers		369
0.55–16.95	Periphyton	Ohio River, Ohio	13–27	5299
18.7	Epiphyton	Lake Glubokoye, USSR	152	176
18.9	Periphyton	East Gallatin R., Montana[b]		209
0.16–19.2	Periphyton	Florida Everglades	82	5212
0.76–21.88	Periphyton	Little Miami River, Ohio	7–71	5299
1–29	Periphyton	Pawnee Reservoir, Nebraska	2–14	2116
10.8–24.3	Periphyton	Oregon coast	14	722
21–36	Periphyton	Nyumba ya Mungu reservoir, Tanzania	28	497
37.5	Periphyton	Outdoor troughs, inorg. enrichment		4788

Table 2-7. Continued

Standing crop	Community type	Locality	Exposure (d)	Reference
40	Periphyton	Artificial stream, sewage enrich.		1221
40	Periphyton	Artificial stream	32	1051
40	Epiphyton	Crescent Pond, Manitoba		2179, 2180
10–49	Epilithon	Two lochs in Scotland		1389
49.5	Periphyton	L. Maarsseveen, The Netherlands, winter		3203
50	Periphyton	Outdoor troughs, enrichment		1223
49.5	Epilithon	White Oak Creek, Tennessee		3412
11.5–56.5	Periphyton	Microcosm		5428
61.5	Epilithon	Lake Superior		1167
65	Periphyton	Donau River, Austria-Slovakia	1 year	1273
15.6–65.7	Periphyton	Artificial indoor stream		3151
66.7	Periphyton	Artificial indoor stream		4170
67.7	Periphyton	Desert stream, Idaho		1187
70.5	Periphyton	Outdoor troughs, inorg. enrichment		1222
44–73	Periphyton	Artificial indoor stream	45–65	3149
82.6	Epilithon	Grand River, Ontario		2846
28–102	Epilithon	Wilson Creek, Kentucky		4741
120	Periphyton	WWTP[c] effluent, Japan		3521
130	Periphyton	WWTP[c] effluent, USA	22	1000
7.3–141.3	Periphyton	Artificial indoor stream		3150
147	Periphyton	Artificial indoor stream		3153
160	Epilithon	Desert stream, Arizona		1360
188	Epiphyton	Silver Spring, Florida		3502
250	Periphyton	Outdoor troughs		1224
290	Epilithon	Desert stream, Arizona		1824
291	Epilithon	Maraekakaho River, New Zealand		370
34.2–518	Epiphyton	Florida Everglades		5210

[a] Periphyton stands for an attached community growing on any kind of artificial substratum
[b] Downstream the sewage outfall
[c] Wastewater treatment plant

$$1 \text{ W m}^{-2} \text{ (Irradiance)} = 5 \ \mu E \text{ m}^{-2} \text{ s}^{-1} \text{ (Quantum irradiance)} = 250 \text{ lux (Illuminance)} \quad (2.22)$$

Conversion factors for the most common light sources are reported by Lüning,[2976] and are appropriate where absolute accuracy is not required.[4643] Full sunlight has a PFD of about 1,700 μE m^{-2} s^{-1} (400–700 nm).[4116]

Photosynthetic PFD is of obvious importance as a major controlling factor for inorganic C uptake by algae.[1103] Another effect of light is its effect on enzyme activation and deactivation. Several enzymes of the Calvin cycle in the chloroplast, e.g. RuBPC, are inactive in darkness but become active following illumination. Other enzymes of glycolysis and of the pentose phosphate pathway (e.g. glucose-6-phosphate dehydrogenase) are active in darkness and inactivated by light and, in some algae, RuBPC becomes inactive again at high light intensities.[827,4861]

Ion transport is often light-dependent and is closely interrelated with changes in membrane potentials.[4615] Concomitant with the light-stimulated uptake of K$^+$ and Na$^+$ in *Hydrodictyon africanum* there is a stimulation of Cl$^-$ uptake. The action spectrum for the responsible Na-K-Cl pump coincides

Table 2-8. Standing crop of attached algal communities (mg chl-*a* m^{-2})

Standing crop	Community type	Locality	Exposure (d)	Reference
0.48–0.8	Periphyton[a]	Artificial indoor stream		3155
0.61–0.96	Epipelon	River Ely, Wales		119
2	Periphyton	Athabasca River, Canada		230
0.17–2.27	Periphyton	Lake Memphremagog, Québec-Vermont	14	731
0.88–2.8	Periphyton	Renggam Stream, Malaysia		2106
5.9	Periphyton	Stream, Tennessee		1238
3.6–6.4	Periphyton	Hart's Run, Kentucky	26	2169
4.5–10.7	Epiphyton	Lake Memphremagog, Québec-Vermont		729
10.5–12.8	Periphyton	Outdoor exp. channels, Michigan		3182
4.5–13.9	Periphyton	Stream, Finland		1235
4.8–15.2	Periphyton	Fox Creek, California		2089
0.2–15.3	Periphyton	Otter Cree, Oklahoma		4412
5.9–16	Periphyton	Wabamun Lake, Alberta	20–163	1701
16.4	Epiphyton	Lake Memphremagog, Québec-Vermont		730
1.4–17.8	Periphyton	Hastings Lake, Alberta		1701
18.5	Periphyton	Valley Creek, Minnesota		5289
22	Periphyton	Columbia River, Washington		937
2.66–22.8	Periphyton	Fleming Creek, Michigan	5–22	3183
24.8	Edaphon	Durant Lake, North Carolina, enrichment	79	630
0.8–26.2	Periphyton	Fire pools, Japan		2645
16.4–33.6	Periphyton	Exp. marsh, Canada, Manitoba		2209
34.7	Periphyton	Silver Spring, Florida		5533
2–36.3	Epiphyton	Wabamun Lake, Alberta	19–163	1701
5.7–41.4	Periphyton	Barnwell creek, California, enrich.	31	2089
0.12–44.7	Periphyton	Tvärmine Zool, Station, Finland		2781
094–47.8	Periphyton	Wye River, Wales	1 month	118
0.2–48	Periphyton	Lake Saimaa, Finland		2482
0.06–50	Periphyton	Outdoor flow-troughs, Canada		472
10–70	Epipelon+ Epilithon+ Epiphyton	Opatovický Pond, Czech Republic		2565
30–70	Periphyton	Arakawa River, Japan		2546
71	Periphyton	Streams, Tennessee	70	3331
49.1–80.4	Periphyton	Artificial stream		117
15.9–97	Epipelon	Inslow Bay, North Carolina		658
104	Periphyton	East Gallalin R., Montana		209
8.4–106	Endolithon	Colorado Plateau, Arizona		298
111	Periphyton	Stream, New Jersey		1381
92–127	Epipelon	Delaware salt marsh		1545
5.8–136.6	Epilithon	New Zealand rivers		369
59–160	Epipelon	Graveline Bay marsh, Mississippi		4836
190	Periphyton	Outdoor flow-trough, Kentucky	26	5044
95–193	Epipelon	Massachusetts marsh		1277
42.8–198.6	Epilithon	Lake Superior		1167
200	Epiphyton	Silver Spring, Florida		3502

Table 2-8. Continued

Standing crop	Community type	Locality	Exposure (d)	Reference
271	Epilithon	Stream, England		3073
78–305	Periphyton	Artificial indoor stream		3151
25–330	Periphyton	Microcosm	32–109	5428
330	Periphyton	Donau River, Austria-Slovakia		1273
335	Periphyton	Desert stream, Idaho		1187
2–370	Epiphelon	Streams and lakes, Spain		3067
30–380	Periphyton	Artificial stream		
420	Periphyton	Stream, Canada		2201
240–465	Periphyton	WWTP[b] effluent, Japan		3521
330–540	Periphyton	Artificial indoor stream	45–65	3149
560	Periphyton	Artificial stream		2202, 5145
680	Periphyton	Tomagawa River, Japan		4934
600–700	Periphyton	WWTP[b] effluent, Norway		5011
880	Periphyton	WWTP[b] effluent, Vermont	40	1000
350–960	Periphyton	Downstream sewage effluent, Utah		280
703–1142	Periphyton	Artificial indoor stream		3152
300–1420	Epilithon	Logan River, Utah		3134
2010	Periphyton	Artificial indoor stream		3153

[a] Periphyton stands for attached communities growing on any kind of artificial substratum
[b] Wastewater treatment plant

Table 2-9. Periphyton standing crop (cells cm^{-2})

Standing crop (density)	Locality	Reference
6730	Wilson Creek, Kentucky	3138
55–11 × 10^3	River Wye System, Wales	118
74–6.8 × 10^4	Oak Creek, Arizona	2573
2.6–9.7 × 10^4	Stream, Finland	1235
6.03–12.05 × 10^4	Ditch, The Netherlands	2523
1.37–2.44 × 105	Two lochs in Scotland	1389
1.5–2.7 × 10^5	Hart's Run, Kentucky	2169
4 × 10^5	Pawnee Reservoir, Nebraska	2116
5 × 10^5	Wheelwright Pond, New Hampshire	4527
9.1–8.35 × 10^5	Lake Memprhemagog, Québec-Vermont	729
5.5 × 10^4–12 × 10^5	New Zealand rivers[a]	369
21 × 10^5	Durant Lake, North Carolina[b]	630
2.2 × 10^6	Susquehanna River, Pennsylvania	2925
2.9 × 10^6	Ohio River, Ohio	5299
3 × 10^6	Outdoor flow-trough, Kentucky	2247
5 × 10^6	Outdoor flow-trough, Kentucky	4746
6.7 × 10^5–6.3 × 10^6	Oak Creek, Arizona	400
7.2 × 10^6	Stream, Tennessee	3331
36 × 10^6	Shetucker River, Connecticut[a]	2539
4.85 × 10^6–2.31 × 10^7	Lake Superior	1167

[a] Epilithon
[b] Edaphon

Table 2-10. Periphyton standing crop—biovolume (μl cm^{-2})

Biovolume	Locality	Exposure (d)	Reference
28	Fox Creek, North California	31	2089
48×10^3	Durant Lake, North Carolina	79	630
15×10^4	Wilson Creek, Kentucky		3138
21.5×10^4	Barnwell Creek, North California	31	2089
26.2×10^4	Stream, Tennessee		3331
2.9×10^4–1.64×10^7	20 New Hampshire Lakes		4745

Table 2-11. Phytoplankton standing crop (mg chl-a m^{-2})

Production	Locality	References
20	Lake Pääjärvi, Finland	2278
15–65	Wisconsin lakes	3045
94	Lake Lovojärvi, Finland	2464
309	McMurdo Sound, Antarctica	3632
10–450	Czech Republic, pond	2565

Table 2-12. Productivity of algae

Species	Productivity	Locality	Reference
	(g DM m^{-2} d^{-1})		
Cladophora glomerata	0.085–2.6	Michigan	4110
Chara sp.	6.8–15.9	Davis, California	5343
Chlorella sp.	3.76–17.8	Laboratory (var. light and temp.)	990
Ulva lactuca	55	Culture, Israel	3416
	(g DM m^{-2} yr^{-1})		
Charophytes	1–91	Europe	2981
Chara sp.	155	Lake Lawrence, Michigan	4096
Chara sp.	262–331	India	1748
Chara fragilis	420	India	1894

Table 2-13. Productivity of attached algal communities (mg DM m^{-2} d^{-1})[a]

Productivity	Locality	Reference
12.6	Sodon Lake, Michigan	3421
0.37–42.5	Florida Everglades	5212
1,424–5,178	Artificial indoor stream	3149
7,294	Outdoor flow-trough, Czech Republic	5198
3,900–21,600	WWTP[b] effluent, Vermont	1000
24,282 (max.)	Florida Everglades (epiphyton)	5210

[a] Periphyton on artificial substrata if not specified
[b] Wastewater treatment plant

Table 2-14. Productivity of attached algal communities

Productivity	Locality	Reference
Epipelic algae **(g C m^{-2} yr^{-1})**		
0.5	Lake Pääjärvi, Finland	2404–5
0.6	Lake Suomunjärvi, Finland	4634
2.0	Lawrence Lake, Michigan	4095
10.1	Lake Ikroavik, Alaska	4685
40	Lake Marion, Canada	1838, 1913
(mg C m^{-2} d^{-1})		
88.7	Michigan, lake	224
109.6	Lake Marion, Canada	1838, 1913
340	Skua Lake, Antarctica	1691
468	Algal Lake, Antarctica	1691
(mg C m^{-2} h^{-1})		
1.71	Priddy Pool, England	2077
1.68–3.68	Lake Pääjärvi, Finland	2406
Epilithic algae **(g C m^{-2} yr^{-1})**		
5.2	Experimental Lakes Area, Canada	4342
(mg C m^{-2} d^{-1})		
231	Little Schultz Creek, Alabama	4784
75–250	Little Schultz Creek, Alabama	2723
274	Eagle Lake, California	2251
731.5 (0–57,600)	Lake Borax, California	5342–3
432–907 (blue-greens)	Little Schulty Creek, Alabama	4785
(mg C m^{-2} h^{-1})		
0.5	Lake Pääjärvi, Finland	2404
4.96	Lake Tahoe, California	2890
1.4–8.2	Stream, Tennessee	3331
8.43	Castle Lake, California	2891
9.04	Donner Lake, California	2891
4.1–12.2	Lake Tahoe, California-Nevada	2892
Epiphytic algae **(g C m^{-2} yr^{-1})**		
37.9	Lawrence Lake, Michigan	4095
67–327	Opatovický pond, Czech Republic	2567
731	Lake Lawrence, Michigan	5355
(mg C m^{-2} d^{-1})		
203	Lake Memphremagog, Vermont	730
1,153	Eagle Lake, California	2251
2,003	Lake Lawrence, Michigan	5355
(mg C m^{-2} h^{-1})		
3.6	Lake Pääjärvi, Finland	2406
4.3–8.9	South Africa	1864
3.3–30.6	Lake Manitoba, Canada	4160
63.9	Priddy Pool, England	2077
Epizoic algae		
84 mg C m^{-2} d^{-1}	Little Schultz Creek, Alabama	4786

Table 2-14.   Continued

Productivity	Locality	Reference
Edaphic algae		
(g C m^{-2} yr^{-1})		
5	Chukchi Sea, Alaska	3102
31	Ythan estuary, Scotland	2727
21–53	Netarts Bay, Oregon	1001
11–60	Florida, Jamaica	620
71	Texas salt marsh	1879
75	Southern Baltic Sea	5273
79	Rija de Arosa, Spain	5139
81	New England estuaries	3077
38–99	Delaware salt marsh	1547
101	Dutch Wadden Sea, tidal flats	657
105	Massachusetts salt marsh	5134
108	Tijuana Estuary, California	5542
115	Wadden Sea	174
30–130	Columbia River estuary, Oregon	3154
143	Lynher River, England	2355
150	Duplin River Marsh, Georgia	3835
28–151 (max. 714)	Graveline Bay Marsh, Mississippi	4836
170	Mugu Lagoon, California	4413
190	Sapelo Island salt marsh, Georgia	3835
200	Mugu Lagoon, California coast	3557
200	Duplin River Estuary, Georgia	3832
213	Boca Ciega Bay, Florida	3833
143–226	Puget Sound, Washington	3635
231	Danish Fjords	1830
56–234	North Inlet Estuary, South Carolina	3800
115–246	Bolsa Bay, California	4156
250	Savin Hill Cove, Massachusetts	1779
50–250	Ems-Dollard Estuary, The Netherlands	841
63–253	Langebaan, South Africa	1344
181–341	California salt marsh	5541
400	Louisiana salt marsh	3278
mg C m^{-2} h^{-1}		
26–59	Bolsa Bay, South California	4156
5–84	Columbia River estuary, Oregon	3154
17–132	Sapelo Island salt marsh, Georgia	3835
1.4–163	Graveline Bay Marsh, Mississippi	4836
14.8–167	North Inlet estuary, South Carolina	3800
20–200	Georgia salt marsh	3832
300	Louisiana salt marsh	3278
Periphytic algae (on artificial substrata)		
(mg C m^{-2} h^{-1})		
0.17–0.80	Tvärminne Zool. Station, Finland	2781
3.52–6.31	Canada, Manitoba	2209
133–250	Artificial chambers	4163

Table 2-15. Productivity of seaweeds

Species	Locality	Productivity	Reference
		(mg C g^{-1} h^{-1})	
jointed calcareous algae	SW U.S.A. and Mexico	0.045 (mean)	2877
(e.g. *Corallina, Jania, Petrocelis*)			
Galaxaura squalida	Canary Islands	0.20	2352
Botryocladia pseudodichotoma	California	0.30	2004
Codim fragile	Mexico	0.57	2877
thick leathery algae	SW U.S.A. and Mexico	0.76 (mean)	2877
(e.g. *Fucus, Sargassum, Macrocystis, Laminaria, Padina*)			
coarsely branched algae	SW U.S.A. and Mexico	1.30 (mean)	2877
(e.g. *Laurentia, Gelidium, Codium, Gigartina, Pterocladia*)			
Gigartina canaliculata	California	1.6	2875
filamentous algae	SW U.S.A. and Mexico	2.47 (mean)	2877
(e.g. *Cladophora, Ceramium, Chaetomorpha*)			
Gelidium robustum	California	2.5	2875
sheet-like algae	SW U.S.A. and Mexico	5.16 (mean)	2877
(e.g. *Ulva, Eneromorpha, Porphyra*)			
Gracilaria verrucosa	Florida	4.10–9.60	2142
Enteromorpha compressa	Finland	0.02–10.8	2445
Cladophora glomerata	Sweden	1.10–12.8	5241
		(mg C g^{-1} d^{-1})	
Enteromorpha compressa	Finland	47	2445
Cladophora glomerata	Sweden	20–180	5241
		(g C $m^{-2}.d^{-1}$)	
Caulerpa prolifera	Canary Islands	1.0	2352
Pterocladia capillacea	California	1.3	2875
Egregia laevigata	California	1.7	2875
Gigartina canaliculata	California	2.0	2875
Sargassum agardianum	California	2.2	2875
Ulva californica	California	2.8	2874
Gelidium pusillum	California	3.1	2875
Nereocystis luetkeana	British Columbia	5.4	2601
Macrocystis integrifolia	British Columbia	6.7	2601
Macrocystis pyrifera	Southern California	0.1–7.0	5010
Macrocystis pyrifera	California	9.5	2301
Fucus vesiculosus	Massachusetts	20	2422
Codium fragile	Connecticut	22	5276
		(g C.$m^{-2}.yr^{-1}$)	
Gigartina exaspertata	Washington	439–758	5213
Laminaria hyperborea	Scotland	377–775	2387
Iridaea cordata	Washington	144–1,012	5213
Laminaria longicruris	Nova Scotia, Canada	1,750	3042
Coralline algae	Hawaii	2,080	2874
Sargassum bed	Curaçao	2,550	5262
Codium fragile	Connecticut	4,700	5276

Table 2-16. Productivity of attached algal communities (mg AFDM m^{-2} d^{-1})

Productivity	Community type	Locality	Reference
1.1	Epipelon	Suomunjärvi Lake, Finland	4634
6.0	Epilithon	Lake Pääjärvi, Finland	2404
11.8	Periphyton[a]	Sodon Lake, Michigan	3420
18	Epipelon	Changing Lake, Antarctica	3897
0.19–23.4	Periphyton	Florida Everglades	5212
16–24	Periphyton	Stream, Tennessee	1237
37.5	Periphyton	Sodon Lake, Michigan	3421
43.4	Periphyton	Outdoor flow-trough, Kentucky	5044
49	Epipelon	Sombre Lake, Antarctica	3897
22–55	Epipelon	Tundra ponds, Alaska	4685
22–58	Periphyton	Fox Creek, California	2089
60	Epipelon	Lake Pääjärvi, Finland	2405
41–104	Periphyton	Renggam Stream, Malaysia	2106
125	Epiphyton	Lake Glubokoye, USSR	176
131	Periphyton	Alkali Lake, Washington	720
42–139	Periphyton	Barnwell Creek, California	2089
147	Epiphyton	Crescent Pond, Manitoba	2179, 2180
148	Periphyton	Falls Lake, Washington	720
110–180	Epiphyton	Nyumba ya Mungu Res. Tanzania	497
200	Periphyton	Stream, Tennessee	2485
282	Periphyton	Red Cedar River, Michigan	2507
25–300	Periphyton	Columbia River, Washington	937
110–320	Epixylon[b]	Nyumba ya Mungu Res., Tanzania	497
30–380	Periphyton	Artificial stream, org. enrich.	1220
262–508	Periphyton	Hart's Run, Kentucky	2169
510	Periphyton	Artificial stream	2485
120–550	Periphyton	Artificial stream, inorg. enrich.	1220
510–580	Periphyton	Artificial stream	2454
140–750	Periphyton	Microcosm	5428
100–1,010	Periphyton	Sedlice Reservoir, Czech Rep.	4557
650–1,020	Periphyton	Nyumba ya Mungu Res., Tanzania	497
430–1,030	Epilithon	Nyumba yz Mungu Res., Tanzania	497
40–1,040	Periphyton	Ohio River, Ohio	5299
1,250	Periphyton	Artificial stream	1051
110–1,390	Periphyton	Little Miami River, Ohio	5299
667–1,622	Periphyton	Artificial stream	3149
710–1,750	Periphyton	Skeleton Creek, Oklahoma	882
11–2,278	Periphyton	Red Cedar River, Michigan	1840
446–1,429	Periphyton	Artificial stream	2484
5,909	Periphyton	WWTP[c] outflow	1000
11,932	Epiphyton	Florida Everglades	5210

[a] Periphyton stands for attached communities growing on any kind of artificial substratum
[b] Epixylon stands for growing on wood
[c] Wastewater treatment plant

Table 2-17. Periphyton productivity (mg chl-a m^{-2} d^{-1})

Productivity	Locality	Reference
0.13–0.25	Hart's Run, Kentucky	2169
0.31	Durant Lake, North Carolina	630
0.15–0.49	Fox Creek, California	2089
0.52	Wabamun Lake, Alberta	1701
0.70	Alberta's lakes	1701
0.18–1.34	Barnwell Creek, California	2089
7.3	Outdoor flow-trough, Kentucky	5044
5–12	Artificial stream	3149
0.031–32	Outdoor flow-troughs, Canada	474

Table 2-18. Productivity of attached algal communities (mg O$_2$ m^{-2} h^{-1})

Productivity	Community type	Locality	Reference
11.5–40.2	Epipelon	Austrian pre-alpine lakes	3422
37–135	Periphyton	Duran Lake, North Carolina	630
140–210	Periphyton	Lake Balaton, Hungary	1329
3,290	Epilithon	Logan River, Utah	3134

with photosystem II.[3994] By contrast, the chloride independent K^+ influx is linked to photosystem I.[3995] The light-stimulated portions of alkali-ion effluxes share several important characteristics with the respective influxes.[218,3993] The light-dependent uptake of HCO_3^- at high pH may depend on a specific, ATP-independent HCO_3^- pump coupled to photosystem I activity.[3992] There is an enhancement by light of sulfate transport across the plasmalemma and tonoplast.

A relationship between irradiance and uptake of inorganic nitrogen[1103,3002,3413,3912-3,4329-30,5356] and phosphorus[3385,3388,5531] has been shown for algae in a number of aquatic systems. Nitrate and nitrite uptake is light-dependent and is linked to photosystem II activity in chloroplast.[1794,3315] Light has a significant effect on algal nitrogen metabolism as light quality with respect to the nature of the products of photosynthesis.[2577,5240] In general, blue light favors the production of amino acids and proteins rather than of carbohydrate with either NO_3^-, NH_4^+, or urea as a nitrogen source.[4861] Nitrogen fixation by blue-green algae is light-dependent with the primary involvement of photosystem I.[1316,2989] Syrett[4861] summarized ways in which light can affect algal nitrogen assimilation. Coccolith formation is also light-dependent[3607-8] as well as the process of calcification in general (see Chapter 5.1.4 for detailed discussion). The recovery of blue-green alga *Anacystis* from manganese deficiency is also light-dependent process.[1607]

Photosynthesis of intertidal species (e.g. *Fucus*) are light-saturated at about 500 μE m^{-2} s^{-1} (full sun approximately 1,700 μE m^{-2} s^{-1}), upper- and mid-sublittoral species (e.g. laminarians) are saturated by about 200 μE m^{-2} s^{-1}, whereas deep-water red algae require only about 100 μE m^{-2} s^{-1} for saturation.[2976] In white light, these values correspond roughly to irradiances of 100, 40, and 20 W m^{-2}, or to illuminances of 25, 10, and 5 klux, respectively.

Richardson et al.[4116] reported that different microalgal classes have significantly different light requirements for growth and photosynthesis. Although there is some variability within each class, dinoflagellates and blue-green algae generally photosynthetize and grow better at low photon flux densities. Diatoms also tend to be able to grow at very low PFD (growth for some species has been reported at less than 1 μE m^{-2} s^{-1}). Comparison of the PFD at which photoinhibition occurs in dinoflagellates and diatoms suggests that the former often experience photoinhibition at comparatively low irradiances. In contrast, diatoms often can tolerate relatively high light environments. This tolerance of a large absolute range of PFD may, in part, explain why diatoms are often associated with spring blooms.[4116] Green algae tend to exhibit higher light compensation points than the other common microalgal classes and can tolerate very high light environments.[4116]

Table 2-19. Phytoplankton productivity

Productivity	Locality	Reference
(g C m^{-2} yr^{-1})		
0.6	Tundra ponds, Alaska	54
2.2	Lake Ikroavik, Alaska	4686
7.0	Schrader Lake, Alaska	2123
11.3	Meratta Lake, Canada, N.W.T.	2410
29	Kattehale Mose, Denmark	3487
24–42	Bothnian Bay, Finland coast	48
43.3	Lawrence Lake, Michigan	4095
64	Saronicos Gulf (Aegean Sea)	266
67	Core Sound, North Carolina	4938
79	Duplin River Marsh, Georgia	3835
84	Store Gribsø, Denmark	3487
91	Borax Lake, California	5343
104	Lake Erken, Sweden	4165
160	Clear Lake, California	1687
77–160	Walter Lakes, Indiana	5347
131–171	Crooked Lake, Indiana	5344
180	Lake Ontario	5182
205	Lake Martin, Indiana	5347
225	North Carolina coast	5447
260	Lake Esron, Denmark	2358
266	Lake Goose, Indiana	5345
9–301	Tomahawk Lagoon, New Zealand	3243
346	North Inlet, S. Carolina	4402
376	Frederiksborg Slotssø, Denmark	3487
290–480	Lake Norrviken, Sweden	34
570	Sylvan Lake, Indiana	5345
640	Lake Victoria, Africa	4887
150–827	Opatovický pond, Czech Republic	2568
(mg C m^{-2} d^{-1})		
1.6	Tundra ponds, Alaska	54
2.5	Lake Peters, Alaska	2123
8.5 (0–170)	Lake Meretta, Canada, N.W.T.	2410
8.04	Marion Lake, Canada, B.C.	989
14	Lake Vanda, Antarctica	1690
19.2	Schrader Lake, Alaska	2123
46.6	Lake Michigan	224
23–66	Ransaren lake, Sweden	4165
80 (0–400)	Kattehale Mose, Denmark	3487
99.3 (5–497)	Lake Lawrence, Michigan	5349
28–133	Lake Tahoe, California-Nevada	1688
158	Brooks Lake, Alaska	1682
194 (24–5,960)	Smith Hole, Indiana	5347
230 (4–680)	Store Gribsø, Denmark	3487
249.3 (10–525)	Borax Lake, California	5342

Table 2-19. Continued

Productivity	Locality	Reference
50–260	Lake Superior	3939
285 (40–2,205)	Lake Erken, Sweden	4165
70–334	Saronicos Gulf (Aegean Sea)	266
356	Eagel Lake, California	2251
402	Greenland Sea, marginal zone	929
438 (2–2,440)	Clear Lake, California	1687
440 (68–1,850)	Pretty Lake, Indiana	5344
462 (0–1,380)	Fureso, Denmark	2358
469	Crooked Lake, Indiana	5344
561 (27–1,708)	Martin Lake, Indiana	5345
200–600	Bothnian Bay, coast, Finland	48
150–700	Lake Huron, U.S.A. - Canada	5182
1,030 (12–4,160)	Frederiksborg Slotsso, Denmark	3487
70–1,030	Lake Michigan, U.S.A.	5182
578–1,103	Waco Reservoir, Texas	2504
60–1,400	Lake Ontario, U.S.A. - Canada	5182
12–1,458	Walter lakes, Indiana	5347
1,564 (9–4,959)	Sylvan Lake, Indiana	5345
1,700 (400–5,000)	Lake Lanao, Philippines	2842
1,750 (1,700–3,800)	Lake Victoria, East Africa	4887
30–4,760	Lake Erie, U.S.A. - Canada	5182
$(mg\ O_2.m^{-2}.d^{-1})$		
150 - >10,000	Czech Republic	2565

By comparison with most vascular plants, unicellular microalgae occupy habitats characterized by very low PFD. Most microalgae live in aquatic environments where light is attenuated exponentially with depth according to the Lambert-Beer Law:[3971,4116]

$$I_d = I_o\ e^{-kd} \qquad (2.23)$$

where d is depth, I_o is the incident radiation upon the surface of the water, I_d is PFD at depth, d, and k is the extinction coefficient. Richardson et al.[4116] reported that clear temperate coastal waters exhibit a k for PAR, 400 to 700 nm of about 0.15, while in the clearest oceanic waters k may be in the region of 0.04.

In benthic microalgal communities, incident light is usually attenuated 100-fold within the upper 1–4 mm, accompanied by major changes in spectral composition.[1332,1858,2370] Mobile cells can change depth position, and thereby optimize light and nutrient utilization.[4212] Microalgal species and photosynthetic bacteria with different spectral light absorption can replace and supplement each other along the intensity and spectral gradient.[2370]

In general, growth appears to be light-saturated at considerably lower irradiances than photosynthesis. Adult thalli of eulittoral species (e.g. *Fucus*) are light-saturated at 150–250 μE m^{-2} s^{-1}, whereas the more bulky species of the upper sublittoral (*Laminaria, Chondrus, Codium*) are saturated at 30–100 μE m^{-2} s^{-1}, and the optimum photon flux density for deep-water red algae seems to be around 10 μE m^{-2} s^{-1}.[2976] Bonin and co-workers,[443] in a comprehensive review, discussed light-shade and light-sun physiological adaptations in phytoplankton species.

The light dependency of photosynthesis is important for calculating primary production rates in pelagic systems.[4615] A mathematical model for the integration of pelagic photosynthesis in space and time, and which includes light inhibition effects, has been developed by Vollenweider.[5176,5178] *In situ* measurements show that the decrease of primary production rates with depth is usually a function of green light intensity, i.e. of the most penetrating light component.[4166] Dring[1144] reported that blue light appears to exert a direct effect on the dark reaction of photosynthesis in brown algae, possibly by activating carbon-fixing enzymes or by stimulating the uptake or transport of inorganic carbon in the plants.

High light intensities often result in release of large proportions of photosynthate by algae, possibly due to membrane damage.[299,1433,2013,3380] The proportion of the photoassimilated carbon released is relatively unaffected by light intensities over an intermediate range where little or no inhibition of photosynthesis takes place.[2013,3380] Relatively small proportions of cellular carbon are lost by algal cells during dark periods following periods of photoassimilation.[2013] High intensity light has been shown to inhibit gas vacuole formation in some blue-green algae.[5249] Hoffmann and Malbrán[2140] reported that some aspects of growth and fertility of *Glossophora kunthii* are controlled by photoperiod.

Light intensity shows direct correlation with specific composition of algal pigmentation.[260,1658,3038] Irradiance level has been positively associated with algal biomass[2089,2987,3350,4283,4726,4729] and can influence community structure[4435,4454,4725-6] and succession.[443]

For other light effects on algal growth and metabolism see also Chapters 4.2.4.6 and 4.3.4.2 and References 133, 443, 476, 1588, 2316-7, 2377, 2976 and 3070.

2.13.1.2 Temperature

Temperature is one of the major factors controlling the rate of photosynthesis in all plants.[982] The effect of temperature on chemical transformations and on transport processes is summarized in an excellent review by Raven and Geider.[4012] In general, growth rate increases exponentially with temperature (Figure 2-34) up to an optimum temperature, then declines rapidly as temperature exceeds this optimum.[1256,3647,4571,4633]

Temperature is probably the most widely measured environmental variable that effects algal growth.[4012] Vogel[5170] suggests that the ease of measurement of temperature has led to an overemphasis on temperature in ecology, in contrast to such less readily measured parameters.

Genotypic variation in the temperature optimum for resource-saturated growth of microalgae has been used to provide envelopes of maximum specific growth rate (μ_m) as a function of temperature. The Q_{10} (ratio of rate at T°C to that at T - 10°C; for further definition see Berry and Raison[346]) value for μ_m for batch-cultured algae with optimal growth temperatures in the range 5–40°C is 1.88; rather higher values (Q_{10} = 2.08 - 2.19) are found, albeit with lower μ values at a given temperature, for continuous cultures.[4012] The typical response for photosynthesis is to increase progressively with increasing temperature, with a Q_{10} of approximately 2.0,[2601,4824] but Q_{10} values may vary. Photosynthesis continues to increase up to an optimum temperature, beyond which it declines rapidly.[982] In some cases, maximum photosynthetic rates occur over a range of several degrees rather than at a single optimum temperature.[3014,3494] A comprehensive review of the effect of temperature on algal photosynthesis has been given by Davidson.[982]

There may be significantly different temperature optima for different species of the same genus. For example, species of the genus *Chlorella* differ markedly in their temperature optima which may be at 25°C, 30°C or 39°C.[2910,4630] Simpson and Eaton[4515] reported the photosynthetic temperature optimum in *Cladophora glomerata* and *Spirogyra* sp. somewhat lower at approximately 20°C. The photosynthetic temperature optimum for *Cladophora glomerata* reported by Bellis and McLarty[301] is 18°C. Vegetative cells of *Cladophopra glomerata* were killed upon freezing or when exposed to temperatures above 30°C. No growth occurred at 5°C and only slight growth was observed at 10°C.[300] Optimum growth of *Tribonema minus,* a dominant alga in blanketing algal mats in the early spring, was obtained at temperatures between 15 and 25°C while optimum growth of *Spirogyra singularis*, a dominant species in algal mats during the summer, was between 20 and 25°C.[1072]

Van Donk and Kilham[5127] reported that temperature alters the maximum growth rate (μ_m), the half-saturation constant (K_s) for growth and the minimum requirements of a cell for a particular resource (the minimum cell quota, Q_o) in several diatoms. Ahlgren[35] reviewed the various formulations used to describe the effects of temperature on algal growth constants and concluded that (1) the effect of temperature on μ_m was essentially linear over the range 0–40°C, (2) K_s and yield at maximum growth rate were independent of temperature, and (3) Q_o had a complex relationship with temperature, usually being higher at low temperature. In P- and N-limited *Scenedesmus* sp., *Asterionella formosa*,[4088] *Monochrysis lutheri*,[1659] and *Oscillatoria agardhii*,[5552] the minimum cell quota increased with decreasing temperatures. These increases in Q_o indicate that suboptimal temperatures increase resource requirements, and thus may also alter optimum ratios. The same may be true for supra-optimal temperatures.[4087,4974]

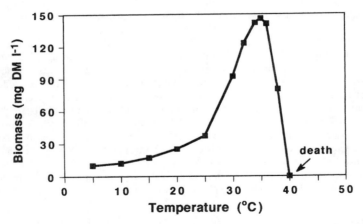

Figure 2-34.   Biomass of *Scenedesmus quadricauda* grown at different temperatures in PAAP medium for 10 days (Vymazal, unpublished results).

Temperature is an important environmental factor determining an algal growth and the geographical distribution of certain algae. Photosynthetic algae occur in the hottest and coldest environments in which autotrophic plants can be found (see Chapters 2.7.4 and 2.7.5).[982] The temperature optima for growth of many marine and freshwater phytoplankters lie in the range 18–25°C, although cold-water forms generally have lower optima. Arctic and Antarctic ice algae achieve net photosynthesis at a constant temperature of -2°C,[3204,3634] and Antarctic soil algae continue to photosynthetize at temperatures as low as -7°C,[981] whereas thermophilic hot spring cyanophytes photosynthetize up to approximately 75°C.[562,723] Some Antarctic diatoms have temperature optima of 4–6°C and will not grow above 7–12°C, whereas an isolate of a tropical diatom will not grow below 13–17°C.[973,2245,3685] Patrick et al.[3685] found that an average temperature of 34 to 38°C resulted in a shift of dominance in the algal flora from diatoms to blue-green algae.

Algal life at extreme temperatures has been reviewed by Brock[563] with special reference to the heat tolerance of thermophilic forms, the upper limit of which seems to be around 74°C (see Chapter 2.7.5). Kol[2554] reviewed the algae inhabiting the surfaces of snow and ice and the physiological characteristics of cryophilic algae have been reviewed by Fogg and Horne.[1432]

Thermophilic algae can be cultured at comparatively low temperatures. For example, *Cyanidium*, which has an upper temperature limit of 55 to 60°C[1114] grows well at 15°C.[268] Cryophilic algae, on the other hand, seem to be much more stenothermic.[4615] The temperature optimum for *Koliella tatrae* is 4°C and the maximum is 10°C.[2097] Lange and Metzner[2701] have measured active photosynthesis below 0°C in lichens, and *Nostoc* sp. and *Prasiola crispa* growing on Antarctic soil can photosynthetize down to -5 and -20°C, respectively.[268] Marre[3075] reported that *Dunaliella salina,* a flagellate capable of growing in concentrated saline media, can under such conditions survive without freezing, and even maintain some motility at temperatures as low as -15°C.

Algae are able to adapt, within limits, to suboptimum temperatures and changes in temperature,[443,2378,4615] but the mechanisms are not well understood. In general, algae with higher optimum temperatures for growth have higher growth rates at their optimum temperatures than algae with lower temperature optima for growth.[973] Goldman and co-workers[1693,1695-7] demonstrated that temperature strongly influences cellular chemical composition and that each species responds somewhat differently. Thompson et al.[4964] reported that strong and consistent patterns of response in the physiological and biochemical parameters of growth rate, chlorophyll-*a* quota, and carbon:chlorophyll-*a* ratios were observed in eight marine phytoplankton species. Thompson et al.[4964] found large changes in relative amounts of certain fatty acids as a result of a 15°C change in growth temperature in the same eight species. Bonin et al.[443] pointed out that blue-green algae are often regarded as organisms favored by high temperature because they are only oxygen-evolving photosynthetic organisms occurring in hot springs[562] and because they are more abundant in tropical than in temperate waters.[1434]

The Q_{10} of active uptake of sulfate ions by *Chlorella* was found to be 3.1 to 4.4.[5102] Another strongly temperature-dependent phenomenon related to ion transport is the formation of coccoliths in *Coccolithus* (= *Emiliania*).[3607] Iodine uptake and accumulation in marine algae is also positively related to temperature.[3644] Respiration is usually temperature dependent and there is a linear increase in O_2 uptake with temperature; temperature-respiration curves may, however, vary.[4645]

The phosphorus requirement of *Scenedesmus* varies with temperature and consumption per unit of biomass was minimum at 25°C.[449] Nitrogen fixation and the excretion of nitrogenous products by blue-green algae are also temperature dependent.[2365] Fogg and Than-Tun[1423] and Fogg and Stewart[1431] obtained Q_{10} values for nitrogen fixation of about 6. Topinka[5005] reported a temperature dependent nitrate (Figure 2-35) and ammonium (Figure 2-36) uptake by *Fucus spiralis*.

Resting stages of many algae possess a greater tolerance to extreme temperatures than do the vegetative stages.[4645] For example, desiccated cysts of freshwater members of the Prasinophyceae remain viable after exposure to 100°C for one hour.[295] A similar heat tolerance is reported for zygote of *Furcilla* although the upper temperature limit of the vegetative flagellate does not exceed 35°C.[294] Brock[563] pointed out that heat tolerance is based, at least partially, on the heat stability of enzymes. The molecular basis for cold tolerance is not clear. Soeder and Stengel[4645] suggested that it may be, from considering data for other groups of organisms, that the synthesis of substances such as free sugars or oligosaccharides prevents the formation of ice crystals in the cell and thus decreases the freezing point of the cytoplasm.

Extensive reviews on the effect of temperature on photosynthesis are those of Berry and Bjork-man,[345] Berry and Raison[346] and Öquist.[350] The effect of temperature on algal growth has been reviewed by Davidson,[982] Ahlgren,[35] Geider,[1586] Li,[2845] and Raven and Geider.[4012]

2.13.1.3 Grazing

Herbivory by aquatic invertebrates can reduce algal biomass,[845,2088–9,2309,2688,3115,4727,4838] and influence the taxonomic and physiognomic structure of algal assemblages.[1945,2309,2088–9,4727,4838]

It is evident that temporal and spatial variations in the abundance and biomass of phytoplankton species are influenced by the grazing activities of a variety of zooplankton and filter-feeding fishes.[3030] This influence is often a consequence of biomass- and size-dependent grazing rates with larger organisms utilizing increasingly broad spectra of particle sizes.[2686,3499,3663]

Malone[3030] summarized that the phytoplankton diet of microzooplankton (< 200 μm), especially protozoans[282–3] and invertebrate larvae,[599,2371,4967] is probably composed mainly of nanoplankton.[3663,4236] Macrozooplankton, such as particle-grazing copepods, appear to selectively remove large particles when presented with multi-sized algal diets[1517,1580,3339,4121] but most copepodes and many other macrozooplank-ton can graze over a broad range of particle sizes.[3030,3857–8] The grazing rates under different environ-mental conditions is discussed in a review by Frost.[1518] Both Frost[1518] and Sournia[4638] stress that no size of class or group of algae should be considered immune from grazing, and a large proportion of any phytoplankton biomass appears available to grazers, though certain cyanophytes may be less palatable than other groups.[153,3850] It should not be assumed that all organic material ingested is incorporated by the grazers.[1262,3299]

The sharp drop in phytoplankton in the summer in temperate zones may not be caused by a depletion of nutrients. Instead, the decline in phytoplankton biomass may be due to a rapid rise in zooplankton[1004,1952] (Figure 2-37). Thus the determination of standing crop (biomass) of a phyto-plankton community may not reflect the actual production because the community may be under constant, intense grazing. Dawson[1006] pointed out that an initial population of 100 cells would produce 6400 cells after only six cell division, but only 1692 cells if 20% were grazed by zooplankton.

Grazing activity by zooplankton might also improve the chances for coexistence among species of planktonic algae. By reducing the total algal biomass, grazing will reduce demand on nutrient resources and therefore reduce the severity of competitive interactions.[973] More direct effect may also be involved because many zooplankton species are known to graze selectively on the basis of prey size.[973] Intense grazing could eliminate some species from the water column. Zooplankton activity will not always increase phytoplankton diversity, however.[3851]

Freshwater grazers are principally crustaceans and rotifers, though protozoa may have significant effects on some diatom populations. Marine grazers include copepods, protozoa (mainly ciliates), and tintinnids together with appendicularians, salps, and pteropods.[4643]

Despite the uncertainty about some of the steps between phytoplankton and zooplankton,[2968] there are many well attested records of the importance of grazing in small water bodies.[3850] It is not uncommon to find mass destruction of vast algal populations by grazing.[1214,3737,5071]

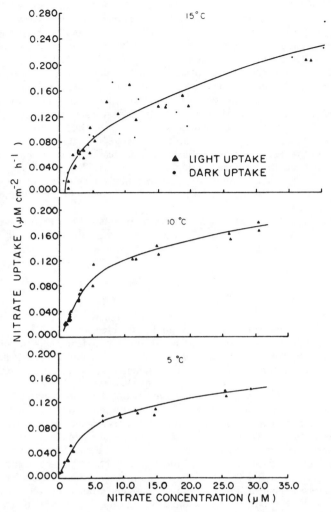

Figure 2-35. Uptake of nitrate by *Fucus spiralis* at 15, 10, and 5°C.[5005] (From Topinka, 1978, in *J. Phycol.*, Vol. 14, p. 243, reprinted with permission of the *Journal of Phycology*.)

Herbivory can play a key process regulating the structure and function of primary producers in many ecosystems. While aquatic herbivory has been studied more thoroughly in the plankton, at least some aspects of the interaction are similar in attached communities.[3138,3140]

The susceptibility of algae to ingestion and digestion varies interspecifically among grazer and algal populations.[3138] Selective feeding by planktonic grazers, such as rotifers and crustaceans, has been related to an alga's size,[2762] rigidity of the cell wall,[385] presence of spines or other projections,[385] and the production of toxic compounds.[153,4680] Benthic grazers, primarily gastropods and insect larvae have not been shown to be as selective, although the growth forms of algal taxa, encompassing parameters such as size and security of attachment, are good indicators of ingestibility.[2465,3676,4838]

Benthic microalgae show many growth forms (including prostrate, apically attached, stalked, filamentous, and motile), each of which would be predicted to have different susceptibilities to herbivory. Filamentous and elevated growth forms are seemingly more susceptible to most types of grazing, whereas motile and prostrate forms are apparently less susceptible.[5044] In a few benthic habitats, herbivory has been shown to effectively remove the large overstory algae, subsequently causing the formation of a new assemblage where adnate forms dominate.[2250,2309,2465,4727]

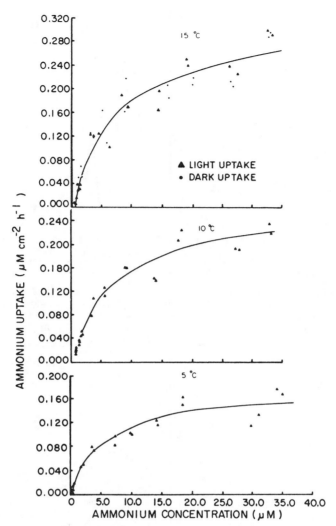

Figure 2-36.   Uptake of ammonium by *Fucus spiralis* at 15, 10, and 5°C.[5005] (From Topinka, 1978, in *J. Phycol.*, Vol. 14, p. 242, reprinted with permission of the *Journal of Phycology*.)

Grazing decreases benthic algal accumulation rates.[925,1814,1945,2688,3115,4727,4729,4838] Qualitative observations suggest that fiddler crabs, snails and herbivorous fish grazing is a major route for loss of epibenthic algal biomass.[1865,2242,2581,3835,5349]

McCormick and Stevenson[3140] found that the effect of snail grazing on epiphytic algae biomass was dependent on the nutrient environment. An overstory of diatoms was susceptible to removal by grazing and was not strongly affected by nutrient enrichment. An understory of *Stigeoclonium* was more resistant to grazing and responded strongly to nutrient enrichment only in the presence of grazers.

Snail grazers may mediate nutrient availability to the understory indirectly by removing overlying cells or by direct nutrient excretion by nutrients.[1495,1823,3140] Hill and Harvey[2090] found that snails had compensatory effects within the periphyton: they diminished biomass and productivity in the loosely attached layer, but stimulated productivity in the tightly attached layer. The reduction of attached algal populations by snails has been extensively studied recently.[2090,2250,2309,3140,4724,4727,4729,5044]

Langeland and Larsson[2704] pointed out that fish predation tends to reduce the individual size of the zooplankton causing less efficient grazing on the phytoplankton. The authors presented a simple box model of the energy transfer in the pelagic system (Figure 2-38). Further information on herbivory fish can be found in References 3861–2.

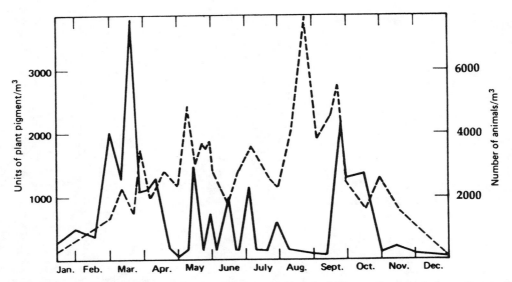

Figure 2-37.   Relationship between density of phytoplankton (solid line, units of plant pigment per cubic meter) and density of zooplankton (dashed line) in oceanic waters off Plymouth, England. When zooplankton counts are high, the phytoplankton counts are usually lower because of grazing.[1952]   (From Harvey et al., 1935, in *J. Mar. Biol. Ass. U.K.,* Vol. 20, p. 408, reprinted with permission of Cambridge University Press.)

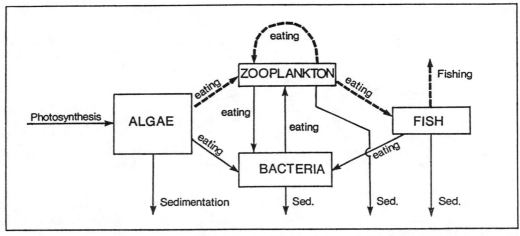

Figure 2-38.   Model of the energy transfer in the pelagic system.[2704]   (Reprinted from *Prog. Wat. Tech.,* Vol. 12, Langeland, A. and P. Larsson, The significance of the predation food chain in lake metabolism, p. 182, © 1980, with kind permission from Pergamon Press Ltd., Headington Hill Hall, Oxford OX3 0BW, U.K.)

Epiphytes always coat aquatic macrophytes. The production of these epiphytes is readily available to grazers, whereas that of the macrophytes apparently remains largely unutilized until the plants become senescent at the end of the growing season.[4577,4644] Macrophyte production should thus contribute significantly to the food supply of herbivores and detrivores only late in the season, whereas epiphyte production could sustain grazers throughout the growing season.[731]

Lubchenko and Gaines[2929] provide an excellent review of much of our knowledge of seaweed-grazer interactions, and it is clear that grazing and predation are powerful biotic forces in the control of species composition and biomass in seaweed communities.[4643] The major animal groups which eat seaweeds are the molluscs, particularly the gastropods and chitons; crustaceans, particularly amphipods; and the sea urchins and fish. Herbivorous fish are rare in temperate waters, but in the tropics 15–25% of fish are algivorous, and the high proportion of toxic algae in the tropics (see also Chapter 2.15) would appear to be an adaptation to high levels of grazing predation.[4643] As in freshwaters, many grazers show considerable species selectivity, and this may be related to seaweed morphology.[4643]

2.13.1.4 Seasonal succession

Seasonal variation in species composition is well known,[647,1082,5310] more rapid changes in abiotic conditions within season due to weather, such as temperature, discharge, and nutrients, may also differentially affect species performance.[3768,3785] A major hypothesis of phytoplankton ecology is that interspecific competition for nutrients determines the species composition and seasonal succession of phytoplankton in water bodies.[1175,2499,3612,4751,4919,4972] The annual changes in phytoplankton abundance and community structure which occur with some degree of predictability in temperate waters are probably the best-studied aspects of algal ecology.[973] During the winter season, with low temperature, low light intensities, and short days, phytoplankton biomass and productivity are generally low in spite of elevated nutrient concentrations. The phytoplankton is dominated by cold-tolerant species of several algal classes which on occasion may account for the surprising amount of production even under an ice cover. Increasing light (higher intensities and longer days) in the early spring is thought to be the principal factor which stimulates the spring outburst of phytoplankton production (Figure 2-39) since temperature often remains low during this period.[973]

In temperate climates there is typically a spring bloom with the largest increase in biomass, followed by a series of smaller variations with a final rise in biomass in the fall, and then a decline.[1004] Arctic regions usually have a single rise in biomass and a subsequent decline in the midsummer months.[1004] In contrast, tropical water bodies show less seasonal variation than temperate ones. Many are eutrophic due to the rapid remineralization occurring at the higher water temperatures. Tropical water bodies in general appear to support a higher biomass of algae, but show less species diversity than temperate ones.[4643]

More detailed explanations for seasonal succession have been suggested in addition to the general involvement of light, nutrients, temperature and grazing pressure (see Chapter 2.13.1.3). For the most comprehensive studies see Round[4210] and Smayda.[4573]

McCormick and Stevenson[3139] pointed out that investigations of algal succession have generally been confined to the study of seasonal changes in species composition, which are largely driven by predictable changes in the abiotic environment during the year.[4210] Within a season, disturbance can initiate a species replacement process in algal assemblages that is similar to that observed during terrestrial plant succession,[2118,2573,3679] although alternative successional pathways (sensu Pickett et al.[3785]) occur.[1887,2240,3227,4725]

Four mechanisms have been posted to cause changes in community composition during succession:[856,3785] facilitation, inhibition, passive tolerance and active tolerance. McCormick and Stevenson[3139] pointed out that successional patterns in benthic algal assemblages appear to result from several processes that defy explanation by a single mechanistic model. Changes in benthic algal species composition are typically observed in association with decreases in diversity during diatom community development,[3508,4740] but increases[2116,2782] and curvilinear patterns[3064,3767] also occur.

Tuchman and Stevenson[5044] found that whether snail grazing enhanced or arrested succession, or changed the direction of succession, was dependent upon growth forms of early and late succession algal species, which varied among habitats and corresponding species assemblages. Community resistance can change with age as a result of structural and/or physiological changes associated with succession.[3770] Sessile communities can become more susceptible to disturbance as vertical structure increases through community development[1210,2116,2982,4641] or as older individuals begin to senesce.[5386] Alternatively, older communities may be more resistant because of stronger attachment.[4640,5449]

Physiological properties of benthic algal communities may also change through time[1887,4628,4728] and differ among vertical strata within a community.[265,2241] The importance of wave disturbance in structuring benthic communities in marine systems is well documented.[1016,4639] Wind induced wave action can also strongly affect attached algal communities in fresh water.[2116,2119,5529]

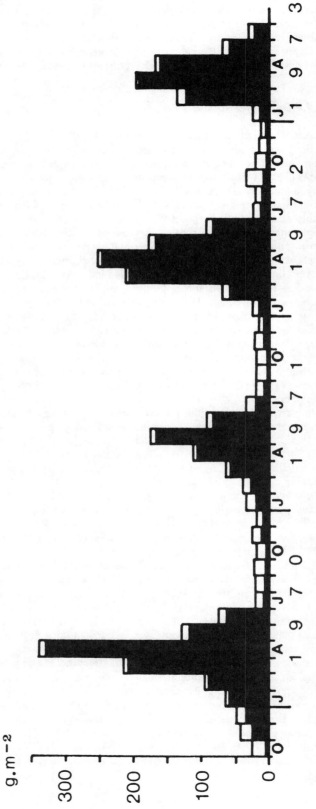

Figure 2-39. Monthly fresh biomass of phytoplankton. Black area = Pyrrophyta (mainly *Peridinium*), white area = other algae. October 1969 to June 1973. Results in g⋅m^{-2}.^{4407} (From Serruya and Berman, 1975, in *J. Phycol.*, Vol. 11, p. 156, reprinted with permission of the *Journal of Phycology*.)

2.13.1.5 pH

The pH dependence of dissociation rates and the ionic state of polar inorganic and organic compounds affects the availability of many algal nutrients such as CO_2, iron[4730] and organic acids.[870] pH also exerts an effect on the electrical charge of the cell wall surface,[1997] on ion transport systems at the plasmalemma, and on the associated membrane potentials.[319]

The pH range at which algae occur is wide.[4615] *Euglena mutabilis*, often found in acid bogs,[1465,2208] can be present in environments with very low pH (< 4) and even as low as 1.8.[2239] *Chlamydomonas acidophila* is common in some highly acidic edaphic-cum-aquatic environments, e.g. peat fields in Czech Republic with pH 1–2.[1465] The optimal pH range of *Zygogonium* sp. extends down to pH 1.0.[2990] Certain algae from alkaline waters have high pH optima; blue-green alga *Spirulina platensis* from Lake Tchad extends from pH 8.5 to pH 11.[819] Certain algae also display a greater range of pH tolerance than required in the environments where they occur.[4615] *Staurastrum pingue* from a slightly acid lake can be cultivated successfully between pH 4 and 11.[4368] Reynolds and Allen[4079] emphasized the role of pH as a factor determining the composition of freshwater phytoplankton communities. Nalewajko and O'Mahony[3386] studied the photosynthesis of algal cultures and phytoplankton following an acid pH shock. They found that phytoplankton in acidifying lakes consists predominantly of species which are tolerant to acid pH for short periods (hours) but cannot grow at these pHs.

Blue-green algae are not found in very acid environments; in those conditions, the only algae found are eukaryotic.[567]

2.13.1.6 Salinity

Salinity is an important environmental parameter for sessile macroalgae and changes in salinity result in physiological and biochemical differences.[443,3340] Algae living in an environment where salinities may change rapidly (such as estuaries, brackish water ponds, salt marshes, or rock pools) have to adapt their physiological processes rapidly to be able to maintain their growth as constant as possible. Likewise, freshwater algae reaching estuaries can survive only if they are somewhat euryhaline.[443] Thus, in many coastal and usually nutrient-rich environments, salinity might be expected to shift the species dominance in a community reaching a variable salinity area, and to be responsible for the peculiar composition of resident populations.[710,794,1115,2127,2244,2900] At low salinity (5 g L^{-1}) *Fucus* demonstrated low photosynthetic O_2 evolution rates, no appreciable organic carbon excretion, high respiratory O_2 consumption, and high total N content.[3343]

Variation in salinity is accompanied by changes in the concentration of major ions that may be energetically coupled to nutrient uptake in algae.[3843,4005] For example, transport of both urea and nitrate were demonstrated to be Na^+-dependent,[4050] silicate uptake by diatoms depends on both K^+ and Na^+,[4828] and both ammonium and amino acid uptake depend on external Na^+ concentrations.[2018,5364] Photosynthesis in some cyanophytes has also been shown to be stimulated by elevated Na^+ concentrations.[3224,3226,4057]

Rueter and Robinson[4222] found that reduction in salinity from normal seawater (33 g L^{-1}) decreased carbon uptake rate but increased nitrate uptake rate in *Fucus distichus* by 50% each. The authors suggested that the stimulation of nitrate uptake at low salinity may be beneficial to plants in estuarine tidal environments in which nitrate is supplied by the fresh water source. The variation in uptake with salinity can also explain the N rich fucoids observed by Munda and Kremer[3343] at low salinities. Nitrogen rich plants may not be an artifact of decreased dry weight of other components as they suggested, but rather a consequence of holding them at a low and non-varying salinity.[4222]

Arrigo and Sullivan[171] found that the effects of salinity on the growth of sea ice microalgae are independent of those elicited by temperature or light, and that the functional relationship between salinity and light or temperature is multiplicative.

Non-marine blue-green algae such as *Aphanothece halophytica* and *Phormidium hypolimneticum* are known to be extremely halotolerant, growing at salinities in excess of 200 g L^{-1}.[454,1124] Herbst and Bradley[2039] reported that algae from two alkaline salt lakes tolerated the salinity up to 150 g L^{-1}.

Further information about the influence of salinity on algal growth can be found in References 443 or 3323.

2.14 UTILIZATION OF ALGAE

Man's use of algae, particularly marine algae, is far more diverse and economically important than generally realized.[10] According to Bonotto[445] there are 5 genera of green seaweeds including 27 species, 40 genera of brown seaweeds including 88 species, and 56 genera of red seaweeds including 344 species, totalling 101 genera and 459 species of seaweeds which are of economic value as human food, as animal food, as manure, for medical or pharmaceutical purposes, or for industrial use. Tseng,[5040] however, listed the total number of economic seaweeds as 107 genera including 493 species. Among the seaweeds utilized by man for various purposes, only 11 genera with less than 20 species are commercially cultivated to any significant extent.[5041] Perhaps only four genera, namely, *Porphyra, Euchemia, Laminaria* and *Undaria*, for which the amount harvested from cultivated sources exceeds that taken from wild populations, can be regarded as truly marine crop plants.[5041] Reviews on utilization of algae are numerous (see References 7, 9, 445, 761, 766, 769, 1004, 1900, 2491, 2790, 3026, 3411, 3424–5, 3524, 4504, 4643, 5040, 5174).

2.14.1 Food

Algae (seaweeds, mainly *Porphyra, Laminaria, Undaria, Gracilaria*) are used worldwide. Approximately 168 species of algae have been reported as commercially important food sources, including 8 blue-green algae, 25 green algae, 54 brown algae and 81 red algae.[10,445,766] Algae are used as a food source mostly in Asia (Japan, China, Philippines, Malaysia). Among non-marine algal species, mostly blue-green algae and coccoid green algae are consumed.[718,3564,5279] In terms of direct consumption as a food, the single most important seaweed is the red alga *Porphyra* (called *nori* in Japan, *zicai* in Chinese, see also Figure 2-44, #8).[10,4041] Other important algae which are used as food are *Euchemia* (known as *gilin cai* in China and as *agar-agar* in Malaysia—different in meaning, however, from the commercial agar-agar) and *Laminaria* (known as *haidai* in China, see also Figure 2-53, #4, 6).[4041]

The principal nutritional value of edible algae comes from their high content of proteins, glycids and fats in many cases, rich concentrations of minerals (e.g. Na and P), trace elements and vitamins (A, B_1, B_2, B_{12}, C, D and E, niacin, panthotenic acid, folic acid) and iodine.[10,636,763,2416,2491,2790]

2.14.2 Agriculture

2.14.2.1 Fertilizer

In a small way algae are used as fertilizers on farmland close to the sea. The larger brown and red algae are used as organic fertilizers; these are usually richer in potassium but poorer in nitrogen and phosphorus than farm manure.[4209,5214] The weed is usually applied direct and ploughed in, but it has also been processed into a seaweed meal for transport inland. A concentrated extract of seaweeds is sold as a liquid fertilizer.[4209]

Probably the widest use of seaweeds in agriculture is as liquid fertilizers.[446,3860,4404,4734] The positive effect of liquid fertilizers is principally explained by the high content of trace elements and growth regulatory substances (particularly cytokinins) found in seaweeds. Calcareous red algae, known as maerl, are collected in the United Kingdom and France to reduce soil acidity.[406-7]

Blue-green algae are widely used in rice fields throughout Asia as nitrogen fixing organisms (so called "algalization"[2491]) in the place of N-rich fertilizers.[1886,4518,4763,5280] Blue-green alga *Tolypothrix tenuis* is grown in cultures and added to rice fields. Watanabe[5278] reported a yield increase up to 20% because of N_2-fixing. Aboul-Fadl et al.[14] reported that inoculation with about 250 g dry mass ha^{-1} of the same species resulted in a 19.5% increase in rice yield as compared with 16.6% produced by a dressing of 25 kg ha^{-1} of ammonium sulfate.

In Japan, rice fields are "fertilized" by water fern *Azolla* which multiplies rapidly and contains symbiotic blue-green alga *Anabaena*, which fixes gaseous N_2. Tzuzimura et al.[5069] reported fixation up to 128 kg N_2 ha^{-1} year^{-1}.

2.14.2.2 Fodder

The use of seaweeds in animal husbandry is a very old and widespread practice in many maritime regions of Europe and North America.[766,4734,5214] Seaweeds are used directly as a feed for livestock,[1884] or in meal powder form to supplement the diets of cattle, pigs, sheep and poultry or in fish cultivation.[10,2079]

2.14.3 Fisheries

Indian coast sardine fisheries yield depends on the mass development of a diatom *Fragilaria oceanica*.[3371] The abundance of phytoplankton is a key for fishpond yields as planktonic algae content of proteins (up to 88%), carbohydrates (up to 58%) and fats (up to 85%) is very high.[636,1703]

2.14.4 Algal Aquaculture

The many uses and functional roles of seaweeds have risen considerable interest in their mariculture and resource management.[3104] The most commonly cultivated seaweeds in aquaculture include *Enteromorpha, Ulva, Monostroma, Laminaria, Macrocystis, Undaria, Chondrus, Euchemia, Gelidium, Pterocladia, Gracilaria, Hypnea, Iridaea, Gigartina, Palmaria,* and *Porphyra*.

Promising is the production of microscopic freshwater algae in a large scale suitable for industrial processing. The most common experiments used green algae *Chlorella* and *Scenedesmus*.[270,636,3361,3374, 4409,4891,5144,5150,5319] Chemical analyses proved that algae are rich in vitamins and proteins. The content of fats and proteins is dependent on cultivation conditions and it is possible to change its amount.[1460]

Further useful information about algal aquaculture can be found in References 270, 1130, 1694, 2243, 3092, 3103, 3410–11, 4248, 5040–1.

2.14.5 Phycological Extracts

The largest commercial use of seaweeds outside the Orient is as a source of colloidal extracts, collectively called phycocolloids.[10,2491,2745,4209,4643] The three major types of phycocolloids—alginic acid, agar and carrageenan—are widely used in food, pharmaceutical, textile, paint, paper, plastics, cosmetic, insectides and other industries.[9,2491,2745,2790]

While all three compounds are cell wall polysaccharides, they differ in their seaweed source, as well as in their commercial uses.[2711,3123,3750] *Alginic acid* (polyuronic polymer consisting of D-mannuronic acid and L-guluronic acid residues[2021] and its salt derivatives (Ca- and Na-alginates) are obtained from brown algae (mainly *Macrocystis, Laminaria* and *Ascophyllum*). *Agar* and *carrageenan* are both galactans, composed of galactose units joined by alternating α-1,3 and β-1,4 glycosidic linkages.[1847,3007] Both polysaccharides are obtained from red algae. Agar primarily comes from a number of species of *Gelidium* and to a less extent, from species of *Gracilaria, Pterocladia, Acanthopeltis* and *Ahnfeltia*.[766] Three-fourths of the demand for agar is as microbiological medium.[9] Agar is the malay word for a gelling substance extracted from *Euchemia,* but now known to be a carrageenan. The term agar is now generally applied to those algal galactans which have agarose, the disaccharide agarobiose, as their repeating unit.[5214] Carrageenan is primarily extracted from *Euchemia* spp. (cultivated in the Philippines) and *Chondrus crispus* (in Canada).[3899,4128] The term carrageenan comes from the name of the small coastal town, Carragheen, in Ireland, where commercial harvests of *Chondrus crispus* (Irish Moss) were made in the late 19th century.[5381] Numerous applications have been developed for carrageenan in food products and processing, in pharmaceutical applications, in cosmetics, in coatings, such as paints and inks, and other products and processes.[5214] Carrageenan has been used in systems studying antibody synthesis on induction of cellular immunity[3419] and to test anti-inflammatory compounds.[3170,5291]

Two other carrageenan-like polymers, especially *furcellaran* (or Danish agar, from the red alga *Furcellaria lumbricalis*) and *funoran* (from *Gloiopeltis* species), are also commercially utilized.[766,2790,3269,5381]

2.14.6 Algae in Medicine

The use of algae for cures or as preventatives for a diversity of medical problems is widespread in certain regions. This is especially true in Asian maritime areas where the sea is intimately involved in daily activities.[4720] Algae may be used as: vermifuges, anesthetics, antipyretics (fever relief), cough remedies, wound healing compounds, thirst quenching remedies, and treatments for gout, gallstone, goitre, hypertension, diarrhea, constipation, dysentery, burns, ulcers, skin disease, lung disease and semen discharge.[2188,3240,3443] Algal extracts have also been used for antitumor treatments.[2434,3366,5514] A review of work on the chemical nature and biological activity of compounds extracted from seaweeds is given by Bhakuni and Silva.[352]

The substance obtained from the red alga *Delesseria sanguinea* can be used as a blood-anti-coagulant compound. The compound is more effective than heparin. Similar effect was found for organic sulfates extracted from the brown algae.[1460] Khan[2491] reported that compounds obtained from algae also serve as substances against a high blood pressure. Pratt et al.[3874] isolated from a green alga *Chlorella* a compound (which they called chlorellin) possessing an antibiotic effect on different bacteria (*Escherichia coli, Staphylococcus aureus, Shigella despenteriae*). Lefévre,[2755] Harder and Opper-mann[1906] and Burkholder et al.[635] found similar compounds but those have not found their use in the medicine. Mason et al.[3094] isolated an antibiotic substance from the blue-green alga *Scytonema hofmanni*. Some other substances which are secreted by or extracted from algae have antibiotic effect.[37,321,407,635,764,1219,1334,1654,1871,2189,2204,2376,2488,2545,4146,4371-3,4719] The compounds in algae which may be useful as antibiotics for killing bacteria, fungi, and viruses, include fatty acids, bromphenols, tannins, phloroglucinol, sulfolipids and terpenods.[1753,1853,1957]

Recently, it has been confirmed that blue-green extracts possess anti-HIV-1 activity.[1753,1853,3702] Patterson et al.[3702] studying approximately 600 strains of cultured blue-green algae, representing some 300 species, found that approximately 10% of the cultures produced substances that caused significant reduction in cytopathic effect normally associated with viral infection. The order Chroococcales proved to be a most prolific producer of antiviral compounds.

2.14.7 Chemical Industry

Seaweeds have been used as a source of chemicals for industry, other than phycocolloids, for centuries. Kelps, particularly *Laminaria*, were harvested in great quantities along the European coast of the Atlantic Ocean during the 18th and 19th centuries to obtain sodium (soda) and potassium (potash) used in the production of soap, glass, porcelain and alum.[10,766,1460] In the early 20th century brown and red algae were used in some countries as a major source of iodine (e.g. red alga *Phyllophira nervosa* contains up to 1.2%).[1460]

2.14.8 Diatomaceous Earth

During Tertiary and Quarternary times, the production of diatoms had been so great in some regions that large sedimentary deposits had been formed. The siliceous cell walls are relatively insoluble and hence these sediments accumulated in marine and freshwater basins and some are relatively uncon-tamined by clay, etc. The natural deposits contain a high proportion of silica (86 to 88% in some American materials) and extremely low loss on ignition (e.g. 4%). When processed it is chemically inert and is mainly used as a filtration aid, abrasive material, as a filler in paints, varnishes, and paper products, and in insulation materials, particularly those for use at high and low temperatures.[4209]

2.14.9 Energy

Although algae are not presently used in the commercial production of energy, it is clear from the literature that seaweeds, in particular, offer significant potential as a renewable source of biomass for energy production, especially the production of methane.[1946,2303,3104,3578,4237,4239]

2.14.10 Algae in Water Treatment

Algae are the key components in wastewater treatment in stabilization ponds.[1728–9,2951–2,3577,3210,3579,3638,5420] In the literature, there have been numerous reports on wastewater treatment or tertiary treatment by means of unicellular algal activity in special reactors.[449,778,1045–6,2027,2721,3080,3108,3859] Periphytic communi-ties have also been employed in water treatment both in rotating discs or in special streams or troughs.[2027,3934,4539,4565,5191,5194,5198,5204,5211]

2.14.11 Algae as Indicator of Water Quality

The use of algae or algal assemblages as indicators of water quality has a long tradition, especially in Europe.[210,846,1061,1376–8,1706,2558–9,2638,4553–6,4558,4561–3,5208–9,5547–8] In the U.S.A., diatoms hava drawn attention as an indicator of pollution.[802,1078,2524,2923,3678,3680,3684,3982,4939]

Algae have been suggested to be good monitoring organisms of heavy metals presence and toxicity in waters,[2380,2453,3231–2,4395,5406] bioassay organisms for eutrophication tests (e.g. algal growth potential),[2955,3085,3132,3846,4986] and bioassay organisms for different types of pollution.[1071,4549,5029,5030,5048,5255–7]

2.15 ALGAE AS NUISANCE FACTORS

Prototheca is the only known alga which is itself a disease-causing organism. This unicellular alga lacks chlorophyll and resembles a yeast, but does not reproduce by budding.[4720] At present, only *P. wickerhamii* and *P. zopfii* are considered as causative organisms for protothecosis.[2427] At least three clinical forms of protothecosis are recognized.[2427] Species of *Prototheca* occur in both fresh and marine waters and are reported from sludge and stagnant waters and ice fields.[2427]

Algae are implicated as causative agents for several instances of dermatitis.[4720] Moikeha et al.[3268] and Moikeha and Chu[3267] have isolated a toxic factor from *Lyngbya majuscula* that causes dermatitis. The best known dermatitis is swimmer's itch reported from Florida, Hawaii, and the Marshall Islands.[1766,3267–8] The diatom *Fragilaria striatula* has been implicated in still another type of dermatitis, "Doggerbach Itch."[1957] Edaphic algae and those present in drying ditches and pond margins are distributed by air currents. Those algae may have an allergic effect.[3143,3145]

Algae are also considered the causative agents of a type of silicosis and a goitre. The algal silicosis (irritation of the lungs caused by fine silica particles) is the result of working with diatomaceous earth, evidently without protection of the lungs from siliceous dust.[4720] Goitre is an enlargement of the thyroid, generally caused by lack of iodine.[1720] However, the goitre can be caused by excess iodine resulting from the ingestion of large amounts of seaweed or seaweed tablets.[2850] The respiratory disease known as "Tingui" or "Tamandare Fever" from northeastern Brazil is considered the result of red tides caused by algae.[4294]

Although algae creating a nuisance may not be the direct result of pollution, nuisance algae result mainly from eutrophication or nutrient enrichment of lentic habitats. Nuisance algal problems are frequently associated with blue-green algae, especially blooms and associated scums.[4323] Scumming gives blue-green algae a competitive advantage in that it enables them to assimilate atmospheric CO_2.[3621]

Nuisance algae cause an economic impact, particularly in relation to treatment of drinking water. Blooms of algae cause filter clogging at water works and are a cause of bad taste and odors due to excessive oil production.[2491,4323] Algal blooms cause the oxygen depletion and byproducts formed during algal decay (e.g. hydroxylamin) are toxic.[2491]

Firm filamentous green algae, e.g. *Rhizoclonium hieroglyphicum* or *Cladophora glomerata* (Figure 2-17, #20, 21), are a nuisance in fish ponds. These algae take up nutrients and retain them as their growing season is relatively long and firm filaments decompose slowly. In addition, during the harvest, filamentous algae clogg nets so that those retain also young and small individuals.[1460] Some filamentous algae cause a significant nuisance to a variety of lake users, especially those along the shoreline (e.g. *Cladophora glomerata*).[4323]

Presence of algae in water can alter its physico-chemical characteristics (pH, dissolved oxygen, alkalinity). Change in these characteristics can influence the behavior of nitrogen, phosphorus and carbon in water. The shift in pH, caused by photosynthetic activity of algae, affects the form of ammonium in water—with rising pH the NH_3 form increases its concentration over NH_4^+ (at pH 9.2, 50% of ammonia is present as NH_3). NH_3-form is very toxic for fishes (gill necrosis) and when significant concentration of ammonia is present (and in fertilized fish ponds it usually is), higher pH may be the cause of fish death.

Silvey et al.[4505] reported a nuisance odor caused by blue-green algae. Many blue-green algae produce 2-methylisoborneol (1,2,7,7-tetramethyl-*exo*-bicycloheptan-2-ol; MIB) and geosmin (*trans*-1,10-dimethyl-*trans*-9-decalol), both of which impart "earthy/musty" tastes and odors.[207,320,1460,2298,3082,3367] Fott[1460] pointed out that fishes from the water reservoirs containing large growths of filamentous blue-green algae *Oscillatoria* and *Phormidium* attain an unpleasant rotten smell.

2.15.1 Algal Toxins

Reich and Aschner[4052] reported a mass death of fish caused by a member of the Prymnesiophyceae (formerly classified as a chrysomonad) *Prymnesium parvum* which can reach a density up to 800,000 cells in 1 mL and be a cause of yellow-brown color of water. Pinto and Silva[3803] reported that edible clams (*Cardium, Tapes, Mytilus*) became poisonous if there was an overproduction of the dinoflagellate alga *Prorocentrum micans* in plankton in Portugal. Abbott and Ballantine[6] reported that species of *Gymnodinium* are poisonous for animals.

It has been known for a long time that blue-green algae can be poisonous for mammals, waterfowl, and fishes or for man. One of the first reports on blue-green algae toxicity is from Australia in 1875.[1460]

The toxic species include e.g. *Microcystis aeruginosa, Aphanizomenon flos-aquae, Anabaena flos-aquae, Coelosphaerium kuetzingianum, Gloeotrichia echinulata, Nodularia spumigea.*[1762,2409] The three most common bloom-forming blue-green algae species are *Microcystis aeruginosa*, (Figure 2-15, #1) *Aphanizomenon flos-aquae* (Figure 2-15, #2) and *Anabaena flos-aquae* (Figure 2-15, #3). Gorham and Carmichael[1767] studied in detail toxins from those three species. The known *Microcystis* toxins are either a fast-acting low-molecular weight alkaloid or a slow-acting, comparatively small polypeptide. The known *Anabaena* and *Aphanizomenon* toxins are fast-acting, low molecular alkaloids. The authors also pointed out that more than 12 species of the blue-green algae belonging to 9 genera have been implicated in animal poisoning.

Stein and Borden[4720] reported that generally the effect of algal toxins is through human consumption of the animals that have eaten toxin-containing algae (fish, shellfish). For toxins produced by dinoflagellates see also chapters 2.7.1.2 and 2.18.9.

Literature on algal toxins is numerous (see References 231, 693–5, 847, 1069, 1230, 1602, 1762–8, 1882, 1901, 1932, 1957, 2273, 2516, 3110, 3289, 3547, 3975, 4317, 4319, 4451, 4655, 4719, 4912, 4987, 5284, 5383).

2.16 EUTROPHICATION

Eutrophication is one of the problems associated with the pollution of surface waters and is mainly imputable to man's activity. It is virtually impossible to make a clear-cut distinction between the problem of eutrophication and the other problems of water pollution, as they are all, at least partially, interrelated. To a certain extent, a line may be drawn in the case of pollutant sources producing obvious toxic effects on aquatic biocenoses, but even here it is not always possible to separate this problem from that of eutrophication.[5177]

Oligotrophy, mesotrophy and eutrophy were introduced as scientific terms by Weber[5298] to describe nutrient conditions in bogs. Oligotrophic bogs were poor in nutrients, and eutrophic bogs rich in nutrients. Mesotrophic bogs had a position between the two extremes. Naumann[3394] related oligotrophy, mesotrophy, and eutrophy to the biological effects of the prevailing nutrient conditions. The composition of the phytoplankton was used as a criterion. Golterman and de Oude[1720] pointed out, however, that there are no absolute definitions of the words "eutrophic" and "oligotrophic." Rodhe[4167] gave some approximate values for phytoplankton production in oligotrophic and eutrophic lakes. He suggested that the annual rates for oligotrophic lakes range between 7 and 25 g C m^{-2} yr^{-1}, naturally eutrophic lakes range between 75 and 250 g C m^{-2} yr^{-1}, and culturally polluted lakes range between 350 and 700 g C m^{-2} yr^{-1}. Likens[2851] pointed out that these values are subject to exceptions and criticism, but they do offer a useful generalization. Vollenweider et al.[5182] gave much broader ranges for oligotrophic waters (Figure 2-40).

Criteria for the terms oligotrophy and eutrophy were intensively discussed. Vollenweider[5177] emphasized that "the term eutrophication . . . may be applied to anything, including external and internal sources, which plays a part in accelerating nutrient loading and increasing the nutrient level, and is directly involved in increasing water productivity."

The present authors prefer to consider eutrophication as a process leading to nutrient rich conditions irrespective of the biological effects. The natural response to increased nutrient concentrations, i.e. increased primary production, may fail. Such a discrepancy reveals a disturbance of the biological system.[33] Golterman and de Oude[1720] defined eutrophication as enhanced plant growth following the increase of the concentration of the plant growth controlling nutrient with P and N as the responsible elements for the present day eutrophication. In general, phosphorus in the form of organic and

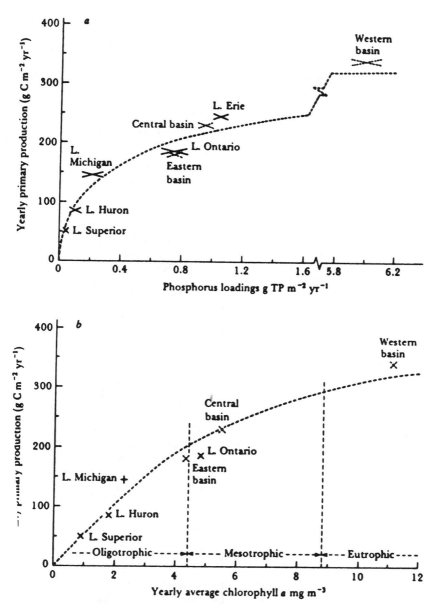

Figure 2-40.   (a) The relationship between annual primary production and annual phosphorus loading and (b) between annual primary production and surface chlorophyll in the Great Lakes. Central, eastern, and western basins refer to Lake Erie.[5182]   (From Vollenweider et al., 1974, in *J. Fish. Res. Board Can.,* Vol. 31, p. 757, reprinted with permission of *Canadian Journal of Fisheries and Aquatic Sciences.*)

inorganic phosphate tends to be the element controlling the rate of plant growth and biomass formation in most temperate lakes.[5181]   In many tropical lakes nitrogen is the controlling factor.[5086]   In heavily loaded lakes with unnaturally high algal densities light may finally become limiting.[2970]

In many inland water bodies and coastal marine regions these habitats have deteriorated as a result of eutrophication causing extensive phytoplankton blooms and shading of bottom-dwelling plants.[463,5349] This has led to an ecosystem with changed food web structure, nutrient cycling, oxygen dynamics and frequently a decline of recreational value and of commercially important fish, crab and mussel populations.[3509]

It seems that in most cases phosphorus limits algal growth and that increased P loading is responsible for the development of nuisance algal problems.[1215,2852,4235,4339,4341,5104] On the other hand, Schindler[4339] concluded that carbon is unlikely to limit the standing crop of phytoplankton in almost any situation.

Primary production as a function of predicted total phosphorus concentration has been reviewed by Golterman[1717] and Jones and Lee.[2366] Vollenweider and Kerekes[5181] showed that in a large number of lakes chlorophyll-*a* was not only related to the phosphate concentration, but to the P-loading, i.e. the amount of phosphate entering a lake per unit surface area and year.[1717]

Several papers have addressed the relation between phosphate loading and biomass, often expressed as chlorophyll-*a* concentration, or as a resulting decrease in light penetration.[36,1085,1714,5180] The relation consists of two steps:[1720]

1. P-concentration as function of P-loading (see Golterman and de Oude[1720] for an excellent review);
2. algal biomass as function of P-concentration.

A simple model relating summer phytoplankton standing crop (determined as chlorophyll-*a*) to the total P concentration at the time of spring overturn was developed by Dillon and Rigler[1085] utilizing data from 19 Ontario lakes together with other data from the literature.[1028,1216,2411,3003,3179,4251,4854] This model was predicted upon a general observation that as nutrient, such as P, increases in concentration concomitant increase in phytoplankton standing crop occurs.[2078] In the literature, many equations describing the relationship between P and chlorophyll concentration have been presented since then (Table 2-20). It can be seen that equations yield different results (for discussion see Golterman and de Oude[1720]). The reasons for this are:

1. differences in sampling (season, depth from a sample is taken), analytical methods for phosphorus concentration determination;
2. analytical problems connected with the extraction of chlorophyll—some green algae and blue-green algae are very resistant toward the extraction with aceton, which was commonly used.

Lambou et al.[2690] provided a comparison of 29 methods commonly used to measure trophic state with respect to their ability to rank 44 test lakes against two standards: total phosphorus, a nutrient or driving force standard and chlorophyll-*a*, an eutrophication response standard. The authors found that the phosphorus loading models had high rank correlation with phosphorus; however, they did not rank the lakes well against chlorophyll.

2.17 ALGAL VIRUSES, PATHOGENIC BACTERIA AND FUNGI

Many freshwater and some marine algal species are infected by parasitic organisms such as fungi, phages, viruses or bacteria.[678,2625,3673,3771,4243,4450,4481]

Viruses that lyse algae have been reported in the literature since the early 1960s. The driving force behind these viral studies was an interest in factors responsible for algal degradation. This interest was motivated by increased concern over eutrophication and associated water blooms.[3081] (See also References 584, 1101, 2775, 3081, 3617.)

The chlorellavorous bacterium "*Vampirovibrio chlorellavorous*" as proposed by Gromov and Mamkaeva[1829] is a predator of the green alga *Chlorella*.[828–9,1828–9,3036] The bacterium has very specific requirements for growth—it seems to grow only by attaching to the cell wall of intact *Chlorella* cells and consuming their cytoplasmic content. In addition, the host range appears to be limited to the genus *Chorella*.[622,828–9,3035] Coder and Goff[829] suggested that this narrow host specificity may be related to cell surface properties.

Peterson et al.[3771] reported the bacterial infection within diatom cells from a benthic community in an Arizona stream. Infected areas appeared as gray rings within a matrix of healthy diatom growth and spread rapidly, eventually covering all benthic substrata and causing algal sloughing within two weeks.

Yamamoto and Suzuki[5515] reported a seasonal distribution of fruiting mycobacteria (gliding bacteria with the unique property of being decomposers of other bacteria and plant debris), which are able to lyse blue-green algae in the water and sediments of eutrophic lakes.

Table 2-20. Relationships between phosphorus (P) and chlorophyll (CHL) concentrations (μg L^{-1}) reported in the literature for lakes

Locality	Relationship	Reference
31 Japanese lakes	log CHL = 1.5831 log P - 1.134	4251
English lakes	log CHL = 0.871 log P + 0.48	2970
Various lakes	log CHL = 1.451 log P - 1.14	1085
Lakes in the United States	log CHL = 1.181 log P - 0.764	3393
143 lakes (worldwide)	log CHL = 1.46 log P - 1.09	2364
Lake Narviken	log CHL = 1.277 log P - 1.009	4608
Friesian lakes (The Netherlands)	log CHL = 0.38 log P + 1.16	1717
British lakes	log CHL = 0.77 log P + 0.75	1717
Lake Erie (mean)	CHL = 0.25 P - 2.1	609
7 lakes in New Zealand	CHL = 0.26 P - 1.68	3129
Lake Washington (several years)	CHL = 0.60 P - 4.22	1216
Lake Minnetonka, Minnesota	CHL = 0.58 P + 4.2	3179
Finger Lake, New York	CHL = 0.574 P - 2.9	3513
IBP Programme (worldwide)	CHL = 0.46 P - 3.87	4340
Bay of Quinte, Ontario	CHL = 0.35 P - 0.04	2305
Holland Marsh, Ontario	CHL = 0.62 P - 1.17	3430
Kawartha Lakes, Ontario	CHL = 0.286 P - 1.21	3431

2.18 ALGAL CLASSES - BRIEF REVIEW

In this chapter the main features of algal classes are summarized. It is necessary to keep in mind that the concept of algal classification is in a state of disorder and there is not much consensus among scientists at present. I decided to include a brief review of algal classes rather than divisions because there is more agreement among phycologists at this level (see Chapter 2.1). From many different classifications I adopted here the classification scheme proposed by South and Whittick[4643] which has received a wide acceptance. As examples, I decided to include mostly freshwater algae as those species rather than oceanic species can be found in wetlands. For most classes I included one figure, for blue-green algae, diatoms and green algae, with respect to their abundance and significance, there are two, three and four figures, respectively.

The reference lists in the end of all previews contain only selected literature (monographs, books, review papers and several very recent papers), especially those that provide additional reference sources. Major xanthophylls are underlined. Size of freshwater species is taken from References 2098 and 2099, size of marine species from References 424 and 1460.

2.18.1 Cyanophyceae

common name: blue-green algae
habitat: freshwater, brackish, marine, terrestrial; examples shown in Figures 2-41 and 2-42
ecology: planktonic, attached
levels of organization: unicellular (coccoid), colonial (aggregations), filamentous (simple, branched), pseudoparenchymatous (parenchymatous)
size: single cells microscopic, colonies and some filamentous species macroscopic (up to several cm), can form attached or floating mats (clusters), algal blooms
cell covering: four-layered, peptidoglycan (murein) principal component, cell wall complete
flagella: absent
thylakoids: free in cytoplasm, unstacked (no chloroplast)

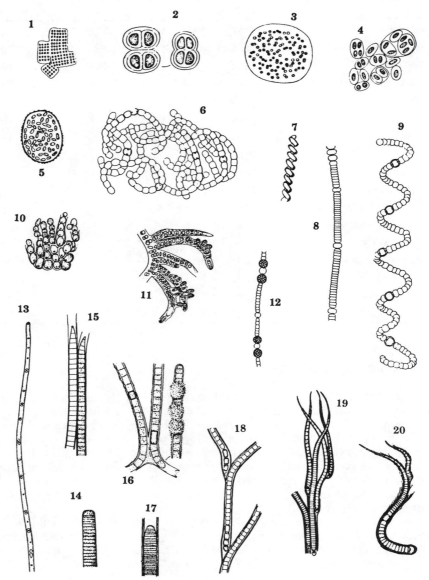

Figure 2-41. Freshwater Cyanophyceae. 1-*Merismopedia tenuissima* (cells 0.5–2 µm in ø); 2-*Chroococcus minutus* (cells < 15 µm in ø); 3-*Microcystis grevillei* (cells < 5 µm in ø); 4-*Gloeothece rupestris* (cells 6–12 × 4.5–5.5 µm); 5-*Aphanothece microscopica* (cells 5–9 × 3–4.5 µm, colonies up to 1 mm); 6-*Nostoc entophytum* (cells 2–3.5 µm wide); 7-*Spirulina maior* (filaments 1.2 µm wide); 8-*Nodularia harveyana* (filaments 4–7 µm wide); 9-*Anabaena spiroides* (cells 6–14 µm wide); 10-*Chamaesiphon polonicus* (cells 4–10 × 3–6 µm); 11-*Stigonema minutum* (filaments up to 30 µm wide); 12-*Anabaena sphaerica* (cells 5–6 µm wide); 13-*Oscillatoria redekei* (cells 6–16 µm long, filaments 1.3–2.3 µm wide); 14-*Oscillatoria limosa* (filaments max. 22 µm wide); 15-*Phormidium autumnale* (filaments (3)–6–10 µm wide); 16-*Scytonema julianum* (filaments 7.5–12 µm wide, with attached bacterial clumps at right); 17-*Lyngbya maior* (filaments 11–16 µm wide); 18-*Tolypothrix tenuis* (filaments 5–15(18) µm wide); 19-*Dichothrix orsiniana* (filaments 10–12 µm wide); 20-*Calothrix parietina* (filaments 10–15 µm wide). Not to scale.[2098] (From Hindák et al., 1975, in *Determination Key for Lower Plants,* SPN Bratislava, reprinted with permission of F. Hindák.)

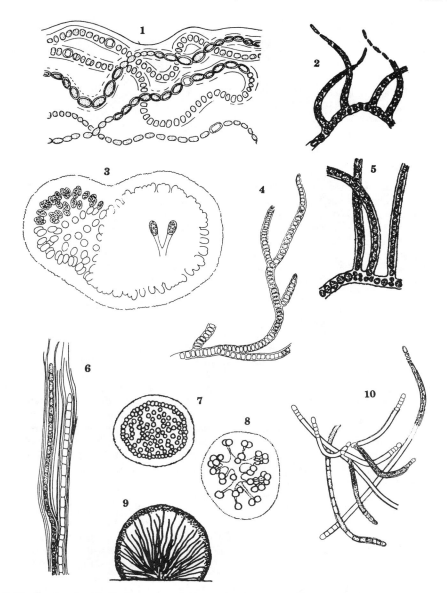

Figure 2-42. Freshwater Cyanophyceae.   1-*Nostoc verrucosum* (cells 3–3.5 μm wide); 2-*Mastigocladus laminosus* (filaments 4–12 μm wide); 3-*Gomphosphaeria naegeliana* (detail of the cells at right, cells 4.5–7 × 1.5 × 5.5 μm, colonies up to 120 μm); 4-*Sommierella cossyrensis* (filaments 10–12 μm wide); 5-*Hapalosiphon fontinalis* (filaments 10–24 μm wide); 6-*Schizothrix friessii* (filaments 3–6 μm wide, up to 3 cm long); 7-*Coelosphaerium kützingianum* (cells 2–5 μm in ø); 8-*Gomphosphaeria lacustris* (cells 2–4 × 1.5–2.5 μm); 9-*Rivularia dura;* 10-*Microchaete tenera* (filaments 5.5–8 μm wide). Not to scale.[2098,4559]   (1, 3, 4, 6, 10 from Hindák et al., 1975, in *Determination Key for Lower Plants,* SPN Bratislava, reprinted with permission of F. Hindák; 2, 5, 7–9 from Sládeček et al., 1973, in *Technical Hydrobiology: A Laboratory Manual,* SNTL Praha, reprinted with permission of V. Sládeček.)

chlorophylls: chlorophyll-*a*
carotenes: β-carotene (γ-carotene, lycopene?)
xanthophylls: antheroxanthin, aphanicin (= canthaxanthin), aphanizophyll, caloxanthin, β-crypto-
        xanthin, isozeaxanthin, lutein, mutatochrome (= flavacin), <u>myxoxanthin</u> (echinenone),
        <u>myxoxanthophyll</u>, nostoxanthin, oscillaxanthin, <u>zeaxanthin</u>

phycobiliproteins: allophycocyanin, C-phycocyanin, C-phycoerythrin, CU-phycoerythrin, phycoerythrocyanin
storage products: cyanophycin granules (arginine and aspartic acid), polyglucose (glycogen-like),
   sulpholipids, myxophycean starch
eyespot: absent
number of genera: approximately 125[55]
number of species: 1,200,[55] 1,500,[2491] 5,000—50% marine[1004]
Literature: References 19, 85–8, 705–6, 727, 1062–3, 1434, 1592–3, 1595, 1598, 2076, 2562–4, 2697,
   4191, 4518, 5478.

Blue-green algae are prokaryotic organisms. While exhibiting an algal morphology, the Cyano-phyceae possess an admixture of bacterial features[1498] (the classification problems will be discussed later in this chapter). The blue-green algae appeared about 3 billion years ago,[1723,3370,4911] and in this context they are often considered to have been the organisms responsible for the early accumulation of oxygen in the earth's atmosphere.[424] Fossil evidence in the form of stromatolites indicates that blue-green algae occupied an important place in the geological record and were the dominant life-form through most of the Proterozoic. Modern stromatolites are believed to resemble closely some of their ancient analogues and are known from marine and freshwater environments.[3742] A stromatolite has been defined by Awzamik and Margulis (in Walter[5259]) as "an organosedimentary structure produced by sediment trapping, binding and/or precipitation as a result of the growth and metabolic activity by microorganisms, principally cyanophytes."

There is a wide range of planktonic and attached blue-green algae occurring in freshwater, brackish, marine and soil habitats. Their distribution is cosmopolitan, with most species occurring throughout the world in both running and standing waters. Several planktonic blue-green algae are characteristical-ly members of water blooms (for discussion see Chapter 2.7.1.2, for toxins produced by blue-green algae see also Chapter 2.15.1). Cyanophyceae are commonly found on moist rocks or soil, forming a brackish crust when dried out (see Chapter 2.7.3). In the marine environment, the picoplankton is composed mostly of small blue-green algae. Blue-green algae are also commonly found in the intertidal zone in tidal pools or as blackish zone on rocks at the high tide marsh.[2745] Blue-green algae may also occur in extreme habitats such as hot springs or hot deserts (see also Chapters 2.7.3.2 and 2.7.5). Blue-green algae are ubiquitous in waters of a great range of salinity and temperature, and they also occur on rocks and in their fissures.[5288] Some soil blue-green algae have been shown to have remained viable for 18–107 years.[673] A number of blue-green algae have been reported from the atmosphere[583] and some have been recovered from house dusts.[343]

In general, blue-green algae seem to be more abundant in neutral or slightly alkaline habitats, e.g. Brock[565] reported that blue-green algae were absent from waters whose pH was less than 4 and 5, whilst certain eukaryotic algae were present. Kallas and Castenholz[2412] pointed out that blue-green algae are apparently unable to occupy habitats with a pH below 4.5–5.0. Mechanisms that have been advanced for dominance of blue-green algae in eutrophic environments include the ability to grow at high pH and low CO_2 concentrations.[2506,4417]

Blue-green algae are mostly obligate photoautotrophs[2165,2702,3225] but nevertheless some may grow heterotrophically[2494] and a light-enhanced uptake of sugars and acetate[76,2121] has been demonstrated in some species. Under certain conditions some blue-green algae are capable of switching over to anoxygenic photosynthesis, presumably having retained this capacity during their evolution from photosynthetic bacteria.[973] In *Oscillatoria limnetica*, anoxygenic photosynthesis is induced within two hours in the light in the presence of H_2S.[3616]

A number of blue-green algae form symbioses with diverse other plants (diatoms, liverworts, water ferns, cycads, angiosperms, and in lichens with fungi), and animals (corals, sea squirts, tortoises, and polar bears).[424,2697] In the partnership with plants the blue-green algae is usually an *Anabaena* or *Nostoc* and contributes fixed nitrogen to the joint venture[4773] (see also chapter 2.7.6). For blue-green symbiotic associations see the discussion by Fogg[1417] or Ahmadjian.[43] On the other hand, blue-green algae can be attached by many different types of cyanophages.[957,3617,4443,4779]

Bold and Wynne[424] pointed out that considerable confusion exists regarding the nomenclature for investments exterior to the cell wall of blue-green algae. There seem to be three types of investment surrounding blue-green algae cell walls: (1) a sheath that is immediately adjacent to the cell wall and visible without staining; (2) a slimy, mucilaginous "shroud" that surrounds the organism (with or without a sheath; this has indefinite, not sharply defined, limits); or (3) similar mucilaginous shrouds with well defined limits.[3084] The available evidence indicates that the outer investments of blue-green algae are composed of pectic acid and mucopolysaccharides.[1186]

The sheath itself is secreted by the cell wall for which it forms a protective envelope. Coccoid cells are rounded and form clusters within a mass of sheath material. Filamentous forms consist of hairlike *trichomes* which are strands of cells attached end-to-end. One, several, or a number of trichomes may share a common tubular sheath. Trichomes and sheath together are then termed a *filament*. In some species each trichome is surrounded by a mucilaginous sheath whereas in other genera more than one trichome may be embedded in the same mucilaginous sheath.[3310] If the trichomes branch, or are arrayed in a fan-shape, then the resulting sheath may have a bush-like or radial form. However, under conditions where there is little need for protection from water movement, sediment particles, sunlight, desiccation or grazing organisms the sheath may develop only weakly or not at all.[4131]

Many filamentous blue-green algae are not enclosed in firm sheaths; the hormogonia of those that are, and some unicellular species, undergo movement when in contact with the substrate. This movement, accomplished without evident organs of locomotion, is called gliding movement and occurs also in some filamentous bacteria.[424] The mechanism is not completely understood; it has been suggested that it is a sort of propulsion caused by the secretion of slime and also by contractile vacuoles on the surface of the cell.[1434,1872-4] The oscillation of certain trichomes (e.g. *Oscillatoria*) is thought to be related to waves of propulsion of the superficial fibrils in the wall.[1873]

Their specialized features include the ability to fix atmospheric nitrogen (which under anaerobic conditions is also associated with specialized cells, the heterocysts (see also Chapter 4.2.4.2), possession of akinetes (Figure 2-43), and the trichome.

The fixation of elemental (i.e. gaseous) nitrogen is biologically of great significance as it makes blue-green algae to be independent of other combined nitrogen sources.[1417,1434,705-6] Nitrogen-fixing blue-green algae are important in all habitats where they occur, and especially in tropical soils, such as rice fields.[4643] An akinete develops from a vegetative cell that become enlarged, darkly pigmented, filled with food reserves (cyanophycin granules) and augments its wall externally by an additional complex investment.[661] After a period of dormancy, the akinete may germinate, giving rise to a vegetative trichome.[424] Many blue-green algae survive periods of unfavorable conditions through akinetes.[973]

Heterocyst is a special type of double-walled pored structure which is produced from ordinary vegetative cells.[1318,1400] The heterocysts have been proven to be a cell modified to provide an anaerobic environment for nitrogen fixation.[4774] Within the cell machinery for growth, carbon fixation and the photosynthetic oxygen evolution[4926] is dismantled and the enzymes for the fixation and assimilation of nitrogen are installed in their place.[1956] These changes have been viewed as a sequence of events directing transformation of a vegetative cell into a heterocyst.[523] However, the morphological expression can be chemically induced without concomitant nitrogenase activity.[4174] (For further details on nitrogen fixation and heterocysts see also Chapter 4.2.4.1.)

Heterocysts occur at definite positions in a filament, either intercalary or terminal. It develops from the vegetative cell which has lost the capacity to divide and become larger in size, lacking reserve carbohydrates and polyphosphates. This stage of development is known as proheterocysts. Proheterocysts synthetize cellulose wall within 12–24 hours,[5064] start accumulating ascorbic acid and reducing TCA, and become heterocysts. The cellulose wall appears to be an essential for proper functioning of the heterocysts by effectively maintaining a reduced state of the protoplast which facilitates N_2 fixation.[2491]

Under anaerobic conditions such as exist in muds, heterotcysts are not formed[4148] and N_2-fixation proceeds in vegetative cells. Under anoxic conditions many heterocystous species will also fix nitrogen.[608,1552] Heterocyst numbers show characteristic fluctuations during the growth cycle, the frequency of these structures being least at the beginning of a period of rapid exponential growth and highest toward its end. Heterocyst frequency shows a negative correlation with cell nitrogen content.[1397] Heterocyst numbers could also be regulated by controlling the concentration of available combined nitrogen in the medium.[1399,3512] Nitrate, glycine, and asparagine gave temporary inhibition of heterocyst formation. Inhibition by ammonia, however, persisted as long as an appreciable concentration remained in the culture medium.[1399]

Ohmori and Hattori,[3520] however, found that less than 4% of the total nitrogen fixed during a relatively short period (5–15 min) was recovered in heterocysts. There was also no positive correlation between nitrogen fixation and heterocyst formation. These results did not support the hypothesis that the heterocysts are the main site for nitrogen fixation in blue-green algae.

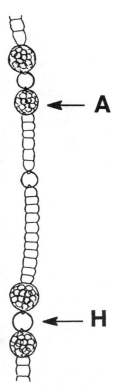

Figure 2-43.   *Anabaena sphaerica* var. *tenuis* with akinetes (A) and heterocysts (H).[2098] (From Hindák et al., 1975, in *Determination Key for Lower Plants,* p. 35, SPN Bratislava, reprinted with permission of F. Hindák.)

Paerl[3618] found that bacteria are attached specifically at the polar regions of heterocysts of *Anabaena circinalis* and *Aphanizomenon flos-aquae*. This algal-bacteria association occurs most frequently during bloom conditions. The author suggested that the attached bacteria assimilate substances released by the heterocysts. Bacterial respiration of organic substances near the N_2 fixing heterocysts may be effective enough to allow for concurrent high rates of photosynthesis and N_2 fixation during blooms.[3618]

In addition to their role in nitrogen fixation, heterocysts have been reported to evoke the formation of akinetes adjacent of them.[5477] They can also germinate to form trichomes[4519,5477] in some cases. Tyagi,[5064] Haselkorn[1956] and Lang and Fay[2696] summarized much that is known about heterocysts.

Planktonic blue-green algae possess flotation devices called gas vacuoles which are unique to prokaryotic organisms.[5253] A single gas vacuole consists of a number of closely packed cylinders with tapered ends (gas vesicles), each 70 nm wide and 200–1000 nm long and limited by a "membrane" of pure protein. Gas vesicles are strong but brittle structures which collapse in an "all-or-nothing" manner under different pressure.[973] The vesicles are essential for providing the vegetative buoyancy in many planktonic species.[2697] Details on gas vacuoles are given in References 396, 1395, 2763, 4674, 5249, 5250 and 5253–4 (see also Chapter 2.7.1.2).

True sexual reproduction is lacking, although there are reports of a low-frequency occurrence of genetic recombination by transformation in some species.[1066,4737] Blue-green algae reproduce vegetative by cell division (binary fission), fragmentation of colonies and filaments (hormogonia) and asexually by akinetes, endospores, exospores and heterocysts.[424] Flagella is lacking, although characteristic gliding occurs in many species.[4643]

The distinctive blue-green color is due to the presence of phycocyanin, one of the accessory pigments known as phycobiliproteins. Phycoerythrin may predominate in some blue-green algae to the extent that they appear purple or red rather than blue-green.[973] Some blue-green algae change the proportion of the phycobiliproteins to make the best use of available wave lengths of light.[4893]

The species concept of blue-green algae remains in a highly confused state.[4643] The taxonomy and nomenclature of the Cyanophyceae are in a state of disorder and there is presently no consensus among experts.[4643]

Cyanophytes (blue-green algae), as prokaryotic phototrophic organisms, were traditionally studied together with eukaryotic algae. This consideration is supported by their functional position in nature since cyanophytes constitute important primary producers in almost all biotopes of the biosphere. In the last decades, two fundamentally different classification schemes of cyanophytes have appeared: F. Drouet's system and the system of R.Y. Stanier's research group.[86] Anagnostidis and Komárek[86] provide a comprehensive analysis of cyanophytes classification. As I regard this problem very important their analysis will be adopted here.

The traditional system (Geitler's conception) of cyanophytes is based mainly on the morphological and cytological, and less on ecophysiological characteristics. The original taxonomic criteria used in the classic works[452,1592,1731,1903,2776,3368,3943,4968,4970] and others, are still accepted without any substantial change. The most important compilations based on these criteria are presented in the monographs of Geitler[1592-4] and Elenkin.[1228] These criteria have been applied with slight modification to all further treatments of this group (e.g. References 486, 801, 1062, 1481, 1513–4, 2238, 2569, 3878, 4690). A number of new valuable techniques of taxonomic evaluation have been developed in recent years and several new traits are at hand. The information obtained is very important for solving the taxonomical problems, but only occasionally useful for species determination. The cyanophyte morphology is more complex than that of any other members of the prokaryotic world. Thus, the morphological attributes, based on criteria of the "Geitler's" system, remain the basis of all revisions and determination keys of cyanophytes in spite of their loose applicability; they must be, however, corrected and modified according to the modern data.

"Drouet's system" (ecophenes approach)[1161–4,1166] introduced a radical reduction in the number of cyanophyte taxa, based on the type-exsiccated form from herbaria. They used the premise (published later) that "blue-green algae (relicts from Archeozoic times) are highly polymorphic, with few strong discontinuities among them and the manifest morphological differences are elicited by the environment both as simple reversible variations, and as less flexible responses"[1440] and that "genetic diversity can be expected within species" (Drouet and Forest in Friedmann and Borowitzka[1502]). Thus, the species sensu Drouet represent a polymorphic cluster of "ecophenes" in a few, broadly conceived genera. Because of the nomenclatural simplicity of Drouet's solution, his system had a great popular appeal, particularly among experimental workers in the USA but it has never been commonly accepted by phycologists. The existence of genetically stable types with narrow genetic variations has been proven experimentally[4550,4675,4684,4816,5287] and it was found that Drouet's interpretation does not correspond to the diversity of cyanophytes in nature or in cultures. Therefore, this classification system cannot be accepted.

The bacteriological approach ("Stanier's system") was proposed by Stanier et al.,[46834] Waterburry and Stanier,[5286–7] Ripka et al.,[4150] Ripka and Cohen-Basin,[4149] and others, who used the strain (clone) as the basic unit of the cyanophyte (cyanobacterial) taxonomy like in bacteriological practice, recommended the acceptance of the bacteriological code of nomenclature for cyanophytes. This system is based exclusively on the evaluation of some of morphological, physiological, cytological and biochemical characters of axenic strains (clones). This proposal soon became popular among the experimental scientists but gave a rise of objections from phycologists.[1724,1599,2825,2828] Anagnostidis and Komárek[86] pointed out that the cultures of cyanophytes are very useful for taxonomy as the additional source of information but they can hardly be the only base of typification for the taxa. The data presented in both phycological and bacteriological papers and the style of work with both natural populations and pure cultures are not contradictory but complementary. The essential differences arise in the interpretation, in different nomenclatural rules, and in the use of conventional terminology. If both approaches were to present the rules in such a way so that they would be comparable, then both methods would be acceptable. However, for those who need to recognize and identify various taxa of aquatic organisms in both nature and culture, the traditional botanical approach remains the only way. The status of species usage, concept, and evolution in the Cyanophyta have been comprehensively reviewed recently by Castenholz.[727]

The fact that phycologists consider cyanophytes (blue-green algae) to be an algal division is expressed in following statements. South and Whittick[4643] wrote: ". . . We regard them (blue-green algae) with the Prochlorophycophyta, as prokaryotic algae and thus, a legitimate part of the science of phycology, even though we do not dispute their affinities with the bacteria. Any comparative account of the algae must of necessity include reference to them; a comprehensive view of algal structure, physiology, ecology, or evolution is otherwise incomplete." Bold and Wynne[424] pointed out: "The authors are impressed by the fact that the blue-green algae contain chlorophyll-*a*, which differs from the chlorophyll of those bacteria which are photosynthetic, and also by the fact that free oxygen is liberated in blue-green algal photosynthesis but not in that of the bacteria." Lee[2745] expressed a very strong criticism of the bacteriological approach: "The Cyanophyceae are most closely related to the prokaryotic bacteria than to eukaryotic algae.[705] This relationship has led to a drive for the recognition of the term blue-green bacteria (cyanobacteria) instead of blue-green algae.[4683-4] Most phycologists, however, regard this suggestion as an imperialistic intrusion into their traditional sphere of influence by expansionist bacteriologists. Phycologists regard any organism with chlorophyll-*a* and a thallus not differentiated into roots, stems and leaves to be an alga. Such a definition covers the Cyanophyceae but not the photosynthetic bacteria." I agree with the aforesaid statements and, therefore, in this book the terms cyanophytes or blue-green algae are used only.

2.18.2 Prochlorophyceae

habitats: freshwater, marine
ecology: planktonic in freshwaters, symbionts in marine waters
levels of organization: unicellular (coccoid), filamentous
size: microscopic (6–30 μm)
cell covering: as in the Cyanophyceae
flagella: absent
thylakoids: free in the cytoplasm, some stacked in pairs or more, no chloroplast
chlorophylls: chlorophyll-*a* and *b*
carotenes: β-carotene
xanthophylls: zeaxanthin, cryptoxanthin
phycobiliproteins: none
storage products: starch-like
number of genera: single genus *Prochloron*
Literature: References 625, 739, 900, 1479, 2829–30, 2836, 2598, 2835, 3298, 3701, 4942–3, 5359.

The division Prochlorophyta (class Prochlorophyceae) was proposed by Lewin[2826-7] and contains the single genus *Prochloron*.[757,2830-1] Establishment of this separate division and evolutionary significance have raised discussion.[109,739,4682,5024] The algae in the Prochlorophyceae are prokaryotic, as are the algae in the Cyanophyceae. However, the prochlorophytes have chlorophylls *a* and *b* and lack phycobiliproteins.[2745]

Prochlorophytes occur as free-living planktonic filaments in freshwater lakes[626] and as symbionts in colonial ascidians, mostly in warm shallow waters of tropical seas.[2830] Although ascidians have been known for some time to harbor photosynthetic symbionts, it was not until 1975 that the symbionts were first brought to the attention of phycologists.[757]

The organisms, now assigned to the genus *Prochoron,* were first described as zooxanthelleae by Smith.[4595] Lewin and Cheng[2833] found that the algae living on *Didemnum* were unidentified coccoid forms similar to *Synechocystis*. Lewin[2823] named his species *Synechocystis didemni* but reported two "anamalous" features for blue-green algae: presence of chlorophyll-*b* and absence of phycobiliprotenis. This was confirmed by Lewin and Withers.[2834] Lewin[2826] proposed a new division, the Prochlorophyta, and in 1977 he established the type genus and species *Prochloron didemni.*[2827]

The structure of the prochlorophytic cell is similar to that of the blue-green algae. Thylakoids are usually grouped in pairs or stacks of many (in contrast to the Cyanophyceae, which have individual thylakoids not touching one another). Some of the cytological structures that occur in the blue-green algae do not occur in the Prochlorophyceae—no gas vacuoles or cyanophycin bodies.[2831] Biochemically, the prochlorophytes are similar to the blue-green algae. The prochlorophytes probably arose from the blue-green algae by the acquisition of chlorophyll-*b* and the loss of phycobiliproteins.[3643] Reproduction is vegetative by constrictive binary fission.

2.18.3 Rhodophyceae

common name: red algae
habitat: freshwater, brackish, marine (majority); examples shown in Figure 2-44
ecology: attached (benthic, periphytic)

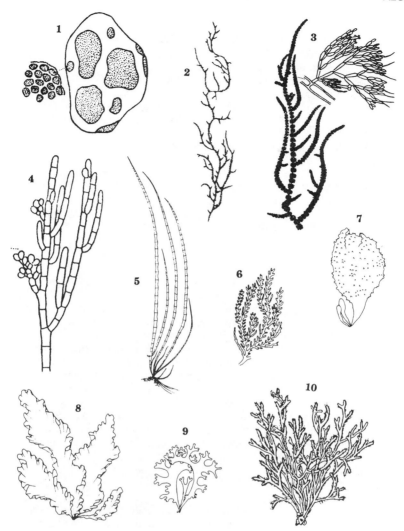

Figure 2-44.   Rhodophyceae.[973,2098,4559]   Freshwater: 1-*Hildebrandia rivularis* (cells 5.2–8.4 μm in ø); 2-*Thorea ramosissima* (up to 50 cm); 3-*Batrachospermum moniliforme* (up to 12 cm); 4-*Audionella* (= *Chantransia*) *chalybea* (up to 10 mm); 5-*Lemanea nodosa* (5–12 cm); Marine: 6-*Gelidium* sp. (ca. 10 cm); 7-*Gigartina* sp. (ca. 3–8 cm); 8-*Porphyra* sp. (ca. 20–40 cm); 9-*Chondrus* sp. (ca. 10–15 cm); 10-*Galaxaura* sp. (ca. 10–15 cm). Not to scale. (1, 3, 4 from Sládeček et al., 1973, in *Technical Hydrobiology: A Laboratory Manual,* SNTL Praha, reprinted with permission of V. Sládeček; 2, 5 from Hindák et al., 1975, in *Determination Key for Lower Plants,* SPN Bratislava, reprinted with permission of F. Hindák; 6–9 from Darley, 1982, in *Algal Biology: A Physiological Approach,* drawings by Andrew J. Lampkin, III, reprinted with permission of Blackwell Scientific Publications and the author; 10 from Hillson, *Seaweeds,* University Park: The Pennsylvania State University Press, 1977, p. 137, Copyright 1977 by The Pennsylvania State University, reproduced with permission of the publisher.)

levels of organization: unicellular (coccoid), filamentous (branched, heterotrichous), pseudoparenchymatous (uniaxal, multiaxal, parenchymatous)
size: microscopic to macroscopic (majority)
cell covering: cellulose, xylans in some, several sulfated polysaccharides (galactan polymers—agar,
carrageenan, funonan, furcellaran, porphyran), some calcified with $CaCO_3$, alginate in
corallinaceae

flagella: absent
number of thylakoids per band: 1 unstacked
number of membranes of CER: CER absent
number of membranes surrounding chloroplast: 2
chlorophylls: chlorophyll-*a,* chlorophyll-*d*
carotenes: α-carotene, β-carotene
xanthophylls: antheraxanthin, β-cryptoxanthin, lutein, neoxanthin, taraxanthin, <u>zeaxanthin</u>
phycobiliproteins: c-, b-, B-, r-, R-allophycocyanin, R-, C-phycocyanin, b-, B-, C-, R-phycoerythrin
eyespot: absent
storage products: floridean starch, oils, outside the chloroplast
number of genera: 400,[55] 675,[2582] 831[1097]
freshwater: 32,[5091] 42[3585]
number of species: 1,500,[55] 4,000,[1004,2095,2745,4387] 4,100,[2582] 5,254[1097]
freshwater: 190,[5091] 200[2491,2745,3585]
Literature: References 8, 438, 835, 1097–8, 1506, 1565, 2095, 2343, 2678, 3423, 3452, 3585, 3935, 4387, 4433, 4921–2, 5470.

The Rhodophyceae are probably one of the oldest groups of eukaryotic algae.[2745] The Rhodophyceae are probably the closest of eukaryotic algae to the Cyanophyta.[1098,1274,3452] In spite of their size and extremely complex life cycles, red algae have a number of characteristics that may be considered primitive including the absence of flagellated cells, simple chloroplast structure and the presence of phycobilin pigments.[973] Most of the red algae are macroscopic seaweeds.

Rhodophycean organisms range from autotrophic, independent plants to complete heterotrophic parasites—nonobligate epiphytes, obligate epiphytes, semiparasites that have some photosynthetic pigments and parasites with no coloration.[2745] Symbiosis and/or parasitism of red algae on other red algae was reviewed by Goff[1679] (see also Chapter 2.7.7). Symbiotic red algae were described also by Apt,[127] Martin and Pocock[3083] and Nonomura.[3450]

Red algae range in morphology from simple filaments to more massive and relatively complex thalli.[1004] Attachment to the substrate for red algae is provided by a variety of devices, e.g. single-celled rhizoids or the holdfast.[424] Freshwater genera are attached to rock or other substrata and they usually are smaller than red seaweeds.[2745,3585] Although the variety of physical and chemical parameters have been monitored in relation to red algae distribution in freshwaters it seems that only stream current is correlated with their occurrence and the Rhodophyceae are usually restricted to permanent flowing waters in tropical and temperate zones.[973,2293,4434,4547] Exceptions, however, occur in lentic[2293] or aerophytic[2678] habitats. The majority of the freshwater Rhodophyceae are blue-green, green, olive-green, gray-blue, brown or black in color rather than red, indicative of the predominance of pigments other than phycoerythrin.[3585]

Red algae are usually smaller and more delicate than the brown algae.[973] Although marine red algae occur at all latitudes, there is a marked shift in their abundance from the equator to colder seas. There are few species in polar and subpolar regions where brown and green algae predominate, but in temperate and tropical regions they far outnumber these groups. The larger species of fleshy red algae occur in cool-temperate areas, whereas in tropical seas the Rhodophyceae (except for massive calcareous forms) are mostly small, filamentous plants.[1006]

All members of Corallinaceae (Gigartinales) and some of the Nemaliales and *Peyssonnelia* deposit $CaCO_3$ extracellularly in the cell walls. Anhydrous calcium carbonate occurs in two crystalline forms, calcite (rhomboidal) and aragonite (orthorhombic). $MgCO_3$ and $SrCO_3$ are commonly deposited along with $CaCO_3$, there being less $MgCO_3$ in aragonite than in calcite.[1097] The 40 genera and 600 species of calcified red algae range from the cold waters of the arctic to the tropical atolls in the Caribbean.[1004] (For further details on calcification in red algae see Chapter 5.1.4.4).

The coralline algae thrive in rock pools and on rocky shores exposed to very strong wave action and swift tidal currents. The red algae that have the highest rates of calcification also have the highest rates of photosynthesis and are usually found in waters less than 20 m deep.[1759] The Rhodophyceae also have the ability to grow at greater depths in the ocean than do members of the other algal classes. They live at depths as great as 200 m. At least a partial explanation for their dominance in deep water is the correlation between the absorption peak of R-phycoerythrin and the blue-green light available at greater depths.[1004,2745]

Red algae are an important source in the food webs of benthic communities.[1004] Red algae are also important to men as direct food (*Palmaria, Porphyra, Eucheumia*)[1004] (see also Chapter 2.14.1). The two most important commercially utilized polysaccharides derived from the Rhodophyceae are agar and carrageenan. Agar obtained commercially from species are often loosely referred to as agarophytes (e.g. *Gelidium, Pterocladia, Gracilaria*). Carrageenan is usually obtained from wild populations of Irish Moss, the name for a mixture of *Chondrus crispus* and the various species of *Gigartina*.[2745] (For the use of agar and carrageenan see Chapter 2.14.5.)

In sexual oogamous reproduction, spermatia are produced which are carried passively by water currents to the female organ, the carpogonium. Asexual reproduction involves spores.[2745]

2.18.4 Chrysophyceae

common name: golden algae
habitat: freshwater (majority), brackish, marine; examples shown in Figure 2-45
ecology: planktonic (majority), attached
levels of organization: unicellular (coccoid, rhizopodal, flagellate), colonial (aggregations, colonial), filamentous (branched), pseudoparenchymatous (parenchymatous)
size: microscopic
cell covering: naked, scales (some composed of silica) or lorica in some
flagella: 1, normally 2, equal or unequal, apical, smooth (acronematic) posterior shorter, hairy (pantonematic) anterior longer
number of thylakoids per band: 3
number of membranes of CER: 2
number of membranes surrounding chloroplast: 4
chlorophylls: chlorophyll-*a*, chlorophyll-*c* ($c_1 + c_2$)
carotenes: β-carotene
xanthophylls: diadinoxanthin, diatoxanthin, dinoxanthin, echinenone, fucoxanthin, lutein
phycobiliproteins: none
eyespot: in the chloroplast
storage products: chrysolaminarin (= leucosin), oils, outside the chloroplast
number of genera: 75,[55] 200[5091]
number of species: 300,[55] 650 (20% marine),[1004] 1,000[2613,5091]
Literature: References 92–3, 481–5, 930, 1632, 2060–2, 2074, 2091, 2613–4, 2728, 2759, 3795, 4272, 4528, 4693, 4712, 4877.

The Chrysophyceae, established by Pascher[3666] and formally typified by Hibberd,[2062] are mainly found in freshwater habitats and most species are planktonic. A main characteristic of the class is the ability to form endogenous, silicified resting stages, termed stomatocysts (statospores).[424,2613]

In 1987, Andersen[92] removed species of the order Synurales out of this class and proposed a new class Synurophyceae, but this classification is still being discussed so for the purpose of this book this order will be treated as a part of the Chrysophyceae. In fact, the Synurophyceae share affinities with the Bacillariophyceae, Chrysophyceae and Phaeophyceae, and it is possible to include them in any of these groups.[92] Because of their silicified outer covering, they have been called "flagellate diatoms"; yet they have many of the characteristics of the Chrysophyceae such as statospores.[2745] Synurophyceae differ from Chrysophyceae first of all in lacking chlorophyll-c_2, eyespot and CER (usually).[2745]

The chrysophytes are golden brown in color and are mostly motile unicells or colonies; a few species exhibit amoeboid, coccoid or filamentous body forms. Most of the species in the Chrysophyceae are freshwater and it has been long believed that they occur in cool environments. Investigations from the tropics[854-5,4878] and from South America,[1192] however, have invalidated this belief. Dop,[1122] Dop et al.,[1123] Gayral and Billard[1585] and Billard and Fresnel[371] have found a rich flora of benthic Chrysophyceae in salt and brackish waters. Most of the Chrysophyceae are found in unpolluted systems and sensitive to changes in the environment.[424,2745]

Photosynthesis is well developed in most members of this class, but colorless forms are not infrequent. Both osmotrophic and phagotrophic modes of heterotrophic nutrition occur. Some forms, such as *Ochromonas*, have both patterns as well as autotrophic nutrition in the same species.[424] Some

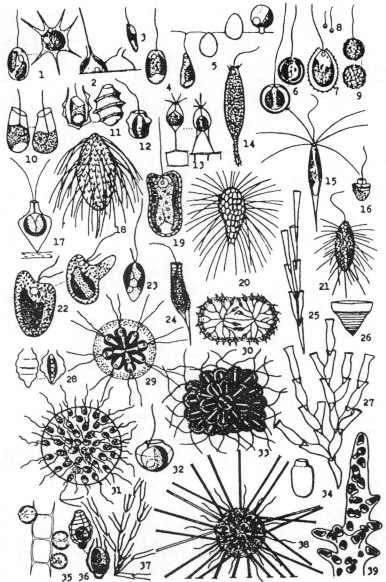

Figure 2-45.   Freshwater Chrysophyceae.[4559] 1-*Chrysamoeba radians;* 2-*Lagynion scherffelii;* 3-*Chromulina nebulosa* (12–16 × 4–6 μm); 4-*C. ovalis* (9–14 × 6–7 μm); 5-*C. rosanofii* (4.5–10 × 4.5–7.5 μm); 6-*Chrysococcus rufescens* (8–12 μm in ø); 7-*C. ornatus;* 8-*C. punctiformis* (2–3 μm in ø); 9-*C. klebsianus;* 10-*Kephyrion ovum* (7 × 4 μm); 11-*K. spirale;* 12-*K. rubri claustri;* 13-*Chrysopyxis stenostoma* (11–22 × 10–16 μm); 14-*Mallomonas insignis;* 15-*M. akrokomos* (23–62 × 4.5–16 μm); 16-*Kephyrion circumvallatum;* 17-*Derepyxis dispar* (18–22 × 10–15 μm); 18-*Mallomonas acaroides* (18–45 × 7–23 μm); 19-*Microglena punctifera* (20–50 × 10–20 μm); 20-*Mallomonas fastigata* (47–100 × 10–30 μm); 21-*M. elegans;* 22-*Ochromonas fragilis* (9–16 × 10–12 μm); 23-*O. ludibunda* (12–18 × 6–13 μm); 24-*Dinobryon suecicum* (cells 19–22 × 4–5 μm); 25-*D. stipitatum* (cells 30–50 × 8–10 μm); 26-*Pseudokephyrion schilleri;* 27-*Dinobryon sertularia* (cells 30–40 × 10–12 μm); 28-*Pseudokephyrion undulatum* (18–25 × 7–12 μm); 29-*Syncrypta volvox* (colonies 20–70 μm in ø); 30-*Uroglena volvox* (colonies 40–400 μm in ø); 31-*Uroglenopsis americana;* 32-*Kephyrion poculum* (7 × 8–9 μm); 33-*Synura petersenii* (colonies 47–57 μm in ø); 34-Pseudokephyrion entzii (9–13 × 7–8 μm); 35-*Chrysosphaera epiphytica* ((7)–15–25 μm long); 36-Pseudokephyrion undulatissimum (10–14 × 7–9 μm); 37-*Dinobryon divergens* (cells 30–65 × 8–11 μm); 38-*Chrysosphaerella longispina* (colonies 35–100 μm in ø); 39-*Hydrurus foetidus* (colonies 1–30(100) cm). Not to scale. (From Sládeček et al., 1973, in *Technical Hydrobiology: A Laboratory Manual,* SNTL Praha, reprinted with permission of V. Sládeček.)

Chrysophyceae have the ability to ingest particles. Under low light conditions, *Dinobryon* can ingest an average of 3 bacterial cells every 5 minutes.[373] Under these conditions, *Dinobryon* obtains about 50% of its total carbon from bacterivory and about 50% from photosynthesis.

Some of the species have a lorica (an envelope around the protoplast but not generally attached to the protoplast as a wall is).[2745] Unlike the loricas of vegetative cells, which are cellulosic, with or without the addition $CaCO_3$ or Fe-compounds, the wall of the statospore is always siliceous.[481]

Reproduction is mainly asexual, although sexual stages are known for an increasing number of species (anisogamy, most often isogamy).[481,1460,2608–13]

2.18.5 Prymnesiophyceae (Haptophyceae)

habitat: freshwater, brackish, marine (majority); examples shown in Figure 2-46
ecology: planktonic (majority), attached
levels of organization: unicellular (flagellate), colonial (aggregations)
size: microscopic
cell covering: organic scales in one to several layers, calcified in many
flagella: 2 equal, anteriorly inserted (acronematic), haptonema
number of thylakoids per band: 3
number of membranes of CER: 2
number of membranes surrounding chloroplast: 4
chlorophylls: chlorophyll-*a,* chlorophyll-*c* ($c_1 + c_2$)
carotenes: β-carotene
xanthophylls: diadinoxanthin, diatoxanthin, echinenone, <u>fucoxanthin</u>
phycobiliproteins: none
eyespot: absent except in one genus (in chloroplast)
storage products: chrysolaminarin, outside the chloroplast
number of genera: 75[3455,4643]
number of species: 200 (50% marine)[1004]
Literature: References 435–6, 439, 581, 585–6, 797, 801, 1804–6, 2065, 2526–7, 2729–32, 3052–3, 3056–60, 3652, 3654, 3794, 4852, 5184.

The class Haptophyceae was created by Christensen[797]—until then the organisms were considered part of the Chrysophyceae. The name Haptophyceae was a descriptive name and not based on a genus in the class; thus Hibberd[2062] later changed the name to Prymnesiophyceae. Dawes[1004] pointed out that much confusion exists as to the classification and placement of genera within the class.

The Prymnesiophyceae are a group of unicellulate flagellates characterized by the possession of a haptonema between two smooth flagella.[2745] A haptonema is a filamentous appendage arising near the flagella but thinner and with different properties and structure. It was mistaken for a third flagellum during earlier studies of phytoflagellates. The functional significance of the organelle is not fully understood but the most conspicuous property of the haptonema is its capacity for surface adhesion, and the cell may then glide autonomously with an attached haptonema.[2731] The participation of the haptonema in phagocytosis has been suggested.[3648–50,4481] This was confirmed by Kawachi et al.[2447] who showed that the haptonema plays an important role as a food capture and transport device in the process of phagocytosis in the marine flagellate *Chrysochromulina hirta*. The reader is referred to Green et al.,[1808] Kawachi et al.[2447] and Gregson et al.[1815–6] for recent reviews on haptonema.

Most of the members of the Prymnesiophyceae have a cell covering consisting of a number of elliptical scales. These are embedded in a mucilaginous substance, and in some organisms a layer of calcified coccolith is outside the scales. Coccoliths are basically organic scales that have $CaCO_3$ (usually calcite) deposits on a surface.[2745] The individual scales (coccoliths) are produced within cisternae of the Golgi apparatus[3049–50,3588,4643] and collectively form a rigid outer skeleton around the cell called the coccosphere.[424] The functional significance of the coccolith cover remain uncertain.[439] (For details on coccolith formation see Chapter 5.1.4.3.)

Braarud et al.[519] created a further subdivision of coccoliths on the basis of their shape. Coccoliths are of two basic types. Holococcoliths have crystals all of the same size and shape, simple rhombohedral calcite crystals forming a covering over an organic template. Heterococcoliths have morphologically diverse crystals, usually of calcite but occasionally aragonite.

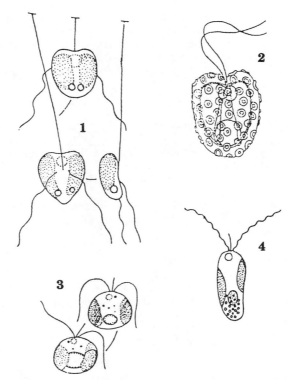

Figure 2-46. Freshwater Prymnesiophyceae.[2098] 1-*Prymnesium czosnowskii* (6.5 × 7.5 µm); 2-*Prymnesium saltans* (19–35 × 7–18 µm); 3-*Chrysochromulina parva* (3–7 × 2.5–5 µm); 4-*Hymenomonas roseola* (18–50 × 10–24 µm). Not to scale. (From Hindák et al., 1975, in *Determination Key for Lower Plants,* SPN Bratislava, reprinted with permission of F. Hindák.)

The light dependency of coccolith formation has been numerously documented.[2528,3794] Recent studies have shown that Ca^{2+} uptake is light dependent but coccolith formation can proceed in the dark, and the quantities of Ca^{2+} taken up by the cell prior to a dark period is critical.[131]

The Prymnesiophyceae are primarily marine organisms, although there are some freshwater species. They make up a major part of the marine nanoplankton and contribute about 45% of the total phytoplankton cells in the middle latitudes of the South Atlantic.[2745] The coccolithophorids (algae with coccoliths) are common in tropical waters because these warm waters have a low partial pressure of CO_2 and are usually saturated or supersaturated with $CaCO_3$, the concentrations being especially high in the upper layers.[2745] Most prymnesiophytes are photosynthetic, but heterotrophic growth is possible, either saprophytic[394,3957] or phagotrophic.[2733,3053,3651] The heterotrophic mode of nutrition would enable these organisms to survive periods when they sink below the euphotic zone.[424] Generally, the Prymnesiophyceae are not completely autotrophic but auxotrophic. The coccolithophorids are among the fastest-growing algae, with growth rates up to 2.25 divisions per day.[2745]

Sexual reproduction occurs and the life history commonly involves alternating motile and nonmotile phases.[4643] Asexual reproduction is by binary cell division or by the release of motile on nonmotile cells. However, there is much plasticity in the mode and type of asexual reproduction.[1004]

2.18.6 Xanthophyceae (Tribophyceae)

common name: yellow-green algae
habitat: freshwater (majority), brackish, marine, terrestrial; examples shown in Figure 2-47

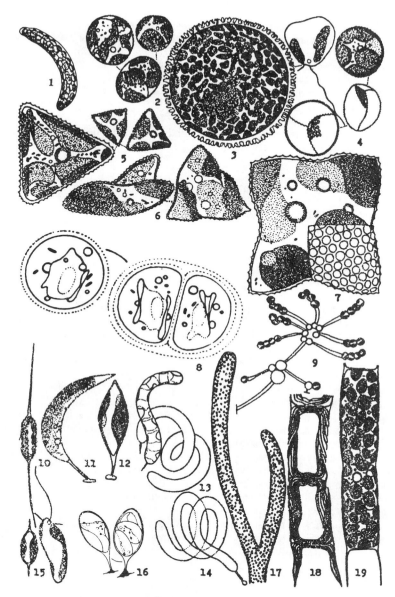

Figure 2-47. Freshwater Xanthophyceae.[2098,4559] 1-*Bumilleriopsis megacystis* (cells 120 × 12 µm); 2-*Chloridella neglecta* (cells 8–14 µm in ø); 3-*Botrydiopsis arhiza* (cells 20–70 µm in ø); 4-*Tetraëdriella subglobosa;* 5-*Goniochloris sculpta* (cells 12–16 µm); 6-*G. torta;* 7-*G. tetragona;* 8-*Heterogloea pyreniger* (cells 10 µm in ø); 9-*Mischococcus confervicola* (cells 5–8(10) µm in ø); 10-*Centritractus belenophorus* (cells 16–17 × 8–15 µm); 11-*Characiopsis lunaris;* 12-*C. acuta* (cells 15–18 × 6–10 µm); 13-*Ophiocytium cochleare* (cells 5–8 µm wide); 14-*O. maius* (cells 2.5 mm × 5–10 µm); 15-*Centritractus dubius;* 16-*Characidiopsis ellipsoidea* (cells 12–15 × ca. 6–8 µm); 17-*Vaucheria geminata* (filaments up to 5–10 cm); 18, 19-*Tribonema viride* (cells 25–120 × 10–15 µm, filaments up to 10 cm or more). Not to scale. (1, 8, 16 from Hindák et al., 1975, in *Determination Key for Lower Plants,* SPN Bratislava, reprinted with permission of F. Hindák; 2–7, 9–15, 17–19 from Sládeček et al., 1973, in *Technical Hydrobiology: A Laboratory Manual,* SNTL Praha, reprinted with permission of V. Sládeček.)

ecology: planktonic, attached
levels of organization: unicellular(coccoid,rhizopodal,flagellate),colonial(aggregations),filamentous
 (simple, branched), pseudoparenchymatous (siphonaceous)
size: microscopic to macroscopic filaments
cell covering: cellulose, glucose and uronic acids, many walls are bipartite
flagella: 2, unequal, apical anterior longer, hairy (pantonematic), posterior shorter, smooth (acronematic)
number of thylakoids per band: 3
number of membranes of CER: 2
number of membranes surrounding chloroplast: 4
chlorophylls: chlorophyll-a, chlorophyll-c ($c_1 + c_2$)
carotenes: β-carotene
xanthophylls: <u>diadinoxanthin</u>, diatoxanthin, b-cryptoxanthin, 5',6'-monoepoxide, 5,6,5',6'-diepoxide, <u>heteroxanthin</u>, lutein, neoxanthin, <u>vaucheriaxanthin ester</u>
phycobiliproteins: none
eyespot: in chloroplast
storage products: chrysolaminarin, oils, outside the chloroplast (carbohydrates, xylans)
number of genera: 60 (15% marine),[1004] 70,[55] 80[5091]
number of species: 360,[55] 400,[2491,5091] 600,[2068] >600[3580]
Literature: References 405, 799, 928, 1107, 1279, 1281–2, 2064, 2068, 3099, 3580–4, 3670, 4134–5, 4495, 4512, 5400.

The Xanthophyceae (= Tribophyceae[2067]) contain primarily freshwater algae with a few marine or brackish representatives but are fairly cosmopolitan.[4900] The Xanthophyceae show a high degree of parallel evolution of body form with the green algae and, in fact, were classified as green algae until the early 1900s. The two groups may be distinguished by a negative iodine test for starch in the Xanthophyceae.[973]

Most xanthophytes are non-motile unicells (coccoid forms), but motile unicells, filaments and siphonaceous forms are also known. *Vaucheria,* a widespread, siphonaceous genus that forms luxuriant, felt-like masses on mud, is often found in estuarine, intertidal habitats.[973] Close to 70% of the species are coccoid unicells found in still waters or on damp soils. About 20% are filamentous species[1004] occurring usually as floating mats with many members possessing a presumed celluloic wall[818] of two overlapping halves. Most of the remaining 10% of the species are coenocytic, and fewer in number are monadoid, rhizopodial, or palmelloid.[2064]

The Xanthophyceae multiply vegetatively by cell division and fragmentation (in filamentous forms), and asexually by motile zoospores (majority of genera) and non-motile aplanospores. In addition, they have the ability to form specialized resting spores—statospores, which are observed especially in brackish and freshwater species. There are only three genera which may reproduce sexually and all three types of gamete production are found.[1004,2745]

Several reports have addressed parasitism in the Xanthopyceae. A zooflagellate parasitic on several genera is thought to play an important role in controlling the numbers of individuals in natural population.[1827] Parasitic rotifers infecting *Vaucheria* were reported by Prud'Homme van Reine[3933] and Davis and Gworek.[998]

2.18.7 Eustigmatophyceae

habitat: freshwater, marine, terrestrial; examples shown in Figure 2-48
ecology: planktonic, attached
levels of organization: unicellular (coccoid)
size: microscopic
cell covering: naked (cellulose?)
flagella: single anterior hairy flagellum (pantonematic) and second basal body (heterokontous)
number of thylakoids per band: 3
number of membranes of CER: 2
number of membranes surrounding chloroplast: 4
chlorophylls: chlorophyll-a, chlorophyll-c

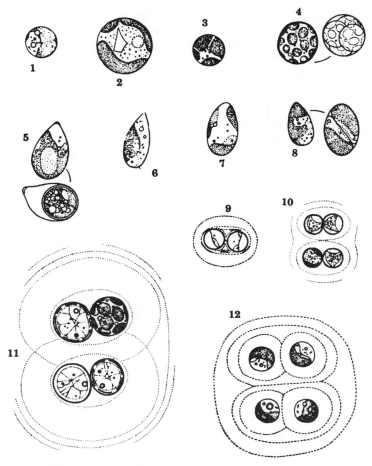

Figure 2-48. Freshwater Eustigmatophyceae.[2098] 1-*Pleurochloris commutata* (5–9(15) μm in ø); 2-*P. magna* (6–12 μm in ø); 3-*P. lobata* (12–15 μm in ø); 4-*P. polychloris* (7–11 μm in ø); 5-*Monodus pyreniger* (12–15 × 9 μm); 6-*M. acuminata* (11 × 2–5 μm); 7-*M. chodatii* (7.8–15 × 7–11.7 μm); 8-*M. subterraneus* (7–9 × 3–5 μm); 9-*Chlorobotrys terrestris* (cells < 4 μm in ø); 10-*C. regularis* (cells 15–20 μm in ø); 11-*C. polychloris* (cells 18–23 μm in ø); 12-*C. simplex* (cells < 7 μm in ø). Not to scale. (From Hindák et al., 1975, in *Determination Key for Lower Plants,* SPN Bratislava, reprinted with permission of F. Hindák.)

carotenes: β-carotene
xanthophylls: antheraxanthin, β-cryptoxanthin, diatoxanthin, 5,6,5′6′-diepoxide, heteroxanthin, 5′6′-monoepoxide, neoxanthin, vaucheriaxanthin ester, violaxanthin, (lutein, diadinoxanthin, zeaxanthin?)
phycobiliproteins: none
eyespot: outside the chloroplast
storage products: oils, leucosin(?), outside the chloroplast
number of genera: 6[424]
number of species: 12[4643]
Literature: References 113, 1464, 2066–7, 2071–3, 2079–80, 2894, 4285, 4495, 5399, 5400.

The class Eustigmatophyceae was created in 1970 by Hibberd and Leedale[2071] out of organisms that were previously classified in the Xanthophyceae on the basis of ultrastructural features of the zoospore.

Hibberd and Leedale[2073] presented the diagnostic features of this class. The Eustigmatophyceae is considered to be quite distinct from the Xanthophyceae and its relatives and not closely related to any of the other algal groups.[973]

The name Eustigmatophyceae was chosen because of the large size of the eyespot in the zoospore. Although only 12 species have been described, the Eustigmatophyceae occur in a relative wide range of habitats—freshwater, marine and terrestrial.[2069] All genera are free-floating except for the sessile genus *Pseudocharaciopsis*. They are green in color though the shade of green is usually more yellow than the grass-green typical for the Chlorophyceae.[2069]

One of the most distinctive features of this class is that the eyespot is located outside the chloroplast, a condition which is known elsewhere only in the Euglenophyceae. All the species which are presently included in the group are coccoid unicells, most of which reproduce by means of zoospores.

2.18.8 Bacillariophyceae

common name: diatoms
habitat: freshwater, brackish, marine, terrestrial; examples shown in Figures 2-49, 2-50 and 2-51
ecology: planktonic, attached
levels of organization: unicellular (coccoid), filamentous (simple, pseudofilamentous), colonial (aggregations)
size: majority microscopic (usually less than 100 μm in diameter or length), some species attain macroscopic sizes (up to 1 mm in length)
cell covering: silica frustule
flagella: 1 (male gametes) apical, hairy (pantonematic), lacks central microtubules
number of thylakoids per band: 3
number of membranes of CER: 2
number of membranes surrounding the chloroplast: 4
chlorophylls: chlorophyll-a, chlorophyll-c ($c_1 + c_2$)
carotenes: α-carotene, β-carotene, ε-carotene
xanthophylls: diadinoxanthin, diatoxanthin, <u>fucoxanthin</u>, neoxanthin
phycobiliproteins: none
eyespot: absent
storage products: chrysolaminarin, oils, lipids (mostly C_{14}, C_{16} and C_{20} fatty acids), outside the chloroplast
number of genera: 200,[55,5091] 240[424]
number of species: 5,000,[55] 6,000,[5091] 12,000[1004]
freshwater: 60 genera with 2000 species[5091]
Literature: References 83, 1102, 1590, 2120, 2584–7, 2793, 2806, 3674, 3682–3, 4207–8, 4214, 4311, 4216, 4444, 4829, 5185, 5332.

Diatoms are perhaps the most ubiquitous of the algal groups. Diatoms comprise the main component of the open-water marine flora and a significant part of the freshwater flora.[2745] The Bacillariophyceae are unicellular, sometimes colonial algae found in almost every habitat as planktonic or attached photosynthetic autotrophs, colorless heterotrophs or photosynthetic symbionts.[123,1004] A few species of colorless diatoms are known, living saprophytically on such substrates as the mucilage of thalli of known algae.[3905] Diatoms known to possess heterotrophic capabilities are more prevalent in areas characterized by high organic content.[2020] Facultative heterotrophy in diatoms was reported by White.[5382] Even though diatoms are autotrophic, a high incidence of auxotrophy occurs in marine littoral species. Investigations have shown that approximately half of the clones isolated required either cobalamine or thiamine or both for growth.[2805,2833] Attached communities composed of diatoms form a typical brownish-green film on substrates (rocks, stones, macrophytes).[2745] Restricted floras are found in soils and aerial habitats[2924] and in areas subject to temporary extremes.[620,1309] Diatoms can also form water blooms.[2804,2810,3162]

Several studies have shown that diatoms diversity is lower in eutrophic than in mesotrophic conditions.[3675,4832,5134] Some diatom species typically thrive in eutrophic conditions and build up large population levels.[4832,5134] Diatoms are frequently employed to characterize aquatic habitats because

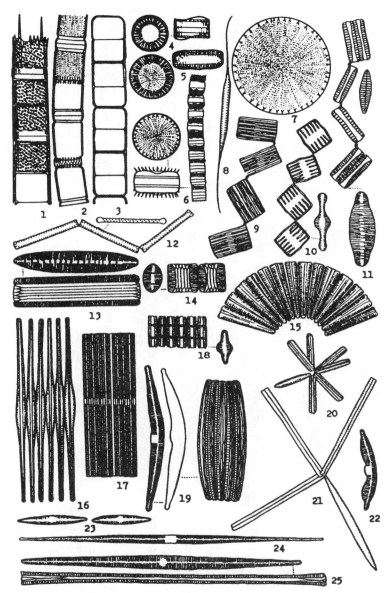

Figure 2-49.   Freshwater Bacillariophyceae.[4559]   1-*Melosira granulata* (cells 21 μm in ø, filaments up to > 1 mm); 2-*M. italica* (cells 3–28 μm in ø, filaments > 1 mm); 3-*M. varians* (cells 8–35 μm in ø, filaments > 1 mm); 4-*Cyclotella meneghiniana* (10–30 μm wide); 5-*Cyclotella comta* (8–50 μm in ø); 6-*Stephanodiscus hantzschii* (8–30 μm in ø); 7-*S. astraea;* 8-*Rhizosolenis longiseta* (70–200 × 4–10 μm); 9-*Tabellaria fenestrata* v. *intermedia* (cells 26–140 × 3–9 μm); 10-*T. flocculosa* (cells 12–50 × 5–16 μm); 11-*Diatoma vulgare* (cells 8–120 × 5–13 μm); 12-*D. tenue* (= *elongatum*) (cells 20–120 × 2–4 μm); 13-*D. hiemale* (cells 12–100 × 6–15 μm); 14-*D. hiemale* v. *mesodon* (cells 12–40 × 6–15 μm); 15-*Meridion circulare;* 16-*Fragilaria crotonensis* (cells 40–170 × 1–3 μm); 17-*F. capucina* (cells 25–100 × 2–5 μm); 18-*F. construens* (cells 7–25 × 5–12 μm); 19-*Ceratoneis arcus;* 20-*Synedra berolinensis;* 21-*S. actinastroides;* 22-*Ceratoneis arcus* v. *amphioxys;* 23-*Synedra vaucheriae* (10–40 × 2–5 μm); 24-*S. acus* (40–500 × 2–5 μm); 25-*S. ulna* (50–600 × 5–9 μm). Not to scale. (From Sládeček et al., 1973, in *Technical Hydrobiology: A Laboratory Manual,* SNTL Praha, reprinted with permission of V. Sládeček.)

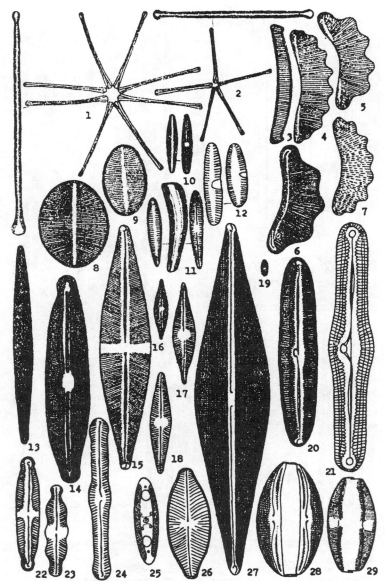

Figure 2-50.   Freshwater Bacillariophycaea.[4559]   1-*Asterionella formosa;* 2-*A. gracillima;* 3-*Eunotia arcus* (25–70 × 3–9 µm); 4-*E. robusta* (10–25 µm wide); 5-*E. robusta* v. *tetraodon;* 6, 7-*E. triodon;* 8-*Cocconeis pediculus* (15–56 × 10–37 µm); 9-*C. placentula* (11–70 × 8–40 µm); 10-*Achnanthes minutissima* (5–40 × 2–4 µm); 11-*Rhoicosphenia curvata* (12–75 × 4–8 µm); 12-*Achnanthes lanceolata* (7–40 × 4–10 µm); 13-*Amphipleura pellucida* (80–140 × 7–9 µm); 14-*Frustulia vulgaris* (50–70 × 10–13 µm); 15-*Stauroneis phoeni-centeron* (45–325 × 16–53 µm); 16-*Navicula cryptocephala* (13–40 × 5–7 µm); 17-*N. rhynchocephala* (35–60 × 10–13 µm); 18-*N. viridula* (40–80 × 10–15 µm); 19-*N. atomus* (6–10 × 2–5 µm); 20-*Pinnularia viridis* (30–170 × 7–30 µm); 21-*P. nobilis* f. *intermedia* (200–300 × 34–50 µm); 22-*P. microstauron* (20–80 × 7–11 µm); 23-*P. microstauron* f. *biundulata;* 24-*P. gibba* (34–140 × 7–14 µm); 25-*Navicula gracilis* (36–60 × 6–10 µm); 26-*N. gastrum* (25–60 × 12–20 µm); 27-*N. cuspidata* (17–37 × 50–170 µm); 28-*Amphora ovalis* (4–140 × 4–63 µm); 29-*A. ovalis* v. *gracilis.* Not to scale. (From Sládeček et al., 1973, in *Technical Hydrobiology: A Laboratory Manual,* SNTL Praha, reprinted with permission of V. Sládeček.)

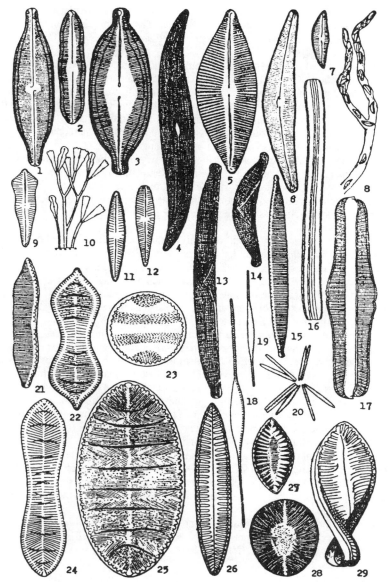

Figure 2-51.   Freshwater Bacillariophyceae.[4559]   1-*Neidium productum* (40–100 × 18–36 μm); 2-*Caloneis silicula* (= *ventricosa*) (2–120 × 6–20 μm); 3-*C. amphisbaena* (36–120 × 20–42 μm); 4-*Gyrosigma acuminatum* (63–200 × 14–20 μm); 5-*Cymbella ehrenbergii* (50–220 × 19–50 μm); 6-*C. lanceolata* (70–210 × 24–34 μm); 7, 8-*C. ventricosa* (10–40 × 5–12 μm); 9-*Gomphonema acuminatum* (20–110 × 5–17 μm); 10-*G. constrictum* (= *truncatum*) (25–65 × 8–14 μm); 11-*G. angustatum* (12–45 × 5–9 μm); 12-*G. olivaceum* (8–70 × 3.5–10 μm); 13-*Epithemia turgida* (60–200 × 15–18 μm); 14-*E. sorex* (15–65 × 8–15 μm); 15-*Nitzschia palea* (20–65 × 2–5 μm); 16-*N. sigmoidea* (160–500 × 8–14 μm); 17-*Rhopalodia gibba* (35–300 × 18–30 μm); 18-*Nitzschia longissima* f. *parva;* 19-*N. acicularis* (50–150 × 3–4 μm); 20-*N. holsatica* (20–55 × 1.5–2 μm); 21-*Hantzschia amphioxys* (20–200 × 5–15 μm); 22-*Cymato-pleura solea* v. *apiculata* (30–300 × 12–40 μm); 23-*C. elliptica* v. *discoidea;* 24-*C. solea* (30–300 × 12–14 μm); 25-*C. elliptica* (5–220 × 40–90 μm); 26-*Surirella biserista* (35–350 × 30–80 μm); 27-*S. turgida* (50–120 × 33–50 μm); 28-*S. ovata* v. *crumena* (15–70 × 8–23 μm); 29-*S. spiralis* (50–200 × 20–80 μm).   Not to scale.   (From Sládeček et al., 1973, in *Technical Hydrobiology: A Laboratory Manual,* SNTL Praha, reprinted with permission of V. Sládeček.)

some taxa or growth forms demonstrate a high habitat-specificity.[2031,2368,2923] Some intertidal diatom species show a cyclic rhythm in the vertical movement toward and away from the surface of estuarine mud.[2183]

Diatoms have an absolute requirement for silicon if cell division is to take place.[2745] In general, diatoms need their cell wall to live, and silicon has been shown to be an absolute requirement for mitosis and frustule formation.[970,974] Silicon is present in the frustule as hydrated amorphous silica $SiO_2 . n H_2O$ (see Chapter 5.10.2 for details).

Diatoms may be separated into two groups—Centrales and Pennales, based on different symmetry and mode of sexual reproduction. Generally speaking, centric and pennate diatoms are often associated with planktonic and benthic communities, respectively. While sometimes useful, this generalization is perhaps beset with more exceptions than most distributional rules.[83] The significant number of fossil diatoms is also known—the earliest types being marine species dating from the Cretaceous.[3682] Freshwater type species appeared in the Oligocene.[424] As a group of organisms there is a greater percentage of extant species of diatoms present in older sediments than other groups of organisms.[5499]

Many planktonic diatoms have regular annual fluctuations in growth that can be attributed to environmental conditions, especially temperature, the growth maximum being mostly in the spring. Once the concentration of dissolved silica (SiO_2) drops to approximately 0.5 mg.l^{-1} most planktonic diatoms cease to grow because of silica limitation. Attached diatoms in standing waters have good growth in spring, as do planktonic diatoms, but do not show as marked a decrease in growth when the concentration of silica in water reaches 0.5 mg.l^{-1} or less.[2745]

The diatom cells are surrounded by a rigid two-part boxlike cell wall composed of silica (SiO_2), called the *frustule*, consisting of two almost equal halves (*valves*), the smaller (*hypotheca*) fitting into the larger (*epitheca*) like a box. Each theca is composed of a more or less flattened plate, with a connecting band attached to the edge of the valve. The two connecting bands are called the *girdle*. The valves of some diatoms may have an opening of fissure running along the apical axis called *raphe*, which is associated with unique gliding motility. The frustules of some diatoms bear special processes or projections, e.g. the *labiate process*, which is an opening through the valve.[424,2745,4643]

Some diatoms are able to glide over the surface of a substrate, leaving a mucilaginous trail. Gliding is restricted to those pennate diatoms with raphe and those centric diatoms with labiate processes.[2745] Jarosh[2314] defined gliding as the active movement of an organism in contact with a solid substratum where there is neither a visible organ responsible for the movement nor a distinct change in the shape of the organism.

Some diatom cells form thick, ornamented walls at different times in their life cycle and become resting spores.[2745] Resting spores constitute the principal method for most benthic and neritic centric diatoms to persist during unfavorable conditions of growth. The protoplasm of the parental frustule darkens and is contracted, and the resting spore develops a very heavy siliceous wall of its own and usually sinks to the bottom. Typically, they consist of only two valves and lack a girdle.[424]

A conspicuous component of diatom growth in both planktonic and sessile forms is the production of extracellular polymeric substances in the form of stalks, tubes, apical pads, adhering films, fibrils and cell coatings. These mucilage structures serve as a variety of functions and their secretion may play a crucial part of their biological success.[1168,2120] (For details on diatom extracellular polymers see an excellent review by Hoagland et al.[2120])

Vegetative reproduction, which predominates, is normally by binary fission, the new valves and girdle form within the parent cell; this process results in a steady decline in cell size in a percentage of the population over several generations.[4643] Cell size is restored following sexual reproduction.[424] Sexual reproduction is isogamous, anisogamous and oogamous.[4643]

Sexuality in diatoms seems to be closely associated with the cell size such that in some species only cells of a size less than a critical level can undergo sexuality.[1140,1594] The product of sexual fusion is the auxospore (zygote), which is not a resting stage but characteristically increases in volume immediately after the gametes have fused.[424]

After the death of diatom cells, the frustules usually dissolve, but under certain circumstances they remain intact and accumulate at the bottom of any water body where the diatom occurs.[4594] Deposits of fossil diatoms, known as diatomaceous earth, are found in various parts of the world (see also Chapter 2.14.8). Stratigraphic analyses of deposits may indicate past environmental events in water bodies.[4203]

2.18.9 Dinophyceae

---

common name: dinoflagellates
habitat: freshwater, brackish, marine, terrestrial; examples shown in Figure 2-52
ecology: planktonic (majority), attached

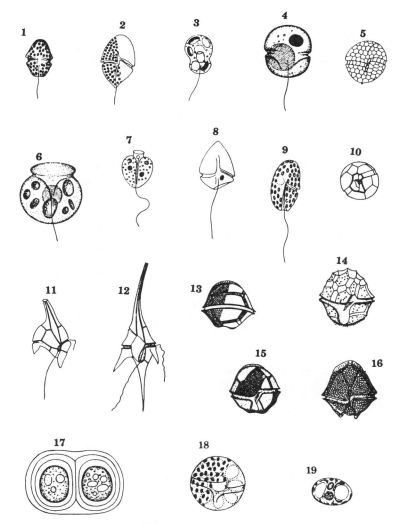

Figure 2-52.   Freshwater Dinophyceae.[2098] 1-*Gyrodinium aeruginosum* (29–36 × 21–28 μm); 2-*G. fuscum* (50–102 × 35–69 μm); 3-*G. bohemicum* (10–16 × 8–10 μm); 4-*G. eurytopum* (18–22 × 16–18 μm); 5-*Woloszynskia neglecta* (30–45 × 25–30 μm); 6-*Amphidinium elenkinii* (10–15 × 10–15 μm); 7-*A. lacustre* (23 × 15 μm); 8-*Katodinium piscinale* (35 × 28 μm); 9-*Heminidium nasutum* (16–35 × 12–22 μm); 10-*Glenodinium palustre;* 11-*Ceratium cornutum* (95–150 × 48–85 μm); 12-*C. hirundinella* (95–450 μm long); 13-*Peridinium willei* (40–83 × 45–70 μm); 14-*P. palatinum* (30–55 × 28–48 μm); 15-*P. palustre* (36–62 × 31–45 μm); 16-*P. biceps* (40–80 × 35–60 μm); 17-*Gloeodinium montanum* (16–60 μm in ø); 18-*Hypnodinium sphaericum* (71–122 μm in ø); 19-*Phytodinium simplex* (42–50 × 30–45 μm). Not to scale. (From Hindák et al., 1975, in *Determination Key for Lower Plants,* SPN Bratislava, reprinted with permission of F. Hindák.)

levels of organization: unicellular (coccoid, rhizopodal, flagellate), colonial (aggragation), filamentous (branched)
size: microscopic
cell covering: naked or cellulosic theca, body scales (rare), mucilaginous substances
flagella: 2, lateral, 1 trailing, 1 girdling, delicate hairs
number of thylakoids per band: 3 (1-3)
number of membranes of CER: 1
number of membranes surrounding the chloroplast: 3

chlorophylls: chlorophyll-*a*, chlorophyll-*c* (c_2)
carotenes: β-carotene
xanthophylls: diadinoxanthin, diatoxanthin, dinoxanthin, fucoxanthin, neoperidinin, peridinin
phycobiliproteins: none
eyespot: various, some highly complex
storage products: starch, oils, outside the chloroplast
number of genera: 100,[55] 120,[5091] 130[1004]
number of species: 230 (30 freshwater),[5091] 900,[55] 1,200 (93% marine)[1004]
Literature: References 424, 898, 1105, 1110–13, 2346, 2888, 3536, 3844, 4287, 4635–6, 4716, 4850, 4915.

The Dinophyceae are commonly known as dinoflagellates—other botanical names include Pyrrhophyceae and Pyrrophyceae. The Dinophyceae are usually biflagellate unicellular organisms, which constitute an important component of marine, brackish and freshwater habitats.[424] They are mainly planktonic; however, some inhabit sand or mud or are symbiotic with other organisms.[1113]

The Dinophyceae are important members of the plankton communities in both fresh and marine waters (although a much greater variety of forms is found in marine members) and as such the photosynthetic species are important primary producers and chemosynthesizers at the base of the food chain.[1004,1113,2745]

Besides photosynthesis, heterotrophic nutrition is well developed in this class—saprophytic, parasitic, symbiotic, and holozoic patterns all being represented.[424,1142] Parasitism is a well-developed mode of nutrition in the Dinophyceae. Both endoparasites (a parasite living within the tissue of its host) and ectoparasites (a parasite living on the outside of its host) are known.[424] Many of the photoautotrophic species, particularly those that are marine, are auxotrophic for various vitamins.[2266] The cell covering, multilayered theca, may consist of two major plates or as many as 100.[424]

Several factors are undoubtedly important in enabling dinoflagellates to grow under varied environmental and nutrient conditions, e.g. vertical migration, competitive exclusion by metabolic rates, utilization of organics, cellular N and P pools, and other adaptive variables.[1113] Pigments are undoubtedly involved in the ability of certain dinoflagellates to adapt to changing light conditions[3883,3885] and they may also be important in the pronounced phototactic response shown by many species.[1453-4]

Dinoflagellates are widely recognized to produce "blooms" or "red tides" in which the concentrations of cells may be so great as to color the ocean locally red, reddish brown, or yellow. Patches up to several square kilometers in extent may be discolored.[2162,3864,4850,4912] Red tides usually last longer than do blooms of nontoxic dinoflagellates because red tides are avoided by macrozooplankton that normally feed on dinoflagellates.[1343]

Some dinoflagellate blooms are associated with the production of toxins,[1882,2346,4316,4318,4715,5008] resulting in fish kills and mortality of other marine organisms; yet some dinoflagellate blooms produce no toxic effects.[1137,5454] This type of poisoning is known by various names: mussel poisoning, paralytic shellfish poisoning, or saxitoxin poisoning.[1825]

Steidenger[4714] distinguished three categories of dinoflagellate toxic blooms: (1) blooms that kill fish but few invertebrates; (2) blooms that kill primarily invertebrates; and (3) blooms that kill few marine organisms but the toxins are concentrated within siphons or digestive glands of filter-feeding bivalve molluscans (clams, mussels, oysters, scallops, etc.), causing paralytic shellfish poisoning (PSP). The most notorious PSP-causing dinoflagellate on the Pacific coast is *Gonyaulax catanella,* its poison being called saxitoxin, a neurotoxin 100,000 times more potent than cocaine.[1295,4715] Interestingly, mussels may become too toxic for human consumption when concentrations of *Gonyaulax catanella* reach only 100–200 cells per mL, but concentrations of at least 20–30,000 cells per mL must be reached before a bloom is apparent.[4320] A number of factors have been suggested as the cause of red tides:[2745] high surface water temperature, wind, light intensity, and nutrients. For further information about red tides and dinoflagellate toxins see also Chapter 2.7.1.2.

The Dinophyceae are the main contributors to marine bioluminescence. The compound responsible for this phenomenon is luciferin, which is oxidized with the aid of the enzyme luciferase, resulting in the emission of light.[2745]

A wide variety of marine invertebrates, including sponges, jellyfish, sea anemones, corals, gastropods, and turbellarians, and some protistans, including ciliates, radiolarians, and foraminiferans, harbor within them golden spherical cells termed zooxanthellae.[424] The translocation of photosynthate from zooxanthellae to the host organism has been studied intensively.[4904-5,4907-8,5183] Values of more than 60% of the carbon fixed in photosynthesis being transferred from the alga to the host have been measured.[4905,5183] (For further information see also Chapter 2.7.6.3.)

The most common mode of reproduction in dinoflagellates is simply by binary fission.[4643] Sexual reproduction is rare (isogamy, anisogamy). Many dinoflagellates are capable of forming encysted stages, which are resting cells formed in response to unfavorable conditions.[4055,4643]

The Dinophyceae have a number of different projectiles, which are fired out of the cell when it is irritated, resulting in a sudden movement of the cell in the opposite direction from the discharge.[2745] Many distinctive cytoplasmic structures are present among dinoflagellates (pusule, ocellus, trichocysts, nematocysts or cnidocysts, peduncle, muciferous bodies). For detailed explanation see the specialized literature.[424,480,1106,3305,3453,5307]

2.18.10 Phaeophyceae

common name: brown algae
habitat: freshwater, brackish, marine (majority, about 99%); examples shown in Figure 2-53
ecology: attached (benthic), free-floating (rare)
levels of organization: filamentous (branched, heterotrichous), pseudoparenchymatous (uniaxal, multiaxal, parenchymatous)
size: microscopic to macroscopic (majority, up to more than 60 m in kelps)
cell covering: two-layered, the inner cellulose, the outer gelatinous layer of pectic material containing alginic acid and fucoidan (sulfated mucopolysaccharides)
flagella: 2 lateral, unequal, longer anterior hairy (pantonematic), shorter posterior smooth (acronematic)
member of thylakoids per band: 3 (2–6)
number of membranes of CER: 2
number of membranes surrounding the chloroplast: 4
chlorophylls: chlorophyll-a, chlorophyll-c ($c_1 + c_2$)
carotenes: α-carotene, β-carotene
xanthophylls: antheraxanthin, diatoxanthin, diadinoxanthin, flavoxanthin, fucoxanthin, lutein, mutatochrome, violaxanthin, zeaxanthin
phycobiliproteins: none
eyespot: in the chloroplast
storage products: laminarin, mannitol, oils, outside the chloroplast
number of genera: 230,[55] 240,[5091] 250,[424] 265[4643,5508–10]
number of species: 1,500[55,1004,2095,2491,4643,5091]
Literature: References 8, 487, 751–3, 816–7, 1016, 2711, 2745, 2789–90, 2885, 2976, 3241, 3334–5, 3423, 3465, 3725, 3750, 4388, 4921–2, 5039, 5041, 5112, 5508–11.

The brown algae are almost exclusively marine in the distribution, freshwater representatives are rare.[485,1004,1121,1616,2095,3936,4344,4692,5426,5508–10] A number of marine forms penetrate into brackish waters, where they often form an important part of the salt marsh flora.[2745] These brackish water plants have almost totally lost the ability to reproduce sexually, and propagate by vegetative means only. Most of the Phaeophyceae grow in the intertidal belt and the upper littoral region.[2745]

Brown algae form a conspicuous intertidal component of temperate rocky shores in both the northern and southern hemisphere. They extend their occurrence from the upper littoral zone into the sublittoral zone, sometimes to depths of 220 m in clear tropical waters.[424] The Phaeophyceae exhibit the greatest diversity in regard to species and morphological expressions in colder temperate to subpolar ocean waters.

Brown algae range in size from microscopic epiphytic branched filaments to the largest of marine plants which exhibit the highest degree of anatomical differentiation in algae (e.g. the giant kelps of the order Laminariales, namely genera *Nereocystis, Pelagophycus, Macrocystis* (see Figure 2-53, #3) which reach up to more than 50 meters in length).[424,4643,5509] A full range of intermediate forms between these two size extremes is represented by the brown algae. No unicellular brown algae are known.[5509] Importance of the larger benthic plants in the marine environment is evident, as they provide habitats (kelp forests, rockweed beds) and food for herbivores.[1004]

Growth occurs by various methods: 1) diffuse growth (generalized, not localized, growth); 2) trichothallic growth (cell division is localized at the base of one or several filaments); and 3) apical growth (with a single cell, groups of apical cells, or marginal apical cells cutting off segments proximately).[424]

Although most members of this class are benthic algae, pelagic species of *Sargassum* have been recognized for centuries, and a free-floating acad of *Ascophyllum nodosum* is present in enclosed habitats

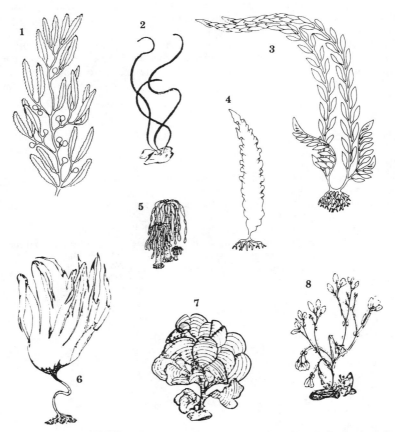

Figure 2-53.   Phaeophyceae.[973,2095] 1-*Sargassum* sp. (ca. 20 cm); 2-*Chorda tomentosa* (up to 4 m); 3-*Macrocystis* sp. (up to > 60 m); 4-*Laminaria* sp. (several m); 5-*Postelsia* sp. (ca. 40 cm); 6-*Laminaria digitata* (up to 2 m); 7-*Padina vickersiae* (ca. 15 cm); 8-*Fucus vesiculosus* (up to 1 m). Not to scale. (1, 3–5 from Darley, 1982, in *Algal Biology: A Physiological Approach,* drawings by Andrew J. Lampkin III, reprinted with permission of Blackwell Scientific Publications and the author; 2, 6–8 after Hillson, *Seaweeds,* University Park: The Pennsylvania State University Press, 1977, pp. 79, 101, 107, 119. Copyright 1977 by The Pennsylvania State University, reproduced with permission of the publisher.)

of the North Atlantic.[789,4642] Most species are attached to the rocks but a few genera are attached to other algae. *Sargassum natans* is a free-floating species growing very abundantly over a large area of the Atlantic Ocean known as Sargasso Sea.[5509]

Asexual reproduction is principally by means of motile zoospores, although a variety of vegetative structures are involved in some species.[4643] Sexual reproduction ranges from isogamous to anisogamous and oogamous.

Kelps are industrially used by man for many purposes (alginates, medical uses, fertilizers, food) (for details see Chapter 2.14).

2.18.11 Raphidophyceae (Chloromonadophyceae)

common name: chloromonads (South and Whittick[4643] pointed out that the name "chloromonads" or Chloromonadophyceae is unacceptable as it is based on the genus *Chloromonas,* which is not a raphidophyte)

habitat: freshwater (majority), brackish, marine; examples shown in Figure 2-54

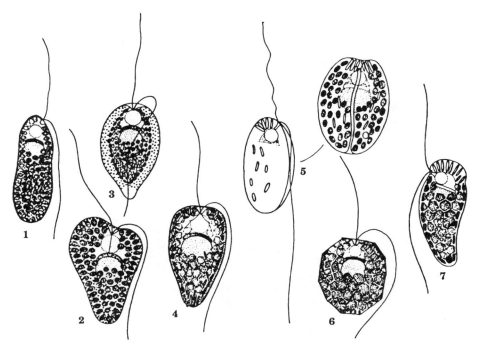

Figure 2-54.   Freshwater Raphidophyceae.[2098]   1-*Vacuolaria virescens* (50–85 × 20–25 *μ*m); 2-*V. viridis* (42–60 × 31–39 *μ*m); 3-*V. penardii* (40–60 × 15–34 *μ*m); 4-*Gonyostomum semen* (40–60(100) × 23–60 *μ*m); 5-*G. octavum* (30–36 × 19–26 *μ*m); 6-*G. latum* (35–40 × 9–13 *μ*m); 7-*Merotrichia bacillata* (40–50 × 9–13 *μ*m). Not to scale. (From Hindák et al., 1975, in *Determination Key for Lower Plants,* SPN Bratislava, reprinted with permission of F. Hindák.)

---

ecology: planktonic, metaphyton
levels of organization: unicellular (flagellate), colonial (aggregations)
size: microscopic
cell covering: naked, soft periplast?
flagella: 2 unequal, anterior longer hairy (pantonematic), posterior shorter smooth (acronematic))
number of thylakoids per band: 3
number of membranes of CER: 2
number of membranes surrounding the chloroplast: 4
chlorophylls: chlorophyll-*a*, chlorophyll-*c* ($c_1 + c_2$)
carotenes: β-carotene
xanthophylls: antheraxanthin, diadinoxanthin, dinoxanthin, epoxide, fucoxanthin, fucoxanthiniol, heteroxanthin, lutein, violaxanthin
storage products: lipid granules, oil, outside the chloroplast
number of species: 50 (50% marine)[1004]
Literature: References 365, 1457, 1461, 2052–9, 2059, 2098, 2551, 3217.

This rather small assemblage of unicellular flagellates has in the past been variously classified as group of uncertain position.[1511,4592]   With the Xanthophyxeae (with which they share similar pigments and flagellation)[1004,4643] in a separate division,[4309] with the Cryptophyceae, or with the Dinophyceae.

There is considerable controversy as to which genera belong to this class.[1632,2057-8]   Systematics on the level of species have also proved to be problematic.[4552,4653]

In freshwaters, the Raphidophyceae are associated with or immediately above the bottom mud in lakes, ponds and bogs with abundant macrophytes[2745,3824,4653,5037] and where the pH is acidic (Drouet and Cohen[1165] found them even at pH < 4.5) or neutral[2058] but they can be found also in alkaline conditions.[1786]   The

Raphidophyceae are common in the marine environment, where they make up a component of algal bloom in red tides in the coastal waters.[3535]

The shape of cells ranges from almost spherical to elongate.[2058] Contractile vacuoles (water expulsion vesicles) are a conspicuous feature of freshwater Raphidophyceae. Reproduction is by longitudal cell division. Fusion has been reported but the process of sexual reproduction is poorly known.[1004]

2.18.12 Cryptophyceae

common name: cryptomonads
habitat: freshwater, brackish, marine; examples shown in Figure 2-55
ecology: planktonic
levels of organization: unicellular (coccoid, flagellate), colonial (aggregations)
size: microscopic
cell covering: periplast
flagella: 2 lateral or apical, equal, hairy
number of thylakoids per band: 2
number of membranes of CER: 2
number of membranes surrounding the chloroplast: 2
chloropylls: chlorophyll-a, chlorophyll-c (c_2)
carotenes: α-carotene, β-carotene, ε-carotene
xanthophylls: <u>alloxanthin</u>, crocoxanthin, diatoxanthin, monadoxanthin, zeaxanthin
phycobiliproteins: 3 spectral types of phycoerythrin and 3 spectral types of phycocyanin
eyespot: in the chloroplast
storage products: starch, oils, between chloroplast envelope and CER
number of genera: 24[1004]
number of species: 100,[2491] 200 (60% marine)[1004]
Literature: References 486, 1110, 1314, 1558–9, 1638, 1978, 2075, 3491, 3493, 4691.

The class Cryptophyceae contains a relatively small group of biflagellate organisms, the cryptomonads, whose asymmetric cells are flattened dosriventrally and bounded by a periplast.[424] The Cryptophyceae occur in both marine and freshwater environments.[2745,3493] The cell body has a dorsiventral shape, with the cells flattened in one plane.[2745] Reproduction is by binary fission, sexual reproduction is not known.

Most forms are photosynthetic, but saprophytic nutrition also occurs.[424,2491] Some of the autotrophic species have been demonstrated to be auxotrophic for various vitamins.[424] Some genera are holozoic.[2491] A few members are heterotrophic.[2745] At least one species is known to live as an endosymbiont within a marine ciliate, the complex called *Mesodinium rubrum*.[2063,3492,4916]

Photosynthetic cryptomonads possess a unique combination of light-harvesting pigments: chlorophylls-a and -c_2, phycocyanin and phycoerythrin.[1983,3517,4219] Peculiar to the cryptomonads, however, is the presence of only one phycobiliprotein in any one species; unlike the red and blue-green algae the cryptomonad biliproteins are not contained in phycobilisomes.[2086] In comparison with other algal groups, the Cryptophyceae appear to be especially light sensitive, often forming the deepest living populations in clear oligotrophic water bodies.[3398]

The Cryptophyceae and other flagellates are present in the water column throughout the winter even below the ice of water bodies. Survival at these extreme low levels depends not only on a highly efficient photosynthetic system but also on slow rates of cell respiration at low water temperatures and reduced winter zooplankton grazing. In the spring after snow and ice melting, cryptomonads suffer from light stress and biomass maximum moves to deeper waters.[2410]

2.18.13 Euglenophyceae

common name: euglenoids
habitat: freshwater, brackish, marine, terrestrial; examples shown in Figure 2-56
ecology: planktonic, metaphyton (creeping)
levels of organization: unicellular (flagellate), colonial (aggregation)
size: microscopic

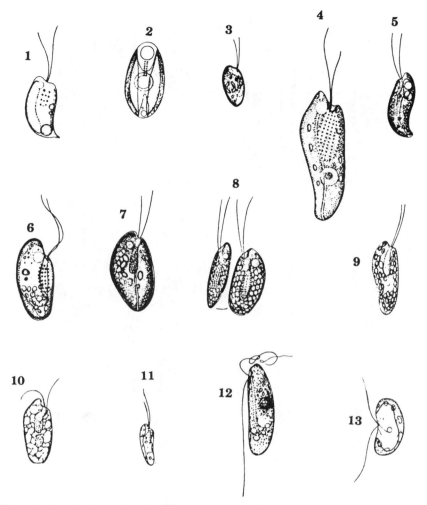

Figure 2-55. Freshwater Cryptophyceae.[2098] 1-*Rhodomonas pusilla* (7–12 × 4–7 μm); 2-*R. rubra* (13–20 × 8–10 μm); 3-*Chroomonas nordstedtii* (12.1–16 × 6.3–8.5 μm); 4-*Cryptomonas curvata* (40–50 × 10–26 μm); 5-*C. reflexa* (26–46 × 11–19 μm); 6-*C. obovata* (26–46 × 13–19 μm); 7-*C. erosa* (13–45 × 6–26 μm); 8-*C. ovata* (14–60 × 6–20 μm); 9-*Chilomonas paramecium* (18–40 × 7–15 μm); 10-*C. oblonga* (20–50 μm long); 11-*Cryptochrysis commutata* (15–19 × 7–10 μm); 12-*Katablepharis phoenikoston* (40 × 10 μm); 13-*Sennia commutata* (23 × 15 μm). Not to scale. (From Hindák et al., 1975, in *Determination Key for Lower Plants*, SPN Bratislava, reprinted with permission of F. Hindák.)

cell covering: proteinaceous pellicle

flagella: 1, 2, or 3 (up to 7), slightly subapical, usually 2, unequal, hairy, 1 often not emergent (pantonematic)

number of thylakoids per band: 3 (sometimes more, up to 12)

number of membranes of CER: 1 (additional CER membr.)

number of membranes surrounding the chloroplast: 3

chlorophylls: chlorophyll-*a,* chlorophyll-*b*

carotenes: β-carotene, γ-carotene

xanthophylls: antherexanthin, astaxanthin ester, canthaxanthin (=euglenanone), β-cryptoxanthin, diadinoxanthin, diatoxanthin, dinoxanthin, echinenone, 3-hydroxyechinenone, lutein, neoxanthin, zeaxanthin

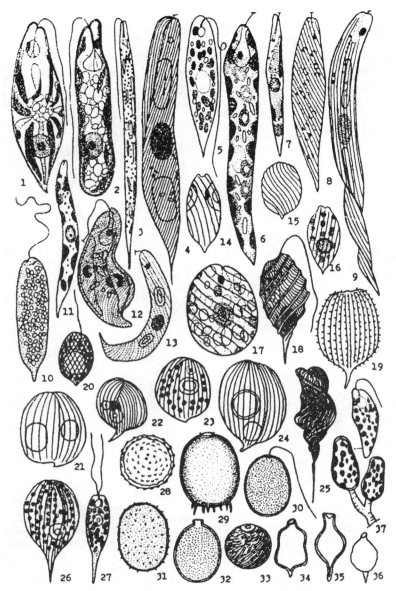

Figure 2-56.   Freshwater Euglenophyceae.[4559]   1-*Euglena viridis* (40–90 × 12–18 μm); 2-*E. gracilis* (35–67 × 6–14(20) μm); 3-*E. acus* (60–311 × 6–29 μm); 4-*E. oxyuris* (106–190 × 17–23 μm); 5-*E. proxima* (60–80 × 15–25 μm); 6-*E. deses* (83–161 × 10–25 μm); 7-*E. mutabilis* ((40)60–121 × 5–8 μm); 8-*E. granulata* (72–112 × ± 27 μm); 9-*E. tripteris* var. *major* (75–205 × 14–22 μm); 10-*E. rubra* (76–200 × 22–60 μm); 11-*E. intermedia* (80–140 × 8-18 μm); 12-*E. pisciformis* ((15)24–35 × 7–11 μm); 13-*E. spirogyra* (80–130 × 10–15 μm); 14-*Lepocinclis cylindrica;* 15-*L. ovum* (21–43 × 13–24 μm); 16-*L. steinii* (21–33 × 8–14 μm); 17-*L. texta* (35–60 × 25–45 μm); 18-*Monomorphina pyrum* v. *costata* (30–55 × 15–21 μm); 19-*Phacus hispidus;* 20-*P. parvulus* (16–30 × 8–12 μm); 21-*P. curvicauda* (20–35 × 18–25 μm); 22-*P. tortuosus;* 23-*P. acuminatus* (25–30 × 8–27 μm); 24-*P. pleuronectes* (16–30 × 8–12 μm); 25-*P. longicauda* v. *tortus* (85–90 × 40–70 μm); 26-*P. caudatus* (31–50 × 15–25 μm); 27-*Eutreptia viridis* (49–90 × 13–16 μm); 28-*Trachelomonas verrucosa* (18–34 μm in ø); 29-*T. armata* (30–59 × 25–30 μm); 30-*T. ovata* (60–64 × 34–42 μm); 31-*T. hispida* (23–36 × 19–25 μm); 32-*T. planctonica* (19–30 × 17–22 μm); 33-*T. regulosa* (14–23 μm in ø); 34-*Strombomonas acuminata* (50–59 × 25–30 μm); 35-*S. fluviatilis* (23–38 × 8–18 μm); 36-*S. planctonica;* 37-*Colacium vesiculosum* (16–35 × 9.4–21 μm).   Not to scale.   (From Sládeček et al., 1973, in *Technical Hydrobiology: A Laboratory Manual,* SNTL Praha, reprinted with permission of V. Sládeček.)

phycobiliproteins: none
eyespot: outside the chloroplast
storage products: paramylon, oil, outside the chloroplast
number of genera: 25,[55] 40,[4643] 45[5091]
number of species: 400,[55,1004] 800,[2747] 930,[5091] 1,000[4643]
Literature: References 611–2, 1109, 2285, 2747–52, 3259, 3439, 4192, 4333, 4349, 4551, 5246, 5481.

Euglenoids occur in most freshwater habitats, particularly in waters contaminated by animal pollution, or decaying organic matter.[611,973] Purer waters have sparse populations of less common euglenoids as planktonic organisms.[2745] Although euglenoids are more common in freshwater environments, some species are found in estuarine and intertidal zones.[1004] Euglenoids are widely distributed also on moist soils and mud. Certain genera of euglenoids have the capacity to encyst and thus to withstand unfavorable environmental conditions.[424] Reproduction is solely by cell division (longitudal furrowing); sexual reproduction is unknown.[4643] The pellicle may be flexible, resulting in the characteristic amoeboid "euglenoid" movement.[1004,3221,4643]

As a group, euglenoids are perhaps the most animal-like of the algae. Many species lack chloroplasts and are obligate heterotrophs, either osmotrophic or phagotrophic. Many species are incapable of utilizing NO_3^- as a nitrogen source.[973]

Lee[2745] pointed out that the Euglenophyceae have a number of modes of nutrition, depending on the species involved. No euglenoid has yet been demonstrated to be fully photoautotrophic—capable of living on a medium devoid of all organic compounds (including vitamins) with CO_2 as a carbon source, nitrates or ammonium salts as a nitrogen source, and light as an energy source. All green euglenoid flagellates so far studied are photoauxotrophic—capable of growing in a medium devoid of organic nutrition, with CO_2, ammonium salts, and light, but needing at least one vitamin (e.g. B_{12}).[973,2265,2745,4643] Some of the green species are facultatively heterotrophic.[2747] None of the green euglenoids is phagotrophic, but some of the colorless ones are.[424]

The euglenoids belong to the acetate flagellates, having the ability to grow photosynthetically in the light or heterotrophically in the dark. In either state, the fixed carbon is used as a source of energy or as building blocks for cell constituents. The substrates that can be used for photosynthetic growth vary from one species to another. The two most commonly used substrates are acetate and ethanol.[870,2745]

Except when they are encysted or in palmelloid phase, euglenoids are flagellate, having two or several flagella. When there are two, one may be nonemergent from the anterior invagination, which consists of a canal and a reservoir.[424] The chlorophyllous members of the Euglenophyceae share with the Chlorophyta the presence of chlorophylls-a and b in the chloroplast.

2.18.14 Chlorophyceae

---

common name: green algae
habitat: freshwater, brackish, marine, terrestrial; examples shown in Figures 2-57, 2-58, 2-59 and 2-60
ecology: planktonic, attached
levels of organization: unicellular (coccoid, flagellate, rhizopodal), colonial (aggregations, coenobial), filamentous (simple, branched, heterotrichous), pseudoparenchymatous (uniaxal, multiaxal, parenchymatous, siphonocladous, siphonous)
size: microscopic to macroscopic
cell covering: some naked, mostly complete cell wall, cellulose, hydroxyproline, glucosides, xylans and mannans, some calcified
flagella: normally 2 (to many), smooth (acronematic), equal, apical, anteriorly inserted
number of thylakoids per band: 2 to many, grana present
number of membranes of CER: CER absent
number of membranes surrounding the chloroplast: 2
chlorophylls: chlorophyll-a, chlorophyll-b
carotenes: α-carotene, β-carotene, γ-carotene
xanthophylls: adonirubin-ester, adonixanthin-ester, antheraxanthin, astaxanthin-ester, cantaxanthin ester, crustaxanthin, echinenone ester, heteroxanthin, lutein, loroxanthin, neoxanthin, pyrenoxanthin, siphonaxanthin, siphonein, violaxanthin, zeaxanthin
phycobiliproteins: none

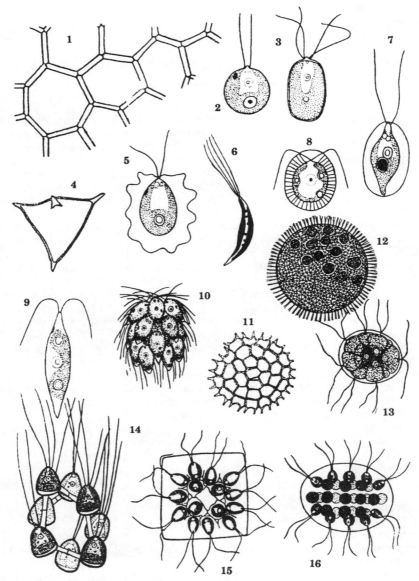

Figure 2-57.   Freshwater Chlorophyceae.[4559] 1-Hydrodictyon reticulatum (cells 1.5 cm, coenobia up to 20 cm); 2-*Chlamydomonas simplex* (cells 12–21 μm in ø); 3-*Carteria klebsii* (cells 15–22 × 7–15 μm); 4-*Tetraëdron regulare* (cells 14–54 μm in ø); 5-*Lobomonas stellata* (cells (6)–13–22 × 9–18 μm); 6-*Spermatozopsis exsultans* (cells 7–9 μm long); 7-*Sphaerellopsis fluviatilis* (cells 14–30 × 10–20 μm); 8-*Haematococcus pluvialis* (cells 63 × 51 μm); 9-*Chlorogonium elongatum* (cells 20–45 × 4–7 μm); 10-*Spondylomorum quaternarium* (16 cells in coenobium, cells 10–26 × 8–15 μm, coenobia up to 50 μm); 11-*Pediastrum boryanum* (8–64 cells in coenobium, outside cells 6–39 × 4–35 μm, inside cells 4–26 × 4–35 μm); 12-*Volvox aureus* (500–1,500 cells in coenobium, cells 5–9 μm in ø, coenobia 500–850 μm in ø); 13-*Pandorina morum* (16 cells in coenobium, cells 9–17 μm in ø, coenobia up to 250 μm in ø); 14-*Chlorcorone bohemica* (8 cells in coenobium, cells 7–12 × 5–9 μm, coenobia up to 50 μm in ø); 15-*Gonium pectorale* (16 cells in coenobium, cells 5–14 × 10 μm, coenobium 70–90 μm in ø); 16-*Eudorina elegans* (32 cells in coenobium, cells (8)–18–24 μm long, coenobium 60–200 μm). Not to scale. (From Sládeček et al., 1973, in *Technical Hydrobiology: A Laboratory Manual,* SNTL Praha, reprinted with permission of V. Sládeček.)

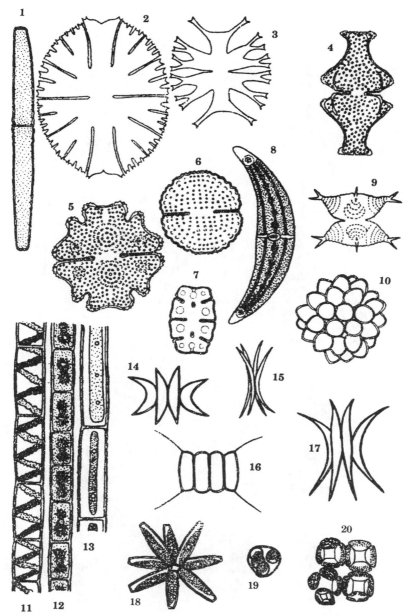

Figure 2-58.   Freshwater Chlorophyceae.[4559]   1-*Pleurotaenium trabecula* (cells 25–50 × 260–660 μm); 2-*Micrasterias rotata* (cells 200–366 × 165–300 μm); 3-*Micrasterias radiata;* 4-*Euastrum insigne* (cells 80–144 × 43–76 μm); 5-*Euastrum verrucosum* (cells 70–120 × 60–103 μm); 6-*Cosmarium formosulum* (cells 40–50 × 34–40 μm); 7-*Euastrum truncatum;* 8-*Closterium ehrenbergii* (cells 240–600 × 45–170 μm); 9-*Staurastrum commutatum;* 10-*Coelastrum microporum* (cells (6)–16–18(27) μm in ø); 11-*Spirogyra porticalis* (up to > 50 cm); 12-*Zygnema stellinum* (cells 27–30 μm wide, filaments up to > 20 cm); 13-*Mougeotia genuflexa* (cells 30–40 μm wide, filaments up to > 20 cm); 14-*Scenedesmus acutus* f. *dimorphus* (cells 7–35 × 2.1–8 μm); 15-*Ankistrodesmus falcatus* (cells 28–80 × 1.2–4.3 μm); 16-*Scenedesmus quadricauda* (cells 6–36 × 2.5–12 μm); 17-*Scenedesmus acuminatus* (cells 10–45(50) × 2.5–7 μm); 18-*Actinastrum hantzschii* (cells 10–35 × 3–6 μm); 19-*Chlorella vulgaris* (cells 2–10 μm in ø); 20-*Crucigenia fenestrata* (coenobia 8–16 μm wide). Not to scale. (From Sládeček et al., 1973, in *Technical Hydrobiolgy: A Laboratory Manual,* SNTL Praha, reprinted with permission of V. Sládeček.)

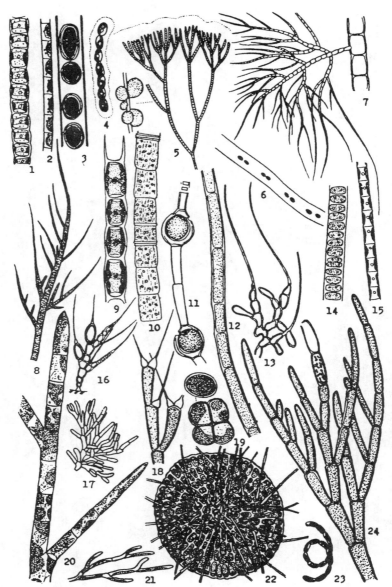

Figure 2-59.   Filamentous freshwater Chlorophyceae.[4559] 1-*Ulothrix zonata;* 2-*Chlorhormidium rivulare;* 3-*Cylindrocapsa involuta;* 4-*Geminella minor;* 5-*Chaetophora elegans;* 6-*Geminella interrupta;* 7-*Draparnaldia glomerata;* 8-*Stigeoclonium flagelliferum;* 9-*Microspora amoena;* 10-*Oedogonium capillare;* 11-*O. crispum;* 12-*Rhizoclonium hieroglyphicum;* 13-*Chaetonema irregulare;* 14-*Ulothrix tenuissima;* 15-*U. tenerrima;* 16-*Bulbochaete nana;* 17-*Microthamnion kützingianum;* 18-*Bulbochaete intermedia;* 19-*Pleurococcus viridis;* 20-*Stigeoclonium* sp.; 21-*Microthamnion strictissimum;* 22-*Coleochaete scutata;* 23-*Gloeotila contorta;* 24-*Cladophora glomerata.* Species macroscopic, some, e.g. *Cladophora* or *Rhizoclonium* up to several meters in length. Not to scale. (From Sládeček et al., 1973, in *Technical Hydrobiology: A Laboratory Manual,* SNTL Praha, reprinted with permission of V. Sládeček.)

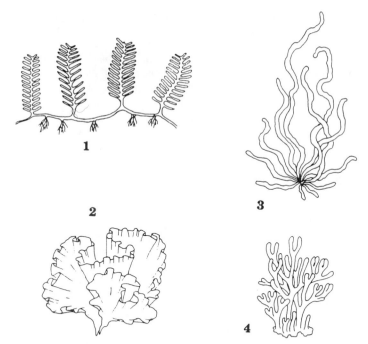

Figure 2-60.   Marine Chlorophyceae.[973]   1-*Caulerpa* sp. (ca. 15 cm, stolons up to 12 per m²); 2-*Ulva* sp. (ca. 30 cm); 3-*Enteromorpha* sp. (ca. 30 cm); 4-*Codium* sp. (ca. 20 cm). (From Darley, 1982, in *Algal Biology: A Physiological Approach,* drawings by Andrew J. Lampkin III, reprinted with permission of Blackwell Scientific Publications and the author.)

eyespot: in chloroplast
storage products: starch (amylose, amylopectin), oil, inside the chloroplast
number of genera: 425,[55] 560[4497]
number of species: 6,500,[55] 7,000 (13% marine),[2095] 8,600,[4497] 20,000[3195]
Literature: References 422, 487, 572, 1280, 1283–5, 1287, 2150, 2402, 2297, 2547, 2560–1, 2901, 2599, 2774, 3191–2, 3195, 3258–9, 3327–8, 3423, 3780, 3789–91, 3931, 4753–4, 4921–2, 4694, 5047, 5111.

The Chlorophyceae comprise one of the major classes of algae, considering the abundance of species and genera and their frequency of occurrence. The Chlorophyceae "sensu lato" include about 560 genera and 8600 species according to Silva,[4497] although probably more than 20,000 species according to Melkonian and Ichimura.[3195] The Chlorophyceae are the dominant class of the division Chlorophyta.

The Chlorophyceae is an extremely large class of algae. Green algae occur in almost any habitat including soil, snow and ice. The Chlorophyceae grow in waters of a great range of salinity and nutrient levels.[424] They also occur in various forms of symbiosis (e.g. with fungi, protozoa, marine invertebrates) but few are entirely parasitic.[3195]

A complete range of morphological types occur in the Chlorophyceae,[424] ranging from naked or walled flagellates, colonial or coenobic organisms, coccoid, sarcinoid or filamentous organisms to parenchymatous or coenocytic organisms. Macroscopic forms include tubular threads, penicillate tufts, stalked discs or branched strands.[3195] The cells are for the most part are uninucleate, but the multi-nucleate (coenocytic) condition characterizes several orders (e.g. Caulerpales) and occur among certain genera of Chlorococcales (e.g. *Hydrodictyon*). Also a wide range of reproduction patterns (vegetative, asexual, sexual-isogamy, anisogamy, oogamy) and life histories occurs.

In many green algae the protoplast fills the cell, while in some, a large, central, aqueous vacuole is present within it. The motile cells of green algae contain contractile vacuoles in their colorless cytoplasm. These serve an osmoregulatory function. Contractile vacuoles are absent in marine species.[424]

Hillis[2093] pointed out that calcareous species of the genus *Halimeda* play an important role in reef construction, and offers support for calling reefs algae rather than coral.

2.18.15 Charophyceae

common name: stoneworts, brittleworts
habitat: freshwater (majority), brackish; examples shown in Figure 2-61
ecology: attached (benthic)
levels of organization: filamentous (uniaxal)
size: macroscopic (up to more than 50 cm)
cell covering: complete cell wall (stages naked), cellulose, many calcified
flagella: male gametes only, 2, equal scaly, anteriorly inserted
number of thylakoids per band: 2–6 or more, grana present
number of membranes of CER: CER absent
number of membranes surrounding the chloroplast: 2
chlorophylls: chlorophyll-*a*, chlorophyll-*b*
carotenes: α-carotene, β-carotene, γ-carotene, lycopene
xanthophylls: astaxanthin, heteroxanthin, lutein, neoxanthin, violaxanthin, zeaxanthin
phycobiliproteins: none
storage products: starch (amylose and amylopectin), inside the chloroplast
number of genera: 7[960,1797]
number of species: 315[5494]
Literature: References 414, 888, 1443–7, 1511, 1790, 1796–7, 2284, 2699, 2700, 2792, 3024, 3257, 3628, 3786–7, 3789, 3919–21, 4158, 4193, 4753, 4883, 4978, 5052, 5174, 5461, 5492–5.

The Charophyceae, commonly known as stoneworts or brittleworts, are a small but unique group of nonvascular hydrophytes with worldwide distribution. The Charophyceae are erect, in still waters, or bend with the current in running waters, and may attain a length of 50 cm or more.[424] They often form high-density "charophyte meadows." They favor oligotrophic calcareous waters to a depth of 15 m[4650,4810] disappearing from water bodies when they become eutrophic.[2768] The Charophyceae are primarily freshwater organisms, although a few species may occur in brackish waters.[4643] One genera, *Lamprothamnium,* is exceptional in that it can survive highly saline conditions, although it requires brackish water to complete its life cycle.[1027]

Many are heavily calcified, with concentrations of plants on the bottom leading to the formation of maerl ($CaCO_3$ and $MgCO_3$ deposits).[2745] (For description of calcification see Chapter 5.1.4.5).

Forsberg[1444] found that the growth rate of axenic *Chara globularis* in the laboratory was reduced at high phosphorus concentrations. Because he also observed luxuriant growth of *Chara* sp. in lakes with total phosphorus concentrations usually lower than 20 $\mu g.L^{-1}$,[1446] he concluded that *Chara* spp. are physiologically sensitive to high phosphorus concentrations. Characeans are often the first submerged macrophytes that disappear during eutrophication.[2695] Melzer et al.[3196] confirmed Forsberg's observation that characeans do not occur at phosphate concentrations greater that 20 $\mu g.l^{-1}$ but the authors assumed light limitation caused by dense growth of phytoplankton in water bodies with P concentrations greater than 20 $\mu g.L^{-1}$ to be responsible for characean disappearance during eutrophication. Blindow[397] found that *Chara* spp. was not inhibited by high P concentration suggesting that the decline of characeans during eutrophication is caused by factors other than phorphorus toxicity.

The macroscopic haploid thallus of charophytes is commonly anchored to a sandy or muddy substrate by well developed rhizoids. The thallus attains its macroscopic size through unlimited apical growth of the main axis. A precise series of transverse divisions in the minute dome-shaped apical cell results in a highly ordered, geometrically-precise thallus in which cylindrical internodal cells of considerable size (may be corticated and even more than 10 cm)[4643] alternate regularly with compact parenchymatous nodal complexes. Lateral branchlets of limited growth arise from these nodes.[1797]

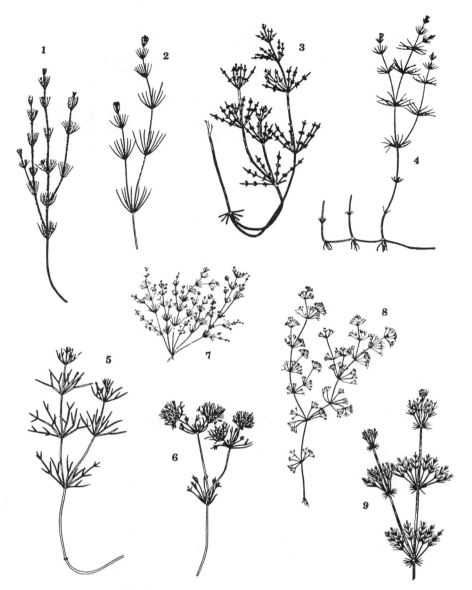

Figure 2-61. Charophyceae.[2098]   1-*Chara hispida* (40–70 cm); 2-*Chara fragilis* (up to 30 cm); 3-*Chara tomentosa* (25–60 cm); 4-*Chara vulgaris* (up to 50 cm); 5-*Nitellopsis obtusa* (= *Tolypellopsis stelligera*) (up to 1 m); 6-*Tolypella intricata* (10–40 cm); 7-*Nitella flexilis* (40–70 cm); 8-*Nitella mucronata* (up to 30 cm); 9-*Lynchothamnus barbatus* (20–30 cm). Not to scale. (From Hindák et al., 1975, in *Determination Key for Lower Plants*, SPN Bratislava, reprinted with permission of F. Hindák.)

The sex organs of the Charophyta are characteristic and unique in the plant kingdom.[424]   Charophytes are also unique among the green algae in that the reproductive cell (termed the oosphere before fertilization and oospore after fertilization) is completely ensheathed by vegetative cells.[424,2768,3310] The reproduction bodies are enclosed within a sterile envelope (otherwise unique in the algae) and the sperm.[4643]   The reproductive organs—antheridia and oogonia—are a large structural complex[3789] and appear in various arrangements along the lateral branchlets or branchlet axils. Reproduction can occur via self-fertilization among bisexual forms in which case the zygotes are genetically identical to each other and to the parent. Allogamy takes place in both unisexual and bisexual forms. Parthenogenesis occurs only rarely.[1797] Asexual zoospores are not formed by charophytes.[2745,4643] Vegetative reproduction occurs through fragmentation and growth of the thallus (especially the nodal complex) and via bulbils. Bulbils are white starchy reproductive structures formed on rhizoids of some charophytes which can often overwinter and initiate protonemal growth when conditions become suitable.[1797]

The presence of chlorophylls-a and b and storage starch in their plastids and their totally aquatic habitat have led some botanists to assign them, as a class or order, to the Chlorophyta. Others, recognizing a significant evolutionary divergence of charophytes from these green algae implied by their complex vegetative and reproductive structures, sperm morphology, and possession of the protonematal stage, have classified them as a separate division, the Charophyta.[424,1797]

2.18.16 Prasinophyceae

habitat: freshwater, brackish, marine; examples shown in Figure 2-62
ecology: planktonic
morphology: unicellular (coccoid, flagellate), colonial (aggregations)
size: microscopic
cell covering: naked, usually scaly (scales organic)
flagella: 1 or 2 unequal, or 4 or more or less equal, scaly, anteriorly inserted
number of thylakoids per band: 2–6, grana present(?)
number of membranes of CER: CER absent
number of membranes surrounding the chloroplast: 2
chlorophylls: chlorophyll-a, chlorophyll-b, chlorophyll-c
carotenes: α-carotene, β-carotene
xanthophylls: antheraxanthin, astaxanthin, lutein, neoxanthin, siphonein, siphonoxanthin, violaxanthin, zeaxanthin
phycobiliproteins: none
eyespot: inside the plastid
storage products: starch, mannitol, inside chloroplast
number of genera: 8–10[3310,3457]
Literature: References 1283, 1286, 2552, 3025, 3051, 3193, 3261–2, 3454, 3456, 3458–9, 4125–6, 4755.

A class of coccoid or flagellae unicells that originally was placed in the order Volvocales of the Chlorophyceae. Most Prasinophyceae are free-living, marine, brackish or freshwater organisms.[3456] Some are symbionts[517,656,1736,2597] (see also Chapter 2.7.6.3). Much remains to be determined regarding the fine structure of the members of this class; at present the taxonomy is tentative, as is the status of the Prasinophyceae.[3653,4643] Asexual reproduction may involve zoospore-forming stages or simple division; sexual reproduction is unknown.[3356,4643]

Unification of these green algae within a class, Prasinophyceae, is defined because of the strong evidence presented by Norris[3454] that the evolutionary series is present among these flagellates.[3356] Derivation of this class, the most primitive of the Chlorophyta, from evolutionary lines in the flagellate chromophytes is suggested by structural and biochemical criteria.[3356]

2.18.17 Glaucophyta

The Glaucophyta include those algae that have endosymbiotic blue-green algae in the cytoplasm instead of chloroplast. Because of the nature of their symbiotic association, they are thought to represent intermediates in the evolution of the chloroplast.[2745] Pascher[3667] coined terms for this association: he called the endosymbiotic blue-green algae cyanelles, the host, a cyanome; and the association between the two, a syncyanosis.

Most of the cyanelles in the Glaucophyta lack a wall and are surrounded by two membranes—the old food vesicle membrane of the cyanome and the plasma membrane of the cyanelle. As evolution progressed, these two membranes became the chloroplast envelope, the various inclusions of the blue-

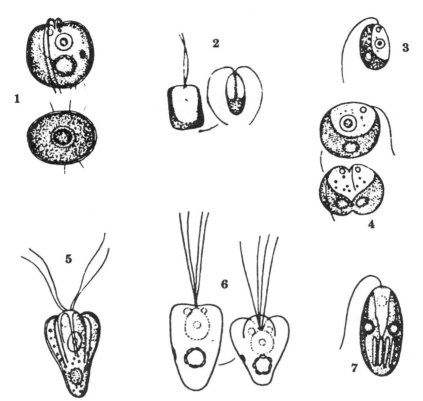

Figure 2-62.   Freshwater Prasinophyceae.[2098]   1-*Tetraselmis* (= *Platymonas*) *cordiformis* (cells 17–21 × 15–19 μm); 2-*Scourfieldia quadrata* (cells 8–10 × 6 μm); 3-*Pedinomonas minor* (cells 4–5 μm long); 4-*Pedinomonas rotunda* (cells 10 μm long); 5-*Pyramimonas tetrarhynchus* (cells 20–28 × 12–18 μm); 6-*Pyramimonas montana* (cells 17–22.5 × ca. 8–12 μm); 7-*Monomastix opistostigma* (cells 6–22 × 3–6 μm). Not to scale. (From Hindák et al., 1975, in *Determination Key for Lower Plants,* SPN Bratislava, reprinted with permission of F. Hindák.)

green algae were lost (polyhedral bodies, cyanophicine granules, polyphosphate bodies), the cyanome cytoplasm took over formation of the storage product (except in the Cryptophyceae), and a pyrenoid was differentiated from proteinaceous bodies in the cyanelle. This endosymbiotic line of evolution probably lead initially to the red algae.[2745]

Skuja[4548] assigned a number of flagellates containing endosymbiotic blue-green algae to the division Glaucophyta (excluding the more recently recognized dinoflagellates containing endosymbiotic blue-green algae "phaeosomes").[4907]

The blue-green component (cyanelle) contains chlorophyll-*a* and phycobiliproteins (as in the Cyanophyceae), however, two of the blue-green algae carotenoids, myxoxanthin and echinenone, are absent.[756]

The extant species include those with naked cyanome (e.g. *Cyanophora*) and their derivatives, those surrounded by a wall (e.g. *Glaucocystis, Gloeochaete*).[1597] South and Whittick[4643] pointed out that the status of this group remains questionable, although Kies[2498] has suggested it forms a natural assemblage with some affinity with the green algae. This conclusion is sustained in the findings of Møestrup,[3260] who provides a summary based on flagellar structure. The Glaucophyta (sensu Møestrup[3260]) could be grouped together with the euglenoids and chlorophytes. Bold and Wynne[424] classified members of this group (*Cyanophora, Glaucocystis*) under "algae of uncertain affinity."

Literature: References 2497, 3218, 3788.

CHAPTER 3

Wetlands are transitional environments.[1278,1750-1,1890,3034,3562] In a spatial context, they lie between dry land and open water—at the coast, around inland lakes and rivers, or as mires draped across the landscape. In an ecological context, wetlands are intermediate between terrestrial and aquatic ecosystems. In a temporal context, most are destined either to evolve into dry land as a result of lowered water tables, sedimentation and plant succession, or to be submerged by rising water-tables associated with relative sea-level rise or climatic change.[3562] Wetlands often form part of a large continuum of community type, and therefore it is difficult to set boundaries.[1751] As a result, in any definition the upper and the lower limits of wetlands excursion are arbitrary boundaries. Consequently, few definitions adequately describe all wetlands. The problem of definition usually arises on the edges of wetlands, toward either wetter or drier conditions.[3246] Because land and water can merge in many ways, it can be frustrating to attempt to define wetlands or determine where wetlands begin or end strictly on the basis of wetness or dryness.[1890]

The term "wetland" is a relatively new one to describe the landscape that many people knew before under different names. Because wetlands occur in many different climate zones, in many different locations and have many different soil and sediment characteristics, they have become an integral part of the landscape since the earliest times, and sometimes of the economy as well. Consequently they have been given various names which add considerably to our confusion.[5440]

Marshes, swamps, and bogs have been well-known terms for centuries, but only relatively recently attempts have been made to group these landscape units under the single term "wetlands." This general term has grown out of a need to understand and describe the characteristics and values of all types of land, and to wisely and effectively manage wetland ecosystems. There is no single, correct, indisputable, ecologically sound definition for wetlands, primarily because of the diversity of wetlands and because the demarcation between dry and wet environments lies along a continuum.[893]

In general terms, wetlands are lands where saturation with the water is the dominant factor determining the nature of soil development and the types of plant and animal communities living in the soil and on its surface. The single feature that most wetlands share is soil or substrate that is at least periodically saturated with or covered by water. The water creates severe physiological problems for all plants and animals except those that are adapted for life in water or in saturated soil.[893]

Wetlands mean different things to different people. To some they are stretches of waterlogged wastelands that harbor pests and disease, and should therefore be drained or filled in order to make better use of the area. To others, wetlands are beautiful serene landscape, usually associated with open waters, and serve as important habitats for waterfowl, fish and other wildlife. The main difficulty in recognizing the true value of wetlands arises from its definition. An all-inclusive definition treating wetlands as "areas of submerged or water-saturated lands, both natural and man-made, permanent or temporary, freshwater or marine," recognizes peat bogs, grass and sedge marshes, swamps, mangroves, tidal marshes, flood plains, shallow ponds and littoral areas of larger water bodies, and paddy fields as wetlands.[1750]

3.1 CLASSIFICATIONS AND DEFINITIONS OF WETLANDS

The definition and classification of wetlands has gone through many stages and a precise wetland definition still has not been developed. Lefor and Kennard[2757] showed that different definitions used for inland wetlands can result from the geologist, soil scientist, biologist, system ecologist, sociologist, economist, political scientist, public health scientist, and lawyer.

Classification of wetlands is fraught with controversy and problems, partly because of the enormous variety of wetland types and their highly dynamic character, and partly because of difficulties in defining their boundaries with any precision. Where, for example, does a wetland end and a deepwater aquatic habitat start? For how long and how intensively does an area have to be flooded, or in any other way saturated with water, for it to be a wetland rather than a terrestrial ecosystem? There are no universally accepted or scientifically precise answers to these questions. The difficulties are compounded by changes over time by which some wetlands may evolve through various stages to become dryland areas.[3034]

Several definition and classification systems have been devised for differing needs and purposes. Most tend to skirt the how-wet-is-wet question by identifying wetlands in terms of soil characteristics and the types of plants these transitional habitats typically support, since shallow standing water or saturated soil soon cause severe problems for all plants except hydrophytes, which are specifically adapted for these conditions. Obviously, wetlands are not continuously dry land. On the other hand, they need not be continuously wet. Many types of wetlands are wet only after heavy rains or during one season of the year.[1890]

Wetland definitions and terms are many and are often confusing. Nevertheless, they are important for both the scientific understanding of these systems and for their proper management. Wetland definitions then, often include three main components:[3246]

1. Wetlands are distinguished by the presence of water.
2. Wetlands often have unique soils that differ from adjacent uplands.
3. Wetlands support vegetation adapted to the wet conditions (hydrophytes), and conversely are characterized by an absence of flooding-intolerant vegetation.

Waksman[5225] has found 90 English terms to describe peatlands, peats and peat-like qualities, and undoubtedly the list has grown since then. In the United States, few terms occur uniformly throughout the country in either scientific or vernacular usage. Unlike Europe, little distinction is made between peat and non-peaty wetlands, perhaps because peatlands comprise a smaller portion of wetlands in the 48 lower states.[5440] The term "marsh" is generally reserved for wetlands dominated by graminoids, grasses or herbs, and "swamp" for wetlands dominated by woody plants, as in the cypress swamps and river bottom hardwood swamps of the South. "Bog" is used to denote ombrogenous mires, but the distinction between all these are not always clear. "Fen" is rarely used, but there are some distinctive additional terms, such as wet meadow, wet prairie, glade and pothole in the northern Plains and pocosin in the Carolinas.[1757,2143,5440]

Gopal et al.[1751] pointed out that the multiplicity of terms used in different languages in different parts of the world for similar types of wetlands are cause for confusion. In English, the general term mire was proposed by Godwin[1671] for all wetlands accumulating significant amounts of peat, and this usage has been followed by Gore.[1757] Ecologically, mires cover several wetland types. The term swamp is sometimes used synonymously with wetland, although North American usage usually refers to wooded or forested wetlands (the palustrine shrub-scrub and forested types of the US FWS classification by Cowardin et al.[893]). The same applies for some terms used in other languages, e.g. "boloto" in Russian. Non-woody wetlands dominated by tall helophytes (e.g. *Phragmites australis, Cyperus papyrus*) and the palustrine emergent wetlands of Cowardin et al.,[893] are often called "swamps" in British usage (e.g. Mason and Standen[3097]). In Figures 3-1 and 3-2, schematic presentation of the SCOPE (Scientific Committee on Problems of the Environment) wetland categories and occurrence of the different SCOPE wetland types along the axes of water regime and nutrients are given.

Definitions vary markedly from marshes of glacial origin, to prairie wetlands, to freshwater marshes of estuarine areas that experience severe fluctuations in water movement under tidal influence. In glaciated regions, marshes are often remnant wetlands of shallow lake systems in which macrovegetation, largely emergent, extends over the entire water surface. Technically, swamp consists of persistent standing water among the vegetation, whereas marshes contain water-saturated sediment with no or little standing water among the vegetation.[5348-9] These definitions, however, are commonly useful only in detailed analyses of successional changes of wetland conditions and biota.[5348]

One of the earliest definitions of wetlands was proposed by Penfound.[3730] Penfound defines five general swamp types and four general marsh types in the southeastern United States. The classification system was relatively simple, dividing wetlands into salt and freshwater swamps or marshes. Penfound emphasized that soil texture is a basic factor in the local distribution of hydric plants, but he did not include soils as part of the classification scheme. Penfound's classification system is straightforward and informative but lacks specificity for regulatory and inventory usage. However, many classification systems developed since 1952 have been based on his system.[1248]

Another early wetland classification was presented by U.S. Fish and Wildlife Service in 1956[4429] in a publication that is frequently referred to as "Circular #39." The classification framework that

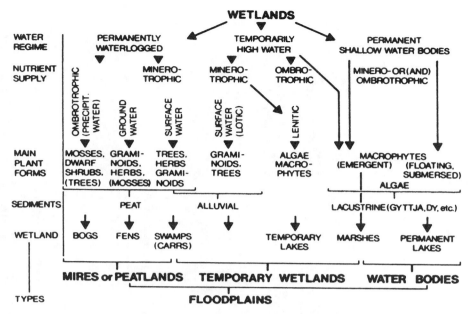

Figure 3-1.   Schematic presentation of the SCOPE wetland categories and their main parameters and features.[1751] (From Gopal et al., 1990, in *Wetlands and Shallow Continental Water Bodies,* Patten, B.C., Ed., p. 14, reprinted with permission of SPB Academic Publishing, The Hague.)

Circular #39 employs places 20 wetland types in one of four major wetland categories: inland fresh, inland saline, coastal fresh and coastal marine. The primary goals of Circular #39 were to delineate the wildlife value of wetlands and to provide a perspective for balanced land use planning.[1248] The term "wetlands" refers to lowlands covered with shallow and sometimes temporary or intermittent waters. They are referred to by such names as marshes, swamps, bogs, wet meadows, potholes, sloughs and river-overflow lands. Shallow lakes and ponds, usually with emergent vegetation as a conspicuous feature, are included in the definition, but the permanent waters of streams, reservoirs, and deep lakes are not included. Neither are water areas that are so temporary as to have little or no effect on the development of moist-soil vegetation.

The International Union for the Conservation of Nature and Natural Resources,[2294] treating wetlands as just "wet areas," adopted this definition: "Wetlands are areas of submerged or water saturated land, whether natural or artificial, permanent or temporary, whether the water is static or flowing, fresh, brackish or salt. Water dominated areas to be considered would include marshes, sloughs, bogs, swamps, fens, peatlands, estuaries, bays, sounds, ponds, lagoons, lakes, rivers, and reservoirs. Where marine or coastal waters are involved, waters up to the depth of 15 m are included."

Another wetland definition was created by Zoltai et al.[5567] for Canadian wetlands: Wetlands is land which has the water table at, near, or above the land surface, or which is saturated for long enough periods to promote wetland or aquatic processes as indicated by poorly drained soils, hydrophilic vegetation and various kinds of biological activity which are adapted to the wet environments. Wetlands include peatlands and areas that are influenced by excess water but which, for climatic, edaphic, or biotic reasons, produce little or no peat. Shallow open water, generally less than 2 m deep, is also included in wetlands.

The classification system developed by Goodwin and Niering[1737] considers only freshwater and inland wetlands. Wetlands were recognized as a site where the water table is near, at, or above the surface of the ground for at least some portion of the growing season. Floodplains are included in this classification as are lakes and ponds where they are ecologically related to specific wetland types. Marshes, swamps and bogs are considered to constitute the major types of wetlands, as in Penfound.[3730]

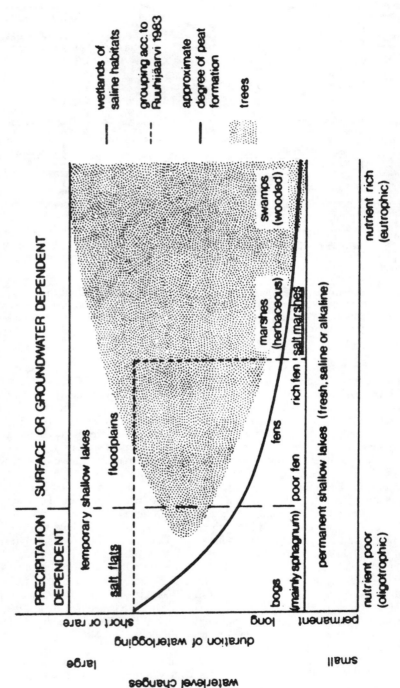

Figure 3-2. Occurrence of the different SCOPE wetland types along the axes of water regime and nutrients. The approximate distribution of peat and trees indicated.[1751] (From Gopal et al., 1990, in *Wetlands and Shallow Continental Water Bodies*, Patten, B.C., Ed., p. 14, reprinted with permission of SPB Academic Publishing, The Hague.)

Zoltai[5566] defined Canadian wetlands as ". . . areas where wet soils are prevalent, having a water table near or above the mineral soil for the most part of the thawed season, supporting a hydrophilic vegetation."

The Corps of Engineers Wetland Delineation Manual[1249] (see also Reference 5341) adopted the following definition: "Corps of Engineers[1321] and the Environmental Protection Agency[1320] jointly define wetlands as: "Those areas that are inundated or saturated by surface or ground water at a frequency and duration sufficient to support, and that under normal circumstances do support, a prevalence of vegetation typically adapted for life in saturated soil conditions. Wetlands generally include swamps, marshes, bogs and similar areas." According to these definitions, wetlands have the following general diagnostic environmental characteristics:

1. Vegetation. The prevalent vegetation consists of macrophytes that are typically adapted to areas having hydrologic and soil conditions described in the definition. Hydrophytic species, due to morphological, physiological, and/or reproductive adaptation(s) have the ability to grow, effectively compete, reproduce, and/or persist in anaerobic soil conditions.
2. Soil. Soils are present and have been classified as hydric, or they possess characteristics that are associated with reducing soil conditions.
3. Hydrology. The area is inundated either permanently or periodically at mean water depths of ≤ 6.6 ft (= 2 m), or the soil is saturated to the surface at some time during the growing season of the prevalent vegetation."

A biological and relatively narrowly conceived definition was adopted during the International Biological Program (IBP):[5339] "A wetland is an area dominated by specific herbaceous macrophytes, the production of which takes place predominantly in the aerial environment above the water level while the plants are supplied with amounts of water that would be excessive for most other higher plants bearing aerial shoots."

Probably the most comprehensive classification of wetlands has been developed for the U.S. Fish and Wildlife Service (US FWS) by Cowardin et al.[893] They define wetlands as lands transitional between terrestrial and aquatic systems where the water table is usually at or near the surface or the land is covered by shallow water. For purposes of this classification wetlands must have one or more of the following three attributes: (1) at least periodically, the land supports predominantly hydrophytes; (2) the substrate is predominantly undrained hydric soil; and (3) the substrate is nonsoil and is saturated with water or covered by shallow water at some time during the growing season of each year.

The term wetland includes a variety of areas that fall into one of five categories: (1) areas with hydrophytes and hydric soils, such as those commonly known as marshes, swamps, and bogs; (2) areas without hydrophytes but with hydric soils—for example flats where drastic fluctuations in water level, wave action, turbidity, or high concentration of salts may prevent the growth of hydrophytes; (3) areas with hydrophytes but nonhydric soils, such as margins of impoundments or excavations where hydrophytes have become established but hydric soils have not yet developed; (4) areas without soils but with hydrophytes such as the seaweed-covered portion of rocky shores; and (5) wetlands without soil and without hydrophytes, such as gravel beaches or rocky shores without vegetation.[893]

The US FWS classification[893] is based on a hierarchical approach analogous to taxonomic classification used to identify plant and animal species. Wetlands are grouped accordingly to ecologically similar characteristics into five ecological systems (Figure 3-3): 1) marine; 2) estuarine; 3) riverine; 4) lacustrine and 5) palustrine. All but the latter include deepwater habitats.

Marine System (Figure 3-4) consists of the open ocean overlying the continental shelf and its associated high-energy coastline. It is mostly a deepwater habitat, with marine wetlands limited to intertidal areas such as beaches, rocky shores and some coral reef.

Estuarine System (Figure 3-5) consists of deepwater tidal habitats and adjacent tidal wetlands that are usually semi-enclosed by land but have open, partly obstructed, or sporadic access to the open ocean, and in which ocean water is at least occasionally diluted by freshwater runoff from the land. The system includes coastal wetlands like salt and brackish tidal marshes, intertidal flats, mangrove swamps, bays, sounds and coastal rivers.

Riverine System (Figure 3-6) includes all wetlands and deepwater habitats contained within a channel, with two exceptions: (1) wetlands dominated by trees, shrubs, persistent emergents, emergent mosses, or lichens, and (2) habitats with water containing ocean-derived salts in excess of 0.05%. This system is restricted to lotic (flowing) freshwater river and stream channels.

Lacustrine System (Figure 3-7) includes wetlands and deepwater habitats with all of the following characteristics: (1) situated in a topographic depression or a dammed river channel; (2) lacking trees,

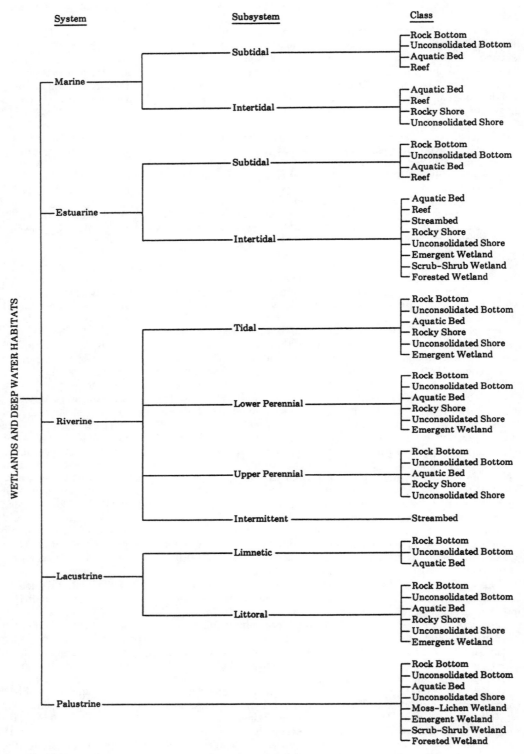

Figure 3-3.   Classification hierarchy of wetlands and deepwater habitats, showing subsystems, and classes. The Palustrine System does not include deepwater habitats.[893]

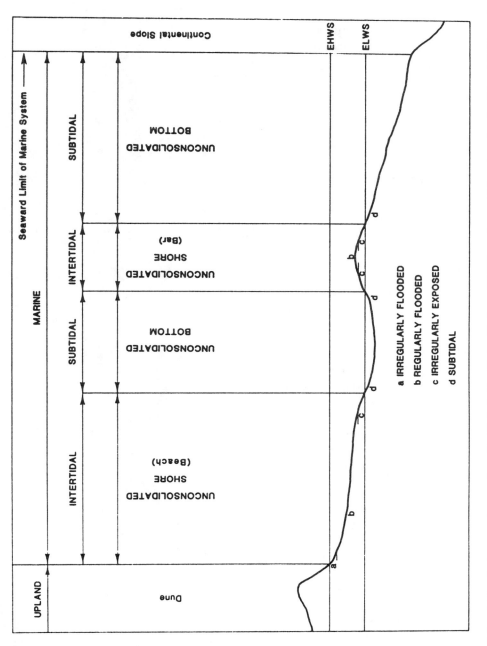

Figure 3-4. Distinguishing features and examples of habitats in the Marine System. EHWS = extreme high water of spring tides; ELWS = extreme low water of spring tides.[893]

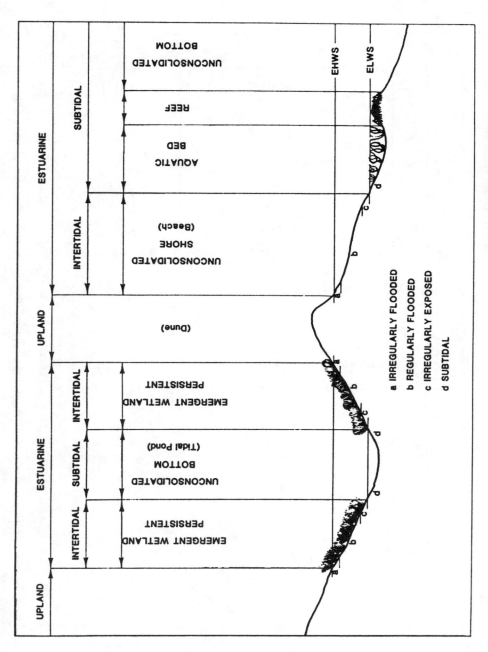

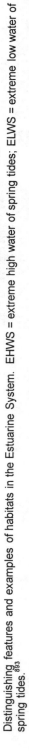

Figure 3-5.   Distinguishing features and examples of habitats in the Estuarine System. EHWS = extreme high water of spring tides; ELWS = extreme low water of spring tides.[893]

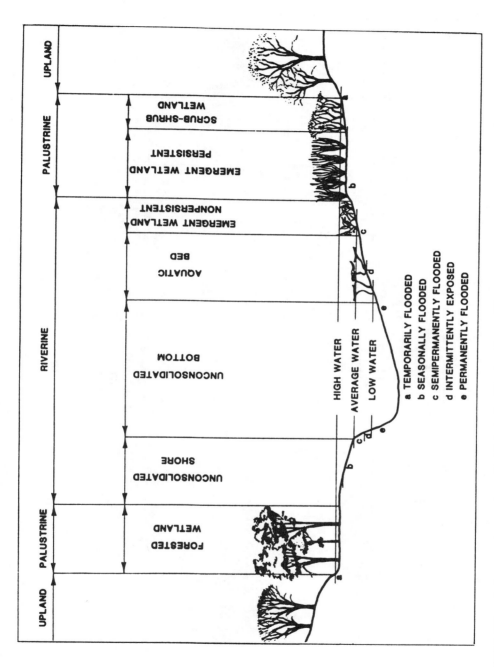

Figure 3-6.   Distinguishing features and examples of habitats in the Riverine System.[893]

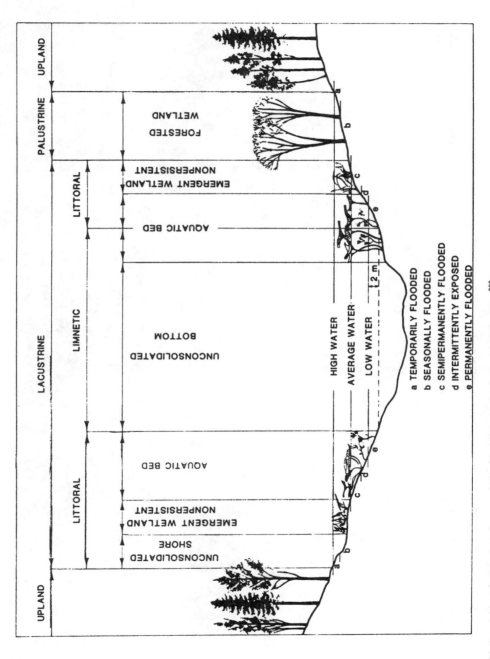

Figure 3-7.   Distinguishing features and examples of habitats in the Lacustrine System.[893]

shrubs, persistent emergents, emergent mosses or lichens with greater than 30% areal coverage; and (3) total area exceeds 8 ha (20 acres). Similar wetland and deepwater habitats totalling less than 8 ha are also included in the Lacustrine System if an active wave-formed or bedrock shoreline feature makes up all or part of the boundary, or if the water depth in the deepest part of the basin exceeds 2 m (6.6 feet) at low water. This system includes lentic (non-flowing) water-bodies such as lakes, reservoirs and deep ponds.

Palustrine System (Figure 3-8) includes all non-tidal wetlands dominated by trees, shrubs, persistent emergents, emergent mosses or lichens, and all such wetlands that occur in tidal areas where salinity due to ocean-derived salts is below 0.05%. It also includes wetlands lacking such vegetation, but with all of the following four characteristics: (1) area less than 8 ha; (2) active wave-formed or bedrock shoreline features lacking; (3) water depth in the deepest part of basin less than 2 m at low water; and (4) salinity due to ocean-derived salts less than 0.05%. This system excludes all deepwater habitats, and includes the majority of inland marshes, bogs, swamps, floodplains, etc.

Deepwater habitats are permanently flooded lands lying below the deepwater boundary of wetlands. Deepwater habitats include environments where surface water is permanent and often deep, so that water, rather than air, is the principal medium within which the dominant organisms live, whether or not they are attached to the substrate. As in wetlands, the dominant plants are hydrophytes; however, the substrates are considered nonsoil because the water is too deep to support emergent vegetation.[5090]

The boundary between wetland and deepwater habitat in the Marine and Estuarine Systems coincides with the elevation of the extreme low water of spring tide; permanently flooded areas are considered deepwater habitats in these systems. The boundary between wetland and deepwater habitat in the Riverine, Lacustrine and Palustrine Systems lies at the depth of 2 m (6.6 feet) below low water; however, if emergents, shrubs, or trees grow beyond this depth at any time, their deepwater edge is the boundary.[893] The 2 m lower limit for inland wetlands was selected because it represents the maximum depth to which emergent plants normally grow.[4384,5323,5427,5558] The 2 m limit is also used by Zoltai et al.[5567] in wetland classification for Canada and by the Corps of Engineering definition.[1249]

The structure of US FWS classification is hierarchical, progressing from systems and subsystems at the most general levels to classes, subclasses, and dominant types. Several subsystems as shown in Figure 3-3 give further definition to the systems. These include the following:

1. Subtidal—the substrate is continuously submerged
2. Intertidal—the substrate is exposed and flooded by tides; this include the splash zone
3. Tidal—for Riverine Systems, the gradient is low and water velocity fluctuates under tidal influence
4. Lower Perennial—Riverine Systems with continuous flow, low gradient, and no tidal influence
5. Upper Perennial—Riverine Systems with continuous flow, high gradient, and no tidal influence
6. Intermittent—Riverine Systems in which water does not flow for part of the year
7. Limnetic—all deepwater habitats
8. Littoral—wetland habitats of Lacustrine Systems which extend from shore to a depth of 2 m below low water or to the maximum extent of nonpersistent emergent plants.

The class of a particular wetland or deepwater habitat describes their general appearance of the ecosystem in terms of either the dominant vegetation or the substrate type. When over 30% cover by vegetation is present, a vegetation class is used (e.g. shrub-scrub wetland). When less than 30% of the substrate is covered by vegetation, then a substrate class is used (e.g. unconsolidated bottom). Further description of the wetlands and deepwater habitats is possible through the use of subclasses, dominance types, and modifiers.[893]

3.2 WETLAND TERMINOLOGY

If classifications and definitions of wetlands are sometimes confusing, the terminology used to describe various wetland types is even more confusing. I am not going to solve the problem of which terms are correct—I just want to show the variability in terminology which is based on different bases: hydrology, soil characteristic, vegetation or nutrient status. It will be shown that the same term could be applied for entirely different wetland type in different parts of the world. The discussion during the IVth International Wetland Conference in Columbus, Ohio, in 1992, clearly showed the need of unification of wetland type terminology.

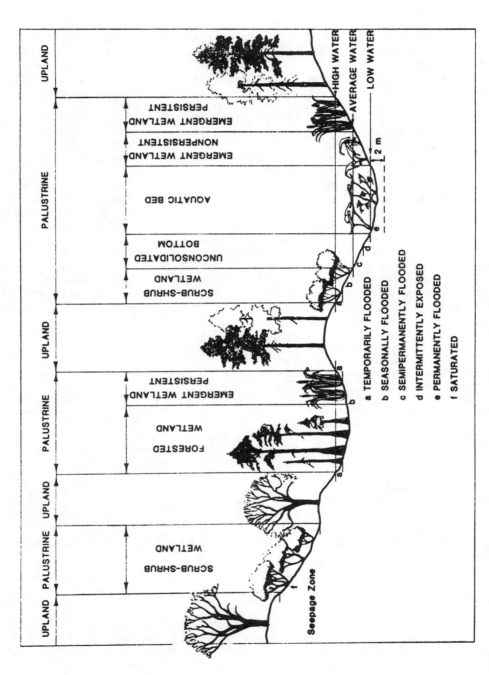

Figure 3-8. Distinguishing features and examples of habitats in the Palustrine System.[893]

3.2.1 Mire

Gore:[1757] The term mire includes all those ecosystems described in English usage such words as swamp, bog, fen, moor, muskeg and peatland. Mire is now an internationally accepted term which includes the generally ombrotrophic types, namely bog (Hochmoor, Weissmoor—in part—and Reisermoor—in part, German; mosse, Swedish) on one hand, and the minerotrophic types such as fen (Niedermoor, Flachmoor and Braunmoor, German; kärr, Swedish) and carr (swamp, North American; Sumpfwald or Bruchmoor, German; lövkärr, Swedish; korpi, Finnish) on the other. The literal translation of Hochmoor and Niedermoor, "high-moor" and "low-moor," are used in much of the older literature in Europe but they tend to be misleading and are falling into disuse.[1757] General words for mire from countries having land of the relevant type include "veen" (Dutch), "Moor" (German), "myr" (Swedish and Norwegian), "myri" (Icelandic), "suo" (Finnish), "boloto" (Russian), "soo" (Estonian), "tourbiere" (French) and "muskeg" (Indian word used in Canada).[1757]

Eurola et al.:[1288] Mire is any peatland or paludified vegetation together with its underlying peat.

Stanek and Worley:[4677] Mire is the term commonly used in the sense of peatland particularly in Europe, New Zealand, and the Soviet Union.

Mitsch and Gosselink:[3246] The term mire is synonymous with any peat-accumulating wetland (European definition).

Masing et al.[3091]: Mires may be defined as unbalanced ecosystems in which an excess of organic matter produced by plants is only partly decomposed and the residues deposited as organic soil, or peat. Mire development is closely related to climate and geomorphology.[3284] Since all mires by definition imply the presence of organic soils, the term has perhaps more pedological and economic significance than ecological meaning. Mires can be subdivided into peat forming bogs (ombrotrophic deriving nutrients from precipitation) and fens (minerotrophic, deriving nutrients from runoff and other non-precipitation sources), and non-peat forming or only moderately peat forming marshes (with vegetation limited to herbaceous forms) and carrs or swamps (with woody plant cover). Many transitional types of mires exist that are not readily classified, in addition to landscape mosaics of various mixtures of bogs, fens, marshes and carrs.

3.2.2 Moor

Moor is a German term meaning peatland. In English, the term moor is also used in the sense of peatland.[4677] A highmoor is a raised bog, while a lowermoor is a peatland in a basin or depression that is not elevated above its perimeter.[3246]

3.2.3 Marsh

Maltby:[3034] Marshes have a number of specific characteristics. They are naturally dominated by reeds, rushes, grasses and sedges. Marshes are sustained by water sources other than direct rainfall, and vary considerably in response to what are often no more than subtle hydrological and chemical differences. There are three major groups: freshwater, tidal salt and tidal freshwater marshes.

Gopal et al.[1751]: Marshes are wetlands dominated by herbaceous emergent vegetation, moderately or nonpeat forming, mainly meso- to eutrophic, and sometimes saline.

Gore[1757]: Marshes are non-peat forming, seasonally waterlogged areas.

Van der Toorn et al.:[5116] Freshwater marshes are defined as follows: vegetation systems consisting of emergent helophytes subject to shallow to deep, intermittent or permanent, freshwater inundation, and usually occurring on the edges of lakes or open water. The nutrient condition is often eutrophic, but varies from eutrophic to mesotrophic with peat formation moderate to absent. The vegetation is usually dominated by perennial monocotyledonous herbs belonging to the following genera: *Arundo, Calamagrostis, Carex, Cladium, Cyperus, Juncus, Phalaris, Phragmites, Scirpus, Typha* and *Zizania*. In this definition, mesotrophic marshes (more or less synonymous with the minerotrophic mores or "rich-fens" of Gore[1757]) are also included because they often occur in combination with eutrophic ones, e.g. zonal gradients in permanently waterlogged situations.

Stanek and Worley[4677] characterize marshes as grassy wet areas, periodically inundated up to a depth of 2 m or less with standing or slowly moving water. Surface water levels fluctuate seasonally, but water remains within the rooting zone of plants during at least part of the growing season. The substratum usually consists of mineral or organic soils with a high mineral content, but there is little peat accumulation. Waters are usually circumneutral to alkaline, and there is a relatively high oxygen saturation. Marshes characteristically show zonal or mosaic surface patterns of vegetation, comprised of unconsolidated grass and sedge sods, frequently interspersed with channels or pools of open water. Marshes may be bordered by peripheral bands of trees and shrubs, but the predominant vegetation consists of a variety of emergent nonwoody plants such as rushes, reeds, reedgrasses and sedges. Where open water areas occur, a variety of submerged and floating aquatic plants flourish.[2325] Zoltai et al.[5567] differentiated the following types of marshes: estuarine, coastal, fluvial, lentic, catchment, and seepage. Other marsh types which have been differentiated include active delta marsh, basin -, channel -, deep -, floodplain -, grass -, inactive delta -, kettle -, marine -, meadow -, seepage track -, shallow basin -, shore -, and streams marsh.

3.2.4 Swamp

Gopal et al.:[1751] Swamps (carrs) are minerotrophic, meso- to eutrophic systems, with or without peat, and with woody shrubs or trees as dominant vegetation.

Maltby:[3034] Swamps generally have saturated soils or are flooded for most, if not all, of the growing season. They are often dominated by a single emergent herb species, or are forested. American authors generally considered swamps to be frequently flooded wetlands which are dominated by woody plants.

Eurola et al.:[1288] Swamp is a mire-type wetland affected by surface water influence.

Gore:[1757] Swamp (Sumpf, German) and marsh imply eutrophic conditions but are less specific words in popular usage. The latter is often confined to wetlands with more or less mineral soils but, apart from the North American usage above, swamp, like wetland, is very widely used to include both mires and marshes.

Mitsch and Gosselink:[3246] Swamp is a wetland dominated by trees or shrubs (U.S. definition). In Europe a forested fen could be called a swamp. In some areas reed grass dominated wetlands are also called swamps.

Stanek and Worley:[4677] Swamp is wet, forested (or treed) minerotrophic peatland where standing to gently flowing waters occur seasonally or persist for long periods on the surface. The waters are circumneutral to moderate acid in reaction, and show little deficiency in oxygen or in mineral nutrients. The substrate consists of mixtures of transported material and organic sediments, or peat deposited in situ. The vegetation cover may consist of coniferous or deciduous trees, tall shrubs, herbs and mosses.[2325] Other swamp types which have been differentiated include: alder swamp, conifer -, flat -, floodplain -, hardwood -, shore -, seepage -, spring -, stream -, and thicket swamp.

Gopal and Masing:[1747] Swamps are recognized as wetlands with dominating shrubs and trees. They are fairly widely distributed in all climatic zones. Their soils may or may not have peat deposits. In areas with thick peat layers underneath the forest cover, swamps are often called forested fens or carrs—terms most commonly used in Europe.

Maltby:[3033] Swamps are flooded throughout most of the growing season and develop in still-water areas, around the margins, and in parts of floodplains. Varied terminology means that they include both forested systems such as the *Cypress* swamps of the United States, the *Melaleuca* forests of New Guinea, or the mangrove forests of the tropical coasts as well as herbaceous systems such as reed-swamps occurring in North America, Europe, Asia, Australia and South America and also the papyrus swamps which are so characteristic of many African wetlands.

3.2.5 Bog

Gopal et al.:[1751] Bogs are typically ombrotrophic, oligotrophic (nutrient-poor), and usually acidic mires, mostly dominated by species of the moss genus *Sphagnum*.

Orme:[3562] Bogs are ombrotrophic, oligotrophic, acid mineral poor, ombrogenous, mostly topogenous areas.

Stanek and Worley:[4677] Bogs are ombrotrophic, wet, nutrient poor, usually strongly acidic peatlands. Peat is usually formed *in situ* under closed drainage and oxygen saturation is very low. Although bogs are usually covered with *Sphagnum*, certain sedges may grow on them. Bogs may be treed or treeless, and they are frequently characterized by a layer of shrubs.[1190,2325,3985] Many landscape types have been differentiated: basin bog, blanket -, climbing -, floating -, maritime -, open -, palsa -, plateau -, quaking -, raised -, spruce -, string -, treed -, valley bog and many more. The term "bog" is still used to mean peatlands in general. The use of the term bog should be restricted to ombrotrophic or extremely oligotrophic peatlands.

Eurola et al.:[1288] Bog is a term used for ombrotrophic mires.

Mitsch and Gosselink:[3246] Bog is a peat-accumulating wetland that has no significant inflows or outflows and supports acidophilic mosses, particularly *Sphagnum*.

Gopal and Masing:[1747] Bogs are usually large expanses of peat covered areas with high acidity (pH below 4.0) and a surface carpet of mosses, chiefly species of *Sphagnum*, often covered by graminoids like *Eriophorum, Trichophorum* and other Cyperaceae, and ericoids, depending upon their developmental stage. They are also ombrotrophic.

Probably the most frequent bogs are blanket bog and raised bog. Blanket bog is formed where the peat covers wide tracts of terrain independent of the details in relief. In such cases, the peat may have expanded beyond the original confines of a lake or pond basin. Raised bog is formed where peat continue to grow upward. This has a distinctive dome-shaped form, and as a result its own peculiar hydrologic system. Eventually, the bog may grow to the point where it is no longer possible for a permanently high water table to be maintained.[3034] Eurola et al.[1288] defined raised bog as a mire complex in which the central part is ombrotrophic, but where only the periphery is minerotrophic and usually supports a supplementary nutrient effect vegetation, frequently characteristic of surface or ground water influence.

3.2.6 Fen

Mitsch and Gosselink:[3246] Fen is a peat-accumulating wetland that receives some drainage from surrounding mineral soil and usually supports marsh-like vegetation.

Eurola et al.:[1288] Fen is the term used for all minerotrophic mires. These may support oligo-, meso-, or eutrophic vegetation.

Gopal et al.:[1751] Fens are minerotrophic, meso- to eutrophic (nutrient rich) mires with predominantly herbaceous plant cover.

Orme:[3562] Fen (treeless fen, tree fen, carr swamp forest) is minerotrophic, eutrophic-mesotrophic, neutral-alkaline, mineral-rich, geogenous, topogenous or doligenous area.

Stanek and Worley:[4677] Fen is a meadow-like, often sedge-rich, peatland on minerotrophic sites, richer in nutrients and less acidic than bogs. In eutrophic fens, *Sphagnum* species are subordinate or absent, whereas *Campylum* spp., *Scorpidium* spp. and *Drepanocladus* species are abundant. Often there is a low shrub cover and sometimes a sparse layer of trees. Fens usually develop in restricted drainage situations where oxygen saturation is relatively low and mineral supply is restricted. Usually very slow internal drainage occurs through seepage down very low gradient slopes, although sheet surface flow may occur during spring melt or periods of heavy precipitation.[1190,2325,4532,4898,5225]

Some authors recognize eutrophic, mesotrophic and oligotrophic fens or rich, intermediate, and poor fens.[964,5430,4533] Several types of fen have been differentiated: basin fen, channel -, collapse -, emergent -, floating -, horizontal -, ladder -, lowland -, polygon -, palsa -, seepage -, shore -, slope -, spring -, stream -, string -, water track fen, and others. Fen should not be used in oligotrophic or ombrotrophic situations where the term bog is more appropriately applied.[4677]

Gopal and Masing:[1747] Fens are areas with slightly to moderately decomposed peat layer, less acidic or somewhat alkaline, with predominately herbaceous vegetation comprised of sedges, graminoids and reeds. Mosses are subordinate or nearly absent. The shrub layer, when developed, is sparse and trees are dwarfed, never forming a closed canopy. They develop in areas with impeded drainage and are supplied with surface or belowground water relatively rich in nutrients.

Fens are often divided into poor and rich, depending upon their nutrient status. Transition bogs are poor fens, and Tansley[4897] called them valley bogs, a term more common in British literature. In the United States, the term fen is rarely used (e.g. References 4374–5, 5117–8). In Africa, and Australia also, fens occur but are often described as sedge meadows or sedge marshes. Though fens can be distinguished from bogs, and poor fens from rich fens, it should be emphasized that all intermediate gradations occur and in bog margins almost a continuum of habitats can be observed. In fact, different mire types can be arranged along two gradients of water and nutrient status of the habitat, as done by Ruuhijärvi.[4231] However, the term fen, like mire, has been used more often in Europe.[1747]

Bogs and fens occur mainly in temperate regions of the world but can be found also in subtropical and tropical regions.[255,478,675,1755–6,1989,1998,2143,2433,3246,3283,3813,4231,4533,4536,4918,4962,5441] The classification of bogs and fens and their plant communities has been a subject of discussion for a long time.[1747] A number of schemes have been proposed for different regions.[522,2001,2325,2905,3089,3284,3576,4231,4533,4897,5068,5567]

3.2.7 Peatland

Mitsch and Gosselink:[3246] Peatland is a generic term of any wetland that accumulates partially decayed plant matter.

Stanek and Worley:[4677] Peatlands are areas having peat-forming vegetation on peat and includes peat originating from the vegetation. The specifications of a peatland vary. In European countries, the peat layer must be 30 cm thick when undrained or more than 20 cm thick when drained.[600,662,1191,2007,4677,5225,5297]

In the Northern Hemisphere, three categories of peatland can be distinguished.[3090,4531] The first type is the topogenic fen, which is supplied with water from its surroundings and represents the late stage of a shallow lake. The soligenic blanket bog, another type, develops in hill slopes with sufficient water supply from the surroundings. The third type, the ombrogenic raised bog, is supplied with water only from precipitation. These different peatland types depend on hydrologic conditions and nutrition, changes which can cause radical changes in the peatland.[2308]

Stanek and Worley[4677] summarized that peat is material constituting peatlands, exclusive of live plant cover, consisting largely of organic residues accumulated as a result of incomplete decomposition of dead plant constituents under conditions of excessive moisture (submergence in water and/or waterlogging). It (1) may contain a variable proportion of transported mineral materials; (2) may form in both base-poor and base-rich conditions, and either as autochthonous peat or allochtonous peat; and (3) may contain usually basal layers or coprogenic elements and comminuted plant remains (such as gyttja) or humus gels (such as dy). The physical and chemical properties of peat are influenced by the nature of plants from which it has originated, by the moisture relations during and following its formation and accumulation, by geomorphological position, and by climatic factors. The moisture content of peat is usually high, the maximum water-holding capacity occurs in *Sphagnum,* being over 10 times its dry mass and over 95% of its volume. Most peats have a high organic content (85% and more). However, in general, peat must have an organic matter content of not less than 30% of the dry mass (about 17% carbon content).[191,662,953,1999,2349,4676,5225,5297]

3.2.8 Shallow Lake

Shallow lakes are difficult to define but the most accepted definition covers all water bodies where the depth is so small in proportion to the area that it allows wind mixing of the whole water column. In general, this depth does not exceed a few meters, more often less than 2 meters. Shallow lakes may or may not support vegetation. Shallow lakes are characterized by mineral soils (or substrata without soils) with little or no peat accumulation. The soils may be slightly acidic to highly alkaline.[1747] Gopal et al.[1751] pointed out that shallow lakes are permanent or intermittent bodies of water, which may be saline, alkaline or fresh, where wind action or turbulent heat transfer (mainly in the tropics) permanently disturb the whole water mass causing high turbidity, unless the lake is densely vegetated.

3.2.9 Floodplain

Maltby:[3034] Floodplains are the flat lands bordering rivers that are subject to periodic flooding. They tend, naturally, to be most expansive along the lower reaches of rivers.

Gopal et al.:[1751] Floodplains are systems of wetlands associated with rivers or streams containing elements belonging to the previous categories fens, marshes, swamps and shallow lakes, and often comprising relatively large areas of riparian forests.

Junk and Welcome:[2386] The specific characteristics of floodplains that distinguishes them from other wetlands is their periodic inundation, which causes a change between a terrestrial and an aquatic phase. Whereas bogs, fens, mires, swamps, shallow lakes and other wetlands are permanently or semipermanently wet areas, large parts of floodplains are, during certain recurrent periods, dry land. Therefore, floodplains can be defined as "areas periodically flooded by freshwater." However, this definition seems too generalized as after heavy rainfall large areas of the earth's surface can be included. Therefore, additional criteria are necessary to characterize the origin of flooding and its biological consequences. The following definition is proposed:[2386] Floodplains are areas of low lying land that are subject to inundation by the lateral overflow of waters from rivers or lakes with which they are associated. The floods further bring about such changes in the physico-chemical environment that biota react by morphological, anatomical, physiological or ethological adaptations, or by change in community structure. The following terms have been differentiated:[5324] fringing floodplains, internal deltas, coastal deltaic floodplains.

Kozlowski:[2579] The term floodplain refers to land adjacent to streams and rivers which naturally cause periodic flooding.

Several terms like river bottom, bottomland, hardwood bottom and alluvial plains are synonymous with floodplains.[1747] Mitsch and Gosselink[3246] defined bottomland as lowland along streams and rivers, usually on alluvial floodplains that are periodically flooded. These are often forested and sometimes called bottomland hardwood forests.

Floodplains are thus wetlands subjected to only periodic flooding, but with large water level changes that are the key to the development of particular plant and animal communities.[2385-6] The floodplain vegetation is generally, though not necessarily, dominated by trees and shrubs. Therefore, it is often called an alluvial, riparian or bottomland forest. Floodplains, in general, represent a complex of diverse habitats such as marshes, swamps and shallow oxbow lakes interspersed on a periodically flooded landscape along river courses as pockets of permanently flooded areas are left behind after floods have receded.[1747]

3.2.10 Pocosin

Pocosins are expansive freshwater wetlands that are confined to the southeastern United States.[4107] Pocosins are palustrine wetlands of the Coastal Plains typically having an overstory of *Pinus serotica* and a dense understory of evergreen, ericaceous shrubs. The hydrology is nonalluvial, fed by rainwater or highly oligotrophic groundwater. Organic soils are the rule and at the pocosins salt water margin underlying mineral soil may actually be below the sea level. Mineral soil surface may also be present where fire has burned away the peat or muck layer. Although the topography appears to be extraordinary flat and featureless, pocosins are commonly slightly dome-shaped and sometimes have lakes in their higher central portions.[3730,4113,4677,4814,5295] Biogeochemistry of pocosins has been extensively studied by Bridgham and Richardson.[543]

3.2.11 Other Terms

Wet prairie—similar to marsh;[3246]

Reedswamp—marsh dominated by *Phragmites* (common reed)—term used particularly in Eastern Europe;[3246]

Wet meadow—grassland with waterlogging soil near the surface but without standing water for most of the year;[3246]

Pothole—shallow marsh-like ponds, particularly as found in the Dakotas, U.S.A.;[3246]

Playa—term used in southeastern United States for marshlike ponds similar to potholes, but with a different geologic origin;[427,3246]

Slough—a swamp or shallow lake system in northern and midwestern United States. A slowly flowing shallow swamp or marsh in southeastern United States;[3246]

Muskeg—large expanses of peatlands or bogs; particularly used in Canada and Alaska.[3246] Stanek and Worley[4677] defined muskegs as a Canadian term frequently applied in ordinary speech to natural and industrial areas covered more or less with *Sphagnum* mosses, sedges, and an open growth of stunted black spruce.[2839,3954]

The meaning of words in languages other than English is not easy resolved, although excellent technical dictionaries dealing specifically with the mires and peat have been prepared, notably by Masing[3088] (German, Estonian, Russian, English, Swedish and Finnish), Bick et al.[357] (German, Polish, English and Russian) and Heikurainen[2000] (English, German, Russian, Swedish and Finnish).

3.3 VEGETATION

Federal Manual[1319] defined hydrophytic vegetation as macrophytic plant life growing in water or soil or on a substrate that is at least periodically deficient in oxygen as a result of excessive water content. Environmental Laboratory[1249] defined hydrophytic vegetation as a sum total of macrophytic plant life that occurs in areas where the frequency and duration of inundation or soil saturation produce permanently or periodically saturated soils of sufficient duration to exert a controlling influence on the plant species present. The vegetation occurring in wetlands may consist of more than one plant community (species association).

In the United States, the FWS in cooperation with CE (Corps of Engineers), EPA (Environmental Protection Agency), and SCS (Soil Conservation Service) has published the "National List of Plant Species That Occur in Wetlands" from a review of the scientific literature and a review by wetland experts and botanists. The list separates vascular plants into four basic groups, commonly called "wetland indicator status," based on a plant species frequency of occurrence in wetlands: (1) *obligate wetland plants* (OBL) that occur almost always (estimated probability >99%) in wetlands under natural conditions (e.g. *Nymphaea, Lemna, Typha, Utricularia*); (2) *facultative wetland plants* (FACW) that usually occur in wetlands (estimated probability 67–99%), but occasionally are found in non-wetlands (e.g. *Salicornia babylonica, Polygonum carnei, Ludwigia maritima*); (3) *facultative plants* (FAC) that are equally likely to occur in wetlands or non-wetlands (estimated probability 34–66%, e.g. *Lobelia inflata, Carex cephalophora, Acer rubrum*); and (4) *facultative upland plants* (FACU) that usually occur in nonwetlands (estimated probability 1–33%, e.g. *Quercus rubra, Potentilla arguta*). If a species occurs almost always (estimated probability > 99%) in nonwetlands under natural conditions, it is considered an *obligate upland plants* (UPL, e.g. *Pinus echinata, Bromus mollis*).[5341]

Federal Manual[1319] stated that an area has hydrophytic vegetation when, under normal circumstances: (1) more than 50% of the composition of the dominant species from the strata are OBL, FACW, and/or FAC species, or (2) a frequency analysis of all species within the community yields a prevalence index value of less than 3.0, where OBL = 1.0, FACW = 2.0, FAC = 3.0, FACU = 4.0 and UPL = 5.0.

Four groups of aquatic macrophytes can be distinguished on a basis of morphology and physiology:[5439]

1) Emergent macrophytes grow on water saturated or submersed soils from where the water table is about 0.5 m below the soil surface to where the sediment is covered with approximately 1.5 m of water (e.g. *Acorus calamus, Carex rostrata, Phragmites australis, Scirpus lacustris, Typha latifolia*);

2) Floating-leaved macrophytes are rooted in submersed sediments in water depths of approximately 0.5 to 3 m and possess either floating or slightly aerial leaves (e.g. *Nymphaea odorata, Nuphar lutea*);

3) Submerged macrophytes occur at all depths within the photic zone. Vascular angiosperms (e.g. *Myriophyllum spicatum, Ceratophyllum demersum*) occur only to about 10 m (1 atm hydrostatic pressure) of water depth and nonvascular macroalgae occur to the lower limit of the photic zone (up to 200 m, e.g. Rhodophyceae);

4) Freely floating macrophytes are not rooted to the substratum; they float freely on or in the water and are usually restricted to nonturbulent, protected areas (e.g. *Lemna minor, Spirodela polyrhiza, Eichhornia crassipes*).

Nutrients are assimilated from the sediments by emergent and rooted floating-leaved macrophytes, and from the water in the free-floating macrophytes.[5349] Various experiments have proven that minerals can be taken up directly by shoot tissues of submerged plants.[132,556,1050,3990,4649,5221] However, there is also no question regarding the uptake capability of nutrients by the roots of these plants.[223] Investigations on quantitative contribution of either the shoots or the roots to the overall nutrition of submerged plants are numerous (e.g. References 554, 556, 686, 688, 1055, 1476, 5221). There is evidence that in submerged plants, oxygen is accumulated in the air spaces during photosynthesis and is later used during the dark phase of respiration. The CO_2 evolved during respiration is then reassimilated in photosynthesis.[1747]

The ability of rooted macrophytes to utilize sediment nutrients may partially account for their greater productivity in comparison with planktonic algae in many lacustrine systems.[5349] The emergent macrophytes act as nutrient pumps and play a key role in seasonal changes in available N, P, and K. The net effect of rooted emergent vegetation is to transfer nutrients from the soil to the surface water in the wetland via leaching and litterfall, especially at the end of the growing season.[2536,3876,4109,4115]

3.3.1 Adaptation of Vegetation to Wetland Conditions

Compared with the vegetation of well-drained soils, wetlands have a world-wide similarity which over-rides climate and is imposed by the common characteristics of a free water supply and the abnormally hostile chemical environment which plant roots must endure.[1278] The hypoxic or anoxic conditions above and below the soil surface, coupled with other factors such as organic matter and temperature, result in several features of the physico-chemical environment of the roots of wetland plants.[1747] For example, the pH is lower, toxic substances produced by the reduction reactions of such ions as ferrous and manganous accumulate, the nitrates and sulfates are reduced to ammonia and sulfides, availability of phosphorus and other nutrients is changed, and the incomplete decomposition of organic matter produces several toxic compounds.[1747] These features have been discussed in detail in the literature[146,916-7,1555,1747,2175,2580,3242,3303,3713,3838,3840,4491] in relation to plants. Thus, plant adaptations are directed both toward the direct effects of flooding, i.e. oxygen stress, and the indirect effects with regard to nutrient supply.[1747] The component species have morphological, anatomical and physiological adaptations which allow them to cope with intermittent flooding, lack of oxygen (anoxia), and the consequent chemical reduction of the soil. Constant flooding limits accessibility to many large herbivores and, particularly in temperate climates, reedswamps and fens are consequently less molded by grazing pressure than the adjacent dry-land vegetation.[1278]

It is not surprising that those plants (and animals) regularly found in wetlands have evolved functional mechanisms to deal with environmental stresses. Adaptations can be broadly classified as those that enable the organism to tolerate the stress and those that enable it to regulate the stress.[3246] Tolerators have functional modifications that enable them to survive, and often to function efficiently, in the presence of the stress. Regulators actively avoid the stress or modify it to minimize its effect. The specific mechanisms for either tolerating or regulating are many and varied. Vascular plants show both structural and physiological adaptations.[3246]

Macrophytes respond to wetland conditions by adopting a variety of morphological and anatomical modifications:[2010]

1. Macrophytization:
   1.1. enlargement of certain organs such as leaves, rhizomes and flowers (e.g. *Nymphaea, Nuphar, Victoria*);
   1.2. emergence of large leaves above the water surface (*Nelumbo*);
   1.3. enlargement of aerial shoots (*Cyperus papyrus, Scirpus*);
   1.4. extension of aerial offshoots and tillers to a large size (*Typha, Echinochloa stagnina, Phragmites, Agrostis semiverticillata*).
2. Microphytization, i.e. the diminution or even total reduction of such organs as leaves and stems:
   2.1. reduction in leaf or frond size in crowded stands (*Wolffia, Lemna, Spirodela, Azolla, Salvinia*), a typical characteristic of the pleustophytes;
   2.2. prevalence of tiny leaves due to increased hydrostatic pressure (*Elodea, Elatine*).
3. Swelling of leaves or stem organs as a result of significant changes in hydrostatic pressure (*Polygonum nodosum, P. acuminatum, P. senegalense, Hippuris, Neptunia, Trapa, Eichhornia, Hydrocotyle*).
4. Hypertrophy of aerenchymatous stems, stolons and rhizomes due to oxygen deficit or anaerobic conditions (*Limnobium, Calla, Oenanthe*).
5. Segmentation of leaf laminae (*Cocomba, Batrachium, Hottonia, Ceratophyllum*).
6. Foliformization of the stems (*Eleocharis acicularis, Juncus bulbosus*).

These morphological adaptations either permanently affect the whole plant habit and are fixed genetically as taxonomic characteristics, or they represent sudden reactions to environmental changes. For example, the development of submerged leaf forms enables some species to survive the hydrophase, although submergence does not provide the optimum ecological conditions for the species.[2010] Another biological adaptation is the formation of vegetative reproductive organs: a dense network of rhizomes or large numbers of underground tubers or corms check the penetration of other species into the stand. The vegetative reproduction enables the plant populations to survive unfavorable seasons, when most or all of their aerial shoots die off.[2010]

Some physiological properties of plants enable settlement in shallow water. The following properties are of great importance:[2010]

1. Rapid growth and the formation of a large biomass and the reaction to favorable life conditions through the maximum exploitation of available resources (*Typha latifolia, Glyceria maxima, Eichhornia crassipes*).
2. A long-term anabiosis (dormancy) of the underground vegetative organs or seeds during unfavorable seasons (long-lasting inundation or dry period) (*Bolboschoenus maritimus, Sagitaria sagittifolia*).

3. Breaking of seed dormancy under conditions favorable for germination and seedling establishment and development (*Oenanthe aquatica*, some species growing on emergent bottom).

Virtually all wetland plants have elaborate structural mechanisms to avoid root anoxia.[915,2175,3246] Rhizosphere oxygenation is considered essential for active root function, and also enables the plants to counteract the effect of soluble phytotoxins, including sulfides and metals, which may be present at high concentrations in anoxic substrata.[145,1555] The main strategy has been the evolution of air spaces (aerenchyma) in roots and stems that allows the diffusion of oxygen from the aerial portions of the plant into the roots[129,139–45,147–51,271,557, 868,891,951,1184,1870,2685,3197,4487–8,4815,4924,5133,5318] (see also Figure 3-9). Moorhead and Reddy[3291] reported that radial oxygen loss from wetland plant roots can range from 100 to 400 mg O_2 m^{-2} h^{-1}. Whereas in normal macrophytes, plant porosity is usually a low 2–7% of volume, in wetland species up to 60% of the plant body is pore space.[3246] The species of woody trees that are successful (as opposed to tolerant) in the wetland habitat are few, and include the mangroves, cypress, tupelo, willow, and a few others. As with herbaceous species, an adequate ventilating system seems to be essential for their growth.[3246]

The magnitude of oxygen diffusion through many wetland plants into the roots is apparently large enough not only to supply the roots but also to diffuse out and oxidize the adjacent anoxic soil.[2227,4924] The extent of oxygenation depends on several factors, including the oxygen demand of biological and chemical processes in the soil substratum and the leakiness of the roots for oxygen, which varies with plant species.[143] Evidence for oxygen diffusion outward from the roots is derived from observations

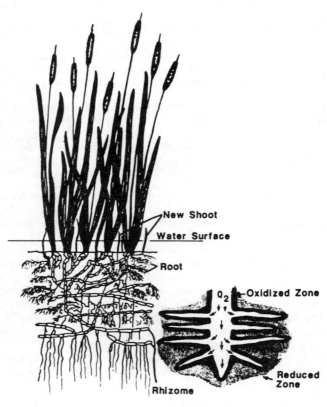

ROOT HAIR ENLARGED

Figure 3-9.   The ability of wetland plants to transport oxygen to support their roots growing in anaerobic substrates.[1890]

of the precipitation of oxidized ferric iron and changes in redox conditions close to root surfaces[772,4917, 4932,5468] and by direct measurement using cylindrical platinum electrodes surrounding the roots.[145,149]

In flood-intolerant species, a reduction in soil aeration has been shown to be accompanied by a rapid increase in glycolysis (the Pasteur effect) as well as by dramatic induction of alcohol dehydrogenase (ADH) activity.[915] Ethanol accumulation resulting from increased glycolysis can readily lead to membrane destruction by lipid solubilization and this will inactivate mitochondrial enzyme activity and further increase the preponderance of glycolytic activity. Flood-tolerant plants differ from intolerant species in not exhibiting this acceleration of glycolysis and show little or no induction or change in the kinetic properties of ADH when subjected to partial anoxia.[915] These tolerant species distinguish themselves from the sensitive species by their much lower rates of respiration, a minimum Pasteur effect, and a partial replacement of ethanol by lactate as the end product of glycolysis.

One adaptation that many wetland plant species share with plants in other stressed environments, especially in drought-stressed environments, is the C_4 biochemical pathway of photosynthesis, formally called the Hatch-Slack-Kortschak pathway, after the discoverers.[1964] It gets its identity from the fact that the first product of CO_2 incorporation is a four-carbon compound, oxaloacetic acid.[235] The first compound resulting from CO_2 incorporation in C_3 plants is a 3-carbon compound phosphoglyceric acid. C_3 plants are much more common that are C_4 plants. Comparison of C_4 and C_3 plants is given in Table 3-1.

Plants with the C_4 pathway can use CO_2 more effectively than other plants. They are able to use it from the atmosphere until its concentration falls below 20 ppm. This is achieved by using phosphoenolpyruvate (PEP), which has a high affinity for CO_2, as the CO_2 acceptor, instead of the ribulose biphosphate (RuBP) acceptor of the conventional pathway. In addition, the malate formed by this carboxylation is nontoxic and can be stored in the cell until it is later decarboxylated and the carbon dioxide is fixed through the normal C_3 pathway (i.e. in Calvin cycle).[3246]

Among the common wetland angiosperms that have been shown to photosynthetize through the C^4 pathway are *Spartina alterniflora, S. foliosa, S. townsendii, Cyperus rotundus, Echinochloa crus-galli, Panicum dichotomiflorum, P. virgatum, Paspalum distichum* and *Sporobolus cryptandrus.*[3246]

Photosynthesis, photorespiration and carbon assimilation in wetland and aquatic plants have been extensively reviewed recently.[468,495–6,1138,2904,2994,3553–4,4276]

The adaptations of wetland plants to flooding have been intensively studied and reported in the literature.[129,234,309,532,783,914–5,892,917,1636,1850,2176,2345,2360, 2448,3302,3696,4384,4498,4569, 4988–9,5140,5451]

3.3.2 Biomass and Productivity of Wetland Plants

3.3.2.1 Biomass

Biomass is the mass of all living organs that can be harvested. It is commonly divided into *belowground* (roots, rhizomes, tubers, etc.) and *aboveground* biomass (all vegetative and reproductive parts above the ground level). The *standing crop* includes also the "standing dead" which is still attached to the plants but cannot be always easily differentiated or segregated. Most biomass values therefore generally refer to the standing crop only. *Litter* refers to only those dead parts which have fallen on the ground.[5187] Biomass, standing crop and litter are most frequently expressed in g m^{-2}, or kg ha^{-1} on dry mass basis.

Total storage of a substance in a particular compartment is called *standing stock*. Nutrient standing stocks in vegetations are computed by multiplying concentrations by biomass per unit area and are expressed as mass per unit area (g.m^{-2}, mg.m^{-2} or kg ha^{-1} on dry mass basis).[2351]

Květ[2656] gives the approximate relationships between dry mass and other units which can be used for macrophyte biomass expression:

1 g dry mass (ash < 10%) = 0.9–1.0 g organic matter = 0.4 g C = 1.5 g CO_2 = 1.07 g O_2.

Wetzel[5349] gives the following values of ash (as the dry mass percentage): emergent species 12% (5–25%), floating-leaved 16% (10–25%), submerged 21% (9–25)%, average for all species 18%.

Energy content of the macrophyte biomass given in the literature varies among authors:

Table 3-1. Photosynthetic and growth characteristics of herbaceous C_3 and C_4 plants.[386,495,1877,2709,3246]

	C_3	C_4
CO_2 compensation concentration at opt. temperature ($\mu l\ l^{-1}$)	30–70	0–10
Transpiration ratio, g H_2O transpired/g DM	450–950	230–350
Theoret. energy required for net CO_2 fixation (CO_2 - ATP - NADPH)	1 : 3 : 2	1 : 5 : 2
Optimum day temperature for net CO_2 fixation (°C)	15–25	30–47
Light saturation of photosynthesis	at intermediate intensities	no saturation at highest intensities
Maximum light energy conversion efficiency	3.5–4.5%	5.0–5.8%
Maximum rate of net photosynthesis (mg CO_2 per dm^2 of leaf surface per hour)	15–40	40–80 (108)
Maximum growth rate (g m^{-2} d^{-1})	19.5	30.3
Dry matter production (g m^{-2} yr^{-1})	medium	high
Primary CO_2 acceptor	RuBP	PEP
First product of photosynthesis	C_3 acids (PGA)	C_4 acids (malate, aspartate)
Chlorophyll-*a* : -*b* ratio	ca 3 : 1	ca 4 : 1
Photosynthesis depression by O_2	Yes (35–40%)	No
CO_2 release in light (apparent photorespiration)	Yes	No
Net photosynthetic capacity	slight - high	high - very high
Redistribution of assimilation products	slow	rapid
Chloroplasts	throughout the leaf mesophyll cells	restricted to concentric rings of mesophyll and bundlesheath cells (Kranz anatomy)

Květ:[2656] aquatic macrophytes: 17.6 kJ g^{-1} DM*
Wetzel:[5349] emergent species: 18.8 (17.6–20.9) kJ g^{-1} DM
        floating-leaved species: 20.0 (19.1–21.5) kJ g^{-1} DM
        submerged species: 19.2 (17.4–21.8) kJ g^{-1} DM
        average for all groups: 19.1 kJ g^{-1} DM
Barko et al.:[224] aquatic macrophytes: 20 kJ g^{-1} AFDM
Cummins and Wuycheck:[932] aquatic monocots: 20 kJ g^{-1} AFDM
Larcher:[2709] vascular plants — shoots: 15.9–18.0 kJ g^{-1} DM
        roots: 13.4–19.7 kJ g^{-1} DM
        seeds: 18.4–21.0 kJ g^{-1} DM

The biomass measurement of floating plants or aboveground parts of emerged, submerged or floating-leaved species usually does not make problems unless the water is very deep. However, it is nearly impossible to harvest satisfactorily all the belowground roots and rhizomes and also differentiate between the dead and live belowground organs.[1205,1621] Vyas et al.[5187] pointed out that these problems have been discussed in many publications but no simple solution is yet available.

Herbaceous plants

In the literature, some estimates of wetland plants biomass have been reported. Whittaker and Likens[5396] listed a biomass range of 3,000–50,000 g m^{-2}, with a mean of 15,000 g m^{-2} for swamp and marsh. In these estimates, evidently herbaceous and woody plants have been put together. Květ[2656] gives the following estimates for biomass range of different herbaceous plant communities (in g m^{-2}): submerged (Euhydatophyta, e.g. *Elodea, Potamogeton*): 50–500; short emergent species (Hydro-

---

*1 Joule = 0.23892 cal.

ochthophyta, e.g. *Eleocharis, Bolboschoenus*): 300–2,000; tall emergent species (Ochthohydrophyta, e.g. *Phragmites, Typha, Glyceria*): 1,800–9,900; and tall sedges and grasses (graminoids, Euochthophyta, e.g. *Carex, Calamagrostis*): 700–3,000. For a comparison, he estimated the maximal phytoplankton biomass in the range of 1 to 2.3 g.m^{-2}. However, the turnover coefficient for phytoplankton is several hundred-fold higher than that for higher plants (for details see Chapter 3.3.2.2). In the Appendix (Table 7-1) a review on herbaceous wetland plants biomass reported in the literature is given.

Roots and rhizomes of submersed macrophytes make up a lesser portion (10–40%) of total plant biomass than is the case for floating-leaved (30–70%) or emergent (30–90%) macrophytes.[329,1195–6,1892,5336,5338,5349] Květ[2656] estimated that belowground biomass makes up 1–17% for submerged plants (i.e. belowground: maximal aboveground biomass ratio 0.01 - 0.2); 29–67% for short emergent species (ratio 0.4 - 2.0); 38–88% for tall emergent species (ratio 0.6 - 7.6); and 50–80% for tall graminoids (ratio 1.0 - 4.0). Appendix Table 7-2 shows that values vary considerably, however.

Woody plants

Mitsch and Gosselink,[3246] summarizing a large data set from the literature sources, reported the range of standing biomass for deepwater swamps (mostly in southeastern United States) of 7,400 to 40,500 g m^{-2} and for riparian forested wetlands throughout the United States of 10,000 to 30,300 g m^{-2} (numbers are given for aboveground biomass). Mitsch and Ewel[3244] reported the range of biomass for various cypress stands in Florida of 6,800 to 19,000 g m^{-2}. Aboveground biomass in the Great Dismal Swamp in Virginia ranged from 19,457 to 34,526 g m^{-2} for various forest communities.[1011,1730] Brown and Lugo[587] reported the biomass for freshwater and saltwater forested wetlands in the range of 19,200 to 37,200 g m^{-2} and 11,900 to 27,900 g m^{-2}, respectively. Richardson,[4101] summarizing the literature data on wetland forests, found the range of 4,650–37,800 g m^{-2} for biomass. Lugo and Snedaker[2953] reported the range of mangrove systems biomass in Florida from 900 (scrub mangroves) to 20,800 (riverine mangroves) g m^{-2}.

3.3.2.2 Productivity

Production is the amount of organic matter, expressed as dry mass, produced per unit area during the whole growing period. The rate of production per unit time (day, month, year) is preferably called *productivity*. However, in current literature, the term "production" with the specification of time interval (daily or annual) refers to the rate as well.[5187] The production is usually differentiated into *gross production* (total amount of organic matter photosynthetized by the plants) and *net production* (gross minus respiration and other losses due to death and predation). Productivity is measured most frequently on dry mass basis in g m^{-2} yr^{-1}, kg ha^{-1} yr^{-1} or g m^{-2} d^{-1}. Less frequently productivity is measured on C basis. Unfortunately, the terms productivity, production, biomass and standing crop have sometimes been used interchangeably in the literature.

Due to strong seasonality of climate and because most of the water bodies are temporary, a large number of macrophytes behave like annuals, i.e. they exhibit a unimodal growth with a single peak of biomass. Thus, the peak standing crop (or the difference between the peak and lowest standing crop if the plants do not die completely) during the growth period can be theoretically considered equivalent to the net production.[5187] However, this often represents an underestimation because substantial amounts of plant parts are lost in death and predation. In submerged and free-floating species with rapid vegetative multiplication, death of individual parts or plant parts accounts for very large losses.[5187] The estimation of belowground production in emergents and other rhizomous plants remains an unsolved problem. In fact, most productivity data are estimates computed by different methods and not the real measurements. In comparison with algal productivity, where daily production rates are measured by different direct techniques, daily production rates in macrophytes are mostly computed from the differences in standing crop, sometimes by simply dividing the observed standing crop or net production by 365 (or shorter period). This, of course, does not provide the real rates as most of the aquatic plants grow for only a few months.

Herbaceous plants

Wetzel[5349] pointed out that on a unit-area basis, the net primary productivity of aquatic macrophytes is among the highest of any community in the biosphere. Emergent macrophytic productivity is the higher, 1,500–4,500 g C m^{-2} yr^{-1}, submersed macrophytic productivity is considerably less (50–1,000 g C m^{-2} yr^{-1}) but often equals or exceeds that of phytoplankton (50–450 g C m^{-2} yr^{-1}).[5349]

Westlake[5336] estimated the following average productivity (g m^{-2} yr^{-1}) of various plant communities: freshwater submerged macrophytes: 600 (temperate climate) and 1,700 (tropical); marine submerged macrophytes (mostly macroalgae): 2,900 (temperate) and 3,500 (tropical); salt marsh: 3,000; reed-swamp (i.e. *Phragmites* marsh) 4,500 (temperate) and 7,500 (tropical). For a comparison, the author also estimated the productivity of microalgae (g m^{-2} yr^{-1}): phytoplankton: 200 (ocean and lake), 300 (coastal) and 600 (polluted lake). For cultivated algae in sub-tropical climate he listed a value of 4,000 g m^{-2} yr^{-1}.

Gopal and Masing,[1747] based on their comprehensive review, listed the following values for primary productivity of various plant groups (in g m^{-2} yr^{-1}): mosses, < 100 to < 1,000; tall emergent species, 500 to 2,000; and short emergent species, 200 to 1,500. The authors pointed out that the tall emergent species (e.g. *Phragmites, Typha*) clearly show higher production in the tropics and subtropics than in temperate regions.

Květ[2656] estimated the net primary productivity as follows (g m^{-2} yr^{-1}): submerged species (e.g. *Elodea, Potamogeton*): 100–500; short emergent species (e.g. *Eleocharis, Bolboschoenus*): 500–3,000; tall emergent species (e.g. *Phragmites, Typha, Glyceria*): 1,400–4,800; and tall sedges and grasses (graminoids, e.g. *Carex, Calamagrostis*): 1,000–3,100. For a comparison, he estimated net primary productivity of phytoplankton in the range of 600–2,400 g m^{-2} yr^{-1}. The author also gives the turnover coefficients (i.e. net primary productivity/maximum season biomass) for these communities: submerged species: 1.1–1.5, short emergent species: 1.05–1.5, tall emergent species: 1.05–1.3, tall sedges and grasses (graminoids): 1.15, and phytoplankton: 450–600.

Whittaker and Likens[5396] listed a net primary productivity range for swamp and marsh 800–6,000, with a mean of 3,000 g m^{-2} yr^{-1}. Lieth[2849] estimated the mean productivity to be 2,000 g m^{-2} yr^{-1} with a range of 800 to 4,000 g m^{-2} yr^{-1}. Richardson[4101] in his review gave the range of productivity values from 700 to 2,700 g m^{-2} yr^{-1} with an overall mean of 1,520 g.m^{-2}.yr^{-1}. Unfortunately, the values are difficult to compare as definitions of "marsh" and "swamp" may differ considerably (see Chapter 3.2).

In the Appendix (Tables 7-3 and 7-4) the annual and daily productivity of herbaceous aquatic and wetland plants from different regions and different wetland ecosystems are given.

Woody plants

Richardson,[4101] summarizing the literature data on wetland forests, reported the range of 480 to 1,570 g m^{-2} yr^{-1}. Conner and Day[858] found seasonally flooded cypress-tupelo areas throughout the southeastern United States to have a production of 950 to 1,780 g m^{-2} yr^{-1}. Productivity of natural swamp forests with nearly year-round flooding ranged from 690 to 1,120 g m^{-2} yr^{-1} and in a constantly flooded, stagnant area, the productivity was 192 g m^{-2} yr^{-1}. The authors reported the stem growth in the range of 253 to 1,250 g m^{-2} yr^{-1}. Conner et al.[860] found net primary productivity in Louisiana swamps to be 887, 1,166 and 1,780 g m^{-2} yr^{-1} for swamps under permanent flooding, natural flooding and controlled flooding, respectively. Gopal and Masing[1747] in their review reported productivity values from forested swamps and floodplains in the range of 712 and 2,758 g m^{-2} yr^{-1}. Conner and Day[857] reported the measured net productivity of bottomland hardwood and baldcypress-water tupelo areas in a Louisiana swamp to be 1,574 and 1,140 g m^{-2} yr^{-1}, respectively. These authors, summarizing the literature data, gave the range for other swamps in the range of 192 to 1,170 g m^{-2} yr^{-1}. Fowler and Hershner[1468] found net primary production of 744 g m^{-2} yr^{-1} (492 for stems and branches and 252 for litterfall) in a tidal freshwater wetland (Cohoke Swamp) in Virginia. The authors, summarizing literature data, reported the aboveground net primary production in depression swamps and riverine swamps in the range of 707–1,030 g m^{-2} yr^{-1} and 971 to > 1,607 g m^{-2} yr^{-1}, respectively. Brown and Lugo[587] reported the litter production in the range of 330 to 450 g m^{-2} yr^{-1}. The annual leaf litter production in the Great Dismal Swamp, Virginia, ranged from 455 to 536 g m^{-2} yr^{-1} in four different communities.[1730] Bates,[240] working in a forested wetland in South Carolina, found the total annual aboveground production of 1,575 g m^{-2} yr^{-1} including a litter production of 500 g m^{-2} yr^{-1}. Mitsch and Gosselink[3246] reported, based on literature data, net primary productivity in deepwater swamps in southeastern United States for litter fall, stems and total aboveground productivity in the range of 120–678 g m^{-2} yr^{-1}, 117–1,230 g m^{-2} yr^{-1} and 387–1,780 g m^{-2} yr^{-1}, respectively.

For riparian forested wetlands throughout the United States, Mitsch and Gosselink[3246] reported the following net primary productivity ranges: litter fall: 412–574 g m^{-2} yr^{-1}, stem growth: 177–914 g m^{-2}

yr^{-1}, and aboveground productivity: 668–1,374 g m^{-2} yr^{-1}. Conner and Day[858] reported the range for litter production in southeastern United States swamps from 120 to 757 g m^{-2} yr^{-1}.

Odum et al.[3507] reported the range of mangrove litterfall production in Florida from 130 (scrub mangroves) to 1,280 (riverine mangroves) g m^{-2} yr^{-1} and annual mangrove biomass production in the range from 380 (scrub mangrove) to 2,740 (basin mangroves) g m^{-2} yr^{-1}. Day et al.[1014] reported the mangrove production rate (woody growth and litterfall) in two localities in Mexico in the range of 1,606 to 2,458 g m^{-2} yr^{-1}. Other reported values for mangrove production rates are e.g. those by Golley et al.[1705] in Puerto Rico (307 g m^{-2} yr^{-1}) and Christensen[795] in Thailand (2,000 g m^{-2} yr^{-1}).

The variability of woody species in forested wetlands, however, is limited as compared with herbaceous plants. The most frequently studied trees in forested wetlands are *Taxodium distichum* (baldcypress), *Acer rubrum* (red maple), *Nyssa aquatica* (water tupelo, water gum), *Nyssa sylvatica* (black gum), *Quercus* spp. (oaks). In addition, most data on biomass and productivity of forested wetlands have come from the southeastern United States only.

Gopal and Masing,[1747] in their comprehensive review, reported a summary of production rates in different wetland communities (Table 3-2). However, all attempts to give "general values" for biomass and productivity in wetlands are, in my opinion, rather misleading as they are mostly based on (1) relatively small data sets, (2) regional studies, and (3) for productivity they are mostly based on estimates and not direct measurements.

3.3.3 Decomposition in Wetlands

Decomposition generally refers to the disintegration of dead organisms into particulate form (or detritus), and the further breakdown of large particles to smaller and smaller particles, until the structure can no longer be recognized and complex organic molecules have been broken down into CO_2, H_2O and mineral components.[3095] In wetland studies, the term decomposition is mostly confined to the breakdown and subsequent decay of dominant macrophytes which leads to the production of detritus.[503,744,992–3,1039,1042,1666–9,1990,2218,2510,2622,2718,3096,3505,5223,5385] Most net annual aboveground production of wetlands is not consumed by herbivores but decomposes on the wetland surface.[1546,3830–1] Rates of decomposition vary in wetlands[744,992,995,1010,2155,3400,3504–5,5378] and the fate of materials released and adsorbed during decomposition depends on the physical and chemical composition of materials[225,569,992,1009,1023,1771] as well as environmental conditions at the site of decomposition.[569,1009,1010,1546,2222,4227,4424,5378]

The decomposition of litter and resultant release of nutrients involve at least two processes.[569,701,1666,3187] An initial loss of soluble materials is attributed to abiotic leaching.[503,569,1041,1666–9,1771] This process is quite rapid (Figure 3-10) and accounts for the majority of mass reduction during the early stages of decomposition.[1041] Nykvist[3490] showed that leaching occurred almost immediately under both aerobic and anaerobic conditions with most of the water-soluble organic substances being released within 6 to 12 months. The rapid initial release of nutrients by leaching has been documented in many marsh plants (Figure 3-11)—up to 30% of nutrients are lost by leaching alone during the first few days of decomposition.[993,3096,3819] In submerged and floating-leaved plants, leaching accounts for up to 50% loss of dry matter within the first 2–3 days.[1742] Released nutrients may be incorporated into the protoplasm of decomposer organisms[1771] where activities such as respiration and denitrification account for additional nutrient losses.[1774,3096] De la Cruz[1041] included one more process which takes place during decomposition: weathering or mechanical fragmentation due to abrasion by wind and ice, wave and tidal action, or animal trampling and incomplete grazing by herbivores.

Accumulation of materials by components of the microfloral and microfaunal detritus community and abiotic factors can cause increased nutrient content of decomposing litter[568,992–3,2125,2219,3937] (see also Figure 3-12). The net effect of abiotic and biotic interactions during decomposition is that some materials are released and presumably used in plant growth at or near the point of release, while other nutrients are immobilized or stored temporarily or permanently at the point of release.[488,4247] Davis and Van der Valk[992] presented a detailed study on the content, flow, and release of phosphorus, nitrogen, potassium and calcium during decomposition of *Scirpus fluviatilis* and *Typha glauca* in an Iowa marsh (Figures 3-13, 3-14, 3-15, and 3-16).

Several types of organisms are involved in aquatic decomposition. Bacteria and fungi are probably the most important and certainly the best known. Protozoans may also be important, but very little research has been performed to elucidate their role in decomposition. The impact of multicellular animals on the decomposition process in aquatic habitats is uncertain.[1670] Bacteria are the most versatile decomposers, both in terms of habitats exploited and substrates metabolized. Bacteria can be either

Table 3-2. Summary of production rates (average range) in different wetland communities.[1747] (After Gopal and Masing, 1990, in *Wetlands and Shallow Continental Water Bodies*, Patten, B.C., Ed., p. 146, reprinted with permission of SPB Academic Publishing bv, The Hague.)

Wetland/community	Production (g m^{-2} yr^{-1})
Bogs	
Sphagnum carpet	100–400
Other sedges and shrubs, also fens	200–1,000
Bog forest	500–1,500
Marshes	
1. Emergent C$_4$ plants, tropical freshwaters[a] (esp. *Cyperus papyrus*)	6,000–9,000
2. Emergent C$_3$ plants, freshwaters (esp. *Phragmites australis, Typha* spp.)	5,000–7,000
3. Emergent, floating C$_3$ plants, subtropical (esp. *Eichhornia crassipes*)	4,000–6,000
4. Phytoplankton	1,500–3,000 (6,000)[b]
5. Submerged plants - tropical[a]	2,000
- temperate	500–1,000
6. Swamps (aboveground) - North America	500–1,600

[a] Few values, [b] enriched, shallow, agitated cultures
1. Production: or seasonal maximum biomass (SMB) x 3 for turnover, or daily growth rate x 365 x 0.6 to allow for translocation (see Thompson,[4961] Thompson et al. [4963])
2. Production: or SMB x 2 (i.e., 0.2 x standing crop aboveground for losses, 0.67 x this aboveground production for underground production)
3. SMB x 2 for leaf losses (cf. Center and Spencer[737])
4. Mostly net production
5. Net photosynthesis
6. Production: or SMB x 1.25 as partial correction for losses

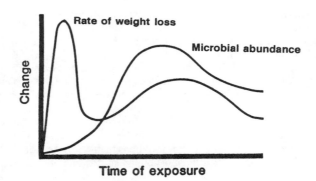

Time of exposure

Figure 3-10.   Generalized trends in rate of weight loss and microbial abundance on detritus in flooded environment. Weight loss is rapid at first because of abiotic leaching; the second smaller peak in weight corresponds to maximum microbial abundance and activity later in decomposition.[1670]   (From Godshalk and Barko, 1985, in *Microbial Processes in Reservoirs,* Gunnison, D., Ed., p. 67, reprinted with permission of Kluwer Academic Publisher, Dordrecht.)

attached or free-living. The interrelationships and interchangeability between these two groups are not clear.[1670]

Generally, the total number of microorganisms is higher in organic soils, especially peat, than in non-organic soils due to the higher proportion of available carbon from organic matter.[1007] The high

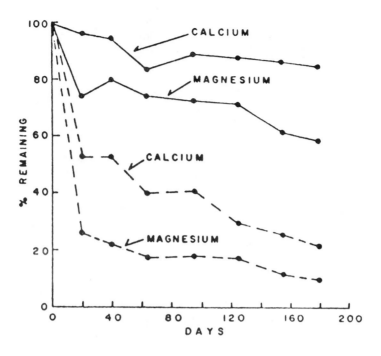

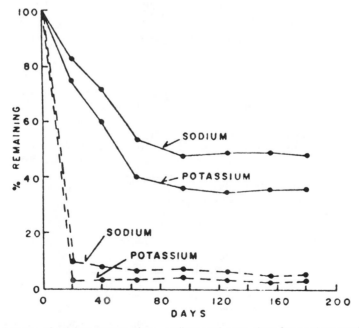

Figure 3-11.   Losses of calcium, magnesium, sodium and potassium from submerged (dashed line, on the surface of the mud beneath about 30 cm of water) and suspended (solid line, about 12 cm above the water surface) bags.[503] (From Boyd, 1970, in *Arch. Hydrobiol.*, Vol. 66, p. 515, reprinted with permission of E. Schweizerbart' sche Verlagsbuchhandlung (Nägele und Obermiller), Stuttgart.)

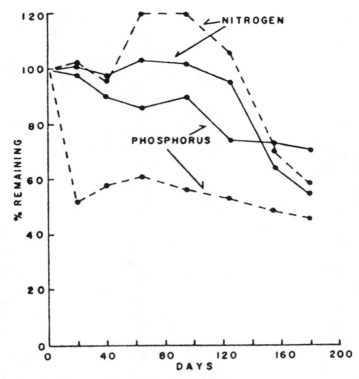

Figure 3-12. Losses of nitrogen and phosphorus from submerged (dashed line, on the surface of the mud beneath about 30 cm of water) and suspended (solid line, about 12 cm above the water surface) bags.[503] (From Boyd, 1970, in *Arch. Hydrobiol.*, Vol. 66, p. 515, reprinted with permission of E. Schweizerbart' sche Verlagsbuchhandlung (Nägele und Obermiller), Stuttgart.)

concentration of C in organic soils and litter provides the substrate necessary for rapid decomposition when N and P are freely available. These higher microorganisms numbers tend to create higher levels of microbial activity than is generally estimated.[2647]

Waterlogged soils, especially organic soils, develop a very thin oxidized layer a few millimeters thick in the surface and many, if good drainage exists, even have some local oxidized areas in the remaining reduced strata.[121] In general, however, the surface layer undergoes aerobic respiration, while the remaining areas undergo anaerobic respiration and fermentation. Anaerobic decomposition is much slower[1667,3233,5450] and is done by obligate anaerobic and some facultative anaerobic bacteria.[744] In addition, anaerobic decomposition generally does not proceed to completion owing to the lack of oxidizing power and adequate levels of alternate electron acceptors (nitrate, oxidized forms of iron and mangan, sulfate) to fuel the process.[225]

Wetland decomposition may be either aerobic or anaerobic.[1401] Aerobic decomposition occurs: (1) in the air, while dead plants are still standing or while dead plant tissues are still attached to living plants; (2) in the water, after dead plant parts are carried by wind or precipitation or simply dropped into the water flowing over a wetland (depending on their size and moisture saturation, these plant parts may either float on the surface, become suspended in the water column, or sink to the bottom); and (3) on the wetland floor, where litter may be stranded or tossed to and from by the tides. The depth and extent of the anaerobic zone depend on the type of rooted aquatic vegetation (since plants differ in their ability to pump oxygen back to soil), the nature of the rhizospheres, the activity of the benthic organisms, and the other physical conditions of the substrate.[1041]

Decomposition is generally higher at locations subjected to flooding than at drier sites (see also Figure 3-11). However, the most rapid processing occurs under aerobic conditions, which in wetlands

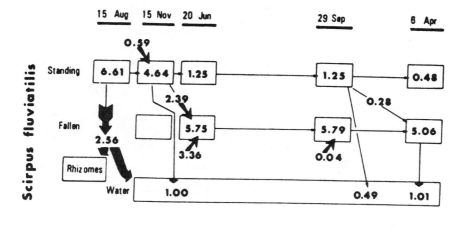

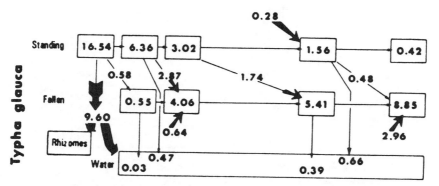

Figure 3-13.   Nitrogen content, flow, and release during decomposition of standing and fallen litter of *Scirpus fluviatilis* and *Typha glauca* at Goose Lake (g m^{-2}) 1974–1976.[992] (From Davis and Van der Valk, 1978, in *Freshwater Wetlands, Ecological Processes and Management Potentials*, Good, R.E., et al., Eds., p. 107, reprinted with permission of Academic Press and the author.)

necessitates an optimal regime of submergence and exposure.[547] Litter decomposition in flooded habitats has traditionally been described as being limited by anaerobic conditions in the water column and the soil.[5450]

A common assumption, made earlier in terrestrial studies,[2331,3546] and widely used in wetland studies, is that decay rate is a constant proportion of the amount remaining, i.e. it decreases logarithmically. Jenny et al.[2331] formulated nonconstant integrative parameters, known as k and k', to denote the decomposition rate of vegetation under steady state conditions. Simply stated, k' equals the percentage loss of the original weight over a specified period of time with a single year usually used. The term k equals the -ln(1 - k') or the negative natural logarithm of the percentage weight remaining after a designated period of time:[744]

$$\ln (W_t/W_0) = -kt \quad \text{or} \quad W_t = W_0\, e^{-kt}, \tag{3.1}$$

where W_0 is the dry mass initially present, W_t is the dry mass remaining at the end of the period of measurement, and t is time in years. Using this expression, the time for 50% decomposition is calculated as 0.693 k^{-1} and for 95% decomposition as 3 k^{-1}.[1742] This equation does not account for rapid leaching losses, however, and therefore large variations are introduced into the calculated rates based on different study periods but similar leaching losses. Louisier and Parkinson,[2918] Godshalk and Wetzel[1669] and Carpenter[700] proposed equations which are more complex. Carpenter[701] compared

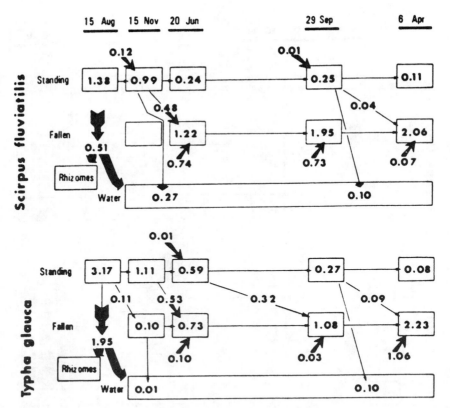

Figure 3-14. Phosphorus content, flow, and release during decomposition of standing and fallen litter of *Scirpus fluviatilis* and *Typha glauca* at Goose Lake (g m^{-2}) 1974–1976.[992] (From Davis and Van der Valk, 1978, in *Freshwater Wetlands, Ecological Processes and Management Potentials,* Good, R.E., et al., Eds., p. 108, reprinted with permission of Academic Press and the author.)

different equations for the same data set, and concluded that the composite exponential equation fits the data more closely.

Godshalk and Barko,[1670] summarizing literature data, listed decay coefficients (k, d^{-1}) for various substrates during aquatic decomposition: phytoplankton 0.0159; submerged, floating vegetation 0.0080; emergent, marsh vegetation 0.0031; deciduous leaves 0.0064; conifer needles 0.0049; and wood 0.00013. Some examples of decomposition constants for various aquatic plants are given in the Appendix (Table 7-5). From this table it is seen, however, that decay coefficients vary greatly.

A relatively small proportion of total nutrients accumulated in aquatic plants is transferred to the herbivorous animals, and therefore, decomposition of dead plants constitutes the most important process whereby the nutrients are released back into the environment.[1742] Though there is evidence to suggest that some nitrogen may be excreted, together with dissolved organic matter, out of growing shoots particularly in submerged and free-floating plants,[1918,2213,2486, 3206–7,3734,5346,5349,5354] decomposition starts with the senescence of plant tissue.[1742] Soon after senescence sets in, water soluble compounds start leaching out of the plant parts in contact with water. In emergent plants, leaching starts after the litter falls in water. During senescence aerial shoots of free-floating and emergent plants become more susceptible to attack by microorganisms (bacteria and fungi) which damage the cell walls. This may result in leaching nutrients by rain water also from the senescent and standing dead organs.[995,1742]

Godshalk and Barko[1670] pointed out that vascular plants show vast differences in their susceptibility to decomposition. Aquatic plants generally decay quickly if they are soft, flaccid, and pithy (floating-leaved plants, e.g. water lilies) or finely dissected (most submersed aquatic plants). Emergent aquatic plants with significant amounts of cellulose and lignin, tough vascular tissues, and protective outer

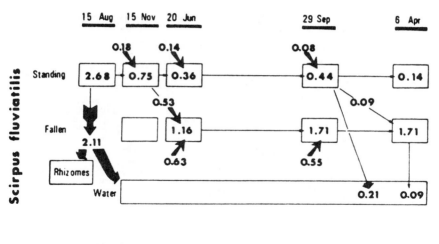

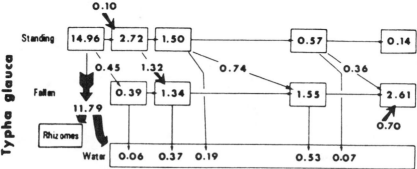

Figure 3-15.   Potassium content, flow, and release during decomposition of standing and fallen litter of *Scirpus fluviatilis* and *Typha glauca* at Goose Lake (g m⁻²) 1974–1976.[992]   (From Davis and Van der Valk, 1978, in *Freshwater Wetlands, Ecological Processes and Management Potentials,* Good, R.E., et al., Eds., p. 109, reprinted with permission of Academic Press and the author.)

layers decay more slowly.[966,1668]   Leaves generally decay faster than petioles and stems.[1276]   The underground portions of aquatic plants, which in emergent species often dominate their biomass, decay less rapidly than the aboveground plant parts.[1859,2221]

It is important to realize that the release of nutrients during the decomposition does not necessarily follow the loss in dry mass. The more labile nutrient elements like sodium and potassium (see Figures 3-11 and 3-12) are lost most rapidly in the early stages of decomposition and other nutrients follow a sequence depending upon the nature of compounds in which they are found.[1742]   Sequence and rate of nutrient release during decomposition are summarized by e.g. Godshalk and Barko[1670] and Gopal.[1742]

The stages of macrophyte decomposition are fairly predictable. Decay is first initiated by the liberation of soluble material. Released substances apparently consist of intracellular cytoplasmic compounds, containing labile organic substances (sugars, fatty acids, amino acids), N, P, and cations (Na, K, Ca, Mg).[225]   Leaching of materials from fresh detritus may explain the rapid decrease in mass weight (more than one third) often observed during initial decomposition of detritus,[699,1276,3764,3831,5054] while later mass weight loss is associated with decomposition of more resistant organic materials such as cellulose and lignin.[225,1670]

Plant tissues having a high N content decompose quickly, releasing N to the environment.[703]   In contrast, tissues with low N levels (high C:N ratios) lose readily soluble (i.e. non-cell wall) nitrogen at the beginning of decomposition, then may accumulate nitrogen.

Factors controlling decomposition include temperature, pH, oxygen content, nutrient availability, water depth, nutrient levels of the water and sediments, the substrate itself (composition, structure, size,

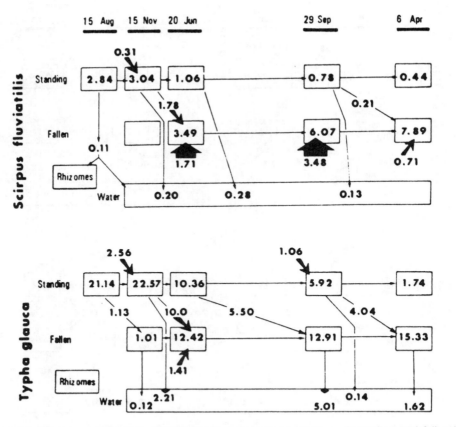

Figure 3-16.   Calcium content, flow, and release during decomposition of standing and fallen litter of *Scirpus fluviatilis* and *Typha glauca* at Goose Lake (g m^{-2}) 1974–1976.[992]   (From Davis and Van der Valk, 1978, in *Freshwater Wetlands, Ecological Processes and Management Potentials,* Good, R.E., et al., Eds., p. 110, reprinted with permission of Academic Press and the author.)

surface area) and mechanical factors.[1670,1742] Bridgham et al.[542] reported that disturbance caused a large increase in decay rates in peatlands.

Further details on decomposition in wetlands can be found in References 90, 743, 1582, 1670, 1667, 2080, 2572, 2996, 3400, 3414, 3505, 3792, 4396, 5154.

3.3.4 Seasonality and Translocation

Bernard[329] found that at winter's minimum aboveground standing crop value, some 74% of the biomass was below ground, but at the summer's maximum aboveground value, only 22% was belowground. Through most of the season, the belowground material made up less than 50% of the total biomass of *Carex rostrata*. Many species of sedge have a considerable standing live component even during winter.[333] Standing live biomass of sedge at the end of a winter was reported to be 13.3–17.4% of the following summer maximum.[328–30,1761,3989] Handoo and Kaul[1892] found that seasonal changes in the aboveground standing crop of *Sparganium ramosum* were inversely related to those in the belowground tissues.

Tissue nitrogen and phosphorus concentrations have been found to be different in plants collected at different times of the year.[513,1611,1892,3433,4114,4180,5060] Vegetation nutrient concentrations tend to be highest early in the growing season, decreasing as the plant matures and senesces.[332,335,500,504,3096,3433] Patterns of seasonal changes in composition vary for both species and nutrients and broad generalization probably cannot be made.[504] Rørslett et al.[4180] found that for *Elodea canadensis*, shoot nitrogen and phosphorus declined rapidly in autumn and the minimum shoot phosphorus concentration was found in winter.

Květ and Ostrý[2659] concluded that it is assumed that at the peak of the growing season in a temperate climate, in late spring and early summer, a significant proportion of available mineral nutrients is contained in the biomass of the vegetation. Consequently, the rhizosphere nutrient pool is expected to be somewhat depleted.

Garver et al.[1575] studied in detail seasonal patterns in accumulation and partitioning of biomass and macronutrients in *Typha* spp. grown in cultivated stands in Minnesota (Figure 3-17). They reported that biomass and nutrient accumulation are in a lag phase during the first 4–8 weeks of growth in the spring. The plants then enter a rapid growth phase in which 47–80% of the total seasonal biomass production and nutrient uptake occurs in a 4–8 week period. During this time, *Typha* leaves account for 60–70% of the biomass and represent the major nutrient sink. As the rate of biomass and nutrient accumulation diminishes, translocation of nutrients and photoassimilate from leaves to rhizomes occurs, with the estimated 40% of leaf nitrogen, 35–44% of leaf phosphorus and 4–38% of leaf potassium translocated to the rhizomes by the end of October. Over the winter, 75% or more of the rhizome biomass, N, P, and K is preserved. Gopal and Sharma[1746] estimated that about 35% N and 50% P of the total maximum content in the leaves of *Typha elephantina* was translocated to the rhizomes during winter and later some of it is used again for new shoot growth.

The rate of heavy metals uptake and cycling depends not only on metal availability, but also on specific plant physiology. The first process which might affect the rate of cycling is an upward transport from roots to shoots.[2632] Higher concentrations of heavy metals in roots, as compared to shoots,[4328,5065,5484] may serve as evidence for a mechanism inhibiting upward transport. The nature of this regulation is not known. Immobilization of metals may occur through the formation of fine crystals on the surface of cell walls in roots, as shown by Malone et al.[3029] for corn plants. However, similar metal precipitates were found by Sharpe and Denny[4425] in leaves of *Potamogeton pectinatus*. Kufel[2623] reported that it is possible that different rates of cycling depend on the role the metal plays in plant physiology. Toxic metals may be immobilized in underground organs, or, conversely, readily excreted from plant tissue. An element that in any way is important for plant functioning (e.g. enzymatic reactions) may be more closely combined into shoot tissue.

Further details on translocation and seasonality in wetland and aquatic plants can be found in References 500, 1442, 1209, 1611 4117, 4582, 4672–3, 5059.

3.3.5 Macrophytes and Algae

Competition for light is probably the most important factor for the balance among phototrophic communities. Phytoplankton has an advantage in light utilization because it is suspended in the water and can shade the other phototrophic communities below (Figure 3-18).[4428] Eutrophication of water bodies results in phytoplankton blooming and reduced water clarity because of light absorption by phytoplankton and increased concentrations of dissolved organic compounds produced by phytoplankton.[4281] Eutrophication therefore restricts depth penetration of benthic microalgae and rooted macrophytes and may ultimately exterminate them.[463,1904] It is quite likely that dense phytoplankton blooms, in addition to shading, inhibit growth of rooted macrophytes and attached algae by generating high pH and thus hampering inorganic carbon uptake and making it more dependent on the light-driven utilization of HCO_3^-.[2994] A similar inhibitory effect of dense epiphytic communities on rooted macrophytes is perhaps also present.[4282]

More light is needed to maintain populations of advanced rooted macrophytes compared with unicellular algae.[4281] Among microalgae, the experimental I_c values typically vary from 1 to 10 μE m^{-2} s^{-1} in a 12 h light/12 h dark cycle.[1587,4116] The mean value for photosynthetic light compensation points from many investigations was 6.7 μE m^{-2} s^{-1},[4116] but when algae were given sufficient time to adapt, recent experiments have suggested I_c values below 1 μE m^{-2} s^{-1}.[1587] Light compensation points for photosynthesis of green shoots of rooted macrophytes are usually between 8 and 50 μE m^{-2} s^{-1}.[2994,5115] Sand-Jensen and Borum[4281] pointed out that growth and biomass development of rooted macrophytes are probably greatly dependent on light availability considering that they are less often nutrient limited, have high light compensation points and light saturation levels for single shoots and extensive self-shading within stands.

It has been demonstrated that epiphytic material absorbs PAR (photosynthetically active radiation) before reaching the leaf surfaces of vascular plants.[614,3781,4277–8,4281,5061] Light attenuation through dense

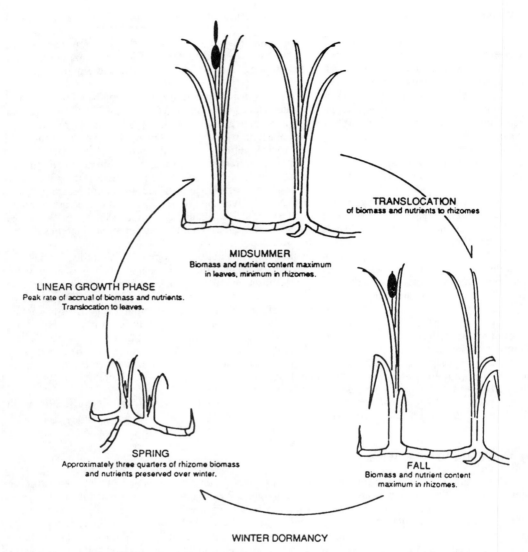

Figure 3-17.   Summary of accumulation and partitioning of biomass and nutrients in *Typha* tissue. Existing *Typha* stands enter a lag growth phase in the spring when large rhizome stores of carbohydrates and nutrients begin to be translocated to the leaves. Plants then enter a rapid growth phase as the leaf canopy develops, concurrent with or closely preceded by peak nutrient uptake. As dry-matter production diminishes, carbohydrate and nutrients are translocated to the rhizomes. Biomass and nutrient content of rhizomes peak at this time, just prior to the onset of dormancy.[1575] (From Garver et al., 1988, in *Aquatic Botany*, Vol. 32, p. 124, reprinted with permission of Elsevier Science Publishers bv, Amsterdam and the authors.)

epiphytic communities is sometimes higher than in the water above the plant.[4277,4279] Twilley et al.[5061] reported that the accumulation of epiphytic material resulted in > 80% attenuation of the incident radiation at the leaf surface. Ozimek et al.[3599] found that clusters of a green filamentous alga *Cladophora glomerata* reduced underwater light by 30–80%, depending on the density of filaments and their compactness and position (distance from macrophytes). Losee and Wetzel[2917] found that geometric configuration of periphyton components also affected PAR attenuation. The algal pigment covering a surface was a major factor influencing light absorption. Sand-Jensen and Borum[4281] pointed out that

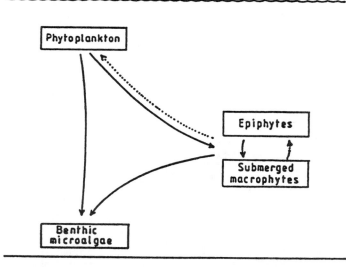

Figure 3-18.   Shading effects among aquatic phototrophic communities of phytoplankton, benthic microalgae and rooted macrophytes with their associated epiphytic algae. Arrows point to functional groups that suffer in competition for light. Arrows with dashed lines show shading effects usually of secondary importance.[4281]   (From Sand-Jensen and Borum, 1991, in *Aquatic Botany,* Vol. 41, p. 154, reprinted with permission of Elsevier Science Publishers bv, Amsterdam and the authors.)

the relationship between epiphytic algal density, epiphyte shading and water chemistry is far from simple and more intricate than the relationship of phytoplankton biomass to total phosphorus in lakes.

The development of epiphytic communities on the leaves of vascular plants may reduce net production through several mechanisms other than (PAR) attenuation, including the reduction of diffusive transport of inorganic carbon, nitrogen and phosphorus. Furthermore, the relation between light reduction and loss of photosynthesis is only linear at PAR levels below the light-saturated level of photosynthesis.[5061]   Reduction of photosynthesis (both O_2 production and ^{14}C incorporation) of submersed plants was observed to correspond with increased epiphytic material associated with nutrient enrichment.[5061]

It has been postulated that many of the losses of submerged plants were attributable to cultural eutrophication[349,463,671,1053,1234,1904,2682,3597,3781,4364] (Figure 3-19). The decrease in macrophyte biomass is associated with the mass appearance of littoral filamentous algae in many water bodies.[3595,3599,3781,4515] The importance of filamentous algae for macrophyte decline has been suggested both for submerged[3599,3781] and emergent[4363] macrophytes. The hypothesized mechanisms whereby nutrient additions could lead to reductions in submerged plants involve the promotion of algal growth, either phytoplanktonic[2388] or epiphytic[3781] and a resulting reduction in light available to the vascular plants.[3733,3781,4274,4281] Simpson and Eaton,[4515] on the basis of experimental studies of the photosynthesis of *Elodea canadensis* and the filamentous algae (*Cladophora glomerata* and *Spirogyra* sp.), suggest that the change in water chemistry induced by plant photosynthesis can put algae at an advantage over a vascular competitor.

On the other hand, rooted macrophytes shade underlying communities of benthic algae, epiphytic algae attached to basal leaves, and phytoplankton within the macrophyte stands[450,4283] (Figure 3-18). Macrophyte shading on phytoplankton is expected to be particularly important in shallow habitats where macrophyte cover is extensive and plants reach the water surface.[4281]   Competition for light between phytoplankton and rooted macrophytes is therefore mutual, but macrophyte shading on phytoplankton is only effective in shallow regions, whereas phytoplankton shading on macrophytes always takes place.[4281]

Changes due to macrophytes in the surrounding water column include a reduction in light availability,[2092] decreased water circulation due to the resistance of foliage to water currents, and a change in the chemical milieu of the surrounding water due to photosynthetic and respiratory processes of the

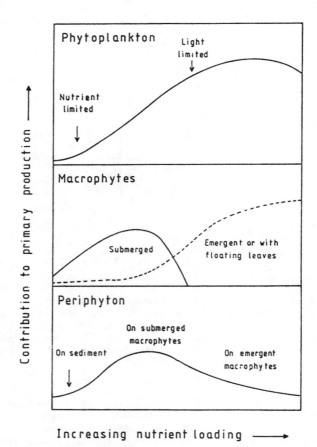

Figure 3-19.   General trends in primary productivity of phytoplankton, macrophytes and attached algae with increasing nutrient loading of relatively small and shallow temperate lake.[4275] (From Sand-Jensen, 1980, in *Vatten,* Vol. 2, p. 110, reprinted with permission of VA-TEKNIK, Vatten, Lund.)

macrophyte community.[1780,2502] These effects themselves vary seasonally as the macrophytes grow, senesce, and enter the detrital phase.[3556]

More factors appear to be in favor of macrophytes reducing (nutrient uptake from water and sediment reducing release to the water, oxidation of surface sediments, shading of phytoplankton, dissolved inorganic carbon uptake and oxygen release to water, release of allelopathic substances, shelter for zooplankton and piscivorous fish, filtrators in macrophyte stands) rather than stimulating (nutrient uptake from sediments and decomposition in water, reduced conditions in dense stands, shelter for planktivorous fish) phytoplankton growth.[4281] The authors, in their excellent paper, pointed out, however, that what is important is the relative magnitude of these factors which can only be evaluated by future experiments.

Sand-Jensen and Borum[4281] showed that nutrient pools in water and sediment are interconnected (Figure 3-20). Nutrient availability is usually much lower in the water column than in the sediment which receives a flux of particle-bound nutrients and carbon from above. Sediment nutrient concentrations and sediment mineralization rates per unit volume are often orders of magnitude above those in the water column. Sediment nutrients are utilized by contact exchange of ions on particles and uptake of dissolved elements.[1056] Conversely, there is usually a diffusive flux of inorganic nutrients from the sediment to the water.[4281] This efflux depends on the presence of rooted macrophytes or microalgae at the sediment surface that can utilize the nutrients and via photosynthesis can oxidize the sediment and convert or immobilize several dissolved elements (iron, manganese, phosphate, ammonium) thereby reducing efflux.[689]

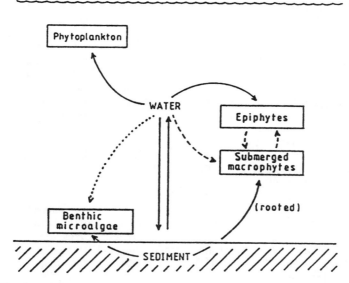

Figure 3-20.   Utilization of nutrient sources in the open waters and in the sediment by phototrophic communities of phytoplankton, benthic microalgae and rooted macrophytes with their associated epiphytic algae. Primary importance of a particular nutrient source is shown by a full line, intermediary importance by a dashed line and small importance by a dotted line.[4281] (From Sand-Jensen and Borum, 1991, in *Aquatic Botany*, Vol. 41, p. 144, reprinted with permission of Elsevier Science Publishers bv, Amsterdam and the authors.)

The half saturation constant for CO_2 uptake by rooted macrophytes is of the order of 100–200 μmol L^{-1},[4280] and this value is about 100-fold lower for most phytoplankton species.[2224,3065] Sand-Jensen and Borum[4281] pointed out, using cited C : N : P ratios[177,4040] and assuming 10-fold lower growth rates for rooted macrophytes,[23,3437,4072] that their requirements per unit time will be 30-fold lower for nitrogen and 50-fold lower for phosphorus compared with microalgae.

3.4 WETLAND SOILS

Soils are complex assemblages of inorganic and organic material at the earth's surface that reflect long-term environmental changes. Specifically, any particular soil is a function of parent material acted on by organisms and climate and conditioned by relief over time.[2330]

Wetland soil is both the medium in which many of the wetland chemical transformations take place and the primary storage of available chemicals for most wetland plants.[3246] It is often described as a *hydric soil.* Hydric soil is a soil that is saturated, flooded, or ponded long enough during the growing season to develop anaerobic conditions that favor the growth and regeneration of hydrophytic vegetation.[5089] Hydric soils may be classified into two broad categories: (1) *mineral soil* or (2) *organic, or peat soil* (also called *histosol*).[1249,3246]

3.4.1 Organic Soils

Cowardin et al.[893] defined organic soils and organic soil materials under either of two conditions of saturation:

1. Are saturated with water for long periods or are artificially drained and, excluding live roots, (a) have 18% or more organic carbon if the mineral fraction is 60% or more clay, (b) have 12% or more organic carbon if the material fraction has no clay, or (c) have a proportional content of organic carbon between 12 and 18% if the clay content of the mineral fraction is between 0 and 60%; or
2. Are never saturated with water for more than a few days and have 20% or more organic carbon.

Item 1 in this definition covers materials that have been called peats or mucks. Item 2 is intended to include what has been called litter or 0 horizon. For further details see Cowardin et al.[893]

Organic soil ("peat") is formed by the accumulation of organic matter where biomass production exceeds decomposition rate.[2351] Typically the development of an organic horizon occurs only when the conditions for bacterial and fungal decomposition are insufficient for the annual deposition of detritus to be completely oxidized.[3659,4979] Long-term flooding, low pH, and cool growing season temperatures all can contribute to low rates of microbial decomposition.[935,3246,4979] Most organic soils occur in wetlands, because rates of decay are slow due to lack of oxygen. Unlike mineral sediment deposition, which depends on inputs of soil material from outside of the wetland, organic soil accumulation depends primarily on the production and decomposition of materials *in situ.*[2351]

Formation of peat consists primarily of three processes: (1) loss of organic matter by leaching or attack by animals and microorganisms; (2) loss of physical structure; and (3) change of chemical state (i.e. the production of new types of molecules by microorganisms and spontaneous chemical reactions).[823] A number of factors can affect the rate of organic matter accumulation:[2007] (1) nature of plant material (different plant species decompose at different rates); (2) climate (influences decomposition rates, depth to permafrost, regional flora and fauna); (3) fire (initial destruction of peat and vegetation, ashes increase available nutrients in peat for several years after fire); (4) geologic factors (glacial readvance, erosion); (5) flooding (can cause anaerobic conditions); and (6) human disturbance (logging, cultivation, drainage, burning, blocking drainages with road fills, etc.).

Rates of peat accumulation in wetlands in Canada, Ireland, and Finland range from about 10 to 100 g dry matter m^{-2} year^{-1}.[3624,4016,5000] Hemond[2028] estimated net peat accumulation in Thoreau's Bog in Massachusetts to be 180 g m^{-2} year^{-1}.

3.4.2 Mineral Soils

Most mineral wetland soils consist of alluvial material: mainly fluvial, lacustrine estuarine or marine (inland parts of coastal plains). Very few wetland soils are on residual parent material. Bottom soils of shallow lakes with limestone in their source area may form an exception. These may be travertine: calcium carbonate precipitated because aquatic vegetation has lowered the CO_2 pressure by photosynthesis.[546] The mineral soils are extremely variable. They may range from soft, semi-liquid, massive or laminated in perennially submerged conditions to consolidated, firm, with a massive, blocky or prismatic structure, in e.g. periodically emergent wetlands. Textures may range from clays to sands and soil profile development may range from absent to very strong.[546] Brinkman and Van Diepen[546] pointed out that in contrast to peat soils, the material of which can only accumulate in wetland conditions, neither the material nor the characteristics of mineral soils of freshwater wetlands are necessarily related to the present wetland conditions. The characteristics of mineral wetland soils depend on their past history, including any periods in which they may have developed above a water table, but the processes currently taking place in these soils are related with the present perennially or periodically water-saturated or flooded conditions.

Mineral soils have been defined by Cowardin et al.[893] as follows: Mineral soil material either

1. is never saturated with water for more than a few days and has < 20% organic carbon by weight; or
2. is saturated with water for long periods or has been artificially drained, and has (a) less than 18% organic carbon by weight if 60% or more of the mineral fraction is clay; (b) less than 12% organic carbon by weight if the mineral fraction has no clay; or (c) a proportional content of organic carbon between 12 and 18% if the clay content of the mineral fraction is between zero and 60%. Soil material that has more organic C than the amounts just given is considered to be organic material.

Organic soils are different from mineral soils in several physiochemical features (Table 3-3).

3.4.3 Wetland Soil Processes

Wetland soils are dominated by anaerobic conditions induced by soil saturation and flooding. Freshwater wetland soils can generally be distinguished from upland, non-wetland soils by two

Table 3-3. Comparison of physical and chemical properties of wetland soils.[414,524,1313,2351,3246,3922,5161]

	Mineral soil	Organic soil
organic content (%)	< 12–20	> 12–20
pH	6.0–7.0	< 6.0
bulk density* (g cm^{-3})	gravel ca 2.1	fibric < 0.09
	sand 1.2–1.8	hemic 0.09–0.20
	clay 1.0–1.6	sapric > 0.20
porosity (%)	gravel 20	fibric > 90
	sand 35–50	hemic 84–90
	clay 40–60	sapric < 84
hydraulic conductivity (m d^{-1})	gravel 100–1,000	fibric > 1.3
	sand 1–100	hemic 0.01–1.3
	clay < 0.01	sapric < 0.01
water holding capacity	low	high
nutrient availability	generally high	often low
cation exchange capacity	low, dominated by major cations	high, dominated by hydrogen ion
mean accretion rate (cm yr^{-1})	0.69	0.12
mean mass accretion (g m^{-2} yr^{-1})	1,680	96

* Bulk density = dry mass of soil material per unit of volume

interrelated characteristics: (1) an abundance of water and (2) accumulation of organic matter. Excess water causes many physical and chemical changes in soils, and wetland hydrologic regimes can range from nearly continuous saturation to infrequent, short-duration flooding (riparian systems). The most significant result of flooding is the isolation of the soil system from atmospheric oxygen, which activates several biological and chemical processes that change the system from aerobic and oxidizing to anaerobic and reducing.[1313] Downward transport of oxygen in flooded soils is extremely difficult due to saturation of the soil with water.[3266,4736] Diffusion of oxygen in an aqueous solution has been estimated at 10,000 times slower than oxygen diffusion through a porous medium such as drained soil.[1555,1812,2777] Evans and Scott[1289] reported that the concentration of oxygen in the water used for saturating the soil decreases to 1/100 of its initial content in 75 minutes.

In flooded soils, oxic metabolism occurs only in a thin oxygenated surface layer while in deeper horizons anoxic processes take place.[121,4736] It has been shown that aerobic activity is very limited several mm below the waterlogged peat surface.[1077] However, De Datta[1026] also adds that, in many flooded soils, there may be small oxidized pockets in the reduced soil matrix and oxidized streaks corresponding to root channels. The oxidized layer is characterized by its brown color and is metabolically aerobic in nature and contains oxidized ions such as Fe^{3+}, Mn^{4+}, NO_3^-, SO_4^{2-}, etc. The reduced subsurface layer is often bluish gray in color, and is metabolically anaerobic with low Eh values and contains reduced products such as ammonia, nitrous oxide, ferrous and manganous salts, sulfides, and the products of anaerobic decomposition of organic matter like aldehydes, alcohols, organic acids, mercaptans, etc.[3263,3837] Anaerobic bacteria function most efficiently at near neutral pH and low redox potential values < -200 mV adjusted to a pH of 7 and given as Eh_7.[121]

Flooding the soil increases the pH of acid soils and decreases the pH of alkali soils, and keeps the soil pH around the neutral point.[3263-4] The pH values of most flooded mineral soils are between 6.7 and 7.2, irrespective of the pH of the drained soils.[3840] However, if the organic matter content of a high-pH soil is low, the pH of the flooded soil may not decrease below 8.0.[2679] After a period of time, an equilibrium is reached and both pH and Eh become stable. In many mineral soils this results in a circumneutral pH, while in soils with high organic content bacterial action may produce organic acids which accumulate and result in a stable pH at a much lower level.[5392]

Once soil is flooded, the oxygen present is quickly consumed by microbial respiration and chemical oxidation.[2226] Subsequently, anaerobic microorganisms use a variety of substances to replace oxygen as the terminal electron acceptor during respiration.[3838] This electron transfer causes significant changes in the valence state of the chemical species used and the overall soil reduction.[1313] Reduction of a saturated soil is a sequential process governed by the laws of thermodynamics.[3838,5051] Nitrate is the first soil component reduced ($NO_3^- \rightarrow NO_2^-$, Eh 220 mV) after oxygen, though this process can proceed before oxygen is completely consumed.[2626] Manganic manganese ($Mn^{4+} \rightarrow Mn^{2+}$, Eh 200 mV) closely follows NO_3^- in the reduction sequence, even before NO_3^- has completely disappeared.[1555,4027,5051] While the preceding reactions can and do overlap, the subsequent sequential reactions of ferric iron to ferrous iron ($Fe^{3+} \rightarrow Fe^{2+}$, Eh 120 mV), sulfate to sulfide ($SO_4^{2-} \rightarrow S^{2-}$, Eh -75 to -150 mV) and carbon dioxide to methane ($CO_2 \rightarrow CH_4$, Eh -250 to -350 mV) will not occur unless the preceding component has been completely reduced.[1313,3246] (The Eh values given are from references 1313 and 3246, some authors, however, have given slightly different values—see References 886, 3836, 3796, 5392.) Bacteria-mediated reduction and oxidation processes in waterlogged soils are summarized in an excellent review by Laanbroek.[2679] For further details see also Chapters 4.1.3, 4.2.3, 4.3.3, 4.4.3, 5.2.2 and 5.3.2.

As long as free dissolved oxygen is present in solution, the redox potential varies little (in the range of +400 to +700 mV). However, it is a sensitive measure of the degree of reduction of wetland soils after oxygen disappears, ranging from +400 mV down to -400 mV.[1555] The greater range of redox potentials for flooded soils versus aerobic soils is important. Natural wetland systems maintain a wider range of redox reactions than upland soils, and their most important function may be as chemical transformers. Wetlands are often the major reducing ecosystems on the landscape and, as such, have great potential for processing nutrients and other materials.[4105]

Periodic or even constant flooding of a soil's surface, characteristic of wetlands, leads to an overall decrease in the activity of soil fauna.[5058] The surface layer of peat has been shown to lack many common forest soil faunal elements and is unable to support them if they are introduced.[1077] In peat, for example, nematodes and earthworms are usually absent due to the unavailability of oxygen.[2578] In addition, most fungi are obligate aerobs, and as most waterlogged wetland ecosystems are anaerobic, fungi members are usually very low and greatly outnumbered by bacteria and actinomycetes.[561]

Gilmour and Gale[1640] presented a review of the chemistry of metals and trace elements in submerged soils. The influence of redox potential on the environmental chemistry of contaminants in soils and sediments is discussed by Gambrell and Patrick.[1556]

Details on wetland and flooded soils can be found in References 152, 546, 621, 1555, 1849, 3292, 3693, 3837–41, 4027, 4871, 5516.

3.5 CHEMICAL MASS BALANCES OF WETLANDS

A quantitative description of the inputs, outputs, and internal cycling of materials in an ecosystem is called an *ecosystem mass balance*.[3447,5375] If the material being measured is one of several elements such as phosphorus, nitrogen, or carbon that are essential for life, then the mass balance is called a *nutrient budget* (Figure 3-21). In wetlands, mass balances have been developed both to describe ecosystem function and to determine the importance of wetland sources, sinks, and transformers of chemicals.[3246] Figures 3-22, 3-23, 3-24 and 3-25 illustrate major chemical storages and flows, biogeochemical pools, and components and processes in wetlands.

Richardson[4105] pointed out that wetlands reportedly maintain unique biogeochemical processes that allow these ecosystems to transform some elements (denitrification of nitrate-nitrogen, reduction of sulfate, etc.), act as a sink or source for others (e.g. carbon), and function as sediment filtering systems in the landscape.[134,228,477,885,1029,2439,3246,3447,3692,4108,4976,5009]

Richardson,[4105] in his excellent paper made several important points which should be stressed here. "The words filter and sink are often used interchangeably in ecosystem level studies, but it is clear from an analysis of their etymology and definitions that these terms are not synonymous." It is suggested to restrict the usage of the term *filter* to the removal of suspended materials and solids from liquids. Thus, by definition a dissolved nutrient (in solution) that is filtered by the wetland will pass through the ecosystem. The term *sink* should be used when we wish to convey the idea that the ecosystem is storing the element or material and that outputs or net yield of material is less than inputs of that material. It is important to keep in mind that a system functioning as a net sink can still release a significant mass of material to downstream ecosystems. This is especially true for wetlands which often receive high loading rates from upstream areas in the watershed. The term *source* should be used when the ecosystem output exceeds the input of the particular material in question. Finally, the word *transformer* is defined as "the thing that changes the form or outward appearance. . . . To change the condition, nature, or function of . . ." Wetlands would be defined as nutrient transformers when processes within the ecosystem result in significant changes in the valence state of an element, change from inorganic to organic forms of the nutrient (or vice versa), or a conversion from the liquid phase

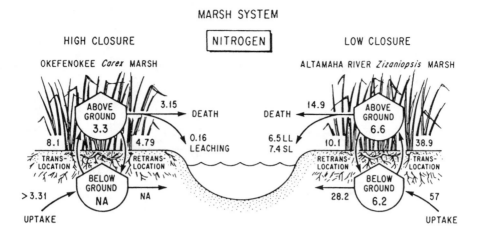

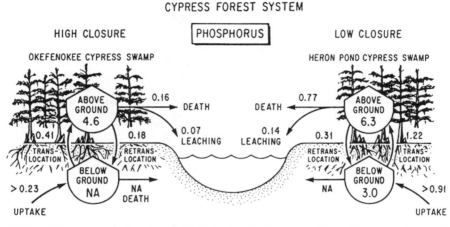

Figure 3-21.   Nitrogen and phosphorus budgets for plant communities of open and closed wetland ecosystems.[2185] (From Hopkinson, 1992, in *Estuaries,* Vol. 15, p. 553, reprinted with permission of the Estuaries.)

to a gaseous phase. It should be noted from these definitions that a wetland can simultaneously function as both transformer and either a sink or source on the landscape."

Richardson[4106] summarized that in general it has been reported that wetlands tend to accumulate nutrients during the growing season, especially N, P and C, and release them during the high spring and fall flows.[2741,2537,3329,4108] The death of wetland vegetation is typically followed by rapid release to the water of 35 to 75% of the plant tissue P and somewhat smaller but substantial amounts of N.[503,992,2536-7]

There has been much discussion as to whether wetlands are nutrient sources, sinks, or transformers. Richardson[4105-6] summarizing a large data set, concluded that:

1. wetlands can function as either a sink or source depending on the element in question, the type of wetland, the loading level, the season of the year, and whether or not the ecosystem is aggrading;
2. wetlands are not efficient sinks for either K or Na;
3. wetlands with organic soils do not retain P as efficiently as forested systems and are often a source rather than a sink;
4. aggrading wetlands are a net sink for C and may have an important role in the global C cycle, a factor not yet recognized by many climatologists; aggrading wetlands export a significant amount of C via hydrologic outflow, especially when compared to terrestrial ecosystems (Figure 3-26);

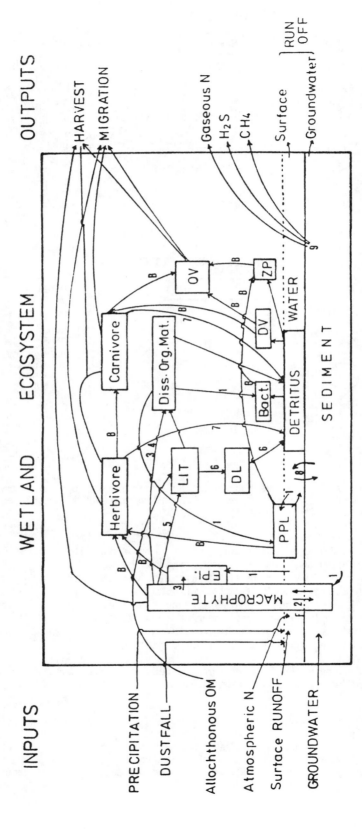

Figure 3-22. Diagrammatic representation of major pathways of nutrient dynamics in a wetland ecosystem. The relative importance of various pools and storages in biota, as well as magnitude of inputs and outputs, varies in different wetland types. Different transfers are: B-transfer of food; F-nitrogen fixation by microbes, including that in the rhizosphere; 1-uptake by primary producers; 2 - translocation between above and belowground organs; 3-excretion; 4-leaching; 5-death; 6-decomposition; 7-excretion by animals; 8-exchanges between sediments and overlying water column; 9-denitrification, fermentation and other anaerobic processes. Several other physicochemical and biological transformations (such as ammonification, nitrification, precipitation, etc.) are not shown. The compartments are: DL-decomposing litter; DV-detritivores; EPI-epiphytes; LIT-litter (fresh); OV-omnivores; PPL-phytoplankton; ZP-zooplankton.[1747] (From Gopal and Masing, 1990, in *Wetlands and Shallow Continental Water Bodies*, Patten, B.C., Ed., p. 181, reprinted with permission of SPB Academic Publishing, The Hague.)

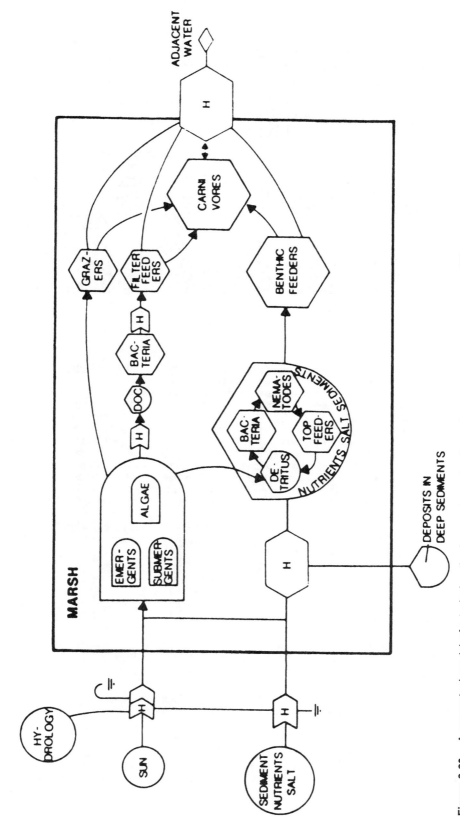

Figure 3-23.   A conceptual model of a typical wetland ecosystem, showing major components and processes.[1770]

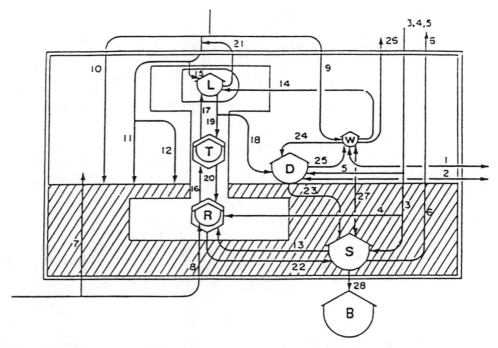

Figure 3-24. Model of major chemical storages and flows in wetlands. <u>Storages</u>: L = aboveground shoots or leaves; T = trunks and branches, perennial aboveground storage; R = roots and rhizomes; W = dissolved and suspended particulates in surface water; D = litter or detritus; S = near-surface sediments; B = deep sediments essentially removed from internal cycling. <u>Flows</u>: 1 and 2 are exchanges of dissolved and particulate material with adjacent waters; 3–5 are nitrogen fixation in sediments, rhizosphere microflora, and litter; 6 is denitrification by sediments (N_2 and N_2O); 7 and 8 are groundwater inputs to roots and surface water; 9 is atmospheric deposition on water; 10 is on land; 11 and 12 are aqueous deposition from the canopy and in stemflow; 13 is uptake by roots; 14 is foliar uptake from surface water; 15 is uptake from rainfall; 16 and 17 are translocations from roots through trunks and stems to leaves; 18 is the production of litter; 19 and 20 are the readsorption of materials from leaves through trunks and stems to roots and rhizomes; 21 is leaching from leaves; 22 is death or sloughing of root material; 23 is incorporation of litter into sediments or peats; 24 is uptake by decomposing litter; 25 is release from decomposing litter; 26 is volatilization of ammonia; 27 is sediment-water exchange; 28 is long-term burial in sediments.[3447]

5. wetlands function as effective transformers of N, P, and C;
6. wetlands maintain biogeochemical processes that are responsible for transforming and releasing significant quantities of dinitrogen.

Studies on nutrient budgets for wetland ecosystems are numerous (see References 886, 902, 922, 1013, 1324–6, 1503, 1633, 1967, 2012, 2029, 2185, 2397, 2428, 2537, 2631, 3246, 3248, 3445–7, 3506, 3876, 3916, 4103, 4105–6, 4109, 4115, 4516, 5097, 5099, 5123, 5153, 5375, 5518).

3.6 WETLAND HYDROLOGY

Wetland hydrology is defined by Environmental Laboratory[1249] as follows. "The term 'wetland hydrology' encompasses all hydrologic characteristics of areas that are periodically inundated or have soils saturated to the surface at some time during the growing season. Areas with evident characteristics of wetland hydrology are those where presence of water has an overriding influence on

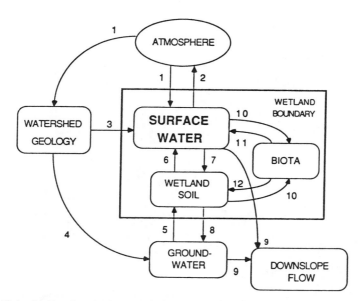

Figure 3-25. Major biogeochemical pools and exchange pathways for a generalized wetland ecosystem. The surface water pool is emphasized. Pathways: 1-precipitation; 2-volatilization; 3-weathering and surface runoff; 4-weathering and infiltration; 5-groundwater discharge; 6 and 7-soil and water exchanges; 8-groundwater recharge; 9-downslope transport; 10-biotic uptake; 11-excretion, leaching, and aboveground decomposition; 12-excretion and belowground decomposition.[4315] (From Schalles, 1989, in *Wetlands Ecology and Conservation: Emphasis in Pennsylvania,* Majumdar, S.K., Ed., p. 76, reprinted with permission of Pennsylvania University Press, Easton.)

characteristics of vegetation and soils due to anaerobic and reducing conditions, respectively. Such characteristics are usually present in areas that are inundated or have soils that are saturated to the surface for sufficient duration to develop hydric soils and support vegetation typically adapted for life in periodically anaerobic soil conditions. Hydrology is often the least exact of the parameters, and indicators of wetland hydrology are sometimes difficult to find in the field. However, it is essential to establish that a wetland area is periodically inundated or has saturated soils during the growing season."

Duever[1173] presented a conceptual hydrologic model (Figure 3-27) which represents a general statement about freshwater wetlands that has relevance both to available field data and feasible management options for both natural systems and systems that have been altered by man. The conceptual model illustrates major system components and their relationships to each other and the dominant external inflows and outflows. The components are functionally, although not physically, isolated from one another.

The hydrology of wetlands creates the unique physiological conditions that make such an ecosystem different from both well-drained terrestrial and deepwater aquatic systems. Hydrologic pathways such as precipitation, surface runoff, groundwater, tides, and flooding rivers transport energy and nutrients to and from wetlands.[3246] Topographic position, stratigraphy, and soil permeability influence both the frequency and duration of inundation and soil saturation.[1249] Areas of lower elevation in a floodplain or marsh have more frequent periods of inundation and/or greater duration than most areas at higher elevations.[1249] Water depth, flow patterns, and duration and frequency of flooding, which are the result of all of the hydrologic inputs and outputs, influence the biochemistry of the soil and are the major factors in the ultimate selection of the biota of wetlands.[3246] Mitsch and Gosselink[3246] pointed out that hydrology is probably the single most important determinant for the establishment and maintenance of specific types of wetlands and wetland processes.

Wetlands are no more passive to their hydrologic conditions than are other ecosystems to their physical environment. Biotic components of wetlands can control their water conditions through a

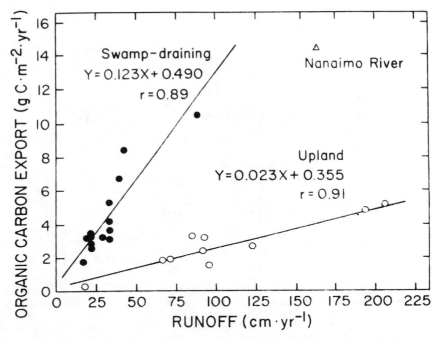

Figure 3-26.   Annual export of organic carbon as a function of annual runoff for various upland (o), swamp-draining (●) and the Nanaimo River (Δ) watersheds.[3330] (From Mulholland and Kuenzler, 1979, in *Limnol. Oceanogr.*, Vol. 24, p. 964, reprinted with permission of American Society of Limnology and Oceanography.)

variety of mechanisms, including peat building, sediment trapping, water shading, and transpiration.[3246] Many marshes and some riparian wetlands accumulate sediments, thereby eventually decreasing the frequency with which they are flooded. Wetland vegetation influences hydrologic conditions by binding sediments to reduce erosion, by trapping sediments, by interrupting water flows, and by building peat deposits.[1770] On the other hand, transpiration rates are higher in areas of abundant plant cover, which may reduce the duration of soil saturation.[1249]

The hydroperiod is the seasonal pattern of the water level of a wetland. It defines the rise and fall of a wetland's surface and subsurface water. Many terms are used to quantitatively describe a wetland's hydroperiod[810,893] (Table 3-4).

The hydroperiod, or hydrologic state of a given wetland can be summarized as being a result of the following factors:[3246]

1. the balance between the inflows and outflows of water
2. surface contours of the landscape
3. subsurface soil, geology, and groundwater conditions.

The first condition defines the water budget of wetland, while the second and the third define the capacity of the wetland to store water. Because water transport is so important to inputs and outputs of nutrients at this level, good water budgets are essential to good nutrient balances.[712] The general balance between water storage and inflows and outflows is expressed as:[2396,3246]

$$\Delta V = P_n + S_i + G_i - ET - S_o - G_o \pm T_{i,o},$$ (3.2)

where

ΔV	= change in volume of water storage in wetland
P_n	= net precipitation
S_i	= surface inflows, including flooding streams
G_i	= groundwater inflows

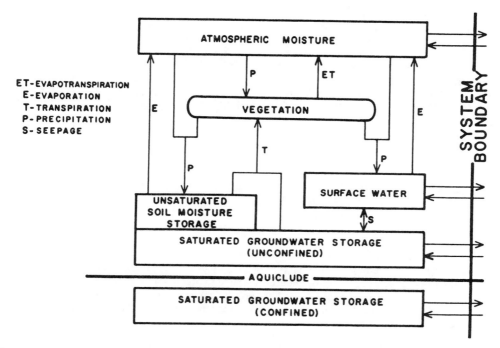

Figure 3-27.   Conceptual model of wetland hydrology.[1173]   (From Duever, 1990, in *Wetlands and Shallow Continental Water Bodies,* Patten, B.C., Ed., p. 62, reprinted with permission of SPB Academic Publishing, The Hague.)

Table 3-4. Wetland hydroperiods[893]

Tidal water regimes (largely determined by oceanic tides)

Subtidal - substrate is permanently flooded with tidal water

Irregularly exposed - the land surface is exposed by tides less often than daily

Regularly flooded - tidal waters alternately floods and exposes the land surface at least once daily

Irregularly flooded - tidal water floods the land surface less often than daily

Nontidal water regimes

Permanently flooded - water covers the land surface throughout the year in all years (vegetation is composed of obligate hydrophytes)

Intermittently exposed - surface water is present throughout the year except the years of extreme drought

Semipermanently flooded - surface water persists throughout the growing season in most years; when surface water is absent, the water table is usually at or very near the land surface

Seasonally flooded - surface water is present for extended periods especially early in the growing season, but is absent by the end of the season in most years; when surface water is absent, the water table is often near the land surface

Saturated - the substrate is saturated to the surface for extended periods during the growing season, but surface water is seldom present

Temporarily flooded - surface water is present for brief periods during the growing season, but the water table usually lies well below the soil surface for most of the season (characteristic vegetation is both upland and wetland)

Intermittently flooded - the substrate is usually exposed, but surface water is present for variable periods without detectable seasonal periodicity

(Artificially flooded - the amount and duration of flooding is controlled by means of pumps and siphons in combination with dams or dikes)

    ET   = evapotranspiration
    S_o = surface outflows
    G_o = groundwater outflows
    $T_{i,o}$ = tidal inflow (+) or outflow (-)

It is beyond the scope of this brief information to discuss all aspects of wetland hydrology; only brief information about precipitation and evapotranspiration will be discussed. For further information about wetland hydrology the reader is referred to following books and papers: References 275, 543, 582, 711–2, 893, 1012, 1173, 1772, 2283, 2397–2401, 2684, 2855, 3245, 3562, 3976, 4482, 4757, 5358, 5161.

3.6.1 Precipitation

The total amount of precipitation that actually reaches the water surface or substrate of a wetland is called *net precipitation* (P_n) and is defined as:[3246]

$$P_n = P - I, \qquad (3.3)$$

where P is total precipitation and I is *interception* (= the amount of precipitation that is retained in the overlying canopy). The amount that actually passes through the vegetation to the water or substrate below is called *throughfall* (TF).[3246] Another term related to precipitation is *stemflow* (= SF, water that passes down the stems of the vegetation). This flow is generally a minor component of the water budget of a wetland. Giving all parameters together yields the most commonly used form for estimating net precipitation in wetlands:[3246]

$$P_n = TF + SF \qquad (3.4)$$

3.6.2 Evapotranspiration

Atmospheric water losses from a wetland occur from the water and soil—*evaporation,* and from emergent portions of plant—*transpiration.* The combination is termed *evapotranspiration.*[2400] Evapotranspiration is an important component of the water balance and ecology of wetland ecosystems.[2684] Koerselman and Beltman[2549] pointed out that evapotranspiration is a major component of the water budget of peatlands. The values of evapotranspiration vary widely, some examples from different regions of the world are given in Table 3-5. A number of workers[318,341,1024,2270,3732,3976,4611,4977] have shown that most aquatic plants enhance water loss; the evapotranspiration to open water evaporation ratio (E/E_o) varies, however (Table 3-6).

The water surface may be permanently or periodically saturated, with periods of shallow standing water. A variable fraction of the surface area may be bare or covered with a litter layer which forms an effective mulch.[2399] This complexity would appear to lead to widely disparate water losses, but in fact does not. Two factors cause a narrow range of values: the energy from vaporization must come from the sun; and many wetlands are normally saturated, and thus capable of realizing their full evaporation potential. The meteorological factors which modify the solar energy utilization for vaporization are well known: wind, relative humidity, temperature and cloud cover.[2399]

A number of predictive equations have been developed to allow estimation of evapotranspiration in wetlands. Meteorological factors together with a cover type are all recognized as controlling factors, but the most useful predictive techniques are usually those which use only the meteorological measurements which are commonly available. The more popular general methods do not account for variation in transpiration between plant species.[2400] Reviews of the major techniques currently in use, have been presented by Chang,[745] Chow,[793] Penman,[3735–6] Knisel,[2541] and Scheffe.[4321] Several direct measurement techniques can be used in wetlands to determine evapotranspiration,[588,2003,2400,3247] using available data, summarized the main features of wetland evapotranspiration.

Table 3-5. Evapotranspiration (ET) in wetlands

Locality	ET (mm d^{-1})	Reference
Quaking fen, The Netherlands	2.5	2549
Subarctic coastal marsh, Canada, Ontario	2.6–3.1	2684
Phragmites australis marsh, Czech Republic	3.2	3890
Carex acutiformis/Spahgnum marsh, The Netherlands	1.0–3.7	2549
Cypress dome, Florida	3.1–3.8	588
Carex diandra marsh, The Netherlands	1.1–3.9	2549
Sedge - grass marsh, Czech Republic	2.2–4.5	3891
Arctic bog, Canada	4.54	4201
Typha latifolia marsh, The Netherlands	0.9–4.7	2549
Willow carr, Czech Republic	2.4–4.8	3891
Freshwater marsh, Florida	5.1	1117
Forested wetlands, Florida	0.9–5.6	587
Typha sp. marsh, Czech Republic	3.2–5.7	3887
Nymphaea lotus marsh, India	2.5–6.0	3976
Tall sedges community, Czech Republic	6.5 (max.)	3888–9
Salvinia molesta marsh, India	2.1–6.8	3976
Phragmites australis marsh, Czech Republic	1.4–6.9	4574
Eichhornia crassipes marsh, India	3.8–10.2	3976
Marsh grasses community, Czech Republic	2.0–10.5	4234
Phragmites australis marsh, Czech Republic	6.9–11.4	2654, 4233

Table 3-6. Ratio of evapotranspiration and evaporation (E/E$_o$) for different wetlands

Plant species/locality	E/E$_o$	Reference
Sedge dominated marsh, Canada, Ontario	0.74–0.85	2684
Nymphaea lotus, India	0.82–1.35	3976
Salvinia molesta, India	0.96–1.39	3976
Carex acutiformis/Sphagnum, The Netherlands	1.65	2549
Carex diandra, The Netherlands	1.68	2549
Typha latifolia, The Netherlands	1.87	2549
Eichhornia crassipes, India	1.30–1.96	3976
Typha latifolia, Poland	1.20–2.40	341
Typha litifolia, Alabama	1.05–2.50	4611
Eichhornia crassipes, Alabama	1.31–2.52	4611
Eichhornia crassipes, Florida	1.50–2.52	1024
Eichhornia crassipes, Texas	3.20–5.30	2270
Eichhornia crassipes, worldwide	2.50–9.80	979, 3732, 4654, 4977

CHAPTER 4

Macronutrients

4.1 CARBON

4.1.1 Global Cycle

Carbon is a key element of life. The exceptional position of carbon among other elements is based on the ability of its atoms to link manifoldly together to form a huge variety of compounds which is essential to the development of living things.[5543]

The global aspects of the carbon cycle and carbon fluxes have been reviewed many times.[428-9,490-1, 1727,1800,5543] The carbon soil cycle has been reviewed extensively by Stevenson.[4738] The global cycle of carbon can be roughly divided into two cycles, a biological cycle and a geological cycle.[5543] The major active reservoirs of carbon are the atmosphere, oceans, terrestrial biota and pedosphere and lithosphere.[428] Figure 4-1 gives a schematic diagram of the global carbon cycle and the main fluxes and Figure 4-2 gives the carbon cycle and fluxes in the oceans, but the estimates, so far reported in the literature, vary considerably (see Zehnder[5543] for discussion).

Within its global biogeochemical cycle, the element carbon occurs in the form of various chemical compounds that are continuously being transformed and moved within two major, mutually interconnected cycles:[1727] (1) the inorganic carbon, or carbonate cycle in which carbon passes through a series of chemical equilibria; and (2) the organic carbon cycle in which carbon passes through the processes of biosynthesis and mineralization of organic matter (Figure 4-3). From the global viewpoint, two interlocking carbon cycles have been recognized, the exogenic cycle (including the carbonate and the organic carbon cycles) with carbon cycled by the atmosphere, hydrosphere, and biosphere, and the endogenic cycle, with carbon cycled by the crust of the earth, the lithosphere and, in part, by the external mantle (Figure 4-4).

Although CO_2 is only a trace gas in our atmosphere, it plays a key role in maintaining life on our planet in its function in photosynthesis. The biological CO_2 cycle is shown in Figure 4-5. CO_2 exchange in plants has been reviewed many times, e.g. by Larcher.[83] In pre-industrial times the CO_2 cycle was essentially balanced. With progressive civilization vast areas of forests were destroyed and decomposition processes and woodburning converted the carbon bound in plants into CO_2. The actual amount of the released CO_2 is controversial, however.[5543]

4.1.2 Carbon in the Environment

C in total biosphere: $0.75 - 7.00 \times 10^{14}$ kg[490]
Earth's crust: 480 mg kg^{-1},[490] 250 mg kg^{-1} (903)
Common minerals: $CaCO_3$ (limestone)
Soils: 20,000 org.C (7,000-500,000) mg kg^{-1}, up to 90% of soil C may be found as either humus or
    $CaCO_3$ (490-1)
Sediments: 29,400 mg kg^{-1} (491)
Air: 164,000 mg m^{-3} as CO_2 (490-1)
Fresh water: 11 (6–19) mg L^{-1} (490-1)
Sea water: 28 mg L^{-1} (variable) (490-1)
Land plants: 45.4% DM[490]
Aquatic and wetland macrophytes: 43–48% DM,[5336] 29.3–48.8% DM[510]
Freshwater algae: 12.2–67.7% DM (Table 4-1)
Marine algae: 28–35% DM,[491] 25–36.1% DM (Table 4-1)
Land animals: 46.5% DM[490]
Marine animals: 40% DM[491]

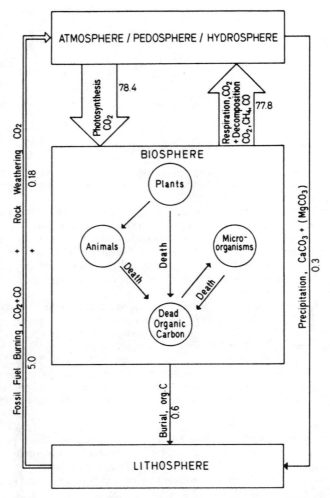

Figure 4-1.   The global carbon cycle. Fluxes are given in 10^{15} g C year^{-1}.[5543] (From Zehnder, 1982, in *The Handbook of Environmental Chemistry,* Vol. 1, Part B, Hutzinger, O., Ed., p. 86, reprinted with permission of Springer-Verlag and the author.)

4.1.2.1 *Aquatic chemistry*

The CO_2 - HCO_3^- - CO_3^{2-} equilibrium system is a major component of the buffering system of most natural waters.[5300-1,5333] The equilibrium (Figure 4-6) depends on the hydrogen ion concentration, amount of excess base, the partial pressure of carbon dioxide in the atmosphere and the temperature:[455,1699,1727,1799,3815,3825]

$$CO_2 + H_2O \overset{pK' = 1.47*}{\Longleftrightarrow} H_2CO_3 \overset{pK' = 6.35*}{\Longleftrightarrow} HCO_3^- + H^+ \overset{pK' = 10.33*}{\Longleftrightarrow} CO_3^{2-} + 2H^+ \qquad (4.1)$$

(* at 25°C)

The first reaction is slow, the second and the third reactions are very fast.[1727,3815] The mechanism of carbon dioxide neutralization depends on pH. At pH < 8 the following reaction takes place immediately:[3815]

$$H_2CO_3 + OH^- \Leftrightarrow HCO_3^- + H_2O \qquad (4.2)$$

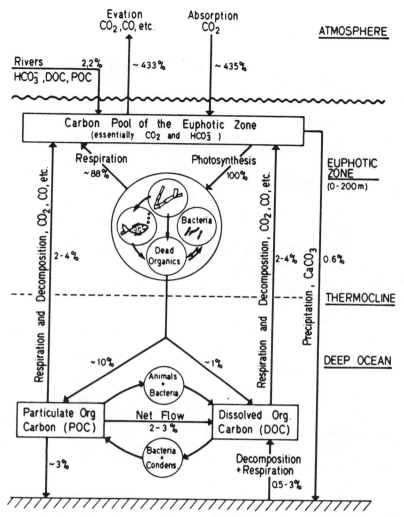

Figure 4-2.   Fate of the carbon in the ocean. The fluxes are given in percent of the net primary production of the euphotic zone.[5543] (From Zehnder, 1982, in *The Handbook of Environmental Chemistry,* Vol. 1, Part B, Hutzinger, O., Ed., p. 90, reprinted with permission of Springer-Verlag and the author.)

Above a pH of 8, the following reactions are dominant:[1727,3815]

$$CO_2 + OH^- \Leftrightarrow HCO_3^- \tag{4.3}$$

and

$$HCO_3^- + OH^- \Leftrightarrow CO_2^{3-} + H_2O \tag{4.4}$$

The equilibrium system is largely affected by changes in pH resulting from algal growth. The extraction of CO_2 from an algal growth system through assimilation into algal biomass at a rate faster than it can be replaced through atmospheric CO_2 diffusion, respiration, fermentation processes, and readjustment of solid carbonate equilibria leads to an increase in pH level.[1699]

Weber and Stumm[5300-1] and Kleijn[2533] showed the effect of $CaCO_3$ saturation on the buffering capacity of natural waters. A water saturated with $CaCO_3$ is considerably more buffered than the same

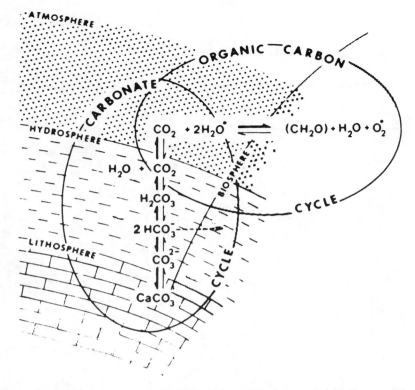

Figure 4-3.   Interaction between the organic and inorganic (carbonate) carbon cycles. Processes promoting precipitation, bold arrows; processes promoting dissolution, fine arrows; incorporation of bicarbonate by plants, dashed arrow.[1727] (From Golubic et al., 1979, in *Biogeochemical Cycling of Mineral-Forming Elements,* Trudinger, P.A., and Swaine, D.J., Eds., p. 30, reprinted with permission of Elsevier Science Publishers bv, Amsterdam and the authors.)

water which is unsaturated. It follows that the water in its heterogenous natural environment is more strongly buffered than the same water studied in the laboratory. Thus, factors other than the homogenous CO_2 - HCO_3^- - CO_3^{2-} equilibrium system help control the buffering capacity of these waters and their role in supplying or denying carbon for the growth of algae is still not fully understood.[1699] Thermodynamically, it would take considerable energy to cause the CO_2 from solid phase $CaCO_3$ to become available for algal growth without addition of H^+. Therefore the utilization of CO_2 would lead to a rise in pH and actually lead to a decrease in available inorganic carbon for growth as in the following reaction which would predominate at a pH greater than 6.5:[1699]

$$Ca^{2+} + 2\ HCO_3^- = CO_2 + CaCO_3 + H_2O \qquad (4.5)$$
$$\text{(algal uptake)} \quad \text{(precipitate)}$$

4.1.3 Carbon Cycling in Wetlands

In recent years, it has been suggested that wetlands may be a vital link in regulating global atmospheric CO_2 concentrations.[135,4103] Wetland peat deposits in the biosphere constitute a major long term sink for carbon.[528] Sjörs[4534-5] calculated that peatlands contain some 300×10^9 tons of C, which represents over 40% of the C content (700×10^9) of the present atmosphere. He suggests that the withdrawal of C from the atmosphere by wetlands has been so slow that the atmosphere has been able to replace these losses through equilibrium with the oceans. On a worldwide scale, the annual

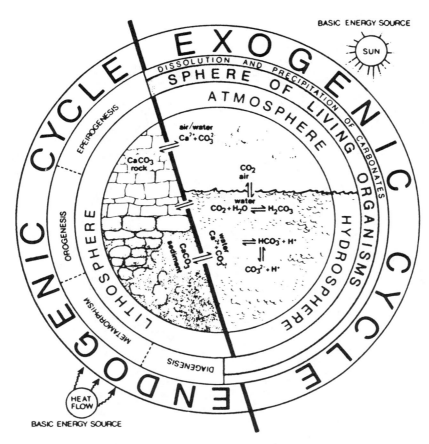

Figure 4-4.   Schematic representation of the exogenic and endogenic cycling of inorganic carbon.[1727] (From Golubic et al., 1979, in *Biogeochemical Cycling of Mineral-Forming Elements,* Trudinger, P.A., and Swaine, D.J., Eds., p. 34, reprinted with permission of Elsevier Science Publishers bv, Amsterdam and the authors.)

accumulation rate of organic matter in wetlands has been estimated to represent 0.135×10^9 t yr^{-1} of C. A larger estimate of 0.21×10^9 t yr^{-1} for C storage was reported by Bramryd[529] and was based on a mean global accumulation rate of 0.7 mm yr^{-1} of organic matter. Armentano and Verhoeven[135] pointed out that differences in climate and chemical properties of histosols explain some of the large regional differences in carbon accumulation rates in wetland sediments. In their excellent review paper[135] they reported that the accretion rate for organic soils in different wetlands in the temperate zone range from 0.20 to 1.22 t C ha^{-1} yr^{-1}.

Temperate zone wetlands which accumulate C under undisturbed conditions have been subjected to extensive drainage activities for agriculture and forestry.[134,136] It has been suggested that wetland organic soils, including tundra and boreal forest areas, sequestered as much as 0.13×10^9 t C yr^{-1} under undisturbed conditions, but after drainage could release as much as 0.21×10^9 t C yr^{-1}.[134] Estimated rates of C release from the recent development of tropical wet organic soils is 0.074 to 0.108×10^9 t C yr^{-1}.[136] Armentano and Verhoeven[135] reported that mean carbon release from organic wetland soils in various regions of the temperate zone varies from 0.56 to 21.94 t C ha^{-1} yr^{-1} due to regional differences in climate and properties of histosols even if cropping patterns are similar.

Armentano and Verhoeven[135] pointed out that the balance in wetlands drained or otherwise altered may be radically changed from the cycle in undisturbed wetlands. When the water table is lowered, oxygen is introduced into deposits lying above the water table, promoting decomposition of the organic matter to its simple inorganic constituents.[1218,4902] Consequences for the carbon cycle are seen largely

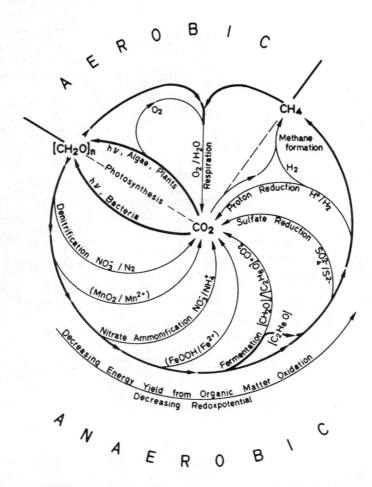

Figure 4-5.   The biological CO_2 cycle.[5543] (From Zehnder, 1982, in *The Handbook of Environmental Chemistry,* Vol. 1, Part B, Hutzinger, O., Ed., p. 104, reprinted with permission of Springer-Verlag and the author.)

as CO_2 release when organic soils are drained.[134] However, drained wetlands probably act as CH_4 sink, but at rates well below the CO_2 releases.[1938]

Richardson[4106] pointed out that the amount of C stored in wetlands and the potential magnitude of the C flux from these ecosystems as a result of drainage suggest that the role of wetlands in global C cycling has been greatly underestimated.

Oxygen is supplied to wetland soils by diffusion through the floodwater water and through transport via the vascular system of wetland plants. Under both conditions, oxygen is supplied more slowly than the potential consumption rate. These conditions result in the development of two distinctly different soil layers: 1) a thin oxidized or aerobic surface layer where oxygen is present, and 2) an underlaying reduced or anaerobic layer in which no free oxygen is present.[3714]

Pathways of carbon during decomposition in sediments are given in Figure 4-7. Typically, in a drained soil, oxygen can be used as an electron acceptor, during aerobic respiration. Upon flooding sequential reduction of electron acceptors occurs as a function of Eh. Oxygen is reduced at Eh of > 300 mV, followed by the reduction of NO_3^- and Mn^{4+} at Eh of about 200 mV. Iron reduction is initiated after all NO_3^- is reduced and is complete at about -100 mV. Sulfate reduction is initiated at Eh of less than 100 mV, followed by CO_2 reduction.[4032] Although redox reactions may be far from equilibrium, they tend to occur in the order of their energy yield. This may be seen in the framework

Table 4-1. Carbon concentration in algae (% dry mass)

Species	Concentration	Locality	Reference
Freshwater			
Cladophora glomerata	26.7–33.8	Clark Fork, Montana	2899
Cladophora glomerata	12.2–40.5	Lake Erie, Michigan-Ohio	2909
Chara sp.	44.6	Davis, California	5343
Anabaenopsis sp.	44.9	Laboratory	3235
Chlamydomonas sp.	45.5	Laboratory	3235
Chlorella sp.	41.1–47.4	Outdoor mass culture	5275
Oikomonas termo	49.1	Laboratory	3235
Chlorella sp.	43.5–49.1	Laboratory mass culture	5275
Stichococcus bacillaris	49.2–50.7	Laboratory	3235
Chlorella pyrenoidosa	47.8–67.7	Laboratory	3235
Chlorella sp.	41.2–67.7	Laboratory	4657
Marine			
Amphipleura rutilans	25.0	Half Moon Bay, California	3235
Macrocystis pyrifera	28.7	Half Moon Bay, California	3235
Navicula torquatum	31.4	Half Moon Bay, California	3235
Egregia menziesii	34.1	Half Moon Bay, California	3235
Gigartina agardhii	35.3	Half Moon Bay, California	3235
Ulva sp.	36.1	Half Moon Bay, California	3235

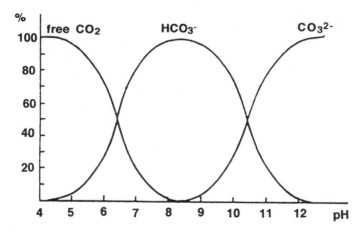

Figure 4-6.   Relationship between pH and percentage abundance of CO_2, HCO_3^- and CO_3^{2-}.

of the biochemical redox cycle (Figure 4-8). The oxidation of organic matter produced in photosynthesis yields energy; the amount of energy depends on the nature of the oxidant, or electron acceptor. The energetically most favorable oxidant is oxygen; after oxygen is depleted there follows a succession of organisms capable of reducing NO_3^-, MnO_2, $FeOOH$, SO_4^{2-}, CO_2 with each oxidant yielding successively less energy for the organism mediating the reaction.[5333]

Pathways of organic carbon decomposition in wetland soil during aerobic, facultative anaerobic and obligate anaerobic respiration are presented in Figure 4-9. Detailed reviews on these pathways were

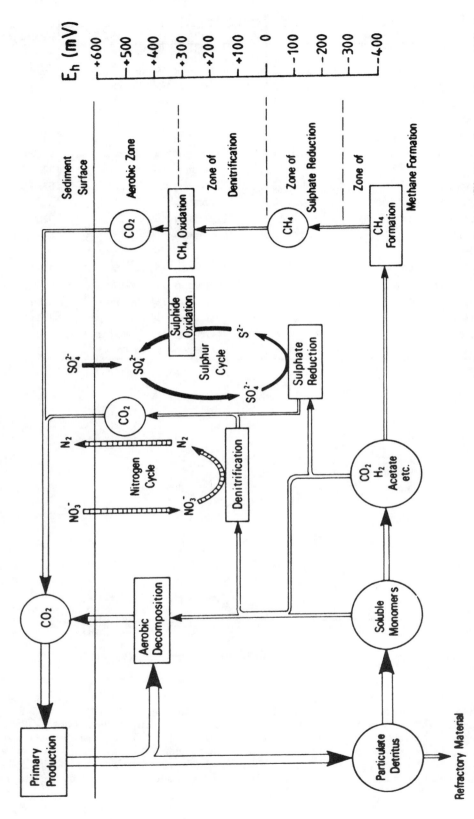

Figure 4-7. Pathways of carbon during decomposition in an aquatic sediment: relationship with sulfur, nitrogen and redox.[1799] (From Grant and Long, 1981, in *Environmental Microbiology*, p. 2, reprinted with permission of Blackie Academic & Professional, an imprint of Chapman & Hall.)

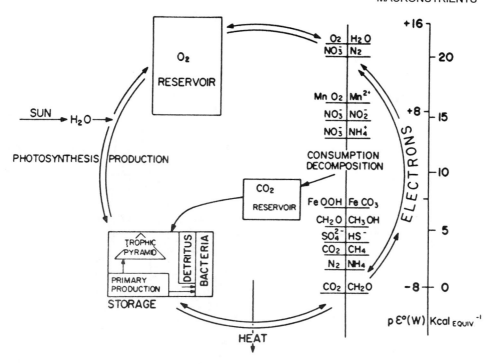

Figure 4-8.   The biochemical redox cycle. Heterotrophs utilize the oxidants which yield the greatest amount of energy, as long as these oxidants are available; then less energy yielding oxidants are utilized.[5333] (From Westall and Stumm, 1980, in *The Handbook of Environmental Chemistry*, Vol. 1, Part B, Hutzinger, O., Ed., p. 49, reprinted with permission of Springer-Verlag and the authors.)

presented by Yoshida[5523] and Fenchel and Jorgensen.[1333] In wetland soils, aerobic respiration occurs primarily in the floodwater, aerobic soil layer (oxygen reduction zone) and in the rhizosphere of wetland plants. Soil organic carbon decomposition in the surface aerobic layer depends on the supply of readily oxidizable organic compounds and steady supply of oxygen. Aerobic decomposition of soil organic C in wetland soils is primary regulated by oxygen supply, since C is usually not limiting.[4032] The oxidation of organic compounds is catalyzed by heterotrophic organisms according to the following reaction:

$$(CH_2O) + O_2 \rightarrow CO_2 + H_2O \qquad (4.6)$$

Oxygen diffusing through floodwater is rapidly consumed by microbial respiration in the first few mm of surface soil. Below this zone obligate aerobes can no longer function. The microbial community shifts to facultative anaerobic bacteria, which utilize NO_3^-, oxidized manganese compounds, and then ferric iron compounds as electron acceptors.[4032]

Anaerobic respiration occurs in the soil zone below the Fe^{3+} reduction zone (Figure 4-9). The process can be carried out in wetland soils by either facultative or obligate anaerobes. It represents one of the major ways in which high molecular mass carbohydrates are broken down to low molecular weight organic compounds, usually as dissolved organic carbon, which are, in turn, available to microbes.[5096] The primary end-products of fermentation are fatty acids such as acetic, butyric, and lactic acids, and the gasses CO_2 and H_2:[3246,4032]

$$C_6H_{12}O_6 \rightarrow 2\ CH_3CHOHCOOH \qquad (4.7)$$
$$\text{(lactic acid)}$$

or

$$C_6H_{12}O_6 \rightarrow 2\ CH_3CH_2OH + 2\ CO_2 \qquad (4.8)$$
$$\text{(ethanol)}$$

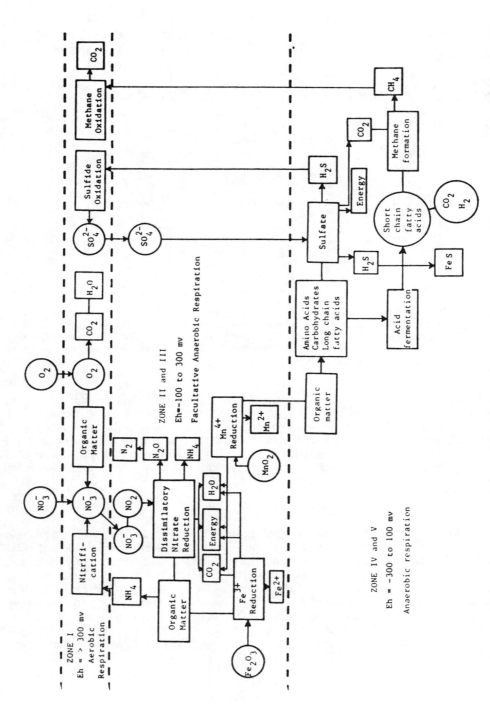

Figure 4-9.   Pathways of organic carbon decomposition in a wetland soil during aerobic, facultative anaerobic and obligate anaerobic respiration.[4032] (From Reddy and Graetz, 1988, in *Ecology and Management of Wetlands*, Vol. 1, Hook, D.D., et al., Eds., p. 308, reprinted with permission of Timber Press, Inc., Portland.)

Acetic acid is the primary acid formed in most flooded soils and sediments. Sulfate-reducing and methanogenic bacteria then utilize the end-products of fermentation and in fact depend on the complex community of fermentative bacteria to supply substrate for their metabolic activities. Both groups play an important role in organic matter decomposition and carbon cycling in wetland soil environments.[4032]

The content of carbon dioxide in the air of flooded soils and within the aquatic plants may reach 20%.[948,3689,4736,5527] Although CO_2 is produced in excess of the oxygen consumed, its total production under the flooded conditions is usually lower than that in well drained soils, as the mineralization rate of organic carbon in anoxia is lower and some carbon is emitted to the atmosphere as methane. Carbon dioxide formed in flooded soils is partly reduced to methane, and is partly emitted to the atmosphere via the soil and/or via the plants.[2081,2172,5283,5513] In the latter case it can be partly or entirely assimilated in the plant shoots.[951] Although the diffusion rate of CO_2 in air is somewhat lower than that of O_2, in water-saturated soils it proceeds 24 times faster than diffusion of O_2 at the same partial pressure gradient. Nevertheless, diffusion through the soil is not efficient enough to remove all the CO_2 produced in the flooded soil, so it is emitted by ebullition together with methane and other gases.[4736] It has been found that ebullition is very important in the unplanted field and in the presence of plant cover, internal transport via the plants dominated.[2172] Wada et al.[5217] found that upward diffusion of CO_2 through the soil and through the floodwater was negligible in comparison to that through the rice plants. Of the total carbon transported from the soil solution to the shoot about 70% was fixed in the shoot under light conditions, but 100% was recovered in the air under dark conditions.

Methane is an important gas evolving from flooded soils. Most of the atmospheric methane is produced biogenetically; rice fields, marshes and lakes are the most important sources.[808,950,1323,4390] The major pathways of methane formation are 1) reduction of electron acceptors and 2) decarboxylation of acetic acid at the expense of hydrogen:[1800,5523]

$$4\ H_2 + CO_2 \rightarrow CH_4 + 2\ H_2O \qquad (4.9)$$

$$CH_3COOH + 3\ H_2 \rightarrow 2\ CH_4 + 2\ H_2O \qquad (4.10)$$

or by a unique aceticlastic reaction:[1800]

$$CH_3COOH \rightarrow CH_4 + CO_2 \qquad (4.11)$$

The relative importance of the two substrates, CO_2 and acetic acid, may depend on the organic loading to the sediment.[2362] In freshwater lake sediments, acetate is the dominant substrate[4802,5457] whereas in marine sediments, CO_2 appears to be dominant as the CH_4 precursor. Methane production requires extremely reduced conditions, with a redox potential more negative than -330 mV, after other terminal electron acceptors (O_2, NO_3^-, Mn^{4+}, Fe^{3+}, SO_4^{2-}) have been used. Methane is produced by a taxonomically diverse group of strictly anaerobic archaebacteria, the methanogens, which use only a limited range of substrates.[1800,2172]

Methane is either released to the atmosphere or is oxidized to CO_2 by methanotrophic bacteria as soon as it enters the oxic zone of an aquatic environment (Figure 4-10). It has been found that in the submerged soil without plants, 35% of the produced CH_4 was emitted to the atmosphere.[2172] The presence of rice plants stimulated methanogenesis in the submerged soil but also enhanced the CH_4 oxidation rates within the rhizosphere, so that only about 23% of the produced CH_4 was emitted.[2172]

Stepniewski and Glinski[4736] summarize that the emission of methane occurs due to ebullition and/or plant-mediated transport.[808,948–50,2127,4390] The contribution of the plant-mediated transport depends on the kind and developmental stage of the plant cover. It has been found that in an unvegetated paddy field CH_4 was emitted almost exclusively by ebullition, whereas in the presence of well-established plant cover (rice, reed, weeds) 60–94% of CH_4 emission was due to plant-mediated transport.[4736]

In general, methane is found at low concentrations in reduced soils if sulfate concentrations are high.[1555,5096] Possible reasons for this phenomenon include (1) competition for substrates that occurs between sulfur and methane bacteria, (2) the inhibitory effects of sulfate or sulfide on methane bacteria, or (3) a possible dependence of methane bacteria on products of sulfur-reducing bacteria.[1555] Valiela[5096] suggested that methane may actually be oxidized to CO_2 by sulfate reducers.

The sulfur cycle is very important in some wetlands for the oxidation of organic carbon. This is particularly true in coastal wetlands (salt marshes and mangroves) where sulfur is abundant.[3246] Sulfur-reducing bacteria require an organic substrate, generally of low molecular weight, as a source of energy in converting sulfate to sulfide. The fermentation process can supply these necessary low molecular mass organic compounds, such as lactate:[5096]

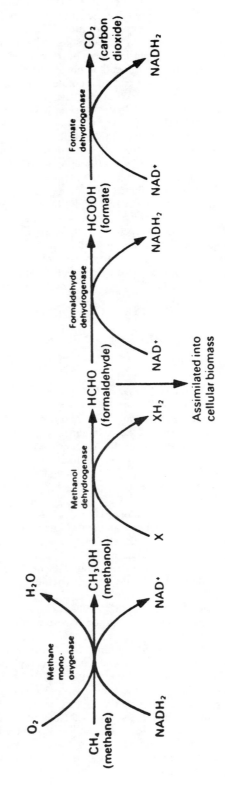

Figure 4-10.   Microbial oxidation of methane.[963]   (From Dalton and Stirling, 1982, in *Phil. Trans. R. Soc. Lond.*, Vol. B 297, p. 484, reprinted with permission of The Royal Society, London and the authors.)

$$2\ CH_3CHOHCOO^- + SO_4^{2-} + 3\ H^+ \rightarrow 2\ CH_3COO^- + 2\ CO_2 + 2\ H_2O + HS^- \qquad (4.12)$$
$$\text{(lactate)} \qquad\qquad\qquad\qquad \text{(acetate)}$$

and

$$CH_3COO^- + SO_4^{2-} \rightarrow 2\ CO_2 + 2\ H_2O + HS^- \qquad (4.13)$$

Methane cycling in aquatic environments has extensively been reviewed by Rudd and Taylor.[4221] Tropical regions (20°N - 30°S) were calculated to release 66 Tg CH_4 per year, 60% of the total wetland emission of 109 Tg year[-1]. Emissions from subtropical and temperate wetlands (45°N-20°N and 30°S-50°S) were calculated relatively low at 5 Tg year[-1]. Northern wetlands (north 45°N) were calculated to release a total of 38 Tg CH_4 year[-1] (34% of total flux); 34 Tg year[-1] from wet soils and 4 Tg year[-1] from relatively dry tundra.[229] Mitsch and Gosseling[3246] reported, based on literature data, the rates of methane production in freshwater and saltwater wetlands in the range of 1–440 mg C m[-2] d[-1] and 0.8 to 109 mg C m[-2] d[-1], respectively. Extensive flux data base and the wetland areas were compiled by Matthews and Fung.[3106]

The global cycle of carbon in wetlands has been presented by Armentano and Verhoeven[135] or Valiela[5096] (Figure 4-11). Carbon balance in local wetlands has been reviewed e.g. by Sjörs,[4535] Valiela et al.[5100] and Richardson.[4105] Two examples of carbon budget and fluxes in a coastal North Carolina wetland and a Mississippi River deltaic marsh are given in Figures 4-12 and 4-13, respectively.

4.1.4 Utilization of Carbon by Algae

4.1.4.1 Uptake and utilization of inorganic carbon

Approximately 90% of all photosynthetic CO_2 fixation is attributed to algae.[444] From planktonic microcells to the massive 50 or more meter kelps, algal cell wall polymers comprise nearly half of the total nonaqueous biomass; between one tenth to one half of all the solar energy converted into chemical bond energy on our planet is stored in the cell walls of algae.[4483]

Most algae are submerged aquatic organisms, and so are exposed to H_2CO_3 and its ions, HCO_3^- and CO_3^{2-}, as well as unhydrated CO_2 available to terrestrial plants.[4001] Water in equilibrium with air at 15°C contains about 10 μM CO_2 dissolved in it; the amount of HCO_3^- and CO_3^{2-} increases with pH. Below pH 5 only free CO_2 is of any importance, between pH 7–9 bicarbonate is most significant and above pH 9.5, carbonate begins to be important. Thus in very acidic waters, only CO_2 is likely to be utilized, whereas at higher pH, bicarbonate may enter the cells and above pH 9.0 calcium will be precipitated resulting in calcium-deficient water.[4209]

It is generally considered that algae use free CO_2 in photosynthesis but at very high pH, i.e. from 9.0 upwards the absence of free CO_2 may be an important ecological factor; the number of species is reduced at this level although many other factors could also be involved.[4209] All algae seem to be able to take up free CO_2 which readily diffuse across cell membranes and is exclusively utilized as a substrate for RuBPC.[4001] (For review on RuBPC see Jensen and Bahr[2333].) Not all algae can take up HCO_3^-.[1857,2356,2993,3572–3,3996] However, there are a large number of algae which have been proved to use HCO_3^-.[264,2276,2356,2759,2993,3572,3992,4945,5017] Under conditions, where photosynthesis is limited by dissolved inorganic carbon supply, the use of HCO_3^- along with CO_2 may give the organism a competitive advantage over organisms that rely on CO_2 only. However, uptake of HCO_3^- at rates sufficient to support net photosynthesis requires active transport of HCO_3^-[3992,3996] and thus energy. Maintenance of the transport apparatus also requires energy.[4011] It is possible that the rate at which inorganic carbon enters the algal cell can limit the rate of photosynthesis under natural conditions.[3992,4339]

Algae can utilize CO_2 from four major sources in a natural water:[1699]

1) from CO_2 diffused from the atmosphere;
2) from the respiration of heterotrophic forms;[2702]
3) from anaerobic fermentation;
4) from bicarbonate alkalinity.

The first three sources provide a direct supply of CO_2. The alkalinity, on the other hand, provides CO_2 by a continual readjustment of the concentrations of the various carbon sources making up the CO_2 - HCO_3^- - CO_3^{2-} system as shown by the following equations:[1699]

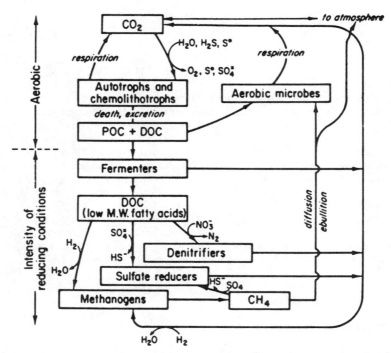

Figure 4-11.   Major transformations of carbon in wetlands.[5096] (From Veliela, 1984, in *Marine Ecologi-cal Processes,* p. 303, reprinted with permission of Springer-Verlag and the author.)

$$2 \, (HCO_3^-) \Leftrightarrow (CO_3^{2-}) + (H_2O) + (CO_2) \tag{4.14}$$

$$(HCO_3^-) + (H_2O) \Leftrightarrow (H_2O) + (CO_2) + (OH^-) \tag{4.15}$$

$$(CO_3^{2-}) + (H_2O) \Leftrightarrow (CO_2) + 2 \, (OH^-) \tag{4.16}$$

Typically, the pH of natural waters at equilibrium with air is about 8.3 where HCO_3^- is the major ion (see Figure 4-6). Thus, as CO_2 is extracted from solution by growing algae at a pH around 8.3, additional carbon dioxide is provided through these reactions.[1699] Both equations (4-14) and (4-15) describe the principal reactions, with the reaction in equation (4-14) being the dominant of the two. As the pH rises, CO_3^{2-} becomes the major carbon species and it, too, can be converted directly to CO_2 by a dydration process as shown in equation (4-16).[1699] This reaction similarly results in a pH rise. It is not uncommon to have pH values as high as 10–11 in active algal systems such as waste stabiliza-tion ponds[267,1729] or in the laboratory.[2993] This rise in pH gives the evidence that the CO_2 supplied from the first three sources mentioned is either unavailable (i.e. diffusion gradients) or insufficient to meet the demands of the growing algae and that a further demand is placed on the bicarbonate alkalinity through the adjustment of the CO_2 - HCO_3^- - CO_3^{2-} system.[1699] Maberly,[2993] testing 35 species of marine macroalgae, reported that three species (*Enteromorpha intestinalis, Ulva lactuca* and *Urospora penicilliformis*) exhibited the highest ability to use HCO_3^- and raised the pH over 10.5 and depleted the concentration of CO_2 effectively to zero, and depleted the concentration of inorganic carbon to less than 50% of that at air-equilibrium. In contrast, the six species, all Rhodophyta, restricted to CO_2 did not raise pH above 9.0 at a CO_2 concentration of about 1.5 μmol L^{-1} and depleted the concentration of inorganic carbon to about 80% of that at air-equilibrium. Obviously, only a portion of the total inorganic carbon can be extracted during intense algal activity in a natural water principally buffered by the CO_2 - HCO_3^- - CO_3^{2-} system. If all the HCO_3^- and CO_3^{2-} were converted to CO_2 and OH^- as described by equations (4-14) through (4-16), the pH would approach a value of 14. Normally metabolic inhibition of algal growth occurs at a pH between 10 and 11, depending on the algal species, thus placing an upper limit on the amount of CO_2 available from HCO_3^- and CO_3^{2-}.[1699]

CREEPING SWAMP SEGMENT ECOSYSTEM

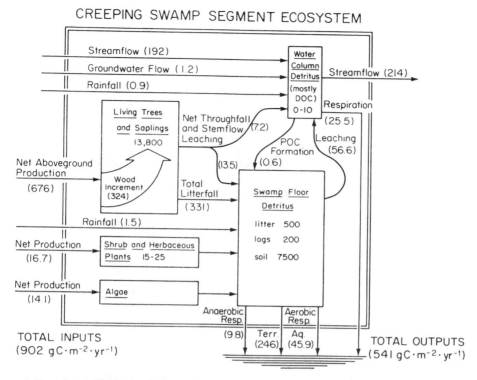

Figure 4-12. A model of organic carbon flux in a swamp ecosystem in coastal North Carolina, USA. Carbon standing stocks are reported as g C per m² and all fluxes as g C m⁻² yr⁻¹.[3329] (From Mulholland, 1980, Report No. 157, p. 79, reprinted with permission of Water Resources Research Institute of The University of North Carolina.)

Inorganic carbon is also produced inside algal cells by decarboxylation reactions in the light as well as in the dark. Some of this is used in carboxylation reactions before it can escape to the medium-reassimilation.[3999]

Most of the available evidence suggests that the pathway of autotrophic CO_2 fixation in algae is almost certainly the same as that for C_3-type higher plants, the reductive pentose phosphate cycle (RPPC = Calvin cycle = photosynthetic carbon reduction cycle) (Figure 4-14). That is, with 3-phosphoglyceric acid (PGA) formed as the first stable product of photosynthesis by carboxylation of ribulose 1,5-biphosphate (RuBP).[4645]

RPPC was first demonstrated in a freshwater green alga, *Chlorella*[238,668-70] (see also References 669 and 4735). It has subsequently been found in every other photosynthetic plant, including the blue-green algae that has been investigated.[2601,3729] It operates only in light with a direct supply of ATP and $NADPH_2$ from photosynthetic electron transport (see Chapter 2.12.1). All the enzymes of the Calvin cycle are located in the plastids.[2886]

The metabolic pathways involved in the photosynthetic fixation of carbon have been extensively studied in the unicellular green algae. Short-term exposure of *Chlorella* and *Scenedesmus* to ¹⁴C-labelled carbon dioxide showed that the first stable product of these reactions was phosphoglyceric acid (3-phosphoglycerate, PGA).[236] PGA is reduced, using $NADPH_2$ and ATP, to two other C_3 compounds, glyceraldehyde-3-phosphate (GAP) and dihydroxyacetone phosphate (DHAP). From then on, two series of reactions take place: GAP and DHAP can be combined to give fructose-1,6-biphosphate (FBP), and these compounds can be interconverted to regenerate RuBP (Figure 4-14). The net effect of the Calvin cycle is:

$$6\ CO_2 + 18\ ATP + 12\ NADPH_2 \rightarrow hexose + 18\ ADP + 18\ PO_4^{3-} + 12\ NADP \qquad (4.17)$$

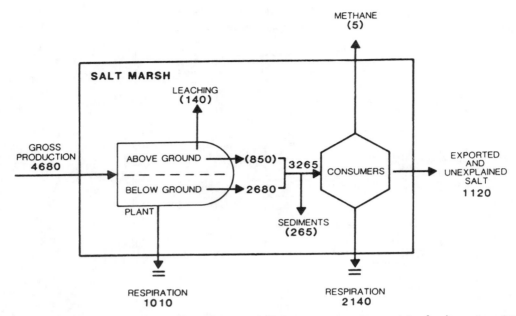

Figure 4-13.   Carbon budget of a Mississippi River deltaic salt marsh. Rates (g C m^{-2} yr^{-1}) are from CO_2 flux measurements, except numbers in parenthesis, which are from other sources.[1770]

Following this common fixation in PGA and FBP, carbon accumulates in various compounds (see Chapter 2.4) that are characteristic of the class to which the alga belongs.[906,2601,4001] The essentially identical nature of the pathway of carbon fixation in all algal groups is not reflected in their storage products[4,643] (see Chapter 2.4).

The photosynthetic reactions show the formation of the carbohydrate product which represents a carbon content, on a dry mass basis, of over 50%.[1699] Generally, most values given in the literature indicate a carbon content of algal cells very close to this value. Ketchum[2479] reported values of 51–56% for various algae grown under continuous illumination. Parsons et al.[2664] showed a range from 15.9 to 53.2% for a variety of marine phytoplankton. Thus, the reported carbon content for various algal species vary (Table 4-1) with lower values occurring in Bacillariophyceae and Charophyceae with respect to their high Si and Ca, contents. Algae growing under continuous illumination and adequate nutrient supply have a carbon content of between 51–56% of the ash-free dry mass,[2481] whilst 49.5–70.17% has been recorded for *Chlorella* grown under varying environmental conditions.[4657]

It is apparent that the pathway for photosynthetic carbon fixation by filamentous green algae such as *Chara, Nitella* or *Spirogyra* is also solely via the C_3 photosynthetic carbon reduction cycle.[3460,4997]

There is evidence that both ATP and NADPH$_2$ required for the photosynthetic carbon reduction cycle come from non-cyclic photophosphorylation (see Chapter 2.12.1), without the need for extra supplies from cyclic or pseudocyclic photophosphorylation or from oxidative phosphorylation.[3997-8,4510,4896]

The rate at which an algal cell fixes CO_2 in photosynthesis is conditioned by a number of external and internal factors and the interaction between them.[3996,3999] Internal factors include the algal species,[576] the stage in the cell cycle for rapidly dividing cells,[3811,4892] and circadian rhythms for slowly dividing cells.[931,2023]

Raven[3996] in his comprehensive review, summarized that external factors controlling the rate of CO_2 fixation include the inorganic carbon supply and pH;[576,3996] the intensity and wavelength of light;[3605,3994] the oxygen level;[5050] the organic and inorganic nutrient supply;[2418–20,4226] and temperature.

The way in which these factors can influence the rate at which CO_2 is fixed include: first, the control of the amount of enzyme present;[4226,4576] second, the control of the activity of the enzyme that is present,[237,3875] and third, control of the supply of substrates CO_2, ATP and NADPH$_2$.[237,3996] (For further information about CO_2 fixation see also References 236, 1628, 2164, 2784, 3996.)

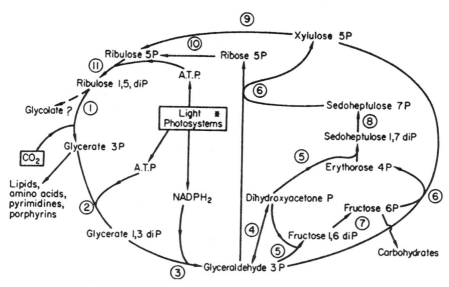

Figure 4-14. The photosynthetic carbon reduction cycle. The enzymes of the cycle, referred to by numbers, are: (1) ribulose biphosphate carboxylase (regulated enzyme), (2) 3-phos-phoglycerate kinase, (3) 3-phosphoglyceraldehyde dehydrogenase, NADP-linked (regulated enzyme), (4) phosphotriose isomerase, (5) fructose diphosphate aldolase, (6) transketolase, (7) fructose-1,6-diphosphate-1-phosphatase (regulated enzyme), (8) sedoheptulose-1,7-diphosphate-1-phosphatase (regulated enzyme), (9) phosphopentose epimerase, (10) phosphoribose isomerase, (11) phosphoribulokinase (regulated enzyme), * = catalytic role of CO_2.[4001] (From Raven, 1974, in *Algal Physiology and Biochemistry*, Stewart, W.D.P., Ed., p. 436, reprinted with permission of Blackwell Scientific Publications, Oxford.)

Since the PCRC was established using algae grown at elevated CO_2 concentrations, there was some question whether the "high affinity" cells grown at air CO_2 concentrations might have a different, more efficient carbon-fixation pathway.[4645] It seems well documented, however, that the pathway for fixation of carbon during photosynthesis in air-adapted microalgae is the same as in those grown at elevated CO_2 concentrations, the C_3 PCRC.[1784,2146,3726]

Algae adapted to air-levels of CO_2 have near-zero CO_2 compensation concentrations and exhibit little or no oxygen inhibition of photosynthesis.[377,1347,2884,4646] The mechanism of photorespiration suppression is believed to function through a CO_2-concentrating mechanism that provides a high intracellular concentration of CO_2, thereby saturating CO_2 binding capacity and suppressing the oxygenase activity of RuBPCO. Algae adapted to air-levels of CO_2 exhibit high levels of carbonic-andydrase activity and this enzyme is thought to have a specific role in the transport and accumulation of inorganic carbon.[204,3308] Patel and Merrett[3672] found that an internal accumulation of inorganic carbon relative to the external medium does not occur in all species, however.

Beardall and Raven[261] pointed out that investigations of carbon assimilation by a number of algae, including blue-green algae, have indicated that, when grown in air-equilibrated medium, the gas exchange characteristics are similar to those of C_4 higher plants. Thus, low CO_2 grown algae lack both overt photorespiration and inhibition of net carbon fixation by normal atmospheric O_2 levels relative to near-zero levels, and have low O_2 compensation concentrations and high affinities for CO_2, these attributes contrast with those of cells grown on 1–5% CO_2.[203–4,262,278–9,344,360,495,839,850,639,1347,2884,3223,4005,4008, 4439–40,5551] Despite having physiological characteristics of C_4 plants, both high CO_2 adapted and low CO_2 adapted algae, including blue-green algae, are biochemically C_3 types[259,2463,4005,4008] (for comparison of C_3 and C_4 plants see also Table 3-1).

When first demonstrated by Badger et al.,[204] the CO_2-concentrating system of the microalgae was proposed to accumulate inorganic carbon via active transport of bicarbonate ions. This was based on the idea that active transport of the charged bicarbonate ion was more easily explained than active transport of CO_2, and also on the observation by Berry et al.[344] and Birmingham and Colman[377] that the apparent $K_{0.5}$ (CO_2) (CO_2 concentration at which the rate of photosynthesis was one half of the maximum) for photosynthesis in air-adapted *Chlamydomonas reinhardtii* decreased with increasing pH. In contrast to those earlier observations, Moroney and Tolbert[3307] found that the pH-dependencies of the apparent $K_{0.5}$ (CO_2) and $K_{0.5}$ (HCO_3^-) were most consistent with CO_2 uptake by *Chlamydomonas reinhardtii*. Shelp and Canvin[4439] reported similar results for air-adapted *Chlorella pyrenoidosa*. Spalding[4645] in his excellent review summarizes that the evidence indicating that CO_2 is the inorganic carbon species crossing the plasmalemma could be interpreted in two ways: CO_2 is actively transported at the plasmalemma or CO_2 passively diffuses across the plasmalemma and active transport (CO_2 or bicarbonate) occurs at an internal membrane, probably at the chloroplast envelope. The author also presented possible mechanisms for inorganic carbon uptake in unicellular and filamentous algae (Figure 4-16).

Hatch and Slack[1963] demonstrated that in sugar cane an alternative mechanism for photosynthetic carbon uptake occurred, involving carboxylation of phosphoenolpyruvate (PEP) to produce oxaloacetate (Figure 4-15). Two enzymes are potentially involved in this carboxylation,[125] PEP carboxylase (PEPC) and PEP carboxykinase (PEPCK); the former utilizes the bicarbonate ion (HCO_3^-) and the latter CO_2. It is, however, an auxiliary mechanism which facilitates reductive carbon assimilation, and all plants in which it is found rely entirely on RuBPC for net carbon fixation.[2709,2940,3199,5559]

PEPCK has been isolated from a variety of algae from the brown algal line including the Phaeophyceae,[5315] Dinophyceae[125] and Bacillariophyceae[2157,2604] whereas PEPC is found in the Bacillariophyceae[988,5095] and Chlorophyceae.[1662] Zimba et al.[5559] pointed out that the principal difference between PEPC and PEPCK involves the phosphorylation of ADP to form ATP by PEPCK and a requirement of $MnCl_2$ or $MgCl_2$ for activation; $MgCl_2$ but not $MnCl_2$ is required for PEPC activation. These enzymes provide an energetically efficient mechanism for coupling anabolic (photosynthetic) reactions with catabolic ones associated with the TCA cycle.[2600]

PEPC activity is, however, extremely low in the Chlorophyceae, Phaeophyceae, Rhodophyceae and Bacillariophyceae.[2603-4] By contrast, PEPCK activity, while almost zero in the Chlorophyceae, amounts for 10% of the activity of RuBPC in the Rhodophyceae, and in some Phaeophyceae its activity reaches 70%.[2603] High values are also reported for the Bacillariophyceae.[2604] There is no evidence for a photosynthetically driven C_4 pathway similar to that operating in higher plants. Zimba et al.[5559] reported that the first product of photosynthetic carbon fixation in the diatom *Amphora micrometra* was 3-phosphoglycerate even though this alga had a PEPC activity that was three times higher than that of RuBPCO.

Colman[850] pointed out that experimental evidence indicates that CO_2 fixation in blue-green algae is mediated by the C_3 pathway and although C_4 acids are among the initial products of photosynthesis, these organisms do not have the enzymatic capacity for the operation of a C_4 pathway. C_4 acid synthesis appears to be required for the biosynthesis of amino acids, necessitated by the absence of a complete TCA cycle in these organisms.

Hassack[1959] first suggested that HCO_3^- as well as CO_2 might provide an exogenous photosynthetic carbon source. Since then, both indirect and direct methods have been used to differentiate between the ability or inability to use HCO_3^- among algae. Indirect methods for microalgae[1330,1699,2933,2939,2949,3573,3575,4889] and macrophytes[130,4228-9,4707] have involved a comparison of rates of net oxygen evolution by plants in CO_2 solutions and in HCO_3^- solutions with the same CO_2 concentrations, or the measurements of pH increase during photosynthesis in solutions of known alkalinity, or electrophysiological measurements. Direct methods involving the uptake of $^{14}CO_2$ and $H^{14}CO_3^-$ established that HCO_3^- is used by various cyanophyte, chlorophyte and charophyte species[203,2452,2930-1,2934,2943,2949,3992,4586] (see also Chapter 5.1.4.5).

Allen and Spence[59] found that the microalgae had considerably greater apparent affinities for HCO_3^- and slightly greater apparent affinities for CO_2 than the macrophyte, while the macrophytes had larger apparent affinities for CO_2 than for HCO_3^- and larger diffusive resistances to CO_2.

Bicarbonate use is the uptake of HCO_3^- ions by active transport, followed by their dehydration within the cytoplasm due to the dehydration activity of carbonic anhydrase, the assimilation of CO_2, and the efflux of OH^- ions.[59,4043] When HCO_3^- is the form of inorganic carbon which enters the algal cell, it must be converted to CO_2 before it can be used in photosynthesis. This dehydration is slow if unhydrated, and photosynthetic cells contain the enzyme carbonic anhydrase, a zinc-containing metalloprotein (see Tsuzuki and Miyachi[5042] for an excellent review), which catalyses the reversible hydration of CO_2:[494,1623,4001,5296]

$$CO_2 + H_2O \Leftrightarrow HCO_3^- + H^+ \tag{4.18}$$

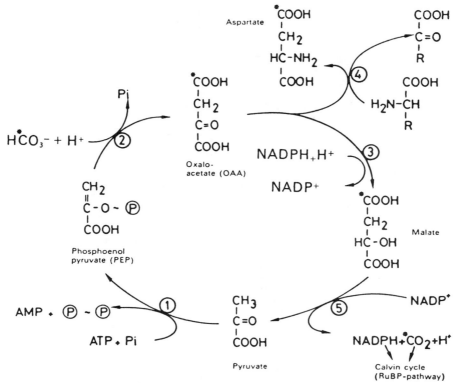

Figure 4-15. CO_2 assimilation of C_4 plants (C_4 pathway).[3199] (From Mengel and Kirkby, 1982, in *Principles of Plant Nutrition*, p. 162, reprinted with permission of International Potash Institute, Basel.)

There is evidence that this enzyme is specifically involved in the use of HCO_3^- in photosynthesis.[2896,3996] Carbonic anhydrase may also be involved in inorganic carbon transport in the cytoplasm, and in preparatory reactions of inorganic carbon prior to the action or RuBPCO.[1788,3996]

Three mechanisms for HCO_3^- assimilation apparently exist: H^+ - HCO_3^- symport, external acidification of HCO_3^- into CO_2, and an increase rate conversion of HCO_3^- into CO_2 by the action of the enzyme carbonic anhydrase.[3907] The ecological significance of photosynthetic HCO_3^- utilization seems to be that it compensates for slow supply CO_2 by diffusion and/or suppresses photorespiration by creating a high intracellular CO_2 concentration.[3907]

In freshwater algae the use of HCO_3^- is quite common both in unicellular microalgae,[258,261,1345-6,4890,5043] blue-green algae[202,2423,2425,5042] as well as filamentous species and Characeae.[2931,2936,2938,2941,3992,4588–90,5234] Prins and Elzenga[3907] pointed out that HCO_3^- utilization occurs also in marine macroalgae or seaweeds, Chlorophyceae as well as Rhodophyceae and Phaeophyceae.[276–7,363,851] The use of HCO_3^- can often be induced. Suppression occurs in growing conditions with high CO_2 (3–5%) and activation in low CO_2 concentrations.[3907] When cultured in high CO_2 concentrations, many microalgae exhibit a lower affinity for HCO_3^-. Björk et al.[382] showed that the level of external carbonic anhydrase activity, as well as the affinity for HCO_3^-, in three common species of *Ulva*, increased when plants were exposed to low concentrations of CO_2.

Transport of HCO_3^- across a membrane is very hard to discriminate from the transport of molecular CO_2, or the undissociated carbonic acids (H_2CO_3).[3907] Prins and Elzenga,[3907] in their superb review on HCO_3^- utilization by aquatic plants, stressed that even now, when there is ample evidence that HCO_3^- is utilized for photosynthesis and that active pumps are involved, there is much debate about which form of dissolved inorganic carbon actually crosses the membrane.[1341,2939–41,3908–11,4588,4648,5232–3,5235]

While photosynthetically driven uptake of inorganic carbon is by far the most effective carboxylating system in photoautotrophic plants, heterotrophic CO_2 fixation also occurs in some algae.[2601,4001]

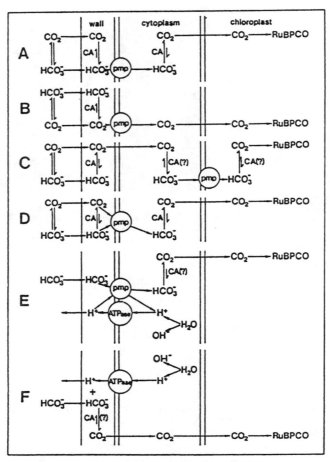

Figure 4-16.   Schematic representation of possible mechanisms for inorganic carbon uptake in unicellular and filamentous green algae. CA = carbonic anhydrase; pmp = inorganic carbon transporter.[4645] (From Spalding, 1989, in *Aquatic Botany*, Vol. 34, p. 196, reprinted with permission of Elsevier Science Publishers bv, Amsterdam.)

Heterotrophic (dark) CO_2 is essential for both autotrophic and heterotrophic growth of algae. In this type of reaction, designated ß-carboxylation, a C_4 dicarboxylic acid is produced by condensation of molecular CO_2 and C_3 acid.[2601] This reaction, which in most species of algae examined is catalyzed by PEPC takes place at comparatively low rates not sufficient to establish any type of C_4 metabolism.[2601] Some algae are capable of utilizing organic carbon substrates either via photosynthetically driven uptake or heterotrophically.[1157] Species lacking photosynthetic pigments are known in most algal classes, and these must of necessity be heterotrophic for carbon; others retain their photosynthetic apparatus and may utilize simple organic substrates, such as acetate, while providing ATP and reductant via photosynthesis.

4.1.4.2 Uptake and utilization of organic carbon

Classically all algae form their cellular carbon solely from carbon dioxide by photosynthesis. However, some are facultative heterotrophs and are able to utilize organic substances as a carbon source. Also there are obligate heterotrophic algae which must obtain at least some organic compounds from their surroundings.[5536]

Compared with bacteria and fungi, heterotrophic algae can utilize only a limited range of substrates.[3408] The ecological significance of heterotrophy must, however, be related to the ability of algal species to compete successfully with other heterotrophs for the low amounts of single substrates that may be present in water.[431] Such would demand that species possess active uptake mechanisms[61,2124,3344] as uptake by simple diffusion might not be sufficient for successful competition. In an ecological context it is also important to assess the potential contribution of an alga's ability to assimilate dissolved compounds to its production of particulate organic carbon.[431] Early work suggested that algae remove organic compounds from their environment by diffusion and only in high substrate concentrations.[61,2126,5503] Following investigations[312,3463] show that some species possess carrier systems for active uptake, but that these have such low affinities for the substances that the heterotrophic potentials are extremely low.

Acetate can be utilized for heterotrophic growth in darkness by a wide range of algae (members of Cryptophyceae, Dinophyceae, Bacillariophyceae, Chrysophyceae, Xanthophyceae, Euglenophyceae, Chlorophyceae).[3408] Acetate is generally oxidized metabolically through intermediates of the TCA cycle. This is functional in most microorganisms except for obligately photolithotrophic Cyanophyta, chemolithotrophs and methylotrophs, all of which have undetectable levels of 2-ketoglutarate dehydrogenase. Organisms growing at the expense of acetate must also possess a glyoxylate cycle to provide carbon skeleton for biosynthetic processes.[3408] In the light, acetate is assimilated into lipids by several algae.[3408]

Organic anions such as lactate and pyruvate, which are generally metabolized in the same way as acetate, can also be utilized by some algae.[431,912,2991,5463] Ethanol is able to support heterotrophic growth of several algae which can also grow at the expense of acetate.[296,1149,2991] The ethanol must first be dehydrogenated to acetate, which then enters the TCA cycle.[3408]

Several algae are able to grow at the expense of higher fatty acids (e.g. propionate, n - butyrate, valerate, caproate) and alcohols (n - propanol, n - butanol, n - amyl alcohol, n - hexanol).[3408] No example of heterotrophic growth in the dark at the expense of glycolate has been established for any alga.[3408] Although Droop and McGill[1159] could detect no assimilation of glycolate in the light by any of a number of marine algae, Palmer and Starr[3631] showed that a freshwater volvocine, Pandorina morum, is able to photoassimilate glycolate as well as acetate. Heterotrophic growth at the expense of glycerol has been demonstrated in several algae from various classes.[3408]

Glucose may be metabolized under either aerobic or anaerobic conditions. Of several pathways used by microorganisms for aerobic dissimilation of glucose, apparently only two, the Embden-Meyerhof-Parnas (EMP) and the pentose-phosphate pathway (PPP), have been demonstrated in algae[3408] (see also Chapter 2.12.2). Helleburst[2016] found that glucose transport capacity of the diatom Cyclotella cryptica increased exponentially over 24 hr after transfer of the cells from light to complete darkness with little simultaneous increase in cell number. The transport system was rapidly inactivated when cells are transferred back to continuous light. Helleburst[2015] found that the glucose transport process in a diatom Cyclotella cryptica is highly specific, and depends on energy metabolism. The Q_{10} for the process is 2.2 (15–25°C). Glucose taken up by cells is almost quantitatively phosphorylated within 10 minutes, either through the transport process itself or by a high affinity kinase system in the cells. Kwon and Grant[2663] found that Dunaliella tertiolecta, a green euryhaline flagellate, is unable to use glucose as a substitute for photosynthetically fixed CO_2 to maintain growth. The authors concluded that the failure of D. tertiolecta to use glucose is due to membrane impermeability, not lack of hexokinase. From other sugars, disaccharides, pentoses, hexoses can serve as a source for the heterotrophic growth of algae.[3408]

Some algae are able to use amino acids as sources of nitrogen during growth in the light (see Chapter 4.7). The ability to use amino acids as sources both of nitrogen and carbon seems to be a rare phenomenon. This is probably due to two factors. In the first place, many algae seem to be rather impermeable to amino acids at physiological pH, although some amides, e.g. asparagine and glutamine, seem to be taken up more readily than the corresponding acids. Secondly, the products of oxidative deamination, e.g. pyruvate, 2-ketoglutarate and oxalacetate, are often unsuitable as carbon sources, at least when supplied exogenously.[3408] One exception among photosynthetic algae is Euglena gracilis: Z strain, which can readily utilize glutamate as a source of both carbon and nitrogen.[2460]

Zajic and Chiu[5536] presented in their excellent review paper a comprehensive list of organic substrates and growth conditions of heterotrophic algae; transport of organic compounds is discussed by Neilson and Lewin.[3408]

4.1.4.3 Organic carbon excretion

The excretion of organic products of photosynthesis by freshwater phytoplankton was first described by Allen[66] and Fogg[1406] and subsequently investigated in detail by Nalewajko,[3379] Fogg and Nalewajko[1425-6] and Fogg et al.[1433] Since then many further studies have revealed that algae may release a variety of organic substances, including glycolate, carbohydrates, polypeptides, free amino acids, lipids, dissolved organic phosphorus, phenolics, DNA, RNA, vitamins, proteins, enzymes, antibiotics, autoinhibitors, toxins and other substances (see References 4, 303, 362, 628, 691, 1317, 1409, 1414, 1427-8, 1430, 1966, 2013, 2017, 2252, 2353, 2365, 2601, 2756, 3230, 3255, 3380, 3382, 3704, 3707, 3869-70, 3872, 4010, 4241, 4609, 4759, 4910, 4998, 5251-2, 5294).

The transfer of these substances between algae and algae and between algae and other organisms has been demonstrated.[247,1918,2702] Nalewajko and Lean[3382] demonstrated that bacteria utilize low molecular weight extracellular metabolites of algal origin and larger molecular weight compounds are formed.

Cells of *Chlorella vulgaris* cultured in a synthetic medium under conditions that are highly favorable for growth produce a substance that tends to retard their further multiplication.[3869-73] Jørgensen[2374] found that the growth of the epiphytic diatoms was inhibited by substances produced by the planktonic diatoms and green algae.

There is good evidence that healthy, actively growing phytoplankton species release a considerable proportion of their photoassimilated carbon into the aquatic environment.[96,1406,1433,2013,2195,4266,4295,4949] A number of investigators of excretion reported excretion rates in exponential growth phase populations.[1641,3381,5294] Other reports, however, indicate highest relative excretion by cultures and natural populations in stationary or declining phases.[1448,1844,3072]

Nalewajko[3380] found that the excretion of planktonic algae was less than 2% of the total carbon fixed during photosynthesis in short-term experiments with dilute cell suspensions under conditions of abundant CO_2 supply and limiting or saturating light intensities. Excretion was increased at high population density, under conditions of limiting CO_2 levels, and at light intensities sufficient to inhibit photosynthesis.

Extracellular excretion of photosynthetic carbon products also varies with light intensity and seems to reach its maximum at light intensities which inhibit photosynthesis slightly.[5293] At even higher light intensities there may be an inhibition of extracellular photosynthate release.[1433] The release of extracellular carbon is also high at very low light intensities.[5292-4] Light quality may also affect the release of extracellular carbon.[4615]

4.2 NITROGEN

4.2.1 Global Cycle

As a main constituent of protein and genetic material, nitrogen is an element of crucial importance for life on this planet.[135] All of the global biogeochemical cycles, the N cycle is the most complicated and the most difficult to understand in a qualitative as well as quantitative sense. In the first place, the nitrogen cycle includes major atmospheric pathways; further, a complex set of bacterial processes is involved in transformations of the various different N compounds that occur in the biosphere.[135] Söderlund and Rosswall[4614] pointed out that the inability of most living species to utilize the molecular form of nitrogen gives rise to the paradox that while nitrogen is one of the most abundant elements in the biosphere it is considered as being in limited supply. The reason for this is that 99.96% of the nitrogen is in the form of N_2 with only 0.04% in a combined form.

In the biosphere it undergoes what is essentially an eight electron shuttle between the most oxidized form at valence +5 (NO_3^-) and the most reduced form at valence -3 (NH_3 or NH_4^+). A number of stable intermediate valence states including NO_2^- (+3), NO (+2), N_2O (+1), N_2 (0), and NH_2OH (-1) have a significant impact on the biosphere.[1800] The main features of the N cycle are shown in Figures 4-17 and 4-18.

Söderlund and Svensson,[4613] in their excellent paper, divided the global nitrogen cycle into three compartments—atmospheric, aquatic (oceanic), and terrestrial. The boundaries are by no means obvious, and hence often have no physical meaning, the relevance only lying in relation to the description of the process under study.[4614] The authors pointed out that in describing the nitrogen cycle as consisting of exchanges between the three reservoirs, the focus is put on the small transfer between

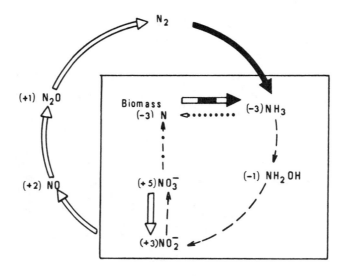

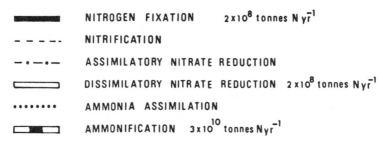

	NITROGEN FIXATION	2×10^8 tonnes $N \, yr^{-1}$
	NITRIFICATION	
	ASSIMILATORY NITRATE REDUCTION	
	DISSIMILATORY NITRATE REDUCTION	2×10^8 tonnes $N \, yr^{-1}$
	AMMONIA ASSIMILATION	
	AMMONIFICATION	3×10^{10} tonnes $N \, yr^{-1}$

Figure 4-17. The nitrogen cycle. The transformations within the box are quantitatively the most significant.[1799] (From Grant and Long, 1981, in *Environmental Microbiology,* p. 128, reprinted with permission of Blackie Academic & Professional, an imprint of Chapman & Hall.)

these systems rather than on the large internal circulation within them. The fluxes within the systems are many times larger than the exchange rates. The circulation of nitrogen within the terrestrial system is at least 10 times larger than its exchange with the aquatic and atmospheric environments. The nitrogen cycle can be treated as four subcycles with different classes of nitrogen compounds:[4613]

1) ammonia/ammonium compounds,
2) NO_x compounds—compounds related to NO/NO_2,
3) N_2/N_2O, and
4) organic nitrogen compounds.

The rationale for this subdivision is based on the small rates of interchange between these classes of compounds within the atmosphere.[4614]

The global cycle is still far from being precisely understood in a quantitative sense.[135] The most comprehensive reviews of global nitrogen cycle and fluxes are those given by Söderlund and Svensson,[4613] Söderlund and Rosswall[4614] (Figure 4-19) and Rosswall;[4195-6] atmospheric nitrogen chemistry has been reviewed by Hauck;[1972] soil cycle has been reviewed by Stevenson[4738] and Bohn et al.;[419] and ecology of the nitrogen cycle has been reviewed by Sprent.[4661] Major nitrogen reservoirs are atmosphere, ocean, pedosphere including biota and lithosphere.[4196] Söderlund and Svensson[4613] listed the most important processes involved in biosphere nitrogen transformations: N_2 fixation, denitrification,

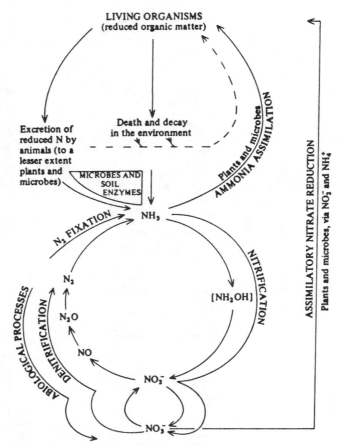

Figure 4-18.   A summary of the major reactions of the nitrogen cycle.[4661] (From Sprent, 1987, in *The Ecology of the Nitrogen Cycle*, p. 6, reprinted with permission of Cambridge University Press, Cambridge.)

nitrification, leaching and runoff, decomposition, dry and wet deposition, combustion, plant uptake, volatilization and reactions in atmosphere and stratosphere.

4.2.2 Nitrogen in the Environment

Earth's crust: 20–25 mg kg⁻¹ (490,903)

Common minerals: $NaNO_3$ (rare)

Soils: 1,000–2,000 (200–5,000) mg kg⁻¹, 90% is usually present as non-basic N bound in humus, NH_4^+ can be fixed by clay minerals, NO_3^- is an important exchangeable anion, especially in arid soils[490-1]

Sediments: 470 mg kg⁻¹ (491)

Air: 9.73×10^{10} μg N_2 m⁻³ (490)

Fresh water: 0.23 mg L⁻¹ as NO_3^-,[490] 0.05 as NO_3^- (0.002–1.8) mg L⁻¹ (491), highly variable, in polluted surface waters > 100 mg L⁻¹ as NO_3^-

Sea water: 0.50 mg L⁻¹ (variable),[490] 0.64 mg L⁻¹ as $NH_3 + NO_3^-$ and 15.5 mg L⁻¹ as N_2 (491)

Land plants: 3.0 (1.2–3.8) % DM[490-1]

Freshwater algae: 0.78–11.25% DM (Table 4-2)

Marine algae: 1.8–2.5% DM,[491] 0.86–5.73% DM (Table 4-2)

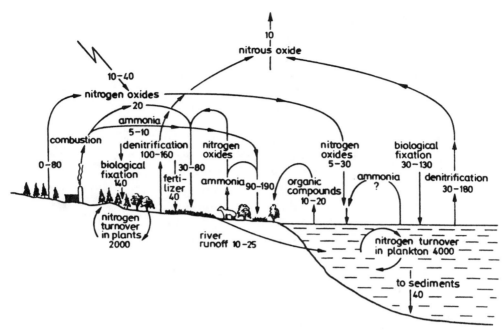

Figure 4-19. A global nitrogen budgets (Tg N yr^{-1}).[4614] (From Söderlund and Rosswall, 1982, in *The Handbook of Environmental Chemistry,* Vol. 1. Part B, Hutzinger, O., Ed., p. 78, reprinted with permission of Springer-Verlag and the authors.)

Land animals: 10.0% DM[490]
Marine animals: 7.5% DM[491]

4.2.2.1 Aquatic chemistry

Sources to waters: N$_2$ fixation, agriculture (fertilizers), municipal wastewaters, minor amounts from pulp and paper production, refining and mining, manufacturing of amines and nitriles.[3287,3815]

Nitrogen is present in the aquatic environments in several oxidation states (from 3- in ammonia to 5$^+$ in nitrate), in both ionic and non-ionic forms. Distribution of forms is mainly influenced by biochemical processes in waters (see Figure 4-17). The main forms are:[1709,3815]

1. Molecular nitrogen (N$_2$)
   N$_2$, dissolved in water in equilibrium with the nitrogen in the atmosphere (about 15 mg per liter at 20°C).[1705]

2. Inorganic nitrogen (NO$_3^-$, NO$_2^-$, NH$_3$ or NH$_4^+$)
   NH$_3$, in equilibrium with NH$_4^+$ and NH$_4$OH, which is produced from organically bound nitrogen, either from allochtonous or from autochthonous sources. At a pH value of 8.0, approximately 95% of the ammonia nitrogen is in the form of NH$_4^+$.[3210] An alkaline pH value shifts the equilibrium equation NH$_3$ + H$_2$O $\Leftrightarrow$ NH$_4^+$ + OH$^-$ towards gaseous ammonia production.[3210] At a pH value of 9.3, the NH$_4^+$ forms about 50% at a temperature of 25°C.[3815] Temperature affects both the equilibrium constant and mass transfer coefficient.[3210] At a pH value of 9.5, and a temperature of 0°C NH$_3$ forms about 20% of total ammoniacal nitrogen, while at a temperature of 30°C it is about 70%.[3815]

Table 4-2. Nitrogen concentration in algae (% dry mass)

Species	Concentration	Locality	Reference
Freshwater			
Tribonema aequale	2.32	Laboratory	1466
Navicula pelliculosa	2.56	Laboratory	1466
Monodus subterraneus	1.22–2.94	Laboratory	1419
Navicula pelicullosa	2.2–3.0	Laboratory	1404
Chara vulgaris	2.43–3.19	Czech Republic	1207
Tribonema aequale	0.97–3.36	Laboratory	1419
Cladophora glomerata	0.78–3.40	Clark Fork, Montana	2899
Chlorella vulgaris	4.30	Laboratory	1466
Chlorella vulgaris	1.89–4.34	Laboratory	1419
Cladophora glomerata	1.2–4.7	Lake Michigan	20
Oikomonas termo	5.09	Laboratory	3235
Enteromorpha prolifera	4.94–5.26	Forfach Loch, Scotland	2107
Chlamydomonas sp.	5.53	Laboratory	3235
Anabaena cylindrica	6.51	Laboratory	1466
Anabaenopsis sp.	6.60	Laboratory	3235
Scenedesmus acutus	6.7–7.1	Cultivation, wastes	5144
Scenedesmus obliquus	6.96–7.14	Laboratory	2593
Chlorella pyrenoidosa	2.6–7.3	Laboratory	1366
Oscillatoria spp.	5.65–7.42	Laboratory	1419
Chlorella vulgaris	8.0	Laboratory	1604
Chlamydomonas geitleri	7.6–8.3	Cultivation, wastes	5144
Chlorella pyrenoidosa	1.38–8.99	Laboratory	3235
Anabaena cylindrica	3.04–9.00	Laboratory	1417
Westiellopsis prolifica	3.1–9.1	Laboratory	3704
Microcystis aeruginosa	3.77–9.15	Laboratory	1610
Cladophora glomerata	1.31–9.90	Lake Erie, Michigan-Ohio	2909
Stichococcus bacillaris	3.20–9.91	Laboratory	3235
Phytoplankton	3.80–10.3	Germany	2607
Chlorella sp.	9.20–10.70	Outdoor mass culture	5275
Chlorella sp.	7.90–10.90	Laboratory mass culture	5275
Chlorella sp.	1.13–11.25	Laboratory	4657
Marine			
Pelvetiopsis limitata	0.86–2.22	Oregon coast	5366
Laminaria saccharina	0.90–2.30	Rhode Island	173
Ascophyllum nodosum	1.10–2.60	Rhode Island	173
Codium fragile	1.64–2.60	Oregon coast	5366
Fucus vesiculosus	1.00–3.00	Rhode Island	173
Amphipleura rutilans	3.05	Half Moon Bay, California	3235
Egregia menziesii	3.68	Half Moon Bay, California	3235
Ulva sp.	3.96	Half Moon Bay, California	3235
Enteromorpha intestinalis	1.38–4.07	Laboratory	385
Macrocystis pyrifera	4.11	Half Moon Bay, California	3235
Porphyra sp.	2.50–4.55	Oregon coast	5366
Gigartina agardhii	4.79	Half Moon Bay, California	3235

Table 4-2. Continued

Species	Concentration	Locality	Reference
Chondrus crispus	1.65–4.80	Rhode Island	173
Codium fragile	2.00–4.80	Rhode Island	173
Ulva lactuca	2.17–5.03	Hong Kong coast	2111
Enteromorpha intestinalis	2.02–5.34	Oregon coast	385,5366
Ulva fenestrata	1.32–5.42	Laboratory	385
Navicula torquatum	5.71	Half Moon Bay, California	3235
Ulva fenestrata	2.44–5.73	Oregon coast	385,5366

NO_2^-, small quantities of nitrite are a normal constituent of water bodies. Larger quantities occur under certain conditions. They are formed during the oxidation of ammonia, but will normally be oxidized to NO_3^-, except in anaerobic conditions.[1709] NO_3^-, the only stable form of a nitrogen compound, unless taken up by algae or bacteria.[1705]

3. Organic nitrogen
   Organic nitrogen, either in dissolved state or as particulate nitrogen.[1705]

4.2.3 Nitrogen Cycling in Wetlands

Nitrogen biogeochemistry in wetlands has been reviewed many times recently.[135,489,1313,1555,2351,3266, 4027,4032,4105–6]

Nitrogen has a complex biogeochemical cycle with multiple biotic/abiotic transformations involving seven valence states (+5 to -3). The major N pools in natural freshwater wetlands are total sediment N (mostly organic N), total plant N, and available inorganic N in sediments.[489] Organic N consists of compounds from amino acids, amines, proteins, and humic compounds with low N content. Inorganic N consists of ammonium N, nitrate and nitrite N; in sediments both nitrate and nitrite N occur in trace quantities. Ammonium N is the predominant form of inorganic N in sediments and is mainly derived through mineralization of organic N. The gaseous forms of N that occur in flooded soils and sediments include ammonia (NH_3), dinitrogen (N_2) and nitrous oxide (N_2O).[4027]

The total sediment pool is the largest, ranging from 100 to 1000 g N m^{-2}. The total plant N pool is roughly an order of magnitude less than total sediment N, while inorganic sediment N is another order of magnitude less than the plant pool.[1313]

The major nitrogen transformations in wetland soils are presented in Figure 4-20. Detailed reviews have been presented by Focht and Verstraete,[1393] Buresh et al.,[624] Savant and De Datta,[4298] Reddy and Graetz[4032] and Reddy and Patrick.[4027] All transformations are affected by the type of microbial metabolism.

4.2.3.1 Ammonification

Ammonification is the process where organic N is converted into inorganic N, especially NH_4-N. In wetland soils, mineralization rates are fastest in the oxygen zone, and decrease with depth as mineralization switches from aerobic to facultative anaerobic and obligate anaerobic microflora.[4027] Since depth of the aerobic zone is usually less than 1 cm, the contribution of aerobic mineralization to the overall N mineralization would be very small, compared to facultative anaerobic and obligate anaerobic mineralization.[4032]

The rate of ammonification in flooded soils is dependent on temperature, pH, C/N ratio of the residue, available nutrients in the soil, and soil conditions such as texture and structure.[4027] Reddy et al.[4039] concluded from the literature data that the rate of aerobic ammonification doubles with a temperature increase of 10°C. The optimum pH range for the ammonification process is between 6.5 and 8.5. Under flood conditions, pH is buffered around neutrality, whereas under well-drained conditions, pH of the soil decreases as a result of nitrate accumulation during mineralization.[3687]

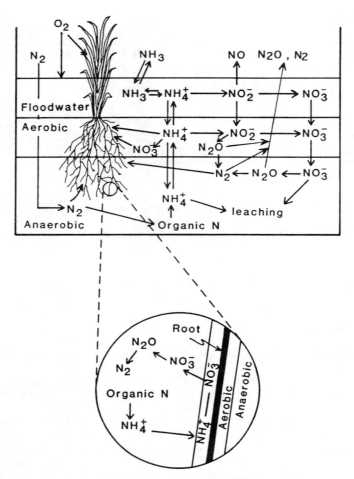

Figure 4-20.   Nitrogen transformations in wetlands.[4027]   (From Reddy and Patrick, 1984, in *CRC Crit. Rev. Environ. Control,* Vol. 13, p. 274, reprinted with permission of CRC Press, Inc., Boca Raton.)

Mineralization of sediment organic N to NH_4^+ is the major process supplying N to wetland plants and responsible for maintenance of high rates of primary productivity and high standing crops of phytoplankton in seashore and estuarine water.[1866,1940,3124,3127,4687] Ammonium N accumulation in wetland soils was found to be rapid during the first two weeks after flooding.[3838] Ammonium N formed during mineralization is rapidly partitioned into: (1) NH_4^+ adsorbed on the cation-exchange complex, and (2) equilibrium NH_4^+ in soil pore water. Ammonium N in pore water moves in two directions, namely (1) upward movement into the overlying aerobic zone and the floodwater, and (2) movement toward plant roots. Rates of NH_4^+ movement is regulated by the concentration gradient established as a result of (1) plant uptake; (2) loss mechanisms in the rhizosphere; and (3) loss mechanisms at the soil-water interface. Other factors governing the overall process include: mineralization rate, concentration of NH_4^+ in the pore water, cation-exchange capacity, types of other cations on the exchange complex and bulk density.[4025,4027]

4.2.3.2 Ammonia volatilization

Ammonia volatilization is a physiochemical process where ammonium N is known to be in equilibrium between gaseous and hydroxyl forms as indicated below:[4027]

$$NH_3 \text{ (aq)} + H_2O = NH_4^+ + OH^- \tag{4.19}$$

Ammonia volatilization can be very active in wetland soils, where the pH of the overlying water is above 7.[3220] Reddy and Patrick[4027] pointed out that losses of NH_3 through volatilization from flooded soils and sediments are insignificant if pH is below 7.5 and very often losses are not serious if the pH is below 8.0.

Volatilization rate is controlled by the NH_4^+ concentration in the floodwater, temperature, wind velocity, solar radiation, the nature and number of aquatic plants, and capacity of the system to change the pH in diurnal cycles (absence of CO_2 increases volatilization).[4024,4027,4032] Several researchers[1335,4665,4799] observed that reaction rate was increased approximately from 1.3 to 3.5 times for each 10°C rise in temperature when the system temperature was between 0 to 30°C. In floodwater containing algae, CO_2 can be depleted during photosynthesis, thereby increasing pH, and enhancing the NH_3 volatilization.[4032]

4.2.3.3 Nitrification

Nitrification is usually defined as the biological oxidation of ammonium to nitrate with nitrite as an intermediate in the reaction sequence. This definition has some limitations where heterotrophic microorganisms are involved,[50] but is adequate for the autotrophic and dominant species.[1972]

Nitrification can occur in the overlying water,[934,4220] in the surface aerobic soil layer[4033,4880] and in the root zone.[4028] The diffusion of oxygen from wetland plants roots to adjacent soil creates an aerobic environment around the roots[2227,2272,3561] (see also Chapter 3.3.1). This aerobic soil envelope around the roots can support nitrification of NH_4^+ diffused from surrounding anaerobic zones.[4032] Nitrification rate in wetland soils depends on the supply of NH_4^+ to the aerobic zone, pH and alkalinity of water, temperature, presence of nitrifying bacteria and the thickness of the aerobic soil layer.[4027,4033,4298]

The nitrifiers derive energy from the oxidation of ammonium N and/or nitrite N. These organisms require O_2 during ammonium N oxidation to nitrite N and nitrite N oxidation to nitrate N. Oxidation of ammonium N to nitrate is a two-step process:[1048,1972,5236]

$$NH_4^+ + 1.5 \, O_2 \rightarrow NO_2^- + 2 \, H^+ + H_2O \tag{4.20}$$

$$NO_2^- + 0.5 \, O_2 \rightarrow NO_3^- \tag{4.21}$$

$$\overline{NH_4^+ + 2 \, O_2 \rightarrow NO_3^- + 2 \, H^+ + H_2O} \tag{4.22}$$

The first step, the oxidation of ammonia to nitrite, is executed by strictly chemolithotrophic bacteria (obligate chemolithotrophs) which are entirely dependent on the oxidation of ammonia for the generation of energy for growth. In soil, species belonging to the genera *Nitrosospira* (*N. briensis*), *Nitrosovibrio* (*N. tenuis*), *Nitrosolobus* (*N. multiformis*) and *Nitrosomonas* (*N. europaea*) have been identified.[1799,1800,4347] *Nitrosomonas europaea* is also found in fresh water.[1799,1800] Three species of the genus *Nitrosococcus* (*N. nitrosus, N. oceanus* and *N. mobilis*) are found in marine environments;[4347] Grant and Long[1799,1800] however, reported *Nitrosococcus nitrosus* to be found in soil. These organisms are all strictly aerobic.[2679] The probable reaction sequence for the oxidation of ammonium to nitrite by Nitroso group bacteria is:[1972]

ammonia (NH_3/NH_4) $\rightarrow$ hydroxylamine (NH_2OH) $\rightarrow$ nitroxyl (NOH) $\rightarrow$ nitrohydroxylamine ($NO_2 . NH_2OH$) $\rightarrow$ nitrite (NO_2^-).

The postulated intermediate compounds NOH and NO_2. NH_2OH, have not been isolated, but their participation in the reaction sequence is consistent with the assumption that two electrons are transferred for each oxidation step between NH_4^+ and NO_2^- (Hauck[1972] and references cited therein).

The second step in the process of nitrification, the oxidation of nitrite to nitrate, is performed by facultative chemolithotrophic bacteria which can also use organic compounds, in addition to nitrite, for the generation of energy for growth. In contrast with the ammonia-oxidizing bacteria, only one species of nitrite-oxidizing bacteria is found in the soil and fresh waters, i.e. *Nitrobacter winogradskyi*.[1799,1800,2679] (Schmidt,[4347] however, reported in addition to *Nitrobacter* also a genus *Nitrospira* to be found in soil and fresh as well as marine waters.) *Nitrococcus mobilis* is found only in marine environments.[1800] Also in contrast to ammonia-oxidizing bacteria, nitrite-oxidizing bacteria, or at least some species, can grow mixotrophically on nitrite and a carbon source, or are even able to grow in the absence of oxygen.[411] An excellent review on the biochemistry of nitrifying microorganisms has been given by Wallace and Nicholas.[5236]

Viable counts of nitrifiers in soils and sediments range from 10^3 to 10^5 organisms per 1 gram, but much larger numbers (10^7 to 10^8 organisms g^{-1}) are generally described in high NH_4^+ environments such as activated sludge.[1799]

In aquatic systems, nitrification rate is influenced by temperature, pH, alkalinity of the water, inorganic C source, microbial population, and ammonium N (substrate concentration).[4027] The optimum temperature for nitrification in pure cultures ranges from 25 to 35°C and in soils ranges from 30 to 40°C.[50,1480,3021] Lower temperatures (below 15°C) have a much more drastic effect on nitrification rate, compared to temperatures between 15 and 35°C.[4027] Reddy et al.[4036] estimated the overall NH_4-N flux from the sediment to the overlying water 4.8 μg cm^{-2} d^{-1} in Lake Apopka, Florida. Discussion on rate equations of nitrification has been given by Reddy and Patrick.[4027]

4.2.3.4 Denitrification

The first anaerobic oxidation process to occur after oxygen depletion is the reduction of nitrate to molecular nitrogen or ammonia. The reduction of nitrate is performed by two different groups of nitrate-reducing bacteria: the denitrifying bacteria which produce N_2O and N_2 as major reduction products, and the nitrate-ammonifying bacteria which produce NH_4^+ as the major end product of the reduction of nitrate. In sediments and soils, both denitrification and nitrate ammonification are observed.[2679]

Different numbers of electrons are used in the reduction of one molecule of nitrate in both nitrate-reducing systems: 5 in the case of denitrification and 8 in the case of nitrate ammonification. Therefore, more organic matter can be oxidized per molecule of nitrate by nitrate-ammonifying bacteria than by denitrifying bacteria.[2679] In addition, nitrate reduction is generally performed by fermentative bacteria that are not dependent on the presence of nitrate for growth under anaerobic conditions. So nitrate-ammonifying bacteria may be favored by nitrate-limited conditions.[2679] Hence, the carbon to nitrate ratio in a system may determine the predominance of nitrate-ammonifying or denitrifying bacteria.[4969] Nitrate ammonification is found in facultatively anaerobic bacteria belonging to the genera *Bacillus, Citrobacter* and *Aeromonas*, or in the members of the Enterobacteriaceae.[833,1800,2998,4600] However, strictly anaerobic bacteria belonging to the genus *Clostridium* are also able to reduce nitrate to ammonia.[1955]

Denitrification is most commonly defined as the biochemical reduction of NO_2^- or NO_3^- to N_2 or gaseous N oxides.[1972,2351] However, the assimilatory reduction of NO_3^- to NH_4^+ and nitrification also produce N oxides (N_2O and/or NO), so that a more precise definition is desirable to keep pace with current knowledge.[1972] From a biochemical viewpoint, denitrification is a bacterial process in which N oxides (in ionic and gaseous forms) serve as terminal electron acceptors for respiratory electron transport. Electrons are carried from an electron donating substrate (usually, but not exclusively, organic compounds) through several carrier systems to a more oxidized N form. The resultant free energy is conserved in ATP, following phosphorylation, and is used by the denitrifying organisms to support respiration.[1972]

Denitrification is illustrated by following equation:[1972]

$$6 \ (CH_2O) + 4 \ NO_3^- \rightarrow 6 \ CO_2 + 2 \ N_2 + 6 \ H_2O \qquad (4.23)$$

This reaction is irreversible, and occurs in the presence of available organic substrate only under anaerobic or oxygen-free conditions (Eh = +350 to +100 mV), where nitrogen is used as an electron acceptor in place of oxygen.[2351,4027] More and more evidence is being provided from pure culture studies that nitrate reduction does occur in the presence of oxygen.[2626] Hence, in waterlogged soils nitrate reduction may also start before the oxygen is depleted.[2679]

Gaseous N production during denitrification also can be depicted by:[1972]

$$4 \ (CH_2O) + 4 \ NO_3^- \rightarrow 4 \ HCO_3^- + 2 \ N_2O + 2 \ H_2O \qquad (4.24)$$

$$5 \ (CH_2O) + 4 \ NO_3^- \rightarrow H_2CO_3 + 4 \ HCO_3^- + 2 \ N_2 + 2 \ H_2O \qquad (4.25)$$

Brodrick et al.[570] found that the majority of denitrification activity was restricted to the top 0–6 cm of the soil, and to the decaying vegetation lying on the soil surface. The peak soil denitrification rates were coincident with the area of peak nitrate plus nitrite concentration, both rates and concentration declining with increasing depth in the soil profile.

Denitrifying ability has been demonstrated in 17 genera of bacteria. Although more abundant than nitrifiers, the denitrifiers are not as diverse as bacteria capable of reducing NO_3^- to NO_2^- or the dissimilatory reduction of NO_3^- to NH_4^+, of which 73 genera have been identified.[1878] Most denitrifying bacteria are chemoheterotrophs. They obtain energy solely through chemical reactions and use organic compounds as electron donors and as a source of cellular carbon.[1972] The genera *Bacillus, Micrococcus* and *Pseudomonas* are probably the most important in soils; *Pseudomonas, Aeromonas* and *Vibrio* in the aquatic environment.[1799] It is not uncommon to isolate 10^6 denitrifiers per g of soil or sediment.[1799] Other denitrifiers are members of the genera *Agrobacterium, Alcaligenes, Azospirillum, Paracoccus,* or *Thiobacillus*.[1353,2543] A list of genera involved in the denitrification process has been given by e.g. Focht and Verstraete.[1393] When oxygen is available, these organisms oxidize a carbohydrate substrate to CO_2 and H_2O.[4027] Aerobic respiration using oxygen as an electron acceptor or anaerobic respiration using nitrogen for this purpose is accomplished by the denitrifier with the same series of electron transport system. This facility to function both as an aerobe and as an anaerobe is of great practical importance because it enables denitrification to proceed at a significant rate soon after the onset of reducing conditions (a redox potential of about 330 mV) without change in microbial population.[1972] Because denitrification is carried out almost exclusively by facultative anaerobic heterotrophs that substitute oxidized N forms for O_2 as electron acceptors in respiratory processes, and because these processes follow aerobic biochemical routes, it can be misleading to refer to denitrification as an anaerobic process. It is rather one that takes place under anaerobic (i.e. anoxic) conditions.[1972]

The actual sequence of biochemical changes from nitrate to elemental gaseous nitrogen is:[1799,1800,1972,3711-2,4614,5491,5569] $2 NO_3^- \rightarrow 2 NO_2^- \rightarrow 2 NO \rightarrow N_2O \rightarrow N_2$. Both N_2O and N_2 contribute to atmospheric nitrogen.[2351,3239]

The quantity of N_2O evolved during denitrification depends upon the amount of nitrogen denitrified and the ratio of N_2 to N_2O produced.[4931] This ratio is also affected by aeration, pH, temperature and nitrate to ammonia ratio in the denitrifying system.[387,1354,1392,1972,4931] If the pH is below 4.5, the denitrification rate is relatively slow and only N_2O is released; at pH > 5, N_2 is the main end product of denitrification under conditions of low redox potential, whereas the relative importance of N_2 decreases if the circumstances are less anaerobic.[135] The rate of N_2O emission increases with increasing moisture content of the soil and with increasing temperature up to 37°C.[1485] Smith et al.[4579] reported that only about 50 mg N m^{-2} yr^{-1} are released as N_2O and estimated that about 5 g N m^{-2} yr^{-1} is released as N_2 through denitrification. Annual nitrous oxide emissions from brackish and freshwater marshes averaged 48 and 55 mg N m^{-2} yr^{-1}, respectively.[4581] Smith et al.[4578] reported the amount of N_2O evolved from stimulated overland flow wastewater treatment system to range from 0.07 to 1.3 mg N m^{-2} d^{-1}, or 0.15 to 2.55 g N m^{-2} yr^{-1}. Seitzinger et al.[4401] reported a ratio of 250 : 1 for N_2 : N_2O being evolved from coastal sediments, whereas Terry and Tate[4931] reported a ratio of 220 : 1 for organic soils. As such, brackish and freshwater marshes release annually an estimated 10 to 12 g N m^{-2} through denitrification.[1326] Armentano and Verhoeven[135] reported the denitrification rates in different wetland types in the range of 0.0024 to 14.3 g N m^{-2} yr^{-1}. Nixon and Lee,[3447] summarizing a large data set reported the range of denitrification rates from 0.027 mg N m^{-2} yr^{-1} in Alaskan tundra[226] up to 110 g N m^{-2} yr^{-1} in several Florida soils.[1785] Several salt marsh studies indicated that denitrification exceeds nitrogen fixation.[1867,2428]

In wetland soils, denitrification occurs in the anaerobic soil below the aerobic soil layer. The rate of denitrification in this zone depends on the supply of NO_3^-, energy source, microflora and temperature. The supply of NO_3^- in wetland soils is derived by nitrification of NH_4^+ in the aerobic zone followed by the downward diffusion of NO_3^- from the aerobic zone to the underlying anaerobic zone.[3692] The flux of NO_3^- from the floodwater and the aerobic soil layer into anaerobic zones is controlled by available carbon supply,[4035,4678] thickness of the aerobic layer,[1274] NO_3^- concentration of the floodwater,[4034] temperature,[4679] and mixing and aeration in the floodwater. Nitrate diffusion rate in wetland soils has been found to be in the range 1.2–1.9 cm d^{-1}, which is about seven times faster than NH_4^+ diffusion.[4027] Rapid diffusion of NO_3^- followed by high denitrification rates is the main cause of low NO_3^- accumulation in the floodwater/aerobic soil layer.[4032] In addition to nitrification, the denitrification also requires the presence of readily available carbon.[4035] Other environmental factors known to influence denitrification rates include absence of O_2, redox potential,[1784,4580] soil moisture,[2616] temperature,[2030,5335] pH,[3712] presence of denitrifiers,[872,2255] soil type[1752] and the presence of overlying and the presence of overlying water.[1025,1247] The process of denitrification and its consequences have been extensively reviewed by Payne.[3712]

Nitrification and denitrification are known to occur simultaneously in flooded soils where both aerobic and anaerobic zones exist such as would be the case in a flooded soil or water bottom containing an aerobic surface layer over an anaerobic layer or in the aerobic root rhizosphere of a swamp plant growing in an anaerobic soil.[3251,3690,3692,3714,4027] By combining these two reactions, a balanced equation occurring in aerobic and anaerobic layers can be written as:[4027]

$$24 \text{ NH}_4^+ + 48 \text{ O}_2 \rightarrow 24 \text{ NO}_3^- + 24 \text{ H}_2\text{O} + 48 \text{ H}^+ \tag{4.26}$$

$$24 \text{ NO}_3^- + 5 \text{ C}_6\text{H}_{12}\text{O}_6 + 24 \text{ H}^+ \rightarrow 12 \text{ N}_2 + 30 \text{ CO}_2 + 42 \text{ H}_2\text{O} \tag{4.27}$$

$$\overline{24 \text{ NH}_4^+ + 5 \text{ C}_6\text{H}_{12}\text{O}_6 + 48 \text{ O}_2 \rightarrow 12 \text{ N}_2 + 30 \text{ CO}_2 + 66 \text{ H}_2\text{O} + 24 \text{ H}^+} \tag{4.28}$$

Denitrification rates are maximized by the coupling of nitrification and denitrification via diffusion across the aerobic-anaerobic soil boundary.[2856,2908,3692] Their significance in geochemical and ecological processes of wetland soils, estuarine and coastal marine sediments was reviewed by Kemp et al.[2459]

Ammonium diffusion and nitrification limit N loss from wetland soils, whereas NO_3^- diffusion and denitrification usually occur at a rapid rate and are not likely to limit the overall process. The ultimate conversion of NH_4^+ to N_2 gas through nitrification and denitrification coupled with NH_4^+ and NO_3^- diffusion was demonstrated by Patrick and Reddy[3692] for wetland soils, and by Jenkins and Kemp[2327] for estuarine processes. Sherr[4445] estimated denitrification of 65 g N m^{-2} yr^{-1} in a Georgia salt marsh. Johnston[2351] summarizing the literature data reported denitrification rates determined by laboratory incubation of wetland soils in the range of 0 to 55 mg N m^{-2} d^{-1} and 0.03 to 0.15 g N m^{-2} yr^{-1}; denitrification rates determined in the field ranged from 0.002 to 0.34 g N m^{-2} yr^{-1}. Sloey et al.[4567] reported that up to 350 mg N m^{-2} d^{-1} may be removed in wetlands.

4.2.3.5 Fixation

In wetland soils, biological N_2 fixation may occur in the floodwater, on the soil surface, in aerobic and anaerobic flooded soils, in the root zone of plants, and on the leaf and stem surfaces of plants.[624] A wide variety of symbiotic (associated with nodulated host plants) and asymbiotic (free-living) organisms can fix nitrogen in wetlands.[2351] To find further details on nitrogen fixation systems the reader is referred to an excellent book by Sprent and Sprent[4662] and other books and papers[1800,3854-5,5168] (see also Chapter 4.2.4.1).

A number of environmental factors influence the rate of nitrogen fixation in flooded soils.[624] The availability and quality of carbon compounds appear to be the primary factor limiting growth of heterotrophic nitrogen-fixing bacteria because these microorganisms must obtain their energy from carbon compounds synthetized by other organisms. Added labile carbon substances have been shown to stimulate nitrogen fixation in flooded soils[4022] and excretion of organic compounds from plant roots helps make the rhizosphere a favorable environment for heterotrophic nitrogen fixation.[3510] Factors that inhibit nitrogen fixation include high ambient nitrogen concentration of inorganic nitrogen, low light intensities (decreases autotrophic nitrogen fixation), high oxygen concentrations (inhibits nitrogenase), high redox potential (fixation is greater under reduced than under oxidized conditions), and high (> 8.0) or low (< 5.0) pH levels[624,3510] (see also Chapters 2.18.1 and 4.2.4.1).

Depending on the population of N_2 fixers and environmental conditions a wide range of N_2 fixation rates in the floodwater have been reported.[624,2351] Johnston,[2351] summarizing literature data reported an average asymbiotic nitrogen fixation rates of 1.33 and 2.50 g N m^{-2} yr^{-1} for temperate and tropical wetland regions, respectively. Armentano and Verhoeven[135] reported rates of N fixation by free-living microorganisms in different wetlands from 0.03 to 7.2 g N m^{-2} yr^{-1}. Nitrogen fixation reaching 3 g N m^{-2} yr^{-1} in rice fields is favored under flooding by providing a photo-oxic water layer and surface soil for phototrophic blue-green algae and anoxic soil suitable for micro-aerobic and anaerobic heterotrophic bacteria.[5217] Whitney et al.[5393] measured nitrogen fixation up to 40 g N m^{-2} yr^{-1} in Sapelo Island salt marsh in Georgia. Nitrogen fixed by blue-green algae of the arctic tundra ranged from 0.0043 to 9.56 g N m^{-2} yr^{-1} with a mean value of 0.13 mg N m^{-2} yr^{-1} (53, 5110). Nixon and Lee[3447] reported the range of nitrogen fixation in different regions of the United States in the range of 0.007 (tundra site) to 39.7 g N m^{-2} yr^{-1} (tall *Spartina alterniflora* marsh in Georgia). Gopal and Masing[1747] reported the nitrogen fixation rate in the range of 0.03 to 40 g N m^{-2} yr^{-1} worldwide. Significant rates of N_2 fixation by the heterotrophic bacteria in the anaerobic soil layer have also been reported. Nitrogen fixation by these bacteria can account for 0.1–0.5 kg N ha^{-1} yr^{-1}, as compared to 25 kg N ha^{-1} yr^{-1} by blue-green algae.[624]

There is a large population of N_2-fixing organisms present in the rhizosphere of wetland plants.[554,4925,5524-5] Nitrogen fixation in the root zone of wetland plants is possible only if N_2 gas is present. Wetland plants have a transport mechanism whereby atmospheric N_2 is transported to the root zone.[5526] In addition, N_2 generated during the denitrification is also potentially available for N_2 fixation.[4032]

4.2.3.6 Ammonium and nitrate diffusion

Ammonium and nitrate diffusion in wetland soils has been comprehensively summarized by Reddy and Patrick.[4027] Ammonium N in the aerobic soil layer and overlying water is derived from: (1) decomposition of organic matter in the water column; (2) mineralization of organic N in the aerobic soil layer; or (3) mass transfer of ammonium N from the anaerobic soil layer to the aerobic soil layer and floodwater.[4027] The latter process contributes a major portion of ammonium N to the overlying aerobic soil layer and floodwater. Mass transfer is accomplished by diffusion, bioturbation, and wave and current stirrings at and near the sediment-water interface.[4027] Rate of ammonium N movement from the anaerobic soil layers governed by Reddy and Patrick:[4027] (1) the concentration gradient established as a result of ammonium-N consumption in the aerobic zone during nitrification and NH_3 volatilization; (2) the ammonium-N regeneration rate in the anaerobic zone; (3) the ammonium-N concentration in the pore water; (4) other cations on the soil exchange complex; (5) cation exchange capacity of the soil; and (6) the relative volume of pore space which is the function of bulk density of the sediment.

Nitrate N in floodwater and in the aerobic soil layer is derived from:[4027] (1) nitrification of ammonium-N in these zones, and (2) input from drainage water effluents and from waste waters when flooded soils are used as a treatment system. In a natural system, oxidation of ammonium-N is the major source of nitrate-N in the floodwater and aerobic surface soil layer. In flooded soils and sediments, nitrate-N present in the floodwater and aerobic soil layer readily diffuses into the anaerobic soil layer as a result of a downward concentration gradient. This diffusion is controlled by (1) available carbon supply in the anaerobic portion of the sediment; (2) thickness of aerobic soil layer; (3) floodwater depth; (4) nitrate-N concentration in the floodwater; (5) temperature; and (6) mixing and aeration in the floodwater.[4027]

Reddy and Patrick,[4027] in their excellent review paper on nitrogen transformations and losses in flooded soils and sediments, presented a schematic representation of the sequential processes functioning in the transport of N from flooded soils and sediments (Figure 4-21). The authors pointed out that the processes shown in the figure will function continuously in flooded soils, shallow water bodies, lake bottoms, and ocean muds.

The literature on nitrogen budget and dynamics in wetlands is voluminous (see e.g. References 135, 489, 2351, 4036, 4105–6). Several local cases are shown in Figures 4-22, 4-23 and 4-24 (see also Figure 4-21).

Johnston[2351] summarized that the nitrogen concentration in wetland soils is in the range of 0.02–65 mg N g DM^{-1}, organic soils average about twice as much as mineral soils. The author reported annual accumulation in wetland soils in the range of 0.9 to 2.7 g N m^{-2} yr^{-1} with the average value of 1.6 g N m^{-2} yr^{-1}. Nichols[3432] reported the nitrogen accumulation in wetland soils 0.1–4.7 and up to 10.0 g N m^{-2} yr^{-1} for moderate to cold climates and warm, highly productive areas, respectively.

4.2.3.7 Nitrogen in wetland plants

Woody plants[2351]

Concentration in % dry mass:
leaves: 2.12 (0.70–4.50)
shrub stems: 0.57 (0.35–0.91)
tree boles and roots: 0.39 (0.08–0.68)
standing stock (g N m^{-2}):
leaves: 7.8 (4.7–12.5)
litter: 12.1 (4.5–24.1)

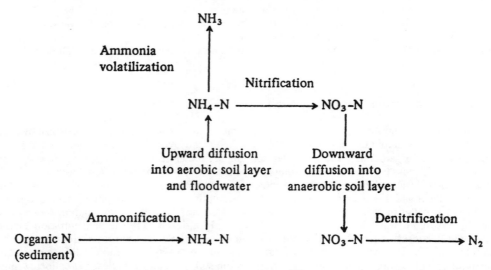

Figure 4-21. Nitrate and ammonium diffusion in wetlands.[4027] (From Reddy and Patrick, 1984, in *CRC Crit. Rev. Environ. Control,* Vol. 13, p. 300, reprinted with permission of CRC Press, Inc., Boca Raton.)

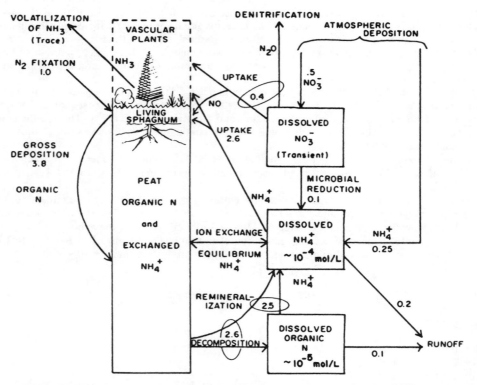

Figure 4-22. Proposed mean annual budget for nitrogen in Thoreau's Bog in g m^{-2}.[2029] (From Hemond, 1983, in *Ecology,* Vol. 64, p. 106, reprinted with permission of the Ecological Society of America.)

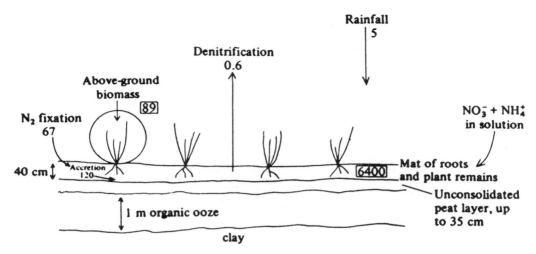

Figure 4-23.   Nitrogen pools (figures in rectangles, kg ha^{-1}) and flows (kg ha^{-1} yr^{-1}) for a freshwater marsh. The dominant macrophyte is *Panicum hemitomon*.[4661] (From Sprent, 1987, in *The Ecology of the Nitrogen Cycle*, p. 108, reprinted with permission of Cambridge University Press, Cambridge.)

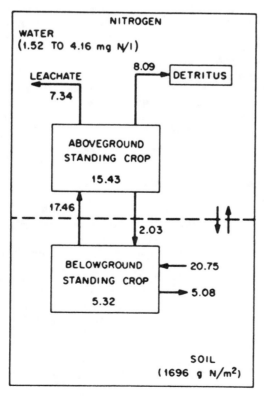

Figure 4-24.   Flow of nitrogen through a *Scirpus fluviatilis* stand. Flows in g m^{-2} yr^{-1} and compartments in g m^{-2} in standing crop.[2537] (From Klopatek, 1978, in *Freshwater Wetlands; Ecological Processes and Management Potentials*, Good, R.E., et al., Eds., p. 207, reprinted with permission of Academic Press and the author.)

Herbaceous plants

Concentration in % dry mass (see Appendix, Table 7-6):
emergent species: aboveground: 0.18–9.5, belowground: 0.1–3.66
floating-leaved species: aboveground: 1.09–5.09, belowground: 0.89–2.90
submerged species: 0.82–6.01 (aboveground)
floating species: 1.30–7.28
bryophytes: 0.56–2.87
Standing stock (g N m^{-2}) (see Appendix, Table 7-7):
emergent species: aboveground: 0.04–205, belowground: 0.92–87
floating-leaved species: 1.34–12.4 (aboveground)
submerged species: 0.66–11.3 (aboveground)
floating species: 0.4–234

Annual nitrogen uptake rates by various macrophytes (see Appendix, Table 7-8) range between 10 and 263 g m^{-2} yr^{-1} for emergent species and between 41 and 611 g N m^{-2} yr^{-1} for floating species. Daily uptake rates range between 18 and 510 mg N m^{-2} d^{-1}.

4.2.4 Nitrogen Sources for Algal Growth

After carbon, oxygen, and hydrogen, nitrogen is the most abundant element in algae without mineralized walls. Nitrogen is of a major importance in amino acids, purines, pyrimidines, porphyrins, amino sugars or amines. The normal requirement of nitrogen in cultures of Chlorophyceae was found to be 6.5–8.3% of the ash-free dry mass, but under conditions of nitrogen starvation this level can be greatly reduced.[2481] In algae growing without nitrogen limitation it may constitute up to 10% of the dry mass.[4861] In Table 4-2, the literature summary of nitrogen concentration in algae is shown.

Nitrogen is available to algae in three basic forms: free nitrogen gas, and as combined inorganic or organic compounds.[4643] In natural habitats the main sources of nitrogen are nitrate and ammonium salts, but in highly polluted waters the organic nitrogen compounds may become important and there may be a relationship between some of the products of animal excretions (ammonia, urea, uric acid and amino acids) and the growth of certain flagellates.[4209]

The assimilation of nitrogen is indirectly connected with photosynthesis since the energy source, hydrogen donors and carbon skeletons are ultimately derived from the latter process. Whatever the origin of the nitrogen, conversion to ammonia takes place, hence the need for hydrogen donors.[1403]

The role of nitrogen in plant (including algae) metabolism has been reviewed many times.[311,642,1402,1407,1519,1841,2470,2592,2841,3156,3363,4750,4860–1,5081–2,5304]

4.2.4.1 Gaseous nitrogen—nitrogen fixation

The ability to reduce or fix nitrogen gas is found only among the prokaryotes, and in the algae exclusively in the Cyanophyceae. Definite proof that a blue-green alga can assimilate or "fix" molecular nitrogen (N$_{2z}$) was for the first time published by Drewes[1143] in 1928 although the suspicion that certain members of this group possessed this property had been current for forty years before this.[1417]

Nitrogen fixation is the conversion of gaseous nitrogen (N$_2$) to ammonia.[1516] Nitrogen fixation requires nitrogenase, an oxygen-sensitive iron-, sulphur- and molybdenum-containing enzyme complex[471] which also brings about the reduction of other substrates containing triple covalent bonds (nitrous oxide, cyanides, isocyanides, cyclopropene, and acetylene). Acetylene reduction to ethylene and the simple separation of substrate and product by gas chromatography has been adopted as the standard assay for nitrogen fixation.[4643]

The reduction of gaseous nitrogen (N$_2$) to ammonia (NH$_3$) takes place very rapidly and for this reason the individual steps in the reaction have not been investigated in detail.[3198] It is supposed that the whole reaction is a 3-step, 2 electrons per step mechanism:[5460]

$$N \equiv N \;\rightarrow\; \underset{\text{diimide}}{HN = NH} \;\rightarrow\; \underset{\text{hydrazine}}{H_2N - N_2H} \;\rightarrow\; NH_3 \qquad (4.29)$$

The major reactions involved in N_2 fixation are shown in Figure 4-25. Stewart et al.[4781] and Stewart and Alexander[4777] showed that the nitrogenase activities of blue-green algae were tied to concentrations of available phosphorus. Nitrogenase activity requires ATP and reduced ferrodoxin.[471] ATP is produced in the heterocyst primarily via photosystem I cyclic photophosphorylation (see Chapter 2.12.2), as is some reduced ferrodoxin. Most ferrodoxin, however, appear to be reduced by NADPH obtained principally from the hexose monophosphate shunt pathway (see Chapter 2.12.2) using carbohydrate imported from adjacent cells.

Ammonia has been long recognized as a key intermediate and probably the first stable product in nitrogen fixation.[1417] Magee and Burris[3020] found it to be the first substance to become labelled with ^{15}N when nitrogen-fixing blue-green algae were provided with N_2 enriched with this isotope. Similar results were obtained with *Westiellopsis prolifica*.[1430] Fogg[1417] reported that it is generally assumed that the common path of entry is by reductive amination of α-ketoglutarate to give glutamate.[1417] However, it has been shown the presence of alanine dehydrogenase,[1985,3407,4382] and high levels of glutamine synthetase.[1073] A variety of other ammonia-incorporation enzymes have also been reported in *Anabaena cylindrica*.[1985] The idea that citrulline is another substance through which ammonia enters organic combination[1423] was based on the finding that this happens in alder root nodules.

Most nitrogen-fixing cyanophytes are filamentous (e.g. genera *Aphanizomenon, Anabaena, Nostoc, Cylindrospermum, Gloeotrichia, Mastigocladus*) and contain large, pale, thick-walled cells called heterocysts (see Figure 2-43), which are formed in the absence of utilizable combined nitrogen. They lack the oxygen-evolving photosystem II apparatus, RuBisCO, and may lack or have reduced amounts of the photosynthetic biliproteins.[4774,5479] The heterocyst walls contain oxygen-binding glycolipids,[2687,5478] and these together with respiratory oxygen consumption maintain the anaerobic conditions necessary for nitrogen fixation.[471]

Ammonia, the first product of fixation[5480] is assimilated via glutamine synthetase[4778] into glutamine[3175] (see also Chapter 4.2.4.3). Nitrogen is reduced to NH_4^+, which is converted to glutamine before export from the heterocyst (for details about heterocysts see Chapter 2.18.1) to adjacent cells of the filament (Figure 4-26). Six electrons are required for each N_2 molecule reduced; these are derived from a low potential reductant such as reduced ferredoxin. ATP is also required, most probably 12 molecules for each N_2 molecule.[3318] In blue-green alga both reduced ferredoxin and ATP can probably be generated either by photochemical reactions or by dark respiratory reactions.[1551]

For many years, it was believed that nitrogen-fixing blue-green algae belonged only to the family Nostocaceae. However, nitrogen fixation has been demonstrated in members of families Scytonemataceae, Rivulariaceae, and Oscillatoriaceae.[3326,5282] Nitrogen fixation has also been demonstrated in several members of the genus *Anabaena*.[69,70,997,1017,1396] Complete listings of nitrogen fixing organisms including blue-green algae are given by e.g. Sprent[4661] and Sprent and Sprent.[4662]

Nitrogen-fixation by blue-green algae relies on oxygenic photosynthesis as a carbon source.[4771,5519] Photosynthetically evolved oxygen can, however, be a formidable barrier to N_2 fixation, because the enzyme complex mediating the latter (i.e. nitrogenase) is readily inactivated by oxygen.[5519] In some blue-green algae, photosynthesis and nitrogen fixation appear to be very sensitive to oxygen; here inhibition begins below atmospheric O_2 concentrations.[4776]

Although it is now known that some non-heterocystous blue-green algae are able to fix N_2 under certain circumstances, e.g. *Lyngbya aestuarii*,[3622] *Plectonema boryanum*,[4775] and marine genus *Trichodesmium*,[1177-9,1675] many investigations have confirmed the essential correctness of the heterocyst hypothesis (see Fogg[1417] for discussion).

In the field, nitrogen fixation by blue-green algae is usually light-dependent,[1176,1673,2192] the rate showing much the same relationship as photosynthesis itself to the light intensity as it varies with depth, with inhibition at the surface in full sunlight, a maximum some way below the surface, and light limitation below this.[2192-3] Low, but statistically significant rates of nitrogen fixation have been recorded below the photic zone[1176,1673,2192] but there is a possibility that this may be carried out by photosynthetic bacteria.

Ammonia suppresses the formation of nitrogenase and also inhibits the formation of heterocysts.[67,1396,1399,2342,3512,4780] Suppression of heterocyst formation by ammonium salts is usually complete but by nitrate may be only partial. Inhibition by urea is complete.[67] Ogawa and Carr[3512] found heterocysts to be still present in *Anabaena variabilis* grown in the presence of 28 mg L^{-1} NO_3-N. For an equal amount of N accumulated, an external concentration of 10 mg L^{-1} ammonium-N was much more inhibitory to nitrogenase in blue-green alga *Calothrix* D764 than 1 mg L^{-1} N.[2292]

Okuda and Yamaguchi[3537] concluded that growth of nitrogen fixing algae in paddy fields was most usually limited by low pH and deficiency in phosphorus. Molybdenum is also likely to be a limiting

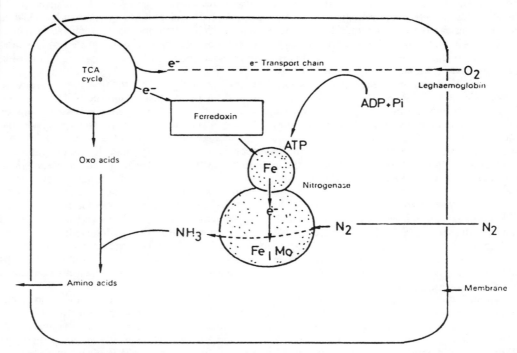

Figure 4-25. The major reactions involved in N₂ fixation. A bacteroid is shown containing nitrogenase, an electron transport chain and a tricarboxylic acid cycle.[3199] (From Mengel and Kirkby, 1979, in *Principles of Plant Nutrition,* p. 153, reprinted with permission of International Potash Institute, Basel.)

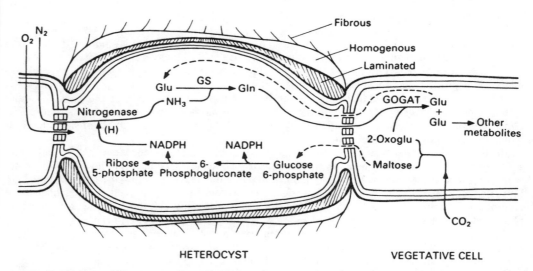

Figure 4-26. Diagram of the flow of carbon and nitrogen between heterocysts and vegetative cells in nitrogen-fixing blue-green algae. GS = glutamine synthetase; GOGAT = glutamine oxoglutarate amino transferase; Glu = glutamic acid; Gln = glutamine. The layered wall of the heterocysts is impermeable to nitrogen, CO₂ and O₂.[1956] (From Haselkorn, 1978, reproduced, with the permission, from *Annual Reviews of Plant Physiology,* Vol. 29, p. 330, © 1978 by Annual Reviews, Inc.)

factor.[3539] Phosphorus-starved cells of *Anabaena flos-aquae* rapidly increase their capacity to reduce acetylene when phosphate is supplied, and this has been proposed as a basis for bioassay of available phosphorus in aquatic ecosystems.[4777,4781] Active nitrogen fixation by blue-green algae has been observed, both at nearly freezing point, in the Antarctic[1431,2191] and in hot springs.[4765]

Ecologically, nitrogen-fixing blue-green algae are of particular interest because of their potential to contribute simultaneously to the carbon and nitrogen input of certain water or soil strata. Large pops of heterocystous blue-green algae and maximum N_2 fixing rates are generally associated, spatially and temporally, with minimal concentrations of combined inorganic N forms.[1188,1673,2194,2197,4782] However, the absence of inorganic forms does not necessarily preclude elemental N fixation.[1181,1674,2192] General accounts of the ecology of nitrogen fixation by blue-green algae are given by Stewart[4763–4] and by Fogg et al.[1434] and more specialized reviews relating to tropical soils by Singh[4518] and Watanabe and Yamamoto,[5281] to temperate soils by Shtina[4461] and Henriksson,[2034] to freshwaters by Fogg[1415] and Stewart,[4762] and to the sea by Stewart.[4767] For further details on nitrogen fixation the reader is referred to References 643, 1018–9, 1191, 1403, 1406, 1408, 1417, 1420, 1429, 1434, 3854–5, 4758, 4760–1, 4763–4, 4766, 4768, 5168 (see also Chapter 4.2.3.5).

4.2.4.2 Combined inorganic nitrogen

Almost all chlorophyll-containing algae studied in culture will grow with either nitrate (NO_3^-), nitrite (NO_2^-), or ammonium (NH_4^+) as a nitrogen source.[4861] Maximum growth rates are generally much the same with either NO_3^- or NH_4^+ as a N source.[2429,2589,2878,4265] Paasche,[3609] however, found that *Dunaliella tertiolecta* grows 10–30% faster on NH_4^+ than on NO_3^-. On the other hand, growth rates of red algae *Goniotrichum* and *Nemalion* were greater when the plants were supplied with nitrate as compared to ammonia.[1506]

It has been known for a long time that nitrate nitrogen must be reduced to the ammoniacal form before incorporating into metabolic reactions.[3873] Now it is clear that whatever the source of inorganic nitrogen it is generally agreed that conversion to NH_4^+ occurs before incorporation into cellular organic compounds.[4861] It has long been known that the uptake of nitrate and ammonium by algal cultures is accompanied by pH changes, the medium becoming alkaline when nitrate is used and acidic when ammonium is taken up.[2589] That such changes in H^+ and OH^- must inevitably accompany the conversion of NO_3^- or NH_4^+ nitrogen into cellular nitrogen was well shown by the equations by Cramer and Myers[911] for *Chlorella* growth:

$$1.0\ NO_3^- + 5.7\ CO_2 + 5.4\ H_2O \rightarrow C_{5.7}H_{9.8}O_{2.3}N + 8.5\ O_2 + 1.0\ OH^- \qquad (4.30)$$

$$1.0\ NH_4^+ + 5.7\ CO_2 + 3.4\ H_2O \rightarrow C_{5.7}H_{9.8}O_{2.3}N + 6.25\ O_2 + 1\ H^+ \qquad (4.31)$$

Syrett[81] pointed out that only in the late 1970s that changes in pH were linked mechanistically to the uptake of the ions by algae.[4009,5078] Pokorný et al.[3827] pointed out that the form of inorganic nitrogen should be considered when photosynthesis is measured as a pH drift and/or O_2 release because pH and O_2 changes are caused not only by carbon dioxide reduction but also by reduction of nitrate. The photosynthetic reduction of nitrate may play an important role in the sudden increase in pH following the addition of nitrate fertilizer to surface water.

Figure 4-27 depicts a much idealized algal cell setting out some of the major features of nitrogen assimilation. The diagram shows all nitrogen sources crossing the cell wall and outer cell membrane (the plasmalemma) to enter the cytoplasm. For all the compounds depicted, there is evidence for active uptake systems dependent on metabolism although there may be passive entry by free diffusion as well.[4861]

The uptake of ammonium, nitrite and nitrate is mostly described by the Michaelis-Menten hyperbolic curve.[832,1022,1129,1175,1259,1264,1896,1923,3001–2,3495,3912,5080,5356]

Principal environmental factors affecting nitrate uptake by phytoplankton are light,[832,1177,3002,4990] the concentration of NO_3^-,[1175,1263,4991] and the concentration of ammonia.[3915] Ammonium uptake is a very complicated and variable process that is strongly influenced by environmental conditions, such as preconditioning N source and N availability.[1129] In Tables 4-3 and 4-4, kinetic constants for ammonium and nitrate uptake, respectively, are given. The reports on kinetic constants for nitrite uptake are rare in comparison to those reported for ammonium and nitrate, e.g. Hanisak and Harlin[1896] reported V_{max} in the range of 3.68 to 8.99 μM NO_2 g DM^{-1} h^{-1} and K_s in the range of 3.06 to 10.39 μM NO_2 for *Codium fragile* var. *tomentosoides*.

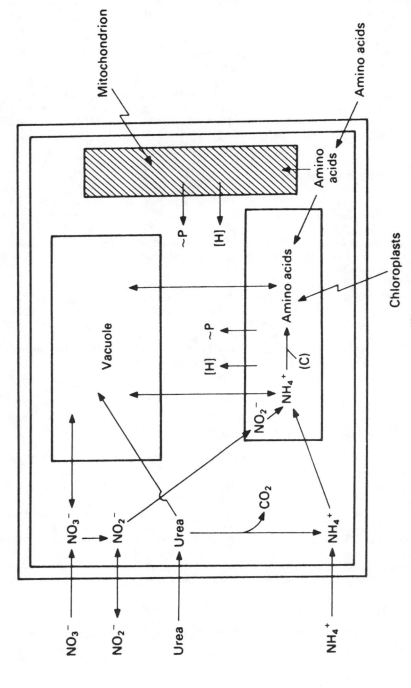

Figure 4-27.   Main features of nitrogen assimilation and partitioning in an algal cell.[461] (From Syrett, 1981, in Physiological Bases of Phytoplankton Ecology, Platt, T., Ed., *Can. Bull. Fish. Aquat. Sci.*, Vol. 210, p. 185, reprinted with permission of Department of Fisheries and Oceans, Ottawa.)

Table 4-3. Kinetic constant values for ammonium uptake by algae

Species	K_s (μM NH_4^+)	V_{max} (μmol NH_4^+ g DM^{-1} h^{-1})	Reference
Freshwater			
Aphanizomenon flos-aquae	2.64–21.4		3352
Ceratium sp.	1.29		3352
Cladophora glomerata		138–438	5197
C. glomerata	17.4–41.9	81–499	2899
Microcystis aeruginosa	5.79–16.2		3352
Oscillatoria agardhii	8–27	720–880	5554,5556
O. agardhii		80–770	5555
Pseudanabaena catenata	0.14		1993
Scenedesmus quadricauda	0.15		1993
Lower Great Lakes phytoplankton	0.2–1.2		3351
Marine			
Codium fragile subsp. tomentosoides	1.40–2.10	13–28	1896
Enteromorpha compressa	18.67	662	2445
Enteromorpha prolifera	2.90–13.4	39.2–188	3495
Fucus spiralis	5.80–9.60		5005
Gracilaria foliifera	1.60	30.6	1047
Hypnea musciformis	16.60		1868
Iridaea cordata	2.50	5.5	1899
Macrocystis pyrifera	5.30	23.8	1868
Neoagardhiella baileyi	2.30–4.90	7.20–38.4	1047
Thalassiosira pseudonana	0.55–0.82		1129
T. pseudonana	0.02–0.66		1261
Ulva lactuca	6.7–35.4	64.3–501	830
U. lactuca	19.3–52.7	177–324	1526
Phytoplankton	0.5–1.57		1361

Hanisak and Harlin[1896] found that the values of V_{max} and K_s for nitrate and nitrite uptake in *Codium fragile* generally increased with temperatures between 6 and 24°C. Values of V_{max} for NH_4^+ uptake also increased with temperature; however, K_s values appeared independent of temperatures in the range of 6–30°C. Uptake rates of NH_4^+ and NO_3^- are also affected by irradiance levels with higher rates of uptake measured in the light as compared to dark.[1047,1896,1920,1923]

La Roche and Harrison[2710] pointed out that the non-linearity of inorganic nitrogen uptake and in particular NH_4^+ uptake is now well documented both in field studies[1635,1700,5365,5368] and in laboratory experiments.[865,867,1049,1698,2206,3126]

Nitrogen uptake capacity of seaweeds (V_{max}) is a direct function of surface area-to-volume ratio (SA/V),[4186] whereas N storage capacity varies approximately inversely with SA/V.[1182,4184] This suggests that the degree of coupling between seaweed growth rates and N supply may also be a function of SA/V. High SA/V species, which have high V_{max} but low storage capacity, would have growth rates highly correlated with N supply, whereas low SA/V species with low V_{max} would store N and have growth rates relatively independent of N availability.[4184] Thus, a seaweed's "functional form"[2876] may determine its ability to buffer nutrient variability.

In comparison with macrophytes, nitrogen uptake rate is only rarely expressed per unit of area. Richardson and Schwegler[4110] reported uptake rate of 55 mg N m^{-2} d^{-1} for *Cladophora glomerata* in a Michigan marsh. In waters with high nitrogen concentration the uptake rate may reach much higher

Table 4-4. Kinetic constant values for nitrate uptake by algae

Species	K_s (μM NO_3^-)	V_{max} (μmol NO_3^- g^{-1}DM h^{-1})	Reference
Freshwater			
Anabaena cylindrica	70		1968
Chlamydomonas reinhardtii	148		5239
Cladophora glomerata	7.3–15.2	14–70.3	2899
Navicula pelliculosa	14.9		5239
Oscillatoria agardhii		0–550	5555
Oscillatoria agardhii	12–21	250–1,170	5555
Oscillatoria agardhii	23–63	380–820	5553
Scenedesmus sp.	2.9–8.4		4085
Phytoplankton	2.75–4.21		832
Phytoplankton	14.49		4992
Phytoplankton	1.15–173		4991
Marine			
Biddulphia aurita	0.91–3.19		5085
Codium fragile subsp. tomentosoides	1.18–7.65	2.80–10.92	1896
Ditylum brightwellii	2.00–2.10		1257
Enteromorpha prolifera	2.31–13.3	75.4–169	3495
Enteromorpha spp.	16.6	129.4	1920
Fucus spiralis	5.60–12.8		5005
Gracilaria foliifera	0.20–7.10	4.3–20.9	1047
Hypnea musciformis	4.90	28.5	1868
Laminaria longicruris	4.10–5.90	7.0–9.6	1923
Laminaria saccharina	1.32–1.47		755
Macrocystis pyrifera	13.1	30.5	1868
Neoagardhiella baileyi	2.10–2.70	13.8–16.2	1047
Thalassiosira pseudonana	0.37–3.70		1129
T. pseudonana	0.47–0.97		1261
Oceanic phytoplankton	0.1–0.7		1264
Oceanic phytoplankton	0.1–0.6		698
Oceanic phytoplankton	0.2–0.4		682

values, however. Davis et al.[1000] reported the uptake rate as high as 1,900 mg N m^{-2} d^{-1} for periphyton growing in wastewater treatment plant effluent. Richardson and Schwegler[4110] reported nitrogen standing stock in *Cladophora glomerata* of 4.3 g N m^{-2}.

4.2.4.3 Assimilation of ammonium

There are several possible pathways of ammonium assimilation. Early works[214,4058,4855] interpreted ammonium assimilation as being consistent with the then accepted mechanism. The key compound was glutamic acid which was formed from NH_4^+ and α-oxoglutaric acid by the reductive reaction catalyzed by glutamic dehydrogenase (GDH) (Figure 4-28, reaction 1).

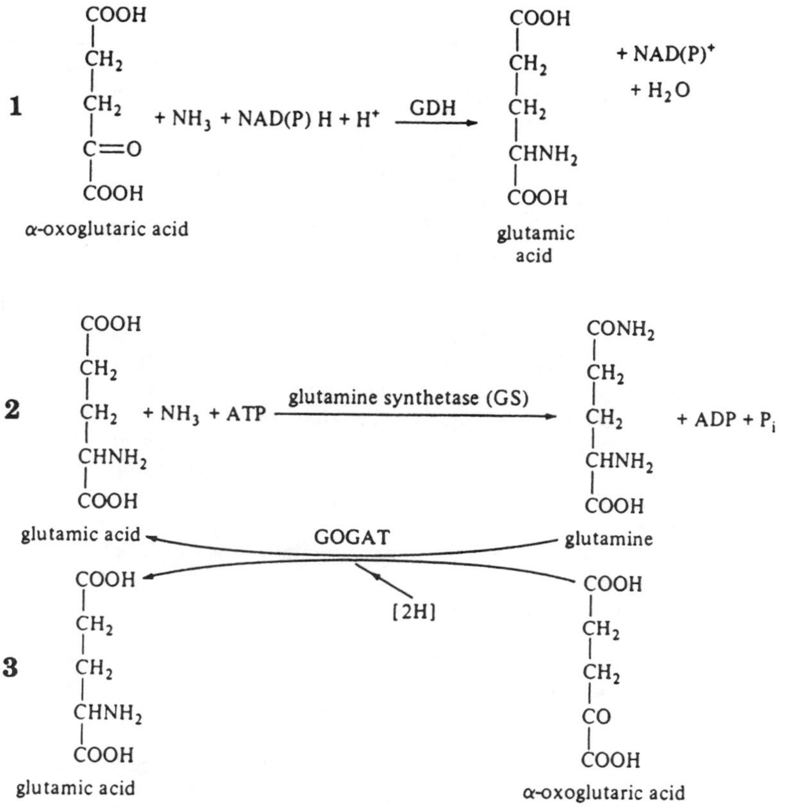

$$1 \quad \begin{array}{c} COOH \\ | \\ CH_2 \\ | \\ CH_2 \\ | \\ C{=}O \\ | \\ COOH \end{array} + NH_3 + NAD(P)H + H^+ \xrightarrow{\text{GDH}} \begin{array}{c} COOH \\ | \\ CH_2 \\ | \\ CH_2 \\ | \\ CHNH_2 \\ | \\ COOH \end{array} \begin{array}{c} + NAD(P)^+ \\ + H_2O \end{array}$$

α-oxoglutaric acid → glutamic acid

$$2 \quad \begin{array}{c} COOH \\ | \\ CH_2 \\ | \\ CH_2 \\ | \\ CHNH_2 \\ | \\ COOH \end{array} + NH_3 + ATP \xrightarrow{\text{glutamine synthetase (GS)}} \begin{array}{c} CONH_2 \\ | \\ CH_2 \\ | \\ CH_2 \\ | \\ CHNH_2 \\ | \\ COOH \end{array} + ADP + P_i$$

glutamic acid → glutamine

GOGAT

glutamic acid ← glutamine

[2H]

$$3 \quad \begin{array}{c} COOH \\ | \\ CH_2 \\ | \\ CH_2 \\ | \\ CHNH_2 \\ | \\ COOH \end{array} \qquad \begin{array}{c} COOH \\ | \\ CH_2 \\ | \\ CH_2 \\ | \\ CO \\ | \\ COOH \end{array}$$

glutamic acid   α-oxoglutaric acid

Figure 4-28. Pathways of incorporation of ammonia. GDH = glutamic dehydrogenase, GOGAT = glutamine oxoglutarate aminotransferase.[4861] (From Syrett, 1981, in *Physiological Bases of Phytoplankton Ecology*, Platt, T., Ed., *Can Bull. Fish. Aquat. Sci.*, Vol. 210, p. 186, reprinted with permission of Department of Fisheries and Oceans, Ottawa.)

At least two forms of GDH exist that preferentially use NADH or NADHP. NADHP-GDH is induced after addition of ammonium to cultures of *Chlorella*[4426] and has a K_m for ammonium sufficiently low (50 μM) to operate in the amination direction.[4981] The induction of this enzyme may require high concentrations of ammonium (e.g. 5 mM[4981]) and therefore is not synthetized by ammonium-grown cells in chemostats.

However, in the early 1970s, work by Brown and his colleagues[575,4927] with bacteria revealed an alternative pathway of NH_4^+ assimilation through the glutamine synthetase–glutamate synthase cycle (GS/GOGAT cycle). Miflin and Lea[3214-5] and Lea and Miflin[2725-6] showed that the pathway could also operate in leaves of higher plants and in blue-green and green algae. Demonstration of GS activity in phytoplankton was then given by several authors.[539,1311,2100,3706,4950] There are also reports on GS isomorphs in algae.[719,1387,4837]

In this pathway, glutamine rather than glutamic acid is the first product of NH_4^+ assimilation (Figure 4-28, reaction 2) and the NH_4^+ incorporated into the amide group of glutamine is then transferred to α-oxoglutaric acid in a reductive reaction (Figure 4-28, reaction 3) from which two molecules of glutamic acid result. The enzyme catalyzing reaction (3) is glutamine-oxoglutarate aminotransferase or glutamate synthase; it is often called GOGAT. By reactions (2) and (3) operating in sequence, NH_4^+ is assimilated into glutamic acid by combination with α-oxoglutaric acid but GDH is not involved. The overall reaction is still reductive, as in reaction (1), but it also requires ATP which reaction (1) does not.[4861] Syrett[4861] pointed out that it is particularly attractive to think of reaction (2) as being the major reaction responsible for the incorporation of NH_4^+ into organic combination because GS has a much higher affinity for ammonium than does GDH.[3215-6] In a number of marine phytoplankters, for example, the apparent K_m (concentration supporting half-maximum velocity of reaction) of GDH for NH_4^+ was 4,500–10,000 μM.[45,1260] Ahmad and Helleburst[40] found K_m of GDH for NH_4^+ 33,000 μM in the soil alga *Stichococcus bacillaris* and Ahmad and Helleburst[38] reported apparent K_m for ammonium of NADPH-GDH isolated from the green alga *Chlorella autotrophica* 20,000 μM. For *Skeletonema*, Falkowski and Rivkin[1311] found a K_m for GDH of 28,000 μM; for GS it was only 29 μM which is similar to the value for higher plant enzymes.[3215] However, Everest and Syrett[1296] reported that NADPH-GDH from *Stichococcus bacillaris* has a K_m for ammonium of about 1,000 μM. According to Miflin and Lea,[3216] the energy cost of ammonium assimilation via the GS/GOGAT cycle is only moderately higher than via the GDH pathway.

Since the discovery of the GS/GOGAT assimilation pathway much work has been done to discover whether NH_4^+ assimilation mainly takes place through GDH or through the GS/GOGAT reactions. One approach is using inhibitors,[3215,3968-9,4136,4138,4778] the second one is to survey algae for the presence of the key enzymes, GDH, GS, and GOGAT, and to estimate their activities.[39,40,1213,2726,3314,4427] These studies can sometimes be misleading because an enzyme activity may appear to be absent or low in cell-free extracts because of failure to establish optimal conditions for assay.[4861] It seems, however, that the major pathway for incorporation of ammonium into amino acids in bacteria, blue-green algae and higher plants is via the GS/GOGAT pathway and that GDH plays only a minor role.[3215-6,4982,5545] In fungi, the GDH pathway may be more significant.[575] In the green algae, the situation is less clear because few detailed studies have been carried out to determine the relative importance of the two pathways.[1296] The GS/GOGAT pathway is the primary route for ammonium incorporation into amino acids in the fronds of several seaweeds.[1981,4354]

Of other pathways of NH_4^+ assimilation, the direct formation of alanine from NH_4^+ and pyruvic acid is significant in some blue-green algae, those that lack GDH activity.[3407] Alanine dehydrogenase is also present in *Chlamydomonas* where, during the life cycle, its activity changes in an inverse way to that of GDH.[2435]

Another subsidiary pathway of NH_4^+ assimilation is that which results in carbamoyl phosphate and hence incorporation into citrulline (and then arginine), pyrimidines (including thiamine), and biotin;[4861] it is probable that in plants the ammonia for carbamoyl phosphate synthesis is derived from the amide group of asparagine of glutamine rather than from free NH_4^+ ions.[284]

Synthesis of amino acids and proteins and their functions in algae have been summarized by Fowden[1466-7] and Morris.[3311]

4.2.4.4 Assimilation of nitrate and nitrite

Assimilation of NO_3^- takes place by reduction of NO_3^- to NO_2^- followed by a reduction of NO_2^- to NH_4^+ which is then converted to organic compounds.[1841,4861] The diagrammatic nitrate conversion is shown in Figures 4-29 and 4-30. It is firmly established[286,1571,2051,2914] that the assimilatory nitrate-reducing system consists of only two metalloproteins, namely nitrate reductase and nitrite reductase, which catalyze the stepwise reduction of nitrate to nitrite and ammonia:[1841]

$$NO_3^- \xrightarrow[2e^-]{\text{Nitrate reductase}} NO_2^- \xrightarrow[6e^-]{\text{Nitrite reductase}} NH_4^+ \qquad (4.32)$$

There are two types of nitrate reductase known in algae. The first, and better known enzyme, is found in eukaryotic algae.[4861] This nitrate reductase complex is similar to that found in fungi and higher plants, consisting of haem (cytochrome *b*-557), flavin adenine dinucleotide (FAD), and molybdenum as a prosthetic group.[285,1339,1644,1841,2472,3372,4618,4620,4861] The best characterized algal enzyme is from *Chlorella*; this enzyme has a molecular mass of about 350,000 and a complex structure.[4861] The enzyme catalyzes the reduction of nitrate to nitrite by reduced pyridine nucleotides[1571,2048,2050,2914] in accordance with the equation:[1841,4861]

$$NO_3^- + NAD(P)H + H^+ \rightarrow NO_2^- + NAD(P)^+ + H_2O \qquad (4.33)$$

Molybdenum appears to be at or close to the site at which NO_3^- is reduced and the enzyme works by molybdenum being reduced by electrons derived from NADH and then reoxidized, possibly to Mo^{VI}, by NO_3^- which is consequently reduced to NO_2^- (Figure 4-31).

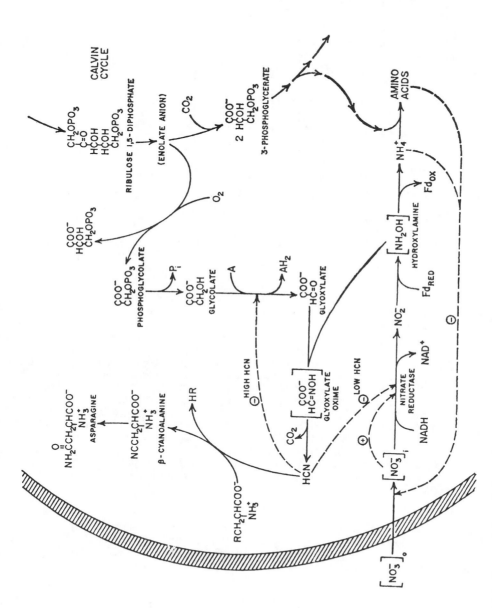

Figure 4-29. Model for the regulation of nitrate assimilation. Solid lines represent metabolic conversions. Dashed lines represent inhibition (-) or activation (+) reactions.[4619] (From Solomonson and Shepar, 1977, reprinted with permission from *Nature*, Vol. 265, p. 373. Copyright (1977) Macmillan Magazines Limited, London.)

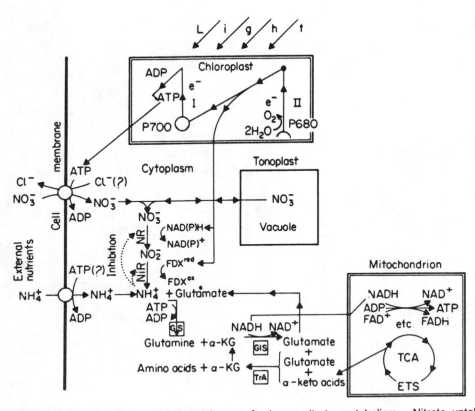

Figure 4-30. A generalized speculative scheme of primary nitrate metabolism. Nitrate uptake is facilitated by a (NO_3^-, Cl^-)-activated ATPase, located in the plasma membrane. Once transported into the cytoplasm, NO_3^- may be stored in a vacuole, or reduced via nitrate reductase (NR) to NO_2^-. Subsequently, NO_2^- is reduced to NH_4^+ via nitrite reductase (NiR). The primary source of reductant cofactors for these processes in light is probably from photosystem I and II. The incorporation of NH_4^+ into amino acids appears to be mediated by glutamine synthetase (GIS)* and glutamate synthase (GS)*. NH_4^+ may inhibit NiR and NR through an unknown intermediary. Although the occurrence of ATPases in animal systems is undisputed, their presence in the algae is not well documented. Other abbreviations: a-KG = a-ketoglutarate; TCA = tricarboxylic acid cycle; ETS = electron transport system; FDX^{red}, FDX^{ox} = reduced and oxidized ferredoxin; TrA = transaminase.[1310] (From Falkowski, 1978, in *Nitrogen in the Environment,* Vol. 2, Nielsen, D.R., and MacDonald, J.G., Eds., p. 145, reprinted with permission of Academic Press.) *Author's note: at present, it is usual to use GS and GOGAT instead of GIS and GS, respectively.

With the higher plant enzyme there is evidence that Mo is contained in a small complex (molecular mass less than 30,000) that can be fairly easily separated from the bulk of the enzyme.[3468] Haem is present as a cytochrome b-557[1841,4620] which is reduced by NADH and reoxidized by NO_3^-; cyanide stops its reoxidization by NO_3^- but not its reduction by NADH.[4618] The pathway of electrons from NAD(P)H to nitrate through nitrate reductase from eukaryotes may be depicted as:[1841]

$$NAD(P)H \rightarrow (FAD \rightarrow cyt\ b\text{-}557 \rightarrow Mo) \rightarrow NO_3^- \qquad (4.34)$$

The pyridine nucleotide specificity of the enzymes differs in different algae. Nitrate reductase is always active with NADH as the electron donor and, in many algae, only with this. Some algae, however, are not able to utilize NADPH.[84,1263,2101,4137,4406]

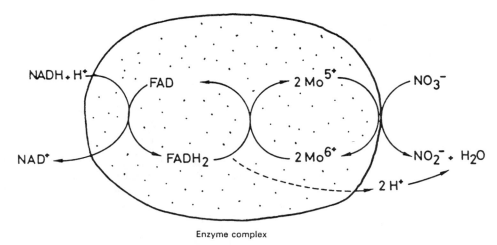

Enzyme complex

Figure 4-31. Nitrate reductase enzyme.[3198] (From Mengel and Kirkby, 1979, in *Principles of Plant Nutrition*, p. 148, reprinted with permission of International Potash Institute, Basel.)

The second type of nitrate reductase is found in prokaryotic cells and, in the algae, in the blue-green algae. The prokaryotic nitrate reductase is simpler and smaller (molecular mass of about 75,000); it also contains molybdenum but not flavin or cytochrome.[2914] The important difference from the enzyme of eukaryotes is that it does not use pyridine nucleotide[1970,3062-3,3565] as electron donor but reduced ferredoxin. It therefore catalyzes the reaction:[1841,4861]

$$NO_3^- + 2Fd_{red} + 2H^+ \rightarrow NO_2^- + 2Fd_{ox} + H_2O \qquad (4.35)$$

The nature of the stimulation by light of nitrate assimilation in photosynthetic cells has been a matter of controversy for a long time.[1841] Nitrate reduction in green cells has previously been considered as a process far removed from the light reactions of photosynthesis and dependent on organic substrates as the immediate source of reducing power. Now it seems to be well established that the photosynthetic reduction of nitrate is, at least in certain aspects, even more directly coupled to the light reactions of photosynthesis than that of carbon dioxide.[2914-5] This is particularly evident for the blue-green algae, where both nitrate and nitrite reductases are tightly bound to chlorophyll-containing membrane fractions and able to utilize photosynthetically generated reducing power via ferredoxin.[3062]

Regulation of nitrate reduction by changes in the activities of the enzymes of the nitrate-reducing pathway is important in the control of the overall process of nitrate assimilation. The level of nitrite reductase in different cells and tissues is usually much higher than that of nitrate reductase and, accordingly, accumulation of nitrite is seldom observed. The reduction of nitrate to nitrite, rather than the further reduction of nitrite to ammonia, appears therefore more likely to be a rate-controlling step in nitrate reduction.[1841] Nitrate reductase levels have been shown to fluctuate in response to changes of environmental conditions such as light, temperature, pH, carbon dioxide and oxygen tensions, water potential, nitrogen source, and other factors,[286,1862,2048,2051,2914,2954,4666,4864] changes that usually also influence the capacity of the organisms to assimilate nitrate.[1841]

The level of nitrate reductase is increased by adding nitrate and decreased by adding ammonium to *Chlorella* cultures.[3313] Eppley et al.[1263] reported that nitrate reductase synthesis and nitrate use began when ammonium was deleted to 0.5–1.0 μM. Hersey and Swift[2042] reported that additions of NH_4^+ (0.5–50 μM) to a dinophyte *Amphidinium carteri* culture decreased the amount of extractable nitrate reductase. It was suggested that a product of ammonium assimilation rather than ammonium *per se* was responsible for decreasing the nitrate reductase increase. Following experiments with amino acids as products of ammonium metabolism showed that it is likely that nitrate reduction in *Chlorella* is controlled by repression of enzyme synthesis and not by feedback inhibition.[4591]

The reduction of NO_2^- to NH_4^+ is catalyzed by ferredoxin nitrite reductase, characteristic of photosynthetic organisms.[1841] (The second type of nitrite reductase, NAD(P)H-nitrite reductase is found in nonphotosynthetic organisms.[1841]) This enzyme appears to be much the same in algae and in leaves of higher plants. It is a small molecule with a molecular mass of 60,000–70,000.[1795,4861] In algae it has been studied by e.g. Hattori and Uesugi,[1971] Grant,[1795] Zumft,[5568] Ho et al.,[2105] Llama et al.[2882] and Kamin and Stein Privella.[2413] The enzyme contains sirohaem which is an iron tetrahydroporphyrin[3353] and where NO_2^- probably attaches, it also contains an iron-sulfur center which participates in electron transport.[2914] The reaction catalyzed is:[1841,4861]

$$NO_2^- + 6\ Fe_{red} + 8\ H^+ \rightarrow NH_4^+ + 6\ Fe_{ox} + 2\ H_2O \qquad (4.35)$$

In the leaves of higher plants, nitrite reductase has been localized definitely in the chloroplast;[1912,2048,2050,2914,3213,3986-7] information about its localization in algal cells is lacking, but some results[2474] suggest a close linkage between nitrite reduction and the photochemical reactions in the chloroplasts.[4861]

Simultaneous presence of NO_3^- with NO_2^- in the medium resulted in lower rates of NO_2^- uptake by marine diatom *Ditylum brightwellii*, although this uptake was not depressed by adding 20 μM NH_4Cl. Inhibition of NO_2^- uptake by NO_3^- appears to be competitive by analogy to kinetics of enzyme inhibition.[1257] Also Harlin and Craige[1923] for *Laminaria longicruris* and Hanisak and Harlin[1896] for *Codium fragile* subsp. *tomentosoides* reported the inhibition of NO_2^- uptake by NO_3^-. On the other hand, assimilation of nitrate by *Chlamydomonas reinhardtii* was inhibited by addition of nitrite.[4935]

4.2.4.5 Effect of ammonium on nitrate uptake and assimilation

It has been known for a long time that ammonium suppresses nitrate uptake in many algal species (Figure 4-32) and also in various higher plants.[241,863,1868,1896,3237,3814,3977,4343,4861-2,5079-80] Suppression of NO_3^- uptake by phytoplankton typically occur in the range 0.5–1.0 μM NH_4^+ range.[1047,1264,1266,3127] This inhibition has been observed to some extent in species of *Gracilaria*,[1047] *Hypnea musciformis*,[1868] *Enteromorpha*[349] but not in *Fucus*[5005] (Figure 4-33). It appears that some seaweeds have the ability to take up NO_3^- and NH_4^+ simultaneously,[374,5006] although NH_4^+ is usually taken up at a higher rate.[1047,1868,1896,5004]

Although NH_4^+ inhibition of NO_3^- assimilation by algae has often been demonstrated both with laboratory cultures and in field experiments[3001-2] it is clear that algae that are highly nitrogen deficient or growing with a limiting nitrogen supply assimilate both NH_4^+ and NO_3^- simultaneously.[367,683,5092]

Ammonium and nitrate assimilation in cultures seem to be mutually exclusive events, sequential in time, as shown for *Chlorella vulgaris*[3313] and for the diatom *Cylindrotheca closterium*.[1793] It was demonstrated many times that in the presence of both ammonium and nitrate in the medium, the NH_4^+ is assimilated first (Figure 4-34), and only when it has gone is NO_3^- utilized.[1950,2950] Preferential uptake of NH_4^+ has been shown often for both marine and freshwater algae.[540,682-3,863,1263,3127-8,3915,4367] In the literature, however, there have been few reports showing preferential assimilation of nitrate; this phenomenon has been reported e.g. for *Pandorina* and *Haematococcus*.[3918,4809]

There are several reasons for this preferential assimilation of ammonium. Active nitrate reductase is not formed in the presence of NH_4^+ nor in the NO_3^- uptake system. And even if active nitrate reductase and an NO_3^- uptake system are present, the addition of NH_4^+ can lead to a rapid cessation of NO_3^- utilization.[4861] There are at least three mechanisms by which nitrate reductase activity can disappear from cells. These include two sorts of reversible inactivation phenomena and an irreversible loss of the enzyme due, presumably, to degradation (see Syrett[4861] for further discussion).

The effect of addition of NH_4^+ to cells assimilating NO_3^- are complex. The first effect appears to be an inhibition of NO_3^- uptake but this is followed by loss of nitrate reductase (and nitrite reductase) activities. The loss of nitrate reductase activity will be partly due to reversible inactivation, and partly due to irreversible loss of enzyme with the rate of proteolytic breakdown of nitrate reductase possibly being greater in the presence of NH_4^+.[2102] At the same time addition of NH_4^+ stops the synthesis of nitrate reductase.

The regulation of the formation of nitrite reductase has received much less attention. Like nitrate reductase, its formation is repressed in the presence of ammonium.[1260,2041,2882,2916,4127] Nitrate does not act either as an inducer or as a repressor of nitrite reductase.[3519] Hattori[1969] found that under light-aerobic conditions, nitrite reducing system was induced by nitrite but not by gaseous nitrogen, ammonia and glutamine. No formation was also observed under the anaerobic conditions or in the dark. Certain algae containing a hydrogenase system have been revealed to be capable of using molecular hydrogen for reduction of nitrate and nitrite.[1969,2468-9]

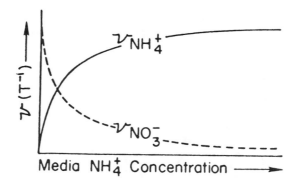

Figure 4-32.   Schematic approximation of the simultaneous rates of NO_3^- and NH_4^+ uptake as a function of NH_4^+ concentration.[3125] (From McCarthy, 1981, in Physiological Bases of Phytoplankton Ecology, Platt, T., Ed., *Can. Bull. Fish. Aquat. Sci.*, Vol. 210, p. 225, reprinted with permission of Department of Fisheries and Oceans, Ottawa.)

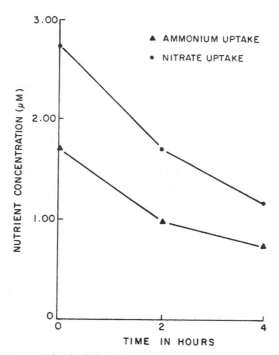

Figure 4-33.   Simultaneous uptake in light of ammonium and nitrate at 15°C by *Fucus spiralis*.[5005] (From Topinka, 1978, in *J. Phycol.*, Vol. 14, p. 244, reprinted with permission of the *Journal of Phycology*.)

Nitrate uptake by algal cells generally stops very quickly when NH_4^+ is added and recommences when NH_4^+ has disappeared because of its assimilation[867,1263,4862] (sse also Figure 4-34). This inhibition is too rapid to be accounted for by the inactivation of nitrate reductase[4861] and there have been several suggestions that it is due to an effect of uptake.[919,4405]

With highly carbon-deficient cells, NH_4^+ does not inhibit NO_3^- uptake; indeed NO_3^- is reduced to NH_4^+ which accumulates.[4862,4935,5078] These findings suggest that it is not NH_4^+ that is the inhibitor but an organic product of NH_4^+ assimilation. Studies using various inhibitors of GS and GOGAT[1388,4136]

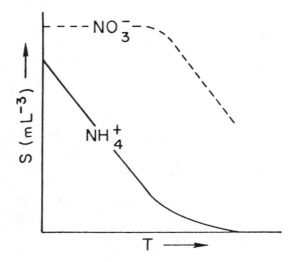

Figure 4-34.   Typical patterns of NO_3^- and NH_4^+ disappearance from culture media resulting from preferential uptake of NH4+.[3125] (From McCarthy, 1981, in Physiological Bases of Phytoplankton Ecology, Platt, T., Ed., *Can. Bull. Fish. Aquat. Sci.,* Vol. 210, p. 223, reprinted with permission of Department of Fisheries and Oceans, Ottawa.)

found that if the inhibitor is organic, it is not glutamine, but something formed from it. Conway et al.[867] suggested that inhibition was caused by the total internal amino acid pool. Another possibility[4009] is that the inhibition results not from an organic product of NH_4^+ assimilation but from the production of protons as NH_4^+ incorporation may be represented as:

$$NH_4^+ + \text{organic C} \rightarrow \text{organic N} + H^+ \tag{4.37}$$

It is unlikely, however, because NO_2^- addition has the same inhibitory effect on NO_3^- assimilation.[4935]

McCarthy et al.[3128] have defined a Relative Preference Index (RPI) useful in studies of this interaction:

$$RPI_{NH4} = (NH_4 \text{ assim. rate}/\Sigma \text{ N assim. rate})/(\text{ambient } NH_4 \text{ conc.}/\Sigma \text{ N conc.}) \tag{4.38}$$

McCarthy et al.[3128] considered all inorganic forms and urea as well. When RPI = 1.0, assimilation is equitable to availability. A value of RPI < 1 implies rejection, while RPI >1 implies preferential utilization.

The effect of NH_4^+ on the relative utilization of NO_3^- and NH_4^+ in estuarine and wetland sediments could be important, since the flux of NH_4^+ from sediments may result in locally high concentrations of NH_4^+.[3495]

4.2.4.6 Effect of light and carbon on nitrate assimilation

Warburg and Negelein[5246] made the fundamental observation that illuminated suspensions of *Chlorella* evolved O_2 in the presence of NO_3^-. These experiments were repeated and extended by others, leading to an early understanding of the photosynthetic nature of nitrate utilization in green microalgae.[441,5131]

As much as 50% of algal carbon is integrally coupled with N-metabolism.[5130] As the assimilation of N into protein requires both energy and organic skeletons, it is not surprising that there are major interactions between N-assimilation and photosynthetic metabolism.[5057] The assimilation of both NH_4^+ and NO_3^- is dependent on photosynthesis, that is assimilation requires light and CO_2; removal of either of these prevents assimilation.[1090,1968,2707,2715,4861,4935,5309] *Chlamydomonas reinhardtii* did not assimilate nitrate or ammonium unless a suitable source of carbon was provided. Suitable sources were CO_2 in light or, in darkness, acetate or the carbon reserves accumulated in nitrogen-starved cells.[4861,4935]

Di Martino Rigano et al.[1090] reported that N-sufficient cells of the red alga *Cyanidium caldarium* obtain carbon skeletons for ammonium assimilation exclusively by photosynthetic reactions. Upon N-limitation they develop the ability, apparently through derepression or activation of regulatory enzyme system(s), to obtain a consistent quantity of additional carbon skeletons and ATP from mobilization of carbon reserves. This enables the N-limited cells to assimilate ammonium not only in light but also in darkness, and at a higher rate than N-sufficient cells.[1090]

Nitrate metabolism is strongly light dependent in green algae.[122,1103,3441,4983] Several steps may be possible targets for light regulation. Nitrate uptake can be stimulated by light,[1386] leading to indirect activation of nitrate reductase. The enzyme nitrate reductase can also be directly activated by light;[190] light can furthermore induce nitrate reductase synthesis.[1356,3979] Nitrate reductase activity was increased both by red and blue light pulses and the effect was partially reversed by far red light in the green marine alga *Ulva lactuca*.[2906] Continuous blue light generally is more effective than continuous red light at stimulating nitrate uptake and metabolism.[190,665] Light also controls nitrite reductase activity in *Chlorella* sp.[4980] Cole and Toetz[832] reported that light limited nitrate uptake by freshwater reservoir phytoplankton below 10–20% of incident light and also at higher light intensities. Possible interactions of light with inorganic nitrogen metabolism are summarized by Syrett.[4861]

Inhibition of nitrite or nitrate uptake at high light intensities has not been observed by most authors.[241,3002,3615] This has led the authors to suggest that the uptake of nitrate as a function of irradiance obeys Michaelis-Menten kinetics. Curtis and Megard[936] reported that nitrite uptake by the green alga *Chlamydomonas* as a function of irradiance did not follow Michaelis-Menten kinetics because of inhibition at high irradiances. The Haldane equation described nitrite uptake better. However, the nitrite uptake as a function of oxygenic photosynthesis followed Michaelis-Menten kinetics.

Solomonson and Shepar[4619] proposed a model (Figure 4-29) for the regulation of nitrate assimilation in which CO_2 fixation and nitrate assimilation are coordinately controlled by the intracellular ratio of $[O_2]/[CO_2]$. This model accounts for the observed effects of O_2, CO_2, and light on nitrate assimilation and provides a rationale for the *in vivo* regulation of nitrate reductase by cyanide.

Losada[2913] and Vega et al.[5147] have emphasized that reduction of a N atom in NO_3^- to a N atom in an amino acid requires 10 electrons. Reduction of one CO_2 molecule to (CH_2O) requires 4 electrons. It is therefore not surprising that it has been repeatedly shown that assimilation of both NH_4^+ and NO_3^- in normally grown algae is dependent on photosynthesis, with little or no uptake occurring in the dark.[4861] However, this is not an obligate process, and nitrogen-starved *Chlamydomonas* with accumulated carbohydrate reserves assimilate both NO_3^- and NH_4^+ in the dark, but does so at even higher rates in the light.[4935] When ammonium-N is assimilated 3.3 carbohydrate carbon atoms are converted to organic nitrogen compounds for every one which is lost as carbon dioxide. When nitrate is assimilated the ratio is 0.75.[4859]

Syret[4861] pointed out that NO_3^- assimilation, like CO_2 assimilation, is a reductive process. As the C:N ratio of phytoplankton is about 5, the ratio of electrons required for CO_2 incorporation to those required for NO_3^- N incorporation by growing cells is about $5 \times 4 : 10$. i.e. 2 : 1. This argument does not necessarily imply that the reductant for NO_3^- assimilation is derived directly from photochemical reactions but it does illustrate that considerable amounts of reductant have to be generated for NO_3^- N assimilation.[4861] For an excellent review of this problem see Syrett.[4861]

4.2.4.7 Organic nitrogen sources

Many algae are capable of using organically combined nitrogen, especially amino acids, urea, and purines, as their sole nitrogen source.[3409] The following summary is mostly based on an excellent review by Antia et al.[116]

Urea

Harvey[1949] was probably the first to conclude from experimental data that urea was assimilated intact by marine phytoplankton. Since than, numerous studies have shown that the capacity to utilize urea as a sole source of N is not universal among unicellular algae.[3124] It has been known for many years that urea, a neutral, small molecule, can move into and out of cells by passive diffusion.

Evidence shows that urea is also taken up by active processes.[116] A summary of K_s values for marine and freshwater phytoplankton is given in Table 4-5.

The first extensive study of the mechanism involved in urea uptake was on the green algae *Chlamydomonas reinhardtii*[2138,5448] and *Chlorella vulgaris*.[2135] Our understanding of urea uptake mechanisms has been extended further by the thorough studies of Rees, Syrett and co-workers[291,2758,4048-9,4864,4866]

Following its uptake, urea is generally accumulated, presumably in an unmodified form, in one or two pools within the algal cell.[958,4049,4866] Depending on the availability and activity of the urea-degrading enzyme present, the intracellular urea is converted to ammonium and bicarbonate. The ammonium so produced is then assimilated primarily by the glutamine synthetase-glutamate synthase pathway (see Chapter 4.2.4.3) commonly used for the assimilation of ammonium supplied as a nutrient.[539,3215,4839] Interestingly, the formation of bicarbonate from urea degradation signifies that the utilization of urea-C is equivalent to an autotrophic CO_2-fixation process.[77,115]

Urea degradation is catalyzed by either urease or ATP:urea amidolase (UALase).[291,342,713,2758,3005,3543,4179,4861,4864]

The enzyme urease is a nickel-ligated metalloprotein with a capacity to catalyze the hydrolysis of urea to ammonia and carbonic acid according to the equation:[4861]

$$NH_2 \cdot CONH_2 + H_2O \rightarrow 2\ NH_3 + CO_2 \qquad (4.39)$$

Its nickel dependence was first identified in the enzyme from jack-bean,[391,1095] and was subsequently shown to be similar for the ureases from several bacteria, fungi, and the microalgae *Anabaena cylindrica*, *Cyclotella cryptica*, *Phaeodactylum tricornutum* and *Tetraselmis subcordiformis*.[1977,3005,3542,4048] In general, the nickel is tightly bound to the apoprotein in these ureases, and is not readily removed by chelators such as EDTA (at physiological pH), but when it is removed by other means the enzyme activity is irreversibly lost.[1977] Cobalt can partially replace nickel in restoring urease activity of *Phaeodactylum* and *Cyclotella*, but other nickel analogues tested lack this ability.[3542,4048]

The enzyme ATP:urea amidolase was first studied in yeast,[4178-9] and was found to be a complex biotin-linked enzyme of high molecular mass catalyzing two distinct enzymatic reactions in accordance with equations:[4861]

$$urea + HCO_3^- + ATP \underset{ureaCA*}{\overset{Mg^{2+}\ K^+}{\Leftrightarrow}} allophanate + ADP + P_i \qquad (4.40)$$

* urea carboxylase

$$allophanate \overset{AH**}{\rightarrow} 2\ NH_3 + 2\ CO_2 \qquad (4.41)$$

** allophanate hydrolase

The overall reaction is thus:[4861]

$$CO(NH_2)_2 + ATP + H_2O \rightarrow CO_2 + 2\ NH_3 + ADP + P_i \qquad (4.42)$$

As a result of their ability to degrade urea to ammonium, most of the tested algae, both marine and freshwater species, have been shown to be capable of efficient growth on urea serving as the sole nitrogen source.[376,659,1070,1147,1150,1265,1842,1928,3409,48725053,5165] Kirk and Kirk[2512-3] found that ammonium, nitrate and the amino acid leucine repressed urea uptake by about 60% in *Chlamydomonas reinhardtii*.

Free amino acids

The mechanism of amino acid transport in algal cells has been reviewed by Raven[4005] and Eddy.[1211] Amino acids appear to be taken up by at least three different transport systems, which correspond to the net charge that they bear at physiological pH values.[2014,2512-4,2813,2878-9,3236,3464,4097,4338,5363]

The best studied example of an acidic amino acid is the transport of glutamate.[1368,2020] A thorough study of the influx of the neutral amino acid phenylalanine has been reported for *Chlorella fusca*.[3723]

Table 4-5. Summary of maximum specific uptake rate (V_m) and half-saturation constant (K_s) values for urea uptake by marine and freshwater phytoplankters reported in the literature.[116] (From Antia et al., 1991, in *Phycologia*, Vol. 30, p. 21, reprinted with permission of Blackwell Scientific Publications, Oxford and the authors.)

Algal species	V_m (h^{-1})	K_s (μM)	Comments	Reference
Marine diatoms				
Phaeodactylum tricornutum	0.005–0.02	1.0	20°C, 24 h N-deplete	4049
	0.002–0.005	0.6	20°C, urea-deplete	4049
	0.003	-	20°C, darkness 18 h N-deplete	4049
Ditylum brightwellii	0.011	0.4	18°C, urea-replete	3124
Lauderia sp.	0.013	1.7	18°C, urea-replete	3124
Thalassiosira weissflogii				
clone Actin	0.024	1.7	18°C, urea-replete	3124
clone T. fluviatilis	0.030	0.5	18°C, urea-replete	3124
Skeletonema costatum	0.015	1.4	18°C, urea-replete	3124
	0.12	0.5	18°C, N-deplete[1]	2206
Thalassiosira pseudonana	0.008	0.4	18°C, urea-replete	3124
	0.08	0.5	18°C, N-deplete[1]	2206
Freshwater chlorophytes				
Chlorella fusca	-	15.0	25°C, N-replete	4865
	0.02	16.5	25°C, N-replete	292
Chlamydomonas reinhardtii	0.67	5.1	25°C, N-deplete	5448
Ankistrodesmus braunii	-	0.8	29.5°C, NO$_3^-$-replete	2514
Chlorella pyrenoidosa	-	0.6	29.5°C, NO$_3^-$-replete	2514
Scenedesmus obliquus	-	0.8	29.5°C, NO$_3^-$-replete	2514
Eudorina elegans	-	0.1	29.5°C, NO$_3^-$-replete	2514
Golenkinia minutissima	-	0.7	29.5°C, NO$_3^-$-replete	2514
Gonium pectorale	-	0.1	29.5°C, NO$_3^-$-replete	2514
Pandorina morum	-	2.6	29.5°C, NO$_3^-$-replete	2514
Pleodorina californica	-	0.5	29.5°C, NO$_3^-$-replete	2514
Volvox carteri	-	0.9	29.5°C, NO$_3^-$-replete	2514
Scenedesmus quadricauda	-	1.2	20°C, N-replete	1993
Freshwater blue-green algae				
Pseudanabaena catenata	-	0.4	20°C, N-replete	1993

[1] Five-minute uptake rate

Active uptake of two neutral amino acids, L-leucine and L-tyrosine, has also been demonstrated for *Chlorella fusca*.[4097] Another neutral amino acid that has been extensive studied is glycine.[2022,2928,3461-3] Uptake of L-lysine and L-arginine by marine diatom *Phaeodactylum tricornutum* has been studied by Flynn and Syrett[1390-1] and Hayward.[1986] Arginine was found to be a good nitrogen source for *Chlorella vulgaris*.[169]

The presence of ammonia in the medium suppressed amino acid uptake rates. However, the suppression of uptake did not show any particular relation to the nitrogenous cell composition of *Platymonas subcordiformis*.[5362] The uptake of amino acids by marine phytoplankton is inversely related to the availability of nitrogen. Under conditions of N deprivation, cell N decreases and uptake rates increase. Conversely, if the cells are grown in the presence of excess N, cell N increases and uptake rates decrease.[3462-3,5367]

Purines

The uptake kinetics of uric acid was studied by Douglas.[1133] Uric acid was taken up by active transport in the prasinomonad *Platymonas convolutae* and accumulated in the cells in both solid and dissolved forms. Birdsey and Lynch[376] studied uric acid uptake by green algae, blue-green algae, euglenoids and red algae; most of the studied green algae utilized uric acid, whereas members of other classes did not.

Guanin is the only other purine for which uptake mechanisms have been studied. Pettersen and Knutsen[3774] found that guanine was accumulated intracellularly in *Chlorella fusca* to concentrations that were 10^5-fold higher than external concentrations, with little or no efflux. Subsequent studies by Pettersen[3773] revealed that *Chlorella* compartmentalized guanine intracellularly between a small cytoplasmic pool and a large vacuolar pool, analogous to the compartmentalization of alanine, arginine and lysine observed in *Platymonas* by Wheeler and Stephens.[5363] Other studies on guanine uptake by algae include those e.g. by Shah and Syrett[4415-6] and Syrett et al.[4867]

Adenine is another purine with five nitrogen atoms per molecule but, unlike guanine, it has not generally served as a good N-source for growth of microalgae.[11,111,1150,4415] However, adenine was reported to be a good N-source for the growth of some algae.[82,659,1212,4668] Studies on xanthine, another purine, include those by Birdsey and Lynch,[376] Ammann and Lynch[82] and Antia and Chorney.[111] While the purines generally serve as a good N-source for microalgal growth, their direct utilization for nucleic acid synthesis is a controversial subject.[116]

Further organic sources of nitrogen include pyrimidines (e.g. uracil), amides (e.g. glutamine, asparagine, acetamide, succinamide), amines, betaines (quaternary ammonium compounds), and nucleic acids.[116,2019,2022]

Kinetic parameters for organic-N compounds are given in Table 4-6.

4.2.4.8 Nitrogen storage

Assimilated nitrogen may be stored as nitrate, ammonium, or low molecular mass organic compounds and used for growth at some future time.[613,754,4184,4643] The Cyanopyceae appear unique in their ability to store nitrogen as structured granules of cyanophycin,[2698] a polypeptide polymer of arginine and aspartic acids. Synthesis of this peptide is unusual as it appears to be non-ribosomal.[4509]

Nitrates accumulate in the vacuoles of marine algae, in *Valonia* to the extent of 2000 times and in *Halicystis* 500 times the nitrate value of seawater.[2310] *Laminaria longicruris* has been reported to concentrate NO_3^- as high as 28,000 fold during periods of N sufficiency.[754] Ammonium and nitrate are accumulated in the vacuolated cells of the marine diatom *Ditylum brightwellii* and these intracellular pools serve as substrate for the assimilatory enzymes.[1260] Nitrite is either not accumulated or is concentrated in a very small cellular compartment.[1260] The proximal mechanism for seaweed's accumulation of N at low light and temperatures may be that N uptake is less limited by light and temperature than is growth.[1183]

4.2.4.9 Nitrogen deficiency

Algal cultures can exist in various states of nitrogen nutrition. The effects of the cellular physiological state on nitrogen uptake rates by phytoplankton were first demonstrated by Syrett[4855] and Harvey.[1950] They showed that NH_4^+ and NO_3^- uptake by batch cultures of nitrogen-starved cells was much more rapid than by normal cells that were nitrogen-replete.

The effects of nitrogen deprivation are well summarized by Syrett[4861] and Turpin.[5056] Some of the more obvious effects are an accumulation of carbon compounds such as polysaccharides and fats,[4861] a reduction in the rate of photosynthesis,[4861,5056] decline in nitrogenous photosynthetic pigments (chlorophylls, phycobilins)[72,1312,1488,2044,3822] or reduction in thylakoid stacking, and absorptivity.[3822,4089-90]

The availability of fixed inorganic nitrogen is a major factor influencing the growth and chemical composition of algae. In unicellular algae nitrogen starvation alters the partitioning of photosynthetically fixed carbon, resulting in an increased accumulation of carbohydrates and lipids.[837,4169,4449,4657,5007]

Table 4-6. Literature summary of K_s values for uptake of amino acids and purines by marine and freshwater phytoplankton species.[116] (From Antia et al., 1991, in *Phycologia*, Vol. 30, p. 33, reprinted with permission of Blackwell Scientific Publications and the authors.)

Algal species	Substrate	K_s (μM)	Comments	Reference
Marine diatoms				
Cyclotella cryptica	Arginine	3.0	N-replete, 20°C	2878–81
	Glutamate	36.0	N-replete, 20°C	2878–81
	Proline	6.0	N-replete, 20°C	2878–81
Melosira nummuloides	Arginine	8.0	N-replete	2014
	Valine	40.0	N-replete	2014
Navicula pavillardi	Glutamate	20.0	N-replete	2811
Navicula pelliculosa	Glutamate	100.0	N-replete	2357
	Aspartate	100.0	N-replete	2357
Nitzschia angularis	Glutamate	20.0	N-replete	2812
Nitzschia laevis	Glutamate	30.0	N-replete	2813
	Alanine	20.0	N-replete	2813
Nitzschia ovalis	Arginine	2.0	N-replete	3468
Phaeodactylum tricornutum	Lysine	2.0	N-replete, 20°C	1390
	Lysine	0.8	N and C-deplete	1390
	Arginine	0.5	N-replete, 20°C	1390
	Glycine	3.0	N-replete, 20°C	2928
	Guanine[1]	0.5	N-replete, 20°C	4415
Other marine algae				
Gymnodinium breve	Glycine	110.0	N-replete, 30°C	201
	Valine	150.0	N-replete, 30°C	201
	Methionine	125.0	N-replete, 30°C	201
Tetraselmis subcordiformis	Glycine	19.0	N-deplete, 20°C	3463
		14.0	N-deplete, 20°C	3463
		6.7	N-deplete, 20°C	3462
Platymonas convolutae	Uric acid[1]	1.5–34.0	H-L affinity syst[2]	1133
Freshwater chlorophytes				
Chlorella fusca	Leucine	2.5	N-replete, 25°C	4097
	Tyrosine	0.4	N-replete, 25°C	4097
	Phenylalanine	5.0	N-replete, 30°C	3723
	Guanine	0.1–1.0	N-replete, 30°C	3723
Chlamydomonas reinhardtii	Arginine	3.9	N-replete, 30°C	2513
Volvox carteri	Arginine	3.8	N-replete, 30°C	2512
Pandorina morum	Leucine	52.0	N-replete, 30°C	2514
Scenedesmus obliquus	Leucine	47.0	N-replete, 30°C	2514
Ankistrodesmus braunii	Leucine	16.0	N-replete, 30°C	2514
Freshwater cyanophytes				
Anabaena variabilis	Leucine	10.8	N-replete, 32°C	4940
	Glutamine	1.1–13.8	N-replete, 30°C	759
	Glutamate	1.4–100	N-replete, 30°C	759
Anacystis nidulans	Leucine	125	N-replete, 35°C	2753

[1] Serves to distinguish purines from the amino acids, [2] High and low affinity systems

In many cases, neutral lipids (primarily triacylglycerols) comprise the bulk of the lipid present in starved cells.[4169] An excellent comparative study by Shifrin and Chisholm[4449] indicated that the lipid contents of 15 chlorophycean strains grown under nitrogen-deficient conditions increased to 130–320% of the values observed for exponential phase cultures. Falkowski et al.[1312] reported that their results with nitrogen limited *Isochrysis galbana* generally support the hypothesis that in nitrogen limited cells, proteins encoded in the nuclear genome are synthetized preferentially over those encoded in the chloroplast. Deliberate nitrogen starvation is used in commercial algal mass culture to increase the yield of valuable lipids.[2044] However, the major problem associated with nitrogen starvation is a rapid, marked reduction in photosynthetic performance. As cells become nitrogen starved, light saturated photosynthetic rates, photosynthetic energy conversion efficiency at the growth irradiance, and maximum quantum yields all decline.[820,4090,4861,5328]

In those eukaryotic algae containing more than one type of chlorophyll, N-limitation often brings about a change in the ratio of chlorophyll-*a* to accessory chlorophyll, but there seems to be no general rule as to whether chlorophyll-*a* or the accessory chlorophyll declines preferentially.[2044,3822] Carotenoid/chlorophyll-*a* ratios increase dramatically under N-limitation.[3822] Herzig and Falkowski[2044] reported that nitrogen limitation led to an overall reduction in pigmentation in *Isochrysis galbana* and a decrease in cellular concentration of reaction centers; however, the optical absorption cross section, normalized to chlorophyll-*a*, increased. Moreover, chlorophyll-*c*/-*a* ratios were higher in nitrogen-limited cells.

Cell division may still occur for some period after the onset of nitrogen starvation, and some investigators therefore have suggested that some of the protein involved in the photosynthetic apparatus may act as a storage reservoir for nitrogen which can be mobilized in times of nitrogen deprivation.[820,4090] Falkowski et al.[1312] summarized that both nitrogen starvation and limitation led to a dramatic decrease in the efficiency of energy transfer from light harvesting complexes to PS II reaction centers.[2044,2496,2556,3886] It is not exactly clear, however, which proteins may serve a storage role or what structural elements are lacking in nitrogen deficient cells which reduce energy transfer efficiency. Moreover, it is not clear what, if any, proteins differentially accumulate or are degraded as cells become nitrogen limited.[2556,3821]

Algae growing under nutrient stress usually show enhanced uptake of NH_4^+ (e.g. 1365, 4187) and comparison of tissue composition and uptake rates may indicate whether algae are nutrient-limited. The tissue concentration of an element below which growth is limited by that element is defined as the critical level and for most algae, that is approximately 2% nitrogen by dry mass.[1895,4187] Syrett and his colleagues in a series of papers reported that nitrogen deprivation of the cells of the marine diatom *Phaeodactylum tricornutum* leads to an increase in the rate at which urea,[4049] nitrate,[920] nitrite,[921] guanine,[4415] methylammonium,[5504] lysine[1390] and arginine[4867] are taken up when supplied to the cells.

Numerous studies have demonstrated that nitrogen-deficient or -starved phytoplankton have the ability to take up ammonium at initially elevated (surge) rates that greatly exceed their growth rate and N uptake rate of nitrogen-sufficient phytoplankton in culture.[825,867,1635,1698,1941,3661–2,4846,5365] In N-limited cells capable of maintaining high rates of NH_4^+ assimilation in the dark, NH_4^+ resupply may cause up to a 40-fold increase in nonphotosynthetic CO_2 fixation.[4367]

In contrast, after a nitrate addition to an N-starved culture there is often but not always a lag of variable duration before nitrate uptake is observed at either elevated, normal, or reduced velocities.[848] Nitrogen starvation may[825,2206,3895,4867] or may not[292] increase urea uptake rates but only a few species have been investigated so far.

Nitrogen-starved cells of *Chlorella vulgaris* assimilate ammonium rapidly and convert it into soluble organic nitrogenous compounds.[1365,4855] The assimilation of ammonia-N was about 4 times as quick as nitrate-N at pH 6.1.[4858] The assimilation of both forms of nitrogen is accompanied by a high respiration rate and a decrease of intracellular carbohydrate. Carbohydrate disappears more rapidly from the ammonium-treated cells.[4856,4859] The rapid assimilation of small quantities of ammonia, nitrite and nitrate-N by N-starved cells of *Chlorella vulgaris* is accompanied by high rates of gas exchange. When all the added nitrogen is assimilated the rates of gas exchange return to the control values.[4857]

Assimilation of nitrogen by nitrogen-deficient or nitrogen-limited cells is limited by the rate of protein synthesis, as suggested by the early work of Syrett.[4858] This has been substantiated by evidence for the accumulation of internal pools of nitrate, ammonium and free amino acids after the addition of nitrogen to nitrogen-limited cultures of phytoplankton[1049,1128] and seaweeds.[1981] Such pools would not accumulate if rates of protein synthesis were equal or greater than rates of membrane transport and subsequent metabolism to amino acids.

4.3 PHOSPHORUS

4.3.1 Global Cycle

Phosphorus is a very important and irreplaceable element in all living systems. As DNA, RNA, phospholipids, ATP, ADP, AMP, and c-AMP it plays a key role in almost all the essential functions of a living cell such as creation, structure, operation and reproduction.[1243,3797-8]

Phosphorus is a macronutrient but its availability is often in the ng g^{-1} range. The effects of phosphorus in nature are, therefore, very profound. Phosphorus discharged by a single person in one year (about 2 kg P) is sufficient for the growth of 1 Mg of plant material,[5104] a fact which served to illustrate the link between urban communities and eutrophication.[3798] (For more information about eutrophication see Chapter 2.16).

Phosphorus is an element which is often limiting for plant growth, since the concentration of plant-accessible phosphorus in soil—free orthophosphate ions PO_4^{3-}—is often very low. Phosphorus as such is not really scarce, but it exists predominantly in nearly insoluble forms, such as apatite and other metal complexes. The soluble inorganic phosphorus liberated by weathering is to a large extent quickly immobilized by iron and aluminum and sometimes calcium in the soil.[3797] In waters, the phosphate ions are not as easily immobilized since the concentrations of iron, aluminum, and also often calcium are much lower in water than in soil. This is one reason why eutrophication poses a problem.[3797]

Global cycle and major reservoirs and fluxes of phosphorus in the environment have been reviewed many times in the literature.[2779,3797-8,4120,4817] Phosphorus soil cycle has been reviewed by Stevenson[4738] and Bohn et al.[419] and the role of phosphorus in biochemistry has been reviewed by Williams.[5443]

The movement of phosphorus can be resolved into three cycles: two biological cycles superimposed on an inorganic cycle.[1243] The primary inorganic cycle—spirale (Figure 4-35)—turns very slowly, its rate of revolution being measured in Gy (10^9 year). The two biocycles turn rapidly in comparison—the land-based cycle (Figure 4-36) is made up of several components and there may be an annual turnover of some of the phosphate, although there are reservoirs of phosphate in the soil that can store it for centuries. In the water-based biocycle (Figure 4-37) phosphate turnover is measured in days and months.[1243]

Emsley,[1243] in his comprehensive review of the phosphorus cycle, pointed out that it is difficult to quantify the amount of phosphate moving through these three cycles although attempts have been made to do this.[3797,4817] The amounts of phosphorus in various compartments of the environments have been presented by various authors in the literature (Figures 4-38, 4-39 and 4-40). The author pointed out that the picture is incomplete because there are areas in which little work has been done as yet, e.g. the phosphorus in wind borne dust and sea spray, the uptake of soil phosphorus by microbes, and the utilization of phosphate by bacteria in water. These parts of the natural cycle are likely to be relatively small, however, and would not significantly alter the data in Figure 4-38.

4.3.2 Phosphorus in the Environment

Earth's crust: 700–1,000 mg kg^{-1} (490, 903)
Common minerals: apatite group minerals, e.g. $3Ca_3(PO_4)_2 \cdot Ca(F,Cl)_2$ (apatite), $Ca_5(PO_4)_3OH$ (hydroxyapatite), $Ca_5F(PO_4)_3$ (fluoroapatite); other minerals, e.g. $CaHPO_4 \cdot 2H_2O$ (brushite), $Al_3(OH)_3(PO_4)_2 \cdot 5H_2O$ (wavellite), $Ca_9(Mg,Fe)H(PO_4)_7$ (whitlockite), $FePO_4 \cdot 2H_2O$ (strengite)[853,1243,1800,3132]
Soils: 800 (35–5,300) mg kg^{-1} (491); 650 mg kg^{-1}, fixed by hydrous oxides of Al and Fe in acid soils, and as various calcium phosphates in alkaline soils[490]
Sediments: 670 mg kg^{-1} (491)
Air: < 180–1,100 ng m^{-3} (491)
Fresh water: 5 μg L^{-1} (490); 20 (1–300) μg L^{-1},[491] highly variable, in polluted surface waters up to >1 mg L^{-1}
Sea water: 70 μg L^{-1} (490, 3815), 60 (60–88) μg L^{-1} (491)
Land plants: 0.23 (0.012–0.3) % DM[490-1]
Freshwater algae: 0.04–0.96% DM (in laboratory culture up to 7.98% DM (Table 4-7)
Marine algae: 0.28–0.4% DM,[491] 0.04–0.86% DM (Table 4-7)
Land animals: 1.7–4.4% DM[490]
Marine animals: 0.4–1.8% DM[490]

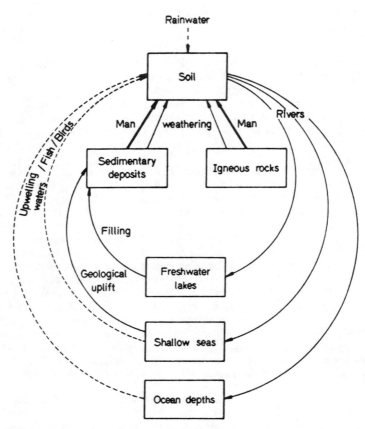

Figure 4-35.   The primary inorganic phosphate cycle - spiral.[1243] (From Emsley, 1980, in *The Hand-book of Environmental Chemistry,* Vol. 1, Part A, Hutzinger, O., Ed., p. 151, reprinted with permission of Springer-Verlag and the author.)

4.3.2.1 Aquatic chemistry

Sources to waters: natural weathering processes, agriculture (fertilizers, wastewaters), brewery, laundry and textile manufacture wastewaters, municipal sewage.[3815,5137]

Phosphorus in aquatic ecosystems occurs as phosphate in organic and inorganic compounds.[1569] Free orthophosphate is the only form of phosphorus believed to be utilized directly by algae and thus represents a major link between organic and inorganic phosphorus cycling in water bodies. (Figure 4-41).

Inorganic phosphate is used by growing algae which are extremely efficient in removing phosphate from solution. Following the death of the algae most of the phosphate is released back into the water. For the sake of convenience a distinction can be made between an "internal" or "metabolic" phosphate cycle:[1709]

$$(PO_4\text{-}P)_{water} \quad \xrightarrow{\text{primary production}} \quad cell\text{-}PO_4 \quad \xrightarrow{\text{mineralization}} \quad (PO_4\text{-}P)_{water} + Org\text{-}P_{water} \tag{4.43}$$

and an "external" phosphate cycle:[1709]

$$(PO_4\text{-}P)_{water} \rightarrow\rightarrow\rightarrow\rightarrow sediments \rightarrow\rightarrow\rightarrow\rightarrow (PO_4\text{-}P)_{water} + Org\text{-}P_{water} \tag{4.44}$$

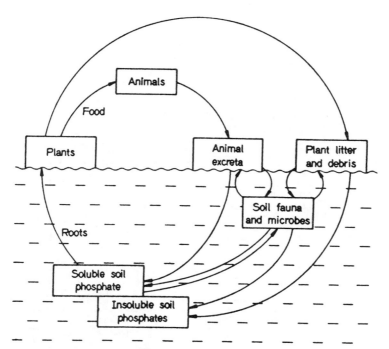

Figure 4-36. The land-based phosphate biocycle.[1243] (From Emsley, 1980, in *The Handbook of Environmental Chemistry*, Vol. 1, Part A, Hutzinger, O., Ed., p. 157, reprinted with permission of Springer-Verlag and the author.)

The first cycle summarizes biological aspects, while the second is a geochemical cycle (Figure 4-41).

In natural waters, orthophosphate occurs in ionic equilibrium, i.e. as:

$$\overset{pK = 2.2}{H_3PO_4 \quad \Leftrightarrow} \quad \overset{pK = 7.2}{H_2PO_4^- \quad \Leftrightarrow} \quad \overset{pK = 12.3}{HPO_4^{2-} \quad \Leftrightarrow \quad PO_4^{3-}} \qquad (4.45)$$

with $H_2PO_4^-$ and HPO_4^{2-} being the predominant species over pH range of 5 to 9.[4819] Another group of inorganic phosphorus compounds are polyphosphates (linearly condensed and cyclic).[853,3815]

Dissolved inorganic phosphates may be incorporated into minerals or sorbed by sediments,[1707-8, 1710-13,1721,3112,4466,5434-6] particulate organic material,[5433] clay particles[3121,4819] or inorganic precipitates of metal hydroxides and oxides.[3121,5387]

Organic compounds containing phosphorus in natural waters are a product of biological processes. Hutchinson[2257] reported that dissolved organic phosphorus forms 30–60% of the total dissolved phosphorus in lakes. Organically bound phosphorus is present e.g. in phospholipids, nucleis acids, nucleoproteins, phosphorylated sugars or organic condensed polyphosphates (coenzymes, ATP, ADP).[853,2166,3815,4311]

A broad spectrum of phosphorus cycling models in the natural waters has been developed. These models range from relatively simple, empirical models of annual, lake-wide average total phosphorus concentration to more detailed seasonal, spatially segmented models describing dynamic interactions among many phosphorus components.[4311]

The first category of models (management models[4311]) includes those designed around the phosphorus loading concept. These models are simply balances of phosphorus inflow, outflow, and sedimentation; the latter process is used to calculate the retention coefficient for phosphorus.[4311] The development and refinement of the mass-balance-based phosphorus model for lakes is attributed to

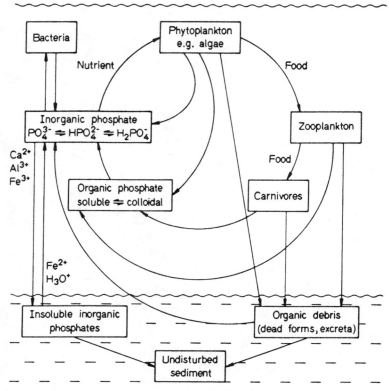

Figure 4-37.   The water-based phosphate biocycle.[1243] (From Emsley, 1980, in *The Handbook of Environmental Chemistry*, Vol. 1, Part A, Hutzinger, O., Ed., p. 161, reprinted with permission of Springer-Verlag and the author.)

Vollenweider.[5176,5178] Further alterations have been given e.g. by Dillon and Rigler,[1086–7] Chapra,[767–8] Lorentzen et al.[2911] and Golterman.[1710,1712] Critical evaluation of these models has been given by Reckhow.[4021] Scavia[4311] pointed out that these models represent the only P-related models that have undergone extensive statistical evaluation with respect to general applicability across the wide range of lakes, they do not reveal much of the structure or function of the ecosystems.

A second category of models, termed eutrophication,[4311] are concerned with stimulation and prediction of seasonal phytoplankton biomass dynamics. This type of model, based on the early work of Riley et al.[4144] was reintroduced in the early 1970s.[773,1092–3,4944] These models focus on one or two groups of phytoplankton and use chlorophyll or dry mass as biomass indicators. Although this class of models has provided important advances in the use of more mechanistic models in management of aquatic resources, they too fall short of providing the detail and focus necessary to explore phosphorus cycling quantitatively.[4311]

The third class of models—ecosystem models[4311]—although fundamentally similar to the eutrophication models, place emphasis on more detailed aspects of ecosystem dynamics. Scavia[4311] pointed out that ecosystem models emphasize more detailed aspects of plankton dynamics than do eutrophication models, e.g. seasonal succession,[368,2764,4312] internal storage of nutrients,[368,2764] grazing,[4312] sediment-water column interactions,[2712,4312] bacteria dynamics,[403,1057] zooplankton vertical migration[4703] and population structure.[4702] Scavia[4311] presented a phosphorus flow diagram within the water column (Figure 4-42).

Models, based on phosphorus as a tracer have been developed in the 1970s.[2734–5,2737,3620] Phosphorus-based models provide the important first step in identifying significant non-living phosphorus components and major pathways of phosphorus flow under steady-state, experimental conditions.[4311] However, because of their steady-state orientation, they do not provide theory for, nor test of,

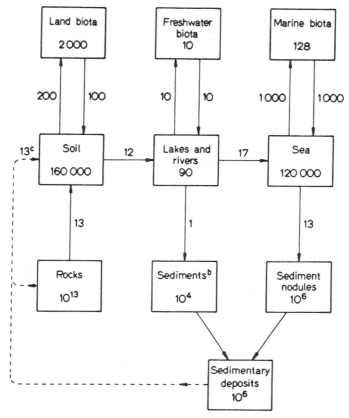

Figure 4-38. The amounts of phosphorus in the various compartments of the environment and the annual natural movements between them (Mtonnes P)[a]. [a] 1 tonne P = ca. 5 tonnes fluoroapatite rock, [b] = lake sediments are the least easy to quantify, [c] = deposits mined by man for fertilizer use.[1243] (From Emsley, 1980, in *The Handbook of Environmental Chemistry,* Vol. 1, Part A, Hutzinger, O., Ed., p. 150, reprinted with permission of Springer-Verlag and the author.)

hypotheses regarding seasonal, dynamic interactions among phosphorus components. Also they are not oriented toward food-web dynamics and consider only radioactive labeled compounds over a relatively short time interval.[4311]

4.3.3 Phosphorus Cycling in Wetlands

The soil phosphorus cycle is fundamentally different from the N cycle. There are no valency changes during biotic assimilation of inorganic P or during decomposition of organic P by microorganisms. Phosphorus has no gaseous phase, and it has a major geochemical cycle.[4738] Soil P primarily occurs in the +5 (oxidized) valency state, because all lower oxidation states are thermodynamically unstable and readily oxidize to PO_4 even in highly reduced wetland soils[2859] Mitsch and Gosselink[3246] pointed out that phosphorus is one of the most important chemicals in ecosystems, and wetlands are no exception. It has been described as a major limiting nutrient in northern bogs,[2002] freshwater marshes,[2537] and southern deepwater swamps.[588,3246]

There are only a few studies that outline the mechanisms controlling phosphorus retention capacity in freshwater wetlands.[901,2217,4104–5,4112] Most of these studies have shown that sediment/peat accumulation is the major long-term phosphorus sink and that natural wetlands are not particularly

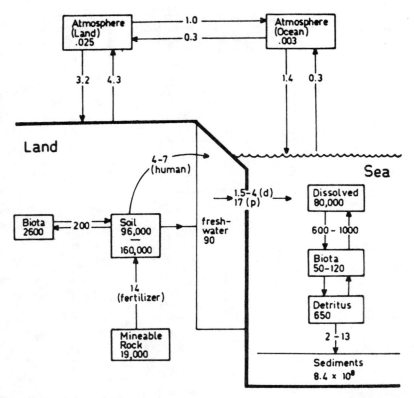

Figure 4-39.   A global phosphorus cycle. Fluxes are Tg P yr^{-1} and reservoirs are in Tg P.[4120] (From Richey, 1983, in SCOPE Report 21, p. 54, reprinted with permission of Scientific Committee on Problems of the Environment, Paris.)

effective as phosphorus sink when compared with terrestrial ecosystems.[4104] The sediment-litter compartment is the major P pool (> 95%) in natural wetlands, with much lower plant pool and little in the overlying water.[1313,5153] Most soil P (> 95%) in peatland systems is in the organic form[4109,4115,5153] with cycling between pools controlled by biological forces (i.e. microbes and plants). The percentage of organic P is generally lower in wetlands with mineral substances.[1313]

Inorganic phosphorus transformations, subsequent complexes, and P retention in wetland soils and sediments are controlled by the interaction of redox potential, pH, Fe, Al, and Ca minerals, organometallic complexes, clay minerals, and the amount of native soil P.[469,1313,1716,2158,2211,2859,3645,3695,4104] In acid soils, inorganic P is adsorbed on hydrous oxides of Fe and Al and may precipitate as insoluble Fe-phosphates (Fe-P) and Al-phosphates (Al-P). Precipitation as insoluble Ca-phosphates (Ca-P) is the dominant transformation at pH greater than 7.0.[215,2859,4837] Adsorption of phosphorus is greater in mineral vs. organic soils.[4103-4,4115] In organic soils P adsorption has been related to either high Al, Fe or Ca levels[402,945,1470,2713] and P sorption capacity of wetland soils may be predicted solely from the oxalate-extractable (amorphous) aluminum content of the soil.[4104] This fact was also confirmed by Walbridge and Struthers.[5226] Phosphorus sorption in wetlands is lower during periods of high water levels and flooding.[637,3432] A large amount of the phosphorus in the soil and sediments are present in the organic fraction. This complicates the system due to the dynamic equilibria between organic and inorganic phosphorus forms.

Richardson and Marshall[4109] found that soil adsorption and peat accumulation (i.e. phosphorus stored in organic matter) control long-term phosphorus sequestration. But microorganisms and small sediments control initial uptake rates, especially during periods of low nutrient concentration and standing surface water. Both biotic and abiotic control mechanisms are thus functional in the peatland, and the proportional effect of each on P transfers is dependent on water levels, the amount of available

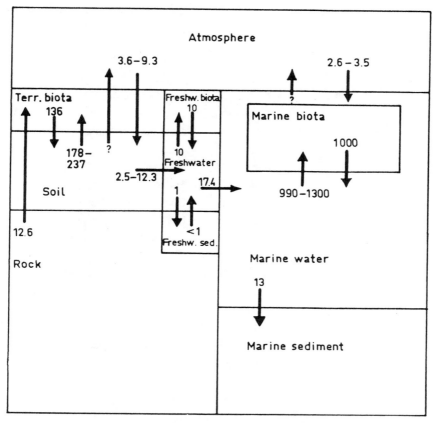

Figure 4-40.   A summary of the global flows (Tg yr^{-1}) of phosphorus.[3797]   (From Pierrou, 1976, in *Ecol. Bull.*, Vol. 22, p. 86, reprinted with permission of the *Ecological Bulletin*.)

P, fluctuating microorganism populations, seasonal changes in P absorption by macrophytes, and P soil adsorption capacity.

The most important retention mechanisms are ligand exchange reactions, where phosphate displaces water or hydroxyls from the surface of Fe and Al hydrous oxides to form monodentate and binuclear complexes within the coordination sphere of the hydrous oxide.[945,1470,2489,3691,4109,4853] The P sorption capacity of an oxidized soil may increase following flooding and reduction due to amorphous ferrous hydroxides, which have a greater surface area and more sorption sites than the more crystalline, oxidized, ferric forms.[2489,3645,3691] Golterman[1718] pointed out that specific extractions[1037,1719] have made it possible to study the distribution of phosphate between the different pools. It appears that even in sediment with about 20% of $CaCO_3$ most of the phosphate was adsorbed onto Fe(OOH), of which the concentration was only 0.5%.[1719]

Although the mechanisms by which P is removed from solutions by sediments is not clearly understood,[402,419,4853] it is thought to be a sorption process rather than a precipitation process. Shukla et al.[4466] attributed P sorption to a gel complex consisting largely of hydrated iron oxide. In soils and sediments exposed to free oxygen the active iron is in the Fe^{3+} form, probably as ferric oxyhydroxide, but under anaerobic conditions most of the active iron is in Fe^{2+} form, with some occurring as ferrous hydroxide gel complex. Faulkner and Richardson[1313] pointed out that the interaction of pH, redox, Fe, Al, adsorption, and precipitation often confounds interpretation of results from P removal studies.

It has been shown[1713,1715] that the adsorption of phosphate by sediments follows the adsorption mechanics:

$$(o - P)_{sed} = A \left[(o - P)_w \right]^B , \qquad (4.46)$$

Table 4-7. Phosphorus concentration in algae (% dry mass)

Species	Concentration	Locality	Reference
Freshwater			
Filamentous algae[a]	0.04–0.08	Nesyt Pond, Czech Republic	2843
Asterionella formosa	0.09–0.17	England, natural	2964
Periphyton	0.11–0.27	New England lake	2466
Navicula pelliculosa	0.18–0.36	Laboratory	3383
Cladophora glomerata	0.18–0.37	Clark Fork, Montana	2899
Chara vulgaris	0.36–0.46	Czech Republic	1207
Periphyton	0.09–0.65	Lake Huron	182
Enteromorpha prolifera	0.51–0.75	Forfach Loch, Scotland	2107
Cladophora glomerata	0.10–0.81	Lake Erie, Michigan-Ohio	2909
Microcystis aeruginosa	0.34–0.83	Laboratory	1610
Cladophora glomerata	0.27–0.96	Lake Michigan, Wisconsin	20
Chlorella vulgaris	1.10	Laboratory	1604
Chlamydomonas geitleri	0.70–1.20	Cultivation (wastewater)	5144
Anabaena flos-aquae	0.40–1.40	Laboratory	1992
Asterionella formosa	0.46–1.40	Laboratory	2964
Chlorella pyrenoidosa	0.94–1.51	Laboratory	4378
Scenedesmus acutus	1.20–1.80	Cultivation (wastewater)	5144
Periphyton	0.4–2.4	WWTPE[b]	1000
Chlorella pyrenoidosa	1.35–2.76	Laboratory	2540
Scenedesmus quadriacauda	0.59–7.98	Laboratory	3383
Marine			
Cladophora intestinalis	0.04	Bermuda	4361
Ulva lactuca	0.11–0.33	Hong Kong coast	2111
Codium frgile	0.28–0.49	Oregon coast	5366
Pelvetiopsis limitata	0.27–0.50	Oregon coast	5366
Ulva fenestrata	0.21–0.51	Laboratory	385
Cladophora sp.	0.07–0.54	Baltic Sea	5243
Enteromorpha intestinalis	0.30–0.56	Laboratory	385
Ulva fenestrata	0.32–0.62	Oregon coast	385,5366
Enteromorpha intestinals	0.37–0.75	Oregon coast	385,5366
Porphyra sp.	0.37–0.86	Oregon coast	5366

[a] *Spirogyra* sp., *Oedogonium* sp., *Cladophora fracta*, *Enteromorpha intestinalis*
[b] Wastewater treatment plant effluent

where $(o - P)_w$ and $(o - P)_{sed}$ are orthophosphate concentrations in the water and the sediment, respectively. Golterman[1715] found values for constants A and B in the range of 0.2–0.3 and 0.3–0.4, respectively, with the correlation coefficient > 0.9. Therefore, an increase of $(o - P)_w$ by a factor 10 is only causing an increase in the sediment phosphate concentration by a factor 2–2.5. The constant A depends on the amount of sediments entering a water body and their adsorption characteristic.

Even though the oxidation state of P is unaffected by redox reactions, redox potential is important because of Fe reduction. The form of Fe-P known as reductant-soluble phosphorus (RS-P) is especially significant in wetland systems. RS-P is a poorly-crystalline Fe compound that is stable under oxidized

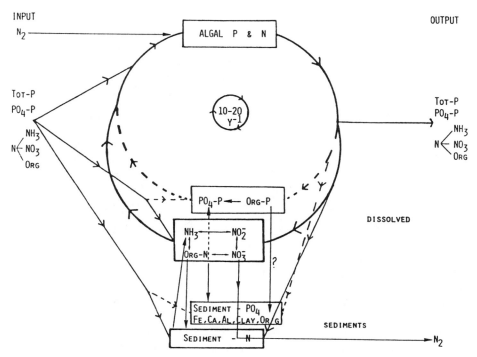

Figure 4-41.   Schematic representations of the N- and P-cycles. The indicated turnover of 10–20 times per year may vary depending on climate, trophic status, etc.[1715] (From Golterman, 1984, in *Verh. Internat. Verein. Limnol.*, Vol. 22, p. 31, reprinted with permission of E. Schweizerbart'sche Verlagsbuchhandlung (Nägele u. Obermiller), Stuttgart.)

conditions, but releases adsorbed and occluded P when Fe^{3+} is reduced to Fe^{2+} following submergence.[1555] This sediment pool may release large amounts of native P into solution under flooded conditions.[2158,3691]

The amount of dissolved inorganic orthophosphate in flooded soils, swamp and marsh sediments, and shallow bodies of water depends on the capacity of the soil or sediment to release orthophosphate-P to a solution low in P and to sorb it from a solution high in P.[3691] The sorption and release of P is affected by, among other factors, oxidation; reduction status of the soil or sediment.[3319-20] Mortimer[3319-20] showed that the disappearance of dissolved oxygen and the subsequent reduction of the sediment resulted in a several-fold increase of dissolved P in freshwater lake. Oxygenation of the sediment reversed this condition and decreased the P concentration. However, the investigations have shown that not only under anaerobic, but also under aerobic conditions essential quantities of phosphate can be released from the sediment only by diffusion-controlled exchange processes.[2228,2542,2742] Although release under aerobic conditions is about 10 times less than under anaerobic conditions,[640,2156] the significance of aerobic P release should not be minimized.[3378] High bacterial activity at high temperatures can cause microanaerobic conditions at the sediment surface and the phosphorus release to aerobic water, in such cases, might be explained by the same process as the anaerobic release.[2156]

Phosphorus release from sediments requires that mechanisms, which transfer phosphorus to the pool of dissolved phosphorus in the pore water, occur simultaneously, or within a short space of time, with processes which can transport the released phosphorus to the overlying water. Important "mobilization" processes are desorption, dissolution, ligand exchange mechanisms, and enzymatic hydrolysis.[469] These processes are affected by a number of environmental factors, of which redox potential, pH and temperature are most important.[469]   Essential transport mechanisms are diffusion, wind-induced turbulence, bioturbation, current and gas convection.[469,2542]

One of the proposed mechanisms for the release of phosphorus from soils upon submergence is the reductive dissolution of Fe(III) and Mn(IV) phosphate minerals.[3685,3695,5432] Anaerobic soils release more phosphate to soil solutions low in phosphate and sorbed more phosphate from soil solutions high in

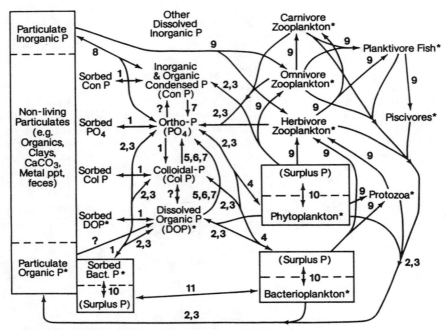

Figure 4-42.   Phosphorus flow-diagram indicating major components and pathways within the water column. * indicates several functional groups or life stages within each component. 1 = sorption, 2 = death, 3 = excretion/egestion, 4 = uptake, 5 = enzymatic release, 6 = photolysis, 7 = hydrolysis, 8 = precipitation, 9 = feeding, 10 = growth, 11 = microbial colonization.[4311] (From Scavia, 1980, in Spec. Rept. #83 of the Great Lakes Res. Div., p. 127, reprinted with permission of the author.)

soluble phosphate than do aerobic soils.[3691] The difference in behavior of phosphate under aerobic and anaerobic conditions is attributed to the change brought about in ferric oxyhydroxide by soil reduction.[3691] The probably greater surface of the gel-like reduced ferrous compounds in anaerobic soils results in more soil phosphate being solubilized when solution phosphate is low and more solution phosphate being sorbed when solution phosphate is high.[3691] Very important is also the capacity of soils and sediments to sorb and release phosphorus under conditions where the P concentration is high in one or both phases.[3691]

The possible release of phosphate from sediments is often calculated by measuring, *in situ,* the increase of the phosphate concentration of the water overlying the sediments or, *in vitro,* by measuring the phosphate concentration in isolated cores.[1718] Release rates found in these ways range between 0 and 100 mg m^{-2} yr^{-1}.[469,1720] A review by French[1483] gives the range of release rates between 10 and 50 mg m^{-2} d^{-1} for aerobic and 0 to 150 mg m^{-2} d^{-1} for anaerobic conditions. The higher release under anaerobic conditions is probably caused by mineralization of organic matter containing phosphate and not by reduction of Fe(OOH)≈P, as is usually believed in the literature.[1718]

Sediment processes control the long-term P removal capability of wetland ecosystems.[251,3432,4109,4111] There is a little direct uptake of phosphate from the water column by emergent wetland vegetation because the soil is the major source of nutrients.[4384] Growing vegetation is a temporary nutrient-storage compartment resulting in seasonal exports following plant death.[503,992,2537,4114-5,4647] The long-term role of emergent vegetation is to transform inorganic P to organic forms. Microorganisms play a definite role in P cycling in wetlands; however, the microbial pool is small.[1313,4109,4112] When phosphate was added to aquatic or marsh ecosystems containing higher plants, microorganisms in the water column took up inorganic P faster than plants.[887,1984] Richardson and Marshall[4109] pointed out that an understanding of the rate and magnitude of soil-plant-water-microorganism interactions in wetlands appears to be essential to determining the processes of controlling the cycling of phosphorus, and in turn the P removal and storage potential of wetland ecosystems. So far, however, interactions including microorganisms have been studied only rarely in wetland ecosystems. Doremus and Clesceri[1125] presented a conceptual model of sediment water phosphorus interactions, including microbial metabolism (Figure 4-43), and recently, Richardson and Craft[4112] have presented a very interesting model of P retention in wetlands based on the concept that P nutrient storage compartments have an operational capacity that is analogous to reservoirs of varying size (shown as buckets and bathtub) with either short-term or long-term storage capacity (Figure 4-44). The size of the short-term storage compartments are drawn in relative scale (soil adsorption, precipitation > plants > periphyton), with the long-term retention capacity being directly related to peat accretion rates. The initial uptake rates in the short-term compartments can be quite high until the reservoirs have reached their capacity, especially in a system that has not received nutrient inputs before. The size and uptake rate of each short-term storage compartment is finite. Once they have reached capacity, they will no longer function as effective storage areas. The fact that short-term storage is equated with long-term storage is often the basis for confusion in the literature concerning the efficiency of wetlands to store P. Short-term processes do control water quality, have fast recycle and uptake rates (i.e., minutes, hours, or days vs. years), but limited storage capacity. The storage and cycling processes are primarily biological and chemical in nature, although the physical factors of high water input rates, increased water depth, and lower retention times for water can override these two processes and decrease storage efficiency. The largest reservoir (bathtub size in Figure 4-44) is the long-term peat or soil compartment. This compartment usually stores more than 80% of the nutrients in the top 20 cm of the strata and is comprised of buried organic P from undecomposed litter from previous years, as well as annual additions from adsorption and precipitation (3), periphyton (1), plant roots (2), and new material from the litter compartment (4). The efficiency of any wetland to store P on a long-term basis is thus determined by the peat or soil accretion rate times the net increase of P stored by these processes each year.[4112]

Johnston,[2351] summarizing the huge literature data, reported phosphorus concentration in wetland soils in the range of 0.001 to 7.0 mg P g[-1], average P concentrations per unit mass being comparable for both organic and mineral soils. Nixon and Lee,[3447] summarizing data from various regions of the United States, reported the range of 0.003 to 20 mg P g[-1], the upper value being reported for a wetland receiving sewage effluent.[518] This value, however, is extremely high as phosphorus concentrations in soils are usually < 1.0 mg P g[-1].[3447] Research to date would suggest that permanent storage of phosphorus is below 1 g m[-2] yr[-1] and usually averages around 0.5 g m[-2] yr[-1].[901,2351,3432,4104,4109,4112] Johnston[2351] reported the average soil standing stock of 64 (16–179) g P.m[-2].

In the literature, numerous models of phosphorus budgets and flows for local wetlands have been reported. Figure 4-45 represents a phosphorus budget for an alluvial cypress swamp in Illinois, Figure

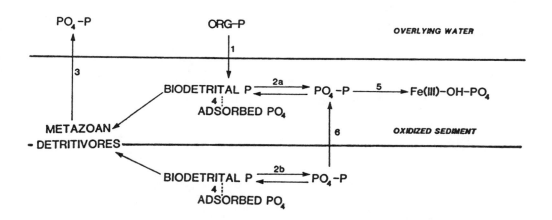

Figure 4-43. A conceptual model of profoundal sediment-water phosphorus interactions. Major processes include: 1 = sedimentation of water column particulate P; 2a, b = cycling of phosphorus within the biodetrital complex; 3 = mobilization of phosphorus by metazoan detrivores; 4 = adsorption of inorganic P onto biodetritus; 5 = adsorption of inorganic P onto ferric hydroxide; 6 = diffusion of deep inorganic P upward.[1125] (From Doremus and Clesceri, 1982, in *Hydrobiologia*, Vol. 91, p. 266, reprinted with permission of Kluwer Academic Publishers, Dordrecht.)

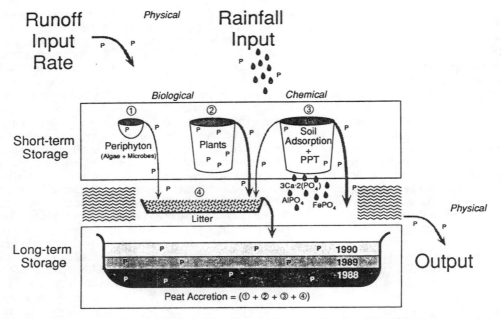

Figure 4-44. A conceptual model of phosphorus retention in wetlands. Only the major reservoirs are shown and no attempt is made to show a complete phosphorus cycle among biotic and abiotic compartments.[41][12] (From Richardson and Craft, 1993, in *Constructed Wetlands for Water Quality Improvement,* Moshiri, G.A., Ed., p. 272, reprinted with permission of CRC Press, Inc., Boca Raton.)

4-46 shows a phosphorus budget for a central Michigan wetland, and Figure 4-47 illustrates phosphorus flow through a *Scirpus fluviatilis* stand in Wisconsin (see also Figure 3-21).

4.3.3.1 Phosphorus in wetland plants

Woody plants[2351]

Concentration in % dry mass:
leaves: 0.19 (0.04–0.60)
shrub stems: 0.06 (0.02–0.08)
tree boles and roots: 0.011 (0.003–0.044)
standing stock (g P m^{-2}):
leaves: 0.89 (0.08–1.7)
wood: 3.14 (0.18–4.28)
litter: 0.96 (0.21–1.94)

Herbaceous plants

Concentration in % dry mass (see Appendix, Table 7-9):
emergent species: aboveground: 0.007–1.15, belowground: 0.02–0.59
floating-leaved species: aboveground: 0.14–0.64, belowground 0.30–0.39 (FD)
submerged species: 0.03–1.05 (aboveground)
floating species: 0.085–2.84
bryophytes: 0.0064–0.61
Standing stock (g P m^{-2}) (see Appendix, Table 7-10):
emergent species: aboveground: 0.019–22.9, belowground: 0.24–7.40
floating-leaved species: 0.076–1.75 (aboveground)

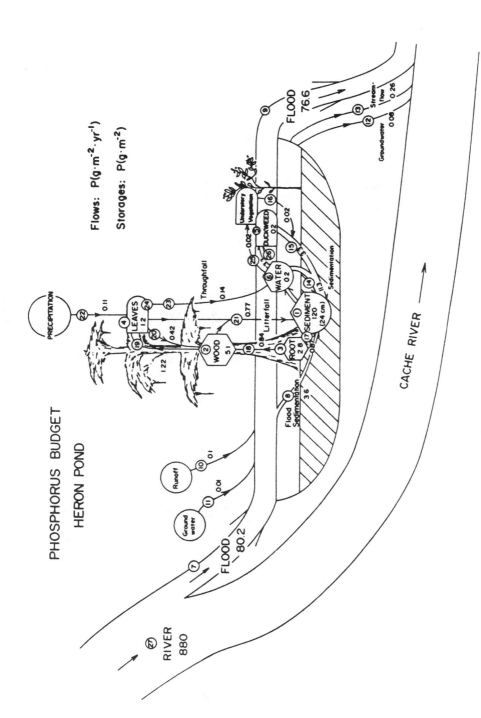

Figure 4-45. Annual phosphorus budget for Heron Pond alluvial cypress swamp in southern Illinois.[3248] Numbers in circles refer to calculations given in Mitsch et al.[3247] (From Mitsch et al., 1979, in *Ecology*, Vol. 60, p. 1120, reprinted with permission of the Ecological Society of America.)

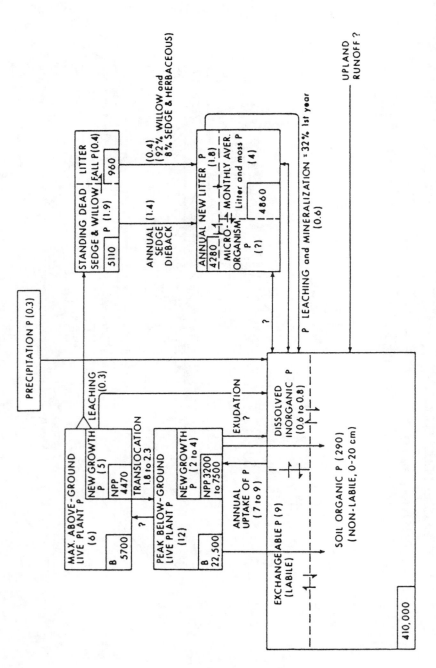

Figure 4-46. The annual productivity, biomass, and phosphorus budget for the sedge-willow cover type in the Houghton Lake Peatland in central Michigan. Estimates of annual net primary productivity are labeled as NPP and reported in kg ha^{-1} yr^{-1}. Biomass (B) and standing mass of dead litter, etc. (unlabeled) are shown in enclosed boxes and are reported in kg ha^{-1}. Uptake, storage, and transfer rates of P are shown in parentheses and are reported in kg ha^{-1} yr^{-1}.[4109] (From Richardson and Marshall, 1986, in *Ecol. Monogr.*, Vol. 56, p. 287, reprinted with permission of the Ecological Society of America.)

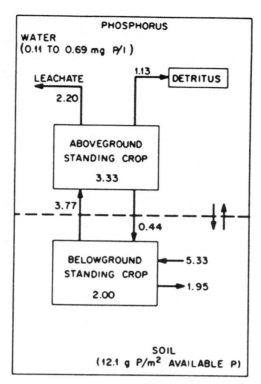

Figure 4-47.   Flow of phosphorus through a *Scirpus flaviatilis* stand. Flows in g m^{-2} yr^{-1} and compartments in g m^{-2} in standing crop.[2537] (From Klopatek, 1978, in *Freshwater Wetlands: Ecological Processes and Management Potentials*, Good, R.E., et al., Eds., p. 207, reprinted with permission of Academic Press and the author.)

submerged species: 0.078–2.3 (aboveground)
floating species: 0.1–34.5 g

Annual phosphorus uptake rates by various macrophytes (see Appendix, Table 7-11) range between 0.77 and 40 g P m^{-2} yr^{-1} for emergent species and between 10.5 and 126 g P m^{-2} yr^{-1} for floating species. Daily uptake rates range between 1.5 and 640 mg P m^{-2} d^{-1}.

4.3.4 Phosphorus in Algal Growth

Uptake of phosphorus by algae and its role in algal metabolism has been extensively reviewed.[735–6, 2592,2634,2636–7,3529,3923] The role of phosphorus in metabolism and photosynthesis was reviewed by Arnon.[155] The phosphorus-deficiency symptoms are summarized by Kuhl.[2637] Much of the following discussion is synthetized from these studies.

Phosphorus plays a significant role in most cellular processes, especially those involved in generating and transforming metabolic energy. It is thus indispensable for the growth and reproduction of living organisms and its metabolic pathways have been extensively studied.[2637] Particular attention has been paid to the role of phosphorylated compounds in the conversion of light energy to biological energy in green plants,[155] an area of research where studies with algae have contributed significantly to our understanding of the processes involved.[2637] Phosphorus plays a key role in many biomolecules, such as nucleic acids, proteins, and phospholipids (the latter are important components of membranes). Its most important role, however, is in energy transfers through ATP and other high energy compounds in photosynthesis and respiration (Figure 4-48) and in "primary" molecules for metabolic pathways.[2886] Phosphorus requirement of the Chlorophyceae is 2–3% of the dry mass[2481] but the phosphorus levels

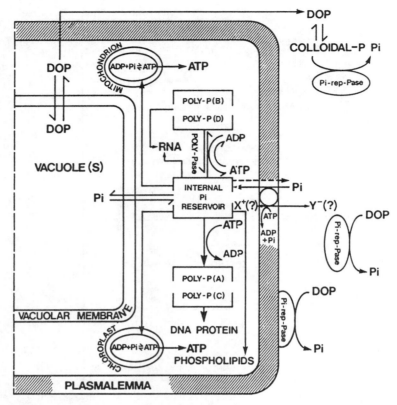

Figure 4-48.   Main features of phosphorus uptake and assimilation in a microbial cell. DOP = dissolved organic phosphate; Pi = inorganic phosphate.[736] (From Cembellia et al., 1984, in *CRC Crit. Rev. Microbiol.*, Vol. 11, p. 15, reprinted with permission of CRC Press, Inc., Boca Raton.)

within algal cells may fluctuate widely depending on whether the algae are growing under phosphorus-limited conditions or not.[2637] Even at very low levels in the environment the algae can accumulate and store phosphorus.[3004] Mackereth[3004] showed that as little as 1 μg L^{-1} of phosphorus was sufficient to provide optimum growth of a diatom *Asterionella formosa* in experiments. Depleted cells can absorb phosphate in the dark whereas replete cells absorb it only in the light.[4209] The concentration of phosphorus in freshwater algae ranges between 0.04 and 7.98% DM, the values > 1% being recorded mostly in laboratory cultures (Table 4-7). The concentration of phosphorus in marine algae is in the range of 0.04 to 0.86% DM (Table 4-7).

The optimum requirements for phosphorus varies widely, e.g. 0.45 μm P l^{-1} for the blue-green alga *Coccochloris peniocystis*, 0.002 μm P l^{-1} for *Asterionella formosa*, whilst 0.005 μg P l^{-1} are reported to inhibit the growth of *Dinobryon* and *Uroglena*.[4209] Chu[803] studied the optimal phosphate concentrations for the growth of diatom and green algae under defined laboratory conditions. Concentrations below 50 μg L^{-1} of phosphorus were limiting, those about 20 mg L^{-1} were inhibitory and 0.1 to 2 mg L^{-1} were optimum. Rodhe[4164] later differentiated three main groups of freshwater algae according to their ability to tolerate phosphate within the ranges below, around, and above 20 μg L^{-1} of phosphorus. Most planktonic algae fall into the groups with low or medium phosphorus tolerance. However, Blum[404] observed saturation of phosphate uptake by *Euglena* at 3.1 mg L^{-1} of phosphorus.

Any form of phosphorus present in (or added to) a water body can be categorized as being immediately available, potentially available, or likely unavailable.[1569] The principal form of phosphorus known to be directly available to plants is orthophosphate phosphorus. It is the only important inorganic phosphorus source for algae.[735,1416,2178,2634,2637,4140,4143]

Potentially available phosphorus forms include a large number of compounds which can be converted to orthophosphate and thus become available to algae.[1569] This category includes a broad spectrum of dissolved inorganic and organic phosphorus sources which are capable of sustaining growth, and utilization of such compounds is relatively widespread among algae[736,804,1119,1950,4723,5415–6] (for review see References 442 and 735). Those compounds include glucose-6-phosphate,[3,2291,2627,2630,3023] glycerophosphate,[2893,3277,3931] AMP,[1473,2627,3164] ATP,[327,602] CMP (cytidine-(3' or 5')-monophosphate),[3164,3801,3929] inorganic pyrophosphate,[3023] inorganic polyphosphate,[1554,4154] and many others.[735] Potentially available phosphorus can appropriately be further categorized into forms available over short periods (hours - days) and forms which become available over long periods (weeks - months).[1569] The conditions necessary for a transformation of potentially available phosphorus to orthophosphates have been widely studied.[137,894–5,2743,2898,5157]

The likely unavailable phosphorus compounds are relatively abundant in water bodies but are probably the least understood of the three categories.[3444] Although very few forms of phosphorus in the environment can be considered as totally unavailable over long periods (years), some forms (e.g. apatite and other minerals) are likely unavailable during their residence in water bodies.[2743,5437]

The availability of organic phosphates (specifically phosphate monoesters) as a potential nutrient source for microalgae is generally contingent upon the ability of the cells to enzymatically cleave the ester linkage joining the P_i group to the organic moiety.[735] This ability is achieved by the activity of phosphomonoesterases (commonly called phosphatases) at the cell surface.[2886] Two groups of these enzymes have been distinguished on the basis of their pH optima, phosphate repressibility, and cellular location. Alkaline phosphatases are phosphate repressible, inducible, and generally located on the cell surface or released into the surrounding water[326,531,1255,1554,2630,2637,5507] (for review see Rivkin and Swift[4154]) and presumably the same transferase is responsible for the transport across the plasmalemma of both orthophosphate and the PO_4^{3-} released from the phosphate esters.[2630] Acid phosphatases are phosphate irrepressible, constitutive, and generally found intracellularly in the cytoplasm. Both types may be found simultaneously in algal cells, with alkaline phosphatases aiding in uptake of organic phosphorus compounds and acid phosphatases playing a crucial role in cleavage and phosphate transfer reactions in metabolic pathways within the cell.[2886]

The essential feature of phosphatases that allows them to participate efficiently in cellular metabolism is the ability to be alternately induced or repressed, depending on metabolic requirements. When external inorganic phosphate concentrations are high, the synthesis of alkaline phosphatases is repressed and cells exhibit little ability to use organic phosphorus compounds.[2886] Alkaline phosphatase is induced by phosphorus limitation[1992] and may be liberated from the cells into the surrounding medium.[3755,4054] Fitzgerald and Nelson[1374] have shown that algae which are phosphorus limited have an alkaline phosphatase activity of 25 times more than algae with sufficient available P. These data were confirmed by Bone[433] who found that the activity of alkaline phosphatase in *Anabaena* varied 20-fold with the lowest values being found in cells containing excess phosphate. The induction of the alkaline phosphatase under phosphate-limited conditions seems to be a characteristic phenomenon which allows this enzyme assay to be used as a measure of phosphorus-limited growth.[2637] The ability to utilize organically bound phosphates is advantageous in aquatic habitats in which such combined phosphate sources are present in concentrations higher than those of free polyphosphate.[4643]

It is necessary to emphasize that the nomenclature applied to the phosphate fractions is sometimes confusing and as it has been pointed out by Cembella et al.[735] terms have sometimes been used carelessly. For example "soluble reactive phosphorus" (SRP) (which passes through a 0.45-μm membrane filter and reacts with the molybdate reagent) is occasionally used synonymously with "low molecular weight inorganic phosphates," or even "orthophosphate," however, interchanging such designations should no longer be considered excusable.[735] Furthermore, the so-called "soluble organic phosphate" fraction may include considerable amounts of pyrophosphate and polyphosphate, which are technically inorganic.[2178]

In the literature, several approaches for phosphorus fractionation in waters have been proposed.[1084,2257,3761–2,4806–7,4901]

Based upon the molybdenum blue technique, Strickland and Parsons[4807] discriminated between the following phosphorus fractions in aquatic ecosystems: 1) soluble inorganic reactive phosphate (orthophosphate); 2) soluble organic reactive phosphate—low molecular mass organics, such as easily hydrolyzable sugar phosphates; 3) soluble organic unreactive phosphate—not readily hydrolyzable; 4) soluble enzyme-hydrolyzable phosphate—sugar phosphate, linear inorganic polyphosphates, etc.; 5) soluble polyphosphate—inorganic and organic polyphosphates; 6) particulate unreactive phosphate; 7) particulate organic unreactive phosphate; and 8) particulate reactive phosphate—including both inorganic and organic phosphorus in particulate form. Unfortunately, the characterization of fractions within this scheme is analytically very complicated.[644,742,4671]

Rigler[4141] pointed out that 60–70% of the total phosphorus is present in suspended form (inorganic + organic), 15–30% is an organic dissolved form and approximately 10% is in the form of inorganic phosphates. Rigler[4142] found that common analytical orthophosphate methods using molybdenum blue method give values 10 to 100% higher. This increase is caused by hydrolysis of organic compounds and polyphosphates or by desorption of PO_4-P particles present in water with acidic pH. Lean and Nalewajko[2737] found this phenomenon also in laboratory cultures of algae. These findings support the idea that molybdenum blue methods measure SRP rather than "true" PO_4-P.

Chamberlain and Shapiro[742] and Walton and Lee[5261] proved that algal growth can be rather correlated with SRP concentration. Cowen and Lee[895] reported that an average amount of biologically available phosphorus is possible to quantify with the following relationship:

$$\text{available P} = \text{TSP} + 0.3\ \text{TPP}, \tag{4.47}$$

where TSP is total soluble phosphorus and TPP is total particulate phosphorus, while Lee et al.[2743] describe the amount of available phosphorus by the equation:

$$\text{available P} = \text{SRP} + 0.2\ \text{TPP} \tag{4.48}$$

Sediments are known sinks for phosphorus and are major reservoirs of phosphates which can potentially be made available to organisms through chemical, biological, and physical processes.[1569] Chemical changes in redox potential cause phosphates to be released when metal precipitates dissolve causing increased phosphate levels in pore water and in water column.[1569] Biological regeneration of phosphates from sediments by benthic macroorganisms and microbes appears to be an important mechanism to return particulate phosphorus to water in dissolved form.[2156,2742,3378,5490] Physical mixing of the upper layers of the sediments results in transport of phosphorus rich pore water to upper layers of the water column.[1569,4699]

A number of investigations have been performed to estimate the available sediment (and particulate) P for natural populations of algae.[782,815,895,1058,1126,1368,1543,1707–8,1710,1721,1996,2522,3847,4147,5437,5473,5532] As a result, some chemically-determined fractions of particulate phosphorus have been shown to be directly correlated with bioassay results on available phosphorus.[1126,1710,5437]

Earlier studies have shown that total reactive P (TRP), measured chemically in unfiltered samples, may be the available fraction of TP for algae in water sample with high content of inorganic seston.[389] Reactive phosphorus (RP) may often vary from 25 to 75% of particulate phosphorus (PP).[2618] The organic content in cultivated mineral soils may be high and more than 50% of TP may be organic-P, mostly as inositol phosphates.[4356] Krogstad and Løvstad[2619] found a linear relationship between available P for blue-green algae and soil reactive phosphorus measured chemically for both cultivated surface soil and uncultivated underground soil. Grobbelaar[1826] reported that more than 84% of adsorbed PO_4-P were available to algae in the Amazon River. Algae were able to extract more PO_4-P from the suspended solids, than could be extracted by distilled water in chemical equilibria or due to small pH shifts. The author suggested that organic compounds excreted by algae were responsible for some of the PO_4-P released from the suspended solids. Sufficient P, adsorbed onto the suspended sediments, is alga-available to produce large quantities of algal biomass. Clasen and Bernhardt[814] reported also that a bloom-forming chrysophyte *Synura uvella* used phosphorus released from the sediment in Wahnbach reservoir in Germany.

4.3.4.1 Orthophosphate uptake

It is unlikely that passive diffusion accounts for more than a small fraction of total inorganic phosphorus (P_i) influx. The role of facilitated diffusion in P_i uptake in microalgae has not been clearly established.[735] The active transport of P_i into algal cells has been demonstrated for many species.[404,2319,2666,2669,4002,4585] Based upon cellular energetics, the two primary bioenergetic candidates for active P_i transport driving forces are high energy nucleotide triphosphates generated by photophosphorylation in chloroplasts and oxidative phosphorylation in mitochondria, and the electron transport system of these organelles.[735] By manipulation of experimental conditions and the use of metabolic inhibitors, it was possible to substantiate the central role of ATP in the transmembrane P_i uptake process (for summary see Cembella et al.[735]).

The evidence that light is generally capable of stimulating P_i uptake[785,2664,2666,4002–3,4155] directly implies photophosphorylation as a major source of ATP for this transport process. In the light, P_i

influx is much less affected by inhibition of oxidative phosphorylation than of photophosphorylation.[2666,4002] As the amount of ATP required for transport represents only a small fraction of the total cellular ATP, and is perhaps below detectable limits, it is difficult to attribute ATP for uptake to a particular pathway. However, current evidence[4004] favors the noncyclic photophosphorylation pathway as the major source.

An increase in endogenous ATP levels on adding phosphorus to phosphorus-deficient cells has been noted by Batterton and Van Baalen[245] and Stewart and Alewander.[4777] The increase occurred both in the light and in the dark indicating that there was a stimulation both of mitochondrial oxidative phosphorylation and photophosphorylation. In the dark, ATP for P_i uptake may be obtained via mitochondrial oxidative phosphorylation, with perhaps a relatively small contribution from substrate level phosphorylation.[735] The enhanced dark uptake effect due to preillumination[4511] indicates that ATP pools saturated through photophosphorylation may be used to power uptake during postillumination darkness. Although P_i transport rates are typically reduced in darkness, an experiment on three P-saturated dinoflagellate species[1020] revealed that uptake was fastest in the dark, with intermediate light intensity usually yielding the slowest uptake.

Orthophosphate uptake by algae is generally described by Michaelis-Menten kinetic model,[181,637,786,1523,3576,3757,4083,4811,4972] but deviation from simple Michaelis-Menten kinetics has been reported.[578,785] Examples of kinetic parameters are given in Table 4-8.

In comparison with macrophytes, nitrogen uptake rate is only rarely expressed per unit of area. Richardson and Schwegler[4110] reported an uptake rate of 12 mg P m^{-2} d^{-1} for *Cladophora glomerata* in a Michigan marsh. In waters with high nitrogen concentration the uptake rate may reach much higher values, however. Davis et al.[1000] reported the uptake rate as high as 160 mg P m^{-2} d^{-1} for periphyton growing in wastewater treatment plant effluent. Richardson and Schwegler[4110] reported nitrogen standing stock in *Cladophora glomerata* of 0.96 g P m^{-2}.

The linearity or nonlinearity of nutrient uptake in time course experiments is a function of the magnitude of the nutrient perturbation relative to the cell density, as well as the relative nutritional status of the cells.[735] Thus, while some researchers have reported linear P_i uptake over incubation periods ranging from minutes to a few hours for laboratory[637,1775,4083] populations, and over 6 hours[2606] and 24 hours[3756] for natural populations, others showed nonlinear uptake in laboratory[188,785,2629] and natural[2736] populations, within the first few hours after nutrient addition.

Cembella et al.[735] pointed out that further complications in interpreting estimates of uptake kinetic parameters based on $^{32}P_i$ or $^{33}P_i$ uptake may occur in oligotrophic areas where ambient P_i concentrations are frequently at or below the limit of detection and one does not accurately know what concentration of substrate (P_i) to use in plotting the uptake vs. substrate concentration curve, and, consequently, determination of K_s (in particular) may be inaccurate.

Although most studies of P_i uptake in microalgae tend to support a monophasic uptake model, at least the confines of environmentally realistic external P_i concentrations (typically < 5 μM = < 155 μg L^{-1}), there is evidence of multiphasic uptake systems worthy of serious consideration, particularly for high P_i concentrations (for discussion see Cembella et al.[735] and references cited therein).

Rapid dark uptake of PO_4^{3-} is a frequently observed characteristic of P-depleted cells,[786,1992,2629] however, P-replete microalgae usually do not assimilate PO_4^{3-} in the dark.[785,1992,2637]

Although Michaelis-Menten kinetics are typically used to describe P_i uptake kinetics as a function of external concentrations, observations that P_i uptake is inversely related to internal P-pool concentrations[73,182,1775,4083-4] are evidence of the coupling of uptake kinetics to cellular nutrient status. As an example, Auer and Canale[182] reported for a green filamentous alga *Cladophora glomerata* that low internal phosphorus levels (0.12% P) were associated with a high maximum uptake rate (164 × 10^{-8} g P mg DM^{-1} h^{-1}) and a high half-saturation constant (91 μg L^{-1} P) while higher levels of internal phosphorus (0.23% P), resulted in lower values of uptake rate (40 × 10^{-8} g P mg DM^{-1} h^{-1}) and half-saturation constant (52 μg L^{-1} P). The authors also reported a variation in value for the half-saturation constant (K_m) for phosphorus uptake with internal phosphorus concentration (Q, in % P): K_s = 246.86 Q_0/Q. The Droop model provided an excellent fit of net specific growth rate and internal phosphorus concentration in *Cladophora glomerata* (Figure 4-49).[182] Smith[4606] reported that light-saturated photosynthesis is a hyperbolic function of cellular phosphorus quota (Q_p), whether Q_p is expressed per unit cellular chlorophyll-*a*, per unit dry mass, or per cell number (Figure 4-50). The author pointed out that the cell quota model is a valid empirical description of the photosynthetic behavior of nutrient-limited cells.

Although the intracellular phosphate levels are frequently considered to be the major factor controlling uptake kinetics,[404,577,1158,1522-3,1992,4002,4083] Perry[3756] demonstrated that P_i uptake rates for P_i-loaded cells from a P_i-limited chemostat tend to reflect the nutritional past history, rather than the instantaneous internal concentrations, at least in the short term.

Table 4-8. Kinetic parameter values for orthophosphate uptake by algae.[735] (Based on Cembella et al., 1983, in *CRC Crit. Rev. Microbiol.*, Vol. 10, p. 337, reprinted with permission of CRC Press, Boca Raton.)

Species	K_s (μM)	V_{max} (μmol cell^{-1}.h^{-1} $\times$ 10^{-9})	Reference
Freshwater			
Anabaena flos-aquae	1.0–2.0		1992
A. flos-aquae	0.96–2.84		3383
Anabaena planctonica	0.06	73.5	4602
Ankistrodesmus braunii	2.27–6.45		5074
Ankistrodesmus falcatus	3.95	32.5	1755
Aphanizomenon flos-aquae	0.06	57.1	4602
Asterionella formosa	1.90–2.80	9.9–13.2	4972
A. formosa	0.60	91.4	4984
A. formosa	0.60	42.9	1775
A. formosa	0.70	15.0	2163
A. formosa	0.34	2.6–16.1	4088
A. formosa	0.06	18.0	4602
Chlorella pyrenoidosa	0.68	48	3489
C. pyrenoidosa	2.10–4.00		2321
Cladophora glomerata	0.24		4183
C. glomerata	0.61		4182
C. glomerata	0.19–1.0		4182
C. glomerata	2.10		183
C. glomerata	0.48–2.78		2899
C. glomerata	0.97–8.06		181
Cryptomonas erosa	0.13	2.2	3297
Cyclotella meneghiniana	0.80	5.5	4972
C. meneghiniana	0.75	5.5	4984
Diatoma elongatum	2.80	8.0	2500
Dinobryon cylindricum	0.72	2.0	2759
Euglena gracilis	0.70–2.80	0.9–150.6	784–6
Fragilaria crotonensis	1.06	3.4	1775
F. crotonensis	0.03	132.3	4602
Microcystic aeruginosa	0.57		3545
M. aeruginosa	1.08		3544
M. aeruginosa	1.23	8.0	2163
Navicula pelliculosa	5.11–11.75		3383
Oscillatoria agardhii	0.29–1.45		5557
Oscillatoria tenuis	0.04	332.3	4602
Pediastrum duplex	0.90–1.48	5.8–51.3	2760
Peridinium triquentum	3.0		1905
Peridinium sp.	6.30	559	2759
Scenedesmus obliquus	2.0–5.0		1992
Scenedesmus quadricauda	0.36–4.18		3383

Table 4-8. Continued

Species	K_s (μM)	V_{max} (μmol cell^{-1}.h^{-1} $\times$ 10^{-9})	Reference
Scenedesmus sp.	0.60	4.8	4083
Selenastrum capricornutum	2.58		3489
S. capricornutum	0.52–4.26		578
Staurastrum luetkemuellerii	0.48		3545
S. luetkemuellerii	0.04–4.10		3455
Stigeoclonium tenue	0.18		4182
S. tenue	0.20		4183
S. tenue	0.48–3.0		4182
Synedra acus	0.06	15.6	4602
Ulothrix zonata	7.91		183
Volvox aureus	0.17–0.96	0.2–2.8	4403
Volvox globator	0.19–4.20	0.1–2.1	4403
Natural phytoplankton			
Jordan River	0.18		1885
Lake Kinneret, Israel	0.21–0.79		1885
Lake Wingra, Wisconsin	0.45		578
Mirror Lake, New Hampshire	0.10–0.27		2759
Lake Memphremagog, Canada	0.01–0.08		4602
Marine			
Agardhiella subulata	0.40		1022
Amphidium carteri	0.01	41.1	1020
Cladophora albida	0.67		1754
Enteromorpha compressa	1.0		2445
Monochrysis lutheri	0.51	28.1	637
Olisthodiscus luteus	1.00–1.98	228–433.2	5001
Prorocentrum minimum	1.96	167	735
Protogonyaulax tamarensis	0.40	299	735
Pyrocystis noctiluca	1.90–2.70	2.9×10^4–10.5×10^4	4155
Symbiodinium microadriaticum[a]	0.01	5.4	1020
Thalassiosira pseudonana[b]	0.58	8.6	1523
T. pseudonana[c]	0.50–0.70	2.8–17.3	3756
Thalassiosira fluviatilis[d]	1.72	250	1523
Natural phytoplankton			
Central North Pacific	0.14–0.18		3757
California current	0.40		3757

[a] = Gymnodinium, [b] neritic, [c] oceanic, [d] = T. weissflogii

4.3.4.2 Factors affecting phosphorus uptake

The evidence regarding light stimulation of P_i uptake is frequently far from unequivocal (Cembella et al.[375] and reference cited therein). In the literature, there are numerous reports on light-enhanced P_i

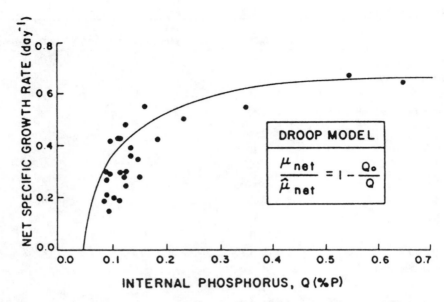

Figure 4-49.   Fit of the Droop formulation to net specific growth rate in *Cladophora*.[182]   (From Auer and Canale, 1982, in *J. Great Lakes Res.*, Vol. 8, p. 97, reprinted with permission of the International Association for Great Lakes Research.)

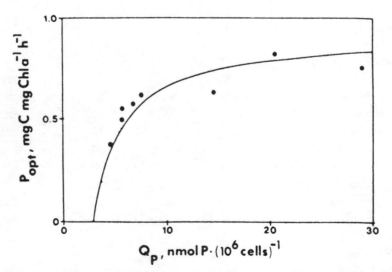

Figure 4-50.   Relationship between light-saturated photosynthesis (P_{opt}, mg C mg chl a^{-1} h^{-1}) and phosphorus quota (Qp) in phosphorus-limited *Scenedesmus quadricauda* ($r^2 = 0.85$).[4606] (From Smith, 1983, in *J. Phycol.*, Vol. 19, p. 308, reprinted with permission of the *Journal of Phycology*.)

uptake.[785–6,1942,2319,2664,2666,4002–3,4155,4511,4585]   By compiling data on natural phytoplankton assemblages, Reshkin and Knauer[4068] described a Michaelis-Menten-type hyperbolic relationship relating P_i uptake rate (V) to percent surface irradiance:

$$V = V_{max} (I/K_L + I),$$ (4.49)

where I is the irradiance and K_L is the half-saturation constant, i.e. the irradiance where $V = 1/2\ V_{max}$. In a study of P_i uptake kinetics in *Euglena gracilis*, Chisholm and Stross[785-6] showed that the influence of light upon uptake was a function of cellular metabolic status. Much more experimental work is needed to determine how the degree of P limitation and/or the degree of light limitation affects the relationship between irradiance and P_i uptake.[735]

From a physicochemical standpoint it is conceivable that temperature could indirectly alter cellular uptake kinetics by affecting ambient water properties, the ionic speciation ratios of transportable P_i, the rate of diffusion into the cellular boundary layer, and the frequency of molecular collision with cell-surface P_i receptors.[735] The influence of temperature upon nutrient uptake kinetics may be expressed as not only a species-specific, but also as a nutrient-specific response. There is a scarcity of data on how a transient in temperature affects the P_i uptake rate.[735]

Although PO_4^{3-} uptake may not be directly coupled with pH, it may stimulate PO_4^{3-} uptake by 1) providing additional carbon substrate for oxidative metabolism, 2) increasing the demand for PO_4^{3-} as a result of photophosphorylation reactions, or 3) providing metabolic energy for PO_4^{3-} uptake and the synthesis of polyphosphate reserves to store assimilated PO_4^{3-}.[2637] P_i uptake as a pH-dependent phenomenon has been reported by several authors.[404,2318,4008] Maximal uptake rates have been reported to appear at approximately neutral pH.[404,2318,4003,5074]

The rate of absorption of phosphate by marine diatom *Nitzschia closterium* was directly dependent on the concentration of both phosphate and nitrate in the medium (Figure 4-51). The ratio in which phosphate and nitrate were absorbed was constant for each concentration of phosphate in the medium regardless of the concentration of nitrate, except that at very low concentrations of nitrate the ratio of absorption increased. The rate of absorption of nitrate was independent of the concentration of phosphate in the medium. The maximum rate of P uptake in N limitation is lower by a factor of about 8 than the rate in P limitation.[4083] Studies of primary productivity in lakes[4605] demonstrated a close linear dependence of *in situ* rates of light-saturated photosynthesis on the concentrations of total phosphorus and total nitrogen in the euphotic zone. On the other hand, Terry[4930] reported for a marine prymnesiophyte *Pavlova lutheri* (formerly *Monochrysis lutheri*, classified as chrysophyte) that short-term phosphate uptake was independent of the nitrate concentration, but the short-term nitrate uptake rate was reduced in the presence of phosphate. The severity of inhibition of nitrate uptake by phosphate was positively correlated with the preconditioning N:P supply ratio and the preconditioning growth rate. For the influence of other factors see the excellent review paper by Cembella et al.[735]

Inorganic phosphate transported across the plasmalemma enters a dynamic intracellular phosphate pool[366,3253,3384,3485] from which it is incorporated into phosphorylated metabolites or stored as luxury phosphorus in vacuoles or in polyphosphate vesicles in microalgae (Figure 4-48).[2886] There are four main organic phosphorus fractions in cells: RNA, DNA, and lipid- and ester-phosphorus.[3384] Phosphorus in DNA is truly segregated while the terminal P groups of ATP turn over very rapidly, in a matter of seconds.[3384] Some of the cytoplasmic phosphate pool may leak back out of the cell and reappear as external phosphate.

There are three major processes by which algae as well as other green plants can incorporate orthophosphate into organic "high-energy" compounds: substrate phosphorylation, oxidative phosphorylation and photophosphorylation.[2636-7] Thus:

$$\text{ADP + orthophosphate} \quad \xrightarrow{\text{energy}} \quad \text{ATP} \qquad (4.50)$$

photo-
oxidative- } phosphorylation
substrate-

By these energy-generating systems adenosine triphosphate (ATP) as the main "high-energy" product is formed by transferring orthophosphate to adenosine diphosphate.[2637] Substrate phosphorylation is coupled directly to the oxidation of the respiratory substrate while the phosphorylation reactions of oxidative phosphorylation are linked to the electron transport system of the mitochondria.[2637] In photophosphorylation light energy is converted into the "energy-rich" phosphate bonds of ATP (see Chapter 2.12.1). The influence of environmental conditions on ATP levels in algal cells has been discussed by Holm-Hansen.[2167]

It has been long known that microalgal cells are capable of "luxury uptake," whereby surplus P is accumulated in excess of that required for immediate growth.[58,2478,3004,4083,4164,4972] Phosphorus-deficient algae of different species commonly possess the ability to incorporate phosphorus extremely rapidly

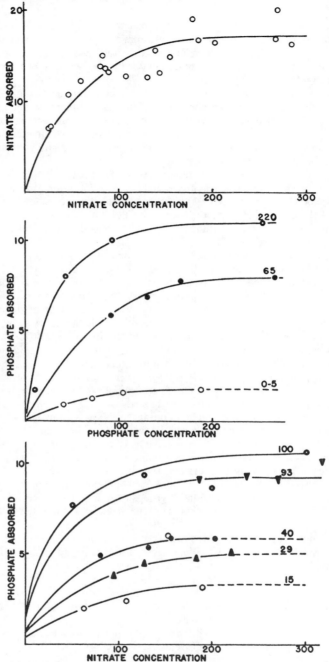

Figure 4-51.   A. The relation between rate of absorption of nitrate by illuminated cultures of *Nitzschia closterium* and the concentration of nitrate in the medium. Abscissa, concentration of nitrate in the medium (gamma N l^{-1}). Ordinate, nitrate absorbed per cell per hour (gamma N × 10^8). B. Relation between the rate of absorption of phosphate by illuminated cultures of *Nitzschia closterium* at 16°C and the concentration of phosphate in the medium, at three different concentrations of nitrate (gamma N l^{-1}). Ordinate, phosphate absorbed per cell per hour (gamma PO$_4$ × 10^8). Abscissa, concentration of phosphate in the medium (gamma PO$_4$ l^{-1}). C. Relation between the rate of absorption by illuminated cultures of *Nitzschia closterium* at 16°C and the concentration of nitrate in the medium at five different phosphate concentrations (gamma PO$_4$ l^{-1}). Ordinate, phosphate absorbed per cell per hour (gamma PO$_4$ × 10^8). Abscissa, concentration of nitrate in the medium (gamma N l^{-1}).[2477] (From Ketchum, 1939, in *Am. J. Bot.,* Vol. 26, p. 402, reprinted with permission of the *American Journal of Botany.*)

if they are provided with this compound.[2637,5284] In most cases the incorporated phosphate exceeds by far the actual requirement of the cells as was noted by early workers.[2478,3004,4379] The excess is built into polyphosphates by the action of polyphosphate kinase:[735,2637]

$$\text{ATP} + (\text{polyphosphate})_n \rightarrow \text{ADP} + (\text{polyphosphate})_{n+1} \tag{4.51}$$

Blum[404] has shown that within one minute of uptake at 25°C over 95% of ^{32}P-labelled phosphate was converted to organic compounds in *Euglena*. Stewart and Alexander[4777] found that in various blue-green algae new polyphosphate bodies were formed within one to two hours of re-adding phosphate to phosphorus-deficient cells. There is some evidence[2669] that polyphosphate synthesis is stimulated by conditions which inhibit ATP synthesis, but on the other hand Simonis[4510] found that light preferentially stimulates uptake into organic compounds.

Polyphosphate is a ubiquitous long-chained linear polymer present in a wide variety of plants and animals, although it is most often found in bacteria and unicellular algae.[1924,2451,4469] Polyphosphates (condensed inorganic phosphates) are normal constituents of algal cells and play an important role in the general metabolism of these organisms, as they do in all organisms.[2637] The natural distribution of these chemical compounds seems to be limited to microorganisms, although there are a few exceptions.[3252,4933] Polyphosphates were first discovered in algae by Sommer and Booth[4624] using *Chlorella*. General reviews of polyphosphates have been given by Schmidt,[4348] Wiame,[5418] Kuhl,[2633-4,2637] Langen,[2705] Harold,[1924] and Bieleski[366] (see also Chapter 2.11.1.3).

Polyphosphate functions primarily as a phosphorus reserve.[4469] It can be formed under several distinct nutritional conditions: (1) restoration of phosphate following a phosphate deficiency;[2335,4471] (2) nutrient imbalance other than phosphorus;[2722,4596,4656,5169] and (3) disturbance of nucleic acid metabolism.[1924] The first detailed investigation to elucidate the conditions of polyphosphate synthesis in algae was carried out by Wintermans.[5462] He found that cells of *Chlorella vulgaris* were able to synthetize polyphosphates from external orthophosphate in the light. The influence of light on polyphosphate synthesis in *Chlorella* was clearly demonstrated by Kuhl.[2635] The stimulation of polyphosphate synthesis by light could be suppressed by inhibitors of photosynthesis, thus indicating the participation of photosynthetic phosphorylation.[2637] Further investigations,[5075-7] however, have demonstrated that polyphosphate formation in the light is influenced by a number of other metabolic conditions as well, including the presence or absence of O_2 or CO_2 and the pH value of the medium. Badour[205] reported that the synthesis of polyphosphates in the light and the synthesis in the dark are markedly influenced by the absence or presence of some cations (K^+, Mg^{2+}, Mn^{2+}) in the surrounding medium.

Rigby et al.[4139] found Ca to stimulate phosphate uptake in *Synechocystis leopoliensis* and considered this element to be a nonspecific activator of an enzyme system for polyphosphate synthesis. The same was noted for *Plectonema boryanum*[4477] and *Anabaena*.[1992] Watanabe et al.[5285] found that during luxury phosphate uptake a raphidophyte *Heterosigma akashiwo* requires Mn for polyphosphate synthesis and that Mn was excreted when the polyphosphate pool was saturated. The exchange of Mn through the cells was rapid, and the authors suggest that charge balance in the cells appears to be preserved by the rapid exchange of other metals in the opposite direction to Mn.

The amount of surplus phosphorus in algae can be separated from essential phosphorus compounds and determined after extracting for 60 minutes with boiling water.[1365,1374] Lin[2853] found that concentrations of a hot-water extractable phosphorus from *Cladophora glomerata* were correlated closely with total dissolved P in ambient Lake Michigan water. Much of it is apparently released as orthophosphate after killing the algae, indicating that it is highly labile.[1375] Most algae store excess phosphate as polyphosphate granules of 30–500 nm diameter,[1924] though some may store ionic phosphate in vacuoles.[1992] The storage compounds are classified into cyclic and linear polyphosphates (Figure 4-52). These two types cannot be easily separated by simple extraction procedures but can be divided into four categories (Figure 4-48): A (acid soluble); B (acid insoluble but soluble in cold alkali at pH 9); C (insoluble in the treatments mentioned before, but extractable with 2N KOH and reprecipitable by neutralizing extract); and D (extractable with 2N KOH but remaining soluble after neutralization).[120,2415,3253-4] Kanai et al.[2415] reported that under normal conditions polyphosphates "A" and "C" are functioning as intermediates transferring phosphate from inorganic orthophosphate to DNA and phosphoprotein. On the other hand, polyphosphates "B" and "D" are essentially P-reservoirs which accumulate in the algal cells in the presence of excess phosphate in the culture medium and are degraded to orthophosphate under P-deficiency. In N limitation, the ratios of fractions A, B, C, and D are quite different from the ratios of P limitations at comparable growth rates. The concentrations of polyphosphate fraction A in N-limited cells are much higher than the levels in P-limited cells, and this fraction becomes more predominant at low growth rates in N limitation.[484]

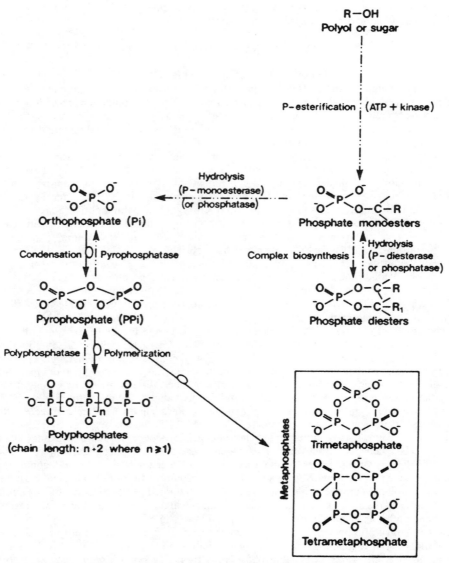

Figure 4-52.   Interconversion of inorganic and organic phosphate storage compounds.[735]   (From
Cembella et al., 1983, in *CRC Crit. Rev. Microbiol.*, Vol. 10, p. 319, reprinted with
permission of CRC Press, Inc., Boca Raton.)

The uptake and storage of a nutrient beyond the immediate metabolic needs of the cells is termed
luxury consumption and presumably provides the alga with a supply of phosphorus when external levels
might otherwise be limiting. It has been suggested, however, that the storage function of polyphosphates is only secondary to their role in regulating the concentration of free phosphate ions in the cell.
The formation of polyphosphates is an important difference between phosphorus metabolism in vascular
plants and algae.[2886]

The induction effect of P_i starvation was shown to increase uptake rates for P-limited cells by one
to two orders of magnitude greater than that for P_i-sufficient cells.[188,786,1775,3756] Intracellular polyphosphate acts as a noncompetitive inhibitor of phosphate uptake in phosphate-limited cultures of the
freshwater unicellular alga *Scenedesmus*.[4083-4] The kinetics could be described by an equation similar
to that for enzyme kinetics under noncompetitive inhibition:[4083]

$$V = V_{max}/(1 + K_s/S) \ (1 + i/K_i) \qquad\qquad (4.52)$$

where V = initial uptake rate, V_{max} = maximum uptake rate at saturating substrate concentration, S = substrate (phosphate ion) concentration, K_s = half-saturation constant for substrate, i = inhibitor concentration, and K_i is the half-saturation constant for inhibitor. Nevertheless, feedback control of P_i uptake as described by Rhee[4083-4] was not evident for certain other species.[637]

Phosphate is concentrated in a striking manner by algal cells, e.g. Krumbholz[2624] reported that phosphates are concentrated by *Euglena* 100,000, *Volvox* 140,000, *Pandorina* 285,000 and *Spirogyra* 850,000 times. Most algae store excess PO_4^{3-} as polyphosphate in cytoplasmic granules of 30–500 nm diameter.[973]

4.3.4.3 Nitrogen/phosphorus ratios

The total phosphorus in a cell fluctuates with changes in the phosphorus supply.[3384] Phytoplankton cells growing in a nutrient-unlimited environment typically show a C:N:P atomic ratio of 106:16:1, the so-called Redfield ratio.[4040-1] Decomposition of this organic matter occurs in the same ratio.[2886] Ryther and Dunstan[4238] reported N:P ratios for coastal phytoplankton between 5:1 and 15:1. Chemostat studies with *Scenedesmus*,[4085] however, suggest that cells are nitrogen-limited up to an N:P ratio of about 30, and are phosphorus-limited above this level (Figure 4-53). The maximum potential for nitrate uptake V_m, is an inverse function of the cellular N:P ratio (Figure 4-54). Under N limitation and for a single N:P ratio in the medium, V_m is an inverse function of both μ and the N-cell quota. Under P-limitation, however, the V_m for nitrate at a single media N:P ratio is independent of the N-cell quota although it is still inversely proportional to μ.[4085] Gerloff et al.[1614] also found that maximum growth of a blue-green alga *Coccochloris peniocystis* occurred in a culture solution with a N:P ratio of approximately 30. Another blue-green alga, *Microcystis aeruginosa*, was found to produce maximum growth in culture solutions with N:P ratios as high as 75:1.[1615] Takamura and Iwakuma[4881] reported internal N:P ratios for phytoplankton and epiphyton from the hypertrophic Lake Kasumigaura in Japan 4.8–38.7 and 6.3–39, respectively. Healey and Hendzel[1995] reported N:P ratio > 20 to indicate P deficiency in several algae in culture. Björnsäter and Wheeler[385] reported tissue N:P 16–24 to optimal for a growth of marine green alga *Ulva fenestrata*. Imbamba[2280] has given mean N:P internal ratios 19.3, 21.0 and 9.9 for benthic green, red and brown algae, respectively. There can be, however, great variation also within the same species as shown by Wallentinus,[5242] who reported N:P internal ratios ranging from 8 to 66 for *Cladophora glomerata*. Benthic marine algae are depleted more in phosphorus and less in nitrogen relative to carbon than phytoplankton, and show an average ratio of 550:30:1.[177]

Wheeler and Björnsäter[5366] found that variations in N:P in five macroalgae from Oregon coastal waters showed a more distinct seasonal pattern than either tissue N or tissue P. The authors suggested

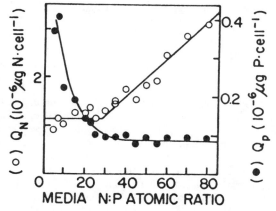

Figure 4-53. Total cell N and P concentration in *Scenedesmus* sp. as a function of N:P in inflow medium. (From Rhee, 1978, in *Limnol. Oceanogr.*, Vol. 23, p. 15, reprinted with permission of the American Society of Limnology and Oceanography.)

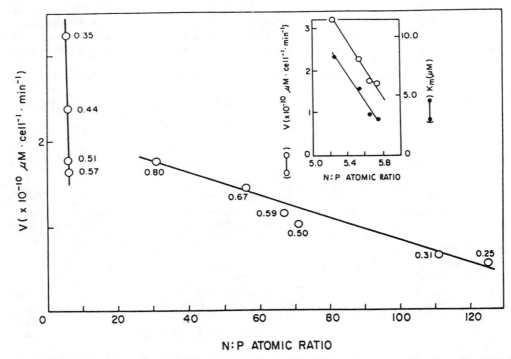

Figure 4-54.   Change in apparent maximum uptake velocity (V) for nitrate as a function of cell N:P in *Scenedesmus* sp.[4085] Values below the ratio 10 were obtained with a N-limited culture and those above it were calculated using data from a P-limited culture.[4083] Numbers next to open circles are relative growth rate (μ/μm). Scale of N:P between 5 and 6 is expanded in insert. Culture conditions for N-limited and P-limited studies were different. (From Rhee, 1978, in *Limnol. Oceanogr.*, Vol. 23, p. 19, reprinted with permission of the American Society of Limnology and Oceanography.)

that tissue N:P ratio for macroalgae may be a good index for evaluating *in situ* nutrient status. The ratios observed were very low, ranging from 0.93 to 2.28 suggesting N limitation.

Large departures from "Redfield" ratios suggest nutrient limitation, and the general response is a decrease in the cell quota of the limiting nutrient. The effects of both phosphorus and nitrogen limitation have been extensively studied in the marine diatom *Thalassiosira pseudonana*.[779,1073,1261,1624,3126,3756,4350,4795]

4.4 SULFUR

4.4.1 Global Cycle

The global sulfur cycle involves exchange between the atmosphere, hydrosphere, pedosphere and lithosphere.[5544] Zehnder and Zinder[5544] and Granat et al.,[1792] in their excellent reviews on the global sulfur cycle, gave the estimates of sulfur fluxes (Figure 4-55) but data in the literature vary.[18,490–1,1486,2295–6,2384,3273] Among major cycle components, the atmosphere is clearly the most dynamic; sulfur in the atmosphere has a mean residence time of only 5.7 days.[5544]

The sulfur cycle bears some similarity to the nitrogen cycle since it also is essentially an eight electron shuttle between the most oxidized valence +6 form (SO_4^{2-}) and the most reduced valence -2 form (S^{2-}). The main reservoirs of SO_4^{2-} are seas and evaporites, whereas those of S^{2-} are sediments, shales, metamorphic, and igneous rocks.[1800] The most important inorganic sulfur compounds in the environment are, according to Zehnder and Zinder:[5544] sulfur trioxide (SO_3), sulfate (H_2SO_4, HSO_4^-,

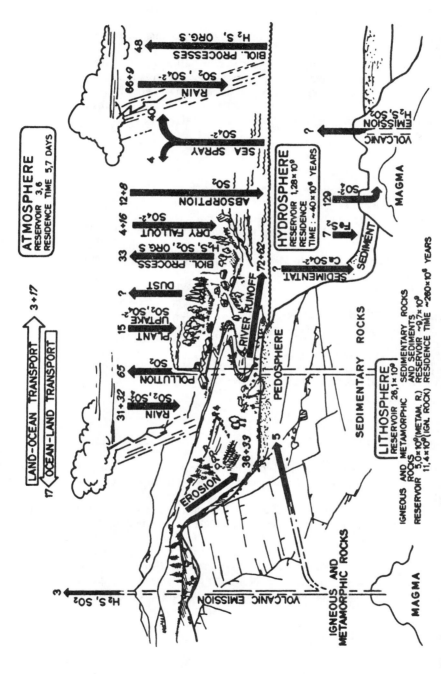

Figure 4-55. The global sulfur cycle. The fluxes shown are given in millions of tonnes sulfur per year (Tg S yr⁻¹). Roman typed numbers denote the transfers as estimated to have prevailed before civilization had a significant influence on the sulfur cycle. The italic numbers give the amounts of what man had added by his various activities.[5544] (From Zehnder and Zinder, 1980, in *The Handbook of Environmental Chemistry, Vol. 1, Part A*, Hutzinger, O., Ed., p. 116, reprinted with permission of Springer-Verlag and the authors.)

SO_4^{2-}), sulfur dioxide (SO_2), sulfite (H_2SO_3, HSO_3^-, SO_3^{2-}), dithionite ($S_2O_4^{2-}$), polythionates ($S_nO_6^{2-}$), thiosulfate ($H_2S_2O_3$, $HS_2O_3^-$, $S_2O_3^{2-}$), elementar sulfur (S_8), disulfide (S_2^{2-}), polysulfides (S_n^{2-}; n = 3–6), and sulfide (H_2S, HS^-, S^{2-}). The biological and biogeochemical sulfur cycles have been reviewed extensively by Trudinger[5033] and Krouse and McCready,[2621] the sulfur soil cycle has been extensively reviewed by Stevenson[4738] and Bohn et al.[419]

The cycling of sulfur is based on the oxidation of H_2S and its ionic forms, biochemical reduction of SO_4^{2-}, and biochemical oxidation and reduction of elemental sulfur (Figure 4-56). Excellent reviews on microbial sulfur transformations have been given by Grant and Long,[1799,1800] Krouse and McCready[2620] and Ralph.[3965]

Sulfur is an important bioelement. A large variety of sulfur-containing compounds are found in living cells:[5544] common amino acids cysteine, methionine; cofactors thiamine, biotin, coenzyme A, lipoic acid, ferredoxin, coenzyme M; sulfate esters chrondroitin sulfate, tyrosin sulfate, choline sulfate; sulfonates cysteic acid, taurine, sulfolipid; and miscellaneous compounds such as dimethyl-ß-propoithetin, phosphatidyl sulfocholine, dimethyl sulfone, elemental sulfur, humin sulfur, diallyl disulfide, propane thial-S-oxide.

4.4.2 Sulfur in the Environment

Earth's crust: 260 mg kg^{-1} (490, 903)
Common minerals: FeS_2 (pyrite), Fe_3S_4 (greigite), (Ni,Fe)S_2 (bravoite), Cu_2S (chalcopyrite), $CuFe_2S_3$ (cubanite), MoS_2 (molybdenite), $CaSO_4$ (anhydride), $CaSO_4 \cdot 2 H_2O$ (gypsum)[491,1800,3287]
Soils: 700 (30–1,600) mg kg^{-1}, up to 90% of soil S may be bound in humus, SO_4^{2-} is a major exchangeable anion in many soils[490-1]
Sediments: 2,200 mg kg^{-1} (491)
Air: 2.3–50 μg m^{-3} (variable)[490-1]
Fresh water: 3.7 (0.2–20) mg L^{-1} (490–1); 25 mg L^{-1} (5218); acid mine drainage up to 63 g L^{-1} (232)
Sea water: 885–905 mg L^{-1} (490–1)
Land plants: 0.34 (0.1–0.9) % DM[490-1]
Freshwater algae: 0.42–1.10% DM (Table 4-9)
Marine algae: 0.8–3.0% DM[491]
Land animals: 0.5% DM[490]
Marine animals: 0.5–1.9% DM[490]

4.4.2.1 Aquatic chemistry

Sources to waters: natural (weathering, volcano emissions, forest fires, microbial decomposition of organic material, sea salt aerosols), anthropogenic emissions (dominate the sulfur cycle in many parts of the world).[3287]

Sulfur exists in natural waters inorganically and organically bound in four oxidation states—S^{2-}, S^0, S^{4+} and S^{6+}.[3815] The dominant sulfur species under the pH and Eh conditions commonly encountered in surface waters are SO_4^{2-}, HSO_4^-, sulfides (H_2S, HS^-, and S^{2-}; the concentration ratio of H_2S and HS^- is approximately 1 around pH 7),[3815] and the elemental sulfur.[5544] SO_4^{2-} together with HCO_3^- and Cl^- are the most frequent anions in natural waters.[3815] In most natural waters, the actual species is variably regulated by the presence of iron; S^{6+} and S^{2-} are the dominant, stable oxidation states, but under reducing conditions, thionates, thiosulfate, polysulfides, and sulfites may be present.[5544]

4.4.3 Sulfur Cycling in Wetlands

In contrast to N and P, there are few studies on S cycling and retention in wetland systems, but available data indicate that most soil S is organic.[715,5421-2,5442] Most studies of S in natural wetland systems have found that carbon-bonded (as opposed to ester-sulfate) organic S is the dominant fraction of the total S soil pool.[252,579,715,5421,5423] Many freshwater wetlands appear to retain S on an annual basis.[252,527,2028]

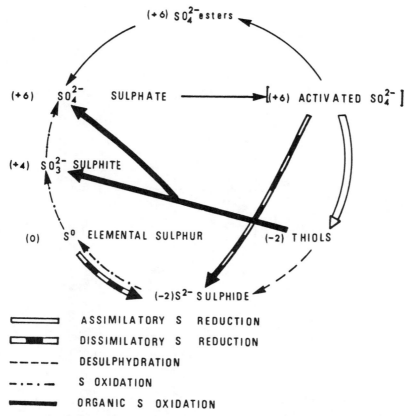

Figure 4-56. The sulfur cycle.[1799] (From Grant and Long, 1981, in *Environmental Microbiology*, p. 148, reprinted with permission of Blackie Academic & Professional, an imprint of Chapman & Hall.)

Table 4-9. Sulfur concentration in algae (% dry mass)

Species	Concentration	Locality	Reference
Freshwater			
Chlorella pyrenoidosa	0.42–0.77	Laboratory	2540
Scenedesmus quadricauda	0.91	Laboratory	2590
Chlorella vulgaris	1.10	Laboratory	1604

The amount of sulfur contained in wetlands varies according to several factors, but especially proximity to the oceans. Some wetland soils, such as mangrove swamps, may consist of well over 1% by dry mass of sulfur, most of it directly contributed by sea water.[715,79,4966] Freshwater wetland soils, which average more organic matter than upland soils, also contain higher quantities of sulfur.[135,715] Although sulfur is emitted in various forms from wetlands, sulfur accumulation also occurs, particularly in wetlands that are accreting peat or muck on a long-term basis. In freshwater sites beyond marine influence, sequestering of organic sulfur is linked to organic matter accumulation.[135]

Sulfur transformations are biologically mediated and, like P and N transformations, are affected by redox and pH interactions.[1313] Major transformations in oxidized environments are assimilatory sulfate reduction, inorganic sulfide and elemental sulfur oxidation, and mineralization of organic S to inorganic SO_4^{2-} (Figure 4-57). When the anaerobic oxidants—nitrate, manganese (IV) oxide and ferric oxide—

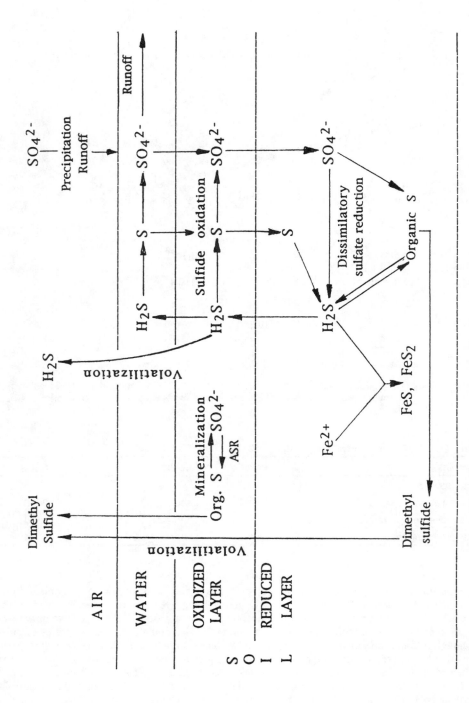

Figure 4-57. Sulfur transformations in wetlands.[1313]   ASR = assimilatory sulfate reduction.

have all been depleted, the reduction of sulfate may start.[2679] Under reducing conditions, dissimilatory sulfate reduction transforms SO_4^{2-} to H_2S (Figure 4-57) during respiration by several genera of strictly anaerobic bacteria:[1555,2681,3246,3778,3838,5224,5544]

$$SO_4^{2-} + lactate \rightarrow H_2S + acetate + CO_2 \qquad (4.53)$$

$$SO_4^{2-} + acetate \rightarrow H_2S + CO_2 \qquad (4.54)$$

Two groups of sulfate-reducing bacteria can be distinguished: a group oxidizing organic compounds to the level of fatty acids and a group oxidizing organic matter completely to carbon dioxide. The latter group seems to be restricted to marine sediments, whereas the first group also occurs in fresh-water environments.[2679] The known strains of microorganisms which carry out sulfate reduction belong to three different genera: *Desulfovibrio, Desulfotomaculum* and *Desulfuromonas*.[5544] Different genera of sulfate-reducing bacteria can coexist at one location.[2680] In neutral to alkaline soils, concentrations of SO_4^{2-} as high as 1,500 mg.l^{-1} may be reduced to zero within some weeks of submergence.[3418]

Sulfate respiration has been shown to be responsible for significant amounts of organic matter decomposition particularly in marine environments where high levels of SO_4^{2-} occur.[4032] Sulfate reduction was identified as the major form of respiration in salt marsh sediments in both Georgia and Massachusetts.[2224] In freshwater environments, sulfate respiration apparently plays a lesser, but important role, in organic matter decomposition.[4032]

The H_2S formed by dissimilatory sulfate reduction can be released to the atmosphere or react with organic matter providing another pathway for converting inorganic S to organic S.[579,580,716] With SO_4^{2-} reduction and sufficient Fe, iron sulfides (FeS, FeS_2) can form; pyrite (FeS_2) formation requires alternating (either temporally or spatially) anaerobiosis with limited aeration.[5108] The reductive reactions in the sulfur cycle have been extensively reviewed by Krouse and McCready.[2620]

Sulfides can be oxidized by both chemoautotrophic and photosynthetic microorganisms to elemental sulfur and sulfates in the aerobic zones of some wetland soils. Certain species of *Thiobacillus* obtain energy from the oxidation of hydrogen sulfide to sulfur, while other species in this genus can further oxidize elemental sulfur to sulfate.[3246] These reactions are summarized as follows:[3246]

$$2 H_2S + O_2 \rightarrow 2 S + 2 H_2O + energy \qquad (4.55)$$

and

$$2 S + 3 O_2 + 2 H_2O \rightarrow 2 H_2SO_4 + energy \qquad (4.56)$$

Thiobacillus and *Beggiatoa* species are the best known representatives of the colorless sulfide-oxidizing bacteria.[2369] Since most of these bacteria are dependent on oxygenic sulfide oxidation for generation of energy, the activity of the colorless sulfide-oxidizing bacteria is more or less restricted to anoxic-oxic interface environments.[2679] The oxidative reactions in the sulfur cycle have been extensively summarized by Ralph.[3965]

Photosynthetic bacteria such as the purple (e.g. *Chromatium*) and green (*Chlorobium*) sulfur bacteria found on salt marshes and mud flats are capable of producing organic matter in the presence of light according to the following equation:[3246,5033,5544]

$$CO_2 + 2 H_2S \xrightarrow{light} CH_2O + S^\circ + H_2O \qquad (4.57)$$

This reaction uses hydrogen sulfide as an electron donor rather than the H_2O used in the more traditional photosynthesis equation, but otherwise the process is the same. This reaction often takes place under anaerobic conditions where hydrogen sulfide is abundant but at the surface of sediments where sunlight is also available.[3246] Zehnder and Zinder[5544] pointed out that elemental sulfur is often an intermediate, while the end product is H_2SO_4.

Despite the small size of the inorganic pool, this fraction is the most important for S cycling, retention, and mobility. Fluxes through the inorganic pool dominate S cycling in wetlands with high SO_4^{2-} inputs.[1313] Wieder and Lang[5422] calculated that 3.5 to 4 times as much inorganic S was processed (through alternating SO_4^{2-} reduction and sulfide/sulfur oxidation) as compared to the organic pool. This has important implications for wetland S cycles because S inputs are primarily SO_4^{2-} from atmospheric deposition and either natural or amended hydrologic sources. Sulfate retention by aerobic, mineral soils is dominated by the same adsorption mechanisms involved in PO_4^{3-} retention.[1313] However, adsorbed SO_4^{2-} is displaced by PO_4^{3-} on the exchange sites, but PO_4^{3-} is not displaced by SO_4^{2-}.[2347,3645]

Gaseous losses from wetlands have been studied only rarely.[728] It seems, however, that H_2S may not be the primary biogenic sulfide gas; the organic gases methyl sulfide and dimethyl sulfide can be of equal or greater importance.[2620]

The sulphur cycle is interesting not because S has been reported to limit plant growth in marshes, but because of its important role in energy transfer. This is a new and still not fully understood role. When oxygen and nitrate are depleted in flooded soils, sulfate can act as a terminal electron acceptor and is reduced to sulfide in the process.[1770] (This gives the marsh its characteristic rotten egg odor.) In anoxic salt marshes sulfate is a major electron acceptor. In fresh marshes where the supply of sulfate is limited, C is reduced to methane instead. The sulfide radical is a form of stored energy that can be tapped by S bacteria in the presence of oxygen or other oxidants.[2225]

In the northeastern Atlantic coast marsh the energy flow through reduced inorganic S compounds was equivalent to 70% of the net belowground primary productivity of the dominant grasses. Apparently most of the stored sulfides are reoxidized annually, by oxygen diffusing into the substrate from the marsh grass roots,[2223] but there is a possibility of soluble sulfides being flushed from the marsh to become a source of biological energy elsewhere. In the marsh cited above, Howarth et al.[2225] estimated that 2.5 to 3.5 moles of reduced S m^{-2} yr^{-1} are exported by pore water exchange with adjacent creeks. This amounts to about 3–7% of the S reduced in the sediment, and as much as 20–40% of net aboveground production.

4.4.3.1 Sulfur in wetland plants

Herbaceous plants

Concentration in % dry mass (see Appendix, Table 7-12):
emergent species: aboveground: 0.08–0.68
floating-leaved species: aboveground: 0.11–0.32
submerged species: aboveground: 0.16–0.92

Standing stock (see Appendix, Table 7-13) ranges in emergent species between 0.09 and 4.04 g S m^{-2}. Sulfur uptake rates in aquatic and wetland plants (see Appendix, Table 7-14) range between 18 and 65 g S m^{-2} yr^{-1} and 7.7 and 23.3 mg S m^{-2} d^{-1}.

4.4.4 The Role of Sulfur in Algal Nutrition

The role of sulfur in biology has been reviewed by Anderson,[98] metabolism of sulfur in plants has been reviewed by Anderson,[98] Thompson[4958] and Wilson.[5552] Nutrition and utilization of sulfur in algae have been reviewed by Schiff,[4330,4334-5] O'Kelley,[3530] Healey,[1991] Raven,[4005] Ikawa et al.[2273] and Møller and Evans.[3274]

Sulfur is generally present in small quantity in all plant cells but is probably not a limiting factor for many algae under normal conditions. Sulfur is incorporated into numerous organic compounds and sulfates are present in the vacuoles. There is an evidence for the connection between divalent sulfur compounds and the assimilation of silica in diatoms.[4209] Sulfur is required by algae for both autotrophic and heterotrophic growth (Figure 4-58). As compared with other macronutrients sulfur uptake and metabolism in algae have been studied only scarcely. In fact, major studies on sulfur assimilation by algae were done more than 20 years ago and present research in this field equals nearly nothing. The sulfur concentration in algae is summarized in Table 4-9.

Since most algae can supply all of their sulfur requirement by reduction of sulfate,[890,2133-7,2665,3530,4329-30,5306] the most abundant form of sulfur in nature, few studies have been made involving the use of other sources of sulfur.[1627,3530,4958] Sulfite supports the growth of two blue-green algae[3866] and thiosulfate the growth of *Chlorella pyrenoidosa*[2136] at rates similar to sulfate. Both are good sources for *Porphyridium cruentum*.[2368] Among the amino acids, methionine and cysteine can act as a sole S source for the growth of *Chlorella pyrenoidosa*.[2136] Methionine can also provide S to several strains of *Chlorella*,[4460] *Anacystis nidulans* and *Anabaena variabilis*.[3866] Shrift[4455] found that *Chlorella vulgaris* was able to utilize either sulfate, D-methionine, or L-methionine as the only source of sulfur for growth. Cystine

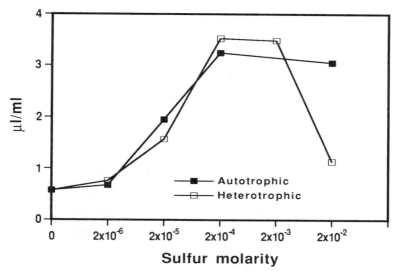

Figure 4-58. Growth of *Chlorella pyrenoidosa* at different concentrations of sulfur. Autotrophic: 2 days growth from 10 μL washed cells. Heterotrophic: 35 days growth from 1 loopful cells.[1305] (From Eyster et al., 1958, in *Trace Elements*, Lamb et al., Eds., p. 164, reprinted with permission of Academic Press.)

did not support growth under the conditions of the experiments. Also Shrift[4458] reported utilization of L-methionine by *Chlorella vulgaris*. The inability of methionine to support the growth of some strains of *Chlorella* is not due to inability to take up the amino acid, but to metabolize it to all required S compounds.[4460] Many studies on algae indicate a special requirement for sulfur in cell division.[3530]

In the literature, there are not many data on sulfate uptake rates. Coughlan[890] found $K_s = 6.9 \times 10^{-2}$ mM in *Fucus serratus*, Jackson and McCandless[2306] found $K_s = 3$ mM in *Chondrus crispus*. Sulfate uptake by the unicellular marine red alga *Rhodella maculata* was reported to be biphasic with K_s of 22 mM for the low affinity system and 63 μM for the high affinity system.[3222] Uptake of sulfate by both *Chlorella pyrenoidosa*[5306] and *Scenedesmus* sp.[2667] is stimulated by light. As with N assimilation, light could be acting by providing energy via photophosphorylation, reductant, or C skeletons.[1991] Sensitivity of sulfate uptake by both these algae to uncouplers shows an energy requirement for uptake and/or reduction. In the case of *Scenedesmus*, stimulation by light is greater in the presence than the absence of CO_2, showing that here the provision of C skeletons is a part of the explanation.

Shrift[4455-7] showed that selenate is a competitive antagonist of sulfate utilization for growth in *Chlorella vulgaris*, inhibiting both the absorption of this ion and subsequent sulfur metabolism. Although methionine partially reversed selenate inhibition, it did not do so competitively. Selenomethionine, on the other hand, competed with the absorption and utilization of methionine but did not competitively interfere with sulfate uptake.[4456] Coughlan[890] reported that in addition to selenate, sulfate uptake was also inhibited by molybdate, tungstate, and especially chromate; the kinetics of this inhibition were not worked out.

Before sulfate can be incorporated into various compounds, it must be activated, since it is a relatively unreactive compound.[4337] Cell-free extracts of *Chlorella pyrenoidosa* reduce labelled sulfate to two major labelled products.[4336] One is thiosulfate with the label virtually confined to the sulfite-S. The sulfide-S is contributed by cellular protein, sulfate presumably being bound to the thiol group of a carrier protein before reduction[2788] (Figure 4-59). The other major product is adenosine-3′-phosphoadenosine-5′-phosphosulfate (PAPS). The enzyme ATP sulfurylase catalyzes the substitution of SO_4^{2-} for two of the phosphate groups of ATP, to form adenosine-5′-phosphosulfate (APS) (Figures 4-59 and 4-60). APS can have another phosphate added (catalyzed by APS kinase) from another ATP to form PAPS (Figure 4-60) which is believed to be the starting point for sulfate ester formation in many systems and for sulfate reduction (like nitrogen, sulfur is incorporated into proteins in its most reduced form).[1799,2886] Conversion of PAPS to thiosulfate requires ATP, Mg, any of several thiol

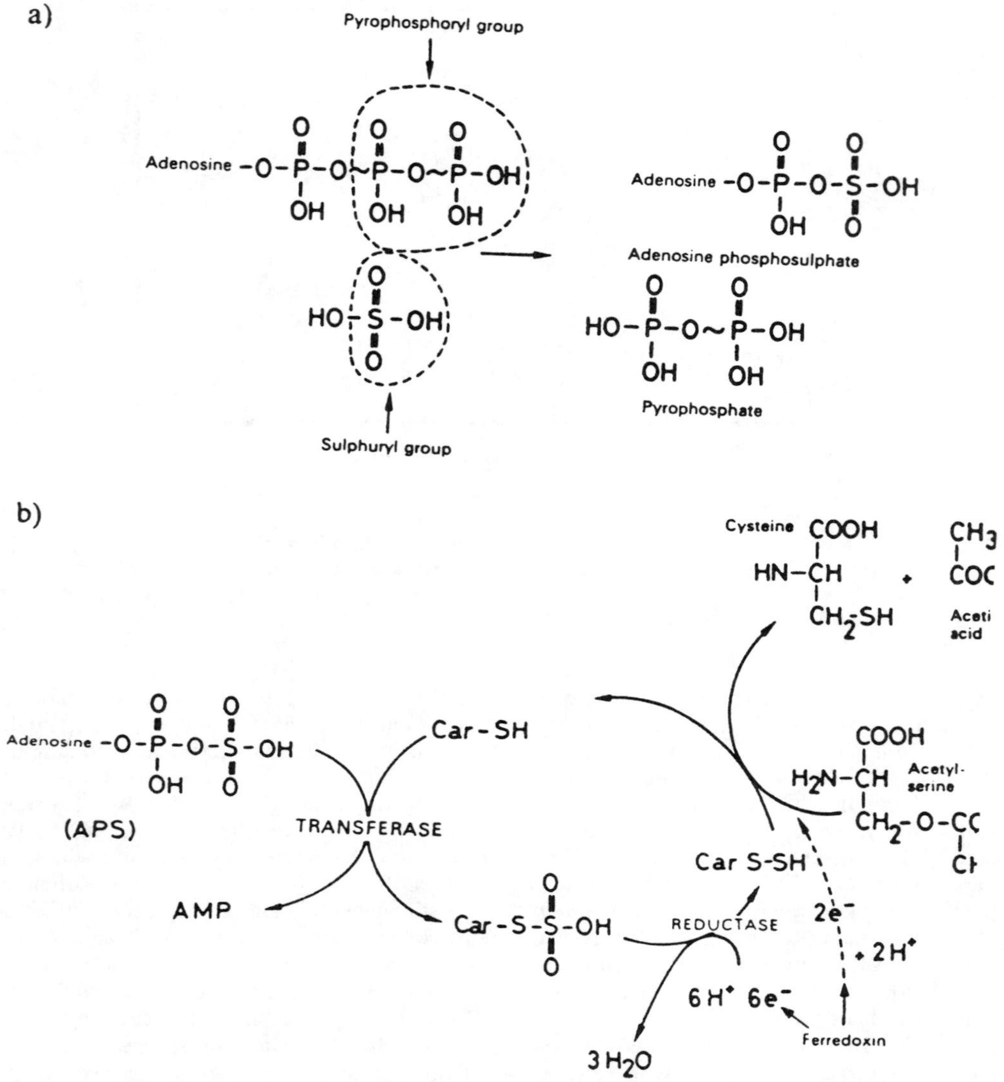

a)

b)

Figure 4-59. Simplified scheme showing the sulfate reduction pathway. a) ATP sulfurylase reaction producing adenosine phosphosulfate. b) Transfer of the sulfhydryl group to the car-SH complex (car = carrier) and reduction steps.[3199] (From Mengel and Kirkby, 1982, in *Principles of Plant Nutrition,* p. 181, reprinted with permission of International Potash Institute, Basel.)

compounds, and two enzymatic components, fractions A and S.[2134] Fraction S appears to be on a side reaction from the mainstream of sulfate reduction to sulfide.[2134] Conversion of PAPS to thiosulfate involves APS as an intermediate and is mediated by fraction A[5036] (Figure 4-60).

A cell-free sulfate-reducing system from a strain of *Chlorella pyrenoidosa* shows maximal activity when fortified with ATP, an ATP-generating system, NADP, an NADP-reducing system and MgCl$_2$.[4336] Thiosulfate is its major product;[2786-8] S-adenosyl methionine,[4329] PAPS,[2135] and related compounds[4332] are also formed.

A large part of the sulfur in most algae is incorporated into protein. Two sulfur-containing amino acids, cysteine and methionine, are very important in maintaining the three-dimensional configuration

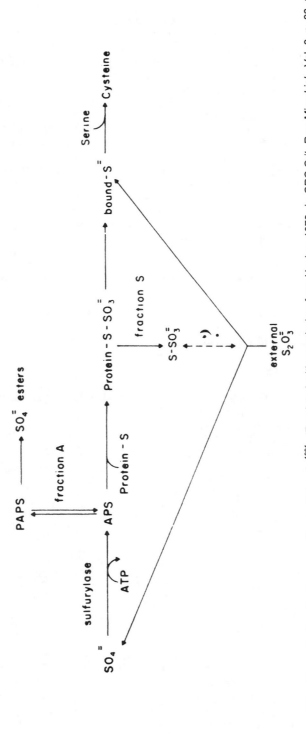

Figure 4-60.   The pathway of sulfate reduction in *Chlorella*.[1991]   (Reprinted with permission from Healey, 1973, in *CRC Crit. Rev. Microbiol.*, Vol. 3, p. 83. Copyright CRC Press, Inc., Boca Raton.)

of proteins through sulfur bridges.[2886] In addition to its occurrence in the amino acids methionine and cysteine, sulfur may also exist in considerable quantity in algae as a component of other materials, such as sulfolipids,[1869,2461] taurine and derivatives (sulfur at the sulfite level of reduction),[2858] and sulfonium compounds.[4331] Sulfur is also part of the biologically important molecules biotin and thiamine and coenzyme A.

Another important group of sulfur-containing compounds in the seaweeds are the sulfated polysaccharides including agar, carageenan, furcellaran, porphyran, L-fucose and several other polysaccharides in the Rhodophyceae, Phaeophyceae and Chlorophyceae.[3122,3498,3750] Sulfated polysaccharides are important in thallus rigidity (e.g. carageenan in red algae) and adhesion (e.g. fucoidan in brown algae) (for more information about sulfated polysaccharides see Chapters 2.4.1.2 and 2.14.5).

Hodson et al.[2136] reported that *Chlorella pyrenoidosa* utilizes thiosulfate for growth as effectively as sulfate, and more effectively than a variety of organic sulfur compounds containing sulfur in various oxidation states. Labels from both sulfur atoms of thiosulfate and from sulfate are incorporated into the cysteine, homocysteine, and glutathione of the soluble pools, and into the methionine and cysteine of protein in the insoluble fraction. Exogenous thiosulfate undergoes early dismutation in which the SO_3-sulfur is preferentially oxidized, and the SH-sulfur is preferentially incorporated in a reduced state.

Incorporation of sulfur from the sulfate in the medium into normal cells of *Scenedesmus* was enhanced by light, relatively most in the case of lipid S and least in the inorganic sulfate fraction; the effects of light were, generally, increased by the presence of CO_2 and nitrogen salts.[2667] The author also suggested that lipid S is formed as a "sink," when a step between sulfite and -SH becomes increasingly rate-limiting in the overall reduction of sulfate. Furthermore, incorporation as SO_4^{2-} and as lipid S may be regulated by more or less independent processes.

Sulfate may tend to be excluded by some algae, but in others it may accumulate.[3530] Marine species *Valonia* and *Halicystis* exclude sulfate almost completely. Freshwater species of *Nitella* accumulate appreciable amounts of sulfate, in addition to 1% of the dry mass in the form of nonvacuolar sulfur.[2115]

4.5 POTASSIUM

4.5.1 Potassium in the Environment

Earth's crust: 21,000 mg kg^{-1} (490)
Common minerals: KCl, micas, feldspars[491]
Soils: 14,000 (80–37,000) mg kg^{-1}, a major exchangeable cation in all but the most alkaline soils, fixed by some clay materials[490–1]
Sediments: 20,000 mg kg^{-1} (491)
Air: 0.3–40,000 ng m^{-3} (491)
Fresh water: 2.2 (0.5–10) mg L^{-1} (491)
Sea water: 380–399 mg L^{-1} (490–1)
Land plants: 1.4 (0.5–3.4) % DM[490–1]
Freshwater algae: 0.04–3.91% DM (Table 4-10)
Marine algae: 3.2–5.2% DM,[491] 1.22–7.45% DM (Table 4-10)
Land animals: 0.74% DM[490]
Marine animals: 0.5–3.0% DM[490]

Sources to waters: natural weathering of alkaline feldspars e.g. $KAlSi_3O_8$ (microcline and orthoclase) and $K(Si_3Al)Al_2O_{10}(OH)_2$ (muscovite) or other minerals, e.g. $KFe_3(OH)_6(SO_4)_2$ (jarosite).[3705,3815]

Aquatic chemistry: potassium occurs in waters primarily as simple cation K^+. In more mineralized waters, ionic associations with HCO_3^- and SO_4^{2-} may be present.[3815]

4.5.2 Potassium in Wetlands

Very limited information is available on the transformation of potassium (as well as calcium and magnesium) under waterlogged situations. Singh and Ram[4517] found an increase in exchangeable potassium upon continuous submergence. Other information about potassium in wetland soils has been given by Mohanty and Patnaik[3265] and Mohanty and Dash.[3266] Klopatek[2537] described the potassium flow through the *Scirpus fluviatilis* stand in Wisconsin (Figure 4-61). The potassium leaching during the decomposition of wetlands plants was described by Boyd[503] (see Figure 3-11), Davis and van der

Table 4-10. Potassium concentration in algae (% dry mass)

Species	Concentration	Locality	Reference
Freshwater			
Scenedesmus quadricauda	0.15–0.38	Laboratory	1612
Chlamydomonas geitleri	0.30–0.50	Cultivation (wastewater)	5144
Microcystis aeruginosa	0.21–0.55	Laboratory	1612
Nostoc muscorum	0.51–0.99	Laboratory	1612
Scenedesmus acutus	0.20–1.30	Cultivation (wastewater)	5144
Chlorella pyrenoidosa	0.04–1.44	Laboratory	4378
Chlorella pyrenoidosa	0.11–1.44	Laboratory	1612
Chlorella vulgaris	1.50	Laboratory	1604
Cladophora glomerata	0.76–1.80	Lake Michigan	20
Chara vulgaris	1.61–2.25	Czech Republic	1207
Enteromorpha prolifera	1.84–2.32	Forfar Loch, Scotland	2107
Filamentous algae[a]	0.06–2.52	Nesyt Pond, Czech Republic	2843
Stigeoclonium tenue	0.44–2.57	Laboratory	1216
Draparnaldia plumosa	0.52–3.91	Laboratory	1216
Marine			
Ascophyllum nodosum	2.31	Canada, Nova Scotia	5528
Ulva lactuca	2.49	Canada, Nova Scotia	5528
Fucus vesiculosus	3.10	Canada, Nova Scotia	5528
Ulva lactuca	1.22–3.14	Hong Kong coast	2111
Chondrus crispus	3.38	Canada, Nova Scotia	5528
Phyllophora membranifolia	3.38	Canada, Nova Scotia	5528
Spongomorpha arcta	4.00	Canada, Nova Scotia	5528
Laminaria longicruris	4.67	Canada, Nova Scotia	5528
Laminaria digitata	4.95	Canada, Nova Scotia	5528
Ahnfeltia plicata	5.50	Canada, Nova Scotia	5528
Halosaccion ramentaceum	5.50	Canada, Nova Scotia	5528
Rhodymenia pallmata	7.11	Canada, Nova Scotia	5528
Phaeodactylum tricornutum	2.18–7.45	Laboratory	1987

[a] *Spirogyra* sp., *Oedogonium* sp., *Cladophora fracta, Enteromorpha intestinalis*

Valk[992] (see Figure 3-15), Chamie and Richardson,[744] Gopal[1743] and Kulshreshtha and Gopal.[2642] Potassium budgets for various wetlands have been reported by e.g. Richardson et al.[4115] and Proctor.[3916]

4.5.2.1 Potassium in wetland plants

Herbaceous plants

Concentration in % dry mass (see Appendix, Table 7-15):
emergent species: aboveground: 0.07–5.86, belowground: 0.39–2.45
floating-leaved species (FD): aboveground: 4.06, belowground: 2.94–3.35
submerged species: 0.28–7.4 (aboveground)
floating species: 2.33–4.0
bryophytes: 0.23–10.3
Standing stock (g K m^{-2}) (see Appendix, Table 7-16):
emergent species: aboveground: 0.1–7.5, belowground: 2.54–67.1
floating-leaved species: aboveground: 0.50–6.2

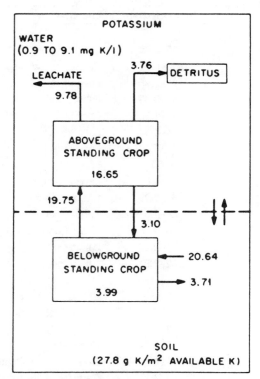

Figure 4-61.   Flow of potassium through a *Scirpus fluviatilis* stand. Flows in g m⁻² yr⁻¹ and compartments in g m⁻² in standing crop.[2537]   (From Klopatek, 1978, in *Freshwater Wetlands:
Ecological Processes and Management Potentials,* Good, R.E., et al., Eds., p. 207,
reprinted with permission of Academic Press and the author.)

submerged species: 1.7–9.9
floating species: 44.1–165

Potassium uptake rates in aquatic and wetlands plants (see Appendix, Table 7-17) range between 252 and 457 g K m⁻² yr⁻¹ and between 59.3 and 333 mg K m⁻² d⁻¹.

4.5.3 Potassium in Algal Nutrition

Potassium is present in many algae in high concentrations relative to the external medium. Its functions include osmotic regulation and the maintenance of the electrochemical environment of the algal cells. It is also a cofactor for variety of enzymes. Potassium is known to be a highly mobile element which gets readily distributed during active growth.[3390]

Potassium is a requirement for algae tested and under low potassium conditions growth and photosynthesis are low and respiration high. Potassium plays an important role as an enzyme activator;[490,3429,3530] many protein synthesis enzymes do not act efficiently in the absence of K⁺, but the way in which K⁺ binds to the enzymes and affects them is not well understood. It is known to bind ionically to pyruvate kinase, which is essential in respiration and carbohydrate metabolism.[361]

Providing sufficient potassium is available the ratio of K to Na within the cell is independent of the ratio in the medium.[4378] Allen[65] reported very low requirements of blue-green algae for potassium. Gerloff and Fishbeck[1612] found requirements for K in 6 algal species varied from critical concentrations lower than the average values established for higher plants (0.25–0.5% dry mass) to values equal to or in excess of higher plant averages (0.8–2.4% dry mass). Gerloff and Krombholz[1611] reported a critical concentration for K to be 10 mg g⁻¹, i.e. 1% dry mass. The potassium concentration in algal biomass is given in Table 4-10.

Sodium may replace potassium, at least in part.[64,1240] The replacement of potassium by rubidium has been reported for a number of algal species.[248,2458,3805] In *Ankistrodesmus braunii*, replacement of K by Rb was considered to be partial at first and then complete following mutative adaptation.[2457] Osretkar and Krauss,[3571] however, found that Rb can not substitute for K in *Chlorella pyrenoidosa*. *Chlamydomonas reinhardtii* continued to grow in the rubidium replacement but lost motility.[248]

In the dark, the inflow of potassium ions to the alga *Hydrodictyon reticulatum* is reduced and at the same time its characteristic selectivity pattern $Li^+ > Cs^+ > Rb^+ > K^+ > Na^+$ is changed into $Rb^+ > Cs^+ > K^+ > Li^+ > Na^+$. This observation suggests that the higher uptake of potassium ions in illuminated cells is not a simple result of more energy available, but rather that a new transport mechanism, with a profoundly different selectivity pattern (preferentially lithium selective), is switched on by the light.[3417]

The uptake of copper by living cells of *Chlorella vulgaris* is associated with the appearance of cellular K in the external medium.[3119] The leakage of K was not a direct exchange for copper. The authors considered that K is released due to a graded response of a barrier, normally of low permeability, to increasing amounts of bound copper.

Schaedle and Jacobson[4314] reported that limited capacity of *Chlorella* to accumulate K could be explained on the basis of the limited ability of the cell to create new negatively charged sites, possibly organic acid anions. Bowen[490] reported concentration factor (CF) for the marine alga *Valonia* about 42. Cushing and Ranticelli[940] found CF for phytoplankton of Columbia River in Washington in the range of 280 to > 3,100. Cushing[939] reported CF for phytoplankton of the same river 9,000.

4.6 MAGNESIUM

4.6.1 Magnesium in the Environment

Earth's crust: 23,000 mg kg^{-1} (490)
Common minerals: $MgCO_3$(magnesite), $CaCO_3 . MgCO_3$(dolomite), $MgSiO_3$(enstatite), $(Mg,Fe)SiO_3$
    (hypersthene), Mg_2SiO_4 (forsterite), $MgSO_4 . 7 H_2O$ (epsomite)[3705,3815]
Soils: 5,000 (400–9,000) mg kg^{-1}, the second most common exchangeable cation in most soils[490-1]
Sediments: 14,000 mg kg^{-1} (491)
Air: 1–11,000 ng m^{-3} (491)
Fresh water: 4.0 (0.4–6) mg L^{-1} (491); 10 mg L^{-1} (5218)
Sea water: 1,290–1,350 mg L^{-1} (490–1)
Land plants: 0.32 (0.1–0.9) % DM[490-1]
Freshwater algae: 0.02–2.52% DM (Table 4-11)
Marine algae: 0.64–2.0% DM,[491] 4.92–7.81% DM (Table 4-11)
Land animals: 0.1% DM[490]
Marine animals: 0.5% DM[490]
    Sources to waters: natural weathering (major source).
    Aquatic chemistry: Magnesium occurs in waters primarily as simple cation Mg^{2+}. Ionic pairs $[MgHCO_3]^+$, $[MgCO_3 (aq)]^0$, $[MgSO_4]^0$ and $[MgOH]^+$ may be present in low concentrations. In natural surface waters these ionic pairs form less than 5% of the total magnesium concentration.[3815]

4.6.2 Magnesium in Wetland Plants

Concentration in dry mass (mg kg^{-1}) (see Appendix, Table 7-18):
emergent species: aboveground: 600–61,000, belowground: 600–7,100
floating-leaved species: aboveground: 2,010–58,900
submerged species: aboveground: 900–41,400
floating species: 4,900–87,000
bryophytes: 470–3,730
Standing stock (g Mg m^{-2}) (see Appendix, Table 7-19):
emergent species: aboveground: 0.003–16.8, belowground: 0.3–11.7
floating-leaved species: aboveground: 0.082–5.36
submerged species: aboveground: 0.3–5.87

Table 4-11. Magnesium concentration in algae (% dry mass)

Species	Concentration	Locality	Reference
Freshwater			
Scenedesmus quadricauda	0.03–0.08	Laboratory	1612
Chlamydomonas geitleri	0.20–0.30	Cultivation (wastewater)	5144
Chara vulgaris	0.30–0.33	Czech Republic	1207
Chlorella pyrenoidosa	0.03–0.37	Laboratory	1612
Nostoc muscorum	0.18–0.46	Laboratory	1612
Stigeoclonium tenue	0.09–0.47	Laboratory	1612
Chlorella vulgaris	0.50	Laboratory	1604
Draparnaldia plumosa	0.02–0.52	Laboratory	1612
Enteromorpha prolifera	0.60–0.68	Forfar Loch, Scotland	2107
Scenedesmus acutus	0.30–0.70	Cultivation (wastewater)	5144
Cladophora glomerata	0.41–1.48	Lake Michigan	20
Chlorella pyrenoidosa	0.26–1.51	Laboratory	4377
Microcystis aeruginosa	0.17–1.56	Laboratory	1612
Filamentous algae[a]	0.68–2.52	Nesyt Pond, Czech Republic	2843
Marine			
Phaeodactlym tricornutum	4.92–7.81	Laboratory	1987

[a] Spirogyra sp., Oedogonium sp., Cladophora fracta, Enteromorpha intestinalis

Magnesium uptake rates in aquatic and wetland plants (see Appendix, Table 7-20) range between 31 and 80 g Mg m^{-2} yr^{-1} and between 2.3–33.8 mg Mg m^{-2} d^{-1}. The changes in magnesium content during decomposition of wetland plants have been summarized by Gopal.[1743]

4.6.3 Magnesium in Algal Nutrition

Magnesium, since it is a constituent of chlorophyll (see Figure 2-9), is obviously an absolute requirement for pigmented algae of all groups and is also necessary for the formation of catalase. Magnesium is an essential cofactor or activator in many reactions, such as nitrate reduction, sulfate reduction, and phosphate transfers (except phosphorylases). It is also important in several carboxylation and decarboxylation reactions, including the first step of carbon fixation, where the enzyme RuBPCO attaches CO_2 to RUBP. Magnesium also activates enzymes[490,3429,3646,4370] involved in nucleid acid synthesis, and binds together the subunits of ribosomes. Magnesium markedly stimulated the Hill reaction in cell-free preparations from *Anabaena variabilis*.[4845] The several means by which Mg may act is summarized by Bidwell:[361] 1) it may link enzyme and substrate together, as, for example, in reactions involving phosphate transfer from ATP; 2) it may alter the equilibrium constant of a reaction by binding with the product, as in certain kinase reactions; 3) it may act by complexing with an enzyme inhibitor; 4) it can form metalloporphyrins, such as chlorophyll; and 5) it can play a role in binding charged polysaccharide chains to one another, since it is a divalent cation.

In comparison with calcium, magnesium seems to be required by microorganisms in relatively large amounts.[3429,4209] Gerloff and Fishbeck[1612] found critical concentrations for 6 algal species equal or only slightly less than in higher plants (0.15–0.30%) with the exception of *Scenedesmus quadricauda* (0.05%). The concentration of magnesium in algal biomass is given in Table 4-11.

Finkle and Appleman[1350] found that Mg deficiency interrupted cell multiplication. The Mg-deficient cells were up to 20-fold larger in volume than those grown in cultures with sufficient Mg. The increases in cell size were paralleled by proportional increases in N-content and dry mass. Finkle and Appleman[1349] demonstrated the cessation of chlorophyll synthesis in Mg-deficient medium. *Chlorella* cells deprived of magnesium become chlorotic, enlarged and extensively vacuolated.[4069] Magnesium-deficient algae can exhibit a number of metabolic disturbances; nitrogen metabolism can be disturbed and there can be a temporary accumulation of carbohydrate material;[3810] an abnormally high quantity of labile phosphate may be produced. Net synthesis of RNA may stop immediately following magnesium withdrawal from a culture while protein synthesis remains unaffected for several hours;[3530] soluble nitrogen compounds including uracil, orotic acid and hypoxanthine may accumulate, along with carbohydrates.[1550]

5.1 CALCIUM

5.1.1 Calcium in the Environment

Earth's crust: 41,000 mg kg^{-1} (490)
Common minerals: $CaCO_3$ (calcite and aragonite), $CaCO_3 . MgCO_3$ (dolomite), $CaSO_4$ (anhydride), $CaSO_4 . 2 H_2O$ (gypsum), $CaAl_2Si_2O_8$ (anorthite), $CaMgSi_2O_6$ (diopside)[3705]
Soils: 13,700–15,000 (7,000–500,000) mg kg^{-1}, highest in limestone soils, lowest in acid soils, the main exchangeable cation in soils of pH 5–8[490–1]
Sediments: 66,000 mg kg^{-1} (491)
Air: 0.5–7,000 ng m^{-3} (491)
Fresh water: 15 (2–120) mg L^{-1} (490–1,5218)
Sea water: 400–412 mg L^{-1} (490–1)
Land plants: 1.8 (0.3–1.4) % DM[490–1]
Freshwater algae: 0–22.6% DM, high values for the Charophyceae (Table 5-1)
Marine algae: 0.4–2.4% DM;[491] 0.25–2.81% DM (Table 5-1), up to 84% DM in calcified species[490,4688]
Land animals: 0.02–8.5% DM[490]
Marine animals: 0.15–2% DM[490]

Sources to freshwater: natural weathering of minerals, e.g. anorthite, industrial wastewaters, especially those where acids are neutralized with lime or limestone.[3815]

Aquatic chemistry: Calcium occurs in waters predominantly as a simple cation Ca^{2+}. Low concentrations of ionic associations such as $[CaHCO_3]^+$, $[CaOH]^+$, $[CaCO_3(aq)]^0$ or $[CaSO_4(aq)]^0$ may be present at pH < 9.[3815] At pH > 9 dissolved ionic association $[CaCO_3 (aq)]^0$ predominates. Ca^{2+} ion is usually the dominant cation in waters with lower total mineralization. With increasing concentration of dissolved compounds the calcium content relatively decreases.

5.1.2 Calcium in Wetlands

The cycling of calcium has not been studied intensively in wetlands[503] (Figure 3-11), Davis van der Valk[992] (Figure 3-16) and Gopal[1743] described the loss of calcium from the biomass during the decomposition of wetland plants. Klopatek[2537] described the calcium flow through the *Scirpus fluviatilis* stand in Wisconsin (Figure 5-1). Richardson et al.[4116] and Proctor[3916] reported calcium budgets for various wetlands.

5.1.2.1 Calcium in wetland plants

Concentration in % dry mass (see Appendix, Table 7-21):
emergent species: aboveground: 0.03–2.48, belowground: 0.01–1.36
floating-leaved species: aboveground: 0.94–4.04
submerged species: 0.28 - >10
floating species: 0.78–9.8
bryophytes: 0.10–57.1
Standing stock (g Ca m^{-2}) (see Appendix, Table 7-22):
emergent species: aboveground: 0.00004–54.3, belowground: 0.24–25.5
floating-leaved species: aboveground: 0.30–12.8
submerged species: aboveground: 1.6–11.3
floating species: 10.5–56.1

Table 5-1. Calcium concentration in algae (% dry mass)

Species	Concentration	Locality	Reference
Freshwater			
Chlorella pyrenoidosa	0.0–0.01	Laboratory	1612
Stigeoclonium tenue	0.03–0.25	Laboratory	1612
Draparnaldia plumosa	0.01–0.26	Laboratory	1612
Chlorella pyrenoidosa	0.006–0.4	Laboratory	2540
Nostoc muscorum	0.01–0.44	Laboratory	1612
Scenedesmus quadricauda	0.01–0.66	Laboratory	1612
Chlamydomonas geitleri	0.20–0.70	Cultivation (wastewater)	5144
Scenedesmus acutus	0.60–0.70	Cultivation (wastewater)	5144
Microcystis aeruginosa	0.03–0.95	Laboratory	1612
Chlorella pyrenoidosa	0.00–1.55	Laboratory	4378
Enteromorpha prolifera	1.62–3.08	Forfar Loch, Scotland	2107
Cladophora glomerata	1.51–4.29	Lake Michigan	20
Filamentous algae[a]	4.15–5.07	Nesyt Pond, Czech Republic	2843
Chara vulgaris	6.63–7.29	Czech Republic	1207
Chara sp.	0.72–22.6	Florida Everglades	5210
Soil			
Nostoc commune	0.98	Tennessee	3829
Marine			
Halosaccion ramentaceum	0.25	Canada, Nova Scotia	5528
Rhodymenia palmata	0.47	Canada, Nova Scotia	5528
Ahnfeltia plicata	0.48	Canada, Nova Scotia	5528
Ulva lactuca	0.85	Canada, Nova Scotia	5528
Fucus vesiculosus	0.98	Canada, Nova Scotia	5528
Laminaria longicruris	1.04	Canada, Nova Scotia	5528
Ascophyllum nodosum	1.10	Canada, Nova Scotia	5528
Laminaria digitata	1.29	Canada, Nova Scotia	5528
Chondrus crispus	1.33	Canada, Nova Scotia	5528
Phyllophora membranifolia	2.34	Canada, Nova Scotia	5528
Spongomorpha arcta	2.40	Canada, Nova Scotia	5528
Phaeodactylum tricornutum	1.48–2.68	Laboratory	1987
Ulva lactuca	0.35–2.81	Hong Kong coast	2111

[a] *Spirogyra* sp., *Oedogonium* sp., *Cladophora fracta, Enteromorpha intestinalis*

Calcium uptake rates in aquatic and wetland plants (see Appendix, Table 7-23) range between 32 and 600 g Ca m^{-2} yr^{-1} and between 9.5 and 109.3 mg Ca m^{-2} d^{-1}.

5.1.3 Calcium in Algal Nutrition

Calcium is one of the elements present in highest concentrations and required in largest amounts by angiosperm plants. (A review on calcium and plant development is given by Hepler and Wayne.[2038]) It seems highly unusual, therefore, that some algae (and fungi and bacteria) have been reported either not to require calcium or to require it only in micronutrient amounts.[3363,3532,4713,5018,5228,5230] In a number

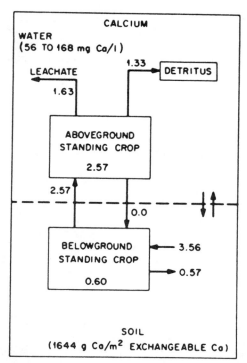

Figure 5-1.   Flow of calcium through a *Scirpus fluviatilis* stand. Flows in g m^{-2} yr^{-1} and compartments in g m^{-2} in standing crop.[2537] (From Klopatek, 1978, in *Freshwater Wetlands: Ecological Processes and Management Potentials*, Good, R.E., et al., Eds., p. 207, reprinted with permission of Academic Press and the author.)

of species of algae studied, however, the calcium requirement has been sufficiently high to allow for its ready demonstration.[310,805,1304,1475,3017,3270,3900-1,5268] Eyster[1304] reported that optimal concentration for algae is 0.03–0.5 mg L^{-1} without chelate and 40 mg L^{-1} with EDTA. In some algae, such as *Chara*, excess calcium is inhibitory; in *Chara* calcium level of 20 mg L^{-1} greatly reduces the rate of photosynthesis.[5352]

For many years calcium was not thought to be an absolute requirement for algae, but it is now certain that where it is required the amount is small and in many experiments it may have been supplied as contaminant, e.g. for *Nitzschia closterium* only 0.5 μg L^{-1} is required.[2264] The calcium requirement of many species is considerably less than that found in natural habitats. Gerloff and Fishbeck[1612] reported very low critical cell concentrations (0.0–0.06%) for 6 algal species. The calcium concentration in algae is given in Table 5-1. There seems to be little evidence for the limitation of production from a direct lack of calcium, although relatively high concentrations seem to be needed by some non-planktonic blue-green algae.[67] Calcium ions undoubtedly play a part in the maintenance of cytoplasmic membranes and in wall structures. It is also a major component of the walls of members of several algal classes (will be discussed later in Chapter 5.1.4).

Allen and Arnon[69] have shown that the nitrogen-fixing blue-green alga *Anabaena cylindrica* required macroquantities of calcium for growth regardless of whether the algae were given molecular nitrogen or nitrate nitrogen (Figure 5-2). Eyster[1303] presented the evidence that calcium enhanced nitrogen fixation in the blue-green alga *Nostoc muscorum* (Figure 5-3). Calcium was also discovered to be involved in the production of motile cells in *Protosiphon*.[3531-2] Kylin and Das[2668] found that additions of Ca above 10^{-3} mM enhance cell multiplication of a green alga *Scenedesmus* and, at least partly in connection with the morphogenetic effect, decrease the average cell size without changes in the percentage of dry matter. The assimilation of inorganic phosphates increased already by 10^{-3} mM Ca^{2+}, with no sign of further interactions over the range tested.

Strontium was found to replace the requirement of *Chlorella* for trace amounts of calcium[5228,5230] and Kevern[2483] found that a green alga *Oocystis eremosphaera* takes up Sr as a substitute for Ca. However, a filamentous green alga *Pithophora oedogonia* and other green algae cannot substitute Sr for Ca for their basic metabolic needs.[5439] O'Kelley and Herndon[3531] reported that studies on *Protosiphon botryoides* point to a critical role for calcium, which cannot be assumed by strontium, in the production of motile cells. Vollenweider[5175] found that Ca could partially substitute for Mg in the growth of blue-green alga *Oscillatoria rubescens* and green alga *Ankistrodesmus falcatus*.

5.1.4 Calcification in Algae

It has long been known that photosynthetic organs of certain plant species develop calcium carbonate incrustation on their outer surfaces. Numerous hypotheses have been proposed to explain the manner in which these incrustations are linked to cellular function.[2942]

Steeman-Nielsen[4704] proposed that some algal cells were able to utilize HCO_3^- directly in exchange for OH^-, and by this exchange process the pH of the bathing solution increased, ultimately causing the precipitation of $CaCO_3$. Spear et al.[4648] and Smith[4587] observed external alkalinization of the bulk medium immediately adjacent to the cell wall of the characean species. Spear et al.[4648] proposed that external alkalinity was due to net passive proton influx. They did not completely reject the possibility of net passive OH^- efflux, but considered this less likely because of the low permeability of the plasmalemma to anions. On the other hand, Smith[4587] suggested that the external alkaline regions were due to the absorption of HCO_3^-, the OH^- being supplied by the reaction:

$$HCO_3^- \rightarrow CO_2 \text{ (fixed photosynthetically)} + OH^- \tag{5.1}$$

Calcification in algae involves the precipitation of $CaCO_3$ around or within algal cells; it occurs in Cyanophyceae, Chlorophyceae, Charophyceae, Prymnesiophyceae, Chrysophyceae, Dinophyceae, Phaeophyceae, and Rhodophyceae.[455,2799,3746,4131] Two forms of calcium carbonate may be deposited: calcite, a rhombohedral crystal, and aragonite, an orthorhombic crystal.[4688] No alga depositing a mixture of the two has been reported.[771,2799,4206] Aragonite is most frequently found in the marine algae, and calcite in the freshwater species, although the marine coralline red algae are notable exceptions in their production of calcite.[4643]

McConnel and Colinvaux[3133] reported that aragonite precipitations on several marine algae contain other inorganic substances in very small amounts which probably represent contaminants. The presence of associated sulfates or phosphates has apparently not been reported in algal skeletons.[2799,3131]

Algal calcification has been a subject of research for over 100 years as can be seen from reviews by Pia[3782] and Lewin.[2799] Reviews of algal calcification have been presented many times.[168,453,455,1726,2623,2799,3526,3740,3745–6,3782,4131]

Figure 5-4 illustrates the approximate relative abundance of modern calcified genera within the algal classes divided between freshwater and the sea.[3746] Most classes are seen to possess only small numbers of genera with calcified species. One class, the Prymnesiophyceae, holds a pre-eminent position, with more than 50% of the genera having calcified species (coccolithophorids). These minute algae are major primary producers in the oceans and must form more biogenic carbonate than any other group of plants.[3746] All genera of the Charophyceae calcify although the degree of mineralization is often slight and sometimes confined to the reproductive organs.[3746]

Goreau[1759] was able to show that marine algae would calcify faster in the light than in the dark. Light-stimulated $CaCO_3$ deposition has also been demonstrated for different species.[456–8,1067,1175,3527,3600–2,3607,3719,4688]

The most acceptable and classical hypothesis[457] for light stimulation of algal calcification (with the possible exclusion of coccolith formation) is that the local (CO_3^{2-}) concentration is increased as a result of CO_2 uptake during photosynthesis or due to alkalinization of the medium due to OH^- extrusion from the cell after HCO_3^- uptake. Light-stimulated OH^- extrusion has been demonstrated by e.g. Lucas and Smith[2942] for *Chara*, Lucas et al.[2949] for *Potamogeton*, Steeman-Nielsen[4704] for *Myriophyllum*, and Ruttner[4228] for *Elodea* and *Vallisneria*. Calcification in these plants appears to be a by-product of the localized pH increase due to the OH^- efflux. Further support for the hypothesis that at least some algal calcification is a by-product of localized increases in (CO_3^{2-}) comes from the observation that, with the exception of coccolithophores, coralline red algae, and some blue-green algae, the deposits are of the crystal type expected to be produced by inorganic precipitation and that these deposits show little organization with respect to the cell outside of which they are precipitated.[460]

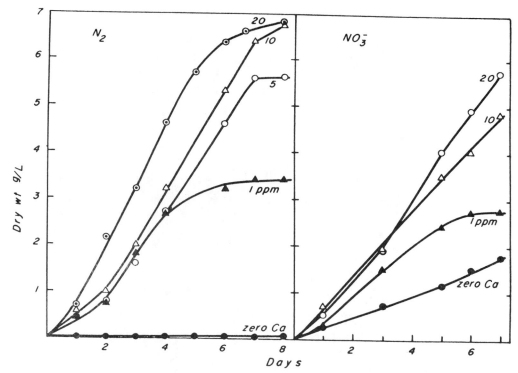

Figure 5-2.   The calcium requirement of *Anabaena cylindrica*. Light intensity 7500 lux. Calcium concentrations (ppm) are shown on the curves.[69] (From Allen and Arnon, 1955, in *Plant Physiol.*, Vol. 30, p. 367, reprinted with permission of the American Society of Plant Physiology.)

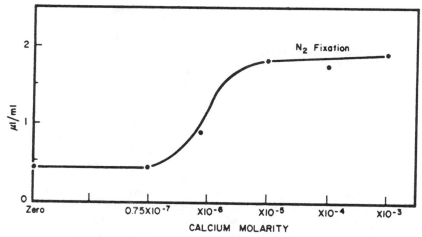

Figure 5-3.   Effect of different levels of calcium on the growth of *Nostoc muscorum* in the absence of nitrate, urea, and ammonium salts but in the presence of nitrogen gas. Culture grown 6 days, while being shaken, bubbled with 5% CO_2 in air, and illuminated with 1500 ft-c cool white fluorescent light. Inoculum: 0.3 μL cells from calcium deficient culture.[1303] (From Eyster, 1964, in *Algae and Man*, Jackson, D.F., Ed., p. 99, reprinted with permission of Plenum Publishing Corporation, New York.)

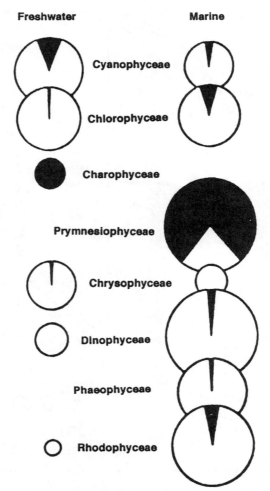

Figure 5-4. Pie charts illustrating the approximate relative proportion of calcified genera to uncalcified genera in the major algal classes. Circle size is approximately proportional to the known abundance of the groups in freshwater and the oceans, respectively. Minor groups are omitted. Calcified genera black, uncalcified white.[3746] (From Pentecost, 1991, in *Calcareous Algae and Stromatolites,* Riding, R., Ed., p. 5, reprinted with permission of Springer-Verlag and the author.)

Pentecost[3746] summarized that most calcifying regimes are characterized by a water pH of 7.8–9.5. Logic dictates that photosynthesizing cells will utilize HCO_3^- rather than CO_2 under these conditions because the concentration of CO_2 will be low, typically 5–10 μM. Uptake of HCO_3^- may be either active or passive, depending upon the sign of the free energy change resulting from the transport of the ion across the diffusion gradient of an electrically charged membrane (for details see Walker[5232]).

To maintain electroneutrality in the external medium, H^+ must also be removed, resulting in an effective loss of H_2CO_3. In an inbounded medium this will be replaced by CO_2 arriving through laminar and turbulent diffusion, with the length of the diffusion path, depending upon the relative water movement. Within the diffusion zone, depletion of H^+ and HCO_3^- results in a pH rise and CO_3^{2-} formation, assuming instantaneous equilibration. Such conditions favor $CaCO_3$ precipitation and increasing the photosynthesis rate will therefore favor calcification.[3746] (For detailed information see Pentecost[3746] and Borowitzka.[455])

There are many algae with as high a rate of photosynthesis as the calcifiers, and growing adjacent to them, which show no calcification.[4688] One theory is that some algae produce on their cell surface substances, such as polyphenols, which inhibit $CaCO_3$ crystal formation.[4082] Phosphate compounds inhibit the process of calcium carbonate precipitation.[595,3964,4506–7]

5.1.4.1 Cyanophyceae

Pentecost and Riding[3747] defined calcification in cyanophytes as "the nucleation of calcium carbonate upon or within the mucilaginous sheath." It is necessary to make this distinction so that trapping and binding of carbonate can be excluded as a calcification process.[3746] However, calcification is unlikely to occur with the complete exclusion of trapping and binding. The problem is illustrated in Figure 5-5, where three calcification processes are shown in addition to binding. It is assumed in this model that the surrounding water is supersaturated with respect to a $CaCO_3$ mineral and that nucleation begins both upon and within the sheath.[3746]

Calcified blue-green algae are individually normally microscopic, but the deposits which they construct, such as oncoids, stromatolites, tufas and reefal fabric, are macroscopic and conspicuous.[4131] The relationship between blue-green algae and carbonate deposits and products of blue-green algae calcification are discussed by Golubic[1722] and Riding.[4131]

Two key features in blue-green algae calcification are: 1) the site of calcification is the enveloping sheath, rather than the cell itself, of the organism, and 2) calcification appears to be proven only under conditions where $CaCO_3$ precipitation is thermodynamically favored.[3743,3748,4131,4621] The presence of the sheath, and also its consistency (whether dense or diffluent), influences calcification because the crystals of $CaCO_3$ nucleate within or upon it, possibly as a result of the attraction of Ca ions by uronic

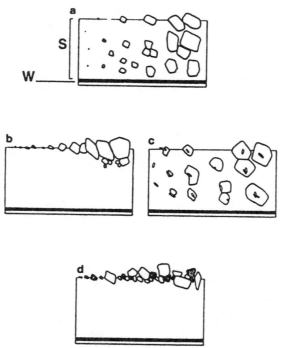

Figure 5-5.   Calcification and carbonate-binding by the cyanophyte sheath. a) Calcification resulting from biogenic nucleation within and upon the sheath. b) Calcification at the sheath surface resulting from the adhesion of seed crystals or heteronuclei originally present in the surrounding water. Note secondary nucleation at right caused by the fragmentation of larger crystals. c) Calcification associated with bacteria upon and within the sheath. d) Carbonate binding at the sheath surface. This process is non-selective and includes non-carbonate grains. Time increases from left to right along the abscissa. S = sheath, W = cell wall.[3746] (From Pentecost, 1991, in *Calcareous Algae and Stromatolites*, Riding, R., Ed., p. 6, reprinted with permission of Springer-Verlag and the author.)

acids.[4131] Prát[3868] was the first who found that blue-green algal incrustations, although a natural and characteristic part of these algae, were not necessary for their growth.

Pentecost[3743] found that $CaCO_3$ content of cyanophyte *Homoeothrix crustacea* was up to 51.2% on wet mass basis. Pentecost[3742] found a calcification of freshwater cyanophyte *Rivularia haematilis* with maximum radial growth rates of 12–14 μm d^{-1} during the summer and minimum rates of less than 2 μm d^{-1} during winter. There was a good correlation of calcification with temperature. Pentecost[3746] pointed out that, at least in some species, during the winter, the calcification is abiotic, while in summer, it results from a combination of abiogenic and biogenic (photosynthetic) processes.

The major blue-green algae genera which produce calcification are *Calothrix, Geitleria, Homoeothrix, Lyngbya, Phormidium, Plectonema, Rivularia, Schizothrix,* and *Scytonema.*[1722,3747]

5.1.4.2 Chlorophyceae

The calcification process is best understood in the tropical green alga *Halimeda*, which deposits needle-like crystals of aragonite outside the cell walls, but within an intracellular space separated from the external seawater by a layer of tightly appressed utricles (Figure 5-6). Although no direct measurements have been made on these spaces the model is supported by several lines of evidence using radioactive tracers and metabolic inhibitors.[455-8] The model can be applied to other aragonite-depositing marine algae in which the presence of densely interwoven filaments allows the development of a long diffusion path from the external seawater to the site of precipitation.[455] Stark et al.[4688] reported that $CaCO_3$ formed 84.4% of dry mass of *Halimeda opuntia*.

A small number of calcified freshwater species occasionally produce significant amounts of calcareous sediments.[3336] The most interesting of these are planktonic species *Phacotus* and *Coccomonas*. Calcification also occurs in epilithic freshwater genera *Cladophora*[4489,4873-5,5490] and *Oocardium*.[1725,5245] The precipitation is thought to be partly the result of photosynthesis and partly by O_2 evasion to the atmosphere.

Sikes[4489] found that onset of carbonate deposition coincided with the appearance of carbonic anhydrase activity in the cells of *Cladophora glomerata*. He suggested that carbonate deposition may be a function of HCO_3^- use as a source of CO_2 for photosynthesis. He also reported increased calcification as *Cladophora* aged (12.3–160 mg Ca g^{-1} dry mass). This reflected increased pectin layer in thickening cell walls (up to 23% of dry mass). Lewin[2830] reported calcification of cell walls of *Chlamydomonas* on agar media.

Further details on calcification in green algae can be found in References 417–8, 459, 1067, 3079, 3133, 3336, 3744, 4688, 5490.

5.1.4.3 Prymnesiophyceae

The coccolithophorids, members of the Prymnesiophyceae, are the only algae in which intracellular calcification occurs[455,4643] (Figure 5-7). In most species the deposition of calcite occurs on an organic base within the cisternae of the Golgi apparatus[2530] to form delicately sculptured calcareous plates, or coccoliths. Coccolith contains polysaccharides with carboxyl and ester-sulfate groups,[1074] and associated proteinaceous materials which are capable of binding Ca^{2+} ions and hence presumably act as nucleating agents in the precipitation process.[455] Calcification proceeds in the light, and in *Emiliania huxleyi*—the most studied coccolithophorid, CO_2 is utilized in photosynthesis while HCO_3^- provides the carbon for calcification.[4490,4643]

Literature dealing with coccolithophorids is voluminous (see e.g. References 393, 451, 1074, 1127, 1489, 1803, 1807, 2289, 2526–8, 2530, 2860, 3054–5, 3600–5, 3658, 3740, 4706, 5126, 5129, 5277, 5334.

5.1.4.4 Rhodophyceae

Coralline red algae are the most widely distributed calcified benthic algae in the seas today. Most vegetative cells of the crustose and non-articulated branched species are calcified to some extent.[3746] Calcification is entirely extracellular[5164] and two phases of mineralization are apparent.[654] A calcification mechanism was proposed by Digby[1081] which involves the photosynthetic oxidation of water

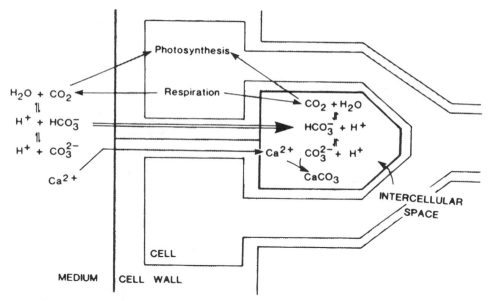

Figure 5-6. *Halimeda.* Schematic representation of the postulated ion fluxes which affects $CaCO_3$ precipitation. Ions diffuse from the sea water to the intracellular space (ICS) via the cell walls of the appressed utricles. During photosynthesis CO_2 is taken up both from the ICS and the external medium and respiratory CO_2 diffuses out of the cell. These CO_2 fluxes result in pH changes in the ICS resulting in $CaCO_3$ precipitation in the light. HCO_3^- uptake for photosynthesis (not shown) requires OH^- efflux from the cell. This would further enhance calcification. H^+ fluxes which may also occur have been omitted.[457] (From Borowitzka and Larkum, 1976, in *J. Exp. Bot.,* Vol. 27, p. 89, reprinted with permission of Oxford University Press.)

followed by export of the hydrogen ions to sea water. Intracellular hydroxyl ions react with bicarbonate and the resulting carbonate ion is exported to the cell wall where it reacts with calcium to form calcite. An alternative model involving localized export of H^+ and OH^- has also been proposed.[3741] There is some evidence for H^+ efflux to seawater[1081] and, if this is a local effect, parallel can be seen with the Charophyceae. However, the mechanisms of deposition are not well understood.[3746] Members of the Corallinaceae deposit calcite among the fibrils of the cell wall.[654] Deposition is linked to photosynthesis.[454-5,3739] It is important to note that Borowitzka[455] rejected Digby's model. Further information on calcification of the Rhodophyceae are given by e.g. Pearse,[3719] Okazaki and Furuya[3525] and Okazaki et al.[3528]

5.1.4.5 Charophyceae

Extracellular deposits of $CaCO_3$ occur on the internodal and some reproduction cells of charophytes. Calcification has been reported in all genera but is most apparent in *Chara.* Regular bands of calcite are often a conspicuous feature of certain ecorticate species and these are known to be associated with alkaline regions of the cell wall where the pH attains values of 9 or more. This high pH is thought to result primarily from a passive influx of H^+ through the plasmalemma.[4013,4648,5234] An active efflux of H^+ occurs in adjacent uncalcified acid bands resulting in the circulation of electric current. This efflux is powered by ATP, the ultimate energy source being provided by photosynthetic reactions.[3746]

An alternative hypothesis of alkaline bands development was proposed by Lucas and his coworkers.[1342,2930,2932,2934,2937,2942-3,2947] According to the theory, alkaline bands develop on the outer surface of *Chara* cells when they photosynthetically fix CO_2, supplied as HCO_3^- (Figure 5-8). These bands originate from the localized export of OH^-, produced during CO_2 fixation. Photosynthetic fixation of the carbon supplied by HCO_3^- results in the production of one OH^- for each HCO_3^- converted to CO_2.[2949] This "waste product" (OH^-) is exported from the cytoplasmic phase, across the plasmalemma, by the

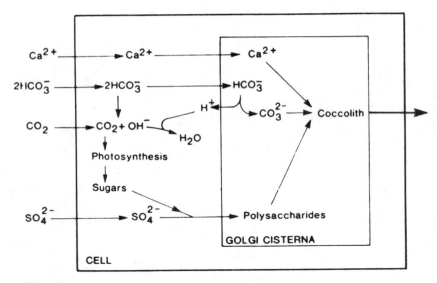

MEDIUM

Figure 5-7.   *Emiliania huxleyi.* Schematic representation of the postulated major ion fluxes from the seawater to the Golgi cisterna where coccolith formation takes place. Ca^{2+} uptake into the cell and the Golgi cisterna is either by diffusion or active transport. For every two molecules of HCO_3^- taken up (possibly by active membrane transport), one is converted to CO_2 in the cytoplasm (possibly by carbonic anhydrase activity) whereas the other is transported to the Golgi cisterna where it is deprotonated to form CO_3^{2-}. The $CaCO_3$ is precipitated in the Golgi cisterna. The so formed proton is extruded from the cisterna. CO_2 may also enter the cell by diffusion and is fixed in photosynthesis. Some of the photosynthetic products are further metabolized together with SO_4^{2-} to form the polysaccharide components of the coccolith. The completed coccolith is expelled from the cell and lodges in the external mucilage.[455] (From Borowitzka, 1982, in *Progress in Phycological Research*, Vol. 1, Round, F.E., and Chapman, D.J., Eds., p. 166, reprinted with permission of Elsevier Science Publishers bv, Amsterdam and the author.)

OH^- efflux system at discrete OH^- efflux sites.[2930,2932,2942,2944] Each alkaline band is formed by the transport of OH^- over a surface,[2931] the length of which is approximately 0.05 cm (cf. 4–5 cm total cell length). Accumulation of OH^- at the cell surface, and its diffusion into the neighboring solution, gives rise to the observed alkaline banding phenomenon.[2930,2942] Under HCO_3^- assimilation conditions, the OH^- bands are spaced at somewhat regular intervals along the length of the internodal cell. The number of bands per cell and their efflux activities depend on the level of exogenous HCO_3^-, the incident light intensity and the length of the actual cell.[2930,2932,2942] Lucas and Dainty[2944] reported that there exist an hierarchy of functional OH^- efflux sites within the plasmalemma. Both HCO_3^- and OH^- transport must work in synchrony.[2936,2944-5] OH^- efflux rather than H^+ influx is thought to be involved because band formation is independent of external pH between 5.5 and 10.[2937] The origin of these bands is completely dependent on light[2932] and the alkalization is prevented by photosynthetic inhibitors.[2942] Extrusion of OH^- and H^+ is spatially separated creating alternatively acid and alkaline bands along the giant characean cells.[2941] A schematic representation of these two alternative methods of alkaline band formation is presented in Figure 5-9.

Further information on calcification of charophytes can be found in References 1341, 2931–2, 2934–6, 2939–41, 2943–6, 3131, 3894, 4013, 4586, 5231, 5234–5.

5.1.4.6 Phaeophyceae

Calcification has only been reported in one genus, *Padina* (Dictyotales); its species are widely distributed in warmer waters but only about half of them calcify with aragonite.[3746]

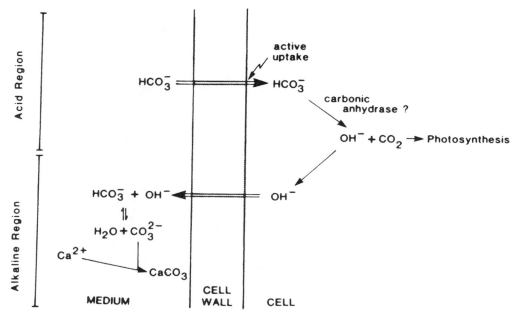

Figure 5-8.   *Chara corallina.* Schematic representation of the major ion fluxes associated with HCO_3^- uptake in the acid region of the cell wall and OH^- efflux in the alkaline regions which results in $CaCO_3$ precipitation if sufficient Ca^{2+} is present.[455] (From Borowitzka, 1982, in *Progress in Phycological Research,* Vol. 1, Round, F.E., and Chapman, D.J., Eds., p. 154, reprinted with permission of Elsevier Science Publishers bv, Amsterdam and the author.)

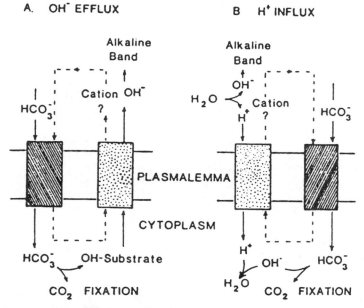

Figure 5-9.   Schematic representation of two modes by which alkaline bands may be formed on the surface of *Chara corallina* cells. HCO_3^- and OH^- (or H^+) transporters exist at spatially separate sites within the plasmalemma.[2937] (From Lucas, 1979, in *Plant Physiol.,* Vol. 63, p. 249, reprinted with permission of the American Society of Plant Physiology.)

5.1.4.7 Other classes

Calcification in other classes is rarely reported. The best documented case of dinophyte calcification is afforded by *Thoracosphaera*, which until recently was classified as a coccolithophorid.[4894] The mineralization is extracellular.

In other classes, calcification has been observed only in freshwater and in usually associated with mucilage coatings. Calcification in the Chrysophyceae has been reported e.g. by Tschermak-Woess[5038] or Tappan.[4899] Calcification in the Bacillariophyceae has been reported by Winsborough and Golubic.[5458]

5.2 IRON

5.2.1 Iron in the Environment

Earth's crust: 41,000 mg kg^{-1},[490] 50,000 mg.kg^{-1} (903)
Common minerals: FeS_2 (pyrite), FeS (troilite), Fe_3O_4 (magnetite), α-Fe_2O_3 (hematite), γ-Fe_2O_3 (maghemite), $2Fe_2O_3 \cdot H_2O$ (limonite), $Fe_2O_3 \cdot n\,H_2O$ (ferrihydrite), α-$FeOOH$ (goethite), γ-$FeOOH$ (lepidocrocite), $FeSO_4 \cdot 7H_2O$ (melanterite), Fe_2SiO_4 (fayalite), $FeCO_3$ (siderite), $KFe_3(SO_4)_2(OH)_6$ (jarosite)[490,2395,2975,3705]
Soils: 2,000–550,000 mg kg^{-1}, availability depends on the oxidation state, this is determined by microbialaction, which in turn depends on external conditions;[490-1] 200–100,000 mg kg^{-1} (4738)
Sediments: 41,000 mg kg^{-1} (491)
Air: 0.51–14,000 ng m^{-3} (491, 2395)
Fresh water: 10–1,400 μg L^{-1} (491); 4–90,000 μg L^{-1} (3287)
Sea water: 2 (0.03–70) μg L^{-1} (491)
Land plants: 140 (7–700) mg kg^{-1} (490–1); 1.4–1,700 mg kg^{-1} (2395)
Freshwater algae: 100–49,510 mg kg^{-1} (Table 5-2)
Marine algae: 13–1,900 mg kg^{-1} (491); 15–15,473 mg kg^{-1} (Table 5-2)
Land animals: 160 mg kg^{-1} (490)
Marine animals: 400 mg kg^{-1} (490)

5.2.1.1 Aquatic chemistry

Sources of iron in waters are municipal wastewater, sewage sludge discharge, continental and volcanic dust flux, base metal mining, and natural weathering.

The primary oxidation states of iron in water are Fe^{2+} (ferrous) and Fe^{3+} (ferric). In most surface waters, Fe^{3+} predominates and, when combined with its salts, is practically insoluble, at least in aerobic conditions.[3287] Forms of dissolved and particulate iron in waters depend on pH, redox potential and the presence of complexation substances of both inorganic and organic origin.[3815]

In reduced environment, Fe^{II} is present as Fe^{2+}, $[FeOH]^+$ or $[Fe(OH)_3]^-$. Solubility of Fe^{II} is limited mostly by the solubility of $FeCO_3$ (s), $Fe(OH)_2$ (s) and $FeS(s)$. The formation of complexes $[FeHCO_3]^\circ$ and $[FeCO_3$ (aq)$]^\circ$ is negligible.[3815] In oxygenated environments the solubility of Fe^{III} is influenced by the solubility of hydrated Fe_2O_3. The presence of inorganic substances which are able to form soluble complexes with Fe^{III} (e.g. $[FeCl]^{2+}$, $[FeCl_2]^+$, $[FeSO_4]^+$, $[Fe(HPO_4)]^+$), increases the solubility of iron in waters. These complexes are formed in waters with low OH$^-$ ions concentration that decrease the formation of hydroxycomplexes.[3815]

The oxidation-reduction cycle is important in controlling the fate of iron in most surface waters. Oxygen concentrations at the water-sediment interface often approach zero. This causes the reduction of Fe^{3+} to soluble Fe^{2+}, which is then transported upward in the water column. The oxygenated water results in reoxidation to the insoluble Fe^{3+}, which settles to the bottom to repeat the cycle.[3287] When Fe^{3+} is formed as part of the reduction-oxidation cycle in the presence of phosphate, a basic iron phosphate—$Fe_2(OH)_3PO_4$—is formed with a Fe : P stoichiometry of 2 : 1.[138] Another possible reaction occurs when Fe^{3+} is formed and hydrolyzed prior to interaction with phosphate; in the scenario, adsorption of phosphate by $Fe(OH)_3$ (s) results in Fe : P ratio > 5:1.[3287]

Table 5-2. Iron concentration in algae (mg kg^{-1} dry mass)

Species	Concentration	Locality	Reference
Freshwater			
Chlamydomonas geitleri	100	Cultivation (wastewater)	5144
Anacystis nidulans	280	Culture	2205
Chlorella vulgaris	400	Laboratory	1604
Oedogonium sp.	700	Lower Swansea Valley, Wales	5027
Enteromorpha prolifera	610–750	Forfar Loch, Scotland	2107
Chara fragilis	787	Tübingen, Germany	2205
Cladophora glomerata	800	Lake Balaton, Hungary	2575
Scenedesmus acutus	200–1,000	Cultivation (wastewater)	5144
Oscillatoria sp.	2,800[b]	Lower Swansea Valley, Wales	5027
Aphanizomenon flos-aquae	3,370	Culture	2205
Cladophora glomerata	2,910–3,940	Lower Swansea Valley, Wales	5027
Ulothrix sp.	7,780[b]	Lower Swansea Valley, Wales	5027
Spirogyra sp.	460–8,930	Lower Swansea Valley, Wales	5027
Cladophora glomerata	697–16,369	Roding River, England	3146
Mougeotia sp.	17,610[a]	Lower Swansea Valley, Wales	5027
Tribonema sp.	9,970–33,920[a]	Lower Swansea Valley, Wales	5027
Chlorella pyrenoidosa	237–34,530	Laboratory	2540
Zygnema sp.	39,850[a]	Lower Swansea Valley, Wales	5027
Microspora sp.	42,310[b]	Lower Swansea Valley, Wales	5027
Coccomyxa sp.	49,510[a]	Lower Swansea Valley, Wales	5027
Soil			
Nostoc commune	3,500	Tennessee	3829
Marine			
Hizikia fusiforme	63–68	Hong Kong	2113
Enteromorpha linza	104–114	Hong Kong	2113
Ascophyllum nodosum	15–120	Great Britain	1455
Endarachne binghamiae	129	Hong Kong	2114
Dermonema frappieri	63–145	Hong Kong	2113
Laminaria saccharina	200	Helgoland, North Sea	2205
Gloiopeltis furcata	38–292	Hong Kong	2113
Fucus vesiculosus	41–360	Great Britain	1455
Ishige okamurai	52–362	Hong Kong	2113
Sargassum hemiphyllum	386	Hong Kong	2114
Ishige foliacea	129–388	Hong Kong	2113
Sargassum vachellianum	451	Hong Kong	2114
Ascophyllum nodosum	51–468	Trondheimsfjorden, Norway	2694
Porphyra suborbiculata	68–485	Hong Kong	2113
Gelidium amansii	314–587	Hong Kong	2113
Fucus vesiculosus	30–790	Sweden, coast	4612
Blue-green algae	946	Penang Island, Malaysia	4526
Fucus vesiculosus	30–967	Great Britain estuaries	604
Fucus vesiculosus	140–1,168	Trondheimsfjorden, Norway	2694

314   ALGAE AND ELEMENT CYCLING IN WETLANDS

Table 5-2. Continued

Species	Concentration	Locality	Reference
Padina arboresceus	176–1,401	Hong Kong	2113
Fucus spp.	56–1,517	Irish Sea, North Sea	3882
Chaetomorpha antennina	480–1,840	Hong Kong	2113
Ahnfeltia plicata	1,840	Helgoland, North Sea	2205
Corallina pilulifera	234–2,518	Hong Kong	2113
Rhizoclonium riparium	2,661–2,846	Hong Kong	2113
Ulva lactuca	90–3,280	Hong Kong coast	2111
Scytosiphon lomentaria	271–3,680	Hong Kong	2113
Porphyra spp.	104–3,800	Irish Sea	3882
Ectocarpus siliculosus	5,067–5,242	Hong Kong	2113
Colpomenia simuosa	1,728–5,497	Hong Kong	2113
Gymnogongrus flabelliformis	199–5,804	Hong Kong	2113
Enteromorpha compressa	1,074–6,184	Hong Kong	2113
Enteromorpha flexuosa	999–7,476	Hong Kong	2113
Caulacanthus okamurai	7,293–8,418	Hong Kong	2113
Enteromorpha sp.	4,735–13,816	Penang Island, Malaysia	4526
Fucus vesiculosus	348–15,473	Penang Island, Malaysia	4526

[a] Adjacent to zinc smelter waste, [b] near zinc smelting waste

5.2.2 Iron in Wetlands

Iron is found in wetlands primarily in its reduced form (ferrous, Fe^{2+}) which is more soluble and more readily available to organisms. When the reduction of nitrate stops by depletion of this electron acceptor, the reduction of ferric oxide starts in waterlogged soils. Ferric oxides are assumed to be one of the most abundant electron acceptors in soils as well as in sediments.[1776,2921,3838] Within a few weeks of submergence 5–50% of active Fe^{3+} present in a soil may be reduced to Fe^{2+}, depending on microbial activity (temperature, organic matter), nitrate content, and crystallinity of the Fe oxides.[3418]

Under natural conditions, the reduction of ferric oxide may proceed chemically by the involvement of sulfide or organic materials such as phenols and carboxylic acids.[2920] So, sulfate-reducing, sulfide-producing bacteria may be indirectly involved in the reduction of ferric oxide.[267] For details on iron reduction processes in waterlogged soils see Lanbroek.[2679]

At neutral pH and at positive redox potentials, oxidized iron is relatively stable.[3694] The direct enzymatic oxidation of Fe^{2+} (and also Mn^{2+}) is confined to a restricted range of organisms and the majority of the bacteria cause precipitation of Fe and/or Mn by indirect means by altering Eh or pH which in turn leads to chemical oxidation and precipitation.[1800] Hence the oxidation of ferrous ions may proceed almost completely chemically under these conditions and it is better to speak of iron-depositing bacteria than of iron-oxidizing bacteria.[2361] Chemolithotrophic oxidation of Fe^{2+} is confirmed for *Thiobacillus ferrooxidans*, *Sulfolobus* spp., and probably *Gallionella* spp.,[1800] Lundgren and Dean[2975] listed also *Leptothrix* and *Siderocapsa* groups and *Metallogenium*. Generally, the following equations describe the biological oxidation of ferrous iron by *Thiobacillus ferroxidans*:[2975]

$$4\ FeSO_4 + 2\ H_2SO_4 + O_2 \rightarrow 2\ Fe_2(SO_4)_3 + 2\ H_2O \tag{5.2}$$

$$2\ Fe_2(SO_4)_3 + 12\ H_2O \rightarrow 4\ Fe(OH)_3 + 6\ H_2SO_4 \tag{5.3}$$

The organism consumes oxygen in the ratio of one mole O_2 per four moles of Fe^{2+} and the oxidized product is Fe^{3+}. Energy (ATP) is derived when the electrons liberated from the iron are coupled to cytochrome c via cytochrome-c reductase. The electron(s) is then transported from cytochrome c to cytochrome a via cytochrome oxidase and is subsequently accepted by oxygen.[2975]

Direct involvement of bacteria in the oxidation of ferrous ions is only likely at low pH.[2679,2975] Several genera of bacteria seem to be involved in the deposition of ferric ions,[1622,2361,5136] among them filamentous, prosthecate and encapsulated bacteria. Only a few of these bacteria have been isolated in pure culture, e.g. filamentous bacteria of the genera *Sphaerotilus* and *Leptothrix*. These bacteria produce sheaths containing ferric oxides (for details see Laanbroek[2679]). Ferrous iron, diffusing to the surface of roots of wetlands, may be oxidized by oxygen leaking from root cells, immobilizing phosphorus and coating roots with an iron oxide, causing a barrier to nutrient uptake.[1555] For further information about microbially-mediated iron transformations see the works by Grant and Long[1799,1800] and especially the excellent paper by Lundgren and Dean.[2975]

Iron, in its reduced form, $Fe(OH)_2$ causes a grey-green coloration of mineral soils instead of its normal red or brown color of oxidized conditions—$Fe(OH)_3$. This gives a relatively easy field check on the oxidized and reduced layers in a mineral soil profile.[3246] This process of soil coloration is called *gleying*, and waterlogged mineral soils are often referred to as *gleys*.[1278]

5.2.2.1 Iron in wetland plants

Concentration in dry mass (mg kg^{-1}) (see Appendix, Table 7-24):
emergent species: aboveground: 19–59,000, belowground: 42–6,900
floating-leaved species: aboveground: 100–650, belowground: 79–6,200
submerged species: aboveground: 80–110,000
floating species: 50–8,440
bryophytes: 231–143,000

Standing stock (see Appendix, Table 7-25) ranges in emergent species between 0.12 and 102 g Fe m^{-2} and between 0.64 and 2,973 g Fe m^{-2} in aboveground and belowground parts, respectively. Iron uptake rates in aquatic and wetland plants (see Appendix, Table 7-26) range between 113 and 80,000 mg Fe m^{-2} yr^{-1}.

5.2.3 Iron in Algal Nutrition

The essentiality of iron for plants has been known for more than a century;[3219] more than 60 years ago, Hopkins and Wann[2184] reported that iron was necessary for the culturing of *Chlorella*. Quantitatively, iron is the most important trace metal for phytoplankton.[2353] It is required in numerous redox reactions and in the synthesis of chlorophyll.[3530,5424] On the other hand, higher concentrations can be toxic, e.g. 80 to 280 mg L^{-1} for *Ankistrodesmus braunii*.[3629] Eyster et al.[1305] have shown that more iron was required for autotrophic growth of *Chlorela pyrenoidosa* than for its heterotrophic growth (Figure 5-10). The critical concentration of iron for heterotrophic growth of *Chlorella pyrenoidosa* has been shown to be 1×10^{-9} M, while for autotrophic growth it is 1.8×10^{-5} M[1302] (see also Figure 5-10). The iron concentration in algal biomass is summarized in Table 5-2; iron concentration factors in algae are shown in Table 5-3.

Iron has been long known to be essential to algae. It is a key element in metabolism, being a constituent of numerous iron-containing enzymes, e.g. peroxidase, nitrate reductase, nitrogenase, catalase, cytochrome *c*, cytochrome oxidase, cytochrome *f*, cytochrome b_6 and photosynthetic pyridine nucleotide reductase.[1251,1303,2171,2274,2353,3753-4,5032,5162,5302-3] Iron plays a parallel role to magnesium in chlorophyll, in being at the center of the cytochrome molecules, which transfer electrons in the respiratory chain and in photosynthesis. The importance of iron in this connection lies in its ability to change valence between Fe^{2+} and Fe^{3+}, but iron is also present in a number of oxidizing systems (such as catalaze) in which it does not change valence.[361] While iron is not part of the chlorophyll molecule, it is required as a cofactor in the synthesis of chlorophylls, at least in *Euglena*.[3530] Other authors also found iron to be necessary for chlorophyll synthesis.[684,2430-1,3892,5132] Reuter et al.[4224] found that blue-green alga *Trichodesmium* NIBB 1067 showed higher chlorophyll-*a* content, oxygen production and nitrogen fixation in the presence of iron as compared with the medium with no added Fe.

Another important iron-containing molecule, involved in photosynthesis and other electron transfers, is ferredoxin,[470,3249-50,4575,4284,4452] a protein that also contains labile sulfur (in addition to the sulfur in amino acids).[2765] In *Scenedesmus* the level of iron is directly related to the extent of hydrogenase activity.[4290] Iron has been demonstrated to be essential to hydrogenase development in *Scenedesmus*.[5517]

A number of investigators have reported that iron is an important algal growth-limiting nutrient in freshwater environments.[1686,4252,4322] In marine environments, iron is believed to assume an even greater

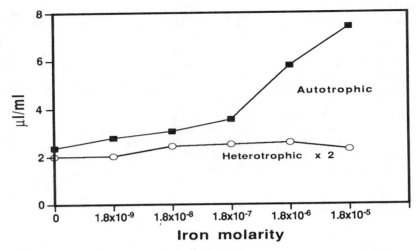

Figure 5-10. Growth of *Chlorella pyrenoidosa* at different concentrations of iron. Autotrophic: 4 days growth from 1.4 μL washed cells. Heterotrophic: 21 days growth from 1 loopful cells.[1305] (From Eyster et al., 1958, in *Trace Elements,* Lamb, C.A., et al., Eds., p. 165, reprinted with permission of Academic Press.)

Table 5-3. Iron concentration factors (CF) in algae

Species	CF	Locality	Reference
Freshwater			
Cladophora glomerata	130–404	Šárecký Brook, Czech Republic	5188
Oedogonium sp.	7,200	Lower Swansea Valley, Wales	5027
Cladophora sp.	7,500–10,000	Lower Swansea Valley, Wales	5027
Oscillatoria sp.	10,000[b]	Lower Swansea Valley, Wales	5027
Ulothrix sp.	14,000[b]	Lower Swansea Valley, Wales	5027
Cladophora glomerata	1,400–60,000	Roding River, England	3146
Mougeotia sp.	74,000[a]	Lower Swansea Valley, Wales	5027
Microspora sp.	76,000[b]	Lower Swansea Valley, Wales	5027
Spirogyra sp.	7,700–80,000	Lower Swansea Valley, Wales	5027
Cladophora glomerata	128,000	Laboratory experiment	5188
Coccomyxa sp.	160,000[a]	Lower Swansea Valley, Wales	5027
Zygnema sp.	160,000[a]	Lower Swansea Valley, Wales	5027
Tribonema sp.	42,000–290,000[a]	Lower Swansea Valley, Wales	5027

[a] Adjacent to zinc smelter waste, [b] near zinc smelting waste

role as an algal growth-limiting nutrient.[530,983,1661,3200] The optimum amount of iron for growth depends on the species investigated and to some extent on the composition of nutrient solution,[1304,1680,2181,3362,5424] especially on the presence of chelating agents. The optimal concentrations reported in the literature vary widely, however.

Iron exists in ionic form in all natural waters although the amount in solution in alkaline waters is very small. Hopkins[2181] has indicated that ionic iron is necessary for the growth of *Chlorella*. However, it is suggested that colloidal iron can be utilized.[1680,4164] Harvey[1948] found that diatoms were able to assimilate ferric hydroxide or ferric phosphate in colloidal or particulate form and to utilize it in their growth. Bitcover and Sealing[381] pointed out that the favorable effect of organic substances on the availability of iron had been noticed as early as 1895, and that citrates and tartrates were used in plant nutrition solutions as early as 1916. Waris[5270] found that EDTA, when added to nutrient solution, makes it possible for desmids to reproduce normally for a long period of time even when other microorganisms are not completely excluded. The author also found that in the presence of EDTA, the pH range within which the reproduction of *Micrasterias* is possible is much widened as compared with purely mineral solution. Although unchelated iron has been reported to stimulate the growth of some algal species,[805] chelated iron is the form most often employed in studies resulting in the enhancement of algal growth[2809,2854,4324,4796] (Figure 5-11). When chelators were found to stimulate growth, their effect was presumed to be due to their ability to solubilize iron, making it more available for algal uptake.[1151,2354] The mechanisms by which chelated iron stimulates algal growth has not been clearly elucidated. In some cases a chelator makes iron more available for uptake by the algae.[102,2809,3000,3353,3619,5229] However, chelators such as EDTA may also affect algal metabolism by reducing the concentration of free metal ions, thus making the metals less available for biological uptake (chelated metals are generally considered unavailable for uptake—see also Chapter 2.11).[60,100,780,4707,4709,4840,4932] It has also been proposed that complexation of iron with a chelator retards the formation of iron-

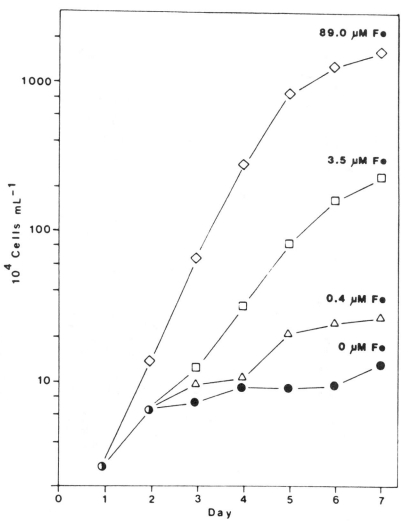

Figure 5-11.   Growth of *Scenedesmus bijugatus* in Bold's basal medium enriched with 0, 0.4, 3.5 and 89.0 μM of EDTA-chelated iron.[4796] (From Storch and Dunham, 1986, in *J. Phycol.*, Vol. 22, p. 112, reprinted with permission of the *Journal of Phycology*.)

phosphorus precipitates.[2548] As a result of this extracellular phenomenon, algal growth may be stimulated in the presence of a chelator, because higher concentrations of soluble phosphorus are available for uptake by the algae. In special cases when bacteria and algae have been grown under iron limitation, they have produced extracellular chelators called siderophores.[3404,3406] Siderophore production is induced under iron-limiting conditions and repressed when the iron concentration is high.

It was suggested that phytoplankton take up iron mainly as the ferrous ion (Fe^{2+}), but ferric ions (Fe^{3+}) may be taken up also.[100,102] Ferrous ions were previously thought to be present in very low amounts, but evidence suggests the they can be formed by (1) oxidation rates ($Fe^{2+} \rightarrow Fe^{3+}$) that are 150 times slower than previous estimates;[102] (2) chelators that can chemically or photochemically reduce Fe^{3+} to Fe^{2+},[2253] and (3) direct photoreduction of Fe^{3+}. Other suggested iron sources have been iron oxides, which, often in associations with colloids, adsorb onto the cell surface and subsequently dissolve, releasing iron for transport into the cell. However, since iron oxide dissolution kinetics are not well understood, the possible importance of this process is unknown.[2886]

The most obvious symptom of iron deficiency is a retardation of growth.[1225,1680,2181,3362,3448,5228-9] Storch and Dunham[4796] found that cell yield and photosynthesis of several planktonic algae were frequently inhibited in the presence of unchelated iron over the range of 3.6 to 53.7 μM iron as $FeCl_3$.

5.3 MANGANESE

5.3.1 Manganese in the Environment

Earth's crust: 950 mg kg^{-1} (490, 903, 3093, 3287)
Common minerals: β-MnO_2 (pyrolusite), MnO (manganosie), $3Mn_2O_3$. $MnSiO_2$ (braunite), Mn_3O_4 (hausmanite), $Mn(OH)_2$ (pyrochroite), MnS (alabandite), $MnFeO_4$ (jacobsite), $MnCO_3$ (rhodochrosite, dialogite), γ-$MnOOH$ (manganite), $K(Mg,Fe^{II})_3(AlSi_3)O_{10}(OH)_2$ (biotite)[491,2257,3076,3287,3705,3815]
Soils: 7–9,200 mg kg^{-1} (2395); 200–3,000 mg kg^{-1} (4849); 1,000 (20–10,000) mg kg^1, may be a major exchangeable cation in very acid soils (490–1); 20–6,000 mg kg^1 (4738)
Sediments: 770 mg kg^{-1} (491)
Air: 0.004–900 ng m^{-3} (491, 2395)
Fresh water: 8 (0.02–130) μg L^{-1} (491), 20 μg L^{-1} (5218), 2–4,800 μg.L^{-1} (3287), up to > 1,000 μg.L^{-1} (3815), particulate manganese accounts for > 90%, and often > 95% of total Mn concentration
Sea water: 0.2 (0.03–21) μg L^{-1} (491)
Land plants: 630 (20–700) mg kg^{-1} (490–1); 1.3–1,840 mg kg^{-1} (2395)
Freshwater algae: 50–25,950 mg kg^{-1} (Table 5-4)
Marine algae: 1–500 mg kg^{-1} (491), 15–15,473 mg kg^{-1} (Table 5-4)
Land animals: 2,000 mg kg^{-1} (490)
Marine animals: 1–60 mg kg^{-1} (490)

Sources to waters: municipal wastewaters, sewage sludge discharge, smelting and refining (iron and steel, nonferrous metals), manufacturing processes (metals, chemicals, pulp and paper), steam electrical production, base metal mining and dressing, atmospheric fallout.[3480]

Aquatic chemistry: Although manganese may exist in oxidation states ranging from -3 to +7, the Mn^{2+} (manganous) and Mn^{4+} (manganic) states are the most important in aqueous systems.[3287] Under neutral and acidic conditions, a simple hydrated cation $[Mn(H_2O)_6]^{2+}$ is the dominant dissolved form. Dissolved complexes $[Mn(OH)]^+$, $[MnCO_3]^+$, $[Mn(OH)_3]^-$, $[MnSO_4]^0$ and at higher chloride concentrations also $[MnCl]^+$, $[MnCl_2]^-$ and $[MnCl_3]^-$ may be present.[3815] Low oxygen concentration at the water-sediment interface causes the reduction of Mn^{4+} to soluble Mn^{2+}, which is then transported upward in the water column. The oxygenated water results in reoxidation of Mn^{2+} to insoluble Mn^{4+}, which settles to the bottom to repeat the cycle.[3287] This, in fact, resembles the iron cycle. Many trace metals sorb to Mn-Fe hydrous oxides in bottom sediments, making these deposits either the primary or secondary site of metal scavenging in water bodies in rivers.[4263]

5.3.2 Manganese in Wetlands

Within the first weeks of submergence almost all active manganese in soil is reduced to Mn^{2+} preceding the reduction of iron. The reduction of Mn is both chemical and biochemical. Acid soils,

Table 5-4. Manganese concentration in algae (mg kg⁻¹ dry mass)

Species	Concentration	Locality	References
Freshwater			
Cladophora glomerata	50	Lake Balaton, Hungary	2575
C. glomerata	235 - > 410	Lake Michigan	20
Cymbella sp.	793	Upper Bee Fork, Missouri	5469
Spirogyra sp.	2,557	Upper Bee Fork, Missouri	5469
Cladophora glomerata	122–10,117	Roding River, England	3146
Mougeotia sp.	13,612	Strother Creek, Missouri	5469
Chlorella pyrenoidosa	241–25,950	Laboratory	2540
Marine			
Dunaliella tertiolecta	3.8	Laboratory	4145
Chlamydomonas sp.	6.8	Laboratory	4145
Endarachne binghamiae	10.7	Hong Kong	2114
Dunaliella primolecta	11.5	Laboratory	4145
Heteromastix longifillis	14.5	Laboratory	4145
Laminaria digitata	20	Canada, Nova Scotia	5528
Olisthodiscus luteus	20	Laboratory	4145
Ascophyllum nodosum	25	Canada, Nova Scotia	5528
Laminaria longicruris	25	Canada, Nova Scotia	5528
Ishige foliacea	10–26	Hong Kong	2113
Porphyra suborbiculata	14–26	Hong Kong	2113
Hemiselmis brunescens	33	Laboratory	4145
Ascophyllum nodosum	9–35	Great Britain	1455
Chlorella salina	48	Laboratory	4145
Fucus vesiculosus	50	Canada, Nova Scotia	5528
Chaetomorpha antennina	16–51	Hong Kong	2113
Asterionella japonica	54	Laboratory	4145
Sargassum hemiphyllum	62	Hong Kong	2114
Tetraselmis tetrathele	62	Laboratory	4145
Monochrysis lutheri	69	Laboratory	4145
Phaeodactylum tricornutum	73	Laboratory	4145
Rhizoclonium riparium	80–87	Hong Kong	2113
Porphyra spp.	14–93	Irish Sea	3882
Ectocarpus siliculosus	109–112	Hong Kong	2113
Blue-green algae	123	Penang Island, Malaysia	4526
Sargassum vachellianum	129	Hong Kong	2114
Scytosiphon lomentaria	16–129	Hong Kong	2113
Fucus vesiculosus	52–130	Great Britain	1455
Colpomenia sinuosa	52–136	Hong Kong	2113
Fucus spp.	33–190	Irish Sea	3882
Enteromorpha sp.	92–221	Penang Island, Malaysia	4526
Gymnogongrus flabelliformis	13–225	Hong Kong	2113
Fucus vesiculosus	108–230	Great Britain estuaries	604
Enteromorpha fluxuosa	74–244	Hong Kong	2113

Table 5-4. Continued

Species	Concentration	Locality	References
Enteromorpha compressa	18–264	Hong Kong	2113
Fucus vesiculosus	90–284	Penang Island, Malaysia	4526
Dermonema frappieri	130–406	Hong Kong	2113
Fucus vesiculosus	65–410	Sweden, coast	4612
Fucus vesiculosus	13–680	Iceland	3341
Caulacanthus okamurai	913–1,704	Hong Kong	2113

high in Mn and organic matter, build up water-soluble Mn concentrations of more than 100 mg L^{-1} immediately after submergence. The level declines thereafter to a stable concentration of around 10 mg L^{-1}. In calcareous soils, the water-soluble Mn may not exceed even 0.5 mg L^{-1}.[3418]

In the waterlogged soils the reduction of manganese (IV) oxide starts when redox potential has become sufficiently low (Eh = 200 mV). Reduction of Mn^{4+} can occur in the presence of low levels of NO_3^-. Although enrichment cultures of bacteria are able to stimulate the reduction of MnO_2 to Mn^{2+},[623] the direct involvement of bacteria in the reduction of MnO_2 has long been doubtful as MnO_2 may be reduced by sulfide or fatty acids produced by anaerobic bacteria.[2315] Laanbroek[2679] pointed out that it has been observed that direct contact between *Clostridium* sp. and MnO_2 was essential for a rapid reduction.[1471] In addition, it has recently been showed by Lovley et al.[2921] that *Alteromonas putrefaciens* is able to grow on hydrogen and MnO_2. Hence, the generation of energy by bacterial reduction of oxidized manganese can no longer be excluded.[2679] Marshall[3076] gave a partial list of microorganisms involved in manganese transformations that shows the range of different morphological types. The list includes conventional bacteria (*Arthrobacter, Pseudomonas*), prosthecate bacteria (*Pedomicrobium, Hyphomicrobium, Metallogenium*), sheathed bacteria (*Leptothrix discophora*), fungi (*Cephalosporium, Cladosporium*), and possible synergistic mixtures (*Corynebacterium-Chromobacterium, Coniotherium-Metallogenium*). Many, but not all, of the microorganisms capable of manganese transformations catalyze similar transformations of iron.[4502] For details on manganese transformations see Marshall.[3076]

Non-biological oxidation of Mn^{2+} may proceed rapidly at pH values > 9.0.[1622,5136] Between pH 7.5 and 9.0, the chemical oxidation of Mn^{2+} is slow, but may be stimulated by hydroxycarboxylic acids. Below pH 7.5, oxidation becomes more and more dependent on microorganisms.[2679] In acid soils with pH < 5.0, there is hardly any oxidation of manganous ions.[3076] It is often very difficult to determine the involvement of bacteria in the oxidation of manganous ions. This is due to the adsorption of Mn^{2+} ions to particulate MnO_2 and the subsequent autoxidation of the adsorbed Mn^{2+} ions.[4194] For further details on mangan oxidation by bacteria see Laanbroek.[2679]

5.3.2.1 Manganese in wetland plants

Concentration in dry mass (mg kg^{-1}) (see Appendix, Table 7-27):
emergent species: aboveground: 13–5,400, belowground: 13–365
floating-leaved species: aboveground: 100–220 (FD), belowground: 86–250 (FD)
submerged species: aboveground: 73–17,100
floating species: 400–13,000
bryophytes: 2–143,000
    Standing stock (see Appendix, Table 7-28) range between 60 and 589 g Mn m^{-2} in aboveground parts of emergent species. Manganese uptake rate in aquatic and wetland plants (see Appendix, Table 7-29) ranges between 9.6 and 30,000 mg Mn m^{-2} yr^{-1}.

5.3.3 Manganese in Algal Nutrition

McHargue[3147] was the first to give definite evidence for the essentiality of manganese, and reported that plants deficient in Mn were stunted in their growth and produced no seeds. The essentiality of

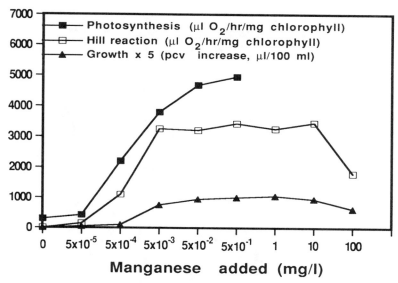

Figure 5-12.   Autotrophic *Chlorella pyrenoidosa*. Photosynthesis, Hill reaction, and growth at various levels of manganese (pcv = packed cell volume).[594] (From Brown et al., 1958, in *Trace Elements,* Lamb, C.A., Ed., p. 145, reprinted with permission of Academic Press.)

manganese for the growth of *Chlorella* was shown originally by Hopkins.[2182] Manganese activated many enzymes and is a constituent of several enzymes.[490,491,3144,3429,3646,4540] Manganese exists in algae in other material than protein. Complexes of Mn and galactosyldiglyceride exist in both green and blue-green algae.[5070]

The rate of photosynthesis is always decreased by manganese deficiency[154,591,594,2467,3804,3807-8,3812] (Figure 5-12). In manganese-deficient cells photosynthesis is lowered but, unlike other mineral deficiencies, there is no immediate effect on respiration, nitrate reduction or the oxidative assimilation of glucose in darkness.[4209]

In 1937 Pirson[3804] discovered that the low photosynthetic activity in manganese-deficient algae could be restored to normal in less than an hour simply by adding manganese salts to the suspension medium (Figure 5-13). The same results were obtained with Mn-deficient cultures of *Ankistrodesmus*.[3812] Homann[2173] found that the time needed to reactivate photosynthesis in Mn-deficient algae varies with each culture, and is often very short when Mn is added not before illumination but during the light period. About 20 years after Pirson's discovery, Kessler[2467] showed that manganese is not required for the photoreduction of CO_2 with hydrogen by adapted green algae. Subsequent studies[156,594,775-7,1303,1306,2173,2553,3807,4122,4651-2] are consistent with Kessler's interpretation that Mn is a catalyst within the oxygen-evolving reactions of photosystem II. Eyster et al.[1306] reported that much higher concentrations of manganese are required for autotrophic growth as compared to heterotrophic growth (Figure 5-14). Brown et al.[594] reported 100- to 1,000-fold higher manganese requirements for *Chlorella* growing in light as compared to the growth in dark. The requirement was demonstrated for green algae as well as for blue-green algae and higher plants.[774,3530] The manganese concentration and concentration factors in algae are summarized in Tables 5-4 and 5-5, respectively.

Arnon et al.[162,163] reported that manganese increases the capacity of isolated chloroplasts to fix carbon dioxide. Glycolate synthesis in *Chlorella* depends upon manganese.[4895] *Chlorella* can grow in the dark in manganese-deficient solution when glucose is supplied, but in the light manganese is essential.[3809] Manganese-deficient *Chlorella* cells were inhibited by high oxygen concentrations that had no effect on healthy cells.[592]

The uptake of manganese follows classical saturation enzyme kinetics (Michaelis-Menten equation) with the regulation by negative feedback control of V_{max}.[4843-4] Sunda and Huntsman[4844] found that the half-saturation constants (K_s) did not vary with the manganese ion concentration of the growth medium. By contrast, the maximum saturating uptake rate (V_{max}) for diatoms *Thalassiosira pseudonana* and *T. oceanica* decreased by about seven-fold with an increase in the free manganese ion concentration in the growth medium.

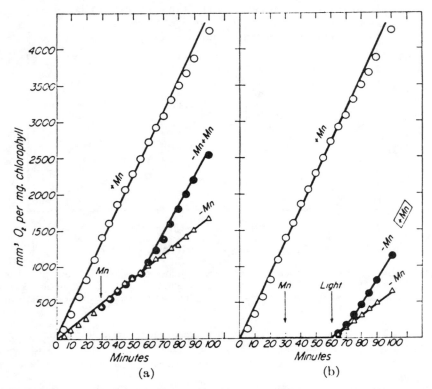

Figure 5-13. Restoration of normal photosynthetic rates in *Scenedesmus obliquus* cells on addition of manganese to -Mn cells. The addition of Mn was made, at the time indicated by arrows, to -Mn cells kept either in light (a) or in the dark (b). -Mn cells which were kept in the dark (b) were exposed to light at the time indicated by the second arrow.[156] (From Arnon, 1958, in *Trace Elements,* Lamb, C.A., Ed., p. 28, reprinted with permission of Academic Press.)

Manganese may be toxic as it behaves as an antimetabolite.[490] At concentrations commonly occurring in surface waters this action can not be achieved, however. Manganese can be inhibitory or toxic in excess amounts; the anionic form was toxic to *Microcystis* at a level of 2 mg L^{-1}.[5149] Manganese was found to be a negative fertility factor in the centric diatom *Ditylum brightwellii* in that auxospores and sperms were formed preferentially in manganese-free medium.[4701]

5.4 ZINC

5.4.1 Zinc in the Environment

Earth's crust: 65–80 mg kg^{-1} (490,903,2129,3093)
Common minerals: ZnS (sphalerite, wurtzite), Zn_2SiO_4 (willemite), $ZnCO_3$ (calamine, smithonite)[3287,4467]
Soils: 10–300 mg kg^{-1} (4849), 40–58 mg kg^{-1} (2129), 50 (10–300) mg kg^{-1} (490), 90 (1–900) mg kg^{-1} (491), 2–50 mg kg^{-1} (4738), 3–762 mg kg^{-1} (2395), in contaminated soils up to 180,000 mg kg^{-1} (2395)
Sediments: 9–470,000 mg kg^{-1} (5192), 95 mg kg^{-1} (491)
Air: 0.002–16,000 ng m^{-3} (491, 2395)
Fresh water: 0.8–330 μg L^{-1} (5192), 15 (0.2–100) μg L^{-1} (491), 1–50 μg L^{-1} (3287)
background levels: 0.5–30 μg L^{-1} (5192)
mine drainage: 6.8–3,840 mg L^{-1} (5192)

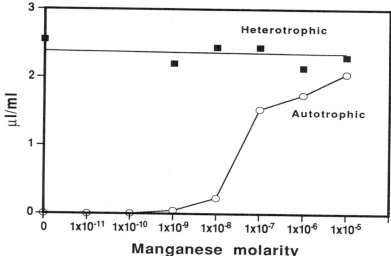

Figure 5-14. Growth of *Chlorella pyrenoidosa* at different concentrations of manganese. Autotrophic: 1 day growth from 85 μL washed cells which had previously been grown in the dark without added manganese. Heterotrophic: 18 days growth from 1 loopful cells.[1305] (From Eyster et al., 1958, in *Trace Elements*, Lamb, C.A., Ed., p. 165, reprinted with permission of Academic Press.)

Table 5-5. Manganese concentration factors (CF) in algae

Species	CF	Locality	Reference
Freshwater			
Cladophora glomerata	3,810–4,260	Šárecký Brook, Czech Republic	5188
C. glomerata	38,000	Laboratory experiment	5188
C. glomerata	2,800–72,000	Roding River, England	3146
Marine			
Enteromorpha intestinalis	18–492	Laboratory	3342
Enteromorpha intestinalis	2,500	Slovenia coast	3342
Fucus vesiculosus	4,600–19,000	South-west England	605
Fucus vesiculosus	19,000	Menai Straits, Wales	1455
Fucus vesiculosus	23,000	Irish Sea	3882
Fucus vesiculosus	23,000–220,000	Bristol Channel	309

Sea water: 4.9 (0.2–48) μg L^{-1} (491)

Land plants: 100 (20–400) mg kg^{-1} (490–1), 1.2–126 mg kg^{-1} (2395), in contaminate sites up to 4,510 mg kg^{-1} (2395)

Freshwater algae: 3.8–219,000 mg kg^{-1} (Table 5-6)

Marine algae: 6–260 mg kg^{-1} (491), 4.6–3,700 mg kg^{-1} (Table 5-6)

Land animals: 160 mg kg^{-1} (490)

Marine animals: 6–1,500 mg kg^{-1} (490)

Sources to waters: manufacturing processes (metals, chemicals, pulp and paper), municipal wastewaters, smelting and refining (iron and steel, nonferrous metals), atmospheric fallout, coal-burning power plants, base metal mining and dressing.[3480]

Aquatic chemistry: Zinc is classified as a borderline metal, forming bonds with both oxygen as well as nitrogen and sulfur donor atoms. Under aerobic conditions, Zn^{2+} is the predominant species at acidic pH, but it is replaced by $Zn(OH)_2$ at pH 8–11, and $Zn(OH)_3^-/Zn(OH)_4^{2-}$ at pH > 11.[5192] Anaerobic conditions lead to the formation of ZnS regardless of pH within the range 1–14. Zinc binds readily

Table 5-6. Zinc concentration in algae (mg kg⁻¹ dry mass)

Species	Concentration	Locality	Reference
Freshwater			
Chara sp.	14.8	Lake Huron	1209
Chara sp.	15	Lake St. Clair marsh, Canada	3346
Cymbella sp.	23	Upper Bee Fork, Missouri	5469
Cladophora glomerata	8.2–23.7	Lake Ontario, Canada	2543
Spirogyra sp.	26	Upper Bee Fork, Missouri	5469
Aphanizomenon flos-aquae	39	Culture	2205
Cladophora rivularis	5.8–84.2	Elsenz River, Germany	355
Chlorella pyrenoidosa	3.8–93	Laboratory	2540
Cladophora glomerata	17–93	Lake Michigan	20
Oedogonium sp.	120	Lower Swansea Valley, Wales	5027
Cladophora sp.	40–160	Lake Huron	1209
C. glomerata	94.6–178.6	Lake Balaton, Hungary	2575
Anacystis nidulans	180	Culture	2205
Cladophora glomerata	62–190	Leine River, Germany	13
Caulerpa prolifera	230	Tübingen, Germany	2205
Scenedesmus acutus	37.2–246	Cultivation (wastewater)	5144
Cladophora sp. + *Hydrodictyon* sp.	262	Strother Creek, Missouri	5469
Not specified	298	Lake Ijseel, The Netherlands	4260
Cladophora glomerata	67–345	Roding River, England	3146
Cladophora glomerata	24–375	New Zealand	4017
Chlamydomonas geitleri	339–460	Cultivation (wastewater)	5144
Cladophora sp.	890–970	Lower Swansea Valley, Wales	5027
Lemanea fluviatilis	990	Derwent Res., North England	1910
Oscillatoria sp.	1,880[b]	Lower Swansea Valley, Wales	5027
Spirogyra sp.	320–1,920	Lower Swansea Valley, Wales	5027
Chara fragilis	1,970	Tübingen, Germany	2205
Ulothrix sp.	3,560[b]	Lower Swansea Valley, Wales	5027
Lemanea sp.	86–3,682	72 sites in Europe	1909
Oscillatoria sp.	3,260–4,030	Strother Creek, Missouri	5469
Mougeotia sp.	4,898	Strother Creek, Missouri	5469
Microspora sp.	9,260[b]	Lower Swansea Valley, Wales	5027
Coccomyxa sp.	19,050[a]	Lower Swansea Valley, Wales	5027
Tribonema sp.	17,440–21,110[a]	Lower Swansea Valley, Wales	5027
Mougeotia sp.	44,940[a]	Lower Swansea Valley, Wales	5027
Zygnema sp.	45,890[a]	Lower Swansea Valley, Wales	5027
Mougeotia sp.	219,000	Caplecleugh Low Level, North England	3703
Marine			
Polysiphonia urceolata	4.6	Rhode Island coast	4389
Laminaria saccharina	9.6	Rhode Island coast	4389
Hizikia fusiforme	11	Hong Kong	2113
Sargassum hemiphyllum	21	Hong Kong	2114

Table 5-6. Continued

Species	Concentration	Locality	Reference
Phycodrys rubens	15.2–22	Rhode Island coast	4389
Ptilota serrata	16.2–23.7	Gulf of Maine	4389
Enteromorpha linza	32–33	Hong Kong	2113
Ascophyllum nodosum	35	Canada, Nova Scotia	5528
Ahnfeltia plicata	38	Helgoland, North Sea	2205
Rhodymenia palmata	41	Canada, Nova Scotia	5528
Laminaria saccharina	46	Helgoland, North Sea	2205
Fucus vesiculosus	49	Canada, Nova Scotia	5528
Endarachne binghamiae	50	Hong Kong	2114
Colpomenia sinuosa	26–52	Hong Kong	2113
Porphyra suborbiculata	11–54	Hong Kong	2113
Corallina officinalis	53–55	Portugal and Spain	4732
Ishige foliacea	36–58	Hong Kong	2113
Scytosiphon lomentaria	53–59	Hong Kong	2113
Ulva lactuca	62	Canada, Nova Scotia	5528
Sargassum vachellianum	63	Hong Kong	2114
Laminaria digitata	64	Canada, Nova Scotia	5528
Chondrus crispus	64	Canada, Nova Scotia	5528
Delesseria sanguinea	72	Portugal and Spain	4732
Chorda filum	85	Portugal and Spain	4732
Enteromorpha compressa	17–86	Hong Kong	2113
Halosaccion ramentaceum	87	Canada, Nova Scotia	5528
Spongomorpha arcta	91	Canada, Nova Scotia	5528
Phyllophora membranifolia	93	Canada, Nova Scotia	5528
Gymnogongrus flabelliformis	46–95	Hong Kong	2113
Laminaria longicruris	97	Canada, Nova Scotia	5528
Gelidium amansii	64–97	Hong Kong	2113
Micromonas squamata	105	Laboratory	4145
Chondrus crispus	145	Portugal and Spain	4732
Ulva lactuca	59–160	Portugal and Spain	4732
Chaetomorpha antennina	22–177	Hong Kong	2113
Porphyra spp.	35–177	Irish Sea	3882
Fucus vesiculosus	16–210	Penang Island, Malaysia	4526
Caulacanthus okamurai	177–212	Hong Kong	2113
Pseudopedinella pyriformis	243	Laboratory	4145
Stichococcus bacillaris	251	Laboratory	4145
Fucus vesiculosus	88–262	Great Britain	1520
Ascophyllum nodosum	82–278	Great Britain	1455
Chlorella salina	301	Laboratory	4145
Enteromorpha flexuosa	28–310	Hong Kong	2113
Phaeodactylum tricornutum	325	Laboratory	4145
Macrocystis integrifolia	10–335	British Columbia coast	5500
Fucus sp.	110–345	Portugal and Spain	4732
Fucus vesiculosus	98–398	Great Britain	1455

Table 5-6. Continued

Species	Concentration	Locality	Reference
Dunaliella primolecta	405	Laboratory	4145
Enteromorpha sp.	19–437	British North Sea coast	4308
Ascophyllum nodosum	59–446	Trondsheimsfjorden, Norway	2694
Fucus spp.	42–450	Irish Sea, North Sea	3882
Hemiselmis brunescens	480	Laboratory	4145
Fucus vesiculosus	216–507	Cardigan Bay, Wales	2286
Fucus vesiculosus	55–666	Trondsheimsfjorden, Norway	2694
Fucus vesiculosus	125–1,080	Sweden, coast	4612
Fucus vesiculosus	199–1,240	South-west England	605
Fucus vesiculosus	85–1,360	Great Britain estuaries	604
Nereocystis luetkeana	7–2,800	British Columbia coast	5500
Ascophyllum nodosum	66–3,700	Norway, coast	1975
Fucus vesiculosus	1,500–3,700	Sorfjørden, Norway	3186
Enteromorpha sp.	950–6,223	Sorfjørden, Norway	3186

[a] Adjacent to zinc smelter waste, [b] near zinc smelting waste

with many organic ligands, particularly in the presence of nitrogen or sulfur donor atoms. Zinc shows variable behavior in binding to suspended particulates, depending on pH and Eh conditions, and the input of anthropogenically derived zinc.[3287]

5.4.2 Zinc in Wetland Plants

Concentration in dry mass (mg kg^{-1}) (see Appendix, Table 7-30):
emergent species: aboveground: 1–337, belowground: 20–440
floating-leaved species: aboveground: 10.6–65, belowground: 21–59
submerged species: 11.5–650 (aboveground)
floating species: 24–3,224
bryophytes: 4–22,300
Standing stock (see Appendix, Table 7-31) ranges between 3.8 and 580 g Zn m^{-2} in aboveground parts of emergent species. Zinc uptake rate in aquatic and wetland plants (see Appendix, Table 7-32) ranges between 7.2 and 3,000 mg Zn m^{-2} yr^{-1}. Blake et al.[390] presented a detailed study on distribution and accumulation of zinc in *Typha latifolia*.

5.4.3 Zinc in Algal Nutrition

Zinc has definitely been established to be required for the growth of both higher plants and algae. The earliest study regarding the essentiality of zinc was the one of Raulin[3991] for *Aspergillus niger*. However, it was not until 1926 that Sommer and Lipman[4623] presented convincing evidence for the necessity of zinc as a plant nutrient. The first demonstration of a zinc requirement in algae was in *Stichococcus bacillaris*.[1225] The essentiality of zinc has been demonstrated in many other species since then and it is now assumed to be universally required by algae.[3530,4093] The isolation and purification of an enzyme, carbonic anhydrase containing 0.33% Zn as a part of its molecule, by Keilin and Mann[2455] offered the first concrete explanation of a mode of action of this element. Zinc was demonstrated to be essential to the mechanism of action of this enzyme, which catalyzes the dehydration of carbonic acid and participates in the elimination and incorporation of carbon dioxide. The discovery of zinc in many highly purified enzymes[5101] has revealed the diversity of its function in protein and carbohydrate metabolism.[2128] Zinc has been established to be a constituent of many metalloproteins and, in addition, it can activate many other enzymes.[490,3646,4540]

Eyster et al.[1305] reported that more zinc is required for the autotrophic growth than for the heterotrophic growth of *Chlorella pyrenoidosa* (Figure 5-15). Stegmann[4713] demonstrated a decrease in the chlorophyll formation and in the photosynthetic activity in cultures deficient in zinc. On the addition of Zn, photosynthesis recovers with a steep decrease in the chlorophyll concentration. Walker[5229] obtained symptoms of zinc deficiency, notably as a relative decrease in dry mass, at concentrations below 10^{-7} mol L^{-1}. Reuter and Morel[4070] reported that zinc-deficient cultures of a diatom *Thalassiosira pseudonana* exhibited reduced silicic acid uptake rates. The disappearance of cytoplasmic ribosomes correlates with zinc deficiencies in *Chlorella*.[3867]

Evidence that copper and zinc compete for the same active sites comes from a study reported by Price and Quigley[3893] in which the growth of *Euglena gracilis* in zinc-limited cultures was markedly higher in media lacking added copper than when copper was present.

5.4.4 Toxicity and Accumulation in Algae

Zinc can act either in a stimulatory mode or in an inhibitory one, depending on its level of availability.[1245] The most important mechanism of the toxic action of zinc is thought to be the "poisoning" of enzymes.[490] De Filippis et al.[1036] found that Zn inhibited NADP-oxidoreductase, thereby significantly lowering the cell's supply of NADPH. Zinc toxicity to algae has been reviewed by Vymazal.[5193]

The division of *Chlamydomonas variabilis* was not altered by up to 12 μmol L^{-1} Zn, although the cell yield obtained during the stationary phase was significantly reduced.[242] The authors developed the equation for the flux of zinc across the cell membrane. If the accumulation of cellular zinc $(Zn)_c$ is a function of the amount of Zn in solution $[(Zn)_b]$, then the flux F (mol cm^{-2} s^{-1}), of Zn across the cell membrane can be described by: $F = d(Zn)_c$ W/dt $S = k_c (Zn^{2+})_b$, where W = mean cellular dry mass (g $cell^{-1}$), S = the mean cellular surface area, assuming a sphere (cm^2 $cell^{-1}$), k_c = the rate constant (1 $cm^{-2} \cdot s^{-1}$).

Toxicity of zinc to algae is influenced by many environmental factors[525,780,1908,3973–4,4301,4437,5407] (see also Chapter 2.11.3.1). Whitton and co-workers[1907,1909,1917,3703,4302–5,4436,5312,5409–10,5412–3] reported the tolerance of algae to zinc from numerous field sites including acid mine drainages. Those algae were e.g. green algae *Hormidium rivulare*, *Ulothrix moniliformis*, *Mougeotia* sp., diatoms *Achnanthes cryptocephala*, *Achnanthes minutissima*, *Amphora veneta*, *Surirella ovata*, *Eunotia exigua*, *Neidium alpinum*, *Pinnularia borealis*, *Caloneis bacillum*, blue-green algae *Chamaesiphon polymorphus*, *Plectonema* sp., *Phormidium autumnale*, *Schizothrix delicatissima*, *Synechococcus aeruginosa*, *Phormidium* sp., *Lyngbya catenata* and euglenoid *Euglena gracilis*. Say and Whitton[4304] found three algal species growing in streams near large active zinc smelters and associated factories at Viviez, near Decazeville, France, with a zinc concentration of 3,800 mg L^{-1}. This concentration is apparently the highest level of zinc anywhere in the literature at which photosynthetic organisms are present. The three species found were: green algae *Hormidium rivulare* and *Hormidium flaccidum* and the diatom *Pinnularia subcapitata*. Kelly and Whitton[2458] found a significant positive relationship between zinc in the green filamentous alga *Stigeoclonium* and that in the water.

Matulová[3107] reported for the green alga *Scenedesmus quadricauda* that the first deleterious effect on the alga was observed at the concentration of 0.002 mg L^{-1} Zn and the lethal effect at the concentration of 0.3 mg L^{-1} Zn. Rachlin and Farran[3945] reported the EC-50$_{96}$ of 2.4 mg L^{-1} for *Chlorella vulgaris*. The concentration of 2.4 mg L^{-1} Zn promoted the growth of *Chlorella vulgaris*, *Euglena viridis* and *Pediastrum tetras* over the three week test period, while 8.7 mg L^{-1} Zn retarded the growth of *Chlorella* and *Euglena*, but not of *Pediastrum*.[840] Jensen et al.[2337] and Rachlin et al.[3946, 3949–50] determined EC-50$_{96}$ for the green alga *Chlorella saccharophila* and diatoms *Navicula incerta* and *Nitzschia closterium* as 7.05, 10.1 and 0.19 mg L^{-1} Zn, respectively. Rosko and Rachlin[4189] found the EC-50$_{96}$ value for *Nitzschia closterium* to be 0.27 mg L^{-1} Zn. Hollibaugh et al.[2159] found Zn to be toxic to marine diatom *Skeletonema costatum* at a concentration of 19.6 μg L^{-1} and to the *Thalassiosira* culture at 65.4 μg L^{-1}. Jensen et al.[2332] found the Zn concentration of 49.7 μg L^{-1} to be inhibitory to the growth of a marine diatom *Skeletonema costatum* and 248.5 μg L^{-1} to be inhibitory to the growth of another marine diatom *Thalassiosira pseudonana*, when these organisms were grown in dialyzing

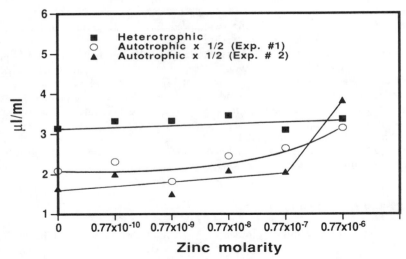

Figure 5-15.   Growth of *Chlorella pyrenoidosa* at different concentrations of zinc. Autotrophic: (upper) 5 days growth from 0.1 μL washed cells; (lower) 3 days growth from 1.4 μL washed cells. Heterotrophic: 35 days growth from 1 loopful cells.[1305] (From Eyster et al., 1958, in *Trace Elements,* Lamb, C.A., Ed., p. 164, reprinted with permission of Academic Press.)

cultures using filtered seawater medium bathed in seawater taken from 30 m in the Trondheim Fjord. Chiaudani and Vighi[780] found that for the green alga *Selenastrum capricornutum* EC-50$_{96}$ was 4.4 μg L^{-1} in AAP medium without EDTA additions and 27 μg L^{-1} in normal AAP medium (300 mg L^{-1} EDTA).

Cushing and Watson[942] reported that killed phytoplankton accumulated significantly more ^{65}Zn than living phytoplankton. This phenomenon is in accordance with experiments of Gutknecht[1854–6] and Bachmann.[196] Gutknecht[1855] hypothesized that the internal chemical changes within the cell as a result of killing increases the number of available sites for cation adsorption by replacement of previously bound other cations. Bachmann[196] explains the differences as either experimental errors, the alteration of the chemical characteristic of the cell by killing, or pH changes due to respiratory activities of the living algae altering the equilibrium. Rose and Cushing[4181] reported that the major site of sorption of zinc-65 in natural, matlike periphyton is on the upper surface of the community. The fact that the zinc-65 is present mainly on the surface and in decreasing amounts within the community indicates that the radionuclides uptake in these communities is largely a surface phenomenon and that there is a diffusion gradient within the periphyton.

Rachlin et al.[3948] observed that Zn caused a reduction in the volume of the intrathylakoidal spaces of the blue-green alga *Plectonema boryanum*. The significance of this finding is not understood, but the result is to bring the thylakoids into closer approximation. This may have some value in facilitating the photosynthetic activity in the metal-stressed cell so that the necessary ATP is manufactured to drive the metabolic machinery required to effect detoxification. Sicko-Goad and Lazinsky,[4473] however, found that relative volume of intrathylakoidal space of *Plectonema boryanum* increased with zinc treatment.

Zinc was taken up by the blue-green alga *Plectonema boryanum*[2336] and sequestered in the cell sectors with polyphosphate bodies. Smaller amounts of zinc were also concentrated in the cell sectors without polyphosphate bodies. The sequestering of Zn in polyphosphate bodies is suggested as serving a dual purpose in cells by providing a storage site for essential metals and acting as a detoxifying mechanism. Jensen et al.[2337] reported that Zn was detected in polyphosphate body cell sectors in *Chlorella saccharophila* starting at 1 mg L^{-1} and in *Navicula incerta* starting at 2.3 mg L^{-1}. Small amounts of Zn were detected in cell sectors away from polyphosphate bodies (in cytoplasmic sectors) in *C. saccharophila* starting at 2.3 mg L^{-1}. No Zn was detected in *Nitzschia closterium* at zinc concentrations up to 2.3 mg L^{-1}.

Although zinc is a micronutrient it is very often found in high concentrations in algal biomass due to accumulation (Tables 5-6 and 5-7).

Table 5-7. Zinc concentration factors (CF) in algae

Species	CF	Locality	Reference
Freshwater			
Ulothrix sp.	730[b]	Lower Swansea Valley, Wales	5027
Oscillatoria sp.	960[b]	Lower Swansea Valley, Wales	5027
Lemanea fluviatilis	1,000	Derwent Res., North England	1910
Cladophora glomerata	1,000	Lake Ontario, Canada	2453
Tribonema sp.	520–1,100[a]	Lower Swansea Valley, Wales	5027
Mougeotia sp.	1,300[a]	Lower Swansea Valley, Wales	5027
Oedogonium sp.	1,500	Lower Swansea Valley, Wales	5027
Coccomyxa sp.	1,700[a]	Lower Swansea Valley, Wales	5027
Microspora sp.	1,900[a]	Lower Swansea Valley, Wales	5027
Cladophora glomerata	2,900	Lake Ontario, Canada	2453
Cladophora glomerata	3,200–3,800	Šárecký Brook, Czech Republic	5188
Zygnema sp.	4,000[a]	Lower Swansea Valley, Wales	5027
Spirogyra sp.	6,600–9,000	Lower Swansea Valley, Wales	5027
Cladophora glomerata	1,100–9,900	Roding River, England	3146
Cladophora sp.	4,000–12,000	Lower Swansea Valley, Wales	5027
Dunaliella sp.	3,880–16,900	Culture	4094
Net plankton	300–19,000	Columbia River, Washington	938
Periphyton	5,000–24,000	Columbia River, Washington	5290
Mougeotia sp.	34,000	Caplecleugh Low Level, North England	3703
Net plankton	3,500–40,000	Columbia River, Washington	5290
Net phytoplankton	8,900–75,600	Columbia River, Washington	940
Cladophora glomerata	100,000	Lake Balaton, Hungary	2575
Phytoplankton	163,750	Columbia River, Washington	939
Cladophora glomerata	314,000	Laboratory experiment	5188
Marine			
Enteromorpha intestinalis	108	Slovenia coast	3342
Enteromorpha intestinalis	70–1,067	Laboratory	3342
Ascophyllum nodosum	1,300	Great Britain	1455
Ecklonia maxima	1,300	Cape of Good Hope	5105
Laminaria digitata	2,455	England	603
Ulva sp.	4,500	Cape of Good Hope	5105
Gigartina rachula	5,300	Cape of Good Hope	5105
Fucus vesiculosus	1,000–6,400	Great Britain	1455
Porphyra capensis	8,500	Cape of Good Hope	5105
Fucus vesiculosus	17,000–20,000	Bristol Channel	3309
Fucus vesiculosus	20,000	Irish Sea	3882
Fucus vesiculosus	7,000–24,000	Norwegian fjord	3186
Fucus vesiculosus	16,600–53,300	Laboratory	604
Fucus vesiculosus	11,000–64,000	England, S-W estuary	605

[a] Adjacent to zinc smelter waste, [b] near zinc smelting waste

5.5 COPPER

5.5.1 Copper in the Environment

Earth's crust: 50–55 mg kg^{-1} (490, 903, 3093, 3287)
Common minerals: CuS_2 (chalcocite), CuS (covellite), Cu_2S (chalcosine), $CuFeS_2$ (chalcopyrite), $CuFe_2S_3$ (cubanite), Cu_2O, (cuprite), $CuCO_3 . Cu(OH)_2$ (malachite), $2CuCO_3 . Cu(OH)_2$ (azurite)[491–2,3287,3815]
Soils: 1–323 mg kg^{-1} (2395), 30 (2–250) mg kg^{-1} (491), strongly absorbed to humus (490), 2–100 mg kg^{-1} (4849), contaminated soils: up to 3,700 mg kg^{-1} (2395)
Sediments: 33 mg kg^{-1} (491), marine sediments: 2–740 mg kg^{-1} (1681,4397)
Air: 0.059–500 ng m^{-3} (492), 0.03–4,900 ng m^{-3} (491,2395), 50–900 ng m^{-3} (2384)
Fresh water: 3 (0.2–30) μg L^{-1} (491), <0.2–135 μg L^{-1} (492), 7 μg L^{-1} (5218), in acid mine drainages < 450 mg L^{-1} (3815)
Seawater: 0.25 (0.05–12) μg L^{-1} (491), 0.034–1.02 μg L^{-1} (492), 0.15 μg L^{-1} (open sea[2131]), 1.0 μg L^{-1} (near-shore waters[2131])
Land plants: 14 (5–15) mg kg^{-1} (490–1), 0.5–33.1 mg kg^{-1} (2395), in contaminated sites up to 1,123 mg kg^{-1} (2395)
Freshwater algae: 0.8–2150 mg kg^{-1} (Table 5-8)
Marine algae: 2–68 mg kg^{-1} (491), 1.39–293 mg kg^{-1} (Table 5-8)
Land animals: 2.4 mg kg^{-1} (490)
Marine animals: 4–50 mg kg^{-1} (491)
    Sources to waters: municipal wastewaters, sewage sludge discharge, manufacturing processes (metals, chemicals, pulp and paper, petroleum products), steam electrical production, smelting and refining of nonferrous metals, base metal mining and dressing, atmospheric deposition.[3479–80]
    Aquatic chemistry: In freshwater at circumneutral pH, most of the inorganic copper in solution is present as complexes with carbonate, nitrate, sulfate, and chloride, rather than as the hydrated divalent cupric ion.[3287] The same distribution is also found in seawater.[2980]
    In some freshwaters, more than 90% of total Cu may be bound to humic acids.[3061] In the presence of equivalent amounts of Co, Cu, Mn, Ni and Zn, humic acids prefer absorption of Cu.[2771,3815,3983] Copper has a strong affinity for hydrous iron and manganese oxides, carbonate materials, clays, and organic matter in bottom sediments.[3287,4263] In surface waters, copper is present mostly in particulate form.[574,3815,5202] Dissolved forms of copper are mostly complex compounds (organic complexes with amino and humic acids and $[CuCO_3(aq)]^0$; simple ionic form forms only a very small portion of the total copper concentration. Higher concentrations of copper may be present only in strongly acid waters.[3815] Aqueous chemistry has been reviewed e.g. by Leckie and Davis.[2739]

5.5.2 Copper in Wetland Plants

Concentration in dry mass (mg kg^{-1}) (see Appendix, Table 7-33):
emergent species: aboveground: <1–1,900, belowground: 1.3–81
floating-leaved species: aboveground: 1.2–9.8, belowground: 1.3–8.1 (FD)
submerged species: 3–206 (aboveground)
floating species: 0–54,500
bryophytes: 4–1,900
    Standing stock (see Appendix, Table 7-34) range between 1.1 and 80,000 g Cu m^{-2} in aboveground parts of emergent species. Zinc uptake rate in aquatic and wetland plants (see Appendix, Table 7-35) ranges between 1.2 and 700 mg Cu m^{-2} yr^{-1}.

5.5.3 Copper in Algal Nutrition

    Sommer[4622] was the first to clearly demonstrate that copper was an essential micronutrient for tomato, sunflower, and flax plants. Since the early work dealing with the relationship of copper to algal growth[1852] it has come to be accepted that all algae have a micronutrient copper requirement.
    Copper is an essential micronutrient, as a constituent of plastocyanin (a protein involved in photosynthetic electron transport[2437]), and as a co-factor for several enzymes. Copper influences the

Table 5-8. Copper concentration in algae (mg kg^{-1} dry mass)

Species	Concentration	Locality	Reference
Freshwater			
Chara sp.	4.7–6.2	Canada, Lake St. Clair	3346
Spirogyra sp.	7.0	Upper Bee Fork, Missouri	5469
Cladophora glomerata	6.4–7.2	Lake Ontario, Canada	2453
Cladophora glomerata	9.1–23	Leine River, Germany	13
Cladophora glomerata	9.5–23.2	Lake Michigan	20
Cladophora rivularis	0.8–31.7	New Zealand	4017
Not specified	32	Lake Ijseel, The Netherlands	4260
Periphyton	22.7–45.5	Elsenz River, Germany	355
Cladophora glomerata	28.7–55.8	Lake Balaton, Hungary	2575
Cladophora sp.	50–60	Lower Swansea Valley, Wales	5027
Chlamydomonas geitleri	1.5–69.7	Cultivation (wastewater)	5144
Cladophora glomerata	4.7–72.4	Roding River, England	3146
Scenedesmus acutus	69.4–80.5	Cultivation (wastewater)	5144
Oedogonium sp.	110	Lower Swansea Valley, Wales	5027
Oscillatoria sp.	173–189	Strother Creek, Missouri	5469
Ulothrix verrucosa	235	Culture (0.01 mM Cu)	1472
Spirogyra sp.	50–290	Lower Swansea Valley, Wales	5027
Mougeotia sp.	299	Strother Creek, Missouri	5469
Oedogonium sp.	300	Culture (0.01 nM Cu)	1472
Chlorella pyrenoidosa	8.2–337	Laboratory	2540
Oscillatoria sp.	340[b]	Lower Swansea Valley, Wales	5027
Draparnardia glomerata	350	Culture (0.01 mM Cu)	1472
Mougeotia sp.	380[a]	Lower Swansea Valley, Wales	5027
Zygnema sp.	460[a]	Lower Swansea Valley, Wales	5027
Ulothrix sp.	480[b]	Lower Swansea Valley, Wales	5027
Coccomyxa sp.	650[a]	Lower Swansea Valley, Wales	5027
Microspora sp.	1,020[b]	Lower Swansea Valley, Wales	5027
Tribonema sp.	400–1,330[a]	Lower Swansea Valley, Wales	5027
Spirogyra cf. *singularis*	1,840	Culture (0.1 mM Cu)	1472
Hormidium sp.	2,150	Culture (0.1 nM Cu)	1472
Marine			
Laminaria saccharina	1.39	Rhode Island coast	4389
Phycodrys rubens	1.5–1.8	Rhode Island coast	4389
Polysiphonia urceolata	2.1	Rhode Island coast	4389
Ptilota serrata	1.8–3.8	Gulf of Maine	4389
Hizikia fusiforme	3.8–4.2	Hong Kong	2113
Sargassum hemiphyllum	4.4	Hong Kong	2114
Sargassum vachellianum	5.0	Hong Kong	2114
Endarachne binghamiae	5.1	Hong Kong	2114
Chorda filum	5.5	Portugal and Spain	4732
Ascophyllum nodosum	6.0	Canada, Nova Scotia	5528
Undaria pinnatifida	1.4–6.9	Suyeong Bay, Korea	2501

Table 5-8. Continued

Species	Concentration	Locality	Reference
Padina arboresceus	3.8–7.4	Hong Kong	2113
Corallina officinalis	7.5–8.0	Portugal and Spain	4732
Rhizoclonium riparium	7.2–8.0	Hong Kong	2113
Fucus vesiculosus	8.0	Canada, Nova Scotia	5528
Delesseria sanguinea	9.0	Portugal and Spain	4732
Fucus vesiculosus	1.6–9.8	Sweden, coast	4612
Scytosiphon lomentaria	6.1–10.0	Hong Kong	2113
Laminaria digitata	11	Canada, Nova Scotia	5528
Laminaria longicruris	12	Canada, Nova Scotia	5528
Corallina pilulifera	4.4–12.2	Hong Kong	2113
Colpomenia sinuosa	7.2–12.9	Hong Kong	2113
Ahnfeltia plicata	13	Canada, Nova Scotia	5528
Porphyra suborbiculata	9.2–13.1	Hong Kong	2113
Chaetomorpha antennina	6.9–16.6	Hong Kong	2113
Gelidium amansii	8.2–17.3	Hong Kong	2113
Porphyra spp.	6.6–19.5	Irish Sea	3882
Chondrus crispus	20	Canada, Nova Scotia	5528
Ectocarpus siliculosus	25	Hong Kong	2113
Ulva lactuca	5.5–26	Portugal and Spain	4732
Rhodymenia palmata	26	Canada, Nova Scotia	5528
Fucus spp.	1.7–28.4	Irish Sea	3882
Fucus sp.	9–31	Portugal and Spain	4732
Monochrysis lutheri	47	Laboratory	4145
Enteromorpha compressa	8.0–51.8	Hong Kong	2113
Stichococcus bacillaris	53	Laboratory	4145
Spongomorpha arcta	55	Canada, Nova Scotia	5528
Dunaliella tertiolecta	57	Laboratory	4145
Ulva lactuca	62	Canada, Nova Scotia	5528
Fucus vesiculosus	35–85	Trondsheimsfjorden, Norway	2694
Hemiselmis virescens	93	Laboratory	4145
Ascophyllum nodosum	6–96	Great Britain	1455
Fucus vesiculosus	7.4–97	Great Britain	1455
Asterionella japonica	105	Laboratory	4145
Fucus vesiculosus	21–107	Sorfjørden, Norway	3186
Ascophyllum nodosum	6–123	Trondheimsfjørden, Norway	2694
Nereocystis luetkeana	3–140	British Columbia coast	5500
Hemiselmis brunescens	188	Laboratory	4145
Heteromastix longifillis	210	Laboratory	4145
Ascophyllum nodosum	3–240	Norway, coast	1975
Macrocystis integrifolia	4–243	British Columbia coast	5500
Enteromorpha flexuosa	7.9–271	Hong Kong	2113
Fucus vesiculosus	4–293	Great Britain estuaries	604

[a] Adjacent to zinc smelter waste, [b] near zinc smelting waste

photosynthetic activity.[3120,4711] Walker[5228] found that *Chlorella* requires Cu when grown in a glucose-urea (or nitrate)-EDTA-salt medium deficient in copper. The copper requirement could not be satisfied by any of 28 elements when ions of these elements were added singly to growth cultures lacking Cu. Nason and McElroy[3390] reported that the most important function of Cu enzymes is to catalyze a direct oxidation of various organic compounds with the atmospheric oxygen.

The concentration of copper in algal biomass is summarized in Table 5-8.

5.5.4 Toxicity and Accumulation in Algae

Copper in water is exceedingly toxic to aquatic biota, in contrast to low toxicity to mammalian consumers of water.[2131] Copper, even though an essential micronutrient, is very toxic to algae and copper sulfate and other copper-containing compounds have been used to control algal blooms in freshwaters since the early 1900s.[233,1362-4,1369-73,3031,3282,3630] Copper inhibits growth as well as photosynthesis of algae.[1525,1851,3530,4957,5394] Toxicity of copper to algae has been reviewed several times.[986,2131,2837] Organic chelators significantly decrease the copper toxicity and also affect uptake of copper.[780,1269,1270,1536,1553,4708] Many studies of copper toxicity to algae have indicated that organically bound copper is not toxic, and Stokes[4790] has also shown that both toxicity to and uptake of copper by *Scenedesmus* can be decreased by addition of EDTA to the medium. The same results are reported for *Selenastrum capricornutum*.[780] On the other hand, Wood[5486] reported that the Cu complexing capacity of some coastal environments may be so great as to reduce the level of biologically available Cu to the point of being nutritionally limiting to phytoplankton. Sunda and Guillard[4840] found that copper toxicity is dependent on the concentration of free Cu^{2+} and not the total copper concentration.

Toxicity effects of copper have been shown to pass through several stages.[4629] First, copper affects the permeability of the plasmalemma, causing loss of K^+ from the cell and changes in cell volume.[1851,3119,3212] Next, Cu^{2+} may be transported to the cytoplasm and then to the chloroplast, where it inhibits photosynthesis[3120,4711] by uncoupling electron transport to $NADP^+$. As the ionic concentration increases, Cu is bound to chloroplast membranes and other cell proteins, causing degradation of chlorophyll and other pigments. At still higher concentrations, copper produces irreversible damage to chloroplast lamellae, preventing photosynthesis and eventually causing death. The toxicity of copper to *Chlorella pyrenoidosa* has been shown to decrease after addition of potassium.[4711]

Concentrations of ionic copper as low as 1 to 2 μg L^{-1} were shown to be poisonous for photosynthesis and growth of unicellular algae.[4708] Gross et al.[1831] found a loss of photosynthetic pigments in *Chlorella* in the presence of copper. The same effect of copper to a natural phytoplankton community dominated by a blue-green alga *Aphanizomenon flos-aquae* has been reported by Whittaker et al.[5394] Shioi et al.[4453] found inhibition of photosystem II by copper in a freshwater green alga *Ankistrodesmus falcatus*.

Copper increases the permeability of the cells.[3119,3212,3671,4249] Copper influences the potassium levels of the cells; when cells of *Chlorella vulgaris* take up Cu from solutions of $CuSO_4$, K is released in amounts which exceed the number of equivalent or divalent copper entering the cells.[3119] Synchronized *Chlorella* cells at the "ripened" stage were prevented from dividing in darkness by cupric ions, especially at pH 6.3.[2417]

Fitzgerald[1370] found a concentration of 50 μg L^{-1} Cu in a commercial preparation to be algicidal to a blue-green alga *Microcystis aeruginosa*, whereas 200 μg L^{-1} as $CuSO_4$ was needed to achieve the same effect. Copper sulfate in low concentrations had no apparent effect on the blue-green alga *Oscillatoria* but concentrations of 1 mg L^{-1} and above had an adverse effect on its growth and resulted in the formation of well defined separation discs followed by marked changes in cell shape and contents.[170] Prescott[3877] gave the amounts required to kill the blue-green alga *Aphanizomenon flos-aquae* as 0.06–0.5 mg L^{-1} Cu. A bloom of the red tides caused by *Gymnodinium breve* was controlled effectively by the application of 180 μg L^{-1} Cu.[4217] Steeman-Nielsen and Wium-Andersen[4708,4710] reported a strong inhibition of photosynthesis of a diatom *Nitzschia palea* in copper concentrations of 1 to 6 μg L^{-1}. Elder and Horne[1227] reported a significant depression of chlorophyll-*a* levels, photosynthesis, and nitrogen fixation at Cu concentrations of 5–10 μg L^{-1} in bioassays using natural algal populations from two California lakes.

Whitton and Shehata[5411] found that the blue-green alga *Anacystis nidulans* is able to build up resistance for copper—the level of strong inhibition is 0.15 mg L^{-1} Cu for a wild type and 0.55 mg L^{-1} for a resistant type. Twiss[5002] demonstrated that strains of *Chlamydomonas acidophila* isolated from acidic, copper-contaminated soils were Cu tolerant; the algistatic Cu concentrations of isolates from Cu-contaminated soils were 20–125 times greater than those of the laboratory strain of *C. acidophila*. Whitton[5404] indicated that some algal species could tolerate copper concentrations exceeding 1 mg L^{-1}

(diatoms *Nitzschia palea, Navicula viridula, Cymbella ventricosa* and *Gomphonema parvulum*) and the diatom *Achnanthes affinis* up to 2 mg[-1].

Several metals low in the electrochemical series are of high toxicity both to the growth and the respiration of *Chlorella vulgaris*.[1960] Copper is unique amongst these in that it is highly toxic when applied to cells under anaerobic conditions, but seldom reduces respiration for many hours when applied at high concentrations in aerated vessels.[1960,3120] Mierle and Stokes[3212] also found that dark, anoxic conditions promoted copper uptake by *Scenedesmus acuminatus*.

Morel et al.[3296] reported that a marine diatom *Skeletonema costatum* was relatively insensitive to cupric ion activity, demonstrating no effect on growth up to $(Cu^{2+}) = 10^{-8.5}$ M. The toxicity of copper was a function of the silicic acid concentration in the medium. The effect was observed in a range of $Si(OH)_4$ concentrations (10^{-5} to 10^{-4} M) above known values for the saturation of silicon uptake kinetics, thus suggesting an influence of copper on silicate metabolism. Reuter and Morel[4070] found that copper toxicity decreased the silicic acid uptake rate. Pace et al.[3614] found that concentrations above 5 mg L[-1] Cu^{2+} are lethal to a marine alga *Dunaliella salina*. Rueter et al.[4223] found that $[^{14}C]CO_2$ fixation in *Oscillatoria* (*Trichodesmium*) *theibautii* decreased in response to copper at low cupric ion activities (10^{-10} M) in test media containing TRIS as a copper chelator and to low concentration (10^{-8} M) of total added copper in filtered seawater. Overnell[3593] found that oxygen evolution in whole cells and chloroplasts (the Hill reaction) of *Chlamydomonas reinhardtii* was suppressed at copper concentrations of 635 to 6,350 μg L[-1]. In work with marine phytoplankton, Overnell[3594] observed a 50% reduction in O_2 evolution in five of the seven test species at copper concentrations ranging from 1.27 to 12.7 mg L[-1].

Sunda and Guillard[4840] demonstrated that growth rate inhibition and Cu content of *Thalassiosira pseudonana* cells were related to Cu^{2+} activity and not to total Cu concentration. Copper inhibited growth rate of *T. pseudonana* at Cu^{2+} activities above 2 ng L[-1] and growth ceased at values above 300 ng L[-1]. Anderson and Morel[94] reported that cells of the dinoflagellate *Gonyaulax tamarensis* were 100% nonmotile at a calculated cupric ion activity of 12 ng L[-1] with 50% of the cells nonmotile at 2 ng L[-1].

Bartlett et al.[224] found that copper at a concentration 300 μg L[-1] was toxic to *Selenastrum capricornutum*. Sublethal effects were found at concentrations of 50–60 μg L[-1], where the lag phase of growth of this alga was extended. Horne and Goldman[2194] found that the addition of 5 μg L[-1] Cu reduced *in situ* nitrogen fixation by 76% in 2 days and by 90 to 95% in 8 days. The addition of 10 μg L[-1] Cu caused a rapid, permanent drop in N_2 fixation. In contrast, rates of photosynthesis were reduced by only 7 to 32%. Erickson[1269] found that population growth and ^{14}C uptake by marine diatom *Thalassiosira pseudonana* displayed inhibition over the entire range of copper added (5–30 μg L[-1]) (Figure 5-16). Growth rate constant decreased with increasing Cu concentration and during the course of growth at each concentration. Correspondingly, mean cell volumes increased with copper concentration and time.

Rachlin and co-workers[3946,3949–50] reported the EC-50[96] values for the diatoms *Nitzschia closterium* and *Navicula incerta* and the green alga *Chlorella saccharophila* 0.016, 10.45 and 0.55 mg L[-1] Cu, respectively. Rosko and Rachlin[4189] found the EC-50[96] value for *Nitzschia closterium* 0.33 mg L[-1] Cu. Fitzgerald[1369] found that a copper concentration of 25 μg L[-1] inhibited phosphorus sorption by 20% and a concentration of 250 μg L[-1] inhibited the sorption completely. Steeman-Nielsen and Kamp-Nielsen[4707] reported that copper affected the cell division in *Chlorella pyrenoidosa*. Steemann-Nielsen et al.[4711] found that copper is bound to the cytoplasmic membrane of the cell, and it is this that prevents cell division. Copper treatment resulted in low proportions of centric diatoms and the complete absence of dinoflagellates 4 days after copper treatment.[4954]

Draparnaldia glomerata accumulates most Cu in the cell wall and has the lowest amount in the chloroplasts, mitochondria and nuclei, whereas *Ulothrix verrucosa* has a more equal distribution between these fractions. *Spirogyra* has more Cu in the great cell organelles while *Hormidium* in the smaller particles, cytoplasm and vacuole content.[1472] The most resistant species to copper were *Mougeotia* sp., *Microspora* cf. *tumidula* and *Hormidium* sp. The very high tolerance of the strain of *Hormidium* seems to be a general feature of this genus, as is its tolerance to other heavy metals.[1916,4307]

The movement of copper into cells is believed to occur mainly by nonmetabolic transport, the plasmalemma is the initial site of copper binding in the sequence leading to intracellular copper uptake.[4791] Thus factors limiting the binding of copper to the plasmalemma or preventing further passage of copper into cells will similarly limit copper toxicity.[4791] Conversely, factors like anaerobic conditions, which were shown to increase copper uptake, have also been observed to increase toxicity.[734] Since copper binds readily to sulfhydryl groups, external binding by S-H groups will remove copper ions that would otherwise bind to sulfhydryl groups on or in the cell.[4791] Potential tolerance mechanisms for copper and algae include exclusion, extracellular binding, precipitation on the outside of the cell membrane, utilization of nonsensitive intracellular sites, and metabolic shunts.[4791]

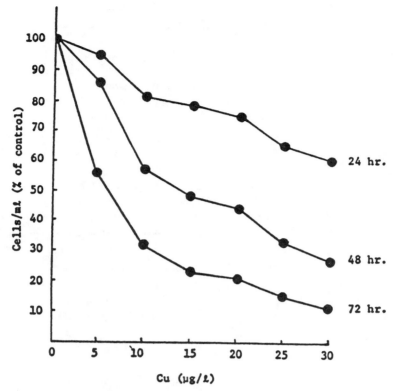

Figure 5-16. Growth of *Thalassiosira pseudonana* in sterile seawater in the presence of copper.[1269] (From Erickson, 1972, in *J. Phycol.*, Vol. 8, p. 319, reprinted with permission of the *Journal of Phycology*.)

Relatively little information is available concerning the physical sites and biochemical forms of copper in algal cells. Silverberg et al.[4501] found intranuclear copper complexes in *Scenedesmus acutiformis*, Nakajima et al.[3373] described a low-molecular-mass copper-binding compound from the cells of *Chlorella vulgaris*.

The similarity of concentration factors from waters of different chemical composition (Table 5-9) and copper content suggests that the amount of copper taken up is a function of the concentration in the water. This has been confirmed experimentally by studies on a number of species, including *Chlorella pyrenoidosa*,[2540] *Scenedesmus acutiformis*,[4794] *Chlorella vulgaris*,[1456,4249] and *Cladophora glomerata*.[2453]

Cells of *Plectonema boryanum* exposed to copper showed the presence of copper in the polyphosphate body cell sectors and to a lesser extent also in the cell sectors without polyphosphate bodies.[2336] Rachlin et al.[3948] found that copper produced a significant increase in cell size and reduction in cellular lipid content in the blue-green alga *Plectonema boryanum*. With copper treatment of the diatom *Fragilaria capucina* there was statistically significant reduction in chloroplast and cytoplasm volumes as well as statistically significant increases in vacuole volume.[4470] Copper treatment also caused a difference in the number of cyanophycin granules in a blue-green alga *Anacystis cyanea*.[4470]

5.6 MOLYBDENUM

5.6.1 Molybdenum in the Environment

Earth's crust: 1.5 mg kg^{-1} (490, 903, 3093, 4920)
Common minerals: MoS_2 (molybdenite), $PbMoO_4$ (wulphenite)[372,491]

Table 5-9. Concentration factors (CF) of copper in algae

Species	CF	Locality	Reference
Freshwater			
Cladophora glomerata	1,000	Lake Erie	4875
Chlorella regularis	1,500	Laboratory	4250
Cladophora glomerata	1,900	Main Duck, Ontario	2453
Cladophora glomerata	2,200	Deadman Bay, Ontario	2453
Cladophora glomerata	2,500	Spokane River, Idaho	1532
Cladophora sp.	1,800–3,500	Lower Swansea Valley, Wales	5027
Scenedesmus acuminatus	4,000	Laboratory	3212
Chlorella vulgaris	700–4,600	Laboratory	1456
Cladophora glomerata	3,720–5,210	Šárecký Brook, Czech Republic	5188
Oedogonium sp.	6,600	Lower Swansea Valley, Wales	5027
Cladophora glomerata	250–9,800	Roding River, England	3146
Ulothrix sp.	10,000[b]	Lower Swansea Valley, Wales	5027
Mougeotia sp.	12,000[a]	Lower Swansea Valley, Wales	5027
Spirogyra sp.	3,300–14,000	Lower Swansea Valley, Wales	5027
Scenedesmus acutiformis	8,700–15,000	Laboratory	4791
Microspora sp.	16,000[b]	Lower Swansea Valley, Wales	5027
Oscillatoria sp.	21,000[b]	Lower Swansea Valley, Wales	5027
Zygnema sp.	35,000[a]	Lower Swansea Valley, Wales	5027
Cladophora glomerata	37,200	Laboratory experiment	5188
Coccomyxa sp.	50,000[a]	Lower Swansea Valley, Wales	5027
Tribonema sp.	13,000–83,000[a]	Lower Swansea Valley, Wales	5027
Cladophora glomerata	10,000–100,000	Lake Balaton, Hungary	2575
Marine			
Fucus vesiculosus	4,500	Irish Sea	3882
Fucus vesiculosus	6,400	Menai Straits, Wales	1455
Ascophyllum nodosum	8,600	Great Britain	1455
Fucus vesiculosus	11,000–14,000	Bristol Channel	3309
Undaria pinnatifida	12,800–15,000	Suyeong Bay, Korea	2501
Fucus vesiculosus	4,800–19,000	Norwegian coast	3186
Sargassum fulvellum	22,800	Suyeong Bay, Korea	2501
Fucus vesiculosus	25,000–27,000	South-west England	605
Fucus vesiculosus	26,000–35,000	Laboratory	604
Fucus vesiculosus	36,000–74,000	Great Britain	4394
Enteromorpha linza	56,000–77,000	Great Britain	4394
Ulva sp.	47,000–86,000	Great Britain	4394
Blidingia minima	71,000–183,000	Great Britain	4394

[a] Adjacent to zinc smelter waste, [b] near zinc smelting waste

Soils: 1.2 (0.1–40) mg kg^{-1} (491) absorbed by humus, especially in alkaline soils[490], 0.013–17.8 mg kg^{-1} (2395), 0.2–5 mg kg^{-1} (4738,4849)
Sediments: 2 mg kg^{-1} (491)
Air: <0.2–10 ng m^{-3} (491, 2395)
Fresh water: 0.1–10 μg L^{-1} (198, 490–1)
Sea water: 0.75–14.0 μg L^{-1} (1988), 10 (4–10) μg L^{-1} (490–1)
Land plants: 0.9 (0.06–3) mg kg^{-1} (490–1), 0.0018–20.5 mg kg^{-1} (2395)

Aquatic and wetland plants: (see Appendix, Table 7-36), standing stock: (see Appendix, Table 7-37)
Marine algae: 0.2–2.0 mg.kg^{-1} (491), 0.23–1.36 mg kg^{-1} (5528)
Land animals: < 0.2 mg kg^{-1} (490)
Marine animals: 0.6–2.5 mg kg^{-1} (490)

5.6.2 Molybdenum in Algal Nutrition

The essentiality of molybdenum for higher plants was established by Arnon and Stout.[159] Previously it was reported that both Mo and V promoted the growth and nitrogen fixation of *Azotobacter*[461] and that molybdenum was necessary for the growth of *Aspergillus niger* and for nitrogen metabolism.[1570,4721] Bortels[462] showed a requirement for molybdenum in nitrogen fixation in blue-green algae and since then molybdenum has been demonstrated as essential for *Chlorella*.[2902,5228]

Like iron, molybdenum can participate in redox reactions by its ability to change valences, in this case between Mo^{5+} and Mo^{6+}. Molybdenum is required in nitrogen metabolism and is necessary for nitrate assimilation and for nitrogen fixation[834,1303,2047,2049,3467,5146,5239] (see also Chapters 4.2.4.1 and 4.2.4.4). It was reported by Bortels[462] and confirmed by Fogg[1399] that Mo has a catalytic effect on N_2 fixation by blue-green algae; along with iron, Mo is a constituent of the nitrogenase enzyme complex.[471,5146] Regarding the essentiality of molybdenum for algae it has been shown that molybdenum-deficient cells of the green alga *Scenedesmus obliquus* fail to assimilate nitrate nitrogen[166,2269] (Figures 5-17 and 5-18). The role of molybdenum as an essential trace element for algae and higher plants in the process of nitrate reduction has been firmly established.[285,2047,2049,3429,5146] The requirement of molybdenum by green alga *Scenedesmus obliquus* was found to be extremely low: to demonstrate marked deficiency the concentration in the nutrient solution had to be reduced to a range between 10^{-9} and 10^{-8} g Mo per liter of nutrient solution.[166] Molybdenum does not affect the growth on ammonium as nitrogen source, but Mo is required for healthy growth of *Anabaena cylindrica* on nitrate or gaseous N_2, the optimal concentrations being about 75 and 200 μg.l^{-1}, respectively[5474] (Figure 5-19).

Photosynthetic rates in cells of a diatom *Navicula pelliculosa* deprived of Mo for 48 hours were significantly lower than in non-deprived cells.[5239] Mo-deficiency in *Anabaena cylindrica* results in a decrease of all the organic nitrogen fractions with the exception of amide.[5475] Mo-deficiency in *Scenedesmus obliquus* resulted in failure of nitrate assimilation and accumulation of starch and a general appearance similar to that of nitrogen-starved cells. The deficiency was also associated with a marked reduction in chlorophyll content.[166] Molybdenum-deficient *Oocystis* cells have been shown to have a low total nitrogen content,[3978] and the nitrate reduction coupled with hydrogenase in *Anabaena* can be stimulated by 0.025 micromolar levels of Mo.[1529]

The concentration of molybdenum in algae is low, the reports in the literature are only rare, however. Young et al.[5528] reported the range of 0.23 to 1.36 mg kg^{-1} for eleven species of marine seaweeds and Kovács et al.[2575] reported the concentration of 0.6 to 2.9 mg kg^{-1} for *Cladophora glomerata* in Lake Balaton in Hungary.

Molybdenum has been shown to be a limiting factor in nature for the growth of algae in Castle Lake, California.[1683-5] Subsequently, it has been reported that Mo stimulated photosynthesis in other lakes.[63,1686]

5.7 SODIUM

5.7.1 Sodium in the Environment

Earth's crust: 33,000 mg kg^{-1} (490)
Common minerals: NaCl (halite), $NaHCO_3$ (nahcolite), Na_2CO_3 . 10 H_2O (soda), Al-Si minerals, e.g.
    $NaAlSi_3O_8$ (albite = Na-feldspar) and $Al_2Si_2O_5(OH)_4$ (kaolinite)[3705,3815,4819]
Soils: 5,000 (150–25,000) mg kg^{-1} (491), a major exchangeable cation in most soils, especially in very alkaline soils[490]
Sediments: 5,700 mg kg^{-1} (491)
Air: 7–7,000 ng m^{-3} (491)
Fresh water: 6.0 (0.7–25) mg.l^{-1} (491)
Sea water: 10.5–10.77 g.l^{-1} (490–1)
Land plants: 1,200 (35–1,500) mg kg^{-1} (490–1)
Freshwater algae: 900–28,500 mg kg^{-1} (Table 5-10)

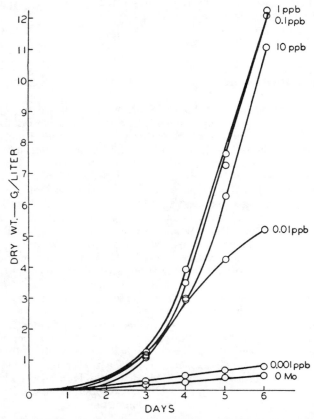

Figure 5-17.   Effect of molybdenum on the dry weight (g dry weight of cells per liter of nutrient solution) of *Scenedesmus obliquus* (10^{-6} g Mo per liter of nutrient solution = 1 ppb).[166] (From Arnon et al., 1955, in *Physiol. Plant,* Vol. 8, p. 544, reprinted with permission of the Physiologia Plantarum.)

Marine algae: 26,000–41,000 mg kg^{-1} (491), 6,100–47,200 (Table 5-10)
Land animals: 4,000 mg kg^{-1} (490)
Marine animals: 4,000–48,000 mg kg^{-1} (490)

   Sources to waters: natural weathering of minerals, industrial wastewaters (production of chemicals), agricultural wastewaters.[3815]

   Aquatic chemistry: Sodium occurs in waters mostly as a simple cation Na$^+$, in more mineralized waters ionic associations such as [NaHCO$_3$]^0, [NaCO$_3$]$^-$ or [NaSO$_4$]$^-$ may be present.[3815]

5.7.2 Sodium in Wetlands

   The sodium cycling in wetlands has not been studied in detail. Several authors reported sodium leaching from decomposing wetland plants[503,995,1743,2642] (see also Figure 3-11). Proctor[3916] reported sodium concentrations in water samples from ombrogenous bogs in Britain and Ireland.

5.7.2.1 Sodium in wetland plants

Concentration in % dry mass (see Appendix, Table 7-38):
emergent species: aboveground: 0.020–1.22, belowground: 0.08–1.18
floating-leaved species: aboveground: 0.009–1.56

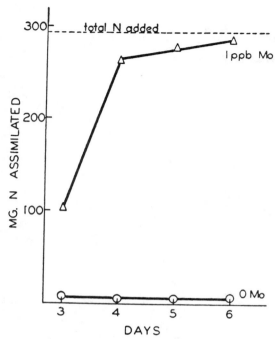

Figure 5-18. Effect of molybdenum on nitrate nitrogen assimilation by *Scenedesmus obliquus*. Ordinate represents mg N assimilated by cells contained in one liter of nutrient solution.[166] (From Arnon et al., 1955, in *Physiol. Plant.*, Vol. 8, p. 547, reprinted with permission of the Physiologia Plantarum.)

submerged species: 0.11–2.87 (aboveground)
floating species: 0.005–0.49
bryophytes: 0.012–0.79
Standing stock (g Na m⁻²) (see Appendix, Table 7-39):
emergent species: aboveground: 0.001–45.7, belowground: 0.31–4.34
floating-leaved species: 0.16–0.92 (aboveground)
submerged species: 0.17–2.2 (aboveground)
Sodium uptake rates in aquatic and wetland plants (see Appendix, Table 7-40) range between 19 and 73 g Na m⁻² yr⁻¹ and between 4.6 and 127.6 mg Na m⁻² d⁻¹.

5.7.3 Sodium in Algal Nutrition

Since sodium and potassium have similar chemical properties, the early experiments dealing with sodium and algal growth were concerned with its possible replacement for potassium.[3530] Benecke[310] described an *Oscillatoria* sp. that grew when potassium in the medium was replaced by sodium. Emerson and Lewis[1239] grew a *Chroococcus* sp. in the presence of sodium and 1 mg L⁻¹ of potassium; however, the *Chroococcus* grew very poorly in the absence of sodium. Allen[64] found 23 strains of blue-green algae that would grow in sodium salt media lacking added potassium.

Sodium is not generally regarded as an absolute requirement for the majority of algae, but the blue-green algae are among the few plants that have an absolute Na requirement.[70] *Anabaena cylindrica* was found to require Na and this could not be substituted for by K, Li, Rb or Cs; 5 mg L⁻¹ of Na allowed optimal growth.[70] This was the first record of the essentiality of Na for an autotrophic plant, but subsequent work suggested that other blue-green algae require Na (in relatively high concentrations).[245,398,596,2589,3160,4783] Requirements of blue-green algae for sodium had also been reported earlier by Benecke,[310] Allen,[64,65] Emerson and Lewis[1239] and Gerloff et al.[1615]

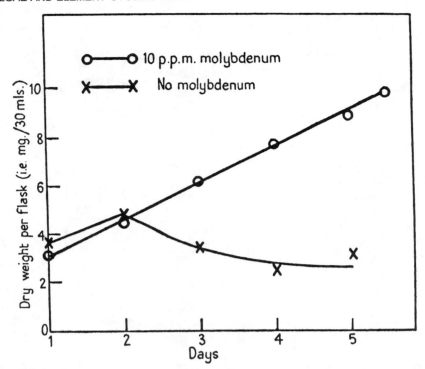

Figure 5-19.   The growth of nitrogen-fixing cultures of *Anabaena cylindrica* in molybdenum-deficient and normal media containing 0.05 g per L of EDTA. Each point is the mean of the dry weights of duplicate samples.[5474] (From Wolfe, 1954, in *Ann. Bot. N.S.,* Vol. 18, p. 305, reprinted with permission of the *Annals of Botany*.)

Sodium is required for photosynthesis,[1566,3018] bicarbonate transport,[3018,4057] urea and nitrate transport,[2708,4050] silicate uptake,[4828] intracellular pH regulation,[2426] alkalotolerance[3226] and affect nitrate reduction in blue-green algae.[5267] Both ammonium and amino acid uptake depend on external Na^+ concentrations.[1391,2018,5364] Photosynthesis in some blue-green algae has also been shown to be stimulated by elevated Na^+ concentrations.[3224,3226,4057] Patel and Merrett[3672] reported that sodium ions increase the affinity of the diatom *Phaeodactylum* for HCO_3^- and even at high HCO_3^- concentrations sodium ions enhance HCO_3^- utilization.

Ward and Wetzel[5267] reported that the blue-green alga *Anabaena cylindrica* with no Na^+ added suffered from decreased rates of acethylene reduction, C assimilation, excretion of organic C as well as lower concentrations of chlorophyll-*a* and particulate organic C compared to cultures supplied with 5, 10, and 50 mg L^{-1} Na. Sodium deficient cells released extracellularly a higher percentage of previously fixed C as anorganic C. No differences in any parameter measured were demonstrable among cultures grown with 5, 10, and 50 mg L^{-1} Na.[5267] Sodium-deficient cultures resulted in enhanced nitrate reductase activity and accumulation of toxic levels of nitrate. Higher levels of Na were required to avoid deficiency symptoms in cultures grown with NO_3^-.[5267] Sodium in the concentration up to 12 mg L^{-1} enhanced growth of the blue-green alga *Anabaena flos-aquae* under nitrate-sufficient but not nitrate-limited conditions.[3541] The authors suggested that the increase in ambient sodium levels in many lakes may provide a competitive advantage to blue-green algae.

Since Scott and Hayward[4380] described a sodium excretion mechanism in *Ulva lactuca*, evidence for "sodium pump" has been reported for many other marine algae.[3009–10,4381] The site of such pumps is no doubt the plasmalemma.[1254] The possibility of a linked K-Na pump has been considered for *Porphyra*[1252–3] and *Nitellopsis*.[961,3010] In *Porphyra*, the presence of an external supply of K or Rb ions is necessary before net Na excretion can be demonstrated, and K accelerates even the passive Na efflux in "Na-free" solutions. Accumulation of K or Rb occurs as Na is excreted.

There is evidence that large amounts of sodium may be inhibitory, which may account for the lack of blue-green algae in marine environments. However, in inland saline lakes they are often abundant.[4209]

Table 5-10. Sodium concentration in algae (% dry mass)

Species	Concentration	Locality	Reference
Freshwater			
Enteromorpha prolifera	0.09–0.13	Forfar Loch, Scotland	2107
Chara vulgaris	0.28	Czech Republic	1207
Chlamydomonas geitleri	0.10–0.40	Cultivation (wastewater)	5144
Scenedesmus acutus	0.10–0.60	Cultivation (wastewater)	5144
Cladophora glomerata	0.48–0.96	Lake Michigan	20
Filamentous algae[a]	1.36–2.85	Nesyt Pond, Czech Republic	2843
Marine			
Ulva lactuca	0.61–1.54	Hong Kong coast	2111
Halosaccion ramentaceum	1.63	Canada, Nova Scotia	5528
Rhodymenia palmata	2.50	Canada, Nova Scotia	5528
Ahnfeltia plicata	2.63	Canada, Nova Scotia	5528
Phyllophora membranifolia	2.79	Canada, Nova Scotia	5528
Laminaria longicruris	2.83	Canada, Nova Scotia	5528
Phaeodactlym tricornutum	1.28–3.07	Laboratory	1987
Laminaria digitata	3.20	Canada, Nova Scotia	5528
Ascophyllum nodosum	3.38	Canada, Nova Scotia	5528
Fucus vesiculosus	3.38	Canada, Nova Scotia	5528
Ulva lactuca	3.51	Canada, Nova Scotia	5528
Chondrus crispus	3.58	Canada, Nova Scotia	5528
Spongomorpha arcta	4.72	Canada, Nova Scotia	5528

[a] Spirogyra sp., Oedogonium sp., Cladophora fracta, Enteromorpha intestinalis

5.8 COBALT

5.8.1 Cobalt in the Environment

Earth's crust: 20–25 mg kg^{-1} (490, 903, 3093, 3287)
Common minerals: CoS_2, $CoAs_2$ (safforite), Co_3S_4 (linnaeite), $CuCo_2S_4$ (carrollite), $(Co,Fe)As_3$ (skutterudite), $Co_3(AsO_4) \cdot 8 H_2O$ (erythrite)[491,3287]
Soils: 1–40 mg kg^{-1} (4849), 0.1–122 mg kg^{-1} (2395), 1–300 mg kg^{-1} (4738), higher in soils derived from basalt or serpentine,[490-1] in contaminated sites up to 520 mg kg^{-1} (2395)
Sediments: 14 mg kg^{-1} (491)
Air: 0.0008–37 ng m^{-3} (491), 0.0001–6.79 ng m^{-3} (2395)
Fresh water: 0.2 (0.04–8) µg L^{-1} (491), 0.5 µg L^{-1} (5218)
Sea water: 0.02 (0.01–4.1) µg L^{-1} (491)
Land plants: 0.5 (0.005–1.0) mg kg^{-1} (490–1), 0.00018–1.7 mg kg^{-1} (2395)
Aquatic and wetland plants: (see Appendix, Table 7-41)
emergent species: 0.01–6.72 mg kg^{-1} (aboveground)
floating-leaved species: 1.0–2.8
submerged species: <1–36.1 mg kg^{-1}
floating species: 0.33–18.6 mg kg^{-1}
bryophytes: 0.7–1,700 mg kg^{-1}
Freshwater algae: 0.04–15.8 mg kg^{-1} (Table 5-11)
Marine algae: 0.02–9 mg kg^{-1} (491) 0.09–12 mg kg^{-1} (Table 5-11)
Land animals: 0.03 mg kg^{-1} (490)
Marine animals: 0.5–5.0 mg kg^{-1} (490)
Sources to waters: municipal sewage, coal gasification, burning of fossil fuels, sea salt sprays.[3287]

Table 5-11. Cobalt concentration in algae (mg kg^{-1} dry mass)

Species	Concentration	Locality	Reference
Freshwater			
Anacystis nidulans	0.04	Culture	2205
Chara sp.	0.26	Lake Huron	1209
Chara fragilis	0.92	Tübingen, Germany	2205
Aphanizomenon flos-aquae	2.1	Culture	2205
Cladophora sp.	1.1–2.6	Lake Huron	1209
Cladophora glomerata	4.3•8.9	Lake Balton, Hungary	2575
Chara sp.	7.4–15.8	Canada, Lake St. Clair	3346
Marine			
Spongomorpha arcta	0.09	Canada, Nova Scotia	5528
Rhodymenia palmata	0.13	Canada, Nova Scotia	5528
Laminaria saccharina	0.19	Helgoland, North Sea	2205
Laminaria digitata	0.25	Canada, Nova Scotia	5528
Chondrus crispus	0.39	Canada, Nova Scotia	5528
Ascophyllum nodosum	0.40	Canada, Nova Scotia	5528
Ahnfeltia plicata	0.49	Canada, Nova Scotia	5528
Laminaria longicruris	0.51	Canada, Nova Scotia	5528
Fucus vesiculosus	0.66	Canada, Nova Scotia	5528
Ulva lactuca	0.69	Canada, Nova Scotia	5528
Ahnfeltia plicata	1.4	Helgoland, North Sea	2205
Fucus vesiculosus	0.1–4.3	Sweden, coast	4612
Phyllophora membranifolia	6.25	Canada, Nova Scotia	5528
Fucus vesiculosus	0.9–7.8	Great Britain estuaries	604
Fucus vesiculosus	1–12	Iceland	3341

Aquatic chemistry: In freshwater, the dominant species are Co^{2+}, $CoCO_3$, $Co(OH)_3$, and CoS. Although lesser amounts of $CoSO_4$ and $CoCl^+$ may also be detected, chloride complexes dominate in seawater.[3287] Hutchinson[2257] pointed out that cobalt cycle, with some exceptions, resembles iron cycle in water bodies.

5.8.2 Cobalt in Algal Nutrition

Cobalt, constituting 4% of vitamin B_{12} (Figure 5-20), appears to be universally essential within the plant kingdom as a micronutrient. Its only known biological function is associated with this vitamin, although organic compounds containing Co are known which reversibly bind oxygen.[617,5263] Cobalt also activates several enzymes,[490,3390,4540] e.g. those which are necessary for nitrogen fixation.

Hutner et al.[2267] were the first to demonstrate an algal requirement for cobalt, as vitamin B_{12} required by *Euglena*; a cobalt requirement by blue-green algal species that could be satisfied using the inorganic form of this element was demonstrated a few years later.[2168] The authors found that cobalt is an essential nutrient for blue-green algae *Nostoc muscorum*, *Calothrix parietina*, *Coccochloris peniocystis* and *Diplocystis aeruginosa*. The quantitative requirement for Co was extremely low, for while at least 0.04 μg L^{-1} Co was ordinarily required for optimal growth; significant increases in biomass and nitrogen content was obtained with concentrations down to 0.02 μg L^{-1} Co. Increases in yield were obtained over a wide Co concentration range. Krauss[2591] reported that cobalt also stimulated the growth of a green alga, *Scenedesmus obliquus*. A number of marine algae either require vitamin

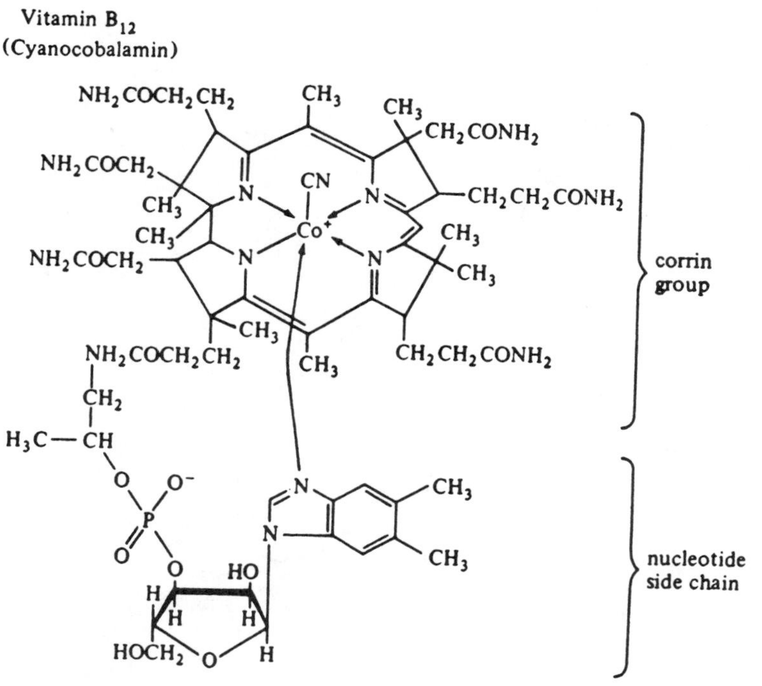

Vitamin B$_{12}$
(Cyanocobalamin)

Figure 5-20.   Chemical structure of vitamin B$_{12}$.

B$_{12}$ or have their growth stimulated by its addition to the culture medium[409,1152,1504–5,1507,3724,3926,4903] (for further details see Chapter 2.9.2.1). Cobalt can partially replace nickel in restoring urease activity of *Phaeodactylum* and *Cyclotella*.[3542,4048]

5.8.3 Toxicity and Accumulation in Algae

Blankenship and Wilbur[393] found that cobalt produced effects on cell division and calcium uptake, reduced cell protein synthesis and increased cell volume in the calcifying alga *Cricosphaera carterae* (now called *Hymenomonas carterae*). At higher cobalt concentrations, the rate of division decreased approximately linearly with concentration until division was blocked almost completely at 100 μM Co (Figure 5-21). The authors also found that cell cobalt can affect the Ca uptake, depending upon the concentration of Co. Ca uptake increased by 53% in 100 μM Co and decreased by 40% in 200 μM Co as compared to cells in 0.09 μM Co.

Cobalt has been found to inhibit transmembrane ion movement and cell division.[1863,4313] Ultrastructurally, Co caused enlargement of the cell vacuole and the appearance of membrane-bound vacuoles containing electron dense bodies.[393] Kumar and Kumar[2643] found growth inhibition of a blue-green alga *Nostoc linckia* to be directly proportional to increasing concentration of cobaltous chloride. Agrawal and Kumar[32] studied the response of a blue-green alga *Anacystis nidulans* to cobalt. They found no significant growth inhibition at Co concentration of 58.9 μg L^{-1}, the growth inhibition was recorded at 589 μg L^{-1} Co, and cell division stopped at Co concentration of 5.9 mg L^{-1}. Chiaudani and Vighi[78] found EC-50$_{96}$ for a green alga *Selenastrum capricornutum* 3.7 μg L^{-1} in the presence of EDTA and 16 μg L^{-1} without EDTA.

Cells exposed to cobalt show the presence of the element in the polyphosphate body cell sectors but it is not detectable in the cell sectors without polyphosphate bodies of the blue-green alga *Plectonema boryanum*.[2336] Rachlin et al.[3948] found that cobalt caused significant decrease in cell size and significant increase in the surface area of the cell's thylakoids of the blue-green alga *Plectonema boryanum*. Cobalt also caused a significant reduction in the volume of the intrathylakoidal spaces, production of extra intracellular membrane whorls, and coalescence of cellular lipid.

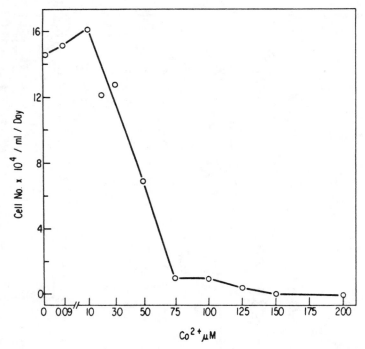

Figure 5-21. Co effects on growth in *Cricosphaera carterae* for 11 concentrations from 0–200 μM. Ordinate shows increase in cell number of log phase cells during 115 h; 18°C.[393] (From Blankenship and Wilbur, 1975, in *J. Phycol.*, Vol. 11, p. 213, reprinted with permission of the *Journal of Phycology*.)

Cobalt tends to accumulate in algae, the values of concentration factors are summarized in Table 5-12.

5.9 BORON

5.9.1 Boron in the Environment

Earth's crust: 10 mg kg^{-1} (491, 3093, 4920)
Common minerals: Na(MgFe)$_3$Al$_6$(BO$_3$)$_3$Si$_6$O$_{18}$(OH)$_4$ (tourmalite), Na$_2$B$_4$O$_7$ (borax), H$_3$BO$_3$ (sassoline)[257,491,3705,3815]
Soils: 2–100 mg kg^{-1} (4738,4849), 20 (2–270) mg kg^{-1} (491), <1–467 mg kg^{-1} (2395)
Sediments: 100 mg kg^{-1} (491)
Air: 4 ng m^{-3} (491)
Fresh water: 15 (7–500) μg L^{-1} (491), < 500 (<10–3,690) μg L^{-1} (3287), 850 mg L^{-1} in Lake Borax in California[5343]
Sea water: 3–5 mg L^{-1} (490–1, 2271, 3287)
Land plants: 50 (11–140) mg kg^{-1} (490–1), 0.3–120 mg kg^{-1} (2395)
Aquatic and wetland plants: (see Appendix, Table 7-42):
emergent species: 1.2–100 mg kg^{-1} (aboveground)
floating leaved species: 8.2–11.3 mg kg^{-1} (aboveground)
submerged species: 0.47–959 mg kg^{-1}
floating species: 1.06–2,567 mg kg^{-1}
bryophytes: 18–10,100 mg kg^{-1} (FD)
Standing stock: emergent species, 0.5–8 g m^{-2} (512,516), floating-leaved species, 0.7–2.9 g m^{-2} (516)
Freshwater algae: 1.07–255 mg kg^{-1} (Table 5-13), concentration factors: 93–10,000 (Table 5-14)
Marine algae: 100–160 mg kg^{-1} (491)

Table 5-12. Cobalt concentration factors (CF) in algae

Species	CF	Locality	Reference
Freshwater			
Cladophora glomerata	2,700–6,270	Šárecký Brook, Czech Republic	5188
C. glomerata	10,000	Lake Balaton, Hungary	2575
C. glomerata	43,330	Laboratory experiment	5188
Phytoplankton	61,700–152,000	Columbia River, Washington	940
Phytoplankton	455,000	Columbia River, Washington	939
Marine			
Enteromorpha intestinalis	1.8–40.4	Laboratory	3342
Enteromorpha intestinalis	960	Slovenia coast	3342

Land animals: 0.5 mg kg^{-1} (490)

Marine animals: 20-50 mg kg^{-1} (490)

Sources to waters: municipal sewage, sewage sludge, industrial wastewaters (e.g. detergent production, photographic industry, glass and china industry).[3287,3815] Natural weathering releases substantially lower amounts than those reported for other elements including Al, Fe, Mn and Pb, and is within an order of magnitude of these reported for Ag and Mo.[5333]

Aquatic chemistry: Boron has two oxidation states, B° and B^{3+}, that form various boranes (hydrides) and organoboron complexes. The elemental form is not normally found in surface waters. Boric acid, H_3BO_3, the most important species in freshwaters, is very weak and moderately soluble, and does not readily dissociate. At neutral pH of natural waters and at acid pH the non-ionic form predominates, while at alkaline pH anion $[B(OH)_4]^-$ predominates. At acid and neutral pH boric acid is also the predominant form in saltwater, and is followed in importance by the borate anion, $B(OH)_4^-$.[3278,3815] Water soluble boron is generally low in acidic soils under high rainfall conditions as a result of leaching.[516]

5.9.2 Boron in Algal Nutrition

Boron has long been recognized as an essential element for higher plants, conclusive evidence for its requirement was provided by Warington[5269] and Sommer and Lipman.[4623] Higher plants, but not animals, bacteria, or fungi have an absolute requirement for boron for normal growth and development, with dicotyledons showing more extensive deficiency symptoms than do the monocots.[4540-1]

Gerloff[1609] pointed out that although the essentiality of boron for higher plants has been recognized for many years, there has been a lack of agreement on the requirements of algae, fungi, and bacteria. Davis et al.[911] reported boron to be essential for *Aspergillus niger* and *Penicillium italicum*; Steinberg[4722] later reported that *Aspergillus niger* did not require the element. Gerretsen and de Hoop[1618] concluded that boron was essential for *Azotobacter chroococcum*. Anderson and Jordan[97] found boron to enhance nitrogen fixation in this bacterium, but they did not consider boron as essential.

Hercinger[2040] was first to report a boron requirement for *Chlorella*, but this was followed by experiments indicating that *Chlorella* had no boron requirement.[4713] McIlrath and Skok[3148] reported boron to be an essential element for *Chlorella vulgaris* on the basis of effects both on cell numbers and cell weights and Bowen et al.[493] reported that several species and strains of *Chlorella* did not require boron. Eyster[1299,1300-1] has shown that the blue-green alga *Nostoc nuscorum* requires boron for growth (Figure 5-22). Boron has been shown to be an essential micronutrient for the growth of cultures of marine and freshwater diatom species.[2801-3] The boron requirement of marine diatoms appears to be greater than that of freshwater diatoms; this may be related to the higher B concentration in the oceans compared with fresh water.[2802] Gerloff[1609] revealed that three species of blue-green algae required boron for nitrogen fixation, but boron was not essential for the growth of three green algae. The blue-green algae adsorbed appreciable amounts of boron, but the green algae did not accumulate significant amounts. A requirement for boron has been demonstrated in *Fucus* embryos[3161] and in several micro-algae.[2814] In *Ulva* and *Dictyota* boron is reported to stimulate reproductive growth specifically.[3391]

Table 5-13. Boron concentration in freshwater algae (mg kg⁻¹ dry mass)

Species	Concentration	Locality	Reference
Chlorella pyrenoidosa	1.07	Laboratory culture	1609
Stigeoclonium tenue	1.42	Laboratory culture	1609
Draparnaldia plumosa	2.65	Laboratory culture	1609
Rhizoclonium sp.	3.4	Alabama, Mississippi	501
Microcystis aeruginosa	3.6	Alabama, Mississippi	501
Euglena sp.	3.8	Alabama, Mississippi	501
Spirogyra sp.	4.2	Alabama, Mississippi	501
Aphanizomenon flos-aquae	4.6	Alabama, Mississippi	501
Chara sp.	6.1	Alabama, Mississippi	501
Mougeotia sp.	8.0	Alabama, Mississippi	501
Oedogonium sp.	8.1	Alabama, Mississippi	501
Nitella sp.	9.8	Alabama, Mississippi	501
Hydrodictyon reticulatum	11.4	Alabama, Mississippi	501
Nostoc muscorum	25	Laboratory culture	1609
Pithophora oedogonia	65.3	Alabama, Mississippi	501
Cladophora glomerata	84.6	Alabama, Mississippi	501
Lyngbya sp.	119.7	Alabama, Mississippi	501
Cladophora glomerata	33.5–247	Lake Balaton, Hungary	2575
Calothrix parietina	255	Laboratory culture	1609

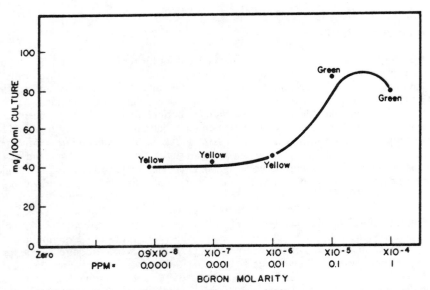

Figure 5-22.   Effect of different levels of boron on the growth of *Nostoc muscorum* in the absence of nitrate, urea, and ammonium slats but in the presence of nitrogen gas. Cultures grown in 5 weeks, while being shaken, bubbled with 5% CO_2 in air, and illuminated with 1,500 ft-c cool white fluorescent light. Inoculum: 0.3 μL cells from boron-deficient culture.[1303] (From Eyster, 1964, in *Algae and Man,* Jackson, D.F., Ed., p. 100, reprinted with permission of Plenum Publishing Corporation, New York.)

Presence of boron in the nutrient medium for *Chlorella* was shown to be salutary, though not essential,[3148] and even its salutory effect has been disputed.[493,3117] Dear and Aronoff[1021] reported no limitation by boron either in cell number or dry mass of *Scenedesmus obliquus*. McBride et al.[3118] also found that boron concentrations, in culture media, from 0.001 to 10 mg L^{-1} did not affect growth of a strain of *Chlorella vulgaris*.

Any theory attempting to explain a function for boron must account for the fact that some species and taxonomic groups have an absolute requirement, while other forms of life have no apparent need.[2814] Lee and Aronoff[2746] proposed that boron is needed where the synthesis is appreciable of phenolic ligands which can complex with borate. They also suggested that boron, as borate, appears to have a role in partitioning metabolism between the glycolytic and pentose-shunt pathway.

Lewin and Chen[2814] found that under conditions of boron deficiency, the concentration of protein, carbohydrate and RNA was depressed below that of control cells of marine pennate diatom *Cylindrotheca fusiformis*; DNA showed no change, while lipids, phenolic compounds, and unaccounted for organic fractions were all increased in concentration. These findings are in general agreement with results found with higher plants. The authors suggested a role of boron as cofactor for some enzymatic reaction (or reactions), rather than as a structural component of molecular constituents.

It appears that boron, under conditions of boron excess, could alleviate nutrient deficiency in some phytoplankters and might cause temporal variations of phytoplankton composition in coastal waters.[4822] Antia and Cheng[112] tested the autotrophic growth of 19 species of marine phytoplankters, from 10 classes of algae, in the presence of boric acid additions of 0–100 mg L^{-1} B. All the growth rates were virtually unaffected by the boron concentrations of 5–10 mg L^{-1}, while 26% of the species were strongly inhibited by 50 mg L^{-1} B and this proportion was increased to 63% species inhibited at 100 mg L^{-1} B. The authors concluded that higher borate concentrations up to 100 mg L^{-1} B are expected to cause species redistribution tending to favor growth of some forms by suppressing that of others. Forsberg et al.,[1449] investigating the influence of synthetic detergents on growth of freshwater species, found that boron concentrations in the range 10–200 μg L^{-1} retarded the growth rate of some species but accelerated that of others.

5.10 SILICON

5.10.1 Silicon in the Environment

Earth's crust: 277,000 mg kg^{-1} (490, 903)
Common minerals: SiO_2 (quartz), $NaAlSi_3O_8$ (albite), $Al_2Si_2O_5(OH)_4$ (kaolinite), $CaAl_2Si_2O_8$ (anorthite), $K(Si_3Al)Al_2O_{10}(OH)_2$ (muscovite), $K(Si_3Al)(MgFe^{II})_3O_{10}(OH)_2$ (biotite)[3705]
Soils: 330,000 (250,000–410,000) mg kg^{-1} (490–1)
Sediments: 245,000 mg kg^{-1} (491)
Air: <0.1–63 μg m^{-3} (491)
Fresh water: 7 (0.5–12) mg L^{-1} (491), 0.2–7 mg L^{-1} (3815)
Sea water: 2.2 (2.2–2.9) mg L^{-1} (491)
Land plants: 200–6,000 mg kg^{-1} (490–1)
Aquatic and wetland plants: 3,300–21,700 mg kg^{-1} (4399)
Diatoms: 10–30% DM[211,2968]
Marine algae: 1,500 mg kg^{-1} (491)
Land animals: 120–6,000 mg kg^{-1} (490)
Marine animals: 70–1,000 mg kg^{-1} (490)
Sources to waters: Wastewaters of inorganic origin may be a source of Si in natural waters. Chemical weathering of Al-Si minerals in the presence of CO_2 and H_2O:[3815]

$$2\ NaAlSi_3O_8 + 2\ CO_2 + 11H_2O = 2\ Na^+ + 2\ HCO_3^- + 4\ H_4SiO_4 + Al_2Si_2O_5(OH)_4 \qquad (5.4)$$
(albite)   (kaolinite)

Aquatic chemistry: In waters with pH < 9, silicon is predominantly present in the form of non-ionic $Si(OH)_4$ and partially in colloidal form.[3815] Ionic forms of silicon, e.g. mononuclear complexes $[SiO(OH)_3]^-$, $[SiO_2(OH)_2]^{2-}$, $[Si(OH)_5]^-$ or $[Si(OH)_6]^{2-}$, may occur in water with pH > 9.[3815]

5.10.2 Silicon in Algal Nutrition

Although silicon is the second most abundant element in the earth's crust, only a few groups of algae have utilized this element for skeletal structures. In addition to the well-known silicified frustule enclosing diatom cells, silicon structures are found in the silicoflagellates (Prymnesiophyceae) and in certain Chrysophyceae, Chlorophyceae, Phaeophyceae, and Xanthophyceae.[972,1521,3613,3655] In virtually all cases examined silicon is found as hydrated amorphous silica, $SiO_2 . n H_2O$, frequently referred to as opal.[2808]

A silicon requirement in diatoms, for valve formation, was proposed first by Bachrach and Lefevre.[199] It is now generally accepted that diatom cells do not divide unless a supply of silicon is available.[2372,2375,2796] The diatoms with their silica frustules have a nutritional requirement for silicon in the form of orthosilicic acid $Si(OH)_4$.[973,2795] In addition to the Si requirement for cell-wall formation, diatoms require small amounts of Si for net DNA synthesis, an unusual requirement which resides in the translation step in the synthesis of DNA polymerase.[973]

The form of silicon utilized by diatoms,[2796] and presumably other siliceous algae, is silicic acid, H_4SiO_4, the un-ionized form of which predominates in the pH range found in most biological systems.[646,4486] Colloidal silica did not support the growth of *Navicula pelliculosa* unless it was first depolymerized with alkali.[2796] Lewin[2796] found that the growth of *Navicula pelliculosa* is proportional to soluble silicon at low concentrations, and is most dense in media containing about 35 mg L^{-1} Si.

Silicon is transported into an intercellular pool via a carrier enzyme located on the cell membrane.[4826-7] The uptake and concentration of silica within the vesicles of the silicalemma require active transport.[2807] The energy requirement appears to be satisfied by respiration, probably involving high energy phosphate,[2795] as uptake rate does not significantly increase in the light.[3415] Early investigations by Lewin[2795,2597] on *Navicula pelliculosa* demonstrated that silicic acid uptake only occurs in living cells grown aerobically. The process is temperature dependent, is linked to energy yielding processes, and in some way requires the presence of water-soluble sulfhydryl groups in the cell membrane of some diatoms[2795,2797] as uptake can be inhibited by the sulfhydryl inhibitor, cadmium chloride, whilst even washing the cells may inhibit silica uptake. Information on the energy requirements associated with silicic acid uptake are provided e.g. by Coombs et al.[878-80] Darley[972] discussed the inactivation of silica deposition.

Experiments with synchronized cultures of diatoms have shown that silicic acid uptake occurs only during cytokinesis and new wall formation within the cell division cycle.[975,2815,4826] Silicon deficiency prevented cell division of marine diatoms almost entirely. When cells did divide in the presence of low silicate, abnormalities in valve structure occurred in some cells.[2161]

This is an absolute requirement for diatoms and for some species of Chrysophyceae and Xanthophyceae, but probably not for other groups.[4209,2529] Klaveness and Guillard[2529] found that concentrations of less than 1 μM silicate greatly decreased the growth rate of *Synura petersenii* and caused morphological changes (Figure 5-23). It seems that chrysophycean algae are much more flexible than diatoms in regard to their silicon requirements.[3613] Silicon metabolism in diatoms has been the focus of much research interest.[646,972,2583,2808,4829] In general, diatoms have an absolute requirement for mitosis and frustule formation.[970,974] The chemistry of the inorganic fraction of the wall has been reviewed by Lewin and Reimann.[2808] The organic cell wall layer is a complex assortment of sugars, lipids, amino acids and uronic acids.[876] Lipids account for 1 to 13% of the cell wall organic matter and differ significantly from cellular lipids.[2436]

Following the division of the parent protoplast into two products such that each is bounded by its own plasura membrane, the formation of new wall components is indicated by the appearance of cytoplasmic vesicles in the region of the new frustule beneath the plasma membrane.[1169,4346] These vesicles fuse laterally, presumably being formed from Golgi-derived vesicles[881,4789] and their formation is followed closely by advancing silica deposition. The common membrane of this vesicle is referred to as the silicalemma,[4056] or "silica deposition vesicle." A rapid deposition of silica, in the form of small spheres or possibly five fibrils, takes place, the silica becoming tightly bound to the silicalemma and somehow assuming the precise shape characteristic of the particular species. The silica become compacted with age.[424] At the completion of silica deposition, a new plasmalemma appears underneath the new cell wall. The silicalemma, possibly including additional material such as the old plasmalemma, continues to surround and adhere to the mature cell wall. The two daughter cells separate when the new valves are complete.[972] Further details have been presented by e.g. Schmid and Schultz,[4346] Schmid,[4345] Volcani[5172] and Dawson.[1008] The processes involved in formation of siliceous walls are shown in Figure 5-24.

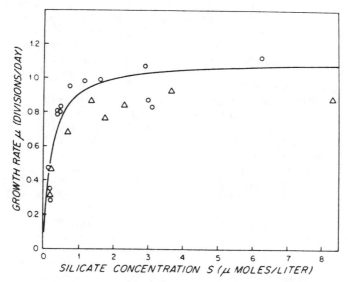

Figure 5-23. Growth rate of *Synura petersenii* as a function of silicate concentration. Δ = data from July-August. o = data from September-October . The curve shown was calculated from the September-October data and 2 more points, not shown (25 and 50 μM, both having 1.12 divisions/day), were included in the calculation.[2529] (From Klaveness and Guillard, 1975, in *J. Phycol.,* Vol. 11, p. 352, reprinted with permission of the *Journal of Phycology.*)

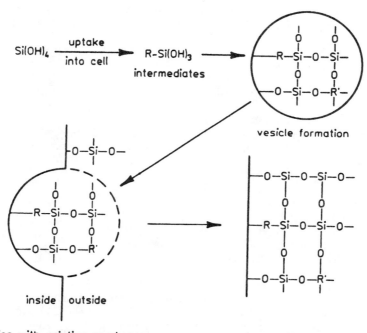

Figure 5-24. Diagrammatic representation of processes involved in formation of siliceous walls.[2005] (From Heinen and Oehler, 1979, in *Biogeochemical Cycling of Mineral-Forming Elements,* Trudinger, P.A., and Swaine, D.J., Eds., p. 435, reprinted with permission of Elsevier Science Publishing bv, Amsterdam.)

Low concentrations of silicate can be utilized by diatoms both in natural habitats and in culture. Owing to the fact that assimilation is directly connected with the formation of new walls, there is a definite low concentration of silicate below which a population cannot live, since cytoplasmic division will proceed without the formation of new walls, e.g. 0.5 μg L^{-1} for *Asterionella formosa* and 0.8 μg L^{-1} for the thicker walled *Melosira italica* var. *subarctica*.[2965] Much higher quantities, at least 25 μg L^{-1}, are required for optimum growth of *Fragilaria crotonensis* and *Nitzschia palea*.[803] For a review on silicic acid as a limiting nutrient under natural conditions see Paasche.[3613]

The silicon uptake follows Michaelis-Menten kinetics[189,865,2500,3415,3610-1,4826,4972] (Figure 5-25). Except at very low external Si(OH)$_4$ concentrations, silicon uptake is likely to be limited by the rate of wall formation rather than by the capacity of the transport system at the cell surface.[3613] Half-saturation constants of Si(OH)$_4$ uptake vary between 0.42 μM for *Skeletonema costatum*[3611] and 7.7 μM for *Asterionella formosa*[4972] (see Paasche[3613] for review). Tilman and Kilham[4972] presented the evidence that the silicon uptake rate depends on phosphate concentration (Figure 5-26). Numerous investigations show that diatoms are able to reduce dissolved silicon to concentrations near the limit of detection.[3613] In some cases, however, a significant amount of Si(OH)$_4$ has been found to remain in the medium after net uptake has ceased.[865,3611] In typical plankton diatoms, silicon constitutes from 10–30% of the dry mass.[211,2968]

The direct effect of silicon starvation on the diatom division cycle is to block cell development either before mitosis or during cell wall formation, depending on the organism and the experimental system. Indirect effect of silicon starvation is a general decline in active metabolism caused by the arresting of further cell development.[972] Silicon starvation of young light-dark synchronized *Cylindrotheca fusiformis* cells inhibited DNA net synthesis, and hence mitosis, while other cell components, including RNA, increased by 75%.[974] Following this report, Sullivan[4825] provided evidence that silicic acid regulates DNA synthesis by stimulating the activity of DNA polymerase at some post-transcriptional level. Silicon deficiency is also accompanied by morphological changes in the cell wall; the valves are thinner than normal. In cells of *Cyclotella cryptica* deprived of silicate, the synthesis of protein, DNA, RNA, chlorophyll, xanthophyll and lipid, as well as photosynthesis, are impaired.[5330] Silicon deficiency promotes lipid accumulation in diatoms.[879,5330]

Germanium, an element which has chemical properties similar to those of silicon is a potent inhibitor of growth in diatoms, although usually not affecting the growth of other algae.[971,974,2802,5330-1] The data suggest that germanic acid acts as a competitive inhibitor of silicic acid metabolism.[972]

5.11 VANADIUM

5.11.1 Vanadium in the Environment

Earth's crust: 135–160 mg kg^{-1} (490, 903, 3093, 3287)
Common minerals: vanadinite, descloizite, partonite, rosceolite, carnotite[3297]
Soils: 0.7–530 mg kg^{-1} (2395), 90–100 (3–500) mg kg^{-1}, absorbed by humus, especially in alkaline soils (490–1)
Sediments: 109 mg kg^{-1} (491), 20–150 mg kg^{-1} (3287)
Air: 0.0015–2,000 ng m^{-3} (491), 0.0006–200 ng m^{-3} (2395)
Fresh water: 0.5 (0.01–20) μg L^{-1} (491), 0.5–50 μg L^{-1} (3287)
Sea water: 2.5 (0.9–2.5) μg L^{-1} (491)
Land plants: 1.6 mg kg^{-1} (490), 0.001–0.5 mg kg^{-1} (491), 0.0005–2.7 mg kg^{-1} (2395)
Aquatic and wetland plants: 2–1,300 mg kg^{-1} (see Appendix, Table 7-43)
Marine algae: 1–16 mg kg^{-1} (491), 1.2–57 mg kg^{-1} (Table 5-15)
Land animals: 0.15 mg kg^{-1} (490)
Marine animals: 0.14–2 mg kg^{-1} (490)

Sources to waters: atmospheric fallout (oil and coal combustion incineration, natural sources—volcanoes, windborne soil particles, sea salt sprays, forest fires, continental biogenic particles), sewage sludge discharge, municipal wastewaters, smelting and refining of nonferrous metals, coal-burning power plants, manufacturing processes (metals, chemicals).[3479-80]

Aquatic chemistry: Vanadium can exist in several oxidation states: V^0, V^+, V^{2+}, V^{3+}, V^{4+}, and V^{5+}. The pentavalent form is the most soluble, and is the primary agent of transport in freshwaters.[3287] Vanadium often appears in association with humic acids.

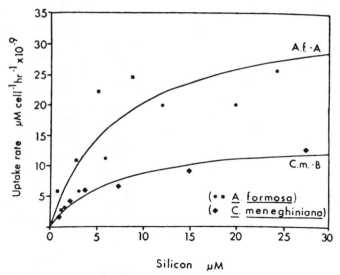

Figure 5-25.   Short-term uptake rates of *Asterionella formosa* (A.f.-A) and *Cyclotella meneghiniana* (C.m.-B) as a function of silicate concentration.[4972] (From Tilman and Kilham, 1976, in *J. Phycol.*, Vol. 12, p. 378, reprinted with permission of the *Journal of Phycology*.)

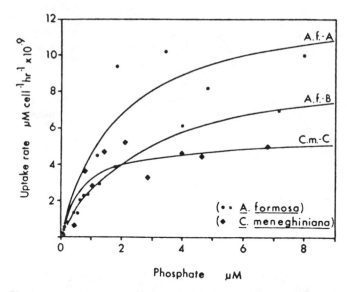

Figure 5-26.   Short-term uptake rates of *Asterionella formosa* (A.f.-A, A.f.-B) and *Cyclotella meneghiniana* (C.m.-C) as a function of phosphate concentration. Experiment A.f.-A performed 2 months prior to experiment A.f.-B.[4972] (From Tilman and Kilham, 1976, in J. *Phycol.*, Vol. 12, p. 378, reprinted with permission of the *Journal of Phycology*.)

5.11.2 Vanadium in Algal Nutrition

Vanadium is one of the most "recent" additions to the list of micronutrients.[154,156,161347,3389] Although earlier papers suggested an essential role of vanadium in algae, the first conclusive evidence was that of Arnon and Wessel[161] for *Scenedesmus obliquus* (Figures 5-27 and 5-28). In terms of the possible

Table 5-14. Boron concentration factors (CF) in freshwater algae

Species	CF	Locality	Reference
Nostoc sp.	93	Laboratory culture	501
Calothrix parietina	870	Laboratory culture	501
Cladophora glomerata	1,000–10,000	Hungary, Lake Balaton	2575

Table 5-15. Vanadium concentration in marine algae (mg kg^{-1} dry mass)

Species	Concentration	Locality	Reference
Chlamydomonas sp.	1.2	Laboratory	4145
Stichococcus bacillaris	2.4	Laboratory	4145
Dunaliella tertiolecta	2.9	Laboratory	4145
Monochrysis lutheri	3.1	Laboratory	4145
Fucus vesiculosus	0–3.4	Sweden, coast	4612
Chlorella salina	3.7	Laboratory	4145
Micromonas squamata	5.7	Laboratory	4145

role of vanadium, Arnon[156] reported that vanadium raises the maximal level of the Hill reaction in photosynthesis by isolated chloroplasts. In *Chlorella*, vanadium is reported to stimulate CO_2 uptake in photosynthesis at low light intensities, possibly by serving as a catalyst for CO_2 reduction.[5266] Vanadium has also been reported to accelerate photosynthesis in both species under high light intensity.[1302] Fries[1510] demonstrated vanadium to be an essential nutrient for several marine algae. O'Kelley[3530] pointed out that experimental difficulties involved in demonstrating the necessity for vanadium result from an affinity of V for Fe and a resulting common contamination of iron compounds with traces of vanadium.

Meisch and Bielig[3184] demonstrated a close relationship between V and Fe in the metabolism of the green algae, *Scenedesmus obliquus* and *Chlorella pyrenoidosa*. Vanadium stimulated the formation of chlorophyll.

5.12 HALOGENS

5.12.1 Halogens in the Environment

5.12.1.1 Chlorine

Earth's crust: 130 mg kg^{-1} (490, 903, 4920)
Common minerals: NaCl (halite)
Soils: 56–1,806 mg kg^{-1} (2395), 100 (8–1,800) mg kg^{-1}, higher in alkaline soils, near the sea and salt
    deserts; a major exchangeable anion in many soils[490-1]
Sediments: 190 mg kg^{-1} (491)
Air: 9–11,000 ng m^{-3} (491)
Fresh water: 7.0 (1–35) mg L^{-1} (491), 15 mg L^{-1} (5218)
Sea water: 19–19.35 g L^{-1} (490–1), Dead Sea in Israel: 127 g L^{-1} (3815)
Land plants: 2,000 (700–27,000) mg kg^{-1} (490–1), 10–7,000 mg kg^{-1} (2395)
Land animals: 2,800 mg kg^{-1} (490)
Marine animals: 5,000–90,000 mg kg^{-1} (490)
    Sources to waters: weathering processes, volcanic activity, industrial wastewaters.[3815]
    Aquatic chemistry: Chlorine occurs predominantly as a simple anion Cl$^-$ but complexes such as $FeCl^+$, $FeCl^{2+}$, or $HgCl_3^-$ may be present. Other forms, e.g. elemental Cl_2 (aq), non-dissociated HClO, anion ClO$^-$ and various amines may also be found.[3815] Chlorides are biochemically and chemically stable and in natural waters they do not undergo any transformations.

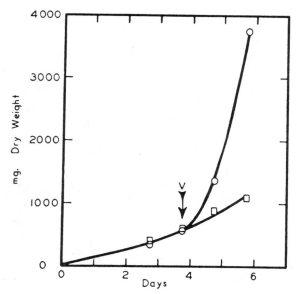

Figure 5-27. Effect of adding vanadium on growth of *Scenedesmus obliquus* in vanadium-deficient cultures. At the time indicated by arrow, 20 micrograms of V (as NH_4VO_3) were added per liter of nutrient solution. Growth is expressed in milligrams dry weight per liter of nutrient solution. One ppm Fe was supplied as $FeCl_3$.[156] (From Arnon, 1958, in *Trace Elements,* Lamb, C.A. et al., Eds., p. 11, reprinted with permission of Academic Press.)

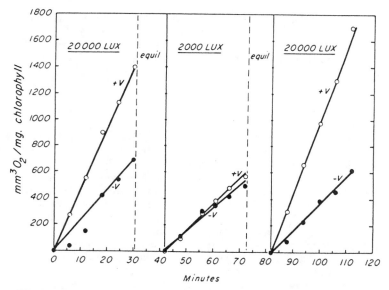

Figure 5-28. Effect of vanadium on the photosynthesis of *Scenedesmus obliquus* at a high and a low light intensity. The same cells were exposed successively to high and to low and again to high light intensity with intervening equilibration periods.[156] (From Amon, 1958, in *Trace Elements,* Lamb, C.A. et al., Eds., p. 25, reprinted with permission of Academic Press.)

5.12.1.2 Bromine

Earth's crust: 0.37 mg kg^{-1} (490), 2 mg kg^{-1} (903)
Common minerals: salt lakes evaporates[491]
Soils: 10 (1–110) mg kg^{-1} (491), < 0.5–515 mg kg^{-1} (2395)
Sediments: 19 mg kg^{-1} (491)
Air: 0.38–2,500 ng m^{-3} (491, 2395)
Fresh waters: 200 μg L^{-1} (490), 14 (0.05–55) μg L^{-1} (491), 1–20 μg L^{-1} (3815), mineral waters up to
        70 mg L^{-1} (3815)
Sea water: 65–67.3 mg L^{-1} (490–1), Dead Sea in Israel: 1.9–4.7 g L^{-1} (3815)
Land plants: 15 mg kg^{-1} (490–1), 0.2–119 mg kg^{-1} (2395)
Marine algae: 220–3,300 mg kg^{-1} (491)
Land animals: 6 mg kg^{-1} (490)
Marine animals: 60–1,000 mg kg^{-1} (490)
    Sources to waters: atmospheric fallout (anthropogenic sources, volcanic activity), fossil mineral waters, industrial wastewaters (pharmaceutical, production of chemicals).[3815]
    Aquatic chemistry: Bromine occurs in waters predominantly as a simple anion Br$^-$.[3815]

5.12.1.3 Iodine

Earth's crust: 0.14 mg kg^{-1} (490), 0.5 mg kg^{-1} (903, 3093)
Common minerals: NaIO$_3$ in evaporates[491,3815]
Soils: 5 (0.1–25) mg kg^{-1}, strongly absorbed by humus,[490–1] 0.06–41 mg kg^{-1} (2395)
Sediments: 16 (?) mg kg^{-1} (491)
Air: 0.08–6,000 ng m^{-3} (2395)
Freshwater: 2 (0.5–7) μg L^{-1} (490–1), < 5 μg L^{-1} (3815), mineral waters of fossil origin: > 135 mg L^{-1}
        (3815)
Sea water: 60 (50–70) μg L^{-1} (490–1)
Land plants: 3–5 mg kg^{-1} (491), 0.005–10.4 mg kg^{-1} (2395)
Marine algae: 7–10,000 mg kg^{-1} (491)
Land animals: 0.43 mg kg^{-1} (490)
Marine animals: 1–150 mg kg^{-1} (490)
    Sources to waters: atmospheric fallout (anthropogenic sources, volcanic activity), fossil mineral waters, industrial wastewaters (pharmaceutical, production of chemicals).[3815]
    Aquatic chemistry: Iodine occurs in waters mostly as a simple anion I$^-$. In surface layers of oceans, due to redox potential around 400 mV and slightly alkalic pH (around 8.0), also elemental iodine, IO$^-$ and IO$_3^-$ may occur.[3815]

5.12.2 Halogens in Algal Nutrition

    Many algae are strong concentrators of halogens. All photosynthesizing algae are believed to require chloride, on the basis of photosynthesis studies indicating such requirement for the Hill reaction, for ATP-formation, and for FMN-catalyzed phosphorylation.[5159] Eyster et al.[1305] demonstrated that *Chlorella pyrenoidosa* required more chloride for autotrophic growth as compared to heterotrophic growth (Figure 5-29). Chloride was designated as a "coenzyme" of photosynthesis specifically concerned with oxygen evolution.[160,5265] This was later confirmed and more thoroughly investigated by Arnon.[157] He found that chloride was necessary for noncyclic photophosphorylation and for the riboflavin phosphate pathway of cyclic photophosphorylation.
    A requirement for chlorine has not been unequivocally demonstrated for algae[3530] although its role in photosynthesis indicates that it is probably essential in most algae.[1022] Chloride ions are presumed to catalyze the electron transfer in the cytochrome chain.[4209]
    The chloride ion is relatively non-toxic, while Cl$_2$, ClO$^-$ and ClO$_3^-$ are very toxic. Toxicity of free chlorine is caused by its combination with the cell membrane which influences its permeability; ClO$_3^-$ behave as antimetabolites.[490]
    There is an apparent bromine requirement in the red alga *Polysiphonia urceolata*.[1507] Bromides were demonstrated in extracts of a number of Rhodophyta.[2673] Sauvageau[4297] claimed to have detected free Br$_2$ in certain cells of *Antithamnion*, which seemed to convert fluorescein to its brominated derivatives, eosin; these cells, however, failed to liberate I$_2$ from iodides.[2674]

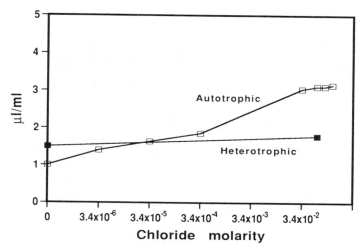

Figure 5-29.   Growth of *Chlorella pyrenoidosa* at different concentrations of chloride (NaCl). Autotrophic: 3 days growth from 0.65 μL washed cells. Heterotrophic: 20 days growth from 1 loopful cells.[1305] (From Eyster et al., 1958, in *Trace Elements,* Lamb, C.A. et al., Eds., p. 173, reprinted with permission of Academic Press.)

Absolute requirements for iodine have been demonstrated in *Polysiphonia urceolata*[1508] and *Ectocarpus fasciculatus.*[3724] Iodine also appears to be necessary for the growth and morphogenesis of *Petalonia fascia*[2229] and *Ectocarpus siliculosus.*[5496]

Phaeophyta are particularly rich in iodine, in the Laminariales and Desmarestiales its concentration may occasionally reach 30,000 times that in sea water.[4432] Some Rhodophyceae, particularly *Ptilota* and some Chlorophyceae, notably *Codium intricatum,* also contain an appreciable amount of iodine.[4432] In fact, brown seaweeds provided the industrial source of the element in the 19th century.[761,1460] The iodine content of freshwater algae has been studied far less frequently; von Fellenberg[5186] reported the iodine concentration of 3.4 mg kg^{-1} in *Vaucheria.* Young plants are generally richer in iodine than are older plants, and those from exposed coasts tend to be richer than those from bays or lochs.[4432]

The literature does not provide much information on the uptake of iodine by algae. There is a body of evidence which suggests that iodine is absorbed not as iodine but as I_2 or in some related oxidation state.[4430,4432] Shaw[4431] found that in certain states of the algae, the uptake of iodine is accompanied by a vigorous burst of respiration. Between 3 and 6 molecules of O_2 are used for each iodine ion taken up. The respiratory quotient is close to unity, which suggests that carbohydrate is used as the substrate.

5.13 SELENIUM

5.13.1 Selenium in the Environment

Earth's crust: 0.05 mg kg^{-1} (490,3093,4920), 0.1 mg kg^{-1} (903, 3287)
Common minerals: associated with sulfide ores[491]
Soils: 0.005–4.0 mg kg^{-1} (2395), 0.2 (0.01–2.0) mg kg^{-1}, absorbed by humus, especially in alkaline soils[490]
Sediments: 0.42 mg kg^{-1} (491), 0.2–30 mg kg^{-1} (946)
Air: 0.0042–30 ng m^{-3} (491, 2395)
Fresh water: 0.2 (0.02–1,000) μg L^{-1} (491), 0.1–5 μg L^{-1} (3287)
Sea water: 0.2 (0.052–0.2) μg L^{-1} (491), < 0.1 μg L^{-1} (4525)
Land plants: 0.2 mg kg^{-1} (490), 0.001–2.03 mg kg^{-1} (2395)
Aquatic and wetland plants (see Appendix, Table 7-44):
emergent species: 0–280 mg kg^{-1} (aboveground), 0–52.3 mg kg^{-1} (belowground)
Freshwater algae: 0–220 mg kg^{-1} (4366)
Marine algae: 0.04–0.1 mg kg^{-1} (491)
Land animals: 1.7 mg kg^{-1} (490)

5.13.1.1 Aquatic chemistry

Sources to waters: coal-burning power plants, smelting and refining of nonferrous metals, municipal wastewaters, sewage sludge discharge, manufacturing processes (metals, chemicals, pulp and paper, petroleum products), atmospheric fallout, base metal mining and dressing.[3480]

Aquatic chemistry: Dissolved inorganic selenium can exist as Se^{2-}, mainly as biselenide (HSe^-); Se^0, as colloidal elemental selenium; as the selenite oxyanion ($HSeO_3^-$ and SeO_3^{2-}); and Se^+, and as the selenate oxyanion (SeO_4^{2-}). Organic Se species are analogous to those of sulfur and include many organic complexes.[3287] Selenite and selenate are the dominant species in most freshwaters, where they typically account for more than 90% of total Se in the water column.[946]

In sediment-water environments Se can exist as selenate (Se^{VI}), selenite (Se^{IV}), elemental Se (Se^0), and selenide (Se^{-II}). Dimethyl selenide (DMSe) and nonvolatile organic Se compounds, produced by biomethylation, are also important chemical species of Se.[3101] For selenium chemistry in sediments see Masscheleyn et al.[3101] and references cited therein; the selenium cycle has been reviewed e.g. by Trudinger et al.[5034] Selenium in higher plants metabolism has been reviewed by Shrift.[4459]

Selenium is of special interest since the element is similar to sulfur, existing in the same valence states in nature, the inorganic forms are structurally very similar to S analogues. A selenium cycle similar to the sulfur cycle is to be found thus:[1800]

$$Se^{2-} \Leftrightarrow Se^0 \Leftrightarrow Se_2O_3^{2-} \Leftrightarrow Se_2O_4^{2-} \tag{5.5}$$

Like S interconversions, some chemical interconversion occurs, but S bacteria have been shown to catalyze these interconversions and there is some evidence of lithotrophic oxidation. Se replaces S in a number of essential organic compounds, and selenomethionine and selenocystine are found in plants growing in high Se areas.[1800]

5.13.2 Selenium in Algal Nutrition

The nutritional importance of selenium for algal growth is becoming increasingly apparent.[3896] Pioneering work by Pintner and Provasoli[3802] documented the stimulatory effects of Se on the growth of three axenic marine *Chrysochromulina* spp. Since then two other studies have shown, using axenic cultures, that *Peridinium cinctum* fa. *westii*[2869] and *Chrysochromulina brevituritta*[5313] have an absolute growth requirement for Se. Other investigations have demonstrated that Se additions to natural and artificial media stimulate the growth of axenic phytoplankton,[5365] axenic macroalgae,[1509] and xenic phytoplankton[2456,2864,2867] cultures.

The specific requirements of *Peridinium glutanense* for low Se concentrations, which increased biomass, growth rates, and oxygen production capacity more than 100 times, were determined by Lindström.[2862,2864,2866,2868] Although Se and S are related chemically,[208] occur in some amino acids, and sometimes follow similar metabolic pathways,[5560] sulfur as sulfate or thisulfate did not preclude the need for Se in *P. glutanense*. However, the suggested role of S in peptide-nucleotide substances, necessary for cell division,[1410] does not exclude Se from being involved in these processes. This may be indicated by the occurrence of deformed giant cells (17% enlargement), which probably were unable to divide when the cultures were Se-depleted.[2863,2865] Similar responses were noted for the marine diatom *Thalassiosira pseudonana*, in which Se deficiency caused reduced cell divisions, reduction in growth rate and increased cell volumes related to elongation of cells,[3896] and aberration of the cell ultrastructure.[1131] Price et al.[3896] tested eleven trace elements for their ability to replace the selenium requirement of *Thalassiosira pseudonana* and all were without effect.

Fries[1509] found that the marine brown alga *Fucus spiralis* and the red alga *Goniotrichum alsidii* increased their growth upon the addition of SeO_3^{2-} and SeO_4^{2-} when cultivated axenically in artificial seawater. In the concentration range 1×10^{-10} to 1×10^{-7} M there were two optima, one at 3.3×10^{-10} M and another at 3.3×10^{-8} M. Organically bound Se had no effect.

Wrench and Measures[5502] observed that decreases in Se^{IV}, but not Se^{VI}, concentration were inversely correlated with increases in phytoplankton biomass and particulate Se in a coastal seawater environment. Selenomethionine and selenocystine may be utilized by some phytoplankton and at least three species grow well using these organic forms of Se as with Na_2SeO_3.[2864,5313] *Chrysochromulina breviturrita* has also been shown to utilize dimethyl selenide (DMSe) as a Se source.[5313] This ability may be relevant to Se cycling in aquatic ecosystems as DMSe and other volatile compounds can be produced by microbial assemblages in lake sediments.[770]

The biochemical basis for the requirement of Se in algae remains unknown.[3896] Several Se-containing enzymes have been isolated from bacteria and a few even from mammals[4670] but these enzyme systems have not been found to contain Se in higher plants.[4597] The essential requirement of Se in mammals and birds is attributed to the Se containing enzyme glutathione peroxidase. Enhanced glutathione peroxidation by cell extracts of *Dunaliella tertiolecta* and *Porphyridium cruentum* was evident when these two algae were cultured in medium containing Na_2SeO_3.[1601] It was concluded, however, that the selenoenzyme, glutathione peroxidase, is absent and that hydroperoxidase dependent oxidation of glutathione is nonenzymatic in nature. By contrast, in *Euglena gracilis* it has been shown that there exist two types of glutathione peroxidases, one of which is a Se-independent form of the enzyme.[3589-90]

Depending on the species involved and the concentration or bioavailability selenium can also act as a toxic agent.[2644,4455,4485] Evidence indicates that selenate and sulfate anions compete with each other at the cell membrane during the absorption process. Unlike most antimetabolites, however, selenate can be metabolized to some degree by the same enzyme systems that reduce sulfate with the formation of organic compounds.[4456] The formation of organoselenium compounds by plants frequently involves the substitution of Se for S in sulfur-containing amino acids (e.g. Se-methionine, Se-cystine, Se-cysteine).[5501] Selenium compounds are metabolized differently than S compounds[2783] and is has been suggested that the replacement of S compounds by dysfunctional seleno-analogues mediates the toxic effect of Se in higher plants.[4455] Other evidence pointing to selenate interference with sulfur metabolism involves the antagonistic action of sulfate and other sulfur metabolites. Sulfate overcomes selenate toxicity in a competitive fashion.[4455] Selenate in a concentration of 2 mM gave a slight and rather irregular inhibition of phosphate uptake by *Scenedesmus*.[2666]

In contrast to the proposed mechanism for Se toxicity, the metabolic basis for a Se growth requirement remains unclear.[1131] While much is known about the biological role of some essential trace metals with a potential for toxicity (e.g. Cu, Zn[3530]) no essential organoselenium compounds have been reported for either algae or higher plants.

CHAPTER 6

Other Elements

6.1 ARSENIC

6.1.1 Arsenic in the Environment

Earth's crust: 1.5–5.0 mg kg^{-1} (490, 903, 3039, 3287, 4920)
Common minerals: As_2S_3, As_4S_4, $FeAs_2$ (arsenopyrite)[491,852]
Soils: 6 (0.1–40) mg kg^{-1} (490–1), 1–50 mg kg^{-1} (4849), 5–10 (0.1–40) mg kg^{-1} (852, 4355), <0.1–95 mg kg^{-1} (2395), in contaminated sites up to 2,470 mg kg^{-1} (2395)
Sediments: 7.7 mg kg^{-1} (491), 0–66,700 mg kg^{-1} (852)
Air: remote regions: < 20 ng m^{-3}, urban areas: <10–160 ng m^{-3}, industrial areas: up to 160 μg m^{-3} (852), 1.5–53 ng m^{-3} (491), 0.007–120 ng m^{-3} (2395)
Fresh water: 1.6–68 μg L^{-1} (2571,5049), 0.5 (0.2–230) μg L^{-1} (491)
Sea water: 2–4 μg L^{-1} (3856), 3.7 (0.5–3.7) μg L^{-1} (491)
Land plants: 0.2 (0.02–7) mg kg^{-1} (490–1), 0.003–1.5 mg kg^{-1} (2395), in contaminated sites up to 6,640 mg kg^{-1} (2395)
Aquatic and wetland plants (see Appendix, Table 7-45):
emergent species: aboveground: 0.04–12 mg kg^{-1}
submerged species: 0.33–3,700 mg kg^{-1}
floating species: 0.5–30 mg kg^{-1}
bryophytes: 0–2,190 mg kg^{-1} (FD)
Freshwater algae: 0.5–182 mg kg^{-1} (Table 6-1)
Marine algae: 1–30 mg kg^{-1} (491), 1.7–190 mg kg^{-1} (Table 6-1)
Land animals: ≤ 0.2 mg kg^{-1} (490)
Marine animals: 0.005–0.3 mg kg^{-1} (490)

6.1.1.1 Aquatic chemistry

Sources to waters: wastewaters, sewage sludge, manufacturing processes (metals, chemicals, pulp and paper, petroleum products), arsenical pesticides, smelting and refining, base metal mining and dressing, coal-fired power plants, steam electrical production, agricultural runoff and land erosion, atmospheric production.[597,667,1357,2383,3480,5019]

Arsenic exists in several oxidation states (+5, +3, 0, -3) in aquatic environments.[106,1337] As^{3-} is found only at low Eh, and the metal is extremely rare.[1337] Under mildly reducing conditions, arsenites predominate, whereas under well-oxygenated conditions and high Eh, arsenates are most common. Oxidation of arsenite to arsenate is slow in most surface waters, increasing at pH extremes. As^{3+} is considered a hard acid and preferentially complexes with oxides and nitrogen, As^{5+}, on the other hand, behaves like a soft acid and preferentially complexes with sulfides.[3287] As^{3+} and As^{5+} also undergo a series of biological transformations in aquatic systems, yielding a large number of compounds.[3022] Like Hg, As is susceptible to methylation by microorganisms such as bacteria and fungi.[897,3116] They produce toxic di- and trimethylarsines.[429]

In sediment-water environments As can exist in three different oxidation states, arsenate (As^V), arsenite and monomethyl arsonic acid (As^{III}), and dimethyl arsinic acid (As^I).[3101] Transformation and fate of As in freshwater systems are governed by the nature and properties of sediments and by limnological conditions.[2237] Arsenic is present in both colloidal and non-colloidal fraction in freshwater sediments.[2235] Arsenic is known to be strongly sorbed on the surfaces of Al and Fe oxides and on the edges of clay minerals[1437,2233,2236,5258] through a ligand exchange mechanism.[2234,3815] As (V) is more

Table 6-1. Arsenic concentration in algae (mg kg^{-1} dry mass)

Species	Concentration	Locality	Reference
Freshwater			
Scenedesmus acutus	0.5–0.7	Cultivation (wastewater)	5144
Chlamydomonas geitleri	1.08–1.15	Cultivation (wastewater)	5144
Phytoplankton	3.2–4.3	Lake Superior	4411
Chara sp.	4.99	Lake Huron	1209
Cladophora glomerata	1.12–6.0	Lake Balaton, Hungary	2575
Cladophora sp.	0.2–21.1	Lake Huron	1209
Nitella hookeri	13–182	New Zealand	4017
Marine			
Fucus serratus	1.7	England	3856
Ahnfeltia plicata	2.0	Canada, Nova Scotia	5528
Ulva lactuca	4.0	Canada, Nova Scotia	5528
Chondrus crispus	5.0	Canada, Nova Scotia	5528
Enteromorpha sp.	< 1–8	Japan	4876
Spongomorpha arcta	8.0	Canada, Nova Scotia	5528
Rhodymenia palmata	10	Canada, Nova Scotia	5528
Enteromorpha compressa	12	England	2359
Ascophyllum nodosum	38	Canada, Nova Scotia	5528
Laminaria digitata	50	Canada, Nova Scotia	5528
Laminaria longicruris	52	Canada, Nova Scotia	5528
Fucus vesiculosus	58	Canada, Nova Scotia	5528
Macrocystis pyrifera	58	New Zealand	5453
Fucus vesiculosus	40–190	Japan	4847

strongly sorbed by sediment components than is As (III).[1848,2160] The sorption of As by the sediments occurs simultaneously with the oxidation of As (III) to As (V).[3566] Therefore, the oxidation plus sorption is referred to as the depletion of As (III) in solution. The depletion of As (III) by the sediments follows first-order kinetics.[3568] Microorganisms and various silicates and calcium carbonates present in the sediments are of minor importance in the oxidation of As (III).[3566-7] Other components which should be considered in the oxidation of As (III) in aquatic environments are organic matter, SO_4^{2-}, NO_3^-, Mn (IV) and Fe (III).[2237]

Biological availability and physiological and toxicological effects of arsenic depend on its chemical form.[3738] Arsenic in the reduced state, As (III) arsenite, is much more toxic, more soluble and mobile, than when in the oxidized state, As (V), arsenite.

6.1.2 Arsenic and Algal Growth

Planas and Healey[3816] showed that concentrations as low as 1 μM arsenate depressed the growth of *Melosira granulata* and *Ochromonas vallesiaca*, whereas concentrations greater than 100 μM were required for a similar growth reduction in *Cryptomonas erosa* and *Anabaena variabilis*. More than 200 μM arsenate was required before growth of *Synechococcus leopoliensis* was affected.[610] Toxicity even varies within one species of algae; growth of *Chlorella vulgaris* was inhibited at 0.8 μM[1052] whereas a different isolate of *Chlorella vulgaris* tolerated more than 0.1 M arsenate.[3016]

Thomas et al.[4955] reported that Cr, Ni, Se, Sb, and As^{3+} were without effect on the morphology of several marine diatoms and silicoflagellates, except for lysis of cells exposed to As^{3+}, at a concentration of 52 μg L^{-1}, well above the concentrations that showed effects with other studied metals (Cu, Zn, Ge, Hg, Cd, Pb).

In many cases, the sensitivity to As depends on the phosphorus levels in the medium. Arsenate is thought to be toxic to cells because it is structurally similar to phosphate and co-transported into cells.[918] Planas and Healey[3816] found that several cultures were much more sensitive to arsenate following depletion of P from the medium. *Chlamydomonas reinhardtii* exposed to 35 μM arsenate, a level much higher than the 10 μM arsenate required to reduce growth rates, showed symptoms of phosphate deficiency.[3816] Similarly, Thiel[4940] found *Anabaena* sp. to be more sensitive to arsenate when phosphate-starved than when phosphate replete.

Creed et al.[918] found that arsenate in concentrations 1–25 μM reduced photosynthesis and cell growth, as reflected by induced lag periods, slower growth rates, and lower stationary cell yields, in *Chlorella vulgaris*. The authors also observed that growth of the phosphorus-limited field strains was stimulated by the addition of arsenic. The cell yield of phosphorus limited *C. vulgaris*, when treated with As, was two times that of the phosphorus-limited control. This pattern was not evident when photosynthesis was used as a measure of cell response.

Arsenate is known to act as an uncoupler of phosphorylation[186,4566] and to inhibit competitively phosphate uptake.[404,652,2320,3816,4200] Arsenate is actively accumulated by some organisms.[1925,2973] In some it is not metabolized further;[1925] in others it is partially reduced to arsenite, some of which is excreted;[395,3799] while in others it is incorporated into polar lipids.[738,2973] The concentration factors of arsenic in algae are shown in Table 6-2.

Some algae are capable of growing in the presence of high concentrations of arsenate. Spring and fall blooms of algae produced chlorophyll concentrations of 73 to 83 mg m^{-3} in the presence of 20–40 μM arsenate.[3816] Arsenic changes the species composition of phytoplankton communities such that sensitive species are replaced by tolerant species.[392,4269] Field studies have shown shifts in species dominance toward smaller species, microflagellates and/or pennate diatoms.[128,4267] These changes in phytoplankton communities can have profound impacts on other trophic levels.[4268]

6.2 CADMIUM

6.2.1 Cadmium in the Environment

Earth's crust: 0.20 mg kg^{-1} (903, 3093, 3287, 4920)
Common minerals: with ZnS
Soils: 0.35 (0.01–2) mg kg^{-1} (491), 0.01–4.0 mg kg^{-1} (2395), in contaminated sites up to 1,781 mg kg^{-1} (2395)
Sediments: 0.17 mg kg^{-1} (491), 0.08–60,000 mg kg^{-1} (5195)
Air: 0.003–620 ng m^{-3} (491, 2395)
Fresh water: 0.1 (0.01–3) μg L^{-1} (491), < 1.0 μg L^{-1} (1493), 0.5 μg L^{-1},[5218] 0.1–0.5 μg L^{-1} (1385), 0.04–20 μg L^{-1} (5195); background levels: 0.05–3.0 μg L^{-1} (256, 521, 1079, 2032); acid mine drainage: 0.14–345,000 μg L^{-1} (5195)
Sea water: 0.11 (<0.01–9.4) μg L^{-1} (variable) (490–1)
Land plants: 0.6 (0.1–2.4) mg kg^{-1} (490–1), 0.006–1.26 mg kg^{-1} (2395), in contaminated sites up to 898 mg kg^{-1} (2395)
Aquatic and wetland plants (mg kg^{-1}) (see Appendix, Table 7-46):
emergent species: aboveground: 0.02–11, belowground: 0.1–16
floating-leaved species: aboveground: 0.41–1.0 (FD), belowground: 0.18–0.76
submerged species: 0.1–32.3

Table 6-2. Arsenic concentration factors (CF) in algae

Species	CF	Locality	Reference
Freshwater			
Cladophora glomerata	100	Lake Balaton, Hungary	2575
Marine			
Fucus serratus	600	England	3856
Enteromorpha compressa	4,000	England	2359
Macrocystis pyrifera	19,000	New Zealand	5453

floating species: 5–10,600
bryophytes: 0.17–89.5
Freshwater algae: 0.08–3,031 mg kg^{-1} (Table 6-3)
Marine algae: 0.3–2.5 mg.kg^{-1} (491), 1.7–190 mg kg^{-1} (Table 6-3)
Land animals: ≤ 0.5 mg kg^{-1} (490)
Marine animals: 0.15–3.0 mg kg^{-1} (490)

Sources to waters: smelting and refining of nonferrous metals, manufacturing processes (chemicals, metals), municipal wastewater, sewage sludge discharge, steam electricity production, base metal mining (mine drainage waters) and dressing, atmospheric fallout.[2771–3,3480] An unknown amount of Cd escapes to the environment wherever zinc escapes.[1383,2024,5485]

Aquatic chemistry: Cadmium is classified as a soft acid, preferentially complexing with sulfides.[3287] Cadmium is relatively mobile in aquatic ecosystems. The most important forms of dissolved cadmium in waters are free ionic Cd^{2+}, inorganic complexes $[CdOH^+]$, $[CdCl^+]$, $[CdSO_4]$, and $[CdCO_3 (aq)]$, and in various organic complexes.[1567,2024,3815] In many freshwaters, the affinity of ligands to complex with cadmium follows the order: humic acids, CO_3^{2-}, OH^-, Cl^-, and SO_4^{2-}.[3287] Sorption to suspended solids such as clay is an important, often dominant fate process in freshwaters. Coprecipitation with hydrous Fe, Al, Mn oxides, and carbonate materials also occurs, and periodically dominates fate processes.[1568,3287,4263]

6.2.2 Toxicity and Accumulation in Algae

The toxicity and accumulation of cadmium in algae have been reviewed several times recently.[192,3962,4629,5026,5195] Cadmium inhibits cell division,[4189] growth,[660,680,1068,3947,5035] biomass,[5035] and motility[1035] and can be lethal for algae.[227,323,864,4398,5026] Trevors et al.[5026] in their excellent review pointed out that cadmium effects are dependent on both the organism used and the toxicity criterion employed, although specific trends in sensitivity are difficult to identify. The authors also pointed out that one well-established trend is that marine algae are usually less sensitive to Cd than are freshwater algae.[323,2639] This is probably due to differences in Cd speciation,[2639] as less Cd^{2+} is available in seawater owing to complexation with chlorides and other anions.[193,4019]

Adshead-Simonsen et al.[26] found that cadmium concentration of 12 μg L^{-1} induced morphological changes in the diatom *Tabellaria flocullosa*. Growth was unaffected at 1.2 μg L^{-1}, but at this concentration cells no longer formed the conventional zig-zag configuration. Thomas et al.[4955] reported that the presence of cadmium caused some disruption of cell separation in the marine diatom *Thalassiosira aestivalis*, so that cell clumps were formed, but the main effect was the formation of elongated, bent, noncolumn-forming cells. The cell elongation as a result of toxic action of cadmium was also reported by Rao and Subramanian[3981] for the diatoms, *Cyclotella meneghiniana, Nitzschia palea* and *Navicula confervacea*. Other cellular abnormalities such as the loss of cross walls,[3981] distorted cells,[4955] swollen chloroplast membranes,[4599] and alterations in cytoplasmic vacuolization and granulation,[4599,4955] have been observed in various diatoms.

Manganese is a competitive inhibitor of cadmium uptake in *Dunaliella salina* and it is possible that these ions are taken up by a common transport system.[4020] The antagonism between cadmium and iron uptake in the marine diatom *Thalassiosira weissflogii* may implicate the iron transport system in cadmium accumulation in this alga.[1939]

Skowronski et al.[4546] found that a significant decrease of cadmium toxicity to the green alga *Stichococcus bacillaris* occurred in both the acidic and alkaline ranges of pH 3–9. In media of pH 3 and 9, cadmium did not affect the dry mass content substantially. Maximum toxicity of cadmium was noticed at pH 6–7.

Rachlin and co-workers[3946-7,3949-50,5271] found the EC-50$_{96}$ values for the green alga *Chlorella saccharophila* and the diatoms *Nitzschia closterium* and *Navicula incerta* to be 0.056–0.11, 0.47 and 2.39–3.01 mg L^{-1} Cd, respectively. Chiaudani and Vighi[780] found that the EC-50$_{96}$ for the green alga *Selenastrum capricornutum* was 3.9 μg L^{-1} and 6.7 μg L^{-1} in AAP medium without and with EDTA, respectively. Bartlett et al.[227] reported that cadmium caused partial growth inhibition of *Selenastrum capricornutum* in EDTA-containing medium at 50 μg L^{-1}, with complete inhibition at 80 μg L^{-1} and algicidal concentration of 650 μg L^{-1}.

Laube et al.[2720] found that the cadmium concentrations of 11.2 and 112.4 mg L^{-1} completely inhibited growth of the blue-green alga *Anabaena* 7120, while concentrations of 11.2 μg L^{-1} and 1.1 mg L^{-1} Cd resulted in an extended lag, and the concentration of 1.12 μg L^{-1} Cd did not affect growth. The cadmium concentration of 2 μg L^{-1} reduced the growth rate of the freshwater diatom *Asterionella*

Table 6-3. Cadmium concentrations in algae (mg kg^{-1} dry mass)

Species	Concentration	Locality	Reference
Freshwater			
Cladophora cladophora	0.29–0.94	Leine River, Germany	13
Chara sp.	0.99	Lake Huron	1209
Cladophora rivularis	0.08–1.37	New Zealand	4017
Cladophora glomerata	1.4	Main Duck, Lake Ontario	2453
Not specified	1.5	IJsselmeer, The Netherlands	4258
Not specified	1.9	IJsselmeer, The Netherlands	4260
Periphyton	1.4–2.0	Elsenz River, Germany	355
Cladophora glomerata	3.9	Deadman Bay, L. Ontario	2453
Chara sp.	2.5–4.0	St. Clair Lake, Canada	3346
Cladophora glomerata	0.19–4.03	Roding River, England	3146
Not specified	5.6	Lake Alpnach, Switzerland	195
Periphyton	10.3–16.8	Rhine River	5171
Oedogonium, Ulothrix	36	Northwestern U.S.A.	5489
Lemanea fluviatilis	9.0–66	Derwent River, North England	1910
Mougeotia sp.	130	Caplecleugh Low Level, England	3703
Lemanea sp.	2–342	River Nent, England	1909
Scenedesmus obliquus	12–3,031	Laboratory	660
Marine			
Ptilota serrata	0.04	Gulf of Maine	4389
Laminaria saccharina	0.22	Rhode Island coast	4389
Phycodrys rubens	0.17–0.34	Rhode Island coast	4389
Chaetomorpha antennina	0.4–0.6	Hong Kong	2113
Sargassum hemiphyllum	0.6	Hong Kong	2114
Endarachne binghamiae	0.6	Hong Kong	2114
Chorda filum	0.7	Portugal and Spain	4732
Gymnogongrus flabelliformis	0.4–0.7	Hong Kong	2113
Enteromorpha compressa	0.6–0.8	Hong Kong	2113
Porphyra spp.	0.05–0.87	Irish Sea	3882
Fucus vesiculosus	1.0	Trondheimsfjorden, Norway	2694
Sargassum vachellianum	1.1	Hong Kong	2114
Enteromorpha flexuosa	0.8–1.2	Hong Kong	2113
Colpomenia sinuosa	0.9–1.4	Hong Kong	2113
Scytosiphon lomentaria	0.5–1.6	Hong Kong	2113
Ascophyllum nodosum	1.5–1.8	Great Britain	1455
Ulva lactuca	0.5–2.0	Portugal and Spain	4732
Fucus vesiculosus	2.10	Great Britain	1455
Polysiphonia urceolata	2.35	Rhode Island coast	4389
Carallina pilulifera	2.2–2.8	Hong Kong	2113
Delesseria sanguinea	4.1	Portugal and Spain	4732
Blue-green algae	4.7	Penang Island, Malaysia	4526
Enteromorpha sp.	0.07–4.8	British North Sea Coast	4308
Chondrus crispus	4.9	Portugal and Spain	4732

Table 6-3. Continued

Species	Concentration	Locality	Reference
Corallina officinalis	4.4–6.2	Portugal and Spain	4732
Fucus vesiculosus	5.6–13.8	Penang Island, Malaysia	4526
Ascophyllum nodosum	<0.7–16	Norway, coast	1975
Enteromorpha sp.	8–16	Penang Island, Malaysia	4526
Fucus vesiculosus	3.8–19.5	Bristol Channel, Great Britain	1520
Fucus spp.	0.4–20.8	Irish Sea	3882
Fucus vesiculosus	3.4–24.5	Sweden, coast	4612
Fucus vesiculosus	1.0–28	Great Britain estuaries	604

formosa by an order of magnitude, and 10 μg L^{-1} caused growth cessation.[864] Cadmium toxicity to the green alga *Ankistrodesmus falcatus* was not observed below 1 mg L^{-1} in an artificial solution compared to 0.5 mg L^{-1} in bay water. In both media virtually no growth was observed above 5 mg L^{-1} Cd. Fennikoh et al.[1336] found that the growth of the green alga *Chlamydomonas reinhardtii* was strongly reduced at Cd concentration of 100 μg L^{-1} with death occurring at 1–10 mg L^{-1} Cd. The growth of *Chlorella pyrenoidosa* was reduced at 1.1 mg L^{-1} Cd but not completely inhibited at 110 mg L^{-1} Cd.[3039] The growth of the green alga *Scenedesmus quadricauda* was inhibited by Cd concentration as low as 6.1 μg L^{-1}.[2525]

Whitton and Shehata[5411] reported that the inhibitory level of Cd for the blue-green alga *Anacystis nidulans* obtained after 25 subcultures in progressively higher Cd concentrations rose from 0.55 to 2.5 mg L^{-1}. Say and Whitton[4304] found the green filamentous alga *Hormidium rivulare* and the diatom *Pinnularia subcapitata* growing in a stream (near a large active zinc smelter at Viviez, near Decazeville, France) with cadmium concentration as high as 345 mg L^{-1}. Say and Whitton[4305] reported a wide variety of algal species, mostly diatoms and green algae, to be present in streams with elevated Cd concentrations.

Overnell[3593] observed that cadmium in a concentration of 4.5 mg L^{-1} inhibited the light-induced oxygen evolution in *Chlamydomonas reinhardtii* but had no effect on the Hill reaction and the modified Mehler reaction. Li[2844] found that for marine diatom *Thalassiosira fluviatilis* the degree of inhibition by cadmium was more severe at low than at high NO$_3^-$ levels but the difference in severity diminished as Cd concentration increased. Rachlin et al.[3947–8,3950] found that, with increasing cadmium concentration, there was a reduction in the growth rate. These results are consistent with the observation that growth reduction is a reasonable determinant of metal toxicity, and that the degree of response of the algae is dependent on the amount of metal that traverses the membrane and has not been bound in the cellular detoxifying mechanisms.[4508] Sarsfield and Mancy[4289] demonstrated that the relative toxic strength of Cd^{2+} is a function of its complex form within the cell. Phthalate, citrate and EDTA complexes of Cd were tested.

Cain et al.[660] found that growth of the green alga *Scenedesmus obliquus* was not significantly affected by Cd concentrations ranging from 0.01 to 1.0 mg L^{-1}. At concentrations above 1 mg L^{-1}, however, growth was inhibited markedly. Rapidly growing, young cultures accumulated less Cd than older cultures approaching stationary growth phase.

Rachlin et al.[3948] studied the ultrastructural changes induced in a blue-green alga *Plectonema boryanum* by cadmium exposure of 100 mg L^{-1}. They reported no change in overall dimensions after a 4-hr exposure. It was concluded that this metal has no apparent effect on cell growth, structural integrity, or growth characteristics under the conditions studied. Exposure to cadmium, however, resulted in an increase in the surface area of the thylakoids of treated cells when compared to the control cells and caused coalescence of cellular lipid inclusions. On contrary, Rachlin et al.[3951] reported that all cadmium concentrations in the range of 13.3 μg L^{-1} to 33.3 mg L^{-1} caused significant reduction of the surface area of the thylakoids of *Anabaena flos-aquae*.

Rachlin et al.[3946] reported a significant enlargement in the polyphosphate bodies of *Plectonema boryanum* as a result of cadmium exposure. It was found that the elemental composition of the polyphosphate bodies of *Anabaena flos-aquae* were changed in that Mg and Ca were no longer present in detectable amounts after cadmium exposure.[3951] Jensen et al.[2336] found that after an exposure of *Plectonema boryanum* to 100 mg L^{-1} of Cd, the metal was sequestered in polyphosphate bodies.

Smaller amounts of cadmium were also sequestered outside the polyphosphate bodies. The authors suggested that the sequestering of Cd and other metals may serve a dual purpose by providing a storage site for essential metals and acting as a detoxifying mechanism. Also the green algae *Ankistrodemus falcatus, Scenedesmus quadricauda* and *Chlorella pyrenoidosa* were found to sequester cadmium in polyphosphate bodies.[638] The fact that cadmium is sequestered in polyphosphate bodies has also been reported by Sicko-Goad and Lazinski[4473] for benthic algae and Rachlin et al.[3951] for the blue-green alga *Anabaena flos-aquae*. In general, cadmium tends to accumulate in algae in large quantities (Table 6-4).

Lazinski and Sicko-Goad[4474] reported a decrease in polyhedral body volume of the blue-green algae *Plectonema boryanum, Microcoleus vaginatus,* and *Schizothrix* sp. exposed to 11.2 μg L^{-1} Cd for three days. In addition, in *Plectonema* and *Schizothrix,* there was a greater relative volume of polyphosphate bodies in Cd treatment. Rosko and Rachlin[4189] reported that cadmium increased the chlorophyll-*a* content per cell of *Chlorella vulgaris* after 33 days of exposure. Conway,[864] however, reported that cadmium reduced the photosynthetic pigment content of *Asterionella formosa.* Cadmium also affected the feedback control of the permease system that transported phosphate or silicate into cells of *A. formosa.*

Burnison et al.[638] reported that changes in ultrastructural morphology were first seen in cells of the green algae *Ankistrodemus falcatus, Scenedesmus quadricauda,* and *Chlorella pyrenoidosa* treated with 30 μg L^{-1} Cd. These changes included dilatation of the endoplasmic reticulum, mitochondrial vacuolation, and granule accumulation. The mitochondrial changes may interfere with the production of ATP. Changes in mitochondrial fine structure has also been observed by Silverberg.[4500] The presence of prominent intramitochondrial-dense granules was the most noticeable feature. Laube et al.[2720] found that 29–34% and 66–71% of the Cd^{2+} bound by the green alga *Ankistrodemus braunii* was associated with the cell wall—cell membrane fraction and the cytoplasm, respectively.

6.3 CHROMIUM

6.3.1 Chromium in the Environment

Earth's crust: 100 mg kg^{-1} (490, 903, 3093, 4261, 4920, 5417)

Common minerals: (Mg,Fe)O(Cr Al,Fe)$_2$O$_3$ (chromite spinel group), depending on the degree of substitution in the Al, Fe, Cr series, the chromites contain from 13% to 65% Cr$_2$O$_3$, PbCrO$_4$ (crocoite)[3287,3815]

Soils: 1–1,384 mg kg^{-1} (2395), 5–1,000 mg kg^{-1} (4849), 70 (5–1,500) mg kg^{-1} (491), 100 (5–3,000) mg kg^{-1}, highest in soils derived from basalt or serpentines[490]

Sediments: < 90 μg g^{-1} (1452,2847,4257,5308), 72 mg kg^{-1} (491)

Air: 0.005–300 ng m^{-3} (491), 0.003–1,100 ng m^{-3} (2395)

Fresh water: 0.18–2 μg L^{-1} (490,1451,3288,5215,5218), 1.0 (0.1–6) μg L^{-1} (491)

Sea water: 0.05–0.5 μg L^{-1} (3172), 0.3 (0.2–50) μg L^{-1} (491)

Land plants: 0.23 (0.03–10) mg kg^{-1} (490–1), 0.013–4.2 mg kg^{-1} (2395)

Aquatic and wetland plants (mg kg^{-1}) (see Appendix, Table 7-47):

emergent species: aboveground: <1.0–17.2

floating-leaved species: < 1.0 (FD)

submerged species: 0.86–122.2

floating species: 0.27–8,600

bryophytes: 0.8–1,330

Freshwater algae: 0.82–67.5 mg kg^{-1} (Table 6-5)

Marine algae: 0.5–13 mg kg^{-1} (491), 0–59 mg kg^{-1} (Table 6-5)

Land animals: 0.075 mg kg^{-1} (490)

Marine animals: 0.2–1 mg kg^{-1} (490)

Sources to waters: manufacturing processes (metals, chemicals, pulp and paper, petroleum products), municipal wastewaters, sewage sludge discharge, smelting and refining of nonferrous metals, base metal mining and dressing, atmospheric deposition.[3480]

Aquatic chemistry: Cr^{3+} is classified as a hard acid and forms relatively strong complexes with oxygen donor ligands. The principal forms in freshwater include CrOH$_2$$^+$, Cr(OH)$_2$$^+$, and Cr(OH)$_4$$^-$.[3287] Cr^{6+}, on the other hand, is water soluble, always existing in solution as a component of a complex anion. The anionic species vary with pH, and may be chromate (CrO$_4$$^{2-}$), hydroxychromate (HCrO$_4$$^-$) or dichromate (Cr$_2O_7$$^{2-}$).[3287]

Table 6-4. Cadmium concentration factors (CF) in algae

Species	CF	Locality	Reference
Freshwater			
Nostoc H	126	Laboratory	1961
Cladophora glomerata	13–920	Roding River, England	3146
Nostoc 586	93–1,140	Laboratory	1961
Scenedesmus obliquus	105–2,700	Laboratory	1961
Scenedesmus obliquus	329–4,940	Laboratory	660
Chlamydomonas sp.	5,480	Laboratory	1961
Cladophora glomerata	2,210–5,550	Šárecký Brook, Czech Republic	5188
Mougeotia sp.	12,000	Northern England	3703
Lemanea fluviatilis	33,000	Northern England	1910
Cladophora glomerata	18,000–49,000	Lake Ontario	2453
Cladophora glomerata	103,900	Laboratory experiment	5188
Lemanea sp.	150,000	River Conway, England	1909
Marine			
Fucus vesiculosus	2,700	Irish Sea	3882
Fucus vesiculosus	4,200–10,000	Norwegian coast	3186
Fucus vesiculosus	10,500	Menai Straits, Great Britain	1455
Fucus vesiculosus	14,000–15,000	Bristol Channel	3309
Fucus vesiculosus	14,500–58,600	Laboratory	604

Adsorption of Cr^{6+} by clays, ferric hydroxide, and ferric and manganese oxides is generally a minor fate process, whereas Cr^{3+} is rapidly adsorbed, at least by clays.[3287] The rate of adsorption of Cr^{3+} increases with pH to the point where the total amount of bound Cr^{3+} exceeds that of Cr^{6+} by 30–300 times.[3287,3980]

6.3.2 Toxicity and Accumulation in Algae

Chromium is an essential micronutrient for human and other mammals involved in the peripheral action of insulin, normal glucose utilization, stimulation of enzyme systems, and possibly in the stabilization of nucleic acids.[2959,4697] Despite much biological requirements, high concentrations of chromium are toxic, mutagenic, carcinogenic, and teratogenic.[2959]

The toxic effects of chromium are known to be valence dependent. Cr^{3+} has a lower toxicity than Cr^{6+}, because of the strong oxidizing power of Cr^{6+}. Also Cr^{6+}, in contrast to Cr^{3+}, is more soluble and passes through biological membranes more readily.[2959,3201,4176] Generally, Cr^{6+} perturbs cell growth and cell cycles. Chromosomal alterations have also been noted.[560] Brochiero et al.[560] reported a toxic action of Cr^{6+} on *Euglena*—the lengthening of the lag-phase, followed by growth whose exponential phase is the same as the control. This response was observed also by Bonaly et al.[432] on *Chlamydomonas* cultured in a synthetic medium containing Cr^{6+}. This response was not found with other toxic metals such as Cr^{3+}, Pb, Ni, Zn or Cd.[220,432]

The toxicity of chromium to algae has been known for a long time. Richards[4098] found total mortality of *Spirogyra insignis* at 2.5 mg L^{-1} while *Oscillatoria limosa* did not succumb until 25 mg L^{-1} level was reached. Hervey[2043] reported an increasing sensitivity to Cr along the species chlorococcal green alga (inhibition at a concentration of 3.2 mg L^{-1}), euglenoid flagellates (0.32 mg L^{-1}), and diatoms (0.032 mg L^{-1}). A chromium dose above 0.5 mg L^{-1} inhibited the growth of *Chlorella*, reduced the protein content, and destroyed chlorophyll.[3449] Similar results were obtained by Romanenko and Velichko[4177] who studied the effect of Cr^{6+} on the green algae *Chlorella pyrenoidosa, Scenedesmus quadricauda* and natural phytoplankton. Bharti et al.[353] described the toxic effect of Cr^{6+} on filamentous

Table 6-5. Chromium concentration in algae (mg kg^{-1} dry mass)

Species	Concentration	Locality	Reference
Freshwater			
Chara fragilis	0.82	Tübingen, Germany	2205
Chara sp.	1.05	Lake Huron	1209
Anacystis nidulans	1.20	Synthetic medium	2205
Chara sp.	1.2–7.9	Canada, St. Clair Lake	3346
Aphanizomenon flos-aquae	9.0	Synthetic medium	2205
Scenedesmus sp.	1.5–11.4	Germany, sewage grown	269
Not specified	12	IJsselmeer, The Netherlands	4258
Not specified	15	IJseelmeer, The Netherlands	4260
Chlorella sp.	1.7–16	Germany, sewage grown	269
Cladophora sp.	4–18.1	Lake Huron	1209
Not specified	20	Haringvliet, The Netherlands	4261
Net phytoplankton	5.4–29.4	Columbia River, Washington	940
Cladophora glomerata	2.3–40	Lake Balaton, Hungary	2575
C. glomerata	12.9–67.5	Lake Michigan	20
Marine			
Laminaria saccharina	0.44	Rhode Island coast	4389
Phycodrys rubens	0.29–0.6	Rhode Island coast	4389
Fucus vesiculosus	1.4	Trondsheimsfjord, Norway	2694
Fucus vesiculosus	0–1.66	Sweden, coast	4612
Polysiphonia urceolata	2.1	Rhode Island coast	4389
Ptilota serrata	0.61–2.3	Gulf of Maine	4389
Stichococcus bacillaris	2.8	Laboratory	4145
Ascophyllum nodosum	2.2–2.8	Great Britain	1455
Ahnfeltia plicata	2.9	Helgoland, North Sea	2205
Dunaliella tertiolecta	3.6	Laboratory	4145
Fucus vesiculosus	1.0–4.0	Trondheimsfjorden, Norway	2694
Fucus vesiculosus	4.5	Great Britain	1455
Asterionella japonica	5.5	Laboratory	4145
Fucus vesiculosus	1.0–5.7	Great Britain estuaries	604
Laminaria saccharina	6.0	Helgoland, North Sea	2205
Dunaliella primolecta	8.4	Laboratory	4145
Monochrysis lutheri	9.4	Laboratory	4145
Ascophyllum nodosum	1–13	Trondheimsfjorden, Norway	2694
Micromonas squamata	13.5	Laboratory	4145
Olisthodiscus luteus	14.7	Laboratory	4145
Blue-green algae	17	Penang Island, Malaysia	4526
Fucus vesiculosus	traces–26	Penang Island, Malaysia	4526
Hemiselmis brunescens	30	Laboratory	4145
Enteromorpha sp.	traces–59	Penang Island, Malaysia	4526

green algae *Ulothrix fimbrinata, Cladophora glomerata* and *Stigeoclonium tenue*. Growth inhibition was evident from a concentration of 0.15 mg L^{-1} upward, and a reduced production of proteins and carbohydrates from a concentration of 0.01 mg L^{-1}. Meisch and Schmitt-Beckmann[3185] found that at the concentrations of 0.1–0.5 mg L^{-1} Cr algal growth and O$_2$ production in *Chlorella* were stimulated, while at concentrations of 0.5–2.0 mg L^{-1} Cr the content of chlorophyll cell numbers and their mass were substantially reduced. Similar results were received by Petria,[3772] who described growth inhibition at a concentration of 20 mg L^{-1} Cr and the inhibition of photosynthesis at 0.5 mg L^{-1} in the alga *Chlorella*.

The toxic effects of Cr^{6+} on the diatom *Nitzschia palea* and the green alga *Chlorella pyrenoidosa* were described by Wium-Andersen.[5467] The growth of *N. palea* was greatly inhibited by the presence of 150 µg L^{-1} Cr. When the initial cell concentration was increased, the inhibitory effect from the addition of 150 µg L^{-1} Cr was overcome after 2 days. The photosynthetic effect of chromium on the growth of *N. palea*, however, was significantly lower than that of copper. A concentration of 5 µg L^{-1} Cu produced a decrease in photosynthesis of about 45%. If chromium was to produce a similar decrease under the same conditions, the concentration must be 650 µg L^{-1}.

Chromium inhibited the photosynthetic capacity (P$_{max}$) of the diatom *Fragilaria crotonensis*, with inhibition being greatest at 2.0 µM Cr^{6+}. Photosynthesis capacity values of the phytoplankton assemblage were also reduced, but the effects were less severe in that 2.0 µM Cr^{6+} reduced P$_{max}$ to 37.5% of the control compared with 5.9% for *F. crotonensis*.[5238]

Mangi et al.[3041] reported that the growth of *Oedogonium* sp., *Hydrodictyon reticulatum*, *Palmella mucosa* and *Palmellococcus prototothecoides* was inhibited by 10 mg L^{-1} of chromium. Patrick et al.[3685] reported inhibition concentrations of Cr^{6+} of 0.26 mg L^{-1} and 0.34 mg L^{-1} for *Nitzschia linearis* and *Navicula seminulum* var. *hustedtii*, respectively. Matulová[3107] reported lethal concentration of 0.3 mg L^{-1} Cr^{6+} for green algae *Scenedesmus quadricauda*, *Chlamydomonas hydra*, *C. eugametos* and *C. moewusii*. No influence was observed up to 0.05 mg L^{-1} Cr.

Chiaudani and Vighi[780] reported EC-50$_{96}$ values for the green alga *Selenastrum capricornutum* to be 3.6 µg L^{-1} and 31 µg L^{-1} of Cr^{3+} in AAP medium without and with EDTA addition, respectively. Aliotta et al.[57] reported EC-50 values for green algae *Chlorella protothecoides*, *C. saccharophila*, *Coenochloris* sp. and *Stichococcus bacillaris* to be 6.6, 13.0, 26.0 and 26.0 mg L^{-1} Cr at pH 3.0 and 3.3, 13.0, 3.3 and 102 mg L^{-1} Cr at pH 6.5, respectively. Patrick[3681] reported that in fresh waters amended with up to 49.5 mg L^{-1} Cr^{3+}, diatoms dominated over blue-green algae, whereas with 376.5 mg L^{-1} Cr^{3+}, blue-green algae were dominated, followed by green algae and greatly reduced numbers of diatoms.

Starý et al.[4697] reported that Cr^{3+} is rapidly cumulated by green algae *Scenedesmus obliquus*, *Chlorella kessleri* and *Chlamydomonas geitleri*, whereas Cr^{6+} is practically not cumulated in algae; this fact was also reported by Schroll.[4362] The cumulation factors increase with the decrease of the radius of algal cells. The authors also found that in the presence of algae, Cr^{6+} was more rapidly reduced into three-valent form (which was then quickly sorbed on the surface of algal cells) than in the control experiments where algal cells were not present. They also concluded that the cumulation of Cr was due predominantly to the chemical sorption on the surface of algal cells. Also Mangi et al.[5238] reported that removal of chromium from solution by filamentous green alga *Oedogonium* sp. was due to adsorption. The values of chromium concentration factors in algae are given in Table 6-6.

6.4 LEAD

6.4.1 Lead in the Environment

Earth's crust: 13–16 mg kg^{-1} (490, 903, 3093, 3287, 4920)
Common minerals: PbS (galena), PbCO$_3$ (cerussite), PbSO$_4$ (anglesite, gelesite), Pb$_5$Cl(PO$_4$)$_3$ (pyromorphite)[491,3815,4465]
Soils: 16 mg kg^{-1} (4465), 35 (2–300) mg kg^{-1} (491), 1.5–286 mg kg^{-1} (2395), in contaminated sites up to 135,000 mg kg^{-1} (2395), 2–30,000 mg kg^{-1} ("geologically polluted"[3471,3473]), strongly absorbed by humus[490]
Sediments: 19 mg kg^{-1} (491), 3–110,000 mg kg^{-1} (freshwater[5205]), 1.0–11,368 mg kg^{-1} (marine[3473]), 2.6–3,500 mg kg^{-1} (freshwater lakes[3473]), 3.9–3,700 mg kg^{-1} (rivers[3473]), background < 40 mg kg^{-1} (1450, 1452, 2769, 2847, 4258–9, 5308)
Air: remote areas 0.049–12.6 ng m^{-3} (3474), urban and rural areas: 0.034–99 µg m^{-3} (3474), 4.4–7,700 ng m^{-3} (5205), 0.19–13,200 ng m^{-3} (491,2395)
Fresh water: 5 µg L^{-1} (average[490,3700]), 0.2 µg L^{-1} (average[5218]), 0.1–2,071 µg L^{-1} (792), 0.09–450 µg L^{-1} (5205), 2–140 µg L^{-1} (2570), 3.0 (0.06–120) µg L^{-1} (491); natural water background: < 0.2–3.0 µg L^{-1} (256, 521, 2032, 5215); acid mine drainage: 70–2,350 µg.l^{-1} (5205); ground water: 0.13–124 µg L^{-1} (792); mineral waters and springs: 0.1–5,000 µg L^{-1} (792); rainwaters: 3–3,000 µg L^{-1} (792, 5205)

Table 6-6. Chromium concentration factors (CF) in algae

Species	CF	Locality	Reference
Freshwater			
Scenedesmus sp.	190–260	Germany, sewage grown	269
Chlorella sp.	210–620	Germany, sewage grown	269
Periphyton	500–4,300	Columbia River, Washington	5290
Cladophora glomerata	2,550–5,390	Sárecky Brook, Czech Republic	5188
Cladophora glomerata	5,650	Laboratory experiment	5188
Net plankton	440–9,400	Columbia River, Washington	5290
Cladophora glomerata	100–10,000	Lake Balaton, Hungary	2575
Chlamydomonas geitleri	50,000	Laboratory	4697
Phytoplankton	57,000	Washington, Columbia River	942
Net phytoplankton	19,000–81,000	Washington, Columbia River	940
Chlorella kessleri	150,000	Laboratory	4697
Scenedesmus obliquus	150,000	Laboratory	4697
Marine			
Fucus vesiculosus	11,000	Menai Straits, U.K.	1455

Sea water: 0.02–8 μg L^{-1} (792), 0.03 (0.03–13) μg L^{-1} (490–1)
Land plants: 2.7 (1–13) mg kg^{-1} (490–1), <0.002–18.8 mg kg^{-1} (2395), in contaminated sites up to 2,714 mg kg^{-1} (2395)
Aquatic and wetland plants (mg kg^{-1}) (see Appendix, Table 7-48):
emergent species: aboveground: 0.04–657, belowground: <0.1–88 (FD)
floating-leaved species: 0–15.6 (aboveground)
submerged species: 0.8–7,750 (aboveground)
floating species: 2.7–25,790
bryophytes: 0.25–15,940
Standing stock (g Pb m^{-2}) (see Appendix, Table 7-49):
emergent species: aboveground: 0.03–5.2, belowground: 0.11–5.30
Freshwater algae: 1–>70,000 mg kg^{-1} (Table 6-7)
Marine algae: 2–40 mg kg^{-1} (491), 0.5–2,200 mg kg^{-1} (Table 6-7)
Land animals: 2.0 mg kg^{-1} (490)
Marine animals: 0.5 mg kg^{-1} (491)
Sources of lead in the environment:
- Natural: windblown dust (more than 50%), volcanic particles (volcanic halogen aerosols and silicate smokes—more than 20%), seasalt sprays, vegetation, forest fire smokes, meteoritic smokes.[3475,3700,4261,5317]
- Man's activity: smelting and refining (nonferrous metals, iron and steel), ore burning, production of paints and dyes, pulp and papers, agricultural pesticides, explosives, water pipes, solder, batteries, antiknock-agent in gasoline, cleaning and duplicating, municipal wastewaters, ammunition industry, steam electrical production.[1450,2026,3480,4465,5205]
Aquatic chemistry: In freshwaters, lead forms a number of complexes of low stability with many of the major anions, including hydroxides, carbonates, sulfides, and less commonly sulfates. Lead also partitions favorably with humic and fulvic acids, forming moderately strong chelates.[3287] In alkaline waters, complexes $[Pb(CO_3)_2]^{2-}$, $[Pb(OH)_2 (aq)]^0$ and $[PbOH]^+$ may be present.[3815] At pH 10, $Pb(OH)^+$ dominates all other species.
Lead undergoes methylation in the environment to form several organic derivatives. The process is mediated by bacteria in sediments to form $(CH_3)_3Pb^+$ and related compounds.[3287] D'Itri,[1094] however, pointed out that whether environmental Pb methylation is a biological or chemical process has been the subject of some controversy. The varying results suggest that the relative importance of biotic versus abiotic $(CH_3)_4Pb$ formation still has not been determined.[1094,4937]

Table 6-7. Lead concentration in algae (mg kg^{-1} dry mass)

Species	Concentration	Locality	Reference
Freshwater			
Spirogyra sp.	1.0	Upper Bee Fork Creek, Missouri	1961
Cymbella sp.	8.0	Upper Bee Fork Creek, Missouri	1961
Cladophora glomerata	9.5	Deadman Bay, Ontario	2453
Cladophora glomerata	12.2	Main Duck, Ontario	2453
Scenedesmus acutus	1.4–12.6	Cultivation (wastewater)	5144
Cladophora sp.	14.9	Vermilion River, Illinois	2769
Filamentous green algae	15	Clearwater Lake, Missouri	1544
Chlamydomonas geitleri	7.0–17.5	Cultivation (wastewater)	5144
Periphyton	10.5–20.4	Elsenz River, Germany	355
Not specified	21	Lake Alpnach, Switzerland	195
Not specified	26	IJsselmeer, The Netherlands	4258
Cladophora glomerata	8–33.5	Lake Balaton, Hungary	2575
Cladophora rivularis	5.8–43.2	New Zealand	4017
Not specified	47	IJsselmeer, The Netherlands	4260
Chara sp.	20.1–47.3	Lake St. Clair, Canada	3346
Cladophora glomerata	5.2–49.2	Leine River, Germany	13
Oedogonium sp.	60	Lower Swansea Valley, Wales	5027
Cladophora rivularis	5.8–84.2	Elsenz River, Germany	355
Cladophora glomerata	10.8–150	Roding River, England	3146
Lemanea fluviatilis	197	Drewent R. Catchment, U.K.	1910
Cladophora sp.	90–230	Lower Swansea Valley, Wales	5027
Cladophora glomerata	8–335	Lake Balaton, Hungary	2575
Cladophora sp.	14.9–347	Vermilion River, Illinois	2769
Anabaena sp.	358	Laboratory	3234
Spirogyra sp.	40–400	Lower Swansea Valley, Wales	5027
Epilithic algae	26–404	Laboratory	3234
Oscillatoria sp.	580[b]	Lower Swansea Valley, Wales	5027
Mougeotia sp.	1,930	Caplecleugh Low Level, U.K.	3703
Microspora sp.	2,160[b]	Lower Swansea Valley, Wales	5027
Ulothrix sp.	2,380[b]	Lower Swansea Valley, Wales	5027
Zygnema sp.	2,600[a]	Lower Swansea Valley, Wales	5027
Coccomyxa sp.	3,230[a]	Lower Swansea Valley, Wales	5027
Oscillatoria sp.	5,483–5,617[c]	Strother Creek, Missouri	5469
Mougeotia sp.	6,190[a]	Lower Swansea Valley, Wales	5027
Mougeotia sp.	9,829[c]	Strother Creek, Missouri	5469
Tribonema sp.	3,680–14,190[a]	Lower Swansea Valley, Wales	5027
Filamentous greens	> 70,000[d]	New Lead Belt, Missouri	1544
Marine			
Laminaria saccharina	1.7	Rhode Island coast	4389
Chorda filum	3.0	Portugal and Spain	4732
Gelidium amansii	0.8–3.1	Hong Kong	2113
Fucus vesiculosus	2.3–3.2	Great Britain	1455
Enteromorpha linza	2.7–3.7	Hong Kong	2113
Polysiphonia urceolata	3.7	Rhode Island coast	4389
Gloiopeltis furcata	1.8–4.4	Hong Kong	2113

Table 6-7. Continued

Species	Concentration	Locality	Reference
Ascophyllum nodosum	4.0–6.0	Norway	4731
Sargassum hemiphyllum	6.0	Hong Kong	2114
Endarachne binghamiae	7.4	Hong Kong	2114
Chorda filum	8.0	Norway	4731
Corallina officinalis	8.0	Portugal and Spain	4732
Fucus spp.	0.5–9.0	Irish Sea	3882
Ptilota serrata	0.9–9.0	Gulf of Maine	4389
Monochrysis lutheri	9.2	Laboratory	4145
Delesseria sanguinea	10.0	Portugal and Spain	4732
Ishige foliacea	6.2–10.3	Hong Kong	2113
Porphyra sp.	0.8–10.5	Irish Sea	3882
Sargassum vachellianum	10.6	Hong Kong	2114
Fucus sp.	5–13	Portugal and Spain	4732
Dunaliella primolecta	16.5	Laboratory	4145
Blue-green algae	17	Penang Island, Malaysia	4526
Ulva lactuca	3–18	Norway, coast	4731
Ulva lactuca	10–18	Portugal and Spain	4732
Stichococcus bacillaris	20.3	Laboratory	4145
Ectocaprus siliculosus	20.9–21.3	Hong Kong	2113
Corallina officinalis	22	Norway	4731
Gymnogongrus flabelliformis	3.1–28.9	Hong Kong	2113
Scytosiphon lomentaria	2.6–31.1	Hong Kong	2113
Fucus vesiculosus	1.8–31.9	Sweden, coast	4612
Gigartina stellata	7–32	Norway, coast	4731
Hemiselmis virescens	32	Laboratory	4145
Laminaria digitata	8–37	Norway, coast	4731
Fucus ceranoides	19–38	Norway, coast	4731
Fucus vesiculosus	1.6–39	Great Britain estuaries	604
Chaetomorpha antennina	4.5–39	Hong Kong	2113
Heteromastix longifillis	50	Laboratory	4145
Fucus vesiculosus	5–50	Penang Island, Malaysia	4526
Enteromorpha compressa	4.9–53.8	Hong Kong	2113
Hemiselmis brunescens	56.5	Laboratory	4145
Enteromorpha sp.	13–59	Penang Island, Malaysia	4526
Phycodrys rubens	1.3–79	Rhode Island coast	4389
Porphyra umbelicalis	0.8–85	Irish Sea	3882
Ascophyllum nodosum	<3–95	Norway, coast	1975
Fucus vesiculosus	0.5–113	British coast	3882
Caulacanthus okamurai	51.6–127.8	Hong Kong	2113
Enteromorpha flexuosa	8.6–137.1	Hong Kong	2113
Fucus vesiculosus	28–163	Sorfjørden, Norway	3186
Fucus vesiculosus	3–202	Norway, coast	4731
Enteromorpha sp.	7–1,200	Norway coast	4731

[a] Adjacent to Zn smelting wastes, [b] near Zn smelting wastes, [c] water receiving a tailing pond effluent, [d] streams receiving effluents from lead mines and mill

Sorption to sediments plays an important role in the fate of lead complexes.[2231] Lead typically complexes with sulfides and Fe-Mn hydrous oxides in sediments.[3157-8,4263] Particulate Pb generally accounts for > 75% of residues in flowing waters and > 50% in standing waters.[4383,5388,5561] Chemical speciation of in natural waters have been reported many times.[792,3294,4492,4818,5565] Lead chemistry from the thermodynamic point of view has been reported by Hem,[2025] Hem and Durum,[2026] Latimer,[2717] Robie and Waldbaum,[4159] Wagman et al.[5220] and Sillén and Martell.[4493-4] The lead cycling in an aquatic ecosystem is shown in Figure 6-1.

6.4.2 Toxicity and Accumulation in Algae

There is no evidence that lead is an essential trace element for the organisms, although the hypothesis that minute amounts may serve some essential function in metabolism has not been carefully examined. On the contrary, all experimental data obtained both *in vitro* and *in vivo* show that the metabolic effects of lead in concentrations as low as 10^{-6} M (207 μg L^{-1}) are of the inhibitory or adverse type.[5484]

Woolery and Lewin[5497] reported that photosynthesis was completely suppressed by 8.5 mg L^{-1} Pb in cells of the marine diatom *Phaeodactylum tricornutum*, whereas respiration was reduced to 25% at the same concentration of lead. No visually observed effects of lead in concentrations up to 10 mg L^{-1} on vegetative morphology or development or reproductive structures of three species of red algae were observed, although in medium with added lead these algae grew more slowly.[4752] Rosko and Rachlin[4189] reported that 32 mg L^{-1} Pb reduced the chlorophyll-*a* content of the cells of the green alga *Chlorella vulgaris* by 18%. This is consistent with the findings of Overnell[3592] of inhibition of O_2 evolution in a green marine alga *Dunaliella tertiolecta* exposed to lead. Irmer[2288] reported a concentration of 207 μg L^{-1} Pb reduced the growth of a green alga *Chlamydomonas reinhardtii* by 47% within 48 hours, chlorophyll biosynthesis by 26% and photosynthetic oxygen evolution by 39%. Similar toxic effects were observed in the green alga *Chlorella fusca*.[2288] Concurrent with these metabolic changes, exposition to 5 μmol L^{-1} Pb (= 1.04 mg L^{-1} Pb) caused drastic ultrastructural changes in *C. reinhardtii*. The chloroplast membranes were damaged, pyrenoids showed signs of dissolution as did the mitochondria, in the nucleus the nucleolus seemed to dissolve, and the nuclear envelope was swollen resulting in a large perinuclear spaces. Rosko and Rachlin[4189] pointed out that Pb, like Zn, exerts its toxicity more in terms of chlorophyll-*a* content than on cell division in *Chlorella vulgaris*.

Sicko-Goad[4470] found a decrease in surface area of thylakoidal membranes in a lead-treated blue-green alga *Anacystis cyanea*. The same phenomenon was reported for lead-treated blue-green alga *Plectonema boryanum*.[4948] Rachlin et al.[3948] reported that lead produced significant increases in cell size and in the surface area of the cell's thylakoids in a blue-green alga *Plectonema boryanum*. Lead enlarged the cells and the electron micrographs showed no obvious cellular distortions, other than a general increase in cell size. Lead also caused coalescence of cellular lipid but caused no extracellular membrane whorls (like Cu, Hg, Cd, Ni, Zn, and Co). The authors found Pb incorporated in the third cell wall layer, fibrils of which consist primarily of glycoaminopeptides.[4262] It is possible that lead (and also Cu, Ni and Co) may be interacting with these peptides. Sicko-Goad[4470] reported that for freshwater diatom *Fragilaria capucina* there were no significant differences in any cellular component values nor any readily observable trends toward increase or decrease with lead treatment. Sicko-Goad and Lazinski[4473-4] reported for *Plectonema boryanum* that lead exposure increased the number of polyphosphate bodies and caused their enlargement. Constant exposure to the metal also resulted in a significant decrease in the polyhedral body (carboxysome) relative volume and number per volume. Exposure of *Scenedesmus quadricauda* to lead during phosphate uptake also resulted in a number of significant morphological changes during polyphosphate degradation. The decrease in the polyhedral body volume in blue-green algae *Plectonema boryanum*, *Microcoleus vaginatus* and *Schizothrix* sp. treated in media with 20.7 μg L^{-1} Pb for 3 days was reported by Lazinski and Sicko-Goad.[2724]

Röderer[4161-2] reported that tetraethyl lead was not toxic to the chrysophycean flagellate *Poterio-ochromonas malhamensis* in darkness, but population growth, mitosis and cytokinesis were inhibited in light. Hessler[2045] found the log phase of unicellular flagellate alga *Platymonas subcordiformis* cells to be more sensitive than stationary phase cells. Sublethal amounts of Pb (2.5 and 10 mg L^{-1}) tended to retard population growth by delaying cell division and daughter cell separation. A lethal amount of lead (60 mg L^{-1}) caused inhibition of growth and cell death. The flagella were shed or altered in a variety of ways, depending on Pb concentration: motility was least affected by low Pb and completely impaired by high Pb. Hessler[2046] suggested that lead probably does not produce mutation in *Platymonas subcordiformis*.

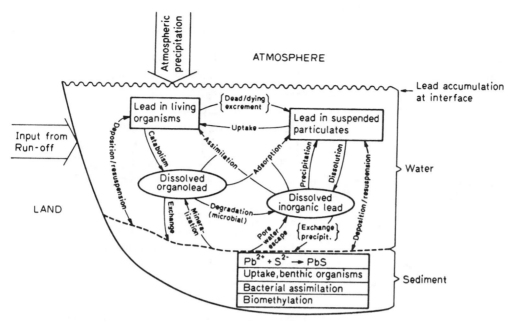

Figure 6-1.   Cycling of lead in an aquatic ecosystem.[4124] (From Rickard and Nriagu, 1978, in *The Biogeochemistry of Lead in the Environment,* Nriagu, J.O., Ed., p. 277, reprinted with permission of Elsevier Science Publishers bv, Amsterdam and the authors.)

Sicko-Goad and Stoermer[4472] reported that cells of a diatom *Diatoma tenue* var. *elongatum* exposed to lead (10.4 μg L^{-1}) in P-deficient medium showed a significant decrease in number of mitochondria with a concomitant increase in their volume and an increase in membranous organelles in the vacuole compared to P-starved and P-sufficient controls. The lead apparently did not affect the phosphate uptake mechanism, since similar numbers of polyphosphate bodies were encountered both for phosphate and phosphate + Pb treatments. Thomas et al.[4955] reported that lead caused some disruption of cell separation in the diatom *Thalassiosira aestivalis*, so that cell clumps were formed, but the main effect was the formation of elongated, bent, out-of-column cells.

Berland et al.[324] reported that concentrations of lead at which toxic effects begin to appear range from 300 nmol L^{-1} (62.2 μg L^{-1}) to 0.5 nmol L^{-1} (0.1 mg L^{-1}) for the marine diatom *Skeletonema costatum*. Whitton[5403] found that all algal species tested were considerably more tolerant of lead than of zinc and copper, with inhibition occurring at 3–60 mg L^{-1} Pb. *Cladophora* was a relatively sensitive genus. The toxic concentration of lead varies with the species involved. Rachlin and co-workers[3946,3949–51] found the EC-50$_{96h}$ concentrations of lead to the blue-green alga *Anabaena flos-aquae,* diatoms *Nitzschia closterium* and *Navicula incerta* and a green alga *Chlorella saccharophila* to be 1.16, 6.1, 10.96 and 63.8 mg L^{-1} Pb, respectively.

Whitton and co-workers[4303–5,5312,5412] reported the growth of some algal species (e.g. blue-green algae *Phormidium autumnale, Aphanocapsa* sp., *Pseudanabaena catenata*, diatoms *Achnanthes minutissima, Navicula tantula, Caloneis bacillum, Cymbella bipartita* or green algae *Mougeotia* sp. and *Hormidium rivulare*) in waters with elevated lead concentrations. Say and Whitton[4305] reported the growth of a blue-green alga *Schizothrix* sp. in close association with anglesite (PbSO$_4$), the concentration of Pb in the sediment having been 110 g kg^{-1}. Many data concerning tolerance and resistance of algae to lead have been summarized by Leland and Kuwabara[2770] and Campbell and Stokes.[676] For further information about lead toxicity to algae see the reviews by Wong et al.[5484] and Vymazal.[5205]

Lead tends to accumulate to high extent in algae (Table 6-8). Lead was found to be sequestered in the polyphosphate bodies of the blue-green alga *Plectonema boryanum*; small amounts of lead were also found in the cell sectors without polyphosphate bodies.[2336] Lead was found in cell sectors with polyphosphate bodies of the green alga *Chlorella saccharophila* at 13.3 mg L^{-1} Pb and diatoms *Navicula incerta* and *Nitzschia closterium* at 10 mg L^{-1} Pb.[2337] Lead was also detectable in the cell wall

Table 6-8. Lead concentration factors (CF) in algae

Species	CF	Locality	Reference
Freshwater			
Oedogonium sp.	570	Lower Swansea Valley, Wales	5027
Microspora sp.	740[b]	Lower Swansea Valley, Wales	5027
Cladophora glomerata	1,000	Lake Balaton, Hungary	2575
Oscillatoria sp.	1,520	Laboratory	1962
Nostoc 586	91–1,560	Laboratory	1961
Nostoc H	226–1,580	Laboratory	1961
Schizothrix calcicola	1,320–1,690	Laboratory	1962
Cladophora sp.	900–2,200	Lower Swansea Valley, Wales	5027
Nostoc sp. 586	1,560–2,400	Laboratory	1962
Cladophora glomerata	2,000–3,400	Šárecký Brook, Prague	5188
Spirogyra sp.	400–3,800	Lower Swansea Valley, Wales	5027
Oscillatoria sp.	5,600[b]	Lower Swansea Valley, Wales	5027
Cladophora glomerata	230–7,500	Roding River, U.K.	3146
Ulothrix sp.	7,600[b]	Lower Swansea Valley, Wales	5027
Cladophora glomerata	16,000	Deadman Bay, Ontario	2453
Tribonema sp.	3,000–18,000[a]	Lower Swansea Valley, Wales	5027
Chlamydomonas sp.	360–18,600	Laboratory	1961
Ulothrix fimbrinata	19,900	Laboratory	1962
Cladophora glomerata	20,000	Main Duck, Ontario	2453
Mougeotia sp.	20,000[a]	Lower Swansea Valley, Wales	5027
Chlamydomonas sp.	16,700–20,100	Laboratory	1962
Cladophora glomerata	25,030	Laboratory	5188
Zygnema sp.	25,000[a]	Lower Swansea Valley, Wales	5027
Coccomyxa sp.	31,000[a]	Lower Swansea Valley, Wales	5027
Nostoc muscorum A	43,800	Laboratory	1962
Marine			
Fucus vesiculosus	2,400	Irish Sea	3882
Fucus vesiculosus	2,900	Menai Straits, Wales	1455
Fucus vesiculosus	3,200–26,000	Norwegian coast	3186
Fucus vesiculosus	37,500–56,600	Laboratory	604

[a] Adjacent to Zn smelting waters, [b] near Zn smelting wastes

sector of *C. saccharophila* at 25 mg L^{-1} and 5.6 mg L^{-1} in *N. closterium*. No lead was detected in the cell margin sector of *N. incerta* at metal concentrations up to 10 mg L^{-1}. The presence of lead in the polyphosphate bodies of various algae has been reported also by Jensen et al.,[2338] Sicko-Goad and Stoermer[2472] and Sicko-Goad and Lazinski.[4473]

6.5 MERCURY

6.5.1 Mercury in the Environment

Earth's crust: 0.05–0.08 mg kg^{-1} (490, 903, 3093, 3287, 4920)
Common minerals: HgS (cinnabar)[491,3287]

Soils: 0.06 (0.01–0.5) mg kg^{-1} (491), 0.071 mg kg^{-1} (903), 0.004–5.8 mg kg^{-1} (2395), in contaminated sites up to 54 mg kg^{-1} (2395), in mineralized areas up to 500 mg kg^{-1} (903)
Sediments: 0.19 mg kg^{-1} (491), 0.3 mg kg-^{1} (903), 0.1–0.5 (0.01–688) mg kg^{-1} (3287), in extreme situations up to 2010 mg kg^{-1} (2519–20)
Air: 0.009–38 ng m^{-3} (491, 2395)
Fresh water: 0.1 (0.0001–2.8) μg L^{-1} (491), 0.01–0.1 μg L^{-1} (3287)
Sea water: 0.03 (0.01–0.22) μg L^{-1} (490–1), <0.01–0.03 μg L^{-1} (3287)
Land plants: 0.015 (0.005–0.02) mg kg^{-1} (490–1), 0.0002–0.086 mg kg^{-1} (2395), in contaminated sites up to 200 mg kg^{-1} (2395)
Aquatic and wetland plants (mg kg^{-1}) (see Appendix, Table 7-50):
emergent species: aboveground: <0.01–2.2, belowground: 0.01–1.47
floating-leaved species: aboveground: 0.01–3.8, belowground: 0.01–0.03 (FD)
submerged species: <0.01–26.4
bryophytes: 0.06–13,500
Freshwater algae: 0.53–25 mg kg^{-1} (Table 6-9)
Marine algae: 0.03 mg kg^{-1} (491), 0.007–20 mg kg^{-1} (Table 6-9)
Land animals: 0.046 mg kg^{-1} (490)

6.5.1.1 Aquatic chemistry

Sources to waters: coal-burning power plants, atmospheric fallout, manufacturing processes (chemicals, metals, petroleum products), municipal wastewaters, sewage sludge discharge, base metal mining and dressing, smelting and refining of nonferrous metals.[3480]

Mercury can exist in three oxidation states in surface waters: Hg^0, Hg^+, and Hg^{2+}. In well-aerated waters Hg^{2+} should dominate, whereas under reducing conditions, Hg^0 may develop.[3287] In anaerobic environments in sediments practically insoluble HgS may be formed.[3815] Under alkaline conditions the solubility of HgS is increased by the formation of the complex anion $[HgS_2]^{2-}$ which is formed in the presence of S^{2-} ion excess.[1308,3815] Ramamoorthy et al.[3967] also reported that Hg^0 may be produced biologically and/or abiologically in microenvironments associated with suspended particulate matter. Aquatic chemistry of mercury has been reviewed e.g. by Gavis and Ferguson.[1584]

The mercury cycle in the aquatic environment is shown in Figure 6-2. Loss of mercury to the atmosphere from the oceans is more important than deposition in sediments. The latter, however, is a better storage reservoir for mercury.[903] The average residence time of mercury in the various reservoirs has been calculated as follows: atmosphere 11 days, ocean 3,200 years, ocean sediments 2.5 × 10^8 years, soils 1,000 years.[903]

Mercury readily binds to a large number of inorganic and organic ligands, the strongest one being with the chloride ion.[3287] Inorganic Hg can be methylated in the environment (see Figure 6-2) to form highly soluble toxic species which are readily absorbed and concentrated by aquatic flora and fauna. The dominant fate process for mercury in both marine and freshwaters is adsorption to suspended solids and sediments.[3287]

Following the original discoveries suggesting methylation of inorganic mercury, it has since been demonstrated on many occasions that mercury added to bottom sediments systems may be converted to methyl mercury.[903] Whereas soluble Hg complexes are adsorbed onto organic or inorganic particulates and removed by sedimentation in aerobic waterways, as the sediments become increasingly anaerobic the precipitated Hg compounds usually are converted into mercuric sulfide (HgS), which reduces the possibility that they will be recycled into the overlying water.[1583] This removal mechanism is absent from aquatic ecosystems that are aerobic year round. Under anaerobic conditions most Hg that is not present as HgS is bound to organic matter.[1094] Under aerobic conditions HgS can be oxidized to sulfate, which is much more soluble, so that Hg^{2+} becomes available for methylation by microorganisms.[1094] The rate of oxidation by physicochemical processes is very slow and depends on the redox potential. Direct biological, enzymatic oxidation may lead to a faster release of Hg^{2+} ions. This suggests that the oxidation of HgS is the rate determining step for the transformation of Hg^{2+} ions to CH_3Hg^+.[1583,2334] Until it is oxidized, Hg bound as HgS demonstrates little methylation even under aerobic conditions. Further details on mercury cycling in the environment and mercury methylation the reader is referred to References 903–5, 1094, 2328–9, 2339–40, 5456, and 5487 and references cited therein.

Table 6-9. Mercury concentration in algae (mg kg⁻¹ dry mass)

Species	Concentration	Locality	Reference
Freshwater			
Cladophora glomerata	0.53–0.68	Leine River, Germany	13
Periphyton	0.50–25	Rhine River	5171
Marine			
Polysiphonia urceolata	0.007	Rhode Island coast	4389
Ptilota serrata	0.027	Gulf of Maine	4389
Laminaria saccharina	0.03	Rhode Island coast	4389
Phycodrys rubens	0.006–0.64	Rhode Island coast	4389
Fucus vesiculosus	0.30–0.71	Portugal coast	1340
Ulva lactuca	0.50–2.20	Portugal coast	1340
Fucus vesiculosus	2.7	Trondheimsfjord, Norway	2694
Gracilaria verrucosa	0.96–4.68	Portugal coast	1340
Fucus spp.	1.8–18	Irish Sea	3882
Ascophyllum nodosum	0.05–20	Norway, coast	1975

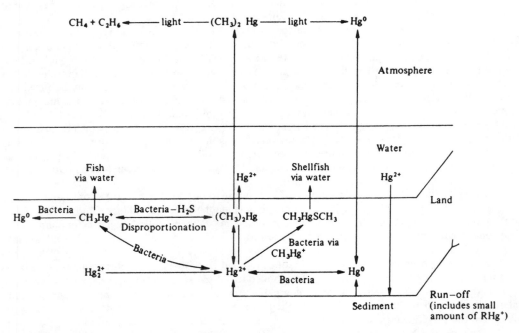

Figure 6-2.   Aquatic mercury cycle.[903]   (From Craig, 1980, in *The Handbook of Environmental Chemistry*, Vol. 1, Part A, Hutzinger, O., Ed., p. 179, reprinted with permission of Springer-Verlag and the author.)

6.5.2 Toxicity and Accumulation in Algae

There are many studies showing the variety of effects Hg can have on phytoplankton.[306,984-5,1897,1937,2414,2544,3484,5003,5072] Mercury is the most toxic heavy metal of the "non-essential" group and is widely distributed in water bodies.[3962] Mercury inhibits or stops algal growth,[31,1032,1897,2263,2414,3484,3962,4092,4800,5563] inhibits photosynthesis,[399,1033,5563] increases cell permeability and loss of potassium ions from the cell,[3962] reduces galactolipid production,[3105] decreases nitrogen fixation by inhibiting nitrogenase activity,[4800] and reduces chlorophyll content and synthesis.[1033,1589,3105,3960–1]

Rice et al.[4094] found the growth inhibition of the diatom *Skeletonema* at mercury concentration 0.5 μg L^{-1}. Zingmark and Miller[5563] reported inhibition of natural phytoplankton population photosynthesis at a concentration of 1.0 μg L^{-1} Hg. The growth of a marine dinoflagellate *Amphidinium carteri* was inhibited at Hg concentration of 1.0 μg L^{-1}. Davies[985] reported giant cell formation of a marine green alga *Dunaliella tertiolecta* as a result of growth unaccompanied by cell division at mercury concentration of 10 μM (= 2 mg L^{-1}). He ascribed this phenomenon to the inhibition of methionine production. A concentration of 1 μg L^{-1} Hg inhibited completely growth of a green alga *Chlorella vulgaris*;[1642] Rai[3959] reported the same effect in *C. vulgaris* to be caused by a concentration of 1 mg L^{-1}.[3959] Raising the initial mercury concentration gradually from 0 to 1.0 mg L^{-1} caused an increasing lag period in the growth of *Chlamydomonas*. In the culture grown in a medium containing 2 mg L^{-1} Hg, growth was completely retarded.[306]

The growth of the blue-green alga *Nostoc linckia* was significantly inhibited at 0.2 mg L^{-1} of mercuric chloride and it was completely inhibited at 1 mg L^{-1}.[2643] There was a linear relationship between the growth inhibition and mercuric chloride concentrations over the range of 0.1–1.0 mg L^{-1}. Nitrate reductase and ammonium uptake activities were stimulated by 11% at 0.2 mg L^{-1} of mercuric chloride.[2643] Stratton et al.[4800] reported that the growth of the blue-green alga *Anabaena inaequalis* was stopped at a mercury concentration of 8 μg L^{-1}, while inhibition of photosynthesis was observed at concentrations > 100 μg L^{-1}. Sick and Windom[4468] found mercury concentrations of 0.03 to 0.35 μg L^{-1} to significantly depress the growth of unicellular marine algae. Lowered cell-division rates were accompanied by a decrease in total cellular nitrogen and lipid. Nuzzi[3484] found mercury, as phenylmercuric acetate, to be inhibitory to three phytoplankton species at concentrations as low as 0.06 μg L^{-1}. Knowles and Zingmark[2544] found that mercury additions inhibited the phytoplankton at a concentration of 0.05 μg L^{-1} (for each 10^4 cells l^{-1}); the most sensitive organism was a chrysophyte *Synura petersenii*.

Glooschenko[1655] found that marine diatom *Chaetoceros costatum* cells in light accumulated ^{203}Hg longer that did non-dividing cells, indicating the possibility of some active uptake of mercury. Hannan and Patouillet[1897] alleged 100 μg L^{-1} methylmercury inhibited growth of *Chaetoceros galvestonensis* but had no lasting effect on growth of other marine diatoms *Cyclotella nana* and *Phaeodactylum tricornutum*. Richardson et al.[4119] reported that green alga *Pediastrum boryanum* is able to survive and produce new colonies when exposed to concentrations of Hg as high as 1 mg L^{-1}. Tompkins and Blinn[5003] reported that the division rates of freshwater diatoms, *Fragilaria crotonensis* and *Asterionella formosa*, were totally inhibited at 74 and 370 μg L^{-1}, respectively. Blinn et al.[399] reported that at least a 40% reduction in photosynthetic activity of phytoplankton occurred at Hg concentration as low as 60 μg.l^{-1} (Figure 6-3). A toxic mercury threshold concentrations of 60 μg.l^{-1} was demonstrated for the summer phytoplankton assemblage, but a distinct threshold concentration was absent for the spring diatom assemblage. Differences in spring and summer phytoplankton populations may suggest subtle differences in Hg sensitivity between phytoplankton assemblages in combination with temperature acting on total community metabolism.[399] Additions of Hg at concentrations 1 μg L^{-1} and higher resulted in reduction of relative growth rates of marine dinoflagellates in batch and continuous cultures.[2449] The effect of Hg on the increase in length of five intertidal thresholds showed that exposure to an average concentration of 100 to 200 μg L^{-1} Hg for 10 days gave a 50% reduction in growth rate.[4808] Even at 5 to 9 μg L^{-1} Hg, a reduction in growth was seen in adults of *Fucus spiralis*.

Methylmercuric chloride produced an inhibition in *Anabaena flos-aquae* at a concentration as low as 1 μg L^{-1}.[4947] Phenylmercuric acetate was more toxic than mercuric chloride but less than methylmercuric chloride. Effects caused by the mercuric compounds included bleaching of individual cells, cell size changes, and destruction of whole cells. Harris et al.[1937] reported that concentration of organomercurial fungicides as low as 1 μg L^{-1} reduced photosynthesis and growth of the marine diatom *Nitzschia delicatissima* and several natural phytoplankton communities from Florida lakes. Fitzgerald[1370] reported algicidal concentrations for HgCl$_2$, CH$_3$COOHgC$_6$H$_5$, and HgClCH$_3$ to be 0.5, 0.1, and 0.1 mg L^{-1}, respectively. Leland et al.[2271] reported the order of toxicity of Hg-compounds: C$_6$H$_5$NO$_3$Hg > HgCl$_2$ > (CH$_3$COO)$_2$Hg > Hg(NO$_3$)$_2$.

Hannerz[1898] concluded that uptake of mercurials by *Oedogonium* sp. and other aquatic plants was primarily by surface adsorption. Surface adsorption was also reported by Glooschenko[1655] for *Chaetoceros costatum*. On the other hand, Burkett[628] found that live *Cladophora glomerata* sorbed more CH$_3$Hg than dead *Cladophora* indicating an active transport.

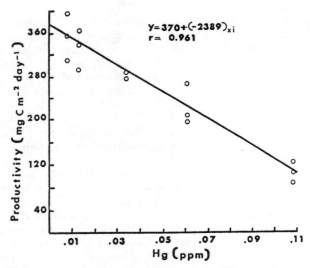

Figure 6-3.   Effect of Hg concentrations (as $HgCl_2$) on *in situ* phytoplankton productivity during spring (3 April 1975).[399]   (From Blinn et al., 1977, in *J. Phycol.*, Vol. 13, p. 60, reprinted with permission of the *Journal of Phycology*.)

The presence of Hg in the cells was found to have a considerable effect upon the mean cell volumes in the cultures of *Isochrysis galbana* causing almost a doubling of size at the highest sublethal concentrations examined; this was probably due to combination of the metal with sulfur-containing compounds which are known to be of importance in the process of cell division.[984]

Mercuric ions induce a greater loss of potassium than does methylmercuric chloride.[4446]   Overnell,[3592] however, observed a more severe potassium efflux with methylmercury than with mercuric ions. De Fillips[1030-1] observed a major loss of K (90%) to occur under both light and dark conditions; 10% loss was light-independent. Interaction of metals with sulfhydryl groups was suggested as a cause of breakdown of permeability barriers of cells eventually leading to efflux of potassium.[1030,3671]

Several authors have reported the ability of algae to release metallic mercury which may then volatilize.[304-5,1034]   Both enzymatic and non-enzymatic mechanisms of mercury release by microorganisms have been proposed.[305,1034]

Starý et al.[4698] found that Hg cumulation by a green unicellular alga *Scenedesmus obliquus*, in contrary to other metals, was pH independent. Havlík et al.[1979] reported that after a 14 days cultivation of the green algae *Chlorella kessleri* and *Scenedesmus obliquus* and the blue-green alga *Microcystis incerta*, 60 to 80% of added mercury was present in algal cells and the portion of mercury washable from the cell surface did not exceed 2%. In the cells of the blue-green alga *Plectonema boryanum* exposed to Hg the significant concentration of this element was present in the polyphosphate body cell sectors and small amounts in the cell sectors away from the polyphosphate bodies.[2336] Rachlin et al.[3948] reported that mercury caused a significant reduction in the volume of the intrathylakoidal spaces of the blue-green alga *Plectonema boryanum*. Mercury also caused the production of extra intracellular whorls and coalescence of cellular lipid. The concentration factors of mercury in algae are summarized in Table 6-10.

6.6 NICKEL

6.6.1 Nickel in the Environment

Earth's crust: 75–80 mg kg^{-1} (490, 903, 3093, 3287)
Common minerals: NiS, $(Fe,Ni)_9S_8$ (pentlandite), $(Ni,Mg)_6(OH)_6(Si_4O_{11})H_2O$ (garnierite), $(Ni,Fe)O(OH) \cdot nH_2O$ (limonite)[3287]

Table 6-10. Mercury concentration factors (CF) in algae

Species	CF	Locality	Reference
Freshwater			
Nostoc H	333–2,630	Laboratory	1961
Schizothrix calcicola	2,150–3,070	Laboratory	1962
Oscillatoria sp.	2,260–3,380	Laboratory	1962
Microcystis incerta[a]	4,000	Laboratory	1980
Nostoc 586	318–4,130	Laboratory	1961
Phytoplankton	4,800	Columbia River, Washington	940
Nostoc sp. 586	1,950–5,120	Laboratory	1962
Scenedesmus obliquus	319–6,540	Laboratory	1961–2
Chlamydomonas sp.	6,780	Laboratory	1961
Navicula pelliculosa	3,460–6,780	Laboratory	1962
Microcystis incerta[b]	9,900	Laboratory	1890
Mougeotia sp.	10,900–12,800	Laboratory	1962
Scenedemsus obliquus[a]	13,000	Laboratory	1980
Chlorella kessleri[a]	15,000	Laboratory	1980
Chlamydomonas sp.	9,060–15,810	Laboratory	1962
Zygnema sp.	9,200–17,100	Laboratory	1962
Pediastrum boryanum	17,700	Culture	4119
Scenedesmus obliquus[b]	21,000	Culture	1980
Phytoplankton[c]	22–24,600	Laboratory	1979
Ulothrix fimbrinata	18,220–25,700	Laboratory	1962
Chlorella pyrenoidosa	24,000–47,300	Laboratory	1962
Chlorella kessleri[b]	63,000	Laboratory	1980
Nostoc muscorum A	43,500–87,500	Laboratory	1962
Marine			
Fucus vesiculosus	11,800–35,500	Portugal coast	1340
Gracilaria verrucosa	48,000–65,000	Portugal coast	1340
Ulva lactuca	22,700–30,500	Portugal coast	1340

[a] Phenylmercury, [b] Methylmercury, [c] Chlorella kessleri, Scenedesmus obliquus, Microcystis incerta

Soils: 40 (10–1,000) mg kg^{-1} (490), 5–500 mg kg^{-1} (4849), 0.2–660 mg kg^{-1} (2395), in contaminated sites up to 26,000 mg kg^{-1} (2395)
Sediments: 52 mg kg^{-1} (491), < 100 mg kg^{-1} (3287)
Air: 1–120 ng m^{-3} (491, 2395)
Fresh water: 10 μg L^{-1} (490), 0.5 (0.02–27) μg L^{-1} (491), 1–3 μg L^{-1} (3287)
Sea water: 0.56 (0.13–43) μg L^{-1} (491)
Land plants: 3 (1–5) mg kg^{-1} (490–1), <0.07–8.2 mg kg^{-1} (2395)
Aquatic and wetland plants: (mg kg^{-1}) (see Appendix, Table 7-51):
emergent species: <1.0–13,500 (aboveground)
floating-leaved species: 0.8–5.21 (aboveground, FD)
submerged species: 4–87.3
floating species: 4–33.9
bryophytes: 1–1,300
Freshwater algae: 3.2–1,070 mg kg^{-1} (Table 6-11)
Marine algae: 0.4–5 mg kg^{-1} (491)
Land animals: 0.8 mg kg^{-1} (490)
Marine animals: 0.4–2.5 mg kg^{-1} (491)

Table 6-11. Nickel concentration in algae (mg kg^{-1} dry mass)

Species	Concentration	Locality	Reference
Freshwater			
Chara sp.	8.62	Lake Huron	1209
Cladophora sp.	13.2	Lake Huron	1209
Chara sp.	7.6–15.7	Lake St. Clair, Canada	3346
Cladophora glomerata	11.9–23.8	Leine River, Germany	13
Oedogonium sp.	70	Lower Swansea Valley, Wales	5027
Cladophora glomerata	13.4–80	Lake Balaton, Hungary	2575
Cladophora glomerata	3.2–99.8	Roding River, England	3146
Cladophora sp.	30–100	Lower Swansea Valley, Wales	5027
Microspora sp.	110[b]	Lower Swansea Valley, Wales	5027
Spirogyra sp.	30–120	Lower Swansea Valley, Wales	5027
Coccomyxa sp.	150[a]	Lower Swansea Valley, Wales	5027
Mougeotia sp.	240[a]	Lower Swansea Valley, Wales	5027
Periphyton	46–265	Vermilion River, Illinois	2769
Tribonema sp.	160–290[a]	Lower Swansea Valley, Wales	5027
Ulothrix sp.	300[b]	Lower Swansea Valley, Wales	5027
Zygnema sp.	700[a]	Lower Swansea Valley, Wales	5027
Oscillatoria sp.	1,070[b]	Lower Swansea Valley, Wales	5027
Marine			
Laminaria digitata	0.30	Canada, Nova Scotia	5528
Ascophyllum nodosum	0.57	Canada, Nova Scotia	5528
Laminaria longicruris	0.86	Canada, Nova Scotia	5528
Fucus vesiculosus	2.0	Canada, Nova Scotia	5528
Chondrus crispus	>2.0	Canada, Nova Scotia	5528
Rhodymenia palmata	>2.0	Canada, Nova Scotia	5528
Spongomorpha arcta	>2.0	Canada, Nova Scotia	5528
Hemiselmis virescens	2.8	Laboratory	4145
Chlorella salina	3.1	Laboratory	4145
Rhizoclonium riparium	4.0	Hong Kong	2113
Dunaliella tertiolecta	4.3	Laboratory	4145
Endarachne binghhamiae	5.2	Hong Kong	2114
Sargassum vachellianum	5.4	Hong Kong	2114
Tetraselmis tetrathele	5.6	Laboratory	4145
Porphyra suborbiculata	1.7–5.8	Hong Kong	2113
Hizikia fusiforme	5.0–5.8	Hong Kong	2113
Phaeodactylum tricornutum	6.2	Laboratory	4145
Ascophyllum nodosum	3.9–6.3	Great Britain	1455
Dunaliella primolecta	6.4	Laboratory	4145
Fucus vesiculosus	2–7	Trondheimsfjorden, Norway	2694
Chlamydomonas sp.	7.1	Laboratory	4145
Ishige foliacea	5.4–7.9	Hong Kong	2113
Chaetomorpha antennina	4.4–8.2	Hong Kong	2113
Fucus vesiculosus	7.1–8.9	Great Britain	1455
Porphyra spp.	0.2–9.6	Irish Sea	3882

Table 6-11. Continued

Species	Concentration	Locality	Reference
Sargassum hemiphyllum	10.2	Hong Kong	2114
Heteromastix longifillis	10.3	Laboratory	4145
Colpomenia sinuosa	7.4–14.3	Hong Kong	2113
Fucus spp.	1.8–18	Irish Sea, North Sea	3882
Enteromorpha flexuosa	5.2–18.5	Hong Kong	2113
Ascophyllum nodosum	1.0–22	Trondheimsfjorden, Norway	2694
Caulacanthus okamurai	18.2–24.4	Hong Kong	2113
Dermonema frappieri	15.0–35.9	Hong Kong	2113
Fucus vesiculosus	4.5–36	Great Britain estuaries	604
Fucus vesiculosus	1.5–67	Sweden, coast	4612

[a] Adjacent to zinc smelter waste, [b] near zinc smelting waste

Sources to waters: municipal wastewaters, sewage sludge discharge, smelting and refining of nonferrous metals, coal-burning power plants, atmospheric fallout, base metal mining and dressing, manufacturing processes (metals, chemicals, pulp and paper, petroleum products).[3480]

Aquatic chemistry: Nickel occurs principally as Ni^{2+} in surface waters, but oxidation states ranging from Ni^- to Ni^{4+} have been reported from time to time.[3287] Nickel is classified as a borderline element between hard and soft acid acceptors in chemical reactions with donor atoms. Under reducing conditions in surface waters, Ni forms insoluble sulfides, provided that sulfur is present in excess. Under aerobic conditions and pH < 9, nickel complexes with hydroxides, carbonates, sulfates, and naturally occurring organic ligands.[3287]

6.6.2 Nickel and Algal Growth

Nickel ions are known to be essential cofactors for four bacterial enzymes,[1977,4936] of which hydrogenase[954,1977] and urease[342,3005] exist in Cyanophyceae. The enzyme urease is a nickel-ligated metalloprotein with a capacity to catalyze the hydrolysis of urea to ammonia and carbonic acid[116,4048] (see also Chapter 4.2.4.7).

Although the growth of one species of cyanophyta has been reported to be dependent on nickel,[5107] the exponential growth rate of nitrogen-fixing bacteria *Anabaena cylindrica* was found to be unaffected, over a range of growth conditions, by omission of nickel from the growth medium.[955] Thus the hydrogenase and urease, and therefore nickel are apparently not essential for the exponential growth of the organism. Daday et al.[956] reported that nickel depletion results in a diversion of ammonia from protein synthesis into cyanophycin, thereby delaying the "*de novo*" synthesis of proteins and enzymes required for the synthesis of active nitrogenase in *Anabaena cylindrica*.

Oliveira and Antia[3542] reported that the inability of the diatom *Cyclotella cryptica* to grow photo-autotrophically on urea in an oceanic-sea-water medium was due to an inadequate concentration of Ni^{2+} ions in the medium; at least 4–5 nM Ni^{2+} was required for measurable growth, while the ocean water contributed only 1.4 nM Ni^{2+} to the growth medium. Interestingly, a wide range of Ni^{2+} concentrations (0.01–10.0 μM) supported good growth of this diatom, while higher Ni^{2+} concentrations were inhibitory to growth, presumably as a result of heavy-metal toxicity.

At 2.0 mg L^{-1} of $NiCl_2$, no significant effect on either lag phase or culture doubling of the blue-green alga *Nostoc linckia* was found, but at 10 mg L^- the growth was totally inhibited.[2643] The authors found also nitrate reductase and nitrate uptake activities inhibition by low concentrations of nickel (< 0.2 mg L^{-1}).

Matulová[3107] reported the first deleterious effect, strong deleterious effect, and lethal effect of Ni to the chlorococcal green algae (*Scenedesmus quadicauda*, *Chlamydomonas eugametos* and *C. hydra*) to be 0.02, 0.05 and 0.1 mg L^{-1}, respectively. Nickel addition of 0.01 mg L^{-1} did not cause any harmful effect. Stokes et al.[4793] found growth inhibition of a laboratory strain of the green alga *Scenedesmus* after 10 days at Ni concentration of 0.5 mg L^{-1}. The natural population of *Scenedesmus* was inhibited by 20% at nickel concentration of 1.5 mg L^{-1}. On the other hand, Hutchinson and Stokes[2263] reported

Chlorella vulgaris growth stimulation to about 140% at nickel concentration of 0.5 mg L^{-1}. Henriksson and Da Silva[2036] reported a significant inhibition of nitrogenase activity in the blue-green alga *Westiellopsis* sp. at Ni concentration of 5 μg L^{-1} while in another blue-green alga, *Nostoc muscorum*, no significant reduction in nitrogenase activity was found at Ni concentration of 125 μg L^{-1}.

Chiaudani and Vighi[780] found the EC-50$_{96}$ for the green alga *Selenastrum capricornutum* 2.5 μg L^{-1} and 13 μg L^{-1} in the medium without and with EDTA, respectively. Skaar et al.[453] reported that kinetic studies of the uptake of ^{63}Ni by the cultures of the diatom *Phaeodactylum tricornutum* revealed that the uptake capacity of the cells depended strongly on their metabolic state. Phosphate-starved cells gave saturation-type uptake curves and had low nickel-binding capacity. This was increased markedly by the addition of phosphate to the cultures. While as little as 0.5 μg L^{-1} of ionic nickel in the medium influenced the nickel uptake, more than 500 μg L^{-1} of added nickel was needed to reduce the growth rate of the alga. Stokes[4790] reported that algae living in waters with elevated Ni concentrations increased their resistance toward this metal. While concentrations 0.5 to 0.75 mg L^{-1} completely inhibited the growth of laboratory cultures, algae isolated from waters with an average Ni concentration about 2.5 mg L^{-1} can grow at Ni concentration of 3 mg L^{-1}.

In Ni-exposed cells of the blue-green alga *Plectonema boryanum* this element was detected in the polyphosphate body cell sectors but not in the cell sectors without polyphosphate bodies.[2336] Rachlin et al.,[3948] in their comprehensive morphometric analysis of the blue-green alga *Plectonema boryanum*, reported that nickel caused significant decrease in cell size, significant increase in the surface area of the cell's thylakoids, and significant reduction in the volume of intrathylakoidal spaces. Nickel also caused the production of extra intracellular membrane whorls and a reduction in cellular lipid content. The concentration factors of nickel in algae are given in Table 6-12.

Table 6-12. Nickel concentration factors (CF) in algae

Species	CF	Locality	Reference
Freshwater			
Microspora sp.	36[b]	Lower Swansea Valley, Wales	5027
Ulothrix sp.	140[b]	Lower Swansea Valley, Wales	5027
Cladophora sp.	380–890	Lower Swansea Valley, Wales	5027
Coccomyxa sp.	1,200[a]	Lower Swansea Valley, Wales	5027
Oedogonium sp.	1,200	Lower Swansea Valley, Wales	5027
Mougeotia sp.	1,600[a]	Lower Swansea Valley, Wales	5027
Spirogyra sp.	250–1,600	Lower Swansea Valley, Wales	5027
Tribonema sp.	1,100–3,000[a]	Lower Swansea Valley, Wales	5027
Cladophora glomerata	1,540–4,700	Šárecký Brook, Czech Republic	5188
Zygnema sp.	6,000[a]	Lower Swansea Valley, Wales	5027
Cladophora glomerata	160–6,500	Roding River, England	3146
Cladophora glomerata	8,770	Laboratory experiment	5188
Oscillatoria sp.	9,100[b]	Lower Swansea Valley, Wales	5027
Cladophora glomerata	10,000	Lake Balaton, Hungary	2575
Marine			
Fucus vesiculosus	2,800	Irish Sea	3882
Fucus vesiculosus	6,800	Menai Straits, Wales	1455

[a] Adjacent to zinc smelter waste, [b] near zinc smelting waste

Appendix

A. Standing crop, productivity, element concentrations, uptake rates, and standing stocks in aquatic and wetland plants

List of abbreviations used in the following tables:

ABG = aboveground biomass or productivity
BLG = belowground biomass or productivity
C = cultivation
E = experimentation
F = fertilizer experiments
FL = floating leaves
GL = green leaves
GR = greenhouse
L = leaves
N = natural
P = petioles (peduncles)
R = roots
RH = rhizomes
S = stems
SBL = submerged leaves
SL = senescent leaves
ST = stolons
T = tops
TI = tips
W = wastewater

Synonyms:

Glyceria maxima = *Glyceria aquatica*
Phragmites australis = *Phragmites communis*
Scirpus lacustris = *Schoenoplectus lacustris*
Sparganium erectum = *Sparganium ramosum*
Sparganium emersum = *Sparganium simplex*

Sums of above- and belowground biomass or productivity may not equal total numbers as figures indicate amounts which may have occurred on different dates.

Some species of the genus *Potamogeton* (e.g. *P. natans*) are sometimes included in the group of floating-leaved species; in the following tables all *Potamogeton* species are included in the group of submerged species.

Species are listed in alphabetical order.

Table 7-1. Standing crop of wetland and aquatic macrophytes (g DM m^{-2})

Plant species	Standing crop			Locality	Reference
	ABG	BLG	Total		
Emergent species					
Acorus calamus	140–996	312		India	1893, 2443
A. calamus	605			New Jersey	3137
A. calamus	819			New Jersey	5370
A. calamus	1,250			Czech Republic	5144
Alisma plantago-aquatica	52–444			Canada, Alberta	5120
Althernanthera philoxeroides	118			South Carolina	3784
A. philoxeroides	800			U.S.A.	505
Bidens laevis	17.4			Virginia	1468
B. laevis	22			South Carolina	3784
B. laevis	282			New Jersey	5376
Bidens sp.	900			Pennsylvania	3136
Bolboschoenus maritimus	167			Czech Republic	2657
B. maritimus	334			Czech Republic	2568
B. maritimus	456	870	1,326	Czech Republic	1201
B. maritimus	480			Czech Republic	3552
B. maritimus	535–613			Czech Republic	2658
B. maritimus	1,659			Czech Republic	3550
Butomus umbellatus	87–94	49	136	India	1893, 2443
Carex acuta	605			Sweden	3306
Carex acutiformis	320–353			The Netherlands	5156
C. acutiformis	630			England	3717
C. acutiformis	692			Poland	2588
C. acutiformis	800			The Netherlands	5155
C. acutiformis	1,140	1,190	2,330	The Netherlands	28
C. acutiformis	1,180	1,560	2,740	Germany	3777
Carex aquatilis	101	402	503	Canada, Alberta	1054
C. aquatilis	327			Canada, Alberta	5120
C. aquatilis	380			Canada, Alberta	176ї
C. aquatilis	806			Canada, Québec	179

Table 7-1. Continued

Plant species	Standing crop			Locality	Reference
	ABG	BLG	Total		
Carex atherodes	319			Canada, NWT	4081
C. atherodes	624			Minnesota	1760
C. atherodes	667			Iowa	5121
C. atherodes	795			Minnesota	1760
Carex diandra	112–197			The Netherlands	5156
C. diandra	350			The Netherlands	5155
C. diandra	350	780	1,130	The Netherlands	28
Carex elata	890			Sweden	3306
C. elata	353–1,205			Czech Republic	2657
Carex gracilis	807			Germany	1160
C. gracilis	950			Czech Republic	3552
C. gracilis + vesicaria	402–732			Czech Republic	2657
Carex lacustris	465			Michigan	1619
C. lacustris	485			Minnesota	1760
C. lacustris	575			Sweden	3306
C. lacustris	1,008			Czech Republic	2011, 2254
C. lacustris	1,037	433	1,470	New York	330
C. lacustris	1,145	575		New York	332
C. lacustris		430		New York	331
C. lacustris		907–1,172		Iowa	5128
Carex lasiocarpa	143–305			Sweden	3306
C. lasiocarpa	270			Sweden	3718
C. lasiocarpa	416			Minnesota	1760
C. lasiocarpa	510			England	3717
C. lasiocarpa	367	1,310	1,677	New York	336
C. lasiocarpa	940	910	1,850	The Netherlands	28
Carex nebraskensis	161			California	4271
C. nebraskensis	214–310			California	3988
C. nebraskensis	474			California	3989
Carex rostrata	12–25	14–17		Sweden	4616
C. rostrata	66	31	97	Sweden	1271
C. rostrata	10–150			Sweden	4529
C. rostrata	89–143			The Netherlands	5156
C. rostrata	114–235	300		Minnesota	329
C. rostrata	165–222			Sweden	2246
C. rostrata	150–328	114–852	1,002	Minnesota	329

Table 7-1. Continued

Plant species	Standing crop			Locality	Reference
	ABG	BLG	Total		
C. rostrata	300			The Netherlands	5155
C. rostrata	320–610			Sweden	3306
C. rostrata	380			Finland	4993
C. rostrata	389			Poland	2588
C. rostrata	420			England	3717
C. rostrata	434			Minnesota	1760
C. rostrata	740			Canada, Alberta	1761
C. rostrata	660	1,430	2,090	Minnesota	336
C. rostrata	975	431	1,406	New York	335
C. rostrata	650	1,350	2,000	The Netherlands	28
Carex stricta	585			New Jersey	2341
C. stricta	736			New York	1639
Carex trichocarpa	944	263	1,294	New York	336
Carex spp.	1–1,205			Czech Republic	2657
Carex spp.	506			Michigan	5329
Carex spp.	667			Iowa	5121
Carex spp.		4,289		Michigan	4114, 5329
Cladium jamaicense	403–803			Florida	1003
C. jamaicense	150–1,000	5,930–8,610		North Carolina	302
C. jamaicense	1,130			Florida	4747
Cladium mariscoides	224	119		New York	338
Cyperus papyrus	9,000–15,000			Central Africa	5338
Cyperus serotinus	6–279	232	447	India	1893, 2443
Cyperus sp.			780	India	3958
Distichlis spicata	670			New Jersey	1733
D. spicata	620–740			Louisiana	5385
D. spicata	280–970			Florida	2622
D. spicata	991			Louisiana	2186–7
D. spicata	950–960	13,990		North Carolina	302
D. spicata		12,400		Delaware	1548
Eleocharis acicularis	118–644			Czech Republic	5148
E. acicularis (F)			1,093–1,905	Czech Republic	5148
Eleocharis atropurpurea			40	India	2312
Eleocharis capitata			78	India	2312
Eleocharis equisetoides	377			South Carolina	516, 3828
Eleocharis palustris	12–153	149	302	India	1893, 2443

Table 7-1. Continued

Plant species	Standing crop ABG	BLG	Total	Locality	Reference
E. palustris	183–447			Canada, Alberta	5120
E. palustris			1,150–1,322	India	3636
Eleocharis plantaginea	408	110	518–719	India	81, 5158
E. plantaginea			2,200	India	3636
Eleocharis quadrangulata	725			South Carolina	516, 3828
E. quadrangulata	881			South Carolina	512
Eleocharis sp.	50–63			Florida	5227
Equisetum debile	88–113	60	148	India	1893, 2443
Equisetum fluviatile	10–21	20		Sweden	4616
E. fluviatile	320			Minnesota	535
E. fluviatile	1,244			Czech Republic	2658
Equisetum limosum	590			Czech Republic	1207
Glyceria grandis	656–673			England	5337
Glyceria maxima	78–208			England	962
G. maxima	826			Czech Republic	3550
G. maxima	0.5–932			Czech Republic	2657
G. maxima	659–970			Czech Republic	2568
G. maxima	1,122			Czech Republic	4232
G. maxima	795–1,220			Czech Republic	3552
G. maxima (W)	1,500	1,400	2,900	Poland	3596
G. maxima	2,690			Czech Republic	5144
Hydrocotyle umbellata	188			South Carolina	516, 3828
Impatiens capensis	22			South Carolina	3784
I. capensis	119			New Jersey	5376
Ipomoea aquatica			251–2,141	India	4408
Juncus articulatus	13	9	22	India	2442
Juncus effusus	800			England	3717
J. effusus			1,592	South Carolina	508
Juncus kraussi	1,790	1,600	2,390	Australia	811–2
Juncus roemerianus	786			North Carolina	5223
J. roemerianus	1,173			North Carolina	4813
J. roemerianus	1,240			Louisiana	2186–7
J. roemerianus	480–1,350	9,930–11,980		North Carolina	302
J. roemerianus	1,959			Louisiana	5385

Table 7-1. Continued

| Plant species | Standing crop | | | Locality | Reference |
	ABG	BLG	Total		
J. roemerianus	2,100			North Carolina	5446
J. roemerianus		4,060–5,140		Florida	2622
J. roemerianus		7,700–12,400		Mississippi	1043
Juncus squarrosus	690			England	3717
Justicia americana	95	94	189	North Carolina	5060
J. americana	2,458			Alabama	500
Ludwigia peploides (N)	500–700			California	4066
L. peploides (C)	1,900			California	4066
Ludwigia uruguayensis	129			South Carolina	3784
Lycopus europeans	76–83	41	117	India	1893, 2443
Lycopus rubellus	82			South Carolina	3784
Lythrum salicaria	1,373			Pennsylvania	3136
L. salicaria	2,104			New Jersey	5370
Oenanthe aquatica	285			Czech Republic	2658
Orontium aquaticum	244			South Carolina	3828
Panicum hemitomon	1,075			South Carolina	516, 3828
Panicum repens	264			India	354
Panicum virgatum	652			Maryland	105
Panicum sp.	74.2			Florida	5227
Papyrus sp.	2,140			Uganda	1577
Peltandra virginica	84			South Carolina	3784
P. virginica	386			New Jersey	5376
Phalaris arundinacea	203–296			Czech Republic	2657
P. arundinacea	566			New Jersey	5370
P. arundinacea	10–683			India	1893
Phragmites australis	182			Finland	4993
P. australis	745			Czech Republic	3550
P. australis	775			Denmark	2714
P. australis	872			Czech Republic	2568
P. australis	934			Iowa	5121
P. australis	942			England	3096
P. australis	990			Louisiana	2186
P. australis	840–1,000			Czech Republic	3552
P. australis	1,000			Finland	2421
P. australis	1,020			Poland	340–1
P. australis	1,115			Czech Republic	2658
P. australis	1,260			Poland	2623

Table 7-1. Continued

Plant species	Standing crop			Locality	Reference
	ABG	BLG	Total		
P. australis	16–1,442			Czech Republic	2657
P. australis	1,451			Maryland	2348
P. australis	1,500–1,600			Sweden	3486
P. australis	1,727			New Jersey	3137
P. australis	960–1,880			Czech Republic	1197
P. australis (F)	740–2,366			Czech Republic	5144
P. australis	2,400			Sweden	383
P. australis	468–4,424	992–4,164		England	410
P. australis	8,147			Florida	627
P. australis	8,650			India	1745
P. australis	780	620	1,400	Denmark	89
P. australis	1,360	574	1,934	Germany	1818
P. australis		1,121–1,565		Iowa	5128
P. australis	1,846	3,144	4,990	South Africa	5455
P. australis	2,050	6,260	8,310	Czech Republic	1201
P. australis	1,170–2,068			Czech Republic	4232
P. australis	2,960			Czech Republic	1206
P. australis	3,250	8,560		Czech Republic	1204
P. australis	265–5,280	824–3,048	1,089–8,328	India	1893, 2443
P. australis (W)	6,334	5,348	11,1682	South Africa	5505
P. australis (C)	3,401	8,886	12,287	Czech Republic	1206
Phragmites karka	10,950			India	1745
P. karka	3,181–7,543	1,460–2,188	4,641–9,730	India	1748
Polygonum arifolium	12.1			South Carolina	3784
P. arifolium	31			Virginia	1468
P. arifolium	200			New Jersey	5376
Polygonum glabrum			1,040–5,300	India	1748, 4300
Polygonum hydropiper	613			Czech Republic	4061
P. hydropiper	98–114	67	183	India	1893, 2443
Polygonum punctatum	14.9			South Carolina	3784
Pontenderia cordata	257			South Carolina	3784
P. cordata	716			South Carolina	516, 3828
Sagittaria falcata	648			Louisiana	2186–7
Sagittaria latifolia	460			Iowa	5121
S. latifolia (W)	12,563			Florida	627
Sagittaria graminea	29			South Carolina	3784

Table 7-1. Continued

Plant species	Standing crop			Locality	Reference
	ABG	BLG	Total		
Sagittaria sagittifolia	14–142	6–73	20–191	India	1893, 2443
Schoenoplectus lacustris	650			Czech Republic	2568
S. lacustris	2,050			Czech Republic	3552
S. lacustris	4,200			Czech Republic	5144
Scirpus acutus	851			Iowa	5121
S. acutus		1,208–1,870		Iowa	5128
Scirpus americanus	150			South Carolina	504
S. americanus	115–185	40–211	396	Canada, Québec	4042
S. americanus	410			South Carolina	516, 3828
Scirpus articulatus	523	159	682	India	354
Scirpus fluviatilis	466			Iowa	5121
S. fluviatilis		1,254–1,424		Iowa	5128
Scirpus lacustris	197–774			India	1893, 2439
S. lacustris	1,330	1,220	2,550	Denmark	89
S. lacustris	1,800			Poland	340–1
Scirpus maritimus	6–71	42	101	India	1893, 2439
Scirpus palustris	365	333	624	India	1893, 2440
Scirpus pungens (W)	394			Florida	627
Scirpus validus	28			South Carolina	3784
S. validus	330			Iowa	5121
S. validus	1,381			South Carolina	516, 3828
Sparganium angustifolium	0.03–0.14			Maine	2249
Sparganium erectum	12–117			England	962
S. erectum	930			Czech Republic	3552
S. erectum	86–976	308–697		India	1893
S. erectum	1,033			Czech Republic	2658
S. erectum	942–1,610			Czech Republic	1202
S. erectum	1,880			Czech Republic	5144
S. erectum	710–985	412–634	1,122–1,619	India	1893, 2443
Sparganium eurycarpum	638			Iowa	5121
S. eurycarpum	770	1,252–1,945		Iowa	5128
S. eurycarpum	1,035	1,280		Michigan	5084
S. eurycarpum	1,054			Iowa	5122
S. eurycarpum (F)	637–1,185	681–1,123		Iowa	3399
S. eurycarpum	1,950			Wisconsin	2861
Spartina alterniflora	649			Virginia	1651

Table 7-1. Continued

Plant species	Standing crop ABG	BLG	Total	Locality	Reference
S. alterniflora	754			Louisiana	2186–7
S. alterniflora	788–1,018			Louisiana	2511
S. alterniflora	1,320			North Carolina	4813
S. alterniflora	1,473	100–250		Louisiana	5385
S. alterniflora	1,592			New Jersey	4664
S. alterniflora	2,410			Virginia	5274
Spartina cynosuroides	808			Louisiana	2186–7
S. cynosuroides	826			Georgia	3501
S. cynosuroides	968			Maryland	105
S. cynosuroides	700–970	6,930–9,260		North Carolina	302
S. cynosuroides	1,250	6,100–8,200		Georgia	4365
S. cynosuroides	1,401			Virginia	5274
S. cynosuroides	2,000			Mississippi	1043
S. cynosuroides		2,000–17,500		Georgia	1548
Spartina patens	640			North Carolina	5223
S. patens	805			Virginia	5274
S. patens	1,376			Louisiana	2186–7
S. patens	2,194			Louisiana	5385
Typha angustata	305–1,565	100–428	405–1,993	India	4419
T. angustata	138–2,068			India	1893
T. angustata	708–6,803	586–4,440	1,808–8,131	India	4420
Typha angustifolia (W)	48–65			Michigan	5385
T. angustifolia	800			Poland	2632
T. angustifolia	1,118			England	3096
T. angustifolia	1,780			Czech Republic	3550, 3552
T. angustifolia	2,560	1,015	3,575	Texas (1983)	2084
T. angustifolia	2,895	401	3,296	Texas (1984)	2084
T. angustifolia	3,710			Czech Republic	1206
T. angustifolia	4,000			Czech Republic	5144
T. angustifolia (C)	3,039	2,070	5,109	Czech Republic	1206
T. angustifolia + latifolia	478–814	3,295–4,889	4,261–6,460	Michigan	4255
Typha domingensis	225–590			Florida	1002–3
T. domingensis	250–600	7,010–11,280		North Carolina	302
T. domingensis	1,483			South Carolina	516, 3828
Typha elephantina	975–2,464	1,542–5,269	2,517–7,733	India	4419
Typha glauca	1,156			Iowa	5121
T. glauca	1,281			Iowa	933
T. glauca	1,309–1,477			New York	334

Table 7-1. Continued

Plant species	Standing crop			Locality	Reference
	ABG	BLG	Total		
T. glauca	1,549	1,167–1,450		Iowa	5128
T. glauca	2,000	1,431	3,431	Iowa	995
T. glauca (F)	1,351–2,343	1,300–1,799		Iowa	3399
Typha latifolia (W)	48–62			Michigan	5385
T. latifolia	181–322			Canada, Alberta	5120
T. latifolia	574			South Carolina	516, 3828
T. latifolia	684			South Carolina	504
T. latifolia	456–848	393–807(R)		Canada, Alberta	2848
T. latifolia	1,070			England	3717
T. latifolia	1,150			Czech Republic	1197, 3552
T. latifolia	1,400			Wisconsin	4582
T. latifolia	1,496			Wisconsin	2538
T. latifolia	147–1,527			Oklahoma	3731
T. latifolia	1,483			South Carolina	3828
T. latifolia	1,400–1,600			Czech Republic	2655
T. latifolia	1,620			Czech Republic	2660
T. latifolia	1,566			New Jersey	2341
T. latifolia	428–2,252			S.E. USA	514
T. latifolia	3,600			Czech Republic	5144
T. latifolia	416	556	972	Nebraska	3171
T. latifolia	378	892	1,250	South Dakota	3171
T. latifolia	404	912	1,316	North Dakota	3171
T. latifolia	730	804	1,534	Oklahoma	3171
T. latifolia (GR)	1,363	2,005	3,368	Florida	4030
T. latifolia	215–1,054	395–2,223	610–3,405	Oregon	3171
T. latifolia	1,336	2,646	3,982	Texas	3171
T. latifolia	488–731	3,402–4,303	4,648–5,970	Michigan	4255
T. latifolia (F)	720–4,371	1,444–3,860	2,164–8,231	Czech Republic	5144
Typha sp.	966			Maryland	2348
Typha sp.	987			New Jersey	3137
Typha sp.	1,297			New Jersey	5370
Typha sp.	1,310			Pennsylvania	3136
Typha sp.	2,338			Maryland	2004
Typha orientalis (GR)	2,304	1,116	3,420	Australia	714
Zizania aquatica	560			Virginia	5274
Z. aquatica	866			New Jersey	5370
Z. aquatica	1,200			New Jersey	2341
Z. aquatica	1,390			New Jersey	3137

Table 7-1. Continued

Plant species	Standing crop			Locality	Reference
	ABG	BLG	Total		
Zizaniopsis miliacea	442			South Carolina	3784
Z. miliacea	1,039			Georgia	3784
Floating-leaved species					
Nelumbo lutea	184			South Carolina	515, 3828
Nelumbo nucifera	200			India	81
N. nucifera	549–813			India	2240
Nuphar advena	84			South Carolina	516, 3828
N. advena	245			Virginia	5274
N. advena	516			New Jersey	3137
N. advena	529	4,799		New Jersey	5370
N. advena	605	1,146		New Jersey	1734
N. advena	1,175			Pennsylvania	3136
Nuphar lutea	36	119	155	North Carolina	5060
Nuphar variegatum	89–235			Canada, Alberta	5120
Nymphaea alba	21–246			India	1893, 2443
Nymphaea odorata	110			Minnesota	535
N. odorata	256			South Carolina	516, 3838
Nymphaea stellata	18–319			India	1893, 2443
Nymphoides peltata	4–112		70–114	India	1893, 2443
N. peltata	228	273	501	The Netherlands	5125
Submerged species					
Batrachium fluitans	160			Czech Republic	1207
Batrachium rionii	90			Czech Republic	4062
Ceratophyllum demersum	500–700			Sweden	1442
C. demersum			15–447	India	354
C. demersum			16–480	India	2440, 2443
Elodea canadensis	450			Czech Republic	3826
Heteranthera dubia	5			Québec-Vermont	687
Hydrilla verticillata			29–470	India	354
H. verticillata			609	India	81
Myriophyllum spicatum	0.43–3.05			Austria	4327
M. spicatum	20–60			Québec-Vermont	687
M. spicatum	220			Wisconsin	21
M. spicatum	402			India	4521
M. spicatum			2–304	Canada, Ontario	686
M. spicatum			644	India	1189

Table 7-1. Continued

Plant species	Standing crop			Locality	Reference
	ABG	BLG	Total		
M. spicatum			115–880	India	2442–3
Myriophyllum verticillatum	0.01–0.22			Maine	2249
M. verticillatum	310–379	360–580	670–959	India	2443
Najas graminea			66–139	India	1672
N. graminea			144–203	India	2443
N. graminea			244	India	81
Najas major			197–469	India	4300
Najas maritima	83–478			Australia	4218
Potamogeton crispus	129			Japan	2277
P. crispus			23–205	India	4245
P. crispus	6,410			Australia	1352
Potamogeton epihydrus	0.02–0.03			Maine	2249
Potamogeton lucens		100–220		Poland	3598
P. lucens			131–380	India	2443
Potamogeton natans	146–295			Canada, Alberta	5120
P. natans	1–89		70–298	India	1893, 2443
P. natans	196–256			Czech Republic	2658
Potamogeton pectinatus	17.5			North Baltic Sea	2446
P. pectinatus	100			Poland	340
P. pectinatus	170			Czech Republic	4062
P. pectinatus	91–297			Canada, Alberta	5120
P. pectinatus			379–445	India	4245
P. pectinatus	78–950			Australia	4218
P. pectinatus	1,313			The Netherlands	5138
Potamogeton perfoliatus	0.85			Hungary	2432
P. perfoliatus	88–986			Australia	4218
P. perfoliatus		82–225		Poland	3598
Potamogeton pusillus			198–279	India	2443
Potomogeton spp.	336			Minnesota	2082
Renunculus aquatilis	12			India	1893
Riccia rhenana	95			Czech Republic	4062
Trapa bispinosa	82–545			India	1893
T. bispinosa			436–994	India	5088
Trapa natans			10–490	India	2443

Table 7-1. Continued

Plant species	Standing crop			Locality	Reference
	ABG	BLG	Total		
T. natans	107			Czech Republic	2658
T. natans	22–500			India	1893
Utricularia intermedia	0.39			Maine	2249
Utricularia vulgaris	1.6–4.0			Maine	2249
Utricularia cf. vulgaris	80			Czech Republic	4062
Utricularia sp.	2.5–6.9			Czech Republic	2565, 2568
Utricularia sp.	143			Florida	5227
Vallisneria americana	2–5			Québec-Vermont	687
Vallisneria spiralis	136–3,632			Australia	4218
V. spiralis			169	India	81
Zannichelia palustris			285–295	India	4414
Floating species					
Azolla pinnata			98	India	740
A. pinnata			70–90	India	4293
Eichhornia crassipes			798	India	81
E. crassipes			963	Japan	3534
E. crassipes (W)			1,060	Czech Republic	5538
E. crassipes			440–1,276	Louisiana	3731
E. crassipes			2,130	Alabama	515
E. crassipes (W)			2,970	Iowa	5498
E. crassipes (F)			3,000	Florida	4037
Lemna gibba			9–150	Czech Republic	4062
Lemna paucicostata			19.2–49.1	Nigeria	3114
Lemna trisulca			93	India	2441
Lemna sp.			75–295	India	2441
Pistia stratiotes			495	India	1189
Spirodela polyrhiza			134	Czech Republic	2658
Spirodela polyrrhiza			43–190	India	2441
Bryophytes					
Sphagnum balticum	440			Finland	5141
Sphagnum fuscum	344			USSR	200
Sphagnum sp.	0.48–8.85			Maine	2249
Sphagnum sp.	54–450			Finland	5141

Table 7-2. Aboveground/belowground biomass ratios of aquatic macrophytes

Plant species	Ratio	Locality	Reference
Emergent species			
Acorus calamus	0.91	New Jersey	5374
Bidens laevis	4.17	New Jersey	5374
Bolboschoenus maritimus	0.53	Czech Republic	1201
Carex acutiformis	0.96	The Netherlands	28
Carex aquatilis	0.25	Canada, Alberta	1054
Carex elata	0.43	Finland	4993
Carex lacustris	2.40	New York	330
Carex lasiocarpa	1.0	Finland	4993
C. lasiocarpa	0.28	New York	336
Carex rostrata	1.0	Finland	4993
C. rostrata	0.22	New York	329
Eleocharis acicularis	5.0	Finland	4993
Eleocharis palustris	0.33	Finland	4993
Eleocharis plantaginea	3.71	India	81, 5158
Equisetum fluviatile	0.33	Finland	4993
Glyceria maxima	1.07	Poland	3596
Impatiens capensis	2.72	New Jersey	5374
Isoëtes lacustris	1.7	Finland	4993
Juncus bulbosus	5.3	Finland	4993
Juncus kraussii	1.12	Australia	811–2
Justicia americana	1.01	North Carolina	5060
Lobelia dortmanna	1.0	Finland	4993
Lycopus europeans	1.85	India	1893, 2443
Lysimachia thyrsiflora	0.67	Finland	4993
Peltandra virginica	0.12	New Jersey	5374
Phragmites australis	2.37	Germany	1818
P. australis	0.59	South Africa	5455
P. australis	0.43	Finland	4993
P. australis	0.38	Czech Republic	1203, 1206
P. australis	0.33	Czech Republic	1201
Polygonum arifolium	6.67	New Jersey	5374
Polygonum punctatum	2.17	New Jersey	5374
Polygonum sagittatum	7.14	New Jersey	5374
Pontenderia cordata	0.77	New Jersey	5374
Sagittaria latifolia	1.06	New Jersey	5374
Scirpus lacustris	0.43	Finland	4993
Scirpus fluviatilis	0.27	New Jersey	5374
Sparganium americanum	0.52	New Jersey	5374
Sparganium spp.	1.73	Finland	4993
Subularia aquatica	5.8	Finland	4993
Typha angustata	3.66	India	4419

Table 7-2. Continued

Plant species	Ratio	Locality	Reference
Typha angustifolia	1.47	Czech Republic	1206
Typha elephantina	0.47	India	4419
Typha latifolia	1.82	New Jersey	5374
T. latifolia	0.44	North Dakota	3171
Zizania aquatica var. *aquatica*	2.44	New Jersey	5374
Floating-leaved species			
Nuphar advena	1.16	New Jersey	5374
Nuphar lutea	0.33	Finland	4993
N. lutea	0.30	North Carolina	5060
Nymphaea spp.	0.33	Finland	4993
Submerged species			
Myriophyllum alterniflorum	5.3	Finland	4993
Myriophyllum verticillatum	0.86	India	2443
Potamogeton gramineus	5.0	Finland	4993
Potamogeton natans	5.0	Finland	4993
Ranunculus reptans	4.0	Finland	4993

Table 7-3. Annual productivity of wetland and aquatic plants (g DM m^{-2} yr^{-1})

Species	Productivity ABG	BLG	Total	Location	Reference
Emergent species					
Acorus calamus	450–830			Czech Republic	2011
A. calamus	712–940			New Jersey	5369
Arundo donax	10,000			Thailand	3511
Bidens laevis	21.3			Virginia	1468
Calamagrostis canadensis	48			Canada, Manitoba	4016
Carex acutiformis	778–790			The Netherlands	5155
Carex aquatilis	340			Canada, Alberta	1761
C. aquatilis	820	210	1,030	Canada, Québec	180
Carex atherodes	432			Canada, NWT	4081
C. atherodes	2,858	548	3,406	Iowa	5121
Carex diandra	288–535			The Netherlands	5156
Carex gracilis	580	250	830	Poland	217
C. gracilis + vesicaria	530–1,390			Czech Republic	2011
Carex lacustris	857	161	1,018	New York	330
C. lacustris	940	130	1,070	Wisconsin	2537
C. lacustris	965	208	1,173	New York	332
C. lacustris	1,186	134	1,320	Wisconsin	2536
Carex lyngbyei	687–1,322			Canada, B.C.	2517
Carex paludosa			710	Germany	1275
Carex rostrata	10–14			Sweden	4617
C. rostrata	515			Canada, Alberta	1761
C. rostrata	57	7	64	Sweden	1271
C. rostrata	116			Canada, Manitoba	4016
C. rostrata	540	260	900	New York	335
C. rostrata	738	180	918	Minnesota	329
C. rostrata	1,917	345	1,262	The Netherlands	5156
C. rostrata	1,080	260	1,340	New York	335
Carex stricta	32			Virginia	1468
Carex subspathacea	80			Hudson Bay	685
Carex sp.	1,340			New Jersey	5372
Cladium jamaicense	802–2,028			Florida	1003
Cladium mariscoides	228	47		New York	338
Cypreus articulatus	923–1677			India	4300

Table 7-3. Continued

Species	Productivity ABG	BLG	Total	Location	Reference
Cyperus papyrus			9,000–15,000	Africa (Equator)	5338
C. papyrus	12,500			Uganda	4960
Distichlis spicata	1,291			Louisiana	5385
D. spicata	1,484			Mississippi	1040
D. spicata	1,967			Louisiana	2187
D. spicata	3,108–3,366			Louisiana	2186
Glyceria maxima			751	Germany	1275
G. maxima	900–1,570			Czech Republic	2011
G. maxima	1,390–2,860			Czech Republic	1195
Heleocharis palustris			138	Germany	1275
Iris pseudacorus			647	Germany	1275
Juncus effusus	1,670	190	1,860	South Carolina	508
Juncus militaris	620	589	1,209	Rhode Island	2145
Juncus roemerianus	796			North Carolina	4813
J. roemerianus	1,360			North Carolina	5223
J. roemerianus	1,697			Mississippi	1040
J. roemerianus	1,806			Louisiana	5385
J. roemerianus	2,156			Georgia	1549
J. roemerianus	3,295			Louisiana	2187
J. roermerianus	3,029–3,794			Louisiana	2186
Lysimachia thyrsiflora			268	Germany	1275
Lythrum salicaria	1,749			Pennsylvania	3136
L. salicaria	2,104			New Jersey	5369
Mentha aquatica			265	Germany	1275
Menyanthes trifiloata			542	Germany	1275
Panicum hemitomon	1,700			Louisiana	4291
Peltandra virginica	144			Virginia	1468
P. virginica	269			Pennsylvania	3136
P. virginica	500–800			New Jersey	5369
Phalaris arunidnacea	800			England	650
Phragmites australis	551–1,080			England	3096
P. australis	183–1,600			Finland	2421, 3486, 4993
P. australis	1,825–2,811			Louisiana	2186
P. australis	265–5,280			India	2442–3
P. australis			1,273	Germany	1275
P. australis	780	620	1,400	Denmark	89

Table 7-3. Continued

Species	Productivity			Location	Reference
	ABG	BLG	Total		
Phragmites karka	3,182–7,543			India	1748
Polygonum glabrum	314–4,793			India	4300
Sagittaria falcata	1,389–1,613			Louisiana	2186
S. falcata	2,310			Louisiana	2187
Sagittaria sp.	628			Pennsylvania	3136
Schoenoplectus lacustris	1,600–3,000			Czech Republic	1195
Scirpus fluviatilis			1,533	Wisconsin	2537
Scirpus lacustris	785			Germany	1275
S. lacustris	1,330	1,220	2,550	Denmark	89
S. lacustris	4,600			World-wide	5336
Scirpus mucronatus	1,696–2,023			India	4300
Sparganium erectum			307	Germany	1275
Sparganium eurycarpum	1,066			Iowa	5121
S. eurycarpum	924–1,448			Iowa	3399
Spartina alterniflora	300			New Jersey	1732
S. alterniflora	63–460			Alabama	1860
S. alterniflora	540–580			Mississippi	1044
S. alterniflora	758–763			Maine	2873
S. alterniflora	973			Georgia	4570
S. alterniflora	1,158			Georgia	4923
S. alterniflora	1,207			Maryland	2348
S. alterniflora	1,296			North Carolina	4813
S. alterniflora	1,332			Virginia	5274
S. alterniflora	1,964			Mississippi	1040
S. alterniflora	2,000			Mississippi	1231
S. alterniflora	1,323–2,645			Louisiana	2511
S. alterniflora	1,381			Louisiana	2187
S. alterniflora	2,523–2,794			Louisiana	2186
S. alterniflora	2,883–3,990			Georgia	3501
S. alterniflora	2,895			Louisiana	5385
S. alterniflora		460–503		North Carolina	4812
S. alterniflora		2,500–3,500		Delaware	5098
Spartina cynosuroides	1,052–1,659			Louisiana	2186
S. cynosuroides	1,134			Louisiana	2187
Spartina patens	805			Virginia	5274
S. patens	1,296			North Carolina	5223
S. patens	1,428			Louisiana	5385
S. patens	1,922			Mississippi	1040

Table 7-3. Continued

Species	ABG	BLG	Total	Location	Reference
	Productivity				
S. patens	3,053–5,509			Louisiana	910
S. patens	4,159			Louisiana	2187
S. patens	4,924–6,163			Louisiana	2186
Typha angustata	1,662–4,577			India	4300
T. angustata	9,339			India	2442
Typha angustifolia	810	1,800	2,610	Denmark	89
T. angustifolia	850	1,800	2,650	USA	1734
T. angustifolia	1,059			Germany	1275
T. angustifolia	1,445			England	3096
T. angustifolia	2,560	2,505	5,065	Texas (1983)	2084
T. angustifolia	2,895	2,314	5,209	Texas (1984)	2084
Typha domingensis	1,483			South Carolina	3828
T. domingensis	1,112–1,580			Malawi	2216
T. domingensis	1,080–2,832			Florida	1002
T. domingensis	1,077–3,035			Florida	1003
Typha elephantina	975–2,464			India	4419
Typha glauca	1,360			Minnesota	536
T. glauca	2,297			Iowa	5121
T. glauca	2,320			Minnesota	108
Typha latifolia	330–418			Oregon	3171
T. latifolia	574			South Carolina	3828
T. latifolia	1,070			England	3717
T. latifolia	520–1,132			South Carolina	508
T. latifolia	1,358			New York	1927
T. latifolia	1,360			Minnesota	536
T. latifolia	1,570	1,050	2,620	New Jersey	2341
T. latifolia	1,600			Czech Republic	2653
T. latifolia	1,657			Czech Republic	3727
T. latifolia	3,338–3,560			Czech Republic	1195
Typha sp.	930			Virginia	5274
Typha sp.	987			New Jersey	3137
Typha sp.	1,240			North Dakota	3171
Typha sp.	730–1,336			Oklahoma	3171
Typha sp.	1,119–1,528			New Jersey	5396
Typha sp.	874–2,064			Pennsylvania	3136
Typha sp.	1,680	1,480	3,160	Minnesota	537
Typha sp.	2,210			Texas	3171
Zizania aquatica			1,450	New Jersey	5373

Table 7-3. Continued

Species	Productivity			Location	Reference
	ABG	BLG	Total		
Z. aquatica	659–1,125			New Jersey	5369
Z. aquatica	1,390			New Jersey	3137
Z. aquatica	605–1,547			Pennsylvania	3136
Floating-leaved species					
Nelumbo lucifera	68–813			India	1748
Nuphar advena	245			Virginia	5274
N. advena	516			New Jersey	3137
N. advena	775			New Jersey	5369
N. advena	1,166–1,188			Pennsylvania	3136
Nuphar lutea	222			North Carolina	5060
N. lutea	200–400			USSR	1099
N. lutea	405			Germany	1275
Nymphaea alba	155–210			India	2443
N. alba	293			Germany	1275
Nymphaea stellata	262			India	1893
Nymphoides peltata	203–305			India	2442
Trapa natans	115–435			India	2440, 2442
Submerged species					
Ceratophyllum demersum	211			Germany	1275
C. demersum	23–565			India	1748
C. demersum	610–960			USSR	1099
Elodea canadensis	326			Germany	1275
Hydrilla verticillata	194–273			India	4414
H. verticillata	104–387			India	2443
H. verticillata	180–523			India	1748
Isoetes savatieri	52–158			Argentina	3402
Myriophyllum spicatum	334			Germany	1275
M. spicatum	710			India	2442
Myriophyllum verticillatum	210–830			USSR	1099
Najas graminea	144–203			India	2443
N. graminea	139–555			India	1748
Najas major	249–799			India	4300
Potamogeton amphibium	167			Germany	1275
Potamogeton crispus	130			Germany	1275
P. crispus	92–276			India	1748
P. crispus	166–361			India	4300
Potamogeton friesii	91			Germany	1275
Potamogeton filiformis	44			United Kingdom	2389

Table 7-3. Continued

Species	Productivity			Location	Reference
	ABG	BLG	Total		
Potomogeton lucens	239			Germany	1275
P. lucens	160–380			India	2443
Potamogeton natans	99			Germany	1275
Potamogeton pectinatus	253			Germany	1275
P. pectinatus	580–830			USSR	1099
Potamogeton perfoliatus	332			Germany	1275
Potamogeton pusillus	198–279			India	2443
Ranunculus circinatus	413			Germany	1275
Utricularia vulgaris	14–39			Austria	2281
Utricularia sp.	370			Malaysia	1534
Utricularia sp. (+ periphyton)	54			Georgia	466
Trapa bispinosa	570			India	80
Vallisneria spiralis	592–709			India	4300
Floating species					
Azolla pinnata			278–400	India	1741, 1747
Eichhornia crassipes			200–291	India	4414
E. crassipes			1,473	Louisiana	3732
E. crassipes			723–2,067	India	1748
E. crassipes			6,520	Alabama	509
Lemna spp.			750–800	Czech Republic	4064
Lemna sp.			>1,000	Israel	3845
Lemna sp. + *Spirodela* sp.			10–802	India	1748
Lemnaceae			750	Czech Republic	4065
Pistia stratiotes			166–394	Argentina	3401
Salvinia natans			266–335	India	2441
Spirodela polyrhiza			12–35	India	2441
S. polyrhiza (W)			1,758–4,400	Louisiana	4246
Bryophytes					
Aulacomnium palustre	5–35			Canada, Manitoba	4016
Fontinalis antipyretica	55			Germany	1275
Hylocomium splendens	80–120			Finland	3028
Hypnum pratense	31			Canada, Manitoba	4016
Pleurotium schreberi	70–110			Finland	3028

Table 7-3. Continued

Species	Productivity			Location	Reference
	ABG	BLG	Total		
P. schreberi	108			Canada, Manitoba	4016
Polytrichum juniperinum	35			Canada, Manitoba	4016
Sphagnum angustifolium	110–140			Estonia	2279
Sphagnum balticum	109			Sweden	27
S. balticum	165			Sweden	4197
S. balticum	190–480			Estonia	2279
S. balticum + majus	210–410			Finland	3625
Sphagnum capillifolium	80–135			England	824
Sphagnum cuspidatum	70–90			Estonia	2279
Sphagnum cuspidum	260			Norway	3722
S. cuspidum	790			England	822
Sphagnum fuscum	70–330			Finland	3625
Sphagnum magellanicum	50–70			Sweden	5237
S. magellanicum	68			England	1438
S. magellanicum	70			Norway	3722
S. magellanicum	95			Sweden	27
S. magellanicum	100			Sweden	965
S. magellanicum	134			Canada	3290
S. magellanicum	250–350			Germany	3591
S. magellanicum	500			USSR	2278
Sphagnum papillosum	35			England	1438
Sphagnum "recurvum"	490–670			Germany	3591
Sphagnum rubellum	70–240			Estonia	2279
S. rubellum	130			England	824
S. rubellum	210–260			Germany	3591
S. rubellum	240–430			England	822
Sphagnum spp.	50			Ireland	1136

Table 7-4. Daily productivity rate of aquatic and wetland plants (g DM m^{-2} d^{-1})

Species	Productivity rate	Location	Reference
Emergent species			
Alisma plantago - aquatica	4.44	Iowa	5120
Althernanthera philoxeroides	17	Florida	505
Carex aquatilis	4	Canada, Alberta	1761
C. aquatilis	15.3	Canada, Québec	180
Carex atherodes	23	Iowa	5121
Carex bigelowii	6	New Hampshire	401
Carex lacustris	8.1	Wisconsin	2536
C. lacustris	15	New York	330
C. lacustris	20.9	New Jersey	2341
Carex nebraskensis	5.3	California	3989
Carex rostrata	6.0	Canada, Alberta	1761
C. rostrata	10.9	Minnesota	329
Carex stricta	23	New Jersey	2341
Carex spp.	0.56–4.16	The Netherlands	5156
Eleocharis palustris	4.97	Iowa	5120
Eleocharis quadrangulata	1.7–18.5	South Carolina	512
Equisetum fluviatile	4.25–6.37	Iowa	5120
Juncus effusus	3.6–14.2	South Carolina	508
Justicia americana	10.1–37.1	Alabama	500
Scirpus americanus	4.67	South Carolina	504
Scirpus validus	4.0	Iowa	5121
Sparganium ramosum	7.33	India	1892
Typha glauca	3.73	Minnesota	536
Typha latifolia	2.70	Iowa	5120
T. latifolia	19.4	South Carolina	504
T. latifolia	49	Grand Canyon	3171
T. latifolia	1.6–52.6	Oklahoma	3731
T. latifolia	0.89–3.14	Oregon	3171
Typha sp.	8.4–10.2	Texas, North Dakota	3171
Typha sp.	20.9	New Jersey	2341
Zizania aquatica	20.9	New Jersey	5373
Floating-leaved species			
Nuphar variegatum	1.37–2.67	Iowa	5120
Submerged species			
Ceratophyllum demersum	5.7	Sweden	1442
Myriophyllum verticillatum	2.8	Sweden	1442
Potamogeton natans	2.27–3.35	Iowa	5120
Potamogeton pectinatus	1.77–2.97	Iowa	5120
Floating species			
Eichhornia crassipes (W)	0.4–12.4	Czech Republic	5538
E. crassipes	12.7–14.6	Louisiana	3731
E. crassipes (W)	14.7	Czech Republic	5537

Table 7-4. Continued.

Species	Productivity rate	Location	Reference
E. crassipes (W)	3.0–15	Czech Republic	2662
E. crassipes	12.5–27.6	Alabama	509
E. crassipes	6.0–33	Florida	5045
E. crassipes (F)	6.3–40	Florida	4037
E. crassipes	40–54	Florida	5534
E. crassipes	60	Subtropic climate	5482
Lemna sp.	14.9	Louisiana	4246
Lemna spp.	3.5	Czech Republic	4064
Pistia stratiotes	10.8	Czech Republic	5537
P. stratiotes	15.3	Florida	3502

Table 7-5. Decomposition constants, k (d^{-1}) and turnover rates for different aquatic and wetland macrophytes

Plant species	k	Time (days) required for 50% decay	Time (days) required for 95% decay	Reference
Emergent species				
Bidens cernua	0.0012	578	2,500	994
Bidens laevis	0.012–0.0069	60–100	250–435	5372
Carex acutiformis	0.0009	770	3,333	5154
Carex diandra	0.0018	385	1,667	5154
Carex riparia	0.0029	240	1,034	5073
Carex rostrata	0.0046	150	652	966
Chara sp. (alga)	0.0084	83	358	239
Cyperus articularis	0.0069	101	437	2642
Distichlis spicata	0.0041	170	732	5385
D. spicata	ca 0.00053	1,300	5,660	5377
Eleocharis dulcis	0.056–0.011	12–63	54–272	3211
Juncus squarrosus	0.0013	533	2,308	2718
Ludwigia leptocarpa	0.005	139	600	2083
Paspulum distichum	0.011–0.022	63–315	272–1,364	3211
Peltandra virginica	0.025–0.020	28–35	120–150	5372
Phragmites australis	0.0031	224	968	3096
Phragmites australis	0.0005	1386	6,000	5124
Phragmites karka	0.0045	154	667	2642
Polygonum glabrum	0.0039	178	769	2642
Sagittaria latfolia	0.0495	14	61	994
Scolochloa festucacea	0.00171	405	1,754	5124
Scirpus fluviatils	0.0018	385	1,667	993
Scirpus mucronatus	0.0044	158	682	2642
Spartina alterniflora	0.023–0.0092	30–75	130–326	5385
Spartina patens	0.0051	135	588	5385
S. patens	ca 0.00069	1,000	4,348	5377
Typha angustata	0.006	116	500	2642
Typha angustifolia	0.0047	147	638	2083
T. angustofolia	0.0019	364	1,579	3096
Typha elephantina	0.0038	180	790	4424
Typha glauca	0.0014	495	2,143	993
Typha glauca (SL)	0.001	63	273	3414
T. glauca (GL)	0.024	29	125	3414
Typha latifolia	0.007	99	429	2085
T. latifolia	0.0104	67	289	5305
T. latifolia	0.0043	160	693	510
Floating-leaved species				
Nelumbo lutea (P)	0.0033	210	909	2083
Nelumbo lutea (L)	0.0108	64	278	2083
Nuphar variegatum	0.035–0.093	7.5–20	32–83	1667

Table 7-5. Continued

Plant species	k	Time (days) required for		Reference
		50% decay	95% decay	
Nymphaea nouchali (L)	0.042	17	71	3355
N. nouchali (RH)	0.014	50	214	3355
Nymphaea odorata	0.009	77	333	2572
Nymphoides indica	0.040	17	75	3355
Submerged species				
Ceratophyllum demersum	0.0213	31	141	239
Ceratophyllum echinatum	0.0659	11	45	2641
Elodea canadensis	0.0912	8	33	239
E. canadensis	0.026	27	115	2085
Hydrilla verticillata	0.020	35	150	2641
Myriophyllum spicatum	0.0315	22	95	239
Najas graminea	0.043	16	70	3355
Najas major	0.0341	29	88	2641
Potamogeton crispus	0.042–0.093	7.5–17	32–71	3211
Potamogeton diversifolius	0.0084	82	356	5472
Potamogeton indicus	0.0095–0.087	8–73	35–316	3211
Potamogeton lucens	0.0525	13	57	2085
Potamogeton nodosus	0.0312–0.0408	14–22	74–96	2082
Potamogeton pectinatus	0.0097	71	309	2641
P. pectinatus	0.082	8.5	37	699
Potamogeton perfoliatus	0.0537	13	56	2085
Vallisneria americana	0.163	4.3	18	699
Floating species				
Azolla pinnata	0.0097	72	309	2641
Eichhornia crassipes (L)	0.038	18	79	4422
E. crassipes (ST)	0.024	29	125	4422
E. crassipes (RH)	0.023	30	130	4422
E. crassipes (R)	0.006	115	500	4422
E. crassipes (ST + L)	0.038–0.025	18–28	79–120	3211
E. crassipes	0.0107	65	280	2218
Lemna minor	0.0109–0.0351	20–64	85–275	2719
L. minor	0.0238	29	126	2641
Salvinia auriculata	0.0277	25	107	2218
Salvinia cucullata	0.0051	136	588	4421
Salvinia molesta	0.0033	210	909	4421
Bryophytes				
Fontinalis antipyretica	0.0138	50	217	239
Mixed tree species leaf litter				
Cypress[a]	0.0016–0.00061	433–1,136	1,875–4,918	1009–10, 5520
Maple-gum[b]	0.0018–0.00049	385–1,414	1,667–6,122	1009–10, 5520
Cedar[c]	0.0013–0.00065	533–1,066	2,307–4,615	1009–10, 5520
Mixed hardwood[d]	0.0011–0.00066	630–1,050	2,727–4,545	1009–10, 5520

Table 7-5. Continued

| Plant species | Time (days) required for | | | Reference |
	k	50% decay	95% decay	
Wetland forest	0.0021[e]	330	1,429	859
	0.0057[f]	122	526	859

[a] Bald cypress (*Taxodium distichum*), Red maple (*Acer rubrum*), Swamp blackgum (*Nyssa sylvatica* var. *biflora*)
[b] Water tupelo (*Nyssa aquatica*), Red maple, Swamp blackgum
[c] Atlantic white cedar (*Chamaecyparis thyoides*), Swamp blackgum, Red maple
[d] Laurel oak (*Quercus laurifolia*), White oak (*Quercus alba*), Sweetgum (*Liquidambar styraciflua*)
[e] dominant species *Taxodium distichum*
[f] dominant species *Nyssa aquatica*

Table 7-6. Nitrogen concentration in aquatic and wetland plants (% dry mass)

Plant species	Concentration		Locality	Reference
	ABG	BLG		
Emergent species				
Acorus calamus	1.80		Poland	340
A. calamus	2.06		New Jersey	5371
A. calamus	1.26–2.92		Czech Republic	1197
A. calamus	2.49		Czech Republic	5143
A. calamus	3.1		Czech Republic	5144
A. calamus	1.26–3.71		Czech Republic	1198
Althernanthera philoxeroides	2.25–3.52		Alabama	500
Bidens laevis	2.22		New Jersey	5371
Bolboscheonus maritimus	1.36–1.79		Czech Republic	1197
B. maritimus	1.91	1.68–2.45	Czech Republic	1201
Carex acutiformis	1.6		Hungary	2574
C. acutiformis (F)	0.7–2.3		The Netherlands	28
Carex aquatilis	1.1		Canada, Québec	178
C. aquatilis	0.82		Canada, Québec	179
C. aquatilis	2.18		Alaska	748
C. aquatilis	2.9–3.3		Alaska	5083
Carex diandra (F)	1.1–3.53		The Netherlands	28
Carex gracilis	1.5		Hungary	2574
C. gracilis	1.3–2.5		Czech Republic	1207
C. gracilis	1.5–2.0		Czech Republic	2659
Carex lacustris	0.6		Wisconsin	2537
C. lacustris	1.0		Canada, Québec	178
C. lacustris	1.3		Iowa	5128
C. lacustris	1.4		Hungary	2574
C. lacustris	1.7		New York	332
C. lacustris	1.72		New York	331
Carex lanuginosa	0.9		Canada, Québec	178
C. lanuginosa	3.27		New York	331
Carex lasiocarpa (F)	0.8–3.0		The Netherlands	28
Carex lyngbyei	1.0		Canada	2518
Carex nebraskensis	0.93–1.97		California	3989
Carex rostrata	1.4		Sweden	4617
C. rostrata	1.47		New York	331
C. rostrata	1.60–1.96		Scotland	2108
C. rostrata	1.1–2.4	0.1	New York	335
C. rostrata (F)	1.1–2.7		The Netherlands	28
C. rostrata	2.2		The Netherlands	5125
Carex tenuiflora	1.7		Canada, Ontario	4568
Carex trisperma	1.2		Canada, Ontario	4568
Carex vesicaria	0.9		Czech Republic	1207

Table 7-6. Continued

Plant species	Concentration ABG	BLG	Locality	Reference
Cladium jamaicense	0.5–0.7		Florida	4747
Dupontia fisheri	1.6		Alaska	748
D. fisheri	2.4–3.0		Alaska	5083
Eleocharis quandrangulata	0.94		South Carolina	512
Eriophorum angustifolium	2.10		Alaska	748
Eriophurum vaginatum	2.30		Alaska	749
Glyceria grandis	1.5–1.7		Canada, Ontario	3347
Glyceria maxima	0.98–2.52		Czech Republic	5143
G. maxima	1.0–1.57		England	651
G. maxima	1.29–1.82		Czech Republic	1197
G. maxima	1.67		Czech Republic	2313
G. maxima	1.05–2.94		Czech Republic	1198
G. maxima (F)	3.11–4.62		Poland	3596
Impatiens capensis	2.08		New Jersey	5371
Iris pseudacorus	S 1.99–2.23	R 1.04–1.44 RH 1.08–1.69	Scotland	2108
Juncus effusus	1.29–1.78		South Carolina	508
J. effusus	S 1.04–1.88	R 0.69–0.82	Scotland	2108
Justicia americana	1.63–3.77		Alabama	500
Papyrus sp.	1.21		Uganda	1577
Peltanda virginica	2.15		New Jersey	5371
Petasites frigidus	2.8–3.5		Alaska	5083
Phalaris arundinaceae	L 2.6–3.0 S 1.16–1.76	R 0.98–1.39	Scotland	2110
Phragmites australis	0.78		South Africa	5455
P. australis	1.05–1.28	0.49–0.61	England	410
P. australis (W)	1.18		South Africa	5505
P. australis	1.23		Poland	340
P. australis	1.24–2.15	0.97–1.05	Czech Republic	1197
P. australis	1.26	0.84–1.29	Czech Republic	1201
P. australis	1.30–2.70		England	73
P. australis	1.40–1.82		England	651
P. australis	1.58–1.94	1.62–1.86	India	2444
P. australis	1.77–2.13		Czech Republic	1197
P. australis	2.07		Scotland	2112
P. australis	2.58		England	3242
P. australis	1.0–2.77	1.05–1.6	Czech Republic	1198
P. australis	2.80		Czech Republic	5144
P. australis	2.91		Czech Republic	5143
P. australis	3.94		England	3096
Polygonum amphibium	S 1.97 L 3.66		Scotland	2108
Polygonum arifolium	1.93		New Jersey	5371
Pontenderia cordata	2.11		New Jersey	5371

Table 7-6. Continued

Plant species	Concentration ABG	Concentration BLG	Locality	Reference
Sagittaria latifolia	1.91		New Jersey	5371
Schoenoplectus lacustris	1.03–1.77		Czech Republic	1197
S. lacustris	1.12		Czech Republic	5143
S. lacustris	1.73		Poland	340
S. lacustris	1.85		England	3242
S. lacustris	0.63–2.59		Czech Republic	1198
Scirpus americanus	0.83–2.72		South Carolina	504
Scirpus lacustris	1.15		South Carolina	504
Sparganium erectum	1.42–2.55		Czech Republic	1197
S. erectum	2.40–2.50	2.40–2.50	India	2444
Sparganium eurycarpum	3.35		New York	331
Sparganium ramosum	1.16–3.15	1.26–2.20	India	1892
Typha angustata	1.96–2.15	1.99–2.13	India	2444
Typha angustifolia	0.86	0.73	Michigan	5084
T. angustifolia	1.15		Czech Republic	1197
T. angustifolia	1.26–1.34		Czech Republic	5143
T. angustifolia	0.8–2.9		Czech Republic	1198
T. angustifolia (W)	9.5		Michigan	5084
Typha domingensis	0.18–0.86	0.65–1.45	Florida	1002
Typha elephantina	1.17	0.91	India	1745
Typha glauca	2.47		New York	331
T. glauca	0.30–3.02	0.42–1.84	Iowa	995
Typha latifolia	0.51–2.4		South Carolina	504
T. latifolia (F)	0.75	0.63	Michigan	5084
T. latifolia	1.18		Poland	340
T. latifolia	1.43		Czech Republic	1197
T. latifolia	2.01		New Jersey	5371
T. latifolia	2.1	1.3	Wisconsin	4582
T. latifolia	0.75–2.25		USA	506
T. latifolia	0.7–2.3		SE USA	3775
T. latifolia	0.64–2.94		Czech Republic	5143
T. latifolia (W)	7.8		Michigan	5084
Typha orientalis (GR)	0.98–2.70		Australia	714
Zizania aquatica	2.12		New Jersey	5371
Floating-leaved species				
Nelumbo nucifera	1.52–1.54		India	2444
Nuphar advena	2.13		New Jersey	5371
Nuphar lutea	L 4.08 P 2.41	R 1.94 RH 1.94	Scotland	2108
N. lutea	3.37		Czech Republic	1207
Nymphaea alba	1.09		India	2444
Nymphaea nouchalli	1.37–5.09	0.89–2.90	India	3355

Table 7-6. Continued

Plant species	Concentration ABG	Concentration BLG	Locality	Reference
Nymphoides peltata	1.8		New Jersey	4133
N. peltata	2.16–2.2		India	2444
N. peltata	2.8		Czech Republic	5143
Submerged species				
Batrachium aquatile	2.70		Czech Republic	1207
Ceratophyllum demersum	2.50–2.80		India	2444
C. demersum	2.11–4.43		Wisconsin	1611
Elodea canadensis	1.81–4.42		England	74
E. canadensis	1.87–4.90		Czech Republic	1207
Elodea spp.	2.10–2.86		Wisconsin	1611
Eriocaulon septangulare	1.98–2.57		Wisconsin	1611
Heteranthera dubia	2.32–3.79		Wisconsin	1611
Isoetes macrospora	2.40–3.70		Wisconsin	467
Littorella uniflora	2.70–3.40		Wisconsin	467
Lobelia dortmana	1.48–1.95		Wisconsin	1611
Myriophyllum spicatum	2.42–2.77		Wisconsin	1611
M. spicatum	3.02		India	2444
Myriophylum verticillatum	1.10–1.56	1.40–1.90	India	2444
Najas graminea	2.10		India	2444
Najas maritima	1.05–1.87		Australia	4218
Potamogeton amplifolius	2.42–3.78		Wisconsin	1611
Potamogeton crispus	2.00–2.10		India	2444
Potamogeton lucens	2.10–2.20		India	2444
Potamogeton epihydrus	2.38–3.19		Wisconsin	1611
Potamogeton natans	1.10–1.20		India	2444
Potamogeton obtusifolius	2.97		Czech Republic	1207
Potamogeton pectinatus	2.20–2.40		India	2444
P. pectinatus	1.22–2.61		Australia	4218
P. pectinatus	4.04–6.01		Scotland	2107
Potamogeton perfoliatus	1.07–2.75		Australia	4218
Potamogeton richardsonii	1.95–3.73		Wisconsin	1611
Potamogeton pusillus	2.10		India	2444
P. pusillus	3.02		Czech Republic	1207
Potamogeton robinsonii	2.12–2.36		Wisconsin	1611
Potamogeton zosteriformis	3.37–3.70		Wisconsin	1611
Potamogeton spp.	2.65		Minnesota	2082
Ranunculus penicillatus	L 4.45 S 2.11	R 2.87	England	717
Trapa natans	0.95		India	2444
Utricularia sp. (+ periphyton)	1.73–3.04		Georgia	466
Vallisneria americana	1.98–3.85 (Total)		Wisconsin	1611
Vallisneria spiralis	0.82–3.24		Australia	4218

Table 7-6. Continued

Plant species	Concentration		Locality	Reference
	ABG	BLG		
Floating species (Total)				
Azolla pinnata	3.85–4.60		Bangladesh	4198
Eichhornia crassipes	1.54–2.05		Alabama	509
E. crassipes	3.02		India	2444
E. crassipes	1.30–3.30		Florida	512
E. crassipes	2.10–3.70		Florida	5045
E. crassipes (W)	T 2.45	R 1.77	Japan	3533
E. crassipes	T 3.11–4.09	R 2.00–2.79	Japan	3534
Lemna gibba	2.70–4.75		England	75
L. gibba + Spirodela polyrhiza	2.97–5.04		Czech Republic	1207
L. gibba + Spirodela polyrhiza	3.29–5.77		Czech Republic	4063,4067
Lemna minor	4.20		Czech Republic	1207
L. minor	4.30		New York	3775
L. minor	2.38–5.31		England	75
Lemna paucicostata (W)	4.21–5.60		Nigeria	3114
L. paucicostata (C, W)	7.28		Nigeria	3114
Lemna polyrhiza	3.94–5.36		England	74
Lemna trisulca	1.60–4.8		England	75
Pistia stratiotes	2.22–5.07		India	2444
Salvinia rotundifolia	1.94–3.85		Florida	29
Spirodela polyrhiza	3.3–4.2		Florida	29
S. polyrhiza	4.75		India	2444
S. polyrhiza	5.04		Czech Republic	4067
Bryophytes				
Fontinalis antipyretica	2.87		Czech Republic	1207
Sphagnum balticum	0.56–1.01		Finland	3625
Sphagnum fuscum	0.66–0.74		Finland	3625
Sphagnum magellanicum	1.40		Czech Republic	1207
Sphagnum majus	0.87–1.06		Finland	3625
Sphagnum recurvum	2.41		Czech Republic	1207
Sphagnum sp.	0.87		Czech Republic	1207
Sphagnum spp.	0.64		Finland	3627

Table 7-7. Nitrogen standing stock in aquatic and wetland plants (g m⁻²)

Plant species	Standing stock ABG	BLG	Locality	Reference
Emergent species				
Acorus calamus	4.3–21.3		Czech Republic	1197–8
Althernanthera philoxeroides	13.5–24.1		Alabama	500
A. philoxeroides	24.0–42.5		Florida	4031
Bolboschoenus maritimus	6.54–11.8		Czech Republic	1197–8
B. maritimus	8.69–15.38		Czech Republic	1201
Carex aquatilis	6.4		Alaska	748
Carex gracilis	9.4		Czech Republic	1207
C. gracilis	1.7–3.1 (winter)		Czech Republic	2659
	6.2–13.7 (summer)			
Carex lacustris	7.71	1.07	Wisconsin	2537
C. lacustris	3.9–18.3	3.3–6.6	New York	332
Carex rostrata	0.18		Sweden	4617
C. rostrata	4.3		The Netherlands	5152
C. rostrata	3.1–14.8	2.4–5.5	New York	335
Carex vesicaria	7.7		Czech Republic	1207
Cladium jamaicense	5.5–8.9		Florida	4747
Eleocharis quadrangulata	8.3		South Carolina	512
Glyceria grandis	45.7–72.0		Canada, Ontario	3347
Glyceria maxima	10.3–44.2		Czech Republic	1197–8
G. maxima	6.6–49.1		Czech Republic	1197
G. maxima	134		Poland	3596
Hydrocotyle umbellata	9–30		Florida	4031
Juncus effusus	26.1		South Carolina	508
Juncus kraussii	24.1	21.1	Australia	811–2
Justicia americana	2.05	1.67	North Carolina	5060
J. americana	12.1–44.3	29	Alabama	500
Papyrus sp.	61.6		Uganda	1577
Phalaris arundinacea	12.31	3.24	Wisconsin	2537
Phragmites australis	22.2		South Africa	5455
P. australis	26	50.9	Czech Republic	1201
P. australis (W)	27	16	The Netherlands	1038
P. australis	28		Czech Republic	2653
P. australis	15.5–38.2	16.4–21.7	India	2444
P. australis	13.7–40.9	35.4–64	Czech Republic	1197–8, 1204
P. australis	0.04–43.3		England	3096
P. australis (W)	75		South Africa	5505
P. australis	88.2		Scotland	2112
Polygonum glabrum	33.2–205	25–87	India	2642
Polygonum sp.	14.2		Alabama	499
Polygonum spp.	8.66	0.92	Wisconsin	2537

Table 7-7. Continued

Plant species	Standing stock		Locality	Reference
	ABG	BLG		
Sagittaria latifolia	21.2		Alabama	499
Schoenoplectus lacustris	12–52.9		Czech Republic	1197–8
Scirpus americanus	1.03–1.66		South Carolina	504
Scirpus fluviatilis	6.6		Iowa	992
S. fluviatilis	15.35	5.32	Wisconsin	2537
Scirpus lacustris (W)	26	32	The Netherlands	1038
Scirpus mucronatus	28–55	16–34	India	2642
Sparganium erectum	18.2–28.7		Czech Republic	1197–8
S. erectum	17.3–23.3	10.2–11.2	India	2444
Sparganium ramosum	3.65–21.1	6.04–13.54	India	1892
Spartina alterniflora	4.7–9.8		Georgia	741
Typha angustata	22.1–54.6	17.7–44.2	India	2444
Typha angustifolia	0.6–13		England	3096
T. angustifolia	9.7–32.7		Czech Republic	1197–8
T. angustifolia	24.5–46.7		Czech Republic	1197
Typha domingensis	2.3–6.1		Florida	1002
Typha elephantina	21.53		India	1745
Typha glauca	16.5		Iowa	992
T. glauca	28.2	3.9	Iowa	995
Typha latifolia	1.24–5.35		South Carolina	504
T. latifolia	9.45	22.07	Wisconsin	2537
T. latifolia	6–12		South Carolina	512
T. latifolia	16.5		Czech Republic	1198
T. latifolia	25.1		Czech Republic	2653
T. latifolia	31		Wisconsin	3876
T. latifolia	50.9		Czech Republic	1197
Typha sp.	4.5		Malawi	2220
Floating-leaved species				
Nelumbo lutea	1.9		Alabama	499–500
Nelumbo nucifera	10.9–12.4		India	2444
Nuphar lutea	1.34	2.17	North Carolina	5060
Nymphaea alba	2.8		India	2444
Nymphaea odorata	4.6		Alabama	499–500
Nymphoides aquaticum	2.8		Alabama	499–500
Nymphoides peltata	2.0–2.05		India	2444
Submerged species				
Betrachium aquatile	2.4		Czech Republic	1207
Batrachium fluitans	6.0		Czech Republic	1207
Ceratophyllum demersum	5.3–5.6		India	2444
Elodea canadensis	8.3–11.3		Czech Republic	1207
Hydrocharis morsus-ranae	4.1–5.5		India	2444
Myriophyllum verticillatum	6.2–6.3	4.1–4.7	India	2444

Table 7-7. Continued

Plant species	Standing stock		Locality	Reference
	ABG	BLG		
Myriophyllum spicatum	0.66		India	2444
Najas graminea	3.1–3.7		India	2444
Potamogeton crispus	5.1–5.4		India	2444
Potamogeton lucens	2.8–3.5		India	2444
Potamogeton natans	2.0–3.5		India	2444
Potamogeton obtusifolius	9.9		Czech Republic	1207
Potamogeton pectinatus	4.4–4.7		India	2444
Potamogeton pusillus	4.0		Czech Republic	1207
P. pusillus	4.2–5.4		India	2444
Potamogeton spp.	5.65		Minnesota	2082
Floating species (Total)				
Eichhornia crassipes	29.2		Japan	3534
E. crassipes	34.8		Alabama	515
E. crassipes (F)	80		Florida	4037
E. crassipes	30–90		Florida	4031
E. crassipes	44–234		India	2641
Lemna minor	0.4–5.0		Florida	4031
Pistia stratiotes	9–25		Florida	4031
Salvinia rotundifolia	1.5–9.0		Florida	4031

Table 7-8. Nitrogen uptake rate by aquatic and wetland plants

Plant species	Rate	Locality	Reference
Emergent species	$(g\ m^{-2}\ yr^{-1})$		
Althernanthera philoxeroides	178	USA	505
Carex lacustris	16	New York	332
Carex rostrata	10	New York	335
Cyperus papyrus	103	Uganda	1578
Justicia americana	229	USA	505
Phragmites australis	22.2	South Africa	5455
P. australis (W)	75	South Africa	5505
P. australis	82.2	Scotland	2112
Scirpus fluviatilis	17.5	Wisconsin	2537
Typha domingensis	11.1–29.3	Florida	1002
Typha latifolia	263	USA	505
Typha spp.	103.2	Florida	4023
Floating species			
Eichhornia crassipes	42	Czech Republic	5538
E. crassipes	198	USA	505
E. crassipes	120	Alabama	509
E. crassipes	250	Florida	4173
E. crassipes (F)	41–357	Florida	4037
E. crassipes (F)	535	Florida	4026
Lemna sp.	611	USA	3426
	$(mg\ m^{-2}\ d^{-1})$		
Eichhornia crassipes (W)	130	Czech Republic	2662
E. crassipes	260–400	Alabama	509
E. crassipes (W)	18–497	Czech Republic	5538
E. crassipes (W)	510	Japan	3533
Lemnaceae	200	Czech Republic	2661
Scirpus americanus	32–34.4	South Carolina	504
Typha latifolia	93–163	South Carolina	504

Table 7-9. Phosphorus concentration in aquatic and wetland plants (% dry mass)

| Plant species | Concentration | | Locality | Reference |
	ABG	BGL		
Emergent species				
Acorus calamus	0.03		Poland	340
A. calamus (W)	0.29		Germany	4399
A. calamus	0.20–0.35		Czech Republic	1197
A. calamus	0.34		Czech Republic	5143
A. calamus	0.23–0.47		Czech Republic	1198
Althernantherra philoxeroidex	0.32–0.36		Alabama	500
Bolboschoenus maritimus	0.28–0.35		Czech Republic	1197
B. maritimus	0.35	0.4–0.52	Czech Republic	1201
Carex acutiformis	0.07		Hungary	2574
Carex aquatilis	0.12		Canada, Québec	178–9
C. aquatilis	0.15		Alaska	748
C. aquatilis	0.2–0.3		Alaska	5083
Carex gracilis	0.12		Hungary	2574
C. gracilis	0.09–0.21		Czech Republic	2659
C. gracilis	0.12–0.38		Czech Republic	1207
Carex lacustris	0.007		Hungary	2574
C. lacustris	0.13		Canada, Québec	178
C. lacustris	0.14		Iowa	5128
C. lacustris	0.16		Wisconsin	2537
C. lacustris	0.17		New York	332
C. lacustris	0.17		Minnesota	2870
C. lacustris	0.25		New York	331
Carex lanuginosa	0.09		Canada, Québec	178
C. lanuginosa	0.45		New York	331
Carex lyngbyei	0.14–0.40		Canada, B.C.	2518
Carex rostrata	0.10		Sweden	4617
C. rostrata	0.18		The Netherlands	5152
C. rostrata	0.14–0.29		Scotland	2108
C. rostrata	0.20–0.30	0.20	New York	335
Carex stricta	0.17		Minnesota	2870
C. stricta (W)	0.22		Germany	4399
Carex tenuiflora	0.15		Canada, Ontario	4568
Carex trisperma	0.15		Canada, Ontario	4568
Carex vesicaria	0.20		Czech Republic	1207
Carex spp.	0.20		Michigan	4114, 5329
Cladium jamaicense	0.022		Florida	4747

Table 7-9. Continued

Plant species	Concentration		Locality	Reference
	ABG	BGL		
C. jamaicense (F)	0.01–0.17	R 0.02–0.12	Florida	4748
Cyperus esculentus (GR)	0.043–0.63		Mississippi	222
Dupontia fisheri	0.09		Alaska	748
D. fisheri	0.2–0.4		Alaska	5083
Eleocharis acicularis	0.24		South Carolina	507
Eleocharis quandrangulata	0.10		South Carolina	512
Eriophorum angustifolium	0.11		Alaska	748
Eriophorum vaginatum	0.40		Alaska	4428
Glyceria grandis	0.13–0.21		Canada, Ontario	3347
Glyceria maxima	0.12		Czech Republic	2313
G. maxima	0.18–0.19		England	651
G. maxima	0.15–0.29		Czech Republic	5143
G. maxima (W)	0.25		Germany	4399
G. maxima	0.18–0.31		Czech Republic	1197
G. maxima	0.23–0.42		Czech Republic	1198
G. maxima (W)	0.50–0.79		Poland	3596
Hydrocotyle sp.	0.18		South Carolina	507
Iris pseudacorus (W)	0.25		Germany	4399
I. pseudacorus	S 0.33–0.43	RH 0.25–0.39 R 0.31–0.59	Scotland	2108
Juncus effusus	0.16–0.30		South Carolina	508
J. effusus	0.27		South Carolina	507
J. effusus	S 0.13–0.31	R 0.10–0.35	Scotland	2108
Justicia americana	0.09–0.21		Alabama	500
Mentha aquatica (W)	0.22		Germany	4399
Panicum hemitomon	0.14		South Carolina	507
Papyrus sp.	0.067		Uganda	1577
Phalaris arundinaceae	L 0.15–0.36 S 0.12–0.35	R 0.08–0.35	Scotland	2110
Phragmites australis	0.01		Poland	340
P. australis	0.09		South Africa	5455
P. australis	0.10		Czech Republic	2653
P. australis (W)	0.12		South Africa	5505
P. australis	0.12–0.14		England	651
P. australis	0.13–0.17		England	73
P. australis (W)	0.14		Germany	4399
P. australis	0.18		England	3096
P. australis	0.17–0.48	0.11–0.19	Czech Republic	1198
P. australis	0.21		Czech Republic	5143
P. australis	0.19–0.28		Czech Republic	1197
P. australis	0.25		Scotland	2112
P. australis	0.21	0.08–0.15	Czech Republic	1201

Table 7-9. Continued

Plant species	Concentration ABG	BGL	Locality	Reference
P. australis	0.16–0.28	0.1–0.11	Czech Republic	1197
Polygonum amphibium	L 0.33 S 0.45		Scotland	2108
Polygonum punctatum	0.32		South Carolina	1574
Pontenderia cordata	0.24		South Carolina	507
Sagittaria lancifolia (F)	0.05–0.45		Florida	1116
Sagittaria latifolia	0.30		South Carolina	507
S. latifolia	0.58		South Carolina	1574
Schoenoplectus lacustris	0.03		Poland	340
S. lacustris	0.23		Czech Republic	5143
S. lacustris	0.23–0.34		Czech Republic	1197
S. lacustris	0.16–0.49		Czech Republic	1198
Scirpus americanus	0.13–0.30		South Carolina	504
S. americanus	0.18		South Carolina	507
Scirpus cyperinus	0.19		South Carolina	1574
S. lacustris (W)	0.20		Germany	4399
S. lacustris	0.23		England	3242
Scirpus validus (GR)	0.046–0.52		Mississippi	222
Sparganium erectum	0.30		Czech Republic	5143
S. erectum (W)	0.39		Germany	4399
S. erectum	0.32–0.48		Czech Republic	1197–8
Sparganium eurycarpum	0.64		New York	331
Sparganium ramosum	0.24–0.50	0.18–0.40	India	1892
Typha angustifolia (F)	0.086–0.157		Michigan	5084
T. angustifolia	0.16		Czech Republic	1198
T. angustifolia	0.18–0.19		Czech Republic	5143
T. angustifolia (W)	0.20		Germany	4399
T. angustifolia	0.15–0.49		Czech Republic	1198
T. angustifolia (W)	1.20		Michigan	5084
Typha domingensis	0.06–0.15	0.04–0.18	Florida	1002
Typha elephantina	0.14	0.39	India	1745
Typha glauca	0.01–0.48	0.14–0.35	Iowa	995
Typha latifolia	0.02		Poland	340
T. latifolia	0.05–0.40		USA	506
T. latifolia (F)	0.078–0.136		Michigan	5084
T. latifolia	0.09–0.31		South Carolina	504
T. latifolia	0.14		South Carolina	507
T. latifolia	0.17–0.34		Czech Republic	5143
T. latifolia	0.22		Czech Republic	1198
T. latifolia	0.37	0.35	Wisconsin	4582
T. latifolia	0.49		Wisconsin	3876
T. latifolia (W)	1.15		Michigan	5084
Typha orientalis (GR)	0.12–0.38		Australia	714

Table 7-9. Continued

Plant species	Concentration		Locality	Reference
	ABG	BGL		
Floating-leaved species				
Brasenia schreberi	0.14		South Carolina	507
Nelumbo lutea	0.19		South Carolina	507
Nuphar advena	0.40		South Carolina	507
Nuphar lutea	L 0.47 P 0.37	R 0.30 RH 0.39	Scotland	2108
Nymphaea odorata	0.18		South Carolina	507
Nymphoides peltata	0.42		New Jersey	4133
N. peltata	0.64		Czech Republic	5143
Submerged species				
Ceratophyllum demersum	0.26		South Carolina	507
C. demersum	0.51–0.75		Wisconsin	1611
Elodea spp.	0.12–0.33		Wisconsin	1611
Eriocaulon septangulare	0.10–0.27 (Total)		Wisconsin	1611
Heteranthera dubia	0.51–0.69		Wisconsin	1611
Isoetes macrospora	0.16–0.33		Wisconsin	467
Littorella uniflora	0.18–0.27		Wisconsin	467
Lobelia dortmania (total)	0.10–0.23 (Total)		Wisconsin	1611
Myriophyllum heterophyllum	0.16		South Carolina	507
M. heterophyllum	0.32–1.05		New Hampshire	2503
Myriophyllum spicatum	0.03		Poland	340
M. spicatum	0.35–0.41		Wisconsin	1611
M. spicatum (TI)	0.14–0.56		Wisconsin	702
M. spicatum (TI)	0.076–0.99		Wisconsin	4352
Najas maritima	0.16		Australia	4218
Najas guadalupensis	0.15		South Carolina	507
Potamogeton amplifolius	0.14–0.22		Wisconsin	1611
Potamogeton diversifolius	0.27		South Carolina	507
Potamogeton epihydrus	0.19–0.33		Wisconsin	1611
Potamogeton pectinatus	0.04		Poland	340
P. pectinatus	0.19		Australia	4218
P. pectinatus	0.21		Wisconsin	699
P. pectinatus	0.46–0.81		Scotland	2107
Potamogeton perfoliatus	0.16		Australia	4218
Potamogeton richardsonii	0.14–0.43		Wisconsin	1611
Potamogeton robinsonii	0.13–0.16		Wisconsin	1611
Potamogeton spp.	0.59		Minnesota	2082
Ranunculus penicillatus	L 0.48 S 0.34	R 0.44	England	717
Utricularia inflata	0.12		South Carolina	507
Utricularia sp. (+ periphyton)	0.045–0.075		Georgia	466
Vallisneria americana	0.24		Wisconsin	699

Table 7-9. Continued

Plant species	Concentration ABG	BGL	Locality	Reference
V. americana	0.37–0.43 (Total)		Wisconsin	1611
Vallisneria spiralis	0.15		Australia	4218
Floating species (Total)				
Eichhornia crassipes	0.19–0.26		Alabama	509
E. crassipes (E)	0.55		Florida	3563
E. crassipes (E)	0.70		Florida	1773
E. crassipes	0.14–0.80		Florida	512
E. crassipes	T 0.70–0.87	R 0.47–0.86	Japan	3534
E. crassipes (W)	T 0.57	R 0.43	Japan	3533
E. crassipes (E)	0.91		Florida	1883
E. crassipes (E)	0.22–1.35		Florida	4038
Lemna gibba (W)	0.72–2.60		Louisiana	927
L. gibba (W)	1.60		Florida	2094
Lemna minor	0.75		New York	3375
Lemna spp. (W)	0.30–1.24		Florida	4847
Salvinia rotundifolia	0.21–0.74		Florida	29
Spirodela oligorhiza (W)	1.1–2.84		Louisiana	926
S. oligorhiza (W)	1.5		Louisiana	2094
Spirodela polyrhiza	0.33–0.68		Florida	29
S. polyrhiza	1.28		Czech Republic	4067
S. polyrhiza (W)	0.085–1.4		Louisiana	4246
S. polyrhiza (W)	1.4		Louisiana	2094
S. polyrhiza + *Lemna gibba*	0.96–2.05		Czech Republic	4063, 4067
S. polyrhiza (W)	0.56–2.10		Louisiana	927
Spirodela punctata (W)	0.61–1.50		Louisiana	927
Bryophytes				
Fontinalis antipyretica	0.61		Czech Republic	1207
Jungermannia vulcanicola	0.17		Japan	4292
Sphagnum balticum	0.035–0.08		Finland	3625
Sphagnum fuscum	0.030–0.042		Finland	3625
Sphagnum magellanicum	0.22		Czech Republic	1207
Sphagnum majus	0.034–0.049		Finland	3625
Sphagnum recurvum	0.23		Czech Republic	1207
Sphagnum sp.	0.18		Czech Republic	1207
Sphagnum spp.	0.0064		Finland	3627

Table 7-10. Phosphorus standing stock in aquatic and wetland plants (g m^{-2})

Plant species	Standing stock AGB	BLG	Locality	Reference
Emergent species				
Acorus calamus	0.5–2.5		Czech Republic	1198
A. calamus	2.1–3.3		Czech Republic	1197
Althernanthera philoxeroides	1.37–3.11		Alabama	500
A. philoxeroides	3.0–5.3		Florida	4031
Bolboschoenus maritimus	1.4–2.4		Czech Republic	1197
B. maritimus	1.59	3.85	Czech Republic	1201
Carex aquatilis	0.33		Alaska	748
Carex gracilis	0.1–1.2	2.4	Czech Republic	2659
Carex lacustris	0.4–1.6	0.8–1.3	New York	332
C. lacustris	1.97	0.24	Wisconsin	2537
Carex rostrata	0.014		Sweden	4617
C. rostrata	0.3		The Netherlands	5152
C. rostrata	0.5–1.9	0.5–1.0	New York	335
Carex vesicaria	1.4		Czech Republic	1207
Cladium jamaicense	0.25		Florida	4747
Eleocharis quadrangulata	1.26		South Carolina	512
Glyceria grandis	5.2–6.8		Canada, Ontario	3347
Glyceria maxima	1.5–7.1		Czech Republic	1198
G. maxima	0.9–8.5		Czech Republic	1197
G. maxima (W)	22.9		Poland	3596
Hydrocotyle umbellata	2.3–7.5		Florida	4031
Juncus effusus	3.02		South Carolina	508
Justicia americana	0.15	0.16	North Carolina	5060
J. americana	0.57–2.78	2.03	Alabama	500
Papyrus sp.	1.43		Uganda	1577
Phragmites australis	0.013–1.96		England	3096
P. australis	2.9		Czech Republic	2653
P. australis	3.2		Wisconsin	3876
P. australis	3.5	2.0	The Netherlands	1038
P. australis	4.37	5.33	Czech Republic	1201
P. australis	1.6–5.3	3.8–7.4	Czech Republic	1198
Schoenoplectus lacustris	2.3–7.2		Czech Republic	1198
S. lacustris	3.8–11.1		Czech Republic	1197
Scirpus americanus	0.12–0.23		South Carolina	504
Scirpus fluviatilis	1.38		Iowa	992
S. fluviatilis	3.33	2.00	Wisconsin	2537
S. lacustris (W)	5.0	5.0	The Netherlands	1038
Sparganium erectum	3.42		Czech Republic	1198
Sparganium ramosum	0.73–2.89	0.85–2.41	India	1892
Spartina alterniflora	0.52	0.74	Georgia	1549

Table 7-10. Continued

Plant species	Standing stock AGB	BLG	Locality	Reference
Typha angustata	0.87–2.30	0.72–1.89	India	2444
Typha angustifolia	0.07–3.21		England	3096
T. angustifolia	1.6–4.6		Czech Republic	1198
T. angustifolia	4.5–6.3		Czech Republic	1197
Typha domingensis	0.16–0.76		Florida	1002
Typha elephantina	3.67		India	1745
Typha glauca	3.17		Iowa	992
T. glauca	3.74	1.39	Iowa	995
Typha latifolia	0.16–0.77		South Carolina	504
T. latifolia	1.60		Czech Republic	2653
T. latifolia	2.49		Czech Republic	1198
T. latifolia	3.20		Wisconsin	3876
T. latifolia	7.7		Czech Republic	1197
Typha sp.	2.46		Malawi	2220
Floating-leaved species				
Nelumbo nucifera	0.076–0.13		India	2444
Nuphar luteum	0.16	0.45	North Carolina	5060
Nymphaea alba	0.282		India	2444
Nymphoides peltata	0.04		India	2444
Submerged species				
Batrachium aquatile	0.45		Czech Republic	1207
Ceratophyllum demersum	0.078–0.13		India	2444
Elodea canadensis	1.9–2.3		Czech Republic	1207
Hydrocharis morsus-ranae	0.19		India	2444
Myriophyllum spicatum	0.15		India	2444
Potamogeton crispus	0.28–0.29		India	2444
Potamogeton lucens	0.13–0.22		India	2444
Potamogeton obtusifolius	1.2		Czech Republic	1207
Potamogeton pectinatus	0.22–0.29		India	2444
Potamogeton pusillus	0.22–0.24		India	2444
P. pusillus	0.5		Czech Republic	1207
Potamogeton spp.	1.29		Minnesota	2082
Floating species (Total)				
Eichhornia crassipes	3.2		Alabama	515
E. crassipes	6.18		Japan	3534
E. crassipes	6–18		Florida	4031
Lemna minor	0.1–1.6		Florida	4031
L. minor	3.3		Illinois	3248
Lemnaceae	34.5		Louisiana	927
Pistia stratiotes	2–5.7		Florida	4031
Salvinia rotundifolia	0.4–2.4		Florida	4031

Table 7-11. Phosphorus uptake rate by aquatic and wetland plants

Plant species	Rate	Locality	Reference
Emergent species	**(g m^{-2} yr^{-1})**		
Althernanthera philoxeroides	20	USA	505
Justicia americana	14	USA	505
Phragmites australis	2.5	South Africa	5455
P. australis (W)	7.6	South Africa	5505
P. australis	9.9	Scotland	2112
Typha domingensis	0.77–3.65	Florida	1002
Typha latifolia	40	USA	505
Floating species			
Eichhornia crassipes (W)	10.5	Czech Republic	5538
E. crassipes	14	Alabama	509
E. crassipes	32	USA	505
E. crassipes (E)	70	Florida	4173
E. crassipes (F)	126	Florida	4026
Lemna sp.	80	USA	3426
	(mg m^{-2} d^{-1})		
Scirpus americanus	1.5–5.3	South Carolina	504
Typha latifolia	2.8–19.3	South Carolina	504
Eichhornia crassipes (W)	20	Czech Republic	2662
E. crassipes (W)	110	Japan	3533
E. crassipes (W)	48–125	Czech Republic	5538
E. crassipes (E)	164	Florida	4038
E. crassipes	330–640	Alabama	509

Table 7-12. Sulfur concentration in aquatic and wetland plants (% dry mass)

Plant species	Concentration ABG	Locality	Reference
Emergent species			
Althernanthera philoxeroides	0.24–0.37	Alabama	500
Eleocharis acicularis	0.28	South Carolina	507
Eleocharis quadrangulata	0.15	South Carolina	507
Hydrocotyle sp.	0.16	South Carolina	507
Juncus effusus	0.16–0.27	South Carolina	508
J. effusus	0.26	South Carolina	507
Justicia americana	0.16–0.25	Alabama	500
Panicum hemitomon	0.23	South Carolina	507
Pontenderia cordata	0.22	South Carolina	507
Sagittaria latifolia	0.15	South Carolina	507
Scirpus americanus	0.55–0.68	South Carolina	504
S. americanus	0.59	South Carolina	507
Typha latifolia	0.08–0.19	South Carolina	504
T. latifolia	0.15	South Carolina	507
Floating-leaved species			
Brasenia schreberi	0.11	South Carolina	507
Nelumbo lutea	0.16	South Carolina	507
Nuphar lutea	0.32	South Carolina	507
Nymphaea odorata	0.14	South Carolina	507
Submerged species			
Ceratophyllum demersum	0.30	South Carolina	507
Isoetes macrospora	0.18–0.34	Wisconsin	467
Littorella uniflora	0.16–0.37	Wisconsin	467
L. uniflora	0.65	Denmark	1380
Lobelia dotrmanna	0.92	Denmark	1380
Myriophyllum heterophyllum	0.24	South Carolina	507
Najas guadalupensis	0.28	South Carolina	507
Potamogeton diversifolius	0.50	South Carolina	507
Utricularia inflata	0.26	South Carolina	507

Table 7-13. Sulfur standing stock in wetland plants (g m^{-2})

Plant species	Standing stock AGB	Locality	Reference
Althernanthera philoxeroidse	1.41–2.44	Alabama	500
Justicia americana	0.80–4.04	Alabama	500
Scirpus americanus	0.27–0.99	South Carolina	504
Typha latifolia	0.09–0.65	South Carolina	504

Table 7-14. Sulfur uptake rate by aquatic and wetland plants

Plant species	Rate	Locality	Reference
	(g m^{-2} yr^{-1})		
Althernanthera philoxeroides	18	U.S.A	505
Justicia americana	20	U.S.A	505
Typha latifolia	25	U.S.A	505
Eichhornia crassipes	25	U.S.A	505
Lemna sp.	65	U.S.A	3426
	(mg m^{-2} d^{-1})		
Typha latifolia	7.7–20	South Carolina	504
Scirpus americanus	13.1–23.3	South Carolina	504

Table 7-15. Potassium concentration in aquatic and wetland plants (% dry mass)

Plant species	Concentration ABG	BLG	Locality	Reference
Emergent species				
Acorus calamus	0.75		Poland	340
A. calamus	1.64		Czech Republic	5143
A. calamus (W)	2.00		Germany	4399
A. calamus	1.85–3.85		Czech Republic	1197–8
Alternanthera philoxeroides	3.03–5.86		Alabama	500
Bolboschoenus maritimus	1.82	1.72	Czech Republic	1201
B. maritimus	1.40–2.69		Czech Republic	1197
Carex acutiformis	0.73		Hungary	2574
Carex aquatilis	0.85		Canada, Québec	179
C. aquatilis	1.2		Alaska	748
C. aquatilis	1.4		Canada, Québec	178
Carex gracilis	1.3		Hungary	2574
C. gracilis	1.1–1.9		Czech Republic	1207
C. gracilis	1.2–1.9		Czech Republic	2659
Carex lacustris	0.7		Minnesota	2870
C. lacustris	0.8		Hungary	2574
C. lacustris	1.4		New York	332
C. lacustris	1.5		Canada, Québec	178
C. lacustris	1.6		Iowa	5128
Carex lanuginosa	1.0		Canada, Québec	178
Carex rostrata	1.3–2.5		New York	335
C. rostrata	1.3–1.4		Scotland	2109
C. rostrata	1.8		The Netherlands	5152
Carex vesicaria	1.6		Czech Republic	1207
Carex stricta	0.5		Minnesota	2870
C. stricta (W)	1.39		Germany	4399
Cladium jamaicense	0.58		Florida	4747
Eleocharis quadrangulata	0.61		South Carolina	512
Glyceria maxima	0.57		Czech Republic	5143
G. maxima	1.10		Czech Republic	2313
G. maxima	1.72–1.80		England	651
G. maxima (W)	2.60		Germany	4399
G. maxima	1.70–3.20		Czech Republic	1197–8
G. maxima (W)	1.94–3.11		Poland	3596
G. maxima	L 2.60 S 3.64	R 1.26	Scotland	2109
Iris pseudacorus (W)	3.54		Germany	4399
I. pseudacorus	S 2.88–3.45	R 0.79–1.22 RH 0.69–0.72	Scotland	2109
Juncus effusus	S 1.05–1.45	R 0.39–0.49	Scotland	2109
Justicia americana	2.35–4.36		Alabama	500
Mentha aquatica (W)	1.05		Germany	4399

Table 7-15. Continued

Plant species	Concentration		Locality	Reference
	ABG	BLG		
Papyrus sp.	2.9		Uganda	1577
Phragmites australis	0.07–0.85		England	73
P. australis	0.42		Poland	340
P. australis (W)	0.81		Germany	4399
P. australis	0.94		Czech Republic	5143
P. australis	1.07–1.71		England	651
P. australis	1.08–1.74		Czech Republic	1197
P. australis (W)	1.13		South Africa	5505
P. australis	1.21		South Africa	5455
P. australis	2.10		Scotland	2112
P. australis	0.07–0.18	0.94–1.19	England	410
P. australis	0.88	0.64–0.82	Czech Republic	1201
P. australis	0.55–2.76	1.11–1.45	Czech Republic	1198
Polygonum amphibium	S 2.36 L 1.61		Scotland	2109
Schoenoplectus lacustris	0.48		Poland	340
S. lacustris	1.15		Czech Republic	5143
S. lacustris	1.36–1.69		Czech Republic	1197
S. lacustris	0.69–2.90		Czech Republic	1198
Scirpus americanus	1.69–3.50		South Carolina	504
Scirpus lacustris (W)	1.03		Germany	4399
Scirpus subterminalis	0.90		Michigan	3206
Sparganium erectum (W)	2.48		Germany	4399
S. erectum	3.60–4.10		Czech Republic	1197
Sparganium ramosum	1.73–3.63	1.35–2.45	India	1892
Typha angustifolia (W)	1.41		Germany	4399
T. angustifolia	1.16–3.20		Czech Republic	1197–8
Typha glauca	0.28–3.27	0.52–1.37	Iowa	995
Typha latifolia	0.39		Poland	340
T. latifolia	1.47		Czech Republic	1197
T. latifolia	1.06–1.94		Czech Republic	5143
T. latifolia	1.60–3.46		South Carolina	504
T. latifolia	0.50–4.50		U.S.A.	506
T. latifolia	3.10	1.20	Wisconsin	4582
Typha orientalis (GR)	2.94–5.78		Australia	714
Floating-leaved species				
Nuphar lutea	P 4.06	R 3.35 RH 2.94	Scotland	2109
Submerged species				
Isoetes macrospora	1.3–7.4		Wisconsin	467
Littorella uniflora	4.0–6.0		Wisconsin	467
Myriophyllum heterophyllum	1.25		South Carolina	507
M. heterophyllum	1.81		New Jersey	4133
M. heterophyllum	3.03		Michigan	3206

Table 7-15. Continued

Plant species	Concentration ABG	Concentration BLG	Locality	Reference
M. heterophyllum	1.1–4.25		New Hampshire	2503
Myriophyllum spicatum	0.28		Michigan	3205
M. spicatum	1.68		New Jersey	4133
M. spicatum	0.97–2.04		Wisconsin	702
Potamogeton pectinatus	3.83–4.27		Scotland	2107
Ranunculus pencillatus	L 3.46 S 5.39	R 4.11	England	717
Utricularia sp. (+ periphyton)	0.48–1.41		Georgia	466
Floating species (Total)				
Lemna paucicostata (W)	3.75–4		Nigeria	3114
Spirodela polyrhiza	2.33		Czech Republic	4067
Spirodela polyrhiza + Lemna gibba	2.57–3.40		Czech Republic	4063,4067
Bryophytes				
Jungermannia vulcanicola	3.98		Japan	4292
Rhynchostegium riparioides	2.34–10.3		England	5311
Sphagnum angustifolium	0.50		Finland	184
Sphagnum balticum	0.34		Finland	184
S. balticum	0.39–0.74		Finland	3625
Sphagnum centrale	0.65		Finland	184
Sphagnum cuspidatum	0.73–0.87		Finland	185
Sphagnum fuscum	0.40		Finland	184
S. fuscum	0.41–0.47		Finland	185
S. fuscum	0.27–0.56		Finland	3625
Sphagnum girgensohnii	0.60		Finland	184
Sphagnum magellanicum	0.44		Finland	184
S. magellanicum	0.43–0.52		Finland	185
S. magellanicum	0.53		Czech Republic	1207
Sphagnum majus	0.33–0.46		Finland	3625
S. majus	0.87		Finland	184
Sphagnum nemoreum	0.39		Finland	184
Sphagnum papillosum	0.40		Finland	184
Sphagnum recurvum	0.55		Czech Republic	1207
Sphagnum rubellum	0.34		Finland	184
S. rubellum	0.33–0.48		Finland	185
Sphagnum russowii	0.36		Finland	184
Sphagnum squarrosum	0.49		Finland	184
Sphagnum tenellum	0.52		Finland	184
Sphagnum sp.	0.67		Czech Republic	1207
Sphagnum spp.	0.23		Finland	3627

Table 7-16. Potassium standing stock in aquatic and wetland plants (g m⁻²)

Plant species	Standing stock		Locality	Reference
	ABG	BLG		
Emergent species				
Acorus calamus	4.4–26.7		Czech Republic	1197–8
Althernanthera philoxeroides	22.4–43.7		Alabama	500
Bolboschoenus maritimus	6.72		Czech Republic	1198
B. maritimus	8.31	9.04	Czech Republic	1201
B. maritimus	10.2–61.8		Czech Republic	1197
Carex aquatilis	2.9		Alaska	748
Carex gracilis	0.4–11.1		Czech Republic	2659
C. gracilis	7.4		Czech Republic	1207
Carex lacustris	7.12	4.87	Wisconsin	2537
C. lacustris	3.4–12.7	4.0–9.2	New York	332
Carex rostrata	3.7–13.6	3.6–7.3	New York	335
C. rostrata	3.0		The Netherlands	5152
Carex vesicaria	10.4		Czech Republic	1207
Eleocharis quandrangulata	6.5		South Carolina	512
Cladium jamaicense	6.5		Florida	4747
Glyceria maxima	7.1–61.8		Czech Republic	1197–8
G. maxima (W)	90.2		Poland	3596
Justicia americana	3.85	2.54	North Carolina	5060
J. americana	44–75		Alabama	500
Papyrus sp.	44		Uganda	1577
Phragmites australis	0.10–11.04		England	3096
P. australis	7.8–37.6	37.3–67.1	Czech Republic	1197–8
P. australis	18.1	40.8	Czech Republic	1201
Schoenoplectus lacustris	15.2–57.3		Czech Republic	1197–8
Scirpus americanus	1.46–2.72		South Carolina	504
Sparganium erectum	38.8		Czech Republic	1198
S. erectum	43.0–67.9		Czech Republic	1197
Sparganium ramosum	5.30–22.3	6.27–10.4	India	1892
Typha angustifolia	0.92–19.9		England	3096
T. angustifolia	10.7–46.8		Czech Republic	1198
T. angustifolia	20.0–66.9		Czech Republic	1197
Typha glauca	30.1		Iowa	995
Typha latifolia	1.79–14.2		South Carolina	504
T. latifolia	16.8		Czech Republic	1198
T. latifolia	52		Czech Republic	1197
Typha sp.	4.5		Malawi	2220
Floating-leaved species				
Nelumbo nucifera	5.7–6.2		India	2444
Nuphar luteum	1.25	3.76	North Carolina	5060
Nymphaea alba	0.50		India	2444

Table 7-16. Continued

Plant species	Standing stock		Locality	Reference
	ABG	BLG		
Nymphoides peltata	0.74		India	2444
Submerged species				
Batrachium aquatile	1.8		Czech Republic	1207
Ceratophyllum demersum	2.8–4.5		India	2444
Elodea canadensis	5.0–9.9		Czech Republic	1207
Hydrocharis morsus ranae	2.7–3.8		India	2444
Myriophyllum spicatum	4.3		India	2444
Polamogeton crispus	3.8–4.3		India	2444
Potamogeton lucens	2.9–4.4		India	2444
Potamogeton obtusifolius	7.0		Czech Republic	1207
Potamogeton pectinatus	4.3–4.6		India	2444
Potamogeton pusillus	1.7		Czech Republic	1207
Floating species (Total)				
Eichhornia crassipes (F)	135		Florida	4037
E. crassipes (E)	165		Florida	4037
Lemnaceae	44.1		Louisiana	927

Table 7-17. Potassium uptake rate in aquatic and wetland plants

Plant species	Uptake rate	Locality	Reference
	$(g\ m^{-2}\ yr^{-1})$		
Althernanthera philoxeroides	322	U.S.A.	505
Eichhornia crassipes	319	U.S.A.	505
Justicia americana	372	U.S.A.	505
Typha latifolia	457	U.S.A.	505
Lemna sp.	252	U.S.A.	3426
	$(mg\ m^{-2}\ d^{-1})$		
Scirpus americanus	119.7–333.1	South Carolina	504
Typha latifolia	59.3	South Carolina	504

Table 7-18. Magnesium concentration in aquatic and wetland plants (mg kg^{-1} dry mass)

| Plant species | Concentration | | Locality | Reference |
	ABG	BLG		
Emergent plants				
Acorus calamus (W)	840		Germany	4399
A. calamus	1,400–2,500		Czech Republic	1197–8
A. calamus	3,000		Czech Republic	5143
Althernanthera philoxeroides	4,000–61,000		Alabama	500
Bolboschoenus maritimus	1,500	600–1,100	Czech Republic	1201
B. maritimus	1,000–1,700		Czech Republic	1197
Carex acutiformis	6,500		Hungary	2574
Carex aquatilis	1,400		Canada, Québec	178
C. aquatilis	2,000		Alaska	748
Carex gracilis	800–1,000		Czech Republic	1207
C. gracilis	1,300–2,100		Czech Republic	2659
C. gracilis	3,300		Hungary	2574
Carex lacustris	900		Iowa	5128
C. lacustris	1,200		New York	332
C. lacustris	1,300		Canada, Québec	178
C. lacustris	1,400		Minnesota	2870
C. lacustris	3,300		Hungary	2574
Carex lanuginosa	1,000		Canada, Québec	178
Carex rostrata	1,000		New York	335
Carex stricta	1,400		Minnesota	2870
C. stricta	2,060		Germany	4399
Carex vesicaria	800		Czech Republic	1207
Eleocharis quadrangulata	700		South Carolina	512
Glyceria maxima	900		Czech Republic	2313
G. maxima	900–2,100		Czech Republic	5143
G. maxima	1,000–1,500		Czech Republic	1197–8
G. maxima (W)	1,460		Germany	4399
G. maxima (W)	1,200–3,000		Poland	3596
Iris pseudacorus (W)	2,580		Germany	4399
Juncus effusus	S 1,040–1,290	R 700–950	Scotland	2109
Justicia americana	3,300–6,700		Alabama	500
Mentha aquatica (W)	1,580		Germany	4399
Phragmites australis	430		South Africa	5455
P. australis	600–1,700	600–700	Czech Republic	1197–8
P. australis	800	600	Czech Republic	1201
P. australis (W)	820		Germany	4399
P. australis (W)	1,020		South Africa	5505
P. australis	1,900		Czech Republic	5143
P. australis	2,290		Scotland	2112
P. australis	1,500–4,100		England	73

Table 7-18. Continued

Plant species	Concentration		Locality	Reference
	ABG	**BLG**		
P. australis	5,100–5,700	6,000–7,100	India	2444
Schoenoplectus lacustris	600–1,500		Czech Republic	1197–8
Scirpus americanus	2,100–3,300		South Carolina	500
Scirpus lacustris (W)	980		Germany	4399
Sparganium erectum (W)	2,100		Germany	4399
S. erectum	1,900–2,900		Czech Republic	1197
S. erectum	4,400		Czech Republic	5143
Sparganium ramosum	2,800–5,000	2,400–3,300	India	1892
Typha angustata	6,000	5,100–6,200	India	2444
Typha angustifolia (W)	1,490		Germany	4399
T. angustifolia	1,300–1,800		Czech Republic	1197–8
T. angustifolia	1,100–2,400		Czech Republic	5143
Typha elephantina	L 8,000	RH 9,000	India	1745
Typha latifolia	1,000–2,100		South Carolina	504
T. latifolia	1,800		Czech Republic	1197
T. latifolia	1,300–4,100		Czech Republic	5143
T. latifolia	3,000	3,400	Wisconsin	4582
Typha orientalis (GR)	2,100–3,100		Australia	714
Floating-leaved species				
Nelumbo nucifera	6,400–6,600		India	2444
N. nucifera	26,200–58,900		India	5087
Nuphar lutea	P 2,010	R 1,440 RH 720	Scotland	2109
N. lutea	2,300		Czech Republic	1207
Nymphaea alba	5,800		India	2444
Nymphaea nouchalli	6,500–13,100		India	5087
Nymphoides peltata	5,000–6,000		India	2444
Submerged species				
Ceratophyllum demersum	1,900–4,900		The Netherlands	348
C. demersum	6,600–6,900		India	2444
Elodea canadensis	>0–3,500		The Netherlands	348
Isoetes macrospora	1,800–3,800		Wisconsin	467
Littorella uniflora	2,300–4,100		Wisconsin	467
Myriophyllum heterophyllum	900–4,300		New Hampshire	2503
Myriophyllum spicatum	4,430–6,890		Wisconsin	702
M. spicatum	6,000–7,000		India	2444
Potamogeton crispus	41,400		India	5087
Potamogeton lucens	6,600–7,300		India	2444
Potamogeton natans	6,900–7,200		India	2444
Potamogeton pectinatus	3,070–5,150		Scotland	2107
P. pectinatus	22,000–24,000		India	2444
Ranunculus penicillatus	L 2,500 S 1,200	R 1,900	England	717

Table 7-18. Continued

Plant species	Concentration ABG	BLG	Locality	Reference
Utricularia sp. (+ periphyton)	1,500–3,600		Georgia	466
Floating species (Total)				
Eichhornia crassipes	10,900–67,000		India	5087
Lemna paucicostata (W)	5,000–6,000		Nigeria	3114
Pistia stratiotes	18,000–87,000		India	5087
Spirodela polyrhiza	10,900–28,000		India	5087
Spirodela polyrhiza + Lemna gibba	4,900–6,700		Czech Republic	4063
Bryophytes				
Rhynchostegium riparioides	756–3,730		England	5311
Sphagnum angustifolium	960		Finland	184
Sphagnum balticum	510		Finland	184
S. balticum	520–730		Finland	3625
Sphagnum centrale	900		Finland	184
Sphagnum cuspidatum	1,028–3,286		Finland	185
Sphagnum fuscum	650		Finland	184
S. fuscum	680–780		Finland	3625
S. fuscum	639–1,115		Finland	185
Sphagnum girgensohnii	910		Finland	184
Sphagnum magellanicum	690		Finland	184
S. magellanicum	696–897		Finland	185
S. magellanicum	1,100		Czech Republic	1207
Sphagnum majus	730–790		Finland	3625
S. majus	1,000		Finland	184
Sphagnum nemoreum	930		Finland	184
Sphagnum papillosum	740		Finland	184
Sphagnum recurvum	500		Czech Republic	1207
Sphagnum rubellum	550		Finland	184
S. rubellum	470–792		Finland	185
Sphagnum russowii	670		Finland	184
Sphagnum squarrosum	880		Finland	184
Sphagnum tenellum	670		Finland	184
Sphagnum sp.	600		Czech Republic	1207
Sphagnum spp.	950		Finland	3627

Table 7-19. Magnesium standing stock in aquatic and wetland plants (g m^{-2})

Plant species	Standing stock		Locality	Reference
	ABG	BLG		
Emergent species				
Acorus calamus	0.2–1.8		Czech Republic	1197–8
Althernanthera philoxeroides	2.3–4.4		Alabama	500
Bolboschoenus maritimus	0.60–1.4	0.69	Czech Republic	1197,1201
Carex aquatilis	0.49		Alaska	748
Carex gracilis	0.7		Czech Republic	1207
C. gracilis	0.2–1.3		Czech Republic	2659
Carex lacustris	0.09–0.34		Wisconsin	2537
C. lacustris	0.4–1.34	0.4–0.6	New York	332
Carex rostrata	0.1–1.6	0.3–0.7	New York	335
Carex vesicaria	0.4		Czech Republic	1207
Eleocharis quadrangulata	0.61		South Carolina	512
Glyceria maxima	0.5–3.5		Czech Republic	1197–8
G. maxima (W)	4.40		Poland	3596
Justicia americana	0.65	0.58	North Carolina	5060
J. americana	4.2–15.0		Alabama	500
Phragmites australis	0.003–1.11		England	3096
P. australis	1.78	3.24	Czech Republic	1201
P. australis	0.7–2.8		Czech Republic	1197–8
P. australis	5.0–11.1	7.2–10.2	India	2444
Schoenoplectus lacustris	0.5–6.3		Czech Republic	1197–8
Scirpus americanus	0.08–0.45		South Carolina	504
Sparganium erectum	2.0–4.5		Czech Republic	1197–8
S. erectum	4.0–5.9	2.06–2.24	India	2444
Sparganium ramosum	0.83–3.67	0.83–3.67	India	1892
Typha angustata	6.27–16.8	5.1–11.7	India	2444
Typha angustifolia	0.032–1.62		England	3096
T. angustifolia	3.5–8.4		Czech Republic	1197
Typha elephantina	0.035		India	1745
Typha latifolia	0.09–1.12		South Carolina	504
T. latifolia	2.04		Czech Republic	1198
T. latifolia	6.32		Czech Republic	1197
Typha sp.	8.59		Malawi	2220
Floating-leaved species				
Nelumbo nucifera	4.64–5.36		India	2444
Nuphar lutea	0.082	0.14	North Carolina	5060
N. lutea	0.60		Czech Republic	1207
Nymphaea alba	1.49		India	2444
Nymphoides peltata	0.48–0.58		India	2444
Submerged species				
Batrachium aquatile	0.3		Czech Republic	1207

Table 7-19. Continued

Plant species	Standing stock ABG	Standing stock BLG	Locality	Reference
Batrachium fluitans	1.0		Czech Republic	1207
Ceratophyllum demersum	1.23–1.67		India	2444
Elodea canadensis	0.4–0.9		Czech Republic	1207
Hydrocharis morsus ranae	1.1–1.56		India	2444
Myriophyllum spicatum	1.08		India	2444
Myriophyllum verticillatum	3.2–3.5		India	2444
Potamogeton crispus	1.50–1.59		India	2444
Potamogeton lucens	1.72–1.89		India	2444
Potamogeton natans	1.3–2.0		India	2444
Potamogeton obtusifolius	0.7		Czech Republic	1207
Potamogeton pectinatus	5.64–5.87		India	2444
Potamogeton pusillus	0.35		Czech Republic	1207
Floating species (Total)				
Eichhornia crassipes	0.97–4.12		India	2641

Table 7-20. Magnesium uptake rate in aquatic and wetland plants

Plant species	Uptake rate	Locality	Reference
	$(g\ m^{-2}\ yr^{-1})$		
Althernanthera philoxeroides	32	U.S.A.	505
Eichhornia crassipes	79	U.S.A.	505
Justicia americana	47	U.S.A.	505
Typha latifolia	31	U.S.A.	505
Lemna sp.	80	U.S.A.	3426
	$(mg\ m^{-2}\ d^{-1})$		
Scirpus americanus	2.3–18.0	South Carolina	504
Typha latifolia	18–33.8	South Carolina	504

Table 7-21. Calcium concentration in aquatic and wetland plants (% dry mass)

Plant species	Concentration ABG	BLG	Locality	Reference
Emergent Species				
Acorus calamus	0.34–0.85		Czech Republic	1197–8
A. calamus	0.57		Poland	340
A. calamus	0.75		Czech Republic	5143
A. calamus (W)	1.23		Germany	4399
A. calamus	0.45–1.4		Canada, Ontario	253
Alisma plantago-aquatica	0.72–2.30		England	74
Althernanthera philoxeroides	0.52–0.62		Alabama	500
Bolboschoenus maritimus	0.11	0.02–0.04	Czech Republic	1201
B. maritimus	0.08–0.63		Czech Republic	1197
Carex acutiformis	0.15–0.33		England	74
C. acutiformis	0.36		Hungary	2574
Carex aquatilis	0.04		Canada, Québec	178
C. aquatilis	0.08		Alaska	748
Carex gracilis	0.16–0.33		Czech Republic	2659
C. gracilis	0.19–0.25		Czech Republic	1207
C. gracilis	0.27		Hungary	2574
Carex lacustris	0.18		Hungary	2574
C. lacustris	0.22		Iowa	5128
C. lacustris	0.25		New York	332
C. lacustris	0.32		Minnesota	2870
C. lacustris	0.44		Canada	178
Carex lanuginosa	0.42		Canada	178
Carex rostrata	0.30		New York	335
Carex stricta	0.30		Minnesota	2870
C. stricta (W)	0.48		Germany	4399
Carex vesicaria	0.18		Czech Republic	1207
Eleocharis quadrangulata	0.20		South Carolina	512
Glyceria maxima	0.12–0.19		Czech Republic	1197–8
G. maxima	0.27		Czech Republic	2313
G. maxima (W)	0.48		Germany	4399
G. maxima	0.43–0.57		Czech Republic	5143
G. maxima (W)	0.71–1.19		Poland	3596
Iris pseudacorus (W)	1.70		Germany	4399
Juncus effusus	S 0.39–0.45	R 0.24–0.29	Scotland	2109
Justicia americana	0.72–2.48		Alabama	500
Mentha aquatica (W)	1.70		Germany	4399
Phragmites australis	0.056		South Africa	5455
P. australis (W)	0.063		South Africa	5505
P. australis	0.10		England	3242
P. australis	0.13		Poland	340

Table 7-21. Continued

Plant species	Concentration ABG	BLG	Locality	Reference
P. australis (W)	0.17		Germany	4399
P. australis	0.14–0.29	0.02–0.06	Czech Republic	1198
P. australis	0.03–0.29	0.01	Czech Republic	1197
P. australis	0.26	0.01–0.03	Czech Republic	1201
P. australis	0.39		Czech Republic	5143
P. australis	0.24–0.77		England	73
Polygonum amphibium	S 0.99 L 1.06		Scotland	2109
Sagittaria safittifolia	0.50–1.55		England	74
Schoenoplectus lacustris	0.07–0.25		Czech Republic	1197
S. lacustris	0.09–0.38		Czech Republic	1198
S. lacustris	0.11		England	3242
S. lacustris	0.16		Poland	340
S. lacustris	0.30		Czech Republic	5143
Scirpus americanus	0.45–0.64		South Carolina	504
Scirpus lacustris (W)	0.40		Germany	4399
Sparganium erectum	0.79		Czech Republic	5143
S. erectum (W)	1.04		Germany	4399
S. erectum	0.18–1.23		Czech Republic	1197
Sparganium ramosum	0.68–1.33	0.43–0.60	India	1892
S. ramosum	0.65–1.64		England	74
Typha angustifolia	0.18–0.73		Czech Republic	1197–8
T. angustifolia	0.87–0.92		Czech Republic	5143
T. angustifolia (W)	1.44		Germany	4399
Typha elephantina	L 1.53	RH 1.36	India	1745
Typha glauca	0.91–1.68	0.56–1.02	Iowa	995
Typha latifolia	0.21		Czech Republic	1197
T. latifolia	0.36		Poland	340
T. latifolia	0.53–0.93		South Carolina	504
T. latifolia	1.0	0.61	Wisconsin	4582
T. latifolia	0.25–1.71		U.S.A.	506
T. latifolia	0.79–1.06		Czech Republic	5143
Typha orientalis (GR)	0.48–1.26		Australia	714
Floating-leaved species				
Nelumbo nucifera	1.62		India	2444
N. nucifera	3.20–3.80		India	5087
Nuphar lutea	P 1.30	R 0.94 RH 0.66	Scotland	2109
Nymphaea alba	1.06		India	2444
Nymphaea nouchalli	2.34–4.04		India	5087
Nymphoides peltata	1.20–1.40		India	2444
Submerged species				
Elodea canadensis	0.73–3.40		England	74,75

Table 7-21. Continued

Plant species	Concentration		Locality	Reference
	ABG	BLG		
Isoetes macrospora	0.44–0.86		Wisconsin	467
Littorella uniflora	1.40–2.20		Wisconsin	467
Myriophyllum heterophyllum	0.55–2.20		New Hampshire	2503
Myriophyllum spicatum	1.49–>10		Wisconsin	702
Potamogeton crispus	4.70–9.66		India	5087
Potamogeton lucens	2.70		India	2444
Potamogeton natans	L 0.43–1.38		England	74
P. natans	1.57		India	2444
Potamogeton obtusifolius	0.28–0.67		England	75
Potamogeton pectinatus	0.99–1.88		Scotland	2107
P. pectinatus	2.60		England	75
P. pectinatus	2.95		India	2444
Potamogeton perforliatus	0.70		England	75
Potamogeton pusillus	2.05		India	2444
Ranunculus penicillatus	L 1.62		England	717
Utricularia sp. (+ periphyton)	0.50–0.65		Georgia	466
Floating species (total)				
Eichhornia crassipes	4.76–7.48		India	5087
Lemna gibba	0.78–1.43		England	74
Lemna minor	1.27–3.55		England	74
Lemna paucicostata	1.03–2.45		Nigeria	3114
Lemna trisulca	0.95–2.02		England	74
Pistia stratiotes	3.12–5.80		India	5087
Spirodela polyrhiza	1.60		Czech Republic	4067
S. polyrhiza	7.56–9.80		India	5087
Spirodela polyrhiza + Lemna gibba	1.05–2.70		Czech Republic	4063
Bryophytes				
Rhynchostegium riparioides	3.8–57.1		England	5311
Scapania undulata	0.42–46.6		EN,FR,GER,IR[a]	5414
Sphagnum angustifolium	0.29		Finland	184
Sphagnum balticum	0.16		Finland	184
S. balticum	0.10–0.12		Finland	3625
Sphagnum centrale	0.24		Finland	184
Sphagnum cuspidatum	0.13–0.76		Finland	185
Sphagnum fuscum	0.19		Finland	184
S. fuscum	0.15–0.21		Finland	3625
S. fuscum	0.18–0.35		Finland	185
Sphagnum gigrensohnii	0.28		Finland	184
Sphagnum magellanicum	0.16		Finland	184
S. magellanicum	0.16–0.24		Finland	195
S. magellanicum	0.29		Czech Republic	1207

Table 7-21. Continued

Plant species	Concentration ABG	BLG	Locality	Reference
Sphagnum majus	0.13		Finland	184
S. majus	0.14–0.15		Finland	3625
Sphagnum nemoreum	0.25		Finland	184
Sphagnum papillosum	0.22		Finland	184
Sphagnum recurvum	0.19		Czech Republic	1207
Sphagnum rubellum	0.18		Finland	184
S. rubellum	0.15–0.18		Finland	185
Sphagnum russowii	0.26		Finland	184
Sphagnum squarrosum	0.21		Finland	184
Sphagnum tenellum	0.12		Finland	184
Sphagnum sp.	0.13		Czech Republic	1207
Sphagnum spp.	0.11		Finland	3627

[a] EN, FR, GER, IR = Spain, France, Germany, Ireland, respectively

Table 7-22. Calcium standing stock in aquatic and wetland plants (g m^{-2})

Plant species	Standing stock ABG	BLG	Locality	Reference
Emergent species				
Acorus calamus	0.30–10.5		Czech Republic	1197–8
Althernanthera philoxeroides	2.37–54.3		Alabama	500
Bolboschoenus maritimus	0.54	0.24	Czech Republic	1201
B. maritimus	3.02		Czech Republic	1198
B. maritimus	0.4–5.5		Czech Republic	1197
Carex aquatilis	2.1		Alaska	748
Carex gracilis	0.5–2.2		Czech Republic	2659
C. gracilis	2.0		Czech Republic	1207
Carex lacustris	4.05	1.15	Wisconsin	2537
C. lacustris	1.1–4.15	0.4–0.9	New York	332
Carex rostrata	0.6–2.8	0.6–0.9	New York	335
Carex vesicarai	1.4		Czech Republic	1207
Eleocharis quadrangulata	1.76		South Carolina	512
Glyceria maxima	0.80–3.5		Czech Republic	1198
G. maxima	0.7–5.3		Czech Republic	1197
Justicia americana	1.25	0.55	North Carolina	5060
J. americana	6.40–19.7		Alabama	500
Phragmites australis	0.00004–2.26		England	3096
P. australis	5.35	0.83	Czech Republic	1201
P. australis	0.2–6.3		Czech Republic	1197
P. australis	1.5–7.4	0.4–1.2	Czech Republic	1198
P. australis	15.5–30	15–25.5	India	2444
Schoenoplectus lacustris	0.5–10.3		Czech Republic	1197–8
Scirpus americanus	0.18–0.97		South Carolina	504
Sparganium erectum	11.0		Czech Republic	1198
S. erectum	7.8–12.0	5–5.5	India	2444
S. erectum	1.9–17.5		Czech Republic	1197
Sparganium ramosum	2.03–10.9	1.61–3.22	India	1892
Typha angustata	10.48–29.05	10.36–24.85	India	2444
Typha angustifolia	0.1–7.21		England	3096
T. angustifolia	0.6–20.6		Czech Republic	1198
T. angustifolia	3.1–29.5		Czech Republic	1197
Typha elephantina	0.045		India	1745
Typha glauca	28.6		Iowa	995
Typha latifolia	2.41		Czech Republic	1198
T. latifolia	7.45		Czech Republic	1197
T. latifolia	0.4–5.46		South Carolina	504
Typha sp.	17.6		Malawi	2220
Floating-leaved species				
Nelumbo nucifera	12.2–12.8		India	2444

Table 7-22. Continued

Plant species	Standing stock		Locality	Reference
	ABG	BLG		
Nuphar luteum	0.30	0.65	North Carolina	5060
Nymphaea alba	2.76		India	2444
Nymphoides peltata	1.18		India	2444
Submerged species				
Batrachium aquatile	6.2		Czech Republic	1207
Ceratophyllum demersum	3.6–5.0		India	2444
Elodea canadensis	1.9–11.3		Czech Republic	1207
Hydrocharis morsus ranae	3.55		India	2444
Myriophyllum spicatum	4.08		India	2444
Myriophyllum verticillatum	6.7–7.0		India	2444
Potamogeton lucens	3.6–6.3		India	2444
Potamogeton natans	2.8–4.6		India	2444
Potamogeton obtusifolius	4.8		Czech Republic	1207
Potamogeton pectinatus	5.7–7.0		India	2444
Potamogeton pusillus	1.6		Czech Republic	1207
P. pusillus	4.2–5.1		India	2444
Floating species (total)				
Eichhornia crassipes	10.5–56.1		India	2641

Table 7-23. Calcium uptake rate in aquatic and wetland plants

Plant species	Uptake rate	Locality	Reference
	$(g\ m^{-2}\ yr^{-1})$		
Althernanthera philoxeroides	32	U.S.A.	505
Eichhornia crassipes	75	U.S.A.	505
Justica americana	102	U.S.A.	505
Typha latifolia	171	U.S.A.	505
Lemna sp.	600	U.S.A.	3426
	$(mg\ m^{-2}\ d^{-1})$		
Scirpus americanus	9.5–37.3	South Carolina	504
Typha latifolia	79.5–109.3	South Carolina	504

Table 7-24. Iron concentration in aquatic and wetland plants (mg kg^{-1} dry mass)

Species	Concentration		Locality	Reference
	ABG	BLG		
Emergent species				
Acorus calamus	280		Czech Republic	356
A. calamus (W)	970		Germany	4399
Carex gracilis	443		Czech Republic	356
Carex husdoni	193		Czech Republic	356
Carex pendula	61		Germany	2205
Carex rostrata	42–84	2,346–6,900	New York	337
Carex stricta (W)	3,800		Germany	4399
Carex vesicaria	520		Czech Republic	356
Eleocharis acicularis	3,600–59,000		Canada, Ontario	3228
Eleocharis quadrangulata	918		South Carolina	512
Glyceria maxima (W)	60.4–258		Poland	3596
G. maxima (W)	1,300		Germany	4399
Iris pseudacorus (W)	1,300		Germany	4399
Juncus effusus	S 100–140	R 1,310–1,492	Scotland	2109
Justicia americana	302–1,644		Alabama	500
Mentha aquatica (W)	3,100		Germany	4399
Phragmites australis	L 64–83 S 19	RH 42–57	Finland	2766
P. australis	135		Czech Republic	356
P. australis (W)	920		Germany	4399
Polygonum amphibium	L 180 S 180		Scotland	2109
Scirpus lacustris	129		Czech Republic	356
S. lacustris (W)	780		Germany	4399
Sparganium erectum	388		Czech Republic	356
S. erectum (W)	3,600		Germany	4399
Typha angustifolia (W)	1,100		Germany	4399
Typha latifolia	138	1,055	Wisconsin	4582
T. latifolia	139		Czech Republic	356
Floating-leaved species				
Nuphar lutea	L 100–650 P 210–580	RH 79–30 R 3,500–6,200	Finland	2766
N. lutea	P 240	R 5,400 RH 160	Scotland	2109
Submerged species				
Ceratophyllum demersum	1,300		The Netherlands	348
Ceratophyllum sp.	82,000–110,000		South Carolina	896
Elodea canadensis	430		Germany	2205
E. canadensis	800		The Netherlands	348
E. canadensis	4,100		England	3111
E. canadensis	9,500–29,000		New Jersey	4133
Isoetes macrospora	80–540		Wisconsin	467
Littorella uniflora	90–540		Wisconsin	467

Table 7-24. Continued

Species	Concentration ABG	BLG	Locality	Reference
Potamogeton pectinatus	820–1,360		Scotland	2107
Myriophyllum heterophyllum	290–12,040		New Hampshire	2503
Myriophyllum spicatum (TI)	308–936		Wisconsin	702
Floating species (Total)				
Eichhornia crassipes	1,760–8,440		India	47
Lemna paucicostata (W)	550–2,400		Nigeria	3114
Salvinia auriculata (GR)	50–750		Germany	2205
Bryophytes				
Fontinalis squamosum	12,070–12,340		Wales	645
F. squamosum	17,000–58,000		Wales	3098
Fontinalis sp.	1,460–2,830		Sweden	4612
Fontinalis spp.	900–48,000		Canada, Ontario	3228
Fontinalis hypnoides + Drepanocladus trichophyllus	11,000		Finland	2766
Philonotis fontana	2,325–11,230		Wales	645
Rhacomitrium aciculare	12,270–17,080		Wales	645
Rhynchostegium riparioides	1,810–143,000		England	5311
Scapania undulata	3,579–17,290		Wales	645
S. undulata	1,120–89,400		EN,IR,GER,FR[a]	5414
Solenostoma crenulata	2,325–14,180		Wales	645
Sphagnum acutifolium	984		Germany	2205
Sphagnum balticum + S. majus	>449		Finland	3626
Sphagnum cuspidatum	402–428		Finland	185
Sphagnum fuscum	231		Finland	3627
S. fuscum	269		Canada, Ontario	1656
S. fuscum	586		Finland	3626
S. fuscum	496–1,457		Finland	185
S. fuscum	382–2,478		Canada, Ontario	1657,1659
S. fuscum	679–2,784		Canada, Québec	1660
S. fuscum	4,182		New York	1533
Sphagnum magellanicum	388–633		Finland	185
S. magellanicum	826		Czech Republic	1207–8
Sphagnum recurvum	3,600		Czech Republic	1207–8
Sphagnum rubellum	163–333		Finland	185
Sphagnum sp.	24,000		Finland	2766
Sphagnum spp.	7,600–40,000		Canada, Ontario	3228

[a] EN, IR, GER, FR = England, Ireland, Germany, and France, respectively

Table 7-25. Iron standing stock in aquatic and wetland plants (g m^{-2})

Species	Standing stock		Locality	Reference
	ABG	BLG		
Emergent species				
Carex rostrata	13–102	670–2,973	New York	337
Eleocharis quadrangulata	0.89		South Carolina	512
Glyceria maxima (W)	1.8		Poland	3596
Justicia americana	0.27	0.64	North Carolina	5060
J. americana	0.12–3.80		Alabama	500
Floating-leaved species				
Nuphar lutea	0.13	0.78	North Carolina	5060

Table 7-26. Iron uptake rate in aquatic and wetland plants (mg m^{-2} yr^{-1})

Species	Uptake rate	Locality	Reference
Alternanthera philoxeroides	4,500	U.S.A.	505
Justicia americana	12,000	U.S.A.	505
Typha latifolia	2,300	U.S.A.	505
Eichhornia crassipes	1,900	U.S.A.	505
Lemna sp.	80,000	U.S.A.	3426
Sphagnum balticum + S. majus	127	Finland	3626
Sphagnum fuscum	113	Finland	3626

Table 7-27. Manganese concentration in aquatic and wetland plants (mg kg^{-1} dry mass)

Species	Concentration ABG	BLG	Locality	Reference
Emergent species				
Acorus calamus	220		Canada, Ontario	253
A. calamus (W)	383		Germany	4399
Alisma plantago-aquatica	75–1,200		England	74
Carex acutiformis	55–460		England	74
Carex gracilis	479		Czech Republic	356
Carex hudsoni	211		Czech Republic	356
Carex rostrata	440–743	280–365	New York	337
Carex stricta (W)	970		Germany	4399
Carex vesicaria	189		Czech Republic	356
Eleocharis acicularis	79–5,400		Ontario	3228
Eleocharis quadrangulata	>400		South Carolina	512
Glyceria maxima (W)	58–172		Poland	3596
G. maxima (W)	586		Germany	4399
Iris pseudacorus (W)	382		Germany	4399
Justicia americana	51–281		Alabama	500
Mentha aquatica (W)	381		Germany	4399
Phragmites australis	L 38–51 S 13–18	RH 13–16	Finland	2766
P. australis	151		Czech Republic	356
P. australis (W)	166		Germany	4399
Sagittaria sagittaria	75–550		England	74
Schoenoplectus lacustris	674		Czech Republic	356
S. lacustris (W)	1,200		Germany	4399
Sparganium erectum	542		Czech Republic	356
S. erectum (W)	604		Germany	4399
Typha angustifolia	642		Czech Republic	356
T. angustifolia (W)	779		Germany	4399
Typha latifolia	243	69.7	Wisconsin	4582
T. latifolia	723		Czech Republic	356
Floating-leaved species				
Nuphar lutea	L 100–110 P 130–220	RH 86–110 R 150–250	Finland	2766
Submerged species				
Ceratophyllum demersum	17,100		The Netherlands	348
Elodea canadensis	700–5,000		England	74
E. canadensis	17,100		The Netherlands	348
Isoetes macrospora	170–430		Wisconsin	467
Littorella uniflora	90–370		Wisconsin	467
Myriophyllum heterophyllum	130–6,420		New Hampshire	2503
Myriophyllum spicatum	345–478		Wisconsin	702
Potamogeton natans	170–3,600		England	74

Table 7-27. Continued

| Species | Concentration | | Locality | Reference |
	ABG	BLG		
Utricularia sp. (+ periphyton)	73–160		Georgia	466
Floating species (Total)				
Lemna gibba	1,400–2,000		England	74
Lemna minor	550–8,500		England	74
Lemna paucicostata (W)	400–663		Nigeria	3114
Lemna trisulca	2,400–13,000		England	74
Bryophytes				
Fontinalis hypnoides + *Drepanocladus trichophyllus*	2,200		Finland	2766
Fontinalis sp.	264–127		Sweden	4612
Fontinalis spp.	94–31,000		Canada, Ontario	3228
Jungermannia vulcanicola	31		Japan	4292
Rhynchostegium riparioides	280–143,000		England	5311
Scapania undulata	2–2,000		Poland	4264
S. undulata	50–16,170		EN,FR,IR,GER[a]	5414
Sphagnum angustifolium	64.7		Finland	184
Sphagnum balticum	52.2		Finland	184
Sphagnum balticum + S. majus	>33		Finland	3626
Sphagnum centrale	135		Finland	184
Sphagnum cuspidatum	60.7–101.8		Finland	185
Sphagnum fuscum	64		Finland	184
S. fuscum	176		Finland	3626
S. fuscum	61.9–192.2		Finland	185
Sphagnum girgensohnii	181		Finland	184
Sphagnum magellanicum	444		Czech Republic	1207–8
S. magellanicum	76.2		Finland	184
S. magellanicum	75.6–90.5		Finland	185
Sphagnum majus	60.7		Finland	184
Sphagnum nemoreum	233.3		Finland	184
Sphagnum papillosum	66.7		Finland	184
Sphagnum recurvum	48.5		Czech Republic	1207–8
Sphagnum rubellum	99.4–101.7		Finland	185
S. rubellum	123.3		Finland	184
Sphagnum rusowii	165.5		Finland	184
Sphagnum squarrosum	92.5		Finland	184
Sphagnum tenellum	27.3		Finland	184
Sphagnum sp.	21–38		Finland	2766
Sphagnum spp.	160–60,000		Canada, Ontario	3228

[a] EN, FR, IR, GER = England, France, Ireland, and Germany, respectively.

Table 7-28. Manganese standing stock in aquatic and wetland plants (g m^{-2})

Species	Standing stock		Locality	Reference
	ABG	BLG		
Emergent species				
Carex rostrata	93–589	16–111	New York	337
Eleocharis quadrangulata	>352		South Carolina	512
Glyceria maxima (W)	213		Poland	3596
Justicia americana	60–310		Alabama	500

Table 7-29. Manganese uptake rate in aquatic and wetland plants (mg m^{-2} yr^{-1})

Species	Uptake rate	Locality	Reference
Alternanthera philoxeroides	2,700	U.S.A.	505
Justicia americana	1,300	U.S.A.	505
Typha latifolia	7,900	U.S.A.	505
Eichhornia crassipes	30,000	U.S.A.	505
Lemna sp.	9,000	U.S.A.	3426
Sphagnum balticum + S. majus	>9.6	Finland	3626
Sphagnum fuscum	33.6	Finland	3626

Table 7-30. Zinc concentration in aquatic and wetland plants (mg kg^{-1} dry mass)

Species	Concentration		Locality	Reference
	ABG	BLG		
Emergent species				
Acorus calamus	34		Czech Republic	356
A. calamus (W)	38		Germany	4399
Carex gracilis	24		Czech Republic	356
Carex hudsoni	33		Czech Republic	356
Carex lacustris	22.5–28.9		Canada, Ontario	3346
Carex pendula	15		Germany	2205
Carex rostrata	14–27	20–24	New York	337
Carex stricta (W)	63		Germany	4399
Carex vesicaria	48		Czech Republic	356
Eleocharis acicularis	43–200		Canada, Ontario	3228
Eleocharis quadrangulata	20		South Carolina	512
Glyceria maxima (W)	73		Germany	4399
G. maxima (W)	72–119		Poland	3596
Iris pseudacorus (W)	50		Germany	4399
Justicia americana	114–287		Alabama	500
Lythrum salicaria	17–26		Canada, Ontario	3346
Mentha aquatica (W)	78		Germany	4399
Phragmites australis	L 14–25 S 35–130	RH 24–25	Finland	2766
P. australis (W)	97		Germany	4399
P. australis	S 217–337 R 158–245	RH 31–38	Denmark	4328
Pontenderia cordata	16–20		Canada	3346
Scirpus lacustris (W)	50		Germany	4399
Scirpus maritimus	15–80	R 30–440	The Netherlands	3587
Sparganium erectum	28		Czech Republic	356
S. erectum (W)	76		Germany	4399
Spartina alterniflora	11–29		Massachusetts	216
Spartina anglica	24–50	R 40–190	The Netherlands	3587
Spartina patens	12–31		Massachusetts	216
Typha angustifolia (W)	43		Germany	4399
Typha latifolia	6.5–17		Canada, Ontario	3346
T. latifolia	31		Czech Republic	356
T. latifolia (E)	1–190	35–360	France	390
Floating-leaved species				
Nuphar lutea	L 35–58 S 52–65	RH 30–59 R 21–26	Finland	2766
Nymphaea odorata	10.6–18.8		Canada	3346
Submerged species				
Ceratophyllum submersum	14.4		Hungary	2575
Elodea canadensis	120		Germany	2205

Table 7-30. Continued

Species	Concentration		Locality	Reference
	ABG	BLG		
Hydrocharis morsus ranae	80.9–321.2		Hungary	2575
Isoetes macrospora	70–240		Wisconsin	467
Littorella uniflora	40–190		Wisconsin	467
Myriophyllum heterophyllum	11.5–32.3		Canada, Ontario	3346
M. heterophyllum	32–332		Washington	944
M. heterophyllum	40–650		New Hampshire	2503
Myriophyllum spicatum	14.3–60.2		Wisconsin	702
M. spicatum	295–349		Hungary	2575
Potamogeton pectinatus	12–128		Hungary	2575
Potamogeton perfoliatus	21.9–208		Hungary	2575
Potamogeton richardsonii	40–182		Washington	944
Floating species (Total)				
Eichhornia crassipes (W)	196–3,224		Czech Republic	5539
Lemna trisulca	73.2		Hungary	2575
Salvinia auriculata	240–880		Germany	2205
Stratiotes aloides	24–79.9		Hungary	2575
Bryophytes				
Drepanocladus trichophyllus + *Fontinalis hypnoides*	100		Finland	2766
Eurhynchium riparoides	3,274		Wales	645
Fontinalis antipyretica	25.2–2,825		Europe	4306
Fontinalis squamosum	2,634–2,841		Wales	645
Fontinalis sp.	94–434		Sweden	4612
Fontinalis spp.	47–820		Canada, Ontario	3228
Jungermannia vulcanicola	13		Japan	4292
Philonotis fontana	977–7,023		Wales	645
Rhynchostegium riparioides	128–22,300		England	5311
Scapania undulata	4–550		Poland	4264
S. undulata	1,950		Wales	3168
S. undulata	777–3,558		Wales	645
Solenostoma crenulata	835–3,419		Wales	645
Sphagnum acutifolium	840		Germany	2205
Sphagnum angustifolium	55.2		Finland	184
Sphagnum balticum	54.2		Finland	184
Sphagnum balticum + *S. majus*	33.4		Finland	3626
Sphagnum centrale	50.0		Finland	184
Sphagnum cuspidatum	58.7–78.2		Finland	185
Sphagnum fuscum	37		Canada, Ontario	1656
S. fuscum	37.3		Finland	3626
S. fuscum	21–57		Canada, Ontario	1657,1659
S. fuscum	60.0		Finland	184
S. fuscum	57.6–70		Finland	185

Table 7-30. Continued

Species	Concentration ABG	BLG	Locality	Reference
S. fuscum	81		New York	1533
S. fuscum	25–148		Canada, Québec	1660
Sphagnum girgensohnii	48.2		Finland	184
Sphagnum magellanicum	54.5		Czech Republic	1207–8
S. magellanicum	54.2–55.9		Finland	185
S. magellanicum	56.8		Finland	184
Sphagnum majus	58.7		Finland	184
Sphagnum nemoreum	54.3		Finland	184
Sphagnum papillosum	61.6		Finland	184
Sphagnum recurvum	36.6		Czech Republic	1207–8
Sphagnum rubellum	42.1		Finland	184
S. rubellum	42.1–52.7		Finland	185
Sphagnum rusowii	39.4		Finland	184
Sphagnum squarrosum	36.5		Finland	184
Sphagnum tenellum	39.3		Finland	184
Sphagnum spp.	47		Finland	2766
Sphagnum spp.	38–190		Ontario	3228

Table 7-31. Zinc standing stock in aquatic and wetland plants (g m^{-2})

Species	Standing stock ABG	BLG	Locality	Reference
Emergent species				
Carex rostrata	3.8–19.8	5.2–10.0	New York	337
Eleocharis quadrangulata	18		South Carolina	512
Glyceria maxima	293		Poland	3596
Justicia americana	50–580		Alabama	500
Phragmites australis	124–259 (Total)		Denmark	2714

Table 7-32. Zinc uptake rate in aquatic and wetland plants (mg m^{-2} yr^{-1})

Species	Uptake rate	Locality	Reference
Althernanthera philoxeroides	600	U.S.A.	505
Justicia americana	3,000	U.S.A.	505
Typha latifolia	600	U.S.A.	505
Eichhornia crassipes	400	U.S.A.	505
Lemna sp.	700	U.S.A.	3426
Sphagnum balticum + S. majus	< 9.5	Finland	3626
Sphagnum fuscum	7.2	Finland	3626

Table 7-33. Copper concentration in aquatic and wetland plants (mg kg^{-1} dry mass)

Plant species	Concentration ABG	BLG	Locality	Reference
Emergent species				
Acorus calamus (W)	4.1		Germany	4399
A. calamus	4.5		Czech Republic	356
Bidens laevis	8.9–36.1		New Jersey	5378
Carex gracilis	1.5		Czech Republic	356
Carex hudsoni	3.0		Czech Republic	356
Carex lacustris	2.0–2.7		Canada	3346
Carex rostrata	3.1–7.3	12–17	New York	337
Carex stricta (W)	5.6		Germany	4399
Carex vesicaria	5.4		Czech Republic	356
Eleocharis acicularis	150–187 (Total)		Connecticut	2008
E. acicularis	7–1,900		Canada, Ontario	3228
Eleocharis quadrangulata	11		South Carolina	512
Glyceria maxima (W)	5.6		Germany	4399
Impatiens capensis	6.6–55.9		New Jersey	5378
Iris pseudacorus (W)	5.7		Germany	4399
Justicia americana	19–44		Alabama	500
Lythrum salicaria	2.0–4.1		Canada	3346
Mentha aquatica (W)	8.5		Germany	4399
Peltandra virginica	15.9–61.4		New Jersey	5378
Phragmites australis	L 2.0 S 1.4–1.5	RH 1.3 - 1.5	Finland	2766
P. australis	2.5		Czech Republic	356
P. australis (W)	4.2		Germany	4399
P. australis	S 3.3–5.3	RH 2.1–3.1 R 14.6–17.7	Denmark	4328
Pontenderia cordata	1.9–2.5		Canada, Ontario	3346
P. cordata	3–24	5–81	Connecticut	2008
Sagittaria subulata	59–153 (Total)		Connecticut	2008
Scirpus lacustris (W)	4.8		Germany	4399
Scirpus maritimus	3–12	R 4–40	The Netherlands	3587
Sparganium erectum	4.7		Czech Republic	356
S. erectum (W)	5.6		Germany	4399
Spartina anglica	2–6	R 3–37	The Netherlands	3587
Typha angustifolia (W)	4.7		Germany	4399
T. angustifolia	7.0		Czech Republic	356
Typha latifolia	< 1.0–4.7		Canada	3346
T. latifolia	8.5		Czech Republic	356
Zizania aquatica v. aquatica	2.5–61		New Jersey	5378
Floating-leaved species				
Nuphar lutea	L 2.3–9.8 P 2.9–7.6	RH 1.3–8.1 R 2.5–6.2	Finland	2766
Nymphaea odorata	1.2–3.0		Canada	3346

Table 7-33. Continued

Plant species	Concentration		Locality	Reference
	ABG	BLG		
Submerged species				
Ceratophyllum submersum	5.2–17.5		Hungary	2575
Hydrocharis morsus ranae	12.9–13.5		Hungary	2575
Myriophyllum heterophyllum	3.3–7.1		Canada, Ontario	3346
M. heterophyllum	7.0–22		Washington	944
M. heterophyllum	30		New Hampshire	2503
Myriophyllum spicatum	6.95–18.6		Wisconsin	702
M. spicatum	49–57		Hungary	2575
Potamogeton crispus (E)	8–57		California	1531
Potamogeton epihydrus	12–206		Connecticut	2008
Potamogeton pectinatus	3.0–20.8		Hungary	2575
P. pectinatus (E)	6–26		California	1531
Potamogeton perfoliatus	17.3–63.7		Hungary	2575
Potamogeton richardsonii	21–26		Washington	944
Floating species (Total)				
Eichhornia crassipes (W)	14–4,064		Czech Republic	5539
Lemna gibba	15–45		England	75
Lemna minor	0–110		England	75
Lemna polyrhiza	3–35		England	75
Lemna trisulca	48.8		Hungary	2575
Lemna valdiviana (E)	75–54,500		Canada	2262
Stratiotes alvides	14.5–41.2		Hungary	2575
Bryophytes				
Fontinalis antipyretica	4.8–119		Europe	4306
Fontinalis squamosum	34–37		Wales	645
Fontinalis sp.	14.4–16.3		Sweden	4612
Fontinalis spp.	10–1,900		Ontario, Canada	3228
Fontinalis hypnoides + Drepanocladus trichophyllus	14		Finland	2766
Jungermannia vulcanicola	27		Japan	4292
Philonotis fontana	10–53		Wales	645
Rhynchostegium riparioides	5.4–157		England	5311
Scapania undulata	7–61		Wales	645
S. undulata	63–202		Wales	3168
S. undulata	9–1,720		EN,IR,FR,GER[a]	5414
Solenostoma crenulata	113–196		Wales	645
Sphagnum angustifolium	8.7		Finland	184
Sphagnum balticum	5.6		Finland	184
Sphagnum balticum + S. majus	5.1		Finland	3626
Sphagnum centrale	9.5		Finland	184
Sphagnum cuspidatum	6.0–11.8		Finland	184
Sphagnum fuscum	6.1		Finland	3626

Table 7-33. Continued

Plant species	Concentration ABG	BLG	Locality	Reference
S. fuscum	6.3		Finland	3627
S. fuscum	8.2		Finland	184
S. fuscum	7.6–8.3		Finland	185
S. fuscum	14		Canada, Ontario	1656
S. fuscum	27		New York	1533
S. fuscum	4–124		Canada, Ontario	1657,1659
S. fuscum	10–138		Canada, Québec	1660
Sphagnum girgensohnii	8.2		Finland	184
Sphagnum magellanicum	5.1		Finland	184
S. magellanicum	7.2		Czech Republic	1207–8
S. magellanicum	4.8–11.4		Finland	185
Sphagnum majus	6.0		Finland	184
Sphagnum nemoreum	5.2		Finland	184
Sphagnum papillosum	9.8		Finland	184
Sphagnum recurvum	7.0		Czech Republic	1207–8
Sphagnum rubellum	7.7		Finland	184
S. rubellum	7.2–7.7		Finland	185
Sphagnum russowii	11.8		Finland	184
Sphagnum squarrosum	8.1		Finland	184
S. squarrosum	50–100		Wales	3168
Sphagnum tenellum	7.3		Finland	184
Sphagnum sp.	26		Finland	2766
Sphagnum spp.	7–140		Germany	4399

[a] EN, IR, FR, GER = England, Ireland, France, and Germany, respectively.

Table 7-34. Copper standing stock in aquatic and wetland plants (g m^{-2})

Plant species	Standing stock ABG	BLG	Locality	Reference
Emergent species				
Carex rostrata	1.1–6.9	2.9–5.8	New York	337
Eleocharis quadrangulata	13		South Carolina	512
Justicia americana	0–80,000		Alabama	500
Phragmites australis	10.2–16.6 (Total)		Denmark	2714

Table 7-35. Copper uptake rate in aquatic and wetland plants (mg m^{-2} yr^{-1})

Plant species	Uptake rate	Locality	Reference
Althernanthera philoxeroides	100	U.S.A.	505
Justicia americana	300	U.S.A.	505
Typha latifolia	700	U.S.A.	505
T. latifolia	ABG 2.0 BLG 5.3	Wisconsin	4582
Eichhornia crassipes	100	U.S.A.	505
Lemna sp.	100	U.S.A.	3426
Sphagnum fuscum	1.2	Finland	3626
Sphagnum balticum + S. majus	< 1.5	Finland	3626

Table 7-36. Molybdenum concentration in aquatic and wetland plants (mg kg^{-1} dry mass)

Plant species	Concentration ABG	Locality	Reference
Emergent species			
Acorus calamus (W)	0.30	Germany	4399
Carex stricta (W)	0.29	Germany	4399
Eleocharis quadrangulata	5.0	South Carolina	512
Glyceria aquatica (W)	0.24	Germany	4399
Iris pseudacorus (W)	0.33	Germany	4399
Mentha aquatica (W)	0.47	Germany	4399
Phragmites australis (W)	0.26	Germany	4399
P. australis	0.10–1.20	Poland	2632
Scirpus lacustris (W)	0.55	Germany	4399
Sparganium erectum (W)	0.24	Germany	4399
Typha angustifolia (W)	0.30	Germany	4399
T. angustifolia	0.40–2.30	Poland	2632
Submerged species			
Ceratophyllum demersum	5.2	Hungary	2575
Hydrocharis morsus ranae	5.4–5.6	Hungary	2575
Myriophyllum spicatum	2.2–6.9	Hungary	2575
Potamogeton pectinatus	1.6–3.7	Hungary	2575
Potamogeton perfoliatus	0.73–31.2	Hungary	2575
Floating species (Total)			
Lemna trisulca	2.2	Hungary	2575
Stratiotes aloides	2.3–6.1	Hungary	2575

Table 7-37. Molybdenum standing stock in aquatic and wetland plants (g m^{-2})

Plant species	Standing stock ABG	BLG	Locality	Reference
Eleocharis quadrangulata	0.004		South Carolina	512
Phragmites australis	0.10	1.80	Poland	2632
Typha angustifolia	0.40	0.64	Poland	2632

Table 7-38. Sodium concentration in aquatic and wetland plants (% dry mass)

Plant species	Concentration		Locality	Reference
	ABG	BLG		
Emergent species				
Acorus calamus	0.026		Czech Republic	5143
A. calamus	0.09–0.24		Czech Republic	1197
A. calamus (W)	0.41		Germany	4399
Bolboschoenus maritimus	0.16–0.66		Czech Republic	1197
Carex rostrata	0.032–0.039		Scotland	2109
Carex stricta (W)	0.22		Germany	4399
Eleocharis quadrangulata	0.27		South Carolina	512
Glyceria maxima	0.02–0.026		Czech Republic	5143
G. maxima	0.03		Czech Republic	2313
G. maxima (W)	0.12		Germany	4399
Iris pseudacorus (W)	0.17		Germany	4399
I. pseudacorus	S 0.28–0.38	RH 0.34–0.37 R 0.7–0.9	Scotland	2108
Juncus effusus	S 0.44–0.50	R 0.39–0.51	Scotland	2108
Mentha aquatica (W)	0.69		Germany	4399
Phragmites australis	0.026		Czech Republic	5143
P. australis	0.03–0.05		Czech Republic	1197
P. australis (W)	0.063		South Africa	5505
P. australis	0.074		South Africa	5455
P. australis	0.084		Scotland	2112
P. australis (W)	0.11		Germany	4399
P. australis	0.08–0.65		England	73
Polygonum amphibium	S 0.084 L 0.077		Scotland	2109
Schoenoplectus lacustris	0.34–0.40		Czech Republic	1197
Scirpus americanus	0.06–0.20		South Carolina	504
Scirpus lacustris	0.14		Czech Republic	5143
S. lacustris (W)	0.63		Germany	4399
Sparganium erectum	0.24		Czech Republic	5143
S. erectum	0.39–0.44		Czech Republic	1197
S. erectum (W)	0.68		Germany	4399
Sparganium ramosum	0.11–0.27	0.08–0.20	India	1892
Typha angustifolia	0.13–0.24		Czech Republic	5143
T. angustifolia	0.25		Czech Republic	1197
T. angustifolia (W)	1.22		Germany	4399
Typha glauca	0.31–0.62	0.33–1.18	Iowa	995
Typha latifolia	0.06–0.21		Czech Republic	5143
T. latifolia	0.24		South Carolina	504
T. latifolia	0.18–0.28		South Carolina	504
T. latifolia	0.30	0.66	Wisconsin	4582
T. latifolia	0.20–1.20		U.S.A.	506
Typha orientalis (GR)	0.17–0.59		Australia	714

Table 7-38. Continued

Plant species	Concentration		Locality	Reference
	ABG	BLG		
Floating-leaved species				
Nelumbo nucifera	0.009–0.012		India	5087
N. nucifera	0.11		India	2444
Nuphar lutea	P 1.56	R 1.36	Scotland	2109
Nymphaea alba	0.10		India	2444
Submerged species				
Ceratophyllum demersum	0.21		India	2444
Hydrocharis morsus ranae	0.11		India	2444
Isoetes macrospora	0.20–0.61		Wisconsin	467
Littorella uniflora	0.41–1.04		Wisconsin	467
Myriophyllum heterophyllum	1.30		New Jersey	4133
M. heterophyllum	1.87		New Jersey	504
M. heterophyllum	0.63–2.87		New Hampshire	2503
Myriophyllum spicatum	1.01		New Jersey	4133
Potamogeton crispus	0.18		India	2444
Potamogeton lucens	0.12		India	2444
Potamogeton natans	0.23		India	2444
Potamogeton pectinatus	1.20–1.44		Scotland	2107
Ranunculus penicillatus	L 1.71 S 0.82	R 1.24	England	717
Utricularia sp. (+ periphyton)	0.46–1.44		Georgia	466
Floating species (Total)				
Eichhornia crassipes	0.013		India	5087
Lemna paucicostata (W)	0.15–0.33		Nigeria	3114
Pistia stratiotes	0.017–0.026		India	5087
Spirodela polyrhiza	0.005–0.01		India	5087
Spirodela polyrhiza	0.49		Czech Republic	4067
Spirodela polyrhiza + Lemna gibba	0.31–0.44		Czech Republic	4063,4067
Bryophytes				
Rhynchostegium riparioides	0.23–0.79		England	5311
Sphagnum angustifolium	0.036		Finland	184
Sphagnum balticum	0.027		Finland	184
Sphagnum centrale	0.026		Finland	184
Sphagnum cuspidatum	0.06–0.1		Finland	185
Sphagnum fuscum	0.012		Finland	184
S. fuscum	0.013–0.04		Finland	185
Sphagnum girgensohnii	0.025		Finland	184
Sphagnum magellanicum	0.03–0.058		Finland	185
S. magellanicum	0.059		Finland	184
S. magellanicum	0.13		Czech Republic	1207
Sphagnum majus	0.061		Finland	184
Sphagnum nemoreum	0.031		Finland	184

Table 7-38. Continued

Plant species	Concentration ABG	BLG	Locality	Reference
Sphagnum papillosum	0.056		Finland	184
Sphagnum recurvum	0.13		Czech Republic	1207
Sphagnum rubellum	0.031		Finland	184
S. rubellum	0.03–0.04		Finland	185
Sphagnum russowii	0.032		Finland	184
Sphagnum squarrosum	0.028		Finland	184
Sphagnum tenellum	0.059		Finland	184

Table 7-39. Sodium standing stock in aquatic and wetland plants (g m^{-2})

Plant species	Standing stock		Locality	Reference
	ABG	BLG		
Emergent species				
Eleocharis quadrangulata	2.44		South Carolina	512
Phragmites australis	0.6–0.9		Czech Republic	1197
P. australis	0.001–2.49		England	3096
P. australis	2.92–5.47	2.87–4.34	India	2444
Scirpus americanus	0.02–0.31		South Carolina	504
Sparganium ramosum	0.31–1.51	0.31–1.51	India	1892
Sparganium erectum	2.61–4.01	1.32–1.89	India	2444
Typha angustata	2.85–4.92	1.61–3.77	India	2444
Typha angustifolia	0.10–9.10		England	3096
Typha glauca	11.9		Iowa	995
Typha latifolia	0.10–1.64		South Carolina	504
Typha sp.	45.7		Malawi	2220
Floating-leaved species				
Nelumbo nucifera	0.83–0.92		India	2444
Nymphoides peltata	0.16–0.17		India	2444
Submerged species				
Batrachium aquatile	0.2		Czech Republic	1207
Ceratophyllum demersum	0.43–0.60		India	2444
Elodea canadensis	1.0–2.2		Czech Republic	1207
Hydrocharis morsus ranae	0.26		India	2444
Myriophyllum spicatum	0.35		India	2444
Myriophyllum verticillatum	0.65–0.95	0.48–0.54	India	2444
Potamogeton lucens	0.17–0.23		India	2444
Potamogeton natans	0.36–0.59		India	2444
Potamogeton obtusifolius	0.8		Czech Republic	1207
Potamogeton pectinatus	0.43–0.49		India	2444
Potamogeton pusillus	0.26–0.36		India	2444
P. pusillus	0.6		Czech Republic	1207

Table 7-40. Sodium uptake rate in aquatic and wetland plants

Plant species	Uptake rate	Locality	Reference
	(g m^{-2} yr^{-1})		
Althernanthera philoxeroides	23	U.S.A.	505
Eichhornia crassipes	26	U.S.A.	505
Justicia americana	19	U.S.A.	505
Typha latifolia	73	U.S.A.	505
Lemna sp.	39	U.S.A.	3426
	(mg m^{-2} d^{-1})		
Scirpus americanus	4.6–16	South Carolina	504
Typha latifolia	39.3–127.6	South Carolina	504

Table 7-41. Cobalt concentration in aquatic and wetland plants (mg kg⁻¹)

Plant species	Concentration	Locality	Reference
Emergent species (ABG)			
Acorus calamus	0.08	Czech Republic	356
A. calamus (W)	0.53	Germany	4399
Carex gracilis	0.30	Czech Republic	356
Carex lacustris	1.5–1.9	Canada	3346
Carex stricta (W)	6.72	Germany	4399
Carex vesicaria	0.01	Czech Republic	356
Glyceria aquatica (W)	0.48	Germany	4399
Glyceria maxima	0.09	Czech Republic	356
Iris pseudacorus (W)	1.10	Germany	4399
Lythrum salicaria	1.4–2.7	Canada	3346
Mentha aquatica (W)	0.54	Germany	4399
Phragmites australis (W)	0.62	Germany	4399
Pontenderia cordata	<1.0–1.4	Canada	3346
Scirpus lacustris	0.13	Czech Republic	356
S. lacustris (W)	5.63	Germany	4399
Sparganium erectum	0.03	Czech Republic	356
S. erectum (W)	1.07	Germany	4399
Typha angustifolia	0.19	Czech Republic	356
T. angustifolia (W)	0.44	Germany	4399
Typha latifolia	<1.0–2.3	Canada	3346
Floating-leaved species (ABG)			
Nymphaea odorata	1.0–2.8	Canada	3346
Submerged species (ABG)			
Ceratophyllum submersum	5.2–30.9	Hungary	2575
Hydrocharis morsus ranae	1.8–32.1	Hungary	2575
Myriophyllum heterophyllum	<1.0–8.7	Canada	3346
Myriophyllum spicatum	15.7–17.5	Hungary	2575
Potamogeton pectinatus	1.2–12.8	Hungary	2575
Potamogeton perfoliatus	2.2–36.1	Hungary	2575
Floating species (Total)			
Lemna trisulca	0.88	Hungary	2575
Stratiotes aloides	0.33–18.6	Hungary	2575
Bryophytes (ABG)			
Rhynchostegium riparioides	4.6–542	England	5311
Scapania undulata	3–1,700	Poland	4264
Sphagnum magellanicum	0.89	Czech Republic	1207–8
Sphagnum recurvum	0.70	Czech Republic	1207–8
Sphagnum sp.	1.0–1.5	Sweden	4612

Table 7-42. Boron concentration in aquatic and wetland plants (mg kg^{-1})

Plant species	Concentration	Locality	Reference
Emergent species (ABG)			
Acorus calamus (W)	56.9	Germany	4399
Carex stricta (W)	21.4	Germany	4399
Eleocharis equisetoides	1.2	South Carolina	516
Eleocharis quadrangulata	3.7	South Carolina	516
E. quandrangulata	9.0	South Carolina	512
Glyceria aquatica (W)	15.0	Germany	4399
Glyceria striata	2.0	South Carolina	516
Hydrocotyle umbellata	7.7	South Carolina	516
Juncus effusus	8.1	South Carolina	516
J. effusus	4.6–51	SE U.S.A.	516
Mentha aquatica (W)	38.3	Germany	4399
Panicum hemitomon	2.3	South Carolina	516
Phragmites australis (W)	8.2	Germany	4399
Pontenderia cordata	7.9	South Carolina	516
Scirpus americanus	2.7	South Carolina	516
Scirpus lacustris (W)	14.6	Germany	4399
Scirpus validus	3.2	South Carolina	516
Sparganium erectum (W)	40.2	Germany	4399
Typha angustifolia (W)	24.5	Germany	4399
Typha domingensis	4.6	South Carolina	516
Typha latifolia	5.2	South Carolina	516
T. latifolia	5.2–100	SE U.S.A.	516
T. latifolia	10.4	Wisconsin	4582
Floating-leaved species (ABG)			
Nelumbo lutea	10.9	South Carolina	516
Nuphar advena	8.2	South Carolina	516
Nymphaea odorata	11.3	South Carolina	516
Submerged species (ABG)			
Ceratophyllum demersum	4.3	South Carolina	516
C. demersum	12.4–29.7	Hungary	2575
C. demersum (W)	41.7–148	Michigan	1645
Hydrocharis morsus ranae	38.4–64.2	Hungary	2575
Myriophyllum heterophyllum	10.6	South Carolina	516
Myriophyllum spicatum	16.2–25.6	Wisconsin	702
Potamogeton berchtoldii (W)	49.6	Michigan	1645
Potamogeton diversifolius	5.3	South Carolina	516
Potamogeton foliosus (W)	29.3	Michigan	1645
Potamogeton pectinatus	52.3–958.5	Hungary	2575
P. pectinatus (W)	279.9	Michigan	1645
Potamogeton perfoliatus	0.47–43.7	Hungary	2575
Utricularia inflata	7.6	South Carolina	516

Table 7-42. Continued

Plant species	Concentration	Locality	Reference
Floating species (Total)			
Lemna minor (W)	781–2,567	Michigan	1645
Lemna trisulca	190.7	Hungary	2575
Stratiotes aloides	1.06–41.2	Hungary	2575
Bryophytes (ABG)			
Scapania undulata	18–10,100	Poland	4264

Table 7-43. Vanadium concentration in aquatic and wetland plants (mg kg^{-1} dry mass)

Species	Concentration	Locality	Reference
Floating species (Total)			
Eichhornia crassipes (W)	8.75	Czech Republic	5539
Bryophytes (ABG)			
Fontinalis sp.	3.7–5.7	Sweden	4612
Scapania undulata	2–1,300	Poland	4264

Table 7-44. Selenium concentration in aquatic and wetland plants (mg kg^{-1} dry mass)

Plant species	Concentration ABG	BLG	Location	Reference
Scirpus maritimus	0–7.7		California	4366
S. maritimus	100–280		California	4366
Typha domingensis	L 17–160	RH 30–89	California	4366
T. domingensis	L 0–0.43	RH 0–1.2	California	4366
Typha latifolia	16.6	52.3	Wisconsin	4582

Table 7-45. Arsenic concentration in aquatic and wetland plants (mg kg^{-1} dry mass)

Plant species	Concentration	Locality	Reference
Emergent species (ABG)			
Scirpus sp.	12	New Zealand	4017
Spartina alterniflora	0.06–0.7	Massachusetts	216
Spartina patens	0.04–0.64	Massachusetts	216
Typha orientalis	8	New Zealand	4017
Submerged species (ABG)			
Ceratophyllum demersum	1.4–971[a]	New Zealand	4017
Ceratophyllum submersum	20.6	Hungary	2575
Elodea canadensis	0.4	Michigan	5327
E. canadensis	3.0–307	New Zealand	4017
Hydrocharis morsus-ranae	107.1	Hungary	2575
Myriophyllum spicatum	0.6–1.6	Michigan	5327
M. spicatum	0.9–19.7	Hungary	2575
Myriophyllum spp.	150–3,700	Canada, N.W.T.	5219
Potamogeton pectinatus	0.33	Hungary	2575
Potamogeton perfoliatus	0.69–8.6	Hungary	2575
Potamogeton sp.	<6.0–178	New Zealand	4017
Potamogeton spp.	1.5–920	N.W.T., Canada	5219
Utricularia spp.	66	N.W.T., Canada	5219
Floating species (Total)			
Eichhornia crassipes (W)	0.5	Czech Republic	5539
Azolla rubra	<2.0	New Zealand	4017
Lemna sp.	2.5–30	New Zealand	4017
Stratiotes aloides	2.33	Hungary	2575
Bryophytes (ABG)			
Scapania undulata	0–2,190	Poland	4264

[a] Geothermal area

Table 7-46. Cadmium concentration in aquatic and wetland plants (mg kg^{-1} dry mass)

Plant species	Concentration ABG	BLG	Locality	Reference
Emergent species				
Acorus calamus	0.04		Czech Republic	356
Bidens laevis	2.5–4.1		New Jersey	5378
Carex hudsoni	0.02		Czech Republic	356
Carex lacustris	< 1.0		Canada	3346
Carex vesicaria	0.12		Czech Republic	356
Eleocharis acicularis	0.4–11.0		Canada, Ontario	3228
Impatiens capensis	2.7–6.0		New Jersey	5378
Lythrum salicaria	< 1.0		Canada	3346
Peltandra virginica	1.5 - 6.1		New Jersey	5378
Phragmites australis	L < 0.1 S < 0.1	RH 0.15–0.16	Finland	2766
P. australis	0.05–0.25	RH 0.10–0.22 R 0.58–1.21	Denmark	4328
Pontenderia cordata	< 1.0		Canada	3346
Scirpus maritimus	0.2–0.7	R 0.4–6.0	The Netherlands	3587
Sparganium erectum	0.04		Czech Republic	356
Spartina anglica	0.1–1.1	R 0.5–16	The Netherlands	3587
Typha angustifolia	0.04		Czech Republic	356
Typha latifolia	0.05		Czech Republic	356
T. latifolia	< 1.0		Canada	3346
Zizania aquatica	0.4–5.3		New Jersey	5378
Floating-leaved species				
Nuphar lutea	L 0.41–0.92 P 0.46–1.00	RH 0.18–0.25 R 0.32–0.76	Finland	2766
Nymphaea odorata	< 1.0		Canada	3346
Submerged species (ABG)				
Ceratophyllum demersum	0.1–1.1		Iraq	5
Elodea canadensis	0.32–32.3		Indiana	3113
Myriophyllum heterophyllum	<1.0–2.2		Canada	3346
Floating species				
Eichhornia crassipes (W)	5–1,180		Czech Republic	5539
E. crassipes (E)	T 36.6–2,326	R 711–10,600	Japan	3345
Bryophytes (ABG)				
Fontinalis antipyretica	0.17–78		Europe	4306
Fontinalis sp.	2.1–2.6		Sweden	4612
Fontinalis spp.	0.4–10.0		Ontario, Canada	3228
Fontinalis hypnoides + *Drepanocladus trichophyllus*	0.90		Finland	2776
Rhynchostegium riparioides	0.50–89.5		England	5311
Sphagnum magellanicum	0.29		Czech Republic	1207–8
Sphagnum recurvum	0.18		Czech Republic	1207–8
Sphagnum sp.	0.54		Finland	2766
Sphagnum spp.	0.5–5.6		Ontario, Canada	3228

Table 7-47. Chromium concentration in aquatic and wetland plants (mg kg^{-1} dry mass)

Plant species	Concentration ABG	Locality	Reference
Emergent species			
Acorus calamus	5.4	Czech Republic	356
Carex gracilis	17.2	Czech Republic	356
Carex hudsoni	5.8	Czech Republic	356
Carex lacustris	<1.0–1.7	Canada	3346
Carex vesicaria	11.7	Czech Republic	356
Glyceria maxima	9.8	Czech Republic	356
Lythrum salicaria	1.5–5.1	Canada	3346
Pontenderia cordata	1.9–2.8	Canada	3346
Scirpus lacustris	3.8	Czech Republic	356
Sparganium erectum	2.7	Czech Republic	356
Typha angustifolia	6.4	Czech Republic	356
Typha latifolia	<1.0–4.4	Canada	3346
T. latifolia	7.2	Czech Republic	356
Floating-leaved species			
Nymphaea odorata	<1.0	Canada	3346
Submerged species			
Ceratophyllum demersum	16.5–122.2	Hungary	2575
Hydrocharis morsus ranae	0.86–1.8	Hungary	2575
Myriophyllum heterophyllum	2.4–12.1	Canada	3346
Myriophyllum spicatum	7.9–65.5	Hungary	2575
Potamogeton pectinatus	5.2–6.4	Hungary	2575
Potamogeton perfoliatus	2.2–42.5	Hungary	2575
Floating species (Total)			
Eichhornia crassipes (W)	2.5	Czech Republic	5539
Lemna gibba (E)	0.58–3,710	Louisiana	4700
Lemna trisulca	3.32	Hungary	2575
Spirodela punctata	0.64–5,610	Louisiana	4700
Spirodela polyrhiza (E)	0.27–8,600	Louisiana	4700
Stratiotes aloides	1.9–5.3	Hungary	2575
Bryophytes			
Fontinalis sp.	0.8–1.8	Sweden	4612
Rhynchostegium riparioides	2.3–1,330	England	5311

Table 7-48. Lead concentration in aquatic and wetland plants (mg kg^{-1} dry mass)

Plant species	Concentration ABG	BLG	Locality	Reference
Emergent species				
Bidens laevis	20.5–446		New Jersey	5378
Carex lacustris	3.2–3.9		Canada	3346
Eleocharis acicularis	52–99 (total)		Connecticut	2008
E. acicularis	6–150		Canada	3228
Glyceria maxima (W)	2.4–13.5		Poland	3596
Impatiens capensis	19.7–457		New Jersey	5378
Lythrum salicaria	4.1–7.9		Canada	3346
Peltandra virginica	158–657		New Jersey	5378
Phragmites australis	0.2–2.8		Poland	2632
P. australis	< 1.0–3.7	R 3.5–9.3 RH < 0.1	Denmark	4328
Pontenderia cordata	3.5–5.0		Canada	3346
P. cordata	7.0–22	11–88	Connecticut	2008
Spartina alterniflora	2.3–12		Massachusetts	216
Spartina patens	1.8–7.8		Massachusetts	216
Typha angustifolia	0.04–0.5		Poland	2632
Typha latifolia	2.6–5.5		Canada	3346
Zizania aquatica var. *aquatica*	25–673		New Jersey	5378
Floating-leaved species				
Nuphar sp.	3.9–15.6		New Jersey	4660
Nymphaea odorata	3.5–4.5		Canada	3346
Nymphaea sp.	0–10.7		New Jersey	4660
Submerged species				
Ceratophyllum demersum	8 (Total)		Connecticut	2008
C. demersum	51.5–52.4		Hungary	2575
Elodea canadensis	5.2–160.9		Indiana	3113
Elodea nuttallii	19–20 (Total)		Connecticut	2008
Hydrocharis morsus ranae	2.2–6.7		Hungary	2575
Myriophyllum heterophyllum	11.5–29.9		Canada	3346
Myriophyllum spicatum	9.8–21.8		Hungary	2575
Myriophyllum farwellii	6.4–116		New Jersey	4660
Myriophyllum sp.	50		Missouri	1544
Myriophyllum sp.	5,000–6,400[a]		Missouri	1544
Potamogeton epihydrus	12–18		Connecticut	2008
Potamogeton pectinatus	1.5–16		Hungary	2575
Potamogeton perfoliatus	0.8–34		Hungary	2575
Potamogeton sp.	23		Missouri	1544
Potamogeton sp.	17.4–118		New Jersey	4660
Potamogeton sp.	148–7,750		Missouri	1544
Utricularia purpurea	23.7–377		New Jersey	4660

Table 7-48. Continued

Plant species	Concentration		Locality	Reference
	ABG	BLG		
Floating species (Total)				
Eichhornia crassipes (W)	2.7–948		Czech Republic	5539
E. crassipes (E)	T 28.9–1,810 R 2,280–25,790		Japan	3345
Lemna trisulca	22.2		Hungary	2575
Stratiotes aloides	2.3–6.1		Hungary	2575
Bryophytes				
Fontinalis antipyretica	0.25–381		Europe	4306
Fontinalis squrrosum	1,301–1,385		Wales	645
Fontinalis sp.	9–20.3		Sweden	4612
Fontinalis spp.	18–290		Canada	3228
Hylocomium splendens	10–100		Sweden	5065
Philonotis fontana	1,232–4,914		Wales	645
Rhynchostegium riparioides	30.9–17,800		England	5311
Scapania undulata	22–3,100		Poland	4264
S. undulata	765–7,806		Wales	645
S. undulata	14,825		Wales	3136
Solenostoma crenulata	6,354–15,940		Wales	645
Sphagnum balticum + *S. majus*	>18.5		Finland	3626
Sphagnum fuscum	5.6		Finland	3627
S. fuscum	23		Canada, Ontario	1659
S. fuscum	78		New York	1533
S. fuscum	23.9		Finland	3626
S. fuscum	20–381		Canada, Ontario	1660
Sphagnum spp.	32–430		Canada	3228

[a] Tailing pond

Table 7-49. Lead standing stock in aquatic and wetland plants (g m^{-2})

Plant species	Standing stock		Locality	Reference
	ABG	BLG		
Glyceria aquatica (W)	5.2		Poland	3596
Phragmites australis	2.53	5.30	Poland	2632
P. australis	ca. 3.8–ca. 9.3 (Total)		Denmark	2714
Typha angustifolia	0.03	0.11	Poland	2632

Table 7-50. Mercury concentration in aquatic and wetland plants (mg kg^{-1} dry mass)

Plant species	Concentration		Locality	Reference
	ABG	BLG		
Emergent species				
Calla palustris	0.04		Finland	2889
Eleocharis acicularis	0.53–1.03		Finland	664
Equisetum fluviatile	0.012		Finland	4288
Glyceria maxima	0.062–0.122		Finland	4288
Phalaris arundinacea	0.079–0.25	0.21–0.45	Canada	3321
Phragmites australis	L 0.01 S 0.01	RH 0.01	Finland	2766
P. australis	0.035		Finland	4288
Polygonum amphibium	0.017		Finland	4288
Sagittaria latifolia	0.054–0.094	0.11–0.21	Canada	3321
Sagittaria sagittifolia	0.02–2.2		Finland	2889
Scirpus lacustris	0.015		Finland	4288
S. lacustris	0.06		Finland	2889
Sparganium angustifolium	0.071–0.21	0.097–0.25	Canada	3321
Sparganium emersum	< 0.01–1.6		Finland	2889
Sparganium erectum	0.052		Finland	4288
S. erectum	0.04–0.06		Finland	2889
Sparganium eurycarpum	0.05–0.16	0.048–0.55	Canada	3321
Sparganium friesii	0.02–0.07		Finland	2889
Sparganium simplex	0.139		Finland	4288
Spartina alterniflora	L 0.04–0.12 S 0.03–0.12	R 0.11–1.47 RH 0.03–0.53	Georgia	5456
Floating-leaved species				
Nuphar lutea	L 0.02 P 0.02	R 0.03 RH 0.01	Finland	2766
N. lutea	0.02–3.1		Finland	2889
N. lutea	FL 0.01–0.08 SL 0.09–0.15		Finland	664
N. lutea	0.011		Finland	4288
N. lutea	FL 3.8		Finland	663
Nuphar sp.	0.022		Canada, B.C.	4484
Nymphaea candida	0.04		Finland	2889
Submerged species				
Ceratophyllum demersum	0.03–4.62		Thailand	4823
C. demersum	0.05–1.10		Finland	4823
C. demersum	0.55		Finland	2889
C. demersum	0.17–0.61		Finland	664
C. demersum	19.2		Finland	663
Elodea canadensis	0.03–0.41		Canada	3321
E. canadensis	0.03–5.9		Finland	2889
Elodea densa	0.01–0.14		Canada	3321
Myriophyllum alterniflorum	<0.01–4.3		Finland	2889

Table 7-50. Continued

Plant species	Concentration		Locality	Reference
	ABG	BLG		
Myriophyllum sp.	0.036–0.13		Canada, B.C.	4484
Myriophyllum sp.	0.035–1.6		Canada	3321
Potamogeton natans	0.02–6.3		Finland	2889
Potamogeton obtusifolius	0.04–1.6		Finland	2889
Potamogeton perfoliatus	0.05–0.64		Finland	664
P. perfoliatus	0.03–6.6		Finland	2889
P. perfoliatus	26.4		Finland	663
Potamogeton sp.	0.015–0.13		Canada, B.C.	4484
Utricularia vulgaris	0.122		Finland	4288
U. vulgaris	0.06–0.20		Finland	2889
U. vulgaris	0.16–0.89		Finland	664
Bryophytes				
Bryum sp.	0.06–0.42		Canada, B.C.	4484
Fontinalis antipyretica	0.31–0.69		Finland	664
Fontinalis hypnoides	0.34		Finland	2889
Fontinalis hypnoides + *Drepanocladus trichophyllus*	0.08		Finland	2766
Jungermannia vulcanicola	40–13,500		Japan	4292
Scapania undulata	< 40–2,400		Japan	4292
Sphagnum sp.	0.12		Finland	2766
Sphagnum sp.	0.11		Canada, B.C.	4484

Table 7-51. Nickel concentration in aquatic and wetland plants (mg kg^{-1} dry mass)

Plant species	Concentration	Locality	Reference
Emergent species (ABG)			
Acorus calamus (W)	1.08	Germany	4399
A. calamus	12.5	Czech Republic	356
Bidens laevis	2.8–27.6	New Jersey	5378
Carex gracilis	19.3	Czech Republic	356
Carex hudsoni	7.5	Czech Republic	356
Carex lacustris	1.0–1.5	Canada	3346
Carex stricta (W)	2.46	Germany	4399
Carex vesicaria	14.3	Czech Republic	356
Eleocharis acicularis	5–1,200	Ontario, Canada	3228
Glyceria maxima (W)	1.99	Germany	4399
G. maxima	13	Czech Republic	356
Impatiens capensis	3.8–48.5	New Jersey	5378
Iris pseudacorus (W)	1.75	Germany	4399
Lythrum salicaria	1.6–6.7	Canada	3346
Mentha aquatica (W)	2.03	Germany	4399
Peltandra virginica	8.2–62.4	New Jersey	5378
Phragmites australis	1.53	Germany	4399
P. australis	8.7	Czech Republic	356
Pontenderia cordata	1.1–1.8	Canada	3346
Scirpus lacustris (W)	1.71	Germany	4399
S. lacustris	5.4	Czech Republic	356
Sparganium erectum (W)	2.27	Germany	4399
S. erectum	5.5	Czech Republic	356
Typha angustifolia (W)	1.86	Germany	4399
T. angustifolia	12.1	Czech Republic	356
Typha latifolia	<1.0–4.7	Canada	3346
T. latifolia	11.4	Czech Republic	356
Zizania aquatica v. aquatica	3.0–53.7	New Jersey	5378
Floating-leaved species (ABG)			
Nymphaea odorata	0.8–5.2	Canada	3346
Submerged species (ABG)			
Ceratophyllum submersum	30.9–87.3	Hungary	2575
Hydrocharis morsus ranae	32.1–33.7	Hungary	2575
Myriophyllum heterophyllum	4.0–10.9	Canada	3346
Myriophyllum spicatum	11.8–65.5	Hungary	2575
Potamogeton pectinatus	9.0–24.0	Hungary	2575
Potamogeton perfoliatus	4.9–36.4	Hungary	2575
Floating species (Total)			
Eichhornia crassipes (W)	6.25	Czech Republic	5539
Lemna trisulca	6.65	Hungary	2575
Stratiotes alvides	4.0–33.9	Hungary	2575

Table 7-51. Continued

Species	Concentration	Locality	Reference
Bryophytes (ABG)			
Fontinalis sp.	4.6–5.7	Sweden	4612
Fontinalis spp.	3–1,300	Canada, Ontario	3228
Rhynchostegium riparioides	13.8–694	England	5311
Scapania undulata	1–150	Poland	4264
Sphagnum spp.	3–180	Canada, Ontario	3228

B. List of algal genera mentioned in the book

The taxonomy of many algal species has been revised. The most important changes concerning species mentioned in the book are (in the following list these genera are marked with an asterisk):

Present name	Previous name
Emiliania huxleyi	*Coccolithus huxleyi* (formerly Chrysophyceae)
Hymenomonas carterae	*Cricosphaera carterae* (formerly Chrysophyceae)
Pavlova lutheri	*Monochrysis lutheri* (formerly Chrysophyceae)
Peridinium cinctum	*Peridinium westii*
Phaeodactylum tricornutum	*Nitzschia closterium* f. *minutissima*
Planktothrix limnetica	*Oscillatoria agardhii*
Poterioochromonas stipitata	*Ochromonas malhamensis*
Protogonyaulax tamarensis	*Gonyaulax tamarensis*
Thalassiosira pseudonana	*Cyclotella nana*
Thalassiosira weissflogii	*Thalassiosira fluviatilis*

If not mentioned, species are classified in the same class.

Cyanophyceae
Aphanizomenon, Anabaena, Anacystis, Aphanothece, Calothrix, Chamaesiphon, Chroococcus, Coccochloris, Coelosphaerium, Cylindrospermum, Dichothrix, Diplocystis, Geitleria, Gloeothece, Gloeotrichia, Gomphosphaeria, Hapalosiphon, Homoeothrix, Lyngbya, Mastigocladus, Merismopedia, Microchaeta, Microcoleus, Microcystis, Nodularia, Nostoc, Oscillatoria,* Phormidium, Plectonema, Pseudanabaena, Rivularia, Schizothrix, Scytonema, Sommierella, Spirulina, Stigonema, Synechococcus, Synechocystis, Tolypothrix, Trichodesmium, Westiellopsis

Prochlorophyceae
Prochloron

Rhodophyceae
Acanthopeltis, Agadhiella, Ahnfeltia, Antithamnion, Asparagopsis, Audionella (= Chantransia), Batrachospermum, Botryocladia, Caulacanthus, Ceramium, Chondrus, Corallina, Cyanidium, Delesseria, Dermonema, Eucheumia, Furcellaria, Galaxaura, Gelidium, Gigartina, Gloiopeltis, Gracilaria, Gymnogongrus, Halosaccion, Harveyella, Hildebrandia, Hypnea, Iridaea, Jania, Laurentia, Lemanea, Lithophyllum, Neoagardhiella, Palmaria, Petrocelis, Peyssonnelia, Phycodrys, Phyllophora, Polysiphonia, Porphyra, Porphyridium, Pterocladia, Ptilota, Rhodymenia, Thorea, Undaria

Chrysophyceae
Chromulina, Chrysamoeba, Chrysococcus, Chrysopyxis, Chrysosphaera, Chrysosphaerella, Derepyxis, Dinobryon, Hydrurus, Kephyrion, Lagynion, Mallomonas, Microglena, Ochromonas,* Olisthodiscus, Poterioochromonas,* Pseudokephyrion, Pseudopedinella, Syncrypta, Synura, Uroglena, Uroglenopsis

Prymnesiophyceae
Chrysochromulina, Emiliania,* Hymenomonas,* Isochrysis, Pavlova,* Prymnesium

Xanthophyceae
Botrydiopsis, Bumilleriopsis, Centritractus, Characidiopsis, Characiopsis, Chloridella, Goniochloris, Heterogloea, Mischococcus, Ophiocytium, Rhizochloris, Tetraëdriella, Tribonema, Vaucheria

Eustigmatophyceae
Chlorobotrys, Monodus, Pleurochloris, Pseudocharaciopsis

Bacillariophyceae
Achnanthes, Amphiphora, Amphipleura, Amphora, Asterionella, Biddulphia, Caloneis, Ceratoneis, Chaetoceros, Cocconeis, Coscinodiscus, Cyclotella,* Cylindrotheca, Cymatopleura, Cymbella, Diatoma, Ditylum, Epithemia, Eunotia, Fragilaria, Frustulia, Gomphonema, Gyrosigma, Hantzschia, Lauderia, Meridion, Melosia, Navicula, Neidium, Nitzschia, Opephora, Phaeodactylum,* Pinnularia, Rhizosolenis, Rhoicosphenia, Rhopalodia, Skeletonema, Stauroneis, Stephanodiscus, Surirella, Synedra, Tabellaria, Thalassiosira*

Dinophyceae
Alexandrium, Amoebophrya, Amphidinium, Ceratium, Dinophysis, Glenodinium, Gloeodinium, Gonyaulax,* Gymnodinium, Gyrodinium, Heminidium, Hypnodinium, Katodinium, Mesodinium, Oxyrrhis, Peridinium,* Phytodinium, Prorocentrum, Protogonyaulax, Pyrocystis, Pyrodinium, Scirppsiella, Symbiodinium, Thoracosphaera, Woloszynskia, Zooxanthella

Phaeophyceae
Ascophyllum, Chorda, Clodostephus, Colpomenia, Cystoseria, Ectocarpus, Egregia, Endarachne, Fucus, Hizikia, Ishige, Laminaria, Macrocystis, Nereocystis, Padina, Pelagophycus, Pelvetiopsis, Postelsia, Sargassum, Scytosiphon

Raphidophyceae
Gonyostomum, Merotrichia, Vacuolaria

Cryptophyceae
Chilomonas, Chroomonas, Cryptochrysis, Cryptomonas, Hemiselmis, Katablepharis, Rhodomonas, Sennia

Euglenophyceae
Colacium, Euglena, Eutreptia, Lepocinclis, Monomorphina, Phacus, Strombomonas, Trachelomonas

Chlorophyceae
Acetabularia, Actinastrum, Ankistrodesmus (Raphidium), Botryococcus, Bulbochaeta, Carteria, Caulerpa, Cephaleuros, Chaetomorpha, Chaetonema, Chaetophora, Chlainomonas, Chlamydomonas, Chlorella, Chlorhormidium, Chlorochytridium, Chlorcorone, Chlorogonium, Cladophora, Closterium, Coccomyxa, Codium, Coelastrum, Coenochloris, Coleochaete, Cosmarium, Crucigenia, Cylindrocapsa, Dictyosphaerium, Draparnaldia, Dunaliella, Enteromorpha, Eremosphaera, Euastrum, Eudorina, Geminella, Gloeotila, Golenkinia, Gonium, Haematococcus, Halicystis, Halimeda, Hormidium, Hydrodictyon, Lobomonas, Micrasterias, Microspora, Microthamnion, Monostroma, Mougeotia, Oedogonium, Oocardium, Oocystis, Palmella, Palmellococcus, Pandorina, Pediastrum, Phacotus, Pithophora, Protosiphon, Prototheca, Pleurococcus, Pleurotaenium, Polytoma, Rhizoclonium, Scenedesmus, Selenastrum, Spermatozopsis, Sphaerellopsis, Sphaerocystis, Spirogyra, Spondylomorum,

Spongomorpha, Staurastrum, Stichococcus, Stigeoclonium, Tetraëdron, Tetraspora, Tetrastrum, Trebouxia, Trentepohlia, Ulothrix, Ulva, Urospora, Valonia, Volvox, Zygnema

Charophyceae
Chara, Lamprothamnium, Lynchothamnus, Nitella, Nitellopsis (= Tolypellopsis), Tolypella

Prasinophyceae
Heteromastix, Micromonas, Monomastix, Pedinomonas, Platymonas (= Tetraselmis), Pyramimonas, Scourfieldia

Glaucophyta
Cyanophora, Glaucocystis, Gloeochaete

References

1. Aach, H.G., Über Wachstum und Zusammensetzung von *Chlorella pyrenoidosa* bei unterschiedlichen Lichtstärken und Nitratmangen, *Arch. Mikrobiol.,* 17, 213, 1952.
2. Aaronson, A., The synthesis of extracellular macromolecules and membranes by a population of the phytoflagellate *Ochromonas danica, Limnol. Oceanogr,* 16, 1, 1971.
3. Aaronson, S., and Patni, N.J., The role of surface and extracellular phosphatases in the phosphorus requirement of *Ochromonas, Limnol. Oceanogr,* 21, 838, 1976.
4. Aaronson, S., de Angelis, B., Frank, O., and Baker, H., Secretion of vitamins and amino acids into the environment by *Ochromonas danica, J. Phycol.,* 7, 215, 1971.
5. Abaychi, J.K., and Al-Obaidy, S.Z., Concentrations of trace elements in aquatic vascular plants from Shatt al arab River, *Iraq. J. Biol. Sci. Res.,* 18, 123, 1987.
6. Abbott, B.C., and Balantine, D., The toxin from *Gymnodinium veneficum* Ballantine, *J. Mar. Bbol. Ass. U.K.,* 36, 169, 1957.
7. Abbott, I.A., The uses of seaweed as food in Hawaii, *Econ. Bot.,* 32, 409, 1978.
8. Abbott, I.A. and Hollenberg, G.J., *Marine Algae of California,* Stanford University Press, Stanford, 1976.
9. Abbott, I.A., and Chapman, F.A., Evaluation of k carrageenan as a substitute for Agar in microbiological media, *Arch. Microbiol.,* 128, 355, 1981.
10. Abbott, I.A., and Cheney, D.P., Commercial uses of algal products: introduction and bibliography, in *Selected Papers in Phycology* II, Rosowski, J.R., and Parker, B.C., Eds., Phycological Society of America, Lawrence, Kansas, 1982, 779.
11. Abeliovich, A., and Dickbuk, S., Adenine inhibition of the synthesis of photosynthetic membranes in the chloroplast of *Scenedesmus obliquus* grown in the dark, *J. Bacteriol.,* 139, 661, 1979.
12. Abercombie, M., Hickman, M., Johnson, M.L., and Thain, M., *The New Penguin Dictionary of Biology,* 8th ed., Penguin Books, Ltd., London, 1990.
13. Abo-Rady, M.D.K., Die Belastung der oberen Leine mit Schwermetallen durch kommunale und industrielle Abwässer, ermittelt anhand von Wasser-, Sediment-, Fisch- und Pflanzenuntersuchungen, Dissertation, University of Göttingen, Germany, 1977.
14. Aboul-Fadl, M., Taha, E.M., Hamissa, M.R., El-Nawawy, A.S., and Sihoukry, A., The effect of the nitrogen fixing blue-green alga, *Tolypothrix tenuis* on the yield of paddy, *J. Mikrobiol. U.A.R.,* 2, 241, 1967.
15. Ackley, S.F., Buck, K.R., and Taguchi, S., Standing crop of algae in the sea ice of the Weddell Sea region, *Deep-Sea Res.,* 26 A, 269, 1979.
16. Acleto, O.C., Las algal marinas del Perú, *Bol. Soc. Peruana Bot.,* 6, 1, 1973.
17. Adamec, J., and Peverly, J.H., Potassium in polyphosphate bodies of *Chlorella pyrenoidosa* (Chlorophyceae) as determined by X-ray microanalysis, *J. Phycol.,* 15, 466, 1979.
18. Adams, D.F., Farwell, S.O., Robinson, E., Pack, M.R., and Bamesberger, W.L., Biogenic sulfur source strengths, *Environ. Sci. Technol.,* 15, 1493, 1981.
19. Adams, D.G., and Carr, N.G., The developmental biology of heterocysts and akinete formation in cyanobacteria, *CRC Crit. Rev. Microbiol.,* 9, 45, 1981.
20. Adams, M.S., and Stone, W., Field studies on photosynthesis of *Cladophora glomerata* (Chlorophyta) in Green Bay, Lake Michigan, *Ecology,* 54, 853, 1973.
21. Adams, M.S., and McCracken, M.D., Seasonal production of the *Myriophyllum* component of the littoral of lake Wingra, Wisconsin, *J. Ecol.,* 62, 457, 1974.
22. Adey, W.H., The effects of light and temperature on growth rates in boreal-subarctic crustose corallines, *J. Phycol.,* 6, 209, 1970.
23. Admiraal, W., Influence of various concentrations of orthophosphate on the division of an estuarine benthic diatom *Navicula arenaria,* in culture, *Mar. Biol.,* 42, 1, 1977.
24. Admiraal, W., Influence of light and temperature on the growth rate of estuarine benthic diatoms in culture, Mar. Biol., 39, 1, 1977.
25. Admiraal, W., The ecology of estuarine diatoms, in *Progress in Phycological Research,* Vol. 3, Round, F.E., and Chapman, D.J., Eds., Biopress, Bristol., 1984, 269.
26. Adshead-Simonsen, P.C., Murray, G.E., and Kushner, D.J., Morphological changes in the diatom, *Tabellaria flocculosa,* induced by very low concentrations of cadmium, *Bull. Environ. Contam. Toxicol.,* 26, 745, 1981.

27. Aerts, R., Wallén, B., and Malmer, N., Growth-limiting nutrients in *Sphagnum* - dominated bogs subject to low and high atmospheric nitrogen supply, *J. Ecol.*, 80, 131, 1992.
28. Aerts, R., de Caluwe, H., and Konings, H., Seasonal allocation of biomass and nitrogen in four *Carex* species from mesotrophic and eutrophic fens as affected by nitrogen supply, *J. Ecol.*, 80, 653, 1992.
29. Agami, M., and Reddy, K.R., Inter-relationships between *Salvinia rotundifolia* and *Spirodella polyrrhiza* at various interaction stages, *J. Aquat. Plant Manage*, 27, 96, 1989.
30. Agardh, C.A., *Systema algarum* I. Lund, 1824.
31. Agrawal, M., and Kumar, H.D., Response of *Chlorella* to mercury pollution, *Ind. J. Ecol.*, 2, 94, 1975.
32. Agrawal, M., and Kumar, H.D., Cobalt toxicity and its possible mode of action in blue-green alga *Anacystis nidulans, Beitr Biol. Pflanzen*, 53, 157, 1977.
33. Ahl, T., Eutrophication in relation to the load of pollution, *Prog. Wat. Res.*, 12, 49, 1980.
34. Ahlgren, G., Response of phytoplankton and primary production to reduced nutrient loading in Lake Norrviken, *Verh. Internat. Verein. Limnol.*, 20, 840, 1978.
35. Ahlgren, G., Temperature functions in biology and their applications to algal growth constants, *Oikos*, 49, 177, 1987.
36. Ahlgren, I., Limnological studies of Lake Norrviken, a eutrophied Swedish lake. 1. Water chemistry and nutrient budget, *Schweiz. Z. Hydrol.*, 29, 53, 1967.
37. Ahling, S., and Hébert, M., The uses of alginates in dentistry, in *Marine Algae in Pharmaceutical Science*, Hoppe, H.A., Levring, T., and Tanaka, Y., Eds., Walter de Gruyter, Berlin, 1979, 631.
38. Ahmad, I., and Helleburst, J.A., Nitrogen metabolism of the marine microalga *Chlorella autotrophica, Plant Physiol.*, 76, 658, 1984.
39. Ahmad, I., and Helleburst, J.A., Effects of methionine sulfoximine on growth and nitrogen assimilation of the marine microalga *Chlorella autotrophica, Mar. Biol.*, 86, 85, 1985.
40. Ahmad, I., and Helleburst, J.A., Pathways of ammonium assimilation in the soil alga *Stichococcus bacillaris* Naeg., *New Phytol.*, 103, 57, 1986.
41. Ahmadjian, V., Lichens, in *Physiology and Biochemistry of Algae*, Lewin, R.A., Ed., Academic Press, New York and London, 1962, 817.
42. Ahmadjian, V., A guide to the algae occurring as lichen symbiosis: isolation, culture, cultural physiology, and identification, *Phycologia*, 6, 127, 1967.
43. Ahmadjian, V., *The Lichen Symbiosis*, John Wiley and Sons, New York, 1993.
44. Ahmadjian, V., and Henriksson, E., Parasitic relationship between two culturally isolated and unrelated lichen components, *Science*, 130, 1251, 1959.
45. Ahmed, S.I., Kenner, R.A., and Packard, T.T., A comparative study of the glutamate dehydrogenase activity in several species of marine phytoplankton, *Mar. Biol.*, 39, 93, 1977.
46. Aitchinson, P.A., and Butt, V.S., The relation between the synthesis of inorganic polyphosphate and phosphate uptake by *Chlorella vulgaris, J. Exp. Bot.*, 24, 497, 1973.
47. Ajmal, M., Khan, R., and Khan, A.U., Heavy metals in water, sediments, fish and plants of river Hindon, U.P., *Hydrobiologia*, 148, 151, 1987.
48. Alasaarela, E., Phytoplankton and environmental conditions in central and coastal areas of the Bothnian Bay, *Ann. Bot. Fennici*, 16, 241, 1979.
49. Albers, D., Reisser, W., and Wiessner, W., Studies on the nitrogen supply of endosymbiotic *Chlorellae* in green *Paramecium bursaria, Plant. Sci. Lett.*, 25, 85, 1982.
50. Alexander, M., Nitrification, in *Soil Nitrogen*, Bartholomew, W.V., and Clark, F.E., Eds., *Am. Soc. Agron.*, Madison, Wisconsin, *Agronomy*, 10, 307, 1965.
51. Alexander, M.C., *An Introduction to Soil Microbiology*, John Wiley & Sons, Inc., New York, 1967.
52. Alexander, V., Interrelationships between the seasonal sea ice and biological regimes, *Cold Regions Sci. Technol.*, 2, 157, 1980.
53. Alexander, V., Billington, M., and Schell, D.M., Nitrogen fixation in arctic and alpine tundra, in *Vegetation and Production Ecology of an Alaskan Arctic Tundra*, Tieszen, L.L., Ed., Springer-Verlag, Berlin, 1978, 539.
54. Alexander, V., Stanley, D.W., Daley, R.J., McRoy, C.P., Primary producers, in *Limnology of Tundra Ponds*, Hobbie, J.E., Ed., Dowden, Hutchinson and Ross, Stroudsburg, Pennsylvania, 1980, 179.
55. Alexopoulos, C.J., and Bold, H.C., *Algae and Fungi*, Macmillan Publishing Company, New York, 1967.
56. Algarra, P., and Niell, F.X., Short-term pigment response of *Corallina elongata* Ellis et Soland to light intensity, *Aquat. Bot.*, 36, 127, 1990.
57. Aliota, G., Pinto, G., and Pollo, A., Observations on tolerance to heavy metals of four green algae in relation to pH , *Estrato da Giornale Botanico Italiano*, 117, 247, 1983.
58. Al Kholy, A.A., On the assimilation of phosphorus in *Chlorella pyrenoidosa, Physiol. Plant.*, 9, 137, 1956.
59. Allen, E.D., and Spence, D.H.N., The differential ability of aquatic plants to utilize the inorganic carbon in fresh waters, *New Phytol.*, 87, 269, 1981.

60. Allen, H.E., Hall, R.H., and Brisbin, T.D., Metal speciation. Effects on aquatic toxicity, *Environ. Sci. Technol.*, 14, 441, 1980.
61. Allen, H.L., Chemo-organotrophic utilization of dissolved organic compounds by planktonic algae and bacteria in a pond, *Int. Rev. Ges. Hydrobiol.*, 54, 1, 1969.
62. Allen, H.L., Primary productivity, chemo-organotrophy, and nutritional interactions of epiphytic algae and bacteria on macrophytes in the littoral of a lake, *Ecol. Monogr.*, 41, 97, 1971.
63. Allen, H.L., Phytoplankton photosynthesis, micronutrient interactions, and inorganic carbon availability in a soft-water Vermont lake, in *Nutrients and Eutrophication*, Likens, G.E., Ed., *Am. Soc. Limnol. Oceanogr. Spec. Symp.*, 1, 63, 1972.
64. Allen, M.B., The cultivation of Myxophyceae, *Arch. Mikrobiol.*, 17, 34, 1952.
65. Allen, M.B., Studies on the nitrogen-fixing blue-green algae. I. The sodium requirement of *Anabaena cylindrica*, *Physiol. Plant.*, 8, 653, 1955.
66. Allen, M.B., Excretion of organic compounds by *Chlamydomonas*, *Arch. Mikrobiol.*, 24, 163, 1956.
67. Allen, M.B., Photosynthetic nitrogen fixation by blue-green algae, *Sci. Mon., N.Y.*, 83, 100, 1956.
68. Allen, M.B., High latitude phytoplankton, *Ann. Rev. Ecol. Syst.*, 2, 261, 1971.
69. Allen, M.B., and Arnon, D.I., Studies on nitrogen-fixing blue-green algae. I. Growth and nitrogen fixation by *Anabaena cylindrica* Lemm., *Plant Physiol.*, 30, 366, 1955.
70. Allen, M.B., and Arnon, D.I., Studies on nitrogen-fixing blue-green algae. II. The sodium requirement by *Anabaena cylindrica*, *Physiol. Plant.*, 8, 653, 1955.
71. Allen, M.B., Dougherty, C., and McLaughlin, J.J.A., Photoreactive pigments in flagellates, *Nature*, 184, 1047, 1959.
72. Allen, M.M., and Smith, A.J., Nitrogen chlorosis in blue-green algae, *Arch. Microbiol.*, 69, 114, 1969.
73. Allen, S.E., and Pearsall, W.H., Leaf analysis and shoot production in *Phragmites*, *Oikos*, 14, 176, 1963.
74. Allenby, K.G., The manganese and calcium contents of some aquatic plants and the water in which they grow, *Hydrobiologia*, 29, 239, 1967.
75. Allenby, K.G., Some analyses of aquatic plants and waters, *Hydrobiologia*, 32, 486, 1968.
76. Allison, R.K., Skipper, H.E., Reid, M.R., Short, W.A., and Hogan, G.L., The assimilation of acetate by *Nostoc muscorum*, *J. Biol. Chem.*, 204, 197, 1953.
77. Allison, R.K., Skipper, H.E., Reid, M.R., Short, W.A., and Hogan, G.L., Studies on the photosynthetic reaction. II. Sodium formate and urea feeding experiments with *Nostoc muscorum*, *Plant Physiol.*, 29, 164, 1954.
78. Alloway, B.J., Ed., *Heavy Metals in Soils*, Blackie and Son Ltd, Glasgow, 1990.
79. Altschuler, Z.S., Schnepfe, M.M., Silber, C.C., and Simon, F.O., Sulfur diagenesis in Everglades peat and origin of pyrite in coal, *Science*, 221, 221, 1983.
80. Ambasht, R.S., Ecosystem study of a tropical pond in relation to primary production of different vegetational zones, *Hidrobiologia* (Bucharest), 12, 57, 1971.
81. Ambasht, R.S., and Ram, K., Stratified primary productive structure of certain macrophytic weeds in a large Indian Lake, in *Aquatic Weeds in South East Asia*, Varshney, C.K., and Rzoska, J., Eds., Dr. W. Junk, The Hague, The Netherlands, 1976, 147.
82. Ammann, E.C., and Lynch, V.H., Purine metabolism by unicellular algae. II. Adenine, hypoxanthine and xanthine degradation by *Chlorella pyrenoidosa*, *Biochim. Biophys. Acta*, 87, 370, 1964.
83. Amspoker, M.C., and Czarnecki, D.B., Bacillariophyceae: introduction and bibliography, in *Selected Papers in Phycology* II, Rosowski, J.R., and Parker, B.C., Eds., Phycological Society of America, Lawrence, Kansas, 1982, 712.
84. Amy, N.K., and Garrett, R.H., Purification and characterization of the nitrate reductase from the diatom *Thalassiosira pseudonana*, *Plant Physiol.*, 54, 629, 1974.
85. Anagnostidis, K., Ed., The 10th Symposium of the International Association for Cyanophyte Research (IAC) report, *Arch. Hydrobiol. Suppl.*, 80 (*Algol. Studies*, 50-53), 1988.
86. Anagnostidis, K., and Komárek, J., Modern approach to the classification system of cyanophytes 1 - Introduction. *Arch. Hydrobiol. Suppl.*, 71 (*Algol. Studies*, 38/39), 291, 1985.
87. Anagnostidis, K., and Komárek, J., Modern approach to the classification system of cyanophytes 3 - Oscillatoriales, *Arch. Hydrobiol. Suppl.*, 80 (*Algol. Studies*, 50-53), 327, 1988.
88. Anagnostidis, K., and Komárek, J., Modern approach to the classification system of cyanophytes 5 - Stigonematales, *Arch. Hydrobiol. Supp.*, 86 (*Algol. Studies*, 59), 1, 1990.
89. Andersen, F.Ø., Primary production in a shallow water lake with special reference to a reed swamp, *Oikos*, 27, 243, 1976.
90. Andersen, F.Ø., Effects of nutrient level on the decomposition of *Phragmites communis* Trin, *Arch. Hydrobiol.*, 84, 42, 1978.
91. Andersen, R.A., The flagellar apparatus of the golden alga *Synura uvella*: four absolute orientations, *Protoplasma*, 128, 94, 1985.
92. Andersen, R.A., Synurophyceae classis nov., a new class of algae, *Am. J. Bot.*, 74, 337, 1987.

93.   Andersen, R.A., and Mulkey, T.J., The occurrence of chlorophylls c_1 and c_2 in the Chrysophyceae, *J. Phycol.*, 19, 289, 1983.

94.   Anderson, D.M., and Morel, F.M.M., Copper sensitivity of *Gonyaulax tamarensis*, *Limnol. Oceanogr.*, 23, 283, 1978.

95.   Anderson, E.H., Studies on the metabolism of the colorless alga *Prototheca zopfii*, *J. Gen. Physiol.*, 28, 297, 1945.

96.   Anderson, G.C., and Zeutschel, R.P., Release of dissolved organic matter by marine phytoplankton in coastal and off-shore areas of the Northeast Pacific Ocean, *Limnol. Oceanogr.*, 15, 402, 1970.

97.   Anderson, G.R., and Jordan, J.V., Boron: A non-essential growth factor for *Azotobacter chroococcum*, *Soil Sci.*, 92, 113, 1961.

98.   Anderson, J.W., *Sulphur in Biology*, Studies in Biology No. 101, Edward Arnold, London, 1978.

99.   Anderson, L.W.J., and Sweeney, B.M., Role of inorganic ions in controlling sedimentation rate of a marine centric diatom *Ditylum brightwelli*, *J. Phycol.*, 14, 204, 1978.

100.   Anderson, D.M., and Morel, F.M.M., Copper sensitivity of *Gonyaulax tamarensis*, *Limnol. Oceanogr.*, 23, 283, 1978.

101.   Anderson, M.A., and Morel, F.M.M., Uptake of Fe (II) by a diatom in oxic culture medium, *Mar. Biol. Lett.*, 1, 263, 1980.

102.   Anderson, M.A., and Morel, F.M.M., The influence of aqueous iron chemistry on the uptake of iron by the coastal diatom *Thalassiosira weissflogii*, *Limnol. Oceanogr.*, 27, 789, 1982.

103.   Anderson, M.A., Morel, F.M.M., and Guillard, R.R.L., Growth limitation of a coastal diatom by low zinc ion activity, *Nature*, 276, 71, 1978.

104.   Anderson, N.S., and Rees, D.A., Porphyran: A polysaccharide with a masked repeating structure, *J. Chem. Soc.*, p. 5880, 1965.

105.   Anderson, R.R., Brown, R.G., and Rappleye, R.D., Water quality and plant distribution along the upper Patuxent River, Maryland, *Chesapeake Sci.*, 9, 145, 1968.

106.   Andreae, M.O., Distribution and speciation of arsenic in natural waters and some marine algae, *Deep-Sea Res.*, 25, 391, 1978.

107.   Andrew, R.W., Biesinger, K.E., and Glass, G.E., Effects of inorganic complexing on the toxicity of copper to *Daphnia magna*, *Water Res.*, 11, 309, 1977.

108.   Andrews, N.J., and Pratt, D.C., The potential of cattails (*Typha* sp.) as an energy source: productivity in managed stands, *J. Minn. Acad. Sci.*, 44, 5, 1978.

109.   Antia, N.J., A critical appraisal of Lewin's prochlorophyta, *Br. Phycol. J.*, 12, 271, 1977.

110.   Antia, N.J., Nutritional physiology and biochemistry of marine cryptomonads and chrysomonads, in *Biochemistry and Physiology of Protozoa*, 2nd ed., Levandovsky, M., and Hunter, H., Eds., Academic Press, New York, 1980, 117.

111.   Antia, N.J., and Chorney, V., Nature of the nitrogen compounds supporting phototrophic growth of the marine cryptomonad *Hemiselmis virescens*, *J. Protozool.*, 15, 198, 1968.

112.   Antia, N.J., and Cheng, J.Y., Culture studies on the effects from borate pollution on the growth of marine phytoplankton, *J. Fish. Res. Board Can.*, 32, 2487, 1975.

113.   Antia, N.J., and Cheng, J.Y., The ketocarotenoids of two marine coccoid members of the Eustigmatophyceae, *Br. Phycol. J.*, 17, 39, 1982.

114.   Antia, N.J., McAllister, C.D., Parsons, T.R., Stephens, K., and Strickland, J.D.H., Further measurements of primary productivity using a large-volume plastic sphere, *Limnol. Oceanogr.*, 8, 166, 1963.

115.   Antia, N.J., Berland, B.R., Bonin, D.J., and Maestrini, S.Y., Effects of urea concentration in supporting growth of certain marine microplanktonic algae, *Phycologia*, 16, 105, 1977.

116.   Antia, N.J., Harrison, P.J., and Oliveira, L., The role of dissolved organic nitrogen in phytoplankton nutrition, cell biology and ecology, *Phycologia*, 30, 1, 1991.

117.   Antoine, S.E., and Benson-Evans, K., The effect of current velocity on the rate of growth of benthic algal communities, *Int. Rev. Ges. Hydrobiol.*, 67, 575, 1982.

118.   Antoine, S.E., and Benson-Evans, K., Phycoperiphyton development on an artificial substrate in the River Wye System, Wales, U.K. Part 1: Biomass variation, *Acta Hydrochim. Hydrobiol.*, 13, 277, 1985.

119.   Antoine, S.E., Esdho, R.T., and Benson-Evans, K., Studies on the bottom sediments and epipelic algae of the River Ely, South Wales, U.K., *Limnologica (Berlin)*, 16, 1, 1984.

120.   Aoki, S., and Miyachi, S., Chromatographic analyses of acid-soluble polyphosphates in *Chlorella* cells, *Plant Cell Physiol.*, 5, 241, 1964.

121.   Aomine, S., A review of research on redox potentials of paddy soils in Japan, *Soil Sci.*, 94, 6, 1962.

122.   Aparicio, P.J., and Azuara, M.P., Wavelength dependence of nitrate release and the effects of different nitrogen sources and CO_2 tensions on *Chlamydomonas reinhardtii* inorganic nitrogen metabolism, in *Blue Light Effects in Biological Systems*, Senger, H., Ed., Springer-Verlag, Berlin, 1984, 196.

123. Apelt, G., Die Symbiose zwischen dem acoelen Turbellar *Convoluta convoluta* und Diatomeen der Gattung *Licmophora, Mar. Biol.,* 5, 165, 1969.

124. Apollonio, S., The chlorophyll content of Arctic sea-ice, *Arctic,* 14, 197, 1961.

125. Appleby, G., Colbeck, J., Holdsworth, E.S., and Wadman, H., Beta carboxylation enzymes in marine phytoplankton and isolation and purification of pyruvate carboxylase from *Amphidinium carterae* (Dinophyceae), *J. Phycol.,* 16, 290, 1980.

126. Applegate, R.D., and McCord, R.C., Productivity of *Oedogonium* in Lake Wingra, Wisconsin, *Am. Midl. Nat.,* 92, 247, 1974.

127. Apt, K.E., Effects of the symbiotic red alga *Hypneocolax stellaris* on its host *Hypnea musciformis* (Hypneaceae, Gigartinales), *J. Phycol.,* 20, 148, 1984.

128. Apte, S.C., Howard, A.G., Morris, R.J., and McCartney, M.J., Arsenic, antimony and selenium speciation during a spring phytoplankton bloom in a closed experimental ecosystem, *Mar. Chem.,* 20, 119, 1986.

129. Arber, A., *Water plants: A Study of Aquatic Angiosperms.* Cambridge University Press, Cambridge, 1920.

130. Arens, K., Physiologisch polarisierter Massenaustausch und Photosynthese bei submersen Wasserpflanzen. I., *Planta,* 20, 621, 1933.

131. Ariovich, D., and Pienaar, R.N., The role of light in the incorporation and utilization of Ca^{++} ions by *Hymenomonas carterae* (Braarud et Fagerl) Braarud (Prymnesiophyceae), *Br. Phycol. J.,* 14, 17, 1979.

132. Arisz, W.H., Influx and efflux by leaves of *Vallisneria spiralis.* I. Active uptake and permeability, *Protoplasma,* 57, 5, 1963.

133. Armbrust, E.V., Chisholm, S.W., and Olson, R.J., Role of light and the cell cycle on the induction od spermatogenesis in a centric diatom, *J. Phycol.,* 26, 470, 1990.

134. Armentano, T.V., Drainage of organic soils as a factor in the world carbon cycle, *BioScience,* 30, 825, 1980.

135. Armentano, T.V., and Verhoeven, J.T.A., Biogeochemical cycles: global, in *Wetlands and Shallow Continental Water Bodies,* Patten, B.C., Ed., SPB Academic Publishers bv, The Hague, The Netherlands, 1990, 281.

136. Armentano, T.V., de la Cruz, A.A., Duever, M., Loucks, O.L., Meijer, W., Mulholland, P.J., Tate, R.L. III., and Whigham, D.F., Recent changes in the global carbon balance of tropical organic soils. U.S. Dept. of Energy, Report No. DOE/CH/10135-1, Holcomb Research Institute, Butler University, Indianapolis, Indiana, 1983.

137. Armstrong, D.E., Perny, J.R., and Flatness, D., Availability of pollutants associated with suspended or settled river sediments which gain access to the Great Lakes, Draft Final Report on EPA Contact No. 68-81-4479, 1979.

138. Armstrong, D.E., Hurley, J.P., Swackhamer, D.L., and Shafer, M.M., Cycles of nutrient elements, hydrophobic organic compounds, and metals in Crystal Lake, in *Sources and fates of Aquatic Pollutants,* Hites, R.A., and Eisenreich, S.J., Eds., American Chemical Society, Washington, D.C., 1987, 491.

139. Armstrong, J., and Armstrong, W., *Phragmites australis*—a preliminary study of soil-oxidizing sites and internal gas transport pathways, *New Phytol.,* 108, 373, 1988.

140. Armstrong, J., and Armstrong, W., Light-enhanced convective through-flow increases oxygenation in rhizomes and rhizosphere of *Phragmites australis* (Cav.) Trin. ex Steud., *New Phytol.,* 114, 121, 1990.

141. Armstrong, J., and Armstrong, W., A convective through-flow of gases in *Phragmites australis* (Cav.) Trin. ex Steud., *Aquat. Bot.,* 39, 75, 1991.

142. Armstrong, J., Armstrong, W., and Beckett, P.M., *Phragmites australis*: a critical appraisal of the ventilating pressure concept and an analysis of resistance to pressurized gas flow and gaseous diffusion in horizontal rhizomes, *New Phytol.,* 110, 383, 1988.

143. Armstrong, W., Oxygen diffusion from the roots of some British bog plants, *Nature,* 204, 801, 1964.

144. Armstrong, W., The oxidizing activity of roots in waterlogged soils, *Physiol. Plant.,* 20, 920, 1967.

145. Armstrong, W., Radial oxygen loss from intact rice roots as affected by distance from the apex, respiration and waterlogging, *Physiol. Plant.,* 25, 192, 1971.

146. Armstrong, W., Waterlogged soils in *Environment and Plant Ecology,* Etherington, J.E., Ed., John Wiley, London, 1975, 181.

147. Armstrong, W., Root aeration in the wetland conditions, in *Plant Life in Anaerobic Environments,* Hook, D.D., and Crawford, R.M.M., Eds., Ann Arbor Science, Ann Arbor, Michigan, 1978, 269.

148. Armstrong, W., Aeration in higher plants, *Adv. Bot. Res.,* 7, 225, 1980.

149. Armstrong, W., and Wright, E.J., Radial oxygen loss from roots: the theoretical basis for the manipulation of flux data obtained by the cylindrical platinum electrode technique, *Physiol. Plant.*, 35, 21, 1975.

150. Armstrong, W., and Beckett, P.M., Internal aeration and the development of stelar anoxia in submerged roots: a multishelled mathematical model combining axial diffusion of oxygen in the cortex with radial losses to the steele, the wall layers and the rhizosphere, *New Phytol.*, 105, 221, 1987.

151. Armstrong, W., Justin, S.H.F.W., Beckett, P.M., and Lythe, S., Root adaptation to soil waterlogging, *Aquat, Bot.*, 39, 57, 1991.

152. Arndt, J.L., and Richardson, J.L., Hydrology, salinity and hydric soil development in a North Dakota prairie-pothole wetland system, *Wetlands*, 8, 93, 1988.

153. Arnold, D.E., Ingestion, assimilation, survival, and reproduction of *Daphnia pulex* fed seven species of blue-green algae, *Limnol. Oceanogr*, 16, 906, 1971.

154. Arnon, D.I., Some recent advances in the study of essential micronutrients for green plants, in *Proc. 8th Internat. Botan. Congress*, Sect. 11, 1954, 73.

155. Arnon, D.I., Phosphorus metabolism and photosynthesis, *Ann. Rev. Plant Physiol.*, 7, 325, 1956.

156. Arnon, D.I., The role of micronutrients in plant nutrition with special reference to photosynthesis and nitrogen assimilation, in *Trace Elements*, Lamb, C.A., Bentley, O.G., and Beattie, J.M., Eds., Academic Press, Inc., New York and London, 1958, 1.

157. Arnon, D.I., Conversion of light into chemical energy in photosynthesis, *Nature*, 184, 10, 1959.

158. Arnon, D.I., and Stout, P.R., The essentiality of certain elements in minute quantity for plants with special reference to copper, *Plant Physiol.*, 14, 371, 1939.

159. Arnon, D.I., and Stout, P.R., Molybdenum as an essential element for higher plants, *Plant Physiol.*, 14, 599, 1939.

160. Arnon, D.I., and Whatley, F.R., Is chloride a coenzyme of photosynthesis?, *Science*, 110, 554, 1949.

161. Arnon, D.I., and Wessel, G., Vanadium as an essential element for green plants, *Nature*, 172, 1039, 1953.

162. Arnon, D.I., Allen, M.B., and Whatley, F.R., Photosynthesis by isolated chloroplasts, *Nature*, 174, 394, 1954.

163. Arnon, D.I., Allen, M.B., and Whatley, F.R., Photosynthesis by isolated chloroplasts. IV. General concept and comparisons of three photochemical reactions, *Biochem. Biophys. Acta*, 20, 449, 1956.

164. Arnon, D.I., Whatley, F.R., and Allen, M.B., Triphosphoxypyridine nucleotide as a catalyst of photosynthetic phosphorylation, *Nature*, 180, 182, 1957.

165. Arnon, D.I., Whatley, F.R., and Allen, M.B., Assimilatory power in photosynthesis. Photosynthetic phosphorylation by isolated chloroplasts is coupled with TPN reduction, *Science*, 127, 1026, 1958.

166. Arnon, D.I., Ichioka, P.S., Wessel, G., Fujiwara, A., and Woolley, J.T., Molybdenum in relation to nitrogen metabolism. I. Assimilation of nitrate nitrogen by *Scenedesmus*, *Physiol. Plant.*, 8, 538, 1955.

167. Arnon, D.I., Losada, M., Nozaki, M., and Tagawa, K., Photoproduction of hydrogen, photofixation of nitrogen and a unified concept of photosynthesis, *Nature*, 190, 601, 1961.

168. Arnott, H.J., and Pautard, F.G.E., Calcification in plants, in *Biological Calcification*, Schraer, H., Ed., Elsevier/North Holland Biomedical Press, Amsterdam, New York, 1970, 375.

169. Arnow, P., Oleson, J.J., and Williams, J.H., The effect of arginine on the nutrition of *Chlorella vulgaris*, *Am. J. Bot.*, 40, 100, 1953.

170. Arora, A., and Gupta, A.B., The effect of copper sulphate on formation of separation disc in *Oscillatoria* spec., *Arch. Hydrobiol.*, 96, 261, 1983.

171. Arrigo, K.R., and Sullivan, C.W., The influence of salinity and temperature covariation on the photophysiological characteristics of Antarctic sea ice microalgae, *J. Phycol.*, 28, 746, 1992.

172. Arvik, J.H., Soil algae of northwest Florida, *Qt. J. Florida Acad. Sci.*, 33, 247, 1970.

173. Asare, S.O., and Harlin, M.M., Seasonal fluctuations in tissue nitrogen for five species of perennial macroalgae in Rhode Island Sound, *J. Phycol.*, 19, 254, 1983.

174. Asmus, R., The measurements on seasonal variation of the activity of primary producers on a sandy tidal flat in the northern Wadden Sea, *Neth. J. Sea Res.*, 16, 389, 1982.

175. Assman, A.V., Role of algal periphyton in the organic production in water reservoirs, *Dokl. Akad. Nauk USSR*, 76, 905, 1951, (in Russian).

176. Assman, A.V., Role of algal periphyton in the organic production in Lake Glukokoye, *Trudy vses. gidrobiol. obshtchestva*, 5, 138, 1953, (in Russian).

177. Atkinson, M.J., and Smith, S.V., C:N:P ratios of benthic marine plants, *Limnol. Oceanogr*, 28, 568, 1983.

178. Auclair, A.N.D., Factors affecting tissue nutrient concentration in a *Carex* meadow, *Oecologia*, 28, 233, 1977.

179. Auclair, A.N., Seasonal dynamics of nutrients in a *Carex* meadow, *Can. J. Bot.*, 60, 1671, 1982.
180. Auclair, A.N.D., Bouchard, A., and Pajaczkowski, J., Plant standing crop and productivity relations in a *Scirpus - Equisetum* wetland, *Ecology*, 57, 941, 1976.
181. Auer, M.T., and Canale, R.P., Ecological studies and mathematical modeling of *Cladophora* in lake Huron. 2. Phosphorus uptake kinetics, *J. Great Lakes Res.*, 8, 84, 1982.
182. Auer, M.T., and Canale, R.P., Ecological studies and mathematical modeling of *Cladophora* in Lake Huron. 3. The dependence of growth rates on internal phosphorus pool size, *J. Great Lakes Res.*, 8, 93, 1982.
183. Auer, M.T., Graham, J.M., Graham, L.E., and Kranzfelder, J.A., Factors regulating the spatial and temporal distribution of *Cladophora* and *Ulothrix* in the Laurentian Great Lakes, in *Periphyton of Freshwater Ecosystems*, Wetzel, R.G., Ed., Dr. W. Junk Publishers, The Hague, The Netherlands, 1983, 135.
184. Aulio, K., Nutrient accumulation in *Sphagnum* mosses. I. A multivariate summarization of the mineral element composition of 13 species from an ombrotrophic raised bogs, *Ann. Bot. Fennici*, 17, 307, 1980.
185. Aulio, K., Nutrient accumulation in *Sphagnum* mosses. II. Intra- and interspecific variation in four species from ombrotrophic and minerotrophic habitats, *Ann. Bot. Fennici*, 19, 93, 1982.
186. Avron, M., and Jagendorf, A.T., Evidence concerning the mechanism of adenosine triphosphate formation by spinach chloroplasts, *J. Biol. Chem.*, 234, 967, 1959.
187. Ax, P., and Apelt, G., Die "Zooxanthellen" von *Convoluta convoluta* (Turbellaria Acoela) entstehen aus Diatomeen. Erster Nachweis einer Endosymbiose zwischen Tieren und Kieselalgen, *Naturwissenschaften*, 52, 444, 196.
188. Azad, H.S., and Borchardt, J.A., Variations in phosphorus uptake by algae, *Environ. Sci. Technol.*, 4, 737, 1970.
189. Azam, F., Hemmingsen, B.B., and Volcani, B.E., Role of silicon in diatom metabolism. V. Silicic acid transport and metabolism in the heterotrophic diatom *Nitzschia alba*, *Arch. Microbiol.*, 97, 103, 1974.
190. Azuara, M.P., and Aparicio, P.J., In vivo blue-light activation of *Chlamydomonas reinhardtii* nitrate reductase, *Plant Physiol.*, 71, 286, 1983.
191. Babel, U., Micromorphology of soil organic matter, in *Soil Components*, Vol. 1. *Organic Components*, Gieseking, I.E., Ed., Springer-Verlag, New York, Heidelberg, Berlin, 1975, 369.
192. Babich, H., and Stotzky, G., Effects of cadmium on the biota: influence of environmental factors, *Adv. Appl. Microbiol.*, 23, 55, 1978.
193. Babich, H., and Stotzky, G., Influence of chloride ions on the toxicity of cadmium to fungi, *Zentralbl. Bakteriol. Mikrobiol. Hyg. Ser. C*, 3, 421, 1982.
194. Babich, H., and Stotzky, G., Heavy metal toxicity to microbe-mediated ecologic processes: a review and potential application to regulatory policies, *Environ. Res.*, 36, 111, 1985.
195. Baccini, P., Untersuchingen über den Schwermetallhaushalt in Seen. *Schweiz. Z. Hydrol.*, 41, 176, 1976.
196. Bachmann, R.W., Zinc-65 in studies of the freshwater zinc cycle, in *Radioecology*, Schultz, V., and Klement, A.W., Eds., Reinhold Publ. Corp., New York, 1963, 485.
197. Bachmann, R.W., and Odum, E.P., Uptake of Zn^{65} and primary productivity in marine benthic algae, *Limnol. Oceanogr.*, 5, 349, 1960.
198. Bachmann, R.W., and Goldman, C.R., The determination of microgram quantities of molybdenum in natural waters, *Limnol. Oceanogr.*, 9, 143, 1964.
199. Bachrach, E., and Lefevre, M., Contribution a l'etude du role de la silice chey les etres vivants. Observations sur la biologie des Diatomées, *J. Physiol. Path. Gen.*, 27, 241, 1929.
200. Bacilevich, N.I., Produktivnost' i biologicheskii krugovorot v mnohovych bolotach yuzhnogo Vacjuganiya, *Rastit. Resursy*, 3, 567, 1967 (in Russian).
201. Baden, D.G., and Mende, T.J., Amino acid utilization by *Gymnodinium breve*, *Phytochemistry* 18, 247, 1979.
202. Badger, M.R., The fluxes of inorganic carbon species during photosynthesis in Cyanobacteria with particulate reference to *Synechococcus* sp., in *Inorganic Carbon Uptake by Aquatic Photosynthetic Organisms*, Lucas, W.J., and Berry, J.A., Eds., Am. Soc. Plant Physiol., Rockville, Maryland, 1985, 39.
203. Badger, M.R., Kaplan, A., and Berry, J.A., A mechanisms for concentrating CO_2 in *Chlamydomonas reinhardtii* and *Anabaena variabilis*, and its role in photosynthetic CO_2 fixation, *Carnegie Inst. of Washington Yearbook*, 77, 251, 1978.
204. Badger, M.R., Kaplan, A., and Berry, J.A., Internal inorganic carbon pool of *Chlamydomonas reinhardtii*. Evidence for a carbon dioxide concentrating mechanism, *Plant Physiol.*, 66, 407, 1980.

205. Badour, A.S.S., Analytisch-chemische Untersuchung des Kaliummangels bei *Chlorella* in Vergleich mit anderen Mangelzuständen, Ph.D. Thesis, University of Göttingen, Germany, 1959.

206. Badour, S.S., Experimental separation of cell division, and silica shell formation in *Cyclotella cryptica, Arch. Mikrobiol.*, 62, 17, 1968.

207. Bafford, R.A., Seagull, R.W., Chung, S.-Y., and Millie, D.F., Intracellular localization of the taste/odor metabolite 2-methylisoborneol in *Oscillatoria limosa* (Cyanophyta), *J. Phycol.*, 29, 91, 1993.

208. Bagnall, K.W., Selenium, tellurium and polonium, in *Comprehensive Inorganic Chemistry*, Vol. 2, Emeleus, J.C., Nyholm, H.J., and Trotman-Dickenson, A.F., Eds., Pergamon Press, Oxford, 1973, 935.

209. Bahls, L.L., Ecology of the diatom community of the upper East Gallatin River, Montana, with in situ experiments of effect of current velocity on features of the Aufwuchs, Ph.D. Dissertation, Montana State University, Bozeman, 1971.

210. Bahls, L.L., Benthic diatom diversity as a measure of water quality, *Proc. Nat. Acad. Sci.*, 38, 1, 1979.

211. Bailey-Watts, A.E., Planktonic diatoms and some marine diatom-silica relations in a shallow eutrophic Scottish loch, *Freshwat. Biol.*, 6, 69, 1976.

212. Bailey, D., Mazurak, A.P., and Rosowski, J.R., Aggregation on soil particles by algae, *J. Phycol.*, 9, 99, 1973.

213. Baillie, P.W., and Welsh, B.L., The effect of tidal resuspension on the distribution of intertidal epipelic algae in an estuary, *Estuar. Coast. Mar. Sci.*, 10, 165, 1980.

214. Baker, J.E., and Thompson, J.F., Assimilation of ammonia by nitrogen-starved cells of *Chlorella vulgaris, Plant Physiol.*, 36, 208, 1961.

215. Banerjee, S., and Mandal, L.N., Inorganic transformation of water soluble phosphates added in fish pond as influenced by the nature of the pond soil, *J. Indian Soc. Soil Sci.*, 13, 167, 1965.

216. Banus, M.D., Valiela, I., and Teal, J.M., Lead, zinc and cadmium budgets in experimentally enriched salt marsh ecosystems, *Estuar. Coast. Mar. Sci.*, 3, 421, 1975.

217. Baradziej, E., Net primary production of two marsh communities near Ispina in the Niepolonice Forest (Southern Poland), *Ekol. Pol.*, 22, 145, 1974.

218. Barber, J., Measurements of the membrane potential and evidence for active transport of ions in *Chlorella pyrenoidosa, Biochem. Biophys. Acta*, 150, 618, 1968.

219. Barber, R.T., White, A.W., and Siegelman, H.W., Evidence for a cryptomonad symbiont in the ciliate, *Cyclotrichium meunieri, J. Phycol.*, 5, 86, 1969.

220. Bariaud, A., Bonaly, J., and Delcourt, A., Divers aspects de l'action toxique du cadmium sur les cellules d'*Euglena gracilis*, in *Actualités de biochimie marine*. Colloque GABIM. La Rochelle, 1978, 135.

221. Barker, H.A., The metabolism of the colorless alga, *Prototheca zopfii* Krüger, *J. Cell. Comp. Physiol.*, 7, 73, 1935.

222. Barko, J.W., and Smart, R.M., The growth and biomass distribution of two emergent freshwater plants, *Cyperus esculentus* and *Scirpus validus,* on different sediments, *Aquat. Bot.*, 5, 109, 1978.

223. Barko, J.W., and Smart, R.M., Sediment-based nutrition of submersed macrophytes, *Aquat. Bot.*, 10, 339, 1981.

224. Barko, J.W., Murphy, P.G., and Wetzel, R.G., An investigation of primary production and ecosystem metabolism in a Lake Michigan dune pond, *Arch. Hydrobiol.*, 81, 155, 1977.

225. Barko, J.W., Gunnison, D., and Carpenter, S.R., Sediment interactions with submersed macrophyte growth and community dynamics, *Aquat. Bot.*, 41, 41, 1991.

226. Barsdate, R.J., and Alexander, V., The nitrogen balance of Arctic tundra: Pathways, rates and environmental implications, *J. Environ. Qual.*, 4, 111, 1975.

227. Bartlett, L.F., Rabe, W., and Funk, W.H., Effect of copper, zinc and cadmium on *Selenastrum capricornutum, Water Res.*, 8, 179, 1974.

228. Bartlett, M.S., Brown, L.C., Hanes, N.B., and Nickerson, N.H., Denitrification in freshwater soil, *J. Environ. Qual.*, 8, 460, 1979.

229. Bartlett, K.B., and Harris, R.C., Review and assessment of methane emissions from wetlands, *Chemosphere*, 26, 261, 1993.

230. Barton, D.R., and Lock, M.A., Numerical abundance and biomass of bacteria, algae and macrobenthos of a large northern river, the Athabasca, *Int. Rev. Ges. Hydrobiol.*, 64, 345, 1979.

231. Barton, L.L., Foster, E.W., and Johnson, G.V., Viability changes in human neutrophils and monocytes following exposure to toxin extracted from *Aphanizomenon flos-aquae, Can. J. Microbiol.*, 26, 272, 1980.

232. Barton, P., The acid mine drainage, in *Sulfur in the Environment*, Part. I. *Ecological Impacts*, Nriagu, J.O., Ed., Wiley, New York, 1978, 313.

233. Bartsch, A.F., Practical methods for control of algae and water weeds, *Public Health Rept.,* 69, 749, 1954.

234. Basilier, K., Investigations on nitrogen fixation in moss communities, in Flower-Ellis, J.G.K., Ed., Int. Biol. Prog., Swedish Tundra Biome Project Tech. Report 16, 83, 1973.

235. Basra, A., and Malik, C.P., Non-photosynthetic fixation of carbon dioxide and possible biological roles in higher plants, *Biol. Bull.,* 60, 357, 1985.

236. Bassham, J.A., Kinetic studies of the photosynthetic carbon reduction cycle, *Ann. Rev. Plant Physiol.,* 15, 101, 1964.

237. Bassham, J.A., The control of photosynthetic carbon metabolism, *Science,* 172, 526, 1971.

238. Bassham, J.A., and Calvin, M., *The Path of Carbon in Photosynthesis,* Prentice Hall, Englewood Cliffs, New Jersey, 1957.

239. Bastardo, H., Laboratory studies on decomposition of littoral plants, *Pol. Arch. Hydrobiol.,* 26, 267, 1979.

240. Bates, R.D., Biomass and primary productivity on an early successional floodplain forest site, in *Proc. Symp. Freshwater Wetlands and Wildlife,* Sharitz, R.R., and Gibbons, J.W., Eds., U.S. Dept. of Energy, 1989, 375.

241. Bates, S.S., Effects of light on nitrate uptake by two species of estuarine phytoplankton, *Limnol. Oceanogr,* 21, 212, 1976.

242. Bates, S.S., Tessier, A., Campbell, P.G.C., and Buffle, J., Zinc adsorption and transport by *Chlamydomonas variabilis* and *Scenedesmus subspicatus* (Chlorophyceae) grown in semicontinuous culture, *J. Phycol.,* 18, 521, 1982.

243. Bates, S.S., Letourneau, M., Tessier, A., and Campbell, P.G., Variation in zinc adsorption and transport during growth of *Chlamydomonas variabilis* (Chlorophyceae) in batch culture with daily additions of zinc, *Can. J. Fish. Aquat. Sci.,* 40, 895, 1983.

244. Batterton, J.C., and Van Baalen, C., Phosphorus deficiency and phosphate uptake in the blue-green alga *Anacystis nidulans, Can. J. Microbiol.,* 14, 341, 1968.

245. Batterton, J.C., and Van Baalen, C., Growth responses of blue-green algae to sodium chloride concentration, *Arch. Mikrobiol.,* 76, 151, 1971.

246. Bauhin, C., *Prodromus theatri botanici.* Basel, 1620.

247. Bauld, J., and Brock, T.D., Algal excretion and bacterial assimilation in hot spring algal mats, *J. Phycol.,* 10, 101, 1974.

248. Baum, L.S., and O'Kelley, J.C., Rubidium substitution for potassium in several species of fresh water algae, *Plant Physiol.,* 41 (suppl.), xxxiv, 1966.

249. Baxter, M., and Jensen, T.E., Uptake of magnesium, strontium, barium, and manganese by *Plectonema boryanum* (Cyanophyceae) with special reference to polyphosphate bodies, *Protoplasma,* 104, 81, 1980.

250. Baxter, M., and Jensen, T.E., A study of methods for in situ X-ray energy dispersive analysis of polyphosphate bodies in *Plectonema boryanum, Arch. Microbiol.,* 126, 213, 1980.

251. Bayley, S.E., The effect of natural hydroperiod fluctuation on freshwater wetlands receiving added nutrients, in *Ecological Considerations in Wetlands Treatment of Municipal Wastewater,* Godfrey, P.J., Kaynor, E.R., Pelczarski, S., and Benforado, J., Eds., Van Nostrand Reinhold Company, New York, 1985, 180.

252. Bayley, S.E., Behr, R.S., and Kelly, C.A., Retention and release of S from a freshwater wetland, *Water Air Soil Pollut.,* 31, 101, 1986.

253. Bayly, I.L., and Freemen, E.A., Seasonal variation of selected cations in *Acorus calamus* L., *Aquat. Bot.,* 3, 65, 1977.

254. Bazzaz, M.B., and Govindjee, Effects of lead chloride on chloroplast reactions, *Environ. Let.,* 6, 175, 1974.

255. Beadle, N-C.W., *The Vegetation of Australia,* Gustav Fischer Verlag, Stuttgart and New York, 1981.

256. Beamish, R.J., Blouw, L.M., and McFarlane, G.A., A fish and chemical study of 109 lakes in the Experimental Lakes Area (ELA), Northwestern Ontario, with appended reports on lake whitefish ageing errors and the Northwestern Ontario baitfish industry, *Fish. Mar. Serv. Res. Dev. Tech. Report* 607, 1976.

257. Bear, F.E., *Chemistry of the Soil,* Amer. Chem. Soc. Monogr. Series, Reinhold, New York, 1964.

258. Beardall, J., Occurrence and importance of HCO_3^- utilization in microscopic algae, in *Inorganic Carbon Uptake by Aquatic Photosynthetic Organisms,* Lucas, W.J., and Berry, J.A., Eds., *Am. Soc. Plant Physiol.,* Rockville, Maryland, 1985, 83.

259. Beardall, J., Photosynthesis and photorespiration in marine phytoplankton, *Aquat. Bot.,* 34, 347, 1989.

260. Beardall, J., and Morris, I., The concept of light intensity adaptation in marine phytoplankton: some experiments with *Phaeodactylum tricornutum, Mar. Biol.,* 37, 377, 1976.

261.   Beardall, J., and Raven, J.A., Transport of inorganic carbon and the "CO_2 concentrating mechanism" in *Chlorella emersonii* (Chlorophyceae), *J. Phycol.*, 17, 134, 1981.

262.   Beardall, J., and Entwisle, L., Evidence for a CO_2 concentrating mechanism in *Botrydiopsis* (Tribophyceae), *Phycologia*, 23,511, 1984.

263.   Beardall, J., Griffiths, H., and Raven, J.A., Carbon isotope discrimination and the CO_2 accumulating mechanism in *Chlorella emersonii, J. Exp. Bot.*, 33, 729, 1982.

264.   Beardall, J., Mukerji, D., Glover, H.E., and Morris, I., The path of carbon in photosynthesis by marine phytoplankton, *J. Phycol.*, 12, 409, 1976.

265.   Bebout, B.M., Paerl, H.W., Crocker, K.M., and Prufert, L.E., Diel interactions of oxygenic photosynthesis and N_2 fixation (acetylene reduction) in a marine microbial community, *Appl. Environ. Microbiol.*, 53, 2353, 1987.

266.   Becacos-Kontos, T., The annual cycle of primary production in the Saronicos Gulf (Aegean Sea) for the period November 1963-October 1964, *Limnol. Oceanogr.*, 13, 485, 1968.

267.   Beck, L.A., Oswald, W.J., and Goldman, J.C., Nitrate removal from agricultural tile drainage by photosynthetic system, Presented before Am. Soc. Civil. Eng., 2nd Nat. Symp. on Sanitary Engineering Research, Development and Design, Cornell University, Ithaca, New York, 1969 (cited by Goldman et al., 1972).

268.   Becker, E.W., Physiologische Untersuchungen zur Photosynthese von Algen unter extremen Temperaturbedingungen, Ph.D. Thesis, Tübingen, 1972.

269.   Becker, E.W., Limitations of heavy metals removal from waste water by means of algae, *Water Res.*, 17, 459, 1983.

270.   Becker, E.W., *Microalgae. Biotechnology and Microbiology,* Cambridge University Press, Cambridge, 1994.

271.   Beckett, P.M., Armstrong, W., Justin, S.H.F.W., and Armstrong, J., On the relative importance of connective and diffusive gas flows in plant aeration, *New Phytol.*, 110, 463, 1988.

272.   Bednar, T., and Holm-Hansen, O., Biotin literation by the lichen alga *Coccomyxa* sp. and by *Chlorella pyrenoidosa, Plant Cell Physiol.*, 5, 297, 1964.

273.   Bednarz, T., and Warkowska-Dratnal, H., Toxicity of zinc, cadmium, lead, copper, and their mixture for *Chlorella pyrenoidosa* Chick, *Acta Hydrobiol.* (Kraków), 25/26, 389, 1983/1984.

274.   Beech, P.L., and Wetherbee, R., Direct observations on flagellar transformation in *Mallomonas splendens* (Synurophyceae), *J. Phycol.*, 26, 90, 1990.

275.   Beeftink, W.G., Salt marshes, in *The Coastline*, Barnes, R.S.K., Ed., John Wiley and Sons, New York, 1977, 93.

276.   Beer, S., and Eshel, A., Photosynthesis of *Ulva* sp. I. Effects of desiccation when exposed to air, *J. Exp. Mar. Biol. Ecol.*, 70, 91, 1983.

277.   Beer, S., and Eshel, A., Photosynthesis of *Ulva* sp. II. Utilization of CO_2 and HCO_3^- when submerged, *J. Exp. Mar. Biol. Ecol.*, 70, 99, 1983.

278.   Beer, S., Spencer, W., and Bowes, G., Photosynthesis and growth of the filamentous blue-green alga *Lyngbya birgei* in relation to its environment, *J. Aquat. Plant. Manage.*, 24, 61, 1986.

279.   Beer, S., Spencer, W., Holbrook, G., and Bowes, G., Gas exchange and carbon fixation properties of the mat-forming cyanophyte *Lyngbya birgei* G.M. Smith, *Aquat. Bot.*, 38, 221, 1990.

280.   Beers, G.D., and Neuhold, J.M., Measurement of stream periphyton on paraffin-coated substrates, *Limnol. Oceanogr.*, 13, 559, 1968.

281.   Beers, J.R., and Stewart, G.L., Microzooplankton and its abundance relative to the larger zooplankton and other seston components, *Mar. Biol.*, 4, 182, 1969.

282.   Beers, J.R., and Stewart, G.L., Numerical abundance and estimated biomass of microzooplankton, in *The Ecology of the Plankton off La Jolla, California in the Period April through September 1967,* Strickland, J.D.H., Ed., University of California Press, Berkeley, 1970, 67.

283.   Beers, J.R., and Stewart, G.L., Microzooplankters in the plankton communities of the upper waters of the eastern tropical Pacific, *Deep Sea Res.*, 18, 861, 1971.

284.   Beevers, L., *Nitrogen Metabolism in Plants,* Edward Arnold, London, 1976.

285.   Beevers, L., and Hageman, R.H., Nitrate reduction in higher plants, *Ann. Rev. Plant Physiol.*, 20, 495, 1969.

286.   Beevers, L., and Hageman, R.H., The role of light in nitrate metabolism in higher plants, in *Photophysiology*, Giese, A.C., Ed., Academic Press, New York, 1972, 85.

287.   Beezley, B.B., Gruber, P.J., and Frederick, S.E., Cytochemical localization of glycolate dehydrogenase in mitochondria of *Chlamydomonas, Plant Physiol.*, 58, 315, 1976.

288.   Behning, A.L., Zur Erforschung der Flussboden der Wolga lebenden Organismen, *Monogr. Biol. Wolga Sta.*, 1, 1, 1924.

289.   Behning, A.L., Das Lebender Wolga. Zugleich eine Einführung in die Fluss-Biologie, in *Die Binnengewasser 5*, Thienemann, A., Ed., Stuttgart, 1928, 1.

290. Behre, K., Die Algensoziologie des Süsswassers (unter besonderer Berücksichtigung der Litoralalgen), Arch. Hydrobiol., 62, 125, 1956.

291. Bekheet, I.A., and Syrett, P.J., Urea-degrading enzymes in algae, Br. Phycol. J., 12, 137, 1977.

292. Bekheet, I.A., and Syrett, P.J., The uptake of urea by Chlorella, New Phytol., 82, 179, 1979.

293. Belay, A., and Fogg, G.E., Photoinhibition of photosynthesis in Asterionella formosa (Bacillariophyceae), J. Phycol., 14, 341, 1978.

294. Belcher, J.H., Reproduction and growth of the mixotrophic flagellate Furcilla stigmatophora (Skuja) Korsh. (Volvocales), Arch. Mikrobiol., 55, 327, 1967.

295. Belcher, J.H., The resistance to desiccation and heat of the asexual cysts of some freshwater Prasinophyceae, Br. Phycol. J., 5, 173, 1970.

296. Belcher, J.H., and Miller, J.D.A., Studies on the growth of Xanthophyceae in pure culture. IV. Nutritional types among the Xanthophyceae, Arch. Mikrobiol., 36, 219, 1960.

297. Bell, R.A., Cryptoendolithic algae of hot semiarid lands and deserts, J. Phycol., 29, 133, 1993.

298. Bell, R.A., Athey, P.V., and Sommerfeld, M.R., Distribution of endolithic algae on the Colorado Plateau of northern Arizona, Southwest. Nat., 33, 315, 1988.

299. Bellin, J.S., and Ronayne, M.E., Effects of photodynamic action on the cell membrane of Euglena, Physiol. Plant., 21, 1060, 1968.

300. Bellis, V.J., Unialgal cultures of Cladophora glomerata (L.) Kütz. 1. Response to temperature, J. Phycol., 4, 19, 1968.

301. Bellis, V.J., and McLarty, D.A., Ecology of Cladophora glomerata (L.) Kütz. in southern Ontario, J. Phycol., 3, 57, 1967.

302. Bellis, V.J., and Gaither, A.C., Seasonality and aboveground and belowground biomass for six salt marsh plant species, J. Elisha Mitchell Sci. Soc., 101, 95, 1985.

303. Belly, R.T., Tansey, M.R., and Brock, T.D., Algal excretion of ^{14}C-labelled compounds and microbial interactions in Cyanidium caldarium mats, J. Phycol., 9, 123, 1973.

304. Ben-Bassat, D., and Mayer, A.M., Volatilization of mercury by algae, Physiol. Plant., 33, 128, 1975.

305. Ben-Bassat, D., and Mayer, A.M., Reduction of mercury chloride by Chlorella: Evidence for a reducing factor, Physiol. Plant., 40, 157, 1977.

306. Ben-Bassat, D., Shelef, G., Gruner, N., and Shuval, H.I., Growth of Chlamydomonas in a medium containing mercury, Nature, 240, 43, 1972.

307. Bender, M.E., Matson, W.R., and Jordan, R.A., On the significance of metal complexing agents in secondary sewage effluents, Environ. Sci. Technol., 4, 520, 1970.

308. Bendix, S.W., Phototaxis, Bot. Rev., 26, 145, 1960.

309. Bendixen, L., Peterson, M.L., Tropism as a basis for flooding tolerance of strawberry clover to flooding conditions, Crop Sci., 2, 223, 1962.

310. Benecke, W., Über Culturbedingungen einiger Algen, Bot. Zeit., 56, 83, 1898.

311. Benes, P., Gjessing, E.T., and Steines, E., Interaction between humus and trace elements in fresh waters, Water Res., 10, 711, 1976.

312. Bennett, M.E., and Hobbie, J.E., The uptake of glucose by Chlamydomonas sp., J. Phycol., 8, 392, 1972.

313. Benson, A.A., and Calvin, M., Carbon dioxide fixation by green plants, Ann. Rev. Plant Physiol., 1, 25, 1950.

314. Benson, A.A., and Muscatine, L., Wax in coral mucus: energy transfer from corals to reef fishes, Limnol. Oceanogr., 19, 810, 1974.

315. Bentley, J.A., Role of plant hormones in algal metabolism and ecology, Nature, 181, 1499, 1958.

316. Bentley, J.A., Plant hormones in marine phytoplankton, zooplankton and sea water, J. Mar. Biol. Ass. U.K., 39, 433, 1960.

317. Bentley-Mowat, J.A., and Reid, S., Survival of marine phytoplankton in high concentrations of heavy metals, and uptake of copper, J. Exp. Mar. Biol. Ecol., 26, 249, 1977.

318. Benton, A.R., Jr., Wesley, P.J., and Rouse, J.W., Jr., Evapotranspiration from water hyacinth (Eichhornia crassipes (Mart.) Solms) in Texas reservoirs, Water Resour. Bull., 14, 919, 1978.

319. Bentrup, F.W., Elektrophysiologie der Zelle, Fortschr. Bot., 33, 51, 1971.

320. Berglend, L., Holtan, H., and Skulberg, O.M., Case studies on off-flavors in some Norwegian lakes, Water Sci. Technol., 15, 199, 1983.

321. Berland, B.R., Bonin, D.J., Cornu, A.L., Maestrini, S.Y., and Marino, J., The antibacterial substances of the marine alga Stichochrysis immobilis (Chrysophyta), J. Phycol., 8, 383, 1972.

322. Berland, B.R., Bonin, D.J., Fiala, M., and Meastrini, S.Y., Importance des vitamines en mer. Consommation et production par les algues et les bactéries. in Actualités de Biochimie marine, Colloque G.A.B.I.M. - C.N.R.S., Marseille, Ed. spéciale du C.N.R.S. Paris, 1976.

323.   Berland, B.R., Bonin, D.J., Kapkov, V.I., Maestrini, S.Y., and Arlhac, D.P., Action toxique de quatre métaux lourds sur la croissance d'algues unicellulaires marines, *C.R. Seance Acad. Sci. Ser. D*, 282, 633, 1976.

324.   Berland, B.R., Bonin, D.J., Guérid-Ancey, O.J., Kapkov, V.I., and Arlhac, D.P., Action de métaux lourds a des doses sublétales sur les caractéristiques de la croissance chez la diatomé *Skeletonema costatum, Mar. Biol.*, 42, 17, 1977.

325.   Berland, B.R., Chretiennot-Dinet, M.-J., Ferrara, R., and Arlhac, D., Action a court terme du mercure sur les populations naturelles phytoplanctoniques et bectériennes d' eaux côtieres de la Méditerranée nord-occidentale, *V^{es} J. Étud. Pollutions*, Cagliari, C.I.E.S.M., 721, 1980.

326.   Berman, T., Alkaline phosphatase and phosphorus availability in Lake Kinneret, *Limnol. Oceanogr.*, 15, 663. 1970.

327.   Berman, T., Chava, S., Kaplan, B., and Wynne, D., Dissolved organic substrates as phosphorus and nitrogen sources for axenic batch cultures of freshwater green algae, *Phycologia*, 30, 339, 1991.

328.   Bernard, J.M., Production ecology of wetland sedges: the genus *Carex, Polsk. Arch. Hydrobiol.*, 20, 207, 1973.

329.   Bernard, J.M., Seasonal changes in standing crop and primary production in a sedge wetland and an adjacent dry old-field in central Minnesota, *Ecology*, 55, 350, 1974.

330.   Bernard, J.M., and MacDonald, J.G., Jr., Primary production and life history of *Carex lacustris, Can. J. Bot.*, 53, 117, 1974.

331.   Bernard, J.M., and Bernard, F.A., Winter standing crop and nutrient contents in five central New York wetlands, *Bull. Torrey Bot. Club*, 104, 57, 1977.

332.   Bernard, J.M., and Solsky, B.A., Nutrient cycling in *Carex lacustris* wetland, *Can. J. Bot.*, 55, 630, 1977.

333.   Bernard, J.M., and Gorham, E., Life history aspects of primary production in sedge wetlands, in *Proc. Symp. Freshwater Wetlands: Ecological Processes and Management Potential*, Good, R.E., Whigham, D.F., and Simpson, R.L., Eds., Academic Press, New York, 1978, 39.

334.   Bernard, J.M., and Fitz, M.L., Seasonal changes in aboveground primary production and nutrient contents in a central New York *Typha glauca* ecosystem, *Bull. Torrey Bot. Club*, 106, 37, 1979.

335.   Bernard, J.M., and Hankinson, G., Seasonal changes in standing crop primary production, and nutrient levels in *Carex rostrata* wetland, *Oikos*, 32, 328, 1979.

336.   Bernard, J.M., and Fiala, K., Distribution and standing crop of living and dead roots in three wetland *Carex* species, *Bull. Torrey Bot. Club*, 113, 1, 1986.

337.   Bernard, J.M., and Bernard, F.A., Seasonal changes in copper, zinc, manganese, and iron levels in *Carex rostrata* Stokes, in *Proc. Symp. Freshwater Wetlands and Wildlife*, Sharitz, R.R., and Gibbons, J.W., Eds., U.S. Department of Energy, 1989, 343.

338.   Bernard, J.M., Seischab, F.K., and Jacoby, G., Life history and production of above- and below-ground structures of *Cladium mariscoides* (Muhl.) Torr. in a western New York fen, *Bull. Torrey Bot. Club*, 112, 288, 1985.

339.   Bernard, J.M., Solander, D., and Kvet, J., Production and nutrient dynamics in *Carex* wetlands, *Aquat. Bot.*, 30, 125, 1988.

340.   Bernatowicz, S., Macrophytes in lake Warniak and their chemical composition, *Ekol. Pol. Acta*, 17, 447, 1969.

341.   Bernatowicz, S., Leszczynski, S., and Tyczynska, S., The influence of transpiration by emergent plants on the water balance in lakes, *Aquat. Bot.*, 2, 275, 1976.

342.   Berns, D.S., Holohan, P., and Scott, E., Urease activity in blue-green algae, *Science*, 152, 1077, 1966.

343.   Bernstein, I.L., and Safferman, R.S., Viable algae in house dust, *Nature*, 227, 851, 1970.

344.   Berry, J., Boynton, J., Kaplan, A., and Badger, M., Growth and photosynthesis of *Chlamydomonas reinhardtii* as a function of CO_2 concentration, *Carnegie Inst. Wash. Year Book*, 75, 423, 1976.

345.   Berry, J.A., and Björkman, O., Photosynthetic response and adaptation to temperature in higher plants, *Ann. Rev. Plant. Physiol.*, 31, 491, 1980.

346.   Berry, J.A., and Raison, J.K., Response of macrophytes to temperature, in *Physiological Plant Ecology*. I. *Response to the Physical Environment*, Lange, O.L., Nobel, P.S., Osmond, C.B., and Ziegler, H., Eds., *Encyclopedia of Plant Physiology New Series*, Vol. 12 A, Springer-Verlag, Berlin, 1981, 277.

347.   Bertrand, D., Vanadium, an oligosynergic element for *Aspergillus niger*, *Ann. Inst. Past.*, 68, 226, 1942.

348.   Best, E.P.H., Seasonal changes in mineral and organic components of *Ceratophyllum demersum* and *Elodea canadensis*, *Aquat. Bot.*, 3, 337, 1977.

349.   Best, E.P.H., de Vries, D., and Reins, A., The macrophytes in the Loosdrecht Lakes: story of their decline in the course of eutrophication, *Verh. Internat. Verein. Limnol.*, 22, 868, 1984.

350. Betzer, N., and Kott, Y., Effect of halogens on algae. II. *Cladophora* sp., *Water Res.*, 3, 257, 1969.

351. Beveridge, T.J., and Murray, R.G.E., Sites of metal deposition in the cell wall of *Bacillus subtilis*, *J. Bacteriol.*, 141, 876, 1980.

352. Bhakuni, D.S., and Silva, M., Biodynamic substances from marine flora, *Bot. Mar.*, 17, 40, 1974.

353. Bharti, A., Saxena, R.P., and Pandey, G.N., Physiological imbalances due to hexavalent chromium in freshwater algae, *Ind. J. Environ. Health*, 21, 234, 1979.

354. Bhatta, K.S., Misra, M.K., and Misra, B.N., Community structure and standing crop biomass of a shallow pond of Orissa, a coastal province of India, *Indian J. Ecol.*, 9, 130, 1982.

355. Bibo, J., Schwermetalluntersuchungen an Wasser, Schwebstoffen, Aufwuchs und *Cladophora rivularis* der Elsenz. M.Sc. Thesis, University of Heidelberg, Germany, 1977.

356. Bican, J., Drbal, K., and Dykyjová, D., Heavy metal retention in aquatic macrophytes, in *Proc. Conf. Macrophytes in Water Management, Water Hygiene and Fishery*, Dum Techniky ČSVTS, České Budějovice, 1982, 39 (in Czech).

357. Bick, W., Robertson, A., Schneider, R., Scheiders, S., Ilnicki, P., *Slovnik torfoznawczy niemiecko-polsko-angielsko-rosyjki (Glossary for Bog and Peat, German-Polish-English-Russian)*.Biblioteczka Widomosci I.M.U.Z. No. 56, Instytut Melioracji i Uzitkow Zielonych, Warsaw, 1976.

358. Bidigare, R.R., Ondrusek, M.E., Kennicutt, M.C., II., Iturriaga, R., Harvey, H.R., Hoham, R.W., and Macko, S.A., Evidence for a photoprotective function for secondary carotenoids of snow algae, *J. Phycol.*, 29, 427, 1993.

359. Bidwell, R.G.S., Photosynthesis and metabolism of marine algae. II. A survey of rate and products of photosynthesis in $C_{14}O_2$, *Can. J. Bot.*, 36, 337, 1958.

360. Bidwell, R.G.S., Photosynthesis and light and dark respiration in fresh-water algae, *Can. J. Bot.*, 55, 809, 1977.

361. Bidwell, R.G.S., *Plant Physiology*, 2nd ed., Macmillan, New York, 1979.

362. Bidwell, R.G.S., Carbon nutrition in plants, in *Plant Physiology: A Treatise*, Vol. 7, Steward, F.C., and Bidwell, R.G.S., Eds., Academic Press, New York, 1983, 287.

363. Bidwell, R.G.S., and McLachlan, J., Carbon nutrition of seaweeds: Photosynthesis, photorespiration and respiration, *J. Exp. Mar. Biol. Ecol.*, 86, 15, 1985.

364. Bidwell, R.G.S., Krotkov, G., and Reed, G.B., Paper chromatography of sugars in plants, *Can. J. Bot.*, 30, 291, 1952.

365. Biecheler, B., Sur une chloromonadine nouvelle d'eau saumatre *Chattonella subsalsa* n. gen., n. sp., *Arch. Zool. Exp. Gén.*, 78, 79, 1936.

366. Bieleski, R.L., Phosphate pools, phosphate transport, and phosphate availability. *Ann. Rev. Plant Physiol.*, 24, 225, 1973.

367. Bienfang, P.K., Steady state analysis of nitrate-ammonium assimilation by phytoplankton, *Limnol. Oceanogr.*, 20, 402, 1975.

368. Bierman, V.J., Jr., Mathematical model of the selective enrichment of blue-green algae by nutrient enrichment, in *Mathematical Modeling of Biochemical Processes in Aquatic Ecosystems*, Canale, R.P., Ann Arbor Science, Ann Arbor, Michigan, 1976, 1.

369. Biggs, B,J.F., Effects of sample storage and mechanical blending on the quantitative analysis of river periphyton, *Freshwat. Biol.*, 18, 197, 1987.

370. Biggs, B.J.F., and Price, G.M., A survey of filamentous algal proliferations in New Zealand rivers, *N. Zealand J. Mar. Freshwat. Res.*, 21, 175, 1987.

371. Billard, C., and Fresnel, J., Nouvelles observations sur le *Pulvinaria feldmannii* (Bourrelly et Magne) comb. nov. (Chrysophycées, Sarcinochrysidales), formant une ceinture sur substrat meuble, *Crypto-gamie; Algol.*, 1, 281, 1980.

372. Bína, J. Ed., *Small Encyclopedia of Chemistry*, Obzor, Bratislava, Slovakia, 1968 (in Czech).

373. Bird, D.F., and Kalff, J., Bacterial grazing by planktonic lake algae, *Science*, 231, 493, 1986.

374. Bird, K.T., Simultaneous assimilation of ammonium and nitrate by *Gelidium nudifrons* (Gelidiales: Rhodophyta), *J. Phycol.*, 12, 238, 1976.

375. Bird, K.T., Habig, C., and De Busk, T., Nitrogen allocation and storage patterns in *Gracilaria tikvahiae* (Rhodophyta), *J. Phycol.*, 18, 344, 1982.

376. Birdsey, E.C., and Lynch, V.H., Utilization of nitrogen compounds by unicellular algae, *Science*, 137, 763, 1962.

377. Birmingham, B.C., and Colman, B., Measurement of carbon dioxide compensation points of freshwater algae, *Plant. Physiol.*, 64, 892, 1979.

378. Bisalputra, T., Plastids, in *Algal Physiology and Biochemistry* Stewart, W.D.P., Ed., Blackwell Scientific Publications, Oxford, 1974, 124.

379. Bishop, C.T., Adams, G.A., and Hughes, E.O., A polysaccharide from the blue-green alga *Anabaena cylindrica*, *Can. J. Chem.*, 32, 999, 1954.

380. Bishop, J.E., *Limnology of a Small Malayan River: Sungai Gombak,* Dr. W. Junk, Publishers, The Hague, The Netherlands, 1973.
381. Bitcover, E.H., and Sieling, D.H., Effect of various factors on the utilization of nitrogen and iron by *Spirodela polyrhiza* (L.) Schleid, *Plant Physiol.,* 26, 290, 1951.
382. Björk, M., Haglund, K., Ramazanov, Z., and Pedersén,M., Inducible mechanisms for HCO$_3^-$ utilization and repression of photorespiration in protoplasts and thalli of three species of *Ulva* (Chlorophyta), *J. Phycol.,* 29, 166, 1993.
383. Björk, S., Ecologic investigations of *Phragmites communis, Folia Limnologica Scandinavica,* 14, 1, 1967.
384. Björkman, O., The effect of oxygen concentration of photosynthesis in higher plants, *Physiol. Plant.,* 19, 618, 1966.
385. Björnsäter, B.R., and Wheeler, P.A., Effect of nitrogen and phosphorus supply on growth and tissue composition of *Ulva fenestrata* and *Enteromorpha interstinalis* (Ulvales, Chlorophyta), *J. Phycol.,* 26, 603, 1990.
386. Black, C.C., Jr., Photosynthetic carbon fixation in relation to net CO$_2$ uptake, *Ann. Rev. Plant Physiol.,* 24, 253, 1973.
387. Blackmer, A.M., and Bremner, J.M., Inhibitory effect of nitrate on reduction of N$_2$O to N$_2$ by soil microorganisms, *Soil Biol. Biochem.,* 10, 187, 1978.
388. Blackmer, A.M., and Bremner, J.M., Stimulatory effect of nitrate on reduction of N$_2$O and N$_2$ by soil microorganisms, *Soil Biol. Biochem.,* 11, 313, 1979.
389. Blakar, I., and Løvstad, Ø., Determination of available phosphorus for phytoplankton in lakes and rivers of southern Norway, *Hydrobiologia,* 192, 271, 1990.
390. Blake, G., Cagnaire-Michard, J., Kirassian, B., and Morand, P., Distribution and accumulation of zinc in *Typha latifolia,* in *Proc. Conf. Aquatic Plants for Watewater Treatment and Resource Recovery* Reddy, K.R., and Smith, W.H., Eds., Magnolia Publishing, Inc., Orlando, Florida, 1987, 487.
391. Blakeley, R.L., and Zerner, B., Jack bean urease: the first nickel enzyme, *J. Mol. Catal.,* 23, 263, 1984.
392. Blanck, H., and Wangberg, S.-A., Validity of an ecotoxicological test system: short-term and long-term effects of arsenate on marine periphyton communities in laboratory systems, *Can. J. Fish. Aquat. Sci.,* 45, 1807, 1988.
393. Blankenship, M.L., and Wilbur, K.M., Cobalt effects on cell division and calcium uptake in the coccolithophorid *Cricosphaera carterae* (Haptophyceae), *J. Phycol.,* 11, 211, 1975.
394. Blankley, W.F., Heterotrophic growth and calcification in coccolithophorids, in *XI Int. Bot. Congr. Abstr,* Seattle, 1969, 16.
395. Blasco, F., Exsorption de l'arsenic par les Chlorelles a la lumiere et a l'obscurite pris en présence des inhibiteurs des groupes SH, *C.R. Acad. Sci. Ser. D.,* 277, 2365, 1973.
396. Blaurock, A.E., and Walsby, A.E., Crystalline structure of the gas vesicle wall from *Anabaena flos-aquae, J. Mol. Biol.,* 105, 183, 1976.
397. Blindow, I., Phosphorus toxicity in *Chara, Aquat. Bot.,* 32, 393, 1988.
398. Blinn, D.W., The influence of sodium on the development of *Ctenocladus circinnatus* Borzi (Chlorophyceae), *Phycologia,* 9, 49, 1970.
399. Blinn, D.W., Tompkins, T., and Zaleski, L., Mercury inhibition on primary productivity using large volume plastic chambers in situ, *J. Phycol.,* 13, 58, 1977.
400. Blinn, D.W., Fredericksen, A., and Korte, V., Colonization rates and community structure of diatoms on three different rock substrata in a lotic system, *Br. Phycol. J.,* 15, 303, 1980.
401. Bliss, L.C., Plant productivity in alpine microenvironments on Mt. Washington, New Hampshire, *Ecol. Monogr,* 36, 125, 1966.
402. Bloom, P.R., Phosphorus adsorption by an aluminium-peat complex, *Soil Sci. Soc. Am. J.,* 45, 267, 1981.
403. Bloomfield, J.A., A mathematical model for decomposition and nutrient cycling in the pelagic zone of Lake George, New York, Ph.D. Thesis, Rensselaer Polytechnic Institute, Troy, New York, 1975.
404. Blum, J.J., Phosphate uptake by phosphate-starved *Euglena, J. Gen. Physiol.,* 49, 1125, 1966.
405. Blum, J.L., Vaucheriaceae. North American Flora, Series II, *The New York Botanical Garden,* 8, 1, 1972.
406. Blunden, G., Binns, W., and Perks, F., Commercial collection and utilization of maerl, *Econ. Bot.,* 29, 140, 1975.
407. Blunden, G., Barwell, C.J., Fidgen, K.J., and Jewers, K., A survey of some British marine algae for anti-influenza virus activity, *Bot. Mar,* 24, 267, 1981.

408.   Blunden, G., Farnham, W., Jephson, N., Barwell, C., Fenn, R., and Plunkett, B., The composition of maerl beds of economic interest in northern Brittany, Cornwall and Ireland, in *Proc. X. Int. Seaweed Symposium,* Levring, T., Ed., Göteborg, Sweden, 1981, 651.

409.   Boalch, G.T., Studies on *Ectocarpus* in culture II. Growth and nutrition of a bacteria-free culture, *J. Mar. Biol. Ass. U.K.,* 41, 287, 1961.

410.   Boar, R.R., Crook, C.E., and Moss, B., Regression of *Phragmites australis* reedswamps and recent changes of water chemistry in the Norfolk Broadland, England, *Aquat. Bot.,* 35, 41, 1989.

411.   Bock, E., Koops, H.-P., and Harms, H., Cell biology of nitrifying bacteria, in *Nitrification,* Prosser, J.I., Ed., Society for General Microbiology, IRL Press, Oxford, 1986, 17.

412.   Boczar, B.A., and Palmisano, A.C., Photosynthetic pigments and pigment-proteins in natural populations of Antarctic sea-ice diatoms, *Phycologia,* 29, 470, 1990.

413.   Bodin, K., and Nauwerck, A., Produktions-biologische Studien über die Moosvegetation eines klaren Gebirgssees, *Schweiz. Z. Hydrol.,* 30, 318, 1969.

414.   Boelter, D.H., Verry, E.S., Peatland and water in the northern lake States, USDA Forest Serv. Gen. Tech. Rep. NC-31, NC For. Exp. Stn., 1977.

415.   Bogorad, L., Chlorophylls, in *Physiology and Biochemistry of Algae,* Lewin, R.A., Ed., Academic Press, New York and London, 1962, 385.

416.   Bogorad, L., Phycobiliproteins and complementary chromatic adaptation, *Ann. Rev. Plant Physiol.,* 26, 369, 1975.

417.   Böhm, E.L., and Goreau, T.F., Composition and calcium binding properties of the water soluble polysaccharides in the calcareous alga *Halimeda opuntia* (L.), *Int. Rev. ges. Hydrobiol.,* 58, 117, 1973.

418.   Böhm, L., Fütterer, D., and Kaminski, E., Algal calcification in some Codiaceae (Chlorophyta): ultrastructure and location of skeletal deposits, *J. Phycol.,* 14, 486, 1978.

419.   Bohn, H.L., McNeal, B.L., and O'Connor, G.A., *Soil Chemistry* 2nd ed. John Wiley and Sons, New York, 1985.

420.   Bohonková, I., Occurrence of mass populations of filamentous algae in ponds near Trebon, M.Sc. Thesis, Pedagogická Fakulta, Ceské Budejovice, Czech Republic, 1977.

421.   Bold, H.C., The cultivation of algae, *Bot. Rev.,* 8, 69, 1942.

422.   Bold, H.C., Cytology of algae, in *Manual of Phycology,* Smith, G.M., Ed., Chronica Botanica, Waltham, Massachusetts, 1951, 203.

423.   Bold, H.C., Some aspects of the taxonomy of soil algae, *Ann. N.Y. Acad. Sci.,* 175, 601, 1975.

424.   Bold, H.C., and Wynne, M.J., *Introduction to the Algae; Structure and Reproduction,* 2nd ed., Prentice-Hall, Inc., Englewood Cliffs, New Jersey, 1985.

425.   Bold, H.C., Alexopoulos, C.J., and Delevoryas, T., *Morphology of Plants and Fungi,* 4th ed., Harper and Row Publishers, New York, 1980.

426.   Bold, H.C., Cronquist, A., Jeffrey, C., Johnson, L.A.S., Margulis, L., Merxmiller, H., Raven, P.H., and Takhtajan, A.L., Proposal (10) to substitute the term "phylum" for "division" for groups treated as plants, *Taxon,* 27, 121, 1978.

427.   Bolen, E.G., Playa wetlands of the U.S. southern High Plains: their wildlife values and challenges for management, in *Proc. Int. Conf. Wetlands: Ecology and Management,* Gopal, B., Turner, R.E., Wetzel, D.F., and Whigham, D.F., Eds., Natl. Inst. of Ecology and Internat. Scientific Publications, Jaipur, India, 1982, 9.

428.   Bolin, B., The carbon cycle, in *The Major Biogeochemical Cycles and Their Interactions,* Bolin, B., and Cook, R.B., Eds., SCOPE Rept. 21, John Wiley and Sons, Chichester, 1983, 41.

429.   Bolin, B., Degens, E.T., Duvigneaud, P., and Kempe, S., The global biogeochemical carbon cycle, in *The Global Carbon Cycle,* Bolin, B., Degens, E.T., Kempe, S., and Ketner, P., Eds., SCOPE Report 13, John Wiley and Sons, Chichester, 1979, 1.

430.   Bollard, E.G., and Butler, G.W., Mineral nutrition of plants, *Ann. Rev. Plant Physiol.,* 17, 77, 1966.

431.   Bollman, R.C., and Robinson, G.G.C., The kinetics of organic acid uptake by three chlorophyta in axenic culture, *J. Phycol.,* 13, 1, 1977.

432.   Bonaly, J., Delcourt, A., and Mestre, J.C., The effects of some metallic ions on the growth of *Chlamydomonas variabilis* and *Euglena gracilis. Mitt. Internat. Verein. Limnol.,* 21, 103, 1978.

433.   Bone, D.H., Relationship between phosphates and alkaline phosphatase of *Anabaena flos-aquae* in continuous culture, *Arch. Mikrobiol.,* 80, 147, 1971.

434.   Boney, A.D., *A Biology of Marine Algae,* Hutchinson, London, 1966.

435.   Boney, A.D., Experimental studies on the benthic phases of Haptophyceae. II. Studies on "aged" benthic phases of *Pleurochrysis scherffellii* E.G. Prings., *J. Exp. Mar. Biol. Ecol.,* 1, 7, 1967.

436.   Boney, A.D., Scale-bearing flagellates: an interim review, *Oceanogr. Mar. Biol. Ann. Rev.,* 8, 281, 1970.

437.   Boney, A.D., *Phytoplankton,* Studies in Biology No. 52, Edward Arnold, London, 1975.

438. Boney, A.D., Taxonomy of red and brown algae, in *Modern Approaches to the Taxonomy of Red and Brown Algae,* Irvine, D.E.G., and Price, J.H., Eds., Academic Press, New York, 1978, 1.

439. Boney, A.D., and Green, J.C., Prymnesiophyceae (Haptophyceae): introduction and bibliography, in *Selected Papers in Phycology* II, Rosowski, J.R., and Parker, B.C., Eds., Phycological Society of America, Lawrence, Kansas, 1982, 705.

440. Boney, A.D., Corner, E.D., and Sparrow, B.E.P., The effects of various poisons on the growth and variability of sporelings of the red alga *Plumaria elegans* (Bonnem) Sch., *Biochem. Pharmacol.,* 2, 37, 1959.

441. Bongers, L.H.J., Aspects of nitrogen assimilation by cultures of green algae, *Meded. Landbouwhogesch. Wageningen,* 56, 1, 1956.

442. Bonin, D.J., and Maestrini, S.Y., Importance of organic nutrients for phytoplankton growth in natural environments: implications for algal species succession, in *Physiological Bases of Phytoplankton Ecology,* Platt,t., Ed., *Can. Bull., Fish. Aquat. Sci.,* 210, 279, 1981.

443. Bonin, D.J., Maestrini, S.Y., and Leftley, J.W., Some processes and physical factors that affect the ability of individual species of algae to compete for nutrient partition, in *Physiological Bases of Phytoplankton Ecology,* Platt, T. Ed., *Can. Bull. Fish. Aquat. Sci.,* 210, 292, 1981.

444. Bonner, J., and Galston, A.W., *Principles of Plant Physiology,* W.H. Freeman and Co., San Francisco, 1952.

445. Bonotto, S., Cultivation of plants. Multicellular plants, in *Marine Ecology,* Vol. 3: *Cultivation,* Part 1, Kinne, O., Ed., John Wiley and Sons, New York, 1976, 467.

446. Booth, E., The manufacture and properties of liquid seaweed extracts, in *Proc. VI. Int. Seaweed Symposium,* Margalef, R., Ed., Madrid, Spain, 1969, 655.

447. Booth, E., Some factors affecting seaweed fertilizers, in *Proc. VIII. Int. Seaweed Symposium,* Fogg, G.E., and Jones, W.E., Eds., Marine Sci. Lab., Menai Bridge, Wales, 1981, 661.

448. Booth, W.E., Algae as pioneers in plant succession and their importance in erosion control, *Ecology,* 22, 38, 1941.

449. Borchardt, J.A., and Azad, H.S., Biological extraction of nutrients, *J. Water Pollut. Control Fed.,* 40, 1739, 1968.

450. Borg, H., Trace metals in Swedish natural freshwaters, *Hydrobiologia,* 101, 27, 1983.

451. Borman, A.H., de Jong, E.W., Huizinga, M., Kok, D.J., Westbroek, P., and Bosch, L., The role of $CaCO_3$ crystallization of an acid Ca^{2+}-polysaccharide associated with coccoliths of *Emiliania hyxleyi,* *Eur. J. Biochem.,* 129, 179, 1982.

452. Bornet, E., and Flahault, C., Revision des Nostocacées hétérocystées. *Ann. Sci. Nat., Bot. Ser.* 7, 3, 323; 4, 343; 5, 51; 7, 177, 1886–1888.

453. Borowitzka, M.A., Algal calcification, *Oceanogr. Mar. Biol.,* 15, 189, 1977.

454. Borowitzka, M.A., Photosynthesis and calcification in the articulated coralline red algae *Amphiroa anceps* and *A. foliacea, Mar. Biol.,* 62, 17, 1981.

455. Borowitzka, M.A., Mechanisms in algal calcification, in *Progress in Phycological Research,* Vol. 1, Round, F.E., and Chapman, C.H., Eds., Elsevier/North Holland Biomedical Press, Amsterdam, New York, 1982, 137.

456. Borowitzka, M.A., and Larkum, A.W.D., Calcification in the green alga *Halimeda.* II. The exchange of Ca^{2+} and the occurrence of age gradients in calcification and photosynthesis, *J. Exp. Bot.,* 27, 864, 1976.

457. Borowitzka, M.A., and Larkum, A.W.D., Calcification in the green alga *Halimeda.* III. The source of inorganic carbon for the photosynthesis and calcification and a model of the mechanism of the calcification, *J. Exp. Bot.,* 27, 879, 1976.

458. Borowitzka, M.A., and Larkum, A.W.D., Calcification in the green alga *Halimeda.* IV. The action of metabolic inhibitors on photosynthesis and calcification, *J. Exp. Bot.,* 27, 894, 1976.

459. Borowitzka, M.A., and Larkum, A.W.D., Calcification in the green alga *Halimeda.* I. An ultrastructure study of thallus development, *J. Phycol.,* 13, 6, 1977.

460. Borowitzka, M.A., Larkum, A.W.D., and Nockolds, C.E., A scanning electron microscope study of the structure and organization of the calcium caron deposits of algae, *Phycologia,* 13, 195, 1974.

461. Bortels, H., Molybdans als Katalysator bei der biologischen Stickstoffbindung, *Arch. Mikrobiol.,* 1, 333, 1930.

462. Bortels, H., Über die Bedeutung des Molybdäns für stickstoffbindende Nostocaceen, *Arch. Mikrobiol.,* 11, 155, 1940.

463. Borum, J., The quantitative role of macrophytes, epiphytes, and phytoplankton under different nutrient conditions in Roskilde Fjord, Denmark, *Proc. Int. Symp. on Aquatic Macrophytes,* Nijmegen, The Netherlands, 1983, 35.

464. Borum, J., Development of epiphytic communities on eelgrass (*Zostera marina*) along a nutrient gradient in a Danish estuary, *Mar. Biol.,* 87, 233, 1985.

465. Borzi, A., *Study algologiei.* Palermo, 1889.
466. Bosserman, R.W., Elemental composition of *Utricularia*—periphytonecosystems from Okefenokee Swamp, *Ecology*, 64, 1637, 1983.
467. Boston, H.L., and Adams, M.S., Productivity, growth and photosynthesis of two small "isoetid" plants, *Littorella uniflora* and *Isoetes macrospora*, *J. Ecol.*, 75, 333, 1987.
468. Boston, H.L., Adams, M.S., and Madsen, J.D., Photosynthetic strategies and productivity in aquatic systems, *Aquat. Bot.*, 34, 27, 1989.
469. Boström, B., Jansson, M., and Forsberg, C., Phosphorus release from lake sediments, *Arch. Hydrobiol. Beih. Ergebn. Limnol.*, 18, 5, 1982.
470. Bothe, H., The role of phytoflavin in photosynthetic reactions, in *Progress in Photosynthesis Research*, Metzner, H.H., and Laupp, B.R., Jr., Eds., Tübingen, 1969, 1482.
471. Bothe, H., Nitrogen fixation, in *The Biology of Cyanobacteria*, Carr, N.G., and Whitton, B.A., Eds., Blackwell Scientific Publications, Oxford, 1982, 87.
472. Bothwell, M.L., All-weather troughs for periphyton studies, *Water Res.*, 17, 1735, 1983.
473. Bothwell, M.L., Phosphorus limitation of lotic periphyton growth rates: an intensive comparison using continuous flow troughs (Thompson River System, British Columbia), *Limnol. Oceanogr*, 30, 527, 1985.
474. Bothwell, M.L., and Jasper, S., A light and dark trough methodology for measuring rates of lotic periphyton settlement and net growth. An evaluation through intersite comparison, in *Periphyton of Freshwater Ecosystems*, Wetzel, R.G., Ed., Dr. W. Junk Publishers, The Hague, The Netherlands, 1983, 253.
475. Bothwell, M.L., Suzuki, K.E., Bolin, M.K., and Hardy, F.J., Evidence of dark avoidance by phototrophic periphytic diatoms in lotic systems, *J. Phycol.*, 25, 85, 1989.
476. Bothwell, M.L., Sherbot, D., Roberge, A.C., and Daley, R.J., Influence of natural ultraviolet radiation on lotic periphyton diatom community growth, biomass accrual, and species composition: short-term versus long-term effects, *J. Phycol.*, 29, 24, 1993.
477. Boto, K.G., and Patrick, W.H., Jr., Role of wetlands in the removal of suspended sediments, in *Wetland Functions and Values: The State of Our Understanding*, Greeson, J.R., and Clark, J.E., Eds., American Water Resources Association, Minneapolis, Minnesota, 1979, 479.
478. Botsch, M.S., and Masing, V., Mire ecosystems in the USSR, in *Ecosystems of the World*, Vol. 4B. *Mires: Swamp, Bog, Fen and Moor*, Gore, A.J.P., Ed., Elsevier Science Publishers, Amsterdam, 1983, 95.
479. Bouck, G.B., Fine structure and organelle associations in brown algae, *J. Cell Biol.*, 26, 523, 1965.
480. Bouck, G.B., and Sweeney, B.M., The fine structure and ontogeny of trichocysts in marine dinoflagellates, *Protoplasma*, 61, 205, 1966.
481. Bourrelly, P., Recherches sur les Chrysophycées. Morphologie, phylogénie, systématique, *Rev. Algol. Mém. Hors-Sér* 1, 1, 1957.
482. Bourrelly, P., Chrysophycées et phylogénie, *Vorträge Gesamtgeb. Bot., Deutsch. Bot. Gesells. N. F.*, 1, 32, 1962.
483. Bourrelly, P., La classification des Chrysophycés, ses problémes, *Rev. Algol.*, 1, 56, 1965.
484. Bourrelly, P., *Recherches sur les Chrysophycées. Morphologie, phylogénie, systématique*, Rev. Algol. Mém. Hors-Sér, No. 1, 1968.
485. Bourrelly, P., *Les algues d'eau douce. Initiation a la systématique. II. Les algues jaunes et brunes. Chrysophycées, Phéophycées, Xanthophycées et Diatomées.* N. Boubée, Paris, 1968.
486. Bourrelly, P., *Les algues d' eau douce. Initiation a la systématique. III. Les algues bleures et rouges, les Eugléniens, Péridiniens et Cryptomonadines,* N. Boubée, Paris, 1970.
487. Bourrelly, P., *Les algues d'eau douce. Initiation a la systématique. I. Les algues vertes,* Rev. ed., N. Boubés, Paris, 1972.
488. Bowden, W.B., Nitrification, nitrate reduction, and nitrogen immobilization in a tidal freshwater marsh sediment, *Ecology*, 67, 88, 1986.
489. Bowden, W.B., The biogeochemistry of nitrogen in freshwater wetlands, *Biogeochemistry*, 4, 313, 1987.
490. Bowen, H.J.M., *Trace Elements in Biochemistry* Academic Press, London, 1966.
491. Bowen, H.J.M., *Environmental Chemistry of the Elements*, Academic Press, London, 1979.
492. Bowen, H.J.M., The cycles of copper, silver and gold, in *The Handbook of Environmental Chemistry* Vol. 1, Part D, *The Natural Environment and the Biogeochemical Cycles*, Hutzinger, O., Ed., Springer-Verlag, Berlin, 1985, 1.
493. Bowen, J.E., Gauch, H.G., Krauss, R.W., and Galloway, R.A., The nonessentiality of boron for *Chlorella, J. Phycol.*, 1, 151, 1965.
494. Bowes, G.W., Carbonic anhydrase in marine algae, *Plant Physiol.*, 44, 726, 1969.

495.   Bowes, G., and Beer, S., Physiological plant processes: Photosynthesis. in *Aquatic Plants for Water Treatment and Resource Recovery*, Reddy, K.R., and Smith, W.H., Eds., Magnolia Publishing, Orlando, Florida, 1987, 311.

496.   Bowes, G., and Salvucci, M.E., Plasticity in the photosynthetic carbon metabolism of submersed aquatic macrophytes, *Aquat. Bot.*, 34, 233, 1989.

497.   Bowker, D.W., and Denny, P., The periphyton communities of Nyumba ya Mungu reservoir, Tanzania, *Biol. J. Linn. Soc.*, 10, 49, 1978.

498.   Boyd, C.E., The elemental composition of several freshwater algae, Ph.D. Thesis, Auburn University, Auburn, Alabama, 1966.

499.   Boyd, C.E., Freshwater plants: A potential source of protein, *Econ. Bot.*, 22, 359, 1968.

500.   Boyd, C.E., Production, mineral nutrient absorption and biochemical assimilation by *Justicia americana* and *Alternanthera philoxeroides*, *Arch. Hydrobiol.*, 66, 139, 1969.

501.   Boyd, C.E., Boron accumulation by native algae, *Am. Midl. Nat.*, 84, 565, 1970.

502.   Boyd, C.E., Seasonal changes in the proximate composition of some common aquatic weeds, *Hyacinth Contr. J.*, 8, 42, 1970.

503.   Boyd, C.E., Losses of mineral nutrients during decomposition of *Typha latifolia*, *Arch. Hydrobiol.*, 66, 511, 1970.

504.   Boyd, C.E., Production, mineral accumulation and pigment concentrations in *Typha latifolia* and *Scirpus americanus*, *Ecology*, 51, 285, 1970.

505.   Boyd, C.E., Vascular aquatic plants for mineral nutrient removal from polluted waters, *Econ. Bot.*, 24, 95, 1970.

506.   Boyd, C.E., Factors influencing shoot production and mineral nutrient levels in *Typha latifolia*, *Ecology*, 51, 296, 1970.

507.   Boyd, C.E., Chemical analyses of some vascular aquatic plants, *Arch. Hydrobiol.*, 67, 78, 1970.

508.   Boyd, C.E., The dynamics of dry matter and chemical substances in a *Juncus effusus* population, *Am. Midl. Nat.*, 86, 28, 1971.

509.   Boyd, C.E., Accumulation of dry matter nitrogen and phosphorus by cultivated water hyacinth, *Econ. Bot.*, 30, 51, 1976.

510.   Boyd, C.E., Chemical composition of wetland plants, in *Proc. Symp. Freshwater Wetlands: Ecological Processes and Management Potential*, Good, R.E., Whigham, D.F., and Simpson, R.L., Eds., Academic Press, New York, 1978, 155.

511.   Boyd, C.E., and Vickers, D.H., Variation in the elemental content of *Eichhornia crassipes*, *Hydrobiologia*, 38, 409, 1971.

512.   Boyd, C.E., and Vickers, D.H., Relationship between production, nutrient accumulation, and chlorophyll synthesis in an *Eleocharis quadrangulata* population, *Can. J. Bot.*, 49, 883, 1971.

513.   Boyd, C.E., and Blackburn, R.D., Seasonal changes in the proximate composition of some common aquatic weeds, *Hyacinth Contr. J.*, 8, 42, 1970.

514.   Boyd, C.E., and Hess, W., Factors influencing shoot production and mineral nutrient level in *Typha latifolia*, *Ecology*, 51, 296, 1970.

515.   Boyd, C.E., and Scarsbrook, E., Influence of nutrient additions and initial density of plants on production of water hyacinth *Eichhornia crassipes*, *Aquat. Bot.*, 1, 252, 1975.

516.   Boyd, C.E., and Walley, W.W., Studies of the biogeochemistry of boron. I. Concentrations in surface waters, rainfall and aquatic plants, *Am. Midl. Nat.*, 88, 1, 1972.

517.   Boyle, J.E., and Smith, D.C., Biochemical interactions between the symbionts of *Convoluta roscoffensis*, *Proc. R. Soc. Lond.*, B 189, 121, 1975.

518.   Boyt, F.L., Bayley, S.E., and Zoltek, J., Jr., Removal of nutrients from treated municipal waste water by wetland vegetation, *J. Water Pollut. Control Fed.*, 49, 789, 1977.

519.   Braarud, T., Deflandre, G., Halldal, P., and Kamptner, E., Terminology, nomenclature, and systematics of the Coccolithophoridae, *Micropaleontology*, 1, 157, 1955.

520.   Bradbeer, C., Transport of vitamin B_{12} in *Ochromonas malhamensis*, *Arch. Biochem. Biophys.*, 144, 184, 1971.

521.   Bradford, G.R., Bair, F.L., and Hunsaker, V., Trace and major element content of 170 High Sierra lakes in California, *Limnol. Oceanogr.*, 13, 526, 1968.

522.   Bradis, E.M., Pro printsipy tipologii bolot SSR, *Ukr. Bot. Zh.*, 30, 681, 1973.

523.   Bradley, S., and Carr, N.G., Heterocyst development in *Anabaena cylindrica*: the necessity of light as an initial trigger and sequential stages of commitment, *J. Gen. Microbiol.*, 101, 291, 1977.

524.   Brady, N.C., *The Nature and Properties of Soils*, 8th ed., Macmillan Publishing Company, New York, 1974.

525.   Bræk, G.S., Jensen, A. and Mohus, A., Heavy metal tolerance of marine phytoplankton. III. Combined effects of copper and zinc ions on cultures of four common species, *J. Exp. Mar. Biol. Ecol.*, 25, 37, 1976.

526. Bræk, G.S., Malnes, D., and Jensen, A., Heavy metal tolerance of marine phytoplankton. IV. Combined effect of zinc and cadmium in growth and uptake in some marine diatoms, *J. Exp. Mar. Biol. Ecol.*, 42, 39, 1980.

527. Braekke, F.H., Hydrochemistry of high altitude catchments in South Norway. 1. Effects of summer droughts and soil-vegetation characteristics, Rep. Norway For. Res. Inst. 36–8, 1981.

528. Bramryd, T., The conservation of peatlands as global carbon accumulation, in *Proc. Int. Symp. Classification of Peats and Peatlands*, International Peat Society, Helsinki, 1979, 297.

529. Bramryd, T., The role of peatlands for the global carbon dioxide balance, in *Proc. 6th Internat. Peat Congress*, Int. Peat Soc., Duluth, Minnesota, Int. Peat Soc., Helsinki, 1980, 9.

530. Brand, L.E., Sunda, W.G., and Guillard, R.R.L., Limitation of marine phytoplankton reproductive rates by zinc, manganese, and iron, *Limnol. Oceanogr.*, 28, 1182, 1983.

531. Brandes, D., and Elston, R.N., An electron microscopical study of the histochemic localization of alkaline phosphatase in the cell wall of *Chlorella vulgaris, Nature*, 274, 1956.

532. Brändle, R., Die Überflutungstoleranz der Seebinse (*Schoenoplectus lacustris* (L.) Palla). IV. Dissimilation und "Energy Charge" in Rhizomgewebeschnitten und Normaxia, Hypoxia and Anoxia, *Bot. Helvetica*, 91, 49, 1981.

533. Bratbak, G., and Thingstad, T.F., Phytoplankton-bacteria interaction: An apparent paradox? Analysis of a model system with both competition and commensalism, *Mar. Ecol. Prog. Ser.*, 25, 23, 1985.

534. Braun, A., *Betrachtungen über die Erscheinung der Versungung in der Nature.* Leipzig, 1851.

535. Bray, J.R., The chlorophyll content of some native and managed plant communities in Central Minnesota, *Can. J. Bot.*, 38, 313, 1960.

536. Bray, J.R., Estimates of energy budgets for a *Typha* (cattail) marsh, *Science,* 136, 1119, 1962.

537. Bray, J.R., Lawrence, D.B., and Pearson, L.C., Primary production in some Minnesota terrestrial communities for 1957, *Oikos,* 10, 39, 1959.

538. Brendemuhl, I., Über die Verbreitung der Erddiatomen, *Arch. Mikrobiol.*, 14, 407, 1949.

539. Bressler, S.L., and Ahmed, S.I., Detection of glutamine synthetase activity in marine phytoplankton: optimization of the biosynthetic assay, *Mar. Ecol. Prog. Ser.*, 14, 207, 1984.

540. Brezonik, P.L., Nitrogen: sources and transformation in natural waters, in *Nutrients in Natural Waters,* Allen, H.E., and Kramer, J.R., Eds., John Wiley, New York, 1972, 1.

541. Briand, F., Trucco, R., and Ramamoorthy, S., Correlation between specific algae and heavy metal binding in lakes, *J. Fish. Res. Board Can.*, 35, 1482, 1978.

542. Bridgham, S.D., Richardson, C.J., Maltby, E., and Faulkner, S.P., Cellulose decay in natural and disturbed peatlands in North Carolina, *J. Environ. Qual.,* 20, 695, 1991.

543. Bridgham, S.D., and Richardson, C.J., Hydrology and nutrient gradients in North Carolina peatlands, *Wetlands*, 13, 207, 1993.

544. Briggs, S.W., Freshwater wetlands, in *Australian Vegetation,* Gooves, R.H., Ed., Cambridge University Press, London, 1981, 335.

545. Bringmann, G., and Kühn, R., Vergleichende abwassertoxikologische Untersuchungen an Bakterien, Algen und Kleinlebewesen, *Gesundheitsingenieur* 80, 115, 1959.

546. Brinkmann, R., and van Diepen, C.A., Mineral soils, in *Wetlands and Shallow Continental Water Bodies,* Patten, B.C., Ed., SPB Academic Publishing bv, The Hague, The Netherlands, 1990, 37.

547. Brinson, M.M., Lugo, A.E., and Browns, S., Primary productivity, decomposition and consumer activity in freshwater wetlands, *Ann. Rev. Ecol. Systemat.*, 12, 123, 1981.

548. Bristol, B.M., On the alga-flora of some desiccated English soils: an important factor of soil biology, *Ann. Bot.*, 34, 35, 1920.

549. Bristol, M., On the retention of vitality by algae from old stored soils, *New Phytol.*, 18, 92, 1919.

550. Bristol Roach, B.M., On the relation of certain soil algae to some soluble carbon compounds, *Ann. Bot.*, 40, 149, 1926.

551. Bristol Roach, B.M., On the carbon nutrition of some algae isolated from soils, *Ann. Bot.*, 41, 509, 1927.

552. Bristol Roach, B.M., The present position of our knowledge of the distribution and function of algae in the soil, Proc. *1st Cong. Soil Sci.*, Washington, DC, 1927, 30.

553. Bristol Roach, B.M., On the influence of light and of glucose on the growth of a soil alga, *Ann. Bot.*, 42, 317, 1928.

554. Bristow, J.M., The structure and function of roots in aquatic vascular plants, in *The Development and Function of Roots,* Torrey, J.G., and Clarkson, D.T., Eds., Academic Press, New York, 1974.

555. Bristow, J.M., Nitrogen fixation in the rhizosphere of freshwater angiosperms, *Can. J. Bot.*, 52, 217, 1974.

556. Bristow, J.M., and Whitcombe, M., The role of roots in the nutrition of aquatic vascular plants, *Am. J. Bot.*, 58, 8, 1971.

557.   Brix, H., Gas exchange through dead clums of reed *Phragmites australis* (Cav.) Trin. ex Steudel, *Aquat. Bot.*, 35, 81, 1989.

558.   Brix, H., Macrophyte-mediated oxygen transfer in wetlands: transport mechanisms and rates, in *Constructed Wetlands for Water Quality Improvement*, Moshiri, G.A., Ed., Lewis Publishers, Boca Raton, Florida, 1993, 391.

559.   Broady, P.A., Qualitative and quantitative observations on green and yellow-green algae in some English soils, *Br. phycol. J.*, 14, 151, 1979.

560.   Brochiero, E., Bonaly, J., and Mestre, J.C., Toxic action of hexavalent chromium on *Euglena gracilis* cells strain Z grown under heterotrophic conditions, *Arch. Environ. Contam. Toxicol.*, 13, 603, 1984.

561.   Brock, T.D., *Principles of Microbial Ecology*, Prentice-Hall, Inc., Englewood Cliffs, New Jersey, 1966.

562.   Brock, T.D., Life at high temperatures, *Science*, 158, 1012, 1967.

563.   Brock, T.D., Microbial growth under extreme environments, *Symp. Soc. Gen. Microbiol.*, 19, 15, 1969.

564.   Brock, T.D., Photosynthesis by algal epiphytes of *Utricularia* in Everglades National Park., *Bull. Mar. Sci.*, 20, 952, 1970.

565.   Brock, T.D., Lower pH limit for the existence of blue-green algae: evolutionary and ecological implications, *Science*, 179, 480, 1973.

566.   Brock, T.D., *Thermophilic Micro-organisms and Life at High Temperatures*, Springer-Verlag, New York, 1978.

567.   Brock, T.D., and Brock, M.L., The algae of Waimangu Cauldron (New Zealand): distribution in relation to pH, *J. Phycol.*, 371, 1970.

568.   Brock, T.C.M., Aspects of the decomposition of *Nymphoides peltata* (Gmel.) O. Knutze (Menyanthaceae), *Aquat. Bot.*, 19, 131, 1984.

569.   Brock, T.C.M., Paffen, B.G.P., and Boon, J.J., The effect of the season and of water chemistry on the decomposition of *Nymphaea alba* L.: Weight loss and pyrolysis mass spectrometry of the particulate matter, *Aquat. Bot.*, 22, 197, 1985.

570.   Brodrick, S.J., Cullen, P., and Maher, W., Denitrification in a natural wetland receiving secondary treated effluent, *Water Res.*, 22, 431, 1988.

571.   Brokaw, C.J., Flagella, in *Physiology and Biochemistry of Algae*, Lewin, R.A., Ed., Academic Press, New York and London, 1962, 595.

572.   Brook, A.J., *The Biology of Desmids, Botanical Monographs*, Vol. 16, University of California Press, Los Angeles, 1981.

573.   Brooks, R.R., and Rumsby, M.G., The biogeochemistry of trace element uptake by some New Zealand bivalves, *Limnol. Oceanogr.*, 10, 521, 1965.

574.   Brown, B.E., Effect of mine drainage on the River Hayle, Cornwall. A. Factors affecting concentrations of copper, zinc and iron in water sediments and dominant invertebrate fauna, *Hydrobiologia*, 52, 221, 1977.

575.   Brown, C.M., Macdonald-Brown, D.S., and Meers, J.L., Physiological aspects of microbial inorganic nitrogen metabolism, *Adv. Microb. Physiol.*, 11, 1, 1974.

576.   Brown, D.L., and Tregunna, E.B., Inhibition of respiration during photosynthesis by some algae, *Can. J. Bot.*, 45, 1135, 1967.

577.   Brown, E.J., and Harris, R.F., Kinetics of algal transient phosphate uptake and the cell quota concept, *Limnol. Oceanogr.*, 23, 35, 1978.

578.   Brown, E.J., Harris, R.F., and Koonie, J.F., Kinetics of phosphate uptake by aquatic microorganisms: deviations from a simple Michaelis-Menten equation, *Limnol. Oceanogr.*, 23, 26, 1978.

579.   Brown, K., Sulphur distribution and metabolism in waterlogged peat, *Soil Biol. Biochem.*, 17, 39, 1985.

580.   Brown, K., Formation of organic sulphur in anaerobic peat, *Soil Biol. Biochem.*,, 18, 131, 1986.

581.   Brown, M.R., Dunstan, G.A., Jeffrey, S.W., Wolkman, J.K., Barett, S.M., and LeRoi, J.-M., The influence of irradiance on the biochemical composition of the prymnesiophyte *Isochrysis* sp. (clone T-ISO), *J. Phycol.*, 29, 601, 1993.

582.   Brown, R.G., Stark, J.R., and Patterson, G.L., Ground-water and surface-water interactions in Minnesota and Wisconsin wetlands, in *The Ecology and Management of Wetlands*, Part 1. *Ecology of Wetlands*, Hook, D.D. et al., Eds., Timber Press, Portland, Oregon, 1988, 176.

583.   Brown, R.M., Jr., Studies on Hawaiian freshwater and soil algae. I. The atmospheric dispersal of algae and fern spores across the island of Oahu, Hawaii, in *Contributions in Phycology*, Parker, B.C., and Brown, R.M., Jr., Eds., Allen Press, Lawrence, Kansas, 1971, 175.

584.   Brown, R.M., Jr., Algal viruses, *Adv. Virus Res.*, 17, 243, 1972.

585. Brown, R.M., Jr., Franke, W.W., Kleinig, H., Falk, H., and Sitte, P., Scale formation in chryso-
      phycean algae. I. Cellulosic and noncellulosic wall components made by the Golgi apparatus, *J. Cell
      Biol.,* 45, 246, 1970.
586. Brown, R.M., Jr., Herth, W., Franke, W.W., and Romanovicz, D., The role of Golgi apparatus in the
      biosynthesis and secretion of a cellulosic glycoprotein in *Pleurochrysis:* a model system for the
      synthesis of structural polysaccharides, in *Biogenesis of Plant Cell Wall Polysaccharides,* Loewus,
      F., Ed., Academic Press, New York, 1973, 207.
587. Brown, S., and Lugo, A.E., A comparison of structural and functional characteristics of saltwater and
      freshwater forested wetlands, in *Proc. Int. Conf. Wetlands: Ecology and Management,* Gopal, B.,
      Turner, R.E., Wetzel, R.G., and Whigham, D.F., Eds., Natl. Inst. of Ecology and Internat. Scientific
      Publications, Jaipur, India, 1982, 109.
588. Brown, S.L., A comparison of the structure, primary productivity, and transpiration of cypress
      ecosystems in Florida, *Ecol. Monogr.,* 51, 403, 1981.
589. Brown, T.A., and Smith, D.G., The effects of silver nitrate on the growth and ultrastructure of the
      yeast *Cryptococcus albidus, Microbios Letters,* 3, 155, 1976.
590. Brown, T.A., and Smith, D.G., Effects of inorganic selenium compounds on growth, cell size, and
      ultrastructure of *Cryptococcus albidus, Microbios Letters,* 10, 55, 1979.
591. Brown, T.E., Comparative studies of photosynthesis and the Hill reaction in *Nostoc muscorum* and
      Chlorella pyrenoidosa, Ph.D. Thesis, Ohio State University, Columbus, Ohio, 1954.
592. Brown, T.E., Physiological aspects of the growth of normal and manganese-deficient *Chlorella* in
      the presence of various atmospheres, *Plant Physiol.,* 36 (suppl.), iii, 1961.
593. Brown, T.E., and Richardson, F.L., The effect of growth environment on the physiology of algae:
      light intensity, *J. Phycol.,* 4, 38, 1968.
594. Brown, T.E., Eyster, H.C., and Tanner, H.A., Physiological effects of manganese deficiency, in
      Trace Elements, Lamb, C.A., Bentley, O.G., and Beattie, J.M., Eds., Academic Press, New York and
      London, 1958, 135.
595. Brown, V., Ducker, S.C., and Rowan, K.S., The effect of orthophosphate concentration on the
      growth of articulated coralline alga (Rhodophyta), *Phycologia,* 16, 125, 1977.
596. Brownell, P.F., and Nicholas, D.J.D., Some effects of sodium on nitrate assimilation and N$_2$ fixation
      in *Anabaena cylindrica, Plant Physiol.,* 42, 915, 1967.
597. Browning, E., *Toxicity of Industrial Metals,* 2nd ed., Butterworths, London, 1969.
598. Broyer, T.C., and Stout, P.R., The macronutrient elements, *Ann. Rev. Plant Physiol.,* 10, 277, 1959.
599. Bruce, J.R., Knight, M., and Parke, M., The rearing of oyster larvae on an algal diet, *J. Mar. Biol.
      Ass. U.K.,* 24, 337, 1940.
600. Brüne, F., *Die Praxis der Moor- und Heidelkultur* Paul Parey, Berlin and Hamburg, 1948.
601. Brunel, J., Prescott, G.W., and Tiffany, L.H., Eds., *The Culturing of Algae,* The Charles F. Kettering
      Foundation, 1950.
602. Bruno, S.F., and McLaughlin, J.J.A., The nutrition of the freshwater dinoflagellate *Ceratium hirundi-
      nella, J. Protozool.,* 24, 548, 1977.
603. Bryan, G.W., The absorption of zinc and other metals by the grown seaweed *Laminaria digitata, J.
      Mar. Biol. Ass. U.K.,* 49, 225, 1969.
604. Bryan, G.W., Brown seaweed, *Fucus vesiculosus,* and the gastropod *Littorina littoralis,* as indicators
      of trace-metal availability in estuaries, *Sci. Total. Environ.,* 28, 91, 1983.
605. Bryan, G.W., and Hummerstone, L.G., Brown seaweed as an indicator of heavy metals in estuaries
      in south-west England, *J. Mar. Biol. Ass. U.K.,* 53, 705, 1973.
606. Bryant, D.A., Glazer, A.N., and Eiserling, F.A., Characterization and structural properties of the
      major biliproteins of *Anabaena* sp., *Arch. Microbiol.,* 110, 61, 1976.
607. Bryant, D.A., Guglielmi, G., Tandeau de Marsac, N., Castets, A.-M., and Cohen-Bazire, G., The
      structure of cyanobacterial phycobilisomes: a model, *Arch. Microbiol.,* 123, 113, 1979.
608. Bryceson, I., and Fay, P., Nitrogen fixation in *Oscillatoria (Trichodesmium) erythraea* in relation
      to bundle formation and trichome differentiation, *Mar. Biol.,* 61, 159, 1981.
609. Brydges, T.J., Chlorophyll a—total phosphorus relationship in Lake Erie, *Proc. 14th Conf. Great
      Lakes Res.,* 1971, 105.
610. Budd, K., and Craig, S.R., Resistance to arsenate toxicity in the blue-green alga *Synechococcus
      leopoliensis, Can. J. Bot.,* 59, 1518, 1981.
611. Buetow, D.E., Ed., *The Biology of Euglena,* Vols. 1 and 2, Academic Press, New York and London,
      1968.
612. Buetow, D.E., Ed., *The Biology of Euglena,* Vols 3 and 4, Academic Press, New York, 1982.
613. Buggeln, R.G., Physiological investigations on *Alaria esculenta* (Laminariales, Phaeophyceae). IV.
      Inorganic and organic nitrogen in the blade, *J. Phycol.,* 14, 156, 1978.

614. Bulthuis, D.A., and Woelkerling, W.J., Biomass accumulation and shading effects of epiphytes on leaves of the seagrass, *Heterozosfera tasmanica*, in Victoria, Australia, *Aquat. Bot.*, 16, 137, 1983.
615. Bunt, J.S., Diatoms of Antarctic sea-ice as agents of primary production, *Nature*, 199, 1255, 1963.
616. Bunt, J.S., Microalgae of the Antarctic pack ice snow, *Proc. SCAR Symp. Antarctic Oceanogr.*, Santiago, Chile, 1967.
617. Bunt, J.S., Uptake of cobalt and vitamin B_{12} by tropical marine macroalgae, *J. Phycol.*, 6, 339, 1970.
618. Bunt, J.S., and Lee, C.C., Data on the composition and dark survival of four ice microalgae, *Limnol. Oceanogr.*, 17, 458, 1972.
619. Bunt, J.S., Owens, O., Van, H., and Hoch, G., Exploratory studies on the physiology and ecology of a psychrophilic marine diatom, *J. Phycol.*, 2, 96, 1966.
620. Bunt, J.S., Lee, C.C., and Lee, F., Primary productivity and related data from tropical and subtropical marine sediments, *Mar. Biol.*, 16, 28, 1972.
621. Buol, S.W., and Robertus, R.A., Soil formation under hydromorphic conditions, in The *Ecology and Management of Wetlands*, Part 1. *Ecology of Wetlands*, Hook, D.D. et al., Eds., Timber Press, Portland, Oregon, 1988, 253.
622. Burczyk, J., Zontek, I., and Szurman, N., Partition of various algal strains in two-polymer phase system of dextran and polyethylene glycol., *Bull. Acad. Polon. Sci., Ser. Sci. Biol.*, 26, 745, 1979.
623. Burdige, D.J., and Nealson, K.H., Microbial manganese reduction by enrichment cultures from coastal marine sediments, *Appl. Environ. Microbiol.*, 50, 491, 1985.
624. Buresh, R.J., Casselman, M.E., and Patrick, W.H., Jr., Nitrogen fixation in flooded soil systems. A review, *Adv. Agron.*, 33, 149, 1980.
625. Burger-Wiersma, T., *Prochlorothrix hollandica*, a filamentous prokaryotic species containing chlorophylls *a* and *b*, *Arch. Hydrobiol. Suppl.*, 92 (*Algol. Studies*, 64), 555, 1991.
626. Burger-Wiersma, T., Veenhuis, M., Korthals, H.J., Wiel, C.C.M., and Mur, L.R., A new prokaryote containing chlorophylls *a* and *b*. *Nature*, 320, 262, 1986.
627. Burgoon, P.S., Reddy, K.R., and De Busk, T.A., Domestic wastewater treatment using plants cultured in gravel and plastic substrates, in *Proc. Int. Conf. Constructed Wetlands for Wastewater Treatment*, Hammer, D.A., Ed., Lewis Publishers, Chelsea, Michigan, 1989, 536.
628. Burkett, R.D., Uptake and release of methylmercury-203 by *Cladophora glomerata, J. Phycol.*, 11, 55, 1975.
629. Burkholder, J.M., and Wetzel, R.G., Microbial colonization on natural and artificial macrophytes in a phosphorus-limited hardwater lake, *J. Phycol.*, 25, 55, 1989.
630. Burkholder, J.M., and Cuker, B.E., Response of periphyton communities to clay and phosphate loading in a shallow reservoir, *J. Phycol.*, 27, 373, 1991.
631. Burkholder, P.R., Some nutritional relationships among microbes of sea sediments and water, in *Symposium on Marine Microbiology*, Oppenheimer, C.H., Ed., C.C. Thomas, Springfield, Illinois, 1963, 133.
632. Burkholder, P.R., and Burkholder, L.M., Vitamin B_{12} in suspended solids and marsh muds collected along the coast of Georgia, *Limnol. Oceanogr.*, 1, 202, 1956.
633. Burkholder, P.R., and Burkholder, L.M., Studies on B vitamins in relation to productivity of Bahia Forforescente, Puerto Rico, *Bull. Mar. Sci. Gulf Carib.*, 8, 201, 1958.
634. Burkholder, P.R., and Mandelli, E.F., Productivity of microalgae in Antarctic sea ice, *Science*, 149, 872, 1965.
635. Burkholder, P., Burkholder, L.M., and Almodovar, L., Antibiotic activity of some marine algae of Puerto Rico, *Bot. Mar.*, 2, 149, 1960.
636. Burlew, J.S., Ed., *Algal Culture: From Laboratory to Pilot Plant.* Carneige Institution of Washington Publ. No. 600, Washington, DC, 1953.
637. Burmaster, D.E., and Chisholm, S.W., A comparison of two methods for measuring phosphate uptake by *Monochrysis lutheri* Droop grown in continuous culture, *J. Exp. Mar. Biol. Ecol.*, 39, 187, 1979.
638. Burnison, G., Wong, P.T.S., Chau, Y.K., and Silverberg, B.A., Toxicity of cadmium to freshwater algae, *Proc. Can. Fed. Biol Soc. Winnipeg*, 18, 182, 1975.
639. Burns, B.D., and Beardall, J., Utilization of inorganic carbon by marine microalgae, *J. Exp. Mar. Biol. Ecol.*, 107, 75, 1987.
640. Burns, N.M., and Ross, C., Oxygen-nutrient relationships within the central basin of Lake Erie, in *Nutrient in Natural Waters*, Allen, H.E., and Kramer, J.R., Eds., Wiley and Sons, New York, 1972, 193.
641. Burris, J.E., Photosynthesis, photorespiration and dark respiration in eight species of algae, *Mar. Biol.*, 39, 371, 1977.
642. Burris, R.H., Nitrogen nutrition, *Ann. Rev. Plant Physiol.*, 10, 301, 1959.
643. Burris, R.H., Biological nitrogen fixation, *Ann. Rev. Plant Physiol.*, 17, 155, 1966.
644. Burton, J.D., Problems in the analysis of phosphorus compounds, *Water Res.*, 7, 291, 1973.

645. Burton, M.A.S., and Peterson, P.J., Metal accumulation by aquatic bryophytes from polluted mine streams, *Environ. Pollut.*, 19, 39, 1979.

646. Busby, W.F., and Lewin, J., Silicate uptake and silica shell formation by synchronously dividing cells of the diatom *Navicula pelliculosa* (Bréb.) Hilse, *J. Phycol.*, 3, 127, 1967.

647. Butcher, R.W., Studies in the ecology of rivers IV. Observations on the growth and distribution of the sessile algae in the River Hull, Yorkshire, *J. Ecol.*, 28, 210, 1940.

648. Butler, M., Haskew, A.E.J., and Young, M.M., Copper tolerance in the green alga, *Chlorella vulgaris, Plant Cell Environ.*, 3, 119, 1980.

649. Butterworth, J., Lester, P., and Nickless, G., Distribution of heavy metals in the Severn Estuary, *Mar. Pollut. Bull.*, 3, 72, 1972.

650. Buttery, B.R., and Lambert, J.M., Competition between *Glyceria maxima* and *Phragmites communis* in the region of Surlingham Broad. The Competition mechanisms, *J. Ecol.*, 53, 163, 1965.

651. Buttery, B.R., Williams, W.T., and Lambert, J.M., Competition between *Glyceria maxima* and *Phragmites communis* in the region of Surlingham Broad. II. The fen gradient, *J. Ecol.*, 53, 183, 1965.

652. Button, D.K., and Dunker, S.S., Biological effects of copper and arsenic pollution, U.S.N.T.I.S. PB Rept. 201648, 1971.

653. Button, K.S., and Hostetter, H.P., Copper sorption and release by *Cyclotella meneghiniana (Bacillariophyceae)* and *Chlamydomonas reinhardtii* (Chlorophyceae), *J. Phycol.*, 13, 198, 1977.

654. Cabioch, J., and Giraud, G., Structural aspects of biomineralization in the coralline algae (calcified Rhodophyceae), in *Biomineralization in Lower Plants and Animals*, Leadbeater, B.S.C., and Riding, R., Eds., Clarendon Press, Oxford, 1986, 141.

655. Cachon, J., Contribution a l'étude des Péridiniens parasites. Cytologie, cycles évolutifs, *Ann. Sci. Nat.*, 12, tom VI, 1, 1964.

656. Cachon, M., and Caram, B., A symbiotic green alga, *Pedinomonas symbiotica* sp. nov. (Prasinophyceae), in the radiolarian *Thalassolampe margarodes, Phycologia,* 18, 177, 1979.

657. Cadée, G., and Hegeman, J., Primary production of the benthic microflora living on tidal flats in the Dutch Wadden sea, *Neth. J. Sea Res.*, 8, 260, 1974.

658. Cahoon, L.B., and Laws, R.A., Benthic diatoms from the North Carolina continental shelf: inner and mid shelf, *J. Phycol.*, 29, 257, 1993.

659. Cain, B.J., Nitrogen utilization in 38 freshwater chlamydomonad algae, *Can. J. Bot.*, 43, 1367, 1965.

660. Cain, J.R., Paschal, D.C., and Hayden, C.M., Toxicity and bioaccumulation of cadmium in the colonial green alga *Scenedesmus obliquus, Arch. Environ. Contam. Toxicol.*, 9, 9, 1980.

661. Caiola, M.G., and de Vecchi, L., Akinete ultrastructure of *Nostoc* species isolated from cycad corolloid roots, *Can. J. Bot.*, 58, 2513, 1980.

662. Cajander, A.K., Studien über die Moore Finnlands, *Acta Forest. Fenn.*, 1913, 1.

663. Cajander, V.-R., and Jääskeläinen, K., Three widely distributed aquatic plants and seston/zooplankton as indicators of mercury contamination in the river Kokemäenjoki, *Ympäristö ja Terveys,* 13, 264, 1982, (in Finnish).

664. Cajander, V.-R., and Ihantola, R., Mercury in some higher aquatic plants and plankton in the estuary of the River Kokemäenjoki, southern Finland, *Ann. Bot. Fennici,* 21, 151, 1984.

665. Calero, F., Ullrich, W.R., and Aparicio, P.J., Regulation by monochromatic light of nitrate uptake in *Chlorella fusca*, in *The Blue Light Syndrome*, Spenger, H., Ed., Springer-Verlag, Berlin, 1980, 411.

666. Callow, J.A., Callow, M.E., and Evans, L.V., Nutritional studies on the parasitic red alga *Choreocolax polysiphoniae, New Phytol.*, 83, 451, 1979.

667. Calvert, C.C., Arsenicals in animal feeds and wastes, in *Arsenical Pesticides*, Woolson, E.A. Ed., CS Symposium Series 7, Am. Chem. Soc., Washington, DC, 1975, 70.

668. Calvin, M., The path of carbon in photosynthesis, *Science,* 135, 879, 1962.

669. Calvin, M., and Benson, A.A., The path of carbon in photosynthesis, *Science,* 107, 476, 1948.

670. Calvin, M., and Bassham, J., *The Photosynthesis of Carbon Compounds*, W.A. Benjamin, Inc., New York, 1962.

671. Cambridge, M.L., Cockburn Sound study technical report on seagrass. Department of Conservation and Environment, Western Australia, Rept. No. 7, 1979.

672. Cameron, R.E., Communities of soil algae occurring in the Sonoran Desert in Arizona, *J. Arizona Acad. Sci.*, 1, 85, 1960.

673. Cameron, R.E., Species of *Nostoc* Vaucher occurring in the Sonoran desert in Arizona, *Trans. Am. Microsc. Soc.*, 81, 379, 1962.

674. Cameron, R.E., Terrestrial algae of southern Arizona, *Trans. Am. Microsc. Soc.*, 133, 212, 1964.

675. Campbell, E.O., Mires of Australia, in *Ecosystems of the World. Mires: Swamp, Bog, Fen and Moor.* Vol. 4B *Regional Studies*, Gore, A.J.P., Ed., Elsevier Science Publishers, Amsterdam, 1983, 153.

676. Campbell, P.G.C., and Stokes, P.M., Acidification and toxicity of metals to aquatic biota, *Can. J. Fish. Aquat. Sci.,* 42, 2034, 1985.

677. Campbell, P.M., and Smith, G.D., Transport and accumulation of nickel ions in the cyanobacterium *Anabaena cylindrica, Arch. Biochem. Biophys,* 244, 470, 1986.

678. Canter, H.M., and Lund, J.W.G., Studies on planktonic parasites. I. Fluctuations in the numbers of *Asterionella formosa* Hass., in relation on fungal epidemics, *New Phytol.,* 47, 238, 1948.

679. Canterford, G.S., Formation and regeneration of abnormal cells of the marine diatom *Ditylum brightwellii* (West) Grunow, *J. Mar. Biol. Ass. U.K.,* 60, 243, 1980.

680. Canterford, G.S., and Canterford, D.R., Toxicity of heavy metals to the marine diatom *Ditylum brightwellii* (West) Grunow: correlation between toxicity and metal speciation, *J. Mar. Biol. Ass. U.K.,* 60, 227, 1980.

681. Caperon, J., and Meyer, J., Nitrogen-limited growth of marine phytoplankton. I. Changes in population characteristics with steady state growth rate, *Deep-Sea Res.,* 19, 601, 1972.

682. Caperon, J., and Meyer, J., Nitrogen-limited growth of marine phytoplankton. 2. Uptake kinetics and their role in nutrient limited growth of phytoplankton, *Deep-Sea Res.,* 19, 619, 1972.

683. Caperon, J., and Ziemann, D.A., Synergistic effects of nitrate and ammonium ion on the growth and uptake kinetics of *Monochrysis lutheri* in continuous culture, *Mar. Biol.,* 36, 73, 1976.

684. Carell, E.F., and Price, C.A., Porphyrins and the iron requirement for chlorophyll formation in *Euglena, Plant Physiol.,* 40, 1, 1965.

685. Cargill, S.M., and Jeffries, R.L., Nutrient limitations of primary production in a sub-arctic salt marsh, *J. Appl. Ecol.,* 21, 657, 1984.

686. Carignan, R., Sediment geochemistry in a eutrophic lake colonized by the submersed macrophyte *Myriophyllum spicatum, Ver. Internat. Verein. Limnol.,* 22, 355, 1984.

687. Carignan, R., Nutrient dynamics in a littoral sediment colonized by the submersed macrophyte *Myriophyllum spicatum, Can. J. Fish. Aquat. Sci.,* 42, 1303, 1985.

688. Carignan, R., and Kalff, J., Phosphorus release by submerged macrophytes: Significance to epiphyton and phytoplankton, *Limnol. Oceanogr,* 27, 419, 1982.

689. Carlton, R.G., and Wetzel, R.G., Phosphorus flux from lake sediments: Effect of epipelic algal oxygen production, *Limnol. Oceanogr,* 33, 562, 1988.

690. Carlucci, A.F., Production and utilization of dissolved vitamins by marine phytoplankton, in *Effect of the Ocean Environment on Microbial Activities,* Colwell, R.R., and Morita, R.Y., Eds., University Park Press, Baltimore, 1974, 449.

691. Carlucci, A.F., and Bowes, P.M., Production of vitamin B_{12}, thiamine, and biotin by phytoplankton, *J. Phycol.,* 6, 351, 1970.

692. Carlucci, A.F., and Bowes, P.M., Vitamin production and utilization by phytoplankton in mixed culture, *J. Phycol.,* 6, 393, 1970.

693. Carmichael, W.W., and Gorham, P.R., Factors influencing the toxicity and animal susceptibility of *Anabaena flos-aquae* (Cyanophyta) blooms, *J. Phycol.,* 13, 97, 1977.

694. Carmichael, W.W., and Gorham, P.R., The mosaic anture of toxic blooms of Cyanobacteria, in *The Water Environment: Algal Toxins and Health,* Carmichael, W.W., Ed., Plenum Press, New York, 1981, 161.

695. Carmichael, W.W., Biggs, D.F., and Gorham, P.R., Toxicology and pharmacological action of *Anabaena flos-aquae* toxin, *Science,* 187, 542, 1975.

696. Caron, D.A., Porter, K.G., and Sanders, R.W., Carbon, nitrogen, and phosphorus budgets for the mixotrophic phytoflagellate *Poterioochromonas malhamensis* (Chrysophyceae) during bacterial ingestion, *Limnol. Oceanogr,* 35, 433, 1990.

697. Carp, E., *Directory of Wetlands of International Importance in the Western Polearctic,* Int. Union for Conservation of Nature and Natural Resources and United Nations Environment Programme, (IUCN-UNEP), Gland, Switzerland, 1980.

698. Carpenter, E,J., and Guillard, R.R.L., Intraspecific differences in nitrate half-saturation constants for three species of marine phytoplankton, *Ecology,* 52, 183, 1971.

699. Carpenter, S.R., Enrichment of Lake Wingra, Wisconsin, by submersed macrophyte decay, *Ecology,* 61, 1145, 1980.

700. Carpenter, S.R., Deacy of heterogenous detritus—a general model, *J. Theor. Biol.,* 90, 539, 1981.

701. Carpenter, S.R., Comparisons of equations for decay of leaf litter in tree—hole ecosystems, *Oikos,* 39, 17, 1982.

702. Carpenter, S.R., and Adams, M.S., The macrophyte tissue nutrient pool of a hardwater eutrophic lake: implications for macrophyte harvesting, *Aquat. Bot.,* 3, 239, 1977.

703. Carpenter, S.R., and Adams, M.S., Effects of nutrient and temperature on decomposition of *Myriophyllum spicatum* L. in a hardwater eutrophic lake, *Limnol. Oceanogr.,* 24, 520, 1979.

704. Carr, N.G., and Pearce, J., Photoheterotrophism in blue-green algae, *Biochem. J.,* 99, 28P, 1966.

705.   Carr, N.G., and Whitton, B.A., Eds., *The Biology of Blue-Green Algae,* Botanical Monographs, Vol. 9, University of California Press, Berkeley, 1973.

706.   Carr, N.G., and Whitton, B.A., Eds., *The Biology of the Cyanobacteria,* University of California Press, Berkeley, 1982.

707.   Carson, J.L., and Brown, R.M., Jr., Studies of Hawaiian freshwater and soil algae. II. Algal colonization and succession on a dated volcanic substrate, *J. Phycol.,* 14, 171, 1978.

708.   Carter, N., A comparative study of the algal flora of two salt marshes: Part 1, *J. Ecol.,* 20, 341, 1932.

709.   Carter, N., A comparative study of the algal flora of two salt marshes: Part 2, *J. Ecol.,* 21, 385, 1933.

710.   Carter, N., New or interesting algae from brackish water, *Arch. Protistenk.,* 90, 1, 1938.

711.   Carter, V., and Novitzki, R.P., Some comments on the relation between ground water and wetlands, in *The Ecology and Management of Wetlands,* Vol. 1. *Ecology of Wetlands,* Hook, D.D. et al., Eds., Timber Press, Portland, Oregon, 1988, 68.

712.   Carter, V., Bedinger, M.S., Novitzky, P., and Wilen, W.O., Water resources and wetlands, in *Proc. Symp. Wetland Functions and Values: The State of Our Understanding,* Greeson, P.E., Clark, J.R., and Clark, J.E., Eds., Am. Water Resour. Ass., Minneapolis, Minnesota, 1979, 344.

713.   Carvajal, N., Fernández, M., Rodríguez, J.P., and Donoso, M., Urease of *Spirulina maxima, Phytochemistry,* 21, 2821, 1982.

714.   Cary, P.R., and Weerts, P.G.J., Growth and nutrient composition of *Typha orientalis* as affected by water temperature and nitrogen and phosphorus supply, *Aquat. Bot.,* 19, 105, 1984.

715.   Casagrande, D.J., Siewfert, K., Berschinski, C., and Sutton, N., Sulfur in peat-forming systems of the Okefenokee Swamp and Florida Everglades: origins of sulfur in coal, *Geochim. Cosmochim. Acta,* 41, 41, 1977.

716.   Casagrande, D.J., Indowu, G., Friedman, A., Rickert, P., and Schlenz, D., H_2S incorporation in coal precursors: origins of sulfur in coal, *Nature,* 282, 599, 1979.

717.   Casey, H., and Downing, A., Levels of inorganic nutrients in *Ranunculus penicillatus* var. *calcareus* in relation to water chemistry, *Aquat. Bot.,* 2, 75, 1976.

718.   Casey, R.P., Lubitz, J.A., Benoit, R., Wissman, B., and Chau, H., Mass culture of *Chlorella, Food Technol.,* 17, 85, 1963.

719.   Casselton, P.J., Chandler, G., Shah, N., Stewart, G.R., and Sumar, N., Glutamine synthetase isoforms in algae, *New Phytol.,* 102, 261, 1986.

720.   Castenholz, R.W., Seasonal changes in the attached algae of freshwater and saline lakes in the Lower Grand Coulee, Washington, *Limnol., Oceanogr.,* 5, 1, 1960.

721.   Castenholz, R.W., The algae of saline and freshwater lakes in the Lower Grand Coulee, Washington, *Res. Stud.,* 28, 125, 1960.

722.   Castenholz, R.W., An evaluation of a submerged glass method of estimating production of attached algae, *Internat. Verein. Limnol. Verh.,* 14, 155, 1961.

723.   Castenholz, R.W., Thermophylic cyanophytes of Iceland and the upper temperature limit, *J. Phycol.,* 5, 360, 1969.

724.   Castenhol, R.W., Thermophilic blue-green algae and the thermal environment, *Bacteriol. Rev.,* 33, 476, 1969.

725.   Castenholz, R.W., The effect of sulfide on the blue-green algae of hot springs. II. Yellowstone National Park, *Microbial Ecol.,* 3, 79, 1977.

726.   Castenholz, R.W., Isolation and cultivation of thermophilic Cyanobacteria, in *The Prokaryotes,* Vol. 1, Starr, M.P., Stolp, H., Trüper, H.G., Baloros, A., and Schlegel, H.G., Eds., Springer-Verlag, Berlin, 1981, 236.

727.   Castenholz, R.W., Species usage, concept, and evolution in the cyanobacteria (blue-green algae), *J. Phycol.,* 28, 737, 1992.

728.   Castro, M.S., and Dierberg, F.E., Biogenic hydrogen sulfide emissions from selected Florida wetlands, *Water Air Soil Pollut.,* 33, 1, 1987.

729.   Cattaneo, A., and Kalff, J., Seasonal changes in the epiphyte community of natural and artificial macrophytes in Lake Memphremagog (Que. & Vt.), *Hydrobiologia,* 60, 135, 1978.

730.   Cattaneo, A., and Kalff, J., Primary production of algae growing on natural and artificial aquatic plants: A study of interactions between epiphytes and their substrate, *Limnol. Oceanogr.,* 24, 1031, 1979.

731.   Cattaneo, A., and Kalff, J., The relative contribution of aquatic macrophytes and their epiphytes to the production of macrophyte beds, *Limnol. Oceanogr.,* 25, 280, 1980.

732.   Cattaneo, A., and Kalff, J., Reply to Gough and Gough, *Limnol. Oceanogr.,* 26, 988, 1981.

733.   Cavalier-Smith, T., Electron and light microscopy of gametogenesis and gamete fusion in *Chlamydomonas reinhardtii, Protoplasma,* 86, 1, 1975.

734.   Cedeno-Maldonado, A., and Swader, J.A., Studies on the mechanism of copper toxicity in *Chlorella, Weed Sci.,* 22, 443, 1974.

735.   Cembella, A., Antia, N.J., and Harrison, P.J., The utilization of inorganic and organic phosphorus compounds as nutrients by eukaryotic microalgae: a multidisciplinary perspective, Part 1, *CRC Crit. Rev. Microbiol.*, 10, 317, 1983.

736.   Cembella, A., Antia, N.J., and Harrison, P.J., The utilization of inorganic and organic phosphorus compounds as nutrients by eukaryotic microalgae: a multidisciplinary perspective, Part 2, *CRC Crit. Rev. Microbiol.*, 11, 12, 1984.

737.   Center, T.D., and Spencer, N.R., The phenology and growth of water hyacinth (*Eichhornia crassipes*) (Mart. Solms.) in a eutrophic North Central Florida Lake, *Aquat. Bot.*, 12, 1, 1981.

738.   Cerbon, J., and Sharpless, N., Arsenic-lipid complex formation during sugar transport, *Biochim. Biophys. Acta*, 126, 292, 1966.

739.   Chadefaud, M., Sur la notation de prochlorophytes, *Rev. Algol. N.S.*, 13, 203, 1978.

740.   Chakravarty, T.K., Contribution of *Azolla pinnata* to the productivity of Hathkatore pond of Bhagalpur (India), *Pol. Arch. Hydrobiol.*, 31, 27, 1984.

741.   Chalmers, A.G., Pools of nitrogen in a Georgia salt marsh, Ph.D. Thesis, University of Georgia, Athens, 1977.

742.   Chamberlain, W., and Shapiro, J., On the biological significance of phosphate analysis; comparison of standard and new methods with a bioassay, *Limnol. Oceanogr.*, 14, 921, 1969.

743.   Chamie, J.P.M., The Effects of Simulated Sewage Effluent upon Decomposition Nutrient Status and Litter Fall in a Central Michigan Peatland, Ph.D. Dissertation, The University of Michigan Press, Ann Arbor, 1976.

744.   Chamie, J.P.M., and Richardson, C.J., Decomposition in northern wetlands, in *Proc. Symp. Freshwater Wetlands, Ecological Processes and Management Potential,* Good, R.E., Whigham, D.F., and Simpson, R.L., Eds., Academic Press, New York, San Francisco, London, 1978, 115.

745.   Chang, J.J., *Climate and Agriculture: an Ecological Survey,* Aldine Publishing Co., Chicago, 1968.

746.   Chantanachat, S., and Bold, H.C., Phycological studies. II. Some algae from arid soils, University of Texas Publ. 6218, Austin, 1962.

747.   Chao, L., and Bowen, C.C., Purification and properties of glycogen isolated from a blue-green alga, *Nostoc muscorum, J. Bact.*, 105, 331, 1971.

748.   Chapin, F.S., III, Van Clieve, K., and Tieszen, L.L., Seasonal nutrient dynamics of tundra vegetation at Barrow, Alaska, *Arctic Alpine Res.*, 7, 209, 1975.

749.   Chapin, F.S., III, Johnson, D.A., and McKendrick, J.D., Seasonal movement of nutrients in plants of differing growth form in an Alaskan USA tundra ecosystem: implications for herbivory, *J. Ecol.*, 68, 189, 1980.

750.   Chapman, A.R.O., Methods for macroscopic algae, in *Handbook of Phycological Methods. Culture Methods and Growth Measurements,* Stein, J.R., Ed., Cambridge University Press, Cambridge, 1973, 87.

751.   Chapman, A.R.O., The ecology of the macroscopic marine algae, *Ann. Rev. Ecol. Syst.*, 5, 65, 1974.

752.   Chapman, A.R.O., Experimental and numerical taxonomy of the Laminariales: a review, in *Modern Approaches to the Taxonomy of Red and Brown Algae,* Irvine, D.E.C., and Price, J.H., Eds., Academic Press, London, 1978, 423.

753.   Chapman, A.R.O., *Biology of Seaweeds: Levels of Organization,* University Park Press, Baltimore, Maryland, 1979.

754.   Chapman, A.R.O., and Craigie, J.S., Seasonal growth in *Laminaria longicruris*: relation with dissolved inorganic nutrients and internal reserves of nitrogen, *Mar. Biol.*, 40, 197, 1977.

755.   Chapman, A.R.O., Markham, J.W., and Lüning, K., Effects of nitrate concentration on the growth and physiology of *Laminaria saccharina* (Phaeophyta) in culture, *J. Phycol.*, 14, 195, 1978.

756.   Chapman, D.J., Pigments of the symbiotic algae (cyanomes) of *Cyanophora paradoxa* and *Glaucocystis nostochinearum* and two Rhodophyceae, *Porphyridium aeruginosa* and *Asterocystis ramosa,* *Arch. Mikrobiol.*, 55, 17, 1966.

757.   Chapman, D.J., and Trench, R.K., Prochlorophyceae: introduction and bibliography, in *Selected Papers in Phycology* II, Rosowski, J.R., and Parker, B.C., Eds., Phycological Society of America, Lawrence, Kansas, 1982, 656.

758.   Chapman, G., and Rae, A.C., Excretion of photosynthate by a benthic diatom, *Mar. Biol.*, 3, 341, 1969.

759.   Chapman, J.S., and Meeks, J.C., Glutamine and glutamate transport by *Anabaena variabilis, J. Bacteriol.*, 156, 122, 1983.

760.   Chapman, R.L., and Good, B.H., Subaerial symbiotic green algae; interactions with vascular plant hosts, in *Algal Symbiosis: A Continuum of Interaction Strategies,* Goff, L.J., Ed., Cambridge University Press, New York, 1982.

761.   Chapman, V.J., *Seaweeds and Their Uses,* 1st ed., Methuen, London, 1950.

762.   Chapman, V.J., *The Algae.* Mac Millan and Company, London, 1962.

763. Chapman, V.J., *Seaweeds and Their Uses,* 2nd ed., Methuen, London, 1970.
764. Chapman, V.J., Seaweeds in pharmaceuticals and medicine: a review, in *Marine Algae in Pharmaceutical Science,* Hoppe, H.A., Levring, T., and Tanaka, Y., Eds., Walter de Gruyter, Berlin, 1979, 139.
765. Chapman, V.J., and Chapman, D.J., *The Algae,* 2nd ed. Mac Millan, New York, 1973.
766. Chapman, V.J., and Chapman, D.J., *Seaweeds and Their Uses*, Chapman and Hall, London, 1980.
767. Chapra, S.C., Comment on "An empirical method of estimating the retention of phosphorus in lakes" by W.B. Kirchner and P.J. Dillon, *Water Resources Res.,* 2, 1033, 1975.
768. Chapra, S.C., Total phosphorus model for the Great Lakes, *J. Environ. Eng. Div., ASCE,* 103 (EE2), 147, 1977.
769. Chase, F.M., Useful algae, *Smithsonian Inst. Ann. Rep.,* 1941, 404, 1941.
770. Chau, Y.K., Wong, P.T.S., Silverberg, B.A., Luxon, P.L., and Bengert, G.A., Methylation of selenium in the aquatic environment, *Science,* 192, 1130, 1976.
771. Chave, K.E., Aspects of the biogeochemistry of magnesium. I. Calcareous marine organisms, *J. Geol.,* 62, 266, 1954.
772. Chen, C.C., Dixon, J.B., and Turner, F.T., Iron coatings on rice roots: mineralogy and quantity influencing factors, *Soil Sci. Soc. Am. J.,* 44, 635, 1980.
773. Chen, C.W., and Orlob, G.T., Ecological stimulation for aquatic environments, in *Systems Analysis and Stimulation in Ecology,* Vol. 3, Patton, B.C., Ed., Academic Press, New York, 1975, 475.
774. Cheniae, G.M., Photosystem II and O_2 evolution, *Ann. Rev. Plant Physiol.,* 21, 467, 1970.
775. Cheniae, G.M., and Martin, I.F., Studies on the function of manganese in photosynthesis, *Brookhaven Symposia in Biology,* 19, 406, 1966.
776. Cheniae, G.M., and Martin, I.F., Site of manganese function in photosynthesis, *Biochim. Biophys. Acta,* 153, 819, 1968.
777. Cheniae, G.M., and Martin, I.F., Photoreactivation of manganese catalyst in photosynthetic oxygen evolution, *Plant Physiol.,* 41, 351, 1969.
778. Chevalier, P., and de la Noüe, J., Efficiency of immobilized hyperconcentrated algae for ammonium and orthophosphate removal from wastewaters, *Biotechnol. Lett.,* 7, 395, 1985.
779. Chiaudani, G., and Vighi, M., The N:P ratio and test with *Selenastrum* to predict eutrophication in lakes, *Water Res.,* 8, 1063, 1974.
780. Chiaudani, G., and Vighi, M., The use of *Selenastrum capricornutum* batch cultures in toxicity studies, *Mitt. Internat. Verein. Limnol.,* 21, 316, 1978.
781. Chiba, Y., and Sasaki, H., Hydrogenase activity in *Scenedesmus* D_3 cultured in a medium containing carrot extract, *Plant Cell Physiol.,* 4, 41, 1963.
782. Chion, C., and Boyd, C., The utilization of phosphorus from muds by phytoplankter, *Scenedesmus dimorphus* and the significance of these findings to the practice of pond fertilization, *Hydrobiologia,* 45, 345, 1974.
783. Chirkova, T.V., The role of anaerobic respiration in the adaptation of some woody plants to temporary anaerobiosis, *Vest. Leningr. Univ. Biol.,* 3, 898, 1973, in Russian.
784. Chisholm, S.W., and Stross, R.G., Light/dark phased cell division in *Euglena gracilis* (Euglenophyceae) in PO_4-limited continuous culture, *J. Phycol.,* 11, 367, 1975.
785. Chisholm, S.W., and Stross, R.G., Phosphate uptake kinetics in *Euglena gracilis* (Z.) (Euglenophyceae) grown on light/dark cycles. I. Synchronized batch cultures, *J. Phycol.,* 12, 210, 1976.
786. Chisholm, S.W., and Stross, R.G., Phosphate uptake kinetics in *Euglena gracilis* (Z.) (Euglenophyceae) grown on light/dark cycles. II. Phased PO_4^-—limited cultures, *J. Phycol.,* 12, 217, 1976.
787. Cho, F., and Govindjee, Low temperature (4–77°K) spectroscopy of *Chlorella;* temperature dependence of energy transfer efficiency, *Biochim. Biophys. Acta,* 216, 139, 1970.
788. Cho, F., and Govindjee, Low temperature (4–77°K) spectroscopy of *Anacystis;* temperature dependence of energy transfer efficiency, *Biochim. Biophys. Acta,* 216, 151, 1970.
789. Chock, J.S., and Mathieson, A.C., Physiological ecology of *Ascophyllum nodosum* (L.) le Jolis and its detached ecad *scorpioides* (Hornemann) Hauck (Fucales, Phaeophyta), *Bot. Mar.,* 22, 21, 1979.
790. Choi, I.C., Phytoplankton production and extracellular release of dissolved organic carbon in the Western North Atlantic, *Deep-Sea Res.,* 19, 731, 1972.
791. Choie, D.D., and Richter, G.W., Lead poisoning: rapid formation of intracellular inclusions, *Science,* 177, 1194, 1972.
792. Chow, T., Lead in natural waters, in *The Biogeochemistry of Lead in the Environment,* Nriagu, J.O., Ed., Elsevier/North Holland Biomedical Press, Amsterdam, 1978, 186.
793. Chow, V.T., Ed., *Handbook of Applied Hydrology,* McGraw-Hill, New York, 1964.
794. Chretiennot, M.-J., Nanoplancton de flaques supralittorales de la région de Marseille. I. Étude qualitative et écologie, *Protistologica,* 10, 469, 1974.

795. Christensen, B., Biomass and primary production of *Rhizophora apiculata* Bl. in a mangrove forest in southern Thailand, *Aquat. Bot.,* 4, 43, 1978.
796. Christensen, E.R., Scherfig, J., and Dixon, P.S., Effects of manganese, copper and lead on *Selenastrum capricornutum* and *Chlorella sigmatophora, Water Res.,* 13, 79, 1979.
797. Christensen, T., Alger, in *Botanik,* II, Nr. 2, *Systematic Botanik,* Böcher, T.W., Lange, M., and Sørensen, T., Eds., Munksgaard, Copenhagen, 1962.
798. Christensen, T., The gross classification of algae, in *Algae and Man,* Jackson, D.F., Ed., Plenum Press, New York, 1964, 59.
799. Christensen, T., *Vaucheria* collections from Vaucher's region, *Biol. Skr,* 16, 1, 1969.
800. Christensen, T., Annotations to a textbook of phycology, *Botanisk Tids.,* 73, 65, 1978.
801. Christensen, T., *Algae. A Taxonomical Survey,* AiO Tryk as, Odense, 1980.
802. Christie, C.E., and Smol, J.P., Diatom assemblages as indicators of lake trophic status in southeastern Ontario lakes, *J. Phycol.,* 29, 575, 1993.
803. Chu, S.P., The influence of the mineral composition of the medium on growth of planktonic algae. II. The influence of the concentration of inorganic nitrogen and phosphate phosphorus, *J. Ecol.,* 31, 109, 1943.
804. Chu, S.P., Utilization of organic phosphorus by phytoplankton, *J. Mar. Biol. Ass. U.K.,* 26, 285, 1946.
805. Chu, S.P., Experimental studies on the environmental factors influencing the growth of phytoplankton, *Sci. Technol. China,* 2, 37, 1949.
806. Chudyba, H., *Cladophora glomerata* and accompanying algae in the Skawa River, *Acta Hydrobiol.,* 7, 93, 1965.
807. Chudyba, H., *Cladophora glomerata* and concomitant algae in the River Skawa. Distribution and conditions of appearance, *Acta Hydrobiol.,* 10, 39, 1968.
808. Cicerone, R.J., and Shetter, J.D., Sources of atmospheric methane: Measurements in rice paddies and a discussion, *J. Geophys. Res.,* 86, 7293, 1981.
809. Clark, J.R., Dickson, K.L., and Cairns, J., Jr., Estimating aufwuchs biomass, in *Methods and Measurements of Periphyton Communities: A Review,* Weitzel, R.L., Ed., American Society for Testing and Materials, S.T.P. 690, Philadelphia, Pennsylvania, 1979, 116.
810. Clark, R.J., and Benforado, J., Eds., *Wetlands of Bottomland Hardwood Forests,* Proc. Workshop Bottomland Forest Wetlands of the Southeastern United States, Elsevier Scientific Publishing Company, New York, 1981.
811. Clarke, P.J., Nitrogen pools in a mangrove-saltmarsh system, *Wetlands,* 3, 85, 1983.
812. Clarke, P.J., Nitrogen pools and soil characteristics of a temperate estuarine wetland in eastern Australia, *Aquat. Bot.,* 23, 275, 1985.
813. Clarke, S.E., Stuart, J., and Sanders-Loehr, J., Induction of siderophore activity in *Anabaena* spp., and its moderation of copper toxicity, *Appl. Environ. Microbiol.,* 53, 917, 1987.
814. Clasen, J., and Bernhardt, H., A bloom of the Chrysophyceae *Synura uvella* in the Wahnbach reservoir as indicator for the release of phosphates from the sediment, *Arch. Hydrobiol. Beih. Ergebn. Limnol.,* 18, 61, 1982.
815. Clasen, J., Bernhardt, H., Hoyer, O., and Wilhelms, A., Phosphate remobilization from the sediment and its influence on algal growth in a lake model, *Arch. Hydrobiol. Beih. Ergebn. Limnol.,* 18, 101, 1982.
816. Clayton, M.N., Studies in the development, life history and taxonomy of the Ectocarpales (Phaeophyta) in southern Australia, *Aust. J. Bot.,* 22, 743, 1974.
817. Clayton, M.N., Phaeophyta, in *Marine Botany: an Australian Perspective,* Clayton, M.N., and King, R.J., Eds., Longman Cheshire, Melbourne, 1981, 104.
818. Cleare, M., and Percival, E., Carbohydrates of the freshwater alga *Tribonema aequale.* II. Preliminary photosynthetic studies with ^{14}C, *Br. phycol. J.,* 8, 181, 1973.
819. Clement, G., and van Landeghem, H., *Spirulina*: ein günstiges Objekt für die Massenkultur von Mikoalgen, *Ber. Dtsch. Bot. Ges.,* 83, 559, 1971.
820. Cleveland, J.S., and Perry, M.J., Quantum yield, relative specific absorption and fluorescence in nitrogen-limited *Chaetoceros gracilis, Mar. Biol.,* 94, 489, 1987.
821. Cloern, J.E., Effects of light and temperature on *Cryptomonas ovata* (Cryptophyceae) growth and nutrient uptake rates, *J. Phycol.,* 13, 389, 1977.
822. Clymo, R.S., The growth of *Sphagnum*: methods of measurement, *J. Ecol.,* 58, 13, 1970.
823. Clymo, R.S., Peat, in *Ecosystems of the World. Mires: Swamp, Bog, Fen and Moor,* Gore, A.J.P., Ed., Elsevier, Amsterdam, 159, 1983.
824. Clymo, R.S., and Reddaway, E.J.F., Growth rate of *Sphagnum rubellum* Wils. on Pennine blanket bog, *J. Ecol.,* 62, 191, 1974.

825.   Cochlan, W.P., and Harrison, P.J., Uptake of nitrate, ammonium, and urea by nitrogen-starved cultures of *Micromonas pusilla* (Prasinophyceae): transient responses, *J. Phycol.*, 27, 673, 1991.

826.   Codd, G.A., and Merrett, M.J., Photosynthetic products of division synchronized cultures of *Euglena*, *Pl. Physiol., Lancaster* 47, 635, 1971.

827.   Codd, G.A., and Stewart, R., Photoinactivation of ribulose biphosphate carboxylase from green algae and cyanobacteria, *FEMS Microbiol. Lett.*, 8, 237, 1980.

828.   Coder, D.M., and Starr, M.P., Antagonistic association of the chlorellavorous bacterium (*"Bdellovibrio chlorellavorus"*) with *Chlorella vulgaris*, *Curr. Microbiol.*, 1, 59, 1978.

829.   Coder, D.M., and Goff, L.J., The host range of the chlorellavorous bacterium (*"Vampirovibrio chlorellavorus"*), *J. Phycol.*, 22, 543, 1986.

830.   Cohen, I., and Neori, A., *Ulva lactuca* biofilters for marine fishpond effluent. 1. Ammonia uptake kinetics and nitrogen content, *Bot. Mar.*, 34, 475, 1991.

831.   Cohen-Bazire, G., and Bryant, D.A., Phycobilisomes: composition and strucute, in *The Biology of Cyanobacteria*, Carr, N.G., and Whitton, B.A., Eds., Blackwell Scientific Publications, Oxford, 1983, 43.

832.   Cole, B., and Toetz, D., Interactions between light and nitrate concentration in controlling nitrate uptake by reservoir phytoplankton, *Arch. Hydrobiol.*, 86, 269, 1979.

833.   Cole, J.A., and Brown, C.M., Nitrite reduction to ammonia by fermentative bacteria: a short circuit in the biological nitrogen cycle, *FEMS Microbiol. Lett.*, 7, 65, 1980.

834.   Cole, J.J., Lane, J.M., Marino, R., and Howarth, R.G., Molybdenum assimilation by cyanobacteria and phytoplankton in freshwater and salt water, *Limnol. Oceanogr.*, 38, 25, 1993.

835.   Cole, K.M., and Sheath, R.G., Eds., *Biology of the Red-Algae*, Cambridge University Press, Cambridge, 1990.

836.   Coleman, A.W., Diversity of plastid DNA configuration among classes of eukaryotic algae, *J. Phycol.*, 21, 1, 1985.

837.   Coleman, L.W., Rosen, B.H., and Schwartzbach, S.D., Environmental control of carbohydrate and lipid synthesis in *Euglena*, *Plant Cell Physiol.*, 29, 423, 1988.

838.   Coleman, J.R., The molecular and biochemical analyses of CO_2-concentrating mechanisms in cyanobacteria and microalgae, *Plant Cell Environ.*, 14, 861, 1991.

839.   Coleman, J.R., and Colman, B., Effect of oxygen and temperature on photosynthetic carbon assimilation in two microscopic algae, *Plant Physiol.*, 65, 980, 1980.

840.   Coleman, R.D., Coleman, R.L., and Rice, E.L., Zinc and cobalt bioconcentration and toxicity in selected algal species, *Bot. Gaz.*, 132, 102, 1971.

841.   Colijn, F., and de Jonge, V.N., Primary production of microphytobenthos in the Ems-Dollard Estuary, *Mar. Ecol. Prog. Ser.*, 14, 185, 1984.

842.   Colin, H., and Guéguen, E., La constitution du principe sucré de *Rhodymenia palmata*, *Comp. rend. acad. sci.*, 190, 163, 1930.

843.   Colin, H., and Guéguen, E., Le sucre des Floridées, *Compt. rend. acad. sci.*, 190, 653, 1930.

844.   Colin, H., and Guéguen, E., Variations saisonnières de la teneur en sucre chez les Floridées, *Compt. rend. acad. sci.*, 190, 884, 1930.

845.   Colletti, P.J., Blinn, D.W., Pickart, A., and Wagner, V.T., Influence of different densities of the mayfly grazer *Heptagenia criddlei* on lotic diatom communities, *J. N. Am. Benthol. Soc.*, 6, 270, 1987.

846.   Collins, G.B., and Weber, C.I., Phycoperiphyton (algae) as indicators of water quality, *Trans. Amer. Micros. Soc.*, 97, 36, 1978.

847.   Collins, M., Algal toxins, *Microbiol. Rev.*, 42, 725, 1978.

848.   Collos, Y., Transient situations in nitrate assimilation by marine diatoms. 4. Nonlinear phenomena and the estimation of the maximum uptake rate, *J. Plankton Res.*, 5, 677, 1983.

849.   Collyer, D.M., and Fogg, G.E., Studies on the fat accumulation by algae, *J. Exp. Bot.*, 6, 256, 1955.

850.   Colman, B., Photosynthesic carbon assimilation and the suppression of photorespiration in the cyanobacteria, *Aquat. Bot.*, 34, 211, 1989.

851.   Colman, B., and Cook, C.M., Photosynthetic characteristics of the marine macrophyte red alga *Rhodymenia palmata*: evidence for bicarbonate transport, in *Inorganic Carbon Uptake by Aquatic Photosynthetic Organisms*, Lucas, W.J., and Berry, J.A., Eds., Am. Soc. Plant Physiol., Rockville, Maryland, 1985, 97.

852.   Committee on Medical and Biologic Effects of Environmental Pollutants, *Arsenic*, Natl. Academy of Sciences, Washington, DC, 1977.

853.   Committee Report: Chemistry of nitrogen and phosphorus in water, *J. Am. Water Works Ass.*, 62, 127, 1970.

854.   Compére, P., *Mallomonas portae-ferreae* Peterfi et Asmund (Chrysophyceae) au lac Tchad, *Bull. Jard. Bot. Nat. Belg.*, 43, 235, 1973.

855.   Compére, P., *Mallomonas bronchartiana,* Chrysophycée nouvelle du lac Tchad, *Bull. Jard. Bot. Nat. Belg.,* 44, 61, 1974.

856.   Connell, J.H., and Slatyer, R.O., Mechanisms of succession in natural communities and their role in community stability and organization, *Am. Natur.,* 111, 1119, 1977.

857.   Conner, W.H., and Day, J.W., Jr., Productivity and composition of a bald cypress—water tupelo site and a bottomland hardwood site in a Louisiana swamp, *Am. J. Bot.,* 63, 1354, 1976.

858.   Conner, W.H., and Day, J.W., Jr., The ecology of forested wetlands in the southeastern United States, in Proc. *Int. Conf. Wetlands: Ecology and Management,* Gopal, B., Tuner, R.E., Wetzel, R.G., and Whigham, D.F., Eds., Natl. Inst. of Ecology and Internat. Scientific Publications, Jaipur, India, 1982, 69.

859.   Conner, W.H., and Day, J.W., Jr., Leaf litter decomposition in three Louisiana freshwater forested wetland areas with different flooding regimes, *Wetlands,* 11, 303, 1991.

860.   Conner, W.H., Gosselink, J.W., and Parrondo, R.T., Comparison of the vegetation of three Louisiana swamp sites with different flooding regimes, *Am. J. Bot.,* 68, 320, 1981.

861.   Conover, J.T., and Sieburth, J.McN., Effect of *Sargassum* distribution on its epibiota and antibacterial activity, *Bot. Mar.,* 6, 147, 1964.

862.   Conover, S.A., Nitrogen utilization during spring blooms of marine phytoplankton in Bedford Basin, Nova Scotia, Canada, *Mar. Biol.,* 32, 247, 1975.

863.   Conway, H.L., Interaction of inorganic nitrogen in the uptake and assimilation by marine phytoplankton, *Mar. Biol.,* 39, 221, 1977.

864.   Conway, H.L., Sorption of arsenic and cadmium and their effects on growth, micronutrient utilization and photosynthetic pigment composition of *Asterionella formosa, J. Fish. Res. Board Can.,* 35, 286, 1978.

865.   Conway, H.L., and Harrison, P.J., Marine diatoms grown in chemostats under silicate or ammonium limitation. IV. Transient response of *Chaetoceros debilis, Skeletonema costatum* and *Thalassiosira gravida* to a single addition of the limiting nutrient, *Mar. Biol.,* 43, 33, 1977.

866.   Conway, H.L., and Williams, S.C., Sorption of cadmium and its effect on growth and the utilization of inorganic carbon and phosphorus of two freshwater diatoms, *J. Fish. Res. Board Can.,* 36, 579, 1979.

867.   Conway, H.L., Harrison, P.J., and Davis, C.O., Marine diatoms grown in chemostats under silicate or ammonium limitation. I. Transient response of *Skeletonema costatum* to a single addition of the limiting nutrient, *Mar. Biol.,* 35, 187, 1976.

868.   Conway, V.M., Studies in the autoecology of *Cladium mariscus* R. Br. III. The aeration of the subterranean parts of the plant, *New Phytol.,* 36, 64, 1937.

869.   Cook, J.R., Adaptations in growth and division in *Euglena* effected by energy supply, *J. Protozool.,* 10, 436, 1963.

870.   Cook, J.R., Influence of light on acetate utilization in green *Euglena, Plant Cell Physiol.,* 6, 301, 1965.

871.   Cook, L.L., and Whipple, S.A., The distribution of edaphic diatoms along environmental gradients of a Louisiana salt marsh, *J. Phycol.,* 18, 64, 1982.

872.   Cooke, J.G., and White, R.E., The effect of nitrate in stream water on the relationship between denitrification and nitrification in a stream-sediment microcosm, *Freshwater Biol.,* 18, 213, 1987b.

873.   Cooke, W.B., Colonization of artificial bare areas by microorganisms, *Bot. Rev.,* 22, 613, 1956.

874.   Cooke, W.J., The occurrence of an endozoic green alga in the marine mollusc *Clinocardium muttallii* (Conrad, 1837), *Phycologia,* 14, 35, 1975.

875.   Coombes, J., and Greenwood, A.J., Compartmentation in the photosynthetic apparatus, in *The Intact Chloroplasts. Topics in Photosynthesis,* Vol. 1, Barber, J., Ed., Elsevier/North Holland, Amsterdam, 1976, 1.

876.   Coombs, J., and Volcani, B.E., Studies on the biochemistry and fine structure of silica shell formation in diatoms. Chemical changes in the wall of *Navicula pelliculosa* during its formation, *Planta,* 82, 280, 1968.

877.   Coombs, J., Halicki, P.J., Holm-Hansen, O., and Volcani, B.E., Studies on the biochemistry and fine structure of silica shell formation in diatoms. Changes in concentration of nucleoside triphosphates during synchronized division of *Cylindrotheca fusiformis* Reimann and Lewin, *Expl. Cell Res.,* 47, 302, 1967.

878.   Coombs, J., Halicki, P.J., Holm-Hansen, O., and Volcani, B.E., Studies on the biochemistry and fine structure of silica shell formation in diatoms. II. Changes in concentration of nucleoside triphosphates in silicon-starvation synchrony of *Navicula pelliculosa* (Bréb.) Hilse, *Expl. Cell Res.,* 47, 315, 1967.

879.   Coombs, J., Darley, W.M., Holm-Hansen, O., and Volcani, B.E., Studies on the biochemistry and fine structure of silica shell formation in diatoms. Chemical composition of *Navicula pelliculosa* during silicon-starvation synchrony, *Plant Physiol.,* 42, 1601, 1967.

880.   Coombs, J., Spanis, C., and Volcani, B.E., Studies on the biochemistry and fine structure of silica shell formation in diatoms. Photosynthesis and respiration in silicon-starvation synchrony of *Navicula pelliculosa, Plant Physiol.*, 42, 1607, 1967.

881.   Coombs, J., Lauritis, J.A., Darley, W.M., and Volcani, B.E., Studies on the biochemistry and fine structure of silica shell formation in diatoms. V. Effects of colchicine on wall formation in *Navicula pelliculosa* (Bréb.) Hilse, *Z. Pflanzenphysiol.*, 59, 124,1968.

882.   Cooper, J.M., and Wilhm, J., Spatial and temporal variation in productivity, species diversity, and pigment diversity of periphyton in a stream receiving domestic and oil refinery effluents, *Southwest. Naturalist*, 19, 413, 1975.

883.   Cooper, P.F., and Findlater, B.C., Eds., *Constructed Wetlands in Water Pollution Control*, Proc. Conf., Pergamon Press, Oxford, 1990.

884.   Copeland, H.F., The kingdoms of organisms, *Quart. Rev. Biol.*, 13, 383, 1938.

885.   Corell, D., Ed., *Watershed Research Perspectives,* Smithsonian Institution, Washington, DC, 1986.

886.   Corell, D.L., and Weller, D.E., Factors limiting processes in freshwater wetlands: an agricultural primary stream riparian forest, in *Proc. Symp. Freshwater Wetlands and Wildlife*, Sharitz, R.R., and Gibbons, J.W., Eds., U.S. Department of Energy, 1989, 9.

887.   Correl, D.L., Faust, M.A., and Severn, D.J., Phosphorus flux and cycling in estuaries, in *Chemistry Biology and the Estuarine System*, Vol. 1. *Estuarine Research*, Cronin, L.E., Ed., Academic Pres, New York, 1975, 108.

888.   Corrillion, R., *Flore et Végetation du Massif Armoricain*, Vol. 4: *Flores des Charophytes (Characées) du Massif Armoricain et des Contrées Voisines d'Europe Occidentale*, Edit. Jouve, Paris, 1975.

889.   Cossa, D., Sorption of cadmium by a population of the diatom *Phaeodactylum tricornutum* in culture, *Mar. Biol.*, 34, 163, 1976.

890.   Coughlan, S., Sulphate uptake by *Fucus serratus, J. Exp. Bot.*, 28, 1207, 1977.

891.   Coult, D.A., Observation on gas movement in the rhizome of *Menyanthes trifoliata* L. with comments on the role of the endodermis, *J. Exp. Bot.*, 15, 205, 1964.

892.   Coults, M.P., and Philipson, J.J., The tolerance of tree roots to waterlogging. II. Adaptation of Sitka spruce and lodgepole pine to waterlogged soils, *New Phytol.*, 80, 71, 1978.

893.   Cowardin, L.M., Carter, V., Golet, F.C., and La Roe, E.T., *Classification of Wetlands and Deepwater Habitats of the United States*, U.S. Fish and Wildlife Service Publ. FWS/OBS-79/31, Washington, DC, 1979.

894.   Cowen, W.F., Algal nutrient availability and limitation in Lake Ontario during IFYGL, Ph.D. Thesis, University of Wisconsin, Madison, 1974.

895.   Cowen, W.F., and Lee, G.F., Phosphorus available in particulate materials transported bu urban runoff, *J. Water Pollut. Control Fed.*, 48, 580, 1976.

896.   Cowgill, U.M., The hydrochemistry of Linsley Pond. II. The chemical composition of the aquatic macrophytes, *Arch. Hydrobiol. Suppl.*, 45, 1, 1974.

897.   Cox, D.P., and Alexander, M., Production of trimethylarsine gas from various arsenic compounds by three sewage fungi, *Bull. Environ. Contam. Toxicol.*, 9, 84, 1973.

898.   Cox, E.R., Ed., *Phytoflagellates, Development in Marine Biology,* Vol. 2, Elsevier/North Holland, New York, 1980.

899.   Cox, E.R., and Hightower, J., Some corticolous algae of McMinn County, Tennessee, U.S.A., *J. Phycol.*, 8, 203, 1972.

900.   Cox, G., and Dwarte, D.M., Freeze-etch ultrastructure of a *Prochloron* species—the symbiont of *Didemnum molle, New Phytol.*, 88, 427, 1981.

901.   Craft, C.B., and Richardson, C.J., Peat accretion and N, P, and organic C accumulation in nutrient-enriched and unenriched Everglades peatlands, *Ecological Applications*, 3, 446, 1993.

902.   Craft, C.B., Broome, S.W., and Seneca, E.D., Exchange of nitrogen, phosphorus and organic carbon between transplanted marshes and estuarine waters, *J. Environ. Qual.*, 18, 206, 1989.

903.   Craig, P.J., Metal cycles and biological methylation. In *The Handbook of Environmental Chemistry* Vol. 1, Part A, *The Natural Environment and the Biogeochemical Cycles*, Hutzinger, O., Ed., Springer-Verlag, Berlin, Heidelberg, New York, 1980, 169.

904.   Craig, P.J., and Morton, S.F., Mercury in Mersey estuary sediments and the analytical procedure for total mercury, *Nature*, 261, 125, 1976.

905.   Craig, P.J., and Morton, S.F., Kinetics and mechanisms of the reaction between methylcobalamin and mercuric chloride, *J. Organometallic Chem.*, 145, 79, 1978.

906.   Craigie, J.S., Storage products, in *Algal Physiology and Biochemistry* Stewart, W.D.P., Ed., University of California Press, Berkeley, 1974, 206.

907.   Craigie, J.S., and McLachlan, J., Excretion of coloured ultraviolet-absorbing substances by marine algae, *Can. J. Bot.*, 42, 23, 1964.

908.   Craigie, J.S., McLachlan, J., Ackman, R.G., and Tocher, C.S., Photosynthesis in algae. III. Distribution of soluble carbohydrates and dimethyl-b-propiothein in marine unicellular Chlorophyceae and Prasinophyceae, *Can. J. Bot.*, 45, 1327, 1967.

909.   Craigie, J.S., McLachlan, K., and Trocher, R.D., Some neutral constituents of the Rhodophyceae with special reference to the occurrence of the floridosides, *Can. J. Bot.*, 46, 605, 1968.

910.   Cramer, G.W., and Day, J.W., Jr., Productivity of the swamps and marshes surrounding Lake Pontchartain, La. in *Environmental Analysis of Lake Pontchartain, Louisiana, Its Surrounding Wetlands, and Selected Land Uses.* Vol. 2., Stone, J.H., Ed., Coastal Ecology Lab., Center for Wetland Resources, Louisiana State University, Baton Rouge, Prepared for U.S. Army Engineers District, New Orleans, 1980, 593.

911.   Cramer, M.L., and Myers, J., Nitrate reduction and assimilation in *Chlorella pyrenoidosa, J. Gen. Physiol.*, 32, 93, 1948.

912.   Cramer, M.L., and Myers, J., Growth and photosynthetic characteristic of *Euglena gracilis. Arch. Mikrobiol.*, 17, 384, 1952.

913.   Crang, R.E., and Jensen, T.E., Incorporation of titanium in polyphosphate bodies of *Anacystis nidulans, J. Cell Biol.*, 67, 80, 1975.

914.   Crawford, R.M.M., The control of anaerobic respiration as a determining factor in the distribution of the genus *Senecio, J. Ecol.*, 54, 403, 1966.

915.   Crawford, R.M.M., Metabolic adaptations to anoxia, in *Plant Life in Anaerobic Environments,* Hook, D.D., and Crawford, R.M.M., Eds., Ann Arbor Science Publishers, Ann Arbor, Michigan, 1978, 119.

916.   Crawford, R.M.M., Physiological responses to flooding, in *Encyclopedia of Plant Physiology,* Vol. 12 B, *Physiological Plant Ecology,* Springer Verlag, Berlin, 1982, 453.

917.   Crawford, R.M.M., Root survival in flooded soils, in *Ecosystems of the World. Mires: Swamp, Bog, Fen and Moor,* Vol. 4A *General Studies,* Gore, A.J.P., Ed., Elsevier Science Publishers, Amsterdam, 1983, 257.

918.   Creed, I.F., Havas, M., and Trick, C.G., Effects of arsenate on growth of nitrogen- and phosphorus-limited *Chlorella vulgaris* (Chlorophyceae) isolates, *J. Phycol.*, 26, 641, 1990.

919.   Cresswell, R.C., and Syrett, P.J., Ammonium inhibition of nitrate uptake by the diatom *Phaeodactylum tricornutum, Plant Sci. Lett.*, 14, 321, 1979.

920.   Cresswell, R.C., and Syrett, P.J., Uptake of nitrate by the diatom *Phaeodactylum tricornutum, J. Exp. Bot.*, 32, 19, 1981.

921.   Cresswell, R.C., and Syrett, P.J., The uptake of nitrate by the diatom *Phaeodactylum tricornutum* between nitrite and nitrate, *J. Exp. Bot.*, 33, 1111, 1982.

922.   Crisp, D.T., Input and output of minerals for an area of Pennine moorland: the importance of precipitation, drainage, peat erosion, and animals, *J. Appl. Ecol.*, 3, 327, 1966.

923.   Crist, R.H., Oberholser, K., Shank, N., and Hguyen, M., Nature of bonding between metallic ions and algal cell walls, *Environ. Sci. Technol.*, 15, 1212, 1981.

924.   Cronquist, A., The divisions and classes of plants, *Bot. Rev.*, 26, 425, 1960.

925.   Cuker, B.E., Grazing and nutrient interactions in controlling the activity and composition of the epilithic algal community of an arctic lake, *Limnol. Oceanogr.*, 28, 133, 1983.

926.   Culley, D.D., Jr., and Epps, E.A., Use of duckweed for waste treatment and animal feed, *J. Water Pollut. Control Fed.*, 45, 337, 1973.

927.   Culley, D.D., Rejmánková, E., Květ, J., and Frye, J.B., Production, chemical quality and use of duckweeds (Lemnaceae) in aquaculture, waste management, and animal feeds, *J. World Maricul. Soc.*, 12, 27, 1981.

928.   Cullinane, J.P., The genus *Vaucheria* in Ireland, *Rev. Algol.*, 11, 249, 1976.

929.   Culver, M.E., and Smith, W.O., Jr., Effects of environmental variation on sinking rates of marine phytoplankton, *J. Phycol.*, 25, 262, 1989.

930.   Cumming, B.F., Smol, J.P., and Birks, H.J.B., Scaled chrysophytes (Chrysophyceae and Synurophyceae) from Adirondack drainage lakes and their relationship to environmental variables, *J. Phycol.*, 28, 162, 1992.

931.   Cumming, B.G., and Wagner, E., Rhythmic processes in plants, *Ann. Rev. Plant Physiol.*, 19, 381, 1968.

932.   Cummins, K.W., and Wuycheck, J.C., Caloric equivalents for investigations in ecological energetics, *Mitt. Internat. Verein. Limnol.*, 18, 1, 1971.

933.   Currier, P.J., Davis, C.B., Van der Valk, A.G., A vegetative analysis of a wetland prairie marsh in northern Iowa, in *Proc. Fifth Midwest Prairie Conference,* Glenn-Lewin, D.C., and Landers, R.Q., Eds., Dept. Botany, Iowa State University, Ames, Iowa, 1978, 65.

934.   Curtis, E.J.C., Durrant, K., and Karman, M.M.I., Nitrification in rivers of the Trent Basin, *Water Res.*, 9, 255, 1975.

935.   Curtis, J.T., *The Vegetation of Wisconsin,* The University of Wisconsin Press, Madison, Wisconsin, 1959.

936.   Curtis, P.J., and Megard, R.O., Interactions among irradiance, oxygen evolution and nitrite uptake by *Chlamydomonas* (Chlorophyceae), *J. Phycol.,* 23, 608, 1987.

937.   Cushing, C.E., Periphyton productivity and radionuclide accumulation in the Columbia River, Washington, U.S.A., *Hydrobiologia,* 29, 125, 1967.

938.   Cushing, C.E., Concentration and transport of ^{32}P and ^{65}Zn by Columbia River plankton, *Limnol. Oceanogr,* 12, 330, 1967.

939.   Cushing, C.E., Trace elements in a Columbia River food web, *Northwest Sci.,* 53, 118, 1979.

940.   Cushing, C.E., and Ranticelli, L.A., Trace element analyses of Columbia River water and phytoplankton, *Northwest Sci.,* 46, 115, 1972.

941.   Cushing, C.E., and Rose, F.L., Cycling of zinc-65 by Columbia River periphyton in a closed lotic microcosm, *Limnol. Oceanogr,* 15, 762, 1970.

942.   Cushing, C.E., and Watson, D.G., Accumulation and transport of radionuclides by Columbia River biota, in *Proc. Conf. Disposal of Radioactive Wastes into Sea, Oceans and Surface Waters,* Vienna, IAEA, 1966, 551.

943.   Cushing, C.E., and Watson, D.G., Accumulation of ^{32}P and ^{65}Zn by living and killed plankton, *Oikos,* 19, 143, 1968.

944.   Cushing, C.E., Jr., and Thomas, J.M., Cu and Zn kinetics in *Myriophyllum heterophyllum* Michx. and *Potamogeton richardsonii* (Ar. Benn.) Rydb., *Ecology,* 61, 1321, 1980.

945.   Cuttle, S.P., Chemical properties of upland peats influencing the retention of phosphate and potassium ions, *J. Soil Sci.,* 34, 75, 1983.

946.   Cutter, G.A., Freshwater systems, in *Occurrence and Distribution of Selenium,* Ihnat, M., Ed., CRC Press, Boca Raton, Florida, 1989, 243.

947.   Dabel, C.V., and Day, F.P., Jr., Structural comparison of four plant communities in the Great Dismal Swamp, Virginia, *Bull. Torrey Bot. Club.,* 104, 352, 1977.

948.   Dacey, J.W.H., Internal winds in water lilies—an adaptation for life in anaerobic sediments, *Science,* 210, 1017, 1980.

949.   Dacey, J.W.H., Pressurized ventilation in the yellow waterlilly, *Ecology,* 62, 1137, 1981.

950.   Dacey, J.W.H., and Klug, M.J., Methane efflux from lake sediments through water lilies, *Science,* 203, 1253, 1979.

951.   Dacey, J.W.H., and Klug, M.J., Tracer studies of gas circulation in *Nuphar:* oxygen-18 and carbon-14 dioxide transport, *Physiol. Plant.,* 56, 361, 1982.

952.   Dacey, J.W.H., and Klug, M.J., Ventilation by floating leaves in *Nymphaea, Am. J. Bot.,* 69, 999, 1982.

953.   Dachnowski, A.P., Quality and value of important types of peat material, U.S. Dept. Agric., *Bur. Plant Ind. Bull.,* 802, 1, 1919.

954.   Daday, A., and Smith, G.D., The effect of nickel on the hydrogen metabolism of the cyanobacterium *Anabaena cylindrica, FEMS Microbiol. Lett.,* 20, 327, 1983.

955.   Daday, A., Mackerras, A.H., and Smith, G.D., The effect of nickel on hydrogen metabolism and nitrogen fixation in the cyanobacterium *Anabaena cylindrica, J. Gen. Microbiol.,* 131, 231, 1985.

956.   Daday, A., Mackerras, A.H., and Smith, G.D., A role for nickel in cyanobacterial nitrogen fixation and growth via cyanophycin metabolism, *J. Gen. Microbiol.,* 134, 2659, 1988.

957.   Daft, M.J., McCord, S.B., and Stewart, W.D.P., Ecological studies on algal-lysing bacteria in fresh waters, *Freshwat. Biol.,* 5, 577, 1975.

958.   Dagestad, D., Lien, T., and Knutsen, G., Degradation and compartmentalization of urea in *Chlamydomonas reinhardii, Arch. Microbiol.,* 129, 261, 1981.

959.   Daiber, F.C., *Conservation of Tidal Marshes,* Van Nostrand Reinhold Company, New York, 1986.

960.   Daily, F.K., Charophyceae, in *Synopsis and Classification of Living Organisms,* Vol. 1, Parker, S.P., Ed., McGraw-Hill, New York, 1982, 161.

961.   Dainty, J., Ion transport across plant cell membranes, *Proc. R. Soc. Edinburgh,* 28, 3, 1960.

962.   Daldorph, P.W.G., and Thomas, J.D., The effect of nutrient enrichment on a freshwater community dominated by macrophytes and molluscs and its relevance to snail control, *J. Appl. Ecol.,* 28, 685, 1991.

963.   Dalton, H., and Stirling, D.I., Co metabolism, *Phil. Trans. R. Soc. Lond.,* B 297, 481, 1982.

964.   Damman, A.W.H., Key to the *Carex* species of Newfoundland by vegetative characteristics, Dept. Forest Publ., 1017, Queen's Printer, Ottawa, 1964.

965.   Damman, A.W.H., Distribution and movement of elements in ombrotrophic peat bogs, *Oikos,* 30, 480, 1978.

966.   Danell, K., and Sjöberg, K., Decomposition of *Carex* and *Equisetum* in a northern Swedish lake: dry weight loss and colonization by macroinvertebrates, *J. Ecol.,* 67, 191, 1979.

967. Danforth, W.F., Substrate assimilation and heterotrophy, in *Physiology and Biochemistry of Algae,* Lewin, R.A., Ed., Academic Press, New York and London, 1962, 99.

968. Daniel, G.F., and Chamberlain, A.H.L., Copper immobilization in fouling diatoms, *Bot. Mar.,* 24, 229, 1981.

969. Danielli, J.F., and Davies, J.T., Reactions at interfaces in relation to biological problems, *Adv. Enzymol.,* 11, 35, 1951.

970. Darley, W.M., Silicon and the division of the diatoms *Navicula pelliculosa* and *Cylindrotheca fusiformis, Proc. N. Am. Paleontol. Conv. Pt.,* G, 994, 1969.

971. Darley, W.M., Silicon requirements for growth and macromolecular synthesis in synchronized cultures of the diatoms, *Navicula pelliculosa*(Brébisson) Hilse and *Cylindrotheca fusiformis*Reimann and Lewin, Ph.D. Thesis, University of California, San Diego, 1969.

972. Darley, W.M., Silicification and calcification, in *Algal Physiology and Biochemistry,* Stewart, W.D.P., Ed., Blackwell Scientific Publications, Oxford, 1974, 655.

973. Darley, W.M., *Algal Biology: A Physiological Approach,* Blackwell Scientific Publications, Oxford, 1982.

974. Darley, W.M., and Volcani, B.E., A silicon requirement for desoxyribonucleic acid synthesis in the diatom *Cylindrotheca fusiformis* Reiman and Lewin, *Exp. Cell Res.,* 58, 334, 1969.

975. Darley, W.M., Sullivan, C.W., and Volcani, B.E., Studies on the biochemistry and fine structure of silica shell formation in diatoms. Division cycle and chemical composition of *Navicula pelliculosa* during light, dark synchronization, *Planta,* 130, 159, 1976.

976. Darley, W.M., Ohlman, C.T., and Wimpee, B.B., Utilization of dissolved organic carbon by natural populations of epibenthic salt marsh diatoms, *J. Phycol.,* 15, 1, 1979.

977. Darley, W.M., Montague, C.L., Plumle, F.G., Sage, W.W., and Psalidas, A.T., Factors limiting edaphic algal biomass and productivity in a Georgia salt-marsh, *J. Phycol.,* 17, 121, 1981.

978. Das, G., Growth and appearance of *Scenedesmus* as influenced by deficient inorganic nutrition, *Sven. Bot. Tidskr,* 62, 457, 1968.

979. Das, R.R., Growth and distribution of *Eichhornia crassipes* and *Spirodela polyrhiza,* Ph.D. Thesis, Banaras Hindu University, Varanasi, India, 1968.

980. Davey, E.W., Morgan, M.J., and Erickson, J., A biological measurements of the copper complexation capacity of seawater, *Limnol. Oceanogr,* 18, 993, 1973.

981. Davey, M.C., The effect of freezing and desiccation on photosynthesis and survival of terrestrial Antarctic algae and cyanobacteria, *Polar Biol.,* 10, 29, 1989.

982. Davidson, I.R., Environmental effects on algal photosynthesis: temperature, *J. Phycol.,* 27, 2, 1991.

983. Davies, A.G., Iron, chelation, and the growth of marine phytoplankton. I. Growth kinetics and chlorophyll production in cultures of the euryhaline flagellate *Dunaliella tertiolecta* under iron-limiting conditions, *J. Mar. Biol. Ass. U.K.,* 50, 65, 1970.

984. Davies, A.G., The growth kinetics of *Isochrysis galbana* in cultures containing sublethal concentrations of mercuric chloride, *J. Mar. Biol. Ass. U.K.,* 54, 157, 1974.

985. Davies, A.G., An assessment of the basis of mercury tolerance in *Dunaliella tertiolecta, J. Mar. Biol. Ass. U.K.,* 56, 39, 1976.

986. Davies, A.G., Pollution studies with marine plankton. Part II. Heavy metals, *Adv. Mar. Biol.,* 15, 381, 1978.

987. Davies, A.G., and Sleep, J.A., Copper inhibition of carbon fixation in coastal phytoplankton assemblages, *J. Mar. Biol. Ass. U.K.,* 60, 841, 1980.

988. Davies, D.D., The central role of phosphoenolpyruvate in plant metabolism, *Ann. Rev. Plant. Physiol.,* 30, 131, 1979.

989. Davies, G.S., Productivity of macrophytes in Marion Lake, British Columbia, *J. Fish. Res. Bd. Can.,* 27, 71, 1970.

990. Davis, E.A., Nitrate reduction by *Chlorella, Plant Physiol.,* 28, 539, 1953.

991. Davis, A.R., Marloth, R.H., and Bishop, C.J., The inorganic nutrition of fungi I. The relation of calcium and boron to growth and spore formation, *Phytopath.,* 18, 949, 1928.

992. Davis, C.B., and van der Valk, A.G., Litter decomposition in prairie glacial marshes, in *Proc. Symp. Freshwater Wetlands: Ecological Processes and Management Potential,* Good, R.E., Whigham, D.F., and Simpson, R.L., Eds., Academic Press, New York, 1978, 99.

993. Davis, C.B., and van der Valk, A.G., The decomposition of standing and fallen litter of *Typha glauca* and *Scirpus fluviatilis, Can. J. Bot.,* 56, 662, 1978.

994. Davis, C.B., and van der Valk, A.G., Mineral release from the litter of *Bidens cernua* L., a mudflat annual at Eagle lake, Iowa, *Verh. Internat. Verein. Limnol.,* 20, 452, 1978.

995. Davis, C.B., and van der Valk, A.G., Uptake and release of nutrients by living and decomposing *Typha glauca* Godr. tissue at Eagle Lake, Iowa, *Aquat. Bot.,* 16, 75, 1983.

996. Davis, E.A., Dedrick, J., French, C.S., Milner, H.W., Myers, J., Smith, J.H.C., and Spoehr, H.A., Laboratory experiments on *Chlorella* culture at the Carnegie Institution of Washington Department of Plant Biology, in *Algal Culture: From Laboratory to Pilot Plant,* Burlew, J.S., Ed., Carnegie Inst. of Washington Publ. 600, Washington, DC, 1953, 105.

997. Davis, E.B., Tischer, R.G., and Brown, L.R., Nitrogen fixation by the blue-green alga *Anabaena flos-aquae* A-37, *Physiol. Plant.,* 19, 823, 1966.

998. Davis, J.S., and Gworek, W.F., A rotifer parasitizing *Vaucheria* in a Florida spring, *Trans. Am. Microsc. Soc.,* 92, 135, 1973.

999. Davis, L.S., Hoffmann, J.P., and Cook, P.W., Seasonal succession of algal periphyton from a wastewater treatment facility, *J. Phycol.,* 26, 611, 1990.

1000. Davis, L.S., Hoffmann, J.P., and Cook, P.W., Production and nutrient accumulation by periphyton in a wastewater treatment facility, *J. Phycol.,* 26, 617, 1990.

1001. Davis, M., and McIntire, C., Effects of physical gradients on the production dynamics of sediment-associated algae, *Mar. Ecol. Prog. Ser.,* 13, 103, 1983.

1002. Davis, S.M., Cattail leaf production, mortality, and nutrient flux in water conservation area 2A, South Florida Water Management District Tech. Publ. 84-8, West Palm Beach, Florida, 1984.

1003. Davis, S.M., Sawgrass and cattail production in relation to nutrient supply in the Everglades, in *Proc. Symp. Freshwater Wetlands and Wildlife,* Sharitz, R.R., and Gibbons, J.W., Eds., U.S. Department of Energy, 1989, 325.

1004. Dawes, C.J., *Marine Botany,* Wiley-Interscience, New York, 1981.

1005. Dawson, E.Y., *How to Know the Seaweeds,* Wm. C. Brown, Dubuque, Iowa, 1956.

1006. Dawson, E.Y., *Marine Botany, an Introduction,* Holt, Rinehart and Winston, New York, 1966.

1007. Dawson, J.E., Organic soils, *Adv. Agron.,* 8, 377, 1956.

1008. Dawson, P.A., Observations on the structure of some forms of *Gomphonema parvulum* Kütz. frustule formation, *J. Phycol.,* 9, 353, 1973.

1009. Day, F.P., Litter decomposition rates in the seasonally flooded Great Dismal Swamp, *Ecology,* 63, 670, 1982.

1010. Day, F.P., Limits on decomposition in the periodically flooded, nonriverine dismal swamp, in *Proc. Symp. Freshwater Wetlands and Wildlife,* Sharitz, R.R., and Gibbons, J.W., Eds., U.S. Department of Energy, 1989, 153.

1011. Day, F.P., and Dabel, C.V., Phytomass budgets for the Dismal Swamp ecosystem, *Va. J. Sci.,* 29, 220, 1978.

1012. Day, F.P., Jr., and Megonigal, P.J., The relationship between variable hydroperiod, production allocation, and belowground organic turnover in forested wetlands, *Wetlands,* 13, 115, 1993.

1013. Day, J.W., Jr., Butler, T.J., and Conner, W.G., Productivity and nutrient export studies in a cypress swamp and lake systems in Louisiana, In *Estuaries Processes,* Wiley, M., Ed., Academic Press, New York, 1977, 255.

1014. Day, J.W., Jr., Conner, W.H., Ley-Lou, F., Day, R.H., and Navarro, A.M., The productivity and composition of mangrove forests, Laguna de Términos, Mexico, *Aquat. Bot.,* 27, 267, 1987.

1015. Day, J.W., Jr., Hall, C.A.S., Kemp, W.M., and Yáñez-Arancibia, A., Eds., *Estuarine Processes,* John Wiley and Sons, New York, 1989.

1016. Dayton, P.K., Competition, disturbance, and community organization: the provision and subsequent utilization of space in a rocky intertidal community, *Ecol. Monogr.,* 41, 351, 1971.

1017. De, P.K., The role of blue-green algae in nitrogen fixation in rice-fields, *Proc. R. Soc. Lond.,* B 127, 121, 1939.

1018. De, P.K., and Sulaiman, M., Fixation of nitrogen in rice soils by algae as influenced by crop, CO_2, and inorganic substances, *Soil Sci.,* 70, 137, 1950.

1019. De, P.K., and Mandal, L.N., Fixation of nitrogen by algae in rice fields, *Soil Sci.,* 81, 453, 1956.

1020. Deane, E.M., and O'Brien, R.W., Uptake of phosphate by symbiotic and free-living dinoflagellates, *Arch. Microbiol.,* 128, 307, 1981.

1021. Dear, J.M., and Aronoff, S., The non-essentiality of boron for *Scenedesmus, Plant Physiol.,* 43, 997, 1968.

1022. De Boer, J.A., Nutrients, in *Biology of Seaweeds,* Lobban, C.S., Wynne, M.J., Eds., The University of California Press, Berkeley, 1981, 386.

1023. De Busk, T.A., and Dieberg, F.E., Effect of nitrogen and fiber content on the decomposition of water hyacinth *Eichhornia crassipes* Mart. (Solms.), *Hydrobiologia,* 118, 199, 1984.

1024. De Busk, T.A., Ryther, J.H., and Williams, L.D., Evapotranspiration of *Eichhornia crassipes* (Mart.) Solms and *Lemna minor* L. in central Florida: relation to canopy structure and season, *Aquat. Bot.,* 16, 31, 1983.

1025. DeBusk, W.F., and Reddy, K.R., Removal of floodwater nitrogen in a cypress swamp receiving primary wastewater effluent, *Hydrobiologia,* 153, 79, 1987.

1026. De Datta, S.K., *Principles and Practices of Rice Production*, John Wiley and Sons, New York, 1981.

1027. De Deckker, P., and Geddles, M.C., Seasonal fauna of ephemeral saline lakes near the Cooron Lagoon, South Australia, *Aust. J. Mar. Freshwater Res.*, 31, 677, 1980.

1028. Deevey, E.S., Limnological studies in Connecticut. 5. A contribution to regional limnology, *Am. J. Sci.*, 238, 717, 1940.

1029. Deevey, E.S., Jr., In defense of mud, *Bull. Ecol. Soc. Amer.*, 51, 5, 1970.

1030. De Filippis, L.F., The effect of heavy metal compounds on the permeability of *Chlorella* cells, *Z. Pflanzenphysiol.*, 92, 39, 1979.

1031. De Filippis, L.F., The effect of sub-lethal concentrations of mercury and zinc on *Chlorella*. V. The counteraction of metal toxicity by selenium and sulphydryl compounds, *Z. Pflanzenphysiol.*, 93, 63, 1979.

1032. De Filippis, L.F., and Pallaghy, C.K., The effect of sublethal concentrations of mercury and zinc on *Chlorella*. I. Growth characteristics and uptake of metals, *Z. Pflanzenphysiol.*, 78, 197, 1976.

1033. De Filippis, L.F., and Pallaghy, C.K., The effect of sublethal concentrations of mercury and zinc on *Chlorella*. II. Photosynthesis and pigment composition, *Z. Pflanzenphysiol.*, 78, 314, 1976.

1034. De Filippis, L.F., and Pallaghy, C.K., The effect of sublethal concentrations of mercury and zinc on *Chlorella*. III. Development and possible resistance to metals, *Z. Pflanzenphysiol.*, 79, 323, 1976.

1035. De Filippis, L.F., Hampp, R., and Ziegler, H., The effects of sublethal concentrations of zinc, cadmium and mercury on *Euglena*. Growth and pigments, *Z. Pflanzenphysiol.*, 101, 37, 1981.

1036. De Filippis, L.F., Hampp, R., and Ziegler, H., The effects of sublethal concentrations of zinc, cadmium and mercury on *Euglena*. II. Respiration, photosynthesis and photochemical activities, *Arch. Microbiol.*, 128, 404, 1981.

1037. de Groot, C.J., and Golterman, H.L., Sequential fractionation of sediment phosphate, *Hydrobiologia*, 192, 143, 1990.

1038. de Jong, J., The purification of wastewater with the aid of rush or reed ponds, in *Biological Control of Water Pollution*, Tourbier, J., and Pirson, R.W., Jr., Eds., The University of Pennsylvania Press, Philadelphia, 1976, 133.

1039. de la Cruz, A.A., A Study of Particulate Organic Detritus in a Georgia Salt Marsh-Estuarine Ecosystem, Ph.D. Dissertation, University of Georgia, Athens, 1965.

1040. de la Cruz, A.A., Primary productivity of coastal marshes in Mississippi, *Gulf Research Reports*, 4, 351, 1974.

1041. de la Cruz, A.A., Production and transport of detritus in wetlands, In *Proc. Nat. Symp. Wetland Functions and Values: The State of Our Understanding*, Greeson, P.E., Clark, J.R., and Clark, J.E., Eds., Am. Water Resources Assoc., Minneapolis, Minnesota, 1979, 162.

1042. de la Cruz, A.A., and Gabriel, B.C., Caloric elemental and nutritive changes in decomposing *Juncus roemarianus* leaves, *Ecology*, 55, 882, 1974.

1043. de la Cruz, A.A., and Hackney, C.T., Energy value, elemental composition, and productivity of belowground biomass of a *Juncus* tidal marsh, *Ecology*, 58, 1165, 1977.

1044. de la Cruz, A.A., and Hackney, C.T., The effects of winter fire on the vegetation structure and primary productivity of tidal marshes in the Mississippi Gulf Coast, Mississippi-Alabama Sea Grant Consortium Publication, MASGP, No. 80-013, 1980.

1045. de la Noüe, J., Lavoie, A., and Walsh, P., Hyperintensive wastewater tertiary treatment by flocculated activated algal sludge, in *Proc. Internat. Conf. Commercial Applications and Implications of Biotechnology* London, 1983, 1005.

1046. de la Noüe, J., Cloutier-Mantha, L., Walsh, P., and Picard, G., Influence of agitation and aeration modes on biomass production by *Oocystis* sp. grown on wastewaters, *Biomass*, 4, 43, 1984.

1047. D'Elia, C.F., and De Boer, J.A., Nutritional studies of two red algae. II. Kinetics of ammonium and nitrate uptake, *J. Phycol.*, 14, 266, 1978.

1048. Delwiche, C.C., The nitrogen cycle, *Sci. Am.*, 223, 137, 1970.

1049. DeManche, J.M., Curl, J.C., Lundy, D.W., and Donaghay, P.L., The rapid response of the marine diatom *Skeletonema costatum* to changes in external and internal nutrient concentrations, *Mar. Biol.*, 53, 323, 1979.

1050. de Marte, J.A., and Hartman, R.T., Studies on absorption of ^{32}P, ^{89}Fe and ^{43}Ca by water-milfoil (*Myriophyllum exalbescens* Fer.), *Ecology*, 55, 188, 1974.

1051. de Nicola, D.M., and McIntire, C.D., Effects of substrate relief on the distribution of periphyton in laboratory streams. I. Hydrology, *J. Phycol.*, 26, 624, 1990.

1052. den Dooren de Jong, L.E., Tolerance of *Chlorella vulgaris* for metallic and non-metallic ions, *Antonie van Leeuwenhoek*, 31, 301, 1965.

1053. de Nie, H.W., The decrease in aquatic vegetation in Europe and its consequences for fish populations, Eur. Inland Fish. Advisory Comm., Occas. Paper No. 19, FAO, Rome, 1987.

1054. Dennis, J.G., Tieszen, L.L., and Vetter, M.A., Seasonal dynamics of above- and below-ground production of vascular plants at Barrow, Alaska, in *Vegetation and Production Ecology of an Alaskan Arctic Tundra,* Tieszen, L.L., Ed., Springer-Verlag, New York, 1978, 113.

1055. Denny, P., Sites of nutrient absorption in aquatic macrophytes, *J. Ecol.,* 60, 819, 1972.

1056. Denny, P., Solute movement in submerged angiosperms, *Biol. Rev.,* 50, 65, 1980.

1057. De Pinto, J.V., Water column decomposition of phytoplankton a necessary part of lake models, in *Perspectives in Aquatic Ecosystem Modeling,* Scavia, D., and Robertson, A., Eds., Ann Arbor Science, An Arbor, Michigan, 1979, 34.

1058. De Pinto, J.V., Young, T.C., and Martin, S.C., Algal-available phosphorus in suspended sediments of lower Great Lakes tributaries, *J. Great Lakes Res.,* 7, 311, 1981.

1059. De Reaumur, R.A., Description des fleures et des graines des divers *Fucus,* et Guelgues autres observations physiques sur ces mémes plantes, *Mém. Acad. Sci. Paris,* 12711, 383, 1711.

1060. Des Abbayes, H., *Traité de lichénologie,* Paul Lechevalier, Paris, 1951.

1061. Descy, J.P., A new approach to water quality estimation using diatoms, *Nova Hedw. Beih.,* 64, 306, 1979.

1062. Desikachary, T.V., *Cyanophyta,* I.C.A.R. Monographs on Algae. New Delhi, 1959.

1063. Desikachary, T.V., Ed., *Taxonomy and Biology of Blue-Green Algae,* Proc. Symp., Madras, India, Bangalore Press, Bangalore, 1972.

1064. De Toni, G.B., *Sylloge algarum omnium hucusque cognitarum.* 1 (*Sylloge Chlorophycearum*), 1–12, I-CXXXIX, Patavii, 1889.

1065. De Toni, G.B., *Sylloge algarum omnium hucusque cognitarum.* 6 (*Sylloge Floridearum 5, Additamenta*), I-XI, Patavii, 1924.

1066. Devilly, C.I., and Houghton, J.A., A study of genetic transformation in *Gloeocapsa alpicola, J. Gen. Microbiol.,* 98, 277, 1977.

1067. Devi Prasad, P.V., and Chowdary, Y.B.K., Effect of some organic carbon sources on the dark calcification of the freshwater green alga *Gloeotaenium, New Phytol.,* 87, 297, 1981.

1068. Devi Prasad, P.V., and Devi Prasad, P.S., Effect of cadmium, lead and nickel on three freshwater green algae, *Water Air Soil Pollut.,* 17, 263, 1982.

1069. Devlin, J.P., Edwards, O.E., Gorham, P.R., Hunter, N.R., Pike, R.K., and Stavric, B., Anatoxin-a, a toxic alkaloid from *Anabaena flos-aquae* NRC-44h, *Can. J. Chem.,* 55, 1367, 1977.

1070. de Vries, P.J.R., and Kamphof, G.J., Growth of some strains of *Stigeoclonium* (Chlorophyta) on nitrate, ammonium, ammonium nitrate and urea, *Br. Phycol. J.,* 19, 349, 1984.

1071. de Vries, P.J.R., and Hotting, E.J., Bioassays with *Stigeoclonium tenue* Kütz. on waters receiving sewage effluents, *Water Res.,* 19, 1405, 1985.

1072. de Vries, P.J.R., and Hillebrand, H., Growth control of *Tribonema minus* (Wille) Hazen and *Spirogyra singularis* Nordstedt by light and temperature, *Acta Bot. Neerl.,* 35, 65, 1986.

1073. de Vries, P.J.R., Torenbeek, M., and Hillebrand, H., Bioassays with *Stigeoclonium* Kütz. (Chlorophyceae) to identify nitrogen and phosphorus limitations, *Aquat. Bot.,* 17, 95, 1983.

1074. de Vrind-de Jong, E.W., Borman, A.H., Thierry, R., Westbroek, P., Gruter, M., and Kamerling, J.P., Calcification in the coccolithophorids *Emiliania huxleyi* and *Pleurochrysis carterae.* II. Biochemical aspects, in *Biomineralization in Lower Plants and Animals,* Leadbeater, B.S.C., and Riding, R., Eds., Clarendon Press, Oxford, 1986, 205.

1075. De Yoe, H.R., Lowe, R.L., and Marks, J.C., Effects of nitrogen and phosphorus on the endosymbiotic load of *Rhopalodia gibba* and *Epithemia turgida* (Bacillariophyceae), *J. Phycol.,* 28, 773, 1992.

1076. Dharmawardene, M.W.N., Haystead, A., and Stewart, W.D.P., Glutamine synthetase of the nitrogen-fixing alga *Anabaena cylindrica, Arch. Mikrobiol.,* 90, 281, 1973.

1077. Dickinson, C.H., Decomposition of litter in soil, in *Biology of Plant Litter* Dickinson, C.H., and Pugh, G.J.F., Eds., Academic Press, New York, 1974, 633.

1078. Dickman, M., A quantitative method for assessing the toxic effects of some water soluble substances, based on changes in periphyton community structure, *Water Res.,* 3, 963, 1969.

1079. Dickson, W., The acidification of Swedish lakes, Inst. of Freshwater Res., Drottningholm, Sweden, Report No. 54, 8, 1975.

1080. Diehn, B., Phototaxis and sensory transduction in *Euglena, Science,* 181, 1009, 1973.

1081. Digby, P.S.B., Growth and calcification in the coralline algae. *Clathromorphum circumscriptum* and *Corallina officinalis* and the significance of pH in relation to precipitation, *J. Mar. Biol. Assoc. U.K.,* 57, 1095, 1977.

1082. Dillard, G.E., The benthic algal communities of a North Carolina Piedmont stream, *Nova Hedw.,* 17, 9, 1969.

1083. Dillenius, J.J., *Histori Muscorum in qua circiter sexcentae veteres et novae ad sua genera relatae describuntus et incomiber genuimis, etc,* Oxonii, 1741.

1084. Dillon, P.J. and Reid, R.A., Input of biologically available phosphorus by precipitation to Precambrian Lakes, in *Atmospheric Pollutants in Natural Waters*, Eisenreich, S., Ed., Ann Arbor Science Publications, Ann Arbor, Michigan, 1981, 183.

1085. Dillon, P.J., and Rigler, F.G., The phosphorus-chlorophyll relationship in lakes, *Limnol. Oceanogr,* 19, 767, 1974.

1086. Dillon, P.J., and Rigler, F.G., A test of a simple nutrient budget model predicting the phosphorus concentration in lake waters, *J. Fish. Res. Board Can.*, 31, 1771, 1974.

1087. Dillon, P.J., and Rigler, F.G., A simple method for predicting the capacity of lake for development based on lake trophic status, *J. Fish. Res. Board Can.*, 32, 1519, 1975.

1088. Dillon, T., and O'Colla, P., The acetolysis of carrageenin, *Nature*, 145, 749, 1940.

1089. Dillon, T., and O'Colla, P.S., The constitution of carrageenin, *Proc. R. Irish Acad.*, B 54, 51, 1951.

1090. Di Martino Rigano, V., Vona, V., Esposito, S., Di Martino, C., and Rigano, C., Carbon skeleton sources for ammonium assimilation in N-sufficient and N-limited cells of *Cyanidium caldarium* (Rhodophyta), *J. Phycol.*, 27, 220, 1991.

1091. Dinsdale, M.T., and Walsby, A.E., The interrelations of cell turgor pressure, gas-vacuolation and buoyancy in a blue-green alga, *J. Exp. Bot.*, 23, 561, 1972.

1092. Di Torro, D.M., Thomann, R.V., and O'Connor, D.J., A dynamic model of phytoplankton population in the Sacramento-San Joaquin Delta, in *Nonequilibrium Systems in Natural Water Chemistry* Gould, R.F., Ed., Advances in Chemistry Series, Vol. 101, 1971, 131.

1093. Di Torro, D.M., O'Connor, D.J., Thomann, R.V., and Mancini, J.L., Phytoplankton-zooplankton-nutrient interaction model for western Lake Erie, in *System Analysis and Stimulation in Ecology*, Patten, B.C., Ed., Academic Press, New York, 1975, 423.

1094. D'Itri, F.M., The biomethylation and cycling of selected metals and metalloides in aquatic sediments, in *Sediments: Chemistry and Toxicity of In-Place Pollutants*, Baudo, R., Giesy, J., and Muntau, H., Eds., Lewis Publishers, Chelsea, Michigan, 1990, 163.

1095. Dixon, N.E., Gazzola, C., Blakeley, R.L., and Zerner, B., Jack bean urease (EC 3.5.1.5): a metalloenzyme. A simple biological role for nickel?, *J. Am. Chem. Soc.*, 97, 4131, 1975.

1096. Dixon, P.S., Perennation, vegetative propagation and algal life histories, with special reference to *Asparagopsis* and other Rhodophyta, *Bot. Gothoburg*, 3, 67, 1965.

1097. Dixon, P.S., *Biology of the Rhodophyta*, Hafner Press, New York, 1973.

1098. Dixon, P.S., Rhodophycophyta, in *Synopsis and Classification of Living Organisms*, Vol. 1, Parker, S.P., Ed., McGraw-Hill, New York, 1982, 61.

1099. Dobrokhotova, K.V., *Water Plants*, Kainar, Alma Ata, Kazakhstan, 1982 (in Russian).

1100. Doddema, H., and Van der Veer, J., *Ochromonas monicis* sp. nov. a particle feeder with bacterial endosymbionts, *Cryptogr. Algol.*, 4, 89, 1983.

1101. Dodds, J.A., Viruses of marine algae, *Experientia*, 35, 4400, 1979.

1102. Dodd, J.J., *Diatoms*, Southern Illinois University Press, Carbondale and Edwardsville, 1987

1103. Dodds, W.K., and Priscu, J.C., Ammonium, nitrate, phosphate, and inorganic carbon uptake in an oligotrophic lake: seasonal variations among light response variables, *J. Phycol.*, 25, 699, 1989.

1104. Dodge, J.D., A review of the fine structure of algal eyespots, *Br. Phycol. J.*, 4, 199, 1969.

1105. Dodge, J.D., Fine structure of the Pyrrophyta, *Bot. Rev.*, 37, 481, 1971.

1106. Dodge, J.D., The ultrastructure of the dinoflagellate pusule: a unique osmo-regulatory organelle, *Protoplasma*, 75, 285, 1972.

1107. Dodge, J.D., *The Fine Structure of Algal Cells*, Academic Press, London, 1973.

1108. Dodge, J.D., Fine structure and phylogeny in the algae, *Sci. Prog.*, 61, 257, 1974.

1109. Dodge, J.D., The fine structure of *Trachelomonas* (Euglenophyceae), *Arch. Protistenk.*, 117, 65, 1975.

1110. Dodge, J.D., The phytoflagellates: fine structure and phylogeny. In *Biochemistry and Physiology of the Protozoa*, Hutner, S.H., and Levandarsky, M., Eds., Academic Press, New York, 1979, 7.

1111. Dodge, J.D., *Marine Dinoflagellates of the British Isles*, H.M.S.O., London, 1982.

1112. Dodge, J.D., *Atlas of Dinoflagellates. A Scanning Electron Microscope Survey*, Farrand Press, London, 1985.

1113. Dodge, J.D., and Steidinger, K.A., Dinophyceae: introduction and bibliography, in *Selected Papers in Phycology* II, Rosowski, J.R., and Parker, B.C., Eds., Phycological Society of America, Lawrence, Kansas, 1982, 691.

1114. Doemel, W.N., and Brock, T.D., The upper temperature limit of *Cyanidium caldarium*, *Arch. Mikrobiol.*, 72, 326, 1970.

1115. Doemel, W.N., and Brock, T.D., Structure, growth, and decomposition of laminated algal-bacterial mats in alkaline hot springs, *Appl. Environ. Microbiol.*, 34, 433, 1977.

1116. Dolan, T.J., Bayley, S.E., Zoltek, J., Jr., and Hermann, A.J., Phosphorus dynamics of a Florida freshwater marsh receiving treated wastewater, *J. Appl. Ecol.*, 18, 205, 1981.

1117. Dolan, T.J., Hermann, A.J., Bayley, S.E., and Zoltek, J., Jr., Evapotranspiration of a Florida, U.S.A., freshwater wetland, *J. Hydrol.*, 74, 355, 1984.

1118. Domozych, D.S., Stewart, K.D., and Mattox, K.R., Comparative aspects of cell wall chemistry in green algae (Chlorophyta), *J. Mol. Evol.*, 15, 1, 1980.

1119. Doonan, B.B., and Jensen, T.E., Physiological aspects of alkaline phosphatase in selected cyanobacteria, *Microbios*, 29, 185, 1980.

1120. Doonan, B.B., Crang, R.E., Jensen, T.E., and Baxter, M., In situ X-ray energy dispersive microanalysis of polyphosphate bodies in *Aureobasidium pullulans*, *J. Ultrastr Res.*, 69, 232, 1979.

1121. Dop, A.J., *Porterinema fluviatile* (Porter) Waern (Phaeophyceae) in the Netherlands, *Acta Bot. Neerl.*, 26, 449, 1979.

1122. Dop, A.J., The genera *Phaeothamnion* Lagerheim, *Tetrachrysis* gen. nov. and *Sphaeridiothrix* Pascher et Vlk (Chrysophyceae), *Acta Bot. Neerl.*, 29, 65, 1980.

1123. Dop, A.J., Kosterman, Y., and van Oers, F., Coccoid and palmelloid benthic Chrysophyceae from the Netherlands, *Acta Bot. Neerl.*, 29, 87, 1980.

1124. Dor, I., and Ehrlich, A., The effect of salinity and temperature gradients on the distribution of littoral microalgae in experimental solar ponds, Dead Sea area, Israel, *Mar. Ecol.*, 8, 193, 1987.

1125. Doremus, C., and Clesceri, L.S., Microbial metabolism in surface sediments and its role in the immobilization of phosphorus in oligotrophic lake sediments, *Hydrobiologia*, 91, 261, 1982.

1126. Dorich, R.A., Nelson, D.W., and Sommers, L.E., Algal availability of sediment phosphorus in drainage water of the Black Creek watershed, *J. Environ. Qual.*, 9, 557, 1980.

1127. Dorigan, J.L., and Wilbur, K.M., Calcification and its inhibition in coccolithophorids, *J. Phycol.*, 9, 450, 1973.

1128. Dortch, Q., Effect of growth conditions on accumulation of internal nitrate, ammonium, and protein in three marine diatoms., *J. Exp. Mar. Biol. Ecol.*, 61, 243, 1982.

1129. Dortch, Q., Thompson, P.A., and Harrison, P.J., Variability in nitrate uptake kinetics in *Thalassiosira pseudonana* (Bacillariophyceae), *J. Phycol.*, 27, 35, 1991.

1130. Doty, M.S., Status of marine agronomy with special reference to the tropics, in *Proc. IX. Int. Seaweed Symp.*, Jensen, A., and Stein, J.R., Eds., Science Press, Princetown, 1979, 35.

1131. Doucette, G.J., Price, N.M., and Harrison, P.J., Effects of selenium deficiency on the morphology and ultrastructure of a coastal marine diatom *Thalassiosira pseudonana* (Bacillariophyceae), *J. Phycol.*, 23, 9, 1987.

1132. Doughetry, R.C., Strain, H.H., Svec, W.A., Uphaus, R.A., and Katz, J.J., The structure, properties and distribution of chlorophyll *c*, *J. Am. Chem. Soc.*, 92, 2826, 1970.

1133. Douglas, A.E., Uric acid utilization in *Platymonas convolutae* and symbiotic *Convoluta roscoffensis*, *J. Mar. Biol. Assoc. U.K.*, 63, 435, 1983.

1134. Douglas, B., The ecology of the attached diatoms and other algae in a small stony stream, *J. Ecol.*, 46, 295, 1958.

1135. Dowd, J.E., and Riggs, D.S., A comparison of estimates of Michaelis-Menten kinetic constants from various linear transformations, *J. Biol. Chem.*, 240, 863, 1965.

1136. Doyle, G.J., Primary production estimates of native blanket bog and meadow vegetation growing on reclaimed peat at Glenamoy, Ireland, in *Proceedings IBP Tundra Biome Symposium. Production and Production Processes*, Bliss, L.C., and Wieloglaski, F.E., Eds., Dublin, 1973, 141.

1137. Dragovich, A., Kelly, J.A., Jr., and Kelly, R.D., Red water bloom of a dinoflagellate *Ceratium furca* Ehr. in Hillsborough Bay, Florida, *Nature*, 207, 1209, 1965.

1138. Drake, B.G., Photosynthesis of salt marsh species, *Aquat. Bot.*, 34, 167, 1989.

1139. Drbal, K., Véber, K., and Zahradník, J., Toxicity and accumulation of copper and cadmium in the alga *Scenedesmus obliquus* LH, *Bull. Environ. Contam. Toxicol.*, 34, 904, 1985.

1140. Drebes, G., On the life history of the marine plankton diatom *Stephanopyxis palmeriana*, *Helgol. Wiss. Meerersunters.*, 13, 101, 1966.

1141. Drebes, G., *Marine phytoplankton. Eine Auswahl der Helgoländer Planktonalgen (Diatomeen, Peridineen)*, Georg Thillne, Stuttgart, 1974.

1142. Drebes, G., Life cycle and host specificity of marine parasitic dinophytes, *Helgolander Meersuntersuchungen*, 37, 1984, 603.

1143. Drewes, K., Über die Assimilation des Luftstickstoffs durch Blaualgen, *Zentr. Bakteriol. Parasitenk. Abt.*, *II.*, 76, 88, 1928.

1144. Dring, M.J., Stimulation of light-saturated photosynthesis in *Laminaria* (Phaeophyta) by blue light, *J. Phycol.*, 25, 254, 1989.

1145. Droop, M.R., On the ecology of flagellates from some brackish and fresh water rockpools in Finland, *Acta Bot. Fennici*, 51, 1, 1953.

1146. Droop, M.R., Phagotrophy in *Oxyrrhis marina* Dujardin, *Nature*, 172, 250.

1147. Droop, M.R., Some new supra-littoral protista, *J. Mar. Biol. Assoc. U.K.*, 34, 233, 1955.

1148. Droop, M.R., Vitamin B_{12} in marine ecology, *Nature,* 180, 1041, 1957.
1149. Droop, M.R., Water-soluble factors in the nutrition of *Oxyrrhis marina. J. Mar. Biol. Ass. U.K.,* 38, 605, 1959.
1150. Droop, M.R., *Haematococcus pluvialis* and its allies. III. Organic nutrition, *Rev. Algol. N.S.,* 4, 247, 1961.
1151. Droop, M.R., Some chemical considerations in the design of synthetic culture media for marine algae, *Bot. Mar.,* 2, 231, 1961.
1152. Droop, M.R., Organic micronutrients, in *Physiology and Biochemistry of Algae,* Lewin, R.A., Ed., Academic Press, New York and London, 1962, 141.
1153. Droop, M.R., Algae and invertebrates in symbiosis, in *Symbiotic Associations,* Nutman, P.S., Mosse, B., Eds., Symp. Soc. Gen. Microbiol., Vol. 13, Cambridge University Press, Cambridge, 1963, 171.
1154. Droop, M.R., Vitamin B_{12} in marine ecology, III. An experiment with a chemostat, *J. Mar. Biol. Ass. U.K.,* 46, 659, 1966.
1155. Droop, M.R., Vitamin B_{12} in marine ecology. IV. The kinetics of uptake, growth and inhibition in *Monochrysis lutheri, J. Mar. Biol. Ass. U.K.,* 48, 689, 1968.
1156. Droop, M.R., Some thoughts on nutrient limitation in algae, *J. Phycol.,* 9, 264, 1973.
1157. Droop, M.R., Heterotrophy of carbon, in *Algal Physiology and Biochemistry,* Stewart, W.D.P., Ed., Blackwell Scientific Publications, Oxford, 1974a, 530.
1158. Droop, M.R., The nutrient status of algal cells in continuous culture, *J. Mar. Biol. Ass. U.K.,* 54, 825, 1974.
1159. Droop, M.R., and McGill, S., The carbon nutrition of some algae: the inability to utilize glycollic acid for growth, *J. Mar. Biol. Ass. U.K.,* 46, 679, 1966.
1160. Droste, M., Above ground standing crop and production of *Carex gracilis* Curt. in a fen, *Arch. Hydrobiol.,* 100, 533, 1984.
1161. Drouet, F., Revision of the classification of the Oscillatoriaceae, *Acad. Nat. Sci. Philadelphia Monogr.* 15, 1, 1968.
1162. Drouet, F., *Revision of the Nostocaceae with the cylindrical trichomes (formerly Scytonemataceae and Rivulariaceae),* Hafner Press, New York, 1973.
1163. Drouet, F., Revision of the Nostocaceae with constricted trichomes, *Beih. Nova Hedw.,* 57, 1, 1978.
1164. Drouet, F., Summary of the classification of blue-green algae, *Beih. Nova Hedw.,* 66, 135, 1981.
1165. Drouet, F., and Cohen, A., The morphology of *Gonyostomum semen* from Woods Hole, Massachusetts, *Biol. Bull.,* 68, 422, 1935.
1166. Drouet, F., and Daily, W., Revision of the coccoid Myxophyceae, *Butler Univ. Bot. Stud.,* 12, 1, 1956.
1167. Drown, D.B., Olson, T.A., and Odlaug, T.O., The response of nearshore periphyton in western Lake Superior to thermal additions, Water Resources Research Center, University of Minnesota, Graduate School, Bulletin No. 77, 1974.
1168. Drum, R., Light and electron microscope observations on the tube-dwelling diatom *Amphipleura rutilans* (Trentepohl) Cleve, *J. Phycol.,* 5, 21, 1969.
1169. Drum, R.W., and Pankratz, H.S., Post mitotic fine structure of *Gomphonema parvulum, J. Ultrastruct. Res.,* 10, 217, 1964.
1170. Drum, R.W., and Pankratz, H.S., Fine structure of an unusual cytoplasmic inclusion in the diatom genus, *Rhopalodia, Protoplasma,* 60, 141, 1965.
1171. Drum, R.W., and Webber, E.E., Diatoms from a Massachusetts salt marsh, *Bot. Mar.,* 11, 70, 1966.
1172. Duckett, J.G., Toth, R., and Soni, S.L., An ultrastructural study of the *Azolla, Anabaena azollae* relationship, *New Phytol.,* 75, 111, 1975.
1173. Duever, M.J., Hydrology, in *Wetlands and Shallow Continental Water Bodies,* Patten, B.C., Ed., SPB Academic Publishing bv, The Hague, The Netherlands, 1990, 61.
1174. Duffer, W.R., and Dorris, T.C., Primary productivity in a southern Great Plains stream, *Limnol. Oceanogr.,* 11, 143, 1966.
1175. Dugdale, R.C., Nutrient limitation in the sea: dynamics, identification, and significance, *Limnol. Oceanogr.,* 12, 685, 1967.
1176. Dugdale, R.C., and Dugdale, V.A., Nitrogen metabolism in Lakes. II. Role of nitrogen fixation in Sanctuary Lake, Pennsylvania, *Limnol. Oceanogr.,* 7, 170, 1962.
1177. Dugdale, R.C., and Goering, J.J., Uptake of new and regenerated forms of nitrogen in primary productivity, *Limnol. Oceanogr.,* 12, 196, 1967.
1178. Dugdale, R.C., Menzel, D.W., and Ryther, J.W., Nitrogen fixation in the Sargasso Sea, *Deep-Sea Res.,* 7, 298, 1961.
1179. Dugdale, R.C., Goering, J.J., and Ryther, J.H., High nitrogen fixation rates in the Sargasso sea and the Arabian Sea, *Limnol. Oceanogr.,* 9, 507, 1964.

1180. Dugdale, R.C., Jones, B.H., Jr., Mac Isaac, J.J., and Goering, J.J., Adaptation of nutrient assimilation, in *Physiological Bases of Phytoplankton Ecology*, Platt, T., Ed., *Can. Bull. Fish. Aquat. Sci.*, 210, 234, 1981.

1181. Dugdale, V.A., and Dugdale, R.C., Nitrogen metabolism in lakes. III. Tracer studies of the assimilation of inorganic nitrogen sources, *Limnol. Oceanogr.*, 10, 53, 1965.

1182. Duke, C.S., Litaker, R.W., and Ramus, J., Seasonal variation in RuBPCase activity and N allocation in the chlorophyte seaweeds *Ulva curvata* (Kütz.) de Toni and *Codium decorticatum* (Woodw.) Howe, *J. Exp. Mar. Biol. Ecol.*, 112, 145, 1987.

1183. Duke, C.S., Litaker, W., and Ramus, J., Effect of temperature, nitrogen supply, and tissue nitrogen on ammonium uptake rates of the chlorophyte seaweeds *Ulva lactuca* and *Codium decorticatum, J. Phycol.*, 25, 113, 1989.

1184. Dunbabin, J.S., Pokorný, J., and Bowmer, K.H., Rhizosphere oxygenation by *Typha domingensis* Pers. in miniature artificial wetland filters used for metal removal from wastewaters, *Aquat. Bot.*, 29, 303, 1988.

1185. Duncan, S.W., and Blinn, D.W., Importance of physical variables on the seasonal dynamics of epilithic algae in a highly shaded canyon stream, *J. Phycol.*, 25, 455, 1989.

1186. Dunn, J.H., and Wolk, C.P., Composition of the cellular envelope of *Anabaena cylindrica, J. Bacteriol.*, 103, 153, 1970.

1187. Dunn, R.W., Seasonal variations in periphyton, chlorophyll a, algal biomass and primary production in a desert stream, M.Sc. Thesis, Idaho State University, Pocatello, 1976.

1188. Duong, T.P., Nitrogen fixation and productivity in a eutrophic hard-water lake: *In situ* and laboratory studies, Ph.D. Dissertation, Michigan State University, East Lansing, 1972.

1189. Durani, P.K., and Rout, D.K., Phytosociology and production ecology of Nandan Kanan lake in Orissa, *Geobios*, 9, 25, 1982.

1190. Du Rietz, G.E., Main units and main limits in Swedish mire vegetation, *Svensk. Bot. Tidskr*, 43, 274, 1949.

1191. Du Rietz, G.E., Die Mineralbodenwasserzeigergrenze als Grundlage einer natürlichen Zweigliederung der nord- und mittrleuropaeischen Moore, *Vegetatio*, 5–6, 571, 1954.

1192. Dürrschmidt, M., Studies on the Chrysophyceae from Rio Cruces, Prov. Valdivia, South Chile by scanning and transmission microscopy, *Nova Hedw.*, 33, 353, 1980.

1193. Duysens, L.N.M., Transfer of excitation energy in photosynthesis, Ph.D. Thesis, University of Utrecht, The Netherlands, 1952.

1194. Dykstra, R.F., Mac Entee, F.J., and Bold, H.C., Some edaphic algae of the Texas coast, *Texas J. Sci.*, 26, 171, 1975.

1195. Dykyjová, D., Production, vertical structure and light profiles in littoral stands of reed bed species, *Hidrobiologia* (Bucharest), 12, 361, 1971.

1196. Dykyjová, D., Ecotypes and ecomorphoses of common reed, *Phragmites communis* Trin., *Preslia* (Prague), 43, 120, 1971 (in Czech).

1197. Dykyjová, D., Accumulation of mineral nutrients in the biomass of reedswamp species, in *Ecosystem Study on Wetland Biome in Czechoslovakia*, Hejný, S., Ed., Czechosl. IBP/PT-PP Report No. 3, Třeboň, 1973, 151b.

1198. Dykyjová, D., Content of mineral macronutrients in emergent macrophytes during their seasonal growth and decomposition, in *Ecosystem Study on Wetland Biome in Czechoslovakia*, Hejný, S., Ed., IBP/PT-PP Rept. No. 3, Třeboň, 163, 1973, b.

1199. Dykyjová, D., Nutrient uptake by littoral communities of helophytes, in *Pond Littoral Ecosystems*, Dykyjová, D., and Květ, J., Eds., Springer-Verlag, Berlin, 1978, 257.

1200. Dykyjová, D., Selective uptake of mineral ions and their concentration factors in aquatic higher plants, *Folia Geobot. Phytotax.*, 14, 267, 1979.

1201. Dykyjová, D., Methods of mineral nutrients pool and ecosystem cycling studies, in *Methods of Ecosystem Studies*, Dykyjová, D., Ed., Academia, Praha, Czech Republic, 1989, 414, (in Czech).

1202. Dykyjová, D. and Ondok, J.P., Biometry and the productive stand structure of coenoses of *Sparganium erectum* L., *Preslia* (Praha), 45, 19, 1973.

1203. Dykyjová, D., and Hradecká, D., Productivity of reed-bed stands in relation to the ecotype, microclimate, and trophic conditions of the habitat, *Pol. Arch. Hydrobiol.*, 20, 111, 1973.

1204. Dykyjová, D., and Hradecká, D., Production ecology of *Phragmites communis*. 1. Relations of two ecotypes to the microclimate and nutrient conditions of habitat, *Folia Geobot. Phytotax.*, 11, 23, 1976.

1205. Dykyjová, D. and Květ, J., Eds., *Pond Littoral Ecosystems: Structure and Functioning*, Springer-Verlag, Berlin, 1978.

1206.   Dykyjová, D., and Véber, K., Experimental hydroponic cultivation of helophytes, in *Pond Littoral Ecosystems: Structure and Functioning*, Dykyjová, D., and Květ, J., Eds., Springer-Verlag, Berlin, 1978, 181.

1207.   Dykyjová, D., and Květ, J., Mineral nutrient economy in wetlands of the Trebon basin Biosphere Reserve, Czechoslovakia, in *Proc. Int. Conf. Wetlands: Ecology and Management*, Gopal, B., Turner, E.R., Wetzel, R.G., and Whigham, D.F., Eds., Internat. Scientific Publ., Jaipur, India, 1982, 325.

1208.   Dykyjová, D., and Drbal, K., Plant chemistry in transient peat and acid bogs, *Preslia* (Praha), 56. 73. 1984, (in Czech).

1209.   Eastbrook, G.F., Burk, D.W., Inman, D.R., Kaufman, P.B., Wells, J.R., Jones, J.D., and Ghosheh, N., Comparison of heavy metals in aquatic plants on Charity Island, Saginaw Bay, Lake Huron, U.S.A., with plants along the shoreline of Saginaw Bay, *Am. J. Bot.*, 72, 209, 1985.

1210.   Ebeling, A.W., Laur, D.R., and Rowley, J.R., Severe storm disturbance and reversal of community structure in a southern California kelp forest, *Mar. Biol.*, 84, 287, 1985.

1211.   Eddy, A.A., Mechanisms of solute transport in selected eukaryotic microorganisms, *Adv. Microb. Physiol.*, 23, 1, 1982.

1212.   Edge, P.A., and Ricketts, T.R., Some notes on the growth and nutrition of *Platymonas striata* Butcher (Prasinophyceae), *Nova Hedw.*, 29, 675, 1978.

1213.   Edge, P.A., and Ricketts, T.R., Studies on ammonium-assimilation enzymes of *Platymonas striata* Butcher (Prasinophyceae), *Planta*, 138, 123, 1978.

1214.   Edmondson, W.T., Secondary production and decomposition, *Verh. Internat. Verein. Limnol.*, 14, 316, 1961.

1215.   Edmondson, W.T., Phosphorus, nitrogen and algae in Lake Washington after diversion of sewage, *Science*, 169, 690, 1970.

1216.   Edmondson, W.T., The present condition of Lake Washington, *Verh. Internat. Verein. Limnol.*, 18, 284, 1972.

1217.   Edwards, R.W., and Owens, M., The effect of plants on river condition, *J. Ecol.*, 48, 151, 1960.

1218.   Egglesmann, R., Peat consumption under influence of climate, soil condition, and utilization, in *Proc. 5th Int. Peat Congress*, Vol. 1, Poznan, Poland, 1976, 233.

1219.   Ehresmann, D.W., Deig, E.F., and Hatch, M.T., Anti-viral properties of algal polysaccharides and related compounds, in *Marine Algae in Pharmaceutical Science*, Hoppe, H.A., Levring, T., and Tanaka, Y., Eds., Walter de Gruyter, Berlin, 1979, 294.

1220.   Ehrlich, G.G., and Slack, K.V., Uptake and assimilation of nitrogen in microbiological systems, *Amer. Soc. Test. Mater. Spec. Tech. Publ.*, 448, 11, 1969.

1221.   Eichenberger, E., On the quantitative assessment of the effects of chemical factors on running water ecosystems, *Schweiz. Z. Hydrol.*, 37, 21, 1975.

1222.   Eichenberger, E., The study of eutrophication of algal benthos by essential metals in artificial rivers, in *Biological Aspects of Freshwater Pollution*, Ravera, O., Ed., Pergamon Press, Oxford, 1979, 111.

1223.   Eichenberger, E., and Wuhrmann, K., Growth and photosynthesis during the formation of benthic algal communities, *Verh. Internat. Verein. Limnol.*, 19, 2035, 1975.

1224.   Eichenberger, E., and Schlatter, A., Effects of herbivorous insects on the production of benthic algal vegetation in outdoor channels, *Verh. Internat. Verein. Limnol.*, 20, 1806, 1978.

1225.   Eilers, H., Zur Kenntnis der Ernährungsphysiologie von *Stichococcus bacillaris*, Näg. *Rec. Trav. Botan. Neerl.*, 23, 362, 1926.

1226.   Eisler, R., Copper accumulations in coastal and marine biota, in *Copper in the Environment*, Part 1. *Ecological Cycling*, Nriagu, J.O., Ed., John Wiley and Sons, New York, 1979, 383.

1227.   Elder, J.F., and Horne, A.J., Copper cycles and $CuSO_4$ algicidal capacity in two California lakes, *Environ. Manage.*, 2, 17, 1978.

1228.   Elenkin, A.A., *Monographia algarum cyanophycearum aquidutcium et terrestrium in fimbus URSS inventarum*. I et II, Izd. Akad. Nauk SSR, Moskva-Leningrad, 1936–1949.

1229.   Eley, J.H., Effect of carbon dioxide concentration on pigmentation in the blue-green alga *Anacystis nidulans*, *Plant Cell Physiol.*, 12, 311, 1971.

1230.   Eleman, T.C., Falconer, I.R., Jackson, A.R.B., and Runnegar, M.T., Isolation, characterization and pathology of the toxin from a *Microcystis aeruginosa* equals *Anacystis cyanea* bloom, *Aust. J. Biol. Sci.*, 31, 209, 1978.

1231.   Eleuterius, L.N., The marshes of Mississippi, *Castanea*, 37, 157, 1972.

1232.   Elliot, P.B., and Bamforth, S.S., Interstitial protozoa and algae of Louisiana salt marshes, *J. Protozool.*, 22, 514, 1975.

1233.   Ellis, R.J., Protein and nucleic acid synthesis by chloroplasts, in *The Intact Chloroplasts. Topics in Photosynthesis*, Vol. 1, Barber, J., Ed., Elsevier/North Holland, Amsterdam, 1976, 335.

1234.   Eloranta, P., Pollution and aquatic flora of waters by sulphite cellulose factory at Mänttä, Finnish Lake District, *Ann. Bot. Fennici*, 7, 63, 1970.

1235. Eloranta, P., and Kunnas, S., The growth and species communities of the attached algae in a river system in Central Finland, *Arch. Hydrobiol.*, 86, 27, 1979.

1236. Elster, H.J., and Motsch, B., Untersuchungen über das Phytoplankton und die organische Urproduktion in einigen Seen des Hochschwarzwalds, im Schluchsee und im Bodensee, *Arch. Hydrobiol. Suppl.*, 28, 291, 1966.

1237. Elwood, J.W., and Nelson, D.J., Periphyton production and grazing rates in a stream measured with a ^{32}P material balance method, *Oikos*, 22, 295, 1972.

1238. Elwood, J.W., Newbold, J.D., Trimble, A.F., and Stark, R.W., The limiting role of phosphorus in a woodland stream ecosystem: effects of P enrichment on leaf decomposition and primary producers, *Ecology*, 62, 146, 1981.

1239. Emerson, R., and Lewis, C.M., The photosynthetic efficiency of phycocyanin in *Chroococcus* and the problem of carotenoid participation in photosynthesis, *J. Gen. Physiol.*, 25, 579, 1942.

1240. Emerson, R., and Lewis, C.M., The dependence of the quantum yield of *Chlorella* photosynthesis on wave length of light, *Am. J. Bot.*, 30, 165, 1943.

1241. Emerson, R.L., Stauffer, J.F., and Umbreit, W.W., Relationship between photophospho-rylation and photosynthesis in *Chlorella*, *Am. J. Bot.*, 31, 107, 1944.

1242. Eminson, D., and Phillips, G., A laboratory experiment to examine the effects of nutrient enrichment on macrophyte and epiphytic growth, *Verh. Internat. Verein. Limnol.*, 20, 82, 1978.

1243. Emsley, J., The phosphorus cycle, in *The Handbook of Environmental Chemistry* Vol. 1, Part A, *The Natural Environment and the Biogeochemical Cycles*, Hutzinger, O., Ed., Springer-Verlag, Berlin, 1980, 147.

1244. Endo, S., Über das Vorkommen von freien Zucker bei einigen Grünalgen und seine Beziehung zur Photosynthese. II. *Cladophora wrightiana* Harvey, *Sci. Repts. Tokyo Kyôiku Bunrika Daigaku*, B 43, 223, 1936., *Sci. Repts.*

1245. Engel, D.W., Sunda, W.G., and Fowler, B.A., Factors affecting trace metals uptake and toxicity to estuarine organisms. I. Environmental parameters, in *Biological Monitoring of Marine Pollutants*, Vernberg, F.J., Calabrese, A., Thurberg, F.P., and Vernberg, W.B., Eds., Academic Press, new York, 1981, 127.

1246. Engler, A., *Syllabus der Pflanzen familien*, Berlin, 1898.

1247. Engler, R.M., and Patrick, W.H., Jr., Nitrate removal from floodwater overlying flooded soils and sediments, *J. Environ. Qual.*, 3, 409, 1974.

1248. Environmental Impact Statement, Phase 1: Freshwater wetlands for wastewater treatment, U.S. EPA 904/9-83-107, Atlanta, Georgia, 1983.

1249. Environmental Laboratory, Corps of Engineers Wetlands Delineation Manual, Tech. Rept. Y-87-1, US Army Engineer Waterways Experiment Station, Vicksburg, Mississippi, 1987.

1250. Epel, B.L., and Butler, W.L., Cytochrome *a*$_3$. Destruction by light, *Science*, 166, 621, 1969.

1251. Epel, B.L., and Butler, W.L., The cytochromes of *Prototheca zopfii*, *Plant Physiol.*, 45, 723, 1970.

1252. Eppley, R.W., Sodium exclusion and potassium retention by the red marine alga, *Porphyra perforata*, *J. Gen. Physiol.*, 41, 901, 1958.

1253. Eppley, R.W., Potassium accumulation and sodium efflux by *Porphyra perforata* tissues in lithium and magnesium sea waters, *J. Gen. Physiol.*, 43, 29, 1959.

1254. Eppley, R.W., Major cations, in *Physiology and Biochemistry of Algae*, Lewin, R.A., Ed., Academic Press, New York and London, 1962, 253.

1255. Eppley, R.W., Hydrolysis of polyphosphates by *Porphyra* and other seaweeds, *Physiol. Plant.*, 15, 246, 1962.

1256. Eppley, R.W., Temperature and phytoplankton growth in the sea, *Fish. Bull.*, 70. 1063, 1972.

1257. Eppley, R.W., and Coatsworth, J.L., Uptake of nitrate and nitrite by *Ditylum brightwellii*—kinetics and mechanisms, *J. Phycol.*, 4, 151, 1968.

1258. Eppley, R.W., and Stricklad, J.D.H., Kinetics of marine phytoplankton growth, in *Advances in Microbiology of the Sea*, Vol. 1, Droop, M.R., and Ferguson-Wood, E.J., Eds., Academic Press, New York, 1968, 23.

1259. Eppley, R.W., and Thomas, W.H., Comparison of half-saturation constants for growth and nitrate uptake of marine phytoplankton, *J. Phycol.*, 5, 375, 1969.

1260. Eppley, R.W., and Rogers, J.N., Inorganic nitrogen assimilation of *Ditylum brightwellii*, a marine plankton diatom, *J. Phycol.*, 6, 344, 1970.

1261. Eppley, R.W., and Renger, E.H., Nitrogen assimilation of an oceanic diatom in nitrogen-limited continuous culture, *J. Phycol.*, 10, 15, 1974.

1262. Eppley, R.W., and Peterson, B.J., Particulate organic matter flux and planktonic new production in the deep ocean, *Nature*, 282, 677, 1979.

1263. Eppley, R.W., Coastworth, J.L., and Solórzano, L., Studies of nitrate reductase in marine phytoplankton, *Limnol. Oceanogr.*, 14, 194, 1969.

1264.   Eppley, R.W., Rogers, J.N., and McCarthy, J.J., Half-saturation constants for uptake of nitrate and ammonium by marine phytoplankton, *Limnol. Oceanogr,* 14, 912, 1969.

1265.   Eppley, R.W., Carlucci, A.F., Holm-Hansen, O., Kiefer, D., McCarthy, J.J., Venrick, E., and Williams, P.M., Phytoplankton growth and composition in shipboard cultures supplied with nitrate, ammonium or urea as the nitrogen source, *Limnol. Oceanogr,* 16, 741, 1971.

1266.   Eppley, R.W., Renger, E.H., Harrison, W.G., and Cullen, J.J., Ammonium distribution in southern California coastal waters and its role in the growth of phytoplankton, *Limnol. Oceanogr,* 24, 495, 1979.

1267.   Epstein, E., Mineral nutrition of plants: mechanisms of uptake and transport, *Ann. Rev. Plant Physiol.,* 7, 1, 1956.

1268.   Epstein, E., *Mineral Nutrition of Plants: Principles and Perspectives,* John Wiley and Sons, New York, 1972.

1269.   Erickson, S.J., Toxicity of copper to *Thalasssiosirapseudonana* in unenriched inshore seawater, *J. Phycol.,* 8, 318, 1972.

1270.   Erickson, S.J., Lackie, N., and Maloney, T.E., A screening technique for estimating copper toxicity to estuarine phytoplankton, *J. Water Pollut. Control Fed.,* 42, 270, 1970.

1271.   Eriksson, F., The macrophytes and their production in Lake Vitalampa. *Scr. Limnol. Upsal.,* 9, 1, 1974.

1272.   Ernst, W., Physiology of heavy metals resistance in plants, in *Proc. Int. Conf. Heavy Metals in the Environment,* Vol. 2, Toronto, 1977, 121.

1273.   Ertl, M., Juris, S., and Tomajka, J., Vorlaufige Angaben über jahreszeitliche Veranderungen und die vertikale Verteilung des Periphytons im mittleren Abschnitt der Donau, *Arch. Hydrobiol. Suppl.,* 44, 34, 1972.

1274.   Erwin, J., and Bloch, K., Polyunsaturated fatty acids in some photosynthetic organisms, *Biochem. Z.,* 388, 496, 1963.

1275.   Esteves, F.A., Die Bedeutung der aquatischen Makrophyten für den Stoffhaushalt des Schöhsees. I. Die Produktion an Biomasse, *Arch. Hydrobiol. Suppl.,* 57, 117, 1979.

1276.   Esteves, F.A., and Barbieri, R., Dry weight and chemical changes during decomposition of tropical macrophytes in Lobo Reservoir-Sao Paulo, Brazil, *Aquat. Bot.,* 16, 285, 1983.

1277.   Estrada, M., Valiela, I., and Teal, J.M., Concentration and distribution of chlorophyll in fertilized plots in a Massachusetts salt marsh, *J. Exp. Mar. Biol. Ecol.,* 14, 47, 1974.

1278.   Etherington, J.R., *Wetland Ecology,* Studies in Biology No. 154, Edward Arnold, London, 1983.

1279.   Ettl, H., Ein Beitrag zur Systematik der Heterokonten, *Bot. Not.,* 109, 411, 1956.

1280.   Ettl, H., *Die Gattung Chlamydomonas Ehrenberg, Beih. Nova Hedw,* 49, 1976.

1281.   Ettl, H., Taxonomische Bermerkungen zu die Xanthophyceen, *Nova Hedw.,* 26, 555, 1977.

1282.   Ettl, H., *Xantophyceae,* Vol. 1, Süssenwasserflora von Mitteleuropa 3, Gustav Fisher Verlag, Stuttgart, 1978.

1283.   Ettl, H., Die taxonomische Abgrenzung der Gattung *Chlorogonium* Ehrenberg (Chlamydomonales, Chlorophyta), *Nova Hedw.,* 33, 709, 1980.

1284.   Ettl, H., *Grundriss der allgemeinen Algologie,* Gustav Fisher, Jena, 1980.

1285.   Ettl, H., *Chlorophyta I. Phytomonadina,* Süsswasserflora von Mitteleuropa 9, Gustav Fisher Verlag, Stuttgart, 1983.

1286.   Ettl, H., and Moestrup, Ø., Light and electron microscopical studies on *Hafniomonas* gen. nov. (Chlorophyceae, Volvocales), a genus resembling *Pyraminomonas* (Prasinophyceae), *Pl. Syst. Evol.,* 135, 177, 1980.

1287.   Ettl, H., and Gärtner, G., *Chlorophyta II. Tetrasporales, Chlorococcales, Gloeodendrales,* Süsswasserflora von Mitteleuropa 10, Gustav Fisher Verlag, Stuttgart, 1988.

1288.   Eurola, S., Hicks, S., and Kaakinen, E., Key to Finnish mire types, in *European Mires,* Moore, P.D., Ed., Academic Press, London, 1984, 11.

1289.   Evans, D.D., and Scott, A.D., A polarographic method of measuring dissolved oxygen in saturated soil, *Soil Sci. Soc. Am. Proc.,* 19, 12, 1955.

1290.   Evans, H.J., and Sorger, G.J., Role of mineral elements with emphasis on the univalent cations, *Ann. Rev. Plant Physiol.,* 17, 47, 1966.

1291.   Evans, L.V., Distribution of pyrenoids among some brown algae, *J. Cell Sci.,* 1, 449., 1966.

1292.   Evans, L.V., Cytoplasmic organelles, in *Algal Physiology and Biochemistry* Stewart, W.D.P., Ed., Blackwell Scientific Publications, Oxford, 1974, 86.

1293.   Evans, L.V., and Holligan, M.S., Correlated light and electron microscope studies on brown algae. I. Localization of alginic acid and sulphated polysaccharides in *Dictyota, New Phytol.,* 71, 1161, 1972.

1294.   Evans, L.V., Callow, J.A., and Callow, M.E., Structural and physiological studies on the parasitic red alga *Holmsella, New Phytol.,* 72, 393, 1973.

1295. Evans, M.H., A comparison of the biological effects of paralytic shellfish poisons from clam, mussel and dinoflagellate, *Toxicon*, 9, 139, 1971.

1296. Everest, S.A., and Syrett, P.J., Evidence for the participation of glutamate dehydrogenase in ammonium assimilation by *Stichococcus bacillaris*, *New Phytol.*, 93, 581, 1983.

1297. Eversole, R.A., Biochemical mutants of *Chlamydomonas reinhardtii*, *Am. J. Bot.*, 43, 404, 1956.

1298. Ewel, K.C., and Odum, H.T., Eds., *Cypress Swamps*, University of Florida Press, Gainesville, 1984.

1299. Eyster, C., Necessity of boron for *Nostoc muscorum*, *Nature*, 170, 755, 1952.

1300. Eyster, C., The microelement nutrition of *Nostoc muscorum*, *The Ohio J. Sci.*, 58, 25, 1958.

1301. Eyster, C., Mineral requirements of *Nostoc muscorum* for nitrogen fixation, in *Proc. IX Int. Bot. Congr., Montreal*, 2, 109, 1959.

1302. Eyster, C., Requirements and function of micronutrients by green plants with respect to photosynthesis, in *Biologistics for Space Systems Symposium*, U.S. Air Force Tech. Doc. Rept. AMRL-TDR-62-116, 1962, 199.

1303. Eyster, C., Micronutrient requirements for green plants, especially algae, in *Algae and Man*, Jackson, D.F., Ed., Plenum Press, New York, 1964, 86.

1304. Eyster, C., Microorganic and microinorganic requirements for algae, in *Algae, Man and the Environment*, Jackson, D.F., Ed., Syracuse University Press, Syracuse, New York, 1968, 27.

1305. Eyster, H.C., Brown, T.E., and Tanner, H.A., Mineral requirements for *Chlorella pyrenoidosa* under autotrophic and heterotrophic conditions, in *Trace Elements*, Lamb, C.A., Bentley, O.G., and Beattie, J.M., Eds., Academic Press, New York and London, 1958, 157.

1306. Eyster, H.C., Brown, T.E., Tanner, H.A., and Hood, S.L., Manganese requirement with respect to growth, Hill-activity and photosynthesis, *Plant Physiol.*, 33, 235, 1958.

1307. Fagerström, T., and Jernelöv, A., Formation of methyl mercury from pure mercuric sulphide in anaerobic organic sediment, *Water Res.*, 5, 121, 1971.

1308. Fagerström, T., and Jernelöv, A., Some aspects of the quantitative ecology of mercury, *Water Res.*, 6, 1193, 1972.

1309. Fairchild, E., and Sheridan, R.P., A physiological investigation of the hot spring diatom *Achnanthes exigua* Grun., *J. Phycol.*, 10, 1, 1974.

1310. Falkowski, P.G., The regulation of nitrate assimilation in lower plants, in *Nitrogen in the Environment*, Vol. 2. *Soil-Plant-Nitrogen Relationship*, Nielsen, D.R., and MacDonald, J.G., Eds., Academic Press, New York, 1978, 143.

1311. Falkowski, P.G., and Rivkin, R.B., The role of glutamine synthetase in the incorporation of ammonium in *Skeletonmema costatum* (Bacillariophyceae), *J. Phycol.*, 12, 448, 1976.

1312. Falkowski, P.G., Sukenik, A., and Herzig, R., Nitrogen limitation in *Isochrysis galbana* (Haptophyceae). II. Relative abundance of chloroplast proteins, *J. Phycol.*, 25, 471, 1989.

1313. Faulkner, S.P., and Richardson, C.J., Physical and chemical characteristics of freshwater wetland soils, in *Proc. Int. Conf. Constructed Wetlands for Wastewater Treatment*, Hammer, D.A., Ed., Lewis Publishers, Chelsea, Michigan, 1989, 41.

1314. Faust, M.A., Structure of the periplast of *Cryptomonas ovata* var. *palustris*, *J. Phycol.*, 10, 121, 1974.

1315. Fawley, M.W., Douglas, C.A., Stewart, K.D., and Mattox, K.R., Light-harvesting pigment-protein complexes of the Ulvophyceae (Chlorophyta): characterization and phylogenetic significance, *J. Phycol.*, 26, 186, 1990.

1316. Fay, P., Photostimulation of nitrogen fixation in *Anabaena cylindrica*, *Biochim. Biophys. Acta*, 216, 353, 1970.

1317. Fay, P., Kumar, H.D., and Fogg, G.E., Cellular factors affecting nitrogen fixation in the blue-green alga *Chlorogloea fritschii*, *J. Gen. Microbiol.*, 35, 351, 1964.

1318. Fay, P., Stewart, W.D.P., Walsby, A.E., and Fogg, G.E., Is the heterocyst the site of nitrogen fixation in blue-green algae? *Nature*, 220, 810, 1968.

1319. Federal Manual for Identifying and Delineating Jurisdictional Wetlands, Cooperative publication of Fish and Wildlife Service, EPA, Dept. of the Army, Soil Conservation Service, 1989.

1320. Federal Register, "40 CFR part 230: Section 404 (b) (1) *Guidelines for Specification of Disposal Sites for Dredged or Fill Material,*" Vol. 45, No. 249, pp. 85352-85353. U.S. Government Printing Office, Washington, DC, 1980.

1321. Federal Register, "Title 33: *Navigation and Navigable Waters:* Chapter III, *Regulatory Programs of the Corps of Engineers,*" Vol. 47, No. 138, p. 31810. U.S. Government Printing Office, Washington, DC, 1982.

1322. Feher, D., Researches on the geographical distribution of soil microflora. Part II. The geographical distribution of soil algae, *Comm. Bot. Inst. Hungar. University of Tech. and Econ. Sciences (Sopron)*, 21, 1, 1948.

1323. Fejitel, T.C., DeLaune, R.D., and Patrick, W.H., Jr., Carbon flow in coastal Louisiana, *Mar. Ecol. Prog. Ser*, 24, 255, 1985.

1324. Fejitel, T.C., DeLaune, R.D., and Patrick, W.H., Jr., Seasonal pore water dynamics and pyrite formation in Barataria Basin marshes, *Soil Sci. Soc. Am. J.*, 52, 59, 1988.

1325. Fejitel, T.C., DeLaune, R.D., and Patrick, W.H., Jr., Biogeochemical control on metal distribution and accumulation in Louisiana sediments, *J. Environ. Qual.*, 17, 88, 1988.

1326. Fejitel, T.C., DeLaune, R.D., and Patrick, W.H., Jr., Carbon, nitrogen and micronutrient dynamics in Gulf coast marshes, in *Proc. Symp. Freshwater Wetlands and Wildlife,* Sharitz, R.R., and Gibbons, J.W., Eds., U.S. Department of Energy, 1989, 47.

1327. Feldhen, C.M., Structure and function of the nuclear envelope, in *Advances in Cell and Molecular Biology,* Vol. 2, DuPraw, E.J., Ed., Academic Press, New York, 1972, 273.

1328. Feldmann, J., Recherches sur la végétation marine de la Méditeranée, *Rev. Algol.*, 10, 1, 1937.

1329. Felföldy, L.J.M., On the chlorophyll content and biological productivity of periphytic diatom communities on the Stony Shores of Lake Balaton, *Ann. Biol. Tihany,* 28, 99, 1961.

1330. Felföldy, L.J.M., On the role of pH and inorganic carbon, sources in the photosynthesis of unicellular algae, *Acta Biol. Acad. Sci. Hungar.*, 13, 207, 1962.

1331. Fenchel, T., Biology of heterotrophic microflagellates. 2. Bioenergetics and growth, *Mar. Ecol. Prog. Ser.,* 8, 225, 1982.

1332. Fenchel, T., and Straarup, B.J., Vertical distribution of photosynthetic pigments and the penetration of light in marine sediments, *Oikos,* 22, 172, 1971.

1333. Fenchel, T.M., and Jørgensen, B.B., Detritus food chains of aquatic ecosystems. The role of bacteria, *Adv. Microbiol. Ecol.,* 1, 1, 1977.

1334. Fenical, W., Halogenation in the Rhodophyta. A review, *J. Phycol.,* 11, 245, 1975.

1335. Fenn, L.B., and Kissel, D.E., Ammonia volatilization from surface of temperature and rate of ammonium nitrogen application, *Soil Sci. Soc. Am. Proc.,* 38, 606, 1974.

1336. Fennikoh, K.B., Hirschfield, H.I., and Kneip, T.J., Cadmium toxicity in planktonic organisms of a freshwater food web, *Environ. Res.,* 15, 357, 1978.

1337. Ferguson, J.F., and Gavis, J., A review of the arsenic cycle in natural waters, *Water Res.,* 6, 1259, 1972.

1338. Ferguson-Wood, E.J., Phytoplankton distribution in the Caribbean region, In *UNESCO Symp. on Investigations and Resources of the Caribbean Sea and Adjacent Regions,* WCNA, 1971, 399.

1339. Fernández, E., and Cárdenas, J., Nitrate reductase from a mutant strain of *Chlamydomonas reinhardtii* incapable of nitrate assimilation, *Z. Naturforsch.,* 38, 439, 1983.

1340. Ferreira, J.G., Factors governing mercury accumulation in three species of marine macroalgae, *Aquat. Bot.,* 39, 335, 1991.

1341. Ferrier, J.M., Apparent bicarbonate uptake and possible plasmalemma proton efflux in *Chara corallina, Plant Physiol.,* 66, 1198, 1980.

1342. Ferrier, J.M., and Lucas, J.W., Plasmalemma transport of OH⁻ in *Chara corallina.* II. Further analysis of the diffusion system associated with OH⁻ efflux, *J. Exp. Bot.,* 30, 705, 1979.

1343. Fiedler, P.C., Zooplankton avoidance and reduced grazing responses of *Gymnodinium splendens* (Dinophyceae), *Limnol. Oceanogr,* 27, 961, 1982.

1344. Fielding, P., Damstra, K., and Branch, G., Benthic diatom biomass, production and sediment chlorophyll in Langeban Lagoon, South Africa, *Estuar. Coast. Shelf Sci.,* 27, 413, 1988.

1345. Findenegg, G.R., Beziehungen zwischen Carboanhydraseaktivität und Aufnahme von HCO_3^- and Cl^- bei der Photosynthese von *Scenedesmus obliquus, Planta,* 116, 123, 1974.

1346. Findenegg, G.R., Ionic transport at the plasmalemma of *Scenedesmus* cells adapted to high and low CO_2 levels, in *Inorganic Carbon Uptake by Aquatic Photosynthetic Organisms,* Lucas, W.J., and Berry, J.A., Eds., *Am. Soc. Plant Physiol.,* Rockville, Maryland, 1985, 155.

1347. Findenegg, G.R., and Fischer, K., Apparent photorespiration of *Scenedesmus obliquus:* increase during adaptation to low CO_2 level, *Z. Pflanzenphysiol.,* 89, 363, 1978.

1348. Findenegg, I., Die Bedeutung kurzwelliger Strahlung für die planktische Primärproduktion, *Ver. Internat. Verein. Limnol.,* 16, 314, 1966.

1349. Finkle, B.J., and Appleman, D., The effect of magnesium concentration on chlorophyll and catalase development in *Chlorella, Plant Physiol.,* 28, 652, 1953.

1350. Finkle, B.J., and Appleman, D., The effect of magnesium concentration on growth of *Chlorella, Plant Physiol.,* 28, 664, 1953.

1351. Finlayson, M., and Moser, M., Eds., *Wetlands,* Facts on File, Oxford, 1991.

1352. Finlayson, C.M., Farrel, T.P., and Griffiths, D.J., Studies of hydrobiology of a tropical lake in northwestern Queensland. III. Growth, chemical composition and potential for harvesting of the aquatic vegetation, *Austr. J. Mar. Freshwater Res.,* 35, 525, 1984.

1353. Firestone, M.K., Biological denitrification, in *Nitrogen in Agricultural Soil,* Stevenson, F.J., Ed., *Am. Soc. Agron.,* Madison, Wisconsin, *Agronomy,* 22, 289, 1982.

1354. Firestone, M.K., Smith, M.S., Firestone, R.B., and Tiedje, J.M., The influence of nitrate, nitrite and oxygen on the composition of the gaseous products of denitrification in soil, *Soil Sci. Soc. Am. J.*, 43, 1140, 1979.

1355. Fischer, F.G., and Dörfel, H., Die Polyuronsäuren der Braualgen, *Hoppe-Seyler's Z. Physiol. Chem.*, 302, 186, 1955.

1356. Fischer, S., and Simmonis, W., Tagesperiodische Schwankungen und lichtinduzierte Zunahme der Nitratreduktase-Aktivität bei Synchronkulturen von *Ankistrodesmus braunii, Z. Pflanzemphysiol.*, 92, 143, 1979.

1357. Fishbein, L., Natural non-nutrient substances in the food chain, *Sci. Total Environ.*, 1, 211, 1972.

1358. Fisher, N.S., and Frood, D., Heavy metals and marine diatoms: influence of dissolved organic compounds on toxicity and selection for metal tolerance among four species, *Mar. Biol.*, 59, 85, 1980.

1359. Fisher, N.S., and Jones, G.J., Heavy metals and marine phytoplankton: Correlation of toxicity and sulfhydryl-binding, *J. Phycol.*, 17, 108, 1981.

1360. Fisher, S.G., Gray, L.J., Grimm, N.B., and Busch, D.E., Temporal succession in a desert stream ecosystem following flash flooding, *Ecol. Monogr.*, 52, 93, 1982.

1361. Fisher, T.R., Carlson, P.R., and Barber, R.T., Some problems in the interpretation of ammonium uptake kinetics, *Mar. Biol. Lett.*, 2, 33, 1981.

1362. Fitzgerald, G.P., Factors affecting the toxicity of copper to algae and fish, *Am. J. Bot.*, 50, 629, 1963.

1363. Fitzgerald, G.P., Evaluation of potassium permanganate as an algicide for water cooling towers, *IEC Prod. Res. Dev.*, 3, 82, 1964.

1364. Fitzgerald, G.P., Laboratory evaluation of potassium permanganate as an algicide for water reservoirs, *Southwest Water Works Assoc. J.*, 45, 16, 1964.

1365. Fitzgerald, G.P., Detection of limiting or surplus nitrogen in algae and aquatic weeds, *J. Phycol.*, 4, 121, 1968.

1366. Fitzgerald, G.P., Field and laboratory evaluations of bioassays for nitrogen and phosphorus with algae and aquatic weeds, *Limnol. Oceanogr.*, 14, 206, 1969.

1367. Fitzgerald, G.P., Aerobic lake muds for the removal of phosphorus from lake waters, *Limnol. Oceanogr.*, 15, 550, 1970.

1368. Fitzgerald, G.P., Evaluations of the availability of sources of nitrogen and phosphorus for algae, *J. Phycol.*, 6, 239, 1970.

1369. Fitzgerald, G.P., Shortcut methods test algicides, *Water Sewage Works*, 121, 85, 1974.

1370. Fitzgerald, G.P., Are chemicals used in algae control biodegradable? *Water Sewage Works*, 122, 82, 1975.

1371. Fitzgerald, G.P., Comparative algicide evaluations using laboratory and field algae, *J. Aquat. Plant Manage.*, 17, 66, 1979.

1372. Fitzgerald, G.P., and Faust, S.L., Bioassays for algicidal vs. algistatic chemicals, *Water and Sewage Works*, 110, 293, 1963.

1373. Fitzgerald, G.P., and Faust, S.L., Factors affecting the algicidal and algistatic properties of copper, *Appl. Microbiol.*, 24, 345, 1963.

1374. Fitzgerald, G.P., and Nelson, T.C., Extractive and enzymatic analysis for limiting or surplus phosphorus in algae, *J. Phycol.*, 2, 32, 1966.

1375. Fitzgerald, G.P., and Faust, S.I., Effect of water sample preservation methods on the release of phosphorus from algae, *Limnol. Oceanogr.*, 12, 332, 1967.

1376. Fjerdingstad, E., The importance of microbiology for estimating the pollution of water, *Nordisk Hygienisk Tidskrift*, 32, 127, 1951.

1377. Fjerdingstad, E., Pollution of streams estimated by benthal phytomicroorganisms, *Int. Rev. Ges. Hydrobiol.*, 49, 63, 1964.

1378. Fjerdingstad, E., Taxonomy and saprobic valency of benthic phytomicro-organisms, *Int. Rev. Ges. Hydrobiol.*, 50, 475, 1965.

1379. Fjerdingstad, E., Kemp. K., Fjerdingstad, E., and Vanggaard, L., Chemical analyses of red snow from East Greenland with remarks on *Chlamydomonas nivalis* (Ban) Wille, *Arch. Hydrobiol.*, 73, 70, 1974.

1380. Fjerdingstad, E., Kemp, K., and Fjerdingstad, E., Trace element analyses and bacteriological investigations in Danish *Lobelia - Isoëtes* lakes, *Arch. Hydrobiol.*, 76, 137, 1975.

1381. Flemer, D.A., Primary productivity of the North Branch of the Raritan River, New Jersey, *Hydrobiologia*, 35, 273, 1970.

1382. Fletcher, J.E., and Martin, W.P., Some effects of algae and molds in the rain-crust of desert soils, *Ecology*, 29, 95, 1948.

1383. Flick, D.F., Kraybill, H.F., and Dimitroff, J.M., Toxic effects of cadmium: a review, *Environ. Res.*, 4, 71, 1971.

1384. Flint, L.H., Notes on the algal food of shrimp and oysters, *Proc. La. Acad. Sci.,* 19, 11, 1956.
1385. Florence, T.M., and Batley, G.E., Determination of the chemical forms of trace metals in natural waters, with special reference to copper, lead, cadmium and zinc, *Talanta,* 24, 151, 1977.
1386. Florencio, F.J., and Vega, J.M., Regulation of the assimilation of nitrate in *Chlamydomonas reinhardtii, Phytochemistry,* 21, 1195, 1982.
1387. Florencio, F.J., and Vega, J.M., Separation, purification and characterization of two isoforms of glutamine synthetase from *Chlamydomonas reinhardii, Z. Natürforsch.,* 38C, 531, 1983.
1388. Flores, E., Guerrero, M.G., and Losada, M., Short-term ammonium inhibition of nitrate utilization by *Anacystis nidulans* and other cyanobacteria, *Arch. Microbiol.,* 128, 137, 1980.
1389. Flower, R.J., An improved epilithon sampler and its evaluation in two acid lakes, *Br. Phycol. J.,* 20, 109, 1985.
1390. Flynn, K.J., and Syrett, P.J., Characteristics of the uptake system for L-lysine and L-arginine in *Phaeodactylum tricornutum, Mar. Biol.,* 90, 151, 1986a.
1391. Flynn, K.J., and Syrett, P.J., Utilization of L-lysine and L-arginine by the diatom *Phaeodactylum tricornutum, Mar. Biol.,* 90, 159, 1986b.
1392. Focht, D.D., The effect of temperature, pH, and aeration on the production of nitrous oxide and gaseous nitrogen—a zero order kinetic model, *Soil Sci.,* 118, 173. 1974.
1393. Focht, D.D., and Verstraete, W., Biochemical ecology of nitrification-denitrification, *Adv. Microbiol. Ecol.,* 1, 124, 1977.
1394. Foerster, J.W., and Schlichting, H.E., Jr., Phyco-periphyton in an oligotrophic lake, *Trans. Amer. Micros. Soc.,* 84, 485, 1965.
1395. Fogg, G.E., The gas vacuoles of the Myxophyceae (Cyanophyceae), *Biol. Rev,* 16, 205, 1941.
1396. Fogg, G.E., Studies on nitrogen fixation by blue-green algae. I. Nitrogen fixation by *Anabaena cylindrica* Lemm., *J. Exp. Biol.,* 19, 78, 1942.
1397. Fogg, G.E., Growth and heterocysts production in *Anabaena cylindrica* Lemm., *New Phytol.,* 43, 164, 1944.
1398. Fogg, G.E., Nitrogen fixation by blue-green algae, *Endeavour,* 6, 172, 1947.
1399. Fogg, G.E., Growth and heterocysts production in *Anabaena cylindrica* Lemm. II. In relation to carbon and nitrogen metabolism, *Ann. Bot. N.S.,* 13, 241, 1949.
1400. Fogg, G.E., Growth and heterocysts production in *Anabena cylibdrica* Lemm. III. The cytology of heterocysts, *Ann. Bot. N.S.,* 15, 23, 1951.
1401. Fogg, G.E., Production of extracellular nitrogenous substances by a blue-green alga, *Proc. R. Soc. Lond.,* 139, 372, 1952.
1402. Fogg, G.E., *The Metabolismus of Algae,* Methuen, London, 1953.
1403. Fogg, G.E., Nitrogen fixation by photosynthetic organisms, *Ann. Rev. Plant Physiol.,* 7, 51, 1956 a.
1404. Fogg, G.E., Photosynthesis and formation of fats in a diatom, *Ann. Bot. N.S.,* 20, 265, 1956.
1405. Fogg, G.E., The comparative physiology and biochemistry of the blue-green algae, *Bact. Rev,* 20, 148, 1956.
1406. Fogg, G.A., Extracellular products of phytoplankton and the estimation of primary production, *Rappt. Proces-Verbaux Reunions, Conseil Per. Internat. Explor. Mer,* 144, 56, 1958.
1407. Fogg, G.E., Nitrogen nutrition and metabolic patterns in algae, *Symp.. Soc. Exp. Biol.,* 13, 106, 1959.
1408. Fogg, G.E., Nitrogen fixation, in *Physiology and Biochemistry of Algae,* Lewin, R.A., Ed., Academic Press, New York and London, 1962, 161.
1409. Fogg, G.E., Extracellular products, in *Physiology and Biochemistry of Algae,* Lewin, R.A., Ed., Academic Press, New York, 1962, 475.
1410. Fogg, G.E., *Algal Cultures and Phytoplankton Ecology,* The University of Wisconsin Press, Madison, Wisconsin, 1966.
1411. Fogg, G.E., The extracellular products of algae, *Oceanogr. Mar. Biol. Ann.Rev.,* 4, 195, 1966.
1412. Fogg, G.E., Observations on the snow algae of the South Orkney Islands, *Phil. Trans. R. Soc. Lond.,* 252 B, 279, 1967.
1413. Fogg, G.E., The physiology of an algal nuisance. *Proc. R. Soc. Lond.,* B 173, 175, 1969.
1414. Fogg, G.E., Extracellular products of algae in freshwater, *Arch. Hydrobiol. Beih. Ergebn. Limnol.,* 5, 1, 1971.
1415. Fogg, G.E., Nitrogen fixation in lakes, *Plant and Soil,* Spec. Vol., 393, 1971.
1416. Fogg, G.E., Phosphorus in primary aquatic plants, *Water Res.,* 7, 77, 1973.
1417. Fogg, G.E., Nitrogen fixation, in *Algal Physiology and Biochemistry* Stewart, W.D.P., Ed., Blackwell Scientific Publications, Oxford, 1974, 560.
1418. Fogg, G.E., *Algal Cultures and Phytoplankton Ecology,* 2nd ed., The University of Wisconsin Press, Madison, 1975.

1419. Fogg, G.E., and Collyer, D.M., The accumulation of lipids by algae, in *Algal Culture: From Laboratory to Pilot Plant*, Burlew, J.S., Ed., Carnegie Inst. of Washington Publ. 600, Washington, DC, 1953, 177.

1420. Fogg, G.E., and Wolfe, M., The nitrogen metabolism of the blue-green algae (Myxophyceae), in *Autotrophic Microorganisms*, Fry, B.A., and Peel, J.L., Eds., Cambridge University Press, London and New York, 1954, 99.

1421. Fogg, G.E., and Westlake, D.F., The importance of extracellular products of algae in freshwater, *Verh. Internat. Verein. Limnol.*, 12, 219, 1955.

1422. Fogg, G.E., and Boalch, G.T., Extracellular products in pure cultures of a brown alga, *Nature*, 181, 789, 1958.

1423. Fogg, G.E., and Than-Thun, Interrelations of photosynthesis and assimilation of elementary nitrogen in a blue-green alga, *Proc. R. Soc.*, B 153, 111, 1960.

1424. Fogg, G.E., and Belcher, J.H., Physiological studies on a plankton "μ-alga," *Verh. Internat. Verein. Limnol.*, 14, 893, 1961.

1425. Fogg, G.A., and Nalewajko, C., Discussion of "The production of glycolate during photosynthesis in *Chlorella*" by C.P. Whittingham and G.G. Pritchard, *Proc. R. Soc. Lond.*, B 157, 381, 1963.

1426. Fogg, G.A., and Nalewajko, C., Glycollic acid as an extracellular product of phytoplankton, *Verh. Internat. Verein. Limnol.*, 15, 806, 1964.

1427. Fogg, G.E., and Watt, W.D., The kinetics of release of extracellular products of photosynthesis by phytoplankton, *Mem. Ist. Ital. Idrobiol.*, 18 (Suppl.), 165, 1965.

1428. Fogg, G.E., and Watt, W.D., Extracellular products of phytoplankton photosynthesis, *Proc. R. Soc. Lond.*, B 162, 517, 1965.

1429. Fogg, G.E., and Stewart, W.D.P., Nitrogen fixation in blue-green algae, *Sci. Progr.*, 53, 191, 1965.

1430. Fogg, G.E., and Pattnaik, H., The release of extracellular nitrogenous products by *Westiellopsis prolifica*, *Phykos*, 5, 58, 1966.

1431. Fogg, G.E., and Stewart, W.D.P., *In situ* determination of biological nitrogen fixation in Antarctica, *Br. Antarct. Survey Bull.*, 15, 39, 1968.

1432. Fogg, G.E., and Horne, A.J., The physiology of antarctic freshwater algae, *Antarctic Res.*, 632, 1968.

1433. Fogg, G.A., Nalewajko, C., and Watt, W.D., Extracellular products of phytoplankton photosynthesis, *Proc. R. Soc. Lond.*, B 162, 517, 1965.

1434. Fogg, G.E., Stewart, W.D.P., Fay, P., and Walsby, A.E., *The Blue-Green Algae*, Academic Press, London and New York, 1973.

1435. Ford, J.E., B$_{12}$ vitamins and growth of the flagellate *Ochromonas malhamensis*, *J. Gen. Microbiol.*, 19, 161, 1958.

1436. Ford, J.E., and Goulden, J.D.S., The influence of vitamin B$_{12}$ on the growth rate and cell composition of the flagellate *Ochromonas malhamensis*, *J. Gen. Microbiol.*, 20, 267, 1959.

1437. Fordham, A.W., and Norrish, K., Arsenate-73 uptake by components of several acidic soils and its implications for phosphate retention, *Aust. J. Soil Res.*, 17, 307, 1979.

1438. Forest, G.I., and Smith, R.A.H., The productivity of a range of blanket-bog types in the Northern Pennines, *J. Ecol.*, 63, 173, 1975.

1439. Forest, H.S., The soil algal community. II. Soviet soil studies, *J. Phycol.*, 1, 164, 1965.

1440. Forest, H.S., and Khan, K.R., The blue-green algae—a program of evaluation of Francis Drouet's taxonomy, in *Taxonomy and Biology of Blue-Green Algae*, Desikachary, T.V., Ed., Bangalore Press, Madras, 1972, 128.

1441. Forest, H.S., Willson, D.L., and England, R.B., Algal establishment on sterilized soil replaced in an Oklahoma prairie, *Ecology*, 40. 475, 1959.

1442. Forsberg, C., Subaquatic macrovegetation in Ösbysjön, Djursholm, *Oikos*, 11, 183, 1960.

1443. Forsberg, C., Phosphorus, a maximum factor in the growth of Characeae, *Nature*, 201, 517, 1964.

1444. Forsberg, C., Nutritional studies of *Chara* in axenic cultures, *Physiol. Plant.*, 18, 275, 1965.

1445. Forsberg, C., Ecological and physiological studies of Charophytes, *Abstr. Uppsala Diss. Sci.*, 53, 1, 1965.

1446. Forsberg, C., Environmental conditions of swedish charophytes, *Symb. Bot. Ups.*, 18, 1, 1965.

1447. Forsberg, C., Sterile germination of oospores of *Chara* and seeds of *Najas marina*, *Physiol. Plant.*, 18, 128, 1965.

1448. Forsberg, C., and Taube, Ö., Extracellular organic carbon from some green algae, *Physiol. Plant.*, 20, 200, 1967.

1449. Forsberg, C., Jinnerot, D., and Davidson, L., The influence of synthetic detergents on the growth of algae, *Vatten*, 23, 2, 1967.

1450. Förstner, U., and Prosi, F., Heavy metal pollution in freshwater ecosystems, in *Biological Aspects of Freshwater Pollution*, Ravera, O., Ed., Pergamon Press, Oxford, 1979, 129.

1451. Förstner, U., and Wittmann, G., *Metal Pollution in Aquatic Environments*, 2nd ed., Springer-Verlag, New York, 1981.

1452. Förstner, U., and Salomons, W., Trace metal analysis on polluted sediments. 1. Assessment of sources and intensities, *Environ. Technol. Lett.*, 1, 494, 1980.

1453. Forward, R.B., Jr., Phototaxis in a dinoflagellate: action spectra as evidence for a two-pigmented system, *Planta*, 111, 167, 1973.

1454. Forward, R.B., Jr., Phototaxis by the dinoflagellate *Gymnodinium splendens* Lebour., *J. Protozool.*, 21, 312, 1974.

1455. Foster, P., Concentrations and concentration factors of heavy metals in brown algae, *Environ. Pollut.*, 10, 45, 1976.

1456. Foster, P.L., Copper exclusion as a mechanism of heavy metal tolerance in a green alga, *Nature*, 269, 322, 1977.

1457. Fott, B., Über den inneren Bau von *Vacuolaria viridis* (Dangeard) Senn., *Arch. Protistenk.*, 84, 242, 1935.

1458. Fott, B., *Algenkunde*, Gustav Fisher, Jena, 1959.

1459. Fott, B., Hologamic and agamic cyst formation in loricate Chrysomonads, *Phykos*, 3, 15, 1964.

1460. Fott, B., *Blue-Green Algae and Algae*, Academia, Praha, 1967, (in Czech).

1461. Fott, B., VIII. Klasse: Chloromonadophyceae, in *Das Phytoplankton des Süsswassers*, Huber-Pestalozzi, G., Ed., Schweizerbart'sche Verlagsbuchhandl., Stuttgart, 1968, 79.

1462. Fott, B., *Studies in Phycology*, Academia, Prague, 1969.

1463. Fott, B., *Algenkunde*, 2nd ed., Gustav Fisher, Jena, 1971.

1464. Fott, B., The phylogeny of the eukaryotic algae, *Taxon*, 23, 446, 1974.

1465. Fott, B., and McCarthy, A.J., Three acidophilic flagellates in pure culture, *J. Protozool.*, 11, 116, 1964.

1466. Fowden, L., A comparison of the composition of some algal proteins. *Ann. Bot. N.S.*, 18, 259, 1954.

1467. Fowden, L., Amino acids and proteins, in *Physiology and Biochemistry of Algae*, Lewin, R.A., Ed., Academic Press, New York and London, 1962, 189.

1468. Fowler, B.K., and Hershner, C., Primary production in Cohoke Swamp, a tidal freshwater wetland in Virginia, in *Proc. Symp. Freshwater Wetlands and Wildlife*, Sharitz, R.R., and Gibbons, J.W., Eds., U.S. Department of Energy, 1989, 365.

1469. Fox, J.L., Odlaug, T.O., and Olson, T.A., The ecology of periphyton in western Lake Superior, *University of Minnesota Bull. Water Resources Res. Center*, 14, 1, 1969.

1470. Fox, R.L., and Kamprath, E.J., Adsorption and leaching of P in acid organic soils and high organic matter sand, *Soil Sci. Soc. Am. Proc.*, 35, 154, 1971.

1471. Francis, A.J., and Dodge, C.J., Anaerobic microbial dissolution of transition and heavy metals oxides, *Appl. Environ. Microbiol.*, 54, 1009, 1988.

1472. Francke, J.A., and Hillebrand, H., Effects of copper on some filamentous Chlorophyta, *Aquat. Bot.*, 8, 285, 1980.

1473. Francko, D.A., Uptake metabolism and release of cAMP in *Selenastrum capricornutum* (Chlorophyceae), *J. Phycol.*, 25, 300, 1989.

1474. Francko, D.A., Modulation of photosynthetic carbon assimilation in *Selenastrum capricornutum* (Chlorophyceae) by cAMP: an electrogenic mechanism?, *J. Phycol.*, 25, 305, 1989.

1475. Frank, T., Kultur und chemische Reizerscheinungen der *Chlamydomonas tingens*, *Bot. Zeit.*, 62, 153, 1904.

1476. Frank, P.A., and Hodgson, R.H., A technique for studying absorption and translocation in submerged plants, *Weeds*, 12, 80, 1974.

1477. Frederick, J.F., Structure of polysaccharides and the biomedical evolution of their synthesis in algae, *Biol. Rev. City Coll. N.Y.*, 17, 33, 1955.

1478. Frederick, J.F., Comparative evolutionary aspects of polyglucoside synthesizing enzymes, *Physiol. Plant.*, 12, 511, 1959.

1479. Frederick, J.F., The a 1,4-glucans of *Prochloron*, a prokaryottic green marine algae, *Phytochem.*, 19, 2611, 1980.

1480. Frederick, L.R., The formation of nitrite from ammonium nitrogen in soils. I. Effect of temperature, *Soil Sci. Soc. Am. Proc.*, 20, 496, 1956.

1481. Frémy, P., Les Cyanophycées des Cotes d'Europe, *Mém. Soc. Nat. Sci. Nat. Mat. Cherbourg*, 41, 1, 1934.

1482. Frémy, P., Les algues perforantes, *Mém. Soc. Nat. Sci. Nat. Mat. Cherbourg*, 42, 275, 1936.

1483. French, M.S., and Evans, L.V., The effects of copper and zinc on growth of the fouling diatoms *Amphora* and *Amphiphora*, *Biofouling*, 1, 3, 1988.

1484. French, R.H., Lake modeling: State of the art, *CRC Crit. Rev. Environ. Control.*, 13, 311, 1983.

1485. Freney, J.R., Denmead, O.T., and Simpson, J.R., Nitrous oxide emission from soils at low moisture contents, *Soil Biol. Biochem.*, 11, 167, 1979.
1486. Freney, J.R., Ivanov, M.V., and Rodhe, H., The sulphur cycle, in *The Major Biogeochemical Cycles and Their Interactions,* Bolin, B., and Cook, R.B., Eds., John Wiley and Sons, Chichester, 1983, 56.
1487. Frenkel, A.W., and Riegler, C., Photoreduction in algae, *Nature,* 167, 1951, 1030.
1488. Fresnedo, O., and Serra, J.L., Effect of nitrogen starvation on the biochemistry of *Phormidium laminosum* (Cyanophyceae), *J. Phycol.*, 28, 786, 1992.
1489. Fresnel, J., Nouvelles observations sur une coccolithacée rare: *Cruciplacolithus neohelis* (McIntyre et Bé) Reinhardt (Prymnesiophyceae), *Protistologica,* 22, 193, 1986.
1490. Freudenthal, H.D., *Symbiodinium* gen. nov. and *Symbiodinium microadriaticum* sp.nov., a zooxanthella: taxonomy, life cycle and morphology, *J. Protozool.*, 9, 45, 1962.
1491. Frey, B.E., Friedel, G., Bass, A.E., and Small, L., Sensitivity of estuarine phytoplankton to hexavalent chromium, *Estuar. Coast. Shelf Sci.*, 17, 181, 1983.
1492. Frey, R., Chitin und Zellulose in Pilzzellwänden, *Ber. Schweiz. Botan. Ges.*, 60, 199, 1950.
1493. Friberg, L., Pucator, M., Nordberg, G.F., and Kjelstrom, T., *Cadmium in the Environment,* CRC Press, Cleveland, 1974.
1494. Friedberg, I., and Avigad, G., Structures containing polyphosphate in *Micrococcus lysodeikticus, J. Bact.,* 96, 544, 1968.
1495. Friedl, F.E., Nitrogen excretion by the freshwater snail *Lymnea stagnalis jagularis* Say, *Comp. Biochem. Physiol.*, 49, 617, 1974.
1496. Friedl, T., New aspects of the reproduction by autospores in the lichen alga *Trebouxia* (Microthamniales, Chlorophyta), *Arch. Protistenk.*, 143, 153, 1993.
1497. Friedmann, I., Light and scanning electron microscopy of the endolithic desert algae habitats, *Phycologia,* 10, 411, 1971.
1498. Friedmann, E.I., Cyanophycota, in *Synopsis and Classification of Living Organisms,* Vol. 1, Parker, S.P., Ed., McGraw-Hill, New York, 1982, 45.
1499. Friedmann, I., and Galun, M., Desert algae, lichens, and fungi, In *Desert Biology,* Brown, G.W., Ed., Academic Press, New York, 1974, 165.
1500. Friedmann, I., and Ocampo, R., Endolithic blue-green algae in the dry valleys: primary producers in the Antarctic desert ecosystem, *Science,* 193, 1247, 1976.
1501. Friedmann, I., Lipkin, Y., and Ocampo-Paus, R., Desert algae of the Negev (Israel), *Phycologia,* 6, 185, 1967.
1502. Friedman, E.I., and Borowitzka, L.J., The symposium on taxonomic concepts in blue-green algae: Towards a compromise with the bacterial code?, *Taxon,* 31, 673, 1982.
1503. Friedman, R.M., and De Witt, C.B., Wetlands as carbon and nutrient reservoirs: a spatial, historical, and societal perspective, in *Wetland Functions and Values: The State of Our Understanding,* Greeson, P.E., Clark, J.R., and Clark, J.E., Eds., American Water Resour. Assoc., Minneapolis, Minnesota, 1979, 175.
1504. Fries, L., *Goniotrichum elegans*: a marine red alga requiring vitamin B_{12}, *Nature,* 183, 558, 1959.
1505. Fries, L., Vitamin requirement of *Nemalion multifidum, Experientia,* 17, 75, 1961.
1506. Fries, L., On the cultivation of axenic red algae, *Physiol. Plant.*, 16, 695, 1963.
1507. Fries, L., *Polysiphonia urceolata* in axenic culture, *Nature,* 202, 110, 1964.
1508. Fries, L., Influence of iodine and bromine on growth of some red algae in axenic culture, *Physiol. Plant.*, 19, 800, 1966.
1509. Fries, L., Selenium stimulates growth of marine macroalgae in axenic culture, *J. Phycol.*, 18, 328, 1982.
1510. Fries, L., Vanadium, an essential element for some marine macroalgae, *Planta,* 154, 393, 1982.
1511. Fritsch, F.E., *Structure and Reproduction of the Algae,* Vol. 1, Cambridge University Press, Cambridge, 1935.
1512. Fritsch, F.E., The role of the terrestrial algae in nature, *Essays in Geobotany,* 13, 195, 1936.
1513. Fritsch, F.E., Present day classification of algae, *Bot. Rev.*, 10, 233, 1944.
1514. Fritsch, F.E., *Structure and Reproduction of the Algae,* Vol. 2, Cambridge University Press, Cambridge, 1945.
1515. Fritz, L., and Nass, M., Development of the endoparasitic dinoflagellate *Amoebophrya ceratii* within host dinoflagellate species, *J. Phycol.*, 28, 312, 1992.
1516. Fritz-Sheridan, R.P., Physiological ecology of nitrogen fixing blue-green algal crusts in the uppersubalpine life zone, *J. Phycol.*, 24, 302, 1988.
1517. Frost, B.W., Effects of size and concentration of food particles on the feeding behavior of the marine planktonic copepod *Calanus pacificus, Limnol. Oceanogr,* 17, 805, 1972.
1518. Frost, B.W., Grazing, in *The Physiological Ecology of Phytoplankton,* Morris, I., Ed., University of California Press, Berkeley, 1980, 465.

1519. Fry, B.A., *The Nitrogen Metabolism in Micro-organisms*, Methuen and Co., Ltd., London, 1955.

1520. Fuge, R., and James, K.H., Trace metal concentrations in *Fucus* from the Bristol Channel, *Mar. Pollut. Bull.*, 5, 9, 1974.

1521. Fuhrman, J.A., Chisholm, S.W., and Guillard, R.R.L., Marine alga *Platymonas* sp. accumulates silicon without apparent requirements, *Nature*, 272, 244, 1978.

1522. Fuhs, W.G., Phosphorus content and rate of growth in the diatoms *Cyclotella nana* and *Thalassiosira fluviatilis*, *J. Phycol.*, 5, 312, 1969.

1523. Fuhs, G.W., Demmerle, S.D., Canelli, E., and Chen, M., Characterization of phosphorus-limited algae (with reflections on the limiting nutrient concept), *Ann. Soc. Limnol. Oceanogr. Spec. Symp.*, 1, 113, 1972.

1524. Fujita, M., and Hashizume, K., Status of uptake of mercury by the freshwater diatom *Synedra ulna*, *Water Res.*, 9, 889, 1975.

1525. Fujita, M., Iwasaki, K., and Takabatake, E., Intracellular distribution of mercury in freshwater diatom *Synedra* cells, *Environ. Res.*, 14, 1, 1977.

1526. Fujita, R.M., The role of nitrogen status in regulating transient ammonium uptake and nitrogen storage by macroalgae, *J. Exp. Mar. Biol. Ecol.*, 92, 283, 1985.

1527. Fujita, Y., and Hattori, A., Formation of phycoerythrin in pre-illuminated cells of *Tolypothrix tenuis* with special reference to nitrogen metabolism, *Plant Cell Physiol.*, 1, 281, 1960.

1528. Fujita, Y., and Hattori, A., Effects of chromatic lights on phycobilin formation in a blue-green alga, *Tolypothrix tenuis*, *Plant Cell Physiol.*, 1, 293, 1960.

1529. Fujita, Y., Ohama, H., and Hattori, A., Hydrogenase activity of cell-free preparation obtained from the blue-green alga, *Anabaena cylindrica*, *Plant Cell Physiol.*, 5, 305, 1964.

1530. Fukushima, H., Studies on cryophytes in Japan, *J. Yokohama Munic. Univ.*, Ser. C, *Natur. Sci.*, 43, 1, 1963.

1531. Fuller, R.H., and Averett, R.C., Evaluation of copper accumulation in the algae of the California aqueduct, *Water Res. Bull.*, 11, 946, 1975.

1532. Funk, W.H., Biological impacts of combined metallic and organic pollution in the Coeur D'Alene-Spokane River drainage system, Washington State University and University of Idaho Report to OWRR, B-044 Wash. and B-015 Idaho, 1973.

1533. Furr, A.K., Schofield, C.L., Grandolfo, M.C., Hofstader, R.A., Guntenmann, W.H., St. John, L.E., Jr., and Lisk, D.J., Element content of mosses as possible indicators of air pollution, *Arch. Environ. Contam. Toxicol.*, 8, 335, 1979.

1534. Furtado, J.I., and Mori, S., Eds., *Tasek Bera. The Ecology of a Freshwater Swamp*, Dr. W. Junk Publishers, The Hague, The Netherlands, 1982.

1535. Gächter, R., Untersuchungen über die Beeinflussung der planktischen Photosynthese durch anorganische metallsalze im eutrophen Alpanachersee und der mesotrophen Horwer Bucht, *Schweiz. Z. Hydrol.*, 38, 97, 1976.

1536. Gächter, R., Lum-Shue-Chan, K., and Chau, Y.K., Complexing capacity of the nutrient medium and its relation to inhibition of algal photosynthesis by copper, *Schweiz. Z. Hydrol.*, 35, 253, 1973.

1537. Gadd, G.M., Accumulation of metals by microorganisms and algae, in *Biotechnology—A Comprehensive Treatise*, Vol. 6b, Rehm, H.-J., Ed., VCH Verlagsgesellschaft, Weinnheim, 1988, 401.

1538. Gadd, G.M., and Griffiths, A.J., Microorganisms and heavy metal toxicity, *Microb. Ecol.*, 4, 303, 1978.

1539. Gaffron, H., Reduction of carbone dioxide with molecular hydrogen in green algae, *Nature*, 143, 204, 1939.

1540. Gaffron, H., Carbon dioxide reduction with molecular hydrogen in green algae, *Am. J. Bot.*, 27, 273, 1940.

1541. Gaffron, H., The effect of specific poisons upon the photoreduction with hydrogen in green algae, *J. Gen. Physiol.*, 26, 195, 1942.

1542. Gaffron, H., Photosynthesis, photoreduction and dark reduction of carbon dioxide in certain algae, *Biol. Revs. Cambridge Phil. Soc.*, 19, 1, 1944.

1543. Gahler, A.R., Sediment-water nutrient interchange, *Proc. Eutrophication-Biostimulation Workshop*, Berkeley, California, 1969, 243.

1544. Gale, N.L., Wixson, B.G., Hardie, M.G., and Jennett, J.C., Aquatic organisms and heavy metals in Missouri's New Lead Belt, *Water Resour. Bull.*, 9, 673, 1976.

1545. Gallagher, J.L., Algal productivity and some aspects of the ecological physiology of the edaphic communities of Canary Creek Tidal Marsh, Ph.D. Thesis, University of Delaware, Newark, 1971.

1546. Gallagher, J.L., Decomposition processes: summary and recommendations, in *Proc. Symp. Wetlands: Ecological Processes and Management Potential*, Good, R.E., Whigham, D.F., and Simpson, R.L., Eds., Academic Press, New York, 1978, 145.

1547. Gallagher, J.L., and Daiber, E.C., Primary production of edaphic algal communities in a Delaware salt marsh, *Limnol. Oceanogr,* 19, 390, 1974.
1548. Gallagher, J.L., and Plumley, F.G., Underground biomass profiles and productivity in Atlantic coastal marshes, *Am. J. Bot.,* 66, 156, 1979.
1549. Gallagher, J.L., Reimold, R.J., Linthurst, R.A., and Pfeiffer, W.J., Aerial production, mortality and mineral accumulation-export dynamics in *Spartina alterniflora and Juncus roemerianus* in plant stands, *Ecology,* 61, 303, 1980.
1550. Galling, G., Analyse des Magnesium-Mangelks bei synchronisierten *Chlorellen, Arch. Mikrobiol.,* 46, 150, 1963.
1551. Gallon, J.R., Nitrogen fixation by photoautotrophs, in *Nitrogen Fixation,* Stewart, W.D.P., and Gallon, J.R., Eds., *Ann. Proc. Phytochem. Soc. Europe,* Vol. 18, Academic Press, London and New York, 1980, 197.
1552. Gallon, J.R., Kurz, W.G.W., and La Rue, T.A., The physiology of nitrogen fixation by a *Gloeocapsa* sp., in *Nitrogen Fixation by Free-Living Micro-Organisms,* Stewart, W.D.P., Ed., *Int. Biol. Programme,* Vol. 6, Cambridge University Press, Cambridge, 1975, 159.
1553. Galloway, R.A., and Krauss, R.W., The differential action of chemical agents, especially polymyxin B, on certain algae, bacteria, and fungi, *Am. J. Bot.,* 46, 40, 1959.
1554. Galloway, R.A., and Krauss, R.W., Utilization of phosphorus sources by *Chlorella,* in *Studies on Microalgae and Photosynthetic Bacteria,* Japanese Soc. Plant Physiol., University of Tokyo Press, Tokyo, 1963, 569.
1555. Gambrell, R.P., and Patrick, W.H., Jr., Chemical and microbiological properties of anaerobic soils and sediments, in *Plant Life in Anaerobic Environments,* Hook, D.D., and Crawford, R.M.M., Eds., Ann Arbor Science Publishers, Ann Arbor, Michigan, 1978, 375.
1556. Gambrell, R.P., and Patrick, W.H., Jr., The influence of redox potential on the environmental chemistry of contaminants in soils and sediments, in *The Ecology and Management of Wetlands,* Part 1. *Ecology of Wetlands,* Hook, D.D. et al., Eds., Timber Press, Portland, Oregon, 1988, 319.
1557. Ganf, G.G., Diurnal mixing and the vertical distribution in a shallow equatorial lake (Lake George, Uganda), *J. Ecol.,* 62, 611, 1974.
1558. Gantt, E., Micromorphology of the periplast of *Chroomonas* sp. (Cryptophyceae), *J. Phycol.,* 7, 177, 1971.
1559. Gantt, E., Photosynthetic cryptophytes, in *Phytoflagellates,* Cox, E.R., Ed., Elsevier/North Holland, New York, 1980, 381.
1560. Gantt, E., Structure and function of phycobilisomes: light harvesting pigment complexes in red and blue-green algae, *Int. Rev. Cytol.,* 466, 45, 1980.
1561. Gantt, E., Ed., *Handbook of Phycological Methods. Development and Cytological Methods,* Cambridge University Press, Cambridge, 1980.
1562. Gantt, E., Phycobilisomes, *Ann. Rev. Plant Physiol.,* 32, 327, 1981.
1563. Gantt, E., Pigmentation and photoacclimation, in *Biology of Red Algae,* Cole, K.H., and Sheath, R.G., Eds., Cambridge University Press, Cambridge, 1990, 203.
1564. Gantt, E., Scott, J., and Lipschultz, C., Phycobiliprotein composition and chloroplast structure in the freshwater red alga *Compsopogon coeruleus* (Rhodophyta), *J. Phycol.,* 22, 480, 1986.
1565. Garbary, D.J., Hansen, G.I., and Scagel, R.F., A revised classification of the Bangiophyceae (Rhodophyta), *Nova Hedw.,* 33, 145, 1980.
1566. Garcia-Gonzales, M., Sanchez-Maeso, E., Quesada, A., and Fernandez-Valiente, E., Sodium requirements for photosynthesis and nitrate assimilation in a mutant of *Nostoc muscorum, J. Plant Physiol.,* 127, 423, 1987.
1567. Gardiner, J., The chemistry of cadmium in natural waters. 1. A study of cadmium complex formation using the cadmium specific-ion electrode, *Water Res.,* 8, 23, 1974.
1568. Gardiner, J., The chemistry of cadmium in natural waters. 2. The adsorption of cadmium on river muds and naturally occurring solids, *Water Res.,* 8, 157, 1974.
1569. Gardner, W.S., and Eadie, B.J., Chemical factors controlling phosphorus cycling in lakes, in *Nutrient Cycling in the Great Lakes: A Summarization of Factors Regulating the Cycling of Phosphorus,* Scavia, D., and Moll, R., Eds., Great Lakes Res. Div. Special Report No. 83, The University of Michigan, Ann Arbor, 1980, 13.
1570. Garner, W.W., Chemical elements essential for nutrition of plants, *U.S. Dept. Agr. Bur. Plant. Industr. Rept.,* 1935.
1571. Garrett, R.H., and Amy, N.K., Nitrate assimilation in fungi, *Adv. Microb. Physiol.,* 18, 1, 1978.
1572. Garrison, D.L., and Buck, K.R., Organism losses during ice melting: a serious bias in sea ice community studies, *Polar Biol.,* 6, 237, 1986.
1573. Garrison, D.L., Buck, K.R., and Fryxell, G.A., Algal assemblages in antarctic pack ice and in ice-edge plankton, *J. Phycol.,* 23, 564, 1987.

1574.   Garten, C.T., Jr., Multivariate perspectives on the ecology of plant mineral element composition, *Am. Nat.*, 112, 534, 1978.

1575.   Garver, E.G., Dubbe, D.R., and Pratt, D.C., Seasonal patterns in accumulation and partitioning of biomass and macronutrients in *Typha* spp., *Aquat. Bot.*, 32, 115, 1988.

1576.   Gauch, H.G., Mineral nutrition of plants, *Ann. Rev. Plant Physiol.*, 8, 31, 1957.

1577.   Gaudet, J.J., Nutrient relationships in the detritus of a tropical swamp, *Arch. Hydrobiol.*, 78, 213, 1976.

1578.   Gaudet, J.J., Uptake, accumulation and loss of nutrients by *Papyrus* in tropical swamps, *Ecology*, 58, 415, 1977.

1579.   Gaudet, J.J., Nutrient dynamics of *Papyrus* swamps, in *Proc. Int. Conf. Wetlands: Ecology and Management*, Gopal, B., Turner, R.E., Wetzel, R.G., and Whigham, D.F., Eds., Natl. Inst. of Ecology and Internat. Scientific Publications, Jaipur, India, 1982, 305.

1580.   Gaudy, R., Feeding four species of pelagic copepods under experimental conditions, *Mar. Biol.*, 25, 125, 1974.

1581.   Gäumann, E., and Jaag, O., Bodenbewohnende Algen als Wuchsstoffspender für bodenbewohnende pflanzenpathogene Pilze, *Phytopathol. Zeitsch.*, 17, 218, 1950.

1582.   Gaur, S., Singhal, P.K., and Hasija, S.K., Relative contribution of bacteria and fungi to water hyacinth decomposition, *Aquat. Bot.*, 43, 1, 1992.

1583.   Gavis, J., Munk and Riley revisited: Nutrient diffusion transport and rates of phytoplankton growth, *J. Mar. Res.*, 34, 161, 1976.

1584.   Gavis, J., and Ferguson, J., The cycling of mercury through the environment, *Water Res.*, 6, 989, 1972.

1585.   Gayral, P., and Billard, C., Synopsis du nouvel ordre des Sarcinochrysidales (Chrysophycées), *Taxon*, 26, 241, 1977.

1586.   Geider, R.J., Light and temperature dependence of the carbon to chlorophyll *a* ratio in microalgae and cyanobacteria: implications for physiology and growth of phytoplankton, *New Phytol.*, 106, 1, 1987.

1587.   Geider, R.J., Osborne, B.A., and Raven, J.A., Light dependence of growth and photosynthesis in *Phaeodactylum tricornutum* (Bacillariophyceae), *J. Phycol.*, 21, 609, 1985.

1588.   Geider, R.J., Osborne, B.A., and Raven, J.A., Growth, photosynthesis and maintenance metabolic cost in the diatom *Phaeodactylum tricornutum* at very low light levels, *J. Phycol.*, 22, 39, 1986.

1589.   Geike, F., Wirkung einiger Quecksilbewr-Verbindungen auf Subwasseralgen, *Z. Pflanzenkrankneiten und Pflanzenschutz*, 84, 84, 1977.

1590.   Geisler, U., Die Variabilität der Schalenmerkmale bei den Diatomeen, *Nova Hedw.*, 19, 623, 1970.

1591.   Geisweid, H.J., and Urbach, W., Sorption of cadmium by the green microalgae *Chlorella vulgaris, Ankistrodesmus braunii* and *Eremosphaera viridis, Z. Pflanzenphysiol.*, 109, 127, 1983.

1592.   Geitler, L., Cyanophyceae, *Pascher's Süssw.-Fl.*, 12, 1, 1925.

1593.   Geitler, L., Cyanophyceae, *Rabenhorst's Krypt.-Fl.*, 14, 1, 1932.

1594.   Geitler, L., Reproduction and life history in diatoms, *Bot. Rev.*, 1, 149, 1935.

1595.   Geitler, L., *Schizophyta, Engler and Prantl's Nat. Pfl.-Fam.*, 2nd ed., 1942.

1596.   Geitler, L., Zur Kenntnis der Bewohner des Oberflächenhäutchens einheimeischer Gewässer, *Biol. Gener.*, 16, 450, 1942.

1597.   Geitler, L., Syncyanosen, in *Ruhland's Handbuch der Pflanzenphysiologie*, Vol. 2, Springer-Verlag, Berlin, 1959, 530.

1598.   Geitler, L., Schizophyzeen, in *Handbuch der Pflanzenanatomie*, Zimmerman, W., and Ozenda, P., Eds., Borntraeger, Berlin, 1969, 1.

1599.   Geitler, L., Einige kritische Bemerkungen zu neuen zusammenfassenden Darstellungen der Morphologie und Systematik der Cyanophyceen, *Plant. Syst. Ecol.*, 132, 153, 1979.

1600.   Gelin, C., Nutrients, biomass and primary productivity of nannoplankton in eutrophic Lake Vombsjön, Sweden, *Oikos*, 26, 121, 1975.

1601.   Gennity, J.M., Bottino, N.R., Zingaro, R.A., Wheeler, A.E., and Irgolic, K.J., A selenium-induced peroxidation of glutathione in algae, *Phytochemistry* 24, 2817, 1985.

1602.   Gentile, J.H., Blue-green algae and green algae toxins, in *Microbiol Toxins*, Kadis, S., Ciegler, A., and Ajl, S.J., Eds., Academic Press, New York, 1971, 27.

1603.   Gentner, S.R., Uptake and transport of iron and phosphate by *Vallisneria spiralis* L., *Aquat. Bot.*, 3, 267, 1977.

1604.   Geoghegan, M.J., Experiments with *Chlorella* at Jealott's Hill, in *Algal Culture: From Laboratory to Pilot Plant*, Burlew, J.S., Ed., Carnegie Inst. of Washington Publ. 600, Washington, DC, 1953, 182.

1605. Gerard, V.A., *In situ* rates of nitrate uptake by giant kelp, *Macrocystis pyrifera* (L.) C. Agardh: tissue difference, environmental effects, and predictions of nitrogen-limited growth, *J. Exp. Mar. Biol. Ecol.*, 62, 211, 1982.

1606. Gerhardt, B., Manganeffecte in photosynthetischen reacktionen von *Anacystis*, *Ber. Dtsch. Botan. Ges.*, 79, 63, 1966.

1607. Gerhardt, B., and Wiessner, W., On the light-dependent reactivation of photosynthetic activity by manganese, *Biochem. Biophys. Res. Commun.*, 28, 958, 1967.

1608. Gerloff, G.C., Comparative mineral nutrition of plants, *Ann. Rev. Plant Physiol.*, 14, 107, 1963.

1609. Gerloff, G.C., The comparative boron nutrition of several green and blue-green algae, *Physiol. Plant.*, 21, 369, 1968.

1610. Gerloff, G.C., and Skoog, F., Cell content of nitrogen and phosphorus as a measure of their availability for growth of *Microcystis aeruginosa, Ecology*, 35, 348, 1954.

1611. Gerloff, G.C., and Krombholz, P.H., Tissue analysis of a measure of nutrient availability for the growth of angiosperm aquatic plants, *Limnol. Oceanogr.*, 11, 529, 1966.

1612. Gerloff, G.C., and Fishbeck, K.A., Quantitative cation requirements of several green and blue-green algae, *J. Phycol.*, 5, 109, 1969.

1613. Gerloff, G.C., Fitzgerald, G.P., and Skoog, F., The isolation, purification, and culture of blue-green algae, *Am. J. Bot.*, 37, 216, 1950.

1614. Gerloff, G.C., Fitzgerald, G.P., and Skoog, F., The mineral nutrition of *Coccochloris peniocystis*, *Am. J. Bot.*, 37, 835, 1950.

1615. Gerloff, G.C., Fitzgerald, G.P., and Skoog, F., The mineral nutrition of *Microcystis aeruginosa, Am. J. Bot.*, 39, 26, 1952.

1616. Gerloff, J., Eine neue Phaeophyceae aus dem Süsswasser: *Pseudobodanella peterfii* nov. gen. et nov. spec., *Rev. Roum. Biol.-Bot.*, 12, 27, 1967.

1617. Gerrard, T.L., Telford, J.N., and Williams, H.H., Detection of selenium deposits in *Escherichia coli* by electron microscopy, *J. Bacteriol.*, 119, 1057, 1974.

1618. Gerretsen, F.C., and de Hoop, H., Boron, an essential micro-element for *Azotobacter chroococcum*, *Plant and Soil*, 5, 349, 1954.

1619. Getz, L.L., Standing crop of herbaceous vegetation in southern Michigan, *Ecology*, 41, 393, 1960.

1620. Ghallab, M.S., Prehistoric civilization in lower Egypt, in *The Nile, Biology of an Ancient River*, Rzóska, J., Ed., Dr. W. Junk Publishers, The Hague, The Netherlands, 1976, 47.

1621. Ghilarov, M.S., Ed., *Methods of Productivity Studies in Root System and Rhizosphere Organisms*, Nauka Leningrad, 1968.

1622. Ghiorse, W.C., Biology of iron- and manganese-depositing bacteria, *Ann. Rev. Microbiol.*, 38, 515, 1984.

1623. Gibbons, B.H., and Edsall, J.T., Kinetic studies of humus carbonic anhydrases B and C, *J. Biol. Chem.*, 239, 2539, 1964.

1624. Gibbons, C.E., Nutrient limitation, *J. Water Pollut. Control Fed.*, 43, 2435, 1971.

1625. Gibbs, M., Respiration, in *Physiology and Biochemistry of Algae*, Lewin, R.A., Ed., Academic Press, New York and London, 1962, 61.

1626. Gibbs, M., Fermentation, in *Physiology and Biochemistry of Algae*, Lewin, R.A., Ed., Academic Press, New York and London, 1962, 91.

1627. Gibbs, M., and Schiff, J.A., Chemosynthesis: the energy relations of chemoautotrophic organisms, in *Plant Physiology A Treatise*, Vol. 1B, Steward, F.C., Eds., Academic Press, New York, 1960, 279.

1628. Gibbs, M., Latzko, M., Harvey, M.J., Plaut, A., and Shain, Y., Photosynthesis in the algae, *Ann. N.Y. Acad. Sci.*, 175, 541, 1970.

1629. Gibbs, S.P. The ultrastructure of the pyrenoids of algae, exclusive of green algae, *J. Ultrastruct. Res.*, 7, 247, 1962.

1630. Gibbs, S.P., The ultrastructure of the chloroplasts of algae, *J. Ultrastruct. Res.*, 7, 418, 1962.

1631. Gibbs, S.P., Nuclear envelope chloroplast relationships in algae, *J. Cell Biol.*, 14, 433, 1962.

1632. Gibbs, S.P., Chu, L.L., and Magnussen, C., Evidence that *Olisthodiscus luteus* is a member of the Chrysophyceae, *J. Phycol.*, 19, 173, 1980.

1633. Giblin, A.E., and Howarth, R.W., Pore water evidence for a dynamic sedimentary iron cycle, *Limnol. Oceanogr.*, 29, 47, 1984.

1634. Giesy, R.M., A light and electron microscope study of interlamellar polyglucoside bodies in *Oscillatoria chalybia, Am. J. Bot.*, 51, 388, 1964.

1635. Gilbert, P.M., and Goldman, J.C., Rapid ammonium uptake by marine phytoplankton, *Mar. Biol. Lett.*, 2, 25, 1981.

1636. Gill, C.J., The flooding tolerance of woody species: a review, *For. Abstr.*, 31, 671, 1970.

1637. Gilliam, J.W., and Skaggs, R.W., Drainage and agricultural development, in *Pocosin Wetlands: an Integrated Analysis of Coastal Plain Freshwater Bogs in North Carolina,* Richardson, J.C., Ed., Hutchinson Ross, Stroudsburg, Pennsylvania, 1981, 109.

1638. Gillot, M.A., and Gibbs, S.P., The cryptomonad nucleomorph: its ultrastructure and evolutionary significance, *J. Phycol.,* 16, 558, 1980.

1639. Gilman, B.A., Wetland Plant Communities Along the Eastern Shoreline of Lake Ontario, M.S. Thesis, Syracuse University, Syracuse, New York, 1976.

1640. Gilmour, J.T., and Gale, P.M., Chemistry of metals and trace elements in a submerged soils, in *The Ecology and Management of Wetlands,* Part 1., *Ecology of Wetlands,* Hook, D.D. et al., Eds., Timber Press, Portland, Oregon, 1988, 279.

1641. Gimmler, H., Ullrich, W., Domanski-Kaden, J., and Urbach, W., Excretion of glycollate during synchronous culture of Ankistrodesmus braunii in the presence of disalicylidene-propanediamine or hydroxypyridine-methane-sulfonate, *Plant Cell Physiol.,* 10, 103, 1969.

1642. Gipps, J.F., and Biro, P., The use of *Chlorella vulgaris* in a simple demonstration of heavy metal toxicity, *J. Biol. Educ.,* 12, 207, 1978.

1643. Gipps, J.F., and Coller, B.A.W., Effects of physical and culture conditions on uptake of cadmium by *Chlorella pyrenoidosa, Aust. J. Mar. Freshwater Res.,* 31, 745, 1980.

1644. Giri, L., and Ramados, C.S., Physical studies on assimilatory nitrate reductase from *Chlorella vulgaris, J. Biol. Chem.,* 254, 11703, 1979.

1645. Glandon, R.P., and McNabb, C.D., The uptake of boron by *Lemna minor, Aquat. Bot.,* 4, 53, 1978.

1646. Glass, A.D.M., *Plant Nutrition. An Introduction to Current Concepts,* Jones and Bartlett Publishers, Boston, 1989.

1647. Glazer, A.N., Phycobilisomes: structure and dynamics, *Ann. Rev. Plant Physiol.,* 36, 173, 1982.

1648. Glazer, A.N., Phycobilisome. A macromolecular complex optimised for light energy transfer, *Biochem. Biophys. Acta,* 768, 29, 1984.

1649. Glazer, A.N., Light harvesting by phycobilisomes, *Ann. Rev. Biophys.,* 14, 47, 1985.

1650. Glazer, A.N., and Bryant, D.A., Allophycocyanin B (l_{max} 671, 618 mm). A new cyanobacterial phycobiliprotein, *Arch. Microbiol.,* 104, 15, 1975.

1651. Gleason, M.L., Drifmeyer, J.E., and Zieman, J.C., Seasonal and environmental variation in Mn, Fe, Cu, and Zn content of *Spartina alterniflora, Aquat. Bot.,* 7, 355, 1979.

1652. Glider, W.V., and Pardy, R.L., Algal-endozoic symbiosis: introduction and bibliography, in *Selected Papers in Phycology* II, Rosowski, J.R., and Parker, B.C., Eds., Phycological Society of America, Lawrence, Kansas, 1982, 761.

1653. Gliwicz, Z.M., and Hillbricht-Ilkowska, A., Efficiency of utilization of nanoplankton primary productivity by communities of filter-feeding animals measured in situ, *Verh. Internat. Verein. Limnol.,* 18, 197, 1975.

1654. Glombitza, K.-W., Antibiotics from algae. in *Marine Algae in Pharmaceutical Science,* Hoppe, H.A., Levring, T., and Tanaka, Y., Eds., Walter de Gruyter, Berlin, 1979, 303.

1655. Glooschenko, W.A., Accumulation of [203]Hg by the marine diatom *Chaetoceros costatum, J. Phycol.,* 5, 224, 1969.

1656. Glooschenko, W.A., and Capobiano, J.A., Metal content of *Sphagnum* mosses from two northern Canadian bog ecosystems, *Water Air Soil Pollut.,* 10, 215, 1978.

1657. Glooschenko, W.A., and De Benedetti, A., Atmospheric deposition of iron from mining activities in northern Ontario, *Sci. Total Environ.,* 32, 73, 1983.

1658. Glooschenko, W.A., Curl, H., and Small, L.F., Jr., Diel periodicity of chlorophyll-*a* concentration in Oregon coastal waters, *J. Fish. Res. Board Can.,* 29, 1253, 1972.

1659. Glooschenko, W.A., Sims, R.A., Gregory, M., and Mayer, T., Use of bog vegetation as a monitor of atmospheric input of metals, in *Atmospheric Pollutants in Natural Waters,* Eisenreich, S.J., Ed., Ann Arbor Science, Michigan, 1981, 389.

1660. Glooschenko, W.A., Holloway, L., and Arafat, N., The use of mires in monitoring the atmospheric deposition of heavy metals, *Aquat. Bot.,* 25, 179, 1986.

1661. Glover, H.E., Effects of iron deficiency on the physiology and biochemistry of *Isochrysis galbana* (Chrysophyceae) and *Phaeodactylum tricornutum* (Bacillariophyceae), *J. Phycol.,* 13, 208, 1977.

1662. Glover, H.E., and Morris, I., Photosynthetic carboxylating enzymes in marine phytoplankton, *Limnol. Oceanogr.,* 23, 510, 1979.

1663. Gnassia-Barelli, M., Romero, M., Laumond, F., and Pesando, D., Experimental studies on the relationship between natural copper complexes and their toxicity to phytoplankton, *Mar. Biol.,* 47, 15, 1978.

1664. Gocke, K., Untersuchingen über Abgabe und Aufnahme von Aminosäuren und Polypeptiden durch Planktonorganismen, *Arch. Hydrobiol.,* 67, 285, 1970.

1665. Godfrey, P.J., Kaynor, E.R., Pelczarski, S., and Benforado, J., Eds., *Ecological Considerations in Wetlands Treatment of Municipal Wastewaters,* Van Nostrand Reinhold Company, New York, 1985.

1666. Godshalk, G.L., and Wetzel, R.G., Decomposition in the littoral zone of lakes, in *Proc. Symp. Freshwater Wetlands: Ecological Processes and Management Potential,* Good, R.E., Whigham, D.F., and Simpson, R.L., Eds., Academic Press, New York, 1978, 131.

1667. Godshalk, G.L., and Wetzel, R.G., Decomposition of aquatic angiosperms. I. Dissolved components, *Aquat. Bot.,* 5, 281, 1978.

1668. Godshalk, G.L., and Wetzel, R.G., Decomposition of aquatic angiosperms. II. Particulate components, *Aquat. Bot.,* 5, 301, 1978.

1669. Godshalk, G.L., and Wetzel, R.G., Decomposition of aquatic angiosperms. III. *Zostera marina* L. and a conceptual model of decomposition, *Aquat. Bot.,* 5, 329, 1978.

1670. Godshalk, G.L., and Barko, J.W., Vegetative succession and decomposition in reservoirs, in *Microbial Processes in Reservoirs,* Gunnison, D., Ed, Dr. W. Junk Publishers, Dordrecht, The Netherlands, 1985, 59.

1671. Godwin, H., The factors which differentiate marsh, fen, bog and heath, *Chronica Botanica,* 6, 11, 1941.

1672. Goel, P.K., Sharma, K.P., and Trivedy, R.K., Ecological observations on some ephemeral ponds around Jaipur, *Acta Limnol. Indica,* 1, 45, 1981.

1673. Goering, J.J., and Neess, J.C., Nitrogen fixation in two Wisconsin lakes, *Limnol. Oceanogr,* 9, 535, 1964.

1674. Goering, J.J., and Dugdale, R.C., Estimates of *in situ* rates of nitrogen uptake by *Trichodesmium* sp. in the tropical Atlantic Ocean, *Limnol. Oceanogr,* 11, 614, 1966.

1675. Goering, J.J., Dugdale, R.C., and Menzel, D.W., Estimates of in situ rates of nitrogen uptake by *Trichodesmium* sp. in the tropical Atlantic Ocean, *Limnol. Oceanogr,* 11, 614, 1966.

1676. Goff, L.J., The biology of *Harveyella mirabilis* (Cryptonemiales, Rhodophyta). V. Host response to parasite infection, *J. Phycol.,* 12, 313, 1976.

1677. Goff, L.J., The biology of *Harveyella mirabilis* (Cryptonemiales, Rhodophyceae). VII. Structure and proposed function of host-penetrating cells, *J. Phycol.,* 15, 87, 1979.

1678. Goff, L.J., The biology of parasitic red algae, in *Progress in Phycological Research,* Round, F.E., and Chapman, D.J., Eds., Elsevier Biomedical Press, Amsterdam, 1982, 289.

1679. Goff, L.J., Ed., *Algal Symbiosis: A Continuum of Interaction Strategies,* Cambridge University Press, London, 1983.

1680. Goldberg, E.D., Iron assimilation by marine diatoms, *Biol. Bull.,* 102, 243, 1952.

1681. Goldberg, E.D., and Arrhenius, G.O.S., Chemistry of Pacific pelagic sediments, *Geochim. Cosmochim. Acta,* 13, 153, 1958.

1682. Goldman, C.R., Primary productivity and limiting factors in three lakes of the Alaska Peninsula, *Ecol. Monogr,* 30, 207, 1960.

1683. Goldman, C.R., Molybdenum as a factor limiting primary productivity in Castle Lake, California, *Science,* 132, 1016, 1960.

1684. Goldman, C.R., Micronutrients limiting factors and their detection in natural phytoplankton populations, *Mem. Ist. Ital. Idrobiol.,* 18 (Suppl.), 121, 1965.

1685. Goldman, C.R., Molybdenum as an essential micronutrient and useful watermass marker in Castle Lake, California, in *Proc. of an IBP Symposium,* Amsterdam, 1966, 229.

1686. Goldman, C.R., The role of minor nutrients in limiting the productivity of aquatic ecosystems, in *Nutrients and Eutrophication,* Likens, G.E., Ed., *Am. Soc. Limnol. Oceanogr Spec. Symp.,* 1, 21, 1972.

1687. Goldman, C.R., and Wetzel, R.G., A study of primary productivity of Clear Lake, Lake County, California, *Ecology,* 44, 283, 1963.

1688. Goldman, C.R., and Carter, R.C., An investigation by rapid carbon-14 bioassay of factors affecting the cultural eutrophication of lake Tahoe, California-Nevada, *J. Water Pollut. Control Fed.,* 37, 1044, 1965.

1689. Goldman, C.R., Mason, D.T., and Wood, B.J.B., Light injury and inhibition in Antarctic freshwater plankton, *Limnol. Oceanogr,* 8, 313, 1963.

1690. Goldman, C.R., Mason, D.T., and Hobbie, J.E., Two antarctic desert lakes, *Limnol. Oceanogr,* 12, 295, 1967.

1691. Goldman, C.R., Mason, D.T., and Wood, B.J.B., Comparative study of the limnology of two small lakes on Ross Islands, Antarctica, in *Antarctic Terrestrial Biology,* Llano, G.A., Ed., American Geophysical Union, Washington, DC, 1972, 1.

1692. Goldman, J.C., Steady state growth of phytoplankton in continuous culture: comparison of internal and external nutrient equations, *J. Phycol.,* 13, 251, 1977.

1693. Goldman, J.C., Temperature effects on phytoplankton growth in continuous culture, *Limnol. Oceanogr,* 22, 932, 1977.

1694. Goldman, J.C., Outdoor algal mass cultures. II. Photosynthetic yield limitations, *Water Res.,* 13, 119, 1979.

1695. Goldman, J.C., Temperature effects on steady state growth, phosphorus uptake, and chemical composition of a marine phytoplankter, *Microb. Ecol.,* 5, 153, 1979.

1696. Goldman, J.C., and McCarthy, J.J., Steady state growth and ammonium uptake of a fast-growing marine diatom, *Limnol. Oceanogr,* 23, 695, 1978.

1697. Goldman, J.C., and Mann, R., Temperature-influenced variations in speciation and chemical composition of marine phytoplankton in outdoor mass cultures, *J. Exp. Mar. Biol. Ecol.,* 46, 29, 1980.

1698. Goldman, J.C., and Gilbert, P.M., Comparative rapid ammonium uptake by four species of marine phytoplankton, *Limnol. Oceanogr,* 27, 814, 1982.

1699. Goldman, J.C., Porcella, D.B., Middlebrooks, E.J., and Toerien, D.F., The effect of carbon on algal growth—its relationship to eutrophication, *Water Res.,* 3, 637, 1972.

1700. Goldman, J.C., Taylor, C.D., and Gilbert, P.M., Non-linear time-course uptake of carbon and ammonium by marine phytoplankton, *Mar. Ecol. Prog. Ser.,* 6, 137, 1981.

1701. Goldsborough, L.G., and Hickman, M., A comparison of periphytic algal biomass and community structure of *Scirpus validus* and on a morphologically similar artificial substratum, *J. Phycol.* 27, 196, 1991.

1702. Gollerbach, M.M., Sovremenoje sostojanie voprosa o roli vodoroslej v pocve, *Sbor. Nac. Rab. Inst. V.L. Kom. SSSR,* 1945, 399, 1946.

1703. Gollerbach, M.M., *Algae, Their Composition, Life and Importance,* Moscow, 1951, (in Russian).

1704. Gollerbach, M.M., and Shtina, E.A., *Soil Algae,* The Academy of Sciences of the U.S.S.R., Moscow, 1969.

1705. Golley, F., Odum, H.T., and Wilson, R., A synoptic study of the structure and metabolism of a red mangrove forest in southern Puerto Rico in May, *Ecology,* 43, 9, 1962.

1706. Golowin, S., Indicator value of bioseston and periphyton for evaluation of pollution degree of flowing waters, *Pol. Arch. Hydrobiol.,* 18, 367, 1971.

1707. Golterman, H.L., Influence of the mud on the chemistry of water in relation to productivity, *Proc. IBP Symp.,* North Holland Publishing, Amsterdam 1967, 297.

1708. Golterman, H.L., Natural phosphate sources in relation to phosphate budgets: a concentration to the understanding of eutrophication, *Water Res.,* 7, 3, 1973.

1709. Golterman, H.L., *Physiological Limnology. An Approach to the Physiology of Lake Ecosystem,* Elsevier Scientific Company, Amsterdam, 1975.

1710. Golterman, H.L., Sediment as source of phosphate for algal growth, in *Interactions Between Sediments and Fresh Water,* Golterman, H.L., Ed., Dr. W. Junk Publishers, Dordrecht, The Netherlands, 1977, 286.

1711. Golterman, H.L., Ed., *Interactions Between Sediments and Fresh Water,* Dr. W. Junk Publishers, Dordrecht, The Netherlands, 1977.

1712. Golterman, H.L., Quantifying the eutrophication process: difficulties caused, for example, by sediments, *Prog. Wat. Res.,* 12, 63, 1980.

1713. Golterman, H.L., Phosphate models, a gap to bridge, *Hydrobiologia,* 72, 61, 1980.

1714. Golterman, H.L., Loading concentration models for phosphate in shallow lakes, *Hydrobiologia,* 91, 169, 1982.

1715. Golterman, H.L., Sediments, modifying and equilibrating factors in the chemistry of freshwater, *Ver. Internat. Verein. Limnol.,* 22, 23, 1984.

1716. Golterman, H.L., The calcium- and iron bound phosphate phase diagram, *Hydrobiologia,* 159, 149, 1988.

1717. Golterman, H.L., Chlorophyll-phosphate relationships, a tool for water management, in *Algae and the Aquatic Environment,* Round, F.E., Ed., Biopress, Ltd., Bristol, 1989, 205.

1718. Golterman, H.L., Biochemical cycles of phosphate in lakes, in *Phosphorus Cycles in Terrestrial and Aquatic Ecosystems,* Tiessen, H., López-Hernández, D., and Salcedo, I.H., Eds., SCOPE and UNEP Workshop, Maracay, Venezuela, 1991, 1.

1719. Golterman, H.L., and Booman, A., The sequential extraction of Ca- and Fe-bound phosphate, *Verh. Internat. Verein. Limnol.,* 23, 904, 1988.

1720. Golterman, H.L., and de Oude, N.T., Eutrophication of lakes, rivers and coastal waters, in *The Handbook of Environmental Chemistry* Vol. 5, Part A, *Water Pollution,* Hutzinger, O., Ed. Springer-Verlag, Berlin, 1991, 80.

1721. Golterman, H.L., Bakels, C.C., and Jacobs-Mögelin, J., Availability of mud phosphates to the growth of algae, *Verh. Internat. Verein. Limnol.,* 17, 467, 1969.

1722. Golubic, S., The relationship between blue-green algae and carbonate deposits, in *The Biology of Blue-Green Algae,* Carr, N.G., and Whitton, B.A., Eds., University of California Press, Berkeley, 1973, 434.

1723. Golubic, S., Taxonomy of extant stromatolite building cyanophytes, in, *Stromatolites,* Walter, M.R., Ed., *Developments in Sedimentology,* 20, Elsevier, New York, 1976, 113.

1724. Golubic, S., Cyanobacteria (blue-green algae) under the bacteriological code? An ecological objection, *Taxon,* 28, 387, 1979.

1725. Golubic, S., and Marcenko, E., Über Konvergenszerscheinungen bei Standdortsformen der Blaualgen unter extremen Lebensbedingungen, *Schweiz. Z. Hydrol.,* 27, 207, 1965.

1726. Golubic, S., and Campbell, S.E., Biogenetically formed aragonite in marine *Rivularia,* in *Phanerozoic Stromatolites,* Monty, C.L.V., Ed., Springer-Verlag, Berlin, 1981, 209.

1727. Golubic, S., Krumbein, W., and Schneider, J., The carbon cycle, in *Biogeochemical Cycling of Mineral-Forming Elements,* Trudinger, P.A., and Swaine, D.J., Eds., Elsevier Scientific Publications, Amsterdam, 1979, 29.

1728. Golueke, C.G., and Oswald, W.J., Harvesting and processing sewage grown algae, *J. Water Pollut. Control Fed.,* 37, 471, 1965.

1729. Golueke, C.G., Oswald, W.J., and Ghee, H.K., Increasing high-rate pond loading by phase isolation, Final Report, SERL, University of California, Berkeley, California, 1962.

1730. Gomez, M.M., and Day, F.P., Jr., Litter nutrient content and production in the Great Dismal Swamp, *Am. J. Bot.,* 69, 1314, 1982.

1731. Gomont, M.M., Monographie des Oscillatoriées (Nostocacées homocystées), *Ann. Sci. Nat. Bot. Ser. 7,* 15, 263; 16, 91, 1892.

1732. Good, R.E., Salt marsh vegetation, Cape May, N.J., *Bull. New Jersey Acad. Sci.,* 10, 1, 1965.

1733. Good, R.E., Salt marsh production and salinity, *Bull. Ecol. Soc. Am. Abstracts,* 53, 22, 1972.

1734. Good, R.E., and Good, N.F., Vegetation and production of the Woodbury Creek Hessian Run freshwater tidal marshes, *Bartonia,* 43, 38, 1975.

1735. Good, R.E., Whigham, D.F., and Simpson, R.L., Eds., *Proc. Symp. Freshwater Wetlands: Ecological Processes and Management Potential,* Academic Press, New York, 1978.

1736. Gooday, G.W., A physiological comparison of the symbiotic alga *Platymonas convolutae* and its free-living relatives, *J. Mar. Biol. Ass. U.K.,* 50, 199, 1970.

1737. Goodwin, R.H., and Niering, W.A., Inland wetlands of the United States: evaluated as potential registered landmarks, Nat. Park. Service, Superintendent of Documents, U.S. Government Printing Office, Washington, DC, 1975.

1738. Goodwin, T.W., Distribution of carotenoids, in *Chemistry and Biochemistry of Plant Pigments,* Goodwin, T.W., Ed., Academic Press, New York and London, 1965, 17.

1739. Goodwin, T.W., The biosynthesis of carotenoids, in *Chemistry and Biochemistry of Plant Pigments,* Godwin, T.W., Ed., Academic Press, New York and London, 1965, 143.

1740. Goodwin, T.W., Carotenoids and biliproteins, in *Algal Physiology and Biochemistry,* Stewart, W.D.P., Ed., Blackwell Scientific Publications, Oxford, 1974, 176.

1741. Gopal, B., Contribution of *Azolla pinnata* R.Br. to the productivity of temporary ponds in Varanasi, *Trop. Ecol.,* 8, 126, 1967.

1742. Gopal, B., Ed., *Ecology and Management of Aquatic Vegetation in the Indian Subcontinent,* Kluwer Academic Publishers, Dordrecht, The Netherlands, 1990.

1743. Gopal, B., Nutrient dynamics of aquatic plant communities, in *Ecology and Management of Aquatic Vegetation in the Indian Subcontinent,* Gopal, B., Ed., Kluwer Academic Publishers, Dordrecht, The Netherlands, 1990, 177.

1744. Gopal, B., and Sharma, K.P., *Water-Hyacinth (Eichhornia crassipes) the Most Troublesome Weed of the World,* Hindasia Publ., New Delhi, 1981.

1745. Gopal, S., and Sharma, K.P., Studies of wetlands in India with emphasis on structure, primary production and management, *Aquat. Bot.,* 12, 81, 1982.

1746. Gopal, B., and Sharma, K.P., Seasonal changes in concentration of major nutrient elements in the rhizomes and leaves of *Typha elephantina* Roxb., *Aquat. Bot.,* 20, 65, 1984.

1747. Gopal, B., and Masing, V., Biology and ecology, in *Wetlands and Shallow Continental Water Bodies,* Vol. 1, Patten, B.C., Ed., SPB Academic Publishing, The Hague, The Netherlands, 1990, 91.

1748. Gopal, B., Sharma, K.P., and Trivedy, R.K., Studies on ecology and production in Indian freshwater ecosystems at primary producer level with emphasis on macrophytes, in *Glimpses of Ecology,* Singh, J.S., and Gopal, B., Eds., Intl. Sci. Publications, Jaipur, India, 1978, 349.

1749. Gopal, B., Turner, R.E., Wetzel, R.G., and Whigham, D.F., Eds., *Proc. 1st Internat. Wetland Conf. Wetlands: Ecology and Management,* Vols. 1+2, Natl. Inst. of Ecology and Internat. Scientific Publications, Jaipur, India, 1982.

1750. Gopal, B., Turner, R.E., Wetzel, R.G., and Whigham, D.F., Introduction, in *Proc. 1st. Internat. Wetland Conf. Wetlands: Ecology and Management*, Gopal, B., Turner, R.E., Wetzel, R.G., and Whigham, D.F., Eds., Natl. Inst. of Ecology and Internat. Scientific Publications, Jaipur, India, 1982, xi.

1751. Gopal, B., Květ, J., Löffler, H., Masing, V., and Patten, B.C., Definition and classification, in *Wetlands and Shallow Continental Water Bodies*, Vol. 1, Patten, B.C., Ed., SPB Academic Publishing, The Hague, The Netherlands, 1990, 9.

1752. Gordon, A.S., Cooper, W.J., and Scheidt, D.J., Denitrification in marl and peat sediments in the Florida Everglades, *Appl. Environ. Microbiol.*, 52, 987, 1986.

1753. Gordon, D.M., and Danishefsky, S.J., Synthesis of a cyanobacterial sulfolipid: confirmation of its structure, stereochemistry, and anti-HIV-1 activity, *J. Am. Chem. Soc.*, 114, 659, 1992.

1754. Gordon, D.M., Birch, P.B., and McComb, A.J., Effects of inorganic phosphorus and nitrogen on the growth on an estuarine *Cladophora* in culture, *Bot. Mar.*, 24, 93, 1981.

1755. Gore, A.J.P., Ed., *Ecosystems of the World. Mires: Swamp, Bog, Fen and Moor. 4A. General Studies*, Elsevier Science Publishers, Amsterdam, 1983.

1756. Gore, A.J.P., Ed., *Ecosystems of the World. Mires: Swamp, Bog, Fen and Moor, 4B. Regional Studies*, Elsevier Science Publishers, Amsterdam, 1983.

1757. Gore, A.J.P., Introduction, in *Ecosystems of the World. Mires: Swamp, Bog, Fen and Moore*, Vol. 4A *General Studies*, Gore, A.J.P., Ed., Elsevier, Amsterdam, 1983, 1.

1758. Goreau, T.F., Problems of growth and calcium deposition in reef corals, *Endeavour*, 20, 32, 1961.

1759. Goreau, T.F., Calcium carbonate deposition by coralline algae and corals in relation to their roles as reef builders, *Ann. N.Y. Acad. Sci.*, 109, 127, 1963.

1760. Gorham, E., and Bernard, J.M., Midsummer standing crops of wetland sedge meadows along a transect from forest to prairie, *J. Minn. Acad. Sci.*, 41, 15, 1975.

1761. Gorham, E., and Somers, M.G., Seasonal changes in the standing crop of two montane sedges, *Can. J. Bot.*, 51, 1097, 1973.

1762. Gorham, P.R., Toxic waterblooms of blue-green algae, *Can. Vet. J.*, 1, 235, 1960.

1763. Gorham, P.R., Laboratory studies on the toxins produced by waterblooms of blue-green algae, *Am. J. Bull. Health*, 52, 2100, 1962.

1764. Gorham, P.R., Toxic algae, in *Algae and Man*, Jackson, D.F., Ed., Plenum Press, New York, 1964, 307.

1765. Gorham, P.R., Toxic waterblooms of blue-green algae, in *Biological Problems of Water Pollution*, 3rd Seminar, U.S. Public Health Service Bull. 999-WP-25, 1965, 37.

1766. Gorham, P.R., and Carmichael, W.W., Phytotoxins from blue-green algae, *Pure Appl. Chem.*, 52, 165, 1979.

1767. Gorham, P.R., and Carmichael, W.W., Toxic substances from freshwater algae, *Prog. Wat. Tech.*, 12, 189, 1980.

1768. Gorham, P.R., McLachlan, J., Hammer, U.T., and Kim, W.M., Isolation and culture of toxic strains of *Anabaena flos-aquae* (Lyngb.) de Bréb., *Verh. Internat. Verein. Limnol.*, 15, 796, 1964.

1769. Gosselin, M., Legendre, L., Therriault, J.-C., and Demers, S., Light and nutrient limitation of sea-ice microalgae (Hudson Bay, Canada Arctic), *J. Phycol.*, 26, 220, 1990.

1770. Gosselink, J.G., *The Ecology of Delta Marshes of Coastal Louisiana: A Community Profile*, U.S. Fish and Wildlife Service, Biol. Services FWS IOBS-84/09, 1984.

1771. Gosselink, J.G., and Kirby, C.J., Decomposition of salt marsh grass, *Spartina alterniflora* Loisel., *Limnol. Oceanogr.*, 19, 825, 1974.

1772. Gosselink, J.G., and Turner, R.E., The role of hydrology in freshwater wetland ecosystem, in *Proc. Symp. Freshwater Wetlands: Ecological Processes and Management Potential*, Good, R.E., Whigham, D.F., and Simpson, R.L., Eds., Academic Press, New York, 1978, 63.

1773. Gossett, D.R., and Norris, W.E., Jr., Relationship between nutrient availability and content of nitrogen and phosphorus in tissues of the aquatic macrophyte, *Eichhornia crassipes* (Mart.) Solms, *Hydrobiologia*, 38, 15, 1971.

1774. Gosz, J.R., Likens, G.E., and Bormann, F.H., Nutrient release from decomposing leaf and branch litter in the Hubbard Brook Forest, New Hampshire, *Ecol. Monogr.*, 43, 173, 1973.

1775. Gotham, I.J., and Rhee, G.-Y., Comparative kinetic studies of phosphate-limited growth and phosphate uptake in phytoplankton in continuous culture, *J. Phycol.*, 17, 257, 1981.

1776. Gotoh, S., and Patrick, W.H., Jr., Transformation of iron in a waterlogged soil as influenced by redox potential and pH, *Soil. Sci. Soc. Am. Proc.*, 38, 66, 1974.

1777. Gough, S.B., and Gough, L.P., Comment on "primary production of algae growing on natural and artificial aquatic plants: a study of interactions between epiphytes and their substrate" (Cattanoe and Kalff), *Limnol. Oceanogr.*, 26, 987, 1981.

1778. Gough, S.B., and Woelkerling, W.J., Wisconsin desmids. 2. Aufwuchs and plankton communities of selected soft water lakes, hard water lakes and calcareous spring pond, *Hydrobiologia*, 49, 3, 1976.

1779. Gould, D., and Gallagher, E., Field measurements of specific growth rate, biomass, and primary production of benthic diatoms of Savin Hill Cove, Boston, *Limnol. Oceanogr.*, 35, 1757, 1990.

1780. Goulder, R., Interactions between rates of production of a freshwater macrophyte and phytoplankton in a pond, *Oikos*, 20, 300, 1969.

1781. Govindjee, and Braun, B.Z., Light absorption, emission and photosynthesis, in *Algal Physiology and Biochemistry* Stewart, W.D.P., Ed., Blackwell Scientific Publications, Oxford, 1974, 346.

1782. Goyer, R.A., Formation of intracellular inclusion bodies in heavy metal poisoning (lead, bismuth, and gold), *Environ. Health Perspect.*, 4, 97, 1973.

1783. Grabowsky, J., and Gantt, E., Photophysical properties of phycobiliproteins from phycobilisomes: fluorescence life times, quantum yields and polarization spectra, *Photochem. Photobiol.*, 28, 39, 1978.

1784. Graetz, D.A., Keeney, D.R., and Aspiras, R.B., The status of lake sediment-water systems in relation to nitrogen transformations, *Limnol. Oceanogr.*, 18, 908, 1973.

1785. Graetz, D.A., Krottje, P.A., Erickson, N.L., Fiskell, J., and Rothwell, D.F., Denitrification in wetlands as a means of water quality improvement, Florida Water Res. Res. Center Publ. No. 48, 1980.

1786. Graffius, J.H., Additions to our knowledge of Michigan Pyrrhophyta and Chloromonadophyta, *Trans. Am. Microsc. Soc.*, 85, 260, 1966.

1787. Graham, D., and Whittingham, C.P., The path of carbon during photosynthesis in *Chlorella pyrenoidosa* at high and low carbon dioxide concentrations, *Z. Pflanzenphysiol.*, 58, 418, 1968.

1788. Graham, D.A., Atkins, C.A., Reed, M.L., Patterson, B.P., and Smillie, R.M., Carbonic anhydrase, photosynthesis and light-induced pH changes, in *Photosynthesis and Photorespiration*, Hatch, M.D., Osmond, C.B., and Slayer, R.O., Eds., Wiley-Interscience, London, 1971, 267.

1789. Graham, L.E., MacEntee, F.J., and Bold, H.C., An investigation on some subaerial green algae, *Texas J. Sci.*, 33, 13, 1981.

1790. Grambast, L.J., Phylogeny of the Charophyta, *Taxon*, 23, 463, 1974.

1791. Gran, H.H., Studies on the biology and chemistry of the Gulf of Maine. II. Distribution of phytoplankton in August 1932. *Biol. Bull.*, 64, 159, 1933.

1792. Granat, L., Rodhe, H., and Hallberg, R.O., The global sulphur cycle. in *Nitrogen, Phosphorus and Sulfur - Global Cycles*, Svensson, B.H., and Söderlund, R., Eds., SCOPE Rept. No. 7, *Ecol. Bull.* (Stockholm), 22, 89, 1976.

1793. Grant, B., Madgwick, J., and Dal Pont, G., Growth of *Cylindrotheca closterium* var. *californica* (Meereschk.) Reimann and Lewin on nitrate, ammonia, and urea, *Austr. J. Mar. Freshwater Res.*, 18, 129, 1967.

1794. Grant, B.R., The action of light on nitrate and nitrite assimilation by the marine chlorophyte, *Dunaliella tertiolecta* Butcher, *J. Gen. Microbiol.*, 48, 379, 1967.

1795. Grant, B.R., Nitrite reductase in *Dunaliella tertiolecta*: isolation and properties, *Plant Cell Physiol.*, 11, 55, 1970.

1796. Grant, M.C., and Proctor, V.W., *Chara vulgaris* and *C. contraria*: Patterns of reproductive isolation for two cosmopolitan species complexes, *Evolution*, 26, 262, 1972.

1797. Grant, M.C., and Sawa, T., Charophyceae: introduction and bibliography, in *Selected Papers in Phycology* II, Rosowski, J.R., and Parker, B.C., Eds., Phycological Society of America, Lawrence, Kansas, 1982, 754.

1798. Grant, W.S., and Horner, R.A., Growth responses to salinity variation in four arctic ice diatoms, *J. Phycol.*, 12, 180, 1976.

1799. Grant, W.D., and Long, P.E., *Environmental Microbiology*, Blackie and Son, Glasgow, 1981.

1800. Grant, W.D., and Long, P.E., Environmental microbiology, in *The Handbook of Environmental Chemistry* Vol. 1, Part D, *The Natural Environment and the Biogeochemical Cycles*, Hutzinger, O., Ed., Springer-Verlag, Berlin, 1985, 125.

1801. Gray, B.H., and Gantt, E., Spectral properties of phycobilisomes and phycobiliproteins from the blue-green alga *Nostoc* sp., *Photochem. Photobiol.*, 21, 121, 1975.

1802. Gray, T.R.G., and Williams, S.T., *Soil Microorganisms*, Hafner, New York, 1971.

1803. Green, J.C., Biomineralization in the algal class Prymnesiophyceae, in *Biomineralization in Lower Plants and Animals*, Leadbeater, B.S.C., and Riding, R., Eds., Clarendon Press, Oxford, 1986, 173.

1804. Green, J.C., and Manton, I., Studies in the fine structure and taxonomy of flagellates in the genus *Pavlova*. I. A revision of *Pavlova gyrans*, the type species, *J. Mar. Biol. Ass. U.K.*, 50, 1113, 1970.

1805. Green, J.C., and Parke, M., A reinvestigation by light and electron microscopy of *Ruttnera spectabilis* Geitler (Haptophyceae), with special reference to the fine structure of the zoids, *J. Mar. Biol. Ass. U.K.*, 54, 539, 1974.

1806. Green, J.C., and Parke, M., New observations upon members of the genus *Chrysotila* Anand, with remarks upon their relationship with the Haptophyceae, *J. Mar. Biol. Ass. U.K.*, 55, 109, 1975.

1807. Green, J.C., and Course, P.A., Extracellular calcification in *Chrysotila lamellosa* (Prymnesiophyceae), *Br. phycol. J.*, 18, 367, 1983.

1808. Green, J.C., Perch-Nielsen, K., and Westbroek, P., Phylum Prymnesiophyta, in *Handbook of the Protoctista*, Margulis, L., Corliss, J.O., Melkonian, M., and Chapman, D.J., Eds., Jones and Bartlett Publishers, Boston, 1990, 293.

1809. Greene, J.C., Miller, W.E., Shiroyama, T., and Merwin, E., Toxicity of zinc to green alga *Selenastrum capricornutum* as a function of phosphorus or ionic strength, *Proc. Biostimul. Nutr. Ass. Workshop*, U.S. EPA 660/3-73-034, 1973, 28.

1810. Greene, J.C., Miller, W.E., Shiroyama, T., Soltero, R.A., and Putnam, K., Use of laboratory cultures of *Selenastrum, Anabaena* and the indigenous isolate *Sphaerocystic* to predict effects of nutrient and zinc interactions upon phytoplankton growth in Long Lake, Washington, *Mitt. Internat. Verein. Limnol.*, 21, 372, 1978.

1811. Greensfield, S.S., Inhibitory effects of inorganic compounds on photosynthesis in *Chlorella*, *Am. J. Bot.*, 29, 121, 1942.

1812. Greenwood, D.J., The effect of oxygen concentration on the decomposition of organic materials in soils, *Plant and Soil*, 14, 360, 1961.

1813. Greeson, P.E., Clark, J.R., and Clark, J.E., Eds., *Proc. Nat. Symp. Wetland Functions and Values: The State of Our Understanding*, American Water Resources Assoc., Minneapolis, Minnesota, 1979.

1814. Gregory, S.V., Plant-herbivore interactions in stream systems, in *Stream Ecology*, Barnes, J.R., and Minshall, G.W., Eds., Plenum Press, New York, 1983, 157.

1815. Gregson, A.J., Green, J.C., and Leadbeater, B.S.C., Structure and physiology of the haptonema in *Chrysochromulina* (Prymnesiophyceae). I. Fine structure of the flagellar/haptonematal root system in *C. acantha* and *C. simplex, J. Phycol.*, 29, 674, 1993.

1816. Gregson, A.J., Green, J.C., and Leadbeater, B.S.C., Structure and physiology of the haptonema in *Chrysochromulina* (Prymnesiophyceae). II. Mechanisms of haptonematal coiling and the regeneration process, *J. Phycol.*, 29, 686, 1993.

1817. Greville, R.K., *Algae Britannica*, Edinburgh, 1830.

1818. Gries, C., and Garbe, D., Biomass, and nitrogen, phosphorus and heavy metal content of *Phragmites australis* during the third growing season in a root zone waste water treatment, *Arch. Hydrobiol.*, 117, 97, 1989.

1819. Grieve, D., and Fletcher, K., Interactions between zinc and suspended sediments in the Fraser River Estuary, British Columbia, *Estuar. Coast. Mar. Sci.*, 5, 415, 1977.

1820. Griffiths, D.W., The pyrenoid, *Bot. Rev.*, 36, 29, 1970.

1821. Grilli, C.M., A light and electron microscopic study of the blue-green algae living either in the coralloid roots of *Macrozamia communis* or isolated in culture, *G. Bot. Ital.*, 108, 161, 1974.

1822. Grilli-Caiola, M., A light and electron microscopic study of blue-green algae growing in the coralloid-roots of *Encephalartos altensteinii* and in culture, *Phycologia*, 14, 25, 1975.

1823. Grimm, N.B., Role of macroinvertebrates in nitrogen dynamics of a desert stream, *Ecology*, 69, 1884, 1988.

1824. Grimm, N.B., Fisher, S.G., and Minckley, W.L., Nitrogen and phosphorus dynamics in hot desert streams of Southwestern U.S.A., *Hydrobiologia*, 83, 303, 1981.

1825. Grindley, J.R., and Sapejka, N., The cause of mussel poisoning in South Africa, *S. Afr. Med. J.*, 43, 275, 1969.

1826. Grobbelaar, J.U., Availability to algae of N and P adsorbed on suspended solids in turbid waters of the Amazon River, *Arch. Hydrobiol.*, 96, 302, 1983.

1827. Gromov, B.V., *Aphelidium tribonemae* Scherffel—parasite of yellow-green algae, *Mikol. Fitopatol.*, 6, 443, 1972 (in Russian).

1828. Gromov, B.V., and Mamkaeva, K.A., Electron microscopic examination of *Bdellovibrio chlorellavorus* parasitism on cells of the green alga *Chlorella vulgaris, Tsitologiya*, 14, 256, 1972 (in Russian).

1829. Gromov, B.V., and Mamkaeva, K.A., Proposal of a new genus *Vampirovibrio* for chlorellavorus bacteria previously assigned to *Bdellovibrio, Mikrobiologiya*, 49, 165, 1980, (in Russian).

1830. Grøntved, J., On the productivity of microbenthos and phytoplankton in some Danish fjords, *Meddr. komm. Danm. Fisk.-og Havunders. (N.S.)*, 3, 55, 1960.

1831. Gross, R.E., Pungo, P., and Dugger, W.M., Observations on the mechanism of copper damage in *Chlorella, Plant Physiol.*, 46, 183, 1970.

1832. Grossi, S.M., Kottmeier, S.T., and Sullivan, C.W., Sea ice microbial communities. III. Seasonal abundance of microalgae and bacteria, *Microb. Ecol.*, 10, 231, 1984.

1833. Grossi, S.M., and Sullivan, C.W., Sea ice microbial communities. V. The vertical zonation of diatoms in an Antarctic fast ice community, *J. Phycol.*, 21, 401, 1985.

1834. Grover, J.P., Influence of cell shape and size on algal competitive ability, *J. Phycol.*, 25, 402, 1989.

1835. Grover, J.P., Phosphorus-dependent growth kinetics of 11 species of freshwater algae, *Limnol. Oceanogr.*, 34, 341, 1989.

1836. Grover, J.P., Algae grown in non-steady continuous culture: population dynamics and phosphorus uptake, *Verh. Internat. Verein. Limnol.*, 24, 516, 1990.

1837. Grover, J.P., Non-steady state dynamics of algal population growth experiments with two chlorophytes, *J. Phycol.*, 27, 70, 1991.

1838. Gruendling, G.K., Ecology of the epipelic algal communities in Marion Lake, British Columbia, *J. Phycol.*, 7, 239, 1971.

1839. Grzenda, A.R., and Ball, R.C., Periphyton production in a warm-water stream, *Mich. Q. Bull.*, 50, 296, 1968.

1840. Grzenda, A.R., Ball, R.C., and Kevern, N.R., Primary production, energetics, and nutrient utilization in a warm water stream, *Tech. Rept. No. 2, Red Cedar River Series*, Inst. Water Res., Michigan State University, East Lansing, 1, 1968.

1841. Guerrero, M.G., Vega, J.M., and Losada, M., The assimilatory nitrate- reducing system and its regulation, *Ann. Rev. Plant Physiol.*, 32, 169, 1981.

1842. Guillard, R.R.L., Organic sources of nitrogen for marine centric diatoms, in *Symposium on Marine Microbiology*, Oppenheimer, C.H., Ed., Thomas, Springfield, Illinois, 1963, 93.

1843. Guillard, R.R.L., B_{12} specificity of marine centric diatoms, *J. Phycol.*, 4, 59, 1968.

1844. Guillard, R.R.L., and Wagnersky, P.J., The production of extracellular carbohydrates by some marine flagellates, *Limnol. Oceanogr.*, 3, 449, 1958.

1845. Guillard, R.R.L., and Cassie, V., Minimum cyanocobalamin requirements of some marine centric diatoms, *Limnol. Oceanogr.*, 8, 161, 1963.

1846. Guillard, R.R.L., and Helleburst, J.A., Growth and the production of extracellular substances by two strains of *Phaeocystis poucheti*, *J. Phycol.*, 7, 330, 1971.

1847. Guiseley, K.B., and Renn, D.W., *Agarose: Purification, Properties and Biomedical Applications*, Marine colloids, FMC Corp., Rockland, Maine, 1975.

1848. Gulens, J., Champ, D.R., and Jackson, R.E., Influence of redox environments on the mobility of arsenic in ground waters, in *Chemical Modelling in Aqueous Systems*, Jenne, E.A., Ed., *Am. Chem. Soc. Symp. Ser.*, 93, 81, 1979.

1849. Gunnison, D., Engler, R.M., and Patrick, W.H., Jr., Chemistry and microbiology of newly flooded soils: relationship to reservoir-water quality, in *Microbial Processes in Reservoirs*, Gunnison, D., Ed., Dr. W. Junk Publishers, Dordrecht, The Netherlands, 1985, 39.

1850. Guppy, H.B., On the postponement of the germination of the seeds of aquatic plants, *Proc. R. Phys. Soc. Edinb.*, 13, 344, 1987.

1851. Gupta, A.B., and Arora, A., Morphology and physiology of *Lyngbya nigra* with reference to copper toxicity, *Physiol. Plant.*, 44, 215, 1978.

1852. Guseva, K.A., Effect of copper on algae, *Mikrobiologiya*, 9, 480, 1940 (in Russian).

1853. Gustafson, K.R., Cardellina II, J.H., Fuller, R.W., Wieslow, O.S., Kiser, R.F., Snader, K.M., Patterson, G.M.L., and Boyd, M.R., AIDS-antiviral sulfolipids from cyanobacteria (blue-green algae), *J. Nat. Cancer Inst.*, 81, 1254, 1989.

1854. Gutknecht, J., Mechanism of radioactive zinc uptake by *Ulva lactuca*, *Limnol. Oceanogr.*, 6, 426, 1961.

1855. Gutknecht, J., Zn-65 uptake by benthic marine algae, *Limnol. Oceanogr.*, 8, 31, 1963.

1856. Gutknecht, J., Uptake and retention of cesium 137 and zinc 65 by seaweeds, *Limnol. Oceanogr.*, 10, 58, 1965.

1857. Guyomarch, C., Influence des pH alcalins et des ions bicarbonates sur la photosynthese de *Chlorella vulgaris* Beij., *Bull. Soc. Scient. Bretagne*, 45, 113, 1970.

1858. Haardt, H., and Nielsen, G.Æ., Attenuation measurements of monochromatic light in marine sediments, *Oceanologia Acta*, 3, 333, 1980.

1859. Hackney, C.T., and De la Cruz, A.A., In situ decomposition of roots and rhizomes of two tidal marsh plants, *Ecology*, 61, 226, 1980.

1860. Hackney, C.T., Stout, J.P., and de la Cruz, A.A., Standing crop and productivity of dominant marsh communities in the Alabama-Mississippi Gulf Coast, Mississippi-Alabama Sea Grant Consortium Publication, MASGP-79-044, 1979.

1861. Häder, D.-P., Ecological consequences of photomovement in microorganisms, *J. Photochem. Photobiol. B: Biol.*, 1, 385, 1988.

1862. Hageman, R.H., Integration of nitrogen assimilation in relation to yield, in *Nitrogen Assimilation of Plants*, Hewitt, E.J., and Cutting, C.V., Eds., Academic Press, London, 1979, 591.

1863. Hagiwara, S., Hayashi, H., and Takashashi, K., Calcium and potassium currents of the membrane and a barnacle muscle fibre in relation to the calcium spike, *J. Physiol.*, 205, 115, 1969.

1864. Haines, D.W., Rogers, K.H., and Rogers, F.E.J., Loose and firmly attached epiphyton: their relative contribution to algal and bacterial carbon productivity in a *Phragmites* marsh, Aquat. Bot., 29, 169, 1987.

1865. Haines, E.B., and Montague, C.L., Food sources of estuarine invertebrates analyzed using $^{13}C/^{12}C$ ratios, *Ecology*, 60, 48, 1979.

1866. Haines, E.B., Nutrient inputs to the coastal zone, in *Estuarine Research*, Vol. 1, Cronin, L., Ed., Academic Press, New York, 1975, 303.

1867. Haines, E.B., Chalmers, A., Hanson, R., and Sherr, B., Nitrogen pools and fluxes on a Georgia salt marsh, in *Estuarine Processes*, Vol. 2, Wiley, M., Ed., Academic Press, New York, 1977, 241.

1868. Haines, K.C., and Wheeler, P.A., Ammonium and nitrate uptake by the marine macrophytes *Hypnea musciformis* (Rhodophyta) and *Macrocystis pyrifera* (Phaeophyta), *J. Phycol.*, 14, 319, 1978.

1869. Haines, T.H., A new sulfolipid in microbes (*Ochromonas danica*), *Diss. Abstr*, 24, 2203, 1964.

1870. Haldemann, C., and Brändle, R., Avoidance of oxygen deficit stress and release of oxygen by stalked rhizomes of *Schoenoplectus lacustris*, *Physiol. Veg.*, 21, 109, 1983.

1871. Hale, R.W., and Pion, R.J., *Laminaria:* an underutilized clinical adjunct, *Clin. Obstet. Gynecol.*, 15, 829, 1972.

1872. Halfen, L.N., Gliding motility of *Oscillatoria*: ultrastructural and chemical characterization of the fibrillar layer, *J. Phycol.*, 9, 248, 1973.

1873. Halfen, L.N., and Castenholz, R.W., Gliding motility in the blue-green alga *Oscillatoria princeps*, *J. Phycol.*, 7, 133, 1971.

1874. Halfen, L.N., and Castenholz, R.W., Energy expenditure for gliding motility in a blue-green alga, *J. Phycol.*, 7, 258, 1971.

1875. Hall, A., Heavy metal co-tolerance in a copper-tolerant population of the marine fouling alga *Ectocarpus siliculosus* (Dillw.) Lyngbye, *New Phytol.*, 85, 73, 1980.

1876. Hall, D.O., and Rao, K.K., *Photosynthesis*, Studies in Biology No. 37, Edward Arnold, London, 1972.

1877. Hall, D.O., and Rao, K.K., *Photosynthesis*, 4th edition, Edward Arnold, London, 1987.

1878. Hall, J.B., Nitrate-reducing bacteria, in *Microbiology*, Schlessinger, D., Ed., Am. Soc. Microbiol., Washington, DC, 1978, 296.

1879. Hall, S.F., and Fisher, F.M., Jr., Annual productivity and extracellular release of dissolved organic compounds by the epibenthic algal communities of a brackish marsh, *J. Phycol.*, 21, 277, 1985.

1880. Halldal, P., Phytoplankton investigation from the Weather Ship M in the Norwegian Sea, 1948–1949, *Hvalradets skrifter* 38, 1, 1953

1881. Halldal, P., Taxes, in *Physiology and Biochemistry of Algae*, Lewein, R.A., Ed., Academic Press, New York, 1962, 583.

1882. Hallegraf, G.M., A review of harmful algal blooms and their apparent global increase, *Phycologia*, 32, 79, 1993.

1883. Haller, W.T., and Sutton, D.L., Effect of pH and high phosphorus concentration on growth of water hyacinth, *Hyacinth Control J.*, 11, 59, 1973.

1884. Hallson, S.V., The uses of seaweed in Iceland, in *Proc. IV. Int. Seaweed Symposium*, Davy de Virville, A., and Feldman, J., Eds., Pergamon Press, Oxford, 1964, 398.

1885. Halman, M., and Stiller, M., Turnover and uptake of dissolved phosphate in freshwater. A study in Lake Kinneret, *Limnol. Oceanogr*, 19, 774, 1974.

1886. Halperin, D.R., De Cano, M.S., De Muele, M.C.Z., and De Caire, G.Z., Algas azueles fijadoras de mitrogenio atmosferico, *Cent. Invest. Biol. Mar, Contr. Tec.* No. 3, 6, 1981.

1887. Hamilton, P.B., and Duthie, H.C., Periphyton colonization of rock surfaces in a boreal forest stream studied by scanning electron microscopy and track autoradiography, *J. Phycol.*, 20, 525, 1984.

1888. Hammer, D.A., (Ed.), *Constructed Wetlands for Wastewater Treatment*, Proc. Conf., Lewis Publishers, Chelsea, Michigan, 1989.

1889. Hammer, D.A., *Creating Freshwater Wetlands*, Lewis Publishers, Chelsea, Michigan, 1992.

1890. Hammer, D.A., and Bastian, R.K., Wetland ecosystems: natural water purifiers? in *Proc. Int. Conf. Constructed Wetlands for Wastewater Treatment*, Hammer, D.A., Ed., Lewis Publishers, Inc., Chelsea, Michigan, 1989, 1.

1891. Hammer, U.T., The succession of bloom species of blue-green algae and some causal factors, *Verh. Internat. Verein. Limnol.*, 15, 829, 1964.

1892. Handoo, J.K., and Kaul, V., Standing crop and nutrient dynamics in *Sparganium ramosum* Hudr. in Kashmir, *Aquat. Bot.*, 12, 375, 1982.

1893. Handoo, J.K., and Kaul, V., Phytosociology and standing crops in four typical wetlands of Kashmir, in *Proc. Int. Conf. Wetlands: Ecology and Management*, Gopal, B., Turner, R.E., Wetzel, R.G., and Whigham, D.F., Eds., Nat. Inst. Ecol. and Int. Sci. Publications, Jaipur, India, 1982, 187.

1894. Haniffa, M.A., and Pandian, T.J., Morphometry, primary productivity and energy flow in a tropical pond, *Hydrobiologia*, 59, 23, 1978.

1895. Hanisak, M.D., Nitrogen limitation of *Codium fragile* ssp. *tomentosoides* as determined by tissue analysis, *Mar. Biol.*, 50, 333, 1979.

1896. Hanisak, M.D., and Harlin, M.M., Uptake of inorganic nitrogen by *Codium fragile* subsp. *tomentoroides* (Chlorophyta), *J. Phycol.*, 14, 450, 1978.

1897. Hannan, P.J., and Patouillet, C., Effect of mercury on algal growth rates, *Biotechnol. Bioeng.*, 14, 93, 1972.

1898. Hannerz, L., Experimental investigations on the accumulation of mercury in water organisms, *Rept. Inst. Freshwater Res., Drottningholm* (Stockholm), 48, 120, 1968.

1899. Hansen, J.E., Population biology if *Iridaea cordata* (Rhodophyta: Gigartinaceae), Ph.D. Thesis, University of California, Santa Cruz, 1976.

1900. Hansen, J.E., Packard, J.E., and Doyle, W.T., Mariculture of red seaweeds, California Grant College Program Report T-CSGCP-002, 1982.

1901. Hansen, P.J., Cembella, A.D., and Moestrup, Ø., The marine dinoflagellate *Alexandrium ostenfeldii*: paralytic shellfish toxin concentration, composition, and toxicity to a tintinnid ciliate, *J. Phycol.*, 28, 597, 1992.

1902. Hansgirg, A., Beiträge zur Kenntnis der bömischen Thermalalgenflora, *Österr Bot. Zeitschr*, 34, 276, 1884.

1903. Hansgirg, A., Synopsis generum subgenerumque Myxophycearum (Cyanophycearum), hucusque cognitorum cum descriptive genesis nov. Dactylococcopsis, *Notarisia*, 3, 548, 1888.

1904. Hansson, L.-A., Effects of competitive interactions on the biomass development of planktonic and periphytic algae in lakes, *Limnol. Oceanogr.*, 33, 121, 1988.

1905. Hanton, J.T., Algal phosphate uptake kinetics: growth rates and limiting phosphate concentrations, M.S. Thesis, University of North Carolina, Chapel Hill, 1969.

1906. Harder, R., and Oppermann, A., Über antibiotische Stoffe bei den Grünalgen *Stichococcus bacillaris* und *Protosiphon botryoides*, *Arch. Mikrobiol.*, 19, 398, 1953.

1907. Harding, J.P.C., and Whitton, B.A., Resistance to zinc of *Stigeoclonium tenue* in the field and the laboratory, *Br. Phycol. J.*, 11, 417, 1976.

1908. Harding, J.P.C., and Whitton, B.A., Environmental factors reducing the toxicity of zinc to *Stigeoclonium tenue*, *Br. phycol. J.*, 12, 17, 1977.

1909. Harding, J.P.C., and Whitton, B.A., Accumulation of zinc, cadmium and lead by field populations of *Lemanea*, *Water Res.*, 15, 301, 1981.

1910. Harding, J.P.C., Burrows, I.G., and Whitton, B.A., Heavy metals in the Derwent Reservoir Catchment, northern England, in *Heavy Metals in Northern England: Environmental and Biological Aspects*, Say, P.J., and Whitton, B.A., Eds., University of Durham, Durham, England, 1981, 73.

1911. Hardy, R.W.F., Burns, R.C., and Parshall, G.W., The biochemistry of N_2 fixation, *Adv. Chem. Ser.*, 100, 219, 1971.

1912. Harel, E., Lea, P.J., and Miflin, B.J., The localization of enzymes of nitrogen assimilation in maize leaves and their activities during greening, *Planta*, 134, 195, 1977.

1913. Hargrave, B.T., Epibenthic algal production and community respiration in the sediments of Marion Lake, *J. Fish. Res. Bd. Canada*, 26, 2003, 1969.

1914. Hargraves, P.E., and Guillard, R.R.L., Structural and physiological observations on some small marine diatoms, *Phycologia*, 13, 163, 1974.

1915. Hargreaves, J.W., and Whitton, B.A., Effect of pH on growth of acid stream algae, *Br. Phycol. J.*, 11, 215, 1976.

1916. Hargreaves, J.W., and Whitton, B.A., Effect of pH on tolerance of *Hormidium rivulare* to zinc and copper, *Oecologia (Berlin)*, 26, 235, 1976.

1917. Hargreaves, J.W., Lloyd, E.J.H., and Whitton, B.A., Chemistry and vegetation of highly acidic streams, *Freshwater Biol.*, 5, 563, 1975.

1918. Harlin, M.M., Transfer of products between epiphytic marine algae and host plants, *J. Phycol.*, 9, 243, 1973.

1919. Harlin, M.M., "Obligate" algal epiphyte *Smithora naiadum* grows on a synthetic substrate, *J. Phycol.*, 9, 230, 1973.

1920. Harlin, M.M., Nitrate uptake by *Enteromorpha* spp. (Chlorophyceae): applications to aquaculture systems, *Aquaculture*, 15, 373, 1978.

1921. Harlin, M.M., Nitrate uptake by *Laminaria longicruris* (Phaeophyceae), *J. Phycol.*, 14, 464, 1978.

1922. Harlin, M.M., and Graigie, J.S., The distribution of photosynthate in *Ascophyllum nodosum* as it relates to epiphytic *Polysiphonia lanosa*, *J. Phycol.*, 11, 109, 1975.

1923. Harlin, M.M., and Craigie, J.S., Nitrate uptake by *Laminaria longicruris* (Phaeophyceae), *J. Phycol.*, 14, 464, 1978.

1924. Harold, F.M., Inorganic polyphosphates in biology: structure, metabolism, and function, *Bact. Rev.*, 30, 772, 1966.

1925. Harold, F.M., and Baarda, J.R., Interaction of arsenate with phosphate-transport systems in wild-type and mutant *Streptococcus faecalis*, *J. Bacteriol.*, 91, 2257, 1966.

1926. Harper, M.A., Movements, In *The Biology of Diatoms*, Werner, D., Ed., Blackwell Scientific Publications, Oxford, 1977, 224.

1927. Harper, R.M., Some dynamic studies of Long Island vegetation, *Plant World*, 21, 33, 1918.

1928. Harris, D.O., Nutrition of *Platymonas caudata* Kofoid, *J. Phycol.*, 5, 205, 1969.

1929. Harris, D.O., An autoinhibitory substance produced by *Platymonas caudata* Kofoid, *Plant Physiol.*, 45, 210, 1970.

1930. Harris, D.O., Growth inhibitors produced by the green algae Volvocaceae, *Arch. Mikrobiol.*, 76, 47, 1971.

1931. Harris, D.O., A model system for the study of algal growth inhibitors, *Arch. Protistenk.*, 113, 230, 1971.

1932. Harris, D.O., and James, D.E., Toxic algae, *Carolina Tips*, 37, 13, 1974.

1933. Harris, G.P., Photosynthesis, productivity and growth: The physiological ecology of phytoplankton, *Arch. Hydrobiol.*, 10 (suppl.), 1, 1978.

1934. Harris, G.P., The measurement of photosynthesis in natural populations of phytoplankton, in *The Physiological Ecology of Phytoplankton*, Morris, I., Ed., University of California Press, 1980, 129.

1935. Harris, G.P., *Phytoplankton Ecology: Structure, Function and Fluctuation*, Chapman and Hall, London, New York, 1986.

1936. Harris, G.P., and Lott, J.N.A., Light intensity and photosynthetic rates in phytoplankton, *J. Fish. Res. Board Can.*, 30, 1771, 1973.

1937. Harris, R.C., White, D.B., and MacFarlane, R.B., Mercury compounds reduce photosynthesis by plankton, *Science*, 170, 736, 1970.

1938. Harris, R.C., Sebacher, D.I., and Day, F.P., Jr., Methane flux in the Great Dismal Swamp, *Nature*, 297, 673, 1982.

1939. Harrison, G.I., and Morel, F.M.M., Antagonism between cadmium and iron in the marine diatom *Thalassiosira weissflogii*, *J. Phycol.*, 19, 495, 1983.

1940. Harrison, W., and Hobbie, J., Nitrogen budget of a North Carolina estuary, *Water Resour. Res. Inst. Univ. N.C.*, 86, 1, 1974.

1941. Harrison, W.G., The time-course of uptake of inorganic and organic nitrogen compounds by phytoplankton from the eastern Canadian Arctic: a comparison with temperate and tropical populations, *Limnol. Oceanogr.*, 28, 1231, 1983.

1942. Harrison, W.G., Azam, F., Renger, E.H., and Eppley, R.W., Some experiments on phosphate assimilation by coastal marine plankton, *Mar. Biol.*, 40, 9, 1977.

1943. Hart, B.A., and Scaife, B.D., Toxicity and bioaccumulation of cadmium in *Chlorella pyrenoidosa*, *Environ. Res.*, 14, 401, 1977.

1944. Hart, B.A., Bertram, P.E., and Scaife, B.C., Cadmium transport by *Chlorella pyrenoidosa*, *Environ. Res.*, 18, 327, 1979.

1945. Hart, D.D., Grazing insects mediate algal interactions in a stream benthic community, *Oikos*, 44, 40, 1985.

1946. Hart, M.R., De Fremey, D., Lyon, C.K., and Kohler, G.O., Processing of *Macrocystis pyrifera* (Phaeophyceae) for fermentation to methane, in *Proc. IX. Int. Seaweed Symposium*, Jensen, A., and Stein, J.R., Eds., Science Press, Princeton, New Jersey, 1979, 493.

1947. Hartshorne, J.N., The function of the eyespot in *Chlamydomonas*, *New Phytol.*, 52, 292, 1953.

1948. Harvey, H.W., The supply of iron to diatoms, *J. Mar. Biol. Ass. U.K.*, 22, 205, 1937.

1949. Harvey, H.W., Nitrogen and phosphorus required for the growth of phytoplankton, *J. Mar. Biol. Ass. U.K.*, 24, 115, 1940.

1950. Harvey, H.W., Synthesis of organic nitrogen and chlorophyll by *Nitzschia closterium*, *J. Mar. Biol. Ass. U.K.*, 31, 477, 1953.

1951. Harvey, H.W., Note on the absorption of organic phosphorus compounds by *Nitzschia closterium* in the dark, *J. Mar. Biol. Ass. U.K.*, 31, 475, 1953.

1952. Harvey, H.W., Cooper, L.H.N., Lebour, M.V., and Russell, F.S., Plankton production and its control., *J. Mar. Biol. Ass. U.K.*, 20, 407, 1935.

1953. Harvey, W.H., Algae, in *Flora Hibernica*, Pt. 3, MacKay, J.T., Ed., Wm. Curry, Dublin, 1836, 157.

1954. Harvey, W.H., *Phycologia Britannica*, 4 vols., London, 1846–1851.

1955. Hasan, S.M., and Hall, J.B., The physiological function of nitrate reductase in *Clostridiumperfringens, J. Gen. Microbiol.,* 87, 120, 1975.

1956. Haselkorn, I.R., Heterocysts, *Ann. Rev. Plant Physiol.,* 29, 319, 1978.

1957. Hashimoto, Y., *Marine Toxins and Other Bioactive Marine Metabolites,* Japan Scientific Societies Press, Tokyo, 1979.

1958. Hasle, G.R., *Navicula endophytica* sp. nov. A pennate diatom with an unusual mode of existence, *Br. Phycol. Bull.,* 3, 475, 1968.

1959. Hassack, C., Über das Vechaltnis von Pflanzen zu Bikarbonaten und über Kalkincrustation, *Tübingen Botanisch Institut Untersuchungen,* 2, 465, 1988.

1960. Hassall, K.A., Uptake of copper and its physiological effects on *Chlorella vulgaris, Physiol. Plant.,* 16, 323, 1963.

1961. Hassett, J.M., Jennett, J.C., and Smith, J.E., Heavy metal accumulation by algae, in *Contaminants and Sediments,* Vol. 2, Baker, R.A., Ed., Ann Arbor Science, Ann Arbor, Michigan, 1980, 409.

1962. Hassett, J.M., Jennett, J.C., and Smith, J.E., Microplate technique for determining accumulation of metals by algae, *Appl. Environ. Microbiol.,* 41, 1097, 1981.

1963. Hatch, M.D., and Slack, C.R., Photosynthesis by sugar cane leaves, *Biochem. J.,* 101, 103, 1966.

1964. Hatch, M.D., and Slack, C.R., Photosynthetic CO_2-fixation pathways, *Ann. Rev. Plant Physiol.,* 21, 141, 1970.

1965. Hatch, M.D., Osmond, C.B., and Slatyer, R.O., Eds., *Photosynthesis and Photorespiration,* Wiley-Interscience, London, 1971.

1966. Hatcher, B.G., Chapman, A.R.O., and Mann, K.H., An annual carbon budget for the kelp *Laminaria longicruris, Mar. Biol.,* 44, 85, 1977.

1967. Hatton, R.S., DeLaune, R.D., and Patrick, W.H., Jr., Sedimentation, accretion and subsidence in marshes of Barataria Bay, Louisiana, *Limnol. Oceanogr,* 28, 494, 1983.

1968. Hattori, A., Light-induced reduction of nitrate, nitrite and hydroxylamine in a blue-green alga, *Anabaena cylindrica, Plant Cell Physiol.,* 3, 355, 1962.

1969. Hattori, A., Effect of hydrogen on nitrite reduction by *Anabaena cylindrica,* in *Studies on Microalgae and Photosynthetic Bacteria,* Japanese Soc. Plant Physiol., Univ. of Tokyo Press, Tokyo, 1963, 485.

1970. Hattori, A., Solubilization of nitrate reductase from the blue-green alga *Anabaena cylindrica, Plant Cell Physiol.,* 11, 975, 1970.

1971. Hattori, A., and Uesugi, I., Purification and properties of nitrite reductase from the blue-green alga *Anabaena cylindrica, Plant Cell Physiol.,* 9, 689, 1968.

1972. Hauck, R.D., Atmospheric nitrogen chemistry, nitrification, denitrification, and their relationships, in *The Handbook of Environmental Chemistry* Vol. 1, Part C, *The Natural Environment and Biogeochemical Cycles,* Hutzinger, O., Ed., Springer-Verlag, Berlin, 1984, 106.

1973. Haug, A., The affinity of some divalentmetals for different types of alginates, *Acta Chem. Scand.,* 15, 1794, 1961.

1974. Haug, A., and Smidorod, O., Strontium, calcium and magnesium in brown algae, *Nature,* 215, 1167, 1967.

1975. Haug, A., Melsom, S., and Omang, S., Estimation of heavy metal pollution in two Norwegian fjord areas by analysis of the brown alga *Ascophyllum nodosum, Environ. Pollut.,* 7, 179, 1974.

1976. Haugstad, M., Ulsaker, L.K., Ruppel, A., and Nilsen, S., The effect of triacantol on growth, photosynthesis and photorespiration in *Chlamydomonas reinhardtii* and *Anacystis nidulans, Physiol. Plant.,* 58, 451, 1983.

1977. Hausinger, R.P., Nickel utilization by microorganisms, *Microbiol. Rev,* 51, 22, 1987.

1978. Hausmann, K., Extrusive organelles of the periplast of the Cryptophyceae during the discharge of ejectosomes, *Arch. Protistenk.,* 122, 222, 1979.

1979. Havlík, B., Starý, J., Prášilová, J., Kratzer, K., and Hanušová, J., Mercury circulation in aquatic biocenoses. Part 1. Mercury (II) - metabolism in phytoplankton, *Acta Hydrochim. Hydrobiol.,* 7, 215, 1979.

1980. Havlík, B., Starý, J., Prášilová, J., Kratzer, K., and Hanušová, J., Mercury circulation in aquatic biocenoses. Part 2. Metabolism of methyl and phenyl mercury in phytoplankton, *Acta Hydrochim Hydrobiol.,* 7, 401, 1979.

1981. Haxen, P.G., and Lewis, O.A.M., Nitrate assimilation in the marine kelp *Macrocystis angustifolia* (Phaeophyceae), *Bot. Mar,* 24, 631, 1981.

1982. Haxo, F., and Strout, P., Nitrogen deficiency and coloration in red algae, *Biol. Bull.,* 99, 360, 1950.

1983. Haxo, F.T., and Fork, D.C., Photosynthetically active accessory pigments of cryptomonads, *Nature,* 184, 1051, 1959.

1984. Hayes, F.R., The role of bacteria in the mineralization of phosphorus in lakes, in *Symposium on Marine Microbiology,* Oppenheimer, C., Ed., Charles C. Thomas, Springfield, Illinois, 1963, 654.

1985.   Haystead, A., Dharmawardene, M.W.N., and Stewart, W.D.P., Ammonia assimilation in a nitrogen-fixing blue-green alga, *Plant Sci. Lett.*, 1, 439, 1973.

1986.   Hayward, J., Studies on the growth of *Phaeodactylum tricornutum* (Bohlin). 1. The effect of certain organic nitrogenous substances on growth, *Physiol. Plant.*, 18, 201, 1965.

1987.   Hayward, J., Studies on the growth of *Phaeodactylum tricornutum*. VI. The relationship to sodium, potassium, calcium and magnesium, *J. Mar. Biol. Ass. U.K.*, 50, 293, 1970.

1988.   Head, P.C., and Burton, J.D., Molybdenum in some ocean and estuarine waters, *J. Mar. Biol. Ass. U.K.*, 50, 439, 1970.

1989.   Heal, O.W., and Perkins, D.F., Eds., *Production Ecology of British Moors and Montane Grasslands*, Ecological Studies No. 27, Springer Verlag, Berlin-Heidelberg-New York, 1978.

1990.   Heald, E.J., The Production of Organic Detritus in a South Florida Estuary, Ph. D. Dissertation, University of Miami, Miami, Florida, 1969.

1991.   Healey, F.P., Inorganic nutrient uptake and deficiency in algae, *CRC Crit. Rev. Microbiol.*, 3, 69, 1973.

1992.   Healey, F.P., Characteristics of phosphorus deficiency in *Anabaena*, *J. Phycol.*, 9, 383, 1973.

1993.   Healey, F.P., Ammonium and urea uptake by some freshwater algae, *Can. J. Bot.*, 55, 61, 1977.

1994.   Healey, F.P., Slope of the Monod equation as an indicator of advantage in nutrient competition, *Microbial Ecol.*, 5, 281, 1980.

1995.   Healey, F.P., and Hendzel, L.L., Indicators of phosphorus and nitrogen deficiency in five algae in culture, *J. Fish. Res. Board Can.*, 36, 1364, 1979.

1996.   Hegemann, D.A., Johnson, A.H., and Keenan, J.D., Determination of algal-available phosphorus on soil and sediment: A review and analysis, *J. Environ. Qual.*, 12, 12, 1983.

1997.   Hegewald, E., Untersuchungen zum Zeta-Potential von Planktonalgen, *Arch. Hydrobiol. Suppl.*, 42, 14, 1972.

1998.   Heikkilä, H., The vegetation and ecology of mesotrophic and eutrophic fens in western Finland, *Ann. Bot. Fennici*, 24, 155, 1987.

1999.   Heikurainen, L., Improvement of forest growth on poorly drained peat soils, *Int. Rev. For. Res.*, 1, 39, 1964.

2000.   Heikurainen, L., *Peatland Terminology for Forestry English, German, Russian, Swedish, Finnish*, I.U.F.R.O. Working Party S 1.05.01 Report, Helsinki, 1977.

2001.   Heikurainen, P., and Pakarinen, P., Mire vegetation and site types, In *Peatlands and Their Utilization in Finland*, Laine, J., Ed., Finnish Peatland Soc., Helsinki, 1982, 14.

2002.   Heilman, P.E., Relationship of availability of phosphorus and cations to forest succession and bog formation in interior Alaska, *Ecology*, 49, 331, 1968.

2003.   Heimburg, K., Hydrology of north-central Florida cypress domes, in *Cypress Swamps*, Ewell, K.C., and Odum, H.T., Eds., University Presses of Florida, Gainesville, Florida, 1984, 72.

2004.   Heine, J.N., Seasonal productivity of two red algae in a central California kelp forest, *J. Phycol.*, 19, 146, 1983.

2005.   Heinen, W., and Oehler, J.H., Evolutionary aspects of biological involvement in the cycling of silica, in *Biogeochemical Cycling of Mineral-Forming Elements*, Trudinger, P.A., and Swaine, D.J., Eds., Elsevier Scientific Publishing, Amsterdam, 1979, 431.

2006.   Heinle, D.R., Flemer, D.A., Ustach, J.F., Murtagh, R.A., and Harris, R.P., The role of organic debris and associated microorganisms in pelagic estuarine food chains, University of Maryland, Water Resour. Res. Center Tech. Rep. 22, College Park, Maryland, 1974.

2007.   Heinselman, M.L., Forest sites, bog processes, and peatland types in the glacial Lake Agassiz region, Minnesota, *Ecol. Monogr.*, 33, 327, 1963.

2008.   Heisey, R.M., and Damman, A.W.H., Copper and lead uptake by aquatic macrophytes in eastern Connecticut, U.S.A., *Aquat. Bot.*, 14, 213, 1982.

2009.   Hejný, S., Ed., *Ecosystem Study on Wetland Biome in Czechoslovakia*, Czechosl. IBP/PT-PP Report No. 3, Třeboň, Czechoslovakia, 1973.

2010.   Hejný, S., and Hroudová, Z., Plant adaptations to shallow water habitats, *Arch. Hydrobiol. Beih. Ergebn. Limnol.*, 27, 157, 1987.

2011.   Hejný, S., Květ, J., and Dykyjová, D., Survey of biomass and net production of higher plant communities in fishponds, *Folia Geobot. Phytotax.*, 16, 73, 1981.

2012.   Hejný, S., Husák, Š., Jeřábková, O., Ostrý, I., and Zákravský, P., Anthropogenic impact on fishpond flora and vegetation, in *Proc. Int. Conf. Wetlands: Ecology and Management*, Gopal, B., Turner, R.E., Wetzel, R.G., and Whigham, D.F., Eds., Natl. Inst. of Ecology and Internat. Scientific Publications, Jaipur, India, 1982, 425.

2013.   Helleburst, J.A., Excretion of some organic compounds by marine phytoplankton, *Limnol. Oceanogr.*, 10, 192, 1965.

2014. Helleburst, J.A., The uptake and utilization of organic substances by marine phytoplankton. *Occas. Publ. Inst. Mar. Sci. University Alaska Collection*, 1, 225, 1970.

2015. Helleburst, J.A., Kinetics of glucose transport and growth of *Cyclotella cryptica* Reimann, Lewin and Guillard, *J. Phycol.*, 7, 1, 1971.

2016. Helleburst, J.A., Glucose uptake by *Cyclotella cryptica*: dark induction and light inactivation of transport system, *J. Phycol.*, 7, 345, 1971.

2017. Helleburst, J.A., Extracellular products, in *Algal Physiology and Biochemistry* Stewart, W.D.P., Ed., Blackwell Scientific Publications, Oxford, 1974, 838.

2018. Helleburst, J.A., Uptake of organic substances by *Cyclotella cryptica* (Bacillariophyceae): effects of ions, ionophores and metabolic and transport inhibitors, *J. Phycol.*, 14, 79, 1978.

2019. Helleburst, J.A., and Guillard, R.R.L., Uptake specificity for organic substances by the marine diatom *Melosira nummuloides, J. Phycol.*, 3, 132, 1967.

2020. Helleburst, J.A., and Lewin, J., Heterotrophic nutrition, in *The Biology of Diatoms*, Werner, D., Ed., University of California Press, Berkeley, 1977, 169.

2021. Helleburst, J.A., and Craigie, J.S., Eds., *Handbook of Phycological Methods: Physiological and Biochemical Methods,* Cambridge University Press, Cambridge, 1978.

2022. Helleburst, J.A., and Ahmad, I., Nitrogen metabolism and amino acid nutritional in the soil alga *Stichococcus bacillaris* (Chlorophyceae), *J. Phycol.*, 25, 48, 1989.

2023. Helleburst, J.A., Terborgh, J., and McLeod, G.C., The photosynthetic rhythm of *Acetabularia crenulata*. II. Measurements of photoassimilation of carbon dioxide and the activities of enzymes of the reductive pentose cycle, *Biol. Bull.*, 133, 670, 1967.

2024. Hem, J.D., Chemistry and occurrence of cadmium and zinc in surface water and groundwater, *Resource Res.*, 8, 661, 1972.

2025. Hem, J.D., Geochemical controls on lead concentrations in stream water and sediments, *Geochim. Cosmochim. Acta*, 40, 599, 1976.

2026. Hem, J.D., and Durum, W.H., Solubility and occurrence of lead in surface waters, *J. Am. Water Works Assoc.*, 65, 562, 1973.

2027. Hemens, J., and Stander, G.J., Nutrient removal from sewage effluents by algal activity, in *Proc. 4th Internat. Conf. Advances in Water Pollution Research*, Pergamon Press, Oxford, 1970, 701.

2028. Hemond, H.F., Biogeochemistry of Thoreau's Bog, Concord, Massachusetts, *Ecol. Monogr.*, 50, 507, 1980.

2029. Hemond, H.F., The nitrogen budget of Thoreau's Bog, *Ecology*, 64, 99, 1983.

2030. Hemond, H.F., and Duran, A.P., Fluxes of N_2O at the sediment-water and water-atmosphere boundaries of a nitrogen-rich river, *Water Resour. Res.*, 25, 839, 1989.

2031. Hendey, N.I., The diagnostic value of diatoms in cases of drowning, *Medicine, Sci., Law*, 13, 23, 1973.

2032. Henriksen, A., and Wright, R.F., Concentrations of heavy metals in small Norwegian lakes, *Wat. Res.*, 12, 101, 1978.

2033. Henriksson, E., Nitrogen fixation by a bacteria-free symbiotic *Nostoc* strain isolated from *Collema*, *Physiol. Plant.*, 4, 542, 1951.

2034. Henriksson, E., Algal nitrogen fixation in temperate regions, *Plant and Soil*, Spec. Vol., 415, 1971.

2035. Henriksson, E., Studies on the physiology of the lichen *Collema*. II. A preliminary report on the isolated fungal partner with special regard to its behavior when growing together with the symbiotic alga, *Svensk. Botan. Tidskr*, 52, 391, 1958.

2036. Henriksson, L.E., and DaSilva, E.J., Effects of some inorganic elements on nitrogen-fixation in blue-green algae and some ecological aspects of pollution, *Z. Allgemeine Mikrobiol.*, 18, 487, 1978.

2037. Hensen, V., Über die Bestimung des Planktons oder des im Meere treibenden Materials an Pflanzen und Thieren, *Ber. Komm. wiss. Unters. Deutsch. Meere, Kiel*, 12–14, 1, 1887.

2038. Hepler, P.K., and Wayne, R.O., Calcium and plant development, *Ann. Rev. Plant Physiol.*, 36, 397, 1985.

2039. Herbst, D.B., and Bradley, T.J., Salinity and nutrient limitations on growth of benthic algae from two alkaline salt lakes in the Western Great Basin (U.S.A.), *J. Phycol.*, 25, 673, 1989.

2040. Hercinger, F., Beiträge zum Wirkungskreislauf der Bors, *Bodenk. u. Pflanzenernähr*, 16, 141, 1940.

2041. Herrera, J., Paneque, A., Maldonado, J.M., Barea, J.L., and Losada, M., Regulation by ammonia of nitrate reductase synthesis and activity in *Chlamydomonas reinhardii, Biochem. Biophys. Res. Comm.*, 48, 996, 1972.

2042. Hersey, R.L., and Swift, E., Nitrate reductase activity of *Amphidinium carteri* and *Cachonina niei* (Dinophyceae) in batch culture: diel periodicity and effects of light intensity and ammonia, *J. Phycol.*, 12, 36, 1976.

2043. Hervey, R.J., Effect of chromium on the growth of unicellular chlorophyceae and diatoms, *Bot. Gaz.*, 111, 1, 1949.

2044. Herzig, R., and Falkowski, P.G., Nitrogen limitation in *Isochrysis galbana* (Haptophyceae). I. Photosynthetic energy conversion and growth efficiencies, *J. Phycol.*, 25, 462, 1989.
2045. Hessler, A., The effects of lead on algae. 1. Effects of Pb on viability and motility of *Platymonas subcordiformis* (Chlorophyta, Volvocales), *Water Air Soil Pollut.*, 3, 371, 1974.
2046. Hessler, A., The effects of lead on algae. II. Mutagenesis experiments on *Platymonas subcordiformis* (Chlorophyta, Volvocales), *Mutation Res.*, 31, 43, 1975.
2047. Hewitt, E.J., The essential nutrient elements: requirements and interactions in plants, in *Plant Physiology* Vol. 3, *Inorganic Nutrition of Plants*, Steward, F.C., Ed., Academic Press, New York and London, 1963, 137.
2048. Hewitt, E.J., Assimilatory nitrate-nitrite reduction, *Ann. Rev. Plant Physiol.*, 26, 73, 1975.
2049. Hewitt, E.J., and Nicholas, D.J.D., Enzymes of inorganic nitrogen metabolism, in *Modern Methods of Plant Analysis*, Vol. 7, Linsken, H.F., Sanwal, B.D., and Tracey, M.V., Eds., Springer Verlag, Berlin, 1964, 67.
2050. Hewitt, E.J., and Notton, B.A., Nitrate reductase systems in eukaryotic and prokaryotic organisms, in *Molybdenum and Molybdenum-containing Enzymes*, Coughlan, M., Ed., Pergamon Press, Oxford, 1980, 273.
2051. Hewitt, E.J., Hucklesby, D.P., and Notton, B.A., Nitrate metabolism, in *Plant Biochemistry* Bonner, J., Ed., Academic Press, New York, 1976, 633.
2052. Heywood, P., Structure and origin of flagellar hairs in *Vacuolaria virescens, J. Ultrastruct. Res.*, 39, 608, 1972.
2053. Heywood, P., Nutritional studies on the Chloromonadophyceae: *Vacuolaria virescens* and *Gonyostomum semen, J. Phycol.*, 9, 156, 1973.
2054. Heywood, P., Mitosis and cytokinesis in the chloromonadophycean alga *Gonyostomum semen, J. Phycol.*, 10, 355, 1974.
2055. Heywood, P., Ultrastructure of mitosis in the chloromonadophycean alga *Vacuolaria virescens, J. Cell Sci.*, 31, 37, 1978.
2056. Heywood, P., Systematic position of the Chloromonadophyceae, *Br. Phycol. J.*, 13, 201 A, 1978.
2057. Heywood, P., Chloromonads, in *Phytoflagellates, Developments in Marine Biology*, Vol. 2, Cox, E.R., Ed., Elsevier/North Holland, New York, 1980, 351.
2058. Heywood, P., Raphidophyceae (Chloromonadophyceae), in *Selected Papers in Phycology* II, Rosowski, J.R., and Parker, B.C., Eds., Phycological Society of America, Lawrence, Kansas, 1982, 719.
2059. Heywood, P., and Godward, M.B.E., Mitosis in the alga *Vacuolaria virescens, Am. J. Bot.*, 37, 423, 1974.
2060. Hibberd, D.J., Observations on the cytology and ultrastructure of *Ochromonas tuberculata* sp. nov. (Chrysophyceae), with special reference to the discobolocysts, *Br. Phycol. J.*, 5, 119, 1970.
2061. Hibberd, D.J., Observations on the cytology and ultrastructure of *Chrysamoeba radians* Klebs (Chrysophyceae), *Br. Phycol. J.*, 6, 207, 1971.
2062. Hibberd, D.J., The ultrastructure and taxonomy of the Chrysophyceae and Prymnesiophyceae (Haptophyceae), a survey with some new observations on the ultrastructure of the Chrysophyceae, *Bot. J. Linn. Soc.*, 72, 55, 1976.
2063. Hibberd, D.J., Observations on the ultrastructure of the cryptomonad endosymbiont of the red water ciliate *Mesodinium rubrum, J. Mar. Biol., Ass. U.K.*, 57, 45, 1977.
2064. Hibberd, D.J., Xantophytes, In *Phytoflagellates, Developments in Marine Biology*, Vol. 2, Cox, E.R., Ed., Elsevier/Nort Holland, New York, 1980 a, 243.
2065. Hibberd, D.J., Prymnesiophytes (=Haptophytes), in *Phytoflagellates, Developments in Marine Biology*, Vol. 2, Cox, E.R., Ed., Elsevier/North Holland, New York, 1980, 273.
2066. Hibberd, D.J., Eustigmatophytes, In *Phytoflagellates, Developments in Marine Biology*, Vol. 2, Cox, E.R., Ed., Elsevier/Nort Holland, New York, 1980 b, 319.
2067. Hibberd, D.J., Notes on taxonomy and nomenclature of the algal classes Eustigmatophyceae and Tribophyceae (synonym Xanthophyceae), *Bot. J. Linn. Soc.*, 82, 93, 1981.
2068. Hibberd, D.J., Xantophyceae, In *Synopsis and Classification of Living Organisms*, Vol. 1, Parker, S.P., Ed., McGraw-Hill, New York, 1982a, 91.
2069. Hibberd, D.J., Eustigmatophyceae: introduction and bibliography, in *Selected Papers in Phycology* II, Rosowski, J.R., and Parker, B.C., Eds., Phycological Society of America, Lawrence, Kansas, 1982, 728.
2070. Hibberd, D.J., Eustigmatophytes, In *Synopsis and Classification of Living Organisms*, Vol. 1, Parker, S.P., Ed., McGraw-Hill, New York, 1982 c, 95.
2071. Hibberd, D.J., and Leedale, G.F., Eustigmatophyceae—a new algal class with unique organization of the motile cell, *Nature*, 225, 758, 1970.
2072. Hibberd, D.J., and Leedale, G.F., A new algal class—the Eustigmatophyceae, *Taxon*, 20, 523, 1971.

2073. Hibberd, D.J., and Leedale, G.F., Observations on the cytology and ultrastructure of the new algal class, Eustigmatophyceae, *Ann. Bot.,* 36, 49, 1972.

2074. Hibberd, D.J., and Chertiennot-Dinet, M.-J., The ultrastructure and taxonomy of *Rhizochromulina marina* gen. et sp. nov., an amoeboid marine chrysophyte, *J. Mar. Biol. Ass. U.K.,* 59, 179, 1979.

2075. Hibberd, D.J., Greenwood, A.D., and Griffiths, H.B., Observations on the ultrastructure of the flagella and periplast in the Cryptophyceae, *Br. Phycol. J.,* 6, 61, 1971.

2076. Hickel, B., Anagnostidis, K., and Komárek, J., Eds., Cyanophyta (Cyanobacteria). Morphology, Taxonomy, Ecology. Proc. 11th Symposium of International Association for Cyanophyte Research, Plön, *Arch. Hydrobiol. Suppl.,* 92 (*Algol. Studies,* 64), 1991.

2077. Hickman, M., The standing crop and primary productivity of the epiphyton attached to Equisetum fluviatile L. in Priddy Pool, North Sommerset, *Br. Phycol. J.,* 6, 51, 1971.

2078. Hickman, M., Phosphorus, chlorophyll and eutrophic lakes, *Arch. Hydrobiol.,* 88, 137, 1980.

2079. Hidu, H., and Ukeles, R., Dried unicellular algae as food for larvae of the hardshell clam, *Mercenaria mercenaria, Proc. Nat. Shellfish Ass.,* 53, 85, 1962.

2080. Hietz, P., Decomposition and nutrients dynamics of reed (*Phragmites australis* (Cav.) Trin. ex Steudel) litter in Lake Neusiedel, Austria, *Aquat. Bot.,* 43, 211, 1992.

2081. Higuchi, T., Gaseous CO_2 transport through the aerenchyma and the intracellular spaces in relation to the uptake of CO_2 by rice roots, *Soil Sci. Plant Nutr.,* 28, 491, 1982.

2082. Hill, B.H., Uptake and release of nutrients by aquatic macrophytes, *Aquat. Bot.,* 7, 87, 1979.

2083. Hill, B.H., The breakdown of macrophytes in a reservoir wetland, *Aquat. Bot.,* 21, 23, 1985.

2084. Hill, B.H., *Typha* productivity in a Texas pond: implications for energy and nutrient dynamics in freshwater wetlands, *Aquat. Bot.,* 27, 385, 1987.

2085. Hill, B.H., and Webster, J.R., Aquatic macrophyte breakdown in an Appalachian river, *Hydrobiologia,* 89, 53, 1982.

2086. Hill, D.R.A., and Rowan, K.S., The biliproteins of the Cryptophyceae, *Phycologia,* 28, 455, 1989.

2087. Hill, R., and Bendall, F., Function of two cytochrome components in chloroplasts: a working hypothesis, *Nature,* 186, 136, 1960.

2088. Hill, W.R., and Knight, A.W., Experimental analysis of the grazing interaction between a mayfly and stream algae, *Ecology,* 68, 1955, 1987.

2089. Hill, W.R., and Knight, A.W., Nutrient and light limitation of algae in two northern California streams, *J. Phycol.,* 24, 125, 1988.

2090. Hill, W.R., and Harvey, B.C., Periphyton response to higher trophic levels and light in shaded stream, *Can. J. Fish. Aquat. Sci.,* 47, 2307, 1990.

2091. Hillard, D.K., and Asmund, B., Studies on Chrysophyceae from some ponds and lakes in Alaska II. Notes on the genera *Dinobryon, Hyalobryon* and *Epipyxis* with the description of new species, *Hydrobiologia,* 22, 331, 1963.

2092. Hillbricht-Ilkowska, A., Kowalczewski, A., and Spodniewska, I., Field experiments on the factors controlling primary production in the lake plankton and periphyton, *Ekologia Polska,* 20, 315, 1972.

2093. Hillis, L., Recent calcified Halimedaceae, in *Calcareous Algae and Stromatolites,* Riding, R., Ed., Springer-Verlag, Berlin, 1991, 167.

2094. Hillman, W.S., and Culley, D.D., Jr., The uses of duckweed, *Am. Sci.,* 66, 442, 1978.

2095. Hillson, C.J., *Seaweeds.* Keystone Books—The Pennsylvania State University Press, University Park and London, 1977.

2096. Hind, G., and Olson, J.M., Electron transport pathways in photosynthesis, *Ann. Rev. Plant Physiol.,* 19, 249, 1968.

2097. Hindák, F., and Komárek, J., Cultivation of the cryosestonic alga *Koliella tatrae* (Kol) Hind., *Biol. Plant.,* 10, 95, 1968.

2098. Hindák, F., Komárek, J., Marvan, P., and Růžička, J., *Determination Key for Lower Plants.* I. *Algae,* SPN Bratislava, 1975 (in Slovak).

2099. Hindák, F., Cyrus, Z., Marvan, P., Javornický, P., Komárek, K., Ettl, H., Rosa, K., Sládečková, A., Popovský, J., Punčochářová, M., and Lhotský, O., *Freshwater Algae,* SPN Bratislava, 1978, (in Slovak).

2100. Hipkin, C.R., and Syrett, P.J., Some effects of nitrogen starvation on nitrogen and carbohydrate metabolism in *Ankistrodesmus braunii, Planta,* 133, 209, 1977.

2101. Hipkin, C.R., Syrett, P.J., and Al-Bassam, B.A., Some characteristics of nitrate reductase in unicellular algae, in *Nitrogen Assimilation in Plants,* Hewitt, E.J., and Cutting, C.V., Eds., Academic Press, London and New York, 1979, 309.

2102. Hipkin, C.R., Al-Bassam, B.A., and Syrett, P.J., The roles of nitrate and ammonium in the regulation of the development of nitrate reductase in *Chlamydomonas reinhardii, Planta,* 150, 13, 1980.

2103. Hirst, E.L., and Rees, D.A., The structure of alginic acid. Part V. Isolation and unambiguous characterization of some hydrolysis products of the methylated polysaccharide, *J. Chem. Soc.*, p. 1182, 1965.

2104. Hirst, E.L., Percival, E., and Wold, J.K., Structural studies of alginic acid, *Chemy. Ind.*, 257, 1963.

2105. Ho, C.-H., Ikawa, T., and Nisizawa, K., Purification and properties of a nitrate reductase from *Porphyra yezoensis* Ueda, *Plant Cell Physiol.*, 17, 417, 1976.

2106. Ho, S.C., Periphyton production in a tropical lowland stream polluted by inorganic sediments and organic wastes, *Arch. Hydrobiol.*, 77, 458, 1976.

2107. Ho, Y.B., Inorganic mineral nutrient level studies on *Potamogeton pectinatus* L. and *Enteromorpha prolifera* in Forfar Loch, Scotland, *Hydrobiologia*, 62, 7, 1979.

2108. Ho, Y.B., Chemical composition studies on some aquatic macrophytes in three Scottish lochs. I. Chlorophyll, ash, carbon, nitrogen and phosphorus, *Hydrobiologia*, 63, 161, 1979.

2109. Ho, Y.B., Chemical composition studies on some aquatic macrophytes in three Scottish lochs. II. Potassium, sodium, calcium, magnesium and iron, *Hydrobiologia*, 64, 209, 1979.

2110. Ho, Y.B., Growth, chlorophyll and mineral nutrient studies on *Phalaris arundinacea* L. in three Scottish lakes, *Hydrobiologia*, 63, 33, 1979.

2111. Ho, Y.B., Mineral element content of *Ulva lactuca* L. with reference to eutrophication in Hong Kong coastal waters, *Hydrobiologia*, 77, 43, 1981.

2112. Ho, Y.B., Mineral composition of *Phragmites australis* in Scottish lochs as related to eutrophication. 1. Seasonal change in organs, *Hydrobiologia*, 85, 227, 1981.

2113. Ho, Y.B., Metals in 19 intertidal macroalgae in Hong Kong waters, *Mar. Pollut. Bull.*, 18, 564, 1987.

2114. Ho, Y.B., Metal levels in three intertidal macroalgae in Hong Kong waters, *Aquat. Bot.*, 29, 367, 1988.

2115. Hoagland, D.R., and Davis, A.R., The composition of the cell sap of the plant in relation to the absorption of ions, *J. Gen. Physiol.*, 5, 629, 1923.

2116. Hoagland, K.D., Short-term standing crop and diversity of periphytic diatoms in a eutrophic reservoir, *J. Phycol.*, 19, 30, 1983.

2117. Hoagland, K.D., and Peterson, C.G., Effects of light and wave disturbance on vertical zonation of attached microalgae in a large reservoirs, *J. Phycol.*, 26, 450, 1990.

2118. Hoagland, K.D., Roemer, S.C., and Rosowski, J.R., Colonization and community structure of two periphyton assemblages, with emphasis on the diatoms (Bacillariophyceae), *Am. J. Bot.*, 69, 188, 1982.

2119. Hoagland, K.D., Zlotsky, A., and Peterson, C.G., The source of algal colonization on rock substrates in a freshwater impoundment, in *Algal Biofouling,* Evans, L.V., and Hoagland, K.D., Eds., Elsevier Scientific Publications, Amsterdam, 1986, 21.

2120. Hoagland, K.D., Rosowski, J.R., Gretz, M.R., and Roemer, S.C., Diatom extracellular polymeric substances: function, fine structure, chemistry, and physiology, *J. Phycol.*, 29, 537, 1993.

2121. Hoare, D.S., Hoare, S.L., and Moore, R.B., The photoassimilation of organic compounds by autotrophic blue-green algae, *J. Gen. Microbiol.*, 49, 351, 1967.

2122. Hoare, D.S., Ingram, L.O., Thurston, E.L., and Walkup, R., Dark heterotrophic growth of an endophytic blue-green alga, *Arch. Mikrobiol.*, 78, 310, 1971.

2123. Hobbie, J.E., Carbon-14 measurements of primary production in two arctic Alaskan lakes, *Verh. Internat. Verein. Limnol.*, 15, 360, 1964.

2124. Hobbie, J.E., and Wright, R.T., Competition between planktonic bacteria and algae for organic solutes, in *Primary Productivity in Aquatic Environments,* Goldman, C.R., Ed., University of California Press, Berkeley, 1965, 177.

2125. Hobbie, J.E., and Lee, C., Microbial production of extracellular material: importance in benthic ecology, in *Marine Benthic Dynamics,* Tenore, K.R., and Coull, B.C., Eds., University of South Carolina Press, Columbia, South Carolina, 1980, 341.

2126. Hobbie, J.E., Raleigh, N.C., and Wright, R.T., A new method for study of bacteria in lakes: description and results, *Mitt. Int. Verein. Limnol.*, 14, 64, 1968.

2127. Hobson, L.A., Some influences of the Columbia river effluent on marine phytoplankton during January 1961, *Limnol. Oceanogr.*, 11, 223, 1966.

2128. Hoch, F.L., and Vallee, B.L., The metabolic role of zinc, in *Trace Elements,* Lamb, C.A., Bentley, O.G., and Beattie, J.M., Eds., Academic Press, New York, 1958, 337.

2129. Hodgson, J.F., Chemistry of micronutrient elements in soils, *Adv. Agronomy*, 15, 119, 1963.

2130. Hodgson, L.M., Photosynthesis of the red alga *Gastroclonium coulteri* (Rhodophyta) in response to changes in temperature, light intensity and desiccation, *J. Phycol.*, 17, 37, 1981.

2131. Hodson, P.V., Borgmann, U., and Shear, H., Toxicity of copper to aquatic biota, in *Copper in the Environment,* Part 2, *Health Aspects,* Nriagu, J.O., Ed., John Wiley and Sons, New York, 1979, 307.

2132. Hodson, R.C., and Thompson, J.F., Metabolism of urea by *Chlorella vulgaris*, *Plant Physiol.*, 44, 691, 1969.

2133. Hodson, R.C., and Schiff, J.A., Studies of sulfate utilization by algae. 8. The ubiquity of sulfate reduction to thiosulfate, *Plant Physiol.*, 47, 296, 1971.

2134. Hodson, R.C., and Schiff, J.A., Studies of sulfate utilization by algae. 9. Fractionation of a cell-free system from *Chlorella* into two activities necessary for the reduction of adenosine-3'-phosphate-5'-phosphosulfate to acid-volatile radioactivity, *Plant Physiol.*, 47, 300, 1971.

2135. Hodson, R.C., Schiff, J.A., Scarsella, A.J., and Levinthal, M., Studies of sulfate utilization by algae. 6. Adenosine-3'-phosphate-5'-phosphosulfate (PAPS) as an intermediate in thiosulfate formation from sulfate by cell-free extracts of *Chlorella*, *Plant Physiol.*, 43, 563, 1968.

2136. Hodson, R.C., Schiff, J.A., and Scarsella, A.J., Studies of sulfate utilization by algae. 7. In vivo metabolism of thiosulfate by *Chlorella*, *Plant Physiol.*, 43, 570, 1968.

2137. Hodson, R.C., Schiff, J.A., and Mather, J.P., Studies od sulfate utilization by algae. 10. Nutritional and enzymatic characterization of *Chlorella* mutants impaired for sulfate utilization, *Plant Physiol.*, 47, 306, 1971.

2138. Hodson, R.C., Williams, S.K., and Davidson, W.R., Jr., Metabolic control of urea catabolism in *Chlamydomonas reinhardii* and *Chlorella pyrenoidosa*, *J. Bacteriol.*, 21, 1022, 1975.

2139. Hofman, L.R., Chemotaxis of *Oedogonium* species, *S. West Nat.*, 5, 111, 1960.

2140. Hoffmann, A.J., and Malbrán, M.E., Temperature, photoperiod and light interactions on growth and fertility of *Grassophora kunthii* (Phaeophyta, Dictyotales), from central Chile, *J. Phycol.*, 25, 129, 1989.

2141. Hoffmann, H.J., and Aitken, J.P., Precambrian biota from the Little Dal Group, McKenzie Mountains, northwestern Canada, *Can. J. Earth Sci.*, 16, 150, 1979.

2142. Hoffmann, W.E., and Dawes, C.J., Photosynthetic rates and primary production by two Florida benthic red algal species from a salt marsh and a mangrove community, *Bull. Mar. Sci.*, 30, 358, 1980.

2143. Hoffstetter, R.H., Wetlands in the United States, in *Ecosystems of the World. Mires: Swamp, Bog, Fen and Moor. 4B. Regional Studies*, Gore, A.J.P., Ed., Elsevier Science Publishers, Amsterdam, 1983, 201.

2144. Hofstetter, A.M., A preliminary report of the algal flora from selected areas of Shelby County, *J. Tennessee Acad. Sci.*, 43, 20, 1968.

2145. Hogeland, A.M., and Killingbeck, K.T., Biomass, productivity and life history traits of *Juncus militaris* Bigel. in two Rhode Island (U.S.A.) freshwater wetlands, *Aquat. Bot.*, 22, 335, 1985.

2146. Hogetsu, D., and Miyachi, S., Operation of the reductive pentose phosphate cycle during the induction period of photosynthesis in *Chlorella*, *Plant Cell Physiol.*, 20, 1427, 1979.

2147. Hoham, R.W., *Chlainomonas kolii* (Hardy et Curl) comb. nov. (Chlorophyta, Volvocales): a revision of the snow alga *Trachelomonas kolii* Hardy et Curl, *J. Phycol.*, 10, 392, 1974.

2148. Hoham, R.W., Optimum temperature and temperature ranges for growth of snow algae, *Arct. Alp. Res.*, 7, 13, 1975.

2149. Hoham, R.W., The life history and ecology of the snow alga *Chloromonas pichinchae* (Chlorophyta, Volvocales), *Phycologia*, 16, 53, 1975.

2150. Hoham, R.W., Unicellular chlorophytes - snow algae, in *Phytoflagellates: Developments in Marine Biology*, Vol. 2., Cox, E.R., Ed., Elsevier/North Holland, New York, 1980, 61.

2151. Hoham, R.W., Snow microorganisms and their interaction with the environment, in *Proc. 57th Ann. Western Snow Conf.*, Shafer, B., Ed., Fort Collins, Colorado, 1989, 31.

2152. Hoham, R.W., Environmental influence of snow algal microbes, in *Proc. 60th Ann. Western Snow Conf.*, Shafer, B., Ed., Jackson Hole, Wyoming, 1992, 78.

2153. Hoham, R.W., and Blinn, D.W., The distribution of algae in an arid region, *Phycologia*, 18, 133, 1979.

2154. Hoham, R.W., Roemer, S.C., and Mullet, J.E., The life history and ecology of the snow-alga *Chloromonas brevispina* comb. nov. (Chlorophyta, Volvocales), *Phycologia*, 18, 55, 1979.

2155. Holding, A.J., Heal, O.W., Maclean, S.F., Jr., and Flanagan, P.W., *Soil Organisms and Decomposition in Tundra*, Internal Biological Program, Tundra Biome Steering Committee, Stockholm, Sweden, 1974.

2156. Holdren, G.C., Jr., and Armstrong, D.E., Factors affecting phosphorus release from intact lake sediment cores, *Environ. Sci. Technol.*, 14, 79, 1980.

2157. Holdsworth, E.S., and Bruck, K., Enzymes concerned with ß-carboxylation in marine phytoplankter. Purification and properties of phosphoenolpyruvate carboxykinase, *Arch. Environ. Microbiol.*, 182, 87, 1977.

2158. Holford, I.C.R., and Patrick, W.H., Jr., Effects of reduction and pH changes on phosphate sorption and mobility in an acid soil. *Soil Sci. Soc. Am. J.*, 43, 292, 1979.

2159.   Hollibaugh, J.T., Siebert, D.L.R., and Thomas, W.H., A comparison of the acute toxicities of ten heavy metals to phytoplankton from Saanich Inlet, B.C., Canada, *Estuar. Coast. Mar. Sci.,* 10, 93, 1980.

2160.   Holm, T.R., Anderson, M.A., Iverson, D.G., and Stanforth, R.S., Heterogenous interactions of arsenic in aquatic systems, in *Chemical Modeling in Aqueous Systems,* Jenne, E.A., Ed., Chem. Soc. Symp. Ser. 93, 711, 1979.

2161.   Holmes, R.W., Light microscope observations on cytological manifestations of nitrate, phosphate, and silicate deficiency in four marine centric diatoms, *J. Phycol.,* 2, 136, 1966.

2162.   Holmes, R.W., Williams, P.M., and Eppley, R.W., Red water in La Jolla Bay, 1964–1966, *Limnol. Oceanogr.,* 12, 503, 1967.

2163.   Holm, N.P., and Armstrong, D.E., Role of nutrient limitation and competition in controlling the population of *Asterionella formosa* and *Microcystis aeruginosa* in semicontinuous culture, *Limnol. Oceanogr.,* 26, 622, 1981.

2164.   Holm-Hansen, O., Assimilation of carbon dioxide, in *Physiology and Biochemistry of Algae,* Lewin, R.A., Ed., Academic Press, New York and London, 1962, 25.

2165.   Holm-Hansen, O., Ecology, physiology and biochemistry of blue-green algae, *Ann. Rev. Microbiol.,* 22, 47, 1968.

2166.   Holm-Hansen, O., Determination of microbial biomass in ocean profiles, *Limnol. Oceanogr.,* 14, 740, 1969.

2167.   Holm-Hansen, O., ATP levels in algal cells as influenced by environmental conditions, *Plant Cell Physiol.,* 11, 689, 1970.

2168.   Holm-Hansen, O., Gerloff, G., and Skoog, F., Cobalt as an essential element for blue-green algae, *Physiol. Plant.,* 7, 665, 1954.

2169.   Holomuzki, J.R., and Stevenson, R.J., Role of predatory fish community dynamics of an ephemeral stream, *Can. J. Fish. Aquat. Sci.,* 49, 2322, 1992.

2170.   Holst, R.W., and Yopp, J.H., Environmental regulation of nitrogenase and nitrate reductase as systems of nitrogen assimilation in the *Azolla mexicana—Anabaena azolae* symbiosis, *Aquat. Bot.,* 7, 369, 1979.

2171.   Holton, R.W., and Myers, J., Cytochromes of a blue-green alga: extraction of a c-type with a strongly negative redox potential, *Science,* 142, 1963.

2172.   Holzapfel-Pschorn, A., Conrad, R., and Seiler, W., Effect of vegetation on the emission of methane from submerged paddy soil, *Plant and Soil,* 92, 223, 1986.

2173.   Homann, P.H., Studies on the manganese of the chloroplast, *Plant Physiol.,* 42, 997, 1967.

2174.   Hongve, D., Skogheim, O.K., Hindar, A., and Abrahamsen, H., Effects of heavy metals in combination with NTA, humic acid and suspended sediment on natural phytoplankton photosynthesis, *Bull. Environ. Contam. Toxicol.,* 25, 594, 1980.

2175.   Hook, D.D., Adaptations to flooding with freshwater, in *Flooding and Plant Growth,* Kozlowski, T.T., Ed., Academic Press, Orlando, Florida, 1984, 265.

2176.   Hook, D.D., and Scholtens, J.R., Adaptation and flood tolerance in the species, in *Plant Life in Anaerobic Environments,* Hook, D.D., and Crawford, R.M.M., Eds., Ann Arbor Science, Ann Arbor, Michigan, 1978, 299.

2177.   Hook, D.D., McKee, W.H., Jr., Smith, H.K., Gregory, J., Burrell, V.G., Jr., De Voe, M.R., Sojka, R.E., Gilbert, S., Banks, R., Stolzy, L.H., Brooks, C., Matthews, T.D., and Shear, T.H., Eds., *Proc. Conf. The Ecology and Management of Wetlands,* Vols 1+2, Timber Press, Portland, Oregon, 1988.

2178.   Hooper, F.F., Origin and fate of organic phosphorus compounds in aquatic systems, in *Environmental Phosphorus Handbook,* Griffith, E.J., Beeton, A., Spencer, J.M., and Mitchell, D.T., Eds., John Wiley and Sons, New York, 1973, 179.

2179.   Hooper, N.M., and Robinson, G.G.C., Primary production of epiphytic algae in a marsh pond, *Can. J. Bot.,* 54, 2810, 1976.

2180.   Hooper-Reid, N.M., and Robinson, G.G.C., Seasonal dynamics of epiphytic algal growth in a marsh pond: productivity, standing crop, and community composition, *Can. J. Bot.,* 56, 2434, 1978.

2181.   Hopkins, E.F., Iron ion concentration in relation to growth and other biological processes, *Bot. Gaz.,* 89, 209, 1930.

2182.   Hopkins, E.F., The necessity and function of manganese in the growth of *Chlorella* sp., *Science,* 72, 609, 1930.

2183.   Hopkins, J.T., The role of water in the behavior of an estuarine mudflat diatom, *J. Mar. Biol. Ass. U.K.,* 46, 617, 1966.

2184.   Hopkins, E.F., and Wann, F.B., Iron requirement for *Chlorella, Bot. Gaz.,* 84, 407, 1927.

2185.   Hopkinson, C.S., A comparison of ecosystem dynamics in freshwater wetlands, *Estuaries,* 15, 549, 1992.

2186. Hopkinson, C.S., Gosselink, J.G., and Parrondo, R.T., Aboveground production of seven marsh plant species in coastal Louisiana, *Ecology*, 59, 760, 1978.

2187. Hopkinson, C.S., Gosselink, J.G., and Parrondo, R.T., Production of coastal Louisiana marsh plants calculated from phenometric techniques, *Ecology*, 61, 1091, 1980.

2188. Hoppe, H.A., Marine algae and their products and constituents in pharmacy, in *Marine Algae in Pharmaceutical Science*, Hoppe, H.A., Levring, T., and Tanaka, Y., Eds., Walter de Gruyter, Berlin, 1979, 25.

2189. Hoppe, H.A., Levring, T., and Tanaka, Y., Eds., *Marine Algae in Pharmaceutical Science*, Walter de Gruyter, Berlin, 1979.

2190. Hori, T., and Green, J.C., The ultrastructure of the flagellar root system of *Isochrysis galbana* (Prymnesiophyta), *J. Mar. Biol. Ass. U.K.*, 71, 137, 1991.

2191. Horne, A.J., The ecology of nitrogen fixation on Signy Island, South Orkney Islands, *Br. Antarct. Survey Bull.*, 27, 1, 1972.

2192. Horne, A.J., and Fogg, G.E., Nitrogen fixation in some English lakes, *Proc. R. Soc. Lond.* B 175, 351, 1970.

2193. Horne, A.J., and Viner, A.B., Nitrogen fixation and its significance in tropical Lake George, Uganda, *Nature*, 232, 417, 1971.

2194. Horne, A.J., and Goldman, C.R., Nitrogen fixation in Clear Lake, Calif. I. Seasonal variation and the role of heterocysts, *Limnol. Oceanogr.*, 17, 678, 1972.

2195. Horne, A.J., Fogg, G.E., and Eagle, D.J., Studies in situ of the primary products of an area of inshore Anctartic Sea, *J. Mar. Biol. Ass. U.K.*, 49, 393, 1969.

2196. Horne, A.J., Javornický, P., and Goldman, C.R., A freshwater "red tide" on Clear Lake, California, *Limnol. Oceanogr.*, 16, 684, 1971.

2197. Horne, A.J., Dillard, J.E., Fujita, D.K., and Goldman, C.R., Nitrogen fixation in Clear Lake Calif. II. Synoptic studies on the autumn *Anabaena* bloom, *Limnol. Oceanogr.*, 17, 693, 1972.

2198. Horner, R.A., History of ice algal investigations, in *Sea Ice Biota*, Horner, R.A., Ed., CRC Press, Boca Raton, Florida, 1985, 1.

2199. Horner, R.A., Ecology of sea ice microalgae, in *Sea Ice Biota*, Horner, R.A., Ed., CRC Press, Boca Raton, Florida, 1985, 103.

2200. Horner, R., and Alexander, V., Algal populations in arctic sea ice. An investigation of heterotrophy, *Limnol. Oceanogr.*, 17, 454, 1972.

2201. Horner, R.R., and Welch, E.B., Stream periphyton development in relation to current velocity and nutrients, *Can. J. Fish. Aquat. Sci.*, 38, 449, 1981.

2202. Horner, R.R., Welch, E.B., and Veenstra, R.B., Development of nuisance periphytic algae in laboratory streams in relation to enrichment and velocity, in *Periphyton of Freshwater Ecosystems*, Wetzel, R.G., Ed., Dr. W. Junk Publishers, The Hague, The Netherlands, 1983, 121.

2203. Horner, R.R., Welch, E.B., Seeley, M.R., and Jacoby, J.M., Responses of periphyton to changes in current velocity, suspended sediment and phosphorus concentration, *Freshwater Biol.*, 24, 215, 1990.

2204. Hornsey, I.S., and Hide, D., The production of antimicrobial compounds by British marine algae. 1. Antibiotic-producing marine algae, *Br. Phycol. J.*, 9, 353, 1974.

2205. Horovitz, C.T., Shock, H.H., and Horovitz-Kisimova, L.A., The content of scandium, thorium, silver, and other trace elements in different plant species, *Plant and Soil*, 40, 397, 1974.

2206. Horrigan, S.G., and McCarthy, J.J., Urea uptake by phytoplankton at various stages of nutrient depletion, *J. Plankton Res.*, 3, 403, 1981.

2207. Hoshiai, T., Seasonal change of ice communities in the sea ice near Syowa Station, Antarctica, in *Polar Oceans*, Dunbar, M.J., Ed., Arctic Institute of North America, Montreal, 1977, 307.

2208. Hosiaisluoma, V., On the ecology of *Euglena mutabilis* on peat bogs in Finland, *Ann. Bot. Fennici*, 12, 35, 1975.

2209. Hosseini, S.M., and Van der Valk, A.G., Primary productivity and biomass of periphyton and phytoplankton in flooded freshwater marshes, In *Proc. Symp. Freshwater Wetlands and Wildlife*, Sharitz, R.R., and Gibbons, J.W., Eds., U.S. Department of Energy, 1989, 303.

2210. Hosseini, S., and Van der Valk, A.G., The impact of prolonged above-normal flooding on metaphyton in a freshwater marsh, in *Proc. Symp. Freshwater Wetlands and Wildlife*, Sharitz, R.R., and Gibbons, J.W., Eds., U.S. Department of Energy, 1989, 317.

2211. Hossner, L.R., and Baker, W.H., Phosphorus transformation in flooded soils, in *The Ecology and Management of Wetlands*, Part 1. *Ecology of Wetlands*, Hook, D.D. et al., Eds., Timber Press, Portland, Oregon, 1988, 293.

2212. Hough, L., Jones, J.K.N., and Wadman, W.H., The polysaccharide components of certain freshwater algae, *J. Chem. Soc.*, 3393, 1952.

2213. Hough, R.A., Light and dark respiration and release of organic carbon in marine macrophytes of the Great Barrier Reef region, *Aust. J. Plant Physiol.*, 3, 63, 1976.

2214. Hough, R.A., and Wetzel, R.G., The release of dissolved organic carbon from submersed aquatic macrophytes: Diel, seasonal, and community relationships, *Verh. Internat. Verein. Limnol.,* 19, 939, 1975.

2215. Hovasse, R., and Teissier, G., Sur la position systématique des Xanthelles, *Bull. Soc. Zool.,* 48, 146, 1923.

2216. Howard-Williams, C., Vegetation and environment in the marginal areas of a tropical African lake (L. Chiwa, Malawi), Ph.D. Thesis, University of London, 1973.

2217. Howard-Williams, C., Cycling and retention of nitrogen and phosphorus in wetlands: a theoretical and applied perspective, *Freshwater Biol.,* 15, 391, 1985.

2218. Howard-Williams, C., and Junk, W.J., The decomposition of aquatic macrophytes in the floating meadow of a Central Amazonian Varzea Lake, *Biogeographica,* 7, 115, 1976.

2219. Howard-Williams, C., and Davis, B.R., The rates of dry matter and nutrient loss from decomposing *Potamogeton pectinatus* in a brackish south-temperate coastal lake, *Freshwater Biol.,* 9, 13, 1979.

2220. Howard-Williams, C., and Howard-Williams, W., Nutrient leaching from the swamp vegetation of lake Chilwa, a shallow African lake, *Aquat. Bot.,* 4, 257, 1978.

2221. Howard-Williams, C., Pickmere, S., and Davies, J., Decay rates and nitrogen dynamics of decomposing watercress (*Nasturtium officinale* R.Br.), *Hydrobiologia,* 99, 207, 1983.

2222. Howarth, R.W., and Fisher, S.G., Carbon, nitrogen and phosphorus dynamics during leaf decay in nutrient enriched stream microecosystems, *Freshwater Biol.,* 6, 221, 1976.

2223. Howarth, R.W., and Teal, J.M., Sulfate reduction in a New England salt marsh, *Limnol. Oceanogr.,* 24, 999, 1979.

2224. Howarth, R.W., and Giblin, A.E., Sulfate reduction in the salt marshes of Sapelo Island, Georgia, *Limnol. Oceanogr.,* 24, 999, 1983.

2225. Howarth, R.W., Giblin, A., Gale, J., Peterson, B.J., and Luther, G.W., III., Reduced sulfur compounds in the pore waters of a New England salt marsh, *Ecol. Bull. (Stockholm),* 35, 135, 1983.

2226. Howeler, R.H., and Bouldin, D.R., The diffusion and consumption of oxygen in submerged soils, *Soil Sci. Soc. Am. J.,* 35, 202, 1971.

2227. Howes, B.L., Howarth, R.W., Teal, J.M., and Valiela, I., Oxidation-reduction potentials in a salt marsh: special patterns and interactions with primary production, *Limnol. Oceanogr.,* 26, 350, 1981.

2228. Hoyer, O., Bernhardt, H., Clasen, J., and Wilhelms, A., In situ studies on the exchange between sediments and water using caissons in the Wahnbach reservoir, *Arch. Hydrobiol. Beih. Ergebn. Limnol.,* 18, 79, 1982.

2229. Hsiao, S.I.C., Life history and iodine nutrition of the marine brown alga, *Petalonia fascia* (O. F. Müll.) Kuntze, *Can. J. Bot.,* 74, 1611. 1969.

2230. Hsiao, S.I.C., Quantitative composition, distribution, community structure and standing stock of sea ice microalgae in the Canadian Arctic, *Arctic,* 33, 768, 1980.

2231. Huang, C.P., Leckie, T.H., and Huang, P.M., Adsorption of inorganic phosphorus by lake sediments, *J. Water Pollut. Control Fed.,* 48, 2754, 1976.

2232. Huang, C.P., Elliott, H.A., and Ashmead, R.M., Interfacial reactions and the fate of heavy metals in solid-water systems, *J. Water Pollut. Control Fed.,* 49, 745, 1977.

2233. Huang, P.M., Retention of arsenic by hydroxy-aluminium on surface of micaceous mineral colloids, *Soil Sci. Soc. Am. Proc.,* 39, 271, 1975.

2234. Huang, P.M., Adsorption processes in soils, in *Handbook of Environmental Chemistry,* Hutzinger, O., Ed., Springer-Verlag, Berlin, 1980, 47.

2235. Huang, P.M., and Liaw, W.K., Distribution and fractionation of arsenic in selected freshwater lake sediments, *Int. Revue Ges. Hydrobiol.,* 63, 533, 1978.

2236. Huang, P.M., and Liaw, W.K., Adsorption of arsenite by lake sediments, *Int. Revue Ges. Hydrobiol.,* 64, 263, 1979.

2237. Huang, P.M., Oscarson, D.W., Liaw, W.K., and Hammer, U.T., Dynamics and mechanisms of arsenite oxidation by freshwater lake sediments, *Hydrobiologia,* 91, 315, 1982.

2238. Huber-Pestalozzi, G., Das Phytoplankton des Süsswassers. Systematik und Biologie I, *Die Binnengewässer* 16, 1, 1938.

2239. Huber-Pestalozzi, G., Euglenophyceen, in *Die Binnengewässer Das Phytoplanktondes Süsswassers,* Vol. 16, Thienemann, A., Ed., 1955, 1.

2240. Hudson, C., and Bourget, E., The effect of light on the vertical structure of epibenthic diatom communities, *Bot. Mar.,* 26, 317, 1983.

2241. Hudson, C., Duthie, H.C., and Paul, B., Physiological modifications related to density increase in periphytic assemblages, *J. Phycol.,* 23, 393, 1987.

2242. Hughes, E.H., Estuarine subtidal food webs analyzed with stable carbon isotope ratios, M.Sc. Thesis, University of Georgia, Athens, 1980.

2243. Huguenin, J.E., An examination of problems and potentials for future large-scale intensive seaweed culture systems, *Aquaculture*, 9, 313, 1976.

2244. Hulburt, E.M., The diversity of phytoplanktic populations in oceanic, coastal, and estuarine regions, *J. Mar. Res.*, 21, 81, 1963.

2245. Hulburt, E.M., and Guillard, R.R.L., The relationship of the distribution of the diatom *Skeletonema tropicum* to temperature, *Ecology*, 49, 337, 1968.

2246. Hultgren, A.B.C., Above-ground biomass variation in *Carex rostrata* Stokes in two contrasting habitats in Central Sweden, *Aquat. Bot.*, 34, 341, 1989.

2247. Humphrey, K.P., and Stevenson, R.J., Response of benthic algae to pulses in current and nutrients during simulations of subscouring spates, *J. N. Am. Benthol. Soc.*, 11, 37, 1992.

2248. Hunt, M.E., Floyd, G.L., and Stout, B.B., Soil algae in field and forest environments, *Ecology*, 60, 360, 1979.

2249. Hunter, M.L., Jr., Jones, J.J., and Witman, J.W., Biomass and species richness of aquatic macrophytes in four Maine (U.S.A.) lakes of different acidity, *Aquat. Bot.*, 24, 91, 1986.

2250. Hunter, R.D., and Russell-Hunter, W.D., Bioenergenic and community changes in intertidal aufwuchs grazed by *Littorina littorea*, *Ecology*, 54, 761, 1983.

2251. Huntsinger, K.R.G., and Maslin, P.E., Contribution of phytoplankton, periphyton, and macrophytes to primary production in Eagle lake, California, *Calif. Fish. Game*, 62, 187, 1976.

2252. Huntsman, S.A., Organic extraction by *Dunaliella tertiolecta*, *J. Phycol.*, 8, 59, 1972.

2253. Huntsman, S.A., and Sunda, W.G., The role of trace metals in regulating phytoplankton growth, in *The Physiological Ecology of Phytoplankton*, Morris, I., Ed., University of California, Berkeley, 1980, 285.

2254. Husák, Š., and Hejný, S., Marginal plant communities of the Nesyt fishpond (South Moravia), *Pol. Arch. Hydrobiol. (Warsaw)*, 20, 461, 1973.

2255. Hussey, M.R., Skinner, Q.D., Adams, J.C., and Harvey, A.J., Denitrification and bacterial numbers in riparian soils of a Wyoming mountain watershed, *J. Range Manage.*, 38, 492, 1985.

2256. Hustedt, F., Die Kieselalgen 1, in *Rabenhorsts Kryptogamen-Flora von Deutschland, Österreich und der Schweiz 7*, Leipzig, 1927–1930.

2257. Hutchinson, G.E., *A Treatise on Limnology*. Vol. I. *Geography, Physics, and Chemistry*, John Wiley and Sons, New York, 1957.

2258. Hutchinson, G.E., *A Treatise on Limnology*. Vol. II. *Introduction to Lake Biology and the Limnoplankton*, John Wiley and Sons, New York, 1967.

2259. Hutchinson, G.E., Eutrophication, *Am. Sci.*, 61, 269,1973.

2260. Hutchinson, G.E., *A Treatise on Limnology*, Vol. III. *Limnological Botany*, John Wiley and Sons, New York, 1975.

2261. Hutchinson, T.C., Comparative studies of the toxicity of heavy metals to phytoplankton and their synergistic interactions, *Water Pollut. Res. Canada*, 8, 68, 1973.

2262. Hutchinson, T.C., and Czyrska, H., Heavy metal toxicity and synergism in floating aquatic weeds, *Verh. Internat. Verein. Limnol.*, 19, 2102, 1974.

2263. Hutchinson, T.C., and Stokes, P.M., Heavy metal toxicity and algal bioassays, in *Water Quality Parameters*, American Soc. for Testing and Materials, Philadelphia, Special Technical Publication No. 573, 1975.

2264. Hutner, S.H., Essentiality of constituents of sea water for growth of a marine diatom, *Trans. N.Y. Acad. Sci. Ser. 2*, 10, 136, 1948.

2265. Hutner, S.H., and Provasoli, L., Comparative biochemistry of flagellates, in *Biochemistry and Physiology of Protozoa*, Vol. 2, Hutner, S.H., and Lwoff, A., Eds., Academic Press, New York and London, 1955, 18.

2266. Hutner, S.H., and Provasoli, L., Nutrition of algae, *Ann. Rev. Plant Physiol.*, 15, 37, 1964.

2267. Hutner, S.H., Provasoli, L., Stockstad, E.L.R., Hoffmann, C.E., Belt, M., Franklin, A.L., and Jukes, T.H., Assay of antipernicious anemia factor with *Euglena*, *Proc. Soc. Exptl. Biol. Med.*, 70, 118, 1949.

2268. Hutner, S.H., Provasoli, L., and Filfus, J., Nutrition of some phagotrophic chrysomonads, *Ann. N.Y. Acad. Sci.*, 56, 852, 1953.

2269. Ichioka, P.S., and Arnon, D.I., Molybdenum in relation to nitrogen metabolism. II. Assimilation of ammonia and urea without molybdenum by *Scenedesmus*, *Physiol. Plant.*, 8, 552, 1955.

2270. Idso, S.B., Evapotranspiration from water hyacinth (*Eichhornia crassipes* (Mart. Solms.) in Texas reservoirs, *Water Resour. Bull.*, 15, 1466, 1979.

2271. Igelsrud, I., Thompson, T.G., and Zwicker, B.M.G., The boron content of sea water and marine organisms, *Am. J. Sci.*, 35, 47, 1938.

2272. Iizumi, H., Hattori, A., and McRoy, C.P., Nitrate and nitrite in interstitial waters of eelgrass beds in relation to the rhizosphere, *J. Exp. Mar. Biol. Ecol.*, 47, 191, 1980.

2273. Ikawa, M., Thomas, V.M., Jr., Buckley, L.J., and Uebel, J.J., Sulfur and the toxicity of the red alga *Ceramium rubrum* to *Bacillus subtilis, J. Phycol.,* 9, 302, 1973.

2274. Ikegami, I., Katoh, S., and Takamiya, A., Light-induced changes of b-type cytochromes in the electron transport chain of *Euglena* chloroplasts, *Plant Cell Physiol.,* 11, 777, 1970.

2275. Ikemori, M., Relation of calcium uptake to photosynthetic activity as a factor controlling calcification in marine algae, *Bot. Mag., Tokyo,* 83, 151, 1970.

2276. Ikemori, M., and Nishida, K., Inorganic carbon source and the inhibition of Diamox on the photosynthesis of marine algae—*Ulva pertusa, Ann. Rep. Noto Marine Lab.,* 7, 1, 1966.

2277. Ikusima, I., Ecological studies on the productivity of aquatic plant communities. I. Measurements of photosynthetic activity, *Bot. Mag.,* 78, 202, 1965.

2278. Ilmavirta, K., and Kotimaa, A.-L., Spatial and seasonal variations in phytoplanktonic primary production and biomass in the oligotrophic lake Pääjärvi, southern Finland, *Ann. Bot. Fennici,* 11, 112, 1974.

2279. Ilomets, M., Growth rate and productivity of the *Sphagnum* carpet in southwestern Estonia, *Bot. J.,* 66, 279, 1981 (in Russian).

2280. Imbamba, S.K., Mineral element content of some benthic marine algae of the Kenya Coast, *Bot. Mar,* 15, 113, 1972.

2281. Imhof, G., and Burian, K., Energy flow studies in a wetland ecosystem (Reed belt of the Lake Neusiedler), Special Publ. of Austrian Acad. Sci. for IBP, Springer-Verlag, Vienna, 1972.

2282. Incoll, L.D., Long, S.P., and Ashmore, M.R., SI units in publications in plant science, *Curr. Adv. Plant Sci.,* 28, 331, 1977.

2283. Ingram, H.A.P., Hydrology, in: *Ecosystems of the World. Mires: Swamp, Bog, Fen and Moor,* Vol. 4A *General Studies,* Gore, J.J.P., Ed., Elsevier Science Publishing Company, Amsterdam, 1983, 67.

2284. Ionescu-Teculescu, U., Characeae associations in the flood zone of the Danube, *Rev. Roum. Biol. Ser. Bot.,* 17, 9, 1972.

2285. Ioriya, T., Notes on some species of colorless Euglenophyceae from Hokkaido, Japan, *Bull. Jap. Soc. Phycol.,* 24, 62, 1976.

2286. Ireland, M.P., Variations in the zinc, copper, manganese and lead content of *Balanus balanoides* in Cardigan Bay, Wales, *Environ. Pollut.,* 7, 65, 1974.

2287. Iriki, Y., and Miwa, T., Chemical nature of the cell wall of the green algae, *Codium, Acetabularia* and *Halicoryne, Nature,* 185, 178, 1960.

2288. Irmer, U., Die Wirkung von Blei auf die Grünalge *Chlamydomonas reinhardtii* Dang. in axenischer Kultur, M.Sc. Thesis, University of Hamburg, Hamburg, 1985.

2289. Isenberg, H.D., Lavine, L.S., Mandell, C., and Wiessfellner, H., Qualitative chemical composition of the calcifying organic matrix obtained from cell-free coccoliths, *Nature,* 206, 1153, 1965.

2290. Ishida, Y., Physiological studies on evolution of dimethylsulfide from unicellular marine algae, *Mem. Coll. Agr. Kyoto Univ.,* 94, 47, 1968.

2291. Islam, R.M., and Whitton, B.A., Phosphorus content and phosphatase activity of the deepwater ricefield cyanobacterium (blue-green alga) *Calothrix* D764, *Microbios,* 69, 7, 1992.

2292. Islam, M.R., and Whitton, B.A., Cell composition and nitrogen fixation by the deepwater ricefield cyanobacterium (blu-green alga) *Calothrix* D764, *Microbios,* 69, 77, 1992.

2293. Israelson, G., The freshwater Florideae of Sweden. Studies of their taxonomy, ecology and distribution, *Symb. Bot. Ups.,* 6, 1, 1942.

2294. IUCN, The Ramsar Conference: Final Act of the International Conference on the Conservation of Wetlands and Waterfowl, Spec. Suppl., *IUNC Bull.,* 2, 1, 1971.

2295. Ivanov, M.V., The global biogeochemical sulfur cycle, in *Some Perspectives of the Major Biogeochemical Cycles,* Likens, G.E., Ed., John Wiley and Sons, New York, 1981, 61.

2296. Ivanov, M.V., and Freney, J.R., *The Global Biogeochemical Sulphur Cycle,* John Wiley and Sons, Chichester, 1983.

2297. Iyengar, M.O.P., and Desikachary, T.V., *Volvocales,* Indian Council of Agriculture Res., New Delhi, 1981.

2298. Izaguirre, G., Hwang, C.J., Kranswer, S.W., and McGuire, M.J., Geosmin and 2-methylisoborneol from cyanobacteria in three water supply systems, *Appl. Environ. Microbiol.,* 43, 708, 1982.

2299. Jackson, D.F., Ed., *Algae and Man,* Plenum Press, New York, 1964.

2300. Jackson, D.F., Ed., *Algae, Man and the Environment,* Syracuse University Press, Syracuse, 1968.

2301. Jackson, G.A., Nutrients and production of giant kelp, *Macrocystis pyrifera,* off Southern California, *Limnol. Oceanogr,* 22, 979, 1977.

2302. Jackson, G.A., and Morgan, J.J., Trace metal-chelator interactions and phytoplankton growth in seawater media: Theoretical analysis and comparison with reported observations, *Limnol. Oceanogr,* 23, 268, 1978.

2303. Jackson, G., Marine biomass production through seaweed aquaculture, in *Biochemical and Photosynthetic Aspects of Energy Production,* San Pietro, A., Ed., Academic Press, New York, 1980, 31.

2304. Jackson, J.E., Jr., and Castenholz, R.W., Fidelity of thermophilic blue-green algae to hot spring habitats, *Limnol. Oceanogr.,* 20, 305, 1975.

2305. Jackson, M.B., Water quality in the Bay of Quinte prior to phosphorus removal at sewage treatment plant, Ontario Ministry of Environment, 1976.

2306. Jackson, S.G., and McCandless, E.L., The effect of sulphate concentration on the uptake and incorporation of [^{35}S] sulphate in *Chondrus crispus, Can. J. Bot.,* 60, 162, 1982.

2307. Jackson, W.A., and Wolk, R.J., Photorespiration, *Ann. Rev. Plant Physiol.,* 21, 385, 1970.

2308. Jacobsen, O.S., Peatlands, in *Wetlands and Shallow Continental Water Bodies,* Vol. 1, Patten, B.C., Ed., SPB Academic Publishing, The Hague, The Netherlands, 1990, 467.

2309. Jacoby, J.M., Grazing effects on periphyton by *Theodoxus fluviatilis* (Gastropoda) in a lowland stream, *J. Freshwater Ecol.,* 3, 265, 1985,

2310. Jacques, A.G., and Osterhout, W.J., The accumulation of electrolytes XI. Accumulation of nitrate by *Valonia* and *Halicystis, J. Gen. Physiol.,* 21, 767, 1938.

2311. Jaenicke, L., Vitamin and coenzyme function: vitamin B$_{12}$ and folic acid, *Ann. Rev. Biochem.,* 33, 287, 1964.

2312. Jain, S.L., Observations on the primary productivity and energetics of the macrophytic vegetation of Gordhan Vilas tank, Udajpur (south Rajasthan), Ph.D. Thesis, University of Udajpur, Udajpur, 1978 (cited by Vas ea 90).

2313. Jakrlová, J., Mineral nutrient uptake by four inundated meadow communities, in *Ecosystem Study on Grassland Biome in Czechoslovakia,* Rychnovská, M., Ed., Czechosl. IBP/PT-PP Report No. 2, Brno, 1972, 51.

2314. Jarosch, R., Gliding, in *Physiology and Biochemistry of Algae,* Lewin, R.A., Ed., Academic Press, New York and London, 1962, 573.

2315. Jauregui, M.A., and Reisenauer, H.M., Dissolution of oxides of manganese and iron by root exudate components, *Soil Sci. Soc. Am. J.,* 46, 314, 1982.

2316. Javornický, P., Light as the main factor limiting the development of diatoms in Slapy Reservoir, *Verh. Internat. Verein. Limnol.,* 16, 701, 1966.

2317. Javornický, P., On the utilization of light by fresh-water phytoplankton, *Arch. Hydrobiol. Suppl. (Algological Studies,* 2/3) 39, 68, 1970.

2318. Jeanjean, R., Phosphorus uptake in *Chlorella pyrenoidosa.* II. Effect of pH and of SH reagents, *Biochemie,* 57, 1229, 1975.

2319. Jeanjean, R., The effect of metabolic poisons on ATP level and on active phosphate uptake in *Chlorella pyrenoidosa, Physiol. Plant.,* 37, 107, 1976.

2320. Jeanjean, R., and Blasco, F., Influence des ions arséniate sur l'absorption des ions phosphate par les Chlorelles, *C.R. Acad. Sci. Ser. D.,* 270, 1897, 1970.

2321. Jeanjean, R., Blasco, F., and Gaudin, C., Etude des mécanismes d'absorption de l'ion phosphate par les Chlorelles, *C.R. Hebd. Seances Acad. Sci. Ser. D Sci. Nat.,* 270, 2946, 1970.

2322. Jeffrey, S.W., Properties of two spectrally different components in chlorophyll *c* preparations, *Biochim. Biophys. Acta,* 177, 456, 1969.

2323. Jeffrey, S.W., Preparation and some properties of crystalline chlorophyll *c*$_1$ and *c*$_2$ from marine algae, *Biochem. Biophys. Acta,* 279, 15, 1972.

2324. Jeffrey, S.W., The occurrence of chlorophyll *c*$_1$ and *c*$_2$ in algae, *J. Phycol.,* 12, 349, 1976.

2325. Jeglum, J.K., Boissonneau, A.N., and Haavisto, V.F., Toward a Wetland Classification for Ontario, Can. For. Serv., Sault Ste. Marie, Ont. Inf. Rep. O-X-215, 1974.

2326. Jenkin, P.M., Oxygen production by the diatom *Coscinodiscus excentricus* Ehr. in relation to submarine illumination in the English Channel, *J. Mar. Biol. Ass. U.K.,* 22, 301, 1937.

2327. Jenkins, M.C., and Kemp, W.M., The coupling of nitrification and denitrification in two estuarine sediments, *Limnol. Oceanogr,* 29, 609, 1984.

2328. Jenne, E.A., Atmospheric and fluvial transport of mercury, in *Mercury in the Environment,* Geol. Survey Prof. Paper 713, U.S. Government Printing Office, Washington, DC, 1970, 40.

2329. Jenne, E.A., Mercury in waters of the Western United States, in *Mercury in the Western Environment,* Buhler, D.R., Ed., A Continuing Education Book, Corvallis, Oregon, 1973, 16.

2330. Jenny, H., Derivation of the State Factor Equations of soil and ecosystems, *Soil Sci. Soc. Am. Proc.,* 25, 385, 1961.

2331. Jenny, H., Gessel, S.P., and Bingham, F.T., Comparative study of decomposition of organic matter in temperate and tropical regions, *Soil Sci.,* 68, 419, 1949.

2332. Jensen, A., Rystad, B., and Melsom, S., Heavy metal tolerance of marine phytoplankton. I. The tolerance of three algal species to zinc in coastal sea waters, *J. Exp. Mar. Biol. Ecol.,* 15, 145, 1974.

2333. Jensen, R.G., and Bahr, J.T., Ribulose 1,5-biphosphate carboxylase-oxygenase, *Ann. Rev. Plant Physiol.*, 28, 379, 1977.

2334. Jensen, S., and Jernelöv, A., Biological methylation of mercury in aquatic organisms, *Nature*, 223, 753, 1969.

2335. Jensen, T.E., and Sicko, L.M., Phosphate metabolism in blue-green algae. I. Fine structure of the "polyphosphate overplus" phenomenon in *Plectonemaboryanum, Can. J. Microbiol.*, 20, 1235, 1974.

2336. Jensen, T.E., Baxter, M., Rachlin, J.W., and Jani, V., Uptake of heavy metals by *Plectonema boryanum* (Cyanophyceae) into cellular components, especially polyphosphate bodies: an X-ray energy dispersive study, *Environ. Pollut., Ser. A*, 27, 119, 1982.

2337. Jensen, T.E., Rachlin, J.W., Jani, V., and Warkentine, B., An X-ray energy dispersive study of cellular compartmentalization of lead and zinc in *Chlorella saccharophila* (Chlorophyta), *Navicula incerta* and *Nitzschia closterium* (Bacillariophyta), *Environ. Exp. Bot.*, 22, 319, 1982.

2338. Jensen, T.E., Sicko, L., Baxter, M., Baxter, M., Warkentine, B., and Jani, V., Heavy metals compartmentalization by algal cells, in *Proc. 4th Annual Meeting of the Electron Microscopy Soc. Am.*, Bailey, G.W., Ed., San Francisco Press, Inc., San Francisco, California, 1984, 294.

2339. Jernelöv, A., Release of methyl mercury from sediments with layers containing inorganic mercury at different depths, *Limnol. Oceanogr.*, 15, 958, 1970.

2340. Jernelöv, A., and Lann, H., Mercury accumulation in food chains, *Oikos*, 22, 403, 1971.

2341. Jervis, R.A., Primary production in the freshwater marsh ecosystem of Troy Meadows, New Jersey, *Bull. Torrey Bot. Club.*, 96, 209, 1969.

2342. Jewell, W.J., and Kulasooriya, S.A., The relation of acetylene reduction to heterocyst frequency in blue-green algae, *J. Exp. Bot.*, 21, 874, 1970.

2343. Johansen, H.W., Morphology and systematics of coralline algae with special reference to *Calliarthron, Univ. Calif. Publ. Bot.*, 35, 165, 1963.

2344. Johansen, J.R., Cryptogamic crusts of semiarid and arid lands of North America, *J. Phycol.*, 29, 140, 1993.

2345. John, C.D., and Greenway, H., Aerobic fermentation and activity of some enzyme in rice roots under anaerobiosis, *Aust. J. Plant Physiol.*, 3, 325, 1976.

2346. Johnsen, G., and Sakshaug, E., Bio-optical characteristics and photoadaptive responses in the toxic and bloom-forming dinoflagellates *Gyrodinium aureolum, Gymnodinium galatheanum*, and two strains of *Prorocentrum minimum, J. Phycol.*, 29, 627, 1993.

2347. Johnson, D.W., and Cole, D.W., Anion mobility in soils: relevance to nutrient transport from forest ecosystems, *Environ. Int.*, 3, 79, 1980.

2348. Johnson, M., Preliminary report on species composition, chemical composition, biomass, and production of marsh vegetation in the upper Patuxent Estuary, Maryland, Chesapeake Biol. Lab., Reference No. 70-130, Solomons, Maryland, 1970.

2349. Johnson, S.W., *Essays on peat, muck and commercial manures*, Brown and Gross, Hartford, 1859.

2350. Johnston, A.M., The acquisition of inorganic carbon by marine macroalgae, *Can. J. Bot.*, 69, 1123, 1991.

2351. Johnston, C.A., Sediment and nutrient retention by freshwater wetlands: effects on surface water quality, *CRC Crit. Rev. Environ. Control*, 21, 491, 1991.

2352. Johnston, C.S., The ecological distribution and primary production in the eastern Canaries, *Int. Rev. Ges. Hydrobiol.*, 54, 473, 1969.

2353. Johnston, C.S., Jones, R.G., and Hunt, R.T., A seasonal carbon budget for a laminarian population in a Scottish sea loch, *Helgol. Wiss. Meeresunters.*, 30, 527, 1977.

2354. Johnston, R., Sea water, the natural medium of phytoplankton. II. Trace metals and chelation, and general discussion, *J. Mar. Biol. Ass. U.K.*, 44, 87, 1964.

2355. Joint, I., Microbial production of an estuarine mudflat, *Estuar. Coast. Shelf Sci.*, 7, 185, 1978.

2356. Joliffe (Thomas), E.A., and Tregunna, E.B., Studies on HC_3^- ion uptake during photosynthesis in benthic marine algae, *Phycologia*, 9, 293, 1970.

2357. Jolley, E.T., and Helleburst, J.A., Preliminary studies on the nutrition of *Naviculapelliculosa*(Bréb.) Hilse, and an associated bacterium *Flavobacterium* sp., *J. Phycol.* 10 (Suppl.), 7, 1974.

2358. Jónasson, P.M., and Mathiesen, H., Measurements of primary production in two Danish eutrophic lakes. Esrom Sø and Furesø, *Oikos*, 10, 137, 1959.

2359. Jones, A.J., The arsenic content of some of the marine algae, *Pharm. J.*, 109, 86, 104, 1922.

2360. Jones, H.E., Comparative studies of plant growth and distribution in relation to waterlogging. II. An experimental study of the relationship between transpiration and the uptake of iron in *Erica cinerea* L. and *E. tetralix, J. Ecol.*, 59, 167, 1971.

2361. Jones, J.G., Iron transformations by freshwater bacteria, *Adv. Microb. Ecol.*, 9, 149, 1986.

2362. Jones, J.G., and Simon, B.M., Interactions of acetogens and methanogens in anaerobic freshwater sediments, *Appl. Environ. Microbiol.*, 49, 944, 1985.

2363. Jones, J.K.N., The structure of the mannan present in *Porphyra umbilicalis, J. Chem. Soc.*, p. 3229, 1950.

2364. Jones, J.R., and Bachmann, R.W., Prediction of phosphorus and chlorophyll levels in lakes, *J. Water Pollut. Control Fed.*, 48, 2175, 1976.

2365. Jones, K., and Stewart, W.D.P., Nitrogen turnover in marine and brackish habitats. III. The production of extracellular nitrogen by *Calothrix scopulorum, J. Mar. Biol. Asso U.K.*, 49, 475,

2366. Jones, R.A., and Lee, G.F., Application of U.S. OECD eutrophication study results to deep lakes, *Prog. Wat. Tech.*, 12, 81, 1980.

2367. Jones, R.F., Extracellular mucilage of the red alga *Porphyridium cruentum, J. Cell. Comp. Physiol.*, 60, 61, 1962.

2368. Jones, R.F., Speer, H.L., and Kury, W., Studies on the growth of the red alga *Porphyridium cruentum, Physiol. Plant.*, 16, 636, 1963.

2369. Jørgensen, B.B., Ecology of the bacteria of the sulphur cycle with special reference to anoxic-oxic interface environments, *Phil. Trans. R. Soc. Lond.*, B 298, 543, 1982.

2370. Jørgensen, B.B., and Marais, D.J.P., Optical properties of benthic photosynthetic communities: Fiber-optic studies of cyanobacterial mats, *Limnol. Oceanogr*, 33, 99, 1988.

2371. Jørgensen, C.B., *Biology of the Suspension Feeding*, Pergamon Press, Oxford, 1966.

2372. Jørgensen, E.G., Effects of different silicon concentrations on the growth of diatoms, *Physiol. Plant.*, 5, 161, 1952.

2373. Jørgensen, E.G., Solubility of the silica in diatoms, *Physiol. Plant.*, 8, 846, 1955.

2374. Jørgensen, E.G., Growth inhibitory substances formed by algae, *Physiol. Plant.*, 9, 712, 1956.

2375. Jørgensen, E.G., Diatom periodicity and silicon assimilation. Experimental and ecological investigations, *Dansk. Bot. Ark.*, 18, 6, 1957.

2376. Jørgensen, E.G., Antibiotic substances from cells and culture solutions of unicellular algae with special reference to some chlorophyll derivatives, *Physiol. Plant.*, 15, 530, 1962.

2377. Jørgensen, E.G., Adaptation to different light intensities in the diatom *Cyclotella meneghiniana* Kütz., *Physiol. Plant.*, 17, 136, 1964.

2378. Jørgensen, E.G., The adaptation of plankton algae. II. Aspects of the temperature adaptation on *Skeletonema costatum, Physiol. Plant.*, 21, 423, 1968.

2379. Joshi, G.V., Studies in photosynthesis in marine plants of Bombay, in *Proc. Seminar Sea, Salt, and Plants,* Krishnamurthy, V., Ed., Catholic Press, Ranchi, India, 1967, 256.

2380. Joubert, G., A bioassay application for quantitative toxicity measurements using the green algae *Selenastrum capricornutum, Water Res.*, 14, 1759, 1980.

2381. Joubert, G., Detailed method for quantitative toxicity measurements using the green algae *Selenastrum capricornutum*, in *Aquatic Toxicology,* Nriagu, J.O., Ed., John Wiley and Sons, New York, 1983, 467.

2382. Joubert, J.J., and Rijkenberg, F.H.J., Parasitic green algae, *Ann. Rev. Phytopathol.*, 9, 45, 1971.

2383. Judson, S., Erosion of the land, *Am. Sci.*, 56, 356, 1968.

2384. Junge, C.E., *Air Chemistry and Radioactivity*, Academic Press, New York and London, 1963.

2385. Junk, W.J., Ecology of swamps on the middle Amazon, in *Ecosystems of the World. Mires: Swamp, Bog, Fen and Moor;* Vol. 4B *Regional Studies,* Gore, A.J.P., Ed., Elsevier Science Publishers, Amsterdam, 1983, 269.

2386. Junk, W.J., and Welcome, R.L., Floodplains, in *Wetlands and Shallow Continental Water Bodies*, Patten, B.C., Ed., SPB Academic Publishing, The Hague, The Netherlands, 1990, 491.

2387. Jupp, B.P., and Drew, E.A., Studies on the growth of *Laminaria hyperborea* (Grunn.) Fosl. I. Biomass and productivity, *J. Exp. Mar. Biol. Ecol.*, 15, 185, 1974.

2388. Jupp, B.P., and Spence, D.H.N., Limitations on macrophytes in a eutrophic lake, Loch Leven. 1. Effect of phytoplankton, *J. Ecol.*, 65, 175, 1977.

2389. Jupp, B.P., Spence, D.H.N., and Britton, R.H., The distribution and production of submerged macrophytes in Loch Leven, Kinross, *Proc. R. Soc. Edinb.*, 74, 195, 1972–1973.

2390. Jurgensen, M.F., Relationship between nonsymbiotic nitrogen fixation and soil nutrient status—a review, *J. Soil Sci.*, 25, 512, 1973.

2391. Jurgensen, M.F., and Davey, C.B., Nitrogen-fixing blue-green algae in acid forest and nursery soils, *Can. J. Microbiol.*, 14, 1179, 1968.

2392. Jussieu, A.L., *Genera plantarum secundum ordines naturales disposita*, Paris, 1789.

2293. Juttner, F., Nor-carotenoids as the major volatile excretion products of *Cyanidium, Z. Naturforsch.*, 34, 186, 1978.

2394. Juttner, F., The algal excretion product, geranylacetone: a protein inhibitor of carotene biosynthesis in *Synechococcus, Z. Naturforsch.*, 34, 957, 1979.

2395. Kabata-Pendias, A., and Pendias, H., *Trace Elements in Soils and Plants*, 2nd ed., CRC Press, Boca Raton, Florida, 1992.

2396. Kadlec, J.A., Nutrient dynamics in wetlands, in *Proc. Conf. Aquatic Plants for Water Treatment and Resource Recovery* Reddy, K.R., and Smith, W.H., Eds., Magnolia Publishing, Orlando, Florida, 1987, 393.

2397. Kadlec, J.A., Effects of deep flooding and drawdown on freshwater marsh sediments, in *Proc. Symp. Freshwater Wetlands and Wildlife,* Sharitz, R.R., and Gibbons, J.W., Eds., U.S. Department of Energy, 1989, 127.

2398. Kadlec, J.A., Effect of depth of flooding on summer water budgets for small diked marshes, *Wetlands*, 13, 1, 1993.

2399. Kadlec, R.H., The hydrodynamics of wetland water treatment systems, in *Proc. Conf. Aquatic Plants for Water Treatment and Resource Recovery* Reddy, K.R., and Smith, W.H., Eds., Magnolia Publishing, Orlando, Florida, 1987, 373.

2400. Kadlec, R.H., Hydrologic factors in wetland water treatment, in *Proc. Int. Conf. Constructed Wetlands for Wastewater Treatment,* Hammer, D.A., Ed., Lewis Publishers, Chelsea, Michigan, 1989, 21.

2401. Kadlec, R.H., Williams, R.B., and Scheffe, R.D., Wetland evapotranspiration in temperate and arid climates, in *Ecology and Management of Wetlands,* Part 1. *Ecology of Wetlands*, Hook, D.D. et al., Eds., Timber Press, Portland, Oregon,1988, 146.

2402. Kadlubowska, J.Z., *Chlorophyta VIII. Conjugatophyceae I. Zygnemales*, Süsswasserflora von Mittelueropa 16, Gustav Fusher Verlag, Stuutgart, New York, 1984.

2403. Kadota, H., and Ishida, Y., Evolution of volatile sulfur compounds from unicellular marine algae, *Bull. Misaki Mar. Biol. Inst., Kyoto Univ.,* 12, 35, 1968.

2404. Kairesalo, T., Measurements of production of epilithiphyton and littoral plankton in Lake Pääjärvi, southern Finland, *Ann. Bot. Fennici,* 13, 114, 1976.

2405. Kairesalo, T., On the production ecology of epipelic algae and littotal plankton in Lake Pääjärvi (southern Finland), *Ann. Bot. Fennici,* 14, 82, 1977.

2406. Kairesalo, T., Comparison of in situ photosynthetic activity of epiphytic, epibenthic and planktonic algal communities in an oligotrophic lake, southern Finland, *J. Phycol.,* 16, 57, 1980.

2407. Kairesalo, T., Gunnarson, K., Jónsson, G.S., and Jónasson, P.M., The occurrence and photosynthetic activity of epiphytes on the tips of *Nitella opacea* Ag. (Charophyceae), *Aquat. Bot.,* 28, 333, 1987.

2408. Kajak, Z., Hillbricht-Ilkowska, A., and Pieczynska, E., The production processes in several Polish Lakes, in *Productivity Problems of Freshwaters,* Kajak, Z., and Hillbricht-Ilkowska, A., Eds., Polish Academy of Sciences, Kraków, 1972, 129.

2409. Kalbe, L., and Tiess, D., Entenmassensterben durch *Nodularia*—Wasserblüte am Kleinen Jasmunder Bodden auf Rügen, *Arch. Exper. Vet.-med.,* 18, 535, 1964.

2410. Kalff, J., and Welch, H.E., Phytoplankton production in Char Lake, a natural polar lake, Cornwallis Is., Northwestern Territories, *J. Fish. Res. Board Can.,* 31, 621, 1974.

2411. Kalff, J., Welch, H.E., and Holmgren, S.K., Pigment cycles in two high-arctic Canadian lakes, *Verh. Internat. Verein. Limnol.,* 18, 250, 1972.

2412. Kallas, T., and Castenholz, R.W., Rapid, transient growth in a blue-green alga at low pH, *J. Phycol. Suppl.,* 16, 22, 1980.

2413. Kamin, H., and Stein Privalle, L., Nitrite reductase, in *Inorganic Nitrogen Metabolism*, Ullrich, W.R., Aparicio, P.J., Syrett, P.J., and Castillo, F., Eds., Springer-Verlag, Berlin, 1987, 112.

2414. Kamp-Nielsen, L., The effect of deleterious concentrations of mercury on the photosynthesis and growth of *Chlorella pyrenoidosa, Physiol. Plant.,* 24, 556, 1971.

2415. Kanai, R., Aoki, S., and Miyachi, S., Quantitative separation of inorganic polyphosphates in *Chlorella* cells, *Plant Cell Physiol.,* 6, 467, 1965.

2416. Kanazawa, A., Vitamins in algae, *Bull. Jap. Sci. Soc. Fish.,* 29, 713, 1963.

2417. Kanazawa, T., and Kanazawa, K., Specific inhibitory effect of copper on cellular division in *Chlorella, Plant Cell Physiol.,* 10, 485, 1969.

2418. Kanazawa, T., Kirk, M.R., and Bassham, J.A., Regulatory effects of ammonia on carbon metabolism in photosynthesizing *Chlorella pyrenoidosa, Biochim. Biophys. Acta,* 205, 401, 1970.

2419. Kanazawa, T., Kanazawa, K., Kirk, M.R., and Bassham, J.A., Regulation of photosynthetic carbon metabolism in synchronously growing *Chlorella pyrenoidosa, Plant Cell Physiol.,* 11, 149, 1970.

2420. Kanazawa, T., Kanazawa, K., Kirk, M.R., and Bassham, J.A., Difference in nitrate reduction in "light" and "dark" stages of synchronously grown *Chlorella pyrenoidosa* and resultant metabolic changes, *Plant Cell Physiol.,* 11, 445, 1970.

2421. Kansanen, A., Niemi, R., and Överlund, K., Pääjärven makrofyytit (Summ.:Macrophytes), *Luonnon Tutkija,* 78, 111, 194, 1974.

2422. Kanwisher, J.W., Photosynthesis and respiration in some seaweeds, in *Some Contemporary Studies in Marine Science,* Barnes, H., Ed., George Allen and Unwin, Ltd., London, 1966, 407.

2423. Kaplan, A., Adaptation to CO_2 levels: induction and the mechanism for inorganic carbon uptake, in *Inorganic Carbon Uptake by Aquatic Photosynthetic Organisms*, Lucas, W.J., and Berry, J.A., Eds., Am. Soc. Plant Physiol., Rockville, Maryland, 1985, 325.

2424. Kaplan, A., Badger, M.R., and Berry, J.A., Photosynthesis and the intracellular inorganic carbon pool in the blue-green algae *Anabaena variabilis*: response to external CO_2 concentration, *Planta*, 149, 219, 1980.

2425. Kaplan, A., Zenvirth, D., Reinhold, L., and Berry, J.A., Involvement of a primary electrogenic pump in the mechanism for bicarbonate uptake by the cyanobacterium *Anabaena variabilis, Plant Physiol.*, 69, 978, 1982.

2426. Kaplan, A., Volokita, M., Zenvirth, D., and Reinhold, L., An essential role for sodium in the bicarbonate transporting system of the cyanobacterium *Anabaena variabilis, FEBS Lett.*, 176, 166, 1984.

2427. Kaplan, W., Prototothecosis and infections caused by morphologically similar green algae, *Pan Am. Health Organ. Sci. Publ.*, 356, 218, 1978.

2428. Kaplan, W., Valiela, I., and Teal, J.M., Denitrification in a salt marsh ecosystem, *Limnol. Oceanogr*, 24, 726, 1979.

2429. Kapp, R., Stevens, S.E., and Fox, J.L., A survey of available nitrogen source for the growth of the blue-green alga, *Agmenellum qaudruplicatum, Arch. Microbiol.*, 104, 135, 1975.

2430. Karali, E.F., Iron, growth and chlorophyll synthesis in *Euglena, Diss. Abstr.*, 24, 1675, 1963.

2431. Karali, E.F., and Price, C.A., Iron, porphyrins and chlorophyl, *Nature*, 198, 708, 1963.

2432. Karpati, I., and Varga, G., Forschungsergebnisse der Laichkrautvegetation in der Balatonbaucht von Keszthely, 1970, *Mitt. Hochsch. Landwirtsch. Keszthely* 12, 1, 1970.

2433. Kashimura, T., and Tachibana, H., The vegetation of the Ozegahara moor and its conservation, in *Ozagahara—Scientific Research of the Highmoor in Central Japan*, Tokyo, 1982, 193.

2434. Kashiwagi, M., Mynderse, J.S., Moore, R.E., and Norton, T.R., Antineoplastic evaluation of Pacific Basin marine algae, *J. Pharm. Sci.*, 69, 734, 1980.

2435. Kates, J.R., and Jones, R.F., Variation in alanine dehydrogenase and glutamine dehydrogenase during the synchronous development of *Chlamydomonas, Biochim. Biophys Acta*, 86, 438, 1964.

2436. Kates, M., and Volcani, B.E., Studies on the biochemistry and fine structure of silica shell formation in diatoms. Lipid components of the cell walls, *Z. Pflanzenphysiol.*, 60, 19, 1968.

2437. Katoh, S., A new copper protein from *Chlorella ellipsoidea, Nature*, 186, 533, 1960.

2438. Kaufman, L.H., Stream aufwuchs accumulation processes: effects of ecosystem depopulation, *Hydrobiologia*, 70, 75, 1980.

2439. Kaul, S., Trisal, C.L., and Kaul, V., Mineral composition of some wetland components in Kashmir, in *Proc. Int. Conf. Wetlands: Ecology and Management*, Part II., Gopal, B., Turner, R.E., Wetzel, R.G., and Whigham, D.F., Eds., Natl. Inst. of Ecology and Internat. Scientific Publications, Jaipur, India, 1982, 83.

2440. Kaul, V., Community architecture biomass and production in some typical wetlands of Kashmir, *Indian J. Ecol.*, 9, 320, 1982.

2441. Kaul, V., and Bakaya, U., The noxious floating lemnid-*Salvinia* aquatic weed complex in Kashmir, in *Aquatic Weeds in South East Asia*, Varshney, C.K., and Rzoska, J., Eds., Dr. W. Junk, The Hague, The Netherlands, 1976, 188.

2442. Kaul, V., Zutshi, D.P., and Vass, K.K., Biomass production of Srinagar lakes, in *Tropical Ecology Emphasizing Organic Production*, Golley, P.M., and Golley, F.B., Eds., University of Georgia, Athens, 1972, 295.

2443. Kaul, V., Trisal, C.L., and Handoo, J.K., Distribution and production of macrophytes in some water bodies in Kashmir, in *Glimpses of Ecology*, Singh, J.S., and Gopal, B., Eds., Intl. Sci. Publications, Jaipur, India, 1978, 310.

2444. Kaul, V., Trisal, C.L., and Kaul, S., Mineral removal potential of some macrophytes in two lakes of Kashmir, *J. Indian Bot. Soc.*, 59, 108, 1980.

2445. Kautsky, L., Primary production and uptake kinetics of ammonium and phosphate by *Enteromorpha compressa* in an ammonium sulfate industry outlet area, *Aquat. Bot.*, 12, 23, 1982.

2446. Kautsky, L., Life cycles of three populations of *Potamogeton pectinatus* L. at different degrees of wave exposure in the Askö area, northern Baltic proper, *Aquat. Bot.*, 27, 177, 1987.

2447. Kawachi, M., Inouye, I., Maeda, O., and Chihara, M., The haptonema as a food-capturing device: observations on *Chrysochromulina hirta* (Prymnesiophyceae), *Phycologia*, 30, 563, 1991.

2448. Kawase, M., and Whitmoyer, R.E., Aerenchyme development in watterloged plants, *Am. J. Bot.*, 67, 18, 1980.

2449. Kayser, H., Waste-water assay with continuous algal cultures: the effect of mercuric acetate on the growth of some marine dinoflagellate, *Mar. Biol.*, 36, 61, 1976.

2450. Keating, K.I., Algal metabolite influence on bloom sequence in eutrophied freshwater ponds, U.S. EPA 600/3-76-081, Washington, DC, 1976.
2451. Keck, K., and Stich, H., The widespread occurrence of polyphosphate in lower plants, *Ann. Bot. N.S.*, 21, 611, 1957.
2452. Keenan, J.D., Bicarbonate utilization in *Anabaena, Physiol. Plant.*, 34, 157, 1975.
2453. Keeney, W.L., Breck, W.G., Vanloon, G.W., and Page, J.A., The dermination of trace metals in *Cladophora glomerata.—C. glomerata* as a potential biological monitor., *Water Res.*, 10, 981, 1976.
2454. Kehde, P.M., The effect of grazing by snails on community structure of periphyton in laboratory streams, M.Sc. Thesis, Oklahoma State University, Stillwater, Oklahoma, 1970.
2455. Keilin, D., and Mann, T., Activity of purified carbonic anhydrase, *Nature*, 153, 107, 1944.
2456. Keller, M., Guillard, R.R.L., Provasoli, L., and Pintner, I.J., Nutrition of some marine ultraplankton clones from the Sarsasso Sea, *Eos*, 65, 898, 1984.
2457. Kellner, K., Die Adaptation von *Ankistrodesmus braunii* an rubidium und kupfer, *Biol. Z.*, 74, 662, 1955.
2458. Kelly, M.G., and Whitton, B.A., Relationship between accumulation and toxicity of zinc in *Stigeoclonium* (Chaetophorales, Chlorophyta), *Phycologia*, 24, 512, 1989.
2459. Kemp, W.M., Wetzel, R.G., Boynton, W.R., D'Elia, C.F., and Stevenson, J.C., Nitrogen cycling in estuarine sediments: Some concepts and research directions, in *Estuarine Comparisons*, Kennedy, V.S., Ed., Academic Press, New York, 1982, 209.
2460. Kempner, E.S., and Miller, J.H., The molecular biology of *Euglena gracilis*. III. General carbon metabolism, *Biochemistry*, 4, 2735, 1965.
2461. Kennedy, G.Y., and Collier, R., The sulpholipids of plants. II. The isolation of sulpholipid from green and brown algae, *J. Mar. Biol. Ass. U.K.*, 43, 613, 1963.
2462. Kentula, M.E., Brooks, R.P., Gwin, S.E., Holland, C.C., Sherman, A.D., and Sifneos, J.C., *An Approach to Improving Decision Making in Wetland Restoration and Creation*, Edited by A.J. Hairston, U.S. EPA, Env. Res. Lab. Corvallis, Oregon, Island Press, Washington, DC, Covelo, California, 1992.
2463. Kerby, N.W., and Raven, J.A., Transport and fixation of inorganic carbon by marine algae, *Adv. Bot. Res.*, 111, 71, 1985.
2464. Keskitalo, J., Phytoplankton pigment concentrations in the eutrophicated lake Lovojärvi, South Finland, *Ann. Bot. Fennici*, 13, 27, 1976.
2465. Kesler, D.H., Periphyton grazing by *Amnicola limosa*: an enclosure-exclosure experiment, *J. Freshwater Ecol.*, 1, 51, 1981.
2466. Kesler, D.H., Periphyton phosphorus concentrations in a small New England lake, *J. Freshwater Ecol.*, 1, 507, 1982.
2467. Kessler, E., On the role of manganese in the oxygen evolving system of photosynthesis, *Arch. Biochem. Biophys.*, 59, 527, 1955.
2468. Kessler, E., Reduction of nitrate with molecular hydrogen in algae containing hydrogenase, *Arch. Biochem. Biophys.*, 62, 241, 1956.
2469. Kessler, E., Stoffwechselphysiologische Untersuchungen an Hydrogenase enhaltenden Grünalgen. II. Dunkel-Reduktion von Nitrat und Nitrit mit molekularem Wasserdtof, *Arch. Mikrobiol.*, 27, 166, 1957.
2470. Kessler, E., Reduction of nitrate by green algae, *Symp. Soc. Exptl. Biol.*, 13, 87, 1959.
2471. Kessler, E., Biochemische Variabilität der Photosynthese: Photoreduktion und verwandte Photosynthesetypen, in *Encyclopedia of Plant Physiology*. Vol. 5. *The Assimilation of Carbon Dioxide*, Part 1., Ruhland, W., Ed., Springer-Verlag, Berlin, 1960, 951.
2472. Kessler, E., Nitrate assimilation by plants, *Ann. Rev. Plant Physiol.*, 15, 57, 1964.
2473. Kessler, E., Hydrogenase, photoreduction and anaerobic growth, in *Algal Physiology and Biochemistry* Stewart, W.D.P., Ed., Blackwell Scientific Publications, Oxford, 1974, 456.
2474. Kessler, E., and Zumft, W.G., Effect of nitrite and nitrate on chlorophyll fluorescence in green algae, *Planta* 111:41, 1973.
2475. Kessler, J.O., The external dynamics of swimming microorganisms, in *Progress in Phycological Research*, Vol. 4, Round, F.E., and Chapman, D.J., Eds., Biopress Ltd., Bristol., 1986, 258.
2476. Kessler, J.O., Hill, N.A., and Häder, D.P., Orientation of swimming flagellates by simultaneously acting external factors, *J. Phycol.*, 28, 816, 1992.
2477. Ketchum, B.H., The absorption of phosphate and nitrate by illuminated cultures of *Nitzschia closterium, Am. J. Bot.*, 26, 399, 1939.
2478. Ketchum, B.H., The development and restoration of deficiencies in the phosphorus and nitrogen composition of unicellular algae, *J. Cell. Comp. Physiol.*, 13, 373, 1939.
2479. Ketchum, B.H., Mineral nutrition of phytoplankton, *Ann. Rev. Plant Physiol.*, 5, 55, 1954.

2480. Ketchum, B.H., and Redfield, A.C., A method for maintaining a continuous supply of marine diatoms by culture, *Biol. Bull.*, 75, 165, 1938.

2481. Ketchum, B.H., and Redfield, A.C., Some physical and chemical characteristics of algae grown in mass culture, *J. Cell. Comp. Physiol.*, 33, 281, 1949.

2482. Kettunen, I., A study of the priphyton of Lake Saimaa, polluted by waste waters of the pulp industry. A method for water pollution control analysis, in *Periphyton in Freshwater Ecosystems*, Wetzel, R.G., Ed., Dr. W. Junk Publishers, The Hague, The Netherlands, 1983, 331.

2483. Kevern, N.R., Strontium and calcium uptake by the green alga, *Oocystis eremosphaeria*, *Science*, 145, 1445, 1964.

2484. Kevern, N.R., and Ball, R.C., Primary productivity and energy relationships in artificial streams, *Limnol. Oceanogr.*, 10, 74, 1965.

2485. Kevern, N.R., Wilhm, J.L., and Van Dyne, G.M., Use of artificial substrate to estimate the production of periphyton, *Limnol. Oceanogr.*, 11, 499, 1966.

2486. Khailov, K.M., and Burlakova, Z.P., Release of dissolved organic matter by marine seaweeds and distribution of their total organic production to inshore communities, *Limnol. Oceanogr.*, 14, 521, 1969.

2487. Khailov, K.M., Burlakova, Z.P., and Lanskaya, L.A., Exogenous metabolites in cultures of *Gymnodinium kovalevskii*, *Mikrobiologiya*, 36, 498, 1963 (in Russian).

2488. Khaleafa, A.F., Kharboush, M.A.M., Metwalli, A., Mohren, A.F., and Serur, A., Antibiotic (fungicidal) action from extracts of some seaweeds, *Bot. Mar.*, 18, 163, 1975.

2489. Khalid, R.A., Patrick, W.H., Jr., and DeLaune, R.D., Phosphorus sorption characteristics of flooded soils, *Soil Sci. Soc. Am. J.*, 41, 305, 1977.

2490. Khan, M., *Fundamentals of Phycology*, The Himachal Times Press, Dehra Dun, India, 1970.

2491. Khan, M., *Algae Today*, Bisher Singh Mahendra Pal Singh, Dehra Dum, India, 1983.

2492. Khan, M.A., Occurrence of a rare euglenoid causing red-bloom in Dal Lake waters of the Kashmir Himalaya, *Arch. Hydrobiol.*, 127, 101, 1993.

2493. Khobot'yev, V.G., Kapkov, V.I., Rukhadze, Y.G., Turrenina, N.V., and Shidlovskaya, N.V., The toxic effect of copper complexes on algae, *Hydrobiol. J.*, 11, 33, 1975.

2494. Khoja, T., and Whitton, B.A., Heterotrophic growth of blue-green algae, *Arch. Mikrobiol.*, 79, 280, 1971.

2495. Khummongkol, D., Canterford, G.A., and Fryer, C., Accumulation of heavy metals in unicellular algae, *Biotechnol. Bioeng.*, 24, 2643, 1982.

2496. Kiefer, D.A., Chlorophyll a fluorescence in marine centric diatomes: Responses of chloroplasts to light and nutrient stress, *Mar. Biol.*, 23, 39, 1973.

2497. Kies, L., Untersuchungen zur Feinstruktur und taxonomischen Einordnung von *Gloeochaeta wittrockiana*, einer apoplastidalen capsalen Alge mit blaugrünen Endosymbionten (Cyanellen), *Protoplasma*, 87, 419, 1976.

2498. Kies, L., Morphology and systematic position of some endocyanomes, in *Endocytology, Endosymbiosis and Cell Biology*, Vol. 1, Schwemmler, W., and Schenk, H.E.A., Eds., Walter de Gruyter and Co., Berlin and New York, 1980, 7.

2499. Kilham, P., A hypothesis concerning silica and the freshwater planktonic diatoms, *Limnol. Oceanogr.*, 16, 10, 1971.

2500. Kilham, S.S., Kott, C.L., and Tilman, D., Phosphate and silicate kinetics for the lake Michigan diatom *Diatoma elongatum*, *J. Great Lakes Res.*, 3, 93, 1977.

2501. Kim, C.Y., and Won, J.H., Concentrations of mercury, cadmium, lead and copper in the surrounding seawater and in seaweeds, *Undaria pinnatifida* and *Sargassum fulvellum*, from Suycong Bay in Busan, *Bull. Korean Fish. Soc.*, 13, 169, 1974.

2502. Kimball, K.D., and Kimball, S.F., Seasonal phytoplankton variations in the shallow Pahlavi Mordab, Iran, *Hydrobiologia*, 55, 49, 1977.

2503. Kimball, K.D., and Baker, A.L., Variations in the mineral content of *Myriophyllum heterophyllum* Michx related to site and season, *Aquat. Bot.*, 14, 139, 1982.

2504. Kimmel, B.L., and Lind, O.T., Factors affecting phytoplankton production in a eutrophic reservoir, *Arch. Hydrobiol.*, 71, 124, 1972.

2505. King, D.L., The role of carbon in eutrophication, *J. Water Pollut. Control Fed.*, 42, 2035, 1970.

2506. King, D.L., Carbon limitation in sewage lagoons, in *Nutrients and Eutrophication: The Limiting-Nutrient Controversy*, Likens, E., Ed., Spec. Symp. Vol. 1, Amer. Soc. Limnol. Oceanogr., Allen Press, Lawrence, Kansas, 1972, 98.

2507. King, D.L., and Ball, R.C., A quantitative measure of Aufwuchs production, *Trans. Am. Micsrosc. Soc.*, 85, 232, 1966.

2508. King, J.M., Some edaphic algae from western Wisconsin, *Trans. Wisconsin Acad. Sci.*, 113, 200, 1975.

2509. King, J.M., and Ward, C.H., Distribution of edaphic algae as related to land usage, *Phycologia,* 16, 23, 1977.

2510. Kirby, C.J., The annual net primary production and decomposition of the salt marsh grass *Spartina alterniflora* in the Barataria Bay Estuary of Louisiana, Ph.D. Thesis, Louisiana State University, Baton Rouge, 1971.

2511. Kirby, C., and Gosselink, J.G., Primary productivity in a Louisiana Gulf Coast *Spartina alterniflora* marsh, *Ecology,* 57, 1052, 1976.

2512. Kirk, M.M., and Kirk, D.L., Carrier-mediated uptake of arginine and urea by *Volvox carteri* f. *magariensis, Plant Physiol.,* 61, 549, 1978.

2513. Kirk, D.L., and Kirk, M.M., Carrier-mediated uptake of arginine and urea by *Chlamydomonas reinhardtii, Plant Physiol.,* 61, 556, 1978.

2514. Kirk, D.L., and Kirk, M.M., Amino acid and urea uptake in ten species of Chlorophyta, *J. Phycol.,* 14, 198, 1978.

2515. Kirk, J.T.O., and Tilney-Bassett, R.A.E., *The Plastids,* W.H. Freeman and Co., London, 1967.

2516. Kirpenko, Y.A., Sirenko, L.A., Orlovskij, V.M., and Lukina, L.F., *Toxin Sinezelenykh Vodoroslej i Organism Zhivotnogo,* Isdatl. Naukova Dumka, Kiev, 1977, (in Russian).

2517. Kistritz, R.U., and Yesaki, I., Primary production, detritus flux, and nutrient cycling in a sedge marsh, Fraser River Estuary, *Wastewater Research Center, Tech. Rep. 17,* 1, 1979.

2518. Kistritz, R.U., Hall, K.J., and Yesaki, I., Productivity, detritus flux, and nutrient cycling in a *Carex lyngbyei* tidal marsh, *Estuaries,* 6, 227, 1983.

2519. Kitamura, S., Hirano, Y., Noguchi, Y., Kojima, T., Kakita, T., and Kumamoto, H., Epidemiological studies on "Minamata Disease" (Supplementary report No. 2), *J. Kumamoto Med. Soc.,* 33 (Suppl. 3), 559, 1959 (in Japanese, cited by D'Itri 90).

2520. Kitamura, S., Kakita, T., Kojo, S., and Kajima, T., Epidemiological studies on Minamata Disease, *J. Kumamoto Med. Soc.,* 34 (Suppl. 3), 477, 1960 (in Japanese, cited by D'Itri 90).

2521. Kjellman, F.R., Phaeophyceae (Fucoideae), In *Die Natürlichen Pflanzenfamilien,* Vol. 1, Engler, A., and Prantl, K.A.E., Eds., Leipzig, 1897.

2522. Klapwijk, S.P., Kroon, J.M.W., and Meijer, R., M.-l., Available phosphorus in lake sediments in The Netherlands, *Hydrobiologia,* 92, 491, 1982.

2523. Klapwijk, S.P., de Boer, T.F., and Rijs, M.J., Effects of agricultural wastewater on benthic algae in ditches in The Netherlands, in *Periphyton in Freshwater Ecosystems,* Wetzel, R.G., Ed., Dr. W. Junk Publishers, The Hague, The Netherlands, 1983, 311.

2524. Klarer, D.M., and Hickman, M., The effect of thermal effluent upon the standing crop of an epiphytic algal community, *Int. Rev. Ges. Hydrobiol.,* 60, 17, 1975.

2525. Klass, E., Rowe, D.W., and Massaro, E.J., The effect of cadmium on population growth of the green alga *Scenedesmus quadricauda, Bull. Environ. Contam. Toxicol.,* 12, 442, 1974.

2526. Klaveness, D., *Coccolithus huxleyi* (Lohman) Kamptner. I. Morphological investigations on the vegetative cell and the process of coccolith formation, *Protistologica,* 3, 335, 1972.

2527. Klaveness, D., *Coccolithus huxleyi* (Lohm.) Kamptner. II. The flagellated cell, abarrant cell types, vegetative propagation and life cycles, *Br. Phycol. J.,* 7, 309, 1972.

2528. Klaveness, D., *Emiliania huxleyi* (Lohmann) Hay and Mohler III. Mineral deposition and origin of the matrix during coccolith formation, *Protistologica,* 12, 217, 1976.

2529. Klaveness, D., and Guillard, R.R.L., The requirement for silicon in *Synura petersenii* (Chrysophyceae), *J. Phycol.,* 11, 349, 1975.

2530. Klaveness, D., and Paasche, E., Physiology of coccolithophorids, in *Biochemistry and Physiology of Protozoa,* Vol. 1, Hutner, S.H., and Levandarsky, M., Eds., Academic Press, New York, 1979, 191.

2531. Klebs, G., Flagellatenstudien. II., *Z. Wiss. Zool.,* 55, 353, 1892.

2532. Klehbahn, H., Beitrage zur Kenntnis der Auxosporenbildung. I. *Rhopalodia gibba, Pringsh. Jahrb. Wiss. Bot.,* 29, 595, 1896.

2533. Kleijn, H.F.W., Buffer capacity in water chemistry, *Int. J. Air Wat. Pollut.,* 9, 401, 1965.

2534. Klein, M., and Cronquist, A., A consideration of the evolutionary and taxonomic significance of some biochemical micromorphological and physiological characters in the thallophytes, *Q. Rev. Biol.,* 42, 105, 1967.

2535. Klemer, A.R., Cyanobacterial blooms: carbon and nitrogen limitation have opposite effects on the buoyancy of *Oscillatoria, Science,* 215, 1629, 1982.

2536. Klopatek, J.M., The role of emergent macrophytes in mineral cycling in a freshwater marsh, in *Mineral Cycling in Southeastern Ecosystems,* Howell, F.G., Gentry, J.B., and Smith, M.H., Eds., ERDA Symp. Series, CONF 740513, 1975, 367.

2537. Klopatek, J.M., Nutrient dynamics of freshwater riverine marshes and the role of emergent macrophytes, in *Proc. Symp. Freshwater Wetlands: Ecological Processes and Management Potential,*

Good, R.E., Whigham, D.F., and Simpson, R.L., Academic Press, New York San Francisco and London, 1978, 195.

2538. Klopatek, J.M., and Stearns, F.W., Primary productivity of emergent macrophytes in a Wisconsin freshwater marsh ecosystem, *Am. Midl. Nat.,* 100, 320, 1978.

2539. Klotz, R.L., Cain, J.R., and Trainor, F.R., Algal competition in an epilithic river flora, *J. Phycol.,* 12, 363, 1976.

2540. Knauss, H.J., and Porter, J.W., The absorption of inorganic ions by *Chlorella pyrenoidosa, Plant Physiol.,* 29, 229, 1954.

2541. Knisel, W.H., Jr., Methods of estimating evaporation and evapotranspiration, in *Hydrological Techniques for Upstream Conservation,* Kunkle, S.H., and Thames, J.L., Eds., FAO of UN, Rome Italy, 1976.

2542. Knoblauch, A.R., Estimation of phosphorus exchange rates in the upper part of the Wahnbach reservoir using balance calculations, *Arch. Hydrobiol. Beih. Ergebn. Limnol.,* 18, 69, 1982.

2543. Knowles, R., Denitrification, *Microbiol. Rev.,* 46, 43, 1982.

2544. Knowles, S.C., and Zingmark, R.G., Mercury and temperature interactions on the growth rates of three species of freshwater phytoplankton, *J. Phycol.,* 14, 104, 1978.

2545. Knübel, G., Larsen, L.K., Moore, R.E., Levine, I.A., and Patterson, G.M.L., Cytotoxic, antiviral indolocarbazoles from a blue-green alga belonging to the Nostocaceae, *J. Antibiotics,* 43, 1236, 1990.

2546. Kobayashi, H., Chlorophyll content in sessile algal communities of a Japanese mountain river. *Bot. Mag.* (Tokyo), 74, 228, 1961.

2547. Koeman, R.P.T., and van den Hoeck, C., The taxonomy of *Ulva* (Chlorophyceae) in the Netherlands, *Br. Phycol. J.,* 16, 9, 1981.

2548. Koenings, J.P., and Hooper, F.F., The influence of colloidal organic matter on iron and iron-phosphorus cycling in an acid bog lake, *Limnol. Oceanogr.,* 21, 684, 1976.

2549. Koerselman, W., and Beltman, B., Evapotranspiration from fens in relation to Penman's potential free water evapotranspiration (E_o) and pan evaporation, *Aquat. Bot.,* 31, 307, 1988.

2550. Kogan, I.G., Anikeeva, I.D., and Vanlina, E.N., Effect of cadmium ions on *Chlorella.* II. Modification of the UV radiation effect, *Genetika* (Moscow), 11, 84, 1975 (in Russian).

2551. Kohata, K., and Watanabe, M., Diel changes in the composition of photosynthetic pigments in *Chattonella antiqua* and *Heterosigma akashiwo* (Raphidophyceae), in *Red Tides: Biology, Environmental Science and Toxicology* Okaichi, T., Anderson, M., and Nemoto, T., Eds., Elsevier, New York, 1988, 329.

2552. Kohata, K., and Watanabe, M., Diel changes in the composition of photosynthetic pigments and cellular carbon and nitrogen in *Pyramimonas parkeae* (Prasinophyceae), *J. Phycol.,* 25, 377, 1989.

2553. Kok, B., and Cheniae, G.M., Kinetics and intermediates of the oxygen evolving step in photosynthesis, in *Current Topics in Bio-Energetics,* Vol. 1, Sanadi, D.R., Ed., Academic Press, New York, 1966, 1.

2554. Kol, E., Kryobiologie, *Die Binnengewässer* 24, 1, 1968.

2555. Kolbe, R.W., Grundlinien einer allgemeinen Ökologie der Diatomeen, *Ergebn. Biol.,* 8, 221, 1932.

2556. Kolber, Z., Zehr, J., and Falkowski, P.G., Effects of growth irradiance and nitrogen limitation on photosynthetic energy conversion on Photosystem II, *Plant Physiol.,* 88, 923, 1988.

2557. Kolkwitz, R., Plankton und Seston, *Bericht der Deutsch. Bot. Ges.,* 30, 334, 1912.

2558. Kolkwitz, R., *Pflanzenphysiologie,* Jena, 1935.

2559. Kolkwitz, R., and Marsson, M., Ökologie der pflanzlichen Saprobien, *Ber. Dtsch. Bot. Ges.,* 26 A, 505, 1908.

2560. Komárek, J., The morphology and taxonomy of crucigenoid algae (Scenedesmaceae, Chlorococcales), *Arch. Protistenk.,* 116, 1, 1974.

2561. Komárek, J., Species concept in coccal green algae, *Arch. Hydrobiol. Suppl. 73, (Algol. Studies, 45),* 437, 1987.

2562. Komárek, J., and Anagnostidis, K., Modern approach to the classification of cyanophytes 2 - Chroococcales, *Arch. Hydrobiol. Suppl.,* 73 (*Algol. Studies,* 43), 157, 1986.

2563. Komárek, J., and Anagnostidis, K., Modern approach to the classification of cyanophytes 4 - Nostocales, *Arch. Hydrobiol. Suppl.,* 82 (*Algol. Studies,* 56), 247, 1989.

2564. Komárek, J., and Hindák, F., Taxonomic review of natural populations of cyanophytes from the *Gomphosphaeria* - comples, *Arch. Hydrobiol. Suppl.,* 80 (*Algol. Studies,* 50-53), 203, 1988.

2565. Komárková, J., Primary production of phytoplankton and periphyton in Opatovický fishpond (South Bohemia) in 1972, in *Ecosystem Study on Wetland Biome in Czechoslovakia,* Hejný, S., Ed., Czechoslovak IBP/PT-PP Report No. 3, Třeboň, Czechoslovakia, 1973, 197.

2566. Komárková, J., and Komárek, J., Comparisons of pelagial and littoral primary production in a South Bohemian fishpond (Czechoslovakia), *Symp. Biol. Hungar.,* 15, 77, 1975.

2567. Komárková, J., and Marvan, P., Primary production and functioning of algae in the fish pond littoral, in *Pond Littoral Ecosystems: Structure and Functioning*, Dykyjová, D., and Květ, J., Eds., Springer-Verlag, Berlin, 1978, 321.

2568. Komárková, J., and Marvan, P., The role of algae in the littoral zone of carp ponds, *Arch. Hydrobiol. Beih. Ergebn. Limnol.*, 27, 239, 1987.

2569. Kondrateva, N.V., *Morphogenesis and the Main Evolutionary Tendencies in Hormogonal Algae*, Izd. Naukova dumka, Kiev, U.S.S.R., 1975.

2570. Kopp, J.F., and Kroner, R.C., Tracing water pollution with an emission spectrograph, *J. Water Pollut. Control Fed.*, 39, 1659, 1967.

2571. Kopp, J.F., and Kroner, R.C., Trace metals in waters of the United States, U.S. Dept. of the Interior, Federal Water Pollut. Control. Admin., Div. Pollut. Surveillance, Cincinnati, Ohio, 1970.

2572. Kormody, E.J., Weight loss of cellulose and aquatic macrophytes in a Carolina Bay, *Limnol. Oceanogr.*, 13, 522, 1968.

2573. Korte, V.E., and Blinn, D.W., Diatom colonization on artificial substrata in pool and riffle zones studies by light and scanning electron microscopy, *J. Phycol.*, 19, 332, 1983.

2574. Kovács, M., Die Bedetung der Balaton-Uferzone für den Umweltschutz am See, *Acta Bot. Acad. Sci. Hungar. (Budapest)*, 22, 85, 1976.

2575. Kovács, M., Nyáry, I., and Tóth, L., The microelement content of some submerged and floating aquatic plants, *Acta Botan. Hungar.*, 30, 173, 1984.

2576. Kowallik, W., Eine Förderne Wirkung von Blaulicht auf die Säureproduktion anaerob gehalterner *Chlorellen*, *Planta*, 87, 372, 1969.

2577. Kowallik, W., Light effects on carbohydrate and protein metabolism in algae, in *Photobiology of Microorganisms*, Halldal, P., Ed., Wiley-Interscience, London and New York, 1970, 165.

2578. Kozlovskaya, L.S., Role of soil organisms in decomposition of organic remains in swamped forest soils, in *The Increase in Productivity of Swamped Forests*, Kozlovskaya, L.S., Ed., Israel Program for Sci. Translations, Jerusalem, 1963, 27.

2579. Kozlowski, T.T., Extent, causes and impacts of flooding, In *Flooding and Plant Growth*, Kozlowski, T.T., Ed., Academic Press, Orlando, 1984, 1.

2580. Kozlowski, T.T., Responses of woody plants to flooding, in *Flooding and Plant Growth*, Kozlowski, T.T., Ed., Academic Press, Orlando, Florida, 1984, 129.

2581. Kraeuter, J.N., and Wolf, P.L., The relationship of marine macroinvertebrates to salt marsh plants, in *Ecology of Halophytes*, Reimond, R.J., Queen, W.H., Eds., Academic Press, New York, 1974, 449.

2582. Kraft, G.T., Rhodophyta: morphology and classification, In *The Biology of Seaweeds*, Lobban, C.S., and Wynne, M.J., Eds., Blackwell Scientific Publications, Oxford, 1981, 6.

2583. Kramer, K., Zur Deutung einiger Schalenstrukturen bei pennaten Diatomeen, *Nova Hedw.*, 35, 75, 1981.

2584. Krammer, K., and Lange-Bartalot, H., *Bacillariophyceae*, Vol. 1, *Naviculaceae*, Süsswasserflora von Mittelueropa 2/1, Gustav Fisher Verlag, Stuttgart, New York, 1986.

2585. Krammer, K., and Lange-Bartalot, H., *Bacillariophyceae*, Vol. 2, *Bacillariaceae, Epithemiaceae, Surirellaceae*, Süsswasserflora von Mitteleuropa 2/2, Gustav Fisher Verlag, Stuttgart, New York, 1988.

2586. Krammer, K., and Lange-Bartalot, H., *Bacillariophyceae*, Vol. 3, *Centrales, Fragilariaceae, Eunotiaceae*, Süsswasserflora von Mitteleuropa 2/3, Gustav Fisher Verlag, Stuttgart, New York, 1991.

2587. Krammer, K., and Lange-Bartalot, H., *Bacillariophyceae*, Vol. 4, *Achnanthaceae, Kritisch Ergänzungen zu Navicula (Lineolatae) und Gomphonema*, Süsswasserflora von Mittelueropa 2/4, Gustav Fisher Verlag, Stuttgart, New York, 1991.

2588. Kraska, M., Net primary production of sedge communities (The Magnocaricion Alliance) of the Slovensky Národní park, *Bull. Soc. Sci. Lett. Poznan*, D-16, 47, 1976.

2589. Kratz, W.A., and Myers, J., Nutrition and growth of several blue-green algae, *Am. J. Bot.*, 42, 282, 1955 b.

2590. Krauss, R.W., Limiting factors affecting the mass culture of *Scenedesmus obliquus* (Turp.) Kütz. in an open system, Ph.D. Thesis, University of Maryland, 1951.

2591. Krauss, R.W., Nutrient supply for large-scale algal cultures, *Sci. Monthly*, 80, 21, 1955.

2592. Krauss, R.W., 1958, Physiology of the fresh-water algae, *Ann. Rev. Plant Physiol.*, 9, 207, 1958.

2593. Krauss, R.W., and Thomas, W.H., The growth and inorganic nutrition of *Scenedesmus obliquus* in mass culture, *Plant Physiol.*, 29, 205, 1954.

2594. Kreger, D.R., Cell walls, in *Physiology and Biochemistry of Algae*, Lewin, R.A., Ed., Academic Press, New York and London, 1962, 315.

2595. Kreger, D.R., and Meeuse, B.J.D., X-ray diagrams of *Euglena* paramylon, of the acid-insoluble glucan of yeast cell walls and of laminarin, *Biochim. Biophys. Acta,* 9, 699, 1952.

2596. Kreger, D.R., and van de Veer, J., Paramylon in a chrysophyte, *Acta Bot. Neerl.,* 19, 401, 1970.

2597. Kremer, B.P., $^{14}CO_2$-fixation by the endosymbiotic alga *Platymonas convolutae* within the turbellarian *Convoluta roscoffensis, Mar. Biol.,* 31, 219, 1975.

2598. Kremer, B.P., *Prochloron.* Neue Kategorie im System der Algen, *Mikrokosmos,* 69, 83, 1980.

2599. Kremer, B.P., Taxonomic implications of algal photoassimilate patterns, *Br. Phycol. J.,* 15, 399, 1980.

2600. Kremer, B.P., Photorespiration and ß-carboxylation in brown algae, *Planta,* 150, 189, 1980.

2601. Kremer, B.P., Carbon metabolism, in *The Biology of Seaweeds,* Lobban, C.S., and Wynne, M.J., Eds., University of California Press, Berkeley, 1981, 493.

2602. Kremer, B.P., Carbon economy and nutrition on the alloparasitic red alga *Harveyella mirabilis, Mar. Biol.,* 76, 231, 1983.

2603. Kremer, B.P., and Küppers, U., Carboxylating enzymes and pathway of photosynthetic carbon assimilation in different marine algae—evidence for the C_4 pathway?, *Planta,* 133, 191, 1977.

2604. Kremer, B.P., and Berks, R., Photosynthesis and carbon metabolism in marine and freshwater diatoms, *Z. Pflanzenphysiol.,* 87, 149, 1978.

2605. Kremer, B.P., and Markham, J.W., Primary metabolic effects of cadmium in the brown alga *Laminaria saccharina, Z. Pflanzenphysiol.,* 108, 125, 1982.

2606. Krempin, D.W., McGrath, S.M., SooHoo, J.B., and Sullivan, C.W., Orthophosphate uptake by phytoplankton and bacterioplankton from the Los Angeles Harbor and Southern California coastal waters, *Mar. Biol.,* 64, 23, 1981.

2607. Krey, J., Eine neue Methode zur quantitativen Bestimmung des Planktons, *Kieler Meeresforsch.,* 7, 58, 1950.

2608. Kristiansen, J., Some cases of sexuality in *Kephyriopsis* (Chrysophyceae), *Bot. Tidsskr,* 56, 128, 1960.

2609. Kristiansen, J., Sexual reproduction in *Mallomonas caudata, Bot. Tidsskr,* 57, 306, 1961.

2610. Kristiansen, J., Observations on some Chrysophyceae from North Wales, *Br. Phycol. J.,* 14, 231, 1979.

2611. Kristiansen, J., Chrysophyceae from some Greek lakes, *Nova Hedw.,* 33, 167, 1980.

2612. Kristiansen, J., Chrysophyceae, in *Synopsis and Classification of Living Organisms,* Parker, S.P., Ed., McGraw-Hill, New York, 1982, 81.

2613. Kristiansen, J., and Takahashi, E., Chrysophyceae: introduction and bibliography, in *Selected Papers in Phycology* II, Rosowski, J.R., and Parker, B.C., Eds., Phycological Society of America, Lawrence, Kansas, 1982, 698.

2614. Kristiansen, J., and Andersen, R.E., Eds., *Chrysophytes: Aspects and Problems,* Cambridge University Press, Cambridge, 1986.

2615. Krock, H.-J., and Mason, D.T., Bioassays of lower trophic levels. A study of toxicity and biostimulation in San Francisco Bay-Delta waters, *University of California, Berkeley Sanit. Eng. Res. Lab., Report No. 71,* 1, 1971.

2616. Kroeckel, L., and Stolp, H., Influence of the water regime on denitrification and aerobic respiration in soil, *Biol. Fertil. Soils,* 2, 15, 1986.

2617. Kroes, H.W., Excretion of mucilage and yellow-brown substances by some brown algae from the intertidal zone, *Bot. Mar.,* 13, 107, 1970.

2618. Krogstad, T., and Løvstad, Ø., Erosion, phosphorus and phytoplankton response in rivers of South-Eastern Norway, *Hydrobiologia,* 183, 33, 1989.

2619. Krogstad, T., and Løvstad, Ø., Available soil phosphorus for planktonic blue-green algae in eutrophic lake water samples, *Arch. Hydrobiol.,* 122, 117, 1991.

2620. Krouse, H.R., and McCready, R.G.L., Reductive reactions in the sulphur cycle, in *Biogeochemical Cycling of Mineral-Forming Elements,* Trudinger, P.A., and Swaine, D.J., Eds., Elsevier Scientific Publishing Company, Amsterdam, 1979, 315.

2621. Krouse, H.R., McCready, R.G.L., Biogeochemical cycling of sulfur, in *Biogeochemical Cycling of Mineral-Forming Elements,* Trudinger, P.A., and Swaine, D.J., Eds., Elsevier Scientific Publishing Company, Amsterdam, 1979, 401.

2622. Kruczynski, W.L., Subrahmanyan, C.B., and Drake, S.H., Studies on the plant community of a North Florida salt marsh. Part 1. Primary production, *Bull. Mar. Sci.,* 28, 316, 1978.

2623. Krumbein, W.F., Calcification by bacteria and algae, in *Biogeochemical Cycling of Mineral-Forming Elements,* Trudinger, P.A., and Swaine, D.J., Eds., Elsevier Scientific Publishing Company, Amsterdam, 1979, 47.

2624. Krumholtz, L.A., *U.S. Atomic Energy Comm. Inform. Serv., ORO-132,* 1, 1954.

2625. Kudoh, S., and Takahashi, M., Fungal control of population changes of the planktonic diatom *Asterionella formosa* in a shallow eutrophic lake, *J. Phycol.*, 26, 239, 1990.

2626. Kuenen, J.G., and Robertson, L.A., Ecology of nitrification and denitrification, in *The Nitrogen and Sulphur Cycles*, Cole, J.A., and Ferguson, S.J., Eds., Cambridge University Press, Cambridge, 1987, 162.

2627. Kuenzler, E.J., Glucose-6-phosphate utilization by marine algae, *J. Phycol.*, 1, 156, 1965.

2628. Kuenzler, E.J., Dissolved organic phosphorus excretion by marine phytoplankton, *J. Phycol.*, 6, 7, 1970.

2629. Kuenzler, E.J., and Ketchum, B.H., Rate of phosphorus uptake by *Phaeodactylum tricornutum*, *Biol. Bull.*, 123, 134, 1962.

2630. Kuenzler, E.J., and Perras, J.P., Phosphatases of marine algae, *Biol. Bull.*, 128, 271, 1965.

2631. Kuenzler, E.J., Mulholland, P.J., Yarbro, L.A., and Smock, L.A., Eds., *Distribution and Budgets of Carbon, Phosphorus, Iron and Manganese in a Floodplain Swamp Ecosystems*, University of North Carolina Water Resources Research Report 157, 1980.

2632. Kufel, I., Lead and molybdenum in reed and cattail-open versus closed type of metal cycling, *Aquat. Bot.*, 40, 275, 1991.

2633. Kuhl, A., Die Biologie der kondensierten anorganischen Phosphate, in *Ergebnisse der Biologie*, Vol. 23, Autrum, H., Ed., Springer-Verlag, Berlin, 1960, 144.

2634. Kuhl, A., Inorganic phosphorus uptake and metabolism, in *Physiology and Biochemistry of Algae*, Lewin, R.A, Ed., Academic Press, New York, 1962, 211.

2635. Kuhl, A., Zur Physiologie der Speicherung kondensierter anorganischer Phosphate in *Chlorella*, *Vortr. Botan. hrsg. Deutsch. Botan. Ges. (N.F.)*, 1, 157, 1962.

2636. Kuhl, A., Phosphate metabolism of green algae, in *Algae, Man and the Environment*, Jackson, D.F., Ed., Syracuse University Press, Syracuse, 1968, 37.

2637. Kuhl, A., Phosphorus, in *Algal Physiology and Biochemistry* Stewart, W.D.P., Ed., Blackwell Scientific Publications, Oxford, 1974, 636.

2638. Kuhn, D.L., Plafkin, J.L., Cairns, J., Jr., and Lowe, R.L., Qualitative characterization of aquatic environments using diatoms life-form strategies, *Trans. Am. Microsc. Soc.*, 100, 165, 1981.

2639. Kuiper, J., Fate and effects of cadmium in marine plankton communities in experimental enclosures, *Mar. Ecol. Prog. Ser.*, 6, 161, 1981.

2640. Kullberg, R.G., Algal distribution in six thermal stream effluents, *Trans. Am. Microscop. Soc.*, 90, 412, 1971.

2641. Kulshreshtha, M., and Gopal, B., Decomposition of freshwater wetland vegetation. I. Submerged and free-floating macrophytes, in *Proc. Int. Conf. Wetlands: Ecology and Management*, Gopal, B., Turner, R.E., Wetzel, R.G., and Whigham, D.F., Eds., Nat. Inst. of Ecology and Internat. Scientific Publications, Jaipur, India, 1982, 259.

2642. Kulshreshtha, M., and Gopal, B., Decomposition of freshwater wetland vegetation. II. Aboveground organs of emergent macrophytes, in *Proc. Int. Conf. Wetlands: Ecology and Management*, Gopal, B., Turner, R.E., Wetzel, R.G., and Whigham, D.F., Eds., Nat. Inst. of Ecology and Internat. Scientific Publications, Jaipur, India, 1982, 279.

2643. Kumar, D., Jha, M., and Kumar, H.D., Heavy-metal toxicity in the cyanobacterium *Nostoc linckia*, *Aquat. Bot.*, 22, 101, 1985.

2644. Kumar, H.D., and Prakash, G., Toxicity of selenium to the blue-green algae, *Anacystis nidulans* and *Anabaena variabilis*, *Ann. Bot.*, 35, 697, 1971.

2645. Kurasawa, H., Studies on the biological production of fire pools in Tokyo. XII. The seasonal changes in the amount of algae attached on the wall of pools, *Misc. Rep. Res. Inst. Nat. Resour.*, 51, 15, 1959.

2646. Kusler, J.A., and Kentula, M.E., Eds., *Wetland Creation and Restoration. The State of the Science*, Island Press, Washington, Covelo, 1990.

2647. Küster, E., Influence of peat and peat extracts on the growth and metabolism of microorganisms, *Int. Peat Congress*, 2, 945, 1968.

2648. Kützing, F.T., *Phycologia generalis*, Leipzig, 1843.

2649. Kützing, F.T., *Species algarum*, Leipzig, 1849.

2650. Kuwabara, J.S., Phosphorus-zinc interactive effects on growth by *Selenastrum capricornutum* (Chlorophyta), *Environ. Sci. Technol.*, 19, 417, 1985.

2651. Květ, J., Growth analysis approach to the production ecology of reedswamp plant communities, *Hidrobiologia*, 12, 15, 1971.

2652. Květ, J., Mineral nutrition in shoots of reed *Phragmites communis*, *Polskie Arch. Hydrobiol.*, 20, 137, 1973.

2653. Květ, J., Growth and mineral nutrients in shoots of *Typha latifolia*, *Symp. Biol. Hungar.*, 15, 113, 1975.

2654. Květ, J., Transpiration of South Moravian *Phragmites communis* littoral of the Nesyt Fishpond, in *Studie CSAV 15*, Květ, J., Ed., Academia Praha, 1973, 143.

2655. Květ, J., Growth analysis of fishpond littoral communities, in *Pond Littoral Ecosystems: Structure and Functioning*, Dykyjová, D., and Květ, J., Eds., Springer Verlag, Berlin, New York, 1978, 198.

2656. Květ, J., Production of organic matter in macrophyte stands, in *Proc. Conf. Macrophytes in Water Management, Water Hygiene and Fishery*, Dům Techniky ČSVTS, České Budějovice, 1982, 73 (in Czech).

2657. Květ, J., and Ondok, J.P., Zonation of higher-plant shoot biomass in the littoral of the Opatovický fishpond, in *Ecosystem Study on Wetland Biome in Czechoslovakia*, Hejný, S., Ed., Czechosl. IBP/PT-PP Rept. No. 3, Třeboň, Czechoslovakia, 1973, 87.

2658. Květ, J., and Husák, Š., Primary data on biomass and production estimates on typical stands of fishpond littoral plant communities, in *Pond Littoral Ecosystems: Structure and Functioning*, Dykyjová, D., and Květ, J., Eds., Springer Verlag, Berlin, New York, 1978, 211.

2659. Květ, J., and Ostrý, I., Mineral nutrient accumulation in the principal plant communities of the Rozmberk fishpond littoral, in *Littoral Vegetation of the Rožmberk Fishpond and its Mineral Nutrient Economy*, Hroudová, Z., Ed., *Studie ČSAV*, 9, 95, 1988.

2660. Květ, J., Svoboda, J., and Fiala, K., Canopy development in stands of *Typha latifolia* L. and *Phragmites communis* Trin. in south Moravia, *Hidrobiologia* 10, 63, 1969.

2661. Květ, J., Rejmánková, E., and Rejmámenk, M., Higher aquatic plants and biological waste water treatment, The outline of possibilities, *Proc. of the Aktiv Jihočeských Vodohospodářů Conference*, 1979, 1 (in Czech).

2662. Květ, J., Leciánová, L., and Véber, K., Experience with the cultivation of water hyacinth in waste-waters, in *Proc. Conf. Macrophytes in Water Management, Water Hygiene and Fishing*, Dum Techniky, Ceské Budejovice, 1982, 101, (in Czech).

2663. Kwon, Y.M., and Grant, B.R., Assimilation and metabolism of glucose by *Dunaliella tertiolecta* I. Uptake by whole cells and metabolism by cell free system, *Plant Cell Physiol.*, 12, 29, 1971.

2664. Kylin, A., The influence of phosphate nutrition on growth and sulfur metabolism of *Scenedesmus*, *Physiol. Plant.*, 17, 384, 1964.

2665. Kylin, A., Sulfate uptake and metabolism in *Scenedesmus* as influenced by phosphate, carbon dioxide, and light, *Physiol. Plant.*, 17, 422, 1964.

2666. Kylin, A., The influence of photosynthetic factors and metabolic inhibitors on the uptake of phosphate in P-deficient *Scenedesmus*, *Physiol. Plant.*, 19, 644, 1966.

2667. Kylin, A., The effect of light, carbon dioxide, and nitrogen nutrition on the incorporation of S from external sulphate into different S-containing fractions in *Scenedesmus*, with special reference to lipid S, *Physiol. Plant.*, 19, 883, 1966.

2668. Kylin, A., and Das, G., Calcium and strontium as micronutrients and morphogenetic factors for *Scenedesmus*, *Phycologia*, 6, 201, 1967.

2669. Kylin, A., and Tillberg, J.E., The relation between total photophosphorylation, levels of ATP, and O_2 evolution in *Scenedesmus* as studied with DCMU and Antimycin A, *Z. Pflanzenphysiol.*, 58, 165, 1967.

2670. Kylin, H., Studien über die Entroichlungs geschichte von *Rhodomela virgata* Kjellm., *Svensk. Bot. Tidsskr*, 8, 33, 1914.

2671. Kylin, H., Zur Biochemie der Meeresalgaen, *Z. physiol. Chem. Hoppe-Seyler's*, 83, 171, 1915.

2672. Kylin, H., Untersuchungen über die Biochemie der Meeresalgen, *Z. physiol. Chem., Hoppe-Seyler's*, 94, 337, 1915.

2673. Kylin, H., Über das Vorkommen von Jodiden, Bromiden und Jodidoxydasen bei den Meeresalgen, *Z. Physiol. Chem. Hoppe-Seyler's*, 186, 50, 1929.

2674. Kylin, H., Über die Blasenzellen bei *Bonnemaisonia*, *Trailliella* und *Antithamnion*, *Z. Bot.*, 23, 217, 1930.

2675. Kylin, H., Zur Biochemie der Cyanophyceen, *Kgl. Fysiograf. Sällskap. i Lund Förh*, 13, 1, 1943b.

2676. Kylin, H., Verwandtschaftliche Beziehungen zwischen den Cyanophyceen und den Rhodophyceen, *Kgl. Fysiograf. Sällskap i Lund Förh*, 13, 1, 1943c.

2677. Kylin, H., Zur Biochemie der Phaeophyceen, *Kgl. Fysiograf. Sällskap. i Lund Förh.*, 14, 1, 1944.

2678. Kylin, H., *Die Gattungen der Rhodophyceen*, CWK Gleerups, Lund, 1956.

2679. Laanbroek, H.J., Bacterial cycling of minerals that affect plant growth in waterlogged soils: a review, *Aquat. Bot.*, 38, 109, 1990.

2680. Laanbroek, H.J., and Pfennig, N., Oxidation of short-chain fatty acids by sulfate-reducing bacteria in freshwater and marine sediments, *Arch. Microbiol.*, 128, 330, 1981.

2681. Laanbroek, H.J., and Veldkamp, H., Microbial interactions in sediment communities. *Phil. Trans. R. Soc. Lond.*, B 297, 533, 1982.

2682. Lachavanne, J.B., Influence de l'eutrophisation des eaux sur les macrophytes des lacs suisses: resultats préliminaires, in *Studies on Aquatic Vascular Plants,* Symoens, J.J., Hooper, S.S., and Compere, P., Eds., Royal Soc. of Belgium, Brussels, 1982, 333.
2683. Laderman, A.D., Ed., *Atlantic White Cedar Wetlands,* Westview Press, Inc., Boilder, 1987.
2684. Lafleur, P.M., Evapotranspiration from sedge-dominated wetland surface, *Aquat. Bot.,* 37, 341, 1990.
2685. Laing, H.E., Respiration of the rhizomes of *Nuphar advenum* and other water plants, *Am. J. Bot.,* 27, 574, 1940.
2686. Lam, R.K., and Frost, B.W., Model of copepod filtering response to changes in size and concentration of food, *Limnol. Oceanogr,* 21, 490, 1976.
2687. Lambein, F., and Wolk, C.P., Structural studies on the glycolipids from the envelope of the heterocyst of *Anabaena cylindrica, Biochemistry* 12, 791, 1973.
2688. Lamberti, G.A., and Resh, V.H., Stream periphyton and insect herbivores: an experimental study of grazing by a caddyfish population, *Ecology,* 64, 1124, 1983.
2689. Lamberti, G.A., Ashkenas, L.R., Gregory, S.V., and Steinman, A.D., Effects of three herbivores on periphyton communities in laboratory streams, *J. N. Am. Benthol. Soc.,* 6, 92, 1987.
2690. Lambou, V.W., Taylor, W.D., Hern, S.C., and Williams, L.R., Comparisons of trophic state measurements, *Water Res.,* 17, 1619, 1983.
2691. Lamouroux, J.V.F., *Dissertations sur plusiuers aspeces de Fucus peu connues ou nouvelles,* Algen, 1805.
2692. Lamouroux, J.V.F., Essai sur les genes de la famille des thalassiophytes non articulées, *Mém. Mus. Nat. Hist. Paris,* 20, 21, 115, 267, 1813.
2693. Lamouroux, J.V.F., *Historie des polypiers Coralligenes flexibles, vulgairement nomees zoophytes,* Caen, 1816.
2694. Lande, E., Heavy metal pollution in Trondheimsfjørden, Norway, and the recorded effects on the fauna and flora, *Environ. Pollut.,* 12, 187, 1977.
2695. Lang, G., Die submersen Makrophyten des Bodensees—1978 im Vergleich mit 1967, *Int. Gawässerschutzkommissionfür den Bodensee,* Bericht No. 26, 1981.
2696. Lang, N.J., and Fay, P., The heterocysts of blue-green algae. II. Details of ultrastructure, *Proc. R. Soc. Lond.,* B 178, 193, 1971.
2697. Lang, N.J., and Walsby, A.E., Cyanophyceae: introduction and bibliography, in *Selected Papers in Phycology* II, Rosowski, J.R., and Parker, B.C., Eds., Phycological Society of America, Lawrence, Kansas, 1982, 647.
2698. Lang, N.J., Simon, R.D., and Wolk, C.P., Correspondence of cyasnophycean granules with structured granules in *Anabaena cylindrica, Arch. Mikrobiol.,* 83, 313, 1972.
2699. Langangen, A., The Charophytes of Iceland, *Astarte,* 5, 27, 1972.
2700. Langangen, A., Ecology and distribution of Norwegian Charophytes, *Norw. J. Bot.,* 21, 31, 1974.
2701. Lange, O.L., and Metzner, H., Lichtabhängiger Kohlenstoff-Einbau in Flechten bei tiefen Temperaturen, *Naturwissenschaften,* 52, 191, 1965.
2702. Lange, W., Cyanophyta—bacteria system: effects of added carbon compounds or phosphate on algae growth at low nutrient concentrations, *J. Phycol.,* 6, 230, 1970.
2703. Lange-Bartalot, H., Differentiating species of diatoms: a better criterion of water pollution than leading indicators, *Arch. Hydrobiol. Suppl.,* 51, 393, 1978.
2704. Langeland, A., and Larsson, P., The significance of the predator food chain in lake metabolism, *Prog. Wat. Tech.,* 12, 181, 1980.
2705. Langen, P., Vorkommen und Bedeteung von Polyphosphaten in Organismen, *Biol. Rdsch.,* 2, 145, 1965.
2706. Lanza, G.R., and Silvey, J.K.G., Interactions of reservoir microbiota: eutrophication-related environmental problems, in *Microbial Processes in Reservoirs,* Gunnison, D., Ed., Dr. W. Junk Publishers, Dordrecht, The Netherlands, 1985, 99.
2707. Lara, C., Romero, J.M., Coronil, T., and Guerrero, M.G., Interactions between photosynthetic nitrate assimilation and CO_2 fixation in cyanobacteria, in *Inorganic Nitrogen Metabolism,* Ullrich, W.R., Aparicio, P.J., Syrett, P.J., and Castillo, F., Eds., Springer-Verlag, Berlin, 1987, 45.
2708. Lara, C., Rodríguez, R., and Guerrero, M.G., Sodium-dependent nitrate transport and energetics of cyanobacteria, *J. Phycol.,* 29, 389, 1993.
2709. Larcher, W., *Physiological Plant Ecology,* Springer-Verlag, Berlin, 1983.
2710. La Roche, J., and Harrison, W.G., Reversible kinetic model for the short-term regulation of methylammonium uptake in two phytoplankton species, *Dunaliella tertiolecta* (Chlorophyceae) and *Phaeodactylum tricornutum* (Bacillariophyceae), *J. Phycol.,* 25, 36, 1989.
2711. Larsen, B., Biosynthesis of alginate, in *Proc. of the X. Int. Seaweed Symposium,* Levring, T., Ed., Göteborg, Sweden, 1981, 7.

2712. Larsen, D.P., and Mercier, H.T., Lake phosphorus loading graphs: An alternative, National Eutrophication Survey, Working Paper No. 174, U.S. E.P.A., Corvalis, Oregon, 1975.

2713. Larsen, J.E., Warren, G.F., and Langston, R., Effect of iron, aluminium and humic acid on phosphorus fixation by organic soils, *Soil Sci. Soc. Am. Proc.*, 24, 438, 1959.

2714. Larsen, V.J., and Schierup, H.-H., Macrophyte cycling of zinc, copper, lead and cadmium in the littoral zone of a polluted and non-polluted lake. II. Seasonal changes in heavy metal content of aboveground biomass and decomposition leaves of *Phragmites australis* (Cav.) Trin., *Aquat. Bot.*, 11, 211, 1981.

2715. Larsson, C.-M., and Larsson, M., Regulation of nitrate utilization in green algae, in *Inorganic Nitrogen Metabolism*, Ullrich, W.R., Aparicio, P.J., Syrett, P.J., and Castillo, F., Eds., Springer-Verlag, Berlin, 1987, 203.

2716. Lathwell, D.J., Bouldin, D.R., and Goyette, E.A., Growth and chemical composition of aquatic plants in twenty artificial wildlife marshes, *New York Fish and Game J.*, 20, 108, 1973.

2717. Latimer, W.M., *Oxidation potentials*, Prentice-Hall, Englewood Cliffs, New Jersey, 1952.

2718. Latter, P.M., and Cragg, J.B., The decomposition of *Juncus squarrosus* leaves and macrobiological changes in the profile of *Juncus* moor, *J. Ecol.*, 55, 469, 1967.

2719. Laube, H.R., and Wohler, J.R., Studies on the decomposition of a duckweed (Lemnaceae) community, *Bull. Torrey Bot. Club*, 100, 238, 1973.

2720. Laube, V.M., McKenzie, C.N., and Kushner, D.J., Strategies of responses to copper, cadmium, and lead by a blue-green alga, *Can. J. Microbiol.*, 26, 1300, 1980.

2721. Lavoie, A., and de la Noüe, J., Hyperconcentrated cultures of *Scenedesmus obliquus*. A new approach for wastewater biological tertiary treatment?, *Water Res.*, 19, 1437, 1985.

2722. Lawry, N.H., and Jensen, T.E., Deposition of condensed phosphate as an effect of varying sulfur deficiency in the cyanobacterium *Synechococcus* sp. (*Anacystis nidulans*), *Arch. Microbiol.*, 102, 1, 1979.

2723. Lay, J.A., and Ward, A.K., Algal community dynamics in two streams associated with different geological regions of the southeastern United States, *Arch. Hydrobiol.*, 108, 305, 1987.

2724. Lazinsky, D., and Sicko-Goad, L., Ultrastructural modification of three blue-green algae following heavy metal exposure, *Micron Microscop. Acta*, 14, 257, 1983.

2725. Lea, P.J., and Miflin, B.J., Alternative route for nitrogen assimilation in higher plants, *Nature*, 251, 614, 1974.

2726. Lea, P.J., and Miflin, B.J., The occurrence of glutamate synthase in algae, *Biochim. Biophys. Res. Commun.*, 64, 856, 1975.

2727. Leach, J.H., Epibenthic algal production in an intertidal mudflat, *Limnol. Oceanogr.*, 15, 514, 1970.

2728. Leadbeater, B.S.C., A fine structural study of *Olisthodiscus luteus* Carter, *Br. Phycol. J.*, 4, 3, 1969.

2729. Leadbeater, B.S.C., Preliminary observations on differences of scale morphology at various stages in the life cycle of *"Apistonema-Syracosphaera"* sensu von Stosch, *Br. Phycol. J.*, 5, 57, 1970.

2730. Leadbeater, B.S.C., Observations on the life history of the Haptophycean alga *Pleurochrysis scherffelli* with special reference to the microanatomy of the different types of motile cells, *Ann. Bot.*, 35, 429, 1971.

2731. Leadbeater, B.S.C., Observations by means of cine photography on the behavior of the haptonema in plankton flagellate of the class Haptophyceae, *J. Mar. Biol. Ass. U.K.*, 51, 207, 1971.

2732. Leadbeater, B.S.C., Fine structural observations on six new species of *Chrysochromulina* from the coast of Norway with preliminary observations on scale production in *C. microcylindrica* sp. nov., *Sarsia*, 49, 65, 1972.

2733. Leadbeater, B.S.C., Phagotrophic feeding in some marine flagellates, *Br. Phycol. J.*, 11, 196, 1976.

2734. Lean, D.R.S., Phosphorus compartments in lake water, Ph.D. Thesis, University of Toronto, Toronto, 1973.

2735. Lean, D.R.S., Phosphorus dynamics in lake waters, *Science*, 179, 678, 1973.

2736. Lean, D.R.S., Phosphorus kinetics in lake water: influence of membrane filter pore size and low pressure filtration, *J. Fish. Res. Board Can.*, 33, 2800, 1976.

2737. Lean, D.R.S., and Nalewajko, C., Phosphate exchange and organic phosphorus excretion by freshwater algae, *J. Fish. Res. Board Can.*, 33, 1312, 1976.

2738. Lean, D.R.S., and Rigler, F.G., A test of the hypothesis that abiotic phosphate complexing influences phosphorus kinetics in epilimnion lake water, *Limnol. Oceanogr.*, 19, 784, 1974.

2739. Leckie, J.O., and Davis, J.A., III., Aqueous environmental chemistry of copper, in *Copper in the Environment*, Part 1., Nriagu, J.O., Ed., John Wiley and Sons, New York, 1979, 90.

2740. Lee, G.B., Wetland soils of the upper midwest, in *Proc. Conf. Wetlands Ecology, Values, Impacts*, De Witt, J.R., and Soloway, E., Eds., Inst. for Environ. Studies, University of Wisconsin, 1977, 12.

2741. Lee, G.F., Bently, E., and Amundson, R., Effects of marshes on water quality, in *Coupling of Land and Water Systems*, Hasler, A.D., Ed., Springer-Verlag, New York, 1975, 105.

2742. Lee, G.F., Sonzogni, W.C., and Spear, R.D., Significance of oxic versus anoxic conditions for Lake Mendota sediment phosphorus release, in *Proc. Internat. Symp. Interactions Between Sediments in Fresh Water,* Golterman, H. Ed., Dr. W. Junk Publishers, The Hague, The Netherlands, 1977, 294.

2743. Lee, G.F., Jones, R.A., and Rast, W., Availability of phosphorus to phytoplankton and its implications for phosphorus management strategies, in *Phosphorus Management Strategies for Lakes,* Loehr, R.C., Ed., Ann Arbor Science, Ann Arbor, Michigan, 1980, 259.

2744. Lee, R.E., Evolution of algal flagellates with chloroplast endoplasmic reticulum from the ciliates, *S. Afr. J. Sci.,* 73, 179, 1977.

2745. Lee, R.E., *Phycology,* 2nd ed., Cambridge University Press, Cambridge, 1989.

2746. Lee, S., and Aronoff, S., Boron in plants: a biochemical role, *Science,* 158, 798, 1967.

2747. Leedale, G.F., *Euglenoid Flagellates,* Prentice-Hall, Englewood Cliffs, New Jersey, 1967.

2748. Leedale, G.F., The euglenoids, in *Oxford Biology Readers* 5, Head, J.J., and Lowenstein, O.E., Eds., Oxford University Press, London, 1971, 1.

2749. Leedale, G.F., How many are the kingdoms of organisms, *Taxon,* 23, 261, 1974.

2750. Leedale, G.F., Envelope formation and structure in the euglenoid genus *Trachelomonas, Br. Phycol. J.,* 10, 17, 1975.

2751. Leedale, G.F., Phylogenetic criteria in euglenoid flagellates, *Biosystems,* 10, 185, 1978.

2752. Leedale, G.F., Euglenophycophyta, in *Synopsis and Classification of Living Organisms,* Vol. 1, Parker, S.P., Ed., McGraw-Hill, New York, 1982, 129.

2753. Lee-Kaden, J., and Simons, W., Amino acid uptake and energy coupling dependent on photosynthesis in *Anabaena nidulans, J. Bacteriol.,* 151, 229, 1982.

2754. Lefeuvre, J.C., *Proc. 3rd Internat Wetlands Conf.,* Muséum National d'Histoire Naturelle, Paris, 1989.

2755. Lefévre, M., L'utilisation des algues d'eau douce par les Cladocéres, *Bull Biol. Fr. et Belg.,* 76, 250, 1942.

2756. Lefévre, M., Extracellular products of algae, in *Algae and Man,* Jackson, D.F., Ed., Plenum Press, New York, 1964, 337.

2757. Lefor, M.W., and Kennard, W.C., *Inland Wetland Definitions,* University of Connecticut Institute of Water Resources Report No. 28, 1977.

2758. Leftley, J.W., and Syrett, P.J., Urease and ATP:urea amidolase activity in unicellular algae, *J. Gen. Microbiol.,* 77, 109, 1973.

2759. Lehman, J.T., Ecological and nutritional studies on *Dinobryon* Ehrenb.: Seasonal periodicity and the phosphate toxicity problem, *Limnol. Oceanogr,* 21, 546, 1976.

2760. Lehman, J.T., Photosynthetic capacity and luxury uptake of carbon during phosphate limitation in *Pediastrum duplex* (Chlorophyceae), *J. Phycol.,* 12, 190, 1976.

2761. Lehman, J.T., Enhanced transport of inorganic carbon into algal cells and its implications for the biological fixation of carbon, *J. Phycol.,* 14, 33, 1978.

2762. Lehman, J.T., and Sandgren, C.D., Species-specific rates of growth and grazing loss among freshwater algae, *Limnol. Oceanogr,* 30, 34, 1985.

2763. Lehmann, H., and Jost, M., Kinetics of the assembly of gas vacuoles in the blue-green alga *Microcystis aeruginosa* Kütz. emend. Elekin, *Arch. Mikrobiol.,* 79, 59, 1971.

2764. Lehman, T.D., Botkin, D.B., and Likens, G.E., The assumptions and rationales of a computer model of phytoplankton population dynamics, *Limnol. Oceanogr,* 10, 343, 1975.

2765. Lehninger, A.L., *Biochemistry* 2nd ed., Worth, New York, 1975.

2766. Lehtonen, J., Effects of acidification on the metal levels in aquatic macrophytes in Espoo, Finland, *Ann. Bot. Fennici,* 26, 39, 1989.

2767. Leitgeb, H., Die Inkrustation der Membran von *Acetabularia, Sitzber math. naturw. Kl. bayer Akad. Wiss. München, Serie 1,* 96, 13, 1887.

2768. Leith, A.R., Calcification of the Charophyte oosporangium, in *Calcareous Algae and Stromatolites,* Riding, R., Ed., Springer-Verlag, Berlin, 1991, 204.

2769. Leland, H.V., and McNurney, J.M., Lead transport in a river ecosystem, in *Proc. Internat. Conf. Transport of Persistent Chemicals in Aquatic Ecosystems,* Ottawa, Canada, 1974, 17.

2770. Leland, H.V., and Kuwabara, J.S., Trace metals, in *Fundamentals of Aquatic Toxicology,* Rand, G.M., and Petrocelli, S.R., Eds., Hemisphere Publishing Corp., New York, 1984, 374.

2771. Leland, H.V., Copenhaver, E.D., and Wilkes, D.J., Heavy metals and other trace elements, *J. Water Pollut. Control Fed.,* 47, 1635, 1975.

2772. Leland, H.V., Wilkes, D.J., and Copenhaver, E.D., Heavy metals and related trace elements, *J. Water Pollut. Control Fed.,* 48, 1459, 1976.

2773. Leland, H.V., Luoma, S.N., and Wilkes, D.J., Heavy metals and related trace elements, *J. Water Pollut. Control Fed.,* 49, 1340, 1977.

2774. Lembi, C.A., Unicellular Chlorophyta, in *Phytoflagellates, Development in Marine Biology*, Vol. 2, Cox, E.R., Ed., Elsevier/North Holland, New York, 5, 1980.

2775. Lemke, P.A., Viruses of eukaryotic microorganisms, *Ann. Rev. Microbiol.*, 30, 105, 1976.

2776. Lemmermann, E., Algen I., *Krypt.-Fl. Mark Brandenburg*, 3, 1, 1907–1910.

2777. Lemon, E., and Kristensen, J., An edaphic expression of soil structure, *Trans. 7th Internat. Congr. Soil Sci.*, 1, 232, 1960.

2778. Leonian, L.H., Effect of auxins from some green algae upon *Phytophora cactorum*, *Bot. Gaz.*, 97, 854, 1936.

2779. Lerman, A., Mackenzie, F.T., and Garrels, R.M., Modelling of geochemical cycles: phosphorus as an example, *Geol. Soc. Am. Mem.*, 142, 205, 1975.

2780. Les, A., and Walker, R.W., Toxicity and binding of copper, zinc, and cadmium by the blue-green alga, *Chroococcus paris*, *Water Air Soil Pollut.*, 23, 129, 1984.

2781. Leskinen, E., Colonization of periphytic organisms on artificial substrata on the southwestern coast of Finland, *Ophelia, Suppl.*, 3, 137, 1984.

2782. Leskinen, E., and Sarvala, J., Community analysis of diatom colonization on artificial substrata in a northern Baltic Sea archipelago: a comparison of methods, *Ann. Bot. Fenn.*, 25, 21, 1988.

2783. Levander, D.A., Selected aspects of the comparative metabolism and biochemistry of selenium and sulfur, in *Trace Elements in Human Health and Disease*, Vol. 2, Pradsad, A.S., Ed., Academic Press, New York, 1976, 135.

2784. Levedahl, B.H., Heterotrophic CO_2 fixation in *Euglena*, in *The Biology of Euglena*, II., Buetow, D.E., Ed., Academic Press, New York and London, 1968, 85.

2785. Levin, E.Y., and Bloch, K., Absence of sterols in blue-green algae, *Nature*, 202, 90, 1964.

2786. Levinthal, M., and Schiff, J.A., Formation of thiosulfate by a cell-free sulfate reducing system from *Chlorella*, *Plant Physiol.*, 38 (suppl.), xii, 1963.

2787. Levinthal, M., and Schiff, J.A., Intermediates in thiosulfate formation by cell-free sulfate-reducing system from *Chlorella*, *Plant Physiol.*, 40, (suppl.), xvi, 1965.

2788. Levinthal, M., and Schiff, J.A., Studies of sulfate utilization by algae. 5. Identification of thiosulfate as a major acid-volatile product formed by a cell-free sulfate reducing system from *Chlorella*, *Plant Physiol.*, 43, 555, 1968.

2789. Levring, T., Ed., *Proc. X. Int. Seaweed Symposium*, Göteborg, Sweden, 1981.

2790. Levring, G.T., Hoppe, H.A., and Schmid, O.J., *Marine Algae. A Survey of Research and Utilization*, Cram, De Gruyter and Co., Hamburg, 1969.

2791. Levy, I., and Gantt, E., Light acclimation in *Porphyridium purpureum* (Rhodophyta): growth, photosynthesis, and phycobilisomes, *J. Phycol.*, 24, 452, 1988.

2792. Levy, L.W., and Strauss, R., Recherches sur la precipitation des carbonales alaclino-terreux chez les Characées, *Hydrobiologia*, 45, 217, 1974.

2793. Lewin, J.C., Obligate autotrophy in *Chlamydomonas moewusii* Gerloff, *Science*, 112, 652, 1950.

2794. Lewin, J.C., Heterotrophy in diatoms, *J. Gen. Microbiol.*, 9, 305, 1953.

2795. Lewin, J.C., Silicon metabolism in diatoms. I. Evidence for the role of reduced sulfur compounds in Si utilization, *J. Gen. Physiol.*, 37, 589, 1954.

2796. Lewin, J.C., Silicon metabolism in diatoms. II. Sources of silicon for growth of *Navicula pelliculosa*, *Plant Physiol.*, 30, 129, 1955.

2797. Lewin, J.C., Silicon metabolism in diatoms. III. Respiration and silicon uptake in *Navicula pelliculosa J. Gen. Physiol.*, 39, 1, 1955.

2798. Lewin, J.C., Silicon metabolism in diatoms. IV. Growth and frustule formation in *Navicula pelliculosa*, *Can. J. Microbiol.*, 3, 427, 1957.

2799. Lewin, J.C., Calcification, in *Physiology and Biochemistry of Algae*, Lewin, R.A., Ed., Academic Press, New York and London, 1962, 457.

2800. Lewin, J.C., and Silification, J.C., in *Physiology and Biochemistry of Algae*, Lewin, R.A., Ed., Academic Press, New York and London, 1962, 445.

2801. Lewin, J.C., The boron requirement of a marine diatom, *Naturwissenschaften*, 52, 70, 1965.

2802. Lewin, J., Boron as a growth requirement for diatoms, *J. Phycol.*, 2, 160, 1966.

2803. Lewin, J.C., Physiological studies of the boron requirement of the diatom, *Cylindrotheca fusiformis* Reimann and Lewin, *J. Exp. Bot.*, 17, 473, 1966.

2804. Lewin, J., Persistent blooms of surf diatoms along the northeast coast, in *The Marine Plant Biomass of the Pacific Northwest Coast*, Krauss, R., Ed., Oregon State University Press, Corvallis, 1977, 81.

2805. Lewin, J.C., and Lewin, R.A., Auxotrophy and heterotrophy in marine littoral diatoms, *Can. J. Microbiol.*, 6, 127, 1960.

2806. Lewin, J.C., and Guillard, R.R.L., Diatoms, *Ann. Rev. Microbiol.*, 17, 373, 1963.

2807. Lewin, J., and Chen, C.-H., Silicon metabolism in diatoms. VI. Silicic acid uptake by a colorless marine diatom *Nitzschia alba* Lewin and Lewin, *J. Phycol.*, 4, 161, 1968.

2808. Lewin, J., and Reimann, B.E.F., Silicon and plant growth, *Ann. Rev. Plant Physiol.*, 20, 289, 1969.
2809. Lewin, J., and Chen, C.H., Available iron: a limiting factor for marine phytoplankton, *Limnol. Oceanogr.*, 16, 670, 1971.
2810. Lewin, J., and Mackas, D., Blooms of surf-zone diatoms along the coast of the Olympic Peninsula, Washington. I. Physiological investigations of *Chaetoceros armatum* and *Asterionella socialis* in laboratory culture, *Mar. Biol.*, 16, 171, 1972.
2811. Lewin, J., and Helleburst, J.A., Heterotrophic nutrition of the marine pennate diatom *Navicula pavillardi* Hustedt, *Can. J. Microbiol.*, 21, 1235, 1975.
2812. Lewin, J., and Helleburst, J.A., Heterotrophic nutrition of the marine pennate diatom *Nitzschia angularis* var. *affinis*, *Mar. Biol. (Berlin)*, 36, 313, 1976.
2813. Lewin, J., and Helleburst, J.A., Utilization of glutamate and glucose for heterotrophic growth by the marine pennate diatom *Nitzschia laevis*, *Mar. Biol. (Berlin)*, 47, 1, 1978.
2814. Lewin, J., and Chen, C.-H., Effects of boron deficiency on the chemical composition of a marine diatom, *J. Exp. Bot.*, 27, 916, 1976.
2815. Lewin, J.C., Reimann, B.E., Busby, W.F., and Volcani, B.E., Silica shell formation in synchronously dividing diatoms, In *Cell Synchrony*, Cameron, I.L., and Padilla, G.M., Eds., Academic Press, New York, 1965, 169.
2816. Lewin, R.A., Extracellular polysaccharides of green algae, *Can. J. Microbiol.*, 2, 665, 1956.
2817. Lewin, R.A., Vitamin-bezonoj de algoj, in *Sciencaj Studoj*, Neergaard, P., Ed., Modersmaalet, Haderslev, Copenhagen, Denmark, 1958, 187.
2818. Lewin, R.A., The isolation of algae, *Rev. Algol.*, 3, 181, 1959.
2819. Lewin, R.A., Phytoflagellates and algae, in *Encyclopedia of Plant Physiology*, Ruhland, W., Ed., Springer-Verlag, Berlin, 1961, 401.
2820. Lewin, R.A., Ed., *Physiology and Biochemistry of Algae*, Academic Press New York and London, 1962.
2821. Lewin, R.A., Biochemistry and physiology of algae: taxonomic and phylogenetic considerations, in *Algae, Man and the Environment*, Jackson, D.F., Ed., Syracuse University Press, Syracuse, 1968, 15.
2822. Lewin, R.A., Auxotrophy in marine littoral diatoms, in *Proc. VII. Int. Seaweed Symp.*, 1972, 316.
2823. Lewin, R.A., A marine *Synechocystis* (Cyanophyta, Chroococcales) epizoic on ascidians, *Phycologia*, 14, 153, 1975.
2824. Lewin, R.A., Ed., *The Genetics of Algae*, Blackwell Scientific Publications, London, 1976.
2825. Lewin, R.A., Naming the blue-green algae, *Nature*, 259, 360, 1976.
2826. Lewin, R.A., Prochlorophyta as a proposed new division of algae, *Nature*, 261, 697, 1976.
2827. Lewin, R.A., Prochloron-type genus of the Prochlorophyta, *Phycologia*, 16, 217, 1977.
2828. Lewin, R.A., Formal taxonomic treatment of cyanophytes, *Int. J. Syst. Bacteriol.*, 29, 411, 1979.
2829. Lewin, R.A., *Prochloron* and the theory of symbiosis, *Ann. N.Y. Acad. Sci.*, 361, 325, 1981.
2830. Lewin, R.A., The prochlorophytes, in *The Prokaryotes*, Starr, M.P., Stolz, H., Trüper, H.G., Balows, A., and Schlegel, G.H., Eds., Springer-Verlag, Berlin, 1981, 256.
2831. Lewin, R.A., *Prochloron*—a status report, *Phycologia*, 23, 203, 1984.
2832. Lewin, R.A., Calcification of cell walls of *Chlamydomonas* (Volvocales, Chlorophyta) on agar media, *Phycologia*, 29, 536, 1990.
2833. Lewin, R.A., and Cheng, L., Associations of microscopic algae with didemnid ascidians, *Phycologia*, 14, 149, 1975.
2834. Lewin, R.A., and Withers, N.W., Extraordinary pigment composition of a prokaryotic alga, *Nature*, 256, 735, 1975.
2835. Lewin, R.A., and Cheng, L., Eds., *Prochloron: A Microbial Enigma*, Chapman and Hall, New York, London, 1989.
2836. Lewin, R.A., Cheng, L., and Lafargue, F., Prochlorophytes in the Caribbean, *Bull. Mar. Sci.*, 30, 744, 1980.
2837. Lewis, A.G., and Cave, W.R., The biological importance of copper in oceans and estuaries, *Oceanogr. Mar. Biol. Ann. Rev.*, 20, 471, 1982.
2838. Lewis, A.G., Whitfield, R.H., and Ramnarine, A., Some particulate and soluble agents affecting the relationship between metal toxicity and organism survival in the calanoid copepod, *Euchaeta japonica*, *Mar. Biol.*, 17, 215, 1972.
2839. Lewis, F.J., and Dowding, E.S., The vegetation and retrogressive changes of peat areas (muskegs) in Central Alberta, *J. Ecol.*, 14, 317, 1926.
2840. Lewis, D.H., and Smith, D.C., Sugar alcohols (polyols) in fungi and green plants. I. Distribution, physiology and metabolism, *New Phytol.*, 66, 1443, 1967.
2841. Lewis, O.A.M., *Plants and Nitrogen*, Studies in Biology No. 166, Edward Arnold, London, 1986.
2842. Lewis, W.M., Jr., Dynamics and succession of the phytoplankton in a tropical lake: Lake Lanao, Philippines, *J. Ecol.*, 66, 849, 1978.

2843. Lhotský, O., and Marvan, P., Biomass production and nutrient standing stocks in some filamentous algae, *Proc. Conf. Biological Problems in Water Management III.*, Košice, Slovakia, 1975, 1.

2844. Li, W.K.W., Kinetic analysis of interactive effects of cadmium and nitrate on growth of *Thalassiosira fluviatilis* (Bacillariophyceae), *J. Phycol.*, 14, 454, 1978.

2845. Li, W.K.W., Temperature adaptation in phytoplankton: cellular and photosynthetic characteristics, in *Primary Productivity in the Sea*, Falkowski, P.G., Ed., Plenum Press, New York, 1980, 259.

2846. Liaw, W.K., and MacCrimmon, H.R., Assessing changes in biomass of riverbed periphyton, *Int. Rev. Ges. Hydrobiol.*, 63, 155, 1978.

2847. Lichtfuss, R., and Brummer, G., Natürlicher Gehalt und anthropogene Anreicherung von Schwermetallen in den Sedimenten von Elbe, Eider, Trave and Schwentine, *Catena*, 8, 251, 1981.

2848. Lieffers, V.J., Growth of *Typha latifolia* in boreal forest habitats as measured by double sampling, *Aquat. Bot.*, 15, 335, 1983.

2849. Lieth, H., Primary productivity of the major vegetation units of the world, in *Productivity of the Biosphere*, Lieth, H., and Whittaker, R.H., Eds., Springer-Verlag, New York, 1975, 203.

2850. Liewendahl, K., and Turula, M., Iodine induced goiter and hypo-thyroidism in a patient with chronic lymphocytic thyroiditis, *Acta Endocrinol.* (Copenhagen), 71, 289, 1972.

2851. Likens, G.E., Eutrophication and aquatic ecosystems, in *Nutrients and Eutrophication: The Limiting-Nutrient Controversy* Likens, G.E., Ed., Spec. Symposia, Vol. 1, Amer. Soc. Limnol. Oceanogr., Inc., Alen Press, Lawrence, Kansas, 1972, 3.

2852. Likens, G.E., Ed., *Nutrients and Eutrophication: The Limiting-NutrientControversy* Spec. Symposia, Vol. 1, Amer. Soc. Limnol. Oceanogr., Inc., Alen Press, Lawrence, Kansas, 1972.

2853. Lin, C.K., Accumulation of water soluble phosphorus and hydrolysis of polyphosphates by *Cladophora glomerata* (Chlorophyceae), *J. Phycol.*, 13, 46, 1977.

2854. Lin, C.K., and Schelske, C.L., Effects of nutrient enrichment, light intensity and temperature on growth of phytoplankton from lake Huron, *Great Lakes Res. Div., Univ. Mich. Publ.* No. 63, 1978.

2855. Linacre, E.T., Swamps, in *Vegetation and Atmosphere*, Vol. 2. *Case Studies*, Monteith, J.L., Ed., Academic Press, London, 1976, 329.

2856. Lindau, C.W., Patrick, W.H., Jr., DeLaune, R.D., Reddy, K.R., and Bolich, P.K., Entrapment of nitrogen-15 dinitrogen during soil denitrification, *Soil Sci. Soc. Am. J.*, 52, 538, 1988.

2857. Lindberg, B., Low-molecular carbohydrates in algae. XI. Investigation of *Porphyra umbilicalis, Acta Chem. Scand.*, 9, 1097, 1955.

2858. Lindberg, B., Methylated taurines and choline sulfate in red alga, *Acta Chem. Scand.*, 9, 1323, 1955.

2859. Lindsay, A.L., *Chemical Equilibria in Soils*, John Wiley and Sons, New York, 1979.

2860. Lindschooten, C., van Bleijswijk, J.D.L., van Emburg, P.R., de Vrind, J.P.M., Kempers, E.S., Westbroek, P., and de Vrind-de Jong, E.W., Role of the light-dark cycle and medium composition on the production of coccoliths by *Emiliana huxleyi* (Haptophyceae), *J. Phycol.*, 27, 1991, 82.

2861. Lindsley, D., Schuck, T., and Stearns, F., Productivity and nutrient content of emergent macrophytes in two Wisconsin marshes, in *Freshwater Wetlands and Sewage Effluent Disposal*, Tilton, D.L., Kadlec, R.H., and Richardson, C.J., Eds., University of Michigan, Ann Arbor, 51, 1976.

2862. Lindström, K., *Peridinium cinctum* bioassays of Se in Lake Erken, *Arch. Hydrobiol.*, 89, 110, 1980.

2863. Lindström, K., Environmental requirements of the dinoflagellate *Peridinium cinctum* fa. *westii*. Abstracts of Uppsala Dissertations from the Faculty of Science, University of Uppsala, Sweden, No. 646, 1982.

2864. Lindström, K., Selenium as a growth factor for plankton algae in laboratory experiments and in some Swedish lakes, *Hydrobiologia*, 101, 35, 1983.

2865. Lindström, K., Effect of temperature, light and pH on growth, photosynthesis and respiration of the dinoflagellate *Peridinium cinctum* fa. *westii* in laboratory cultures, *J. Phycol.*, 20, 212, 1984.

2866. Lindström, K., Selenium and algal growth, in *Proc. Third Internat. Symp. on Industrial Uses of Selenium and Tellurium*, Selenium-Tellerium Development Association, Inc., Stockholm, 1984, 441.

2867. Lindström, K., Selenium requirement of the dinoflagellate *Peridiopsis borgei* (Lemm.), *Int. Rev. Ges. Hydrobiol.*, 70, 77, 1985.

2868. Lindström, K., Nutrient requirements of the dinoflagellate *Peridinium gatunense, J. Phycol.*, 27, 207, 1991.

2869. Lindström, K., and Rodhe, W., Selenium as a micronutrient for the dinoflagellate *Peridinium cinctum* fa. *westii, Mitt. Internat. Verein. Limnol.*, 21, 168, 1978.

2870. Linn, J.G., Goodrich, R.D., Meiske, J.C., and Staba, E.J., *Aquatic Plants from Minnesota.* Part 4. *Nutrient Composition,* Water Resources Res. Center, University of Minnesota Bull. 56, 1973.

2871. Linnaues, C., Species plantarum, in *Carl Linnaeceus' Species Plantarum*, Stearn, W.T., Ed., Royal Society London, 1953.

2872. Linnaeus, C., *Genera Plantarum*, 5th Ed., Holmiae, 1954.

2873. Linthurst, R.A., An evaluation of biomass, stem density, net aerial primary production (NAPP) and NAPP estimation methodology for selected estuarine angiosperms in Maine, Delaware, and Georgia, M.Sc. Thesis, North Carolina State University, Raleigh, North Carolina, 1977.

2874. Littler, M.M., The productivity of Hawaiian fringing-reef crustose corallinaceae and an experimental evaluation of productivity methodology, *Limnol. Oceanogr,* 18, 946, 1973.

2875. Littler, M.M., and Murray, S.N., The primary productivity of marine macrophytes from a rocky intertidal community, *Mar. Biol. (Berl.),* 27, 131, 1974.

2876. Littler, M.M., and Littler, D.S., The evolution of thallus form and survival strategies in benthic marine macroalgae: field and laboratory tests of a functional form model, *Am. Nat.,* 116, 25, 1980.

2877. Littler, M.M., and Arnold, K.E., Primary productivity of marine macroalgal functional-form groups from southwestern North America, *J. Phycol.,* 18, 307, 1982.

2878. Liu, M.S., and Helleburst, J.A., Utilization of amino acids as nitrogen sources, and their effects on nitrate reductase in the marine diatom *Cyclotella cryptica, Can. J. Microbiol.,* 20, 1119, 1974.

2879. Liu, M.S., and Helleburst, J.A., Uptake of amino acids by the marine centric diatom *Cyclotella cryptica, Can. J. Microbiol.,* 20, 1109, 1974.

2880. Liu, M.S., and Helleburst, J.A., Effects of salinity and osmolarity of the medium on amino acid metabolism in *Cyclotella cryptica, Can. J. Bot.,* 54, 938, 1976.

2881. Liu, M.S., and Helleburst, J.A., Regulation of proline metabolism in the marine centric diatom *Cyclotella cryptica, Can. J. Bot.,* 54, 949, 1976.

2882. Llama, M.J., Macarulla, J.M., and Serra, J.L., Characterization of the nitrate reductase activity in the diatom *Skeletonema costatum, Plant Sci. Lett.,* 14, 169, 1979.

2883. Lloyd, F.E., *The Carnivorous Plants,* Chronica Botanica, Waltham, Massachusetts, 1942.

2884. Lloyd, N.D.H., Canvin, D.T., and Culver, D.A., Photosynthesis and photorespiration in algae, *Plant Physiol.,* 59, 936, 1977.

2885. Lobban, C.S., and Wynne, M.J., Eds., *The Biology of Seaweeds,* Blackwell Scientific Publications, Oxford, 1981.

2886. Lobban, C.S., Harrison, P.J., and Duncan, M.J., *The Physiological Ecology of Seaweeds,* Cambridge University Press, Cambridge, London, 1985.

2887. Lobban, C.S., Chapman, D.J., and Kremer, B.P., *Experimental Phycology. A Laboratory Manual,* Cambridge University Press, Cambridge, 1988.

2888. LoCicero, V.R., Ed., *Proc. of the 1st Internat. Conf. on Toxic Dinoflagellate Blooms,* Massachusetts Science and Technology Foundation, Wakefield, Massachusetts, 1975.

2889. Lodenius, M., Aquatic plants and littoral sediments as indicators of mercury pollution in some areas in Finland, *Ann. Bot. Fennici,* 17, 336, 1980.

2890. Loeb, S.L., An in situ method for measuring the primary productivity and standing crop of the epilithic periphyton community in lentic systems, *Limnol. Oceanogr,* 26, 394, 1981.

2891. Loeb, S.L., and Reuter, J.E., The epilithic periphyton community: a five-lake comparative study of community productivity, nitrogen metabolism and depth-distribution of standing crop, *Verh. Internat. Verein. Limnol.,* 21, 346, 1981.

2892. Loeb, S.L., Reuter, J.E., and Goldman, C.R., Littoral zone production of oligotrophic lakes: The contribution of phytoplankton and periphyton, in *Periphyton of Freshwater Ecosystems,* Wetzel, R.G., Ed., Dr. W. Junk Publishers, The Hague, The Netherlands, 1983, 161.

2893. Loeblich, A.R., III, Aspects of the physiology and biochemistry of the Pyrrophyta, *Phykos,* 5, 216, 1966.

2894. Loeblich, A.R., III, and Loeblich, L.A., Division Eustigmatophyta, in *CRC Handbook of Microbiology,* 2nd ed., Vol. II. *Fungi, Algae, Protozoa and Viruses,* Laskin, A.I., and Lechevalier, H.A., Eds., CRC Press, Boca Raton, Florida, 1978, 481.

2895. Loeblich, A.R., III, and Sherley, J.L., Observations on the theca of the motile phase of free-living and symbiotic isolates of *Zooxanthella microadriatica* (Freudenthal) com. nov., *J. Mar. Biol. Ass. U.K.,* 59, 195, 1979.

2896. Loeblich, L.A., Growth limitation of *Dunaliella salina* by CO_2 at high salinity, *J. Phycol.,* 6 (suppl.), 9, 1970.

2897. Löffler, H., Human uses, in *Wetlands and Shallow Continental Water Bodies,* Vol. 1, Patten, B.C., Ed., SPB Academic Publishing, The Hague, The Netherlands, 1990, 17.

2898. Logan, T.J., Chemical extraction as an index of bioavailability of phosphate in Lake Erie Basin suspended sediments, Lake Erie Wastewater Management Study, Buffalo, 1978.

2899. Lohman, K., and Priscu, J.C., Physiological indicators of nutrient deficiency in *Cladophora* (Chlorophyta) in the Clark Fork of the Columbia River, Montana, *J. Phycol.,* 28, 443, 1992.

2900. Lohmann, H., Untersuchungen zur Feststellung des vollständigen Gehaltes des Meeres an Plankton. *Wiss. Meeresunters. Abt. Kiel.,* 10, 131, 1908.

2901. Lokhorst, G.M., Taxonomic studies on the marine and brackish-water species of *Ulothrix* (Ulotrichales, Chlorophyceae) in Western Europe, *Blumea*, 24, 191, 1978.

2902. Loneragan, J.F., and Arnon, D.I., Molybdenum in the growth and metabolism of *Chlorella*, *Nature*, 174, 459, 1954.

2903. Long, S.P., and Mason, C.F., *Saltmarsh Ecology*, Blackie, Glasgow and London, 1983.

2904. Longstreth, D.J., Photosynthesis and photorespiration in freshwater emergent and floating plants, *Aquat. Bot.*, 34, 287, 1989.

2905. Lopatin, V.D., "Gladkoe" boloto (Torfyanayazalezh i bolotnye flatsii), *Ukr. Zap. Leningradsk. Gos. Univ. ser. Geogr. Nauk.*, 9, 95, 1954.

2906. López-Figueroa, F., and Rüdiger, W., Stimulation of nitrate net uptake and reduction by red and blue light and reversion by far-red light in the green alga *Ulva rigida*, *J. Phycol.*, 27, 389, 1991.

2907. Lorch, D., and Weber, A., Accumulation, toxicity and localization of lead in cryptogams: experimental results, *Symp. Biol. Hungar.*, 29, 51, 1985.

2908. Lorenz, J.S., and Biesboer, D.D., Nitrification, denitrification, and ammonia diffusion in a cattail marsh, in *Aquatic Plants for Water Treatment and Resource Recovery* Reddy, K.R., and Smith, W.H., Eds., Magnolia Publishing Co., Orlando, Florida, 1987, 525.

2909. Lorenz, R.C., and Herdendorf, C.E., Growth dynamics of *Cladophora glomerata* in western Lake Erie in relation to some environmental factors, *J. Great Lakes Res.*, 8, 42, 1982.

2910. Lorenzen, H., Temperatureinflüsse auf *Chlorella pyrenoidosa* unter besonderer Berücksichtigung der Zellentwicklung, *Flora*, 153, 554, 1963.

2911. Lorenzen, M.W., Smith, D.J., and Kimmel, L.V., A long-term phosphorus model for lakes: Application to Lake Washington, in *Modeling Biochemical Processes in Aquatic Ecosystems*, Canale, R.P., Ed., Ann Arbor Science, Ann Arbor, Michigan, 1976, 75.

2912. Lorimer, G.H., The carboxylation and oxygenation of ribulose 1,5-biphosphate: the primary events in photosynthesis and photorespiration, *Ann. Rev. Plant Physiol.*, 32, 349, 1981.

2913. Losada, M., The photosynthetic assimilation of nitrate, *Proc. Conf. of Federation of European Societies of Plant Physiology* Santiago, 1980, 112.

2914. Losada, M., and Guerrero, M.G., The photosynthetic reduction of nitrate and its regulation, in *Photosynthesis in Relation to Model Systems*, Barber, J., Ed., Elsevier Scientific Publications, Amsterdam, 1979, 365.

2915. Losada, M., Guerrero, M.G., and Vega, J., The assimilatory reduction of nitrate, in *Biochemistry and Physiology of Nitrogen and Sulfur Metabolism*, Bothe, H., and Trebst, A., Eds., Springer Verlag, Berlin, 1981.

2916. Losada, M., Paneque, A., Aparicio, P.J., Vega, J.M., Cárdenas, J., and Herrera, J., Inactivation and repression by ammonium of the nitrate reducing system in *Chlorella*, *Biochem. Biophys. Res., Commun.*, 38, 1009, 1970.

2917. Losee, R.F., and Wetzel, R.G., Selective light attenuation by the periphytic complex, in *Periphyton of Freshwater Ecosystems*, Wetzel, R.G., Ed., Dr. W. Junk Publishers, The Hague, The Netherlands, 1983, 89.

2918. Louisier, J.D., and Parkinson, D., Litter decomposition in a cool temperate deciduous forest, *Can. J. Bot.*, 54, 419, 1976.

2919. Love, J., and Percival, E., The polysaccharides of the green seaweed *Codium fragile*. Part II. The water soluble sulphated polysaccharides, *J. Chem. Soc.*, p. 3338, 1964.

2920. Lovley, D.R., Organic matter mineralization with the reduction of ferric iron: a review, *Geomicrobiol. J.*, 5, 375, 1987.

2921. Lovley, D.R., and Phillips, E.J.P., Organic matter mineralization with reduction of ferric iron in anaerobic sediments, *Appl. Environ. Microbiol.*, 51, 683, 1986.

2922. Lovley, D.R., Phillips, E.J.P., and Lonergan, D.J., Hydrogen and formate oxidation coupled to dissimilatory reduction of iron or manganese by *Alteromonas putrefaciens*, *Appl. Environ. Microbiol.*, 55, 700, 1989.

2923. Lowe, R.L., Environmental Requirements and Pollution Tolerance of Freshwater Diatoms, Environmental Protection Agency, Environment Monitoring Service, EPA-670/4-74-005, Cincinnati, 1974.

2924. Lowe, R.L., and Collins, G.B., An aerophilous diatom community from Hocking County, Ohio, *Trans. Am. Microsc. Soc.*, 92, 492, 1973.

2925. Lowe, R.L., and Gale, W.F., Monitoring river periphyton with artificial benthic substrates, *Hydrobiologia*, 69, 235, 1980.

2926. Lowe, R.L., Rosen, B.H., and Fairchild, G.W., Endosymbiotic blue-green algae in freshwater diatoms: an advantage in nitrogen-poor habitats, *J. Phycol.*, (suppl.), 20, 24, 1984.

2927. Lowry, R.J., Sussman, S., and von Böventer, B., Physiology of the cell surface of *Neurospora ascospores*. III. Distinction between the adsorptive and entrance phases of cation uptake, *Mycologia*, 49, 609, 1957.

2928. Lu, M., and Stephens, G.C., Demonstration of net influx of free amino acids in *Phaeodactylum tricornutum* using high performance liquid chromatography, *J. Phycol.*, 20, 584, 1984.

2929. Lubchenko, J., and Gaines, S.D., A unified approach to marine plant-herbivore interactions. I. Populations and communities, *Ann. Rev. Ecol. Syst.*, 12, 405, 1981.

2930. Lucas, W.J., Analysis of the diffusion symmetry developed by the alkaline and acid bands which form at the surface of *Chara corralina* cells, *J. Exp. Bot.*, 26, 271, 1975.

2931. Lucas, W.J., Photosynthetic fixation of ^{14}carbon by internodal cells of *Chara corallina*, *J. Exp. Bot.*, 26, 331, 1975.

2932. Lucas, W.J., The influence of light intensity on the activation and operation of the hydroxyl efflux system of *Chara corallina*, *J. Exp. Bot.*, 26, 347, 1975.

2933. Lucas, W.J., Plasmalemma transport of HCO_3^- and OH^- in *Chara corallina*: non-antiporter system, *J. Exp. Bot.*, 27, 19, 1976.

2934. Lucas, W.J., The influence of Ca^{2+} and K^+ on $H^{14}CO_3^-$ influx in internodal cells of *Chara corallina*, *J. Exp. Bot.*, 27, 32, 1976.

2935. Lucas, W.J., Plasmalemma transport of HCO_3^- and OH^- in *Chara corallina*: inhibitory effect of ammonium sulphate, *J. Exp. Bot.*, 28, 1307, 1977.

2936. Lucas, W.J., Analogue inhibition of the active HCO_3^- transport site in the characean plasma membrane, *J. Exp. Bot.*, 28, 1321, 1977.

2937. Lucas, W.J., Alkaline band formation in *Chara corallina* due to OH^- efflux or H^+ influx? *Plant Physiol.*, 63, 248, 1979.

2938. Lucas, W.J., Control and synchronization of HCO_3^- and OH^- transport during photosynthetic assimilation of exogenous HCO_3^-, in *Plant Membrane Transport: Current Conceptual Issues*, Spanswick, R.M., Lucas, W.J., and Dainty, J., Eds., Elsevier, Amsterdam, 1980, 317.

2939. Lucas, W.J., Mechanisms of acquisition of exogenous bicarbonate by internodal cells of *Chara corallina*, *Planta*, 156, 181, 1982.

2940. Lucas, W.J., Photosynthetic assimilation of exogenous HCO_3^- by aquatic plants, *Ann. Rev. Plant Physiol.*, 34, 71, 1983.

2941. Lucas, W.J., Bicarbonate utilization by *Chara*: a re-analysis, in *Inorganic Carbon Uptake by Aquatic Photosynthetic Organisms*, Lucas, W.J., and Berry, J.A., Eds., Am. Soc. Plant Physiologists, Rockville, Maryland, 1985, 228.

2942. Lucas, W.J., and Smith, F.A., The formation of alkaline and acid regions at the surface of *Chara corallina* cells, *J. Exp. Bot.*, 24, 1, 1973.

2943. Lucas, W.J., and Dainty, J., HCO_3^- influx across the plasmalemma of *Chara corallina*. Divalent cation requirements, *Plant Physiol.*, 60, 862, 1977.

2944. Lucas, W.J., and Dainty, J., Spatial distribution of functional OH^- carriers along a Characean internodal cell: determined by the effect of cytochalasin B on $H^{14}CO_3^-$ assimilation, *J. Membrane Biol.*, 32, 75 1977.

2945. Lucas, W.J., and Alexander, J.M., Sulfhydryl group involvement in plasmalemma transport of HCO_3^- and OH^- in *Chara corallina*, *Plant Physiol.*, 65, 274, 1980.

2946. Lucas, W.J., and Nuccitelli, R., HCO_3^- and OH^- transport across the plasmalemma of *Chara*. Spatial resolution obtained using extracellular vibrating probe, *Planta*, 150, 120, 1980.

2947. Lucas, W.J., Ferrier, J.M., and Dainty, J., Plasmalemma transport of OH^- in *Chara corallina*. Dynamics of activation and deactivation, *J. Membrane Biol.*, 32, 49, 1977.

2948. Lucas, W.J., Spanswick, R.M., and Dainty, J., HCO_3^- influx across the plasmalemma of *Chara corallina*. Physiological and biophysical influence of 10 mM K^+, *Plant Physiol.*, 61, 487, 1978.

2949. Lucas, W.J., Tyree, M.T., and Petrov, A., Characterization of photosynthetic ^{14}carbon assimilation by *Potamogeton lucens* L., *J. Exp. Bot.*, 29, 1409, 1978.

2950. Ludwig, C.A., The availability of different forms of nitrogen to a green alga (*Chlorella*), *Am. J. Bot.*, 25, 448, 1938.

2951. Ludwig, H.F., and Oswald, W.J., Role of algae in sewage oxidation ponds, *Sci. Monthly*, 74, 3, 1952.

2952. Ludwig, H.F., Oswald, W.J., Gotaas, H.B., and Lynch, V., Algal symbiosis in oxidation ponds. I. Growth characteristics of *Euglena gracilis* cultured in sewage, *Sewage Ind. Wastes*, 23, 1337, 1951.

2953. Lugo, A.E., and Snedakers, S.C., The ecology of mangroves, *Ann. Rev. Ecol. Systematics*, 5, 39, 1974.

2954. Lui, N.S.T., and Roels, O.A., Nitrogen metabolism of aquatic organisms. II. The assimilation of nitrate, nitrite, and ammonia by *Biddulphia aurita*, *J. Phycol.*, 8, 259, 1972.

2955. Lukavsky, J., The evaluation of algal growth potential by cultivation on solid media, *Water Res.*, 17, 549, 1983.

2956. Lukavský, J., First record of cryoseston in the Bohemian Forest Mts. (Šumava), *Algol. Studies*, 69, 83, 1993.

2957. Lukešová, A., Three filamentous green algae isolated from soils, *Arch. Protistenk.*, 139, 69, 1991.

2958. Lukešová, A., Revision of the genus *Follicularia*Mill. (Chlorophyceae), *Arch. Protistenk.*, 143, 87, 1993.

2959. Luli, G.W., Talnagi, J.W., Strohl, W.R., and Pfister, R.M., Hexavalent chromium-resistant bacteria isolated from river sediments, *Appl. Environ. Microbiol.*, 46, 846, 1983.

2960. Lund, J.W.G., The marginal algae of certain ponds, with special reference to the bottom deposits, *J. Ecol.*, 30, 245, 1942.

2961. Lund, J.W.G., Observations on soil algae. I. The ecology, size and taxonomy of British soil diatoms, Part I, *New Phytol*, 44, 196, 1945.

2962. Lund, J.W.G., Observations on soil algae. I. The ecology, size and taxonomy of British soil diatoms, Part 2, *New Phytol*, 45, 56; 1946.

2963. Lund, J.W.G., Observations on soil algae. II. Notes on groups other than diatoms, *New Phytol.*, 46, 35, 1947.

2964. Lund, J.W.G., Studies on *Asterionella formosa* Hass. II. Nutrient depletion and the spring maximum. Part II: Discussion, *J. Ecol.*, 38, 15, 1950.

2965. Lund, J.W.G., The seasonal cycle of the plankton diatom *Melosira italica* (Ehr.) Kütz. subsp. *subarctica* O. Müll., *J.Ecol.*, 42, 151, 1954.

2966. Lund, J.W.G., Buoyancy in relation to the ecology of the freshwater phytoplankton, *Br. Phycol. Bull.*, 1, 1, 1959.

2967. Lund, J.W.G., Soil algae, in *Physiology and Biochemistry of Algae*, Lewin, R.A., Ed., Academic Press, New York and London, 1962, 759.

2968. Lund, J.W.G., The ecology of the freshwater phytoplankton, *Biol. Rev.*, 40, 231, 1965.

2969. Lund, J.W.G., Soil algae, In *Soil Biology*, Burges, A., and Raw, F., Eds., Academic Press, 1967, 129.

2970. Lund, J.W.G., Primary production, *Water Treatment Exam.*, 19, 332, 1970.

2971. Lund, J.W.G., and Talling, J.F., Botanical limnological methods with special reference to the algae, *Bot. Rev.*, 23, 489, 1957.

2972. Lund, J.W.G., Mackereth, F.J.H., and Mortimer, C.H., Changes in depth and time of certain chemical and physical conditions and of the standing crop of *Asterionella formosa* Hass in the North Basin of Windermere in 1947, *Phil. Trans. R. Soc. Lond*, B 246, 255, 1963.

2973. Lunde, B., The synthesis of fat and water soluble arseno organic compounds in marine and limnetic algae, *Acta Chem. Scand.*, 27, 1586, 1973.

2974. Lundell, D.J., and Glazer, A.N., Allophycocyanin B. A comon b sub-unit in *Synechococcus* allophycocyanins (l_{max} 670 nm) and allophycocyanin (l_{max} 650 nm), *J. Biol. Chem.*, 256, 12600, 1981.

2975. Lundgren, D.G., and Dean, W., Biogeochemistry of iron, in *Biogeochemical Cycling of Mineral-Forming Elements*, Trudinger, P.A., and Swaine, D.J., Eds., Elsevier Scientific Publishing Company, Amsterdam, 1979, 211.

2976. Lüning, K., Light, in *Biology of Seaweeds*, Lobban, C.S., and Wynne, M.J., Eds., University of California Press, Berkeley, 1981, 326.

2977. Lüning, K., *Seaweeds: Their Environment, Biogeography and Ecophysiology* John Wiley and Sons, New York, 1990.

2978. Lüning, K., Day and night kinetics of growth rate in green, brown, and red seaweeds, *J. Phycol.*, 28, 794, 1992.

2979. Luntz, A., Untersuchungen über die Phototaxis I., *Z. vergleich. Physiol.*, 14, 68, 1931.

2980. Luther, G.W., Wilk, Z., Ryans, R.A., and Meyerson, A.L., On the speciation of metals in the water column of a polluted estuary, *Mar. Pollut. Bull.*, 17, 535, 1986.

2981. Luther, H., On the life forms and aboveground and underground biomass of aquatic macrophytes: a review, *Acta Bot. Fennici*, 20, 1, 1983.

2982. Luttenton, M.R., and Rada, R.G., Effects of disturbance on epiphytic community architecture, *J. Phycol.*, 22, 320, 1986.

2983. Lwoff, A., *L'évolution physiologique. Étude des pertes de fonction chez les microorganisms*, Hermann Co., Paris, 1943.

2984. Lwoff, A., and Dusi, H., La pyrimidine et la thiazole, facteurs de croissance pour le flagellé *Polytoma coeca, Compt. Rend. Acad. Sci.*, 205, 630, 1937.

2985. Lwoff, A., and Dusi, H., Le thiazole, facteur de croissance pour les flagellés *Polytoma caudatum* et *Chilomonas paramoecium, Compt. Rend. Acad. Sci.*, 205, 756, 1937.

2986. Lwoff, A., and Dusi, H., Le thiazole, facteur de croissance pour le flagellé *Polytoma ocellatum, Compt. Rend. Acad. Sci.*, 205, 882.

2987. Lyford, J.H., and Gregory, S.V., The dynamics and structure of periphyton communities in three Cascade Mountain streams, *Verh. Internat. Verein. Limnol.*, 19, 1610, 1975.

2988. Lyngbye, H.C., *Tentamen hydrophytologiae Danicae,* Copenhagen, 1819.

2989. Lyne, R.L., and Stewart, W.D.P., Emerson enhancement of carbon fixation but not of acetylene reduction (nitrogenase activity) in *Anabaena cylindrica, Planta,* 109, 27, 1973.

2990. Lynn, R., and Brock, T.D., Notes on the ecology of a species of *Zygogonium* (Kütz) in Yellowstone National Park, *J. Phycol.,* 5, 181, 1969.

2991. Lynn, R.L., and Starr, R.C., The biology of the acetate flagellate *Diplostauron elegans* Skuja, *Arch. Protistenk.,* 112, 283, 1970.

2992. Lyon, J.G., *Practical Handbook for Wetland Identification and Delineation,* CRC Press, Boca Raton, 1993.

2993. Maberly, C.S., Exogenous sources of inorganic carbon for photosynthesis by marine macroalgae, *J. Phycol.,* 26, 439, 1990.

2994. Maberly, S.C., and Spence, D.H.N., Photosynthetic inorganic carbon use by freshwater plants, *J. Ecol.,* 71, 705, 1983.

2995. Maberly, S.C., and Spence, D.H.N., Photosynthesis and photorespiration in freshwater organisms: amphibious plants, *Aquat. Bot.,* 34, 267, 1989.

2996. Maccubbin, A.E., and Hodson, R.E., Mineralization of detrital lignocelluloses by salt marsh sediment microflora, *Appl. Environ. Microbiol.,* 40, 735, 1980.

2997. MacEntee, F.J., Schreckenberg, G., and Bold, H.C., Some observations on the distribution of edaphic algae, *Soil Sci.,* 114, 717, 1972.

2998. MacFarlane, G.T., and Herbert, R.A., Nitrate dissimilation by *Vibrio* spp. isolated from estuarine sediments, *J. Gen. Microbiol.,* 128, 2463, 1982.

2999. Macias, F.M., and Eppley, R.W., Development of EDTA media for the rapid growth of *Chlamydomonas mundana, J. Protozool.,* 10, 143, 1963.

3000. Maciasr, F.M., Effect of pH of the medium on the availability of chelated iron for *Chlamydomonas mundana, J. Protozool.,* 12, 500, 1965.

3001. MacIsaac, J.J., and Dugdale, R.C., The kinetics of nitrate and ammonium uptake by natural populations of marine phytoplankton, *Deep-Sea Res.,* 16, 45, 1969.

3002. MacIsaac, J.J., and Dugdale, R.C., Interaction of light and inorganic nitrogen in controlling nitrogen uptake in the sea, *Deep-Sea Res.,* 19, 209, 1972.

3003. Mackenthun, K.M., Keup, L.E., and Stewart, R.K., Nutrients and algae in Lake Sabasticook, Maine, *J. Water Pollut. Contr. Fed.,* 40, 72, 1968.

3004. Mackereth, F.J., Phosphorus utilization by *Asterionella formosa* Hass., *J. Exp. Bot.,* 4, 296, 1953.

3005. Mackerras, A.H., and Smith, G.D., Urease activity of the cyanobacterium *Anabaena cylindrica, J. Gen. Microbiol.,* 132, 2749, 1986.

3006. Mackie, I.M., and Percival, E., Polysaccharides from the green seaweed *Caulerpa filiformis.* II. A glucan of amylopectin type, *J. Chem. Soc.,* 2381, 1960.

3007. Mackie, W., and Preston, R.D., Cell wall and intracellular region polysaccharides, in *Algal Physiology and Biochemistry* Stewart, W.D.P., Ed., University of California Press, Berkeley, 1974, 40.

3008. MacRobbie, E.A.C., Ion uptake, in *Algal Physiology and Biochemistry* Stewart, W.D.P., Ed., Blackwell Scientific Publications, Oxford, 1974, 676.

3009. MacRobbie, E.A.C., and Dainty, J., Sodium and potassium distribution and transport in the seaweed *Rhodymenia palmata, Physiol. Plant.,* 11, 782, 1958.

3010. MacRobbie, E.A.C., and Dainty, J., Ion transport in *Nitellopsis obtusa, J. Gen. Physiol.,* 42, 335, 1958.

3011. Maddux, W.S., and Jones, R.F., Some interactions of temperature, light intensity and nutrient concentration during the continuous culture of *Nitzschia closterium* and *Tetraselmis* sp., *Limnol. Oceanogr.,* 9, 79, 1964.

3012. Madigan, M.T., and Brock, T.D., Adaptation by hot spring phototrophs to reduced light intensities, *Arch. Mikrobiol.,* 113, 111, 1977.

3013. Madlener, J.C., *The Sea Vegetable Book,* Clarkson N. Potter, New York, 1977.

3014. Madsen, T.V., and Maberly, S.C., A comparison of air and water as environments for photosynthesis by the intertidal alga *Fucus spiralis* (Phaeophyta), *J. Phycol.,* 26, 24, 1990.

3015. Madsen, T.V., and Sand-Jensen, K., Photosynthetic carbon assimilation in aquatic macrophytes, *Aquat. Bot.,* 41, 5, 1991.

3016. Maeda, S., Nakashima, S., and Takeshita, T., Bioaccumulation of arsenic by freshwater algae and the application to the removal of inorganic arsenic from an aqueous phase. Part II. *Chlorella vulgaris* isolated from arsenic-polluted environment, *Sep. Sci. Technol.,* 20, 153, 1985.

3017. Maertens, H., Das Wachstum von Blaualgen in mineralischen Nährlösungen, *Beitr. Biol. Pflanzen,* 12, 439, 1914.

3018. Maeso, E.S., Pinas, F.F., Gonzales, M.G., and Valiente, E.F., Sodium requirements for photosynthesis and its relationship with dinitrogen fixation and the external CO_2 concentration in cyanobacteria, *Plant Physiol.*, 85, 585, 1987.

3019. Maestrini, S.Y., and Bonin, D.J., Competition among phytoplankton based on inorganic macronutrients, in *Physiological Bases of Phytoplankton Ecology*, Platt, T., Ed., *Can. Bull. Fish. Aquat. Sci.*, 210, 264, 1981.

3020. Magee, W.E., and Burris, R.H., Fixation of N_2 and utilization of combined nitrogen by *Nostoc muscorum*, *Am. J. Bot.*, 41, 777, 1954.

3021. Mahendrappa, M.K., Smith, R.L., and Christiansen, A.T., Nitrifying organisms affected by climatic region in western United States, *Soil Sci. Soc. Am. Proc.*, 30, 60, 1966.

3022. Maher, W.A., Arsenic in the marine environment of south Australia, in *The Biological Alkylation of Heavy Metals*, Craig, P.J., and Glocking, F., Eds., Royal Society of Chemistry, London, 1988, 120.

3023. Mahoney, J.B., and McLaughlin, J.J.A., The association of phytoflagellate blooms in Lower New York Bay with hypertrophication, *J. Exp. Mar. Biol. Ecol.*, 28, 53, 1977.

3024. Maier, E.X., The Dutch Charophyta, *Wet. Meded. Knnv.*, 93, 1, 1972.

3025. Maiwald, M., A comparative ultrastructural study of *Pyramimonas montana* Geitler and *Pyramimonas* spec., *Arch. Protistenk.*, 113, 334, 1971.

3026. Major, A., *The Book of Seaweed*, Gordon and Cremonesi Publishers, London and New York, 1977.

3027. Majumdar, S.K., Brooks, R.P., Brenner, F.J., and Tiner, R.W., Jr., Eds., *Wetlands Ecology and Conservation: Emphasis in Pennsylvania*, The Pennsylvania Academy of Science, Easton, Pennsylvania, 1989.

3028. Mäkinen, A., and Pakarinen, P., Comparison of some forest and bog plants in heavy metal surveys, *Ympäristö ja Terveys*, 8, 170, 1977, (in Finnish).

3029. Malone, C., Koeppe, D.E., and Miller, R.J., Localization of lead accumulated by corn plants, *Plant Physiol.*, 53, 388, 1974.

3030. Malone, T.C., Algal size, in *The Physiological Ecology of Phytoplankton*, Morris, I., Ed., University of California Press, Berkeley, 1980, 433.

3031. Maloney, T.E., and Palmer, C.M., Toxicity of six chemical compounds in thirty cultures of algae, *Water Sewage Works*, 103, 509, 1956.

3032. Maltby, E., *Waterlogged Wealth*, Earthscan, Russell Press, Nottingham, 1986.

3033. Maltby, E., Global wetlands—history, current status and future, in *Proc. Symp. The Ecology and Management of Wetlands*, Vol. 1 *Ecology of Wetlands*, Hook, D.D., Ed., Timber Press, Wiltshire, Oregon, 1988, 3.

3034. Maltby, E., Wetlands and their values, in *Wetlands*, Finlayson, M., and Moser, M., Eds., Facts on File, Oxford, 1991, 8.

3035. Mamkaeva, K.A., Observations on the lysis of cultures of the genus *Chlorella*, *Mikrobiologiya*, 35, 853, 1966, (in Russian).

3036. Mamkaeva, K.A., and Rybal'chenko, O.V., Ultrastructural features of *Bdellovibrio chlorellavorus*, *Mikrobiologiya*, 48, 159, 1979, (in Russian).

3037. Mandelli, E.F., The inhibitory effects of copper on marine phytoplankton, *Mar. Sci.*, 14, 47, 1969.

3038. Mandelli, E.F., The effect of growth illumination on the pigmentation of a marine dinoflagellate, *J. Phycol.*, 8, 367, 1972.

3039. Mang, S., and Broda, E., Influence of cadmium on the growth of *Chlorella*, *Naturwissenschaften*, 63, 295, 1976.

3040. Mang, S., and Tromballa, H.W., Aufnahme von Cadmium durch *Chlorella fusca*, *Z. Pflanzenphysiol.*, 90, 293, 1978.

3041. Mangi, J., Schmidt, K., Pankow, J., Gaines, L., and Turner, P., Effect of chromium on some aquatic plants, *Environ. Pollut.*, 16, 285, 1978.

3042. Mann, K.H., Seaweeds: Their production and strategy for growth, *Science*, 182, 975, 1973.

3043. Manners, D.J., Mercer, G.A., Stark, J.R., and Ryley, J.F., Studies on the metabolism of the protozoa. The molecular structure of a starch-type polysaccharide from *Polytoma uvella*, *Biochem. J.*, 97, 530, 1965.

3044. Manners, D.J., Pennie, I.R., and Ryley, J.F., The reserve polysaccharides of *Prototheca zopfii*, *Biochem. J.*, 104, 32P, 1967.

3045. Manning, W.M., and Juday, R.E., The chlorophyll content and productivity of some lakes in northeastern Wisconsin, *Trans. Wis. Acad. Sci. Arts Lett.*, 33, 363, 1941.

3046. Manny, B.A., The relationship between organic nitrogen and the carotenoid to chlorophyll a ration in five freshwater phytoplankton species, *Limnol. Oceanogr.*, 14, 69, 1969.

3047. Manny, B.A., Seasonal changes in organic nitrogen content of net- and nanoplankton in two hardwater lakes, *Arch. Hydrobiol.*, 71, 103, 1972.

3048. Manton, I., Plant cell structure, in *Contemporary Botanical Thought*, McLeod, A.M., and Cobley, L.S., Eds., Oliver and Boyd, Edinburgh, 1961, 171.

3049. Manton, I., Observations on scale production in *Prymnesium parvum*, *J. Cell Sci.*, 1, 375, 1966.

3050. Manton, I., Further observations on the fine structure of *Chrysochromulina chiton* with special reference to the haptonema, "peculiar" Golgi structure and scale production, *J. Cell Sci.*, 2, 265, 1967.

3051. Manton, I., Tubular trichocysts in a species of *Pyramimonas* (*P. grossi* Parke), *Österr. Bot. Z.*, 116, 378, 1969.

3052. Manton, I., Preliminary observations on *Chrysochromulina mactra* sp. nov., *Br. Phycol. J.*, 7, 21, 1972.

3053. Manton, I., Observations on the biology and micro-anatomy of *Chrysochromulina megacylindrica* Leadbeater, *Br. Phycol. J.*, 7, 235, 1972.

3054. Manton, I., Functional parallels between calcified and uncalcified periplasts, in *Biomineralization of Lower Plants and Animals*, Leadbeater, B.S.C., and Riding, R., Eds., Clarendon Press, Oxford, 1986, 157.

3055. Manton, I., and Leedale, G.F., Observations on the microanatomy of *Crystallolithus hyalinus* Gaarder and Markali, *Arch. Mikrobiol.*, 47, 115, 1963.

3056. Manton, I., and Oates, K., Fine-structural observations on *Papposphaera* Tangen from the southern hemisphere and on *Pappomonas* gen. nov. from South Africa and Greenland, *Br. Phycol. J.*, 10, 93, 1975.

3057. Manton, I., and Oates, K., *Polycrater galapagensis* gen. et sp. nov., a putative coccolithophorid from the Galapagos Islands with an unusual aragonite periplast, *Br. Phycol. J.*, 15, 95, 1980.

3058. Manton, I., and Sutherland, J., Further observations on the genus *Pappomonas* Manton et Oates with special reference to *P. virgulosa* sp. nov. from West Greenland, *Br. Phycol. J.*, 10, 337, 1975.

3059. Manton, I., Sutherland, J., and Oates, K., Arctic coccolithophorids: two species of *Turrisphaera* gen. nov. from West Greenland, Alaska and the North-West passage, *Proc. R. Soc. Lond.*, B 194, 179, 1976.

3060. Manton, I., Sutherland, J., and Oates, K., Arctic coccolithophorids: *Wigwamma arctica* gen. et sp. nov. from Greenland and Arctic Canada, *W. annulifera* sp. nov. from South Africa and S. Alaska and *Calciarus alaskensis* gen. et sp. nov. from S. Alaska, *Proc. R. Soc. Lond.*, B 197, 145.

3061. Mantoura, R.F.C., Dickson, A., and Riley, J.P., The complexation of metals with humic materials in natural waters, *Estuar. Coast. Mar. Sci.*, 6, 387, 1978.

3062. Manzano, C., Candau, P., Gómez-Moreno, C., Relimpio, A.M., and Losada, M., Ferredoxin-dependent photosynthetic reduction of nitrate and nitrite by particles of *Anacystis nidulans*, *Mol. Cell. Biochem.*, 10, 161, 1976.

3063. Manzano, C., Candau, P., and Guerrero, M.G., Affinity chromatography of *Anacystis nidulans* ferredoxin-nitrate reductase and NADP reductase on reduced ferredoxin-sepharose, *Anal. Biochem.*, 90, 408, 1978.

3064. Marcus, M.D., Periphytic community response to chronic nutrient enrichment by a reservoir discharge, *Ecology*, 61, 389, 1980.

3065. Marcus, Y., Havel, E., and Kaplan, A., Adaptation of the cyanobacterium *Anabaena variabilis* to low CO_2 concentrations in their environment, *Plant Physiol.*, 71, 208, 1983.

3066. Marcus, Y., Volokita, M., and Kaplan, A., The location of the transporting system for inorganic carbon and the nature of the form translocated in *Chlamydomonas reinhardtii*, *J. Exp. Bot.*, 35, 1136, 1984.

3067. Margalef, R., Méthode d'extraction des pigments dans l'étude de la végétation benthique, *Ann. Sta. Centr. Hydrobiol. Appl.*, 8, 99, 1960.

3068. Margulis, L., and Schwartz, K.V., *Five Kingdoms. An Illustrated Guide to the Phyla of Life on Earth*, W.H. Freeman and Company, San Francisco, 1982.

3069. Marin, B., Matzke, C., and Melkonian, M., Flagellar hairs of *Trichodesmium* (Prasinophyceae): ultrastructural types and intrageneric variation, *Phycologia*, 32, 213, 1993.

3070. Markager, S., Light absorption and quantum yield for growth in five species of marine macroalgae, *J. Phycol.*, 29, 54, 1993.

3071. Markager, S., and Sand-Jansen, K., Heterotrophic growth of *Ulva lactuca* (Chlorophyceae), *J. Phycol.*, 26, 670, 1990.

3072. Marker, A.F.H., Extracellular carbohydrate liberation in the flagellates *Isochrysis galbana* and *Prymnesium parvum*, *J. Mar. Biol. Ass. U.K.*, 45, 755, 1965.

3073. Marker, A.F.H., The benthic algae of some streams in southern England I. Biomass of the epilithon in some small streams, *J. Ecol.*, 64, 343, 1976.

3074. Marker, A.F.H., and Casey, H., The population and production dynamics of benthic algae in an artificial recirculating hard-water stream, *Phil. Trans. R. Soc. Lond.*, B 298, 265, 1982.

3075. Marre, E., Temperature, in *Physiology and Biochemistry of Algae*, Lewin, R.A., Ed., Academic Press, New York and London, 1962, 541.

3076. Marshall, K.C., Biogeochemistry of manganese minerals, in *Biogeochemical Cycling of Mineral-Forming Elements*, Trudinger, P.A., and Swaine, D.J., Eds., Elsevier Scientific Publishing, Amsterdam, 1979, 253.

3077. Marshall, N., Oviatt, C.A., and Skauen, D.M., Productivity of the benthic microflora of Shoal estuarine environment in Southern New England, *Int. Rev. Ges. Hydrobiol.*, 56, 947, 1971.

3078. Marshall, S.M., and Orr, A.P., The photosynthesis of diatom cultures in the sea, *J. Mar. Biol. Ass. U.K.*, 15, 321, 1928.

3079. Marszalek, D.S., Skeletal ultrastructure of sediment producing green algae. in *Scanning Electron Microscopy. I.*, Johari, O., and Corvin, T., Eds., Illinois Institute of Technology Research Institute, Chicago, 1971, 273.

3080. Martin, C., de la Noüe, J., and Picard, G., Intensive cultivation of freshwater microalgae on aerated pig manure, *Biomass*, 7, 245, 1985.

3081. Martin, E.L., and Benson, R.L., Algal viruses, pathogenic bacteria and fungi: introduction and bibliography, in *Selected Papers in Phycology* II, Rosowski, J.R., and Parker, B.C., Eds., Phycological Society of America, Lawrence, Kansas, 1982, 793.

3082. Martin, J.F., Izaguirre, G., and Waterstrat, X., A planktonic *Oscillatoria* species from Mississippi catfish ponds that produces the off-flavor compound 2-methylisoborneol, *Water Res.*, 25, 1447, 1991.

3083. Martin, M.T., and Pocock, M.A., South African parasitic Florideae and their hosts. 2. Some South African parasitic Florideae, *J. Linn. Soc. Lond. Bot.*, 55, 45, 1953.

3084. Martin, T.C., and Wyatt, J.T., Extracellular investments in blue-green algae with particular emphasis on the genus *Nostoc, J. Phycol.*, 10, 204, 1974.

3085. Marvan, P., Přibil, S., and Lhotský, O., Eds., *Algal Assays and Monitoring Eutrophication*, E. Schweizerbart'sche Verlagsbuchhandlung, Stuttgart, 1979.

3086. Marvan, P., Komárek, J., Ettl, H., and Komárková, J., Structural elements. Principal populations of algae. Spatial distribution, in *Pond Littoral Ecosystems: Structure and Functioning*, Dykyjová, D., and Květ, J., Eds., Springer-Verlag, Berlin, 1978, 296.

3087. Marvan, P., Komárek, J., Ettl, H., and Komárková, J., Dynamics of algal communities, in *Pond Littoral Ecosystems: Structure and Functioning*, Dykyjová, D., and Kvet, J., Eds., Springer-Verlag, Berlin, 1978, 314.

3088. Masing, V., *Saksa-inglese-rootsi-soome-eesti-veneroote-aduslik oskenssönaslik* (German-English-Swedish-Finnish-Estonian-Russian Mire Science Dictionary), Riikl. Ülik., Tartu, 1960.

3089. Masing, V., Mire typology in the Estonian SSR, in *Some Aspects of Botanical Research in the Estonian SSR*, Masing, V., Ed., Valgus Publ., Tartu, Estonia, 1975, 123.

3090. Masing, V., Ed., *Peatland Ecosystems, Estonian Contribution to the IBP No. 91*, 1982, Tallin "Valgus," 1982.

3091. Masing, V., Svirezhev, Y.M., Löffler, H., and Patten, B.C., Wetlands in the biosphere, in *Wetlands and Shallow Continental Water Bodies*, Vol. I, Patten, B.C., Ed., SPB Academic Publishing, The Hague, The Netherlands, 1990, 313.

3092. Masojídek, J., and Šetlík, I., *Progress in Biotechnology of Photoautotrophic Microorganisms*, Book of Conf. Abstracts, České Budějovice, 1993.

3093. Mason, B., *Principles of Geochemistry*, 3rd ed., John Wiley and Sons, New York, 1966.

3094. Mason, C.P., Edwards, K.R., Carlson, R.E., Pignatello, J., Gleason, F.K., and Wood, J.M., Isolation of chlorine-containing antibiotic from the freshwater cyanobacterium *Scytonema hofmanni, Science*, 215, 400, 1982.

3095. Mason, C.F., *Decomposition*, Studies in Biology No. 74, Edward Arnold, Ltd., London, 1977.

3096. Mason, C.F., and Bryant, R.J., Production, nutrient content and decomposition of *Phragmites communis* Trin. and *Typha angustifolia, J. Ecol.*, 63, 71, 1975.

3097. Mason, C.F., and Standen, V., Aspects of second decay production, in *Ecosystems of the World*, Vol. 4B, *Mires: Swamp, Bog, Fen and Moor; Regional Studies*, Gore, A.J.P., Ed., Elsevier Science Publishers, Amsterdam, 1983, 47.

3098. Mason, C.F., and Macdonald, S.M., Metal concentration in mosses and otter distribution in a rural Welsh river receiving mine drainage, *Chemosphere*, 17, 1159, 1988.

3099. Massalski, A., and Leedale, G.F., Cytology and ultrastructure of the Xanthophyceae. I. Comparative morphology of the zoospores of *Bumilleria sicula* and *Tribonema vulgare, Br. Phycol. J.*, 4, 159, 1969.

3100. Massalski, A., Laube, V.M., and Kushner, D.J., Effects of cadmium and copper on the ultrastructure of *Ankistrodesmus braunii* and *Anabaena* 7120, *Microb. Ecol.*, 7, 183, 1981.

3101. Masscheleyn, P.H., DeLaune, R.D., and Patrick, W.H., Jr., Arsenic and selenium chemistry as affected by sediment redox potential and pH, *J. Environ. Qual.*, 20, 522, 1991.

3102.   Matheke, G.E.M., and Horner, R., Primary productivity of the benthic microalgae in the Chukchi Sea near Barrow, Alaska, *J. Fish. Res. Board Can.*, 31, 1779, 1974.

3103.   Mathieson, A.C., Seaweed aquaculture, *Mar. Fish. Rev.*, 37, 2, 1975.

3104.   Mathieson, A.C., and North, W.J., Algal aquaculture: introduction and bibliography, in *Selected Papers in Phycology* II, Rosowski, J.D., and Parker, B.C., Eds., Phycological Society of America, Lawrence, Kansas, 1982, 773.

3105.   Matson, R.S., Mustoe, G.E., and Chang, S.B., Mercury inhibition on lipid biosynthesis in freshwater algae, *Environ. Sci. Technol.*, 6, 158, 1972.

3106.   Matthews, E., and Fung, I., Methane emissions from natural wetlands: Global distribution, area, and environmental characteristics of sources, *Global Biogeochem. Cycles*, 1, 68, 1987.

3107.   Matulová, D., Toxicity of selected heavy metals to algae and bacteria, *Water Mgmt.*, B 29, 148, 1979 (in Czech).

3108.   Matusiak, K., Studies on the purification of waste water from nitrogen fertilizer industry by intensive algal cultures. I. Growth of *Chlorella vulgaris* in wastes, *Acta Microbiol. Pol.*, 25, 233, 1976.

3109.   Matzku, S., and Broda, E., Die Zinkaufnahme in das Innere von *Chlorella, Planta*, 92, 29, 1970.

3110.   May, V., and McBarron, E.J., Occurrence of the blue-green alga *Anabaena circinalis* Rabenh, in New South Wales and toxicity to mice and honey bees, *J. Aust. Inst. Agric. Sci.*, 39, 264, 1973.

3111.   Mayer, A.M., and Gorham, C., The iron and manganese content of plants present in the natural vegetation of the English Lake District, *Ann. Bot. N.S.*, 15, 247, 1951.

3112.   Mayer, T., Williams, J.D.H., Modified procedures for determining the forms of phosphorus in freshwater sediments, *Can. Nat. Water Res. Inst. Burlington Tech. Bull.*, 119, 1, 1981.

3113.   Mayes, R.A., McIntosh, A.W., and Anderson, V.L., Uptake of cadmium and lead by rooted aquatic macrophytes (*Elodea canadensis*), *Ecology*, 58, 1176, 1977.

3114.   Mbagwu, I.G., and Adeniji, H.A., The nutritional content of duckweed (*Lemna paucicostata* Hegelm.) in the Kainji Lake area, Nigeria, *Aquat. Bot.*, 29, 357, 1988.

3115.   McAuliffe, J.R., Resource depression by a stream herbivore: effects on distribution and abundances of other grazers, *Oikos*, 42, 327, 1984.

3116.   McBride, B.C., and Wolfe, R.S., Biosynthesis of dimethylarsine by methanobacterium, *Biochemistry* 10, 4312, 1971.

3117.   McBride, L.J., The boron requirement of *Chlorella vulgaris:* progress report, U.S. At. Energy Comm. Report ANL 6790 (Argonne National Lab.), Lemont, Illinois, 232, 1962.

3118.   McBride, L., Chorney, W., and Skok, J., Growth of *Chlorella* in relation to boron supply, *Bot. Gaz.*, 132, 10, 1971.

3119.   McBrien, D.C.H., and Hassall, K.A., Loss a cell potassium by *Chlorella vulgaris* after contact with toxic amounts of copper sulphate, *Physiol. Plant.*, 18, 1059, 1965.

3120.   McBrien, D.C.H., and Hassall, K.A., The effect of toxic doses of copper upon respiration, photosynthesis and growth of *Chlorella vulgaris, Physiol. Plant.*, 20, 113, 1967.

3121.   McCallister, D.L., and Logan, T.J., Phosphate adsorption-desorption characteristics of soils and bottom sediments in the Maumee River basin of Ohio, *J. Environ. Anal.*, 7, 87, 1978.

3122.   McCandless, E.L., Polysaccharides of the seaweed, in *The Biology of Seaweeds*, Lobban, C.S., and Wynne, M.J., Eds., University of California Press, Berkeley, 1981, 559.

3123.   McCandless, E., and Craige, J., Sulfated polysaccharides in red and brown algae, *Ann. Rev. Plant Physiol.*, 30, 41, 1979.

3124.   McCarthy, J.J., The uptake of urea by marine phytoplankton, *J. Phycol.*, 8, 216, 1972.

3125.   McCarthy, J.J., The kinetics of nutrient utilization, in *Physiological Bases of Phytoplankton Ecology*, Platt, T., Ed., *Can. Bull. Fish. Aquat. Sci.*, 210, 211, 1981.

3126.   McCarthy, J.J., and Goldman, J.C., Nitrogenous nutrition of marine phytoplankton in nutrient depleted waters, *Science*, 23, 670, 1979.

3127.   McCarthy, J.J., Taylor, W.R., and Taft, J.L., The dynamics of nitrogen and phosphorus cycling in the open waters of the Chesapeake Bay, in *Marine Chemistry in the Coastal Environment*, Church, T.M., Ed., ACS Symp. Ser. 18, 1975, 664.

3128.   McCarthy, J.J., Taylor, W.R., and Taft, J.L., Nitrogenous availability of the plankton in the Chesapeake bay. I. Nutrient availability and phytoplankton preference, *Limnol. Oceanogr.*, 22, 996, 1977.

3129.   McColl, R.H.S., Chemistry and trophic status of 7 New Zealand lakes, *N.Z. Mar. Freshwater Res.*, 6, 399, 1972.

3130.   McCombie, A.M., Actions and interactions of temperature, light intensity and nutrient concentration of the green algae, *Chlamydomonas reinhardtii, J. Fish. Res. Board Can.*, 17, 871, 1960.

3131.   McConnaughey, T., Calcification in *Chara corallina*: CO_2 hydroxylation generates protons for bicarbonate assimilation, *Limnol. Oceanogr.*, 36, 619, 1991.

3132. McConnell, D., Biogeochemistry of phosphate minerals, in *Biogeochemical Cycling of Mineral-Forming Elements*, Trudinger, P.A., and Swaine, D.J., Eds., Elsevier Scientific Publications, Amsterdam, The Netherlands, 1979, 163.

3133. McConnell, D., and Colinvaux, L.H., Aragonite in *Halimeda* and *Tydemania* (order Siphonales), *J. Phycol.*, 3, 198, 1967.

3134. McConnell, W.J., and Sigler, W.F., Chlorophyll and productivity in a mountain river, *Limnol. Oceanogr*, 4, 335, 1959.

3135. McConville, M.J., and Wetherbee, R., The bottom-ice microalgal community from annual ice in the inshore waters of east Antarctica, *J. Phycol.*, 19, 431, 1983.

3136. McCormick, J.F., The natural features of Tinicum Marsh, with particular emphasis on the vegetation, in *Two Studies of Tinctum Marsh, Delaware and Philadelphia Counties, Pa.*, McCormick, J.F., Grant, R.R., Jr., and Patrick, R., Eds., Conservation Foundation, Washington, DC, 1970, 1.

3137. McCormick, J., and Ashbaugh, T., Vegetation of a section of Oldman Creek tidal marsh and related areas in Salem and Glocester Counties, New Jersey, *Bull. New Jersey Acad. Sci.*, 17, 31, 1972.

3138. McCormick, P.V., and Stevenson, R.J., Effects of snail grazing on benthic algal community structure in different nutrient environments, *J. N. Am. Benthol. Soc.*, 8, 162, 1989.

3139. McCormick, P.V., and Stevenson, R.J., Mechanisms of benthic algal succession in lotic environments, *Ecology*, 72, 1835, 1991.

3140. McCormick, P.V., and Stevenson, R.J., Grazer control of nutrient availability in the periphyton, *Oecologia*, 86, 288, 1991.

3141. McCracken, M.D., Gustafson, T.D., and Adams, M.S., Productivity of *Oedogonium* in Lake Wingra, Wisconsin, *Am. Midl. Nat.*, 92, 247, 1974.

3142. McCree, K., Test of current definitions of photosynthetically active radiation against leaf photosynthesis data, *Agric. Meteorol.*, 10, 443, 1972.

3143. McElhenney, T.R., Bold, H.C., Brown, R.M., Jr., and McGovern, J.P., Algae, a cause of inhalant allergy in children, *Ann. Allergy*, 20, 739, 1962.

3144. McElroy, W.D., and Nason, A., Mechanism of action of micronutrient elements in enzyme systems, *Ann. Rev. Plant Physiol.*, 5, 1, 1954.

3145. McGovern, J.P., Hayward, T.J., and McElhenney, T.R., Airborne algae and their allergenicity. II. Clinical and laboratory multiple correlation studies with four genera, *Ann. Allergy*, 24, 146, 1966.

3146. McHardy, B.M., and George, J.J., The uptake of selected heavy metals by the green alga *Cladophora glomerata*, *Symp. Biol. Hungar.*, 29, 3, 1985.

3147. McHargue, J.S., The role of manganese in plants, *J. Am. Chem. Soc.*, 44, 1592, 1922.

3148. McIlrath, W.J., and Skok, J., Boron requirement of *Chlorella vulgaris*, *Bot. Gaz.*, 119, 231, 1958.

3149. McIntire, C.D., Some effects of current velocity on periphyton communities in laboratory streams, *Hydrobiologia*, 27, 559, 1966.

3150. McIntire, C.D., Physiological-ecological studies on benthic algae in laboratory streams, *J. Water Pollut. Control. Fed.*, 40, 1940, 1968.

3151. McIntire, C.D., Structural characteristics of benthic algal communities in laboratory streams, *Ecology*, 49, 520, 1968.

3152. McIntire, C.D., Periphyton assemblages in laboratory streams, in *River Ecology*, Whitton, B.A., Ed., University of California Press, Berkeley, 1975, 403.

3153. McIntire, C.D., and Phinney, H.K., Laboratory studies of periphyton production and community metabolism in lotic environments, *Ecol. Monogr.*, 35, 237, 1965.

3154. McIntire, C.D., and Amspoker, M.C., Effects of sediment properties on benthic primary production in the Columbia River estuary, *Aquat. Bot.*, 24, 249, 1986.

3155. McIntire, C.D., Garrison, R.L., Phinney, H.K., and Warren, C.E., Primary production in laboratory streams, *Limnol. Oceanogr*, 9, 92, 1964.

3156. McKee, H.S., *Nitrogen Metabolism in Plants*, Clarendon Press, Oxford, 1962.

3157. McKee, J.D., Wilson, T.P., Long, D.T., and Owen, R.M., Geochemical partitioning of Pb, Zn, Cu, Fe, and Mn across the sediment-water interface in large lakes, *J. Great Lakes Res.*, 15, 46, 1989.

3158. McKee, J.D., Wilson, T.P., Long, D.T., and Owen, R.M., Pore water profiles and early diagenesis of Mn, Cu, and Pb in sediments from large lakes, *J. Great Lakes Res.*, 15, 68, 1989.

3159. McKnight, D.M., and Morel, F.M.M., Release of weak and strong copper-complexing agents by algae, *Limnol. Oceanogr*, 24, 823, 1979.

3160. McLachlan, J., The culture of *Dunaliella tertiolecta* Butcher—a euryhaline organism, *Can. J. Microbiol.*, 367, 1960.

3161. McLachlan, J., Effects of nutrients on growth and development of embryos of *Fucus edentatus* Pyl. (Phaeophyceae, Fucales), *Phycologia*, 16, 329, 1977.

3162. McLachlan, J., and Lewin, J., Observations on surf phytoplankton blooms along the coasts of South Africa, *Bot. Mar.*, 24, 553, 1981.

3163. McLachlan, J., and Chen, L.C., Formation of adventive embryos from rhizoidal filaments in sporelings of four species of *Fucus, Can. J. Bot.,* 50, 1841, 1972.

3164. McLaughlin, J.J.A., Euryhaline chrysomonads: nutrition and toxicogenesis in *Prymnesium parvum,* with notes on *Isochrysis galbana* and *Monochrysis lutheri, J. Protozool.,* 5, 75, 1958.

3165. McLaughlin, J.J.A., and Zahl, P.A., Axenic Zooxanthellae from various invertebrate hosts, *Ann. N.Y. Acad. Sci.,* 77, 55, 1959.

3166. McLaughlin, J.J.A., and Zahl, P.A., Endozoic algae, in *Symbiosis,* Vol. 1, Henry, S.M., Ed., Academic Press, New York, 1966, 257.

3167. McLean, R.O., The tolerance of *Stigeoclonium tenue* Kütz. to heavy metals in South Wales, *Br. Phycol. J.,* 9, 91, 1974.

3168. McLean, R.O., and Jones, A.K., Studies of tolerance to heavy metals in the flora of the Rivers Ystwyth and Clarach, Wales, *Freshwater Biol.,* 5, 431, 1975.

3169. McLean, M.W., and Williamson, F.B., Cadmium accumulation by the marine red alga *Porphyra umbilicalis, Physiol. Plant.,* 41, 268, 1977.

3170. McMillan, R.M., MacIntyre, D.E., and Gordon, J.L., Stimulation of human platelets by carrageenans, *J. Pharm. Pharmacol.,* 31, 148, 1979.

3171. McNaughton, S.J., Ecotype function in the *Typha* community type. *Ecol. Monogr.,* 36, 297, 1966.

3172. Mearns, A.J., Oshida, P.S., Sherwood, M.J., Young, D.R., and Reish, D.J., Chromium effects on coastal organisms, *J. Water Pollut. Control Fed.,* 48, 1929, 1976.

3173. Meeks, J.C., Chlorophylls, in *Algal Physiology and Biochemistry,* Stewart, W.D.P., Ed., Blackwell Scientific Publications, Oxford, 1974, 161.

3174. Meeks, J.C., and Castenholz, R.W., Growth and photosynthesis in an extreme thermophille, *Synechococcus lividus* (Cyanophyta), *Arch. Mikrobiol.,* 78, 25, 1971.

3175. Meeks, J.C., Wolk, C.P., Lockau, W., Schilling, N., Shaffer, P.W., and Chien, W.-S., Pathways of assimilation of [^{13}N]N and ^{13}NH$_4^+$ by cyanobacteria with and without heterocysts, *J. Bacteriol.,* 134, 125, 1978.

3176. Meeuse, B.J.D., Storage products, in *Physiology and Biochemistry of Algae,* Lewin, R.A., Ed., Academic Press, New York and London, 1962, 289.

3177. Meeuse, B.J.D., and Kreger, D.R., X-ray diffraction of algal starches, *Biochim. Biophys. Acta,* 35, 26, 1959.

3178. Meeuse, B.D., Andries, M., and Wood, J.A., Floridean starch, *J. Exp. Bot.,* 11, 129, 1960.

3179. Megard, R.O., Phytoplankton, photosynthesis and phosphorus in Lake Minnetonka, Minnesota, *Limnol. Oceanogr,* 17, 68, 1972.

3180. Meguro, H., Plankton ice in the Antarctic Ocean, *Antarct. Res.,* 14, 1192, 1962.

3181. Meguro, H., Ito, K., and Fukushima, H., Ice flora (Bottom Type): A mechanism of primary production in polar seas and the growth of diatoms in sea ice, *Arctic,* 20, 114, 1967.

3182. Meier, P.G., and Dilks, D.W., Periphytic oxygen production in outdoor experimental channels, *Water Res.,* 18, 1137, 1984.

3183. Meier, P.G., O'Connor, D., and Dilks, D., Artificial substrata for reducing periphytic variability on replicated samples, in *Periphyton of Freshwater Ecosystems,* Wetzel, R.G., Ed., Dr. W. Junk Publishers, The Hague, The Netherlands, 1983, 283.

3184. Meisch, H.-U., and Bielig, H.-J., Effect of vanadium on growth, chlorophyll formation and iron metabolism in unicellular green algae, *Arch. Microbiol.,* 105, 77, 1975.

3185. Meisch, H.-U., and Schmitt-Beckmann, J., Influence of tri- and hexavalent chromium in two *Chlorella* strains, *Z. Pflanzenphysiol.,* 94, 231, 1979.

3186. Melhuus, A., Seip, K.L., Seip, H.M., and Myklestad, S., A preliminary study of the use of benthic algae as biological indicators of heavy metal pollution in Sørjorden, Norway, *Environ. Pollut.,* 15, 101, 1978.

3187. Meliello, J.M., Naiman, R.J., Aber, J.D., and Linkins, A.E., Factors controlling mass loss and nitrogen dynamics of plant litter decaying in Northern Streams, *Bull. Mar. Sci.,* 35, 341, 1984.

3188. Melkonian, M., The flagellar root system of zoospores of the green alga *Chlorosarcinopsis* (Chlorosarcinales) as compared with the *Chlamydomonas* (Volvocales), *Plant Syst. Evol.,* 128, 79, 1977.

3189. Melkonian, M., Structure and significance of cruciate flagellar root systems in green algae: Zoospores of *Ulva lactuca* (Ulvales, Chlorophyceae), *Helgol. Wiss. Meeresunters.,* 32, 425, 1979.

3190. Melkonian, M., Ultrastructural aspects of basal body associated fibrous structure in green algae: a critical review, *BioSystems,* 12, 85, 1980.

3191. Melkonian, M., Flagellar roots, mating structure and gametic fusion in the green alga *Ulva lactuca* (Ulvales), *J. Cell Sci.,* 46, 149, 1980.

3192. Melkonian, M., Structural and evolutionary aspects of the flagellar apparatus in green algae and land plants, *Taxon,* 31, 255, 1982.

3193. Melkonian, M., Ed., *Algal Cell Motility* Chapman and Hall, New York and London, 1991.

3194.   Melkonian, M., and Robenek, H., Eyespot membranes in newly released zoospores of the green alga *Chlorosarcinopsisgelatinosa*(Chlorosarcinales) and their fate during zoospore settlement, *Protoplasma,* 104, 129, 1980.

3195.   Melkonian, M., and Ichimura, T., Chlorophyceae: introduction and bibliography, in *Selected Papers in Phycology,* Rosowski, J.R., and Parker, B.C., Eds., Phycological Society of America, Lawrence, Kansas, 1982, 747.

3196.   Melzer, A., Haber, W., and Kohler, A., Floristisch-ökologische Charakterisierung und Gliederung der Osterseen (Oberbayern) mit Hilfe von submersen Makrophyten, *Mitt. flor.-soz. Arbeitsgem. N.F.,* 19/20, 139, 1977.

3197.   Mendelssohn, I.A., McKee, K.L., and Patrick, W.H., Jr., Oxygen deficiency in *Spartina alterniflora* roots: metabolic adaptation to anoxia, *Science,* 214, 439, 1981.

3198.   Mengel, K., and Kirkby, E.A., *Principles of Plant Nutrition,* 2nd ed., Internat. Potasch Institute, Bern, Switzerland, 1979.

3199.   Mengel, K., and Kirkby, E.A., *Principles of Plant Nutrition,* 3rd ed., Internat. Potasch Institute, Basel, Switzerland, 1979.

3200.   Menzel, D.W., and Ryther, J.H., Nutrients limiting the production of phytoplankton in the Sargasso Sea with special reference to iron, *Deep-Sea Res.,* 7, 276, 1960.

3201.   Mertz, W., Chromium occurrence and function in biological systems, *Physiol. Rev.,* 49, 163, 1969.

3202.   Metting, B., The systematics and ecology of soil algae, *Bot. Rev.,* 47, 195, 1981.

3203.   Meulemans, J.H., and Heinis, F., Biomass and production of periphyton attached to dead reed stems in Lake Maarsseveen, in *Periphyton of Freshwater Ecosystems,* Wetzel, R.G., Ed., Dr. W. Junk Publishers, The Hague, The Netherlands, 1983, 169.

3204.   Michel, C., Legendre, L., Therriault, J.-C., and Demers, S., Photosynthetic responses of Arctic sea-ice microalgae to short-term acclimation, *Polar Biol.,* 9, 437, 1989.

3205.   Mickle, A.M., The comparative mineral nutrition of nuisance aquatic plants, Ph.D. Thesis, University of Wisconsin, Madison, 1975.

3206.   Mickle, A.M., and Wetzel, R.G., Effectiveness of submersed angiosperm-epiphyte complexes on exchange of nutrients and organic carbon in littoral systems. I. Inorganic nutrients, *Aquat. Bot.,* 4, 303, 1978.

3207.   Mickle, A.M., and Wetzel, R.G., Effectiveness of submersed angiosperm-epiphyte complexes on exchange of nutrients and organic carbon in littoral systems. II. Dissolved organic carbon, *Aquat. Bot.,* 4, 317, 1978.

3208.   Mickle, A.M., and Wetzel, R.G., Effectiveness of submersed angiosperm-epiphyte complexes on exchange of nutrients and organic carbon in littoral systems. III. Refractory organic carbon, *Aquat. Bot.,* 6, 339, 1979.

3209.   Middlebrooks, E.J., and Porcella, D.B., Rational multivariate algal growth kinetics, *J. Sanit. Eng. Div. Proc. Am. Soc. Civ. Eng.,* 97 (SAI), 135, 1971.

3210.   Middlebrooks, E.J., and Pano, A., Nitrogen removal in aerated lagoons, *Water Res.,* 17, 1369, 1983.

3211.   Middleton, B.A., van der Valk, A.H., Williams, R.L., Mason, D.H., and Davis, C.B., Litter decomposition in an Indian monsoonal wetland overgrown with *Paspalum distichum, Wetlands,* 12, 37, 1992.

3212.   Mierle, G., and Stokes, P.M., Heavy metal tolerance and metal accumulation by planktonic algae, in *Trace Substances in Environmental Health,* Vol. 10, Hemphill, D.D., Ed., University of Missouri, Columbia, 1976, 113.

3213.   Miflin, B.J., The location of nitrite reductase and other enzymes related to amino acid biosynthesis in the plastids of root and leaves, *Plant Physiol.,* 54, 550, 1974.

3214.   Miflin, B.J., and Lea, P.J., Glutamate synthase and its role in nitrogen assimilation in plants, *Plant Physiol.,* 56 (suppl.), 88, 1974.

3215.   Miflin, B.J., and Lea, P.J., The pathway of nitrogen assimilation in plants, *Phytochemistry,* 15, 873, 1976.

3216.   Miflin, B.J., and Lea, P.J., Ammonium assimilation, in *The Biochemistry of Plants,* Vol. 5, Stumpf, P.K., and Conn, E.E., Eds., Academic Press, New York, 1980, 169.

3217.   Mignot, J.-P., Compléments a l' étude des Chloromonadines: ultrastructure de *Chattonella subsalsa* Biecheler flagellé d'eau saumatre, *Protistologica,* 12, 279, 1976.

3218.   Mignot, J.-P., Joyon, L., and Pringsheim, E.G., Quelques particularitiés structurales de *Cyanophora paradoxa* Korsch. protozaire flagellé, *J. Protozool.,* 16, 138, 1969.

3219.   Miguel, P., De la culture artificielle des diatomées, *Le Diatomiste,* 1, 93, 1890.

3220.   Mikkelsen, D.S., De Datta, S.K., and Obcemea, W.N., Ammonia volatilization losses from flooded rice soils, *Soil Sci. Soc. Am. J.,* 42, 725, 1978.

3221.   Mikolajczyk, E., and Kuznicki, L., Body concentration and ultrastructure of *Euglena, Acta Protozool.,* 20, 1, 1981.

3222.   Millard, P., and Evans, L.V., Sulphate uptake in the unicellular marine alga *Rhodomella maculata*, *Arch. Microbiol.*, 131, 165, 1982.

3223.   Miller, A.G., and Colman, B., Evidence for HCO_3^- transport by the blue-green alga (Cyanobacterium) *Coccochloris peniocystis*, *Plant Physiol.*, 65, 397, 1980.

3224.   Miller, A.G., and Canvin, D.T., Distinction between HCO_3 and CO_2^- dependent photosynthesis in the cyanobacterium *Synechococcus leopoliensis* on the selective response of HCO_3^- transport to Na^+, *FEBS Lett.*, 187, 29, 1985.

3225.   Miller, A.G., Cheng, K.H., and Colman, B., The uptake and oxidation of glycolic acid by blue-green algae. *J. Phycol.*, 7, 97, 1971.

3226.   Miller, A.G., Turpin, D.H., and Canvin, D.T., Na^+ requirement for growth, photosynthesis and pH regulation in the alkalotolerant cyanobacterium *Synechococcus leopoldiensis*, *J. Bacteriol.*, 159, 100, 1984.

3227.   Miller, A.R., Lowe, R.L., and Rotenberry, J.T., Succession of diatom communities on sand grains, *J. Ecol.*, 75, 693, 1987.

3228.   Miller, G.E., Wille, I., and Hitching, G.G., Patterns of accumulation of selected metals in members of the soft-water macrophyte flora of central Ontario lakes, *Aquat. Bot.*, 15, 53, 1983.

3229.   Miller, J.D.A., and Fogg, G.E., Studies on the growth of Xanthophyceae in pure culture II. The relations of *Monodus subterraneus* to organic substances, *Arch. Mikrobiol.*, 30, 1, 1958.

3230.   Miller, R.M., Meyer, C.M., and Tanner, H.A., Glycolate excretion and uptake by *Chlorella, Plant Physiol.*, 38, 184, 1963.

3231.   Miller, W.E., Greene, J.C., and Shiroyama, T., Use of algal assays to define trace-element limitation and heavy metal toxicity, in *Proc. Symp. Terrestrial and Aquatic Ecological Studies of the Northwest*, EWSC Press, Eastern Washington State College, Cheney, Washington, 1976, 317.

3232.   Miller, W.E., Greene, J.C., and Shiroyama, T., The *Selenastrum capricornutum* Printz algal assay bottle test. Experimental design, application, and data interpretation protocol, U.S. EPA-600/9-78-018, Corvallis, Oregon, 1978.

3233.   Mills, A.L., and Alexander, M., Microbial decomposition of species of freshwater planktonic algae, *J. Environ. Qual.*, 3, 423, 1974.

3234.   Milne, J.B., and Dickman, M., Lead concentrations in algae and plants grown over lead contaminated sediments taken from snow dumps in Ottawa, Canada, *J. Environ. Sci. Health.*, A 12, 173, 1977.

3235.   Milner, H.W., The chemical composition of algae, in *Algal Culture: From Laboratory to Pilot Plant*, Burlew, J.S., Ed., Carnegie Inst. of Washington Publ. 600, Washington, DC, 1953, 285.

3236.   Ming, L., and Stephens, G.C., Uptake of free amino acids by the diatom *Melosira mediocris*, Hydrobiologia, 128, 187, 1985.

3237.   Minotti, P.L., Williams, D.G., and Jackson, W.A., The influence of ammonium on nitrate reduction in wheat seedlings, *Planta*, 86, 267, 1969.

3238.   Minshall, G.W., Autrophy in stream ecosystems, *BioScience*, 28, 767, 1978.

3239.   Minzoni, F., Bonetto, C., and Golterman, H.L., The nitrogen cycle in shallow water sediment systems of rice fields. Part 1. The denitrification process, *Hydrobiologia*, 159, 189, 1988.

3240.   Misra, A., and Sinha, R., Algae as drug plants in India, in *Marine Algae in Pharmaceutical Science*, Hoppe, H.A., Levring, T., and Tanaka, Y., Eds., Walter de Gruyter, Berlin, 1979, 237.

3241.   Misra, J.N., *Phaeophyceae in India*, Indian Council of Agricultural Res., New Delhi, 1966.

3242.   Misra, R., Edaphic factors in the distribution of aquatic plants in the English lakes, *J. Ecol.*, 26, 411, 1938.

3243.   Mitchell, S.F., Primary production in a shallow eutrophic lake dominated alternately by phytoplankton and by submerged macrophytes, *Aquat. Bot.*, 33, 101, 1989.

3244.   Mitsch, W.J., and Ewel, K.C., Comparative biomass and growth of cypress in Florida wetlands, *Am. Midl. Nat.*, 101, 417, 1979.

3245.   Mitsch, W.J., and Rust, W.G., Tree growth responses to flooding in a bottomland forest in northeastern Illinois, *Forest Sci.*, 30, 499, 1984.

3246.   Mitsch, W.J., and Gosselink, J.G., *Wetlands*, Van Nostrand Reinhold Company, New York, 1986.

3247.   Mitsch, W.J., Dorge, C.L., and Weimhoff, J.R., Forested Wetlands for Water Resource Management in Southern Illinois, Res. Rept. No. 132, Illinois University, Water Resources Center, Urbana, Illinois, 1977.

3248.   Mitsch, W.J., Dorge, C.L., and Weimhoff, J.R., Ecosystem dynamics and a phosphorus budget of an alluvial cypress swamp in southern Illinois, *Ecology*, 60, 1116, 1979.

3249.   Mitsui, A., Crystallization and some properties of algal ferredoxins, *Plant Physiol.*, 46, (suppl.), xxxix, 1970.

3250.   Mitsui, A., and Arnon, D.I., Crystalline ferredoxin from a blue-green alga, *Nostoc* sp., *Physiol. Plant.*, 25, 135, 1971.

3251. Mitsui, S., *Inorganic Nutrition, Fertilization and Amelioration for Lowland Rice*, Yokenda, Tokyo, 1954.

3252. Miyachi, S., Inorganic polyphosphate in spinach leaves, *J. Biochem.*, 50, 367, 1961.

3253. Miyachi, S., and Tamiya, H., Distribution and turnover of phosphate compounds in growing *Chlorella* cells, *Plant Cell Physiol.*, 2, 405, 1961.

3254. Miyachi, S., Kanai, R., Mihara, S., Miyachi, S., and Aoki, S., Metabolic roles of inorganic polyphosphates in *Chlorella* cells, *Biochim. Biophys. Acta*, 93, 625, 1964.

3255. Moebus, K., Johnson, K.M., and Sieburth, McN., Re-hydration of desiccated intertidal brown algae: release of dissolved organic carbon and water uptake, *Mar. Biol.*, 26, 127, 1974.

3256. Moeller, R.E., Burkholder, J.M., and Wetzel, R.G., Significance of sedimentary phosphorus to a rooted submersed macrophyte (*Najas flexilis*) and its algal epiphytes, *Aquat. Bot.*, 32, 261, 1988.

3257. Moestrup, Ø., The fine structure of mature spermatozoids of *Chara corallina*, with special reference to microtubules and scales, *Planta*, 93, 295, 1970.

3258. Moestrup, Ø., Observations on the fine structure of spermatozoids and vegetative cells of the green alga *Golenkinia*, *Br. Phycol. J.*, 7, 169, 1972.

3259. Moestrup, Ø., On the phylogenetic validity of the flagellar apparatus in green algae and other chlorophyll *a* and *b* containing plants, *BioSystems* 10, 117, 1978.

3260. Moestrup, Ø., Flagellar structure in algae. A review with new observations particularly on the Chrysophyceae, Phaeophyceae (Fucophyceae), Euglenophyceae and Recketia, *Phycologia*, 21, 427, 1982.

3261. Moestrup, Ø., and Thomsen, H.A., An ultrastructural study of the flagellate *Pyramimonas orientalis* with particular emphasis on Golgi apparatus activity and the flagellar apparatus, *Protoplasma*, 81, 247, 1974.

3262. Moestrup, Ø., and Walne, P.L., Studies on scale morphogenesis in the Golgi apparatus of *Pyramimonas tetrarhynchus* (Prasinophyceae), *J. Cell Sci.*, 36, 437, 1979.

3263. Mohanty, S.K., Effect of submergence on nutrient dynamics, yield and nutrient uptake by rice in different soil types, Ph.D. Thesis, Utkal University, Bhubaneswar, India, 1969.

3264. Mohanty, S.K., and Patnaik, S., Effect of submergence on the physico-chemical and chemical changes in different rice soils. I. Kinetics of pH, Eh, C and N, *Acta Agron. Acad. Sci. Hungary*, 24, 446, 1975.

3265. Mohanty, S.K., and Patnaik, S., Effect of submergence on the chemical changes in different rice soils. III. Kinetics of K, Ca, and Mg, *Acta Agron. Acad. Sci. Hungary*, 26, 187, 1977.

3266. Mohanty, S.K., and Dash, R.N., The chemistry of waterlogged soils, in *Proc. Int. Conf. Wetlands-Ecology and Management*, Gopal, B., Turner, R.E., Wetzel, R.G., and Whigham, D.F., Eds., Int. Sci. Publications, Jaipur, India, 1982, 389.

3267. Moikeha, S.N., and Chu, G.W., Dermatitis-producing alga *Lyngbya majuscula* Gomont in Hawaii. II. Biological properties of the toxic factor, *J. Phycol.*, 7, 8, 1971.

3268. Moikeha, S.N., Chu, G.W., and Berger, R.L., Dermatitis-producing alga *Lyngbya majuscula* in Hawaii. I. Isolation and chemical characterization of the toxic factor, *J. Phycol.*, 7, 4, 1971.

3269. Moirano, A.L., Sulphated seaweed polysaccharides, in *Food Colloids*, Graham, H.D., Ed., Avi Publ. Co., New Haven, Connecticut, 1977, 347.

3270. Molish, H., Die Ernährung der Algen, *Sitz. Akad. Wiss. Wien. Math.-Nat. Kl.*, 104, 783, 1895.

3271. Molish, H., Über die Symbiose der beiden Lebermoose *Blasia pusilla* L. und *Cavicularia densa* St. mit *Nostoc*, *Sci. Rep. Tohoku Imperial Univ. (4th Ser. Biol., Sendai)*, 1, 169, 1924.

3272. Mollenhauer, H.H., and Morré, D.J., Golgi apparatus and plant secretion, *Ann. Rev. Plant Physiol.*, 17, 27, 1966.

3273. Möller, D., Estimation of global man-made sulfur emissions, *Atmos. Environ.*, 18, 19, 1984.

3274. Moller, M.F., and Evans, L.V., Sulphate activation in the unicellular red alga *Rhodella*, *Phytochemistry*, 15, 1623, 1976.

3275. Monahan, T.H., Lead inhibition in *Hormotila blennista* (Chlorophyceae, Chlorophyta), *Phycologia*, 12, 247, 1973.

3276. Monahan, T.H., Lead inhibition of Chlorophycean microalgae, *J. Phycol.*, 12, 358, 1976.

3277. Monahan, T.J., Effects of organic phosphate on the growth and morphology of *Scenedesmus obtusiusculus* (Chlorophyceae), *Phycologia*, 16, 133, 1977.

3278. Moncreiff, C.A., Filamentous algal mat communities in the Atchafalaya River delta, M.S. Thesis, Louisiana State University, Baton Rouge, 1983.

3279. Monod, J., *La Croissance des Cultures Bacteriennes*, Herman and Cie, Paris, 1942.

3280. Montejano, G., Gold, M., and Komárek, J., Freshwater epiphytic ctanoprocaryotes from Central Mexico I. and *Xenococcus*, *Arch. Protistenk.*, 143, 237, 1993.

3281. Moore, B.G., and Tischer, R.G., Biosynthesis of extracellular polysaccharides by the blue-green alga *Anabaena flos-aquae*, *Can. J. Microbiol.*, 11, 877, 1965.

3282. Moore, G.E., and Kellerman, K.F., A method of destroying or protecting the growth of algae and certain pathogenic bacteria in water supplies, *U.S. Dept. Agric. Bur. Plant Ind. Bull.*, 64, 1, 1904.

3283. Moore, P.D., Ed., *European Mires*, Academic Press, London, 1984.

3284. Moore, P.D., and Bellamy, D.J., *Peatlands*, Springer-Verlag, New York, 1974.

3285. Moore, J.F., and Gover, R.A., Lead-induced inclusion bodies: composition and probable role in lead metabolism, *Environ. Health Perspect.*, 7, 121, 1974.

3286. Moore, J.W., Seasonal succession of algae in rivers. 1. Examples from the Avon, the large slow-flowing river, *J. Phycol.*, 12, 342, 1976.

3287. Moore, J.W., *Inorganic Contaminants of Surface Waters. Research and Monitoring Priorities*, Springer-Verlag, New York, 1991.

3288. Moore, J.W., and Ramamoorthy, S., *Heavy Metals in Natural Waters. Applied Monitoring and Impact Assessment*, Springer-Verlag, Berlin, 1984.

3289. Moore, R.E., Toxins from blue-green algae, *BioScience*, 27, 797, 1977.

3290. Moore, T.R., Growth and net production of *Sphagnum* at five fen sites, subarctic eastern Canada, *Can. J. Bot.*, 67, 1203, 1989.

3291. Moorehead, K.K., and Reddy, K.R., Oxygen transport through selected aquatic macrophytes, *J. Environ. Qual.*, 17, 138, 1988.

3292. Moorhead, K.K., and Cook, A.E., A comparison of hydric soils, wetlands, and land use in coastal North Carolina, *Wetlands*, 12, 99, 1992.

3293. Morel, A., and Smith, R.C., Relation between total quanta and total energy for aquatic photosynthesis, *Limnol. Oceanogr.*, 19, 591, 1974.

3294. Morel, F., and Morgan, J., A numerical method for computing in aqueous chemical systems, *Environ. Sci. Technol.*, 6, 58, 1972.

3295. Morel, F.M.M., Kinetics of nutrient uptake and growth in phytoplankton, *J. Phycol.*, 23, 137, 1987.

3296. Morel, N.M.L., Reuter, J.G., and Morel, F.M.M., Copper toxicity to *Skeletonema costatum* (Bacillariophyceae), *J. Phycol.*, 14, 43, 1978.

3297. Morgan, K.C., Studies on the autoecology of the freshwater flagellate, *Cryptomonas erosa* Skuja, Ph.D. Thesis, McGill University, Montreal, 1976.

3298. Moriarty, D.J.W., Muramic acid in the cell walls of *Prochloron*, *Arch. Microbiol.*, 210, 191, 1979.

3299. Moriarty, D.J.W., and Moriarty, C.M., The assimilation of carbon from phytoplankton by two herbivorous fishes: *Tilapia nilotica* and *Haplochromis nigripinnis*, *J. Zool.*, 171, 41, 1973.

3300. Morimoto, H., and James, T.W., Effects of growth rate on the DNA content of *Astasia longa* cells, *Exp. Cell Res.*, 58, 55, 1969.

3301. Morimura, Y., Synchronous culture of *Chlorella*. 2. Changes in content of various vitamins during the course of the algal life cycle, *Plant Cell Physiol.*, 1, 63, 1959.

3302. Morisset, C., Structural and cytoenzymological aspects of the mitochondria in excited roots of oxygen-deprived *Lycopersicum* cultivated *in vitro*, in *Plant Life in Anaerobic Environments*, Hook, D.D., and Crawford, R.M.M., Eds., Ann Arbor Science Publications, Ann Arbor, Michigan, 1978, 497.

3303. Morisset, C., Raymond, P., Mocquot, B., and Pradet, A., Plant adaptation to hypoxia and anoxia, *Bull. Soc. Bot. Fr. Actual. Bot.*, 129, 73, 1982.

3304. Morita, H., Characterization of starch and related polysaccharides by differential thermal analysis, *Anal. Chem.*, 28, 64, 1956.

3305. Mornin, L., and Francis, D., The fine structure of *Nematodinium armatum*, a naked dinoflagellate, *J. Microsc.* (Paris), 6, 759, 1967.

3306. Mörnsjö, T., Studies on vegetation and development of a peatland in Scania, South Sweden, *Opera Bot.*, 24, 1, 1969.

3307. Moroney, J.V., and Tolbert, N.E., Inorganic carbon uptake by *Chlamydomonas reinhardtii*, *Plant Physiol.*, 77, 253, 1985.

3308. Moroney, J.V., Husic, H.D., and Tolbert, N.E., Effect of carbonic anhydrase inhibitors on inorganic carbon accumulation by *Chlamydomonas reinhardtii*, *Plant Physiol.*, 79, 177, 1985.

3309. Morris, A.W., and Bale, A.J., The accumulation of cadmium, copper, manganese and zinc by *Fucus vesiculosus* in the British Channel, *Estuar. Coast. Mar. Sci.*, 3, 153, 1975.

3310. Morris, I., *An Introduction to Algae*, Hutchinson and Co., London, 1967.

3311. Morris, I., Nitrogen assimilation and protein synthesis, in *Algal Physiology and Biochemistry*, Stewart, W.D.P., Ed., Blackwell Scientific Publications, Oxford, 1974, 583.

3312. Morris, I., Ed., *The Physiological Ecology of Phytoplankton*, University of California, Berkeley, 1980.

3313. Morris, I., and Syrett, P.J., The development of nitrate reductase in *Chlorella* and its repression by ammonium, *Arch. Mikrobiol.*, 47, 32, 1963.

3314. Morris, I., and Syrett, P.J., The effect of nitrogen starvation on the activity of nitrate reductase and other enzymes in *Chlorella, J. Gen. Microbiol.,* 38, 21, 1965.

3315. Morris, I., and Ahmed, J., The effect of light on nitrate and nitrite assimilation by *Chlorella* and *Ankistrodesmus, Physiol. Plant.,* 22, 1166, 1968.

3316. Morris, I., and Darley, W.M., Physiology and biochemistry of algae: introduction and bibliography, in *Selected Papers in Phycology II.,* Rosowski, J.R., and Parker, B.C., Eds., Phycological Society of America, Lawrence, Kansas, 1982, 278.

3317. Morris, O.P., and Russell, G., Effect of chelation on toxicity of copper, *Mar. Pollut. Bull.,* 4, 159, 1973.

3318. Mortenson, L.E., and Thorneley, R.N.F., Structure and function of nitrogenase, *Ann. Rev. Biochem.,* 48, 387, 1979.

3319. Mortimer, C.H., The exchange of dissolved substances between mud and water in lakes, *J. Ecol.,* 29, 280, 1941.

3320. Mortimer, C.H., The exchange of dissolved substances between mud and water in lakes, *J. Ecol.,* 30, 147, 1942.

3321. Mortimer, D.C., Freshwater aquatic macrophytes as heavy metal monitors - The Ottawa River experience, *Environ. Monit. Assess.,* 5, 311, 1985.

3322. Moshiri, G.A., Ed., *Constructed Wetlands for Water Quality Improvement,* CRC Press, Inc., Boca Raton, Florida, 1993.

3323. Mossisch, T.D., Effects of salinity on the distribution of *Caloglossa leprieurii* (Rhodophyta) in the Brisbane River, Australia, *J. Phycol.,* 29, 147, 1993.

3324. Moss, B., Morphogenesis, in *Algal Physiology and Biochemistry* Stewart, W.D.P., Ed., Blackwell Scientific Publications, Oxford, 1974, 788.

3325. Moss, B., *Ecology of Freshwaters,* Blackwell Scientific Publications, Oxford, 1980.

3326. Moyse, A., Coudere, D., and Garnier, J., The effect of temperature upon the growth and photosynthesis of *Oscillatoria subbrevis* (Cyanophyceae), *Rev. Cytol. Biol. Veg.,* 18, 293, 1957.

3327. Mrozinska, T., *Zielenice (Chlorophyta): Edogoniowce (Oedogoniales).* Panstwowe Wydawnictwo Naukowe Warszawa, Kraków, 1984.

3328. Mrozinska, T., *Chlorophyta VI. Oedogoniophyceae: Oedogoniales.* Süsswasserflora von Mitteleuropa 14, Gustav Fisher Verlag, Stuttgart, New York, 1985.

3329. Mulholland, P.J., Organic carbon cycling and export, in *Distribution and Budgets of Carbon, Phosphorus, Iron and Manganese in a Floodplain Swamp Ecosystems,* Kuenzler, E.J., Mulholland, - P.J., Yarbro, L.A., and Smock, L.A., Eds., University of North Carolina Water Resources Research Report 157, 1980, 15.

3330. Mulholland, P.J., and Kuenzler, E.J., Organic carbon export from upland and forested wetland watersheds, *Limnol. Oceanogr,* 24, 960, 1974.

3331. Mulholland, P.J., Elwood, J.W., Palumbo, A.V., and Stevenson, R.J., Effect of stream acidification on periphyton composition, chlorophyll, and productivity, *Can. J. Fish. Aquat. Sci.,* 43, 1846, 1986.

3332. Müller, C., On the productivity and chemical composition of some benthic algae in hard-water streams, *Verh. Internat. Verein. Limnol.,* 20, 1457, 1978.

3333. Müller, C., Uptake and accumulation of some nutrient elements in relation to the biomass of an epilithic community, in *Periphyton of Freshwater Ecosystems,* Wetzel, R.G., Ed., Dr. W. Junk Publishers, The Hague, The Netherlands, 1983, 147.

3334. Müller, D.G., Sexual reproduction in British *Ectocarpus siliculosus* (Phaeophyta), *Br. Phycol. J.,* 12, 131, 1976.

3335. Müller, D.G., and Gassmann, G., Sexual hormone specificity in *Ectocarpus* and *Laminaria* (Phaeophyceae), *Naturwissenschaften,* 67, 462, 1980.

3336. Müller, G., and Oti, M., The occurrence of calcified planktonic green algae in freshwater carbonates, *Sedimentology* 28, 897, 1981.

3337. Müller, K.W., and Payer, H.D., The influence of pH on the cadmium-repressed growth of the alga *Coelastrum proboscideum, Physiol. Plant.,* 45, 415, 1979.

3338. Müller, O.F., *Animalcula Infusoria,* Copenhagen, 1786.

3339. Mullin, M.M., Some factors affecting the feeding of marine copepods of the genus *Calanus, Limnol. Oceanogr,* 8, 239, 1963.

3340. Munda, I.M., Observations on variations in form and chemical composition of *Fucus ceranoids* L., *Nova Hedw.,* 8, 403, 1964.

3341. Munda, I.M., Trace metal concentrations in some Icelandic seaweeds, *Bot. Mar.,* 21, 261, 1978.

3342. Munda, I.M., Salinity dependent accumulation of Zn, Co and Mn in *Scytosiphon lomentaria* (Lyngb.) Link and *Enteromorpha intestinalis* (L.) Link from the Adriatic Sea, *Bot. Mar.,* 27, 371, 1984.

3343. Munda, I.M., and Kremer, B.P., Chemical composition and physiological properties of fucoids under conditions of reduced salinity, *Mar. Biol.,* 42, 9, 1977.

3344. Munro, A.L.S., and Brock, T.D., Distinction between bacterial and algal utilization of soluble substances in the sea, *J. Gen. Microbiol.*, 51, 35, 1968.

3345. Muramoto, S., and Oki, Y., Removal of some heavy metals from polluted water by water hyacinth (*Eichhornia crassipes*), *Bull. Environ. Contam. Toxicol.*, 30, 170, 1983.

3346. Murdoch, A., and Capobianco, J., Study of selected metals in marshes on Lake St. Clair, Ontario, *Arch. Hydrobiol.*, 84, 87, 1978.

3347. Murdoch, A., and Capobianco, J.A., Effects of treated effluent on a natural marsh, *J. Water Pollut. Control Fed.*, 51, 2243, 1979.

3348. Murkin, H.R., Stainton, M.P., Boughen, J.A., Pollard, J.B., Titman, R.D., Nutrient status of wetlands in the Interlake region of Manitoba, Canada, *Wetlands*, 11, 105, 1991.

3349. Murphy, M.J., Siegel, L.M., Tove, S.R., and Kamin, H., Siroheme: a new prosthetic group participating in six-electron reduction reactions catalyzed by both sulfite and nitrite reductases, *Proc. Natl. Acad. Sci. U.S.A.*, 71, 612, 1974.

3350. Murphy, M.L., and Hall, J.D., Varied effects of clear-cut logging on predators and their habitat in small streams of the Cascade Mountains, Oregon, *Can. J. Fish. Aquat. Sci.*, 38, 137, 1981.

3351. Murphy, T.P., Ammonia and nitrate uptake in the Lower Great Lakes, *Can. J. Fish. Aquat. Sci.*, 37, 1365. 1980.

3352. Murphy, T.P., and Brownlee, B.G., Blue-green algal ammonia uptake in hypertrophic prairie lakes, *Can. J. Fish. Aquat. Sci.*, 38, 1040, 1981.

3353. Murphy, T.P., Lean, D.R.S., and Nalewajko, C., Blue-green algae: their excretion of iron-selective chelators enables them to dominate other algae, *Science*, 192, 900, 1976.

3354. Murray, A.D., and Kidby, D.K., Sub-cellular location of mercury in yeast grown in the presence of mercuric chloride, *J. Gen. Microbiol.*, 86, 66, 1975.

3355. Murty, K.S.N., An ecological study of lake Kondakarla. Ph.D. Thesis, Andhra University, Waltair, India, 1987.

3356. Muscatine, L., Symbiosis of hydra and algae. III. Extracellular products of the algae, *Comp. Biochem. Physiol.*, 16, 77, 1965.

3357. Muscatine, L., Glycerol excretion by symbiotic algae from corals and *Tridacna* and its control by the host, *Science*, 156, 516, 1967.

3358. Muscatine, L., Endosymbiosis of Cnidarians and algae, in *Coelenterate Biology*, Muscatine, M., and Lenhoff, H.M., Eds., Academic Press, New York, 1974, 359.

3359. Muscatine, L., and Neckelmann, N., Regulation and numbers of algae in the *Hydra-Chlorella* symbioses, *Ber. Dtsch. Bot. Ges.*, 94, 571, 1981.

3360. Muscatine, L., Pool, R.R., and Trench, R.K., Symbiosis of algae and invertebrates: aspects of the symbiont surface and the host-symbiont interface, *Trans. Am. Microsc. Soc.*, 94, 450, 1975.

3361. Myers, J., Culture conditions and the development of the photosynthetic mechanism III, *J. Gen. Physiol.*, 29, 419, 1946.

3362. Myers, J., Culture conditions and the development of the photosynthetic mechanism. V. Influence of the composition of the nutrient medium, *Plant Physiol.*, 22, 590, 1947.

3363. Myers, J., Physiology of the algae, *Ann. Rev. Microbiol.*, 5, 157, 1951.

3364. Myers, J., and Kratz, W.A., Relations between pigment content and photosynthesis in a blue-green alga, *J. Gen. Physiol.*, 39, 11, 1955.

3365. Myklestad, S., and Eide, I., Exchange of heavy metals in *Ascophyllum nodosum* (L.) *in situ* by means of transplanting experiments, *Environ. Pollut.*, 16, 277, 1978.

3366. Mynderse, J.S., Moore, R.E., Kashiwagi, M., and Norton, T.R., Antileukemia activity of Oscillatoriaceae: isolation of debromoaplysiatoxin from *Lyngbya*, *Science*, 196, 538, 1977.

3367. Naes, H., Aarnes, H., Utkilen, H.C., Nilsen, S., and Skulberg, O.M., Effect of photon fluence rate and specific growth rate on geosmin production of the cyanobacterium *Oscillatoria brevis* (Kütz.) Gom., *Appl. Environ. Microbiol.*, 49, 1538, 1985.

3368. Nägeli, C., *Gattungen einzellen Algen*, Zürich, 1849.

3369. Nägeli, C., Sphaerocrystalle in *Acetabularia*, *Botan. Mitt. von C.Nägeli*, 1, 206, 1863.

3370. Nagy, L.A., Transvaal stromatolite: First evidence for the diversification of cells about 2.2×10^9 years ago, *Science*, 183, 514, 1974.

3371. Nair, V.R., and Subrahmanyan, R., The diatom *Fragilaria oceanica*, an indicator of abundance of the Indian oil sardien, *Sardinella longiceps*, *Current Sci.*, 24, 41, 1955.

3372. Nakagawa, H., Yonemura, Y., Yamamoto, H, Sato, T., Ogura, N., and Sato, R., Spinach nitrate reductase: purification, molecular weight and subunit composition, *Plant Physiol.*, 77, 124, 1985.

3373. Nakajima, A., Horikoshi, T., and Sakaguchi, T., Uptake of copper ion by green microalgae, *Agric. Biol. Chem.*, 43, 1455, 1979.

3374. Nakamura, H., Report on the present situation of the Microalgae Research Institute of Japan, *Rep. Microalgae Res. Inst. Jap.*, 2, 1, 1961.

3375. Nakani, D.V., and Korsak, M.N., Effect of chromium, cadmium, and zinc on the rate of photosynthesis in short-term experiments, *Biol. Nauki* (Moscow), 19, 84, 1976 (in Russian).

3376. Nakayama, T.O.M., Carotenoids, in *Physiology and Biochemistry of Algae*, Lewin, R.A., Ed., Academic Press, New York and London, 1962, 409.

3377. Nakano, Y., Okamoto, K., Toda, S., and Fuwa, K., Toxic effects of cadmium on *Euglena gracilis* grown in zinc deficient and zinc sufficient media, *Agric. Biol. Chem.*, 42, 901, 1978.

3378. Nalepa, T.F., White, D.S., Pringle, C.M., and Quigley, M.A., The biological component of phosphorus exchange and cycling in lake sediments, in *Nutrient Cycling in the Great Lakes: A Summarization of Factors Regulating the Cycling of Phosphorus*, Scavia, D., and Moll, R., Eds., Great Lakes Res. Div. Spec. Report No. 83, The University of Michigan, Ann Arbor, Michigan, 1980, 93.

3379. Nalewajko, C., Extracellular products of phytoplankton, Ph.D. Thesis, University of London, 1962.

3380. Nalewajko, C., Photosynthesis and excretion in various planktonic algae, *Limnol. Oceanogr.*, 11, 1, 1966.

3381. Nalewajko, C., and Marin, L., Extracellular production in relation to growth of four planktonic algae and of phytoplankton populations from Lake Ontario, *Can. J. Bot.*, 47, 405, 1969.

3382. Nalewajko, C., and Lean, D.R.S., Growth and excretion in planktonic algae and bacteria, *J. Phycol.*, 8, 361, 1972.

3383. Nalewajko, C., and Lean, D.R.S., Phosphorus kinetics-algal growth relationships in batch cultures, *Mitt. Internat. Verein. Limnol.*, 21, 184, 1978.

3384. Nalewajko, C., and Lean, R.D.S., Phosphorus, in *The Physiological Ecology of Phytoplankton*, Morris, I., Ed., University of California Press, Berkeley, 1980, 235.

3385. Nalewajko, C., and Lee, K., Light stimulation of phosphate uptake in marine phytoplankton, *Mar. Biol.*, 74, 9, 1983.

3386. Nalewajko, C., and O'Mahony, M.A., Photosynthesis of algal cultures and phytoplankton following an acid pH shock, *J. Phycol.*, 25, 319, 1989.

3387. Nalewajko, C., Chowdhuri, N., and Fogg, G.E., Excretion of glycollic acid and the growth of a planktonic *Chlorella*, in *Studies on Microalgae and Photosynthetic Bacteria*, Jap. Soc. Plant Physiol., Tokyo Pres, Tokyo, 1963, 171.

3388. Nalewajko, C., Paul, B., Le, K., and Shear, H., Light history, phosphorus status, and the occurrence of light stimulation or inhibition of phosphorus uptake in Lake Superior phytoplankton and bacteria, *Can. J. Fish. Aquat. Sci.*, 43, 329, 1986.

3389. Nason, A., The metabolic role of vanadium and molybdenum in plants and animals, in *Trace Elements*, Lamb, C.A., Bentley, O.G., and Beattie, J.M., Eds., Academic Press, New York and London, 1958, 269.

3390. Nason, A., and McElroy, W.D., Modes of action of the essential mineral elements, in *Plant Physiology*, Vol. 3. *Inorganic Nutrition of Plants*, Steward, F.C., Ed., Academic Press, New York, 1963, 451.

3391. Nasr, A.H., and Beckheet, L.A., Effect of certain trace elements and soil extract on some marine algae, *Hydrobiologia*, 36, 53, 1970.

3392. Nathanielsz, C.P., and Staff, I.A., A mode of entry of blue-green alga into the apogeotropic roots of *Macrozamia communis*, *Am. J. Bot.*, 62, 232, 1975.

3393. National Eutrophication Survey, The relationship of phosphorus and nitrogen to the trophic state of northwest and north-central lakes and reservoirs, Working paper No. 23, N.E.S. Pacific NW Environ. Res. Lab., Corvallis, Oregon, 1974.

3394. Naumann, E., Några synpunkte anående planktons ökologi. Med sarskild hänsyn till fytoplankton, *Sv. Bot. Tidskr*, 13, 129, 1919.

3395. Nauman, E., Untersuchungen über einige sub- und elitorale Algenassociation unseren Seen, *Ark. Bot.*, 19, no. 16, 1925.

3396. Naumann, E., *Limnologische Termonologie*, Berlin, 1931.

3397. Nauwerck, A., Die Beziehungen zwischen Zooplankton und Phytoplankton in See Erkur, *Symb. Bot. Upsal.*, 17, 1, 1963.

3398. Nauwerck, A., Das Phytoplankton des Latnjajaure 1954-55. *Schweitz. Z. Hydrol.*, 30, 188, 1968.

3399. Neely, R.K., and Davis, C.B., Nitrogen and phosphorus fertilization of *Sparganium eurycarpum* Engelm. and *Typha glauca* Godr. stands. I. Emergent plant production, *Aquat. Bot.*, 22, 347, 1985.

3400. Neely, R.K., and Davis, C.B., Nitrogen and phosphorus fertilization of *Sparganium eurycarpum* Engelm. and *Typha glauca* Godr. stands. II. Emergent plant decomposition, *Aquat. Bot.*, 22, 363, 1985.

3401. Neiff, A.P., and Neiff, J.J., El pleuston de *Pistia stratiotes* de la laguna Barranqueras (Chaco, Argentina), *Ecosur*, 4, 69, 1977.

3402. Neiff, J.J., Contribucion al conocimiento de la distribution y biomass de hidrofitas en le lago Mascardi (Rio Negro, Argentina), *Rev. Asoc. Cienc. Nat. Lit.*, 4, 129, 1973.

3403. Neilands, J.B., Hydroxamic acids in nature, *Science*, 156, 1443, 1967.

3404. Neilands, J.B., Microbial iron transport compounds (siderochromes), in *Inorganic Biochemistry*, Eichhorn, G.L., Ed., Elsevier, New York, 1973, 167.

3405. Neilands, J.B., Iron and its role in microbial physiology, in *Microbial Iron Metabolism*, Neilands, J.B., Ed., Academic Press, New York, 1974, 3.

3406. Neilands, J.B., Iron absorption and transport in microorganisms, *Ann. Rev. Nutr.*, 1, 27, 1981.

3407. Neilson, A.H., and Doudoroff, M., Ammonia assimilation in blue-green algae, *Arch. Mikrobiol.*, 89, 15, 1973.

3408. Neilson, A.H., and Lewin, R.A., The uptake and utilization of organic carbon by algae: an essay in comparative biochemistry (including the addena by N.J. Antia), *Phycologia*, 13, 227, 1974.

3409. Neilson, A.H., and Larsson, T., The utilization of organic nitrogen for growth of algae: physiological aspects, *Physiol. Plant.*, 48, 542, 1980.

3410. Neish, I.C., Role of mariculture in the Canadian seaweed industry, *J. Fish. Res. Bd. Can.*, 33, 1007, 1976.

3411. Neish, I.C., Principles and perspectives of the cultivation of seaweeds in enclosed systems, in *Actas Primer Symposium Sobre Algas Marinas Chilenas*, Santelices, B., Ed., Subsecretaria de Pesca, Ministerio de Economia, Chile, 1979, 59.

3412. Nelson, D.J., Kevern, N.R., Wilhm, J.L., and Griffith, N.A., Estimates of periphyton mass and stream bottom area using phosphorus-32, *Water Res.*, 3, 367, 1969.

3413. Nelson, D.M., and Conway, H.L., Effects of the light regime on nutrient assimilation by phytoplankton in the Baja California and northwest Africa upwelling system, *J. Mar. Res.*, 37, 301, 1979.

3414. Nelson, J.W., Kadlec, J.A., and Murkin, H.R., Seasonal comparisons of weight loss for two types of *Typha glauca* Godr. leaf litter, *Aquat. Bot.*, 37, 299, 1990.

3415. Nelson, D.M., Goering, J.J., Kilham, S.S., and Guillard, R.R.L., Kinetics of silica acid uptake and rates of silica dissolution in the marine diatom *Thalassiosirapseudonana*, *J. Phycol.*, 12, 246, 1976.

3416. Neori, A., Cohen, I., and Gordin, H., *Ulva lactuca* biofilters for marine fishpond effluents. 2. Growth rate, yield, and C:N ratio, *Bot. Mar.*, 34, 483, 1991.

3417. Nešpurková, L., Rybová, R., and Janáček, K., Parallel pathways of potassium transport in the alga *Hydrodictyon reticulatum*. Effects of calcium, *Gen. Physiol. Biophys.*, 6, 263, 1987.

3418. Neue, H.E., and Mamaril, C.P., Zinc, sulfur, and other micronutrients in wetland soils, in *Proc. Conf. Wetland Soils: Characterization, Classification, and Utilization*, Int. Rice Res. Inst., Los Baños, Philippines, 1985, 307.

3419. Neveu, P.J., and Thierry, D., Effects of carrageenan—a macrophage toxic agent—on antibody synthesis and on delayed hypersensitivity in guinea pig, *Int. J. Immunopharm.*, 4, 175, 1982.

3420. Newcombe, C.L., Attachment materials in relation to water productivity, *Trans. Am. Microsc. Soc.*, 58, 355, 1949.

3421. Newcombe, C.L., A quantitative study of attachment materials in Sodon Lake, Michigan, *Ecology*, 31, 204, 1950.

3422. Newrkla, P., and Gunatilaka, A., Benthic community metabolism of three Austrian pre-alpine lakes of different trophic conditions and its oxygen dependency, *Hydrobiologia*, 92, 531, 1982.

3423. Newton, L., *A Handbook of the British Seaweeds*, The Trustees of the British Museum, London, 1931.

3424. Newton, L., *Seaweed Utilization*, Sampson Low, London, 1951.

3425. Newton, L., Uses of seaweeds, *Vitas in Botany*, 2, 325, 1963.

3426. Ngo, V., Boosting pond performance with aquaculture, *Operations Forum*, 4, 20, 1987.

3427. Nicholas, D.J.D., The function of trace metals in the nitrogen metabolism in plants, *Ann. Bot. N.S.*, 21, 587, 1957.

3428. Nicholas, D.J.D., Minor mineral nutrients, *Ann. Rev. Plant Physiol.*, 12, 63, 1961.

3429. Nicholas, D.J.D., Inorganic nutrient nutrition of microorganisms, in *Plant Physiology, A Treatise*. Vol. III. *Inorganic Nutrition of Plants*, Steward, F.C., Ed., Academic Press, New York, 1963, 363.

3430. Nicholls, K.H., Nutrient-phytoplankton relationship in the Holland Marsh, Ontario, *Ecol. Monogr.*, 46, 179, 1976.

3431. Nicholls, K.H., The phytoplankton of the Kawartha Lakes (1972–1976), in *Kawartha Lakes Water Management Study*, Ontario Ministry of Environment and Ontario Ministry of Natural Resources, 1977, 29.

3432. Nichols, D.S., Capacity of natural wetlands to remove nutrients from wastewater, *J. Water Pollut. Control Fed.*, 55, 495, 1983.

3433. Nichols, D.S., and Keeney, D.R., Nitrogen nutrition of *Myriophyllum spicatum*: variation of plant tissue nitrogen concentration with season and site in Lake Wingra, *Freshwater Biol.*, 6, 137, 1976.

3434. Nichols, H.W., Culture and development of *Hildebrandia rivularis* from Denmark and North America, *Am. J. Bot.*, 51, 180, 1965.

3435. Nichols, J.M., and Adams, D.G., Akinetes, in *The Biology of Cyanobacteria*, Carr, N.G., and Whitton, B.A., Eds., Blackwell Scientific Publications, Oxford, 1982, 389.

3436. Nieboer, E., and Richardson, D.H.S., The replacement of the nondescript term "heavy metal" by a biologically and chemically significant classification of metal ions, *Environ. Pollut.* B 1, 3, 1980.

3437. Nielsen, S.L., and Sand-Jensen, K., Variation in growth rates of submerged rooted macrophytes, *Aquat. Bot.,* 39, 109, 1991.

3438. Nienhuis, P.H., Variability in the life cycle of *Rhizoclonium riparium* (Roth) Harv. (Chlorophyceae: Cladophorales) under Dutch estuarine conditions, *Hydrobiol. Bull.,* 8, 172, 1974.

3439. Nigon, V., and Heizmann, P., Morphology, biochemistry and genetics of plastid development in *Euglena gracilis, Internat. Rev. Cytol.,* 53, 212, 1978.

3440. Nilsen, S., and Johnsen, O., Effect of CO_2, O_2 and diamox on photosynthesis and photorespiration in *Chlamydomonas reinhardtii* (green alga) and *Anacystis nidulans* (cyanobacterium, blue-green alga), *Physiol. Plant.,* 56, 273, 1982.

3441. Ninnemann, H., Photoregulation of eukaryotic nitrate reductase, in *Blue Light Response, Phenomena and Occurrence in Plants and Microorganisms*, Vol. 1, Senger, H., Ed., CRC Press, Boca Raton, Florida, 1987, 17.

3442. Nishitani, L., Hood, R., Wakeman, J., and Crew, K.K., Potential importance of an endoparasite of *Gonyaulax* in paralytic shellfish poisoning outbreaks, in *Seafood Toxins,* Ragelis, E.P., Ed., Am. Chem. Soc., Washington, DC, 1984, 139.

3443. Nisiwaza, K., Pharmaceutical studies on marine algae in Japan, in *Marine Algae in Pharmaceutical Science,* Hoppe, H.A., Levring, T., and Tanaka, Y., Eds., Walter de Gruyter, Berlin, 1979, 243.

3444. Nissenbaum, A., Phosphorus in marine and non-marine humic substances, *Geochim. Cosmochim. Acta,* 43, 1973, 1979.

3445. Nixon, S.W., Between coastal marshes and coastal waters—a review of twenty years of speculation and research on the role of salt marshes in estuarine productivity and water chemistry, in *Estuarine and Wetland Processes,* Hamilton, P., and MacDonald, K.B., Eds., Plenum Press, New York, 1980, 437.

3446. Nixon, S.W., and Oviatt, C.A., Ecology of a New England salt marsh, *Ecol. Monogr.,* 43, 463, 1973.

3447. Nixon, S.W., and Lee, V., Wetlands and Water Quality. A Regional Review of Recent Research in the United States on their Role of Freshwater and Saltwater Wetlands as Sources, Sinks, and Transformers of Nitrogen, Phosphorus, and Various Heavy Metals, Tech. Report Y-86-2, prepared by University of Rhode Island for U.S. Army Engineer Waterways Exp. Station, Vicksburg, 1986.

3448. Noack, K., and Pirson, A., Die Wirkung von Eisen und mangan auf die Stickstoffassimilation von *Chlorella, Ber. Dtsch. Botan. Ges.,* 57, 442, 1939.

3449. Nollendorfa, A., Pakalne, D., and Upitis, V., Little known trace elements in *Chlorella* culture. Chromium, *Latv. PSR Zinat. Akad. Vestis,* 7, 33, 1972.

3450. Nonomura, A., Development of *Janczewskia morimotoi* (Ceramiales) on its host *Laurancie nipponica* (Ceramiales, Rhodophyceae), *J. Phycol.,* 15, 154, 1979.

3451. Nordin, R.N., and Blinn, D.W., Analysis of a saline tallgrass prairie ecosystem. IV. Preliminary investigations of soils algae, *Proc. North Dakota Acad. Sci.,* 25, 8, 1972.

3452. Norris, J.N., and Krugens, P., Marine Rhodophyceae: introduction and bibliography, in *Selected Papers in Phycology* II, Rosowski, J.R., and Parker, B.C., Eds., Phycological Society of America, Lawrence, Kansas, 1982, 663.

3453. Norris, R.E., Unarmored marine dinoflagellates, *Endeavour,* 25, 124, 1966.

3454. Norris, R.E., Prasinophytes, In *Phytoflagellates,* Cox, E.R., Ed., Elsevier/North Holland, New York, 1980, 85.

3455. Norris, R.E., Prymnesiophyceae, In *Synopsis and Classification of Living Organisms*, Vol. 1, Parker, S.P., Ed., McGraw-Hill, New York, 1982, 86.

3456. Norris, R.E., Prasinophyceae: introduction and bibliography, in *Selected Papers in Phycology* II, Rosowski, J.R., and Parker, B.C., Eds., Phycological Society of America, Lawrence, Kansas, 1982, 740.

3457. Norris, R.E., Prasinophyceae, In *Synopsis and Classification of Living Organisms,* Vol. 1., Parker, S.P., Ed., McGraw-Hill, New York, 1982, 162.

3458. Norris, R.E., and Pienaar, R.N., Comparative fine-structural studies on five marine species of *Pyramimonas* (Chlorophyta, Prasinophyceae), *Phycologia,* 17, 41, 1978.

3459. Norris, R.E., Hori, T., and Chihara, M., Revision of the genus *Tetraselmis* (Class Prasinophyceae), *Bot. Mag. Tokyo,* 93, 317, 1980.

3460. Norris, S., Norris, R.E., and Calvin, M., A survey of the rates and products of short-term photosynthesis in plants of nine phyla, *J. Exp. Bot.,* 6, 64, 1955.

3461. North, B.B., and Stephens, G.C., Uptake and assimilation of amino acids by *Platymonas, Biol. Bull.,* 133, 391, 1967.

3462. North, B.B., and Stephens, G.C., Dissolved amino acids and *Platymonas nutrition,* in *Proc. VI. Int. Seaweed Symp.,* Margalef, R., Ed., Subsecretaria de la Marina Mercante, Madrid, 1969, 263.

3463. North, B.B., and Stephens, G.C., Uptake and assimilation of amino acids by *Platymonas.* II. Increased uptake in nitrogen-deficient cells, *Biol. Bull.,* 140, 242, 1971.

3464. North, B.B., and Stephens, G.C., Amino acid transport in *Nitzschia ovalis* Arnott., *J. Phycol.,* 8, 64, 1972.

3465. North, W.J., Ed., *The Biology of Giant Kelp Beds (Macrocystis) in California, Beih. Nova Hedw.,* 32, 1971.

3466. Norton, T.A., and Mathieson, A.C., The biology of unattached seaweeds, *Prog. Phycol. Res.,* 2, 2, 333, 1983.

3467. Notton, B.A., and Hewitt, E.J., Incorporation of radioactive molybdenum into protein during nitrate reductase formation, an effect of molybdenum on nitrate reductase and diaphorase activities of spinach (*Spinacea oleracea* L.), *Plant Cell Physiol.,* 12, 465, 1971.

3468. Notton, B.A., and Hewitt, E.J., Structure and properties of higher plant nitrate reductase, especially *Spinacea oleracea,* in *Nitrogen Assimilation of Plants,* Hewitt, E.J., and Cutting, C.V., Ed., Academic Press, London and New York, 1979, 227.

3469. Novichkova-Ivanova, L.N., Principal trends and problems of soil algology in the U.S.S.R., in *Algae, Man and the Environment,* Jackson, D.F., Ed., Syracuse University Press, Syracuse, New York, 1968, 359.

3470. Novitzki, R.P., Hydrologic characteristics of Wisconsin's wetlands and their influence on floods, stream flow, and sediment, in *Wetlands Functions and Values: The State of Our Understanding,* Greeson, P.E., Clark, J.R., and Clark, J.E., Eds., American Water Resour. Assoc., Minneapolis, Minnesota, 1979, 377.

3471. Nriagu, J.O., Ed., *The Biogeochemistry of Lead in the Environment,* Elsevier/North Holland Biomedical Press, 1978.

3472. Nriagu, J.O., Properties and the biogeochemical cycle of lead, in *The Biogeochemistry of Lead in the Environment,* Part A: *Ecological Cycles,* Nriagu, J.O., Ed., Elsevier/North-Holland Biomedical Press, 1978, 1.

3473. Nriagu, J.O., Lead in soils, sediments and major rock types, in *The Biogeochemistry of Lead in the Environment,* Part A: *Ecological Cycles,* Nriagu, J.O., Ed., Elsevier/North Holland Biomedical Press, 1978, 15.

3474. Nriagu, J.O., Lead in the atmosphere, in *The Biogeochemistry of Lead in the Atmosphere,* Part A: *Ecological Cycles,* Nriagu, J.O., Ed., Elsevier/North Holland Biomedical Press, 1978, 137.

3475. Nriagu, J.O., Global inventory of natural and anthropogenic emissions of trace metals to the atmosphere, *Nature,* 279, 409, 1979.

3476. Nriagu, J.O., Ed., *Copper in the Environment,* John Wiley and Sons, New York, 1979.

3477. Nriagu, J.O., The global copper cycle, in *Copper in the Environment,* Part 1: *Ecological Cycling,* Nriagu, J.O., Ed., John Wiley and Sons, New York, 1979, 1.

3478. Nriagu, J.O., Ed., *Zinc in the Environment,* John Wiley and Sons, New York, 1980.

3479. Nriagu, J.O., A global assessment of natural sources of atmospheric trace metals, *Nature,* 338, 47, 1989.

3480. Nriagu, J.O., and Pacyna, J.M., Quantitative assessment of worldwide contamination of air, water and soil by trace metals, *Nature,* 333, 134, 1988.

3481. Nultsch, W., and Häder, D.-P., Photomovement of motile microorganisms, *Photochem. Photobiol.,* 29, 429, 1979.

3482. Nultsch, W., and Häder, D.-P., Photomovement of motile microorganisms II., *Photochem. Photobiol.,* 47, 837, 1988.

3483. Nusch, E.A., Assessment of detrimental biological effects of organic chelators and heavy metal complexes by means of bioassays using bacteria, algae, protozoa, lower metazoa and fish, *Z. Wasser Abwasser Forsch.,* 10, 49, 1977.

3484. Nuzzi, R., Toxicity of mercury to phytoplankton, *Nature,* 237, 38, 1972.

3485. Nyberg, H., The effect of detergents and phosphate starvation on phosphorus fractions in *Nitzschia actinastriodes* (Bacillariophyceae), *Ann. Bot. Fennici,* 18, 37, 1981.

3486. Nybom, C., En ekoligisk undersökning av den högre vattenvegetationen i några Ålandska sjöar, Manuscript, Department of Biology, Åbo Akademi, Finland, 1976.

3487. Nygaard, G., On the productivity of five Danish waters, *Verh. Internat. Verein. Limnol.,* 12, 123, 1955.

3488. Nygaard, K., and Tobiesen, A., Bacteriovory in algae: a survival strategy during nutrient limitation, *Limnol. Oceanogr.* 38, 273, 1993.

3489. Nyholm, N., A mathematical model for growth of phytoplankton, *Mitt. Internat. Verein. Limnol.,* 21, 193, 1978.

3490. Nykvist, N., Leaching and decomposition of litter. 1. Experiments on leaf litter of *Fraxinus excelsior* *Oikos*, 10, 190, 1959.

3491. Oakley, B.R., and Bisalputra, T., Mitosis and cell division in *Cryptomonas* (Cryptophyceae), *Can. J. Bot.*, 55, 2789, 1977.

3492. Oakley, B.R., and Taylor, F.J.R., Evidence for a new type of endosymbiotic organization in a population of the ciliate *Mesodinium rubrum* from British Columbia, *BioSystems*, 10, 361, 1978.

3493. Oakley, B.R., and Santore, U.J., Cryptophyceae: introduction and bibliography, in *Selected Papers in Phycology* II, Rosowski, J.R., and Parker, B.C., Eds., Phycological Society of America, Lawrence, Kansas, 1982, 682.

3494. Oates, B.R., and Murray, S.N., Photosynthesis, dark respiration and desiccation resistance of the intertidal seaweeds *Hesperophycus harveyanus* and *Pelvetia fastigiata* f. *gracilis*, *J. Phycol.*, 19, 371, 1983.

3495. O'Brien, M.C., and Wheeler, P.A., Short term uptake of nutrients by *Enteromorpha prolifera* (Chlorophyceae), *J. Phycol.*, 23, 547, 1987.

3496. Obukowitz, M., Schaller, M., and Kennedy, G.S., Ultrastructure and phenolic histochemistry of the *Cycas revoluta-Anabaena* symbiosis, *New Phytol.*, 87, 751, 1981.

3497. Ochiai, E.-I., *General Principles of Biochemistry of the Elements*, Plenum Press, New York, 1987.

3498. O'Colla, P.S., Mucilages, in *Physiology and Biochemistry of Algae*, Lewin, R.A., Ed., Academic Press, New York and London, 1962, 337.

3499. O'Connors, H.B., Particle-size modification by two size classes of the estuarine copepod *Acartia clausi*, *Limnol. Oceanogr.*, 21, 300, 1976.

3500. Odum, E.P., *Fundamentals of Ecology*, Saunders, Philadelphia, 1971.

3501. Odum, E.P., and Fanning, M.E., Comparison of the productivity of *Spartina alterniflora* and *S. cynosuroides* in Georgia coastal marshes, *Bull. Ga. Acad. Sci.*, 31, 1, 1973.

3502. Odum, H.T., Trophic structure and productivity of Silver Springs, Florida, *Ecol. Monogr.*, 27, 55, 1957.

3503. Odum, H.T., and Hoskins, C.M., Metabolism of a laboratory stream microcosm, *Publ. Inst. Mar. Sci. Univ. Tex.*, 4, 115, 1957.

3504. Odum, W.E., The importance of tidal freshwater wetlands in coastal zone management, in *Coastal Zone 78: Symp. on Tech., Environ., Socioeconomic and Regulatory Aspects of Coastal Zone Mgmt.*, Am. Soc. Civil. Engng., Minneapolis, 1978, 1196.

3505. Odum, W.E., and Heywood, M.A., Decomposition of intertidal freshwater marsh plants, in *Proc. Symp. Freshwater Wetlands: Ecological Processes and Management Potential*, Good, R.E., Whigham, D.F., and Simpson, R.L., Eds., Academic Press, New York, 1978, 89.

3506. Odum, W.E., Fisher, J.S., and Pickral, J.C., Factors controlling the flux of particulate organic carbon from estuarine wetlands, in *Ecological Processes in Coastal and Marine Systems*, Livingston, R.J., Ed., Plenum Press, New York, 1979, 69.

3507. Odum, W.E., McIvor, C.C., and Smith, T.J., III., The ecology of the mangroves of South Florida: a community profile, U.S. Fish and Wildlife Service, Office of Biol. Services, FWS/OBS 81/24, Washington, DC, 1982.

3508. Oemke, M.P., and Burton, T.M., Diatom colonization dynamics in a lotic system, *Hydrobiologia*, 139, 153, 1986.

3509. Officer, C.B., Biggs, R.B., Toft, J.L., Gronin, L.E., Tyler, M.A., and Boynton, W.R., Chesapeake bay anoxia: origin, development and significance, *Science*, 223, 22, 1984.

3510. Ogan, M.T., Factors affecting nitrogenase activity associated with marsh grasses and their soils from eutrophic lakes, *Aquat. Bot.*, 17, 215, 1983.

3511. Ogawa, M., Yoda, K., and Kira, T., A preliminary survey of the vegetation of Thailand, *Nature and Life SE Asia*, 1, 121, 1961.

3512. Ogawa, R.E., and Carr, J.F., The influence of nitrogen on heterocyst production in blue-green algae, *Limnol. Oceanogr.*, 14, 342, 1969.

3513. Oglesby, R.T., Schaffner, W.R., and Mills, E.L., Nitrogen, phosphorus and eutrophication in the Finger Lakes, Tech. Rep. No. 94, Cornell University, Ithaca, New York, 1975.

3514. Ogren, W.L., Photorespiration: pathways, regulation, and modification, *Ann. Rev. Plant Physiol.*, 35, 415, 1984.

3515. O'hEocha, C., Phycobilins, in *Physiology and Biochemistry of Algae*, Lewin, R.A., Ed., Academic Press, New York and London, 1962, 421.

3516. O'hEocha, C., Pigments of the red algae, *Oceanogr. Mar. Biol. Ann. Rev.*, 9, 61, 1971.

3517. O'hEocha, C., and Raftery, M., Phycoerythrins and phycocyanins of cryptomonads, *Nature*, 184, 1049, 1959.

3518. Ohki, K., and Fujita, Y., Photoregulation of phycobilisome structure during complementary chromatic adaptation in the marine cyanophyte *Phormidium* sp. C86, *J. Phycol.*, 28, 803, 1992.

3519. Ohmori, K., and Hattori, A., Induction of nitrate and nitrite reduction in *Anabaena cylindrica, Plant Cell Physiol.*, 11, 873, 1970.

3520. Ohmori, M., and Hattori, A., Nitrogen fixation and heterocysts in the blue-green alga *Anabaena cylindrica, Plant Cell Physiol.*, 12, 961, 1971.

3521. Ohtake, H., Aiba, S., and Sudo, R., Growth and detachment of periphyton in an effluent from the secondary treatment plant of wastewaters, *Jap. J. Limnol.*, 39, 163, 1978.

3522. Ojala, A., The influence of light quality on growth and phycobiliprotein/chlorophyll a fluorescence quotients of some species of freshwater algae in culture, *Phycologia*, 32, 22, 1993.

3523. Ojala, A., Effects of temperature and irradiance on the growth of two freshwater photosynthetic cryptophytes, *J. Phycol.*, 29, 278, 1993.

3524. Okazaki, A., *Seaweeds and Their Uses in Japan*, Tokai University Press, Tokyo, 1971.

3525. Okazaki, M., and Furuya, K., Studies on calcium carbonate deposition of algae - IV. Initial calcification sites of calcareous red alga *Galaxaura fastigiata* Decaisne, *Bot. Mar.*, 25, 511, 1982.

3526. Okazaki, M., and Furuya, K., Mechanisms of algal calcification, *Jap. J. Phycol.*, 33, 328, 1985.

3527. Okazaki, M., Ikawa, T., Furuya, K., Nisizawa, K., and Miwa, T.M., Studies on calcium carbonate deposition of a calcarous red alga *Serraticardia maxima, Bot. Mag. Tokyo*, 83, 193, 1970.

3528. Okazaki, M., Ichikawa, K., and Furuya, K., Studies on the calcium carbonate deposition of algae: 4. Initial calcification site of *Galaxaura fastigiata, Bot. Mar.*, 25, 511, 1982.

3529. O'Kelley, J.C., Mineral nutrition of algae, *Ann. Rev. Plant Physiol.*, 19, 89, 1968.

3530. O'Kelley, J.C., Inorganic nutrients, in *Algal Physiology and Biochemistry*, Stewart, W.D.P., Ed., Blackwell Scientific Publications, Oxford, 1974, 610.

3531. O'Kelley, J.C., and Herndon, W.R., Effect of strontium replacement for calcium on production of motile cells in *Protosiphon, Science*, 130, 718, 1959.

3532. O'Kelley, J.C., and Herndon, W.R., Alkaline earth elements and zoospore release and development in *Protosiphon botryoides, Am. J. Bot.*, 43, 796, 1961.

3533. Oki, Y., Potential utilization of water hyacinth in Japan, *Proc. 9th Asian-Pacific Weed Sci. Soc. Conf. Suppl.*, Manila, Philippines, 1983, 588.

3534. Oki, Y., Nakagawa, K., and Nogi, M., Production and nutrient removal potentials of *Eichhornia crassipes* in Japan, *Proc. 8th Asian-Pacific Weed Sci. Soc. Conf.*, 1981, 133.

3535. Okiachi, T., Marine environmental studies on outbreaks of red tides in neritic waters, *J. Oceanogr. Soc. Jap.*, 39, 267, 1983.

3536. Okiachi, T., Anderson, D.M., and Nemoto, T., Eds., *Proc. Int. Symp. Red Tides—Biology Environmental Science and Toxicity* Elsevier, New York, Amsterdam and London, 1989.

3537. Okuda, A., and Yamaguchi, M., Nitrogen-fixing microorganisms in paddy soils. II. Distribution of blue-green algae in paddy soils and the relationship between the growth of them and soil properties, *Soil Plant Food*, 2, 4, 1956.

3538. Okuda, A., and Yamaguchi, M., Nitrogen-fixing microorganisms in paddy soils. VI. Vitamin B$_{12}$ activity in nitrogen-fixing blue-green algae, *Soil Plant Food.*, 6, 76, 1960.

3539. Okuda, A., Yamaguchi, M., and Nioh, I., Nitrogen-fixing microorganisms in paddy soils. X. Effect of molybdenum on the growth and the nitrogen assimilation of *Tolypothrix tenuis, Soil Sci. and Plant Nutr.*, 8, 35, 1962.

3540. Olenin, A.S., Neistadt, M.I., and Tyuremov, S.N., On the principles of classification of peat species and deposits in the USSR., in *Proc. 4th Internat. Peat Congr.*, Otaniemi, Finland, 1972, 41.

3541. Oleson, D.J., and Makarewicz, J.C., Effect of sodium and nitrate on growth of *Anabaena flos-aquae* (Cyanophyta), *J. Phycol.*, 26, 593, 1990.

3542. Oliveira, L., and Antia, N.J., Evidence of nickel and iron requirement for autotrophic growth of a marine diatom with urea serving as nitrogen source, *Br. Phycol. J.*, 19, 125, 1984.

3543. Oliveira, L., and Antia, N.J., Some observations on the urea-degrading enzyme of the diatom *Cyclotella cryptica* and the role of nickel in its production, *J. Plankton Res.*, 8, 235, 1986 a.

3544. Olsen, Y., Evaluation of competitive ability of *Staurastrum luetkemuellerii*(Chlorophyceae) and *Microcystis aeruginosa* (Cyanophyceae) under P-limitation, *J. Phycol.*, 25, 486, 1989.

3545. Olsen, Y., Vadstein, O., Andersen, T., and Jensen, A., Competition between *Staurastrum luetkemuellerii*(Chlorophyceae) and *Microcystis aeruginosa*(Cyanophyceae) under varying modes of phosphate supply, *J. Phycol.*, 25, 499, 1989.

3546. Olson, J.S., Energy storage and the balance of producers and decomposers in ecological systems, *Ecology*, 44, 322, 1963.

3547. Olson, T.A., Toxic plankton, in *Proc. of Inservice Training Course in Water Works Problems*, University of Michigan School of Public Health, Ann Arbor, 1951, 95.

3548. Oltmanns, F., Über die Cultur und Lebensbedingungen der Meeresalgen, *Jahrb. Wiss. Bot.*, 23, 349, 1892.

3549. Oltmanns, F., *Morphologie und Biologie der Algen*, Vols. 1+2, 2nd ed., Jena, 1922.

3550. Ondok, J.P., Average shoot biomass in monospecific helophyte stands of the Opatovický fishpond, in *Ecosystem Study on Wetland Biome in Czechoslovakia,* Hejný, S., Ed., Czechosl. IBP/PT-PP Report No. 3, Třeboň, Czechoslovakia, 1973, 83.

3551. Ondok, J.P., Models of mineral cycles, succession and ecosystem management, in *Methods of the Ecosystem Studies,* Dykyjová, D., Ed., Academia, Praha, 1989, 661, (in Czech).

3552. Ondok, J.P., and Dykyjová, D., Assessment of shoot biomass of dominant reed-beds in Trebon basin methodological aspects. in *Ecosystem Study on Wetland Biome in Czechoslovakia,* Hejný, S., Ed., Czechosl. IBP/PT-PP Rept. No. 3, Třeboň, Czechoslovakia, 1973, 79.

3553. Ondok, J.P., and Pokorný, J., Modelling photosynthesis of submersed macrophyte stands in habitats with limiting inorganic carbon. 1. Model description, *Photosynthetica,* 21, 543, 1987.

3554. Ondok, J.P., and Pokorný, J., Modelling photosynthesis of submersed macrophyte stands in habitats with limiting inorganic carbon. 2. Application to a stand of *Elodea canadensis* Michx., *Photosynthetica,* 21, 555, 1987.

3555. O'Neil (Morin), J.A., The colonization of periphyton on Myriophyllum heterophyllum Michx., M.Sc. Thesis, University of New York, Binghamton, 1981.

3556. O'Neil Morin, J., and Kimball, K.D., Relationship of macrophyte-mediated changes in the water column to periphyton composition and abundance, *Freshwater Biol.,* 13, 403, 1983.

3557. Onuf, C.P., Quammen, M.L., Shaffer, G.P., Peterson, C.H., Chapman, J.W., Cermak, J., and Holmes, R.W., An analysis of the values of central and southern California coastal wetlands, in *Wetland Functions and Values: The State of Our Understanding,* Greeson, P.E., Clark, J.R., and Clark, J.E., Eds., American Water Resour. Assoc., Minneapolis, Minnesota, 1979, 186.

3558. Oppenheimer, D.R., Seasonal changes in epipelic diatoms along an intertidal shore, Berrow Flats, Somerset, *J. Mar. Biol. Assoc. U.K.,* 71, 579, 1991.

3559. Öquist, G., Changes in pigment composition and photosynthesis induced by iron-deficiency in the blue-green alga *Anacystis nidulans, Physiol. Plant.,* 25, 188, 1971.

3560. Öquist, G., Effects of low temperature on photosynthesis, *Plant Cell Environ.,* 6, 281, 1983.

3561. Oremland, R.S., and Taylor, B.F., Diurnal fluctuations of O_2, N_2, and CH_4 in the rhizosphere of *Thalassia testudinum, Limnol. Oceanogr,* 22, 566, 1977.

3562. Orme, A.R., Wetland morphology, hydrodynamics and sedimentation, in *Wetlands: A Threatened Landscape,* Williams, M., Ed., Basil Blackwell, Oxford, 1990, 42.

3563. Ornes, W.H., and Sutton, D.L., Removal of phosphorus from static sewage effluent by water hyacinth, *Hyacinth Control J.,* 13, 56, 1975.

3564. Ortega, M.M., Estudio de las algas comestibles des Valle de Mexico. I., *Rev. Lat.-Amer. Microbiol.,* 14, 85, 1972.

3565. Ortega, T., Castillo, F., and Cárdenas, J., Photolysis of water coupled to nitrate reduction by *Nostoc muscorum* subcellular particles, *Biochem. Biophys. Res. Commun.,* 71, 885, 1976.

3566. Oscarson, D.W., Huang, P.M., and Liaw, W.K., The oxidation of arsenic by aquatic sediments. *J. Environ. Qual.,* 9, 700, 1980.

3567. Oscarson, D.W., Huang, P.M., and Liaw, W.K., The role of manganese in the oxidation of arsenite by freshwater lake sediments, *Clays and Clay Minerals,* 29, 219, 1981.

3568. Oscarson, D.W., Huang, P.M., and Liaw, W.K., The kinetics and components involved in the oxidation of arsenite by freshwater lake sediments, *Verh. Internat. Verein. Limnol.,* 21, 181, 1981.

3569. Oschman, J.L., Development of the symbiosis of *Convoluta ruscoffensis* (Gruff) and *Platymonas* sp., *J. Phycol.,* 2, 105, 1966.

3570. Oschman, J.L., Structure and reproduction of the algal symbionts of *Hydra viridis, J. Phycol.,* 3, 221, 1967.

3571. Osretkar, A., and Krauss, R.W., Growth and metabolism of *Chlorella pyrenoidosa* Chick during substitution of Rb for K., *J. Phycol.,* 1, 23, 1965.

3572. Österlind, S., Inorganic carbon sources of green algae. I. Growth experiments with *Scenedesmus quadricauda* and *Chlorella pyrenoidosa, Physiol. Plant.,* 3, 353, 1950.

3573. Österlind, S., Inorganic carbon sources of green algae. III. Measurements of photosynthesis in *Scenedesmus quadricauda* and *Chlorella pyrenoidosa, Physiol. Plant.,* 4, 242, 1951.

3574. Österlind, S., Inorganic carbon sources of green algae. IV. Photoactivation of some factor necessary for bicarbonate assimilation, *Physiol. Plant.,* 4, 514, 1951.

3575. Österlind, S., Inorganic carbon sources of green algae. VI. Further experiments concerning photoactivation of bicarbonate assimilation, *Physiol. Plant.,* 5, 403, 1952.

3576. Osvald, H., Die Hochmoortypen Europas, Veröff. Geobot. Inst. ETH, *Stift. Rübel, Zürich,* 3, 707, 1925.

3577. Oswald, W.J., and Gotaas, H.B., Photosynthesis in sewage treatment, *Trans. Am. Soc. Civ. Eng.,* 122, 73, 1957.

3578. Oswald, W., and Benemann, P., A critical analysis of bioconversion with microalgae, in *Biological Solar Energy Conversion*, Mitsui, A., Myachi, S., San Petro, A., and Tamura, S., Eds., Academic Press, New York, 1977, 379.

3579. Oswald, W.J., Gotaas, H.B., Ludwig, H.F., and Lynch, V., Algae symbiosis in oxidation ponds. III. Photosynthetic oxygenation, *Sewage Ind. Wastes*, 25, 692, 1953.

3580. Ott, D.W., Tribophyceae (Xanthophyceae): introduction and bibliography, in *Selected Papers in Phycology*, Rosowski, J.R., and Parker, B.C., Eds., Phycological Society of America, Lawrence, Kansas, 1982, 723.

3581. Ott, D.W., and Brown, R.M., Jr., Light and electron microscopical observations on mitosis in *Vaucheria litorea* Hofman ex. C. Agarhd, *Br. Phycol. J.*, 7, 361, 1972.

3582. Ott, D.W., and Brown, R.M., Jr., Developmental cytology of the genus *Vaucheria*. II. Sporogenesis in *Vaucheria fontinalis* (L.), Christensen, *Br. Phycol. J.*, 9, 333, 1974.

3583. Ott, D.W., and Brown, R.M., Jr., Developmental cytology of the genus *Vaucheria*. III. Emergence, settlement and germination of the mature zoospore of *Vaucheria fontinalis* (L.) Christensen, *Br. Phycol. J.*, 10, 49, 1975.

3584. Ott, D.W., and Brown, R.M., Jr., Developmental cytology of the genus *Vaucheria*. IV. Spermatogenesis, *Br. Pycol. J.*, 13, 69, 1978.

3585. Ott, F.D., and Sommerfeld, M.R., Freshwater Rhodophyceae: introduction and bibliography, In *Selected Papers in Phycology* II, Rosowski, J.R., and Parker, B.C., Eds., Phycological Society of America, Lawrence, 1982, 671.

3586. Otte, A.M., and Bellis, V.J., Edaphic diatoms of a low salinity estuarine marsh system in North Carolina—a comparative floristic study, *J. Elisha Mitchell Sci. Soc.*, 101, 116, 1985.

3587. Otte, M.L., Bestebroer, S.J., van der Linden, J.M., Rozema, J., and Broekman, R.A., A survey of zinc, copper, and cadmium concentrations in salt marsh plants along the Dutch coast, *Environ. Pollut.*, 72, 175, 1991.

3588. Outka, D.E., and Williams, D.C., Sequential coccolith morphogenesis in *Hymenomonas carterae*, *J. Protozool.*, 18, 285, 1971.

3589. Overbaugh, J.M., and Fall, R., Detection of glutathione peroxidases in some microalgae, *FEMS Microbiol. Lett.*, 13, 371, 1982.

3590. Overbaugh, J.M., and Fall, R., Characterization of a selenium independent glutathione peroxidase from *Euglena gracilis*, *Plant Physiol.*, 77, 437, 1985.

3591. Overbeck, F., and Happach, H., Über das Wachstum und den Wasserhaushalt einiger Hochmoorsphagnen, *Flora*, 144, 335, 1957.

3592. Overnell, J., The effect of heavy metals on photosynthesis and loss of cell potassium in two species of marine algae, *Dunaliella tertiolecta* and *Phaeodactylum tricornutum*, *Mar. Biol.*, 29, 99, 1975.

3593. Overnell, J., The effect of some heavy metal ions on photosynthesis in a freshwater alga, *Pestic. Biochem. Physiol.*, 5, 19, 1975.

3594. Overnell, J., Inhibition of marine algal photosynthesis by heavy metals, *Mar. Biol.*, 38, 335, 1976.

3595. Ozimek, T., Aspects of the ecology of a filamentous algae in a eutrophicated lake, *Hydrobiologia*, 191, 23, 1990.

3596. Ozimek, T., and Klekot, L., *Glyceria maxima* (Hartm.) Holmb. in ponds supplied wiwater, *Aquat. Bot.*, 7, 231, 1979.

3597. Ozimek, T., and Kowalczewski, A., Long-term changes of the submerged macrophytes in eutrophic Lake Mikolajske (north Poland), *Aquat. Bot.*, 19, 1, 1984.

3598. OzimeK, T., Prejs, A., and Prejs, K., Biomass and distribution of underground parts of *Potamogeton perfoliatus* L. and *P. lucens* L. in Mikolajske Lake, Poland, *Aquat. Bot.*, 2, 309, 1976.

3599. Ozimek, T., Pieczynska, E., and Hankiewicz, A., Effects of filamentous algae on submerged macrophyte growth: a laboratory experiment, *Aquat. Bot.*, 41, 309, 1991.

3600. Paasche, E., Coccolith formation, *Nature*, 193, 1094, 1962.

3601. Paasche, E., The adaptation of the carbon-14 method for the measurement of coccolith production in *Coccolithus huxleyi*, *Physiol. Plant.*, 16, 186, 1963.

3602. Paasche, E., A tracer study of the inorganic carbon uptake during coccolith formation and photosynthesis in the coccolitrophorid *Coccolithus huxleyi*, *Physiol. Plant. Suppl.*, 3, 1, 1964.

3603. Paasche, E., The effect of 3(p-chlorophenyl)1,1-dimethylurea (CMU) on photosynthesis and light-independent coccolith formation in *Coccolithus huxleyii*, *Physiol. Plant.*, 18, 138, 1965.

3604. Paasche, E., Adjustment to light and dark rates of coccolith formation, *Physiol. Plant.*, 19, 271, 1966.

3605. Paasche, E., Action spectrum of coccolith formation, *Physiol. Plant.*, 19, 770, 1966.

3606. Paasche, E., Biology and physiology of coccolithophorides, *Ann. Rev. Microbiol.*, 22, 71, 1968.

3607. Paasche, E., The effect of temperature, light intensity, and photoperiod on coccolith formation, *Limnol. Oceanogr.*, 13, 178, 1968.

3608. Paasche, E., Light-dependent coccolith formation in two forms of *Coccolithus pelagicus*. With remarks on the ^{14}C zero-thickness counting efficiency of coccolithophorids, *Arch. Mikrobiol.*, 67, 199, 1969.

3609. Paasche, E., Effect of ammonium and nitrate on growth, photosynthesis and ribulosediphosphate carboxylase content of *Dunaliella tertiolecta, Physiol. Plant.*, 25, 294, 1971.

3610. Paasche, E., Silicon and the ecology of marine plankton diatoms. II. *Thalassiosira pseudonana* (*Cyclotella nana*) grown in a chemostat with silicate as the limiting nutrient, *Mar. Biol.*, 19, 117, 1973.

3611. Paasche, E., Silicon and the ecology of marine plankton diatoms. II. Silicate-uptake kinetics of five diatom species, *Mar. Biol.*, 19, 262, 1973.

3612. Paasche, E., Growth of the plankton diatom *Thalassiosira nordenskioeldii* Cleve at low silicate concentrations, *J. Exp. Biol. Ecol.*, 18, 173, 1975.

3613. Paasche, E., Silicon, in *The Physiological Ecology of Phytoplankton*, Morris, I., Ed., University of California Press, Berkeley, 1980, 259.

3614. Pace, F., Ferrara, R., and Del Carratore, G., Effects of sub-lethal doses of copper sulphate and lead nitrate on growth and pigment composition of *Dunaliella salina* Teod., *Bull. Environ. Contam. Toxicol.*, 17, 679, 1977.

3615. Packard, T.T., The light dependence of nitrite reductase in marine phytoplankton, *Limnol. Oceanogr.*, 18, 466, 1973.

3616. Padan, E., Facultative anoxygenic photosynthesis in cyanobacteria, *Ann. Rev. Plant Physiol.*, 30, 27, 1979.

3617. Padan, E., and Shilo, M., Cyanophages—Viruses attacking blue-green algae, *Bacteriol. Rev.*, 37, 343, 1973.

3618. Paerl, H.W., Specific associations of the blue-green algae *Anabaena* and *Aphanizomenon* with bacteria in freshwater bloom, *J. Phycol.*, 12, 431, 1976.

3619. Paerl, H.W., Feasibility of [55]Fe autoradiography as performed on N_2-fixing *Anabaena* spp. populations and associated bacteria, *Appl. Environ. Microbiol.*, 43, 210, 1982.

3620. Paerl, H.W., and Lean, D.R.S., Visual observations of phosphorus movement between algae, bacteria, and abiotic particles in lake water, *J. Fish. Res. Board Can.*, 33, 2805, 1976.

3621. Paerl, H.W., and Kellar, P.E., Significance of bacterial-*Anabaena* (Cyanophyceae) associations with respect to N_2 fixation in freshwater, *J. Phycol.*, 14, 254, 1978.

3622. Paerl, H.W., Prufert, L.E., and Ambrose, W.W., Contemporaneous N_2 fixation and oxygenic photosynthesis in the non-heterocystous mat-forming cyanobacterium *Lyngbya aestuarii, Appl. Environ. Microbiol.*, 57, 3086, 1991.

3623. Pagenkopf, G.K., Russo, R.C., and Thurston, R.V., Effect of complexation on toxicity of copper to fishes, *J. Fish. Res. Board Can.*, 31, 462, 1974.

3624. Pakarinen, P., Bogs as peat-producing ecosystems, *Intern. Peat Soc. Bull.*, 7, 51, 1975.

3625. Pakarinen, P., Production and nutrient ecology of the *Sphagnum* species in southern Finnish raised bog, *Ann. Bot. Fennici*, 15, 15, 1978.

3626. Pakarinen, P., Distribution of heavy metals in the *Sphagnum* layer of bog hummock and hollows, *Ann. Bot. Fennici*, 15, 287, 1978.

3627. Pakarinen, P., and Tolonen, K., Nutrient contents of *Sphagnum* moses in relation to bog water chemistry in northern Finland, *Lindbergia*, 4, 27, 1977.

3628. Pal, B.P., Kundu, B.C., Sundaralingam, V.S., and Venkataraman, G.S., *Charophyta*, Indian Council of Agric. Research, New Delhi, 1962.

3629. Palamar-Mordvyntseva, H.M., The need of *Ankistrodesmus braunii* Brunnth. in the field, *Ukr. Bot. Zh.*, 25, 21, 1968 (in Russian).

3630. Palmer, C.M., and Maloney, T.E., Preliminary screening for potential algicides, *Ohio J. Sci.*, 55, 1, 1955.

3631. Palmer, E.G., and Starr, R.C., Nutrition of *Pandorina morum, J. Phycol.*, 7, 85, 1971.

3632. Palmisano, A.C., and Sullivan, C.W., Sea ice microbial communities (SIMCOs). I. Distribution, abundance, and primary production of ice microalgae in McMurdo Sound in 1980, *Polar Biol.*, 2, 171, 1983.

3633. Palmisano, A.C., Kottmeier, S.T., Moe, R.L., and Sullivan, C.W., Sea ice microbial communities. IV. The effect of light perturbation on microalgae at the ice seawater interface in McMurdo Sound, Antarctica, *Mar. Ecol. Prog. Ser.*, 21, 37, 1985.

3634. Palmisano, A.C., SooHoo, S.B., and Sullivan, C.W., Effects of four environmental variables on photosynthesis—irradiance relationships in Antarctic sea-ice microalgae, *Mar. Biol.*, 94, 299, 1987.

3635. Pamatmat, M.M., Ecology and metabolism of a benthic community on an intertidal sandflat, *Int. Rev. Ges. Hydrobiol.*, 53, 211, 1968.

3636. Pandey, S.B., *Eleocharis palustris*, in Ecological studies of noxious weeds common in India and America which are becoming increasing problem in the Upper Gangetic Plains. Final Tech. Repory, Banaras Hindu University, Varanasi, 1969 (cited by Vyas ea 90).

3637. Pankow, H., Die Bindung von Luftstickstoff durch zwei weitere Blaualgen—Arten: *Fischeriella muscicola* und *F. major, Naturwissenschaften,* 48, 185, 1964.

3638. Pano, A., and Middlebrooks, E.J., Ammonia nitrogen removal in facultative wastewater stabilization ponds, *J. Water Pollut. Control Fed.,* 54, 344, 1982.

3639. Papenfuss, G.F., Proposed names for the phyla of algae, *Bull. Torrey Bot. Club.,* 73, 217, 1946.

3640. Papenfuss, G.F., Classification of the algae, in *A Century of Progress in the Natural Sciences, 1853–1953,* California Academy of Science, San Francisco, 1955, 115.

3641. Pardy, R.L., Some factors affecting the growth and distribution of the algal endosymbionts of *Hydra viridis, Biol. Bull.,* 147, 105, 1974.

3642. Pardy, R.L., Phycozoans, phycozoology, phycozoologists, in *Algal Symbiosis: A Continuum of Interaction Strategies,* Goff, L.J., Ed., Cambridge University Press, Cambridge, 1983.

3643. Pardy, R.L., Lewin, R.A., and Lee, K., The *Prochloron* symbiosis. in *Algal Symbiosis: A Continuum of Interaction Strategy,* Goff, L.J., Ed., Cambridge University Press, Cambridge, 1983, 91.

3644. Parekh, J.M., Bhalala, J., Talreja, S.T., and Doshi, Y.A., Uptake of I-131 by marine algae and the factors influencing iodine enrichment by seaweeds from seawater, *Curr. Sci.,* 38, 268, 1969.

3645. Parfit, R.L., Anion adsorption by soils and soil minerals, *Adv. Agron.,* 30, 1, 1978.

3646. Parisi, A.F., and Vallee, B.L., Zinc metalloenzymes characteristics and significance in biology and medicine, *Amer. J. Clin. Nutr.,* 22, 1222, 1969.

3647. Park, R.A., Scavia, D., and Clesceri, N.L., CLEANER: The lake George model, in *Ecological Modelling in a Resource Management Framework,* Russell, C.S., Ed., Resources for the Future, Inc., Washington, DC, 1975, 49.

3648. Parke, M., Manton, I., and Clarke, B., Studies on marine flagellates. II. Three new species of *Chrysochromulina, J. Mar. Biol. Ass. U.K.,* 34, 579, 1955.

3649. Parke, M., Manton, I., and Clarke, B., Studies on marine flagellates. III. Three further species of *Chrysochromulina, J. Mar. Biol. Ass. U.K.,* 35, 387, 1956.

3650. Parke, M., Manton, I., and Clarke, B., Studies on marine flagellates. IV. Morphology and microanatomy of a new species of *Chrysochromulina, J. Mar. Biol. Ass. U.K.,* 37, 209, 1958.

3651. Parke, M., and Adams, I., The motile (*Crystallolithushyalinus* Gaarder and Markali) and nonmotile phases in the life history of *Coccolithus pelagicus* (Wallich) Schiller, *J. Mar. Biol. Ass. U.K.,* 39, 262, 1960.

3652. Parke, M., and Green, J.C., Haptophyta, in Parke, M., and Dixon, P.S., *Check-list of British marine algae—third revision, J. Mar. Biol. Ass. U.K.,* 56, 527, 1976.

3653. Parke, M., and Green, J.C., Chlorophyta, Prasinophyceae, in Parke, M., and Dixon, P.S., *Check-list of British marine algae - third revision, J. Mar. Biol. Ass. U.K.,* 56, 537, 1976.

3654. Parke, M., Green, J.C., and Manton, I., Observations on the fine structure of zoids of the genus *Phaeocystis* (Haptophyceae), *J. Mar. Biol. Ass. U.K.,* 51, 927, 1971.

3655. Parker, B.C., Occurrence of silica in brown and green algae, *Can. J. Bot.,* 47, 537, 1969.

3656. Parker, B.C., and Bold, H.C., Biotic relationships between soil algae and other microorganisms, *Am. J. Bot.,* 46, 185, 1961.

3657. Parker, B.C., Bold, H.C., and Deason, T.R., Facultative heterotrophy in some chlorococcacean algae, *Science,* 133, 761, 1961.

3658. Parker, S.B., Skarnulis, A.J., Westbroek, P., and Williams, R.J.P., The ultrastructure of coccoliths from the marine alga *Emiliania huxleyi* (Lohmann) Hay and Mohler: an ultra-high resolution electron microscope study, *Proc. R. Soc. Lond.,* B 219, 111, 1983.

3659. Parker, W.B., The use of hydric soil to assist in worldwide wetland inventories, in *Proc. Conf. Ecology and Management of Wetlands.* Vol. 2. *Management, Use and Value of Wetlands,* Hook, D.D. et al., Eds., Timber Press, Wilshire, Oregon, 1988, 82.

3660. Parry, G.D.R., and Hayward, J., The uptake of ^{65}Zn by *Dunaliella tertiolecta* Butcher, *J. Mar. Biol. Ass. U.K.,* 53, 915, 1973.

3661. Parslow, J.S., Harrison, P.J., and Thompson, P.A., Development of rapid ammonium uptake during starvation of batch and chemostat cultures of the marine diatom *Thalassiosira pseudonana, Mar. Biol.,* 83, 43, 1984.

3662. Parslow, J.S., Harrison, P.J., and Thompson, P.A., Saturated uptake kinetics: transient response of the marine diatom *Thalassiosira pseudonana* to ammonium, nitrate, silicate or phosphate starvation, *Mar. Biol.,* 83, 51, 1984.

3663. Parsons, T.R., and LeBrasseur, R.J., The availability of food to different trophic levels in the marine food chain, in *Marine Food Chains,* Steele, J.H., Ed., University of California Press, Berkeley, 1970, 325.

3664. Parsons, T.R., Stephans, K., and Strickland, J.D.H., On the chemical composition of eleven species of marine phytoplankton, *J. Fish. Res. Board Can.,* 18, 1001, 1961.

3665. Pascher, A., *Süsswasserflora Mitteleuropas,* Vols. 1–12, Fisher, Jena, 1913–1932.

3666. Pascher, A., Über Flagellaten und Algen, *Ber. Deutsch. Bot. Ges.,* 32, 136, 1914.

3667. Pascher, A., Über Symbiosen von Spaltpilzen und Flagellaten, *Ber. Dtsch. Bot. Ges.,* 32, 339, 1914.

3668. Pascher, A., Über Gruppenbildung und Geschlechtswechsel bei den Gameten einer Chlamydomonadine (*Chlamydomonas paupera*), *Jb. wiss. Bot.,* 75, 551, 1931.

3669. Pascher, A., Über einen neuen einzelligen Organismus mit Eibefruchtung, *Beih. Bot. Zbl.,* 48, 446, 1931.

3670. Pascher, A., Heterokonten, in *Kryptogamen-Flora von Deutschland, Österreich und der Schweiz,* 2nd ed., Vol. XI., Rabenhorst, L., Ed., Akademische Verlagsgesellschaft, Leipzig, 1937–1939, 481.

3671. Passow, H., Rothstein, A., and Clarkson, T.W., The general pharmacology of the heavy metals, *Pharmacol. Rev,* 13, 185, 1961.

3672. Patel, B.N., and Merrett, M.J., Inorganic-carbon uptake by the marine diatom *Phaeodactlym tricornutum, Planta,* 169, 222, 1986.

3673. Paterson, R.A., Infestation of chytridiaceous fungi on phytoplankton in relation to certain environmental factors, *Ecology,* 41, 416, 1960.

3674. Patrick, R., Factors effecting the distribution of diatoms, *Bot. Rev,* 14, 473, 1948.

3675. Patrick, R., Diatom communities in estuaries, in *Estuaries,* Lauff, G.H., Ed., Am. Assoc. Adv. Sci., Washington, DC, 1967, 311.

3676. Patrick, R., Benthic stream communities, *Am. Sci.,* 58, 546, 1970.

3677. Patrick, R., *Report on Water Quality Criteria,* National Academy of Sciences, National Academy of Engineering, Washington, DC, 1971.

3678. Patrick, R., Use of algae, especially diatoms, in the assessment of water quality, in *Biological Methods for the Assessment of Water Quality, ASTM STP,* 528, 76, 1973.

3679. Patrick, R., The formation and maintenance of benthic diatom communities, *Proc. Am. Phil. Soc.,* 120, 475, 1976.

3680. Patrick, R., Ecology of freshwater diatoms and diatom communities, in *The Biology of the Diatoms,* Werner, D., Ed., Blackwell Scientific Publications, Oxford, 1977, 284.

3681. Patrick, R., Effects of trace metals in the aquatic ecosystems, *Am. Sci.,* 66, 185, 1978.

3682. Patrick, R., and Riemer, C.W., *The Diatoms of the United States exclusive of Alaska and Hawaii,* Vol. 1. *Fragilariaceae, Eunitiaceae, Achnanthaceae, Naviculaceae,* Monogr. Acad. Nat. Sci. Phila, 13, 1966.

3683. Patrick, R., and Riemer, C.W., *The Diatoms of the United States exclusive of Alaska and Hawaii,* Vol. 2., Part 1. *Entomoneidaceae, Cymbellaceae, Gomphonemaceae, Epithemiaceae,* Monogr. Acad. Nat. Sci. Phila., 13, 1975.

3684. Patrick, R., Hohn, M.H., and Wallace, J.H., A new method for determining the pattern of the diatom flora, *Not. Nat. Acad. Sci. Philadelphia,* 259, 1, 1954.

3685. Patrick, R., Cairns, J., Jr., and Schmeyr, A., The relative sensitivity of diatoms, snails and fish to twenty common constituents of industrial waters, *Progress. Fish Culturist,* 30, 137, 1968.

3686. Patrick, W.H., Jr., Extractable iron and phosphorus in a submerged soil at controlled redox potentials, *Trans. 8th Int. Congr. Soil Sci.,* 4, 605, 1964.

3687. Patrick, W.H., Jr., and Wyatt, R., Soil nitrogen loss as a result of alternate submergence and drying, *Soil Sci. Soc. Am. Proc.,* 28, 647, 1964.

3688. Patrick, W.H., Jr., and Mahapatra, I.C., Transformation and availability of nitrogen and phosphorus in waterlogged soils, *Advance,* 20, 323, 1968.

3689. Patrick, W.H., Jr., and Mikkelsen, D.S., Plant nutrient behavior in flooded soils, in *Fertilizer Technology and Use,* 2nd ed., Soil Science Soc. of Am., Madison, 1971, 187.

3690. Patrick, W.H., Jr., and DeLaune, R.D., Characterization of the oxidized and reduced zones in flooded soils, *Soil Sci. Soc. Am. Proc.,* 36, 573, 1972.

3691. Patrick, W.H., Jr., and Khalid, R.A., Phosphate release and sorption by soils and sediments— effect of aerobic and anaerobic conditions, *Science,* 186, 53, 1974.

3692. Patrick, W.H., Jr., and Reddy, K.R., Nitrification—denitrification reactions in flooded soils and water bottoms: dependence on oxygen supply and ammonium diffusion, *J. Environ. Qual.,* 5, 469, 1976.

3693. Patrick, W.H., Jr., and Reddy, C.N., Chemical ranges in rice soils, in *Proc. Conf. Soils and Rice,* Int. Rice Res. Inst., Los Baños, Philippines, 1978, 361.

3694. Patrick, W.H., Jr., and Henderson, R.E., A method for controlling redox potential in packed soil cores, *Soil Sci. Soc. Am. J.,* 45, 35, 1981.

3695. Patrick, W.H., Jr., Gotoh, S., and Williams, B.G., Strengite dissolution in flooded soils and sediments, *Science,* 179, 564, 1973.

3696.   Patriquin, D.G., and Keddy, C., Nitrogenase activity (acetylene reduction) in a Nova Scotia salt marsh. Its association with angiosperms and the influence of some edaphic factors, *Aquat. Bot.*, 4, 227, 1978.

3697.   Patten, B.C., Introduction and overview, in *Wetlands and Shallow Continental Water Bodies*, Patten, B.C., Ed., SPB Academic Publishing, The Hague, The Netherlands, 1990, 3.

3698.   Patten, B.C., Ed., *Wetlands and Shallow Continental Water Bodies*, SPB Academic Publishing bv, The Hague, The Netherlands, 1990.

3699.   Patten, B.C., Gopal, B., Jørgensen, S.E., Koryavov, P.P., Květ, J., Löffler, J., Svirezhev, Y., and Tundisi, J., *Ecosystem Dynamics in Freshwater Wetland and Shallow Water Bodies*, Proc. Meet. in Tallin, John Wiley, New York, 1984.

3700.   Patterson, C.C., Contaminated and natural lead environments of man, *Arch. Environ. Health*, 11, 344, 1965.

3701.   Patterson, G.M.L., and Withers, N.W., Laboratory cultivation of *Prochloron*, a tryptophan auxotroph, *Science*, 217, 1034, 1982.

3702.   Patterson, G.M.L., Baker, K.K., Baldwin, C.L., Bolis, C.M., Caplan, F.R., Larsen, L.K., Levine, I.A., Moore, R.E., Nelson, C.S., Tschappat, K.D., Tuang, G.D., Boyd, M.R., Cardellina II, J.H., Collins, R.P., Gustafson, K.R., Snader, K.M., Weislow, O.W., and Lewin, R.A., Antiviral activity of cultured blue-green algae (Cyanophyta), *J. Phycol.*, 29, 125, 1993.

3703.   Patterson, G., Whitton, B.A., Chemistry of water, sediment and algal filaments in groundwater draining an old lead-zinc mine, in *Heavy Metals in Northern England: Environmental and Biological Aspects*, Say, P.J., and Whitton, B.A., Eds., University of Durham, Durham, England, 1981, 65.

3704.   Pattnaik, H., Growth and nitrogen fixation by *Westiellopsis prolifica* Janet., *Ann. Bot. N.S.*, 30, 231, 1966.

3705.   Paul, E.A., and Huang, P.M., Chemical Aspects of Soil., in *The Handbook of Environmental Chemistry* Vol. 1, Part A, *The Natural Environment and the Biogeochemical Cycles*, Hutzinger, O., Ed., Springer-Verlag, Berlin, 1980, 69.

3706.   Paul, J.H., and Cooksey, K.E., Regulation of asparaginase, glutamine synthetase and glutamate dehydrogenase in response to medium nitrogen concentration in euryhaline *Chlamydomonas* species, *Plant Physiol.*, 68, 1364, 1981.

3707.   Paulsen, B.S., Vieira, A.H., and Klaveness, D., Structure of extracellular polysaccharides produced by a soil *Cryptomonas* sp. (Cryptophyceae), *J. Phycol.*, 28, 61, 1992.

3708.   Pavoni, M., Die Bedeutung des Nannoplanktons in Vergleich zum Netzplankton, *Schweiz. Zeitschrift Hydrol.*, 25, 219, 1963.

3709.   Payne, M., Sur la gélose et les nids de salangane, *Compt. Rend. Acad. Sci.*, 49, 521, 1859.

3710.   Payne, N.F., *Technique for Wildlife Habitat Management of Wetlands*, McGraw-Hill, New York, 1992.

3711.   Payne, W.J., Reduction of nitrogenous oxides by microorganisms, *Bacteriol. Rev.*, 37, 409, 1973.

3712.   Payne, W.J., *Denitrification*, John Wiley and Sons, New York, 1981.

3713.   Pearsall, W.H., The soil complex in relation to plant communities. III. Moorlands and bogs, *J. Ecol.*, 26, 298, 1938.

3714.   Pearsall, W.H., The investigation of wet soils and its agricultural implications, *Emp. J. Agro.*, 18, 289, 1950.

3715.   Pearsall, W.H., and Bengry, R.P., The growth of *Chlorella* in darkness and in glucose solution, *Ann. Bot. N.S.*, 14, 365, 1940.

3716.   Pearsall, W.H., and Bengry, R.P., Growth of *Chlorella* in relation to light intensity, *Ann. Bot. N.S.*, 14, 485, 1940.

3717.   Pearsall, W.H., and Gorham, E., Production ecology. I. Standing crops of natural vegetation, *Oikos*, 7, 193, 1956.

3718.   Pearsall, W.H., and Newbould, P.J., Production ecology I. Standing crops of natural vegetation in the sub-arctic, *J. Ecol.*, 45, 593, 1957.

3719.   Pearse, V.B., Radioisotopic study of calcification in the articulated coraline alga *Bossiella orbigniana*, *J. Phycol.*, 8, 88, 1972.

3720.   Peary, J.A., and Castenholz, R.W., Temperature strains of a thermophilic blue-green algae, *Nature*, 202, 720, 1964.

3721.   Peat, S., and Turvey, J.R., Polysaccharides of marine algae, *Fortschr. Chem. Org. Naturstoffe*, 23, 1, 1965.

3722.   Pedersen, A., Growth measurements of five *Sphagnum* species in south Norway, *Norwegian J. Bot.*, 22, 277, 1975.

3723.   Pedersen, A.G., and Knutsen, G., Uptake of L-phenylalanine in synchronous *Chlorella fusca*. Characterization of the uptake system, *Physiol. Plant.*, 32, 294, 1974.

3724.   Pedersen, M., Marine brown algae requiring vitamin B_{12}, *Physiol. Plant.*, 22, 977, 1969.

3725. Pedersen, P.M., Phaeophyta: life histories, in *The Biology of Seaweeds,* Lobban, C.S., and Wynne, M.J., Eds., University of California Press, Berkeley, 1981, 194.

3726. Pedersen, T.A., Kirk, M., and Bassham, J.A., Light-dark transients in levels of intermediate compounds during photosynthesis in air-adapted *Chlorella, Physiol. Plant.,* 19, 219, 1966.

3727. Pelikán, J., Svoboda, J., and Květ, J., On some relations between the production of *Typha latifolia* and muskrat population, *Zool. listy* 19, 303, 1970.

3728. Pellegrino, F., Wong, D., Alfano, R.R., and Zilinskas, B.A., Fluorescence relaxation kinetics and quantum yield from the phycobilisomes of the blue-green alga *Nostoc* sp. measured as a function of single picosecond pulse intensity, *Photochem. Photobiol.,* 34, 691, 1981.

3729. Pelroy, R.A., and Bassham, J.A., Photosynthetic and dark carbon metabolism in unicellular blue-green algae, *Arch. Mikrobiol.,* 86, 25, 1972.

3730. Penfound, W.T., Southern swamps and marshes, *Bot. Rev,* 18, 413, 1952.

3731. Penfound, W.T., Production of vascular aquatic plants, *Limnol. Oceanogr,* 1, 92, 1956.

3732. Penfound, W.T., and Earle, T.T., The biology of the water hyacinths, *Ecol. Monogr,* 18, 447, 1948.

3733. Penhale, P., Macrophyte-epiphyte biomass and productivity in an eelgrass (*Zostera marina*) community, *J. Exp. Mar. Biol. Ecol.,* 26, 211, 1977.

3734. Penhale, P.A., and Smith, W.O., Jr., Excretion of dissolved carbon by eelgrass (*Zostera marina*) and its epiphytes, *Limnol. Oceanogr,* 22, 400, 1977.

3735. Penman, H.L., Natural evapotranspiration from open-water, bare soil and grass, *Proc. R. Soc. Acad.,* 193, 120, 1948.

3736. Penman, H.L., Estimating evaporation, *Trans. Am. Geogr. Union,* 37, 43, 1956.

3737. Pennington, W., The control of the numbers of freshwater phytoplankton by small invertebrate animals, *J. Ecol.,* 29, 204, 1941.

3738. Penrose, W.R., Arsenic in the marine and aquatic environments: analysis, occurrence, and significance, *CRC Crit. Rev. Environ. Cont.,* 4, 465, 1974.

3739. Pentecost, A., Calcification and photosynthesis in *Corallina officinalis* L. using the $^{14}CO_2$ method, *Br. Phycol. J.,* 13, 383, 1978.

3740. Pentecost, A., Calcification in plants, *Int. Rev. Cytol.,* 62, 1, 1980.

3741. Pentecost, A., Photosynthetic plants as intermediary agents between environmental bicarbonate and carbonate deposits, in *Inorganic Carbon Uptake by Aquatic Photosynthetic Organisms,* Lucas, W.J., and Berry, J.A., Eds., *Am. Soc. Plant Physiol.,* Rockville, Maryland, 1985, 459.

3742. Pentecost, A., Growth and calcification of the freshwater cyanobacterium *Rivularia haematites, Proc. R. Soc. Lond.,* B 232, 125, 1987.

3743. Pentecost, A., Growth and calcification of the cyanobacterium *Homoeothrix crustacea, J. Gen. Microbiol.,* 134, 2665, 1988.

3744. Pentecost, A., Observations on growth rates and calcium carbonate deposition in the green alga *Gongrosira, New Phytol.,* 110, 249, 1988.

3745. Pentecost, A., Growth and calcification of *Calothrix*—dominated oncolites from Northern England, in *5th Int. Conf. Biomineralization: Origin, Evolution and Modern Aspects of Biomineralization in Plants and Animals,* Crick, R.E., Ed., Plenum Press, New York, 1989, 443.

3746. Pentecost, A., Calcification processes in algae and cyanobacteria, in *Calcareous Algae and Stromatolites,* Riding, R., Ed., Springer-Verlag, Berlin, 1991, 3.

3747. Pentecost, A., and Riding, R., Calcification in cyanobacteria, in *Biomineralization of Lower Plants and Animals,* Leadbeater, B.S.C., and Riding, R., Eds., Clarendon Press, Oxford, 1986, 73.

3748. Pentecost, A., and Bauld, J., Nucleation of calcite on the sheaths of cyanobacteria using a simple diffusion cell, *Geomicrobiol. J.,* 6, 129, 1988.

3749. Percival, E., Marine algal carbohydrates, *Oceanogr. Mar. Biol. Ann. Rev,* 6, 137, 1968.

3750. Percival, E., The polysaccharides of green, red and brown seaweeds: their basic structure, biosynthesis and function, *Br. Phycol. J.,* 14, 103, 1979.

3751. Percival, E.G.J., and Johnston, R., Confirmation of 1,3-linkage in carrageenin and isolation of l-galactose derivatives from a resistant fragment, *J. Chem. Soc.,* p. 1994, 1950.

3752. Percival, E., and McDowell, R.H., *Chemistry and Enzymology of Marine Algal Polysaccharides,* Academic Press, New York and London, 1967.

3753. Perini, F., Kamen, M.D., and Schiff, J.A., Iron-containing proteins in *Euglena.* I. Detection and characterization, *Biochim. Biophys. Acta,* 88, 74, 1964.

3754. Perini, F., Schiff, J.A., and Kamen, M.D., Iron-containing proteins in *Euglena.* II. Functional localization, *Biochim. Biophys. Acta,* 88, 91, 1964.

3755. Perry, M.J., Alkaline phosphatase activity in subtropical central North Pacific waters using a sensitive fluorometric method, *Mar. Biol.,* 15, 113, 1972.

3756. Perry, M.J., Phosphate utilization by an oceanic diatom in phosphorus-limited chemostat culture and in oligotrophic waters of the central North Pacific, *Limnol. Oceanogr,* 21, 88, 1976.

3757. Perry, M.J., and Eppley, R.W., Phosphate uptake by phytoplankton in the central North Pacific Ocean, *Deep-Sea Res.,* 28, 39, 1981.

3758. Peters, G.A., The *Azolla - Anabaena* relationship. III. Studies on metabolic capabilities and a further characterization of the symbiont, *Arch. Microbiol.,* 103, 113, 1975.

3759. Peters, G.A., and Mayne, B.C., The *Azolla - Anabaena azollae* relationship. II. Localization of nitrogenase activity as assayed by acetylene reduction, *Plant Physiol.,* 53, 820, 1974.

3760. Peters, G.A., Evans, W.R., and Toia, R.E., Jr., *Azolla - Anabaena azollae* relationship. IV. Photosynthetically driven, nitrogenase-catalyzed H_2 production, *Plant Physiol.,* 58, 119, 1976.

3761. Peters, R.H., Concentrations and kinetics of phosphorus fractions along the trophic gradient of Lake Memphremagog, *J. Fish. Res. Board Can.,* 36, 970, 1979.

3762. Peters, R.H., Phosphorus availability in Lake Memphremagog and its tributaries, *Limnol. Oceanogr,* 26, 1150, 1981.

3763. Petersen, J.B., Algal collected by E.Hulten on the Swedish Kamtschatka—expedition 1920–1922, especially from the hot springs, *Kgl. Danske Videsk. sels. Biol. Med.,* 20, 1, 1946.

3764. Petersen, R.C., and Cummins, K.W., Leaf processing in woodland stream, *Freshwater Biol.,* 4, 343, 1974.

3765. Petersen, J.B., The aerial algae of Iceland, in *The Botany of Iceland,* Vol. 2, Rosenvinger, L.K., and Warming, E., Eds., 1928, 327.

3766. Petersen, J.B., Studies on the biology and taxonomy of soil algae, *Dansk. Bot. Arkiv,* 8, 1, 1935.

3767. Peterson, C.G., Influence of flow regime on development and desiccation response of lotic diatom communities, *Ecology,* 68, 946, 1987.

3768. Peterson, C.G., and Hoagland, K.D., Effects of wind-induced turbulence and algal mat development on epilithic diatom succession in a large reservoir, *Arch. Hydrobiol.,* 118, 47, 1990.

3769. Peterson, C.G., and Stevenson, R.J., Resistance and resilience of lotic algal communities: importance of disturbance timing and current, *Ecology,* 73, 1445, 1992.

3770. Peterson, C.G., Hoagland, K.D., and Stevenson, J.R., Timing of wave disturbance and the resistance and recovery of a freshwater epilithic microalgal community, *J. N. Am. Benthol. Soc.,* 9, 54, 1990.

3771. Peterson, C.G., Dudley, T.L., Hoagland, K.D., and Johnson, L.M., Infection, growth, and community-level consequences of a diatom pathogen in a Sonoran Desert stream, *J. Phycol.,* 29, 442, 1993.

3772. Petria, V., Effect of chromium salts from water sediments on physiological processes in the alga *Chlorella vulgaris, Rev. Roum. Biol., Ser. Biol. Veg.,* 23, 55, 1978.

3773. Pettersen, R., Control by ammonium of intercompartmental guanine transport in *Chlorella, Z. Pflanzenphysiol.,* 76, 213, 1975.

3774. Pettersen, R., and Knutsen, G., Uptake of guanine by synchronized *Chlorella fusca.* Characterization of the transport system in autospores, *Arch. Microbiol.,* 96, 233, 1974.

3775. Peverly, J.H., Element accumulation and release by macrophytes in a wetland stream, *J. Environ. Qual.,* 14, 137, 1985.

3776. Peverly, J.H., Adamec, J., and Parthasarathy, M.V., Association of potassium and some other monovalent cations with occurrence of polyphosphate bodies in *Chlorella pyrenoidosa, Plant Physiol.,* 62, 120, 1978.

3777. Pfadenhauer, J., and Lütke Twenhöven, F., Nährstoffokölogie von *Molinia caerulea* and *Carex acutifotmis* auf baumfreien Niedermooren des Alpenvorlandes, *Flora,* 178, 157, 1986.

3778. Pfennig, N., and Widdel, F., The bacteria of the sulphur cycle, *Phil. Trans. R. Soc. Lond.,* B 298, 433, 1982.

3779. Pfitzer, E., Untersuchungen über Bau and Entwicklung der Bacillariaceen (Diatomaceen), *Hansteins Bot. Abh.,* 2, 1, 1871.

3780. Philipose, M.T., *Chlorococcales,* Indian Council of Agric. Research, New Delhi, 1967.

3781. Phillips, G.L., Emnison, D., and Moss, B., A mechanism to account for macrophyte decline in progressively eutrophicated freshwaters, *Aquat. Bot.,* 4, 103, 1978.

3782. Pia, J., Die Kalkbindung durch Pflanzen, *Botan. Centr. Beih.,* A 52, 1, 1934.

3783. Pickering, D.C., and Puia, I.L., Mechanism for uptake of zinc by *Fontinalis antipyretica, Physiol. Plant.,* 22, 653, 1969.

3784. Pickett, J., McKellar, H., and Kelley, J., Plant community composition, leaf mortality, and above-ground production in a tidal freshwater marsh, in *Proc. Symp. Freshwater Wetlands and Wildlife,* Sharitz, R.R., and Gibbons, J.W., Eds., U.S. Dept. of Energy, 1989, 351.

3785. Pickett, S.T.A., Collins, S.L., and Armesto, J.J., Models, mechanisms and pathways of succession, *Bot. Rev,* 53, 335, 1987.

3786. Pickett-Heaps, J.D., Ultrastructure and differentiation in *Chara* sp. III. *Aust. J. Biol. Sci.,* 21, 255, 1968.

3787. Pickett-Heaps, J.D., Ultrastructure and differentiation in *Chara* (*vibrosa*). IV. Spermatogenesis, *Aust. J. Biol. Sci.*, 21, 655, 1968.
3788. Pickett-Heaps, J.D., Cell division in *Cyanophora paradoxa*, *New Phytol.*, 71, 561, 1972.
3789. Pickett-Heaps, J.D., *Green Algae—Structure, Reproduction and Evolution in Selected Genera*, Sinauer Associates Publishers, Sunderland, Massachusetts, 1975.
3790. Pickett-Heaps, J.D., Cell division and evolution of branching in *Oedocladium* (Chlorophyceae), *Cytobiologie*, 14, 319, 1977.
3791. Pickett-Heaps, J.D., and Marchant, H.J., The phylogeny of the green algae: A new proposal, *Cytobiosis*, 6, 255, 1972.
3792. Pieczynska, E., Sources and fate of detritus in the shore zone of lakes, *Aquat. Bot.*, 25, 153, 1986.
3793. Pieczynska, E., and Szczepanska, W., Primary production in the littoral of several Mazurian lakes, *Verh. Internat. Verein. Limnol.*, 16, 372, 1966.
3794. Pienaar, R.N., The rhythmic production of body covering components in the haptophycean flagellate *Hymenomonas carterae*, in *The Mechanisms of Mineralization in the Invertebrates in Marine Science*, Vol. 5, Watabe, N., and Wilbur, K.M., Eds., University of South Carolina Press, Columbia, 1976, 203.
3795. Pienaar, R.N., Chrysophytes, in *Phytoflagellates, Developments in Marine Biology,* Vol. 2, Cox, E.R., Ed., Elsevier/North Holland, New York, 1980, 213.
3796. Pierce, G.J., Wetland soils, in *Wetlands Ecology and Conservation: Emphasis in Pennsylvania*, Majumdar, S.K., Brooks, R.P., Brenner, F.J., and Tiner, R.W., Jr., Eds., University of Pennsylvania Press, 1989, 65.
3797. Pierrou, U., The global phosphorus cycle, in *Nitrogen, Phosphorus and Sulfur—Global Cycles*, Svensson, B.H., and Söderlund, R., Eds., SCOPE Report No. 7, *Ecol. Bull. (Stockholm)*, 22, 75, 1976.
3798. Pierrou, U., The phosphorus cycle: quantitative aspects and the role of man, in *Biogeochemical Cycling of Mineral-Forming Elements*, Trudinger, P.A., and Swaine, D.J., Eds., Elsevier Scientific Publishing, Amsterdam, 1979, 205.
3799. Pilson, M.E.Q., Arsenate uptake and reduction by *Pocillopora verrucosa*, *Limnol. Oceanogr.*, 19, 339, 1974.
3800. Pinckney, J.L., and Zingmark, R.G., Modeling the annual production of intertidal benthic microalgae in estuarine ecosystems, *J. Phycol.*, 29, 396, 1993.
3801. Pintner, I.J., and Provasoli, L., Nutritional characteristics of some chrysomonads, in *Symposium on Marine Microbiology,* Oppenheimer, C.H., Ed., Charles C. Thomas, Springfield, Illinois, 1963, 114.
3802. Pintner, I.J., and Provasoli, L., Heterotrophy in subdued light of 3 *Chrysochromulina* species, *Bull. Misaki. Mar. Biol. Inst. Kyoto Univ.*, 12, 25, 1968.
3803. Pinto, J.S., and Silva, E.S., The toxicity of *Cardium edule* L. and its possible relation to the Dinoflagellate *Prorocentrum micans* Ehr., *Notas Estudios Inst. Biol. Maritima*, 2, 1, 1956.
3804. Pirson, A., Ernährungs-und stoffwechselphysiologische untersuchung an *Fontinalis* and *Chlorella*, *Z. Bot.*, 31, 193, 1937.
3805. Pirson, A., Über die Wirkung alkalionen auf Wachstum und Stoffwechsel von *Chlorella*, *Planta*, 29, 231, 1939.
3806. Pirson, A., Functional aspects in mineral nutrition of green plants, *Ann. Rev. Plant Physiol.*, 6, 71, 1955.
3807. Pirson, A., Manganese and its role in photosynthesis, in *Trace Elements*, Lamb, C.A., Bentley, O.G., and Beattie, J.M., Eds., Academic Press, New York, 1958, 81.
3808. Pirson, A., Photosynthese und mineralische Faktoren, in *Handbuch der Pflanzenphysiologie,* Vol. V., Part 2, Ruhland, W., Ed., Springer-Verlag, Berlin, 1960, 123.
3809. Pirson, A., and Bergmann, L., Manganese requirement and carbon source in *Chlorella*, *Nature*, 176, 209, 1955.
3810. Pirson, A., and Badour, S.S.A., Kennzeichung von Mineralsalzmangelzustanden bei Grünalgen mit analytischenchemischen Methodik. I. Kohlenhydratspiegel, organischen Stickstoff und Chlorophyll bei Kalimangel im Vergleich mit Magnesium- und Manganmangel, *Flora Jena*, 150, 243, 1961.
3811. Pirson, A., and Lorenzen, H., Synchronized dividing algae, *Ann. Rev. Plant Physiol.*, 17, 439, 1966.
3812. Pirson, A., Tichy, C., and Wilhelmi, G., Stoffwechsel und Mineralsalzernährung einzelliger Grünalgen. I. Vergleichende Untersuchungen an Mangelkulturen von *Ankistrodesmus*, *Planta*, 40, 199, 1952.
3813. Pisano, E., The magellanic tundra complex, in *Ecosystems of the World, Mires: Swamp. Bog, Fen and Moor* Vol. 4B *Regional Studies*, Gore, A.J.P., Ed., Elsevier Science Publishers, Amsterdam, 1983, 295.
3814. Pistorius, E.K., Funkhouser, E.A., and Voss, H., Effect of ammonium and ferricyanide on nitrate utilization by *Chlorella vulgaris*, *Planta*, 141, 279, 1978.

3815.   Pitter, P., *Hydrochemistry* SNTL Praha, Czech Republic, 1981 (in Czech).
3816.   Planas, D., and Healley, F.P., Effects of arsenate on growth and phosphorus metabolism of phyto-plankton, *J. Phycol.*, 14, 337, 1978.
3817.   Planter, M., Elution of mineral components out of dead reed *Phragmites communis* Trin., *Pol. Arch. Hydrobiol.*, 17, 357, 1970.
3818.   Platt, T., Ed., Physiological Bases of Phytoplankton Ecology, *Can. Bull. Fish. Aquat. Sci.*, 210, 1981.
3819.   Platt, T., and Jassby, A.D., The relationship between photosynthesis and light for natural assemblages of coastal marine phytoplankton, *J. Phycol.*, 12, 421, 1976.
3820.   Platt, T., Gallegos, C.L., and Harrison, W.G., Photoinhibition of photosynthesis in natural assem-blages of marine phytoplankton, *J. Mar. Res.*, 38, 687, 1980.
3821.   Plumley, F.G., and Schmidt, G.W., Nitrogen-dependent regulation of photosynthetic gene expression, *Proc. Nat. Acad. Sci. U.S.A.*, 86, 2678, 1989.
3822.   Plumley, F.G., Douglas, S.E., Switzer, A.B., and Schmidt, G.W., Nitrogen-dependent biogenesis of chlorophyll-protein complexes, in *Photosynthesis*, Allan R. Liss, Inc., 1989, 311.
3823.   Poff, N.L., Voelz, N.J., and Ward, J.V., Algal colonization under four experimentally-controlled current regimes in a high mountain stream, *J. N. Am. Benthol. Soc.*, 9, 303, 1990.
3824.   Poisson, R., and Hollande, A., Considérations sur la cytologie, la mitose et les affinités des Chloro-monadines. Étude de *Vacuolaria virescens* Cienk. *Annls. Sic. Nat. (Zool.) Ser. II*, 5, 147, 1943.
3825.   Pokorný, J., Photosynthesis of higher submersed plants. Measurement of the gas exchange in an aquatic habitat, in *Methods of the Ecosystem Studies*, Dykyjová, D., Ed., Academia, Praha, Czech Republic, 1989, 365.
3826.   Pokorný, J., Květ, J., Ondok, J.P., Toul, Z., and Ostry, I., Production-ecological analysis of a plant community dominated by *Elodea canadensis* Michx., *Aquat. Bot.*, 19, 263, 1984.
3827.   Pokorný, J., Pešlová, J., and Chromek, J., Photosynthetic reduction of nitrate and its methodological and ecological implications, *Photosynthetica*, 22, 232, 1988.
3828.   Polisini, J.M., and Boyd, C.E., Relationship between cell-wall fractions, nitrogen and standing crop in aquatic macrophytes, *Ecology*, 53, 484, 1972.
3829.   Pollard, A.L., and Smith, P.B., The adsorption of manganese by algal polysaccharides, *Science*, 114, 413, 1951.
3830.   Polunin, N.V.C., Processes contributing to the decay of reed (*Phragmites australis*) litter in fresh water, *Arch. Hydrobiol.*, 94, 182, 1982.
3831.   Polunin, N.V.C., Processes in the decay of reed (*Phragmites australis*) litter in freshwater, in *Proc. Int. Conf. Wetlands: Ecology and Management*, Gopal, B., Turner, R.E., Wetzel, R.G., and Whig-ham, Eds., Internat. Scientific Publications, Jaipur, India, 1982, 293.
3832.   Pomeroy, L.R., Algal productivity in salt marshes of Georgia, *Limnol. Oceanogr*, 4, 386, 1959.
3833.   Pomeroy, L.R., Primary productivity of Boca Ciega Bay, Florida, *Bull. Mar. Sci. Gulf Caribb.*, 10, 1, 1960.
3834.   Pomeroy, R.L., and Wiegert, R.G., Eds., *The Ecology of a Salt Marsh*, Springer-Verlag, New York, 1981.
3835.   Pomeroy, L.R., Darley, W.M., Dunn, E.L., Gallagher, J.L., Haines, E.B., and Whitney, D.M., Primary production, in *The Ecology of a Salt Marsh*, Pomeroy, L.R., and Wiegert, R.G., Eds., Springer-Verlag, Berlin, 1981, 39.
3836.   Ponnamperuma, F.N., The chemistry of submerged soils in relation to the growth and yield of rice, Ph.D. Thesis, Cornell University, Ithaca, New York, 1955.
3837.   Ponnamperuma, F.N., Dynamic aspects of flooded soils and the nutrition of the rice plant, in *The Mineral Nutrition of Rice Plant*, Johns Hopkins Press, Baltimore, Maryland, 1964, 295.
3838.   Ponnamperuma, F.N., The chemistry of submerged soils, *Adv. Agron.*, 24, 29, 1972.
3839.   Ponnamperuma, F.N., Electrochemical changes in submerged soils and the growth of rice, in *Soils and Rice*, Internat. Rice Res. Inst., Los Baños, Philippines, 1978, 421.
3840.   Ponnamperuma, F.N., Effects of flooding on soil., in *Flooding and Plant Growth*, Kozlowski, T.T., Ed., Academic Press, New York, 1984, 9.
3841.   Ponnamperuma, F.N., Chemical kinetics of wetland rice soils relative to soil fertility, in *Proc. Conf. Wetland Soils: Characterization, Classification, and Utilization*, Int. Rice Res. Inst., Los Baños, Philippines, 1985, 71.
3842.   Pool, R.R., Jr., The establishment of the *Hydra-Chlorella* symbiosis: recognition of potential algal symbionts, *Ber. Deutsch. Bot. Ges.*, 94, 565, 1981.
3843.   Poole, J.R., Energy coupling for membrane transport, *Ann. Rev. Plant Physiol.*, 29, 437, 1978.
3844.   Popovský, J., and Pfiester, L.A., *Dinophyceae (Dinoflagellida)*, Süsswasserflora von Mitteleuropa 6, Gustav Fisher Verlag, Jena, Stuttgart, 1990.
3845.   Porath, D., Hepher, B., and Koton, A., Duckweed as an aquatic crop: evaluation of clones for aquaculture, *Aquat. Bot.*, 7, 273, 1979.

3846. Porcella, D.B., Grau, P., and Radimsky, J., Provisional algal assay procedure, University of California, Berkeley, SERL Report No. 70, 1970.
3847. Porcella, D.B., Adams, V.D., and Cowen, P.A., Sediment-water microcosms for assessment of nutrient interactions in aquatic ecosystems, in *Biostimulation and Nutrient Assessment*, Middlebrooks, E.J., Falkenborg, D.H., and Maloney, T.E., Eds., Ann Arbor Science, Ann Arbor, Michigan, 1976, 56.
3848. Porter, G., Tredwell, C.J., Searle, G.F.W., and Barber, J., Picosecond timensolved energy transfer in *Porphyridium aurentum, Biochem. Biophys. Acta*, 501, 232, 1978.
3849. Porter, K.G., Selective effects of grazing by zooplankton on the phytoplankton of Fuller Pond, Kent, Connecticut, Ph.D. Thesis, Yale University, Hartford, Connecticut, 1973.
3850. Porter, K.G., Selective grazing and differential digestion of algae by zooplankton, *Nature*, 244, 179, 1973.
3851. Porter, K.G., The plant-animal interface in freshwater ecosystems, *Am. Sci.*, 65, 159, 1977.
3852. Porter, K.G., Phagotrophic phytoflagellates in microbial food webs, *Hydrobiologia*, 159, 89, 1988.
3853. Porterfield, W.M., References to the algae in the Chinese classics, *Bull. Torrey Bot. Club.*, 49, 297, 1922.
3854. Postgate, J., Relevant aspects of the physiological chemistry of nitrogen fixation, *Symp. Soc. Gen. Microbiol.*, 21, 287, 1971.
3855. Postgate, J., *Nitrogen Fixation*, Studies in Biology No. 92, Edward Arnold, London, 1978.
3856. Portman, J.E., and Riley, J.P., Determination of arsenic in sea water, marine plants and silicate and carbonate sediments, *Anal. Chim. Acta*, 31, 509, 1964.
3857. Poulet, S.A., Grazing of *Pseudocalanus minutus* on naturally occurring particulate matter, *Limnol. Oceanogr*, 18, 564, 1973.
3858. Poulet, S.A., Seasonal grazing of *Pseudocalanus minutus* on particles, *Mar. Biol.*, 25, 109, 1974.
3859. Pouliot, Y., and de la Noüe, J., Utilisation des micro-algues pour le traitement tertiare des eaux usees, *Proc. 7th Symp. sur le Traitement des Eaux Usees*, Montreal, 1984, 131.
3860. Povolny, M., The effect of the steeping of peat-cellulose flowerpots (Jiffypots) in extracts of seaweeds on the quality of tomato seedlings, in *Proc. VIII. Int. Seaweed Symposium*, Fogg, G.E., and Jones, W.E., Eds., Marine Sci. Lab., Menai Bridge, Wales, 1981, 730.
3861. Power, M.E., and Matthews, W.J., Algae-grazing minnows (*Campostoma anomalum*), piscivorous bass (*Micropterus* spp.) and the distribution of attached algae in a small prairie margin stream, *Oecologia (Berl.)*, 60, 328, 1983.
3862. Power, M.E., Matthews, W.J., and Stewart, A.J., Grazer control of algae in an Ozark Mountain stream: effects of short-time exclusion, *Ecology*, 69, 1894, 1988.
3863. Prager, J.C., Burke, J.M., Marchisotto, J., and McLaughlin, J.J.A., Mass culture of a tropical dinoflagellate and the chromatographic analysis of extracellular polysaccharides, *J. Protozool.*, 6 (Suppl.), 19, 1959.
3864. Prakash, A., and Taylor, R., A "red water" bloom of *Gonyaulax acatenella* in the Strait of Georgia and its relation to paralutic shellfish toxicity, *J. Fish. Res. Bd. Can.*, 23, 1265, 1966.
3865. Prakash, A., Rashid, M.A., Jensen, A., and Subba Rao, D.V., Influence of humic substances on the growth of marine phytoplankton: diatoms, *Limnol. Oceanogr.*, 18, 516, 1973.
3866. Prakash, G., and Kumar, H.D., Studies on sulphur-selenium antagonism in blue-green algae. 1. Sulphur nutrition, *Arch. Mikrobiol.*, 77, 196, 1971.
3867. Prask, J.A., and Plocke, D.J., The role of zinc in the structural integrity of the cytoplasmic ribosomes of *Euglena gracilis, Plant Physiol.*, 48, 150, 1971.
3868. Prát, S., The culture of calcareous Cyanophyceae, *Studies Plant Physiol. Lab. Charles Univ. Prague*, 3, 86, 1925.
3869. Pratt, R., Influence of the size of the inoculum on the growth of *Chlorella vulgaris* in freshly prepared culture medium, *Am. J. Bot.*, 27, 52, 1940.
3870. Pratt, R., Studies on *Chlorella vulgaris*. V. Some properties of the growth-inhibitor formed by *Chlorella* cells, *Am. J. Bot.*, 29, 142, 1942.
3871. Pratt, R., Studies on *Chlorella vulgaris*. VI. Retardation of photosynthesis by a growth-inhibiting substance from *Chlorella vulgaris, Am. J. Bot.*, 30, 32, 1943.
3872. Pratt, R., and Fong, J., Studies on *Chlorella vulgaris*. II. Further evidence that *Chlorella* cells form a growth-inhibiting substance, *Am. J. Bot.*, 27, 431, 1940.
3873. Pratt, R., and Fong, J., Studies on *Chlorella vulgaris*. III. Growth of *Chlorella* and changes in the hydrogen-ion and ammonium-ion concentrations in solutions containing nitrate and ammonium nitrogen, *Am. J. Bot.*, 27, 735, 1940.
3874. Pratt, R., Oneto, J.F., and Pratt, J., Studies on *Chlorella vulgaris*. X. Influence of the age of the culture on the accumulation of chlorellin, *Am. J. Bot.*, 32, 405, 1945.

3875. Preiss, J., and Kosuge, T., Regulation of enzyme activity in photosynthetic systems, *Ann. Rev. Plant Physiol.*, 21, 433, 1970.

3876. Prentki, R.T., Gustafson, T.D., and Adams, M.S., Nutrient movement in lakeshore marshes, in *Proc. Conf. Freshwater Wetlands: Ecological Processes and Management Potentials*, Good, R.E., Whigham, D.F., and Simpson, R.L., Eds., Academic Press, New York, 1978, 169.

3877. Prescott, G.W., Objectionable algae with reference to killing of fish and other animals, *Hydrobiologia*, 1, 1, 1948.

3878. Prescott, G.W., Algae of the Western Great Lakes area, *Cranbrook Inst. Sci. Bull.*, 31, 1, 1951.

3879. Prescott, G.W., A guide to the literature on ecology and life histories of the algae, *Bot. Rev.*, 22, 167, 1956.

3880. Prescott, G.W., *The Algae: A Review*, Houghton Mifflin, Boston, 1968.

3881. Prescott, G.W., *How to Know the Fresh-water Algae*, Wm. C. Brown, Dubuque, Iowa, 1970.

3882. Preston, A., Jefferies, D.F., Dutton, J.W.R., Harvey, B.R., and Steele, A.K., British Isles coastal waters: The concentrations of selected heavy metals in sea water, suspended matter and biological indicators—a pilot survey, *Environ. Pollut.*, 3, 69, 1972.

3883. Prézelin, B.B., The role of peridinin-chlorophyll a-proteins in the photosynthetic light adaptation of the marine dinoflagellate *Glenodinium* sp., *Planta*, 130, 225, 1976.

3884. Prézelin, B.B., Light reactions in photosynthesis, *Can. Bull. Fish. Aquat. Sci.*, 210, 1, 1981.

3885. Prézelin, B.B., and Alberte, R.S., Photosynthetic characteristics and organization of chlorophyll in marine dinoflagellates, *Proc. Natl. Acad. Sci. U.S.A.*, 75, 1801, 1978.

3886. Prézelin, B.B., Samuelsson, G., and Matlick, H.A., Photosystem II photoinhibition and altered kinetics of photosynthesis during nutrient-dependent high-light photoadaptation in *Gonyaulax polyedra*, *Mar. Biol.*, 93, 1, 1986.

3887. Přibáň, K., and Šmíd, P., Evapotranspiration in littoral macrophyte stands, in *Proc. Conf. Macrophytes is Water Management, Water Hygiene and Fishery*, Dům Techniky ČSVTS České Budějovice, 1982, 65 (in Czech).

3888. Přibáň, K., and Ondok, J.P., Seasonal measurements of radiation and heat balances in Mokré louky near Trebon, in *Proc. Conf. Water and Energy Balance in Meadow and Forest Ecosystems*, Nitra, Slovakia, 1979, 32 (in Czech).

3889. Přibáň, K., and Ondok, L.P., The daily and seasonal course of evapotranspiration from a central European sedge-grass marsh, *J. Ecol.*, 68, 547, 1980.

3890. Přibáň, K., and Ondok, J.P., Heat balance components and evapotranspiration from a sedge-grass marsh, *Folia Geobot. Phytotax.*, 20, 41, 1985.

3891. Přibáň, K., and Ondok, J.P., Evapotranspiration of a willow carr in summer, *Aquat. Bot.*, 25, 203, 1986.

3892. Price, C.A., and Carell, E.F., Control by iron of chlorophyll formation and growth in *Euglena gracilis*, *Plant Physiol.*, 39, 862, 1964.

3893. Price, C.A., Quigley, J.W., A method for determining quantitative zinc requirement for growth, *Soil Sci.*, 101, 11, 1966.

3894. Price, G.D., Badger, M.R., Bassett, M.E., and Whitecross, M.I., Involvement of plasmalemmasomes and carbonic anhydrase in photosynthetic utilization of bicarbonate in *Chara corallina*, *Aust. J. Plant Physiol.*, 12, 241, 1985.

3895. Price, N.M., and Harrison, P.J., Uptake of urea C and urea N by the coastal marine diatom *Thalassiosira pseudonana*, *Limnol. Oceanogr.*, 33, 528, 1988.

3896. Price, N.M., Thompson, P.A., and Harrison, P.J., Selenium: An essential element for growth of the coastal marine diatom *Thalassiosira pseudonana* (Bacillariophyceae), *J. Phycol.*, 23, 1, 1987.

3897. Priddle, J., The production ecology of benthic plants in some Antarctic lakes. I. In situ production studies, *J. Ecol.*, 68, 141, 1980.

3898. Prince, H.H., and D'Itri, F.M., Eds., *Coastal Wetlands*, Proc. 1st. Great Lakes Coastal Wetlands Colloquium, East Lansing, Lewis Publishers, Chelsea, Michigan, 1985.

3899. Pringle, J.D., and Sharp, G.J., Multispecies resource management of economically important marine plant communities of eastern Canada, *Helgol. Meeresunters.*, 33, 711, 1980.

3900. Pringsheim, E.G., Über das Ca-Bedürfnis einiger Algen, *Planta*, 2, 555, 1926.

3901. Pringsheim, E.G., Die Kultur von *Micrasterias* und *Volvox*, Abdruck aus *Arch. Protistenk.*, 72, 1, 1930.

3902. Pringsheim, E.G., Über farblose diatomen, *Arch. Mikrobiol.*, 16, 18, 1951.

3903. Pringsheim, E.G., On the nutrition of *Ochromonas*, *Q. J. Microsc. Sci.*, 93, 71, 1952.

3904. Pringsheim, E.G., Phagotrophie, in *Handbuch der Pflanzenphysiologie*, Vol. 11, Ruhland, W., Ed., Springer-Verlag, Berlin, 1959, 179.

3905. Pringsheim, E.G., Zur Physiologie der farblosen Diatomee *Nitzschia putrida*, *Arch. Mikrobiol.*, 55, 60, 1967.

3906. Pringsheim, E.G., and Pringsheim, O., Experimental elimination of chromatophores and eye-spot in *Euglena gracilis, New Phytol.,* 51, 65, 1952.

3907. Prins, H.B.A., and Elzenga, J.T.M., Bicarbonate utilization: function and mechanism, *Aquat. Bot.,* 34, 59, 1989.

3908. Prins, H.B.A., Snel, J.F.H., Helder, R.J., and Zanstra, P.E., Photosynthetic bicarbonate utilization in the aquatic angiosperms *Potamogeton* and *Elodea, Hydrobiol. Bull.,* 13, 106, 1979.

3909. Prins, H.B.A., Snel, J.F.H., Helder, R.J., and Zanstra, P.E., Photosynthetic HCO_3^- utilization and OH^- excretion in aquatic angiosperms. Light induced pH changes at the leaf surface, *Plant Physiol.,* 66, 818, 1980.

3910. Prins, H.B.A., O'Brien, J., and Zanstra, P.E., Bicarbonate utilization in aquatic angiosperms, pH and CO_2 concentrations at the leaf surface, in *Studies on Aquatic Vascular Plants,* Symoens, J.J., Hooper, S.S., and Compere, P., Eds., R. Bot. Soc. Belgium, Brussels, 1982, 112.

3911. Prins, H.B.A., Snel, J.F.H., Zanstra, P.E., and Helder, R.J., The mechanism of bicarbonate assimilation by the polar leaves of *Potamogeton* and *Elodea.* CO_2 concentrations at the leaf surface, *Plant Cell Environ.,* 5, 207, 1982.

3912. Priscu, J.C., A comparison of nitrogen and carbon metabolism in the shallow and deep-water phytoplankton populations of a subalpine lake: Response to photosynthetic photon flux density, *J. Plankton Res.,* 6, 733, 1984.

3913. Priscu, J.C., Photon dependence of inorganic nitrogen transport by phytoplankton in perennially ice-covered antarctic lakes, *Hydrobiologia,* 172, 173, 1989.

3914. Pritchard, G.G., Griffin, W.J., and Whittingham, C.P., The effect of carbon dioxide concentration, light intensity and iso-nicotinyl hydrazide on the photosynthetic production of glycolic acid by *Chlorella, J. Exp. Bot.,* 13, 176, 1962.

3915. Procházková, L., Blažka, P., and Králová, M., Chemical changes involving nitrogen metabolism in water and particulate matter during primary production experiment, *Limnol. Oceanogr,* 15, 797, 1970.

3916. Proctor, M.C.F., Regional and local variation in the chemical composition of ombrogenous mire waters in Britain and Ireland, *J. Ecol.,* 80, 719, 1992.

3917. Proctor, V.W., Preferential assimilation of nitrate ion by *Haematococcus pluvialis, Am. J. Bot.,* 44, 141, 1957.

3918. Proctor, V.W., Some controlling factors in the distribution of *Haematococcus pluvialis, Ecology,* 38, 457, 1957.

3919. Proctor, V.W., Storage and germination of *Chara* oospores, *J. Phycol.,* 3, 90, 1967.

3920. Proctor, V.W., The nature of charophyte species, *Phycologia,* 14, 97, 1975.

3921. Proctor, V.W., Historical biogeography of *Chara* (Charophyta): an appraisal of the Braun-Wood classification plus a falsifiable alternative for future consideration, *J. Phycol.,* 16, 218, 1980.

3922. Protection of public water supplies from ground-water contaminations, Center for Environmental Research, U.S. EPA Seminar Publ. 625/4-85/016, 1985.

3923. Provasoli, L., Nutrition and ecology of protozoa and algae, *Ann. Rev. Microbiol.,* 12, 279, 1958.

3924. Provasoli, L., Growth factors in unicellular marine algae, in *Perspectives in Marine Biology,* Buzzati-Treviso, A.A., Ed., University of California Press, Berkeley, 1958, 385.

3925. Provasoli, L., Organic regulation of phytoplankton fertility, in *The Sea,* Vol. 2, Hill, M.N., Ed., Wiley-Interscience, New York, 1963, 165.

3926. Provasoli, L., Growing marine seaweeds, *Proc. Int. Seaweed Symp.* IV, Devirville, A.D., and Feldman, J., Eds., Pergamon Press, Oxford, 4, 9, 1964.

3927. Provasoli, L., Nutritional relationships in marine organisms, in *Fertility of the Sea,* Vol. 2, Costlow, J.D., Ed., Gordon and Beach, New York, 1971, 369.

3928. Provasoli, L., and Pintner, I.J., Artificial media for fresh-water algae: problems and suggestions, in *The Ecology of Algae,* Pymatuning Symposia in Ecology, Special Publication No.2, Tryon, C.A., Jr., and Hartman, R.T., Eds., University of Pittsburgh Press, Pittsburgh, 1960, 84.

3929. Provasoli, L., and McLaughlin, J.J.A., Limited heterotrophy of some photosynthetic dinoflagellates, in *Symposium on Marine Microbiology,* Oppenheimer, C.H., Ed., Charles C. Thomas, Springfield, Illinois, 1963, 105.

3930. Provasoli, L., and Carlucci, F., Vitamins and growth regulators, in *Algal Physiology and Biochemistry* Stewart, W.D.P., Ed. Blackwell Scientific Publications, Oxford, 1974, 741.

3931. Provasoli, L., and Pintner, I.J., Bacteria induced polymorphism in an axenic laboratory strain of *Ulva lactuca* (Chlorophyceae), *J. Pycol.,* 16, 196, 1980.

3932. Provasoli, L., Yamatsu, L.T., and Manton, I., Experiments on the resynthesis of symbiosis in *Convoluta roscoffensis* with different flagellate cultures, *J. Mar. Biol. Ass. U.K.,* 48, 465, 1968.

3933. Prud'Homme Van Reine, W.F., Gallen in *Vaucheria* - draden (Algen - Xanthophyceae), *Gorteria* (Leiden), 6, 89, 1972.

3934. Przytocka-Jusiak, M., Blaszczyk, M., Kosinska, E., and Blisz-Konarzewska, A., Removal of nitrogen from industrial wastewaters with the use of algal rotating discs and denitrification packed bed reactors, *Water Res.*, 18, 1077, 1984.

3935. Pueschel, C.M., An ultrastructural survey of the diversity of crystalline, proteinaceous inclusions in red algal cells, *Phycologia*, 31, 489, 1992.

3936. Pueschel, C.M., and Stein, J.R., Ultrastructure of a freshwater brown alga from western Canada, *J. Phycol.*, 19, 209, 1983.

3937. Puriveth, P., Decomposition of emergent macrophytes in a Wisconsin marsh, *Hydrobiologia*, 72, 231, 1980.

3938. Putman, E.W., and Hassid, W.Z., Structure of galactosylglycerol from *Iridaea laminarioides*, *J. Am. Chem. Soc.*, 76, 2221, 1954.

3939. Putnam, H.D., and Olson, T.A., Studies on the productivity and plankton of lake Superior, Rep. School Public Health, University of Minnesota, 1961.

3940. Qualls, R.G., The role of leaf litter nitrogen immobilization in the nitrogen budget of a swamp stream, *J. Environ. Qual.*, 13, 640, 1984.

3941. Quillet, M., Sur le Métabolisme glucidique des Algues brunes. Présence de petites quantités de laminarine chez de nombreuses nouvelles espèces, réparties dans tout le groupe des Phéophycées, *Comp. Rend. Acad. Sci.*, 246, 812, 1958.

3942. Quispel, A., Ed., *The Biology of Nitrogen Fixation*, North Holland Res. Monogr. Frontiers in Biology, 33, North-Holland, 1974.

3943. Rabenhorst, L., *Flora europeae algarum aquare dulcis et submarine*, sect. 2, Leipzig, 1865.

3944. Rabinovich, E.I., *Photosynthesis and Related Processes*, Vol. 1, Interscience, New York, 1945.

3945. Rachlin, J.W., and Farran, M., Growth response of the green alga *Chlorella vulgaris* to selective concentrations of zinc, *Water Res.*, 8, 575, 1974.

3946. Rachlin, J.W., Jensen, T.E., and Warkentine, B., The growth response of the green alga (*Chlorella saccharophila*) to selected concentrations of the heavy metals Cd, Cu, Pb, and Zn., in *Trace Substances in Environmental Health* XVI, Hemphill, D.D., Ed., University of Missouri, Columbia, 1982, 145.

3947. Rachlin, J.W., Warkentine, B., and Jensen, T.E., The growth responses of *Chlorella saccharophila*, *Navicula incerta* and *Nitzschia closterium* to selected concentrations of cadmium, *Bull. Torrey Bot. Club.*, 109, 129, 1982.

3948. Rachlin, J.W., Jensen, T.E., Baxter, M., and Jani, V., Utilization of morphometric analysis in evaluating response of *Plectonema boryanum* (Cyanophyceae) to exposure to eight heavy metals, *Arch. Environ. Contam. Toxicol.*, 11, 323, 1982.

3949. Rachlin, J.W., Jensen, T.E., and Warkentine, B., The growth response of the diatom *Navicula incerta* to selected concentrations of the metals: cadmium, copper, lead and zinc, *Bull. Torrey Bot. Club.*, 110, 217, 1983.

3950. Rachlin, J.W., Warkentine, B., and Jensen, T.E., The response of the marine diatom *Nitzschia closterium* to selected concentrations of the divalent cations Cd, Cu, Pb and Zn, in *Trace Substances in Environmental Health* XVII, Hemphill, D.D., Ed., University of Missouri, Columbia, 1983, 72.

3951. Rachlin, J.W., Jensen, T.E., and Warkentine, B., The toxicological response of the alga *Anabaena flos-aquae* (Cyanophyceae) to cadmium, *Arch. Environ. Contam. Toxicol.*, 13, 143, 1984.

3952. Rachlin, J.W., Jensen, T.E., and Warkentine, B., Morphometric analysis of the response of *Anabaena flos-aquae* and *Anabaena variabilis* (Cyanophyceae) to selected concentrations of zinc, *Arch. Environ. Contam. Toxicol.*, 14, 395, 1985.

3953. Rader, R.B., and Richardson, C.J., The effect of nutrient enrichment on algae and macroinvertebrates in the Everglades: a review, *Wetlands*, 12, 121, 1992.

3954. Radforth, N.W., Suggested classification of muskeg for the engineer, *Eng. J.*, 35, 1, 1952.

3955. Ragan, M.A., Chemical constituents of seaweeds, in *The Biology of Seaweeds*, Lobban, C.S., and Wynne, M.J., Eds., University of California Press, Berkeley, 1981, 589.

3956. Ragan, M.A., and Chapman, D.J., *A Biochemical Phylogeny of the Protists*, Academic Press, New York, 1978.

3957. Rahat, M., and Spira, Z., Specificity of glycerol for dark growth of *Prymnesium parvum*, *J. Protozool.*, 14, 45, 1967.

3958. Rai, D.N., and Datta Munshi, J.S., Ecological characteristics of chaurs of North Bihar, *Int. J. Ecol. Environ. Sci.*, 7, 89, 1981.

3959. Rai, L.C., Mercury toxicity to *Chlorella vulgaris*. I. Reduction of toxicity by ascorbic acid and reduced glutathione (GSH), *Phykos*, 18, 105, 1979.

3960. Rai, L.C., and Dey, R., Environmental effects on the toxicity of methylmercuric chloride to *Chlorella vulgaris*, *Acta hydrochi, Hydrobiol.*, 8, 319, 1980.

3961. Rai, L.C., and Khatoniar, N., Response of *Chlorella* to mercury pollution, *Ind. J. Environ. Health*, 22, 113, 1980.

3962. Rai, L.C., Gaur, J.P., and Kumar, H.D., Phycology and heavy-metal pollution, *Biol. Rev.*, 56, 99, 1981.

3963. Rains, D.W., Mineral metabolism, in *Plant Biochemistry*, 3rd ed., Bonner, J., and Varner, J.E., Eds., Academic Press, New York, 1976, 561.

3964. Raistrick, B., The influence of foreign ions on crystal growth from solution. I. The stabilization of the supersaturation of calcium carbonate solutions by anions possessing O-P-O-P-O chains, *Disc. Faraday Soc.*, 5, 234, 1949.

3965. Ralph, B.J., Oxidative reactions in the sulphur cycle, in *Biogeochemical Cycling of Mineral-Forming Elements*, Trudinger, P.A., and Swaine, D.J., Eds., Elsevier Scientific Publishing, Amsterdam, 1979, 369.

3966. Ramamoorthy, S., and Kushner, D.J., Heavy metal binding components of river water, *J. Fish. Res. Board Can.*, 32, 1755, 1975.

3967. Ramamoorthy, S., Cheng, T.C., and Kushner, D.J., Mercury speciation in water, *Can. J. Fish. Aquat. Sci.*, 40, 85, 1983.

3968. Ramos, J.L., Flores, E., and Guerrero, M.G., Glutamine synthetase—glutamate synthase: the pathway of ammonium assimilation in *Anacystis nidulans*, in *Proc. Conf. of Federation of European Societies of Plant Physiology*, Santiago, 1980, 579.

3969. Ramos, J.L., Guerrero, M.G., and Losada, M., Photosynthetic production of ammonia by blue-green algae, in *Proc. Conf. of Federation of European Societies of Plant Physiology*, Santiago, 1980, 581.

3970. Ramus, J., The production of extracellular polysaccharide by the unicellular red alga *Porphyridium aerugineum*, *J. Phycol.*, 8, 97, 1972.

3971. Ramus, J., The capture and transduction of light energy, in *The Biology of Seaweeds*, Lobban, C.S., and Wynne, M.J., Eds. University of California Press, Berkeley, 1981, 458.

3972. Ramus, J., and Robins, D.M., The correlation of Golgi activity and polysaccharide secretion in *Porphyridium*, *J. Phycol.*, 11, 70, 1975.

3973. Rana, B.C., and Kumar, H.D., The toxicity of zinc to *Chlorella vulgaris* and *Plectonema boryanum* and its protection by phosphate, *Phykos*, 13, 60, 1974.

3974. Rana, B.C., and Kumar, H.D., Effects of toxic waste and waste water components on algae, *Phykos*, 13, 67, 1974.

3975. Ransom, R.E., Nerad, A., and Meier, P.G., Acute toxicity of some blue-green algae to the protozoan *Paramecium caudatum*, *J. Phycol.*, 14, 114, 1978.

3976. Rao, A.S., Evapotranspiration rates of *Eichhornia crassipes* (Mart.) Solms, *Salvinia molesta* D.S. Mitchell and *Nymphaea lotus* (L.) Willd. Linn. in a humid tropical climate, *Aquat. Bot.*, 30, 215, 1988.

3977. Rao, K.P., and Rains, D.W., Nitrate absorption by barley. i. Kinetics and energetics, *Plant Physiol.*, 57, 55, 1976.

3978. Rao, K.V.M., The effect of molybdenum on the growth of *Oocystis marssonii* Lemm., *Ind. J. Plant Physiol.*, 6, 142, 1963.

3979. Rao, L.V.M., Datta, N., Sopory, S.K., and Mukherjee, G.S., Phytochrome mediated induction of nitrate reductase activity in etiolated maize leaves, *Physiol. Plant.*, 50, 208, 1980.

3980. Rao, V.M., and Sastri, M.N., Determination of chromium in natural waters - a review. *J. Sci. Ind. Res.*, 41, 607, 1982.

3981. Rao, V.N.R., and Subramanian, S.K., Metal toxity tests on growth of some diatoms, *Acta Bot. Indica*, 10, 274, 1982.

3982. Raschke, R.L., Diatom (Bacillariophyta) community response to phosphorus in the Everglades National Park, USA, *Phycologia*, 32, 48, 1993.

3983. Rashid, M.A., Absorption of metals on sedimentary and peat humic acids, *Chem. Geol.*, 13, 15, 1974.

3984. Raspopov, I.M., Vegetation der grossen seichten Seen in Nordwesten der USSR und ihre Produktion, *Arch. Hydrobiol.*, 86, 242, 1979.

3985. Ratcliffe, D.A., Mires and bogs, in *The Vegetation of Scotland*, Burnett, J.H.L., Eds., Oliver and Noyd, London, 1964, 426.

3986. Rathnam, C.K.M., and Das, V.S.R., Nitrate metabolism in relation to the aspartate-type C-4 pathway of photosynthesis in *Eleusine coracana*, *Can. J. Bot.*, 52, 2599, 1974.

3987. Rathnam, C.K.M., and Edwards, G.E., Distribution of nitrate-assimilating enzymes between mesophyll protoplast and bundle sheath cells in leaves of three groups of C-4 plant, *Plant Physiol.*, 57, 881, 1976.

3988. Ratliff, R.D., Livestock grazing not detrimental to meadow wildflower, *U.S. For. Ser. Res. Note PSW*, 70, 1, 1972.

3989. Ratliff, R.D., and Westfall, S.E., Biomass trends in a Nebraska sedge meadow, Sierra National Forest, California, *Aquat. Bot.*, 30, 109, 1988.

3990. Rattray, M.R., Howard-Williams, C., and Brown, J.M.A., Sediment and water as sources of nitrogen and phosphorus for submerged rooted aquatic macrophytes, *Aquat. Bot.*, 40, 225, 1991.

3991. Raulin, J., Etudes chimiques sur la vegetation, *Ann. Sci. Nat. Bot.* 5, Ser. 11, 92, 1869.

3992. Raven, J.A., The mechanism of photosynthetic use of bicarbonate by *Hydrodictyon africanum*, *J. Exp. Bot.*, 19, 193, 1968.

3993. Raven, J.A., The linkage of light-stimulated Cl influx to K and Na influxes in *Hydrodictyon africanum*, *J. Exp. Bot.*, 19, 233, 1968.

3994. Raven, J.A., Action spectra for photosynthesis and light-stimulated ion transport processes in *Hydrodictyon africanum*, *New Phytol.*, 68, 45, 1969.

3995. Raven, J.A., Effects of inhibitors on photosynthesis and the active influxes of K and Cl in *Hydrodictyon africanum*, *New Phytol.*, 68, 1089, 1969.

3996. Raven, J.A., Exogenous inorganic carbon sources in plant photosynthesis, *Biol. Rev.*, 45, 167, 1970.

3997. Raven, J.A., The role of cyclic and pseudocyclic photophosphorylation in photosynthetic $^{14}CO_2$ fixation in *Hydrodictyon africanum*, *J. Exp. Bot.*, 21, 1, 1970.

3998. Raven, J.A., Cyclic and non-cyclic photophosphorylation as energy sources for active K influx in *Hydrodictyon africanum*, *J. Exp. Bot.*, 22, 420, 1971.

3999. Raven, J.A., Endogenous inorganic carbon sources in plant photosynthesis. 1. Occurrence of the dark respiratory pathway in illuminated green cells, *New Phytol.*, 71, 227, 1972.

4000. Raven, J.A., Photosynthetic electron flow and photophosphorylation, in *Algal Physiology and Biochemistry* Stewart, W.D.P., Ed., Blackwell Scientific Publications, Oxford, 1974, 391.

4001. Raven, J.A., Carbon dioxide fixation, in *Algal Physiology and Biochemistry*, Stewart, W.D.P., Ed., Blackwell Scientific Publications, Oxford, 1974, 434.

4002. Raven, J.A., Energetics of active phosphate influx in *Hydrodictyon africanum*, *J. Exp. Bot.*, 25, 221, 1974.

4003. Raven, J.A., Phosphate transport in *Hydrodictyon africanum*, *New Phytol.*, 73, 421, 1974.

4004. Raven, J.A., Transport in algal cells, in *Encyclopedia of Plant Physiology*, New Series, Vol. 2, *Transport in Plants II*, Lütge, U., and Pitman, M.G., Eds., Springer-Verlag, New York, 1976, 129.

4005. Raven, J.A., Nutrient transport in microalgae, *Adv. Microb. Physiol.*, 21, 47, 1980.

4006. Raven, J.A., The energetics of freshwater algae: energy requirements for biosynthesis and volume regulations, *New Phytol.*, 92, 1, 1982.

4007. Raven, J.A., Physiology of inorganic C acquisition and implications for resource use efficiency by marine phytoplankton: relation to increased CO_2 and temperature, *Plant Cell Environ.*, 14, 779, 1991.

4008. Raven, J.A., and Glidewell, S.M., C_4 characteristics of photosynthesis in the C_3 alga *Hydrodictyon africanum*, *Plant Cell Environ.*, 1, 185, 1978.

4009. Raven, J.A., and De Michelis, M.I., Acid-base regulation during nitrate assimilation in *Hydrodictyon africanum*, *Plant Cell Environ.*, 2, 245, 1979.

4010. Raven, J.A., and Beardall, J., Respiration and photorespiration, *Can. Bull. Fish. Aquat. Sci.*, 210, 55, 1981.

4011. Raven, J.A., and Lucas, W.J., Energy costs of carbon acquisition, in *Inorganic Carbon Uptake by Aquatic Photosynthetic Organisms*, Lucas, W.J., and Berry, J.A., Eds., Am. Soc. Plant Physiol., Rockville, Maryland, 1985, 305.

4012. Raven, J.A., and Geider, R.J., Temperature and algal growth, *New Phytol.*, 110, 441, 1988.

4013. Raven, J.A., Smith, F.A., and Walker, N.A., Biomineralization in the Charophyceae sensu lato, in *Biomineralization of Lower Plants and Animals*, Leadbeater, B.S.C., and Riding, R., Eds., Clarendon Press, Oxford, 1986, 125.

4014. Ray, J., *Synopsis methodica stirpium Britannicorum*, London, 1696.

4015. Reader, R.J., Primary production in northern bog marshes, in *Proc. Symp. Freshwater Wetlands: Ecological Processes and Management Potential*, Good, R.E., Whigham, D.F., and Simpson, R.L., Eds., Academic Press, New York, 1978, 53.

4016. Reader, R.J., and Stewart, J.M., The relationship between net primary production and accumulation for a peatland in southeastern Manitoba, *Ecology*, 53, 1024, 1972.

4017. Reay, P.F., The accumulation of arsenic from arsenic-rich natural waters by aquatic plants, *J. Appl. Ecol.*, 9, 557, 1972.

4018. Reazin, G.H., Jr., On the dark metabolism of a golden-brown alga *Ochromonas mathamensis*, *Am. J. Bot.*, 41, 771, 1954.

4019. Rebhun, S., and Ben-Amotz, A., The distribution of cadmium between the marine alga *Chlorella stigmatophora* and sea water medium, *Water Res.*, 18, 173, 1984.

4020. Rebhun, S., and Ben-Amotz, A., Antagonistic effect of manganese to cadmium toxicity in the alga *Dunaliela salina*, *Mar. Ecol. Prog. Ser.*, 42, 97, 1988.

4021. Reckhow, K.H., Empirical lake models for phosphorus development, applications, limitations, and uncertainty, in *Perspectiveson Lake EcosystemModeling,* Scavia, D., and Robertson, A., Eds., Ann Arbor Science, Ann Arbor, Michigan, 1979, 183.

4022. Reddy, K.R., and Patrick, W.H., Jr., Nitrogen fixation in flooded soil, *Soil Sci.,* 128, 80, 1979.

4023. Reddy, K.R., and Bagnall, L.O., Biomass production of aquatic plants used in agricultural drainage water treatment, in *Proc. 2nd Int. Gas. Res. Conf.,* Govt. Inst., Inc., Rockville, Maryland, 1981, 376.

4024. Reddy, K.R., and Graetz, D.A., Use of shallow reservoirs and flooded soil systems for wastewater treatment: Nitrogen and phosphorus transformations, *J. Environ. Qual.,* 10, 113, 1981.

4025. Reddy, K.R., and Rao, P.S.C., Nitrogen and phosphorus fluxes from a flooded organic soil, *Soil Sci.,* 136, 300, 1983.

4026. Reddy, K.R., and Tucker, J.C., Productivity and nutrient uptake of water hyacinth, *Eichhornia crassipes.* I. Effect of nitrogen source, *Econ. Bot.,* 37, 237, 1983.

4027. Reddy, K.R., and Patrick, W.H., Jr., Nitrogen transformations and loss in flooded soils and sediments, *CRC Crit. Rev. Environ. Control.,* 13, 273, 1984.

4028. Reddy, K.R., and Patrick, W.H., Jr., Fate of fertilizer nitrogen in the rice root zone, *Soil Sci. Soc. Am. J.,* 50, 649, 1986.

4029. Reddy, K.R., and Smith, W.H., Eds., *Aquatic Plants for Water Treatment and Resource Recovery* Magnolia Publishers, Orlando, Florida, 1987.

4030. Reddy, K.R., and Portier, K.M., Nitrogen utilization by *Typha latifolia*L. as affected by temperature and rate of nitrogen application, *Aquat. Bot.,* 27, 127, 1987.

4031. Reddy, K.R., and DeBusk, W.F., Nutrient storage capabilities of aquatic and wetland plants, in *Aquatic Plants for Water Treatment and Resource Recovery* Reddy, K.R., and Smith, W.H., Eds., Magnolia Publishing Inc., Orlando, Florida, 1987, 337.

4032. Reddy, K.R., and Graetz, D.A., Carbon and nitrogen dynamics in wetland soils, in *Ecology and Management of Wetlands,* Vol. 1. *Ecology of Wetlands,* Hook, D.D., et al., Eds., Timber Press, Portland, Oregon, 1988, 307.

4033. Reddy, K.R., Patrick, W.H., Jr., and Phillips, R.E., Ammonium diffusion as a factor in nitrogen loss from flooded soils, *Soil Sci. Soc. Am. J.,* 40, 528, 1976.

4034. Reddy, K.R., Sacco, P.D., and Graetz, D.A., Nitrate reduction in an organic soil-water system, *J. Environ. Qual.,* 9, 283, 1980.

4035. Reddy, K.R., Rao, P.S.C., and Jessup, R.E., The effect of carbon mineralization on denitrification kinetics in mineral and organic soils, *Soil Sci. Soc. Am. J.,* 46, 62, 1982.

4036. Reddy, K.R., Jessup, R.E., and Rao, P.S.C., Nitrogen dynamics in a eutrophic lake sediment, *Hydrobiologia,* 159, 177, 1988.

4037. Reddy, K.R., Agami, M., and Tucker, J.C., Influence of nitrogen supply rates on growth and nutrient storage by water hyacinth (*Eichhornia crassipes*) plants, *Aquat. Bot.,* 36, 33, 1989.

4038. Reddy, K.R., Agami, M., and Tucker, J.C., Influence of phosphorus on growth and nutrient storage by water hyacinth (*Eichhornia crassipes* (Mart.) Solms) plants, *Aquat. Bot.,* 37, 355, 1990.

4039. Reddy, K.R., Khaleel, R., Overcash, M.R., and Westerman, P.W., A nonpoint source model for land areas receiving animal wastes. 1. Mineralization of organic nitrogen, *Trans. ASAE,* 22, 863, 1979.

4040. Redfield, A.C., On the proportions of organic derivatives in sea water and their relation to the composition of plankton, in *James Johnstone Memorial Volume,* University of Liverpool Press, Liverpool, 1934, 177.

4041. Redfield, A.C., Ketchum, B.H., and Richards, F.A., The influence of organisms on the composition of seawater, in *The Sea,* Hill, M.N., Ed., Wiley-Interscience, New York, 1963, 26.

4042. Reed, A., Use of freshwater tidal marsh in the St. Lawrence Estuary by great snow geese, in *Proc. Conf. Freshwater Wetlands and Wildlife,* Sharitz, R.R., and Gibbons, J.W., Eds., U.S. Dept. of Energy, 1989, 605.

4043. Reed, M.L., and Graham, D., Carbonic anhydrase in plants: distribution, properties and possible physiological functions, *Prog. Phytochem.,* 7, 47, 1981.

4044. Reed, R.H., and Collins, J.C., The kinetics of Rb$^+$ and K$^+$ exchange in *Porphyra purpurea, Plant Sci. Lett.,* 20, 281, 1981.

4045. Reed, R.H., and Moffat, L., Copper toxicity and copper tolerance in *Enteromorpha compressa* (L.) Grev., *J. Exp. Mar. Biol. Ecol.,* 68, 85, 1983.

4046. Reed, R.H., and Gadd, G.M., Metal tolerance in eukaryotic and prokaryotic algae, in *Heavy Metal Tolerance in Plants: Evolutionary Aspects,* Shaw, J.A., Ed., CRC Press, Boca Raton, Florida, 1989, 105.

4047. Rees, D.A., and Samuel, J.W.B., The structure of alginic acid, Part VI. *J. Chem. Soc,* C., 2295, 1967.

4048. Rees, T.A.V., and Bekheet, I.A., The role of nickel in urea assimilation by algae, *Planta,* 156, 385, 1982.

4049.   Rees, T.A.V., and Syrett, P.J., The uptake of urea by the diatom *Phaeodactylum, New Phytol.*, 82, 169, 1979.

4050.   Rees, T.A.V., Cresswell, R.C., and Syrett, P.J., Sodium-dependent uptake of nitrate and urea by a marine diatom, *Biochim. Biophys. Acta*, 596, 141, 1980.

4051.   Reese, W.D., *"Chlorochytrium"* a green alga endophytic in Musci, *Bryologist*, 84, 75, 1981.

4052.   Reich, K., and Aschner, M., Mass development and control of the phytoflagellate *Prymnesium parvum* in fish ponds in Palestine, *Pales. J. Bot.*, 4, 14, 1947.

4053.   Reichardt, W., Influence of methylheptenone and related phytoplankton noncarotenoids on heterotrophic aquatic bacteria, *Can. J. Microbiol.*, 27, 144, 1981.

4054.   Reichardt, W., Overbeck, J., and Steubing, L., Free dissolved enzymes in lake waters, *Nature*, 216, 1345, 1967.

4055.   Reid, P.C., Dinoflagellate cyst distribution around the British Isles, *J. Mar. Biol. Ass. U.K.*, 52, 939, 1972.

4056.   Reimann, B.E.F., Lewin, J.C., and Volcani, B.E., Studies on the biochemistry and fine structure of silica shell formation in diatoms. II. The structure of the cell wall of *Navicula pelliculosa* (Bréb.). Hilse, *J. Phycol.*, 2, 74, 1966.

4057.   Reinhold, L., Volokita, M., Zenvirth, D., and Kaplan, A., Is HCO_3^- transport in *Anabaena* a Na^+ symport? *Plant Physiol.*, 76, 1090, 1984.

4058.   Reisner, G.S., Gering, R.K., and Thompson, J.F., The metabolism of nitrate and ammonia by *Chlorella, Plant Physiol.*, 35, 48, 1960.

4059.   Reisser, W., Host-symbiont interaction in *Paramecium bursaria.* Physiological and morphological features and their evolutionary significance, *Ber. Deutsch. Bot. Ges.*, 94 557, 1981.

4060.   Reiter, M.A., and Carlson, R.E., Current velocity in streams and the composition of benthic algal mats, *Can. J. Fish. Aquat. Sci.*, 43, 1156, 1986.

4061.   Rejmánek, M., and Velasquez, J., Communities of emerged fishpond shores and bottoms, in *Pond Littoral Ecosystems,* Dykyjová, D., and Kvet, J., Eds., Springer Verlag, Berlin, 1978, 206.

4062.   Rejmánková, E., Biomass production and growth rate of duckweeds (*Lemna gibba* and *L. minor*), in *Ecosystem Study on Wetland Biome in Czechoslovakia*, Hejný, S., Ed., Czechosl. IBP/PT-PP Report No. 3, Třeboň, Czechoslovakia, 1973, 101.

4063.   Rejmánková, E., Growth, production and nutrient uptake of duckweeds in fishponds and in experimental culture, in *Pond Littoral Ecosystems: Structure and Functioning*, Dykyjová, D., and Kvet, J., Eds., Springer-Verlag, Berlin, 1978, 278.

4064.   Rejmánková, E., On the production ecology of duckweeds. *Proc. Int. Workshop on Aquatic Macrophytes*, Illmitz, Austria, 1981, 1.

4065.   Rejmánková, E., The role of duckweeds (Lemnaceae) in small wetland water bodies of Czechoslovakia, in *Proc. Int. Conf. Wetlands: Ecology and Management*, Gopal, B., Turner, R.E., Wetzel, R.G., and Whigham, D.F., Eds., Natl. Inst. of Ecology and Internat. Scientific Publications, Jaipur, India, 1982, 397.

4066.   Rejmánková, E., Ecology of creeping macrophytes with special reference to *Ludwigia peploides* (H.B.K.) Raven, *Aquat. Bot.*, 43, 283, 1992.

4067.   Rejmánková, E., and Hapala, P., The importance od duckweeds in pond management, in *Proc. Conf. Macrophytes in Water Management, Water Hygiene and Fishing*, Dům Techniky ČSVTS, České Budějovice, 1982, 93.

4068.   Reshkin, S.J., and Knauer, G.A., Light stimulation of phosphate uptake in natural assemblages of phytoplankton, *Limnol. Oceanogr.*, 24, 1121, 1979.

4069.   Retovský, R., and Klášterská, I., Study of the growth and development of *Chlorella* populations in the culture as a whole. V. The influence of $MgSO_4$ on autospore formation, *Folia Microbiol.*, 6, 115, 1961.

4070.   Reuter, J.G., and Morel, F.M.M., The interaction between zinc deficiency and copper toxicity as it affects the silicic acid uptake mechanisms in *Thalassiosira pseudonana, Limnol. Oceanogr.*, 26, 67, 1981.

4071.   Reuter, J.E., Loeb, S.L., and Goldman, R.C., Inorganic nitrogen uptake by epilithic periphyton in an N-deficient lake, *Limnol. Oceanogr.*, 31, 149, 1986.

4072.   Reynolds, C.R., The response of phytoplankton communities to changing lake environments, *Schweiz. Z. Hydrol.*, 49, 220, 1987.

4073.   Reynolds, C.S., The breaking of the Shropshire meres: some recent investigations, *Bull. Shropshire Conservat. Trust*, 10, 9, 1967.

4074.   Reynolds, C.S., The ecology of the planktonic blue-green algae in the North Shropshire meres, *Field Studies*, 3, 409, 1971.

4075.   Reynolds, C.S., Growth, gas-vacuolation and buoyancy in a natural population of a blue-green alga, *Freshwater Biol.*, 2, 87, 1972.

4076.   Reynolds, C.S., The phytoplankton of Crose Mere, Shropshire, *Br. Phycol. J.*, 8, 153, 1973.
4077.   Reynolds, C.S., Phytoplankton periodicity of some North Shropshire meres, *Br. Phycol. J.*, 8, 301, 1973.
4078.   Reynolds, C.S., *The Ecology of Freshwater Phytoplankton,* Cambridge University Press, Cambridge, 1984.
4079.   Reynolds, C.S., and Allen, S.E., Changes in the phytoplankton of Oak Mere following the introduction of base-rich water, *Br. Phycol. J.*, 3, 451, 1968.
4080.   Reynolds, C.S., and Walsby, A.E., Water blooms, *Biol. Rev.,* 50, 437, 1975.
4081.   Reynolds, H.W., Hanson, R.M., and Peden, D.G., Diets of the Slave River Lowland Bison herd, Northwestern Territories, Canada, *J. Wildl. Manage.,* 42, 581, 1978.
4082.   Reynolds, R.C., Jr., Polyphenol inhibition of calcite precipitation in Lake Powel, *Limnol. Oceanogr.,* 23, 585, 1978.
4083.   Rhee, G.-Y., A continuous culture study of phosphate uptake growth rate and polyphosphate in *Scenedesmus* sp., *J. Phycol.,* 9, 495, 1973.
4084.   Rhee, G.-Y., Phosphate uptake under nitrate limitation by *Scenedesmus* sp. and its ecological implications, *J. Phycol.,* 10, 470, 1974.
4085.   Rhee, G.-Y., Effects of N:P atomic ratios and nitrate limitation on algal growth, cell composition and nitrate uptake, *Limnol. Oceanogr.,* 23, 10, 1978.
4086.   Rhee, G.-Y., Continuous culture in phytoplankton ecology, *Adv. Aquat. Microbiol.,* 2, 151, 1980.
4087.   Rhee, G.-Y., Effects of environmental factors and their interactions on the phytoplankton growth, in *Advanced in Microbial Ecology,* Vol. 6, Marshall, K.C., Ed., Plenum Press, New York, 1982, 33.
4088.   Rhee, G.-Y., and Gotham, I.J., The effect of environmental factors on phytoplankton growth: temperature and the interactions of temperature with nutrient limitation, *Limnol. Oceanogr.,* 26, 635, 1981.
4089.   Rhiel, E., Morschel, E., and Wehrmeyer, W., Correlation of pigment deprivation and ultrastructural organization of thylakoid membranes in *Cryptomonas maculata* following nutrient deficiency, *Protoplasma,* 129, 62, 1985.
4090.   Rhiel, E., Krupinskak, K., and Wehrmeyer, W., Effect of nitrogen starvation on the function and organization of the photosynthetic membranes in *Cryptomonas maculata* (Cryptophyceae), *Planta,* 169, 361, 1986.
4091.   Ribelin, B.W., and Collier, A.W., Ecological considerations of detrital aggregates in the salt marsh, in *Ecological Processes in Coastal and Marine Systems,* Livingstone, R.J., Ed., Plenum Press, New York, 1979, 47.
4092.   Rice, H.V., Leighty, D.A., and McLeod, G.C., The effects of some trace metals on marine phytoplankton, *CRC Crit. Rev. Microbiol.,* 3, 27, 1973.
4093.   Rice, T.R., Review of zinc in ecology, in *Proc. Nat. Symp. on Radioecology,* Schultz, V., and Klement, A.W., Eds., Reinhold Publ. Corp., New York, 1961, 619.
4094.   Rice, T.R., Accumulation of radionuclides by aquatic organisms, in *Studies of the Fate of Certain Radionuclides in Estuarine and Other Aquatic Environments,* Saho, J.J., and Bedrosian, P.H., Eds., U.S. Public Health Service Publ. No. 999-R-3, 1963, 35.
4095.   Rich, P.H., and Wetzel, R.G., Detritus in the lake ecosystem, *Ann. Nat.,* 112, 57, 1978.
4096.   Rich, P.H., Wetzel, R.G., and Thuy, N.V., Distribution, production and role of aquatic macrophytes in a southern Michigan marl lake, *Freshwat. Biol.,* 1, 3, 1971.
4097.   Richards, L., and Thurston, C.F., Uptake of leucine and tyrosine and their intracellular pools in *Chlorella fusca* var. vacuolata, *J. Gen. Microbiol.,* 121, 39, 1980.
4098.   Richards, O., Killing organisms with chromium, *Physiol. Zool.,* 9, 246, 1936.
4099.   Richardson, B., Ed., *Proc. Conf. Wetland Values and Management,* Freshwater Society, Navarre, 1981.
4100.   Richardson, B., Orcutt, D.M., Schwertner, H.A., Martinez, C.L., and Wickline, H.E., Effects of nitrogen limitation on the growth and composition of unicellular algae in continuous culture, *Appl. Microbiol.,* 18, 245, 1969.
4101.   Richardson, C.J., Primary productivity values in fresh water wetlands, in *Proc. Nat. Symp. Wetland Functions and Values: The State of Our Understanding,* Greeson, P.E., Clark, J.R., and Clark, J.E., Eds., American Water Resources Assoc., Minneapolis, Minnesota, 1979, 131.
4102.   Richardson, C.J., Ed., *Pocosins Wetlands: An Integrated Analysis of Coastal Plain Freshwater Bogs in Northern Carolina,* Hutchinson Ross Publishing Company, Stroudsburg, Pennsylvania, 1981.
4103.   Richardson, C.J., Pocosins: Ecosystem processes and the influence of man on system response, in *Pocosin Wetlands: An Integrated Analysis of Coastal Plain Freshwater Bogs in North Carolina,* Richardson, C.J., Ed., Hutchinson Ross Publishing Company, Stroudsburg, Pennsylvania, 1981, 3.
4104.   Richardson, C.J., Mechanisms controlling phosphorus retention capacity of freshwater wetlands, *Science,* 228, 1424, 1985.

4105.   Richardson, C.J., Freshwater wetlands: transformers, filters, or sinks, in *Proc. Symp. Freshwater Wetlands and Wildlife*, Sharitz, R.R., and Gibbons, J.W., Eds., U.S. Dept. of Energy, 1989, 26.
4106.   Richardson, C.J., Biogeochemical cycles: regional, in *Wetlands and Shallow Continental Water Bodies*, Patten, B.C., Ed., SPB Academic Publishing, The Hague, The Netherlands, 1990, 259.
4107.   Richardson, C.J., Pocosins: an ecological perspective, *Wetlands*, 11, 335, 1991.
4108.   Richardson, C.J., and Nichols, D.S., Ecological analysis of freshwater management criteria in wetland ecosystems, in *Ecological Considerations in Wetlands Treatment of Municipal Wastewaters*, Godfrey, P.J., Kaynor, E.R., Pelczarski, S., and Benforado, J., Eds., Van Nostrand Reinhold Company, New York, 1985, 351.
4109.   Richardson, C.J., and Marshall, P.E., Processes controlling movement, storage, and export of phosphorus in a fen peatlands, *Ecol. Monogr.*, 56, 279, 1986.
4110.   Richardson, C.J., and Schwegler, B.R., Algal bioassay and gross productivity experiments using sewage effluent in a Michigan wetland, *Water Resour. Bull.*, 22, 111, 1986.
4111.   Richardson, C.J., and Davis, J.A., Natural and artificial wetland ecosystems: ecological opportunities and limitations, in *Aquatic Plants for Water Treatment and Resource Recovery*, Reddy, K.R., and Smith, W.H., Eds., Magnolia Publishing, Orlando, Florida, 1987, 819.
4112.   Richardson, C.J., and Craft, C.B., Efficient phosphorus retention in wetlands: fact or fiction?, in *Proc. Int. Conf. Constructed Wetlands for Water Quality Improvement*, Moshiri, G.A., Ed., CRC Press, Inc., Boca Raton, Florida, 1993, 271.
4113.   Richardson, C.J., and Gibbons, J.W., Pocosins, Carolina bays, and mountain bogs, in *Biodiversity of the Southeastern United States/Lowland Terrestrial Communities*, Martin, W.H., Boyce, S.G., and Echternacht, A.C., Eds., John Wiley and Sons, New York, 1993, 257.
4114.   Richardson, C.J., Kadlec, J.A., Wentz, W.A., Chamie, J.P.M., and Kadlec, R.H., Background ecology and the effects of nutrient additions on a central Michigan wetland, in *Proc. 3rd Wetland Conf.*, Lefor, M.W., Kennard, W.C., and Helfgott, T.B., Eds., The University of Connecticut, 1976, 34.
4115.   Richardson, C.J., Tilton, D.L., Kadlec, J.A., Chamie, J.P.M., and Wentz, W.A., Nutrient dynamics of northern wetland ecosystems, In *Proc. Symp. Freshwater Wetlands: Ecological Processes and Management Potential*, Good, R.E., Whigham, D.F., and Simpson, R.L., Eds., Academic Press, New York, 1978, 217.
4116.   Richardson, K., Beardall, J., and Raven, J.A., Adaptation of unicellular algae to irradiance: an analysis of strategies, *New Phytol.*, 93, 157, 1983.
4117.   Richardson, M., *Translocations in Plants*, Studies in Biology No. 10, Edward Arnold, London, 1968.
4118.   Richardson, M., Microbodies (glyoxysomes and peroxisomes) in plants, *Sci. Progr.*, 61, 41, 1974.
4119.   Richardson, T.R., Millington, W.F., and Miles, H.M., Mercury accumulation in *Pediastrum boryanum* (Chlorophyceae), *J. Phycol.*, 11, 320, 1975.
4120.   Richey, J.E., The phosphorus cycle, in *The Major Biogeochemical Cycles and Their Interactions*, Bolin, B., and Cook, R.B., Eds., SCOPE Report No. 21, John Wiley and Sons, Chichester, 1983, 51.
4121.   Richman, S., and Rogers, J.N., The feeding of *Calanus helgolandicus* on synchronously growing populations of the marine diatom *Ditylum brightwellii, Limnol. Oceanogr*, 14, 701, 1969.
4122.   Richter, G., Die Auswirkung von Mangan-Mangel auf Wachstum und Photosynthese bei der Blaualge *Anacystis nidulans, Planta*, 57, 202, 1961.
4123.   Richter, O., Zur Physiologie der Diatomeen, *Sitz. Akad. Wiss. Wien (Math.-Nat. Kl.)*, 115, 27, 1906.
4124.   Rickard, D.T., and Nriagu, J.O., Aqueous environmental chemistry of lead, in *The Biogeochemistry of Lead in the Environment*, Nriagu, J.O., Ed., Elsevier/North Holland Biomedical Press, 1978, 219.
4125.   Ricketts, T.R., The cultural requirements of the Prasinophyceae, *Nova Hedw.*, 25, 683, 1974.
4126.   Ricketts, T.R., Nitrate and nitrite reductases in *Platymonas striata*, Butcher (Prasinophyceae), *Br. Phycol. J.*, 13, 167, 1978.
4127.   Ricketts, T.R., and Edge, P.A., Nitrate and nitrite reductases in *Platymonas striata*, Butcher (Prasinophyceae), *Br. Phycol. J.*, 13, 167, 1978.
4128.   Ricohermoso, M., and Deveau, L.E., Review of commercial propagation of *Eucheuma*, in *Proc. IX. Int. Seaweed Symposium*, Jensen, A., and Stein, J.R., Eds., Science Press, Princeton, New Jersey, 1979, 525.
4129.   Riding, R., Cyanoliths (cyanoids): oncoids formed by calcified cyanophytes, in *Coated Grains*, Peryt, T.M., Ed., Springer-Verlag, Berlin, 1983, 276.
4130.   Riding, R., Ed., *Calcareous Algae and Stromatolites*, Springer-Verlag, Berlin, 1991.
4131.   Riding, R., Calcified cyanobacteria, in *Calcareous Algae and Stromatolites*, Riding, R., Ed., Springer-Verlag, Berlin, 1991, 55.
4132.   Ried, A., Über die Wirkung des blauen Lichts auf den photosynthetischen Gasaustausch von *Chlorella, Planta*, 87, 333, 1969.
4133.   Riemer, D.N., and Toth, S.J., A survey of the chemical composition of aquatic plants in New Jersey, *N. J. Agric. Exp. Stat. Rutgers Univ. Bull.*, 820, 1, 1968.

4134. Rieth, A., Über eine bemerkenswerte *Botrydium* - Form aus Kuba, *Beitr. Biol. Pflanzen*, 52, 337, 1976.

4135. Rieth, A., *Xanthophyceae,* Süsswasserflora von Mitteleuropa 4, Gustav Fisher Verlag, Stuutgart, New York, 1980.

4136. Rigano, C., Di Martino Rigano, V., Vona, V., and Fuggi, A., Glutamine synthetase activity, ammonia assimilation and control of nitrate reduction in unicellular red alga *Cyanidium caldarium, Arch. Microbiol.,* 121, 117, 1979.

4137. Rigano, C., Vona, V., Di Martino Rigano, V., and Fuggi, A., Nitrate reductase and glutamate dehydrogenase of the red alga *Porphyridium aerugineum, Plant Sci. Lett.,* 15, 203, 1979.

4138. Rigano, C., Di Martino Rigano, V., Fuggi, A., and Vona, V., Control of the assimilatory nitrate reduction in a unicellular alga and the possible effector, in *Proc. Conf. of Federation of European Societies of Plant Physiology*, Santiago, 1980, 595.

4139. Rigby, C.H., Craig, S.R., and Budd, K., Phosphate uptake by *Synechococcus leopoliensis* (Cyanophyceae): enhancement by calcium ions, *J. Phycol.,* 16, 389, 1980.

4140. Rigler, F.H., A tracer study of the phosphorus cycle in the lake water, *Ecology,* 37, 550, 1956.

4141. Rigler, F.H., The phosphorus fraction and the turnover time of inorganic phosphorus in different type of lakes, *Limnol. Oceanogr,* 9, 511, 1964.

4142. Rigler, F.H., Radiobiological analyses of inorganic phosphorus in lakewater, *Verh. Internat. Verein. Limnol.,* 16, 465, 1966.

4143. Rigler, F.H., A dynamic view of phosphorus cycle in lakes, in *Environmental Chemistry Handbook,* Griffith, N.W., et al., Eds., John Wiley and Sons, New York, 1973, 539.

4144. Riley, G.A., Stommel, H., and Bumpus, D.F., Quantitative ecology of the plankton on the western north Atlantic, *Bull. Bingham Oceanogr. Coll.,* 12, 1, 1949.

4145. Riley, J.P., and Roth, I., The distribution of trace elements in some species of phytoplankton grown in culture, *J. Mar. Biol. Ass. U.K.,* 51, 63, 1971.

4146. Rinehart, K.L., Jr., Shaw, P.D., Shield, L.S., Gloer, J.B., Harbour, G.C., Koker, M.E.S., Samain, D., Schwartz, R.E., Tymiak, A.A., Weller, D.L., Carter, G.T., Munro, M.H.G., Hughes, R.G., Jr., Renia, H.E., Swynenberg, E.B., Stringfellow, D.A., Vava, J.J., Coats, J.H., Zurenko, G.E., Kuentzel, S.L., Li, L.H., Bakus, G.J., Brunsca, R.C., Craft, L.L., Young, D.N., and Conner, J.L., Marine natural products as sources of antiviral, antimicrobial, and antineoplastic agents, *Pure Appl. Chem.,* 53, 795, 1981.

4147. Ripl, W., and Lindmark, G., The impact of algae and nutrient composition on desiment exchange dynamics, *Arch. Hydrobiol.,* 86, 46, 1979.

4148. Rippka, R., and Stanier, R.Y., The effects of anaerobiosis on nitrogenase synthesis and heterocyst development by nostocean cyanobacteria, *J. Gen. Microbiol.,* 105, 83, 1978.

4149. Rippka, R., and Cohen-Bazire, G., The Cyanobacteriales: a legitimate order based on the type strain *Cyanobacterium stanieri?, Ann. Microbiol. (Inst. Pasteur),* 134B, 21, 1983.

4150. Rippka, R., Deruelles, J., Waterbury, J.B., Herdman, M., and Stanier, R.Y., Generic assignments strain histories and properties of pure cultures of cyanobacteria, *J. Gen. Microbiol.,* 111, 1, 1979.

4151. Ris, H., and Singh, R.N., Electron microscope studies on blue-green algae, *J. Biophys. Biochem. Cytol.,* 9, 63, 1961.

4152. Rittenberg, S.C., The obligate autotroph-the demise of a concept, *Antonie van Leeuwenhoek.,* 38, 457, 1972.

4153. Rivkin, R.B., Effects of lead on growth of the marine diatom *Skeletonema costatum, Mar. Biol.,* 50, 239, 1979.

4154. Rivkin, R.B., and Swift, E., Characterization of alkaline phosphatase and organic phosphorus utilization in the oceanic dinoflagellate *Pyrocystis noctiluca, Mar. Biol. (Berl.),* 61, 1, 1980.

4155. Rivkin, R.B., and Swift, E., Phosphate uptake by the oceanic dinoflagellate *Pyrocystis noctiluca, J. Phycol.,* 18, 113, 1982.

4156. Riznyk, R.Z., Edens, J.I., and Libby, R.C., Production of epibenthic diatoms in a southern California impoundment estuary, *J. Phycol.,* 14, 273, 1978.

4157. Robb, D.A., and Pierpoint, W.S., Eds., *Metals and Micronutrients: Uptake and Utilization by Plants,* Academic Press, London, 1983.

4158. Robert, D., Localization cytochimique en microscopic électronique des constituants nucléaires au cours de la spermigenése chez le *Chara vulgaris, Ann. Sci. Nat. Bot. Paris 13ᵉ* ser 1, 67, 1979.

4159. Robie, R.A., and Waldbaum, D.R., Thermodynamic properties of minerals and related substances at 298.15 K (25.0 C) and one atmosphere (1.013 bars) pressure and at higher temperatures, *U.S. Geol. Surv. Bull.,* 1259, 1, 1968.

4160. Robinson, G.G.C., and Pip, E., The application of a nuclear track autoradiographic technique to the study of periphyton photosynthesis, in *Periphyton of Freshwater Ecosystems*, Wetzel, R.G., Ed., Dr. W. Junk Publishers, The Hague, The Netherlands, 1983, 267.

4161. Röderer, G., On the toxic effects of tetraethyl lead and its derivatives on the chrysophyte *Poterio-chromonas malhamensis*. I. Tetraethyl lead, *Environ. Res.*, 23, 371, 1980.

4162. Röderer, G., On the toxic effects of tetraethyl lead and its derivatives in the Chrysophyte *Poterio-chromonas malhamensis*. II. Triethyl lead, diethyl lead and anorganic lead, *Environ. Res.*, 25, 361, 1981.

4163. Rodgers, J.H., Jr., Dickson, K.L., and Cairns, J., Jr., A chamber for in situ evaluations of periphyton productivity in lotic systems, *Arch. Hydrobiol.*, 84, 389, 1978.

4164. Rodhe, W., Environmental requirements of fresh water plankton algae: Experimental studies in the ecology of phytoplankton, *Symb. Bot. Uppsal.*, 10, 1, 1948.

4165. Rodhe, W., Primär produktion und Seetypen, *Verh. Internat. Verein. Limnol.*, 13, 121, 1958.

4166. Rodhe, W., Standard correlation between pelagic photosynthesis and light, *Mem. Ist. Ital. Idrobiol.*, 18 (suppl.), 365, 1965.

4167. Rodhe, W., Crystallization of eutrophication concepts in northern Europe, in *Eutrophication: Causes, Consequences and Correctives*, Nat. Acad. Sci./Nat. Res. Council Publ. 1700, 1969, 50.

4168. Rodhe, W., Vollenweider, R.A., and Nauwerck, A., The primary production and standing crop of phytoplankton, in *Perspectives in Marine Biology*, Buzzati-Traverso, A.A., Ed., University of California Press, Berkeley, 1958, 299.

4169. Roessler, P.G., Environmental control of glycero-lipid metabolism in microalgae: commercial implications and future research directions, *J. Phycol.*, 26, 393, 1990.

4170. Rodgers, J.H., Jr., and Harvey, R.S., The effect of current on periphyton productivity as determined using carbon-14, *Water Resour. Bull.*, 12, 1109, 1976.

4171. Rogers, L.J., and Gallon, J.R., Eds., *Biochemistry of the Algae and Cyanobacteria*, Clarendon Press, Oxford, 1988.

4172. Rogers, H.J., and Perkins, H.R., *Cell Walls and Membranes*, E. and F.N. Spon, London, 1968.

4173. Rogers, H.H., and Davis, D.E., Nutrient removal by water hyacinth, *Weed Sci.*, 20, 423, 1972.

4174. Rogerson, A.C., Modifiers of heterocyst repression and spacing and formation of heterocysts without nitrogenase in the cyanobacterium *Anabaena variabilis*, *J. Bacteriol.*, 140, 213, 1979.

4175. Roll, H., Zur Terminologie des Periphytons, *Arch. Hydrobiol.*, 35, 59, 1939.

4176. Rollinson, C.L., Problems of chromium reactions, in *Andrews Radioactive Pharmaceuticals*, U.S. Atomic Energy Commission Conference 65111, Oak Ridge, Tennessee, 1966.

4177. Romanenko, V.I., and Velichko, I.A., Influence of chromium ions on bacteria and algae, *Biol. Inland Waters*, 21, 12, 1974.

4178. Roon, R.J., and Levenberg, B., An adenosine triphosphate-dependent, avidin-sensitive enzymatic cleavage of urea in yeast and green algae, *J. Biol. Chem.*, 243, 5213, 1968.

4179. Roon, R.J., and Levenberg, B., Urea amidolase. I. Properties of the enzyme from *Candida utilis*, *J. Biol. Chem.*, 247, 4107, 1972.

4180. Rørslett, B., Berge, D., and Johansen, W., Mass invasion of *Elodea canadensis* in a mesotrophic, south Norwegian lake—impact on water quality, *Verh. Internat. Verein. Limnol.*, 22, 2920, 1985.

4181. Rose, F.L., and Cushing, C.E., Periphyton: autoradiography of zinc-65 adsorption, *Science*, 168, 576, 1970.

4182. Rosemarin, A.S., Phosphorus nutrition of the potentially competing filamentous algae, *Cladophora glomerata* (L.) Kütz. and *Stigeoclonium tenue* (Agardh) Kütz. from Lake Ontario, *J. Great Lakes Res.*, 8, 66, 1982.

4183. Rosemarin, A.S., Direct examination of growing filaments to determine phosphate growth kinetics in *Cladophora glomerata* (L.) Kütz. and *Stigeoclonium tenue* (Agardh) Kütz., in *Periphyton of Freshwater Ecosystems*, Wetzel, R.G., Ed., Dr. W. Junk Publishers, The Hague, The Netherlands, 1983, 111.

4184. Rosenberg, G., and Ramus, J., Ecological growth strategies in the seaweeds *Gracilaria foliifera* (Rhodopyceae) and *Ulva* sp. (Chlorophyceae): soluble nitrogen and reserve carbohydrates, *Mar. Biol.*, 66, 251, 1982.

4185. Rosenberg, G., and Ramus, J., Ecological growth strategies in the seaweeds *Gracilaria foliifera* (Rhodophyceae) and *Ulva* sp. (Chlorophyceae): photosynthesis and antenna composition, *Mar. Ecol. Prog. Ser.*, 8, 233, 1982.

4186. Rosenberg, G., and Ramus, J., Uptake of inorganic nitrogen and seaweed surface are: volume ratios, *Aquat. Bot.*, 19, 65, 1984.

4187. Rosenberg, G., Probyn, T.A., and Mann, K.H., Nutrient uptake and growth kinetics in brown seaweeds: Response to continuous and single additions of ammonium, *J. Exp. Mar. Biol. Ecol.*, 80, 125, 1984.

4188. Rosko, J.J., and Rachlin, J.W., The effect of copper, zinc, cobalt and manganese of the growth of the marine diatom *Nitzschia closterium*, *Bull. Torrey Bot. Club*, 102, 100, 1975.

4189. Rosko, J.J., and Rachlin, J.W., The effect of cadmium, copper, mercury, zinc and lead on the cell division, growth, chlorophyll content of the chlorophyte *Chlorella vulgaris, Bull. Torrey Bot. Club.,* 104, 226, 1977.

4190. Rosowski, J.R., and Parker, B.C., Eds., *Selected Papers in Phycology I,* Department of Botany, University of Nebraska, Lincoln, 1971.

4191. Rosowski, J.R., and Parker, B.C., Eds., *Selected Papers in Phycology II,* Phycological Society of America, Lawrence, Kansas, 1982.

4192. Rosowski, J.R., Vadas, R.L., and Kugrens, P., Surface configuration of the lorica of the euglenoid *Trachelomonas* as revealed with scanning electron microscopy, *Am. J. Bot.,* 62, 48, 1975.

4193. Ross, M.M., Morphology and physiology of germination of *Chara gymnopitys* A. Br. I. Development and morphology of the sporeling, *Aust. J. Bot.,* 7, 1, 1959.

4194. Rosson, R.A., Tebo, B.M., and Nealson, K.H., Use of poisons in determination of microbial manganese binding rates in seawater, *Appl. Environ. Microbiol.,* 47, 740, 1984.

4195. Rosswall, T., The biogeochemical nitrogen cycle, in *Some Perspectives of the Major Biogeochemical Cycles,* Likens, G.E., Ed., SCOPE Report No., 17, John Wiley, Chichester, 1981, 25.

4196. Rosswall, T., The nitrogen cycle, in *The Major Biogeochemical Cycles and Their Interactions,* Bolin, B., and Cook, R.B., Eds., SCOPE Report No. 21, John Wiley and Sons, Chichester, 1983, 46.

4197. Roswall, T., Flower-Ellis, J.G.K., Johansson, L.G., Jonsson, S., Ryden, B.E., and Sonesson, M., Stordalen (Abisko), Sweden, in *Structure and Function of Tundra Ecosystems,* Roswall, T., and Heal, O.W., Eds., *Ecol. Bull.,* 20, 265, 1975.

4198. Rother, J.A., and Whitton, B.A., Mineral composition of *Azolla pinnata* in relation to composition of floodwaters in Bangladesh, *Arch. Hydrobiol.,* 113, 371, 1988.

4199. Rothstein, A., Cell membrane as a site of action of heavy metals, *Fed. Proc. Fed. Am. Soc. Exp. Biol.,* 18, 1026, 1959.

4200. Rothstein, A., Interactions of arsenate with the phosphate-transporting system of yeast, *J. Gen. Physiol.,* 46, 1075, 1963.

4201. Roulet, N.T., and Woo, M.K., Wetland and lake eutrophication in the low arctic, *Arct. Alp. Res.,* 18, 195, 1986.

4202. Round, F.E., A note on some communities of the littoral of lakes, *Arch. Hydrobiol.,* 52, 398, 1956.

4203. Round, F.E., The late-glacial and post-glacial diatom succession in the Kentmere valley deposit. I. Introduction, methods and flora, *New Phytol.,* 58, 98, 1957.

4204. Round, F.E., A note on some diatom communities in calcareous springs and streams, *J. Linn. Soc. Bot.,* 55, 662, 1957.

4205. Round, F.E., The algal flora of Massom's slack, Freshfield, Lancashire, *Arch. Hydrobiol.,* 54, 462, 1958.

4206. Round, F.E., *The Biology of Algae,* St. Martine Press, New York, 1965.

4207. Round, F.E., Benthic marine diatoms, *Oceanogr. Mar. Biol. Ann. Rev.,* 9, 83, 1971.

4208. Round, F.E., The formation of girdle, intercalary band and septa in diatoms, *Nova Hedw.,* 23, 449, 1972.

4209. Round, F.E., *The Biology of the Algae,* 2nd ed., Edward Arnold Publishers, London, 1973.

4210. Round, F.E., *The Ecology of Algae,* Cambridge University Press, Cambridge, 1981.

4211. Round, F.E., *Proc. 9th Internat. Diatom Symp.,* Biopress Ltd., Bristol and Koeltz Scientific Books, Koenigstein, 1988.

4212. Round, F.E., and Eaton, J.W., Persistent, vertical migration rhythms in benthic microflora. III. The rhythm of epipelic algae in a freshwater pond, *J. Ecol.,* 54, 609, 1966.

4213. Round, F.E., and Hickman, M., Phytobenthos sampling and estimation of primary production, in *IBP Handbook No. 16,* Holme, N.A., and McIntire, Eds., Blackwell Scientific Publications, Oxford, 1971, 169.

4214. Round, F.E., and Crawford, R.M., The lines of evolution of the Bacillariophyta, *Proc. R. Soc. Lond.,* B 211, 237, 1981.

4215. Round, F.E., and Chapman, D.J., Eds., *Progress in Phycological Research,* Vols. 1–8, Elsevier Biomedical Press Amsterdam, (Vols. 1+2), Biopress Ltd., Bristol. (Vols. 3–8), 1982–1992.

4216. Round, F.E., Crawford, R.M., and Mann, D.G., *Diatoms,* Cambridge University Press, Cambridge, 1990.

4217. Rounsfell, G.A., and Evans, J.E., Large-scale experimental test of copper sulphate as a control for the Florida red tide, U.S. Dept. of Interior, Fish and Wildlife Serv. Spec. Rept. Fish. 270, 1958.

4218. Royle, R.N., and King, R.J., Aquatic macrophytes in Lake Lindell, New South Wales: biomass, nitrogen and phosphorus status, and changing distribution from 1981 to 1987, *Aquat. Bot.,* 41, 281, 1991.

4219. Rowan, K.S., *Photosynthetic Pigments of Algae,* Cambridge University Press, Cambridge, 1989.

4220. Ruane, R.J., and Krenkel, P.A., Nitrification and other factor affecting nitrogen in the Holston River, *J. Water Pollut. Control Fed.,* 50, 2016, 1978.
4221. Rudd, J.W.M., and Taylor, C.D., Methane cycling in aquatic environments, *Adv. Aquatic Microbiol.,* 2, 77, 1980.
4222. Rueter, J.G., and Robinson, D.H., Inhibition of carbon uptake and stimulation of nitrate uptake at low salinities in *Fucus distichus* (Phaeophyceae), *J. Phycol.,* 22, 243, 1986.
4223. Rueter, J.G., McCarthy, J.J., and Carpenter, E.J., The toxic effect of copper on *Oscillatoria*(*Trichodesmium*) *theibautii, Limnol. Oceanogr.,* 24, 558, 1979.
4224. Rueter, J.G., Ohki, K., and Fujita, Y., The effect of iron nutrition on photosynthesis and nitrogen fixation in cultures of *Trichodesmium* (Cyanophyceae), *J. Phycol.,* 26, 30, 1990.
4225. Ruokolahti, C., and Rønnberg, O., Seasonal variation in chlorophyll *a* content of *Fucus vesiculosus* in a northern Baltic archipelago, *Ann. Rev. Fennici,* 25, 385, 1988.
4226. Russell, G.K., and Gibbs, M., Regulation of photosynthetic capacity in *Chlamydomonas mundana, Plant Physiol.,* 41, 885, 1966.
4227. Russell, G.K., and Gibbs, M., Evidence for the participation of the reductive pentose phosphate cycle in photoreduction and the oxyhydrogen reaction, *Plant Physiol.,* 43, 649, 1968.
4228. Ruttner, F., Zur Frage der Karbonatassimilation der Wasserpflanzen. Vol. 1. Die beiden Haupt-typen der Kohlenstoffaufnahme, *Österr. Bot. Zeitschr.,* 94, 265, 1947.
4229. Ruttner, F., Zur Frage der Karbonatassimilation der Wasserpflanzen. Vol. 2. Das Verhalten von *Elodea canadensis* and *Fontinalis antipyretica* in Losaungen von Natrium- bzw. Kalciumkarbonat, *Österr. Bot. Zeitschr,* 95, 208, 1948.
4230. Ruttner, F., *Fundamentals of Limnology,* University of Toronto Press, Toronto, 1953.
4231. Ruuhujäarvi, R., The Finnish mire types and their regional distribution, in *Ecosystems of the World, Mires: Swamp, Bog, Fen and Moor.* Vol. 4B. *Regional Studies,* Gore, A.J.P., Ed., Elsevier Science Publishers, Amsterdam, 1983, 47.
4232. Rychnovská, M., Water relations, water balance, transpiration, and water turnover in selected reedswamps communities, in *Pond Littoral Ecosystems: Structure and Functioning,* Dykyjová, D., and Kvet, J., Eds., Springer-Verlag, Berlin, 1978, 246.
4233. Rychnovská, M., and Šmid, P., Preliminary evaluation of transpiration in two *Phragmites* stands. in *Ecosystem Study on Wetland Biome in Czechoslovakia,* Czechoslovak IBP/PT-PP Report No. 3, Hejný, S., Ed., Třeboň, Czechoslovakia, 1973, 111.
4234. Rychnovská, M., Květ, J., Glosser, J., and Jakrlová, J., Plant water relations in three zones of grassland. *Acta Scientierum Naturalium* (Brno), 6, 1, 1972.
4235. Ryding, S.-O., and Rast, W., Eds., *The Control Of Eutrophication of Lakes and Reservoirs,* UNESCO, Paris, 1989.
4236. Ryther, J.H., Photosynthesis and fish production in the sea, *Science,* 166, 72, 1969.
4237. Ryther, J.H., Fuels from marine biomass, *Oceanus,* 22, 48, 1979.
4238. Ryther, J.H., and Dunstan, W.H., Nitrogen, phosphorus and eutrophication in the coastal marine environment, *Science,* 171, 1008, 1971.
4239. Ryther, J.H., De Boer, J.A., and Lapointe, B.E., Cultivation of seaweeds for hydrocolloids, waste treatment and biomass energy conversion, in *Proc. IX. Int. Seaweed Symposium,* Jensen, A., and Stein, J.R., Eds., Science Press, Princeton, New Jersey, 1979, 1.
4240. Rzewuska, E., and Wernikowska-Ukleja, E., Research on the influence of heavy metals on the development of *Scenedesmus quadricauda* (Turp.) Bréb. I. Mercury, *Arch. Hydrobiol.,* 21, 109, 1974.
4241. Rzhanova, G.N., Extracellular nitrogen-containing compounds of two nitrogen-fixing species of blue-green algae, *Mikrobiologiya,* 36, 639, 1967 (in Russian).
4242. Saboski, E.M., Effects of mercury and tin on frustular ultrastructure of the marine diatom *Nitzschia liebethrutti, Water Air Soil Pollut.,* 8, 461, 1977.
4243. Safferman, R.S., and Morris, M.E., Algal virus: isolation, *Science,* 140, 679, 1963.
4244. Sagan, L. (Margulis, L.), On the origin of mitosin cells, *J. Theor. Biol.,* 14, 225, 1967.
4245. Sahai, R., and Sinha, A.B., Productivity of submerged macrophytes in polluted regions of the eutrophic lake, Ramgarh (Uttar Pradesh), in *Aquatic Weeds in South East Asia,* Varshney, C.K., and Rzoska, J., Eds., Dr. W. Junk, The Hague, The Netherlands, 1976, 131.
4246. Said, M.Z., Culley, D.D., Standifer, L.C., Epps, E.A., Myers, M.W., and Boney, S.A., Effect of harvest rate, waste loading and stocking density on the yield of duckweeds, *Proc. World Maricult. Soc.,* 10, 769, 1979.
4247. Sain, P., Decomposition of wild rice (*Zizania aquatica*) straw in two natural lakes of northwestern Ontario, *Can. J. Bot.,* 62, 1352, 1984.
4248. Saito, Y., *Seaweed Aquaculture in the Northwest Pacific,* FAO Technical Conference on Aquaculture, Kyoto, Japan, FAO, FOR: AQ/Conf./76/R. 14, 1976.

4249. Sakaguchi, T., Horikashi, T., and Nakajima, A., Uptake of copper ion by *Chlorella regularis, J. Agric. Chem. Jap.*, 51, 497, 1977.

4250. Sakaguchi, T., Tsuji, T., Nakajima, A., and Horikoshi, T., Accumulation of cadmium by green microalgae, *Eur. J. Appl. Microbiol. Biotechnol.*, 8, 207, 1979.

4251. Sakamoto, M., Primary production by phytoplankton community in some Japanese lakes and its dependence on lake depth, *Arch. Hydrobiol.*, 62, 1, 1966.

4252. Sakamoto, M., Chemical factors involved in the control of phytoplankton production in the experimental lakes area, northwestern Ontario, *J. Fish. Res. Board Can.*, 28, 203, 1971.

4253. Sakevich, A., Volatile growth-inhibiting metabolites of blue-green algae, *Gidrobiol. Zhur.*, 9, 25, 1973.

4254. Saks, N.M., Stoner, R.J., and Lee, J.J., Autrophic and heterotrophic nutritional budget at salt marsh epiphytic algae, *J. Phycol.*, 12, 443, 1976.

4255. Sale, P.J.M., and Wetzel, R.G., Growth and metabolism of *Typha* species in relation to cutting treatments, *Aquat. Bot.*, 15, 321, 1983.

4256. Salisbury, J.L., and Floyd, G.L., Molecular, enzymatic and ultrastructural characterization of the pyrenoid of the scaly green monad *Micromonas squamata, J. Phycol.*, 14, 362, 1978.

4257. Salomons, W., Voorlopige baseline voor Cd, Zn, Ni, Pb, Cu, and Cr in Nederlandse sedimenten, *Verslag van onderzoek*, R 1790, 9, 1982.

4258. Salomons, W., *Trace Metal Cycling in a Polluted Lake: Ijsselmeer The Netherlands,* Delft Hydraulic Laboratory, The Netherlands, 1983.

4259. Salomons, W., Impact of atmospheric trace metals on the hydrological cycle, in *Toxic Metals in the Air,* Nriagu, J.O., and Davidson, C., Eds., John Wiley and Sons, New York, 1984, 1.

4260. Salomons, W., and Mook, W.G., Biogeochemical processes affecting metal concentrations in lake sediments (Ijsselmeer, The Netherlands), *Sci. Tot. Environ.*, 16, 217, 1980.

4261. Salomons, W., and Förstner, U., *Metals in the Hydrocycle,* Springer-Verlag, Berlin, 1984.

4262. Salton, M.R.J., *The Bacterial Cell Wall,* Elsevier Science Publishing, New York, 1964.

4263. Samanidou, V., and Fytianos, K., Partitioning of heavy metals into selective chemical fractions in sediments from rivers in northern Greece, *Sci. Total Environ.*, 67, 279, 1987.

4264. Samecka-Cymerman, A., Contents of As, V, Ge and other elements in aquatic liverwort *Scapania undulata* (L.) Dum. growing in the sudety mountains, *Pol. Arch. Hydrobiol.*, 38, 79, 1991.

4265. Samejima, H., and Myers, J., On the heterotrophic growth of *Chlorella pyrenoidosa, J. Gen. Microbiol.*, 18, 107, 1958.

4266. Samuel, S., Shah, N.H., and Fogg, G.E., Liberation of extracellular products of photosynthesis by tropical phytoplankton, *J. Mar. Biol. Ass. U.K.*, 51, 793, 1971.

4267. Sanders, J.G., Arsenic geochemistry in Chesapeake bay: dependence upon anthropogenic inputs and phytoplankton species composition, *Mar. Chem.*, 17, 329, 1985.

4268. Sanders, J.G., Direct and indirect effects of arsenic on the survival and fecundity of estuarine zooplankton, *Can. J. Fish. Aquat. Sci.*, 43, 694, 1986.

4269. Sanders, J.G., and Cibik, S.J., Adaptive behavior of euryhaline phytoplankton communities to arsenic stress, *Mar. Ecol. Prog. Ser.*, 22, 199, 1985.

4270. Sanders, R.W., and Porter, K., Phagotrophic phytoflagellates, *Adv. Microb. Ecol.*, 10, 167, 1988.

4271. Sanderson, H.R., Herbage production on high Sierra Nevada meadows, *J. Range Manage.*, 20, 255, 1976.

4272. Sandgren, C.D., Characteristics of sexual and sexual resting cyst (statospore) formation in *Dinobryon cylindricum* Imhof (Chrysophyta), *J. Phycol.*, 17, 199, 1981.

4273. Sandgren, C.D., Ed., *Growth and Reproductive Strategies of Freshwater Phytoplankton,* Cambridge University Press, Cambridge, 1988.

4274. Sand-Jensen, K., Effect of epiphytes on eelgrass photosynthesis, *Aquat. Bot.*, 3, 55, 1977.

4275. Sand-Jensen, K., Balancen mellem autotrofe komponenter i tempererede søer med forskellig næringsbelastning, *Vatten*, 2, 80, 1980.

4276. Sand-Jensen, K., Environmental variables and their effect on photosynthesis of aquatic plant communities, *Aquat. Bot.*, 34, 5, 1989.

4277. Sand-Jensen, K., and Søndergaard, M., Phytoplankton and epiphyte development and the shading effect on submerged macrophytes in lakes of different nutrient status, *Int. Rev. Ges. Hydrobiol.*, 66, 529, 1981.

4278. Sand-Jensen, K., and Borum, J., Regulation of growth of eelgrass (*Zostera marina* L.) in Danish coastal waters, *Mar. Tech. Soc. J.*, 17, 15, 1983.

4279. Sand-Jensen, K., and Borum, J., Epiphyte shading and its effect on diel metabolism of *Lobelia dortmanna* L. during the spring bloom in a Danish lake, *Aquat. Bot.*, 20, 109, 1984.

4280. Sand-Jensen, K., and Gordon, D.M., Differential ability of marine and freshwater macrophytes to utilize HCO_3^- and CO_2, *Mar. Biol.*, 80, 247, 1984.

4281. Sand-Jensen, K., and Borum, J., Interactions among phytoplankton, periphyton and macrophytes in temperate freshwaters and estuaries, *Aquat. Bot.*, 41, 137, 1991.

4282. Sand-Jensen, K., Revsbech, N.P., and Jørgensen, B.B., Microprofiles of oxygen in epiphyte communities on submerged macrophytes, *Mar. Biol.*, 89, 55, 1985.

4283. Sand-Jensen, K., Møller, J., and Oleson, B.H., Biomass regulation of microbenthic algae in Danish lowland streams, *Oikos*, 53, 332, 1988.

4284. San Pietro, A., Plant and algal ferredoxins, *Proc. 8th Internat. Congr. Biochem.*, Tokyo, Abstr. Col. XIII-3, 559, 1967.

4285. Santos, L.M., and Leedale, G.F., First report of a Golgi body in a uniflagellate eustigmatophycean zoospore, *Phycologia*, 31, 119, 1992.

4286. Saraiva, M.C., and Fraizier, A., Contamination per le ^{51}Cr et le ^{109}Cd de culture de l'algue *Dunaliella bioculata*, *Mar. Biol.*, 29, 343, 1975.

4287. Sarjeant, W.A.S., *Fossil and Living Dinoflagellates*, Academic Press, New York, 1974.

4288. Särkkä, J., Hattula, M.-L., Janatuinen, J., and Paasivirta, J., Chlorinated hydrocarbons and mercury in aquatic vascular plants of Lake Päijänne, Finland, *Bull. Environ. Contam. Toxicol.*, 20, 361, 1978.

4289. Sarsfield, L.J., and Mancy, K.H., The properties of cadmium complexes and their effect on toxicity to a biological systems, in *Proc. Symp. Biological Implications of Metals in the Environment*, Drucker, H., and Wildung, R.E., Eds., Energy Res. and Develop. Admin., Richland, Washington, 1977, 335.

4290. Sasaki, H., Effects of culture conditions on hydrogenase activity of *Scenedesmus* D$_3$, *Plant Cell Physiol.*, 7, 231, 1966.

4291. Sasser, C.E., Peterson, G.W., Fuller, D.A., Abernethy, R.K., and Gosselink, J.G., Environmental monitoring program. Louisiana off-shore oil port pipeline. 1981 Annual Report, Coastal Ecol. Lab. Center for Wetland Resources, Louisiana State University, Baton Rouge, 1982.

4292. Satake, K., Soma, M., Seyama, H., and Uehiro, T., Accumulation of mercury in the liverwort *Jungermannia vulcanicola* Steph. in an acid stream Kashiranashigawa in Japan, *Arch. Hydrobiol.*, 99, 80, 1983.

4293. Satapathy, K.B., and Chand, P.K., Studies on the ecology of *Azolla pinnata* R.Br. of Orissa, *J. Indian Bot. Sci.*, 63, 44, 1984.

4294. Sato, S., Paranagua, M.N., and Eskinazi, E., On the mechanism of red tide of *Trichodesmium* in Recife Northeastern Brazil, with some considerations of the relation to the human disease "Tamandare Fever," *Trab. inst. Oceanogr. Univ. Recife*, 5/6, 7, 1963/1964.

4295. Saunders, G.W., The kinetics of extracellular release of soluble organic matter by plankton, *Verh. Internat. Verein. Limnol.*, 18, 140, 1972.

4296. Sauvageau, C., Sur la sexualité hétérogamique d'une Laminaire (*Saccorhiza bulbosa*), *C.R. Acad. Sci. Paris*, D 161, 769, 1915.

4297. Sauvageau, C., Sur quelques algues floridées renfermant du brome a l'état libre, *Bull. Sta. Biol. Arcachon*, 23, 5, 1926.

4298. Savant, N.K., and De Datta, S.K., Nitrogen transformation in wetland rice soils, *Adv. Agron.*, 35, 241, 1982.

4299. Saxena, J., and Howard, P., Environmental transformations of alkylated and inorganic forms of certain metals, *Adv. Appl. Microbiol.*, 21, 185, 1977.

4300. Saxena, M., Dry matter production of freshwater and marsh plants around Jaipur, *Limnologica*, 17, 127, 1986.

4301. Say, P.J., and Whitton, B.A., Influence of zinc on lotic plants. II. Environmental effects on toxicity of zinc to *Hormidium rivulare*, *Freshwater Biol.*, 7, 377, 1977.

4302. Say, P.J., and Whitton, B.A., Changes in flora down a stream showing a zinc gradient, *Hydrobiologia*, 76, 255, 1980.

4303. Say, P.J., and Whitton, B.A., Chemistry and plant ecology of zinc rich streams in the northern Pennines, in *Heavy Metals in Northern England: Environmental and Biological Aspects*, Say, P.J., and Whitton, B.A., Eds., University of Durham, Durham, England, 1981, 55.

4304. Say, P.J., and Whitton, B.A., Chimie et ecologie de la vegetation de cours d'eau en France a fortes teneurs en zinc. 1. Massif Central, *Annls. Limnol.*, 18,3, 1982.

4305. Say, P.J., and Whitton, B.A., Chemistry and plant ecology of zinc-rich streams in France. 2. The Pyrénées, *Annls. Limnol.*, 18, 19, 1982.

4306. Say, P.J., and Whitton, B.A., Accumulation of heavy metals by aquatic mosses. 1. *Fontinalis antipyretica* Hedw., *Hydrobiologia*, 100, 245, 1983.

4307. Say, P.J., Diaz, B.M., and Whitton, B.A., Influence of zinc on lotic plants. I. Tolerance of *Hormidium* species to zinc, *Freshwater Biol.*, 7, 357, 1977.

4308. Say, P.J., Burrows, I.G., and Whitton, B.A., *Enteromorpha* as a monitor of heavy metals in estuaries, *Hydrobiologia*, 195, 119, 1990.

4309. Scagel, R.F., Bandoni, J.R., Rouse, G.E., Schofield, W.B., Stein, J.R., and Taylor, T.M.C., *An Evolutionary Survey of the Plant Kingdom*, Wadsworth, Belmont, California, 1966.

4310. Scagel, R.F., Bandoni, J.R., Maze, J.R., Rouse, G.E., Schofield, W.B., and Stein, J.R., *Nonvascular Plants: An Evolutionary Survey*, Wadsworth, Belmont, California, 1982.

4311. Scavia, D., Conceptual model of phosphorus cycling, in *Nutrient Cycling in the Great Lakes: A Summarization of Factors Regulating the Cycling of Phosphorus*, Scavia, D., and Moll, R., Eds., Great Lakes Res. Div. Spec. Report No. 83, The University of Michigan, Ann Arbor, Michigan, 1980, 119.

4312. Scavia, D., An ecological model of Lake Ontario, *Ecol. Modelling*, 8, 49, 1980.

4313. Schade, A.L., Cobalt and bacterial growth, with special reference to *Proteus vulgaris*, *J. Bact.*, 58, 811, 1949.

4314. Schaedle, M., and Jacobson, L., Ion absorption and retention by *Chlorella pyrenoidosa*. I. Absorption of potassium, *Plant Physiol.*, 40, 214, 1965

4315. Schalles, J.F., The chemical environment of wetlands, in *Wetlands Ecology and Conservation: Emphasis in Pennsylvania*, Majumdar, S.K., Brooks, R.P., Brenner, F.J., and Tiner, R.W., Jr., Eds., The Pennsylvania Academy of Sci., 1989, 75.

4316. Schantz, E.J., Biochemical studies in purified *Gonyaulax catenella* poison, in *Animal Toxins*, Russel, F.E., and Saunders, P.R., Eds., Pergamon Press, Elmsford, New York, 1967.

4317. Schantz, E.J., Algal toxins, in *Properties and Products of Algae*, Zajic, J.E., Ed., Plenum Press, New York, 1970, 83.

4318. Schantz, E.J., The dinoflagellate toxins, in *Microbial Toxins*, Vol. VII. *Algal and Fungal Toxins*, Kadis, S., Ciegler, A., and Ajl, S.J., Eds., Academic Press, New York, 3, 1971.

4319. Schantz, E.J., Poisonous dinoflagellates, in *Biochemistry and Physiology of Protozoa*, Vol. 1, Hutner, S.H., and Levandovsky, M., Eds., Academic Press, New York, 1979.

4320. Schantz, E.J., Lynch, J.M., Vayvada, G., Matsumoto, K., and Rapoport, H., The purification and characterization of the poison produced by *Gonyaulax catenella* in axenic culture, *Biochemistry* 3, 1191, 1966.

4321. Scheffe, R.D., Estimation and prediction of summer evapotranspiration from a northern wetland, M.S. Thesis, The University of Michigan, Ann Arbor, 1978.

4322. Schelske, C.L., Iron, organic matter and other factors limiting primary productivity in a marl lake, *Science*, 136, 45, 1962.

4323. Schelske, C.L., and Lowe, R.L., Algal nuisances and indicators of pollution: introduction and bibliography, In *Selected Papers in Phycology* II, Rosowski, J.R., and Parker, B.C., Eds., Phycological Society of America, Lawrence, 1982, 799.

4324. Schelske, C.L., Hooper, F.F., and Haert, L., Responses of a marl lake to chelated iron and fertilizer, *Ecology*, 43, 646, 1962.

4325. Schenck, R.C., Copper deficiency and toxicity in *Gonyaulax tamarensis* (Lebour), *Mar. Biol. Letters*, 5, 13, 1984.

4326. Schenkman, R.P.F., *Hypnea musciformis* (Rhodophyta): ecological influence on growth, *J. Phycol.*, 25, 192, 1989.

4327. Schiemer, F., and Prosser, M., Distribution and biomass of submerged macrophytes in Neusiedlersee, *Aquat. Bot.*, 2, 289, 1976.

4328. Schierup, H.-H., and Larsen, V.J., Macrophyte cycling of zinc, copper, lead and cadmium in the littoral zone of a polluted and non-polluted lake. I. Availability, uptake and translocation of heavy metals in *Phragmites australis* (Cav.) Trin., *Aquat. Bot.*, 11, 197, 1981.

4329. Schiff, J.A., Studies on sulfate utilization by *Chlorella pyrenoidosa* using sulfate - S^{35}; the occurrence of S-adenosyl methionine, *Plant Physiol.*, 34, 73, 1959.

4330. Schiff, J.A., Sulfate reduction by cell-free extracts of *Chlorella* and by intact cells, *Plant Physiol.*, 37 (suppl.), xiii, 1962.

4331. Schiff, J.A., Sulfur, in *Physiology and Biochemistry of Algae*, Lewin, R.A., Ed., Academic Press, New York, 1962, 239.

4332. Schiff, J.A., Studies of sulfate utilization by algae. II. Further identification of reduced compounds formed from sulfate by *Chlorella*, *Plant Physiol.*, 39, 176, 1964.

4333. Schiff, J.A., The development, inheritance and origin of the plastid in *Euglena*, *Adv. Morphogen.*, 10, 265, 1973.

4334. Schiff, J.A., Pathways of assimilation sulfate reduction in plants and microorganisms, in *Sulfur in Biology*, Elliot, K., and Whelan, J., Eds., Ciba Found. Symp. 72 (New Series), Excerpta Medica, Amsterdam, 1980, 49.

4335. Schiff, J.A., Reduction and other metabolic reactions of sulfate, in *Encyclopedia of Plant Physiology*, Vol. 15, Läuchli, A., and Bieliski, R.L., Eds., Springer-Verlag, New York, 1983, 382.

4336. Schiff, J.A., and Levinthal, M., Studies of sulfate utilization by algae. IV. Properties of a cell-free sulfate-reducing system from *Chlorella, Plant Physiol.,* 43, 547, 1968.

4337. Schiff, J.A., and Hodson, R.C., Pathways of sulfate reduction in algae, *Ann. N.Y. Acad. Sci.,* 175, 555, 1970.

4338. Schimpf, C., and Parthier, B., Modulation of amino acid uptake by *Euglena* cells after preincubation with substrate, *Biochem. Physiol. Pflanzen,* 182, 129, 1987.

4339. Schindler, D.W., Carbon, nitrogen, and phosphorus and the eutrophication of freshwater lakes, *J. Phycol.,* 7, 321, 1971.

4340. Schindler, D.W., Biochemical evolution of phosphorus limitation in nutrient-enriched lakes of the Precambrian Shield, in *Environmental Biogeochemistry* Vol. 2, *Metals Transfer and Ecological Mass Balances,* Nriagu, J.O., Ed., Ann Arbor Science Publications, Ann Arbor, Michigan, 1976, 647.

4341. Schindler, D.W., The evolution of phosphorus limitation in lakes, *Science,* 195, 260, 1977.

4342. Schindler, D.W., Frost, V.E., and Schmidt, R.V., Production of epilithiphyton in two lakes of the Experimental Lakes Area, northwestern Ontario, *J. Fish. Res. Board Can.,* 30, 1511, 1973.

4343. Schloemer, R.H., and Garrett, R.H., Uptake of nitrate by *Neurospora crassa, J. Bacteriol.,* 118, 270, 1974.

4344. Schloesser, R.W., and Blum, J.L., *Sphacelaria lacustris* sp. nov., a freshwater brown alga from Lake Michigan, *J. Phycol.,* 16, 201, 1980.

4345. Schmid, A.M., The development of structure in the shells of diatom, *Beih. Nova Hedw.,* 64, 219, 1979.

4346. Schmid, A.M., and Schultz, D., Wall morphogenesis in diatoms: deposition of silica by cytoplasmic vesicles, *Protoplasma,* 100, 267, 1979.

4347. Schmidt, E.L., Nitrification in soils, in *Nitrogen in Agricultural Soil,* Stevenson, F.J., Ed., Am. Soc. Agron., Madison, Wisconsin, *Agronomy,* 22, 253, 1982.

4348. Schmidt, G., The biochemistry of inorganic pyrophosphates and metaphosphates, in *Phosphorus Metabolism,* Vol. 1, McElroy, W.D., and Glass, B., Eds., Johns Hopkins Press, Baltimore, Maryland, 1951, 443.

4349. Schmidt, G., and Lyman, H., Inheritance and synthesis of chloroplast and mitochondria of *Euglena gracilis,* in *The Genetics of Algae,* Lewin, R.A., Ed., University of California Press, Berkeley, 1976, 257.

4350. Schmidt van Dorp, A.D., De eutrofiering van ondiepe meren in Gijnland, Technische Dienst Hoogheemraadschap van Rijnland, Leiden, The Netherlands, 1978.

4351. Schmiedeberg, J.E.O., Tagblatt der 58. Versammlung deutscher Naturforscher und Ärzte in Strassbo-ug, 1885, 231.

4352. Schmitt, M.R., and Adams, M.S., Dependence of rates of apparent photosynthesis on tissue phospho-rus concentration in *Myriophyllum spicatum* L., *Aquat. Bot.,* 11, 379, 1981.

4353. Schmitz, F., Systematische Übersicht der bisher bekannten Gattungen der Florideen, *Flora* (Jena), 72, 435, 1889.

4354. Schmitz, K., and Riffarth, W., Carrier-mediated uptake of L-leucine by the brown alga *Giffordia mitchelliae, Z. Pflanzenphysiol.,* 67, 311, 1980.

4355. Schneider, R.F., The impact of various heavy metals on the aquatic environment, U.S. EPA Water Quality Office Tech. Rep. 2, 1971.

4356. Schnitzer, M., and Khan, S.U., Soil organic matter, in *Development in Soil Science,* 8, Elsevier, New York, 1978.

4357. Schopf, J.W., Pre-Cambrian micro-organisms and evolutionary events prior to the origin of vascular plants, *Biol. Rev.,* 45, 319, 1970.

4358. Schopf, J.W., Are the oldest "fossils" fossils?, *Origins Life,* 7, 19, 1976.

4359. Schopf, J.W., The evolution of the earliest cells, *Sci. Am.,* 239, 110, 1978.

4360. Schopf, J.W., and Walter, M.R., Origin and early evolution of Cyanobacteria: the geological evidence, in *The Biology of Cyanobacteria,* Carr, N.G., and Whitton, B.A., Eds., Blackwell Scientific Publications, Oxford, 1982, 543.

4361. Schramm, W., and Booth, W., Mass bloom of the alga *Cladophora prolifera* in Bermuda: productivity and phosphorus accumulation, *Bot. Mar,* 24, 419, 1981.

4362. Schroll, H., Determination of the absorption of Cr^{6+} and Cr^{3+} in an algal culture of *Chlorella pyrenoidosa* using ^{51}Cr, *Bull. Environ. Contam. Toxicol.,* 20, 721, 1978.

4363. Schröder, R., Das Achilfsterben am Bodensee - Untersee. Beobachtungen, Untersuchungen und Gegenmassnahmen, *Arch. Hydrobiol.* (Suppl.), 76, 53, 1987.

4364. Schröder, R., and Schröder, H., Change in the composition of the submerged macrophyte community in Lake Constance. A multiparameter analysis with various environmental factors, *Mem. Ist. Ital. Idrobiol.,* 40, 25, 1982.

4365. Schubauer, J.P., and Hopkinson, C.S., Above- and belowground emergent macrophyte production and turnover in a coastal marsh ecosystem, Georgia, *Limnol. Oceanogr.,* 29, 1052, 1984.

4366. Schuler, C.A., Anthony, R.-G., and Ohlendorf, H.M., Selenium in wetlands and waterfowl foods at Kesterson Reservoir, California, 1984, *Arch. Environ. Contam. Toxicol.,* 19, 845, 1990.

4367. Schuler, J.F., Diller, V.M., and Kersten, H.J., Preferential assimilation of ammonium ion by *Chlorella vulgaris, Plant Physiol.,* 28, 299, 1953.

4368. Schulle, H., Ökologische und physiologische Untersuchungen an Planktonalgen des Titisees, Ph.D. Thesis, Freiburg, Germany, 1968.

4369. Schulz-Baldes, M., and Lewin, R.A., Lead uptake in two marine phytoplankton organisms, *Biol. Bull.,* 150, 118, 1976.

4370. Schüte, K.H., *The Biology of the Trace Elements: Their Role in Nutrition,* International Monographs, London, 1964.

4371. Schwimmer, D., and Schwimmer, M., Algae and medicine, in *Algae and Man,* Jackson, D.F., Ed., Plenum Press, New York, 1964, 368.

4372. Schwimmer, M., and Schwimmer, D., *The Role of Algae and Plankton in Medicine,* Grune and Stratton, New York, 1955.

4373. Schwimmer, M., and Schwimmer, D., Medical aspects of phycology, in *Algae, Man and the Environment,* Jackson, D.F., Syracuse University Press, Syracuse, 1968, 279.

4374. Schwintzer, C.R., Nutrient and water levels in a small Michigan bog with high tree mortality, *Am. Midl. Nat.,* 100, 441, 1978.

4375. Schwintzer, C.R., Vegetation and nutrient status of northern Michigan bogs and conifer swamps with a comparison to fens, *Can. J. Bot.,* 59, 842, 1981.

4376. Scott, B.D., and Jitts, H.R., Photosynthesis of phytoplankton and zooxanthellae on a coral reef, *Mar. Biol.,* 41, 307, 1977.

4377. Scott, G.S., Lichen terminology, *Nature,* 179, 234, 1957.

4378. Scott, G.T., The mineral composition of *Chlorella pyrenoidosa* grown in culture media containing varying concentrations of calcium, magnesium, potassium and sodium, *J. Cell. Comp. Physiol.,* 21, 327, 1943.

4379. Scott, G.T., The mineral composition of phosphate deficient cells of *Chlorella pyrenoidosa* during the restoration of phosphate, *J. Cell. Comp. Physiol.,* 26, 35, 1945.

4380. Scott, G.T., and Hayward, H.R., Metabolic factors influencing the sodium and potassium distribution in *Ulva lactuca, J. Gen. Physiol.,* 36, 659, 1953.

4381. Scott, G.T., and Hayward, H.R., Sodium and potassium regulation in *Ulva lactuca* and *Valonia macrophysa,* in *Electrolytes in Biological Systems,* Shanes, A.M., Ed., Am. Physiol. Soc., Washington, DC, 1955, 35.

4382. Scott, W.E., and Fay, P., Phosphorylation and amination in heterocysts of *Anabaena variabilis, Br. Phycol. Bull.,* 7, 283, 1972.

4383. Scoullos, M.J., and Hatzianestis, J., Dissolved and particulate trace metals in a wetland of international importance: Lake Mikri Prespa, Greece, *Water, Air Soil Pollut.,* 44, 307, 1989.

4384. Sculthorpe, C.D., *The Biology of Aquatic Vascular Plants,* Edward Arnold Publishers, London, 1967.

4385. Scutt, J.E., Autoinhibitor production by *Chlorella vulgaris, Am. J. Bot.,* 51, 581, 1964.

4386. Searle, G.F.W., Barber, J., Porter, G., and Tredwell, C.J., Picosecond time-resolved energy transfer in *Porphyridium cruentum.* Part II. In the isolated light harvesting complex (phycobilisomes), *Biochem. Biophys. Acta,* 501, 246, 1978.

4387. Searles, R.B., The strategy of the red algal life history, *Am. Nat.,* 115, 113, 1980.

4388. Sears, J.R., and Wilce, R.T., Sublittoral benthic marine algae of southeastern Cape Cod and adjacent islands: *Pseudolithoderma paradoxum* sp. nov. (Ralfsiaceae, Ectocarpales), *Phycologia,* 12, 75, 1973.

4389. Sears, J.R., Pecci, K.J., and Cooper, R.A., Trace metal concentrations in offshore, deep-water seaweeds in the western North Atlantic Ocean, *Mar. Pollut. Bull.,* 16, 325, 1985.

4390. Sebacher, D.I., Harris, R.C., and Barlett, K.B., Methane emissions to the atmosphere through aquatic plants, *J. Environ. Qual.,* 14, 40, 1985.

4391. Sebrell, W.H., Jr., and Haris, R.S., *The Vitamins,* Vol. 2, 2nd ed., Academic Press, New York, 1968.

4392. Sedláček, J., Kallquist, T., and Gjessing, E.T., Effect of aquatic humus on uptake and toxicity of cadmium to *Selenstrum capricornutum,* in *Aquatic and Terrestrial Humic Materials,* Christman, R.F., and Gjessing, E.T., Eds., Ann Arbor Science Publishers, Ann Arbor, Michigan, 1983, 495.

4393. Seeliger, U., Ed., *Coastal Plant Communities of Latin America,* Academic Press, San Diego, 1992.

4394. Seeliger, U., and Edwards, P., Correlation coefficients and concentration factors of copper and lead in sea water and benthic algae, *Mar. Pollut. Bull.,* 8, 16, 1977.

4395. Seeliger, U., and Cordazzo, O., Field and experimental evaluation of *Enteromorpha* sp. as a quali-quantitative monitoring organism for copper and mercury in estuaries, *Environ. Poll.,* 29, 197, 1982.

4396.   Segal, D.S., Jones, R.H., and Sharitz, R.R., Release of NH$_4$-N, NO$_3$-N, and PO$_4$-P from litter in two bottomland hardwood forests, *Am. Midl. Nat.*, 123, 160, 1990.

4397.   Segar, D.A., and Pellenberg, R.E., Trace metals in carbonate and organic rich sediments, *Mar. Pollut. Bull.*, 4, 138, 1973.

4398.   Segot, M., Codomier, L., and Combaut, G., Action de quatre métaux lourds (cadmium, cuive, mercure, plomb) sur la croissance d'*Asparagopsis armata* Harvey (Rhodophycée: Bonnemaisoniale) en culture, *J. Exp. Mar. Biol. Ecol.*, 66, 41, 1983.

4399.   Seidel, K., Reinigung von Gewässern durch höhere Pflanzen, *Naturwissenschaften*, 53, 289, 1966.

4400.   Seifer, S., Dayton, S., Novic, B., and Muntwyler, E., The estimation of glycogen with the anthrone reagent, *Arch. Biochem. Biophys.*, 25, 191, 1950.

4401.   Seitzinger, S., Nixon, S., Pilson, M.E.Q., and Burke, S., Denitrification and N$_2$O production in nearshore marine sediments, *Geochim. Cosmochim. Acta*, 44, 1853, 1980.

4402.   Sellner, K.G., and Zingmark, R.G., Interpretation of the 14C method of measuring the total annual production of phytoplankton in a South Carolina estuary, *Bot. Mar.*, 19, 119, 1976.

4403.   Senf, W.H.H., Hunchberger, R.A., and Roberts, K.E., Temperature dependence of growth and phosphorus uptake in two species of *Volvox* (Volvocales, Chlorophyta), *J. Phycol.*, 17, 323, 1981.

4404.   Senn, T.L., and Skelton, J., The effect of Norwegian seaweed on metabolic activity of certain plants, in *Proc. VI. Int. Seaweed Symposium*, Margalef, R., Ed., Madrid, Spain, 1969, 731.

4405.   Serra, J.L., Llama, M.J., and Cadenas, E., Nitrate utilization by the diatom *Skeletonema costatum*. II. Regulation of nitrate uptake, *Plant Physiol.*, 62, 991, 1978.

4406.   Serra, J.L., Llama, M.J., and Cedenas, E., Characterization of the nitrate reductase activity in the diatom *Skeletonema costatum*, *Plant Sci. Lett.*, 13, 41, 1978.

4407.   Serruya, C., and Berman, T., Phosphorus, nitrogen and the growth of algae in Lake Kinneret, *J. Phycol.*, 11, 155, 1975.

4408.   Seshavatharam, V., and Venu, P., Some observations on the ecology of Kolleru lake, *Int. J. Ecol. Environ. Sci.*, 7, 35, 1981.

4409.   Šetlík, I., Perspectives of algal production for food and feed purposes, *Naša Voda*, 6, 115, 1959, (in Slovak).

4410.   Šetlík, I., Berková, E., and Kubín, S., Irradiation and temperature dependence of photosynthesis in some chlorococcal strains, *A. Rep. Lab. Algol. Trebon* 1968, 134, 1969.

4411.   Seydel, I.S., Distribution and circulation of arsenic through water organisms and sediments of Lake Michigan, *Arch. Hydrobiol.*, 71, 17, 1972.

4412.   Seyfer, J.R., and Wilhm, J., Variation with stream order in species composition, diversity, biomass, and chlorophyll of periphyton in Otter Creek, Oklahoma, *Southwest. Natur.*, 22, 455, 1977.

4413.   Shafer, G., and Onuf, C., Reducing the error in estimating annual production of benthic microflora: hourly to monthly rates, patchiness in space and time, *Mar. Ecol. Prog. Ser.*, 26, 221, 1985.

4414.   Shah, J.D., and Abbas, S.G., Seasonal variation in frequency, density, biomass and rate of production of some aquatic macrophytes of the river Ganges at Bhagalpur (Bihar), India, *Trop. Ecol.*, 20, 127, 1979.

4415.   Shah, N., and Syrett, P.J., Uptake of guanine by the diatom *Phaeodactylum tricornutum*, *J. Phycol.*, 18, 579, 1982.

4416.   Shah, N., and Syrett, P.J., The uptake of guanine and hypoxanthine by marine microalgae, *J. Mar. Biol. Ass. U.K.*, 64, 545, 1984.

4417.   Shapiro, J., Carbon dioxide and pH: effect on species succession of algae, *Science*, 182, 306, 1973.

4418.   Sharitz, R.R., and Gibbons, J.W., Eds., *Proc. Symp. Freshwater Wetlands and Wildlife*, U.S. Department of Energy, 1989.

4419.   Sharma, K.P., and Gopal, B., Studies on stand structure and primary production in *Typha* species, *Int. J. Ecol. Environ. Sci.*, 3, 45, 1977.

4420.   Sharma, K.P., and Pradhan, V.N., Study on growth and biomass of underground organs of *Typha angustata* Bory & Chaub., *Hydrobiologia*, 98, 147, 1983.

4421.   Sharma, K.P., and Goel, P.K., Studies on decomposition of two species of *Salvinia*, *Hydrobiologia*, 131, 57, 1986.

4422.   Sharma, K.P., and Goel, P.K., Decomposition of water hyacinth, *Eichhornia crassipes* (Mart.) Solms., *Int. J. Ecol. Environ. Sci.*, 13, 13, 1987.

4423.   Sharma, B., and Ahler, R.C., Nitrification and nitrogen removal, *Water Res.*, 11, 897, 1977.

4424.   Sharma, D.P., and Gopal, B., Decomposition and nutrient dynamics in *Typha elephantina* Roxb. under different water regimes, in *Proc. Int. Conf. Wetlands: Ecology and Management*, Gopal, B., Turner, R.E., Wetzel, R.G., and Whigham, D.F., Eds., International Scientific Publications, Jaipur, India, 1982, 321.

4425.   Sharpe, V., and Denny, P., Electron microscopy studies of the absorption of lead in the leaf tissue of *Potamogeton pectinatus* L., *J. Exp. Bot.*, 27, 1155, 1976.

4426. Shatilov, V.R., and Kretovich, V.L., Glutamate dehydrogenase from *Chlorella*: forms, regulation and properties, *Mol. Cell. Biochem.*, 15, 201, 1977.

4427. Shatilov, V.R., Sofin, A.V., Zabrodina, T.M., Mutuskin, A.A., Pshenova, K.V., and Kretovich, V.L., Ferredoxin-dependent glutamate synthase from *Chlorella*, *Biokhimiya*, 43, 1492, 1978.

4428. Shaver, G.R., and Chapin, F.S., III, Response to fertilization by various plant growth forms in an Alaskan tundra: nutrient accumulation and growth, *Ecology*, 61, 622, 1980.

4429. Shaw, S.P., and Fredine, C.G., *Wetlands of the United States. Their Extent and Their Value for Waterfowl and Other Wildlife*, U.S. Dept. of Interior, Fish and Wildlife Service, Circular 39, Washington, 1956.

4430. Shaw, T.I., The mechanism of iodine accumulation by the brown sea weed *Laminaria digitata*. I. uptake of I^{131}, *Proc. R. Soc.*, B 150, 356, 1959.

4431. Shaw, T.I., The mechanism of iodine accumulation by the brown sea weed *Laminaria digitata*. II. Respiration and iodine uptake, *Proc. R. Soc.*, B 152, 109, 1960.

4432. Shaw, T.I., Halogens, in *Physiology and Biochemistry of Algae*, Lewin, R.A., Ed., Academic Press, New York and London, 1962, 247.

4433. Sheath, R.G., The biology of freshwater red algae, in *Progress in Phycological Research*, Vol. 3, Round, F.E., and Chapman, D.J., Eds., Biopress Ltd., Bristol., 1984.

4434. Sheath, R.G., and Hymes, B.J., A preliminary investigation of the fresh-water red algae in streams of southern Ontario, Canada, *Can. J. Bot.*, 58, 1295, 1980.

4435. Sheath, R.G., and Burkholder, J.M., Characteristics of soft-water stream in Rhode Island. II. Composition and seasonal dynamics of macroalgal communities, *Hydrobiologia*, 128, 109, 1985.

4436. Shehata, F.H.A., and Whitton, B.A., Field and laboratory studies on the blue-green algae from aquatic sites with high levels of zinc, *Verh. Internat. Verein. Limnol.*, 21, 1466, 1981.

4437. Shehata, F.H.A., and Whitton, B.A., Zinc tolerance in strains of the blue-green alga *Anacystis nidulans*, *Br. Phycol. J.*, 17, 5, 1982.

4438. Shelef, G., Oswald, W.J., Goldman, J.C., Sobsey, M., Harrison, J.E., Gee, H., and Halperin, R., Kinetics of algal systems in waste treatment. Ammonia-nitrogen as a growth limiting factor and other pertinent topics, Sanitary Engineering Research laboratory, University of California, Berkeley, 1970.

4439. Shelp, B.J., and Canvin, D.T., Utilization of exogenous inorganic carbon species in photosynthesis in *Chlorella pyrenoidosa*, *Plant Physiol.*, 65, 774, 1980.

4440. Shelp, B.J., and Canvin, D.T., Photorespiration and oxygen inhibition of photosynthesis in *Chlorella pyrenoidosa*, *Plant Physiol.*, 65, 780, 1980.

4441. Shelp, B.J., and Canvin, D.T., Photorespiration in air and high CO_2 grown *Chlorella pyrenoidosa*, *Plant Physiol.*, 68, 1500, 1981.

4442. Sheridan, R.P., A qualitative and quantitative study of plastoquinone A in two thermophilic blue-green algae, *J. Phycol.*, 8, 47, 1972.

4443. Sherman, L.A., and Brown, R.M., Jr., Cyanophages and viruses of eukaryotic algae, in *Comprehensive Virology*, Fraenkel-Conrat, H., and Wagner, R.R., Eds., 12, 145, 1978.

4444. Shero, B.R., Parker, M., and Stewart, K.M., The diatoms, productivity and morphometry of 43 lakes in New York State, U.S.A., *Int. Rev. Ges. Hydrobiol.*, 63, 365, 1978.

4445. Sherr, B.F., The ecology of denitrifying bacteria in salt marsh soils—an experimental approach, Ph.D. Thesis, University of Georgia, Athens, 1977.

4446. Shieh, Y.J., and Barber, J., Uptake of mercury by *Chlorella* and its effects on potassium regulation, *Planta*, 109, 49, 1973.

4447. Shields, L.M., and Durrell, L.W., Algae in relation to soil fertility, *Bot. Rev.*, 30, 92, 1964.

4448. Shields, L.M., Mitchell, C., and Drouet, F., Alga - and lichen - stabilized surface crusts as soil nitrogen sources, *Am. J. Bot.*, 44, 489, 1957.

4449. Shifrin, N.S., and Chisholm, S.W., Phytoplankton lipids: interspecific differences and effects of nitrate, silicate, and light-dark cycles, *J. Phycol.*, 17, 374, 1981.

4450. Shilo, M., Lysis of blue-green algae by myxobacter, *J. Bacteriol.*, 104, 453, 1970.

4451. Shilo, M., Toxigenic algae, in *Progress in Industrial Microbiology*, Hockenhull, D.J.D., Ed., Churchill Livingston, Edinburgh and London, 1972, 233.

4452. Shin, M., and Arnon, D.I., Enzyme mechanisms of pyridine nucleotide reduction in chloroplasts, *J. Biol. Chem.*, 240, 1405, 1965.

4453. Shioi, Y., Tamai, H., and Sasa, T., Inhibition of photosystem II in the green alga *Ankistrodesmus falcatus* by copper, *Physiol. Plant.*, 44, 434, 1978.

4454. Shortreed, K.S., and Stockner, J.G., Periphyton biomass and species composition in a coastal rainforest stream in British Columbia: effects of environmental changes caused by logging, *Can. J. Fish. Aquat. Sci.*, 40, 1887, 1983.

4455. Shrift, A., Sulfur-selenium antagonism. 1. Antimetabolite action of selenate on the growth of *Chlorella vulgaris*, *Am. J. Bot.*, 41, 223, 1954.

4456.   Shrift, A., Sulfur-selenium antagonism. 2. Antimetabolite action of selenomethionine on the growth of *Chlorella vulgaris, Am. J. Bot.,* 41, 345, 1954.

4457.   Shrift, A., Biological activities of selenium compounds, *Bot. Rev.,* 24, 550, 1958.

4458.   Shrift, A., Methionine transport in *Chlorella vulgaris, Plant Physiol.,* 41, 405, 1966.

4459.   Shrift, A., Aspects of selenium metabolism in higher plants, *Ann. Rev. Plant Physiol.,* 20, 475, 1969.

4460.   Shrift, A., and Sproul, M., Sulfur nutrition and the taxonomy of *Chlorella, Phycologia,* 3, 85, 1963.

4461.   Shtina, E.A., Algae solorum caespitose - podzolensium regionis Kirovskensis, *Trudy Botan. Int. im. V.L.Komarova Akad. Nauk S.S.S.R., Ser. II, Sporovyje. Rasteniya,* 12, 36, 1959 (in Russian).

4462.   Shtina, E.A., Participation of algae in processes of terrestrial algae of soil formation, *Izv. Akad. Nauk SSSR, Biol. Ser.,* 1, 72, 1964.

4463.   Shtina, E.A., Some regularities in the distribution of blue-green algae in soils, in *Biology of the Cyanophyta,* Fedorova, V.D., and Telichenko, M.M., Eds., Moscow University Press, 1959, 21 (in Russian).

4464.   Shubert, L.E., Soil, in *Biology of Cyanobacteria,* Carr, N.G., and Whitton, B.A., Eds., University of California Press, Berkeley, 1982, 216.

4465.   Shukla, S.S., and Leland, H.V., Heavy metals: a review of lead, *J. Water Pollut. Control Fed.,* 45, 1319, 1973.

4466.   Shukla, S.S., Syers, J.K., Williams, J.D.H., Armstrong, D.E., and Harris, R.F., Sorption of inorganic phosphate by lake sediments, *Soil Sci. Soc. Am. Proc.,* 35, 244, 1971.

4467.   Shuman, L.M., Zinc in soils, in *Zinc in the Environment,* Part 1. *Ecological Cycling,* Nriagu, J.O., Ed., John Wiley and Sons, New York, 1980, 39.

4468.   Sick, L.V., and Windom, H.L., Effects of environmental levels of mercury and cadmium on rates of metal uptake and growth physiology of selected genera of marine phytoplankton, in *Proc. Symp. Mineral Cycling in Southeastern Ecosystems,* Publ. Energy Research and Development Administration, Augusta, Georgia, 1974, 239.

4469.   Sicko, L.M., Physiological and cytological aspects of phosphate metabolism in the blue-green alga *Plectonema boryanum,* Ph.D. Thesis, City University of New York, New York, 1974.

4470.   Sicko-Goad, L., A morphometric analysis of algal response to low dose, short-term heavy metal exposure, *Protoplasma,* 110, 75, 1982.

4471.   Sicko-Goad, L., and Jensen, T.E., Phosphate metabolism in blue-green algae. II. Changes in phosphate distribution during starvation and the "polyphosphate overplus" phenomenon in *Plectonema boryanum, Am. J. Bot.,* 63, 183, 1976.

4472.   Sicko-Goad, L., and Stoermer, E.F., A morphometric study of lead and copper effects on *Diatoma tenue* var. *elongatum* (Bacillariophyta), *J. Phycol.,* 15, 316, 1979.

4473.   Sicko-Goad, L., and Lazinsky, D., Accumulation and cellular effects of heavy metals in benthic and planktonic algae, *Micron,* 22, 289, 1981.

4474.   Sicko-Goad, L., and Lazinsky, D., Synergistic effects of phosphorus nutrient status and lead exposure in three algae, *Micron Microscop. Acta,* 14, 261, 1983.

4475.   Sicko-Goad, L., and Lazinsky, D., Quantitative ultrastructural changes associated with lead-complex luxury phosphate uptake and polyphosphate utilization, *Arch. Environ. Contam. Toxicol.,* 15, 617, 1986.

4476.   Sicko-Goad, L., Stoermer, E.F., and Ladewski, B.G., A morphometric method for correcting phytoplankton cell volume estimates, *Protoplasma,* 93, 147, 1977.

4477.   Sicko-Goad, L., Jensen, T.E., and Ayala, R.P., Phosphate metabolism in blue-green bacteria. V. Factors affecting phosphate uptake in *Plectonema boryanum, Can. J. Microbiol.,* 24, 105, 1978.

4478.   Sicko-Goad, L., Ladewski, B.G., and Lazinsky, D., Synergistic effects of nutrients and lead on the quantitative ultrastructure of *Cyclotella* (Bacillariophyceae), *Arch. Environ. Contam. Toxicol.,* 15, 291, 1986.

4479.   Sieburth, J. McN., Studies on algal substances in the sea. III. Production of extracellular organic matter by littoral marine algae, *J. Exp. Mar. Biol. Ecol.,* 3, 290, 1969.

4480.   Sieburth, J. McN., and Conover, J.T., *Sargassum* tannin, an antibiotic which retards fouling, *Nature,* 208, 52, 1965.

4481.   Sieburth, J. McN., Johnson, P.W., and Hargraves, P.E., Ultrastructure and ecology of *Aureococcus anophagefferens* gen. et sp. nov. (Chrysophyceae): the dominant picoplankter during a bloom in Narraganset Bay, Rhode Island, summer 1985, *J. Phycol.,* 24, 416, 1988.

4482.   Siegel, D.I., A review of the recharge-discharge function of wetlands, in The *Ecology and Management of Wetlands,* Vol. 1. *Ecology of Wetlands,* Hook, D.D. et al., Eds., Timber Press, Portland, Oregon, 1988, 59.

4483.   Siegel, B.Z., and Siegel, S.M., The chemical composition of algal cell walls, *CRC Crit. Rev. Microbiol.,* 3, 1, 1973.

4484. Siegel, S.M., Siegel, B.Z., Lipp, C., Kruckberg, A., Towers, G., and Warren, H., Indicator plant-soil mercury patterns in a mercury-rich mining area of British Columbia, *Water Air Soil Pollut.*, 25, 73, 1985.

4485. Sielicki, M., and Burnham, J.C., The effect of selenite on the physiological and morphological properties of the blue-green alga *Phormidium luridum* var. *olivacea*, *J. Phycol.*, 9, 509, 1973.

4486. Siever, R., and Scott, R.A., Organic geochemistry of silica, in *Organic Geochemistry* Breger, L.A., Ed., Macmillan Co., New York, 1963, 579.

4487. Sifton, H.B., Air space tissue in plants. I., *J. Bot. Rev.*, 11, 108, 1945.

4488. Sifton, H.B., Air space tissue in plants. II., *J. Bot. Rev.*, 23, 303, 1957.

4489. Sikes, C.S., Calcification and cation sorption of *Cladophora glomerata* (Chlorophyta), *J. Phycol.*, 14, 325, 1978.

4490. Sikes, C.S., Roer, R.D., and Wilbur, K.M., Photosynthesis and coccolith formation: inorganic carbon sources and net inorganic reaction of deposition, *Limnol. Oceanogr.*, 25, 248, 1980.

4491. Sikora, L.J., and Keeney, D.R., Further aspects of soil chemistry under anaerobic conditions, in *Ecosystems of the World, Mires: Swamp, Bog, Fen and Moor.* Vol. 4A. *General Studies*, Gore, A.J.P., Ed., Elsevier Science Publishers, Amsterdam, 1983, 247.

4492. Sillén, L.G., The physical chemistry of seawater, *Publ. Am. Ass. Adv. Sci.*, 67, 549, 1961.

4493. Sillén, L.G., and Martell, A.E., *Stability Constants of Metal Ion Complexes*, 2nd ed., Chem. Soc. (London) Spec. Publ., 17, 1964.

4494. Sillén, L.G., and Martell, A.E., *Stability Constants of Metal Ion Complexes.* 1st suppl, Chem. Soc. (London) Spec. Publ., 25, 1971.

4495. Silva, P.C., Review of the taxonomic history and nomenclature of the yellow-green algae, *Arch. Protistenk.*, 121, 20, 1979.

4496. Silva, P.C., *Names of Classes and Families of Living Algae*, Bohn, Scheltema and Holkema, Utrecht and Dr. W. Junk Publishers, The Hague, 1980.

4497. Silva, P.C., Thallobionta, in *Synopsis and Classification of Living Organisms*, Vol. 1, Parker, S.P., Ed., McGraw-Hill, New York, 1982, 59.

4498. Silver, W.S., and Jump, A., Nitrogen fixation associated with vascular aquatic macrophytes, in *Nitrogen Fixation by Free-living Microorganisms*, Stewart, W.D.P., Ed., Cambridge University Press, London, 1975, 121.

4499. Silverberg, B.A., Ultrastructural localization of lead in *Stigeoclonium tenue* (Chlorophyceae, Ulotrichales) as demonstrated by cytochemical and X-ray microanalysis, *Phycologia*, 14, 265, 1975.

4500. Silverberg, B.A., Cadmium-induced ultrastructural changes in the mitochondria of freshwater green algae, *Phycologia*, 15, 155, 1976.

4501. Silverberg, B.A., Stokes, P.M., and Ferstenberg, L.B., Intracellular complexes in a copper tolerant green alga, *J. Cell Biol.*, 69, 210, 1976.

4502. Silverman, M.P., and Ehrlich, H.L., Microbial formation and degradation of minerals, in *Advances in Applied Microbiology*, Vol. 6, Umbreit, W.W., Ed., Academic Press, New York, 1964, 153.

4503. Silvester, W.B., and McNamara, P.J., The infection process and ultrastructure of the *Gunnera - Nostoc* symbiosis, *New Phytol.*, 77, 135, 1976.

4504. Silvethorne, W., and Sørensen, P.E., Marine algae as an economic resources, *Mar. Tech. Soc. Tans. of the 7th Ann. Mar. Sci. Conf.*, 7, 523, 1971.

4505. Silvey, J.K., Henley, D.E., and Wyatt, J.T., Planktonic blue-green algae: growth and odor-production studies, *J. Am. Water Works Ass.*, 64, 35, 1972.

4506. Simkiss, K., Phosphates as crystal poisons of calcification, *Biol. Rev.*, 39, 487, 1964.

4507. Simkiss, K., The inhibitory effect of some metabolites on the precipitation of calcium carbonate from artificial and natural sea water, *J. Cons. Int. Explor. Mer.*, 29, 6, 1964.

4508. Simkiss, K., Metal ions in cells, *Endeavour*, 3, 2, 1979.

4509. Simon, R.D., The effect of chloramphenicol on the production of cyanophycin granule polypeptide in the blue green alga *Anabaena cylindrica*, *Arch. Mikrobiol.*, 92, 115, 1973.

4510. Simonis, W., Zyklische und nichtzyklische Photophosphorylierung *in vivo*, *Ber. Dt. Bot. Ges.*, 80, 395, 1967.

4511. Simonis, W., and Gimmler, H., Effect of inhibitors of photophosphorylation on light-induced dark incorporation of ^{32}P into *Ankistrodesmus braunii*, in *Progress in Photosynthetic Research*, Vol. 3, Metzner, H., Ed., H. Laupp, Jr., Tübingen, Germany, 1969, 1155.

4512. Simons, J., *Vaucheria* species from estuarine areas in the Netherlands, *Neth. J. Sea Res.*, 9, 1, 1975.

4513. Simons, J., The Dutch *Vaucheria* species. *Wet. Meded. K. Ned. Natuurhist. Ver.*, 120, 1, 1977.

4514. Simpson, F.B., and Neilands, J.B., Siderochromes in cyanophyceae: Isolation and characterization of schizokinen from *Anabaena* sp., *J. Phycol.*, 12, 44, 1976.

4515.   Simpson, P.S., and Eaton, J.W., Comparative studies of the photosynthesis of the submerged macrophyte *Elodea canadensis* and the filamentous algae *Cladophora glomerata* and *Spirogyra* sp., *Aquat. Bot.,* 24, 1, 1986.

4516.   Simpson, R.L., and Whigham, D.F., Seasonal patterns of nutrient movement in a freshwater tidal marsh, in *Proc. Conf. Freshwater Wetlands: Ecological Processes and Management Potentials,* Good, R.E., Whigham, D.F., and Simpson, R.L., Eds., Academic Press, New York, 1978, 243.

4517.   Singh, B., and Ram, P., Effect of alternate submergence and drying on the release of potassium in soils growing rice, *Bull. Indian Soc. Soil Sci.,* 10, 129, 1976.

4518.   Singh, R.N., *Role of Blue-green Algae in Nitrogen Economy of Indian Agriculture,* Indian Council Agr. Res., New Delhi, 1961.

4519.   Singh, R.N., and Tiwari, D.N., Frequent heterocyst germination in the blue-green alga *Gloeotrichia ghosei* Singh, *J. Phycol.,* 6, 172, 1970.

4520.   Singh, S.P., and Yadava, V., Cadmium induced inhibition of nitrate uptake by *Anacystis nidulans:* interaction with other divalent cations, *J. Gen. Appl. Microbiol.,* 29, 297, 1983.

4521.   Singh, S.P., Pant, M.C., Sharma, A.P., Sharma, P.C., and Purohit, R., Limnology of shallow water zones of lakes in Kumaun Himalaya (India), in *Proc. Int. Conf. Wetlands: Ecology and Management,* Gopal, B., Turner, R.E., Wetzel, R.G., and Whigham, D.F., Eds., Natl. Inst. of Ecology and Internat. Scientific Publications, Jaipur, India, 1982, 39.

4522.   Sinha, A.B., and Sahai, R., Contribution to the ecology of Indian aquatics. I. Seasonal changes in the biomass and rate of production of two perennial submerged macrophytes (*Hydrilla* and *Najas* sps.) of Ramgarh lake of Gorakhpur, *Trop. Ecol.,* 14, 19, 1973.

4523.   Sirenko, L.A., Stetsenko, N.M., Arendarchuk, V.V., and Kuz'menko, M.I., Role of oxygen conditions in the vital activity of certain blue-green algae, *Microbiology,* 37, 199, 1968.

4524.   Sirodot, M.S., *Les Batrachospermes es-organization functions, development, classification,* Paris, 1884.

4525.   Siu, K.W.M., and Berman, S.S., The marine environment, in *Occurrence and Distribution of Selenium,* Ihnat, M., Ed., CRC Press, Boca Raton, Florida, 1989, 263.

4526.   Sivalingam, P.M., Biodeposited trace metals and mineral content studies of some tropical marine algae, *Bot. Mar.,* 21, 327, 1978.

4527.   Siver, P.A., Comparison of attached diatom communities on natural and artificial substrates, *J. Phycol.,* 13, 402, 1977.

4528.   Siver, P.A., and Hamer, J.S., Seasonal periodicity of Chrysophyceae and Synurophyceae in a small New England lake: implications for paleolimnological research, *J. Phycol.,* 28, 186, 1992.

4529.   Sjöberg, K., and Danell, K., Effects of permanent flooding on *Carex - Equisetum* wetlands in northern Sweden, *Aquat. Bot.,* 15, 275, 1983.

4530.   Sjörs, H., On the relation between vegetation and electrolytes in north Swedish mire waters, *Oikos,* 2, 241, 1950.

4531.   Sjörs, H., *Nordisk Vsekstgeogrfi,* 2nd ed., Bonniers, Stockholm, 1956 (in Swedish).

4532.   Sjörs, H., Bogs and fens in the Hudson's Bay lowlands, *Arctic,* 12, 1, 1959.

4533.   Sjörs, H., Bogs and fens on Attawapiskat River, Northern Ontario, *Nat. Mus. Can. Bull.,* 186, 45, 1963.

4534.   Sjörs, H., Peat on earth: multiple use or conservation? *Ambio,* 9, 303, 1980.

4535.   Sjörs, H., The zonation of northern peatlands and their importance for the carbon balance of the atmosphere, in *Proc. Int. Conf. Wetlands: Ecology and Management,* Part II., Gopal, B., Turner, R.E., Wetzel, R.G., and Whigham, D.F., Eds., Natl. Inst.of Ecology and Internat. Scientific Publications, Jaipur, India, 1982, 11.

4536.   Sjörs, H., Mires of Sweden, in *Ecosystems of the World, Mires: Swamp, Bog, Fen and Moor.* Vol. 4B, *Regional Studies,* Gore, A.J.P., Ed., Elsevier Science Publishers, Amsterdam, 1983, 69.

4537.   Skaar, H., Ophus, E., and Gullvag, B.M., Lead accumulation within nuclei of moss leaf cells, *Nature,* 241, 215, 1973.

4538.   Skaar, H., Rystad, B., and Jensen, A., The uptake of [63]Ni by the diatom *Phaeodactylum tricornutum, Physiol. Plant.,* 32, 353, 1974.

4539.   Skadovskij, S.N., *Periphyton Community in the Function of Bioeliminator* Moscow University Press, Moscow, 1961 (in Russian).

4540.   Skholnik, M.J., *Micronutrients in the Plant Life,* Nauka, Leningrad, 1974 (in Russian).

4541.   Skok, J., The role of boron in the plant cell, in *Trace Elements,* Lamb, C.A., Bentley, O.G., and Beattie, J.M., Eds., Academic Press, New York and London, 1958, 227.

4542.   Skowronski, T., Energy-dependent transport of cadmium by *Stichococcus bacilaris, Chemosphere,* 13, 1379, 1984.

4543.   Skowronski, T., Uptake of cadmium by *Stichococcus bacillaris, Chemosphere,* 13, 1385, 1984.

4544. Skowronski, T., Adsorption of cadmium on the green microalga *Stichococcus bacillaris*, *Chemosphere*, 15, 69, 1986.

4545. Skowronski, T., Influence of some physico-chemical factors on cadmium uptake by the green alga *Stichococcus bacillaris*, *Appl. Microbiol. Biotechnol.*, 24, 423, 1986.

4546. Skowronski, T., Szubinska, S., Pawlik, B., Jakubowski, M., Bilewicz, R., and Cukrowska, E., The influence of pH on cadmium toxicity to the green alga *Stichococcus bacillaris* and on the cadmium forms present in the culture medium, *Environ. Pollut.*, 74, 89, 1991.

4547. Skuja, H., Comments on fresh-water Rhodophyceae, *Bot. Rev.*, 4, 665, 1938.

4548. Skuja, H., Glaucophyta, in *Syllabus der Pflanzenfamilien*, 12, Melchoir, H., and Werdermann, F., Eds., Borntraeger, Berlin, 1954, 56.

4549. Skulberg, O.M., Algal cultures as a means to assess the fertilizing influence of pollution, in *Proc. 3rd Internat. Conf. on Water Pollution Research*, Water Pollution Control Federation, Washington, DC, 1966, section 1, paper No. 6.

4550. Skulberg, O.M., and Skulberg, R., Planktic species of *Oscillatoria* (Cyanophyceae) from Norway—characterization and classification, *Arch. Hydrobiol. Suppl.*, 71 (*Algol. Studies*, 38/39), 157, 1985.

4551. Skvortzov, B.V., and Noda, M., On Brazilian and European species of genus *Vacuolaria*, *J. Jap. Bot.*, 43, 69, 1968.

4552. Skvortzov, B.V., and Noda, M., Flagellates of clean and polluted waters. III. New taxa of colorless flagellates with the flagellum-genus, *Conradimena* gen. nov. and genus *Peranemopsis* Lackey (Peranemaceae, Euglenophyta). *Sci. Rep. Niigata Univ., Ser. D (Biol.)*, 11, 29, 1974.

4553. Sládeček, V., Zur Ermittlung des Indikations-Gewichtes in der biologischen Gewässeruntersuchung, *Arch. Hydrobiol.*, 60, 241, 1964.

4554. Sládeček, V., System of water quality from the biological point of view, *Arch. Hydrobiol. Beih.*, 7, 1, 1973.

4555. Sládeček, V., The reality of three British biotic indices, *Water Res.*, 7, 995, 1973.

4556. Sládeček, V., Diatoms as indicators of organic pollution, *Acta hydrochim. Hydrobiol.*, 14, 555, 1986.

4557. Sládeček, V., and Sládečková, A., Determination of the periphyton productivity by means of the glass slide method, *Hydrobiologia*, 23, 125, 1964.

4558. Sládeček, V., and Perman, J., Saprobic sequence within the genus *Euglena*, *Hydrobiologia*, 57, 57, 1978.

4559. Sládeček, V., Ottová, V., and Sládečková, A., *Technical Hydrobiology. A Laboratory Manual*, SNTL Praha, Czech Republic, 1973 (in Czech).

4560. Sládečková, A., Limnological investigation methods for the periphyton (*Aufwuchs*) community, *Bot. Rev.*, 28, 241, 1962.

4561. Sládečková, A., Periphyton as indicators of the groundwater quality, *Verh. Internat. Verein. Limnol.*, 18, 1011, 1972.

4562. Sládečková, A., Periphyton as indicator of the reservoir water quality. III. Biomonitoring techniques, *Arch. Hydrobiol. Beih. Ergebn. Limnol.*, 33, 775, 1990.

4563. Sládečková, A., and Sládeček, V., Periphyton as indicator of the reservoir water quality. II. Pseudoperiphyton, *Arch. Hydrobiol. Beih. Ergebn. Limnol.*, 9, 177, 1977.

4564. Sládečková, A., Vymazal, J., and Sládeček, V., *Technical Hydrobiology. Laboratory Guide*. SNTL Praha, Czech Republic, 1982 (in Czech).

4565. Sládečková, A., Marvan, P., and Vymazal, J., The utilization of periphyton in waterworks pretreatment for nutrient removal from enriched influents, in *Periphyton of Freshwater Ecosystems*, Wetzel, R.G., Ed., Dr. W. Junk Publishers, The Hague, The Netherlands, 1983, 299.

4566. Slater, E.C., Uncouplers and inhibitors of oxidative phosphorylation, in *Metabolic Inhibitors*, Hochster, R.M., and Quastel, H.H., Eds., Academic Press, New York, 1963, 503.

4567. Sloey, W.E., Spangler, F.L., and Fetter, C.W., Jr., Management of freshwater wetlands for nutrient assimilation, in *Proc. Conf. Freshwater Wetlands: Ecological Processes and Management Potential*, Good, R.E., Whigham, D.F., and Simpson, R.L., Eds., Academic Press, New York, 1978, 321.

4568. Small, E., Ecological significance of four critical elements in plant of raised *Sphagnum* peat bogs, *Ecology*, 53, 498, 1972.

4569. Small, E., Water relationship of plants raised in *Sphagnum* peat bogs, *Ecology*, 53, 726, 1973.

4570. Smalley, E.A., The role of two invertebrate populations, *Littorina irrorata* and *Orchelium fidicenium* in the energy flow of a salt marsh ecosystem, Ph.D. Thesis, University of Georgia, Athens, 1958.

4571. Smayda, T.J., Experimental observations on the influence of temperature, light, and salinity on cell division of the marine diatom *Detonula confervacea* (Cleve) Gran., *J. Phycol.*, 5, 150, 1969.

4572. Smayda, T.J., Bioassay of the growth potential of the surface water of lower Narragansett Bay over an annual cycle using the diatom *Thalassiosira pseudonana* (oceanic clone 13-1), *Limnol. Oceanogr.*, 19, 889, 1974.

4573. Smayda, T.J., Phytoplankton species succession, in *The Physiological Ecology of Phytoplankton*, Morris, I., Ed., University of California Press, Berkeley, 1980, 493.

4574. Šmíd, P., Evaporation from a reedswamp, *J. Ecol.*, 63, 299, 1975.

4575. Smillie, R.M., Isolation of phytoflavin, a flavoprotein with chloroplast ferredoxin activity, *Plant Physiol.*, 40, 1124, 1965.

4576. Smillie, R.M., Enzymology of *Euglena*, in *The Biology of Euglena*, II., Buetow, D.E., Ed., Academic Press, New York and London, 1968, 1.

4577. Smirnov, N.N., Some data about food consumption of plant production of bogs and fens by animals, *Verh. Internat. Verein. Limnol.*, 13, 363, 1958.

4578. Smith, C.J., Chen, R.L., and Patrick, W.H., Jr., Nitrous oxide emission from simulated overland flow wastewater treatment system, *Soil Bull. Biochem.*, 13, 275, 1981.

4579. Smith, C.J., DeLaune, R.D., and Patrick, W.H., Jr., Nitrate reduction in *Spartina alterniflora* marsh soil, *Soil Sci. Soc. Am. J.*, 46, 748, 1982.

4580. Smith, C.J., Wright, M.F., and Patrick, W.H., Jr., The effect of soil redox potential and pH on the reduction and production of nitrous oxide, *J. Environ. Qual.*, 12, 186, 1983.

4581. Smith, C.J., DeLaune, R.D., and Patrick, W.H., Jr., Nitrous oxide emission from Gulf coast wetlands, *Geochim. Cosmochim. Acta*, 47, 1805, 1983.

4582. Smith, C.S., Adams, M.S., and Gustavson, T.D., The importance of belowground mineral element stores in cattails (*Typha latifolia* L.), *Aquat. Bot.*, 30, 343, 1988.

4583. Smith, D.B., and Cook, W.H., Fractionation of carrageenin, *Arch. Biochim. Biophys.*, 45, 232, 1853.

4584. Smith, D.C., What can lichens tell us about real fungi? *Mycologia*, 70, 915, 1978.

4585. Smith, F.A., Active phosphate uptake by *Nitella translucens*, *Biochim. Biophys. Acta*, 126, 94, 1966.

4586. Smith, F.A., Rate of photosynthesis in Characean cells. II. Photosynthetic $^{14}CO_2$ fixation and ^{14}C-bicarbonate uptake by Characean cells, *J. Exp. Bot.*, 19, 207, 1968.

4587. Smith, F.A., The mechanism of chloride transport in Characean cells, *New Phytol.*, 69, 903, 1970.

4588. Smith, F.A., Historical perspective on HCO_3^- assimilation, in *Inorganic Carbon Uptake by Aquatic Photosynthetic Organisms*, Lucas, W.J., and Berry, J.A., Eds., Am. Soc. Plant Physiol., Rockville, Maryland, 1985, 1.

4589. Smith, F.A., Biological occurrence and importance of HCO_3^- utilization systems: macroalgae (Charophytes), in *Inorganic Carbon Uptake by Aquatic Photosynthetic Organisms*, Lucas, W.J., and Berry, J.A., Eds., Am. Soc. Plant Physiol., Rockville, Maryland, 1985, 111.

4590. Smith, F.A., and Walker, N.A., Photosynthesis by aquatic plants: effects of unstirred layers in relation to assimilation of CO_2 and HCO_3^- and to carbon isotropic discrimination, *New Phytol.*, 86, 245, 1980.

4591. Smith, F.W., and Thompson, J.F., Regulation of nitrate reductase in *Chlorella vulgaris*, *Plant Physiol.*, 48, 224, 1971.

4592. Smith, G.M., *The Fresh-Water Algae of the United States*, 2nd ed., McGraw-Hill Book Company, New York, 1950.

4593. Smith, G.M., *Manual of Phycology. An Introduction to the Algae and Their Biology*, Chronica Botanica Company, Waltham, 1951.

4594. Smith, G.M., *Cryptogamic Botany*, Vol. 1, *Algae and Fungi*, McGraw-Hill, New York, 1955.

4595. Smith, H.G., On the presence of algae in certain Ascidiaceae, *Am. Mag. Nat. Hist. (ser. 10)*, 15, 615, 1935.

4596. Smith, I.W., Wilkinson, J.F., and Duguid, J.P., Volutin production in *Aerobacter aerogenes* due to nutrient imbalance, *J. Bacteriol.*, 68, 450, 1954.

4597. Smith, J., and Shrift, A., Phytogenetic distribution of glutathione peroxidase, *Comp. Biochem. Physiol.*, 63B, 39, 1979.

4598. Smith, J.H.C., and Benitez, A., Chlorophylls: analysis in plant materials, in *Modern Methods of Plant Analysis*, Paech, K., and Tracey, M.V., Eds., Springer-Verlag, Berlin, 1955, 142.

4599. Smith, M.A., The effects of heavy metals on the cytoplasmic fine structure of *Skeletonema costatum* (Bacillariophyceae), *Protoplasma*, 116, 14, 1983.

4600. Smith, M.S., and Zimmerman, K., Nitrous oxide production by non-denitrifying soil nitrate reducers, *Soil Sci. Soc. Am. J.*, 45, 865, 1981.

4601. Smith, P.F., Mineral analysis of plant tissues, *Ann. Rev. Plant Physiol.*, 13, 81, 1962.

4602. Smith, R.E.H., and Kalff, J., Size-dependent phosphorus uptake kinetics and cell quota in phytoplankton, *J. Phycol.*, 18, 275, 1982.

4603. Smith, R.E.H., and Kalff, J., Competition for phosphorus among co-occurring freshwater phytoplankton, *Limnol. Oceanogr.*, 28, 448, 1983.

4604. Smith, S.V., Carbon dioxide dynamics: a record of organic carbon production, respiration, and calcification in the Eniwetok reef flat community, *Limnol. Oceanogr.*, 18, 106, 1973.

4605. Smith, V.H., The nutrient dependence of primary productivity in lakes, *Limnol. Oceanogr.*, 24, 1051, 1979.
4606. Smith, V.H., Light and nutrient dependence of photosynthesis by algae, *J. Phycol.*, 19, 306, 1983.
4607. Smith, V.H., Low nitrogen to phosphorus ratios favor dominance by blue-green algae in lake phytoplankton, *Science*, 221, 669, 1983.
4608. Smith, V.H., and Shapiro, J., Phosphorus-chlorophyll relation in individual lakes. Importance of lake restoration strategies, *Environ. Sci. Technol.*, 15, 444, 1981.
4609. Smith, W.O., Jr., The extracellular release of glycolic acid by a marine diatom, *J. Phycol.*, 10, 30, 1974.
4610. Smith, W.O., and Nelson, D.M., Importance of ice edge phytoplankton production in the Southern Ocean, *BioScience*, 36, 251, 1986.
4611. Snyder, R.L., and Boyd, C.E., Evapotranspiration by *Eichhornea crassipes* (Mart.) Solms and *Typha latifolia* L., *Aquat. Bot.*, 27, 217, 1987.
4612. Söderlund, S., Forsberg, Å., and Pedersen, M., Concentrations of cadmium and other metals in *Fucus vesiculosus* L. and *Fontinalis dalecarlica* Br. Eur. from the northern Baltic Sea and the southern Bothnian Sea, *Environ. Pollut.*, 51, 197, 1988.
4613. Söderlund, R., and Svensson, B.H., The global nitrogen cycle, in *Nitrogen, Phosphorus and Sulphur—Global Cycles,* Svensson, B.H., and Söderlund, R., Eds., SCOPE Report NO. 7, *Ecol. Bull. (Stockholm)*, 22, 23, 1976.
4614. Söderlund, R., and Rösswall, T., The nitrogen cycle, in *The Handbook of Environmental Chemistry* Vol. 1. Part B. *The Natural Environment and the Biogeochemical Cycles*, Hutzinger, O., Ed., Springer-Verlag, Berlin, 1982, 61.
4615. Soeder, C., and Stengel, E., Physico-chemical factors affecting metabolism and growth rate, in *Algal Physiology and Biochemistry,* Stewart, W.D.P., Ed., Blackwell Scientific Publications, Oxford, 1974, 714.
4616. Solander, D., Biomass and shoot production of *Carex rostrata* and *Equisetum fluviatile* in unfertilized and fertilized subarctic lakes, *Aquat. Bot.*, 15, 349, 1983.
4617. Solander, D., Production and nutrient content of macrophytes in a natural and a fertilized subarctic lake, Ph.D. Thesis, University of Uppsala, Sweden, 1983.
4618. Solomonson, L.P., Structure of *Chlorella* nitrate reductase, in *Nitrogen Assimilation of Plants*, Hewitt, E.J., and Cutting, C.V., Academic Press, New York and London, 1979, 199.
4619. Solomonson, L.P., and Spehar, A.M., Model for the regulation of nitrate assimilation, *Nature*, 265, 373, 1977.
4620. Solomonson, L.P., and Barber, M.J., Structure-function relationships of assimilatory nitrate reductase, in *Inorganic Nitrogen Metabolism,* Ullrich, W.R., Aparicio, P.J., Syrett, P.J., and Castillo, F., Eds., Springer-Verlag, Berlin, 1987, 71.
4621. Somers, G.F., and Brown, M., The affinity of trichomes of blue-green algae for calcium, *Estuaries*, 1, 17, 1978.
4622. Sommer, A.L., Copper as an essential for plant growth, *Plant Physiol.*, 6, 339, 1931.
4623. Sommer, A.L., and Lipman, C.B., Evidence on the indispensable nature of zinc and boron for higher green plants, *Plant Physiol.*, 1, 231, 1926.
4624. Sommer, A.L., and Booth, T.E., Meta- and pyrophosphate within the algal cell, *Plant Physiol.*, 13, 199, 1938.
4625. Sommer, U., The role of r- and K-selection in the succession of phytoplankton in Lake Constance, *Acta Oecol./Oecol. Gener.*, 2, 237, 1981.
4626. Sommer, U., Nutrient competition between phytoplankton species in multispecies chemostat experiment, *Arch. Hydrobiol.*, 96, 399, 1983.
4627. Søndegaard, M., and Sand-Jensen, K., Total autotrophic production in oligotrophic Lake Kalgaard, Denmark, *Verh. Internat. Verein. Limnol.*, 20, 667, 1978.
4628. Sørensen, J., Jørgensen, T., and Brandt, S., Denitrification in stream epilithon: seasonal variation in Galeak and Rabis Beak, *FEMS Microbiol. Ecol.*, 53, 345, 1988.
4629. Sorentino, C., The effects of heavy metals on phytoplankton: a review, *Phykos*, 18, 149, 1979.
4630. Sorokin, C., Tabular comparative data for the low- and high-temperature strains of *Chlorella, Nature*, 184, 613, 1959.
4631. Sorokin, C., Kinetic studies of temperature effects on the cellular level, *Biochim. Biophys. Acta*, 38, 197, 1960.
4632. Sorokin, C., Injury and recovery of photosynthesis. The capacity of cells of different developmental stages to regenerate their photosynthetic activity, *Physiol. Plant.*, 13, 20, 1960.
4633. Sorokin, C., and Krauss, R.W., Effects of temperature and illuminance on *Chlorella* growth uncoupled from cell division, *Plant Physiol.*, 37, 37, 1962.

4634. Sorsa, K., Primary production of epipelic algae in Lake Suomunjärvi, Finnish North Karelia, *Ann. Bot. Fennici*, 16, 351, 1979.

4635. Sournia, A., Catalogue des especes et taxons infraspécifique de dinoflagellés marins actuels. I. Dinoflagellés libres, *Beih. Nova Hedw.*, 48, 1, 1973.

4636. Sournia, A., Catalogue des especes et taxons infraspécifique de dinoflagellés marins actuels publiés dupluis la révision de J. Schiller. III. Complément, *Rev. Algol. N.S.*, 13, 3, 1978.

4637. Sournia, A., Ed., *Phytoplankton Manual*, UNESCO, Paris, 1981.

4638. Sournia, A., Morphological bases of competition and succession, in *Physiological Bases of Phytoplankton Ecology*, Platt, T., Ed., *Can. Bull. Fish. Aquat. Sci.*, 210,339, 1981.

4639. Sousa, W.P., Disturbance in marine intertidal boulder fields: the nonequilibrium maintenance of species diversity, *Ecology*, 60, 1225, 1979.

4640. Sousa, W.P., The responses of a community to disturbance: the importance of successional age and species' life histories, *Oecologia (Berlin)*, 45, 72, 1980.

4641. Sousa, W.P., Disturbance and patch dynamics on rocky intertidal shores, in *The Ecology of Natural Disturbance and Patch Dynamics*, Pickett, S.T.A., and White, P.S., Eds., Academic Press, Orlando, Florida, 1985, 101.

4642. South, G.R., and Hill, R.D., Studies on marine algae of Newfoundland. I. Occurrence and distribution of free-living *Ascophyllum nodosum* in Newfoundland, *Can. J. Bot.*, 48, 1697, 1970.

4643. South, G.R., and Whittick, A., *Introduction to Phycology*, Blackwell Scientific Publications, Oxford, 1987.

4644. Sozska, G.J., Ecological relations between invertebrates and submerged macrophytes in the lake littoral, *Ekol. Pol.*, 23, 393, 1975.

4645. Spalding, M.H., Photosynthesis and photorespiration in freshwater green algae, *Aquat. Bot.*, 34, 181, 1989.

4646. Spalding, M.H., Spreitzer, R.J., and Ogren, W.L., Carbonic anhydrase-deficient mutant of *Chlamydomonas reinhardtii* requires elevated carbon dioxide concentration for photoautotrophic growth, *Plant Physiol.*, 73, 268, 1983.

4647. Spangler, F., Fetter, C.W., and Sloey, W.E., Phosphorus accumulation discharge cycles in marshes, *Water Res, Bull.*, 13, 1191, 1977.

4648. Spear, D.G., Barr, J.K., and Barr, C.E., Localization of hydrogen ion and chloride ion fluxes in *Nitella*, *J. Gen. Physiol.*, 54, 397, 1969.

4649. Spence, D.H.N., The macrophytic vegetation of freshwater locks, swamps and associated fens, in *The Vegetation of Scotland*, Burnett, M.H., Ed., Oliver and Boyd, Edinburgh, 1964, 306.

4650. Spence, D.H.N., The zonation of plants in freshwater lakes, *Adv. Ecol. Res.*, 12, 37, 1982.

4651. Spencer, D., and Possingham, J.V., The effect of manganese deficiency of photophosphorylation and the oxygen evolving sequence in spinach chloroplasts, *Biochem. Biophys. Acta*, 52, 379, 1961.

4652. Spencer, D., and Possingham, J.V., Manganese as a functional component of chloroplasts, *Aust. J. Biol. Sci.*, 15, 58, 1962.

4653. Spencer, L.B., A study of *Vacuolaria virescens* Cienkowski, *J. Phycol.*, 7, 274, 1971.

4654. Spencer, N., Hyacinth spreading havoc over the world, *Florida Times—Union and Jacksonville Journal* No. 296, October 22, 1972. (in Rao 1988).

4655. Spiegelstein, M., Reich, K., and Bergmann, F., The toxic principles of *Ochromonas* and related Chrysomonadina, *Verh. Internat. Verein. Limnol.*, 17, 778, 1969.

4656. Spitznagel, J.K., and Sharp, D.G., Magnesium and sulfate ions as determinants in the growth and reproduction of *Mycobacterium bovis*, *J. Bacteriol.*, 78, 453, 1959.

4657. Spoehr, H.A., and Milner, H.W., The chemical composition of *Chlorella*; effect of environmental conditions, *Plant Physiol.*, 24, 120, 1949.

4658. Spoehr, H.A., Smit, J.H.C., Strain, H.H., Milner, H.W., and Hardin, G.J., Fatty acid antibacterials from plants, *Carneige Inst. Wash. Publ.* No. 586, 1, 1949.

4659. Sponsler, O.L., New data on cellulose space lattice, *Nature*, 125, 633, 1930.

4660. Sprenger, M., and McIntosh, A., Relationship between concentrations of aluminium, cadmium, lead and zinc in water, sediments and aquatic macrophytes in six acidic lakes, *Arch. Environ. Contam. Toxicol.*, 18, 225, 1989.

4661. Sprent, J.I., *The Ecology of the Nitrogen Cycle*, Cambridge University Press, Cambridge, 1987.

4662. Sprent, J.I., and Sprent, P., *Nitrogen Fixing Organisms. Pure and Applied Aspects*, Chapman and Hall, London, 1990.

4663. Spruit, C.J.P., Photoreduction and anaerobiosis, in *Physiology and Biochemistry of Algae*, Lewin, R.A., Ed., Academic Press, New York, 1962, 47.

4664. Squiers, E.R., and Good, R.E., Seasonal changes in the productivity, caloric content and chemical composition of a population of saltmarsh cordgrass (*Spartina alterniflora*), *Chesapeake Science*, 15, 63, 1974.

4665. Srinath, E.C., and Loehr, R.C., Ammonium desorption by diffused aeration, *J. Water Pollut. Control Fed.*, 46, 1939, 1974.

4666. Srivastava, H.S., Regulation of nitrate reductase activity in higher plants, *Phytochemistry*, 19, 725, 1980.

4667. Stabenau, H., Localization of enzymes of glycolate metabolism in the alga *Chlorogonium elongatum, Plant Physiol.*, 54, 921, 1974.

4668. Stacey, J.L., and Casselton, P.J., Utilization of adenine but not nitrate as nitrogen source by *Prototheca zopfii, Nature*, 211, 862, 1966.

4669. Stackhouse, J., *Nereis Britannica*, Bath, 1801.

4670. Stadtman, T.C., Biological function of selenium, *Trends Biochem. Sci.*, 5, 2083, 1980.

4671. Stainton, M.P., Errors of the molybdenum blue methods for determining orthophosphate in freshwater, *Can. J. Fish. Aquat. Sci.*, 37, 472, 1980.

4672. Stake, E., Higher vegetation and nitrogen in a rivulet in central Sweden, *Schweiz. Hydrol. Z.*, 29, 107, 1967.

4673. Stake, E., Higher vegetation and phosphorus in a small stream in central Sweden, *Schweiz. Z. Hydrol.*, 30, 353, 1968.

4674. Staley, J.T., The gas vacuole: an early organelle of prokaryote motility?, *Orig. Life*, 10, 111, 1980.

4675. Stam, W.T., and Holleman, H.C., The influence of different salinities on growth and morphological variability of a number of *Phormidium* strains (Cyanophyceae) in culture, *Acta Bot. Neerl.*, 24, 379, 1975.

4676. Stanek, W., Annotates bibliography of peatland forestry, Environment Canada Libraries Bibliography Ser. 76/1, Ottawa, 1976.

4677. Stanek, W., and Worley, I.A., A terminology of origin peat and peatlands, in *Proc. Int. Symp. Peat Utilization,* Fuchsam, C.H., and Spiragelli, S.A., Eds., Bemidji State University, Bemidji, Minnesota, 1983, 75.

4678. Stanford, G., Van der Pol, R.A., and Dzienia, S., Denitrification rates in relation to total and extractable soil carbon, *Soil Sci. Soc. Am. Proc.*, 39, 290, 1975.

4679. Stanford, G., Dzienia, S., and Van der Pol, R.A., Effects of temperature on denitrification in soils, *Soil Sci. Soc. Am. Proc.*, 39. 284, 1975.

4680. Stangenberg, M., The effects of *Microcystis aeruginosa* Kg. extracts on *Daphnia longispina* O.F. Müller and *Eucypris virens* Jurine, *Hydrobiologia*, 32, 81, 1968.

4681. Stanier, R.Y., Photosynthetic mechanisms in bacteria and plants: Development of a unitary concept, *Bact. Rev.*, 25, 1, 1961.

4682. Stanier, R.Y., and Cohen-Bazire, G., Phototrophic prokaryotes: Cyanobacteria. *Ann. Rev. MIcrobiol.*, 31, 225, 1977.

4683. Stanier, R.Y., Kunisawa, R., Mandel, M., and Cohen-Bazire, G., Purification and properties of unicellular blue-green algae (order Chroococcales), *Bacteriol. Rev*, 35, 171, 1971.

4684. Stanier, R.Y., Sistrom, W.R., Hansen, T.A., Whitton, B.A., Castenholz, R.W., Pfennig, N., Gorlenko, W.N., Kondratieva, E.N., Eimhjellen, K.E., Whittenburg, R.L., Gherna, R.L., and Trupper, H.G., Proposal to place the nomenclature of the cyanobacteria (blue-green algae) under the rules of Nomenclature of Bacteria, *Int. J. Syst. Bacteriol.*, 28, 335, 1978.

4685. Stanley, D.W., Productivity of epipelic algae in tundra ponds and a lake near Barrow, Alaska, *Ecology*, 57, 1015, 1976.

4686. Stanley, D.W., A carbon flow model of epipelic algae productivity in Alaskan tundra ponds, *Ecology*, 57, 1034, 1976.

4687. Stanley, D., and Hobbie, J., Nitrogen cycling in the Chowan River, *Water Resour. Res. Inst. Univ. N.C.*, 121, 1, 1977.

4688. Stark, L.M., Almodovar, L., and Krauss, R.W., Factors affecting the rate of calcification in *Halimeda opuntia* (L.) Lamouroux and *Halimeda discoidea* Decaisne, *J. Phycol.*, 5, 305, 1969.

4689. Starks, T.L., Shubert, L.E., and Trainor, F.R., Ecology of soil algae: a review, *Phycologia*, 20, 65, 1981.

4690. Starmach, K., Cyanophyta - Sinice, *Flora Sladkowodna Polski*, 2, 1, 1966.

4691. Starmach, K., Cryptophyceae, Dinophyceae, Raphidophyceae, *Flora Sladkowodna Polski*, 4, 1, 1974, (in Polish).

4692. Starmach, K., Phaeophyta - Brunatnice, Rhodophyta - Krasnorosty, *Flora Sladkowodna Polski*, 14, 1, 1977, (in Polish).

4693. Starmach, K., Chrysophyceae - Zlotowiciowce, *Flora Sladkowodna Polski*, 5, 1, 1980, (in Polish).

4694. Starmach, K., *Chrysophyceae und Haptophyceae,* Süsswasserflora von Mitteleuropa 1, Gustav Fisher Verlag, Stuttgart, New York, 1985.

4695. Starr, R.C., The culture collection of algae at the University of Texas in Austin, *J. Phycol.*, 14 (Suppl.), 47, 1978.

4696. Starý, J., and Kratzer, K., The cumulation of toxic metals on alga, *J. Environ. Anal. Chem.*, 12, 65, 1982.

4697. Starý, J., Havlík, B., Zeman, A., Kratzer, K., Prášilová, J., and Hanušová, J., Chromium (III) and chromium (VI) circulation in aquatic biocenoses, *Acta Hydrochim. Hydrobiol.*, 10, 251, 1982.

4698. Starý, J., Havlík, B., Kratzer, K., Prášilová, J., and Hanušová, J., Cumulation of zinc, cadmium, and mercury on the alga *Scenedesmus obliquus*, *Acta Hydrochim. Hydrobiol.*, 11, 401, 1983.

4699. Stauffer, R.E., and Lee, G.F., The role of thermocline migration in regulating algal blooms, in *Modelling the Eutrophication Processes*, Middlebrooks, E.J., et al., Eds., Ann Arbor Science Publishers, Ann Arbor, Michigan, 1973, 73.

4700. Staves, R.P., and Knaus, R.M., Chromium removal from water by the species of duckweed, *Aquat. Bot.*, 23, 261, 1985.

4701. Steele, J.H., Notes on some theoretical problems in production ecology, in *Primary Productivity in Aquatic Environments*, Goldman, R.C., Ed., (*Mem. Ist. Ital. Idrobiol.* 18 suppl.), University of California Press, Berkeley, 1965, 383.

4702. Steele, J.H., and Frost, B.W., The structure of plankton communities, *Phil. Trans. Royal Soc. Lond.*, B 280, 485, 1977.

4703. Steele, J.H., and Mullin, M.M., Zooplankton dynamics, in *The Sea*, Vol. 6, *Marine Modeling*, Goldberg, E.D., McCave, I.W., O'Brien, J.J., and Steele, J.D., Eds., John Wiley and Sons, New York, 1977, 857.

4704. Steeman-Nielsen, E., Photosynthesis of aquatic plants with special reference to carbon source, *Dansk. Bot. Ark.*, 12, 1, 1947.

4705. Steeman-Nielsen, E., On a complication in marine productivity work due to the influence of ultraviolet light, *J. Cons. Int. Explor. Mar.*, 24, 130, 1964.

4706. Steeman-Nielsen, E., The uptake of free CO_2 and HCO_3^- during photosynthesis of plankton algae with special reference to the coccolithophorid *Coccolithus huxleyi*, *Physiol. Plant.*, 19, 232, 1966.

4707. Steeman-Nielsen, E., and Kamp-Nielsen, L., Influence of deleterious concentrations of copper on the growth of *Chlorella pyrenoidosa*, *Physiol. Plant.*, 23, 828, 1970.

4708. Steeman-Nielsen, E., and Wium-Andersen, S., Copper ions as poison in the sea and in freshwater, *Mar. Biol.*, 6, 93, 1970.

4709. Steeman-Nielsen, E., and Wium-Andersen, S., The influence of Cu on photosynthesis and growth in diatoms, *Physiol. Plant.*, 24, 480, 1971.

4710. Steeman-Nielsen, E., and Wium-Andersen, S., Influence of copper on photosynthesis of diatoms, with special reference to an afternoon depression, *Verh. Internat. Verein. Limnol.*, 18, 78, 1972.

4711. Steeman-Nielsen, E., Kamp-Nielsen, L., and Wium-Andersen, S., The effect of deleterious concentrations of copper on the photosynthesis of *Chlorella pyrenoidosa*, *Physiol. Plant.*, 22, 1121, 1969.

4712. Štefanová, I., and Kalina, T., New and rare species of silica-scaled Chrysophytes from Eastern Bohemia (Czechoslovakia), *Arch. Protistensk.*, 142, 167, 1992.

4713. Stegmann, G., Die Bedeutung der Spurenelemente für *Chlorella*, *Z. Bot.*, 35, 385, 1940.

4714. Steidinger, K.A., Phytoplankton ecology: a conceptual review based on eastern Gulf of Mexico research. *CRC Crit. Rev. Microbiol.*, 3, 49, 1973.

4715. Steidinger, K.A., and Joyce, E.A., Florida red tides, *Florida Dept. Natl. Resour. Mar. Res. Lab. Educ. Ser.* 17, 1973.

4716. Steidinger, K.A., and Cox, E.R., Free-living dinoflagellates, in *Phytoflafellates, Developments in Marine Biology*, Vol. 2, Cox, E.R., Ed., Elsevier/North Holland, New York, 1980, 407.

4717. Stein, J.R., Ed., *Handbook of Phycological Methods*, Cambridge University Press, Cambridge, 1973.

4718. Stein, J.R., and Brooke, R.C., Red snow from Mt. Seymour, British Columbia, *Can. J. Bot.*, 42, 1183, 1964.

4719. Stein, J.R., and Borden, C.A., Algae in medicine: introduction and bibliography, in *Selected Papers in Phycology II*, Rosowski, J.R., and Parker, B.C., Eds., Phycological Society of America, Lawrence, Kansas, 1982, 788.

4720. Stein, J.R., and Borden, C.A., Causative and beneficial algae in human disease conditions: a review, *Phycologia*, 23, 485, 1984.

4721. Steinberg, R.A., Relation of accessory growth substances to heavy metals including molybdenum, in the nutrition of *Aspergillus niger*, *J. Agr. Res.*, 52, 438, 1936.

4722. Steinberg, R.A., Specificity of potassium and magnesium for the growth of *Aspergillus niger*, *Am. J. Bot.*, 33, 210, 1946.

4723. Steiner, M., Zur Kenntnis des Phosphatkreislaufes in Seen, *Naturwissenschaften*, 26, 723, 1938.

4724. Steinman, A.D., Effects of herbivore size and hunger level on periphyton communities, *J. Phycol.*, 27, 54, 1991.

4725. Steinman, A.D., and McIntire, C.D., Effects of current velocity and light energy on the structure of periphyton assemblages in laboratory streams, *J. Phycol.*, 22, 352, 1986.

4726. Steinman, A.D., and McIntire, C.D., Effects of irradiance on the community structure and biomass of algal assemblages in laboratory streams, *Can. J. Fish. Aquat. Sci.*, 44, 1640, 1987.

4727. Steinman, A.D., McIntire, C.D., Gregory, S.V., Lamberti, G.A., and Ashkenas, L.R., Effects of herbivore type and density on taxonomic structure and physiognomy of algal assemblages in laboratory streams, *J. N. Am. Benthol. Soc.*, 6, 175, 1987.

4728. Steinman, A.D., McIntire, C.D., and Lowry, R.R., Effects of irradiance and age on chemical constituents of algal assemblages in laboratory streams, *Arch. Hydrobiol.*, 114, 45, 1988.

4729. Steinman, A.D., McIntire, C.D., Gregory, S.V., and Lamberti, G.A., Effects of irradiance and grazing on lotic algal assemblages, *J. Phycol.*, 25, 478, 1989.

4730. Stengel, E., Zustandsänderungen verschiedener Eisenverbindungen in Nährlösungen für Algen, *Arch. Hydrobiol. Suppl.*, 38, 151, 1970.

4731. Stenner, R.D., and Nickless, G., Distribution of some heavy metals in organisms in Hardangerfjord and Skjerstadfjord, Norway, *Water Air Soil Pollut.*, 3, 279, 1974.

4732. Stenner, R.D., and Nickless, G., Heavy metals in organisms of the Atlantic coast of south-west Spain and Portugal, *Mar. Pollut. Bull.*, 6, 89, 1975.

4733. Stephenson, W.A., *Seaweed in Agriculture and Horticulture*, Faber and Faber, London, 1968.

4734. Stephenson, W.A., *Seaweed in Agriculture and Horticulture*, 3rd ed., Rateaver Publ., Valley, California, 1974.

4735. Stepka, W., Benson, A.A., and Calvin, M., The path of carbon in photosynthesis: II. Amino acids, *Science*, 108, 304, 1948.

4736. Stepniewski, W., and Glinski, J., Gas exchange and atmospheric properties of flooded soils, in *The Ecology and Management of Wetlands*, Part 1. *Ecology of Wetlands*, Hook, D.D. et al., Eds., Timber Press, Portland, Oregon, 1988, 269.

4737. Stevens, S.E., and Porter, R.D., Transformation in *Agmenellum quadruplicatum*, *Proc. Natl. Acad. Sci. U.S.A.*, 77, 6052, 1980.

4738. Stevenson, F.J., *Cycles of Soil. Carbon, Nitrogen, Phosphorus, Sulfur, Micronutrients*, John Wiley and Sons, New York, 1986.

4739. Stevenson, R.J., Effects of current and conditions simulating autogenically changing microhabitats on benthic diatom immigration, *Ecology*, 64, 1514, 1983.

4740. Stevenson, R.J., How currents on different sides of substrates in streams affect mechanisms of benthic algal accumulation, *Int. Rev. Ges. Hydrobiol.*, 69, 241, 1984.

4741. Stevenson, R.J., Benthic algal community dynamics in a stream during and after a spate, *J. N. Am. Benthol. Soc.*, 9, 277, 1990.

4742. Stevenson, R.J., and Lowe, R.L., Sampling and interpretation of algal patterns for water quality assessments, in *Rationale for Sampling and Interpretation of Ecological Data in the Assessment of Freshwater Ecosystems*, Isom, B.G., Ed., ASTM STP 894, American Society for Testing and Materials, Philadelphia, 1986, 118.

4743. Stevenson, R.J., and Peterson, C.G., Variation in benthic diatom (Bacillariophyceae) immigration with habitat characteristics and cell morphology, *J. Phycol.*, 25, 120, 1989.

4744. Stevenson, R.J., and Peterson, C.G., Emigration and immigration can be important determinants of benthic diatom assemblages in streams, *Freshwater Biol.*, 26, 279, 1991.

4745. Stevenson, R.J., Singer, R., Roberts, D.A., and Boylen, C.W., Patterns of epipelic algal abundance with depth, trophic status, and acidity in poorly buffered New Hampshire Lakes, *Can. J. Fish. Aquat. Sci.*, 42, 1501, 1985.

4746. Stevenson, R.J., Peterson, C.G., Kirschtel, D.B., King, C.C., and Tuchman, N.C., Density-dependent growth, ecological strategies, and effects of nutrients and shading on benthic diatom succession in streams, *J. Phycol.*, 27, 59, 1991.

4747. Steward, K.K., and Ornes, W.H., The autoecology of sawgrass in the Florida Everglades, *Ecology*, 56, 162, 1975.

4748. Steward, K.K., and Ornes, W.H., Mineral nutrition of sawgrass (*Cladium jamaicense* Crantz) in relation to nutrient supply, *Aquat. Bot.*, 16, 349, 1983.

4749. Steward, F.C., Ed., *Plant Physiology, A Treatise*, Vol. 3, *Inorganic Nutrition of Plants*, Academic Press, New York and London, 1963.

4750. Steward, F.C., and Pollard, J.K., Nitrogen metabolism in plants: ten years in retrospect, *Ann. Rev. Plant Physiol.*, 8, 65, 1957.

4751. Stewart, F.M., and Levin, B.R., Partitioning of resources and the outcome of interspecific competition: a model and some general considerations, *Am. Nat.*, 107, 171, 1973

4752. Stewart, J.G., Effects of lead on the growth of four species of red algae, *Phycologia*, 16, 31, 1977.

4753. Stewart, K.D., and Mattox, K.R., Comparative cytology, evolution and classification of the green algae with some consideration of the origin of other organisms with chlorophylls *a* and *b*, *Bot. Rev.*, 41, 104, 1975.

4754. Stewart, K.D., and Mattox, K.R., Structural evolution in the flagellated cells of green algae and land plants, *BioSystems,* 10, 145, 1978.

4755. Stewart, K.D., and Mattox, K., Phylogeny of phytoflagellates, in *Phytoflagellates,* Cox, E.R., Ed., Elsevier/North Holland, New York, 1980, 433.

4756. Stewart, K.D., and Mattox, K.R., The case for a polyphyletic origin if mitochondria: morphological and molecular comparisons, *J. Mol. Ecol.,* 21, 54, 1984.

4757. Stewart, R.E., and Kantrud, H.A., Classification of natural ponds and lakes in the glaciated prairie region, *U.S. Bur. Sport Fish Wildl. Resour. Pub.* 92, 1971.

4758. Stewart, W.D.P., Fixation of elemental nitrogen by marine blue-green algae, *Ann. Bot. N.S.,* 26, 439, 1962.

4759. Stewart, W.D.P., Liberation of extracellular nitrogen by two nitrogen-fixing blue-green algae, *Nature,* 200, 1020, 1963.

4760. Stewart, W.D.P., Nitrogen turnover in marine and brackish habitats. 1. Nitrogen fixation, *Ann. Bot. N.S.,* 29, 229, 1965.

4761. Stewart, W.D.P., *Nitrogen Fixation in Plants,* Athlone Press, London, 1966.

4762. Stewart, W.D.P., Nitrogen input into aquatic ecosystems, in *Algae, Man and The Environment,* Jackson, D.F., Ed., Syracuse University Press, Syracuse, New York, 1968, 53.

4763. Stewart, W.D.P., Biological and ecological aspects of nitrogen fixation by free-living microorganisms, *Proc. R. Soc. Lond.,* B 172, 367, 1969.

4764. Stewart, W.D.P., Algal fixation of atmospheric nitrogen, *Plant and Soil.,* 32, 555, 1970.

4765. Stewart, W.D.P., Nitrogen fixation by blue-green algae in Yellowstone thermal areas, *Phycologia,* 9, 261, 1970.

4766. Stewart, W.D.P., Physiological studies on nitrogen-fixing blue-green algae, *Plant and Soil,* Spec. Vp., 377, 1971.

4767. Stewart, W.D.P., Nitrogen fixation in the sea, in *Fertility of the Sea,* Costlow, J.D., Ed., Gordon and Breach Science Publishers, London, 1971, 537.

4768. Stewart, W.D.P., Nitrogen fixation in photosynthetic microorganisms, *Ann. Rev. Microbiol.,* 27, 283, 1973.

4769. Stewart, W.D.P., Nitrogen fixation, in *The Biology of Blue-green Algae,* Carr, N.G., and Whitton, B.A., Eds., University of California Press, Berkeley, 1973, 260.

4770. Stewart, W.D.P., Ed., *Algal Physiology and Biochemistry* Blackwell Scientific Publications, Oxford, 1974.

4771. Stewart, W.D.P., Blue-green algae, in *The Biology of Nitrogen Fixation,* Quispel, A., Ed., North Holland Research Monographs, Frontiers in Biology, 33, 1974, 202.

4772. Stewart, W.D.P., Ed., *Nitrogen Fixation by Free-living Micro-organisms,* Cambridge University Press, Cambridge, 1975.

4773. Stewart, W.D.P., A botanical ramble among the blue-green algae, *Br. Phycol. J.,* 12, 89, 1977.

4774. Stewart, W.D.P., Some aspects of structure and function in N_2-fixing cyanobacteria, *Bacteriol. Rev.,* 37, 32, 1980.

4775. Stewart, W.D.P., and Lex, M., Nitrogenase activity in the blue-green alga *Plectonema boryanum* Strain 594, *Arch. Mikrobiol.,* 73, 250, 1970.

4776. Stewart, W.D.P., and Pearson, H.W., Effects of aerobic and anaerobic conditions on growth and metabolism of blue-green algae, *Proc. R. Soc. Lond.,* B 175, 293, 1970.

4777. Stewart, W.D.P., and Alexander, G., Phosphorus availability and nitrogenase activity in aquatic blue-green algae, *Freshwater Biol.,* 1, 389, 1971.

4778. Stewart, W.D.P., and Rowell, P., Effects of L-methionine-DL-sulphoximine on the assimilation of newly fixed NH_3, acetylene reduction and heterocyst production in *Anabaena cylindrica, Biochem. Biophys. Res. Comm.,* 65, 846, 1975.

4779. Stewart, W.D.P., and Daft, M.J., Microbial pathogens of cyanophycean blooms, in *Adv. Aquat. Microbiol.,* 1, 177, 1977.

4780. Stewart, W.D.P., Fitzgerald, G.P., and Burris, R.H., Acetylene reduction by nitrogen-fixing blue-green algae, *Arch. Mikrobiol.,* 62, 336, 1968.

4781. Stewart, W.D.P., Fitzgerald, G.P., Burris, R.H., Acetylene reduction assay for determination of phosphorus availability in Wisconsin lakes, *Proc. Nat. Acad. Sci. U.S.A.,* 66, 1104, 1970.

4782. Stewart, W.D.P., Mague, T., Fitzgerald, G.P., and Burris, R.H., Nitrogenase activity in Wisconsin lakes of differing degrees of eutrophication, *New Phytol.,* 70, 497, 1971.

4783. Stiles, W., Other elements, in *Handbuch der Pflanzenphysiologie,* Vol. IV, Ruland, W., Ed., Springer-Verlag, Berlin, 1958, 599.

4784. Stock, M.S., and Ward, A.K., The establishment of a bedrock epilithic community in a small stream: microbial (algal and bacterial) metabolism and physical structure, *Can. J. Fish. Aquat. Sci.,* 46, 1874, 1989.

4785.   Stock, M.S., and Ward, A.K., Blue-green algal mats in a small stream, *J. Phycol.*, 27, 692, 1991.
4786.   Stock, M.S., Richardson, T.D., and Ward, A.K., Distribution and primary productivity of the epizoic macroalga *Boldia erythrosiphon* (Rhodophyta) in a small Alabama stream, *J. N. Am. Benthol. Soc.*, 6, 168, 1987.
4787.   Stockner, J.G., and Shortreed, K.R.S., Autotrophic production in Carnation Creek, a coastal rainforest stream on Vancouver Island, British Columbia, *J. Fish. Res. Board Can.*, 33, 1553, 1976.
4788.   Stockner, J.G., and Shortreed, K.R.S., Enhancement of autotrophic production by nutrient addition in a coastal rainforest stream on Vancouver Island, *J. Fish. Res. Bd. Can.*, 35, 28, 1978.
4789.   Stoermer, E.F., Pankratz, H.S., and Bowen, C.C., Fine structure of the diatom *Amphipleura pellucida*. II. Cytoplasmic fine structure and frustule formation, *Am. J. Bot.*, 52, 1067, 1965.
4790.   Stokes, P.M., Uptake and accumulation of copper and nickel by metal-tolerant strains of *Scenedesmus*, *Verh. Internat. Verein. Limnol.*, 19, 2128, 1975.
4791.   Stokes, P.M., Copper accumulation in freshwater biota, in *Copper in the Environment*, Part 1. *Ecological Cycling*, Nriagu, J.O., Ed., John Wiley and Sons, New York, 1979, 357.
4792.   Stokes, P.M., Responses of freshwater algae to metals, in *Progress in Phycological Research*, Vol. 2, Round, F.E., and Chapman, D.J., Eds., Elsevier, Amsterdam, 1983, 87.
4793.   Stokes, P.M., Hutchinson, T.C., and Krauter, K., Heavy metal tolerance in algae isolated from contaminated lakes near Sudbury, Ontario, *Can. J. Bot.*, 51, 2155, 1973.
4794.   Stokes, P.M., Maler, T., and Riordan, A.R., A low molecular weight copper binding protein in a copper tolerant strain of *Scenedesmus acutiformis*, in *Trace Substances in Environmental Health*, Vol. 11, Hemphill, D.D., Ed., University of Missouri, Columbia, 1977, 146.
4795.   Storch, T.A., and Dietrich, G.A., Seasonal cycling of algal nutrient limitation in Chatauqua Lake, New York, *J. Phycol.*, 15, 399, 1979.
4796.   Storch, T.A., and Dunham, V.L., Iron-mediated changes in the growth of Lake Erie phytoplankton and axenic algal cultures, *J. Phycol.*, 22, 109, 1986.
4797.   Strain, H.H., Cope, B.T., and Svec, W.A., Procedures for the isolation, identification, estimation and investigation of the chlorophylls, in *Methods in Enzymology*, XXIII., San Pietro, A., Ed., Academic Press, New York and London, 1971, 452.
4798.   Straškraba, M., Share of the littoral region in the productivity of two fishponds in southern Bohemia, *Rozpr. České Akad. Věd., Mat.-Přír Věd Series*, 73, 1, 1963.
4799.   Stratton, F.E., Ammonia nitrogen losses from streams, *J. Sanit. Eng. Div. Proc. Am. Soc. Civ. Eng.*, 94, 1085, 1968.
4800.   Stratton, G.W., and Corke, C.T., The effect of cadmium ion on the growth, photosynthesis and nitrogenase activity of *Anabaena inaequalis*, *Chemosphere*, 8, 227, 1979.
4801.   Stratton, G.W., Huber, A.L., and Corke, C.T., Effect of mercuric ion on the growth, photosynthesis and nitrogenase activity of *Anabaena inaequalis*, *Appl. Environ. Microbiol.*, 38, 537, 1979.
4802.   Strayer, R.F., and Tiedje, J.M., Kinetic parameters of the conversion of the methane precursors to methane in a hypereutrophic lake sediment, *Appl. Environ. Microbiol.*, 36, 330, 1978.
4803.   Strickland, J.D.H., Measuring the production of marine phytoplankton, *Bull. Fish. Res. Board Can.*, 167, 1, 1960.
4804.   Strickland, J.D.H., Research on the marine planktonic food web at the Institute of Marine Resources: a review of the past seven years of work, *Oceanogr. Mar. Biol. Rev.*, 10, 349, 1972.
4805.   Strickland, J.D.H., and Solórzano, L., Determination of monoesterase hydrolyzable phosphate and phosphomonoesterase activity in sea water, in *Some Contemporary Studies in Marine Science*, Barnes, H., Ed., Allen and Unwin, London, 1966, 665.
4806.   Strickland, J.D.H., and Parsons, T.R., A manual of seawater (with special reference to common micronutrients and to particulate organic material), *Fish. Res. Board Can. Bull.*, 125, 1965.
4807.   Strickland, J.D.H., and Parsons, T.R., *A Practical Handbook of Seawater Analysis*, 2nd ed., *Fish. Res. Board Can. Bull.*, 167, Queen's Printer, Ottawa, 1972.
4808.   Strömgren, T., The effect of lead, cadmium, and mercury on the increase in length of five intertidal Fucales, *J. Exp. Mar. Biol. Ecol.*, 43, 107, 1980.
4809.   Stross, R.G., Nitrate preference in *Haematococcus* as controlled by strain, age of inoculum, and pH of the medium, *Can. J. Microbiol.*, 9, 33, 1963.
4810.   Stross, R.G., Density and boundary regulation of the *Nitella* meadow in Lake George, New York, *Aquat. Bot.*, 6, 285, 1979.
4811.   Stross, R.G., and Pemrick, S.M., Nutrient uptake kinetics in phytoplankton: a basis for niche separation, *J. Phycol.*, 10, 164, 1974.
4812.   Stroud, L.M., Net primary production of belowground material and carbohydrate patterns of two height forms of *Spartina alterniflora* in two North Carolina marshes, Ph.D. Thesis, North Carolina State University, Raleigh, North Carolina, 1976.

4813. Stroud, L.M., and Cooper, A.W., Color-infrared aerial photographic interpretation and net primary productivity of a regularly flooded North Carolina salt marsh, Water Resources Inst. University of North Carolina Report No. 14, Raleigh, 1968.

4814. Stubbs, J., A challenging opportunity in forest management: wetland forest, *Forest Farmer,* 21/11, 6, 10, 1972.

4815. Studer, C., and Brändle, R., Sauerstoffkonzum und Versorgung der Rhizome von *Acorus calamus* L., *Glyceria maxima* (Hartman) Holmberg, *Mempanthes trifoliata* L., *Phalaris arundinacea* L., *Phragmites comunis* Trin. und *Typha latifolia* L., *Bot. Helvetica,* 94, 23, 1984.

4816. Stulp, B.K., *Morphological and Molecular Approaches to the Taxonomy of the Genus* Anabaena *(Cyanophyceae, Cyanobacteria),* Groningen, The Netherlands, 1983.

4817. Stumm, W., The acceleration of the hydrogeochemical cycling of phosphorus, *Water Res.,* 7, 131, 1973.

4818. Stumm, W., and Bilinski, H., Trace metals in natural waters; difficulties of their interpretation arising from our ignorance of their speciation, in *Proc. Int. Conf. Advances in Water Pollution Research,* Jenkins, S.D., Ed., Jerusalem, 1972, 39.

4819. Stumm, W., and Morgan, J.J., *Aquatic Chemistry,* Wiley-Interscience, New York, 1970.

4820. Stuve, J., and Galle, P., Role of mitochondria in the handling of gold by the kidney, *J. Cell Biol.,* 44, 667, 1970.

4821. Styron, C.E., Hagan, T.M., Campbell, D.R., Harvin, J., Whittenburg, N.K., Baughman, G.A., Bransford, M.E., Saunders, W.H., Williams, O.C., Woodle, C., Cixon, N.K., and McNeill, C.R., Effect of temperature and salinity on growth and uptake of ^{65}Zn and ^{137}Cs for six marine algae, *J. Mar. Biol. Ass. U.K.,* 56, 13, 1976.

4822. Subba Rao, D.V., Effect of boron on primary production of nanoplankton, *Can. J. Fish. Aquat. Sci.,* 38, 52, 1981.

4823. Suckcharoen, S., *Ceratophyllum demersum* as an indicator of mercury contamination in Thailand and Finland, *Ann. Bot. Fennici,* 16, 173, 1979.

4824. Sukenik, A., Bennett, J., and Falkowski, P.G., Light-saturated photosynthesis—limitation by electron or carbon fixation? *Biochim. Biophys. Acta,* 891, 205, 1987.

4825. Sullivan, C.W., A silicic acid requirement for DNA polymerase, thymidylate kinase and DNA synthesis in the marine diatom *Cylindrotheca fusiformis,* Ph.D. Thesis, University of California, San Diego, California, 1971.

4826. Sullivan, C.W., Diatom mineralization of silica acid. I. Si(OH)$_4$ transport characteristics in *Navicula pelliculosa, J. Phycol.,* 12, 390, 1976.

4827. Sullivan, C.W., Diatom mineralization of silica acid. II. Regulation of Si(OH)$_4$ transport rates during the cell cycle of *Navicula pelliculosa, J. Phycol.,* 13, 86, 1977.

4828. Sullivan, C.W., and Volcani, B.E., Synergistically stimulated (Na$^+$, K$^+$)-adenosine triphosphatase from plasma membranes of a marine diatom, *Proc. Nat. Acad. Sci. U.S.A.,* 71, 4376, 1974.

4829. Sullivan, C.W., and Volcani, B.E., Silicon in the cellular metabolism of diatoms, in *Silicon and Siliceous Structures in Biological Systems,* Simpson, T.L., and Volcani, B.E., Eds., Springer-Verlag, New York, 1981, 15.

4830. Sullivan, C.W., Palmisano, A.C., Kottmeier, S., Grossi, S.M., Moe, R., and Taylor, G.T., The influence of light on development and growth of sea ice microbial communities (SIMCO) in McMurdo Sound, Antarctica, *Antarct. J. U.S.,* 18, 177, 1983.

4831. Sullivan, M.J., Diatom communities from a Delaware salt marsh, *J. Phycol.,* 11, 384, 1975.

4832. Sullivan, M.J., Long-term effects of manipulating light intensity and nutrient enrichment on the structure of a salt marsh diatom community, *J. Phycol.,* 12, 205, 1976.

4833. Sullivan, M.J., Edaphic diatom communities associated with *Spartina alterniflora* and *S. patens* in New Jersey, *Hydrobiologia,* 52, 207, 1977.

4834. Sullivan, M.J., Diatom community structure: taxonomic and statistical analysis of a Mississippi salt marsh, *J. Phycol.,* 14, 468, 1978.

4835. Sullivan, M.J., Distribution of edaphic diatoms in a Mississippi salt marsh: a canonical correlation analysis, *J. Phycol.,* 18, 130, 1982.

4836. Sullivan, M.J., and Montcreiff, C.A., Primary productivity of edaphic algal communities in a Mississippi salt marsh, *J. Phycol.,* 24, 49, 1988.

4837. Sumar, N., Casselton, P.J., McNally, S.F., and Stewart, G.R., Occurrence of isoenzymes of glutamine synthetase in the alga *Chlorella kessleri, Plant Physiol.,* 74, 204, 1984.

4838. Sumner, W.T., and McIntire, C.D., Grazer-periphyton interactions in laboratory streams, *Arch. Hydrobiol.,* 93, 135, 1982.

4839. Summons, R.E., and Osmond, C.B., Nitrogen assimilation in the symbiotic marine alga *Gymnodinium microadriaticum*: direct analysis of ^{15}N incorporation by GC-MS methods, *Phytochemistry* 20, 575, 1981.

4840. Sunda, W.G., and Guillard, R.R.L., The relationship between cupric ion activity and the toxicity of copper to phytoplankton, *J. Mar. Res.,* 34, 511, 1976.

4841. Sunda, W.G., and Lewis, J.M., Effect of complexation by natural organic ligands on the toxicity of copper to a unicellular alga *Monochrysis lutheri, Limnol. Oceanogr,* 23, 870, 1978.

4842. Sunda, W.G., and Gillepsie, P.A., The response of a marine bacterial clone to cupric ion and its use to estimate cupric activity in seawater, *J. Mar. Res.,* 37, 761, 1979.

4843. Sunda, W.G., and Huntsman, S.A., Regulation of cellular manganese and manganese transport rates in the unicellular alga *Chlamydomonas, Limnol. Oceanogr.,* 30, 71, 1985.

4844. Sunda, W.G., and Huntsman, S.A., Relationships among growth rate, cellular manganese concentrations and manganese transport kinetics in estuarine and oceanic species of the diatom *Thalassiosira, J. Phycol.,* 22, 259, 1986.

4845. Susor, W.A., and Krogmann, D.W., Triphosphopyridine nucleotide photoreduction with cell-free preparations of *Anabaena variabilis, Biochim. Biophys. Acta,* 120, 67, 1966.

4846. Suttle, C.A., and Harrison, P.J., Rapid ammonium uptake by freshwater phytoplankton, *J. Phycol.,* 24, 13, 1988.

4847. Sutton, D.L., and Ornes, W.H., Phosphorus removal from static sewage effluent using duckweed, *J. Environ. Qual.,* 4, 367, 1975.

4848. Svedelius, N., Über den bau und die Entwicklung der Florideengattung *Martenisa,Vet. Akad. Handl. Bot.,* 43, Stockholm, 1908.

4849. Swaine, D.J., The trace-element content of soils, Commonwealth Bureau of Soil Science, Commonwealth Agricultural Bureaux, England, Tech, Commun. 48, 1955.

4850. Sweeney, B.M., Red tides, *Nat. Hist.,* 85, 78, 1976.

4851. Swift, D., Vitamins and phytoplankton growth, in T*he Physiological Ecology of Phytoplankton,* Morris, I., Ed., University of California Press, Berkeley, 1980, 329.

4852. Swift, E., 5th., Taylor, W.R., The effect of pH on the division rate of the coccolithophorid, *Crisosphaera elongata, J. Phycol.,* 2, 121, 1966.

4853. Syers, J.K., Harris, R.F., and Armstrong, D.E., Phosphate chemistry in lake sediments, *J. Environ. Qual.,* 2, 1, 1973.

4854. Sylvester, R.O., and Anderson, G.C., A lake's response to its environment, *J. Sanit. Eng. Div. Proc. Am. Soc. Div. Eng.,* 90, 1, 1964.

4855. Syrett, P.J., The assimilation of ammonia by nitrogen-starved cells of *Chlorella vulgaris.* Part 1. The correlation of assimilation with respiration, *Ann. Bot. N.S.,* 17, 1, 1953.

4856. Syrett, P.J., The assimilation of ammonia by nitrogen-starved cells of *Chlorella vulgaris.* Part 2. The assimilation of ammonia to other compounds, *Ann. Bot. N.S.,* 17, 21, 1953.

4857. Syrett, P.J., The assimilation of ammonia and nitrate by nitrogen-starved cells of *Chlorella vulgaris.* I. The assimilation of small quantities of nitrogen, *Physiol. Plant.,* 8, 924, 1955.

4858. Syrett, P.J., The assimilation of ammonia and nitrate by nitrogen-starved cells of *Chlorella vulgaris.* II. The assimilation of large quantities of nitrogen, *Physiol. Plant.,* 9, 19, 1956.

4859. Syrett, P.J., The assimilation of ammonia and nitrate by nitrogen-starved cells of *Chlorella vulgaris.* III. Difference of metabolism dependent on the nature of the nitrogen source, *Physiol. Plant.,* 9, 28, 1956.

4860. Syrett, P.J., Nitrogen assimilation, in *Physiology and Biochemistry of Algae,* Lewin, R.A., Ed., Academic Press, New York and London, 1962, 171.

4861. Syrett, P.J., Nitrogen metabolism of microalgae, in *Physiological Bases of Phytoplankton Ecology,* Platt, T., Ed., *Can. Bull. Fish. Aquat. Sci.,* 210, 182, 1981.

4862. Syrett, P.J., and Morris, I., the inhibition of nitrate assimilation by ammonium in *Chlorella, Biochim. Biophys. Acta,* 67, 566, 1963.

4863. Syrett, P.J., and Wong, H.A., The fermentation of glucose by *Chlorella vulgaris, Biochem. J.,* 89, 308, 1963.

4864. Syrett, P.J., and Leftley, J.W., Nitrate and urea assimilation by algae, in *Perspectives in Experimental Biology,* Vol. 2, *Botany,* Sunderland, N., Ed., Pergamon Press, Oxford, 1976, 221.

4865. Syrett, P.J., and Bekheet, I.A., The uptake of thiourea by *Chlorella, New Phytol.,* 79, 291, 1977.

4866. Syrett, P.J., and Peplinska, A.M., The effect of nickel and nitrogen deprivation on the metabolism of urea by the diatom *Phaeodactylum tricornutum, Br. Phycol. J.,* 23, 387, 1988.

4867. Syrett, P.J., Flynn, K.J., Molloy, C.J., Dixon, G.K., Peplinska, A.M., and Cresswell, R.C., Effects of nitrogen deprivation on rates of uptake of nitrogenous compounds by the diatom *Phaeodactylum tricornutum* Bohlin, *New Phytol.,* 102, 39, 1986.

4868. Sze, P., *A Biology of the Algae,* Wm. C. Brown Publishers, Dubuque, 1986.

4869. Taasen, J.P., Remarks on the epiphytic diatom flora of *Dumontia incrassata* (Müll.) Lamour. (Rhodophyceae), *Sarsia,* 55, 129, 1974.

4870.   Täckholm, V., Ancient Egypt, landscape, flora and agriculture, in *The Nile, Biology of an Ancient River,* Rzóska, J., Ed., Dr. W. Junk Publishers, The Hague, 1976, 51.

4871.   Tadano, T., and Yoshida, S., Chemical changes in submerged soils and their effect on rice growth, in *Soils and Rice,* Internat. Rice Res. Inst., Los Baños, Philippines, 1978, 399.

4872.   Tadros, M.G., and Johansen, J.R., Physiological characterization of six lipid-producing diatoms from the south-eastern United States, *J. Phycol.,* 24, 445, 1988.

4873.   Taft, C.E., History of *Cladophora* in the Great Lakes, in *Cladophora in the Great Lakes,* Shear, H., and Konasewich, D.E., Eds., Great Lakes Research Advisory Board, Int. Joint Commission, Windsor, Ontario, 1975, 9.

4874.   Taft, C.E., and Kishler, W.J., Algae from western Lake Erie, *Ohio J. Sci.,* 68, 80, 1968.

4875.   Taft, C.E., and Kishler, W.J., *Cladophora* as related to pollution and eutrophication in western Lake Erie, Water Resour. Center Report No. 332X, 339X, Ohio State University, Columbus, 1973.

4876.   Tagawa, S., and Kojima, Y., Arsenic content and its seasonal variation in seaweed, *J. Shimonoseki Univ. Fish.,* 25, 67, 1976.

4877.   Takahashi, E., *Electron Microscopical Studies on the Synuraceae (Chrysophyceae) in Japan - Taxonomy and Ecology,* Tokai University Press, Tokyo, 1978.

4878.   Takahashi, E., and Hayakawa, T., The Synuraceae (Chrysophyceae) in Bangladesh, *Phykos,* 18, 129, 1979.

4879.   Takahashi, M., Shimura, S., Yamaguchi, Y., and Fujita, P., Photoinhibition of phytoplankton photosynthesis as a function of exposure time, *J. Oceanogr. Soc. Jap.,* 27, 43, 1971.

4880.   Takai, Y., and Uehara, Y., Nitrification and denitrification in the surface layer of submerged soils, *J. Sci. Soil Manure,* 44, 463, 1973.

4881.   Takamura, N., and Iwakuma, T., Nitrogen uptake and C:N:P ratio of epiphytic algae in the littoral zone of Lake Kasumigaura, *Arch. Hydrobiol.,* 121, 161, 1991.

4892.   Takashima, K., Ishikawa, I.S., and Hase, E., Further notes on the growth and chlorophyll formation of *Chlorella protothecoides, Plant Cell Physiol.,* 5, 321, 1964.

4883.   Takatori, S., and Imahori, K., Light reactions in the control of oospore germination in *Chara delicatula, Phycologia,* 10, 221, 1971.

4884.   Talarico, L., R-phycoerythrin from *Audouinella saviana* (Nemaliales, Rhodophyta). Ultrastructural and biochemical analysis of aggregates and subunits, *Phycologia,* 29, 292, 1990.

4885.   Talling, J.F., Photosynthetic characteristic of some freshwater plankton diatoms in relation to underwater radiation, *New Phytol.,* 56, 29, 1957.

4886.   Talling, J.F., Photosynthesis under natural conditions, *Ann. Rev. Plant Physiol.,* 12, 133, 1961.

4887.   Talling, J.F., The photosynthetic activity of phytoplankton in East African lakes, *Int. Rev. Ges. Hydrobiol.,* 50, 1, 1965.

4888.   Talling, J.F., The underwater light climate as a controlling factor in the production ecology of freshwater phytoplankton, *Mitt. Internat. Verein. Limnol.,* 19, 214, 1971.

4889.   Talling, J.F., The depletion of carbon dioxide from lake water by phytoplankton, *J. Ecol.,* 64, 79, 1976.

4890.   Talling, J.F., Inorganic carbon reserves of natural waters and ecophysiological consequences of their photosynthetic depletion: microalgae, in *Inorganic Carbon Uptake by Aquatic Photosynthetic Organisms,* Lucas, W.J., and Berry, J.A., Eds., Am. Soc. Plant Physiol., Rockville, Maryland, 1985, 403.

4891.   Tamiya, H., Mass culture of algae, *Ann. Rev. Plant Physiol.,* 8, 309, 1957.

4892.   Tamiya, H., Synchronous cultures of algae, *Ann. Rev. Plant Physiol.,* 17, 1, 1966.

4893.   Tandeau de Marsac, N., Occurrence and nature of chromatic adaptation in cyanobacteria, *J. Bacteriol.,* 130, 82, 1977.

4894.   Tangen, K., Brand, L.E., Blackwelder, P.L., and Guillard, R.R.L., *Thoracosphaera heimii* is a dinophyte. Observations on its morphology and life cycle, *Mar. Micropaleontol.,* 7, 193, 1982.

4895.   Tanner, H.A., Brown, T.E., Eyster, C., and Treharne, R.W., A manganese-dependent photosynthetic process, *Biochem. Biophys. Res. Commun.,* 3, 205, 1960.

4896.   Tanner, W., Loffler, W., and Kandler, O., Cyclic photophosphorylation *in vivo* and its relation to photosynthetic CO_2 fixation, *Plant Physiol.,* 44, 422, 1969.

4897.   Tansley, A.G., *The British Islands and Their Vegetation,* Cambridge University Press, Cambridge, 1939.

4898.   Tansley, A.G., *The British Islands and Their Vegetation,* 2nd ed., Cambridge University Press, London, 1949.

4899.   Tappan, H., *The Paleobiology of Plant Protists,* Freeman, San Francisco, 1980.

4900.   Tarapchak, S.J., Studies on the Xanthophyceae of the Red Lake Wetlands, Minnesota, *Nova Hedw.,* 23, 1, 1972.

4901. Tarapchak, S.J., Measurements of phosphate-phosphorus in lake water, in *Nutrient Cycling in the Great Lakes: A Summarization of Factors Regulating the Cycling of Phosphorus,* Scavia, D., and Moll, R., Eds., Great Lakes Res. Div. Spec. Report No. 83, The University of Michigan, Ann Arbor, 1980, 1.

4902. Tate, R.L., III, Microbial oxidation of organic matter of histosols, *Adv. Microb. Ecol.,* 4, 169, 1980.

4903. Tatewaki, M., and Provasoli, L., Vitamin requirements of three species of *Antithamnion, Bot. Mar.,* 6, 193, 1964.

4904. Taylor, D.L., The nutritional relationship of *Anemonia sulcata* (Pennat) and its dinoflagellate symbiont, *J. Cell Sci.,* 4, 751, 1969.

4905. Taylor, D.L., Some aspects of the regulation and maintenance of algal numbers in zooxanthellae-coelenterate symbioses, with a note on the nutritional relationship in *Anemonia sulcata, J. Mar. Biol. Ass. U.K.,* 49, 1057, 1969.

4906. Taylor, D.L., Ultrastructure of the "zooxanthella" *Endonidium chattonii in situ, J. Mar. Biol. Ass. U.K.,* 51, 227, 1971.

4907. Taylor, D.L., The cellular interaction of algal-invertebrate symbiosis, *Adv. Mar. Biol.,* 11, 1, 1973.

4908. Taylor, D.L., Nutrition of algal-invertebrate symbiosis. I. Utilization of soluble organic nutrients by symbiont-free hosts, *Proc. R. Soc. Lond.,* B 186, 357, 1974.

4909. Taylor, D.L., Symbiotic marine algae: taxonomy and biological fitness, in *Symbiosis in the Sea,* Vernberg, W.B., Ed., University of South Carolina Press, Columbia, 1974, 245.

4910. Taylor, D.L., Nutrition of algal-invertebrate symbiosis. II. Effects of exogenous nitrogen sources on growth, photosynthesis, and the rate of excretion by algal symbionts *in vivo* and *in vitro, Proc. R. Soc. Lond.,* B 201, 401, 1978.

4911. Taylor, D.L., Evolutionary impact of intracellular symbiosis, *Ber. Dtsch. Bot. Ges.,* 94, 583, 1981.

4912. Taylor, D.L., and Seliger, H.H., Eds., *Toxic Dinoflagellate Blooms, Development in Marine Biology,* Vol. 1, Elsevier/North Holland, New York, 1979.

4913. Taylor, F.J.R., Parasitism of the toxin-producing dinoflagellate *Gonyaulax catanella* by the endoparasitic dinoflagellate *Amoebophrya ceratii, J. Fish. Res. Board Can.,* 25, 2241, 1968.

4914. Taylor, F.J.R., Basic biological features of phytoplankton cells, in The *Physiological Ecology of Phytoplankton,* Morris, I., Ed., University of California Press, Berkeley, 1980, 3.

4915. Taylor, F.J.W. On dinoflagellate evolution, *BioSystems,* 13, 65, 1980.

4916. Taylor, F.R., Blackbourn, D.J., and Blackbourn, J., The red-water ciliate *Mesodinium rubrum* and its "incomplete symbiosis:" a review including new ultrastructural observations, *J. Fish. Res. Bd. Can.,* 28, 391, 1971.

4917. Taylor, G.J., Crowder, A.A., and Rodden, R., Formation and morphology of an iron plaque on the roots of *Typha latifolia* L. grown in solution culture, *Am. J. Bot.,* 71, 666, 1984.

4918. Taylor, J.A., The peatlands of Great Britain and Ireland, in *Ecosystems of the World, Mires: Swamp, Bog, Fen and Moor.* Vol. 4B. *Regional Studies,* Gore, A.J.P., Ed., Elsevier Science Publishers, Amsterdam, 1983, 1.

4919. Taylor, P.A., and Williams, P.J.L., Theoretical studies on the coexistence of competing species under continuous-flow conditions, *Can. J. Microbiol.,* 21, 90, 1975.

4920. Taylor, S.R., Abundance of chemical elements in the continental crust: a new table, *Geochim. Cosmochim. Acta,* 28, 1273, 1964.

4921. Taylor, W.R., *Marine Algae of the Northeastern Coast of North America,* The University of Michigan Press, Ann Arbor, 1957.

4922. Taylor, W.R., *Marine Algae of the Eastern Tropical and Subtropical Coast of the Americas,* The University of Michigan Press, Ann Arbor, 1960.

4923. Teal, J.M., Energy flow in the salt marsh ecosystem of Georgia, *Ecology,* 43, 614, 1962.

4924. Teal, J.M., and Kanwisher, J.W., Gas transport in the marsh grass *Spartina alterniflora, J. Exp. Bot.,* 17, 355, 1966.

4925. Teal, J.M., Valiela, I., and Berla, D., Nitrogen fixation by rhizosphere and free-living bacteria in salt marsh sediments, *Limnol. Oceanogr.,* 24, 126, 1979.

4926. Tel-Or, E., and Stewart, W.D.P., Photosynthetic components and activities of nitrogen-fixing isolated heterocysts of *Anabaena cylindrica, Proc. R. Soc. Lond.,* B 198, 61, 1977.

4927. Tempest, D.W., Meers, J.L., and Brown, C.M., Synthesis of glutamate in *Acrobacter aerogenes* by a hitherto unknown route, *Biochem. J.,* 117, 405, 1970.

4928. Terlizzi, D.E., Jr., and Karlander, E.P., Soil algae from a Maryland serpentine formation, *Soil Biol. Biochem.,* 11, 205, 1979.

4929. Terry, K.L., Nitrate uptake and assimilation in *Thalassiosira weissflogii* and *Phaeodactylum tricornutum:* Interactions with photosynthesis and with the uptake of other ions, *Mar. Biol.,* 69, 21, 1982.

4930.   Terry, K.L., Nitrate and phosphate uptake interactions in a marine prymnesiophyte, *J. Phycol.*, 18, 79, 1982.

4931.   Terry, R.E., and Tate, R.L., The effect of nitrate on nitrous oxide reduction in inorganic soils and sediments, *Soil Sci. Soc. Am. J.*, 44, 744, 1980.

4932.   Tessanow, U., and Baynes, Y., Experimental effects of *Isoetes lacustris* L. on the distribution of Eh, pH, Fe and Mn in lake sediments, *Verh. Internat. Verein. Limnol.*, 20, 2358, 1978.

4933.   Tewari, K.K., and Singh, M., Acid soluble and acid insoluble inorganic polyphosphates in *Cuscuta reflexa, Phytochemistry* 3, 341, 1964.

4934.   Tezuka, Y., Watanabe, Y., Hayashi, H., Fukunoga, S., and Aizaki, M., Changes in the standing crop of sessile microbes caused by organic pollution of the Tomagawa River, *Jap. J. Ecol.*, 24, 43, 1974.

4935.   Thacker, A., and Syrett, P.J., The assimilation of nitrate and ammonium by *Chlamydomonas reinhardii, New Phytol.*, 71, 423, 1972.

4936.   Thauer, R.K., The new nickel enzymes from anaerobic bacteria, *Naturwissenschaften*, 70, 60, 1983.

4937.   Thayer, J.S., and Brinkman, F.E., The biological methylation of metals and metalloids, *Adv. Organometal. Chem.*, 20, 313, 1982.

4938.   Thayer, G.W., Phytoplankton production and the distribution of nutrients in a shallow unstratified estuarine system near Beaufort, NC, *Chesapeake Sci.*, 12, 240, 1971.

4939.   Theriot, E., Håkansson, H., and Stoermer, E.F., Morphometric analysis of *Stephanodiscus alpinus* (Bacillariophyceae) and its morphology as an indicator of lake trophic status, *Phycologia*, 27, 485, 1988.

4940.   Thiel, T., Transport of leucine in the cyanobacterium *Anabaena variabilis, Arch. Microbiol.*, 149, 466, 1988.

4941.   Thiel, T., Phosphate transport and arsenate resistance in the cyanobacterium *Anabaena varuabilis, J. Bacteriol.*, 170, 1143, 1988.

4942.   Thinh, L., Photosynthetic lamallae of *Prochloron* (Prochlorophyta) associated with the Ascidian *Diplosoma virens* (Hartmeyer) in the vicinity of Townsville, *Aust. J. Bot.*, 26, 617, 1978.

4943.   Thinh, L., *Prochloron* (Prochlorophyta) associated with the ascidian *Trididemnum cyclops* Michaelson, *Phycologia*, 18, 77, 1979.

4944.   Thomann, R.V., Di Toro, D.M., Winfield, R.P., and O'Connor, D.J., Mathematical modeling of phytoplankton in Lake Ontario. 1. Model development and verification, U.S. E.P.A. 600/3-75-005, Corvalis, Oregon, 1975.

4945.   Thomas, E.A., and Tregunna, E.B., Bicarbonate ion assimilation in photosynthesis by *Sargassum muticum, Can. J. Bot.*, 46, 411, 1968.

4946.   Thomas, D.J., and Grill, E.V., The effect of exchange reactions between Fraser River sediment and seawater on dissolved Cu and Zn concentrations in the Strait of Georgia, *Estuar. Coast. Mar. Sci.*, 5, 421, 1977.

4947.   Thomas, D.L., Montes, J.G., Spectrophotometrically assayed inhibitory effects of mercuric compounds on *Anabaena flos-aquae* and *Anacystis nidulans* (Cyanophyceae), *J. Phycol.*, 14, 494, 1978.

4948.   Thomas, D.M., and Goodwin, T.W., Nature and distribution of carotenoids in the Xanthophyta (Heterokontae), *J. Phycol.*, 1, 118, 1965.

4949.   Thomas, J.P., Release of dissolved organic matter from natural populations of marine phytoplankton, *Mar. Biol.*, 11, 311, 1971.

4950.   Thomas, T.E., Turpin, D.H., and Harrison, P.H., g-Glutamyl transferase activity in marine phytoplankton, *Mar. Ecol. Prog. Ser.*, 14, 219, 1984.

4951.   Thomas, W.H., Nutrient requirements and utilization: Algae, in *Metabolism*, Altman, P.L., and Ditmer, D.S., Eds., *Fed. Am. Soc. Exp. Biol.*, 1968, 210.

4952.   Thomas, W.H., Observations on snow algae in California, *J. Phycol.*, 8, 1, 1972.

4953.   Thomas, W.H., and Dodson, A.N., On nitrogen deficiency in tropical Pacific Oceanic phytoplankton. II. Photosynthetic and cellular characteristics of a chemostat-grown diatom, *Limnol. Oceanogr.*, 17, 515, 1972.

4954.   Thomas, W.H., and Seibert, D.L.R., Effects of copper on the dominance and the diversity of algae: controlled ecosystem pollution experiment, *Bull. Mar. Sci.*, 27, 23, 1977.

4955.   Thomas, W.H., Hollibaugh, J.T., and Seibert, D.L.R., Effects of heavy metals on the morphology of some marine phytoplankton, *Phycologia*, 19, 202, 1980.

4956.   Thomas, W.H., Hollibaugh, J.T., Seibert, D.L.R., and Wallace, G.T., Jr., Toxicity of a mixture of ten metals to phytoplankton, *Mar. Ecol. Prog. Ser.*, 2, 213, 1980.

4957.   Thomas, W.H., Holm-Hansen, O., Seibert, D.L.R., Azam, F., Hodson, R., and Takahashi, M., Effect of copper on phytoplankton standing crop and productivity: controlled ecosystem pollution experiment, *Bull. Mar. Sci.*, 27, 34, 1977.

4958.   Thompson, J.F., Sulfur metabolism in plants, *Ann. Rev. Plant Physiol.*, 18, 59, 1967.

4959. Thompson, E.W., and Preston, R.D., Proteins in the cell walls of some green algae, *Nature*, 213, 684, 1967.

4960. Thompson, K., Swamp development in the headwaters of the White Nile, in *The Nile: Biology of an Ancient River*, Rzoska, J., Ed., Dr. W. Junk Publishers, The Hague, The Netherlands, 1976, 177.

4961. Thompson, K., The primary productivity of African wetlands with particular reference to the Okavango Delta, in *Proc. Symp. Okavango Delta and its Future Utilization*, Botswana Society, Gaborone, Botswana, 1976, 67.

4962. Thompson, K., and Hamilton, A.C., Peatlands and swamps of the African continent, in *Ecosystems of the World, Mires: Swamp, Bog, Fen and Moor*, Vol. 4B. *Regional Studies*, Gore, A.J.P., Ed., Elsevier Science Publishers, Amsterdam, 1983, 331.

4963. Thompson, K., Shewry, P.R., and Woolhouse, H.W., Papyrus swamp development in the Upemba Basin, Zaire: Studies of population structure in *Cyperus papyrus* stands, *Bot. J. Linn. Soc.*, 78, 299, 1979.

4964. Thompson, P.A., Guo, M., and Harrison, P.J., Effects of variation in temperature. I. On the biochemical composition of eight species of marine phytoplankton, *J. Phycol.*, 28, 481, 1992.

4965. Thornber, J.P., and Barber, J., Photosynthetic pigments and models for their organization *in vivo* in *Photosynthesis in Relation to Model Systems*, Barber, J., Ed., Elsevier, New York, 1979, 27.

4966. Thornton, I., and Gigioli, M.E.C., Mangrove swamps of Keneba. Lower Gambia River Basin, *J. Appl. Ecol.*, 2, 257, 1965.

4967. Thorson, G., Reproductive and larval ecology of marine bottom invertebrates, *Biol. Rev.*, 25, 1, 1950.

4968. Thuret, G., Essay de classification des Nostochinées. *Ann. Sci. Nat.*, 6, *Bot.*, 1, 372, 1875.

4969. Tiedje, J.M., Sextone, A.J., Myrold, D.D., and Robinson, J.A., Denitrification: Ecological niches, competition and survival, *Antonie van Leeuwenhoek*, 48, 569, 1982.

4970. Tilden, J., The Myxophyceae of North America, in *Minnesota Algae* I., Minneapolis, 1910.

4971. Tilden, J.E., *The Algae and Their Life Relations*, The University of Minnesota Press, Minneapolis, 1935.

4972. Tilman, D., and Kilham, S.S., Phosphate and silicate growth and uptake kinetics of the diatoms *Asterionella formosa* and *Cyclotella meneghiniana* in batch and semicontinuous culture, *J. Phycol.*, 12, 375, 1976.

4973. Tilman, D., and Kilham, S.S., Phytoplankton community ecology: The role of limiting nutrients, *Ann. Rev. Ecol. Syst.*, 13, 349, 1982.

4974. Tilman, D., Mattson, M., and Langer, S., Competition and nutrient kinetics along a temperature gradient: an experimental test of a mechanistic approach to niche theory, *Limnol. Oceanogr.*, 26, 1020, 1981.

4975. Tilton, D.L., and Schwegler, B.R., The values of wetland habitat in the Great Lakes basin, in *Wetland Functions and Values: The State of Our Understanding*, Greeson, P.E., Clark, J.R., and Clark, J.E., Eds., American Water Resour. Assoc., Minneapolis, Minnesota, 1979, 267.

4976. Tilton, D.L., Kadlec, R.H., and Richardson, C.J., Eds., *Proc. Nat. Symp. Freshwater Wetlands and Sewage Effluent Disposal*, University of Michigan, Ann Arbor, 1976.

4977. Timmer, C.E., and Weldom, L.W., Evapotranspiration and pollution of water by water hyacinth, *Water Hyacinth Control.*, 6, 34, 1967.

4978. Tindall, D.R., Sawa, T., and Hotchkis, A.T., *Nitellopsis bulbillifera* in North America, *J. Phycol.*, 1, 147, 1965.

4979. Tiner, R.W., Jr., and Veneman, P.L.M., *Hydric Soils of New England*, University of Massachusetts Cooperation Extension Communications Center, 1987.

4980. Tischner, R., Zur Induktion der Nitrat- and Nitritreduktase in vollsynchronous *Chlorella* - Kulturen, *Planta*, 132, 285, 1976.

4981. Tischner, R., Evidence for the participation of NADP-glutamate dehydrogenase in the ammonium assimilation of *Chlorella sorokiniana*, *Plant Sci. Lett.*, 34, 73, 1984.

4982. Tischner, R., Regulation of ammonium incorporation, in *Inorganic Nitrogen Metabolism*, Ullrich, W.R., Aparicio, P.J., Syrett, P.J., and Castillo, F., Eds., Springer-Verlag, Berlin, 1987, 126.

4983. Tischner, R., and Lorenzen, H., Nitrate uptake and nitrate reduction in synchronous *Chlorella*, *Planta*, 146, 287, 1979.

4984. Titman, D., and Kilham, P., Sinking in freshwater phytoplankton: some ecological implications of cell nutrient status and physical mixing processes, *Limnol. Oceanogr.*, 21, 409, 1976.

4985. Toerien, D.J., Wrigley, T.J., Artificially established reed beds: A way to upgrade sewage effluents, *Water, Sewage and Effluent*, May, 20, 1984.

4986. Toerien, D.F., Huang, C.H., Radimsky, J., Pearson, E.A., and Scherfig, J., Provisional algal assay procedures, Final Report, EPA Project No. 16010 DQB, University of California, Berkeley, Sanitary Engineering Research Laboratory, SERL Rept. No. 71-6, 1971.

4987. Toerien, D.E., Scott, W.E., and Pitout, M.J., *Microcystis* toxins: isolation, identification, implications, *Water S.A.,* 2, 160, 1976.

4988. Toetz, D.W., Diurnal uptake of nitrate and ammonium by a *Ceratophyllum* periphyton community, *Limnol. Oceanogr,* 16, 819, 1971.

4989. Toetz, D.W., Uptake and translocation of ammonium by a freshwater hydrophytes, *Ecology,* 55, 199, 1974.

4990. Toetz, D.W., Diel periodicity in uptake of nitrite and nitrate by reservoir phytoplankton, *Hydrobiologia,* 49, 49, 1976.

4991. Toetz, D., Varga, L.P., and Loughran, E.D., Half-saturation constants for uptake of nitrate and ammonia by reservoir plankton, *Ecology,* 54, 903, 1973.

4992. Toetz, D., Varga, L., and Huss, B., Observations on uptake of nitrate and ammonia by reservoir phytoplankton, *Arch. Hydrobiol.,* 79, 182, 1977.

4993. Toivonen, H., and Lappalainen, T., Ecology and production of aquatic macrophytes in the oligotrophic, mesohumic lake Suomunjärvi, eastern Finland, *Ann. Bot. Fennici,* 17, 69, 1980.

4994. Tolbert, N.E., Glycolate pathway, in *Photosynthetic Mechanism in Green Plants*, N.A.S. - N.R.C., Washington, DC, 1963, 648.

4995. Tolbert, N.E., Microbodies - peroxisomes and glyoxysomes. *Ann. Rev. Plant Physiol.,* 22, 45, 1971.

4996. Tolbert, N.E., Photorespiration, in *Algal Physiology and Biochemistry*, Stewart, W.D.P., Ed., Blackwell Scientific Publications, Oxford, 1974, 474.

4997. Tolbert, N.E., and Zill, L.P., Photosynthesis by protoplasm extruded from *Chara* and *Nitella, J. Gen. Physiol.,* 37, 575, 1954.

4998. Tolbert, N.E., and Zill, L.P., Excretion of glycolic acid by algae during photosynthesis, *J. Biol. Chem.,* 222, 895, 1956.

4999. Tolbert, N.E., and Zill, L.P., Excretion of glycolic acid by *Chlorella* during photosynthesis, in *Research in Photosynthesis*, Gaffron, H., Ed., Interscience, New York, 1957, 228.

5000. Tolonen, K., Peat as a renewable resources, in *Proc. Int. Symp. Classification of Peat and Peatlands,* Internat. Peat Society, Helsinki, 1979, 17.

5001. Tomas, C.R., *Olisthodiscus luteus* (Chrysophyceae). III. Uptake and utilization of nitrogen and phosphorus, *J. Phycol.,* 15, 5, 1979.

5002. Tomita, G., and Rabinowitch, E., Excitation energy transfer between pigments in photosynthetic cells, *Biophys. J.,* 2, 483, 1962.

5003. Tompkins, T., and Blinn, D.W., Effect of mercury on growth rate of *Fragilaria crotonensis* Kitton and *Asterionella formosa* Hass., *Hydrobiologia,* 49, 111, 1976.

5004. Topinka, J., An investigation of nitrogen uptake and nitrogen status in *Fucus spiralis* L., Ph.D. Thesis, University of Massachusetts, Amherst, 1975.

5005. Topinka, J., Nitrogen uptake by *Fucus spiralis* (Phaeophycea), *J. Phycol.,* 14, 241, 1978.

5006. Topinka, J., and Robbins, J.V., Effects of nitrate and ammonium enrichment on growth and nitrogen physiology in *Fucus spiralis, Limnol. Oceanogr,* 21, 659, 1976.

5007. Tornabene, T.G., Holzer, G., Lien, S., and Burris, N., Lipid composition of the nitrogen starved green alga *Neochloris oleoabundans, Env. Microb. Technol.,* 5, 435, 1983.

5008. Torpey, J., and Ingle, R.M, The red tide (Revision of 1955), *Fla. B. Conserv. Mar. Lab. Educ. Ser.* 1, 1966.

5009. Tourbier, J., and Pirson, R.W., Eds., *Biological Control of Water Pollution*, University of Pennsylvania Press, Philadelphia, 1976.

5010. Towle, D.W., and Pearse, J.S., Production of the giant kelp *Macrocystis*, estimated by in situ incorporation of ^{14}C in polyethylene bags, *Limnol. Oceanogr,* 18, 155, 1973.

5011. Traaen, T.S., Biological effects of primary, secondary and tertiary sewage treatment in lotic analog recipients, *Verh. Internat. Verein. Limnol.,* 19, 2064, 1975.

5012. Traaen, T.S., and Lindstrøm, E.-A., Influence of current velocity on periphyton distribution, in *Periphyton of Freshwater Ecosystems*, Wetzel, R.G., Ed., Dr. W. Junk Publishers, The Hague, The Netherlands, 1983, 97.

5013. Trainor, F.R., Temperature tolerance of algae in dry soil, *Phycol. News Bull.,* 15, 3, 1962.

5014. Trainor, F.R., *Introductory Phycology*, John Wile and Sons, New York, 1978.

5015. Trainor, F.R., Survival of algae in soil after high temperature treatment, *Phycologia,* 22, 201, 1983.

5016. Trefry, J.H., and Presley, B.J., Heavy metal transport from the Mississippi River to the Gulf of Mexico, in *Marine Pollutant Transfer*, Windom, H.L., and Duce, R.A., Eds., Lexington, Massachusetts, 1976, 39.

5017. Tregunna, E.B., and Thomas, E.A., Measurement of inorganic carbon and photosynthesis in seawater by pCO_2 and pH analysis, *Can. J. Bot.,* 46, 481, 1968.

5018. Trelease, S.F., and Selsam, M.E., Influence of calcium and magnesium on the growth of *Chlorella, Am. J. Bot.,* 26, 339, 1939.

5019. Tremearne, T.H., and Jacob, K.D., Arsenic in natural phosphates and phosphate fertilizers, *USDA Tech. Bull. No. 781,* Washington, 1941.

5020. Trench, R.K., The physiology and biochemistry of zooxanthellae symbiotic with marine coelenterates. I. Assimilation of photosynthetic products by two marine coelenterates. II. Liberation of fixed ^{14}C by zooxanthellae in vitro. III. The effect of homogenates of host tissues on excretion of photosynthetic products in vitro by zooxanthellae from two marine coelenterates, *Proc. R. Soc. Lond.,* B 177, 225, 1971.

5021. Trench, R.K., Of "leaves that crawl:" functional chloroplasts in animal cells, *Symp. Soc. Exp. Biol.,* 29, 229, 1975.

5022. Trench, R.K., The cell biology of plant - animal symbiosis, *Ann. Rev. Plant Physiol.,* 30, 485, 1979.

5023. Trench, R.K., Cellular and molecular interactions in symbioses between dinoflagellates and marine invertebrates, *Pure Appl. Chem.,* 53, 819, 1981.

5024. Trench, R.K., Physiology, ultrastructure and biochemistry of cyanelle, *Prog. Phycol. Res.,* 1., 257, 1982.

5025. Trench, R.K., Colley, N.J., and Fitt, W.K., Recognition phenomena in symbioses between marine invertebrates and "zooxanthellae" uptake, sequestration and persistence, *Ber. Dtsch. Bot. Ges.,* 94, 529, 1981.

5026. Trevors, J.T., Stratton, G.W., and Gadd, G.M., Cadmium transport, resistance and toxicity in bacteria, algae and fungi, *Can. J. Microbiol.,* 32, 447, 1986.

5027. Trollope, D.R., and Evans, B., Concentrations of copper, iron, lead, nickel and zinc in freshwater algal blooms, *Environ. Pollut.,* 11, 107, 1976.

5028. Tromballa, H.W., Electrogenicity of potassium transport in *Chlorella, Z. Pflanzenphysiol.,* 96, 123, 1980.

5029. Trottier, D.M., and Hendricks, A.C., The use of *Stigeoclonium subsecundum* (Chlorophyceae) as a bioassay organism. I. Attachment of the algal cells to glass slides, *Water Res.,* 10, 909, 1976.

5030. Trottier, D.M., Hendricks, A.C., and Cairns, J., Jr., The use of *Stigeocloniumsubsecundum* (Chlorophyceae) as a bioassay organism. III. Response to intermittent chlorination, *Water Res.,* 12, 185, 1978.

5031. Troxler, R.F., and Bogorad, L., Studies on the formation of phycocyanin, porphyrins and a blue phycobilin by wild-type and mutant strains of *Cyanidium caldarium, Plant Physiol.,* 41, 491, 1966.

5032. Trubachev, I.N., On the role of ascorbic acid, H_2O_2 and iron in assimilation of nitrate by *Chlorella, Fiziol. Rast.,* 15, 658, 1968 (in Russian).

5033. Trudinger, P.A., The biological sulfur cycle, in *Biogeochemical Cycling of Mineral-Forming Elements,* Trudinger, P.A., and Swaine, D.J., Eds., Elsevier Scientific Publishing Company, Amsterdam, 1979, 293.

5034. Trudinger, P.A., Swaine, D.J., and Skyring, G.W., Biogeochemical cycling of elements-general considerations, in *Biogeochemical Cycling of Mineral-Forming Elements,* Trudinger, P.A., and Swaine, D.J., Eds., Elsevier Scientific Publishing Company, Amsterdam, 1979, 1.

5035. Truhaut, R., Ferard, J.F., and Jouany, J.M., Cadmium IC_{50} determination on *Chlorella vulgaris* involving different parameters, *Ecotoxicol. Environ. Saf.,* 4, 215, 1980.

5036. Tsang, M.L., Goldsmith, E.E., and Schiff, J.A., Adenosine-5'-phosphosulfate (APS^{35}) as an intermediate in the conversion of adenosine-3'-phosphate-5'-phosphosulfate ($PAPS^{35}$) to acid-volatile radioactivity, *Plant Physiol.,* 47 (suppl.), 20, 1971.

5037. Tschermak-Woess, E., Das sogenannte Alveolarplasma und die Schleimbildung bei *Vacuolaria virescens, Österr. Bot. Z.,* 101, 328, 1954.

5038. Tschermak-Woess, E., Zur Kenntnis von *Tetrasporopsis fuscescens, Plant Syst. Evol.,* 133, 121, 1980.

5039. Tseng, C.K., Commercial cultivation, in *The Biology of Seaweeds,* Lobban, C.S., and Wynne, M.J., Eds., University of California Press, Berkeley, 1981, 680.

5040. Tseng, C.K., Marine phycoculture in China, in *Proc. X. Int. Seaweed Symposium,* Levring, T., Ed., Walter de Gruyter, Berlin-New York, 1981, 123.

5041. Tsuda, R.T., Marine benthic algae of Guam. I. Phaeophyta, *Micronesica,* 8, 87, 1972.

5042. Tsuzuki, M., and Miyachi, S., The function of carbonic anhydrase in aquatic photosynthesis, *Aquat. Bot.,* 34, 85, 1989.

5043. Tsuzuki, M., Miyachi, S., and Berry, J.A., Intracellular accumulation of inorganic carbon and its active species taken by *Chlorella vulgaris* 11h, in *Inorganic Carbon Uptake by Aquatic Photosynthetic Organisms,* Lucs, W.J., and Berry, J.A., Am. Soc. Plant Physiol., Rockville, Maryland, 1985, 53.

5044. Tuchman, N.C., and Stevenson, R.J., Effects of selective grazing by snails on benthic algal succession, *J. N. Am. Benthol. Soc.,* 10, 430, 1991.

5045. Tucker, C.S., and De Busk, T.A., Seasonal variation in the nitrate content of water hyacinth (*Eichhornia crassipes* [Mart.] Solms), *Aquat. Bot.,* 15, 419, 1983.

5046. Tung, H.F., and Shen, T.C., Studies of the *Azolla pinnata - Anabaena azolae* symbiosis, *New Phytol.*, 87, 743, 1981.

5047. Tupa, D., An investigation of certain chaetophoracean algae, *Beih. Nova Hedw.*, 46, 1973.

5048. Turbak, S.C., Olson, S.B., and McFeters, G.A., Comparison of algal assay systems for detecting waterborne herbicides and metals, *Water Res.*, 20, 91, 1986.

5049. Turekian, K.K., The oceans, streams and atmosphere, in *Handbook of Geochemistry*, Wedepohl, K.-H., Ed., Springer-Verlag, Berlin, 1969, 297.

5050. Turner, J.S., and Brittain, E.G., Oxygen as a factor in photosynthesis, *Biol. Rev.*, 37, 130, 1962.

5051. Turner, F.T., and Patrick, W.H., Jr., Chemical changes in waterlogged soils as a result of oxygen depletion, *Trans. 9th Int. Cong. of Soil Sci.*, 4, 53, 1968.

5052. Turner, F.R., An ultrastructural study of plant spermatogenesis: spermatogenesis in *Nitella*, *J. Cell Biol.*, 37, 370, 1968.

5053. Turner, M.F., Nutrition of some marine microalgae with special reference to vitamin requirements and utilization of nitrogen and carbon sources, *J. Mar. Biol. Ass. U.K.*, 59, 535, 1979.

5054. Turner, R.E., Community plankton respiration in a salt marsh estuary and the importance of macrophytic leachates, *Limnol. Oceanogr.*, 23, 442, 1978.

5055. Turpin, D.H., Physiological mechanisms in phytoplankton resource competition, in *Growth and Reproductive Strategies of Freshwater Phytoplankton*, Sandgren, C.D., Ed., Cambridge University Press, New York, 1988, 316.

5056. Turpin, D.H., Effects of inorganic N availability on algal photosynthesis and carbon metabolism, *J. Phycol.*, 27, 14, 1991.

5057. Turpin, D.H., Elrifi, I.R., Birch, D.G., Weger, H.G., and Holmes, J.J., Interactions between photosynthesis, respiration, and nitrogen assimilation in microalgae, *Can. J. Bot.*, 66, 2083, 1988.

5058. Tusneem, M.E., and Patrick, W.H., Jr., Nitrogen transformation in waterlogged soils, *Louisiana State University Agric. Exp. Station Bull.*, 6, Baton Rouge, 1972.

5059. Twilley, R.R., Brinson, M.M., and Davis, G.J., Phosphorus absorption, translocation, and secretion in *Nuphar luteum*, *Limnol. Oceanogr.*, 22, 1022, 1977.

5060. Twilley, R.R., Blanton, L.R., Brinson, M., and Davis, G.J., Biomass production and nutrient cycling in aquatic macrophyte communities of the Chowan River, North Carolina, *Aquat. Bot.*, 22, 231, 1985.

5061. Twilley, R.R., Kemp, W.M., Staver, K.W., Stevenson, J.C., and Boynton, W.R., Nutrient enrichment of estuarine submersed vascular plant communities. 1. Algal growth and effects on production of plants and associated communities, *Mar. Ecol. Prog. Ser.*, 23, 179, 1985.

5062. Twiss, M.R., Copper tolerance of *Chlamydomonas acidophila* (Chlorophyceae) isolated from acidic, copper-contaminated soils, *J. Phycol.*, 26, 655, 1990.

5063. Twiss, M.R., and Nalewajko, C., Influence of phosphorus nutrition on copper toxicity to three strains of *Scenedesmus acutus* (Chlorophyceae), *J. Phycol.*, 28, 291, 1992.

5064. Tyiagi, V.V.S., The heterocysts of blue-green algae (Myxophyceae), *Biol. Rev.*, 50, 247, 1975.

5065. Tyler, G., Heavy metals pollute nature may reduce productivity, *Ambio*, 1, 52, 1972.

5066. Tyler, M.A., and Seliger, H.H., Annual subsurface transport of a red tide dinoflagellate in its bloom area: water circulation patterns and organism distribution in the Chesapeake Bay, *Limnol. Oceanogr.*, 23, 227, 1978.

5067. Tyler, M.A., and Seliger, H.H., Selection for a red tide organism: physiological responses to the physical environment, *Limnol. Oceanogr.*, 26, 310, 1981.

5068. Tyurmenov, S.N., *Torfyanye mestorozhdenija*, Nedra, Moscow, 1976 (in Russian).

5069. Tzuzimura, F.I., Ikeda, F., and Tsukamoto, K., Studies on with reference to the green manure for rice fields, *J. Sci. Soil Manure Japan*, 28, 275, 1957.

5070. Udel'nova, T.M., Kondrateva, E.N., and Boichenko, E.A., Iron and manganese content in various photosynthetic microorganisms, *Mikrobiologiya*, 37, 197, 1968 (in Russian).

5071. Uhlmann, D., Über den Einfluss von Planktonorganismen auf ihr Milieu, *Int. Rev. Hydrobiol.*, 46, 115, 1961.

5072. Ukeles, R., Growth of pure cultures of marine phytoplankton in the presence of toxicants, *Appl. Microbiol.*, 10, 532, 1962.

5073. Úlehlová, B., Decomposition process in the fish pond littoral, in *Pond Littoral Ecosystems*, Dykyjová, D., and Kvĕt, J., Eds., Springer-Verlag, Berlin, 1978, 341.

5074. Ullrich-Eberius, C.I., and Simonis, W., Der Einfluss von Natrium- und Kaliumionen auf die Phosphataufnahme bei *Ankistrodesmus braunii*, *Planta*, 93, 214, 1970.

5075. Ullrich, W.R., Die Wirkung von O_2 and CO_2 auf die ^{32}P-Markierung der Polyphosphate von *Ankistrodesmus braunii* bei der Photosynthese, *Ber. Dtsch. Bot. Ges.*, 83, 435, 1970.

5076. Ullrich, W.R., Zur Wirkung von Sauerdtoff auf die ^{32}P-Markierung von Polyphosphaten und organischen Phosphaten bei *Ankistrodesmus* in Licht, *Planta*, 90, 272, 1970.

5077. Ullrich, W.R., Der Einfluss von CO_2 und pH auf die ^{32}P-Markierung von Polyphosphaten und organischen Phosphaten bei *Ankistrodesmus braunii* im Licht, *Planta*, 102, 37, 1972.
5078. Ullrich, W.R., Die Nitritaufnahme bei Grünalgen und ihre Regulation durch äussere Faktoren, *Ber. Dtsch. Bot. Ges.*, 92, 273, 1979.
5079. Ullrich, W.R., Uptake and reduction of nitrate: Algae and fungi, in *Encyclopedia of Plant Physiology. NS, Vol. 15. Inorganic Plant Nutrition*, Läuchli, A., and Bieleski, R.L., Eds., Springer-Verlag, Berlin, 1983, 376.
5080. Ullrich, W.R., Nitrate and ammonium uptake in green algae and higher plants: mechanism and relationship with nitrate metabolism, in *Inorganic Nitrogen Metabolism*, Ullrich, W.R., Aparicio, P.J., Syrett, P.J., and Castillo, F., Eds., Springer-Verlag, Berlin, 1987, 30.
5081. Ullrich, W.R., Aparicio, P.J., Syrett, P.J., and Castillo, F., Eds., *Inorganic Nitrogen Metabolism*, Springer-Verlag, Berlin, 1987.
5082. Ullrich, W.R., Rigano, C., Fuggi, A., and Aparicio, P.J., Eds., *Inorganic Nitrogen in Plants and Microorganisms*, Springer-Verlag, Berlin, 1990.
5083. Ulrich, A., and Gersper, P.L., Plant nutrition limitations of tundra plant growth, in *Vegetation and Production Ecology of an Alaskan Arctic Tundra*, Tieszen, L.L., Ed., Ecological Studies 29, Springer-Verlag, New York, 1978, 457.
5084. Ulrich, K.E., and Burton, T.M., An experimental comparison of the dry matter and nutrient distribution patterns of *Typha latifolia* L., *Typha angustifolia* L., *Sparganium eurycarpum* Engelm. and *Phragmites australis* (Cav.) Trin. ex Steudel, *Aquat. Bot.*, 32, 129, 1988.
5085. Underhill, P.A., Nitrate uptake kinetics and clonal variability in the neritic diatom *Biddulphia aurita*, *J. Phycol.*, 13, 170, 1977.
5086. UNEP, Chapter 4: Inland water—changes in water quality, in The World Environment 1972–1982, A report by the UNEP, Tycooly, Dublin, Ireland, 1982.
5087. Unni, K.S., Seasonal variation in chemical constituents of some aquatic plants, *J. Bombay Nat. Hist. Soc.*, 69, 242, 1970.
5088. Unni, K.S., Seasonal changes in growth rate and organic matter production of *Trapa bispinosa* Roxb., *Trop. Ecol.*, 25, 125, 1984.
5089. United States Department of Agriculture (USDA)—Soil Conservation Service (SCS), Hydric Soils of the United States, *USDA-SCS nat. Bull. No. 430-5-9*, Washington, 1985.
5090. United States Soil Conservation Service (U.S. SCS), Soil taxonomy: a basic system of soil classification for making and interpreting soil surveys, U.S. SCS Agric. Handbook 436, 1975.
5091. Urban, Z., and Kalina, T., *System and Evolution of Lower Plants*, SPN, Prague, 1980, (in Czech).
5092. Urhan, O., Beiträge zur Kenntis der Stickstoffassimilation von *Chlorella* und Scenedesmus, *Jahrb. Wiss. Bot.*, 75, 1, 1932.
5093. Usher, H.D., and Blinn, D.W., Influence of various exposure periods on the biomass and chlorophyll *a* of *Cladophora glomerata* (Chlorophyta), *J. Phycol.*, 26, 244, 1990.
5094. Utermöhl, H., Limnologische Phytoplankton - Studien, *Arch. Hydrobiol. Suppl.*, 5, 1, 1925.
5095. Utter, M.F., and Kolenbrander, H.M., Formation of oxaloacetate by CO_2 fixation on phosphoenolpyruvate, in *The Enzymes*, Boyer, P.D., Ed., Academic Press, New York, 1972, 117.
5096. Valiela, I., *Marine Ecological Processes*, Springer-Verlag, New York, 1984.
5097. Valiela, I., Teal, J.M., and Sass, W., Nutrient retention in salt marsh plots experimentally fertilized with sewage sludge, *Estuar. Coast. Mar. Sci.*, 1, 261, 1973.
5098. Valiela, I., Teal, J.M., and Persson, N.Y., Production and dynamics of experimentally enriched salt marsh vegetation: belowground biomass, *Limnol. Oceanogr.*, 21, 247, 1976.
5099. Valiela, I., Teal, J.M., Volkmann, S., Shafer, D., and Carpenter, E.J., Nutrient and particulate fluxes in a salt marsh ecosystem: tidal exchange and inputs by precipitation and groundwater, *Limnol. Oceanogr.*, 23, 798, 1978.
5100. Valiela, I., Howes, B., Howarth, R., Giblin, A., Foreman, K., Teal, J.M., and Hobbie, J.E., Regulation of primary production and decomposition in a salt marsh ecosystem, In *Proc. Int. Conf. Wetlands: Ecology and Management*, Gopal, B., Turner, R.E., Wetzel, R.G., and Whigham, D.F., Eds., Natl. Inst. of Ecology and Internat. Scientific Publications, Jaipur, India, 1982, 151.
5101. Vallee, B.L., Zinc and metalloenzymes, in *Advances in Protein Chemistry*, Anson, M.L., Ed., Academic Press, New York, 1955, 317.
5102. Vallee, B.L., and Jeanjean,R., Le systeme de transport de SO_4^{2-} chez *Chlorella pyrenoidosa* et sa regulation. I. Etude cinetique de la permeation, *Biochim. Biophys. Acta*, 150, 599, 1968.
5103. Vallee, B.L., and Ulmer, D.D., Biochemical effects of mercury, cadmium and lead, *Ann. Rev. Biochem.*, 41, 91, 1972.
5104. Vallentyne, J.R., *The Algal Bowl. Lakes and Man*, Dept. Environ., Fish. and Mar. Serv. Miscel. Spec. Publ. 22, Ottawa, 1974.

5105. van As, D., Fourie, H.O., and Vleggaar, C.M., Accumulation of certain trace elements in the marine organisms from the Sea around the Cap of Good Hope, in *Radioactive Contamination of the Marine Environment*, Internat. Atomic Energy Agency, vienna, 1973, 615.

5106. van Baalen, C., and Marler, J.E., Characteristics of marine blue-green algae with uric acid as nitrogen source, *J. Gen. Microbiol.*, 32, 457, 1963.

5107. van Baalen, C., and O'Donnell, R., Isolation of nickel-dependent blue-green alga, *J. Gen. Microbiol.*, 105, 351, 1978.

5108. van Breeman, N., Redox processes of iron and sulfate involved in the formation of acid sulfate soils, in *Iron in Soils and Clay Minerals*, Stucki, J.W., Goodman, B.A., and Schwertmann, U., Eds., D. Reidel Publishing, Dordrecht, The Netherlands, 1988, 825.

5109. Vance, B.D., Composition and succession of cyanophycean water bloom, *J. Phycol.*, 1, 81, 1965.

5110. van Cleve, K., and Alexander, V., Nitrogen cycling in tundra and boreal ecosystems, *Ecol. Bull. (Stockholm)*, 33, 375, 1979.

5111. van den Hoeck, C., Chlorophyta: morphology and classification, in *The Biology of Seaweeds*, Lobban, C.S., and Wynne, M.J., Eds., University of California Press, Berkeley, 1981, 86.

5112. van den Hoeck, C., and Jahns, H.M., *Algen. Einführung in die Phykologie*, Thieme Verlag, Stuttgart, 1978.

5113. van der Ben, C., Ecophysiologie de quelques Desmidiées. II. Perturbations experimentales de la nitrate-respiration, *Hydrobiologia*, 38, 165, 1971.

5114. van der Berg, C.M.G., Wong, P.T.S., and Chau, Y.K., Measurement of complexing materials excreted from algae and their ability to ameliorate copper toxicity, *J. Fish. Res. Board Can.*, 36, 901, 1979.

5115. van der Bijl, L., Sand-Jensen, K., and Hjermind, A.L., Photosynthesis and canopy structure of a submerged plant, *Potamogeton pectinatus*, in a Danish lowland stream, *J. Ecol.*, 77, 947, 1989.

5116. van der Toorn, J., Verhoeven, J.T.A., and Simpson, R.L., Fresh water marshes, in *Wetlands and Shallow Continental Water Bodies*, Patten, B.C., Ed., SPB Academic Publishing, The Hague, The Netherlands, 1990, 445.

5117. van der Valk, A.G., Floristic composition and structure of fen communities in northern Iowa, *Proc. Iowa Acad. Sci.*, 82, 113, 1975.

5118. van der Valk, A.G., Zonation, competitive displacement and standing drop of northern Iowa fen communities, *Proc. Iowa Acad. Sci.*, 83, 83, 1976.

5119. van der Valk, A.G., Ed., *Northern Prairie Wetlands*, Iowa State University Press, Ames, 1989.

5120. van der Valk, A.G., and Bliss, L.C., Hydrarch succession and the primary production of oxbow lakes in Central Alberta, *Can. J. Bot.*, 49, 1177, 1971.

5121. van der Valk, A.G., and Davis, C.B., Primary production of prairie glacial marshes, in *Proc. Conf. Freshwater Wetlands: Production Processes and Management Potential*, Good, R.E., Whigham, D.F., and Simpson, R.L., Eds., Academic Press, New York, 1978, 21.

5122. van der Valk, G.A., and Davis, C.B., The role of seed banks in the vegetation of prairie glacial marshes, *Ecology*, 59, 322, 1978.

5123. van der Valk, A.G., Davis, C.B., Baker, J.L., and Beer, C.E., Natural fresh water wetlands as nitrogen and phosphorus traps for land runoff, in *Proc. Symp. Wetland Functions and Values: The State of Our Understanding*, Greeson, P.E., Clark, J.R., and Clark, J.E., Eds., American Water Resources Assoc., Minneapolis, Minnesota, 1979, 457.

5124. van der Valk, A.G., Rhymer, J.M., and Murkin, H.R., Flooding and the decomposition of litter of four emergent plant species in a prairie wetland, *Wetlands*, 11, 1, 1991.

5125. van der Velde, G., Giesen, T.G., and van der Heijden, L., Structural, biomass and seasonal changes in biomass of *Nymphoides peltata* (Gmel.) O. Kuntze (Menyanthaceae), a preliminary study, *Aquat. Bot.*, 7, 279, 1979.

5126. van der Wal, P., de Jong, L., Westbroek, P., de Bruijn, W.C., and Mulder-Stapel, A.A., Polysaccharide localization, coccolith formation and Golgi dynamics in the coccolithophorid *Hymenomonas carterae*, *J. Ultrastruct. Res.*, 85, 1139, 1983.

5127. van Donk, E., and Kilham, S.S., Temperature effects on silicon and phosphorus-limited growth and competitive interactions among three diatoms, *J. Phycol.*, 26, 40, 1990.

5128. van Dyke, G.D., Aspects relating to emergent vegetation dynamics in a deep marsh, North Central Iowa, Ph.D. Thesis, Iowa State University, Ames, 1972.

5129. van Emburg, P.R., Coccolith formation in *Emiliania huxleyi*, Ph.D. Thesis, Leiden University, The Netherlands, 1989.

5130. Vanlerberghe, G., Schuller, K.A., Smith, R.G., Feil, R., Plaxton, W.C., and Turpin, D.H., Relationship between NH_4^+ assimilation rate and in vivo phosphoenol pyruvate carboxylase activity: regulation of anaplerotic carbon flow in the green alga *Selenastrum minutum*, *Plant Physiol.*, 94, 284, 1990.

5131. van Niel, C.B., Allen, M.B., and Wright, B.E., On the photochemical reduction of nitrate by algae, *Biochim. Biophys. Acta,* 12, 67, 1953.

5132. van Noort, D., Ribonucleic acid and chlorophyll synthesis in *Chlorella vulgaris* during recovery from iron deficiency, *Diss. Abstr,* 25, 1545, 1964.

5133. van Raalte, M.H., On the oxygen supply of rice roots, *Ann. Bot. Gard. Buit.,* 51, 43, 1941.

5134. van Raalte, C.D., Valiela, I., and Teal, J.M., Production of epibenthic salt marsh algae: light and nutrient limitation, *Limnol. Oceanogr,* 21, 862, 1976.

5135. van Raalte, C.D., Valiela, I., and Teal, J.M., The effects of fertilization on the species composition of salt marsh diatoms, *Water Res.,* 10, 1, 1976.

5136. Van Veen, W.L., Mulder, E.G., and Deinema, M.H., The *Sphaerotilus - Leptothrix* group of bacteria, *Microbiol. Rev.,* 42, 329, 1978.

5137. van Wazer, R., The compounds of phosphorus, in *Environmental Phosphorus Handbook,* Griffith, E., Beeton, A., Spencer, J., and Mitchell, D., Eds., John Wiley and Sons, New York, 1973, 169.

5138. van Wijk, R.J., Ecological studies on *Potamogeton pectinatus* L. I. General characteristics, biomass production and life cycles under field conditions, *Aquat. Bot.,* 31, 211, 1988.

5139. Varela, M., and Penas, E., Primary production of benthic microalgae in an intertidal sand flat of the Ria de Arosa, NW Spain, *Mar. Ecol. Prog. Ser.,* 25, 111, 1985.

5140. Vartapetian, B.B., Andreeva, I.N., and Nuritdinov, N., Plant cells under oxygen stress, in *Plant Life in Anaerobic Environments,* Hook, D.D., and Crawford, R.M.M., Eds., Ann Arbor Science, Ann Arbor, Michigan, 1978, 13.

5141. Vasander, H., Plant biomass and production in virgin, drained and fertilized sites in a rised bog in southern Finland, *Ann. Bot. Fennici,* 19, 103, 1982.

5142. Vaucher, J.P., *Histoire des Conferves d'eau douce,* Geneve, 1803.

5143. Vavruška, A., Determination of nutrients in the most widely spread aquatic, littoral and swamp plants with regard to their utilization for composting, *Práce VÚRH Vodňany,* 6, 41, 1966, (in Czech with English summary).

5144. Véber, K., and Zahradník, J., Tertiary treatment by means of autotrophic microorganisms and higher plants, Academia Praha, *Studie ČSAV,* 24, 1, 1986. (in Czech)

5145. Veenstra, R.B., Effects of nutrients and velocity upon periphytic algal populations in laboratory streams, M.S. Thesis, Department of Civil Engineering, University of Washington, Seattle, 1982.

5146. Vega, J.M., Herrera, J., Aparicio, P.J., Paneque, A., and Losada, M., Role of molybdenum in nitrate reduction by *Chlorella, Plant Physiol.,* 48, 294, 1971.

5147. Vega, J.M., Gotor, C., and Menacho, A., Enzymology of the assimilation of ammonium by the green alga *Chlamydomonas reinhardtii,* in *Inorganic Nitrogen Metabolism,* Ullrich, W.R., Aparicio, P.J., Syrett, P.J., and Castillo, F., Eds., Springer-Verlag, Berlin, 1987, 132.

5148. Velásquez, J., Communities and production of ecology of *Eleocharis acicularis* (L.) R. et Sch., Thesis, Inst. of Botany, Czechoslovak Academy of Sciences, Průhonice, 1985.

5149. Velichko, I.M., The role of iron and manganese in the vital activities of blue-green algae of the genus *Microcystis, Mikrol. Lem. Selskochoz. Med. Respub. Mezhvedom SB,* 4, 11, 1968 (in Russian).

5150. Venkataraman, G.S., *The Cultivation of Algae,* Indian Council of Agricultural Research, New Delhi, 1969.

5151. Vergara, J.J., and Niell, F.X., Effects of nitrate availability and irradiance on internal constituents in *Corallina elongata* (Rhodophyta), *J. Phycol.,* 29, 285, 1993.

5152. Verhoeven, J.T.A., Nutrient dynamics in mesotrophic fens under the influence of eutrophicated ground water, in *Proc. Int. Symp. Aquatic Macrophytes,* Nijmegen, The Netherlands, 1983, 241.

5153. Verhoeven, J.T.A., Nutrient dynamics in minerotrophic peat mires, *Aquat. Bot.,* 25, 117, 1986.

5154. Verhoeven, J.T.A., and Arts, H.H.M., *Carex* litter decomposition and nutrient release in mires with different water chemistry, *Aquat. Bot.,* 43, 365, 1992.

5155. Verhoeven, J.T.A., Van Beek, S., Dekker, S., and Strom, W., Nutrient dynamics in small mesotrophic fens surrounded by cultivated land.I. Productivity and nutrient uptake by the vegetation in relation to the flow of eutrophicated ground water, *Oecologia,* 60, 25, 1983.

5156. Verhoeven, J.T.A., Schmitz, M.B., and Pons, T.L., Comparative demographic study of *Carex rostrata* Stokes, *C. diandra* Schrank and *C. acutiformis* Ehrn. in fens of different nutrient status, *Aquat. Bot.,* 30, 95, 1988.

5157. Verhoff, F.H., and Heffner, M.R., Rates of availability of total phosphorus in river waters, *Environ. Sci. Technol.,* 13, 844, 1979.

5158. Verma, K.R., Pandey, D., and Ambasht, R.S., Productive status of marsh zone vegetation of Gujar lake (khetasarai, Jaipur), India, *Int. J. Ecol. Environ. Sci.,* 7, 29, 1981.

5159. Vernon, L.P., Mechanism of oxygen evolution of photosynthesis, in *Biologistics for Space Systems,* U.S. Air Force Doc. Rept. AMRL-TDR-62-116), 1962, 131.

5160. Veroy, R.L., Montaño, N., de Guzman, M.L.B., Laserna, F.C., and Cajipe, G.J.B., Studies on the binding of heavy metals to algal polysaccharides from Philippine seaweeds. I. Carrageenan and the binding of lead and cadmium, *Bot. Mar.*, 23, 59, 1980.

5161. Verry, E.S., and Boelter, D.H., Peatland hydrology, in *Proc. Symp. Wetland Function and Values: The State of Our Understanding*, Greeson, P.E., Clark, J.R., and Clark, J.E., Eds., American Water Resources Assoc., Minneapolis, Minnesota, 1979, 389.

5162. Verstreate, D.R., Storch, T.A., and Dunham, V.L., A comparison of the influence of iron on the growth and nitrate metabolism of *Anabaena* and *Scenedesmus*, *Physiol. Plant.*, 50, 47, 1980.

5163. Vesk, M., and Jeffrey, S.W., Effect of blue-green light on photosynthetic pigments and chloroplast structure in unicellular marine algae from six classes, *J. Phycol.*, 13, 280, 1977.

5164. Vesk, M., and Borowitzka, M.A., Ultrastructure of tetrasporogenesis in the coralline alga *Haliptilon cuvieri* (Rhodophyta), *J. Phycol.*, 20, 501, 1984.

5165. Vieira, A.A.H., and Klaveness, D., The utilization of organic nitrogen compounds as sole nitrogen source by some freshwater phytoplankton, *Nordic J. Bot.*, 6, 93, 1986.

5166. Vilenkin, B.Y., and Pertsov, N.A., Cell division of alga at the interface in different hydrodynamic regimes of streaming, *Biophysics* (English translation of *Biofizika*), 28, 494, 1983.

5167. Vincent, W.F., and Goldman, C.R., Evidence for algal heterotrophy in Lake Tahoe, California, Nevada, *Limnol. Oceanogr,* 25, 89, 1980.

5168. Virtanen, A.I., and Miettinen, J.K., Biological nitrogen fixation, in *Plant Physiology*, Vol. 3, *Inorganic Nutrition of Plants*, Steward, F.C., Ed., Academic Press, New York and London, 1963, 539.

5169. Voelz, H., Voelz, U., and Ortigoza, R.O., The "polyphosphate overplus" phenomenon in *Myxococcus xanthus* and its influence on the architecture of the cell., *Arch. Mikrobiol.*, 55, 371, 1966.

5170. Vogel, S., *Life in Moving Fluids: the Physical Biology of Flow*, Willard Grant Press, Boston, Massachusetts, 1981.

5171. Vogt, G., and Kittelberger, F., Studie zur Aufnahme und Anreicherung von Schwermetallen in typischen Algenassoziationen des Rheimes zwischen Germersheim und Gernsheim, in *Fisch und Umwelt*, Vol. 3, Reichenbach-Klinke, H., Ed., Gustav Fischer, Stuttgart, 1977, 15.

5172. Volcani, B.E., Cell wall formation in diatoms: morphogenesis and biochemistry, in *Silicon and Siliceous Structures in Biological Systems*, Simpson, T.L., and Volcani, B.E., Eds., Springer-Verlag, New York, 1981, 157.

5173. Volesky, B., Zajic, J.E., and Knettig, E., Algal products, in *Properties and Products of Algae*, Zajic, J.E., Ed., Plenum Press, 1970, 49.

5174. Volkova, S.P., Lunevskii, V.Z., Spridonov, N.A., Vinokurov, M.G., and Berestovskij, G.N., Chemical composition of calcium channels in cells of the charophyte algae, *Bio. Fizika*, 25, 537, 1980, (in Russian).

5175. Vollenweider, R.A., Ökologische Untersuchungen von praktischen Algen auf experimentelles Grundlage, *Schweiz. Z. Hydrol.*, 12, 194, 1950.

5176. Vollenweider, R.A., Calculation models of photosynthesis-depth curves and some implications regarding day rate estimates in primary production measurements, *Mem. Ist. Ital. Idrobiol.*, (Suppl.) 18, 425, 1965.

5177. Vollenweider, R.A., Scientific fundamentals of the eutrophication of lakes and flowing waters, with particular reference to phosphorus and nitrogen as factors in eutrophication, OECD/DAS/C 51/68, Paris, 1968.

5178. Vollenweider, R.A., Possibilities and limits of elementary models concerning the budget of substances in lakes, *Arch. Hydrobiol.*, 66, 1, 1969.

5179. Vollenweider, R.A., Ed., *A Manual on Methods for Measuring Primary Production in Aquatic Environments*, I.B.P. Handbook No. 12, Blackwell Scientific Publications, Oxford, 1974.

5180. Vollenweider, R.A., Input-output models; with special reference to the phosphorus loading concept in limnology, *Schweiz. Z. Hydrol.*, 37, 53, 1975.

5181. Vollenweider, R.A., and Kerekes, J., Eutrophication in waters: monitoring, assessment and control, OECD Report, Paris, 1982.

5182. Vollenweider, R.A., Munavar, M., and Stadelman, P., A comparative review of phytoplankton and primary production in the Laurentian Great Lake, *J. Fish. Res. Bd. Can.*, 31, 739, 1974.

5183. von Holt, C., and von Holt, M., Transfer of photosynthetic products from zooxanthellae to coelenterate hosts, *Comp. Biochem. Physiol.*, 24, 75, 1968.

5184. von Stosch, H.A., Haptophyceae, in *Vegetative Fortpflanzung, Parthenogenese und Apogamie bei Algen*, Ruhland, W., Ed., *Encycl. Plant Physiol.*, 18, 646, 1967.

5185. von Stosch, H.A., An amended terminology of the diatom girdle, *Nova Hedw. Beih.*, 53, 1, 1975.

5186. von Fellenberg, T., Das vorkommen, der Kreislauf und der Stoffwechsel des Jods, *Ergeb. Physiol.*, 25, 176, 1926.

5187. Vyas, L.N., Sharma, K.P., Sankhla, S.K., and Gopal, B., Primary production and energetics, in *Ecology and Management of Aquatic Vegetation in the Indian Subcontinent*, Gopal, B., Ed., Kluwer Academic Publishers, Dordrecht, The Netherlands, 1990, 149.

5188. Vymazal, J., Heavy metal content in Šárecký Brook, Praha, *Water Technol. - Econ. Inf.*, 26, 187, 1984 (in Czech).

5189. Vymazal, J., Short-term uptake of heavy metals by periphyton algae, *Hydrobiologia*, 119, 171, 1984.

5190. Vymazal, J., Occurrence of heavy metals in surface waters and their removal by means of periphyton, in *Proc. Ministry of Water Management Workshop*, Praha, Czech Republic, 1984, 17 (in Czech).

5191. Vymazal, J., The use of periphyton for phosphorus, nitrogen and heavy metal removal from polluted waters, Ph.D. Thesis, Prague Institute of Chemical Technology, Praha, Czech Republic, 1985.

5192. Vymazal, J., Occurrence and chemistry of zinc in freshwaters—its toxicity and bioaccumulation with respect to algae: a review. Part 1. Occurrence and chemistry of zinc in freshwaters, *Acta Hydrochim. Hydrobiol.*, 13, 627, 1985.

5193. Vymazal, J., Occurrence and chemistry of zinc in freshwaters—its toxicity and bioaccumulation with respect of algae: a review. Part 2. Toxicity and bioaccumulation with respect to algae, *Acta Hydrochim. Hydrobiol.*, 14, 83, 1986.

5194. Vymazal, J., The use of periphyton for nutrient removal from waters, *Br. Phycol. Soc. Newsletter*, 21, 1, 1986.

5195. Vymazal, J., Toxicity and accumulation of cadmium with respect to algae and cyanobacteria: a review, *Toxic. Assess.*, 2, 387, 1987.

5196. Vymazal, J., Zn uptake by *Cladophora glomerata*, *Hydrobiologia*, 148, 97, 1987.

5197. Vymazal, J., Ammonium uptake and biomass interaction in *Cladophora glomerata* (Chlorophyta), *Br. Phycol. J.*, 22, 163, 1987.

5198. Vymazal, J., The use of periphyton communities for nutrient removal from polluted streams, *Hydrobiologia*, 166, 225, 1988.

5199. Vymazal, J., *Cladophora glomerata*—an indicator of heavy metal pollution of aquatic environments, in *Proc. 8th Conf. Czechoslovak Limnological Society*, 1988, 217. (in Czech).

5200. Vymazal, J., The time course of heavy metal uptake by periphytic alga, in *Proc. 8th Conf. Czechoslovak Limnological Society*, 1988, 220 (in Czech).

5201. Vymazal, J., The influence of pH on heavy metal uptake by *Cladophora glomerata*, *Br. Phycol. J.*, 23, 297, 1988.

5202. Vymazal, J., Size fraction of heavy metals in waters, *Acta Hydrochim. Hydrobiol.*, 17, 309, 1989.

5203. Vymazal, J., Heavy metals accumulation by *Cladophora glomerata*, *Water Manage.*, B 39, 220, 1989 (in Czech).

5204. Vymazal, J., Use of periphyton for nutrient removal from waters, in *Constructed Wetlands for Wastewater Treatment*, Hammer, D.A., Ed., Lewis Publishers, Chelsea, Michigan, 1989, 558.

5205. Vymazal, J., Toxicity and accumulation of lead with respect to algae and cyanobacteria: a review, *Acta Hydrochim. Hydrobiol.*, 18, 513, 1990.

5206. Vymazal, J., Uptake of lead, chromium, cadmium and cobalt by *Cladophora glomerata*, *Bull. Environ. Contam. Toxicol.*, 44, 468, 1990.

5207. Vymazal, J., Uptake of heavy metals by *Cladophora glomerata*, *Acta Hydrochim. Hydrobiol.*, 18, 657, 1990.

5208. Vymazal, J., and Sládečková, A., Periphyton—an indicator of wastewater treatment efficiency, *Water Manage. Tech. Econ. Inform.*, (Prague), 29, 316, 1987 (in Czech).

5209. Vymazal, J., and Sládečková, A., Wastewater treatment efficiency evaluation by means of periphytic communities, in *Proc. Conf. Technical Hydrobiology—Biology of Treatment Processes*, DT ČS VTS, Pardubice, Czech Republic, 1987, 151 (in Czech).

5210. Vymazal, J., and Richardson, C.J., Species composition, biomass and nutrient content of periphyton in the Florida Everglades, *J. Phycol.*, 1994, in press.

5211. Vymazal, J., Matulová, D., and Sládečková, A., The use of periphyton for nutrient removal from waters, in *Proc. Conf., Vegetation in Water Treatment and its Application*, Žáková, Z., Květ, J., Lhotský, O., and Marvan, P., Eds., VÚV Brno, Czech Republic, 1987, 145 (in Czech).

5212. Vymazal, J., Craft, C.B., and Richardson, C.J., Periphyton response to nitrogen and phosphorus additions in Florida Everglades, *Algol. Studies*, 73, 75, 1994.

5213. Waaland, J.R., Growth of Pacific Northwest marine algae in semi-closed culture, in *The Marine Plant Biomass of the Pacific Northwest Coast*, Krauss, R.W., Ed., Oregon State University Press, 1977, 117.

5214. Waaland, J.R., Commercial Utilization, in *The Biology of Seaweeds*, Lobban, C.S., and Wynne, J.M., Eds., University of California Press, Berkeley, 1981, 726.

5215. Wachs, B., Die Bioindikation von Schwermetallen in Fliessgewässern, *Münchener Beitr Abwasser-, Fischerei- und Flussbiol.*, 34, 301, 1982.

5216.   Wacker, W.E.C., Nucleic acids and metals. III. Changes in nucleic acid, protein, and metal content as a consequence of Zn deficiency in *Euglena gracilis, Biochemistry* 1, 859, 1962.

5217.   Wada, T., Yokoyama, T., and Takai, Y., Absorption of CO_2 by rice roots from soil solution of the submerged soil. 1. Transfer of CO_2 from rice rhizosphere to rice shoot and the air, *Jpn. J. Soil Sci. Plant Nutr.,* 54, 215, 1983.

5218.   Wagemann, R., Some theoretical aspects of stability and solubility of inorganic arsenic in the freshwater environment, *Wat. Res.,* 12, 139, 1978.

5219.   Wagemann, R., Snow, N.B., Rosenberg, D.M., and Lutz, A., Arsenic in sediments, water and aquatic biota from lakes in the vicinity of Yellowknife, Northwest Territories, Canada, *Arch. Environ. Contam. Toxicol.,* 7, 169, 1978.

5220.   Wagman, D.D., Evans, W.H., Parker, V.B., Hallow, I., Bailey, S.M., and Schumm, R.H., Selected values of chemical thermodynamic properties, *Natl. Bur. Stand. Tech. Note,* 270–3, 1, 1969.

5221.   Waisel, Y., Agami, M., and Shapira, Z., Uptake and transport of ^{86}Rb, ^{32}P, ^{36}Cl and ^{32}Na by four submerged hydrophytes, *Aquat. Bot.,* 13, 179, 1982.

5222.   Waite, T.D., and Mitchell, R., The effect of nutrient fertilization on the benthic alga *Ulva lactuca, Bot. Mar.,* 15, 151, 1972.

5223.   Waits, E.D., Net primary productivity of an irregularly flooded North Carolina salt marsh, Ph.D. Thesis, North Carolina State University, Raleigh, 1967.

5224.   Wake, L.V., Christopher, R.K., Rickard, A.D., Andersen, J.E., and Ralph, B.J., A thermodynamic assessment of possible substrates for sulfate-reducing bacteria, *Aust. J. Biol. Sci.,* 30, 155, 1977.

5225.   Waksman, S.A., The parts of New Jersey and their utilization. Part. A: Nature and origin of peat composition and utilization, N.J. Dept. Cons. Dev. and Rutgers University, New Jersey, *Agric. Exp. Station Bull.,* 55, 1942.

5226.   Walbridge, M.R., and Struthers, J.P., Phosphorus retention in non-tidal palustrine forested wetlands of the mid-Atlantic region, *Wetlands,* 13, 84, 1993.

5227.   Walker, D.R., Flora, R.D., Rice, R.G., and Scheidt, D.J., Response of the Everglades marsh to increased nitrogen and phosphorus loading, National Park Service, South Florida Research Center, Everglades national Park, Homestead, Florida, 1980.

5228.   Walker, J.B., Inorganic micronutrient requirements of *Chlorella*. I. Requirements for calcium (or strontium), copper, and molybdenum, *Arch. Biochem. Biophys.,* 46, 1, 1953.

5229.   Walker, J.B., Inorganic micronutrients requirement of *Chlorella*. II. Quantitative requirements for iron, manganese and zinc, *Arch. Biochem. Biophys.,* 53, 1, 1954.

5230.   Walker, J.B., Strontium inhibition of calcium utilization by a green algae, *Arch. Biochem. Biophys.,* 60, 264, 1956.

5231.   Walker, N.A., The transport systems of Charophyte and Chlorophyte giant algae and their integration into modes of behavior, in *Plant Membrane Transport: Current Conceptual Issues,* Spanswick, R.M., Lucas, W.J., and Dainty, J., Eds., Elsevier North-Holland, Amsterdam, 1980, 287.

5232.   Walker, N.A., The uptake of inorganic carbon by freshwater plants, *Plant Cell Environ.,* 6, 323, 1983.

5233.   Walker, N.A., The carbon species taken up by *Chara*: a question of unstirred layers, in *Inorganic Carbon Uptake by Aquatic Photosynthetic Organisms,* Lucas, W.J., and Berry, J.A., Eds., Am. Soc. Plant Physiol., Rockville, Maryland, 1985, 31.

5234.   Walker, N.A., and Smith, F.A., Circulating electric currents between acid and alkaline zones associated with HCO_3^- assimilation in *Chara, J. Exp. Bot.,* 28, 1190, 1977.

5235.   Walker, N.A., Smith, F.A., and Cathers, I.R., Bicarbonate assimilation by freshwater charophytes and higher plants. II. Membrane transport of bicarbonate ions is not proven, *J. Membr. Biol.,* 57, 51, 1980.

5236.   Wallace, W., and Nicholas, D.J.D., The biochemistry of nitrifying microorganisms, *Biol. Rev.,* 44, 359, 1969.

5237.   Wallén, B., Falkengren-Grerup, U., and Malmer, N., Biomass, productivity and relative rate of photosynthesis of *Sphagnum* at different water levels on a South Swedish peat bog, *Holarctic Ecol.,* 11, 70, 1988.

5238.   Wallen, D.G., The toxicity of chromium (VI) to photosynthesis of the phytoplankton assemblage of Lake Erie and the diatom *Fragilaria crotonensis* Kitton, *Aquat. Bot.,* 38, 331, 1990.

5239.   Wallen, D.G., and Cartier, L.D., Molybdenum dependence, nitrate uptake and photosynthesis of freshwater plankton algae, *J. Phycol.,* 11, 345, 1975.

5240.   Wallen, D.G., and Geen, G.H., Light quality in relation to growth, photosynthetic rates and carbon metabolism in two species of marine plankton-algae, *Mar. Biol.,* 10, 34, 1971.

5241.   Wallentinus, I., Productivity studies on Baltic macroalgae, *Bot. Mar.,* 21, 365, 1978.

5242. Wallentinus, I., Environmental influences on benthic macrovegetation in the Trosa Askö area, northern Baltic proper. III. On the significance of chemical constituents in some macroalgal species, Askö Laboratory Mimeogr. Rep., 1979.

5243. Wallentinus, I., Productivity studies on *Cladophora glomerata* (L.) Kützing in the northern Baltic proper. in *Proc. 10th Eur. Symp. Mar. Biol. Population Dynamics of marine Organisms in Relation to Nutrient Cycling in Shallow Waters,* Part 2, Persoone, G., and Jaspers, E., Eds., Universa Press, Wetteren, 1981, 631.

5244. Wallentinus, I., Comparisons of nutrient uptake rates for baltic macroalgae with different thallus morphologies, *Mar. Biol.,* 80, 215, 1984.

5245. Wallner, J., *Oocardium stratum* Näg., eine wichtige tuffbildende Alge Südbayerns, *Planta,* 20, 287, 1933.

5246. Walne, P.L., Euglenoid flagellates, in *Phytoflagellatyes,* Cox, E.R., Ed., Elsevier/North Holland, New York, 1980, 165.

5247. Walsby, A.E., The permeability of blue-green algal gas-vacuole membrane to gas, *Proc. R. Soc. Lond.,* B 173, 235, 1969.

5248. Walsby, A.E., The nuisance algae: curiosities in the biology of planktonic blue-green algae, *Water Treatment and Examination,* 19, 359, 1970.

5249. Walsby, A.E., The pressure relationships of gas-vacuoles, *Proc. R. Soc. Lond.,* B 178, 301, 1971.

5250. Walsby, A.E., Structure and function of gas vacuoles, *Bacteriol. Rev.,* 36, 1, 1972.

5251. Walsby, A.E., The extracellular products of *Anabaena cylindrica* Lemm. I. Isolation of macromolecular pigment-peptide complex and other components, *Br. Phycol. J.,* 9, 371, 1974.

5252. Walsby, A.E., The extracellular products of *Anabaena cylindrica* Lemm. II. Fluorescent substances containing serine and threonine; and their role in extracellular pigment formation, *Br. Phycol. J.,* 9, 383, 1974.

5253. Walsby, A.E., The gas vacuoles of blue-green algae, *Sci. Am.,* 237, 90, 1979.

5254. Walsby, A.E., and Klemer, A.R., The role of gas vacuoles in the microstratification of a population of *Oscillatoria agardhii* var. *isothrix* in Deming Lake, Minnesota, *Arch. Hydrobiol.,* 74, 375, 1974.

5255. Walsh, G.E., and Alexander, S.V., A marine algal bioassay method: results with pesticides and industrial wastes, *Water Air Soil Pollut.,* 13, 45, 1980.

5256. Walsh, G.E., and Garnas, R.L., Determination of bioactivity of chemical fractions of liquid wastes using freshwater and saltwater algae and crustacea, *Environ. Sci. Technol.,* 17, 182, 1983.

5257. Walsh, G.E., Duke, K.M., and Foster, R.B., Algae and crustaceans as indicators of bioactivity of industrial wastes, *Water Res.,* 16, 879, 1982.

5258. Walsh, L.M., and Keeney, D.R., Behavior and phototoxicity of inorganic arsenicals in soils, in *Arsenical Pesticides,* Worelson, E.A., Ed., Am. Chem. Soc., Washington, 1975, 35.

5259. Walter, M.R., Ed., *Stromatolites,* Elsevier, New York, 1976.

5260. Walter, M.R., Introduction, in *Stromatolites,* Walter, M.R., Ed., Elsevier, Amsterdam, 1976, 1.

5261. Walton, C.P., and Lee, G.F., A biological evaluation of the molybdenum blue method for orthophosphate analysis, *Verh. Internat. Verein. Limnol.,* 18, 676, 1972.

5262. Wanders, J.B.W., The role of benthic algae in the shallow reef of Curaçao (Netherlands Antilles). II. Primary productivity of the Sargassum beds on the north-east coast submarine plateau, *Aquat. Bot.,* 2, 327, 1976.

5263. Wang, B., and Schaeffer, W.P., Structure of an oxygen-carrying cobalt complex, *Science,* 166, 1404, 1969.

5264. Warburg, O., and Negelein, E., Über die Reduktion der Salpetersäure in grünen Zellen, *Biochem. Z.,* 110, 66, 1920.

5265. Warburg, O., and Lüttgens, W., Photochemische Reduktion des Chinons in grünen Zellen und Granula, *Biokhimiya,* 11, 303, 1946.

5266. Warburg, O., Krippahl, G., and Bucholz, W., Wirkung von Vanadium auf die Photosynthese, *Z. Naturf.,* 10 B, 422, 1955.

5267. Ward, A.K., and Wetzel, R.G., Sodium: some effects on blue green algal growth, *J. Phycol.,* 11, 357, 1975.

5268. Waren, H., Nahrungsphysiologische Versuche an *Micrasterias rotata,* Soc. Sci. Fennica *(Helsingfors), Commentat Biol.* II, 8, 1, 1926.

5269. Warington, K., The effect of boric acid and borax on the broad bean and certain other plants, *Ann. Bot.,* 37, 630, 1923.

5270. Waris, H., The significance for algae of chelating substances in the nutrient solutions, *Physiol. Plant.,* 6, 538, 1953.

5271. Warkentine, B.E., and Rachlin, J.W., A test of proposed organizational framework for the ordering of algal toxicity responses, *Bull. Torrey Bot. Club,* 113, 12, 1986.

5272. Warming, E., *Handbuch der systematischen Botanik,* Berlin, 1850.

5273. Wasmund, N., Ecology and bioproduction in the microphytobenthos of the chain of shallow inlets (Boddens) south of the Darss-Zingst Peninsula (southern Baltic Sea), *Int. Rev. ges. Hydrobiol.*, 71, 153, 1986.

5274. Wass, M.L., and Wright, T.D., Coastal wetlands of Virginia. Interim report to the Governor and General Assembly. *Virginia Inst. Mar. Sci. Rpt. in Appl. Mar. Sci. and Ocean Eng.*, 10, 1, 1969.

5275. Wassing, E.C., Kok, B., van Oorschot, J.L.P., The efficiency of light-energy conversion in *Chlorella* cultures as compared with higher plants, in *Algal Culture: From Laboratory to Pilot Plant*, Burlew, J.S., Ed., Carnegie Inst. of Washington Publ. 600, Washington, DC, 1953, 55.

5276. Wassman, E.R., and Ramus, J., Primary-production measurements for the green seaweed *Codium fragile* in Long Island Sound, *Mar. Biol.*, 21, 289, 1973.

5277. Watabe, N., Crystallographic analysis of the coccolith of *Coccolithus huxleyi*, *Calcif. Tiss. Res.*, 1, 114, 1967.

5278. Watanabe, A., Production of some amino acids by the atmospheric nitrogen - fixing blue-green algae, *Arch. Biochem. Biophys.*, 34, 50, 1951.

5279. Watanabe, A., Studies on application of Cyanophyta in Japan, *Schweiz. Z. Hydrol.*, 32, 566, 1970.

5280. Watanabe, A., Practical significance of algae in Japan - Nitrogen fixation by algae, in *Advance of Phycology in Japan*, Tokida, J., and Hirose, H., Eds., Dr. W. Junk Publishers, The Hague, The Netherlands, 1975, 255.

5281. Watanabe, A., and Yamamoto, Y., Heterotrophic nitrogen fixation by the blue-green alga *Anabaenopsis circularis*, *Nature*, 214, 788, 1967.

5282. Watanabe, A., Nishigaki, S., and Konishi, C., Effects of nitrogen-fixing blue-green algae on the growth of rice plants, *Nature*, 168, 748, 1951.

5283. Watanabe, I., Nitrogen fixation by non-legumes in tropical agriculture with special reference to wetland rice, *Plant Soil*, 90, 343, 1986.

5284. Watanabe, M.M., Kaya, K., and Takamura, N., Fate of the toxic cyclic heptapeptides, the microcystins, from blooms of *Microcystis* (Cyanobacteria) in a hypertrophic lake, *J. Phycol.*, 28, 761, 1992.

5285. Watanabe, M., Takamatsu, T., Kohata, K., Kunugi, M., Kawashima, M., and Koyama, M., Luxury phosphate uptake and variation of intracellular metal concentrations in *Heterosigma akashiwo* (Raphidophyceae), *J. Phycol.*, 25, 428, 1989.

5286. Waterbury, J.B., and Stanier, R.Y., The unicellular cyanobacteria which reproduce by budding, *Arch. Mikrobiol.*, 115, 249, 1977.

5287. Waterbury, J.B., and Stanier, R.Y., Patterns of growth and development in pleurocapsalean Cyanobacteria, *Microb. Rev.*, 42, 2, 1978.

5288. Waterbury, J.B., and Stanier, R.Y., Isolation and growth of Cyanobacteria from marine and hypersaline environments, in *The Prokaryotes*, Vol. 1, Starr, M.P., Stolp, H., Trüper, H.G., Balows, A., Schlegel, H.G., Eds., Springer-Verlag, Berlin, 1981, 221.

5289. Waters, T.F., Notes on the chlorophyll method of estimating the photosynthetic capacity of stream periphyton, *Limnol. Oceanogr.*, 6, 486, 1961.

5290. Watson, D.G., Cushing, C.E., Coutant, C.C., and Templeton, W.L., Cycling of radionuclides in Columbia River biota, in *Trace Substances in Environmental Health* IV., Hemphill, D.D., Ed., University of Missouri, Columbia, Missouri, 1970, 144.

5291. Watt, J., and Marcus, R., Experimental ulcerative disease of the colon in animals, *Gut*, 14, 506, 1973.

5292. Watt, W.D., Release of dissolved organic material from the cells of phytoplankton populations, *Proc. R. Soc. Lond.*, B 164, 521, 1966.

5293. Watt, W.D., Extracellular release of organic matter from two freshwater diatoms, *Ann. Bot.*, 33, 427, 1969.

5294. Watt, W.D., and Fogg, G.E., The kinetics of extracellular glycollate production by *Chlorella pyrenoidosa*, *J. Exp. Bot.*, 17, 117, 1966.

5295. Weakley, A.S., and Schafale, M.P., Classification of pocosins of the North Coastal Plains, *Wetlands*, 11, 355, 1991.

5296. Weaver, C.I., and Wetzel, R.G., Carbonic anhydrase levels and internal lacunar CO_2 concentrations in aquatic macrophytes, *Aquat. Bot.*, 8, 173, 1980.

5297. Weber, C.A., Über Torf, Humus und Moor, *Abhandl. Naturw. Ver. Bremen*, 17, 456, 1903.

5298. Weber, C.A., Aufbau und Vegetation der Moore Norddeutschlands, *Beiblatt Bot. Jahrbuch.*, 90, 19, 1907.

5299. Weber, C.I., and McFarland, B.H., Effects of exposure time, season, substrate type, and planktonic populations on the taxonomic composition of algal periphyton on artificial substrates in the Ohio and Little Miami Rivers, Ohio, in *Ecological Assessment of Effluent Impacts on Communities of Indigenous Aquatic Organisms*, Bates, J.M., and Weber, C.I., Eds., American Society for Testing and Materials, Philadelphia, Pennsylvania, 1981, 166.

5300. Weber, W.J., and Stumm, W., Buffer systems of natural fresh waters, *J. Chem. Engng. Data,* 8, 464, 1963.
5301. Weber, W.J., and Stumm, W., Mechanism of hydrogen ion buffering in natural waters, *J. Am. Wat. Wks. Ass.,* 55, 1553, 1963.
5302. Webster, D.A., and Hackett, D.P., Respiratory chain of colorless algae. I. Chlorophyta and Euglenophyta, *Plant Physiol.,* 40, 1091, 1965.
5303. Webster, D.A., and Hackett, D.P., Respiratory chain of colorless algae. II. Cyanophyta, *Plant Physiol.,* 41, 599, 1966.
5304. Webster, G.C., Nitrogen metabolism, *Ann. Rev. Plant Physiol.,* 6, 43, 1955.
5305. Webster, J.R., and Simmons, G.M., Jr., Leaf breakdown and invertebrate colonization on a reservoir bottom, *Verh. Internat. Verein. Limnol.,* 19, 1587, 1978.
5306. Wedding, R.T., and Black, M.K., Uptake and metabolism of sulfate by *Chlorella.* I. Sulfate accumulation and active sulfate, *Plant Physiol.,* 35, 72, 1950.
5307. Wedemayer, G.J., Wilcox, L.W., and Graham, L.E., *Amphidinium cryophilum* sp.nov. (Dinophyceae) a new freshwater dinoflagellate. I. Species description using light and scanning electron microscopy, *J. Phycol.,* 18, 13, 1982.
5308. Wedepohl, K.H., Environmental influence on the chemical composition of shales and clays, *Phys. Chem. Earth,* 8, 305, 1970.
5309. Weger, H.C., and Turpin, D.H., Mitochondrial respiration can support NO_3^- and NO_2^- reduction during photosynthesis. Interaction between photosynthesis, respiration, and N assimilation in the N-limited green alga *Selenastrum minutum, Plant Physiol.,* 89, 409, 1989.
5310. Wehr, J.D., Analysis of seasonal succession of attached algae in a mountain stream, the North Alouette River, British Columbia, *Can. J. Bot.,* 59, 1465, 1981.
5311. Wehr, J.D., and Whitton, B.A., Accumulation of heavy metals by aquatic mosses. 2. *Rhynchostegium riparioides, Hydrobiologia,* 100, 261, 1983.
5312. Wehr, J.D., and Whitton, B.A., Aquatic cryptogams of natural acid springs enriched with heavy metals: Kootenay Paint Pots, British Columbia, *Hydrobiologia,* 98, 97, 1983.
5313. Wehr, J.D., and Brown, L.M., Selenium requirement of a bloom-forming planktonic alga from softwater and acidified lakes, *Can. J. Fish. Aquat. Sci.,* 42, 1783, 1985.
5314. Wehrmeyer, W., Organization and composition and rhodophycean phycobilisomes, in *Photosynthetic Prokaryotes: Cell Differentiation and Function,* Papageourgiou, G.C., and Packer, L., Eds., Elsevier Science Publishing, Amsterdam, 1983, 1.
5315. Weidner, M., and Küppers, U., Phosphoenolpyruvate-carboxykinase und Ribulose-1,5-diphosphate-carboxylase von *Laminaria hyperborea* (Gunn.) Fosl.: Das Verteilungsmuster der Enzymaktivataten im Thallus, *Planta,* 114, 365, 1973.
5316. Weinmann, G., Gelöste Kohlenhydrate und andere organische Stoffe in naturlichen Gewässern und in Kulturen von *Scenedesmus quadricauda, Arch. Hydrobiol.* (Suppl.), 37, 164, 1970.
5317. Weisel, C.P., The atmospheric flux of elements from the oceans, Ph.D. Thesis, University of Rhode Island, 1981.
5318. Weisner, S.E.B., Factors affecting the internal oxygen supply of *Phragmites australis* (Cav.) Trin. ex Steudel in situ, *Aquat. Bot.,* 31, 329, 1988.
5319. Weiss, H., Beitrag zur Frage einer Algenmassenzucht in Deutschland, *Zentralbl. Bakter Parasit. Int. und Hygiene II. Abstr,* 107, 230, 1953.
5320. Weiss, R.L., Fine structure of the snow alga (*Chlamydomonas nivalis*) and associated bacteria, *J. Phycol.,* 19, 200, 1983.
5321. Weitzel, R.L., Ed., *Methods and Measurements of Periphyton Communities: A Review,* ASTM Special Tech. Publ., 690, American Society for Testing and Materials, Philadelphia, 1979.
5322. Weitzel, R.L., Periphyton measurements and applications, in *Methods and Measurements of Periphyton: A Review,* Weitzel, R.L., Ed., ASTM STP 690, Am. Soc. for Testing and Materials, Philadelphia, 1979, 3.
5323. Welch, P.S., *Limnology,* 2nd ed, McGraw-Hill, New York, 1952.
5324. Welcomme, R.L., *Fisheries Ecology of Floodplains Rivers,* Longman, London, 1979.
5325. Weller, M.W., *Freshwater Marshes. Ecology and Wildlife Management,* 2nd ed., University of Minnesota Press, Minneapolis, 1987.
5326. Weller, M.W., and Fredrikson, L.H., Avian ecology of a managed glacial marsh, *Living Bird,* 12, 269, 1974.
5327. Wells, J.R., Kaufman, P.B., and Jones, J.D., Heavy metal concentration in some macrophytes from Saginaw Bay (Lake Huron, U.S.A.), *Aquat. Bot.,* 9, 185, 1980.
5328. Welschmeyer, N.A., and Lorenzen, C.J., Chlorophyll-specific photosynthesis and quantum efficiency at subsaturating light intensities, *J. Phycol.,* 17, 283, 1981.

5329. Wentz, W.A., The effects of sewage effluents on the growth and productivity of peatland plants, Ph.D. Thesis, University of Michigan, Ann Arbor, 1975.

5330. Werner, D., Die Kieselsäure im Stoffwechsel von *Cyclotellacryptica*Reimann, Lewin and Guillard, *Arch. Mikrobiol.*, 55, 278, 1966.

5331. Werner, D., Hemmung der Chlorophyllsynthese und der NADP$^+$-abhängigen Glycerinaldehyd-3-Phosphate-Dehydrogenase durch Germaniumsäure bei *Cyclotellacryptica,Arch. Mikrobiol.*, 57, 51, 1967.

5332. Werner, D., Ed., *The Biology of Diatoms*, University of California Press, Berkeley and Los Angeles, 1977.

5333. Westall, J., and Stumm, W., The hydrosphere, in *The Handbook of Environmental Chemistry* Vol. 1, Part A. *The Natural Environmental and Biogeochemical Cycles*, Hutzinger, O., Ed., Springer-Verlag, Berlin, 1980, 17.

5334. Westbroek, P., van der Wal, P., van Emburg, P.R., de Vrind-de Jong, E.W., and de Bruijn, W.C., Calcification in the coccolithophorids *Emilianiahuxleyi*and *Pleurochrysiscarterae*.I. Ultrastructural aspects, in *Biomineralizationin Lower Plants and Animals*, Leadbeater, B.S.C., and Riding, R., Eds., Clarendon Press, Oxford, 1986, 189.

5335. Westermann, P., and Ahring, B.K., Dynamics of methane production, sulfate reduction, and denitrification in a permanently waterlogged alder swamp, *Appl. Environ. Microbiol.*, October, 2554, 1987.

5336. Westlake, D.F., Comparisons of plant production. *Biol. Rev.*, 38, 385, 1963.

5337. Westlake, D.F., The biomass productivity of *Glyceria grandis*. I. Seasonal changes in biomass, *J. Ecol.*, 57, 745, 1966.

5338. Westlake, D.F., Primary production of freshwater macrophytes, in *Photosynthesis and Productivity in Different Environments*, Cooper, J.P., Ed., Cambridge University Press, London, 1975, 189.

5339. Westlake, D.F., Květ, J., and Szczepanski, A., Eds., *Ecology of Wetlands*, IBP-Wetlands Synthesis Volume. Cambridge University Press, London, in press.

5340. Wetland Systems in Water Pollution Control, Proc. Int. Conf. in Sydney, University of New South Wales, Australia, 1992.

5341. Wetland Training Institute, *Field Guide for Wetland Delineation: 1987 Corps of Engineers Manual*, WTI 92-2, Polesville, 1991.

5342. Wetzel, R.G., Primary productivity of periphyton, *Nature*, 197, 1026, 1963.

5343. Wetzel, R.G., A comparative study of the primary productivity of higher aquatic plants, periphyton and the phytoplankton in a large shallow lake, *Int. Res. Ges. Hydrobiol.*, 49, 1, 1964.

5344. Wetzel, R.G., Productivity and nutrient relationship in mud lakes of northern Indiana, *Verh. Internat. Verein. Limnol.*, 16, 321, 1966.

5345. Wetzel, R.G., Variations in productivity of Goose and hypereutrophic Sylvan lakes, Indiana, *Invest. Indiana Lakes Streams*, 7, 147, 1966.

5346. Wetzel, R.G., Factors influencing photosynthesis and excretion of dissolved organic matter by aquatic macrophytes in hard-water lakes, *Verh. Internat. Verein. Limnol.*, 17, 72, 1969.

5347. Wetzel, R.G., Productivity investigations of interconnected lakes. I. The eight lakes of the Oliver and Walters chains, northeastern Indiana, *Hydrobiol. Stud.*, 3, 91, 1973.

5348. Wetzel, R.G., Foreword and introduction, in *Proc. Nat. Symp. Freshwater Wetlands: Ecological Processes and Management Potentials*, Good, R.E., Whigham, D.F., and Simpson, R.L., Eds., Academic Press, New York, San Francisco, London, 1978, xiii.

5349. Wetzel, R.G., *Limnology*, 2nd ed, Saunders College Publishing, Philadelphia, 1983.

5350. Wetzel, R.G., Ed., *Periphyton of Freshwater Ecosystems*, Dr. W. Junk Publishers, The Hague, The Netherlands, 1983.

5351. Wetzel, R.G., Attached algal-substrata interactions: fact or myth, and when and how?, in *Periphyton of Freshwater Ecosystems*, Wetzel, R.G., Ed., Dr. W. Junk Publishers, The Hague, The Netherlands, 1983, 207.

5352. Wetzel, R.G., and McGregor, D.L., Axenic cultures and nutritional studies of aquatic macrophytes, *Am. Midl. Nat.*, 80, 52, 1968.

5353. Wetzel, R.G., and Westlake, D.F., Periphyton, in *A Manual on Methods for Measuring Primary Production in Aquatic Environments*, Vollenweider, R.A., Ed., IBP Handbook No, 12, Blackwell, Oxford, 1969, 39.

5354. Wetzel, R.G., and Manny, B.A., Secretion of dissolved organic carbon and nitrogen by aquatic macrophytes, *Verh. Internat. Verein. Limnol.*, 18, 162, 1972.

5355. Wetzel, R.G., Rich, P.H., Miller, M.C., and Allen, H.L., Metabolism of dissolved and particulate detrital carbon in a temperate hard-water lake, *Mem. Ist. Ital. Idrcbiol.*, 29, (Suppl.), 185, 1972.

5356. Whalen, S.C., and Alexander, V., Influence of temperature and light on rate of inorganic nitrogen transport by algae in an arctic lake, *Can. J. Fish. Aquat. Sci.*, 41, 1310, 1984.

5357. Whaley, W.G., Dauwalder, M., and Kephart, J.E., Assembly, continuity and exchanges in certain cytoplasmic membrane systems, in *Origin and Continuity of Cell Organelles*, Reinert, J., and Ursprung, Z., Eds., Springer-Verlag, Berlin, 1971, 1.

5358. Wharton, C.H., Odum, H.T., Ewel, K., Duever, M., Lugo, A., Boyt, R., Bartholomew, E., De Bellevue, P., Brown, S., and Duever, L., *Forested Wetlands of Florida—Their Management and Use*, Center for Wetlands, University of Florida, Gainesville, 1976.

5359. Whatley, J.M., The fine structure of *Prochloron, New Phytol.*, 79, 309, 1977.

5360. Whatley, J.M., Ultrastructure of plastid inheritance: green algae to angiosperms, *Biol. Rev.*, 57, 527, 1982.

5361. Wheeler, A.E., Zingaro, R.A., Irgolic, K., and Bottino, N.R., The effect of selenate, selenite and sulphate on the growth of six unicellular green algae, *J. Exp. Mar. Biol. Ecol.*, 57, 181, 94.

5362. Wheeler, P.A., Effect of nitrogen source on *Platymonas* (Chlorophyta) and composition and amino acid uptake rate, *J. Phycol.*, 13, 301, 1977.

5363. Wheeler, P.A., and Stephens, G.C., Metabolic segregation of intracellular free amino acids in *Platymonas* (Chlorophyta), *J. Phycol.*, 13, 193, 1977.

5364. Wheeler, P.A., and Helleburst, J.A., Uptake of concentration of alkylamines by a marine diatom, Effects of H^+ and K^+ and implications for the transport and accumulation of weak bases, *Plant Physiol.*, 67, 367, 1981.

5365. Wheeler, P.A., and McCarthy, J.J., Methylamine uptake by Chesapeake Bay phytoplankton: evaluation of the use of the ammonium analogue for field uptake measurements, *Limnol. Oceanogr.*, 27, 1129, 1982.

5366. Wheeler, P.A., and Björnsäter, B.R., Seasonal fluctuations in tissue nitrogen, phosphorus, and N:P for five macroalgal species common to the Pacific Northwest coast, *J. Phycol.*, 28, 1, 1992.

5367. Wheeler, P.A., North, B.B., and Stephens, G.C., Amino acid uptake by marine phytoplankton, *Limnol. Oceanogr.*, 19, 249, 1974.

5368. Wheeler, P.A., Gilbert, P.M., and McCarthy, J.J., Ammonium uptake and incorporation by Chesapeake Bay phytoplankton: short-term uptake kinetics, *Limnol. Oceanogr.*, 27, 1113, 1982.

5369. Whigham, D.F., Structure function of a freshwater tidal-marsh ecosystem, *Nat. Geogr. Soc. Res. Reports*, 15, 725, 1983.

5370. Whigham, D.F., and Simpson, R.L., Ecological studies on the Hamilton Marshes. Progress report for the period June 1974 - January 1975, Rider College, Biology Dept., Lawrenceville, New Jersey, 1975.

5371. Whigham, D.F., Simpson, R.L., The potential use of freshwater tidal marshes in the management of water quality in the Delaware River, in *Biological Control of Water Pollution*, Tourbier, J., and Pierson, R.W., Jr., Eds., The University of Pennsylvania Press, Philadelphia, 1976, 173.

5372. Whigham, D.F., and Simpson, R.L., Sewage spray irrigation in a Delaware River freshwater tidal marsh, in *Freshwater Wetlands and Sewage Disposal*, Tilton, D.L., Kadlec, R.H., and Richardson, C.J., Eds., University of Michigan, Ann Arbor, Michigan, 1976, 121.

5373. Whigham, D.F., and Simpson, R.L., Growth, mortality, and biomass partitioning freshwater tidal wetland populations of wild rice (*Zizania aquatica* var. *aquatica*), *Bull. Torrey Bot. Club*, 104, 347, 1977.

5374. Whigham, D.F., and Simpson, R.L., The relationship between aboveground and belowground biomass of freshwater tidal wetland macrophytes, *Aquat. Bot.*, 5, 355, 1978.

5375. Whigham, D.F., and Bayley, S.E., Nutrient dynamics in freshwater wetlands, in *Proc. Symp. Wetlands Functions and Values: The State of Our Understanding*, Greeson, P.E., Clark, J.R., and Clark, J.E., Eds., American Water Resources Assoc., Minneapolis, 1979, 468.

5376. Whigham, D.F., McCormick, J., Good, R.E., and Simpson, R.L., Biomass and primary production in freshwater tidal wetlands of the Middle Atlantic Coast, in *Proc. Symp. Freshwater Wetlands: Ecological Processes and Management Potential*, Good, R.E., Whigham, D.F., and Simpson, R.L., Eds., Academic Press, New York, 1978, 3.

5377. Whigham, D.F., O'Neill, J., and McWethy, M., The effect of three marsh management techniques on the ecology or irregularly flooded Chesapeake Bay wetlands, vegetation and water quality studies, Chesapeake Bay Center for Environmental Studies, Smithsonian Institute Mimeo Final Report to Maryland Dept. of Natural Resources Coastal Zone Unit, 1982.

5378. Whigham, D.F., Simpson, R.L., Good, R.E., and Sickels, F.A., Decomposition and nutrient - metal dynamics of litter in freshwater tidal wetlands, in *Proc. Symp. Freshwater Wetlands and Wildlife*, Sharitz, R.R., and Gibbons, J.W., Eds., U.S. Department of Energy, 1989, 167.

5379. Whigham, D.F., Good, R.E., and Květ, J., Eds., *Wetland Ecology and Management. Case Studies*, Kluwer Academic Publishers, Dordrecht, 1990.

5380. Whigham, D.F., Dykyjová, D., and Hejný, S., Eds., *Wetlands of the World I: Inventory, Ecology and Management.* Handbook of vegetation science 15/2, Kluwer Academic Publishers, Dordrecht, The Netherlands, 1993.

5381. Whistler, R.L., *Industrial Gums, Polysaccharides and Their Derivatives,* 2nd ed., Academic Press, London and New York, 1973.

5382. White, A.W., Growth of two facultatively heterotrophic marine centric diatoms, *J. Phycol.,* 10, 292, 1974.

5383. White, A.W., Dinoflagellate toxins as probable cause of an Atlantic hering (*Clupea harengus* var. *harengus*) kill, and pteropods as apparent vector, *J. Fish. Res. Bd. Can.,* 34, 2421, 1977.

5384. White, E.B., and Boney, A.D., Experiments with some endophytic and endozoic *Achrochaetium* species, *J. Exp. Mar. Biol. Ecol.,* 3, 246, 1969.

5385. White, D.A., Weiss, T.E., Trapani, J.M., and Thien, L.B., Productivity and decomposition of the dominant salt marsh plants in Louisiana, *Ecology,* 59, 751, 1978.

5386. White, P.S., Pattern, process, and natural disturbance in vegetation, *Bot. Rev.,* 45, 229, 1979.

5387. White, W.S., and Wetzel, R.G., Nitrogen, phosphorus, particulate and colloidal carbon content of sedimenting seston of a hard water lake, *Verh. Internat. Verein. Limnol.,* 19, 330, 1975.

5388. Whitehead, N.E., Huynh-Ngoc, L., and Aston, S.R., Trace metals in two north Mediterranean rivers, *Water, Air Soil Pollut.,* 42, 7, 1988.

5389. Whitford, L.A., Ecological distribution of freshwater algae, in *Pymaturing Symp. in Ecology,* Pittsburgh, 1960, 1.

5390. Whitford, L.A., and Schumacher, G.J., Effect of current on mineral uptake and respiration by a freshwater alga, *Limnol. Oceanogr,* 6, 423, 1961.

5391. Whitford, L.A., and Schumacher, G.J., Effect of current on respiration and mineral uptake in *Spirogyra* and *Oedogonium, Ecology,* 45, 168, 1964.

5392. Whitlow, T.H., and Harris, R.W., Flood tolerance in plants: A state of the art review, U.S. Army Waterways Exp. Station. Environ. and Water Quality Operational Studies Tech. Rep. E-79-2, 1979.

5393. Whitney, D.M., Chalmers, A.G., Haines, E.B., Hanson, R.B., Pomeroy, L.R., and Sheer, B., The cycles of nitrogen and phosphorus, in *The Ecology of a Salt Marsh,* Pomeroy, L.R., and Wiegert, R-.G., Eds., Springer-Verlag, Berlin, 1981, 163.

5394. Whittaker, J., Barica, J., King, H., and Buckley, M., Efficacy of copper sulphate in the suppression of *Aphanizomenon flos-aquae* blooms in prairie lakes, *Environ. Pollut.,* 15, 185, 1978.

5395. Whittaker, R.H., New concept of kingdoms of organisms, *Science,* 163, 150, 1969.

5396. Whittaker, R.H., and Likens, G.E., The biosphere and man, in *Primary Productivity of the Biosphere,* Leith, H., and Whittaker, R.H., Eds., Springer-Verlag, New York, 1975, 305.

5397. Whittick, A., The life history and phenology of *Callithamnion corymbosum* (Rhodophyta: Ceramiaceae) in Newfoundland, *Can. J. Bot.,* 56, 2497, 1978.

5398. Whittingham, C.P., and Pritchard, G.G., The production of glycollate during photosynthesis in *Chlorella, Proc. R. Soc. Lond.,* B 157, 366, 1963.

5399. Whittle, S.J., The major chloroplast pigments of *Chlorobotrys regularis* (West) Bohlin (Eustigmatophyceae) and *Ophiocytium majus* Naegeli (Xanthophyceae), *Br. Phycol. J.,* 11, 111, 1976.

5400. Whittle, S.J., and Casselton, P.J., The chloroplast pigments of some green and yellow-green algae, *Br. Phycol. J.,* 4, 55, 1969.

5401. Whittle, S.J., and Casselton, P.J., The chloroplast pigments of the algal classes Eustigmatophyceae and Xanthophyceae. II. Xanthophyceae, *Br. Phycol. J.,* 10, 192, 1975.

5402. Whitton, B.A., Extracellular products of blue-green algae, *J. Gen. Microbiol.,* 40, 1, 1965.

5403. Whitton, B.A., Toxicity of zinc, copper, and lead to Chlorophyta from flowing waters, *Arch. Mikrobiol.,* 72, 353, 1970.

5404. Whitton, B.A., Toxicity of heavy metals to freshwater algae: a review, *Phykos,* 9, 116, 1970.

5405. Whitton, B.A., Ed., *River Ecology,* University of California Press, Berkeley, 1975.

5406. Whitton, B.A., Plants as Indicators of River Water Quality, in *Biological Indicators of Water Quality,* James, A., Ed., John Wiley and Sons, New York, 1979, Chapter 5.

5407. Whitton, B.A., Zinc and plants in river and streams, in *Zinc in the Environment,* Vol. II. *Health Effects,* Nriagu, J.O., Ed., John Wiley and Sons, New York, 1980, 363.

5408. Whitton, B.A., and Say, P.J., Heavy metals, in *River Ecology,* Whitton, B.A., Ed., Blackwell Scientific Publications, Oxford, 1975, 286.

5409. Whitton, B.A., and Diaz, B.M., Chemistry and plants of streams and rivers with elevated zinc, in *Trace Substances in Environmental Health XIV,* Hemphill, D.D., Ed., University of Missouri, Columbia, 1980, 457.

5410. Whitton, B.A., and Diaz, B.M., Influence of environmental factors on photosynthetic species composition in highly acidic waters, *Verh. Internat. Verein. Limnol.,* 21, 1459, 1981.

5411. Whitton, B.A., and Shehata, F.H.A., Influence of cobalt, nickel, copper and cadmium on the blue-green alga *Anacystis nidulans, Environ. Pollut.,* 27, 275, 1982.

5412. Whitton, B.A., Gale, N.L., and Wixson, B.G., Chemistry and plant ecology of zinc-rich wastes dominated by blue-green algae, *Hydrobiologia,* 83, 331, 1981.

5413. Whitton, B.A., Say, P.J., Wehr, J.D., Use of plants to monitor heavy metals in rivers, in *Heavy Metals in Northern England: Environmental and Biological Aspects,* Say, P.J., and Whitton, B.A., Eds., University of Durham, Durham, England, 1981, 135.

5414. Whitton, B.A., Say, P.J., and Jupp, B.P., Accumulation of zinc, cadmium and lead by the aquatic liverwort *Scapania, Environ. Pollut,* B 17, 299, 1982.

5415. Whitton, B.A., Potts, M., Simon, J.W., and Grainger, S.L.J., Phosphatase activity of the blue-green alga (cyanobacterium) *Nostoc commune* UTEX 584, *Phycologia,* 29, 139, 1990.

5416. Whitton, B.A., Grainger, S.L.J., Hawley, G.R.W., and Simon, J.W., Cell-bound and extracellular phosphatase activities of cyanobacterial isolates, *Microb. Ecol.,* 21, 85, 1991.

5417. WHO Report: Micropollutants in river sediments, EURO Reports and Studies 61, Regional Office for Europe, WHO, Copenhagen, 1980.

5418. Wiame, J.M., Accumulation de l'acide phosphorique (Phytine, Polyphosphates), in *Hanbuch Pflanzenphysiologie,* Vol. 9, Ruhland, W., Ed., Springer-Verlag, Berlin, 1958, 136.

5419. Widra, A., Metachromatic granules of microorganisms, *J. Bact.,* 78, 664, 1959.

5420. Wiedemen, V.E., Heterotrophic nutrition of waste-stabilization pond algae, in *Properties and Products of Algae,* Zajic, J.E., Ed., Plenum Press, New York, 1970, 107.

5421. Wieder, R.K., and Lang, G.E., Fe, Al, Mn, and S chemistry of *Sphagnum* peat in four peatlands with different metal and sulfur input, *Water Air Soil Pollut.,* 29, 309, 1986.

5422. Wieder, R.K., and Lang, G.E., Cycling of inorganic and organic sulfur in peat from Big Run Bog, West Virginia, *Biogeochemistry,* 5, 221, 1988.

5423. Wieder, R.K., Lang, G.E., and Granus, V.A., An evaluation of wet chemical methods for quantifying sulfur fractions in freshwater wetland peat, *Limnol. Oceanogr,* 30, 1109, 1985.

5424. Wiessner, W., Inorganic micronutrients, in *Physiology and Biochemistry of Algae,* Lewin, R.A., Ed., Academic Press, New York and London, 1962, 267.

5425. Wilbert, N., Ökologische Untersuchung der Aufwuchs- und Planktonciliaten eines eutrophen Weihers, *Arch. Hydrobiol. Suppl.,* 35, 411, 1969.

5426. Wilce, R.T., *Pleurocladia lacustris* in Arctic America, *J. Phycol.,* 2, 57, 1966.

5427. Wilen, B.D., Status and trends of inland wetlands and deepwater habitats in the United States, in *Proc. Symp. Freshwater Wetlands and Wildlife,* Sharitz, R.R., and Gibbons, J.W., Eds., U.S. Department of Energy, 1989, 719.

5428. Wilhm, J.L., and Long, J., Succession in algal mat communities at three nutrient levels, *Ecology,* 50, 645, 1969.

5429. Wille, J.N.F., Conjugatae, in *Die natürlichen Pflanzenfamilien,* Vol. 1, abt. 2, Engler, A., and Prantl, K.A.E., Eds., Leipzig, 1897, 1.

5430. Wille, J.N.F., Chlorophyceae, in *Die natürlichen Pflanzenfamilien,* Vol. 2, abt. 2, Engler, A., and Prantl, K.A.E., Eds., Leipzig, 1909, 1.

5431. Wille, J.N.F., Chlorophyceae, in *Die natürlichen Pflanzenfamilien,* Vol. 1 abt. 2, Engler, A., and Prantl, K.A.E., Eds., Leipzig, 1911, 1.

5432. Willett, I.R., The reductive dissolution of phosphated ferrihydride and strengite, *Aust. J. Soil Res.,* 23, 237, 1985.

5433. Williams, E.G., Scott, N.M., and McDonald, M.J., Soil properties and phosphate sorption, *J. Sci. Food Agric.,* 9, 551, 1958.

5434. Williams, J.D.H., Syers, J.K., Harris, R.F., and Armstrong, D.E., Adsorption and desorption of inorganic phosphorus by lake sediments in a 0.1 M NaCl system, *Environ. Sci. Technol.,* 4, 517, 1970.

5435. Williams, J.D.H., Syers, J.K., Shukla, S.S., and Harris, R.F., Levels of inorganic and total phosphorus in lake sediments as related to other sediment parameters, *Environ. Sci. Technol.,* 5, 1113, 1971.

5436. Williams, J.D.H., Syers, J.K., Harris, R.F., and Armstrong, D.E., Fractionation of inorganic phosphate in calcareous lake sediments, *Soil Sci. Soc. Am. Proc.,* 35, 250, 1971.

5437. Williams, J.D.H., Shear, H., and Thomas, R.L., Availability to *Scenedesmus quadricauda* on different forms of phosphorus in sedimentary materials in the Great Lakes, *Limnol. Oceanogr,* 25, 1, 1980.

5438. Williams, J.L., Reproduction in *Dictyota dichotoma, Ann. Bot.,* 11, 545, 1898.

5439. Williams, L.G., Relative strontium and calcium uptake by green algae, *Science,* 146, 1488, 1964.

5440. Williams, M., Understanding wetlands, in *Wetlands: A Threatened Landscape,* Williams, M., Ed., Basil Blackwell, Ltd., Oxford, 1990, 1.

5441. Williams, M., Ed., *Wetlands: A Threatened Landscape,* Basil Blackwell, Oxford, 1990.

5442. Williams, M.C., Bagley, C.E., Jr., Dillard, M.O., Fox, C.E., Suddereth, L., Williams, D.J., McLean, M.R., and Albers, L.E., Acid input to an urban freshwater marsh, in *Freshwater Wetlands and Wildlife*, Sharitz, R.R., and Gibbons, J.W., Ed., U.S. Department of Energy, 1989, 1237.

5443. Williams, R.J.P., Phosphorus biochemistry, in *Phosphorus in the Environment: its Chemistry and Biochemistry* Excerpta Medica, Amsterdam, 1979, 95.

5444. Williams, R.B., Division rates of salt marsh diatoms in relation to salinity and cell size, *Ecology,* 45, 877, 1964.

5445. Williams, R.B., Unusual motility of tube-dwelling pennate diatoms, *J. Phycol.,* 1, 145, 1965.

5446. Williams, R.B., and Murdoch, M.B., Annual production of *Spartina alterniflora* and *Juncus roemerianus, Assoc. of Southeastern Biologist Bull.,* 15, 59, 1966.

5447. Williams, R.B., and Murdoch, M.B., Phytoplankton production and chlorophyll concentration in the Beaufort Channel, North Carolina, *Limnol. Oceanogr,* 11, 73, 1966.

5448. Williams, S.K., and Hodson, R.C., Transport of urea at low concentrations in *Chlamydomonas reinhardii, J. Bacteriol.,* 130, 266, 1977.

5449. Williams, S.L., Disturbance and recovery of a deep-water Caribbean seagrass bed, *Mar. Ecol. Prog. Ser,* 42, 63, 1988.

5450. Williams, S.T., and Gray, T.R.G., Decomposition of litter on the soil surface, in *Biology of Plant Litter Decomposition,* Dickinson, C.H., and Pugh, G.J.F., Eds., Academic Press, New York, 1974, 611.

5451. Williams, W.T., and Barber, D.A., The functional significance of aerenchyma in plants, *Soc. Exp. Biol. Symp.,* 15, 132, 1961.

5452. Wilson, L.G., Metabolism of sulfate: sulfate reduction, *Ann. Rev. Plant Physiol.,* 13, 201, 1962.

5453. Wilson, S.H., and Fieldes, M., Studies in spectrographic analysis. II. Minor elements in a sea-weed (*Macrocystis pyrifera*), *N. Z. J. Sci. Technol.,* 23B, 47, 1942.

5454. Wilton, J.W., and Barham, E.G., A yellow-water bloom of *Gymnodinium flavum* Kofoid & Swezy, *J. Exp. Mar. Biol. Ecol.,* 2, 167, 1968.

5455. Wiltshire, G.H., Productivity of reed beds round a silt-laden dam, Proj. No. NP 14/106/3/1/CSIR, 1981.

5456. Windom, H., Gardner, W., Stephens, J., and Taylor, F., The role of methylmercury production in the transfer of mercury in a salt marsh ecosystem, *Estuar. Coast. Mar. Sci.,* 4, 579, 1976.

5457. Winfrey, M.R., and Zeikus, J.G., Anaerobic metabolism of immediate methane precursors in Lake Mendota sediments, *Appl. Environ. Microbiol.,* 37, 244, 1979.

5458. Winsborough, B.M., and Golubic, S., The role of diatoms in stromatolite growth: two examples from modern freshwater settings, *J. Phycol.,* 23, 195, 1987.

5459. Winter, G., Über die Assimilation des Luftstickstoffs durch endophytische Blaualgen, *Beitr. Biol. Pfl.,* 23, 295, 1935.

5460. Winter, H.C., and Burris, R.H., Nitrogenase, *Ann. Rev. Biochem.,* 45, 409, 1976.

5461. Winter, U., and Kirst, G.O., Turgor pressure regulation in *Chara aspera* (Chlorophyta): the role of sucrose accumulation in fertile and sterile plants, *Phycologia,* 31, 240, 1992.

5462. Wintermans, J.F.G.M., Polyphosphate formation in *Chlorella* in relation to photosynthesis, *Mededel. Landbouwhogeschool Wageningen,* 55, 69, 1955.

5463. Wise, D.L., Carbon nutrition and metabolism of *Polytomella caeca, J. Protozool.,* 6, 19, 1959.

5464. Wisser, S.A., A study on the decomposition of *Cyperus papyrus* in the swamps of Uganda, in natural peat deposits as well as in the presence of various additions, *East Afr. Agric. For. J.,* 29, 268, 1964.

5465. Wiszniewki, J., Remarques relatives aux recherches recentes sur le psammon d'eaux, *Arch. Hydrobiol. Ryb.,* 13, 7, 1947.

5466. Withers, N.W., Alberte, R.A., Lewin, R.A., Thornber, J.P., Britton, G., and Goodwin, T.W., Photosynthetic unit size, carotenoids and chlorophyll protein composition of *Prochloron* sp. a prokaryoric green alga, *Proc. Natl. Acad. Sci. U.S.A.,* 75, 2301, 1978.

5467. Wium-Andersen, S., The effect of chromium on the photosynthesis and growth of diatom and green algae, *Physiol. Plant.,* 32, 308, 1974.

5468. Wium-Andersen, S., and Andersen, J.M., The influence of vegetation on the redox profile of the sediment of Grande Langrø, a Danish *Lobelia* lake, *Limnol. Oceanogr,* 17, 948, 1970.

5469. Wixson, B.G., Ed., *The Missouri Lead Studies,* Final Report NSF Grant 74/22953-A01, University of Missouri, Rolla, Missouri, 1977.

5470. Woelkerling, Wm.J., *The Corraline Red Algae,* Oxford University Press, Oxford, London, 1988.

5471. Woelkerling, W.J., and Gough, S.B., Wisconsin desmids. 3. Desmid community composition and distribution in relation to lake type and water chemistry, *Hydrobiologia,* 51, 3, 1976.

5472. Wohler, J.R., Robertson, D.B., and Laube, H.R., Studies on the decomposition of *Potamogeton diversifolius, Bull. Torrey Bot. Club,* 102, 76, 1975.

5473. Wolf, A.M., Baker, D.E., Pionke, H.B., and Kunishi, H.M., Soil test for estimating labile, soluble, and algal-available phosphorus in agricultural soil., *J. Environ. Qual.*, 14, 341, 1985.

5474. Wolfe, M., The effect of molybdenum upon the nitrogen metabolism of *Anabaena cylindrica*. I. A study of the molybdenum requirement for nitrogen fixation and for nitrate and ammonia assimilation, *Ann. Bot. N.S.*, 18, 299, 1954.

5475. Wolfe, M., The effect of molybdenum upon the nitrogen metabolism of *Anabaena cylindrica*, II. A more detailed study of the action of molybdenum in nitrate assimilation, *Ann. Bot. N.S.*, 18, 309, 1954.

5476. Wolk, C.P., Heterocysts germination under defined conditions, *Nature*, 205, 201, 1965.

5477. Wolk, C.P., Physiological basis of the pattern of vegetative growth in a blue-green alga, *Proc. Natl. Acad. Sci. USA*, 57, 1246, 1967.

5478. Wolk, C.P., Physiological chemistry of blue-green algae, *Bacteriol. Rev.*, 37, 32, 1973.

5479. Wolk, C.P., Heterocysts, in *The Biology of Cyanobacteria*, Carr, N.G., and Whitton, B.A., Eds., Blackwell Scientific Publications, Oxford, 1982, 359.

5480. Wolk, C.P., Thomas, J., Shaffer, P.W., Austin, S.M., and Galonsky, A., Pathway of nitrogen metabolism after fixation of ^{13}N-labeled nitrogen gas by the cyanobacterium *Anabaena cylindrica, J. Biol. Chem.*, 251, 5027, 1976.

5481. Wolken, J.J., *Euglena. An Experimental Organism for Biochemical and Biophysical Studies*, Appleton-Century Crofts, New York, 1967.

5482. Wolverton, B.C., and McKown, M.M., Water hyacinths for removal of phenols from polluted waters, *Aquat. Bot.*, 2, 191, 1976.

5483. Wong, P.K., and Chang, L., Effects of copper, chromium and nickel on growth, photosynthesis and chlorophyll *a* synthesis of *Chlorella pyrenoidosa* 251, *Environ. Pollut.*, 72, 127, 1991.

5484. Wong, P.T.S., Silverberg, B.A., Chau, Y.K., and Hodson, P.V., Lead and the aquatic biota, in *The Biogeochemistry of Lead in the Environment*, Part B. *Biological Effects*, Nriagu, J.O., Ed., Elsevier/North Holland Biomedical Press, New York, 1978, 279.

5485. Wong, P.T.S., Mayfield, C.I., and Chau, Y.K., Cadmium toxicity to phytoplankton and microorganisms, in *Cadmium in the Environment*, Part 1., Nriagu, J.O., Ed., John Wiley and Sons, New York, 1980, 571.

5486. Wood, A.M., Available copper ligands and the apparent bioavailability of copper to natural phytoplankton assemblages, *Sci. Total Environ.*, 28, 51, 1983.

5487. Wood, J.M., Biological cycles for toxic elements in the environment, *Science*, 183, 1049, 1974.

5488. Wood, J.M., and Wang, H.K., Microbial resistance to heavy metals, *Environ. Sci. Technol.*, 15, 582, 1983.

5489. Wood, K.G., Trace element pollution in streams of northwestern U.S.A., *Verh. Internat. Verein. Limnol.*, 19, 1641, 1975.

5490. Wood, K.G., Photosynthesis of *Cladophora* in relation to light and CO_2 limitation; $CaCO_3$ precipitation, *Ecology*, 56, 479, 1975.

5491. Wood, M., *Soil Biology*, Blackie and Son, Ltd., Glasgow, 1989.

5492. Wood, R.D., Characeae in Australia, *Nova Hedw.*, 22, 1, 1971.

5493. Wood, R.D., and Imahori, K., *A Revision of the Characeae*, Vol. 1, Verlag von J. Cramer, Weinheim, 1964.

5494. Wood, R.D., and Imahori, K., *A Revision of the Characeae*, Vol. 2, Verlag von J. Cramer, Weinheim, 1965.

5495. Wood, R.D., and Mason, R., Characeae of New Zealand, *New Zealand J. Bot.*, 15, 87, 1977.

5496. Woolery, M.L., and Lewin, R.A., Influence of iodine on growth and development of the brown alga *Ectocarpus siliculosus* in axenic culture, *Phycologia*, 12, 131, 1973.

5497. Woolery, M.L., and Lewin, R.A., The effects of lead on algae. IV. Effects on respiration and photosynthesis of *Phaeodactylum tricornutum* (Bacillariophyceae), *Water Air Soil Pollut.*, 6, 25, 1977.

5498. Wooten, J.N., and Dodd, J.D., Growth of water hyacinth in treated sewage effluent, *Econ. Bot.*, 30, 29, 1976.

5499. Wornardt, W.W., Jr., Diatoms, past, present, future, in *Proc. First Conf. on Planktonic Microfossils*, Brönnimann, P., and Renz, H.H., Eds., Leiden, Brill., 1969, 690.

5500. Wort, D.J., The seasonal variation in chemical composition of *Macrocystis integrifolia* and *Nereocystis luetkeana* in British Columbia coastal waters, *Can. J. Bot.*, 33, 323, 1955.

5501. Wrench, J.J., Selenium metabolism in the marine phytoplankters *Tetraselmis tetrathele* and *Dunaliella minuta*, *Mar. Biol.*, 49, 231, 1978.

5502. Wrench, J.J., and Measures, C.I., Temporal variations in dissolved selenium in a coastal ecosystem, *Nature*, 290, 431, 1982.

5503. Wright, R.T., and Hobbie, J.E., The uptake of organic solutes in lake water, *Limnol. Oceanogr,* 10, 22, 1965.

5504. Wright, S.A., and Syrett, P.J., The uptake of methylammonium and dimethylammonium by the diatom, *Phaeodactylum tricornutum, New Phytol.,* 95, 189, 1983.

5505. Wrigley, T.J., and Toerien, D.F., The ability of an artificially established wetland system to upgrade oxidation pond effluent to meet water quality criteria, *Water SA,* 14, 171, 1988.

5506. Wylie, P.A., and Schlichting, H.E., A floristic survey of corticolous subaerial algae in North Carolina, *J. Elisha Mitchell Sci. Soc.,* 89, 179, 1975.

5507. Wynne, D., Alterations in activity of phosphatases during the *Peridinium* bloom in Lake Kinneret, *Physiol. Plant.,* 40, 219, 1977.

5508. Wynne, M.J., Phaeophyta: morphology and classification: In *The Biology of Seaweeds,* Lobban, C.S., and Wynne, M.J., Eds., University of California Press, Berkeley, 1981, 52.

5509. Wynne, M.J., Phaeophyceae: introduction and bibliography, in *Selected Papers in Phycology,* Rossowski, J.R., and Parker, B.C., Eds., Phycological Society of America, Lawrence, Kansas, 1982 a, 731.

5510. Wynne, M.J., Phaeophyceae, in *Synopsis and Classification of Living Organisms,* Vol. 1, Parker, S.P., Ed., McGraw-Hill, New York, 1982 b, 115.

5511. Wynne, M.J., and Loiseaux, S., Recent advances in life history studies of the Phaeophyta, *Phycologia,* 15, 435, 1976.

5512. Yamakana, G., and Glazer, A.N., Dynamic aspects of phycobilisome structure: phycobilisome turnover during nitrogen starvation in *Synechococcus, Arch. Microbiol.,* 124, 39, 1980.

5513. Yamakawa, T., and Yamada, Y., Assimilation and absorption of carbon dioxide through roots during the life of rice plants. Part 1. Studies on the absorption of carbon dioxide through roots of rice plants, *Jpn. J. Soil Sci. Plant Nutr,* 54, 292, 1983.

5514. Yamamoto, I., Nagumo, T., Yagi, K., Tominaga, H., and Aoki, M., Antitumor effects of seaweeds. I. Antitumor effects of extracts from *Sargassum* and *Laminaria, Japan J. Exp. Med.,* 44, 543, 1974.

5515. Yamamoto, Y., and Suzuki, K., Distribution and algal-lysing activity of fruiting mycobacteria in Lake Suwa, *J. Phycol.,* 26, 457, 1990.

5516. Yamane, I., Electrochemical changes in rice soils, in *Soils and Rice,* Internat. Rice Res. Inst., Los Baños, Philippines, 1978, 381.

5517. Yanagi, S., and Sasa, T., Changes in hydrogenase activity during the synchronous growth of *Scenedesmus obliquus* D$_3$, *Plant Cell Physiol.,* 7, 593, 1966.

5518. Yarbro, L.A., Phosphorus cycling in the Creeping Swamp floodplain ecosystem and exports from the Creeping Swamp watershed, Ph.D. Thesis, University of North Carolina, Chapel Hill, 1979.

5519. Yates, M.G., Physiological aspects of nitrogen fixation, in *Recent Developments in Nitrogen Fixation,* Newton, W., Postgate, J.R., and Rodrigues-Barucco, C., Eds., Academic Press, London, 1977, 219.

5520. Yates, R.F., and Day, F.P., Jr., Decay rates and nutrient dynamics in confined and unconfined leaf litter in the Great dismal swamp, *Am. Midl. Nat.,* 110, 37, 1983.

5521. Yentsch, C.M., and Taylor, D.L., *Red Tides,* Harvard University Press, Boston, Massachusetts, 1982.

5522. Yentsch, C.S., and Ryther, J.H., Relative significance of the net phytoplankton and nanoplankton in the water of Vineyard Sound, *Journal du Conseil Int. pour l' Exploration de la Mer,* 24, 231, 1959.

5523. Yoshida, T., Microbial metabolism of flooded soils, in *Soil Biochemistry* Vol. 3, Paul, E.A., and McLaren, A.D., Eds., Marcel Dekker, New York, 1975, 83.

5524. Yoshida, T., and Ancajas, R.R., Nitrogen fixing activity in upland and flooded rice fields, *Soil Sci. Soc. Am. Proc.,* 37, 42, 1973 a.

5525. Yoshida, T., and Ancajas, R.R., The fixation of atmospheric nitrogen in the rice rhizosphere, *Soil Biol. Biochem.,* 5, 153, 1973 b.

5526. Yoshida, T., and Broadbent, F.E., Movement of atmospheric nitrogen in rice plants, *Soil Sci.,* 120, 288, 1975.

5527. Yoshida, T., and Nishijama, Y., Quantitative measurement of nitrous oxide evolution from soil environment, *Jpn. J. Soil Sci. Plant Nutr,* 54, 100, 1983.

5528. Young, E.G., and Langille, W.M., The occurrence of inorganic elements in marine algae of the Atlantic provinces of Canada, *Can. J. Bot.,* 36, 301, 1958.

5529. Young, O.W., A limnological investigation of periphyton in Douglas Lake, Michigan, *Trans. Am. Microsc. Soc.,* 64, 1, 1945.

5530. Young, R.G., and Lisk, D.J., Effect of copper and silver ions on algae, *J. Water Pollut. Control Fed.,* 44, 1643, 1972.

5531. Young, T.C., and King, D.L., Interacting limits to algal growth: Light, phosphorus, and carbon dioxide availability, *Water Res.,* 14, 409, 1980.

5532. Young, T.C., and De Pinto, J.V., Algal-availability of particulate phosphorus from diffuse and point sources in the lower Great Lakes basin, *Hydrobiologia,* 91, 111, 1982.

5533. Yount, J.L., Factors that control species numbers in Silver Spring, Florida, *Limnol. Oceanogr.*, 1, 286, 1956.

5534. Yount, J.L., and Grossman, R.A., Eutrophication control by plant harvesting, Part 2., *J. Water Pollut. Control Fed.*, 42, R 173, 1970.

5535. Zajic, J.E., Ed., *Properties and Products of Algae,* Plenum Press, New York, 1970.

5536. Zajic, J.E., and Chiu, Y.S., Heterotrophic culture of algae, in *Properties and Products of Algae,* Zajic, J.E., Ed., Plenum Press, New York, 1970, 1.

5537. Žáková, Z., The experience with the use water hyacinth for water treatment in South Moravia, in *Proc. Conf. Vegetation in Water Treatment and its Application,* Žáková, Z., Květ, J., Lhotský, O., and Marvan, P., Eds., VÚV Brno, Czech Republic, 1987, 83.

5538. Žáková, Z., Cultivation and possibilities of practical use of water hyacinth (*Eichhornia crassipes*) in ČSFR, in *Controlled Cultivation of Aquatic and Wetland Plants,* Čížková-Končalová, H., and Husák, S., Eds, Botanical Institute, Academy of Science of the Czech Republic, Třeboň, 1992, 102.

5539. Žáková, Y., and Véber, K., Biological base of water hyacinth cultivation and use in ČSFR, Práce a studie ČSAV, Academia Praha, Czech Republic, 1991.

5540. Zálužanský ze Zálužan, A., *Methodi herbariae libri tres,* Prague, 1592.

5541. Zedler, J.B., Algal mat productivity comparisons in a salt marsh, *Estuaries,* 3, 122, 1980.

5542. Zedler, J., Winfield, T., and Mauriello, D., Primary productivity in a southern California estuary, in *Coastal Zone 1978. Symposium on Technical, Environmental, Socioeconomic, and Regulatory Aspects of Coastal Zone Management,* Vol. 11, Am. Soc. Civil Engineers, New York, 1978, 649.

5543. Zehnder, A.J.B., The carbon cycle, in *The Handbook of Environmental Chemistry* Vol. 1, Part B, *The Natural Environment and the Biogeochemical Cycles,* Hutzinger, O., Ed., Springer-Verlag, Berlin, 1982, 83.

5544. Zehnder, A.J.B., and Zinder, S.H., The sulfur cycle, In *The Handbook of Environmental Chemistry* Vol. 1, Part A, *The Natural Environment and the Biogeochemical Cycles,* Hutzinger, O., Ed., Springer-Verlag, Berlin, Heidelberg, New York, 1980, 105.

5545. Zehr, J.P., and Falkowski, P.G., Pathway of ammonium assimilation in marine diatom determined with the radiotracer, *J. Phycol.,* 24, 588, 1988.

5546. Zeitschel, B., Why study phytoplankton? in *Phytoplankton Manual,* Sournia, A., Ed., UNESCO, Paris, 1978.

5547. Zelinka, M., and Marvan, P., Zur Präzisierung der biologischen Klassification der Reinheit fliessender Gewässer, *Arch. Hydrobiol.,* 57, 389, 1961.

5548. Zelinka, M., and Sládeček, V., *Hydrobiology for Water Management,* SNTL, Praha, Czechoslovakia, 1964, (in Czech).

5549. Zelitch, I., Organic acids and respiration in photosynthetic tissues, *Ann. Rev. Plant Physiol.,* 15, 121, 1964.

5550. Zelitch, I., *Photosynthesis, Photorespiration, and Plant Productivity* Academic Press, New York and London, 1971.

5551. Zenwirth, D., Volokita, M., and Kaplan, A., Photosynthesis and inorganic carbon accumulation in the acidophilic alga *Cyanidioschyzon merolae, Plant Physiol.,* 77, 237, 1985.

5552. Zevenboom, W., Growth and nutrient uptake kinetics of *Oscillatoria agardhii.* Ph.D. Thesis, University of Amsterdam, The Netherlands, 1980.

5553. Zevenboom, W., and Mur, L.R., Nitrate uptake by nitrogen-limited *Oscillatoria agardhii,* in *Growth and Nutrient Uptake Kinetics of* Oscillatoria agardhii, Zevenboom, W., Ed., Ph.D. Thesis, University of Amsterdam, The Netherlands, 1980, 47.

5554. Zevenboom, W., and Mur, L.R., Ammonium uptake by nitrogen-limited *Oscillatoria agardhii,* in *Growth and Nutrient Uptake Kinetics of* Oscillatoria agardhii, Zevenboom, W., Ed., Ph.D. Thesis, University of Amsterdam, The Netherlands, 1980, 63.

5555. Zebenboom, W., and Mur, L.R., Simultaneous short-term uptake of nitrate and ammonium by *Oscillatoria agardhii* grown in nitrate- or light-limited continuous culture, *J. Gen. Microbiol.,* 126, 355, 1981.

5556. Zevenboom, W., and Mur, L.R., Ammonium-limited growth and uptake by *Oscillatoria agardhii* in chemostat cultures, *Arch. Microbiol.,* 129, 61, 1981.

5557. Zevenboom, W., Knip, K.M., and Mur, L.R., Influence of the nature of the growth-limitation on some physiological properties of *Oscillatoria agardhii* grown in continuous culture, in *Growth and Nutrient Uptake Kinetics of* Oscillatoria agardhii, Zevenboom, W., Ed., Ph.D. Thesis, University of Amsterdam, The Netherlands, 1980, 109.

5558. Zhadin, V.I., and Gerd, S.V., *Fauna and Flora of the Rivers, Lakes and Reservoirs of the U.S.S.R.,* Oldbourne Press, London, 1963.

5559. Zimba, P.V., Sullivan, M.J., and Glover, H.E., Carbon fixation in cultured marine benthic diatoms, *J. Phycol.,* 26, 306, 1990.

5560. Zingaro, R.A., and Cooper, W.C., *Selenium*, Van Nostrand Reinhold Company, New York, 1974.
5561. Zingde, H.V., and Pucci, A.E., Cu, Cd, Pb, and Zn in tributaries to Blanca Bay, Argentina, *Mar. Pollut. Bull.*, 17, 230, 1986.
5562. Zingde, M.D., Rokade, M.A., and Mandalia, A.V., Heavy metals in Mindhola River estuary, India, *Mar. Pollut. Bull.*, 10, 538, 1988.
5563. Zingmark, R.G., and Miller, T.G., The effects of mercury on the photosynthesis and growth of estuarine and oceanic phytoplankton, in *Physiological Ecology of Estuarine Organisms*, Vernberg, F.J., Ed., Belle W. Baruch Library of Marine Science, University of South Carolina Press, Columbia, 1975, 45.
5564. Zinn, J.A., and Copeland, C., *Wetland Management*, Congressional Res. Service, The Library of Congress, Washington, 1982.
5565. Zirino, A., and Yamamoto, S., A-pH dependent model for the chemical speciation of copper, zinc, cadmium and lead in seawater, *Limnol. Oceanogr.*, 17, 661, 1972.
5566. Zoltai, S.C., An outline of the wetland regions of Canada, in *Proc. Workshop on Canadian Wetlands*, Rubec, C.D.A., and Pollet, F.C., Eds., Environment Canada, Ecological Land Classification Series No. 12, Saskatoon, 1979, 1.
5567. Zoltai, S.C., Pollett, F.C., Jeglum, J.K., and Adams, G.D., Development of wetlands classification for Canada, *Proc. 4th North America For. Soil Conf.*, Quebec City, 1975, 497.
5568. Zumft, W.G., Ferredoxine: nitrite oxidoreductase from *Chlorella*. Purification and properties, *Biochim. Biophys. Acta*, 276, 363, 1972.
5569. Zumft, W.G., Gotzmann, D.J., Frunzke, K., and Viebrock, A., Novel terminal oxidoreductases of anaerobic respiration (denitrification) from *Pseudomonas*, in *Inorganic Nitrogen Metabolism*, Ullrich, W.R., Aparicio, P.J., Syrett, P.J., and Castillo, F., Eds., Springer-Verlag, Berlin, 1987, 61.

Subject Index

(includes bacteria and macrophyte species and genera, excluding Appendix); TI = Taxonomical Index

Abioseston 31
Accretion rate 189
Acer rubrum 168
Acetate 60, 221
    uptake 114
Acids
    acetic, 211
    alginic (see Alginic acid)
    amino (see Amino acid)
    fatty 209
    free amino 252
    fulvic 369
    higher fatty 221
    humic 330, 369
    nucleic 51, 271, 298
    organic 61
    oxaloacetic 171
    phosphoglyceric 171
    silicic 348
Acorus calamus 168
Active transport 54–55, 274, 310
ADP 259
Adsorption 52
Aerenchyma 170
Aerial shoots 169
Aeromonas 230–231
Aerophytic algae 42
Agar 1, 17–18, 51, 105, 119, 121, 294
Aggregations 10
Agrobacterium 231
Agrostis semiverticillata 169
Air-adapted microalgae 217
Akinetes 28, 115–116
Alcaligenes 231
Alkaline bands 309
Alcohol dehydrogenase 171
Alginates 18, 119
Alginic acid 1, 17–18, 64, 105, 135
Alkaloids 108
Alteromonas putrefaciens 320
Aluminum 262
Amines 51
Amino acids 51, 60, 68, 92, 221, 244
    uptake 255
Ammonia 107, 225, 227, 237, 248
    assimilation 242
    diffusion 233–234
    incorporation 243
    uptake 239, 249, 377
    uptake constants 241
    volatilization 228
Ammonification 192, 227
Amphisema 9
Amylopectin 27, 145–146
Amylose 145–146
Anabiosis 169
Angiosperms 171

Anisogamy 29, 132, 135–136, 145 (see also Reproduction (sexual))
Anoxia 169–170
Antheridium 6, 29, 147
Anthoceros 46
Antibiotics 61, 106
Antimetabolites 354, 357
Aplanospores 28, 126
APS 291
Aquaculture 105
Arabinose 60
Aragonite 123, 304, 308
Archegonium 6
Arsenic 359–361
    and algal growth 360
    aquatic chemistry 359
    concentration factors 361
    concentration in algae 360
    in the environment 359
    sources to waters 359
Arthrobacter 320
Arundo 163
Ascomycetes 46
Aspergillus niger 326, 337, 345
Astaxanthin 44
ATP 21, 51, 59, 69–72, 74, 215–216, 237, 243, 259, 271, 274–275, 279, 291, 309, 314, 354
    :urea amidolase 252
Attached algae 80, 128
    productivity 87–89, 91
    standing crop 81–86
Aufwuchs 36 (see also Periphyton)
Autocolony 28
Autospores 28
Autotrophic 61, 124, 128
Auxospores 132
Auxotrophy 59, 61, 124, 128
Azolla 33, 46, 104, 169
Azospirillum 231
Azotobacter 337
Azobacter chroococcum 345

Bacillus 231
Bacillus coagulans 45
Bacteria 18, 56, 62, 106, 110, 175, 180, 190, 221, 244, 252, 281, 345, 357, 359, 369
    aerobic 72
    ammonia-oxidizing 230
    anaerobic 72, 189, 232, 320
    attached 116, 175–176
    autotrophic 72
    colorless sulfide-oxidizing 289
    denitrifying 230
    facultative anaerobic 178, 209, 230
    facultative chemolithotrophic 229
    fermentative 211, 230
    filamentous 115, 315

Vegetation (continued)
  hydrophytic 167–168
Vibrio 231
Victoria 169
Viruses 106, 110
Vitamins 59, 70, 104–105, 134, 141
  B$_{12}$ 51, 57, 59, 61, 342–343
Volatilization (see Ammonia)
Volvocales 148

Wastewater treatment 106
Water bloom 33, 114, 128 (see also Blooms)
Water gum 175
Waterlogging 166
Water quality 107
Water fern 104, 114
Water tupelo 174–175
Wet meadow 152, 153, 167
Wet prairie 152, 167
Willow 170
Wolffia 33, 169

Xanthophylls 21, 23, 113, 118, 120–121, 123,
  126–128, 134–135, 137, 139, 141, 146, 221, 348
  structure 23

Xanthophytes 44 (see also Xanthophyceae (TI))
Xylans 17, 119, 141
Xylose 60

Zinc 51, 322–329
  accumulation 327
  aquatic chemistry 323
  concentration factors 329
  concentration in algae 325
  deficiency 327
  in algal nutrition 326
  in the environment 322
  in wetlands plants 326
  sources to waters 323
  toxicity 327
Zizania 163
Zoochlorellae 47
Zooflagellate 126
Zooplankton 31–32, 97, 100, 138, 192
Zoospore 9, 28, 126–127
Zooxanthellae 47–48, 134
Zygospores 30
Zygote 29, 44, 132, 147

SI = Subject Index